Food processing technology

Related titles:

Physical properties of foods and food processing systems
(ISBN 978-1-85573-272-8)

'… an excellent choice as a textbook.' *Food Technology*

A standard introduction to physical properties of foods such as density, rheology, the diffusion of heat, diffusion and mass transfer.

Principles and practices for the safe processing of foods
(ISBN 978-1-85573-362-6)

'Readers will want to have this book, not just because it is such a comprehensive text on safe processing, but because it is so full of sound advice.' *Food Engineering*

A standard guide on safe process design and operation, both for students and the food industry.

Benders' dictionary of nutrition and food technology Eighth edition
(ISBN 978-1-84569-051-9)

'This book is certainly comprehensive and covers all aspects of food and nutrition science.' *British Nutrition Foundation Bulletin*

A comprehensive reference for key terms and definitions in food science.

Details of these books and a complete list of Woodhead's titles can be obtained by:

- visiting our web site at www.woodheadpublishing.com
- contacting Customer Services (e-mail: sales@woodheadpublishing.com; fax: +44 (0) 1223 832819; tel.: +44 (0) 1223 499140 ext. 130; address: Woodhead Publishing Limited, 80 High Street, Sawston, Cambridge CB22 3HJ, UK)

Woodhead Publishing Series in Food Science, Technology and Nutrition:
Number 174

Food processing technology

Principles and practice

Third edition

P. J. Fellows

CRC Press
Boca Raton Boston New York Washington, DC

WOODHEAD PUBLISHING LIMITED

Oxford   Cambridge   New Delhi

Published by Woodhead Publishing Limited,
80 High Street, Sawston, Cambridge CB22 3HJ, UK
www.woodheadpublishing.com

Woodhead Publishing India Private Limited, G-2, Vardaan House, 7/28 Ansari Road, Daryaganj,
New Delhi – 110002, India
www.woodheadpublishingindia.com

Published in North America by CRC Press LLC, 6000 Broken Sound Parkway, NW,
Suite 300, Boca Raton, FL 33487, USA

First edition 1988, Ellis Horwood Ltd
Second edition 2000, Woodhead Publishing Limited and CRC Press LLC
Reprinted 2002, 2003, 2005, 2007, 2008
Third edition 2009, Woodhead Publishing Limited and CRC Press LLC
Reprinted 2011
© 2009, Woodhead Publishing Limited
The author has asserted his moral rights.

British Library Cataloguing in Publication Data
A catalogue record for this book is available from the British Library.

Library of Congress Cataloging in Publication Data
A catalog record for this book is available from the Library of Congress.

Woodhead Publishing ISBN 978-1-84569-216-2 (print)
Woodhead Publishing ISBN 978-1-84569-634-4 (online)
ISSN 2042-8049 Woodhead Publishing Series in Food Science, Technology and Nutrition (print)
ISSN 2042-8057 Woodhead Publishing Series in Food Science, Technology and Nutrition (online)
CRC Press ISBN 978-1-4398-0821-4
CRC Press order number: N10075

The publishers' policy is to use permanent paper from mills that operate a sustainable forestry policy,
and which has been manufactured from pulp which is processed using acid-free
and elemental chlorine-free practices. Furthermore, the publishers ensure that the text paper and
cover board used have met acceptable environmental accreditation standards.

Typeset by Godiva Publishing Services Limited, Coventry, UK.
Printed by Lightning Source.

Contents

Acknowledgements

I am indebted to the large number of people who have given freely of their time and experience, provided me with information, checked the text and given support during this revision of *Food Processing Technology*. My particular thanks to Dr Mike Lewis of the Department of Food Biosciences, Reading University, for his support and technical editing skills; to Dr Mike Dillon, Director, Grimsby Institute of Food and Fisheries, for his assistance with the section on quality assurance; and to Sarah Whitworth of Woodhead Publishing for her ideas, suggestions and administrative support. My thanks also to the many companies that responded positively to my requests for information about their equipment and products; and finally, but not least, my special thanks to Wen for her constructive thoughts, encouragement and forbearance at my long hours away from family life over many months.

Every effort has been made to trace and acknowledge ownership of copyright. The publishers will be glad to hear from any copyright holder whom it has not been possible to contact.

Peter Fellows

For Wen and Paddy

Introduction

A brief history of food processing

Archaeological and ethnographic evidence indicates that the first food processing had its origins in hunter–gatherer societies that used heating over open fires or in boiling water to make meat, roots and vegetables more palatable. However, because of their lifestyle they did not need to preserve foods to any significant extent. The gradual change to agricultural societies necessitated storage and preservation of foods and by 3000–1500 BC, the Egyptians had developed processing techniques, including sun drying to preserve fish and poultry, fermentation to produce alcohol, and cereal grinding and ovens to bake leavened bread. These were slowly adopted by pastoral societies throughout the Middle East and then further afield, to preserve foods against times of famine, to improve their eating quality and to give a more varied diet. By 1500 BC, all of the main food plants that are used today, except sugar beet, had been cultivated somewhere in the world.

During the ensuing thousand years similar food technologies may have developed independently in many places, with local variations due to differences in climate, crops or food preferences. Early processes developed in China include tofu (soybean curd) and roasted dried millet and dried beef as military rations. In Japan, rice wine (saki) was developed, salt made from dried seaweed was used to preserve foods, and soya was processed to soy sauce and miso (soy paste) to flavour foods. In Europe, the first water-powered flour mills and the first commercial bakeries were developed by the Romans, and in India, the manufacture of sugar from cane had developed in the Indus Valley, both by 100 BC (Trager 1995).

In the first millennium AD, the comparative isolation of different civilisations began to change, and first travellers and then traders began to exchange ideas and foods across the world. For example in AD 400, the Vandals introduced butter to southern Europe, which began to replace olive oil. By AD 600, Jewish merchants had established the spice trade with the Orient. Technological developments during this period included the use of maltose from germinating grains as a sweetener, the development of kumiss (fermented mare's milk) and ales made from fermented millet in China. By AD 700, the first written law, which established regulations for the production of dairy products and preservation of foods, was encoded in China.

By the turn of the second millennium, a rapid expansion of trade and exchange of foods and technologies took place with European explorers and military expeditions: for example in 1148, knights returning from the second crusade brought sugar to Britain from the Middle East; Marco Polo brought noodles from China; in the 1500s, the Portuguese brought cloves from the East Indies for use in preserves and sauces, and to disguise spoiled meat. Spanish conquistadors discovered sun-dried llama, duck and rabbit, which were eaten uncooked in Peru; and they returned with foods that had never been seen before in Europe, including avocado, papaya, tomato, cacao, vanilla, kidney beans and later (1539) potatoes. At the same time, the Portuguese introduced chilli peppers and cayenne from Latin America to India, where they were used to prepare spiced dishes.

As societies developed, specialisation took place and trades evolved (e.g. millers, cheese-makers, bakers, brewers and distillers). Small variations in raw materials or processing methods gave rise to thousands of distinctive local varieties of cheeses, beers, wines and breads. These were the forerunners of present day food industries, and some foods have been in continuous production for nearly 800 years. During this period, mechanical processing equipment using water, wind and animal power was developed to reduce the time and labour involved in processing; for example, the Domesday Book (1086) in England lists nearly 6000 water-powered flour mills, one for every 400 inhabitants.

In countries with a temperate climate, processing techniques were developed to preserve food through winter months, including salting and smoking of meats and fish, fermentation to produce vinegar which was also used to preserve meat and vegetables, and boiling fruits or vegetables to reduce the moisture content and produce jams or chutneys. Ice from mountains had been used to refrigerate fruits and vegetables by the Romans, and later (1626) in England, Francis Bacon published his ideas on freezing chickens by stuffing them with snow. The growth of towns and cities gave impetus to the development of preservation technologies and the extended storage life allowed foods to be transported from rural areas to meet the needs of urban populations. During the 1600–1700s, the slave trade helped change food supplies, eating habits, agriculture and commerce. Ships returning from delivering slaves to Brazil took maize, cassava, sweet potato, peanuts and beans to Africa, where they remain staple foods. Cocoa from West Africa was brought to Europe and in 1725 the first chocolate company began operation in Britain. At this time, in Massachusetts, USA, more than 60 distilleries produced rum from molasses that was supplied by slave traders. The rum provided the capital needed to buy African slaves, who were then sold to West Indian sugar planters. A similar circular trade existed in salted cod fish and slaves between Britain, America, Africa the Caribbean and Latin America (Kurlansky 1997, 2002).

The scale of operation by food processing businesses accelerated during the Industrial Revolution in the eighteenth century, but there was an almost total absence of scientific understanding. The processes were still based on craft skills and experience, handed down within families that held the same trades for generations. By the late 1700s, the first scientific discoveries were being made: chlorine was used to purify water and citric acid was used to flavour and preserve foods. At this time, the first 'new' food process was developed in France after Napoleon Boneparte offered a prize of 12 000 francs to invent a means of preserving food for long periods, for military and naval forces. Nicholas Appert, a Parisian brewer and pickler, opened the first vacuum bottling factory (cannery) in 1804, boiling meat and vegetables and sealing the jars with corks and tar, and he won the prize in 1809.

In the nineteenth century, the pace of scientific understanding increased: Russian chemist, Gottlieb Iorchoff, demonstrated that starch breaks down to glucose and a Dutch chemist, Johann Mulder, created the word 'protein'. Technological advances in canning and refrigeration accelerated at an unprecedented rate. In 1810, the first patent for a tin-plated steel container was issued in Britain, and in 1849 a can-making machine was developed in the USA that enabled two unskilled workers to make 1500 cans per day, compared with 120 cans per day that could be made previously by two skilled tinsmiths. In 1861 a canner in Baltimore reduced the average processing time from 6 hours to 30 minutes by raising the temperature of boiling water to 121 °C with calcium chloride; and in 1874, a pressure-cooking retort using live steam was invented, leading to rapid expansion of the industry. In 1858 the first mechanical refrigerator using liquid ammonia was invented in France and in 1873, the first successful refrigeration compressor was developed in Sweden. The pasteurisation process, named after French chemist and microbiologist Louis Pasteur, was developed in 1862. Towards the end of the nineteenth century, scientific understanding had started the change away from small-scale craft-based industry, and by the start of the twentieth century, the food industry as we now know it was becoming established. In 1929, the merger of Lever Brothers and the Margarine Union formed the world's first multi-national food company.

Technological advances gathered speed in all areas of food technology as the twentieth century progressed. For example, 'instant' coffee was invented in 1901, the first patent for hydrogenating fats and oils was issued in 1903, transparent 'cellophane' wrapping was patented in France in 1908, the same year that the flavour enhancer, monosodium glutamate was isolated from seaweed. In 1923 dextrose was produced from maize, and widely used in bakery products, beverages and confectionery.

The widespread introduction of electricity revolutionised the food industry and prompted the manufacture of new specialist food processing machinery. For example in 1918, the Hobart Company developed the first electric dough mixer, electric food cutters and potato peelers. Most food processing at this time supplied staples (e.g. dried foods, sugar, cooking oil) and foods that were further prepared in the home or in catering establishments (e.g. canned meat and vegetables). The impetus for development of some of these foods came from military requirements during World War I. Later a 'luxury' market included canned tropical fruits, confectionery and ice cream. After World War II, development of a wide range of ready-to-eat meals, snackfoods and convenience foods began. Again these developments had been partly stimulated by the need to preserve foods for military rations. From the 1950s, food science and technology were taught at university level, and the scientific underpinning from this and work at food research institutions has created new technologies, products and packaging that have resulted in several thousand new foods being developed for sale each year.

The food industry today

The aims of the food industry today, as in the past, are fourfold:

1 To extend the period during which a food remains wholesome (the shelf-life) by preservation techniques which inhibit microbiological or biochemical changes and thus allow time for distribution, sales and home storage.
2 To increase variety in the diet by providing a range of attractive flavours, colours, aromas and texture in food.

3 To provide the nutrients required for health.
4 To generate income for the manufacturing company and its shareholders.

Each of these aims exists to a greater or lesser extent in all food processing, but a given product may emphasise some more than others. For example, the aim of freezing is to preserve organoleptic and nutritional qualities as close as possible to the fresh product, but with a shelf-life of several months instead of a few days or weeks. In contrast, sugar confectionery and snackfoods are intended to provide variety in the diet, and a large number of shapes, flavours, colours and textures are produced from basic raw materials. All food processing involves a combination of procedures to achieve the intended changes to the raw materials. Each of these unit operations has a specific, identifiable and predictable effect on a food and the combination and sequence of operations determine the nature of the final product.

In industrialised countries the market for processed foods is changing, and consumers no longer require a shelf-life of several months at ambient temperature for the majority of their foods. Changes in family lifestyle, and increased ownership of refrigerators, freezers and microwave ovens are reflected in demand for foods that are convenient to prepare, are suitable for frozen or chilled storage, or have a moderate shelf-life at ambient temperatures. There is also an increasing demand by consumers for foods that more closely resemble the original raw materials and have a 'healthy' or 'natural' image, and have fewer synthetic additives, or have undergone fewer changes during processing. Manufacturers have responded to these pressures by reducing or eliminating synthetic additives from products and substituting them with natural or 'nature-equivalent' alternatives; and by introducing new ranges of low-fat, sugar-free or low-salt products in nearly all subsectors. Functional foods, especially foods that contain probiotic micro-organisms and cholesterol-reducing ingredients, have shown a dramatic increase in demand, and products containing organic ingredients are also now widely available. Consumer pressure has also stimulated improvements to processing methods to reduce damage caused to organoleptic and nutritional properties, and led to the development of novel 'minimal' processes.

Trends that started during the 1960–1970s, and have accelerated during the last 30 years, have caused food processors to change their operations in four key respects: first is increasing investment in capital intensive, automated equipment to reduce labour and energy costs, and improve product quality; secondly, investment in computer control of processing operations, warehousing and distribution logistics to meet more stringent requirements for traceability, food safety and quality assurance; thirdly, high levels of competition and slower growth in food markets in industrialised countries have prompted mergers or take-overs of competitors; and fourthly, there has been a shift in power and control of food markets from manufacturers to large retail companies. Changes to the food industry during the last quarter of the twentieth century are reviewed by Anon (2001).

The changes in technology have been influenced by substantial increases in the costs of both energy and labour, and by public pressure and legislation to reduce negative environmental effects of processing. Food processing equipment now has increasingly sophisticated levels of microprocessor control to reduce processing costs, enable rapid change-over between shorter production runs, to improve product quality and to provide improved records for management decisions. The automation of entire processes, from reception of materials, through processing and packaging to warehousing is a reality. This has allowed producers to generate increased revenue and market share from products that have higher quality and value. Food processing has now become a global industry, dominated by a relatively few multinational conglomerates. A 2005 report (Anon 2005) noted that 30 companies now account for a third of the world's processed food and five companies control 75% of the international grain trade. Two companies dominate sales of half the world's bananas and three trade 85% of the world's tea.

A substantial increase in the power and influence of large retailing companies, especially in the USA and Europe during the past 20 years, has caused food companies to form international strategic alliances that enable them to develop pan-regional economies of scale. These companies are now focusing on development of new markets and are either buying up or forming alliances with local competitors in South East Asia, India, Eastern Europe and Latin America. Global sourcing of raw materials has been a feature of some industries from their inception, but this has now expanded to many more sectors to reduce costs and ensure continuity of supply. These developments have in turn prompted increased consumer awareness of both ethical purchasing issues, such as employment and working conditions in suppliers' factories, and the environmental impact of international transportation of foods by air. There has also been a resurgence of consumer interest in locally distinctive foods and 'fair-traded' foods in some European countries.

Rapid development of global production and distribution (global value chains, GVCs) has been made possible by developments in information and communications technologies. Growth in the buying power of major retail companies, particularly in Europe and the USA, enabled them to drive down prices paid to food processors. As a result, transnational processing corporations (TNCs) have adopted a series of strategies to increase their competitiveness, including mergers and acquisitions with food manufacturers in other countries. TNCs have established tightly integrated global-scale systems in widely separated locations and have reduced the need for highly skilled, highly paid workforces. This makes it possible for companies to move their operations to new countries, often in the developing world, where unskilled and lower-paid workers can be employed. These developments enable food production to be coordinated between distant sites, and suppliers can now be called upon to transfer goods across the world at short notice. These developments have been assisted by international agreements to remove tariff and non-tariff barriers, privatisation and deregulation of national economies to create 'free' markets in trade and foreign investment. The Uruguay Round of the General Agreement on Tariffs and Trade (GATT) held from 1986 to 1994 expanded the principle of 'free' trade in key areas, including agriculture, where countries must reduce subsidies paid to producers and also reduce tariffs on imported goods (Hilary 1999). Agreements related to investment under the World Trade Organization (WTO) extended the scope of GATT negotiations to include services and intellectual property (The General Agreement on Trade in Services), foreign direct investment and copyright, trademarks, patents and industrial designs. This was facilitated by changes introduced by the International Monetary Fund and World Bank that opened up investment opportunities in many developing countries and helped the creation of GVCs. Saul (2005) describes the rise of free trade from the mid-nineteenth century, with globalisation reaching its high point in the mid-1990s with the establishment of the WTO. However, he believes that globalisation had begun to falter within five years, with the Asian financial crisis of 1997–98 underlining its inherent instability, and it is presently characterised by stasis. Whether this is the case or not, the effects of globalisation are likely to remain a significant influence on large-scale food industries for many years.

About this book

This book aims to introduce students of food science and technology or biotechnology to the wide range of processing techniques that are used in food processing. It shows how knowledge of the properties of foods is used to design processing equipment and to control processing conditions on an industrial scale, to achieve the aims of making attractive, saleable, safe and nutritious products and extending the shelf-life of foods. It is a comprehensive, yet basic text, offering an overview of most unit operations (Fig. I.1). It provides details of the processing equipment, operating conditions and the effects of processing on micro-organisms that contaminate foods, the biochemical properties of foods and their sensory and nutritional qualities. It collates and synthesises information from a wide range of sources, combining food processing theory and calculations, and results of scientific studies, with descriptions of commercial practice. Where appropriate, references are given to related topics of food microbiology, nutrition, fundamentals of food engineering, mathematical modelling of food processing operations, biochemical and physical properties of foods, food analyses, and business operations, including quality assurance, marketing and production management.

The book is divided into five parts in which unit operations are grouped according to the nature of the heat transfer that takes place:

- Part I describes some important basic concepts;
- Part II describes unit operations that take place at ambient temperature or involve minimum heating of foods;
- Part III includes operations that heat foods to preserve them or to alter their eating quality;
- Part IV describes operations that remove heat from foods to extend their shelf-life with minimal changes in nutritional quality or sensory characteristics; and
- Part V describes post-processing operations, including packaging and distribution logistics.

In each chapter, the theoretical basis of the unit operation is first described. Formulae required for calculation of processing parameters and sample problems are given where appropriate, and sources of more detailed information are indicated. Details of the equipment used for commercial food production, and developments in technology that relate to cost savings, environmental improvement or improvement in product quality are described. Finally the effects of each unit operation on sensory characteristics and nutritional properties of selected foods, and the effects on micro-organisms are described. The book describes each topic in a way that is accessible without an advanced mathematical background, while providing references to more detailed or more advanced texts. The book is therefore suitable for students studying nutrition, hospitality management/ catering or agriculture, as an additional perspective on their subject areas.

This third edition has been substantially rewritten, updated and extended to include the many developments in food technology that have taken place since the second edition was published in 2000. Nearly all unit operations have undergone significant developments, and these are reflected in the large amount of additional material in each chapter. In particular, advances in microprocessor control of equipment, 'minimal' processing technologies, genetic modification of foods, functional foods, developments in 'active' or 'intelligent' packaging, and storage and distribution logistics are described. Additionally, sections on the composition of foods and food-borne micro-organisms are included for the first time.

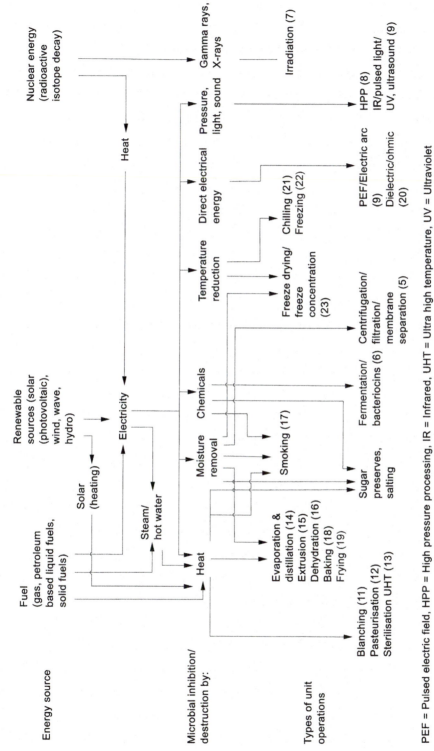

Fig. I.1 Types of processing and their preservative effects (numbers in parentheses indicate chapter numbers).

PEF = Pulsed electric field, HPP = High pressure processing, IR = Infrared, UHT = Ultra high temperature, UV = Ultraviolet

References

ANON, (2001), 25 years of FEI, *Food Engineering and Ingredients*, May, pp. 12–13, 16, 18, 20.

ANON, (2005), Power hungry: six reasons to regulate global food corporations, Action Aid, 19 May, available at www.actionaid.org/docs/power_hungry.pdf.

HILARY J., (1999), Globalisation and employment, Panos Briefing Paper No 33, May, Panos Institute, London.

KURLANSKY, M., (1997), *Cod: A Biography of the Fish That Changed the World*, Penguin, New York.

KURLANSKY, M., (2002), *Salt: A World History*, Penguin, New York.

SAUL, J.R., (2005), *The Collapse of Globalism*, Atlantic Press, London.

TRAGER, J., (1995), *The Food Chronology*, Aurum Press, London.

Part I

Basic principles

1

Properties of food and processing

Abstract: Knowledge of the biochemical composition of foods is important to understand their physical, sensory and nutritional properties, as well as the changes that take place during processing. This chapter describes the important macromolecules in foods and their derivatives, the importance of water in foods, and micro-components including minerals, vitamins, pigments, antioxidants and preservative chemicals. There follows a description of the physical properties of foods, including surface activity, water activity, acids/bases and redox potential, and the principles of heat transfer, moisture management and fluid flow. The chapter summarises important groups of micro-organisms and their effects on food quality and safety, and concludes with a description of how processing controls biochemical and physical changes and micro-organisms to produce the required sensory and nutritional qualities, safety and shelf-life of foods.

Key words: biochemical properties, micro-organisms, food spoilage and safety, heat transfer, moisture management, fluid flow, shelf-life.

Knowledge of the structure and composition of foods, their chemical, sensory and nutritional properties, as well as the types of micro-organisms that are likely to be present in foods, is a necessary prerequisite to understanding how unit operations are used to preserve foods or alter their eating qualities. In this part, the properties of foods are first described, followed by a description of changes to foods that determine their shelf-life and safety, and then an outline of the mechanisms that are used to process and preserve foods. Finally, a summary is given of the effects of processing on the properties of foods and how these are controlled. These aspects are expanded in subsequent chapters that describe individual unit operations.

1.1 Properties of foods

1.1.1 Composition

Foods are composed of chemicals, and for food manufacturers it is the chemical composition that determines all aspects of their products, from the suitability of raw materials for use in particular products and processes, to the sensory characteristics and nutritional value of the processed foods, as well as wider issues such as food safety and quality assurance, traceability, product development and labelling. In this section, the chemical components of foods are divided into first the macromolecular components (carbohydrates, lipids, proteins), then water, which is a major component of many foods, and finally the micro-components (vitamins, minerals, natural colourants, flavours, toxins and additives). Further information is available in a number of food chemistry textbooks, including Damodaran *et al.* (2007), Owusu-Apenten (2004) and Belitz *et al.* (2004). Details of the composition of individual foods are given in a number of publications (e.g. Roe *et al.* 2002) or in on-line databases (e.g. Anon 2008a; Braithwaite *et al.* 2006). The issue of food authentication has become increasingly significant over recent years, but it is beyond the scope of this book and details are given by for example Ashurst and Dennis (1996).

Macromolecules

Carbohydrates

'Carbohydrate' is the generic term for a wide variety of chemicals that form the major part of the dry matter in plants. The simplest forms are 'monosaccharides' (or 'simple sugars') that have between three and nine carbon atoms, and cannot be further broken down by hydrolysis. Some simple sugars have a sweet taste and they are absorbed and metabolised quickly in the body to release energy. Monosaccharides contain both hydroxyl groups and carbonyl groups. They are classified into 'aldoses' if the carbonyl group is an aldehyde (e.g. pentose (5 carbon atoms), or hexose (6 carbon atoms)) or 'ketoses' if it is a ketone (e.g. the corresponding 5- and 6-carbon molecules are pentulose, hexulose). When the position of the aldehyde or ketone group on the molecule allows it to react with oxidants (i.e. act as a reducing agent as for example in Maillard reactions (section 1.2.2)), the sugar is known as a 'reducing sugar'. Examples include glucose, fructose and arabinose, and the disaccharides lactose and maltose (below), but not sucrose, which is a non-reducing sugar.

Monosaccharides also contain 'chiral' carbon atoms (atoms that can exist in two different spatial configurations that are the mirror image of each other). Glucose is the most abundant aldose and the two forms are D- and L-glucose, with the D-form occurring naturally in foods. Other common naturally occurring monosaccharides are D-mannose, D-galactose and D-xylose. The L-form is less common in nature, but two examples are L-arabinose and L-galactose, which are units in polymeric carbohydrates (below). The only ketose found naturally in foods is D-fructose (fruit sugar), which forms ≈55% of high-fructose corn syrup and ≈40% of honey.

Reaction between the carbonyl and hydroxyl groups in the molecule forms monosaccharides into a ring structure. Common types have five-membered (furanose) rings or six-membered (pyranose) rings (Fig. 1.1). Further details of the structure and orientation of monosaccharide molecules are given by BeMiller and Whistler (1996) and the important reactions of monosaccharides are described by Davis and Fairbanks (2002).

Oligosaccharides contain 2–20 monosaccharides joined by glycosidic bonds. Disaccharides have two monosaccharides linked together by a glycosidic bond and the following are the most important in foods:

- Sucrose (table sugar) formed from glucose and fructose. The main commercial sources are sugar cane and sugar beet. It is used in a wide variety of forms, including brown sugar that contains molasses, white (or crystalline) sugar, and icing or fondant sugar, in which the crystals are ground to a smaller size. Sucrose is highly soluble, and concentrated sugar solutions are used in processing and also sold as syrups (e.g. maple

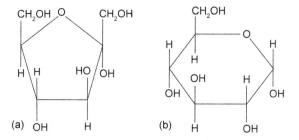

Fig. 1.1   Structure of (a) furanose ring and (b) pyranose ring.

syrup), which have high osmolality. As a result they are used as humectants and preservatives as well as sweeteners. In freezing (Chapter 22, section 22.1.2), sugars are concentrated in cellular liquids as water freezes, and form a glassy state (section 1.4.1) that acts as a cryprotectant and reduces the damage to cellular tissues from freezing, to protect the texture of the food.

- Lactose (milk sugar) is formed from glucose and galactose. It is fermented to lactic acid by lactic acid bacteria in yoghurt and other fermented milk products (Chapter 6, section 6.4.3). The production of the digestive enzyme lactase is genetically controlled, and deficiencies in lactase production increase with age (after about 6 years old) and with ethnicity, leading to a syndrome known as 'lactose intolerance'. Here lactose passes to the large intestine where it undergoes anaerobic bacterial fermentation that results in bowel irritation and gas production leading to diarrhoea and bloating.
- Maltose consists of two glucose molecules and is formed by hydrolysis of starch (below). Commercially it is produced by malting grains, especially barley, using β-amylase that is either naturally occurring or produced by *Bacillus* spp.

Few other oligosaccharides occur naturally, but they are formed by hydrolysis of polysaccharides. An example of a trisaccharide is raffinose made up of galactose and sucrose molecules, and a tetrasaccharide is stachyose formed from sucrose with two galactose units. These oligosaccharides are non-digestible and found in many types of beans. Like lactose, they pass to the large intestine and are anaerobically fermented by bacteria, causing flatulence when eaten.

Polysaccharides contain more than 20 monosaccharide units joined by different types of glycosidic bonds. They are the most abundant form of carbohydrate, the most important in plants being starch, cellulose, pectin and a range of gums (e.g. guar, locust bean, xanthan and carrageenan). In animal tissues, glycogen is a branched glucose polymer that has glucose molecules linked by $\alpha(1-4)$ glycosidic bonds, which is stored in the liver and muscle tissue as an instant source of energy. The molecular structure of polysaccharides enables the individual units to form hydrogen bonds with water, and as a result polysaccharide particles readily hydrate, swell and dissolve either partially or completely. They are therefore used to control water in food systems and viscosity (below) and to influence the physical and functional properties of foods (section 1.1.2). They are also cryostabilisers (i.e. they do not depress the freezing point or increase osmolality as do cryoprotectants), which produce a matrix that increases the viscosity of solutions and restricts the mobility of water. They may also restrict ice crystal growth by absorption to nuclei (see Chapter 22, section 22.1.1).

Starch is the main carbohydrate found in plant seeds and tubers. It exists in the form of granules, each of which consists of several million molecules. It occurs in two forms: (1) α-

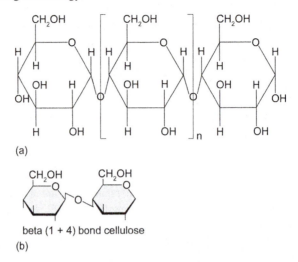

(a)

beta (1 + 4) bond cellulose

(b)

Fig. 1.2   Structure of polysaccharides: (a) $\alpha(1\text{–}4)$ bonds in starch and glycogen; (b) $\beta(1\text{–}4)$ bonds in cellulose. (n is the number of repeating glucose units.)

amylose, in which 500–20 000 D-glucose molecules are linked by $\alpha(1\text{–}4)$ glycosidic bonds (Fig. 1.2a) in linear, helical chains; and (2) amylopectin, in which highly branched chains of up to 30 glucose molecules are linked through $\alpha(1\text{–}4)$ bonds and connected to each other through $\alpha(1\text{–}6)$ branch points. Amylopectin is therefore a much larger molecule than amylose, typically containing one to two million residues, with a mass that is four or five times that of amylose. Starch granules produce low-viscosity pumpable slurries in cold water, and thicken due to gelatinisation when heated to $\approx 80\,^{\circ}\text{C}$. Details of starch gelatinisation are given by Palav and Seetharaman (2006), Xie *et al.* (2006) and Maaruf *et al.* (2001).

Different types of starches have different uses, for example: waxy maize starch produces clear, cohesive pastes; potato starch is used in extruded cereal and snackfood products (Chapter 15, section 15.3) and in dry soup or cake mixes; and rice starch produces opaque gels for baby foods.

Starches can be modified to improve their functional properties (e.g. improved gelatinisation, increased solubility, increased or decreased viscosity, freeze/thaw stability, enhanced clarity and sheen, improved gel strength and reduced syneresis, and changes to paste cohesiveness). Modification also enables starches to withstand conditions of high shear, high temperatures and/or acidic conditions in particular processing applications. Types of modification include crosslinking polymer chains, depolymerisation and pregelatinisation to produce 'instant' or cold water soluble starches. Starch can also be hydrolysed into simpler carbohydrates by acids or enzymes, and the extent of conversion is quantified by the 'dextrose equivalent' (DE), which relates to the fraction of the glycosidic bonds that are broken. Starch is used to produce a very large number of derivative ingredients including modified starches, maltodextrin (DE 10–20), used as a filler and thickener, corn syrups (DE 30–70), which are viscous solutions used as sweeteners and thickeners, dextrose (DE 100), glucose, and high-fructose corn syrup sweetener, made by treating dextrose with glucose isomerase. Details are given by BeMiller and Whistler (1996). Starch is broken down to oligosaccharides and glucose by amylases in animal digestive systems. First amylose and amylopectin are hydrolysed into small fragments by α-amylase, which only breaks internal $\alpha(1\text{–}4)$ glycosidic bonds to produce three different oligosaccharides: maltose, maltotriose (a trisaccharide) and a group of dextrins that contain the branch points from amylopectin. Secondly, the oligosaccharides are hydrolysed by maltase to produce glucose. Microbial enzymes are used commercially to produce a wide range of starch degradation products that are used as ingredients and as substrates for fermentations (see also Chapter 6, section 6.5).

Cellulose is composed of unbranched linear chains of 1000–10 000 D-glucose molecules, linked together by β(1–4) glycosidic bonds (Fig. 1.2b). The linear structure promotes hydrogen bonding that holds together nearby cellulose molecules to form a three-dimensional structure of microfibres, which in turn interact to form cellulose fibres. A typical fibre contains about 500 000 cellulose molecules, and the large number of hydrogen bonds creates a crystalline structure that has a high tensile strength. As a result, cellulose is a stiff material that is used by plants as a structural molecule to support leaves and stems. In contrast to other more amorphous polymeric carbohydrates, the crystalline structure also makes cellulose insoluble and resistant to enzymic breakdown, a property that also makes cellulose suitable as a packaging film (Chapter 25, section 25.2.4). Additional details of the properties of cellulose are given by Chaplin (2008a).

Hemicelluloses have amorphous branched structures composed of a variety of sugars, including xylose, arabinose and mannose, which can become highly hydrated to form gels. Cellulose or hemicelluloses cannot be enzymatically digested by humans or animals and they form dietary fibre that passes through the intestine essentially unchanged (ruminant herbivores digest cellulose using cellulase- and hemicellulase-producing bacteria in their fore-stomachs or large intestines). Further information is given by Carpi (2008), Cui (2008) and Bowen (2006).

Cellulose is modified to impart specific functional properties: for example microcrystalline cellulose (MCC) is made by hydrolysis of cellulose and separation of the constituent microcrystals. The powdered form is used as a flavour carrier and anti-caking agent and the colloidal form has properties that are similar to gums (Appendix A.1). Carboxymethylcellulose (CMC) is produced by the reaction of cellulose with chloroacetic acid to substitute polar carboxyl groups for hydroxyl groups. This makes the cellulose soluble and more chemically reactive. The functional properties of CMC depend on how many hydroxyl groups are involved in the substitution reaction and the chain length of the cellulose backbone. It produces a wide range of viscous stable solutions and is used to stabilise protein solutions such as egg albumin before drying or freezing, and to prevent casein precipitation in milk products. Methylcelluloses and hydroxypropylmethylcelluloses (HPMC) are cold water soluble, produce heat-reversible gels and have surface active properties that can be used to stabilise emulsions and foams (section 1.1.2). They are also used to reduce the amount of fat in products by both providing fat-like properties and reducing fat absorption in fried foods (Chapter 19, section 19.4.1). Their gel structure provides a barrier to oil and moisture and acts as a binding agent.

Polysaccharide gums (hydrocolloids) are used at concentrations of 0.25–1% to thicken aqueous solutions, form gels, or to modify and control the properties of liquid foods or the texture of semi-solid foods (Appendix A.1) (also section 1.1.2). The extent to which these linear or branched polysaccharides increase viscosity depends on their molecular weight, shape and flexibility of the hydrated molecules. Also the presence of electrically charged groups causes mutual repulsion of molecular chains that extends them to create higher viscosities (e.g. alginates, carrageenans and xanthan gums). When they are used to form gels, the polysaccharide molecules come out of solution to form a three-dimensional network that is joined together by hydrogen bonding, hydrophobic van der Waals associations, ionic cross-linking or covalent bonding. Examples of gels include aspic, restructured fruit pieces, jams and other preserves. The selection of a hydrocolloid for a particular application is complicated; each gum has one or more particular properties that may be used as a starting point but selection depends on many factors, including for example the required strength or rheology of the gel (section 1.1.2), the pH, ionic strength, temperature and other ingredients in the food. Details of hydrocolloid use are given by Hoefler (2004) and Phillips and Williams (2000).

Guar and locust bean (or carob) gums are thickening agents, whose main component is a galactomannan consisting of a chain of β-D-mannopyranosyl units joined by (1–4) bonds with α-D-galactopyranosyl branches. Locust bean gum also reacts with xanthan and carrageenan to form rigid gels, and guar gum is frequently used in combination with other gums. Xanthan gum is composed of chains that are identical to cellulose, but with trisaccharide side-chains of mannopyranosyl and glucuronopyranosyl units. It has unusual properties: being soluble in both hot and cold water; producing a high viscosity at low concentrations with no change in viscosity from 0 to 100 °C; having stability in acidic foods and exposure to freezing/thawing; being compatible with salt; and having the ability to stabilise suspensions and emulsions.

Carrageenans are a group of over 100 types of sulphated galactans that are derived from seaweed. The basic structure has chains of D-galactopyranosyl units that have alternating (1–3)-α-D- and (1–4)-β-D-glycosidic linkages with sulphate groups esterified to the hydroxyl groups. The three basic types are named 'kappa', 'iota' and 'lambda', which form gels with potassium or calcium ions. Kappa-type gels are strong and brittle, whereas iota-type gels are softer and more resilient, and have good freeze/thaw stability. Lambda-type carrageenans are soluble and non-gelling. Carrageenans are used to form weak thixotropic pourable gels, especially in milk products such as chocolate milk, where they prevent the cocoa particles from settling out. In other applications they form gels that are clear/turbid, rigid/elastic or thermally reversible/heat stable. They are also used to stabilise freeze/thaw cream and air bubbles in ice cream (see overrun, section 1.1.2 and Chapter 3, section 3.2.2) as well as improving the water holding capacity of meat products and meat analogues. Details of carrageenans are given by Chaplin (2008b).

Alginates are the sodium salt of alginic acid, a polyuronic acid composed of units of β-D-mannopyranosyluronic acid and α-DL-gulopyranosyluronic acid. They are used both as thickeners and to form gels with calcium ions (e.g. in reformed fruit pieces that are stable during pasteurisation, or dessert gels that do not require refrigeration). Propylene glycol alginates are less sensitive to acidity and calcium ions, and are used to thicken or stabilise dairy products and salad dressings.

Pectins are a group of poly α-D-galactopyranosyluronic acids that have differing amounts of methyl ester groups along the chains. High methoxyl (HM) pectins have more than half the carboxyl groups as methyl esters and form gels in the presence of high sugar concentrations and acid (e.g. jams, jellies, marmalades). Low methoxyl (LM) pectins have less than half the carboxyl groups as methyl esters and form gels with calcium ions, thus requiring less sugar as, for example, in diabetic preserves. Treatment of pectins with ammonia and methanol produces amidated LM pectins. There are a large variety of different types of pectins for different applications and further information is available from suppliers (e.g. Anon 2005).

Gum arabic (or gum acacia) is one of a group of tree exudates that is used as an emulsifying agent and emulsion stabiliser in foods. Others, including gum karaya, gum ghatti and gum tragacanth, have diminishing use because of increased costs and restricted availability. Gum arabic has two fractions: highly branched arabinogalactan chains with side chains of galactopyranosyl units, and another fraction that has proteins as an integral part of the structure. It is highly soluble and produces a low-viscosity solution that does not affect the product viscosity. It is used in the bakery, soft drink and confectionery industries to stabilise flavour emulsions and essential oils. In sugar confectionery, it also prevents sucrose crystallisation and emulsifies fat ingredients, prevents surface 'bloom' and is used as a glaze for pan-coated confectionery (Chapter 24, section 24.2.3). Flavourings stabilised with gum arabic can also be spray-dried to produce non-hygroscopic powders, in which the flavour is protected against oxidation and volatilisation (see also Chapter 16, section 16.2.1). They are used in dry cake, beverage and soup mixes.

Lipids
The distinction between 'fats' and 'oils' is based solely on whether a lipid is solid or liquid at room temperature and the terms are used interchangeably. Important lipids in food processing include for example vegetable oils from coconut, cottonseed, olive, palm/palm kernel, peanut (groundnut) and soybean. Animal sources of lipids include lard and the fats contained in meats and dairy products (butter, cheese, cream). The properties of lipids, including their composition, crystalline structure, melting points and association with non-lipid ingredients, are important influences on the functional properties of foods. Changes to lipids during processing can have either beneficial or adverse effects on both the sensory properties and nutritional value of foods (section 1.4). Dietary lipids are also the subject of ongoing research in relation to obesity, toxicity and diseases such as cardiovascular disease and diabetes (see also Chapter 19, section 19.4.3). The nutritional aspects of lipids and health are described by O'Neil and Nicklas (2007), Gunstone (2006), Akoh and Min (2002) and Gurr (1999).

Fats are composed of mono-, di- and tri-esters of glycerol with fatty acids (monocarboxylic acids). The older terms 'mono-', 'di-' and 'triglycerides' have been replaced by 'monoacylglycerols' 'diacylglycerols' and 'triacylglycerols' respectively. Glycerol is a trihydric alcohol containing three hydroxyl (–OH) groups that can combine with up to three fatty acids to form a wide variety of monoacylglycerols, diacylglycerols or triacylglycerols (Fig. 1.3). A monoacylglycerol has one fatty acid unit per molecule of glycerol, which may be attached to carbon atom 1 or 2 on the glycerol molecule; a diacylglycerol has two fatty acids as either the 1,2 form or the 1,3 form depending on how they are attached to the glycerol molecule. Triacylglycerols are the main constituents of vegetable oils and animal fats, and their structure involves a chain of carbon atoms with a carboxyl group (–COOH) at one end. The triacylglyceride structural formula (Fig. 1.3d showing olive oil) consists of two radicals of oleic acid and one of palmitic acid attached to the glycerol molecule (the vertical chain of carbon atoms).

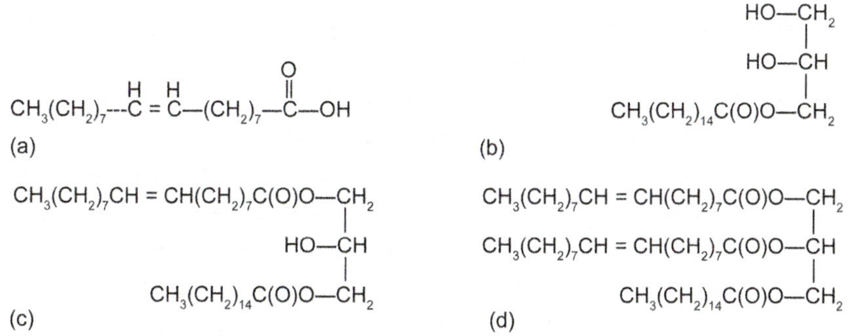

Fig. 1.3   Components of lipids: (a) oleic acid, (b) 1-monoacylglyceride, (c) 1,3-diacyglyeceride, (d) triacylglyceride.

Saturated fatty acids (SFAs) have no double bonds between the carbon atoms, monounsaturated fatty acids (MUFAs) have only one double bond, and polyunsaturated fatty acids (PUFAs) have more than one double bond. Nawar (1996) has reviewed different forms of nomenclature of lipids and in this book the common names for fats and fatty acids are normally used (Table 1.1). The numbers at the beginning of the scientific names indicate the locations of the double bonds. By convention, the carbon atom of the carboxyl group is number one and Greek numbers such as 'di, tri, tetra, penta, and hexa' are used to describe the length of carbon chains. For example, linoleic acid is 9,12-octadecadienoic acid, which indicates that there is an 18-carbon atom chain (octa deca) with two double bonds (di en) located at carbon atoms 9 and 12, with carbon atom 1 being a carboxyl group (oic acid). The structural formula corresponds to:

$$CH_3CH_2CH_2CH_2CH_2CH=CHCH_2CH=CHCH_2CH_2CH_2CH_2CH_2CH_2CH_2COOH$$

which can be abbreviated as:

$$CH_3(CH_2)_4CH=CHCH_2CH=CH(CH_2)_7COOH$$

Fatty acids can also be written as for example C18:2, which indicates that the fatty acid consists of an 18-carbon atom chain with two double bonds. Although this could describe

Table 1.1   Chemical names and descriptions of some common fatty acids

Common name	Carbon atoms	Double bonds	Scientific name	Sources
Butyric acid	4	0	Butanoic acid	Butterfat
Caproic acid	6	0	Hexanoic acid	Butterfat
Caprylic acid	8	0	Octanoic acid	Coconut oil
Capric acid	10	0	Decanoic acid	Coconut oil
Lauric acid	12	0	Dodecanoic acid	Coconut oil
Myristic acid	14	0	Tetradecanoic acid	Palm kernel oil
Palmitic acid	16	0	Hexadecanoic acid	Palm oil
Palmitoleic acid	16	1	9-Hexadecenoic acid	Animal fats
Stearic acid	18	0	Octadecanoic acid	Animal fats
Oleic acid	18	1	9-Octadecenoic acid	Olive oil
Ricinoleic acid	18	1	12-Hydroxy-9-octadecenoic acid	Castor oil
Vaccenic acid	18	1	11-Octadecenoic acid	Butterfat
Linoleic acid	18	2	9,12-Octadecadienoic acid	Grape seed oil
α-Linolenic acid (ALA)	18	3	9,12,15-Octadecatrienoic acid	Flaxseed (linseed) oil
γ-Linolenic acid (GLA)	18	3	6,9,12-Octadecatrienoic acid	Borage oil
Arachidic acid	20	0	Eicosanoic acid	Peanut oil, fish oil
Gadoleic acid	20	1	9-Eicosenoic acid	Fish oil
Arachidonic acid (AA)	20	4	5,8,11,14-Eicosatetraenoic acid	Liver fats
EPA	20	5	5,8,11,14,17-Eicosapentaenoic acid	Fish oil
Behenic acid	22	0	Docosanoic acid	Rapeseed (canola) oil
Erucic acid	22	1	13-Docosenoic acid	Rapeseed oil
DHA	22	6	4,7,10,13,16,19-Docosahexaenoic acid	Fish oil
Lignoceric acid	24	0	Tetracosanoic acid	Small amounts in most fats

From Zamora (2005)

a number of possible fatty acids that have this chemical composition, in common use it shows the naturally occurring fatty acid (i.e. linoleic acid).

Double bonds are 'conjugated' when they are separated from each other by one single bond (e.g. (–CH=CH–CH=CH–)) so 9,11-conjugated linoleic acid has the formula:

$$CH_3(CH_2)_5CH=CH-CH=CH(CH_2)_7COOH$$

Double bonds enable 'configurational isomers' to exist, which are described by Latin prefixes 'cis' (on the same side) and 'trans' (on the other side) and indicate the orientation of hydrogen atoms to the double bond. Naturally occurring fatty acids mostly have the cis-configuration.

Fats crystallise below their melting point in a two stage process: nucleation followed by crystal growth. Details of this process are described in Chapter 22 (section 22.1.1) in relation to water, but fats also crystallise in a similar way. The size and shape of crystals depend on the temperature but are also influenced by minor components in the oil (e.g. free fatty acids) and by stirring. Lipids can also be polymorphic; adopting different crystal morphologies and changing from one form to another without melting. These transformations take place to produce the more stable forms of the crystals. The factors that determine the type of polymorphic form include temperature and rate of cooling, the purity of the oil and the presence of crystalline nucleii. Details are given in Chapter 24 (section 24.1.1) of polymorphism in cocoa butter used in chocolate manufacture.

Omega-3 (ω3) and omega-6 (ω6) fatty acids are unsaturated 'essential fatty acids' (EFAs) that cannot be created from other fatty acids in human metabolism. They are polyunsaturated and described by the terms 'n–3 PUFAs' and 'n–6 PUFAs' respectively. Greek letters (α to ω) are used to identify the location of the double bonds, with the α-carbon atom being closest to the carboxyl group (carbon atom 2), and the ω-carbon atom being the last in the chain. Using the example of linoleic acid, this is an ω6 fatty acid because it has a double bond that is six carbon atoms away from the ω carbon atom. Similarly, α-linolenic acid is an ω3 fatty acid because it has a double bond three carbon atoms away from the ω carbon atom (Fig. 1.4). The classification can be found by subtracting the highest double-bond location in the scientific name from the number of carbon atoms in the fatty acid chain.

The composition of fatty acids in a lipid depends on the type of animal or crop, and whether a crop has been selectively bred to achieve a particular ratio of fatty acids. For example, rapeseed (canola) is selectively bred to contain less than 2% erucic acid and other crops are bred to contain high levels of monounsaturated or polyunsaturated fatty acids (e.g. oleic types of safflower oil have about 78% monounsaturated, 15% polyunsaturated and 7% saturated fatty acids) (Table 1.2) (see also Chapter 19, section 19.3).

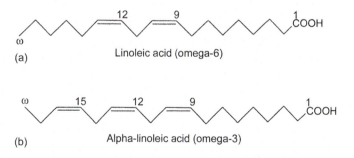

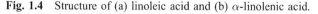

Fig. 1.4   Structure of (a) linoleic acid and (b) α-linolenic acid.

Table 1.2 Fatty acid composition of some common edible fats and oils

Oil or fat	Ratio: unsaturated/ saturated	Saturated					Monounsaturated	Polyunsaturated	
		Capric acid C10:0	Lauric acid C12:0	Myristic acid C14:0	Palmitic acid C16:0	Stearic acid C18:0	Oleic acid C18:1	Linoleic acid (ω6) C18:2	α-Linolenic acid (ω3) C18:3
Almond	9.7	–	–	–	7	2	69	17	–
Beef tallow	0.9	–	–	3	24	19	43	3	1
Butterfat (cow)	0.5	3	3	11	27	12	29	2	1
Cocoa butter	0.6	–	–	–	25	38	32	3	–
Coconut	0.1	6	47	18	9	3	6	2	–
Cod liver	2.9	–	–	8	17	–	22	5	–
Cottonseed	2.8	–	–	1	22	3	19	54	1
Flaxseed	9.0	–	–	–	3	7	21	16	53
Grapeseed	7.3	–	–	–	8	4	15	73	–
Lard (pork fat)	1.2	–	–	2	26	14	44	10	–
Maize (corn)	6.7	–	–	–	11	2	28	58	1
Olive	4.6	–	–	–	13	3	71	10	1
Palm	1.0	–	–	1	45	4	40	10	–
Palm kernel	0.2	4	48	16	8	3	15	2	–
Palm olein	1.3	–	–	1	37	4	46	11	–
Peanut (groundnut)	4.0	–	–	–	11	2	48	32	–
Rapeseed	15.7	–	–	–	4	2	62	22	10
Safflower[a]	10.1	–	–	–	7	2	13	78	–
Sesame	6.6	–	–	–	9	4	41	45	–
Soybean	5.7	–	–	–	11	4	24	54	7
Sunflower[a]	7.3	–	–	–	7	5	19	68	1
Walnut	5.3	–	–	–	11	5	28	51	5

Percent by weight of total fatty acids

Adapted from Zamora (2005)

[a] Not high-oleic variety.

Palm olein is the liquid oil obtained by fractionation of palm oil after crystallisation at controlled temperatures.

Note: there is considerable variation in composition between varieties of crops or animals. Average values are used. Percentages may not equal 100 owing to rounding and other constituents not listed.

Phospholipids

Phospholipids, including lecithin and cephalin, are natural emulsifiers that consist of an alcohol, such as glycerol, fatty acids and a phosphoric acid compound. Lecithin has a mixture of diacylglycerides of stearic, palmitic and oleic acids that are linked to the choline ester of phosphoric acid. Phospholipids are present in cell walls and other membranes where the fatty acid tails are oriented toward each other and the phosphate groups form the outer surfaces of the membrane. These semipermeable bilipid layers regulate the type of molecules that can pass through the membrane. Lecithin is widely used as an emulsifier in products such as chocolate (Chapter 24) and bakery products (Chapter 18).

Sterols (e.g. cholesterol) have a similar structure to steroid hormones such as testosterone, progesterone and cortisol. Cholesterol is produced by the liver and has a role in the organisation and permeability of cell membranes. Cholesterol derivatives in the skin are also converted to vitamin D when the skin is exposed to ultraviolet light. However, a high level of cholesterol in the blood is considered to be a risk factor for cardiovascular disease (see below), and it is recommended that levels are reduced by exercise, reduced calorie diets without hydrogenated fats and increased proportions of polyunsaturated fatty acids.

Phytosterols in vegetables have a similar structure to cholesterol, but have different side chains. They are of interest because they lower blood cholesterol levels. Fully saturated sterols are named 'stanols' and are being developed as a component of functional foods (Chapter 6, section 6.3).

Hydrogenation and fat substitutes

Unsaturated fats are more susceptible to rancidity and hydrogenation is used to decrease the number of double bonds in unsaturated fats and thus reduce the potential for rancidity. As a result, liquid vegetable oil becomes a solid fat, known as 'shortening', which is used in many processes, especially baking (Chapter 18). The oils are heated with pressurised hydrogen gas in the presence of metal catalysts and the hydrogen is incorporated into the fatty acid molecules and they become more saturated. Oleic acid (C18:1) and linoleic acid (C18:2) are both converted to stearic acid (C18:0) when fully saturated. However, fully saturated fats are too waxy and solid, and manufacturers therefore partially hydrogenate the oils until they have the required consistency for the particular application. As a side effect, the process causes a large percentage of naturally occurring *cis* double bonds to be converted to *trans* double bonds. *Trans* fatty acids have been implicated in health concerns: specifically, polyunsaturated *trans* fatty acids have shapes that are not recognised by digestive enzymes, and when they are incorporated into cell membranes, they create denser membranes that alter the normal functions of the cell. These fats also raise the level of low-density lipoproteins (LDL or 'bad' cholesterol), reduce levels of high-density lipoproteins (HDL or 'good' cholesterol), and raise levels of triacylglycerols in the blood, which increases the risk of coronary heart disease (also Chapter 19, section 19.4.3). Further information on the structure, properties, preparation and health effects of edible oils is given in a number of publications including Strayer (2006), O'Brien (2004) and Gunstone *et al.* (1995).

Fat substitutes are created using sucrose instead of glycerol, with up to eight fatty acids. The molecule is too large to be metabolised and passes through the body unchanged. Polyglycerol (or glyceran) fatty acid esters have the general structure R–$(OCH_2–CH(OR)–CH_2O)_n$–R (where R represents fatty acids and the average value of $n = 3$). They are metabolised like fats but polymerised glycerol is not digested and is excreted in urine.

Proteins

Proteins have a large number of functions in biological systems. They form:

- structural components of cells (e.g. collagen, elastin);
- thousands of enzymes that control and regulate metabolic activity;
- muscle tissue (e.g. myosin, actin);
- hormones (e.g. insulin);
- transfer proteins (e.g. serum albumin, haemoglobin);
- antibodies (e.g. immunoglobulins);
- storage proteins (e.g. albumen, seed proteins); and
- protective proteins (e.g. allergens, toxins) (Damodaran 1996).

Proteins are highly complex polymers that are made up from 20 amino acids. It is the sequence of amino acids and the way in which they are linked using substituted amide bonds, together with different three-dimensional structural forms, that produce the very large number of different proteins and enzymes.

Amino acids are the units from which proteins are formed. An amino acid consists of a central (α) carbon atom to which are bound an amino (NH_2) group, a hydrogen atom, a carboxyl (COOH) group and a side chain (R). Each amino acid has a similar yet unique structure. The different side chains vary in their complexity and properties and produce amino acids that have different physicochemical properties. Amino acids can be classified by the chemical nature of the side chains into two groups: polar (or hydrophilic) side chains that interact with water; and non-polar (or hydrophobic) types that do not react with water. Polar amino acids are: arginine, asparagine, aspartic acid (or aspartate), glutamine, glutamic acid (or glutamate), histidine, lysine, serine and threonine. Non-polar amino acids are: alanine, cysteine, glycine, isoleucine, leucine, methionine, phenylalanine, proline, tryptophan, tyrosine and valine. Amino acids can also be grouped into those that have ionisable side chains (e.g. arginine, aspartate, cysteine, glutamate, histidine, lysine and tyrosine). These amino acids contribute to the charge exhibited by peptides and proteins. Both the amino group and the carboxyl group of each amino acid are ionisable and the acid dissociation constants (pK_a values – see section 1.1.3) of amino acid side chains are important for the activity of enzymes and the stability of proteins. This is because ionisation alters their physical properties such as solubility and lipophilicity. At physiological pH amino acids exist as 'zwitterions' that have a negatively charged carboxyl group and a positively charged amino group. Examples of pK_a values of amino acid side chains are given in Table 1.3.

The α-carbon of amino acids is chiral, which produces two optically active forms, designated D- and L-forms. The L-forms are most common, although some peptides contain both D- and L-amino acids. Proteins are also chiral and consist of only L-amino

Table 1.3   Approximate acid dissociation constants of amino acid side chains

Amino acid	pK_a	Functional group
Arginine	12.0	Guanidinium
Aspartate and glutamate	4.4	Carboxyl
Cysteine	8.5	Sulphydryl
Histidine	6.5	Imidazole
Lysine	10.0	Amine
Tyrosine	10.0	Phenol

Adapted from Gorga (2007)

acids, which is important in understanding their function (e.g. some enzymes bind one stereoisomer of a compound with a thousand times greater affinity than the other).

Peptides, polypeptides and proteins

A covalent peptide bond is formed by dehydration between the carbon atom in the carboxyl group of one amino acid to the nitrogen atom of the amino group of another amino acid (Fig. 1.5). Peptides are formed by joining amino acids together via amide bonds. Amides are made by condensing together a carboxylic acid and an amine. Small peptides (containing fewer than 25 amino acids) are 'oligopeptides' and longer peptides are 'polypeptides'. Peptides have a polarity, with a free amino group on the amino-terminal residue and a free carboxyl group on the carboxyl-terminal residue (Fig. 1.6), both of which are ionisable groups.

There may also be ionisable groups in the side chains of some amino acids. The overall charge on a peptide (or protein) is the sum of the charges of each ionisable group, and depends on its amino acid content and the pH of the solution. When the pH of a solution equals the pK_a of an ionisable group, the group exists as an equal mixture of its acidic form and the conjugate base. The pH at which a zwitterion is neutral is known as the 'isoelectric point'. If the pH is less than the pK_a, the acid form predominates, whereas a pH greater than the pK_a enables the base to predominate. The further the pH is from the pK_a the more unbalanced are the acid and base groups.

A polypeptide chain forms a protein and has a series of amino acid residues linked by peptide bonds, with a 'backbone' made up by the repeated sequence of three atoms of

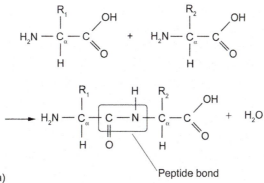

(a)

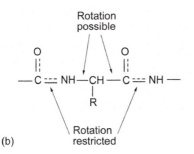

(b)

Fig. 1.5   (a) Formation of a peptide bond between two amino acid residues; (b) rotation of peptide bond.

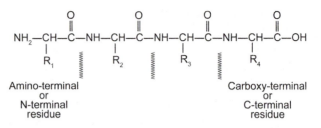

Fig. 1.6   A tetrapeptide showing a free amino group on the amino-terminal residue and free carboxyl group on the carboxyl-terminal residue of an amino acid.

each amino acid residue (the amide N, the α carbon and the carbonyl carbon). Because the bond between the carbonyl carbon and the nitrogen is a partial double bond, rotation around this bond is restricted and the peptide unit has a rigid structure. Rotation in the peptide backbone is restricted to bonds involving the α-carbon, and the chain can rotate to create three-dimensional structures in proteins. There are four levels of structure that can be described as follows:

1   The primary structure (the sequence of amino acid residues in the polypeptide chain, which is determined by the gene that encodes it).
2   The secondary structure (formed by hydrogen bonds between backbone atoms in a chain, producing two types of stable structures: α-helices and β-sheets).
3   The tertiary structure (the arrangement of α-helices, β-sheets and random coils along a polypeptide chain. The polypeptide folds so that the side chains of non-polar amino acids are within the structure and the side chains of polar residues are exposed on the outer surface. The tertiary structure of some proteins is stabilised by disulphide bonds between cysteine residues).
4   The quaternary structure (the spatial organisation of chains if there are more than one polypeptide chain in a complex protein).

Not all proteins show a quaternary structure and in many the polypeptides fold independently into a stable tertiary structure with the folded units associating with each other to form the final structure. In contrast, quaternary structures are stabilised by non-covalent interactions including hydrogen bonding, van der Waals interactions and ionic bonding. These intricate three-dimensional structures are unique to each protein and it is these that allow proteins to function. Most globular proteins consist of a core composed mainly of hydrophobic residues surrounded by a skin of mainly hydrophilic residues. Disulphide bonds are formed by the oxidation of thiol (–SH) groups in cysteine residues (Fig. 1.7) and they can occur within a single polypeptide chain where they stabilise the tertiary structures, or between two chains, where they stabilise quaternary structures.

Proteins can be grouped according to their structural organisation into 'globular' proteins that have spherical or ellipsoidal shapes due to folding of polypeptide chains (e.g. enzymes), and 'fibrous' proteins that have twisted linear polypeptide chains (e.g. structural proteins) or are formed by the linear aggregation of globular proteins (e.g. muscle tissues). Proteins that form complexes with non-protein materials (known as 'prosthetic groups') are known as 'conjugated' proteins and these may be grouped into the following:

• Nucleoproteins (e.g. ribosomes in which prosthetic group is a nucleic acid).
• Glycoproteins (e.g. ovalbumin, κ-casein in which prosthetic group is a sugar or chain of sugars).

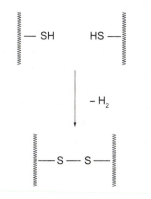

Fig. 1.7 Disulphide bond formation.

- Phosphoproteins (e.g. α- and β-caseins in which the prosthetic group is phosphate).
- Lipoproteins (e.g. egg yolk, blood plasma proteins in which the prosthetic group is a lipid).
- Metalloproteins (e.g. haemoglobin, myoglobin and some enzymes in which the prosthetic group is a metal ion, such as Fe^{2+}, Cu^{2+}, Zn^{2+}).

Some authors also describe flavoproteins (in which prosthetic group is a flavin, e.g. flavin adenine dinucleotide, FAD) and chromoproteins (where the prosthetic group is a pigment group). These complexes can become dissociated under some types of processing conditions and alter the activity of the protein.

The wide variety of protein configurations is due to the large number of different sequences of amino acid residues, which by convention is written with the amino terminus on the left and the carboxyl terminus on the right. Amino acid sequences can be written using either a three-letter code or a one-letter code (Table 1.4). For example the code for a small eight-residue peptide is written as: Asp-Ile-Glu-Phe-Arg-Val-Leu-His. Further information on the structure and function of proteins is given by Gorga (2007), Petsko and Ringe (2004) and Damodaran (1996).

The use of proteins in food processing is described by Yada (2004) and detailed descriptions of the interactions between proteins, lipids and other food components are given in Gaonkar and McPherson (2006).

Enzymes are proteins that range in size from molecular weights of 12 000 to 1 million, that catalyse thousands of highly specific biochemical reactions needed by living organisms. The specificity is due to a limited number of substrates that can bind

Table 1.4 Codes for amino acids

Amino acid	3 letter code	1 letter code	Amino acid	3 letter code	1 letter code
Alanine	Ala	A	Leucine	Leu	L
Arginine	Arg	R	Lysine	Lys	K
Asparagine	Asn	N	Methionine	Met	M
Aspartate	Asp	D	Phenylalanine	Phe	F
Cysteine	Cys	C	Proline	Pro	P
Histidine	His	H	Serine	Ser	S
Isoleucine	Ile	I	Threonine	Thr	T
Glutamine	Gln	Q	Tryptophan	Trp	W
Glutamate	Glu	E	Tyrosine	Tyr	Y
Glycine	Gly	G	Valine	Val	V

stereospecifically into the active site of the enzyme before catalysis can occur. Enzymes accelerate reactions by factors of 10^3 to 10^{11} times that of non-enzyme catalysed reactions (Whitaker 1996). There are six main types of enzymes in food processing:

1   Oxidoreductases (e.g. catalase) that are involved in oxidation/reduction reactions.
2   Transferases (e.g. glucokinase) that remove groups from substrates and transfer them to acceptor molecules (excluding hydrogen and water).
3   Hydrolases (e.g. lipases, proteases) that break covalent bonds and add water into the bonds.
4   Lyases (e.g. pectin lyase) that remove groups from the substrates (not by hydrolysis) to leave a double bond or add groups to double bonds.
5   Isomerases (e.g. glucose isomerase) cause isomerisation of a substrate (isomers have the same molecular formula but a different structural formula and properties).
6   Ligases (previously 'sythetases') that catalyse covalent linking of two molecules with the breaking of a pyrophosphate bond (Whitaker 1996).

The rate of enzyme activity is determined by the enzyme concentration, the substrate concentration and any substrate inhibitors or activators, the substrate pH, temperature and water activity. Further details are given in food biochemistry textbooks.

Water
More than 70% of the Earth's surface is covered with water, and it is estimated that the hydrosphere (including the atmosphere, oceans, lakes, rivers, glaciers, snowfields and groundwater) contains about 1.36 billion cubic kilometres of water, mostly in liquid form (Pidwirny 2006). Water is essential for life and is the major constituent of almost all life forms, with most animals and plants containing >60% water by volume. The chemical structure of water (two hydrogen atoms covalently bound to an oxygen atom) causes the molecules to have unique electrochemical properties. The two hydrogen atoms are slightly positively charged and the oxygen atom is negatively charged. This separation between negative and positive charges produces a polar molecule that has an electrical charge on its surface. The net interaction between the covalent bond and the attraction and repulsion between the positive and negative charges produces a 'V'-shaped molecule. The V-shape is important because it allows different configurations of water to be formed. For example, each water molecule can form bonds with four other water molecules to form a tetrahedral arrangement or the ordered lattice structure of ice in Fig. 1.8(a). Water molecules also form hydrogen bonds that allow it to exist as a liquid, and together with the molecular polarity, this causes water molecules to be attracted to each other, forming strong cohesive forces between the molecules that produce a high surface tension and make water adhesive and elastic and allow it to move and flow (Anon 2008b). Water has a maximum density at ≈4 °C. When water changes phase to ice the molecular arrangement leads to an increase in volume and it expands by about 9% (Table 1.5), producing a decrease in density. This makes it the only substance that does not have a maximum density when it is solidified. Changes in density in liquid water and water vapour also result in thermal convection (Chapter 10, section 10.1.2). Changes in phase are described in more detail in Chapters 10 and 14 in relation to water vapour and steam, and in Chapter 22 (section 22.1.1) in relation to ice.

   The molecular polarity, together with a high dielectric constant (Table 1.5) and small molecular size, allows water molecules to bind with other molecules, which makes it a powerful solvent that is able to dissolve a large number of chemical compounds (a 'universal solvent') especially polar and ionic compounds (see also cleaning in Chapter 27, section 27.1.1).

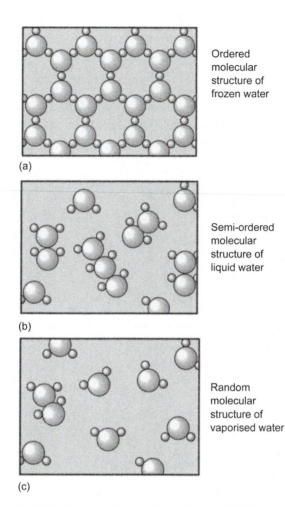

(a)

Ordered
molecular
structure of
frozen water

(b)

Semi-ordered
molecular
structure of
liquid water

(c)

Random
molecular
structure of
vaporised water

Fig. 1.8   Structure of water (from Pidwirny 2006).

When a substance, such as salt or sugar, or some gases (especially oxygen and carbon dioxide), dissolves in water the molecules separate from each other and become surrounded by water molecules to form a solution. The dissolved substance is the 'solute', and the liquid is the 'solvent'. Substances that readily dissolve in water are termed 'hydrophilic' and are composed of ions or polar molecules that attract water molecules through their electrical charge. Water molecules surround each ion and dissolve them, or they form hydrogen bonds with polar substances to dissolve them. Different types of ions and ionic groups on organic molecules can reduce or increase the mobility of water molecules in dilute solutions. For example positively charged, small multivalent ions (e.g. Na^+, Ca^{2+}, Mg^{2+} and Al^{3+}) have a net structure-forming effect (i.e. solutions are less fluid than pure water), whereas larger positively charged monovalent ions and negatively charged ions (e.g. NH_4^+, IO_3^-, Cl^-, NO_3^-) have a net structure-breaking effect (i.e. solutions are more fluid than pure water) (Fennema 1996).

Molecules that contain mostly non-polar bonds (especially hydrocarbons that contain C–H bonds) are usually insoluble in water and are termed 'hydrophobic'. Water molecules are less attracted to such molecules and so do not surround and dissolve them.

Table 1.5   Physical properties of water

Property	Value
Molar mass	18.015
Molar volume (moles litre^{-1})	55.5
Phase transition properties:	
Boiling point (°C at 101.3 kPa)	100.00
Melting point (°C at 101.3 kPa)	0.00
Triple point (K)	273.16
Pressure at the triple point (Pa at 0.01 °C)	611.73
Critical temperature (°C)	373.99
Critical pressure (MPa)	22.06
Critical volume (m^3 kmol^{-1})	0.056
Latent heat of vaporisation at 100 °C (kJ kg^{-1})	2270
Latent heat of fusion at 0 °C (kJ kg^{-1})	334
Vapour pressure (kPa at 20 °C)	2.338
Heat capacity (kJ kg^{-1} K^{-1})	
Water	4.182
Ice	2.108
Water vapour	1.996
Thermal conductivity at 20 °C (W m^{-1} K^{-1})	0.5984
Thermal conductivity (ice) at 0 °C (W m^{-1} K^{-1})	2.240
Thermal diffusivity at 20 °C (m^2 s^{-1})	1.4×10^{-7}
Temperature at maximum density (°C)	3.98
Density (kg m^{-3})	
0 °C (solid)	916.8
0 °C (liquid)	998.4
4 °C	1000
20 °C	998.2
40 °C	992.2
60 °C	983.2
80 °C	971.8
100 °C	958
Dielectric constant (permittivity) at 20 °C	80.20
Viscosity (centipoise) at 20 °C	1.00
Surface tension (N m^{-1}) at 20 °C	73×10^{-3}
Speed of sound (m s^{-1})	1480
Refractive index (relative to air)	
Ice; 589 nm	1.31
Water; 430–490 nm	1.34
Water; 590–690 nm	1.33
Electrical conductivity (s m^{-1})	
Ultra-pure water	5.5×10^{-6}
Drinking water	0.005–0.05
Seawater	5

Adapted from Fennema (1996), Ambrose (2008) and Anon (2008c)

However, it is simplistic to regard non-polar molecules as not being attracted to water. For example, a droplet of oil forms a thin film on the surface of water instead of remaining as a droplet. This indicates both that molecular attraction takes place between the oil and water, and that an individual oil molecule is attracted to a water molecule by a force that is greater than the attraction between two oil molecules. It is the attraction between water molecules that is stronger than both the oil–oil and water–oil attractions, and this keeps the substances separate as two phases (see also emulsions below and Chapter 3, section 3.2.2). Surface active agents (e.g. phospholipids) have hydrophobic and hydrophilic portions. When placed in an aqueous environment, the hydrophobic portions stick together, as do the hydrophilic parts to form a very stable arrangement as a lipid bilayer (for example in cytoplasmic membranes) (see also detergents in Chapter 27, section 27.1.1).

Water also has unique hydration properties for biological molecules, especially proteins and nucleic acids, and this, together with associations of water with hydrophobic non-polar groups of proteins, determine their three-dimensional structures, and hence their functions in solution. Fennema (1996) describes the properties of proteins during different stages in a hydration sequence (see also water activity, section 1.1.2). Hydration also forms gels that can undergo reversible gel–sol phase transitions that underlie many cellular mechanisms.

Water is miscible with many liquids (e.g. ethanol) in all proportions, where it forms a single homogeneous liquid. As a gas, water vapour is completely miscible with air. Water forms an 'azeotrope' with many solvents (an azeotrope is a mixture of two or more pure compounds in a ratio that cannot be changed by simple distillation. This is because when an azeotrope is boiled, the vapour has the same ratio of constituents as the original mixture of liquids). Each azeotrope has a characteristic boiling point that is either less than the boiling points of any of its constituents (a positive azeotrope such as 95.6% ethanol and 4.4% water (by weight) or greater than the boiling point of any of its constituents (a negative azeotrope, such as 20.2% hydrogen chloride and 79.8% water (by weight)). Further details are given in Chapter 14, (section 14.2.1).

Water has several other unique physical properties (Table 1.5):

- Water is a tasteless, odourless and colourless liquid over the temperature range 0–100 °C at ambient pressure.
- Water is transparent and only strong UV light is slightly absorbed.
- Water has the second highest specific heat capacity after ammonia, and a high latent heat of vaporisation, both of which are due to extensive hydrogen bonding between its molecules. The high specific heat (the energy required to change the temperature of a substance (Chapter 10, section 10.1.2)) means that it can absorb large amounts of heat before it becomes hot and it releases heat slowly when it cools. It also conducts heat more easily than any liquid except mercury. For these reasons it is widely used for process heating and cooling.
- Pure water has a low electrical conductivity, but it increases significantly when a small amount of ionic material is dissolved in it (see also microwave heating (Chapter 20, section 20.1) and ohmic heating (Chapter 20, section 20.2)). Because the electrical current is transported by the ions in solution, the conductivity increases as the concentration of ions increases.
- Above certain temperature (the 'critical temperature') water (and other substances) cannot be liquefied, no matter how high a pressure is applied. The pressure needed to liquefy water at the critical temperature is the 'critical pressure'. Fluids above the critical temperature are 'supercritical' fluids that have properties of both liquid and gas and are very different from those of the liquid (see Chapter 5, section 5.4.3). The 'triple point' of water is described in Chapter 10 (section 10.1.1).

Water can exist in a number of different forms in foods: it can be physically entrapped within a matrix of molecules, usually macromolecules such as gels and cellular tissues. This water does not flow from food tissues, even when they are cut, but it is easily removed during dehydration (Chapter 16, section 16.1) and easily frozen (Chapter 22, section 22.1). Removing the entrapment capability (i.e. the 'water-holding capacity') has substantial effects on food quality, including syneresis of gels, thaw exudates from frozen foods (Chapter 22, section 22.3.3) and post-mortem changes in meat (Chapter 21, section 21.3.1).

'Bound water' is used to refer to water that has limited mobility and different properties from physically entrapped water, although Fennema (1996) and others have suggested that the term should be discontinued because it is too imprecise, and instead describe it as water with 'hindered mobility'. This water is chemically bound within foods and is not available as a solvent, and does not readily freeze or evaporate. Further descriptions are given in section 1.1.2 (water activity).

Micro-components

This section describes a diverse range of chemicals that are found in foods in small quantities and includes naturally occurring vitamins, minerals, pigments and antioxidants, and added preservatives and colourants.

Vitamins

Vitamins comprise 13 organic compounds that are essential micronutrients, necessary for normal metabolism in animals, but are either not synthesised in the body or are synthesised in inadequate quantities. As a result, vitamins must be obtained from the diet. They act as enzyme precursors or coenzymes, as components of antioxidant defence systems, as factors involved in genetic regulation, and in other specialised functions (e.g. the role of vitamin A in vision). Health conditions that result from vitamin deficiency are recognised for all vitamins and, for some, excessive intake can also lead to disease. In addition, some vitamins also act in foods as reducing agents, free radical scavengers, flavour precursors or reactants in browning reactions (section 1.4.2). Gregory (1996) gives detailed descriptions of different vitamins, and the effects of processing on vitamin retention or activity are described in subsequent chapters. The nutritional functions and Recommended Dietary Allowance (RDA) of different vitamins are described by Anon (2006), and fortification of foods is described by Mejia (1994) and Wirakartakusumah and Hariyadi (1998). The nature and properties of vitamins and main sources in foods are summarised in Appendix A.2 and changes caused by processing are described in section 1.4.3.

The stability of vitamins in foods is summarised in Fig. 1.9, but it should be noted that this overview does not take into account the stability of different forms of each vitamin, and some can show great differences (e.g. synthetic folic acid used in food fortification is very stable to oxidation, whereas naturally occurring tetrahydrofolic acid, which has the same nutritional properties, is very susceptible). Details of the structures and functions of vitamins are given by Gregory (1996).

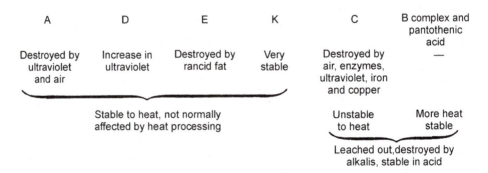

Fig. 1.9   Stability of vitamins in foods.

Minerals

Minerals are found in nearly all foods, but their concentration in plant sources can vary substantially depending on the local soil conditions. They are grouped into 'major' minerals (>100 mg/day required in the diet) and trace elements (<100 mg/day required in the diet). Unlike organic food components, minerals cannot be destroyed by heat, oxidation, etc. and the main sources of loss are leaching or during milling of cereals or legumes. Minerals are absorbed from foods to different extents, with, for example, nearly all ingested sodium being absorbed whereas about 10% of iron is absorbed. Unlike carbohydrates, lipids and proteins, mineral absorption is affected by other food components, specifically phytic acid, tannins and oxalates that are present in foods such as legumes and leaves. Acidic foods increase the solubility of minerals, and enhance absorption. Miller (1996) describes factors that influence the absorption of minerals from foods, including the chemical form of the mineral, the presence of ligands that form chelates with minerals and the redox activity of the food. The general functions of minerals in the body are to maintain electro-neutrality across cell membranes to maintain water balance; to contribute to the structural integrity of the skeleton and the activity of many proteins; and to act as cofactors to many enzymes. There are recommended dietary intakes for 12 minerals and a summary of the functions of 16 important minerals in foods and in the body is given in Appendix A.3.

Other minerals are less well documented. Arsenic is essential in trace amounts; its deficiency depresses growth and impairs reproduction. Boron may affect the metabolism of calcium and magnesium, membrane function and prevents some forms of cardiovascular disease. Its deficiency is linked to osteoporosis and arthritis and may be related to vitamin D production. Chlorine is essential in maintaining cellular fluid and electrolyte balances and its deficiency can cause hair and tooth loss, poor muscular contraction and impaired digestion. Cobalt is an integral part of vitamin B_{12} and deficiency may lead to pernicious anaemia, retarded growth and nervous disorders. Fluorine is incorporated into bones and teeth and may increase the deposition of calcium. However, at high levels, it is toxic and has adverse effects on many enzyme systems. Gallium controls brain chemistry and may have anti-tumour activity. Lithium controls aggressiveness. Silicon is needed for production of connective tissues (tendons, cartilage, blood vessels, nails, skin and hair) and with calcium to makes bones. Tin supports hair growth and can enhance reflexes. Deficiency symptoms include baldness, reduced response to loud noises and reduced haemoglobin synthesis. Vanadium is required for development of bones, cartilage and teeth and for cellular metabolism. Deficiency may be linked to reproductive problems and kidney disease. Lanthanum, praeseodymium, neodymium, thulium, samarium, europium and ytterbium are each involved in enhanced cell growth and extended lifespan (Clark 2006; Anon undated). Details of the chemical, functional and nutritional properties of minerals are given by Miller (1996).

Colourants and pigments

Colourants are any chemicals that impart colour to foods by reflecting or emitting energy at wavelengths that can stimulate the retina in the eye (i.e. visible light in the electromagnetic spectrum; Fig. 20.1 in Chapter 20). Black and intermediate shades of grey are also regarded as colours. Pigments are naturally occurring substances in plant or animal cells that impart colour (Table 1.6). In plants the main ones are chlorophyll, carotenoids, betalaines, and flavonoids or other phenols. Chlorophyll has a magnesium ion in the molecule and when this is displaced by heat or acids, it forms pheophytin and pyropheophytin and the colour of the pigment changes to olive-brown. Carotenoids are

Table 1.6 Naturally occurring pigments in foods

Pigment	Typical source	Oil or water soluble	Stability to the following			
			Heat	Light	Oxygen	pH change
Anthocyanins	Fruits	Water soluble	High	High	High	Low
Betalaines	Beetroot	Water soluble	Moderate	High	High	High
Bixin	Seed coat of *Bixa orellana*	Oil soluble	Moderate to low	Low	High	–
Canxanthin		Oil soluble	Moderate	Moderate	Moderate	Moderate
Caramel	Heated sugar	Water soluble	High	High	High	High
Carotenes	Leaves	Oil soluble	Moderate to low	Low	Low	High
Chlorophylls	Leaves	Water soluble	High	High	High	Low
Cochineal[a]	Insect (*Dactylopius coccus*)	Water soluble	High	High	–	Moderate to high
Curcumin	Turmeric	Water soluble	Low	Low	Low	–
Norbixin	See Bixin	Water soluble	Moderate to low	Low	High	–
Oxymyoglobin	Animals	Water soluble	Low	–	High	Low
Polyphenols	Tea leaf	Water soluble	High	High	High	High
Quinones	Roots, bark	Water soluble	High	Moderate	–	Moderate
Xanthophylls	Fruits	Water soluble	Moderate	High	High	Low

[a] As aluminium lake.
From the data of Zapsalis and Beck (1985) and Coultate (1984)

yellow-orange pigments found in plants and marine algae. They are nutritionally important as precursors for vitamin A (especially β-carotene) and also act as antioxidants. Natural carotenoids include α-carotene (in carrots), lycopene (in tomatoes), capsanthin (in red peppers and paprika), bixin (in annatto) and astaxanthin (pink colour in salmon, lobster and shrimps). β-Carotene is present in all green vegetables but the chlorophyll masks the colour. Natural and synthetic carotenoids are used as food colourants. Anthocyanins are a flavonoid subgroup of phenolic compounds and have a wide spectrum of colours from purple, through blue and magenta to red and orange. They are relatively unstable pigments and are easily degraded by heat, oxygen, light or changes in pH. Betalaines are a group of pigments that contain red betacyanins and yellow betaxanthins and are most commonly found in beetroot and amaranth. Unlike anthocyanins, they are not affected by pH and are more resistant to heat, but the colour is affected by oxygen and light. Details of plant pigments are given by von Elbe and Schwartz (1996).

In animal tissues they are the different forms of haem pigments (haemoglobin and the different forms of myoglobin). Synthetic colourants (or food dyes) are water-soluble chemicals, and 'lakes' are oil-soluble dyes that are extended on a substratum of aluminium that are approved for use in foods by regulatory authorities (Anon 2007a, 2002, and Appendix A4).

Antioxidants

Antioxidants are compounds that protect cells against damage caused by reactive oxygen species (e.g. superoxide, peroxyl radicals and hydroxyl radicals) produced by oxidation of fats or in the body by metabolic activity. An imbalance between antioxidants and reactive oxygen species results in oxidative stress, leading to cellular damage that has been linked to cancer, aging, atherosclerosis and neurodegenerative diseases. Antioxidants in foods include β-carotene, vitamins C and E, selenium and some polyphenolic compounds (e.g. flavonoids such as flavonols, flavones, flavanones, isoflavones, catechins, anthocyanidins and chalcones). Further information on the antioxidant activity of flavenoids is given by Buhler and Miranda (2000). Antioxidants are also added to foods to prevent oxidative rancidity, including vitamins C and E, spice extracts from cloves (*Eugenia caryophyllata*), cinnamon (*Cinnamomum zeylanicum*) and rosemary (*Rosmarinus officinalis*), and synthetic antioxidants, including tertiary butyl hydroxy quinone (TBHQ), butylated hydroxy anisole (BHA), butylated hydroxy toluene (BHT) and propyl gallate (PG) (Anon 2007a, 2002, and Appendix A4).

Preservatives

Preservatives are compounds that have antimicrobial (and sometimes antioxidant) activity and include ethanol, salt, sugar, grape seed extract, acetic, citric and ascorbic acids, and synthetic antimicrobial compounds including the following:

- Benzoates (sodium benzoate, benzoic acid, hydroxybenzoate alkyl esters), effective against yeasts and bacteria, less so against moulds, and used in soft drinks, fruit juices, ketchup, salad dressings.
- Epoxides (ethylene and propylene oxides) effective against all micro-organisms and spores, used in some low-moisture foods and to sterilise aseptic packaging materials (Chapter 13, section 13.2.3).
- Nitrites and nitrates (sodium or potassium nitrite or nitrate), which are effective against pathogenic bacteria, including *Clostridium* spp., and react with haem pigments to produce pink nitrosomyoglobin in cured meats. There is ongoing debate over the risk of toxic nitrosamine production in cured meats.
- Propionates (propionic acid, sodium, calcium or potassium propionate) effective against moulds, and a few bacteria, and used in bakery products and cheeses.
- Sorbates (sodium or potassium sorbate, sorbic acid) effective against moulds in particular and also yeasts, and used in cheeses, cakes, salad dressings, fruit juices, wines and pickles.
- Sulphites (sodium sulphite, sulphur dioxide, sodium bisulphite, potassium hydrogen sulphite) more effective against insects and bacteria (Gram-negative) than against moulds, yeasts and Gram-positive bacteria. They have antioxidant action and prevent enzymic browning, used in dried fruits, wines and fruit juices.

Other preservatives include sodium erythorbate, sodium diacetate, sodium succinate/succinic acid, disodium ethylenediaminetetraacetic acid (EDTA) and sodium dehydroacetate (Anon 2007a, 2002, and Appendix A4). Antibacterial chemicals (e.g. nicin) are described in Chapter 6 (section 6.7). Details of the methods of action of preservatives are given by Lindsay (1996a).

Natural toxicants

Although all chemicals in foods have the potential to become toxic when eaten in excess, there are a number of plant chemicals that can cause toxicity when eaten in smaller amounts. These include protease-inhibiting proteins in legumes, potatoes and cereals and haemagglutinin proteins in beans, both of which can cause impaired food utilisation and growth. Others include thioglycosides in brassicas that cause thyroid enlargement, cyanogenic glucosides in cassava and some pulses that can cause cyanide poisoning, and favism (haemolytic anaemia) from fava beans. There are also a variety of natural carcinogens in celery, mushrooms, Brussels sprouts and some herbs and spices described by Pariza (1996).

1.1.2 Physical properties

Density and specific gravity

The density of a material is equal to its mass divided by its volume and has SI units of $kg\,m^{-3}$. Examples of the density of solid foods and other materials used in food processing are shown in Table 1.7 and examples of densities of liquids are shown in Table 1.8. The density of materials is not constant and changes with temperature (higher temperatures reduce the density of materials) and pressure. This is particularly important in fluids where differences in density cause convection currents to be established. Knowledge of the density of foods is important in separation processes (Chapter 5), and differences in density can have important effects on the operation of size reduction and mixing equipment (Chapters 3 and 4).

The density of liquids or a solid piece of material is a straightforward measure of mass/volume at a particular temperature, but for particulate solids and powders there are two forms of density: the density of the individual pieces and the density of the bulk of material, which also includes the air spaces between the pieces. This latter measure is termed the 'bulk density' and is 'the mass of solids divided by the bulk volume'. The fraction of the volume that is taken up by air is termed the 'porosity' (ϵ) and is calculated by:

$$\epsilon = V_a/V_b \qquad\qquad \boxed{1.1}$$

where V_a (m^3) = volume of air and V_b (m^3) = volume of bulk sample.

The bulk density of a material depends on the solids density and the geometry, size and surface properties of the individual particles. Examples of bulk densities of particulate foods are shown in Table 1.7 and bulk density is discussed in relation to spray-dried powders in Chapter 16 (section 16.2.1).

The density of liquids can be expressed as 'specific gravity' (SG), a dimensionless number that is found by dividing the mass (or density) of a liquid by the mass (or density) of an equal volume of pure water at the same temperature:

$$SG = \text{mass of liquid/mass water} \qquad\qquad \boxed{1.2}$$

$$SG = \text{density of liquid/density water} \qquad\qquad \boxed{1.3}$$

If the specific gravity of a liquid is known at a particular temperature, its density can be found using:

$$\rho_L = (SG)_\theta \cdot \rho_w \qquad\qquad \boxed{1.4}$$

where ρ_L = liquid density (kg m^{-3}) and ρ_w = density of water (kg m^{-3}), each at temperature θ (°C).

Table 1.7   Densities of foods and other materials

Material	Density (kg m^{-3})	Bulk density (kg m^{-3})	Temperature (°C)
Solids			
Aluminium	2640	–	0
Copper	8900	–	0
Stainless steel	7950	–	20
Concrete	2000	–	20
Grapes	1067	368	–
Tomatoes	–	672	–
Lemons/oranges	–	768	–
Fresh fruit	865–1067	–	–
Frozen fruit	625–801	–	–
Fresh fish	967	–	–
Frozen fish	1056	–	–
Water (0 °C)	1000	–	0
Ice (0 °C)	916	–	0
Ice (−10 °C)	933	–	−10
Ice (−20 °C)	948	–	−20
Fat	900–950	–	20
Salt	2160	960	–
Sugar (granulated)	1590	800	–
Sugar (powdered)	–	480	–
Starch	1500	–	–
Wheat	1409–1430	790–819	–
Wheat flour	–	480	–
Barley	1374–1415	564–650	–
Oats	1350–1378	358–511	–
Rice	1358–1386	561–591	–
Gases			
Air	1.29	–	0
Air	0.94	–	100
Carbon dioxide	1.98	–	0
Carbon dioxide	1.46	–	100
Nitrogen	1.30	–	0

Adapted from data of Earle (1983), Lewis (1990), Milson and Kirk (1980), Peleg (1983) and Mohsenin (1970)

Specific gravity is widely used instead of density in brewing and other alcoholic fermentations (Chapter 6, section 6.4.3), where the term 'original gravity (OG)' is used to indicate the specific gravity of the liquor before fermentation (for example '1072' or '72' refers to a specific gravity of 1.072).

The density of gases depends on their pressure and temperature (Table 1.7). Pressure is often expressed as 'gauge pressure' when it is above atmospheric pressure, or as 'gauge vacuum' when it is below atmospheric pressure. Pressure is calculated using the ideal gas equation as follows:

$$PV = nRT \qquad \boxed{1.5}$$

where P (Pa) = absolute pressure, V (m^3) = volume, n (k mole) = number of k moles of gas, R = the gas constant (8314.4 J kmol^{-1} K^{-1}) and T (K) = temperature.

This equation is useful for calculation of gas transfer in applications such as modified atmosphere storage or packaging (Chapter 21, section 21.2.5), cryogenic freezing

Table 1.8   Properties of fluids

	Thermal conductivity $(W\,m^{-1}\,{}^\circ K^{-1})$	Specific heat $(kJ\,kg^{-1}\,{}^\circ K^{-1})$	Density $(kg\,m^{-3})$	Dynamic viscosity $(N\,s\,m^{-2})$	Temperature $({}^\circ C)$
Air	0.024	1.005	1.29	1.73×10^{-5}	0
	0.031	1.005	0.94	2.21×10^{-5}	100
Carbon dioxide	0.015	0.80	1.98		0
Oxygen		0.92		1.48×10^{-3}	20
Nitrogen	0.024	1.05	1.30		0
Refrigerant 12	0.0083	0.92			
Water	0.57	4.21	1000	1.79×10^{-3}	0
	0.68	4.21	958	0.28×10^{-3}	100
Sucrose solution (60%)				6.02×10^{-2}	20
Sucrose solution (20%)	0.54	3.8	1070	1.92×10^{-3}	20
Sodium chloride solution (22%)	0.54	3.4	1240	2.7×10^{-3}	2
Acetic acid	0.17	2.2	1050	1.2×10^{-3}	20
Ethanol	0.18	2.3	790	1.2×10^{-3}	20
Rape-seed oil			900	1.18×10^{-1}	20
Maize oil		1.73			20
Olive oil	0.168			8.4×10^{-2}	29
Sunflower oil		1.93			20
Whole milk	0.56	3.9	1030	2.12×10^{-3}	20
				2.8×10^{-3}	10
Skim milk			1040	1.4×10^{-3}	25
Cream (20% fat)			1010	6.2×10^{-3}	3
Locust bean gum (1% solution)				1.5×10^{-2}	
Xanthan gum (1% solution)			1000		

From Earle (1983), Lewis (1990) and Peleg and Bagley (1983)

Sample problem 1.1
Calculate the amount of oxygen (kmole) that enters through a polyethylene packaging material in 24 h at 23 °C, if the pack has a surface area of 750 cm² and an oxygen permeability of 120 ml m⁻² per 24 h at 23 °C and 85% relative humidity (see Chapter 25, Table 25.2) (atmospheric pressure $= 10^5$ Pa).

Solution to sample problem 1.1
The volume of oxygen entering through the polyethylene:

$$V = 120 \times \frac{750}{100^2}$$

$$= 9.0 \text{ cm}^3$$

Using Equation 1.5,

$$n = 10^5 \times 9 \times 10^{-6}/(8314 \times 296)$$

$$= 3.66 \times 10^{-7} \text{ kmole}$$

(Chapter 22, section 22.1) and the permeability of packaging materials (Chapter 25, section 25.1).

The density of gases and vapours is also referred to as 'specific volume', which is 'the volume occupied by unit mass of gas or vapour' and is the inverse of density. This is used for example in the calculation of the amount of vapour that must be handled by fans during dehydration (Chapter 16) or by vacuum pumps in freeze drying (Chapter 23, section 23.1) or vacuum evaporation (Chapter 14, section 14.1). Further details are given by Toledo (1999a) and Lewis (1990).

When air is incorporated into liquids (e.g. cake batters, ice cream, whipped cream) it creates a foam and the density is reduced. The amount of air that is incorporated is referred to as the 'overrun' and is calculated using equation 1.6:

$$\text{Overrun} = \frac{\text{volume of foam} - \text{volume of liquid}}{\text{volume of liquid}} \times 100 \qquad \boxed{1.6}$$

Typical overrun values are 95–105% for ice cream and 100–120% for whipped cream (also Chapter 3, section 3.2).

Viscosity

Viscosity is an important characteristic of liquid foods in many areas of food processing. For example the characteristic mouthfeel of food products such as tomato ketchup, cream, syrup and yoghurt depends on their viscosity (or 'consistency'). The viscosity of many liquids changes during heating/cooling or concentration and this has important effects on, for example, the power needed to pump these products. Viscosities of some common fluids in food processing are shown in Table 1.8.

Viscosity may be thought of as a liquid's internal resistance to flow. A liquid can be envisaged as having a series of layers and when it flows over a surface, the uppermost layer flows fastest and drags the next layer along at a slightly lower velocity, and so on through the layers until the one next to the surface is stationary. The force that moves the liquid is known as the shearing force or 'shear stress' and the velocity gradient is known as the 'shear rate'. If shear stress is plotted against shear rate, most simple liquids and gases show a linear relationship (line A in Fig. 1.10) and these are termed 'Newtonian' fluids. Examples include water, most oils, gases, and simple solutions of sugars and salts. Where the relationship is non-linear (lines B–E in Fig. 1.10), the fluids are termed 'non-Newtonian'. Further details are given by Nedderman (1997). For all liquids, viscosity decreases with an increase in temperature but for most gases it increases with temperature (Lewis 1990).

Many liquid foods are non-Newtonian, including emulsions and suspensions, and concentrated solutions that contain starches, pectins, gums and proteins. These liquids often display Newtonian properties at low concentrations but as the concentration of the solution is increased, the viscosity increases rapidly, and there is a transition to non-Newtonian properties (Rielly 1997). Non-Newtonian fluids can be classified broadly into the following types:

- Pseudoplastic fluid (line B in Fig. 1.10) – viscosity decreases as the shear rate increases (e.g. emulsions, and suspensions such as concentrated fruit juices and purées).
- Dilatant fluid (line C in Fig. 1.10) – viscosity increases as the shear rate increases. This behaviour is less common but is found with liquid chocolate and cornflour suspension.
- Bingham or Casson plastic fluids (lines D and E in Fig. 1.10). There is no flow until a

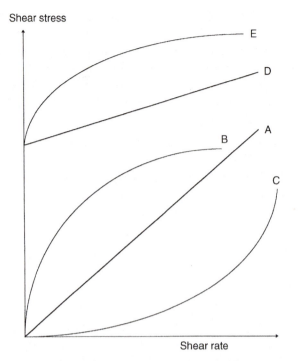

Fig. 1.10   Changes in viscosity of (A) Newtonian fluid and different types of non-Newtonian fluids: (B) pseudoplastic fluid; (C) dilitant fluid; (D) Bingham plastic fluid and (E) Casson plastic fluid (after Lewis 1990).

critical shear stress is reached and then the shear rate is either linear (Bingham type) or non-linear (Casson type) (e.g. tomato ketchup).

- Thixotropic fluid – the structure breaks down and viscosity decreases with continued shear stress (e.g. most creams).
- Rheopectic fluid – the structure builds up and viscosity increases with continued shear stress (e.g. whipping cream).
- Viscoelastic material – has viscous and elastic properties exhibited at the same time. When a shear stress is removed the material never fully returns to its original shape and there is a permanent deformation (e.g. dough, cheese, gelled foods).

The measurement of viscosity is complicated by the range of terms used to describe it. The simplest is the ratio of shear stress to shear rate, which is termed the 'dynamic viscosity' ($kg\,m^{-1}\,s^{-1}$). This is related to another term, 'kinematic viscosity' ($m^2\,s^{-1}$), as follows:

$$kinematic\ viscosity = \frac{dynamic\ viscosity}{density} \qquad [1.7]$$

Other terms, including 'relative viscosity', 'specific viscosity' and 'apparent viscosity', together with descriptions of methods of measuring viscosity, are given by Toledo (1999b) and Lewis (1990).

Surface activity
A large number of foods comprise two or more immiscible components, which have a boundary between the phases (Table 1.9) (see emulsions in Chapter 3, section 3.2.2). The

Table 1.9 Examples of colloidal food systems

Dispersed phase	Continuous phase	Name	Example
Liquid	Gas	Fog, mist, aerosol	Sprays (e.g. spray drying)
Solid	Gas	Smoke, aerosol	Carbon dioxide fog, smoke
Gas	Liquid	Foam	Whipped cream, aeration of fermentation media, hydrogenation of oils (hydrogen in oil)
Liquid	Liquid	Emulsion	Cream, mayonnaise, margarine and low-fat spreads, salad cream, sausagemeat and cakes
Solid	Liquid	Sol, colloidal solution, gel suspension	Chocolate drinks, fruit juice
Gas	Solid	Solid foam	Meringue, ice cream

From Lewis (1990)

phases are known as the 'dispersed' phase (the one containing small droplets or particles) and the 'continuous' phase (the phase in which the droplets or particles are distributed).

One characteristic of these systems is the very large surface area of the dispersed phase that is in contact with the continuous phase. In order to create the increased surface area, a considerable amount of energy needs to be put into the system using, for example, a high-speed mixer or an homogeniser. Droplets are formed when new surfaces are created. To understand the reason for this it is necessary to know the forces acting in liquids: within the bulk of a liquid the forces acting on each individual molecule are equal in all directions and they cancel each other out. However, at the surface the net attraction is towards the bulk of the liquid and as a result, the surface molecules are 'pulled inwards' and are therefore in a state of tension (produced by surface tension forces). This causes liquid droplets to form into spheres because this shape has the minimum surface area for the particular volume of liquid.

Emulsions
Chemicals that reduce the surface tension in the surface of a liquid are known as 'surfactants' 'emulsifying agents' or 'detergents'. By reducing the surface tension, they permit new surfaces to be produced more easily when energy is put into the system (e.g. by homogenisers) and thus enable larger numbers of droplets to be formed. There are naturally occurring surfactants in foods, including alcohols, phospholipids and proteins, and these are sometimes used to create food emulsions (e.g. using egg in cake batters). However, synthetic chemicals have more powerful surface activity and only require very small amounts to create emulsions. Details of synthetic emulsifying agents are given by Anon (2002) and in Chapter 3 (section 3.2.2) and Appendix A4.

Surface active agents contain molecules that are polar (or 'hydrophilic') at one end and non-polar (or 'lipophilic') at the other. In emulsions, the molecules of emulsifying agents become oriented at the surfaces of droplets, with the polar end in the aqueous phase and the non-polar end in the lipid phase.

In detergents, the surface active agents reduce the surface tension of liquids to both promote wetting (spreading of the liquid) and to act as emulsifying agents to dissolve fats. The detergent molecules have a lipophilic region of long chain fatty acids and a hydrophilic region of either a sodium salt of carboxylic acid (soapy detergents) or the sodium salt of an alkyl or aryl sulphonate (anionic detergents). Enzymes may also be added to detergents to remove proteins, and other ingredients may include polyphosphates (to soften water and keep dirt in suspension), sodium sulphate or sodium silicate (to make detergent powder fee-flowing) and sodium perborate (bleaching agent). Further details are given in Chapter 27, section 27.1.1).

Foams

Foams are two-phase systems that have gas bubbles dispersed in a liquid or a solid, separated from each other by a thin film. In addition to food foams (Table 1.9), foams are also used for cleaning equipment (Chapter 27, section 27.1.1). The main factors needed to produce a stable foam are:

- a low surface tension to allow the bubbles to contain more air and prevent them contracting;
- gelation or insolubilisation of the bubble film to minimise loss of the trapped gas and to increase the rigidity of the foam; and
- a low vapour pressure in the bubbles to reduce evaporation and rupturing of the film.

In food foams, the structure of the foam may be stabilised by freezing (e.g. ice cream), by gelation (e.g. setting gelatin in marshmallow), by heating (e.g. cakes, meringues) or by the addition of stabilisers such as proteins or gums (section 1.1.1).

Water activity

Deterioration of foods by micro-organisms can take place rapidly, whereas enzymic and chemical reactions take place more slowly during storage. In either case the water content is a important factor controlling the rate of deterioration. The moisture content of foods can be expressed (Lewis 1990) either on a wet-weight basis:

$$m = \left(\frac{\text{mass of water}}{\text{mass of sample}}\right) \times 100 \qquad \boxed{1.8}$$

$$m = \left(\frac{\text{mass of water}}{\text{mass of water} + \text{mass of solids}}\right) \times 100 \qquad \boxed{1.9}$$

or on a dry-weight basis:

$$m = \frac{\text{mass of water}}{\text{mass of solids}} \qquad \boxed{1.10}$$

The dry-weight basis is more commonly used for processing calculations, whereas the wet-weight basis is frequently quoted in food composition tables. It is important, however, to note which system is used when expressing a result. Wet-weight basis is used throughout this text unless otherwise stated.

A knowledge of the moisture content alone is not sufficient to predict the stability of foods. Some foods are unstable at a low moisture content (e.g. peanut oil deteriorates if the moisture content exceeds 0.6%), whereas other foods are stable at relatively high moisture contents (e.g. potato starch is stable at 20% moisture). It is the *availability* of water for microbial, enzymic or chemical activity that determines the shelf-life of a food, and this is measured by the water activity (a_w) of a food, also known as the relative vapour pressure (RVP).

Examples of unit operations that reduce the availability of water in foods include those that physically remove water (dehydration (Chapter 16), evaporation (Chapter 14, section 14.1) and freeze drying or freeze concentration (Chapter 23)) and those that immobilise water in the food (e.g. by the use of humectants in 'intermediate-moisture' foods (Esse

Table 1.10   Moisture content and water activity of foods

Food	Moisture content (%)	Water activity	Degree of protection required
Ice (0 °C)	100	1.00[a]	
Fresh meat	70	0.985	
Bread	40	0.96	Package to prevent moisture loss
Ice (−10 °C)	100	0.91[a]	
Marmalade	35	0.86	
Ice (−20 °C)	100	0.82[a]	
Wheat flour	14.5	0.72	
Ice (−50 °C)	100	0.62[a]	Minimum protection or no packaging
Raisins	27	0.60	required
Macaroni	10	0.45	
Cocoa powder	5.0	0.40	
Boiled sweets	3.0	0.30	
Biscuits	5.0	0.20	Package to prevent moisture uptake
Dried milk	3.5	0.11	
Potato crisps	1.5	0.08	

[a] Vapour pressure of ice divided by vapour pressure of water.
Adapted from Troller and Christian (1978), van den Berg (1986) and Brenndorfer et al. (1985)

and Saari 2004) and by formation of ice crystals in freezing (Chapter 22, section 22.1.1)). Examples of the moisture content and a_w of foods are shown in Table 1.10 and the effect of reduced a_w on food stability is shown in Table 1.11.

Water in food exerts a vapour pressure, and the size of the vapour pressure depends on:

- the amount of water present;
- the temperature; and
- the concentration of dissolved solutes (particularly salts and sugars) in the water.

Water activity is defined as 'the ratio of the vapour pressure of water in a food to the saturated vapour pressure of water at the same temperature':

$$a_w = \frac{P}{P_0} \hspace{4cm} \boxed{1.11}$$

where P (Pa) = vapour pressure of the food and P_0 (Pa) = vapour pressure of pure water at the same temperature. a_w is related to the moisture content by a number of equations, including:

$$\frac{a_w}{M(1 - a_w)} = \frac{1}{M_1 C} + \frac{C - 1}{M_1 C} a_w \hspace{3cm} \boxed{1.12}$$

where a_w = water activity, M (% dry weight basis) = moisture, M_1 = moisture (dry weight basis) of a monomolecular layer and C = a constant.

Detailed derivations of other equations for calculation of water activity in high- and low-moisture foods are described by Toledo (1999c). A proportion of the total water in a food is strongly bound to specific sites (e.g. hydroxyl groups of polysaccharides, carbonyl and amino groups of proteins, and hydrogen bonding) (section 1.1.1). When all sites are (statistically) occupied by adsorbed water the moisture content is termed the 'Brunauer–

Table 1.11   The importance of water activity in foods

a_w	Phenomenon	Examples
1.00		Highly perishable fresh foods
0.95	*Pseudomonas, Bacillus, Clostridium perfringens* and some yeasts inhibited	Foods with 40% sucrose or 7% salt, cooked sausages, bread
0.90	Lower limit for bacterial growth (general), *Salmonella, Vibrio parahaemolyticus, Clostridium botulinum, Lactobacillus*, and some yeasts and fungi inhibited	Foods with 55% sucrose, 12% salt, cured ham, medium-age cheese. Intermediate-moisture foods ($a_w = 0.90 - 0.55$)
0.85	Many yeasts inhibited	Foods with 65% sucrose, 15% salt, salami, mature cheese, margarine
0.80	Lower limit for most enzyme activity and growth of most fungi; *Straphlococcus aureus* inhibited	Flour, rice (15–17% water) fruit cake, sweetened condensed milk, fruit syrups, fondant
0.75	Lower limit for halophilic bacteria	Marzipan (15–17% water), jams
0.70	Lower limit for growth of most xerophilic fungi	
0.65	Maximum velocity of Maillard reactions	Rolled oats (10% water), fudge, molasses, nuts
0.60	Lower limit for growth of osmophilic or xerophilic yeasts and fungi	Dried fruits (15–20% water), toffees, caramels (8% water), honey
0.55	Deoxyribonucleic acid becomes disordered (lower limit for life to continue)	
0.50		Dried foods ($a_w = 0$–0.55), spices, noodles
0.40	Minimum oxidation velocity	Whole egg powder (5% water)
0.30		Crackers, bread crusts (3–5% water)
0.25	Maximum heat resistance of bacterial spores	
0.20		Whole milk powder (2–3% water), dried vegetables (5% water), cornflakes (5% water)

Emmett–Teller (BET) monolayer value' (Fennema 1996). Typical examples include gelatin (11%), starch (11%), amorphous lactose (6%) and whole spray-dried milk (3%). The BET monolayer value therefore represents the moisture content at which the food is most stable. At moisture contents below this level, there is a higher rate of lipid oxidation and at higher moisture contents, Maillard browning and then enzymic and micro-biological activities are promoted (Fig. 1.11).

The effect of a_w on microbiological and selected biochemical reactions is shown in Fig. 1.11 and Table 1.11. Almost all microbial activity is inhibited below $a_w = 0.6$, most fungi are inhibited below $a_w = 0.7$, most yeasts are inhibited below $a_w = 0.8$ and most bacteria below $a_w = 0.9$. The interaction of a_w with temperature, pH, oxygen and carbon dioxide, or chemical preservatives has an important effect on the inhibition of microbial growth. When any one of the other environmental conditions is sub-optimal for a given micro-organism, the effect of reduced a_w is enhanced. This permits the combination of several mild control mechanisms that result in the preservation of food without substantial loss of nutritional value or sensory characteristics (Table 1.12) (see also the 'hurdle effect' in section 1.3.1). Further details are given by Alzamora *et al.* (2003).

Enzymic activity virtually ceases at a_w values below the BET monolayer value. This is due to the low substrate mobility and its inability to diffuse to the reactive site on the enzyme. Chemical changes are more complex. The two most important that occur in foods that have a low a_w are Maillard browning and oxidation of lipids. The a_w that causes the maximum rate of browning varies with different foods. However, in general, a

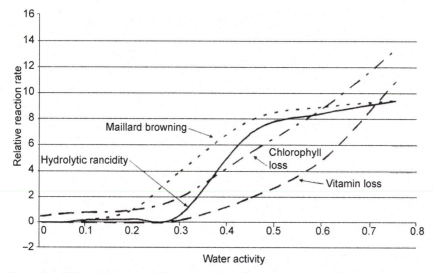

Fig. 1.11   Effect of water activity on chemical changes to foods (Esse and Saari 2004).

low a_w restricts the mobility of the reactants and browning is reduced. At a higher a_w, browning reaches a maximum. Water is a product of the condensation reaction in browning and at higher moisture levels, browning is slowed by 'end product inhibition'. At high moisture contents, water dilutes the reactants and the rate of browning falls.

Oxidation of lipids occurs at low a_w values owing to the action of free radicals. Above the BET monolayer value, antioxidants and chelating agents (which sequester trace metal catalysts) become soluble and reduce the rate of oxidation. At higher a_w values the catalytic activity of metals is reduced by hydration and the formation of insoluble hydroxides but, at high a_w values, metal catalysts become soluble, and the structure of the food swells to expose more reactive sites (Fig. 1.11).

The movement of water vapour from a food to the surrounding air depends upon both the food (its moisture content and composition) and the condition of the air (temperature and humidity). At a constant temperature, the moisture content of food changes until it comes into equilibrium with water vapour in the surrounding air. The food then neither gains nor loses weight on storage under those conditions. This is termed the 'equilibrium moisture content' of the food and the relative humidity of the storage atmosphere is known as the 'equilibrium relative humidity'. When different values of relative humidity

Table 1.12   Interaction of a_w, pH and temperature in selected foods

Food	pH	a_w	Shelf life	Notes
Fresh meat	>4.5	>0.95	days	Preserve by chilling
Cooked meat	>4.5	0.95	weeks	Ambient storage when packaged
Dry sausage	>4.5	<0.90	months	Preserved by salt and low a_w
Fresh vegetables	>4.5	>0.95	weeks	'Stable' while respiring
Pickles	<4.5	0.90	months	Low pH maintained by packaging
Bread	>4.5	>0.95	days	
Fruit cake	>4.5	<0.90	weeks	Preserved by heat and low a_w
Milk	>4.5	>0.95	days	Preserved by chilling
Yoghurt	<4.5	<0.95	weeks	Preserved by low pH and chilling
Dried milk	>4.5	<0.90	months	Preserved by low a_w

Adapted from Anon (2007b)

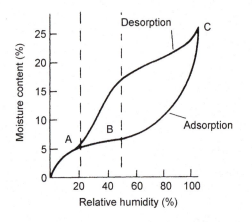

Fig. 1.12   Water sorption isotherm.

versus equilibrium moisture content are plotted, a curve known as a 'water sorption isotherm' is obtained (Fig. 1.12).

Each food has a unique set of sorption isotherms at different temperatures. The precise shape of the sorption isotherm is caused by differences in the physical structure, chemical composition and extent of water binding within the food, but all sorption isotherms have a characteristic shape, similar to that shown in Fig. 1.12. The first part of the curve, to point A, represents monolayer water, which is very stable, unfreezable and not removed by drying. The second, relatively straight part of the curve (AB) represents water adsorbed in multilayers within the food and solutions of soluble components. The third portion (above point B) is 'free' water condensed within the capillary structure or in the cells of a food. It is mechanically trapped within the food and held by only weak forces. It is easily removed by drying and easily frozen, as indicated by the steepness of the curve. Free water is available for microbial growth and enzyme activity, and a food which has a moisture content above point B on the curve is likely to be susceptible to spoilage.

The sorption isotherm indicates the a_w at which a food is stable and allows predictions of the effect of changes in moisture content on a_w and hence on storage stability. It is used to determine the rate and extent of drying (Chapter 16), the optimum frozen storage temperatures (Chapter 22) and the moisture barrier properties required in packaging materials (Chapter 25, section 25.1).

The rate of change in a_w on a sorption isotherm differs according to whether moisture is removed from a food (desorption) or whether it is added to dry food (absorption) (Fig. 1.12). This is termed a 'hysteresis loop'. The difference is large in some foods (e.g. rice) and is important for example in determining the protection required against moisture uptake.

1.1.3   Biochemical properties
Acids, bases and pH

Acids are substances that release hydrogen ions into solution. Many food acids are only partially dissociated (i.e. 'weak' acids such as those containing a carboxyl group, –COOH). They dissociate to give a hydrogen ion (H^+) in solution. Substances that reduce the number of hydrogen ions in solution are known as 'bases'. Some (e.g. ammonia), combine directly with hydrogen ions, whereas others (e.g. sodium hydroxide) create hydroxyl (OH^-) ions that then combine with H^+ ions to make H_2O and hence indirectly reduce the number of H^+ ions. Many bases found in foods are partially dissociated and are termed 'weak' bases (e.g. compounds that contain an amino group, $-NH_2$).

Positively charged hydrogen ions can move from one water molecule to another and pH is a measure of the activity of dissolved H^+ ions. The formula for calculating pH is:

$$pH = -\log_{10} \alpha_H{}^+ \tag{1.13}$$

where $\alpha_H{}^+$ denotes the activity of H^+ ions, and is dimensionless. In dilute solutions, activity is approximately equal to the numeric value of the H^+ concentration in moles per litre (also known as 'molarity') and denoted as $[H^+]$. Therefore, pH is defined as:

$$pH = -\log_{10} [H^+] \tag{1.14}$$

Pure water at 25 °C dissociates into H^+ and OH^- ions that have equal concentrations of 10^{-7} moles/litre. This is defined as 'neutral' and corresponds to pH = 7.0. (NB: when pure water is exposed to air, it absorbs carbon dioxide, which partially reacts with water to form carbonic acid and H^+, thereby lowering the pH to ≈ 5.7). The pH of a solution is obtained by comparing unknown solutions with one of known pH (i.e. pure water). For example, lemonade with a $[H^+]$ concentration of 0.0050 moles per litre, has

$$pH \approx -\log_{10}(0.0050) \approx 2.3.$$

Conversely, a solution of pH = 8.2 has an $[H^+]$ concentration of $10^{-8.2}\,\mathrm{mol\,l^{-1}}$, or $\approx 6.31 \times 10^{-9}\,\mathrm{mol\,l^{-1}}$. Thus, its hydrogen activity $\alpha_H{}^+$ is $\approx 6.31 \times 10^{-9}$.

Solutions in which the concentration of H^+ exceeds that of OH^- (acids) have a pH value lower than 7.0 and conversely, solutions in which OH^- exceeds H^+ (bases or alkalis) have a pH value greater than 7.0 (the terms 'weak' and 'strong' acids or bases do not refer to pH, but describe the extent to which an acid or base ionises in solution). The pH scale is an inverse logarithmic representation of hydrogen ion (H^+) concentration (i.e. each pH unit is a factor of 10 different to the next higher or lower unit, so for example a change in pH from 2 to 4 represents a 100-fold decrease in H^+ concentration). Examples of pH values of foods and other materials are given in Table 1.13.

Weak acids do not dissociate completely and an equilibrium is reached between the hydrogen ions and the conjugate base. To calculate the pH it is necessary to know the equilibrium constant of the reaction for each acid.

$$K_a = \frac{[\text{hydrogen ions}][\text{acid ions}]}{[\text{acid}]} \tag{1.15}$$

An example of a calculation is given in sample problem 1.2.

In plant and animal foods, pH is controlled by buffer chemicals, including proteins and amino acids, carboxylic acids, weak organic acids and phosphate salts. Synthetic buffering chemicals, including sodium salts of gluconic, acetic, citric and phosphoric acids, are used in foods and are listed in Appendix A4. Further details of the properties of water are given by Anon (2008c) and Fennema (1996).

Redox potential
Biochemical processes involve redox reactions where an electron is transferred to or from a molecule or ion to change its oxidation state. The oxidation state of atoms, ions or molecules is the number of electrons they have compared with the number of protons, and is denoted by a '$+$' or '$-$' sign (e.g. $O_2{}^-$ has an oxidation state of -1). When an atom

Table 1.13   Approximate pH values of foods and other materials

Material	Approximate pH	Material	Approximate pH
Ammonia	11.5	Mango, ripe	3.40–4.80
Apple, eating	3.30–4.00	Lye (NaOH)	13.5
Apricot	3.30–4.80	Mango, green	5.80–6.00
Artichoke	5.50–6.00	Maple syrup	5.15
Asparagus	6.00–6.70	Melons, honeydew	6.00–6.67
Avocado	6.27–6.58	Milk, cow	6.40–6.80
Banana	4.50–5.20	Milk, *acidophilus*	4.09–4.25
Bean	5.60–6.50	Milk, condensed	6.33
Beetroot	5.30–6.60	Molasses	4.90–5.40
Blackberry	3.85–4.50	Olives, ripe	6.00 -7.50
Bleach	12.5	Orange juice	3.30–4.15
Blueberry	3.12–3.33	Marmalade, orange	3.00–3.33
Bread, white	5.00–6.20	Oysters	5.68–6.17
Cantaloupe	6.13–6.58	Papaya	5.20–6.00
Carp	6.00	Parsnip	5.30–5.70
Carrots	5.88–6.40	Peaches	3.30–4.05
Cauliflower	5.60	Peanut butter	6.28
Cheese, Camembert	7.44	Pears, Bartlett	3.50–4.60
Cheese, Cheddar	5.90	Peas, strained	5.91–6.12
Cheese, cottage	4.75–5.02	Pineapple	3.20–4.00
Cherry	4.01–4.54	Plums, damson	2.90–3.10
Coconut, fresh	5.50–7.80	Pomegranate	2.93–3.20
Crab meat	6.50–7.00	Potato	5.40–5.90
Cream, 40%	6.44–6.80	Raspberries	3.22–3.95
Cucumber	5.12–5.78	Rhubarb	3.10–3.40
Cucumbers, dill pickles	3.20–3.70	Salmon, boiled	5.85–6.50
Egg white	7.96	Sardine	5.70–6.60
Egg yolk	6.10	Shrimp	6.50–7.00
Gooseberry	2.80–3.10	Soy sauce	4.40–5.40
Grape, concord	2.80–3.00	Soybean curd (tofu)	7.20
Grapefruit	3.00–3.75	Spinach	5.50–6.80
Guava, nectar	5.50	Squid	6.00–6.50
Herring	6.10	Sturgeon	6.20
Honey	3.70–4.20	Strawberries	3.00–3.90
Jam, fruit	3.50–4.50	Sweet potatoes	5.30–5.60
Ketchup	3.89–3.92	Tangerine	3.32–4.48
Lemon juice	2.00–2.60	Tomatoes	4.30–4.90
Lettuce	5.80–6.15	Vinegar	2.40–3.40
Lime juice	2.00–2.35	Worcester sauce	3.63–4.00
Lychee	4.70–5.01		

Adapted from Anon (2007b)

or ion gives up an electron its oxidation state increases, and the recipient of the negatively charged electron has its oxidation state decreased. Oxidation and reduction therefore always occur together with one atom or ion being oxidised when the other is reduced. The paired electron transfer is a redox reaction and the redox potential (E_h) (or 'oxidation reduction potential' (ORP)) is a measure in mV of the capacity of a compound to donate electrons in an aqueous medium. A simple example is the formation of sodium chloride: when atomic sodium reacts with atomic chlorine, the sodium donates one electron and the oxidation state becomes +1. Chlorine accepts the electron and its oxidation state is reduced to −1. The attraction between the differently charged Na^+ and Cl^- ions causes them to form an ionic bond. The loss of electrons is 'oxidation', and the gain of electrons

Sample problem 1.2

In a $0.1 \, \text{mol} \, \text{l}^{-1}$ solution of methanoic acid, the equilibrium reaction between methanoic acid and its ions can be expressed as:

$$HCOOH(aq) \leftrightarrow H^+ + HCOO^-$$

and the equilibrium constant for HCOOH $(K_a) = 1.6 \times 10^{-4}$ (it is assumed that the water does not provide any hydrogen ions). Calculate the pH of the solution.

Solution to sample problem 1.2

Using Equation 1.15, the acidity constant of methanoic acid is equal to:

$$K_a = \frac{[H^+][HCOO^-]}{[HCOOH]}$$

As an unknown amount of the acid has dissociated, [HCOOH] is reduced by this amount and $[H^+]$ and $[HCOO^-]$ are each increased by this amount. Therefore, [HCOOH] may be replaced by $0.1 - x$, and $[H^+]$ and $[HCOO^-]$ replaced by x, giving:

$$1.6 \times 10^{-4} = \frac{x^2}{0.1 - x}$$

Solving this for x yields 3.9×10^{-3}, which is the hydrogen ion concentration after dissociation. Therefore the pH is $-\log_{10} (3.9 \times 10^{-3})$, or ≈ 2.4.

is 'reduction' and correspondingly the atom or molecule which loses electrons is known as a 'reducing agent' and that which accepts the electrons is an 'oxidising agent'.

Redox reactions are essential for life in living organisms and many enzyme-catalysed reactions are oxidation–reduction reactions. The ability of micro-organisms to carry out oxidation–reduction reactions depends on the redox potential of the growth medium or food. Strictly aerobic micro-organisms can only grow at positive E_h values, whereas strict anaerobes require negative E_h values. In microbial fermentations (Chapter 6, section 6.4.3), for example, the redox potential is used to give information about the metabolism taking place in aerobic or anaerobic cultures to indicate the physiological state of microbial cultures. Redox potential measurements are used to monitor and control the dissolved oxygen concentration, and redox-sensitive pigments can be used to indicate microbial numbers (e.g. Kuda *et al.* 2004). In foods the redox potential represents the sum of all the compounds present that influence oxidation–reduction reactions. It also affects the solubility of nutrients, especially mineral ions. Antioxidants (section 1.1.1) are also known as redox-active compounds and Halvorsen *et al.* (2006) report measurements of redox-active compounds in 1113 foods. Both animal and plant pigments are sensitive to redox potential and changes can therefore affect the colour of foods. For example, Mellican *et al.* (2003) found that development of off-colours in foods is caused by oxidation-reduction interactions between ferric iron and polyphenols that contained *ortho*-dihydroxyl groups.

1.1.4   Sensory characteristics

There are a number of definitions of 'quality' of foods, which are discussed by Cardello (1998). To the consumer, the most important quality attributes of a food are its sensory characteristics (texture, flavour, aroma, shape and colour). These determine an individual's preference for specific products, and small differences between brands of similar products can have a substantial influence on acceptability.

Colour and appearance

The appearance of a food is a combination of its colour and geometric attributes (e.g. shape, size, texture, surface properties, gloss and translucency). The colour (or 'chromatic attributes') of a food depends on three factors: lightness (whether the colour is closer to black or to white); hue (the perceived colour – red, green, etc.) and saturation (the vividness or purity of a colour). These factors depend in part on the type of light that is used to view the food. Visible light is a small portion of the electromagnetic spectrum (Chapter 20, Figure 20.1) and contains different wavelengths from $\approx$ 380 to 770 nm. For accurate assessment of colour, the viewing light must be standardised and the International Commission on Illumination (CIE) has produced standard illuminants to assess colour by both human observers and instrumental methods (Anon 2000).

A viewed object also contributes to the perception of colour by modifying the light from the light source. Different pigments in foods absorb some wavelengths of the light and reflect or transmit others (e.g. a red object reflects red wavelengths and absorbs all other wavelengths). The third component in assessing colour is the observer. Because different people perceive colour differently, the CIE developed a 'standard observer' to best represent the average spectral response of human observers. This approach is still used but developments in machine vision systems (Chapter 2, section 2.3.3) are replacing human assessment.

There have been a number of systems developed to characterise colours, with the $L^*a^*b^*$ colour system widely used to assess the colour of foods. It can be represented diagrammatically (Fig. 1.13) where the vertical L^* axis represents 'lightness' from 0 to

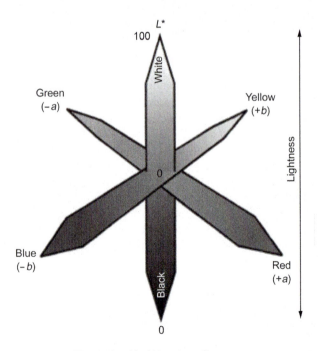

Fig. 1.13 $L^*a^*b^*$ colour diagram.

100. The two horizontal axes at right angles to each other are represented by $a*$ and $b*$. These use the fact that a colour cannot be both red and green, or blue and yellow. The $a*$ axis is green at one end (represented by $-a$), and red at the other ($+a$). The $b*$ axis is blue at one end ($-b$), and yellow ($+b$) at the other. The centre of each axis is 0 for both $a*$ and $b*$ and where the axes cross, the colour is neutral (grey, black or white). The $L*a*b*$ system is used to produce exact colour specifications for foods, for dyes (including those in plastic packaging), for printing inks and paper (e.g. 'Paper Type 1 is $115 \, \mathrm{g \, m^{-2}}$ gloss coated white and is described by $L*$ 95, $a*$ 0, $b*$ 0). Other systems include the Ostwald colour system, which matches colours to a set of standard samples, and the Munsell system and Newton Colour Circle, which both divide hue into 100 equal divisions around a colour circle and assign a unit of radial distance to each perceptible difference in saturation (named units of 'chroma'). Perpendicular to the plane formed by hue and saturation is the brightness scale of 'value' from 0 (black) to 10 (white). A point in the colour space is specified by hue, value and chroma in the form $H \, V/C$. The human eye has three different types of colour-sensitive cones, and the colours are defined by numerical values that are found by combining a set of three primary colours (blue, green and red). Further details are given by Nave (2008), and the pigments found in foods are described in section 1.1.1. The effects of processing on the colour of foods is described in subsequent chapters and by MacDougall (2002).

Taste and flavour
The flavour of foods is second only to appearance in shaping food choices, and people rank flavour as a major reason for selecting particular foods. There is a difference between taste and flavour: taste is the sensations of sweetness, sourness, saltiness, bitterness (or astringency) and 'umami' (the savoury taste of glutamate found in protein foods and monosodium glutamate). Some of these attributes can be detected in very low thresholds in foods (Table 1.14). Taste occurs on the tongue with 10 000 taste buds (or lingual papillae) located on the back, sides and the tip of the tongue, and also on the palate and in the throat. Although there are small differences in sensation on different parts of the tongue, all taste buds can respond to all types of taste. Each taste bud has clusters of 50–100 taste receptor cells that are excited by different chemical stimuli. Sweet, umami and bitter sensations are produced by a signal through a G-protein coupled receptor, whereas salty and sour sensations are produced by ion channels (Heller 2007). Flavour results from the interaction between the senses of taste and smell, with up to 80% of perception resulting from smell. This occurs when odours from foods reach olfactory receptors in the nasal cavity via inhalation through nostrils and from the back of the mouth as food is chewed and swallowed. The perceived flavour of foods is influenced by the rate at which flavour compounds are released during chewing, and hence is closely associated with the texture of foods and the rate of breakdown of food structure during mastication (Clark 1990).

Humans can detect ≈20 000 different odours, with ten intensities of each. Other sensations, including the characteristic 'heat' of chilli and the 'bite' of peppermint, as well as the texture, temperature and appearance of a food, each add to the experience of flavour. Fresh foods contain complex mixtures of volatile compounds, which give characteristic flavours and aromas, some of which are detectable at extremely low concentrations (Table 1.14). Research into flavour release is described by Wyeth and Kilcast (1991). Details of different aspects of taste and flavour are described by a number of authors (This 2006, Spillane 2006, Taylor 2002, Takeoka *et al.* 2001, Seabourne 1999, Lindsay 1996b, Rouseff 1990).

Table 1.14   Detection thresholds for common food components

Compound	Taste or odour	Threshold
Taste compounds		
Hydrochloric acid	Sour	0.0009 N
Citric acid	Sour	0.0016 N
Lactic acid	Sour	0.0023 N
Sodium chloride	Salty	0.01 M
Potassium chloride	Bitter/salty	0.017 M
Sucrose	Sweet	0.01 M
Glucose	Sweet	0.08 M
Sodium saccharine	Sweet	0.000023 M
Quinine sulphate	Bitter	0.000008 M
Caffeine	Bitter	0.0007 M
Odour compounds		
Citral	Lemon	0.000003 mg$\,l^{-1}$
Limonene	Lemon	0.1 mg$\,l^{-1}$
Butyric acid	Rancid butter	0.009 mg$\,l^{-1}$
Benzaldehyde	Bitter almond	0.003 mg$\,l^{-1}$
Ethyl acetate	Fruity	0.0036 mg$\,l^{-1}$
Methyl salicylate	Wintergreen	0.1 mg$\,l^{-1}$
Hydrogen sulphide	Rotten eggs	0.00018 mg$\,l^{-1}$
Amyl acetate	Banana oil	0.039 mg$\,l^{-1}$
Safrol	Sassafras	0.005 mg$\,l^{-1}$
Ethyl mercaptan	Rotten cabbage	0.00000066 mg$\,l^{-1}$

From Cardello (1998), Lafarge *et al.* (2008)

Texture

The texture of foods is mostly determined by the moisture and fat contents, and the types and amounts of structural carbohydrates (cellulose, starches and pectic materials) and proteins that are present. Detailed information on the textural characteristics of food is given by Kilcast (2004), McKenna (2003) and Lewis (1990). The effect of food composition and structure on texture is described by Wilkinson *et al.* (2000).

The texture of foods has a substantial influence on consumers' perception of 'quality' and during chewing (or 'mastication'), information on the changes in texture of a food is transmitted to the brain from sensors in the mouth, from the sense of hearing and from memory, to build up an image of the textural properties of the food. This may be seen as taking place in a number of stages:

1 An initial assessment of hardness, ability to fracture and consistency during the first bite.
2 A perception of chewiness, adhesiveness and gumminess during chewing, the moistness and greasiness of the food, together with an assessment of the size and geometry of individual pieces of food.
3 A perception of the rate at which the food breaks down while chewing, the types of pieces formed, the release or absorption of moisture and any coating of the mouth or tongue with food.

These various characteristics have been categorised (Table 1.15) and used to assess and monitor the changes in texture that affect the quality of foods.

Rheology is the science of deformation of objects under influence of applied forces. When a material is stressed it deforms, and the rate and type of deformation characterise its rheological properties. The rheological properties of solid foods are described in more

Table 1.15   Textural characteristics of foods

Primary characteristic	Secondary characteristic	Popular terms
Mechanical characteristics		
Hardness		Soft→firm→hard
Cohesiveness	Brittleness	Crumbly, crunchy, brittle
	Chewiness	Tender, chewy, tough
	Gumminess	Short, mealy, pasty, gummy
Viscosity		Thin, viscous
Elasticity		Plastic, elastic
Adhesiveness		Sticky, tacky, gooey
Geometrical characteristics		
Particle size and shape		Gritty, grainy, coarse
Particle shape and orientation		Fibrous, cellular, crystalline
Other characteristics		
Moisture content		Dry→moist→wet→watery
Fat content	Oiliness	Oily
	Greasiness	Greasy

Adapted from Szczesniak (1963)

detail in Chapter 3 (section 3.1.1). A large number of different methods have been used to assess the texture of food, including texture profiling by sensory methods using taste panels, quantitative descriptive analysis (QDA) (Fig. 1.14), described by Clark (1990), and empirical methods in which measurements of the forces needed to shear, penetrate, extrude, compress or cut a food are related to a textural characteristic.

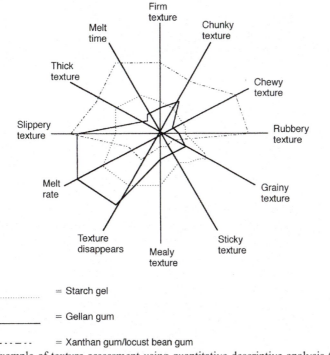

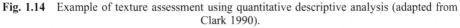

.... = Starch gel

_____ = Gellan gum

— · · — · · = Xanthan gum/locust bean gum

Fig. 1.14   Example of texture assessment using quantitative descriptive analysis (adapted from Clark 1990).

Examples of these instrumental methods include the Brabender system to measure dough texture or the viscosity of starch pastes, cone or rod penetrometers to measure the yield stress of margarine and spreads or the hardness of fruits, the General Foods Texturometer which simulates mastication by compressing foods using a plunger, the quantitative descriptive analysis of textural characteristics of foods, and the Instron Universal Testing machine, which measures stress and strain forces by compression or extension. Chemical methods include measurement of starch or pectin content, and microscopic methods include electron microscopy of emulsions or the flesh structure of meat and fish. These methods are reviewed by Kilcast (1999), Rosenthal (1999) and Lawless and Heyman (1998). Further information on food texture and viscosity is given by Kilcast and McKenna (2003), Bourne (2002) and Lewis (1990). Methods for the sensory evaluation of foods are described by, among others, Lawless and Heyman (1998).

1.1.5   Nutritional quality

This section provides a brief summary of the main nutrients in foods and their functions. Details of nutritional aspects of food components are given in nutrition texts, including Eastwood (2003), Warldaw (2003) and Gibney et al. (2002).

Macronutrients

Macronutrients can be defined as the chemical elements (carbon, hydrogen, nitrogen, oxygen, phosphorus and sulphur) or the classes of chemical compounds (carbohydrates, proteins, fats, water and atmospheric oxygen) that humans consume in the largest quantities (also section 1.1.1). Calcium, sodium, magnesium and potassium are sometimes included as macronutrients because they are required in relatively large quantities compared with other vitamins and minerals, or they may be termed 'macrominerals'. Macronutrients provide energy and are needed for growth, metabolism and other body functions. Carbohydrates and proteins provide $4 \, \text{kcal} \, \text{g}^{-1}$ ($\sim 16.7 \, \text{kJ} \, \text{g}^{-1}$) and fat provides $9 \, \text{kcal} \, \text{g}^{-1}$ ($\sim 37.7 \, \text{kJ} \, \text{g}^{-1}$). Between 45 and 65% of calorie intake should come from carbohydrate, 20–35% from fat and 10–35% from protein to enable the functions shown in Table 1.16 to take place.

Micronutrients

Micronutrients include dietary minerals (or 'microminerals' or 'trace elements') and vitamins (section 1.1.1) that are needed in very small quantities (<100 mg/day). An alternative method classifies these nutrients as either Type I or Type II, based on the way in which the body responds to a nutrient deficiency. A Type I response is specific physical signs of deficiency caused by reduced intake of the nutrient but there is no effect on growth or body weight (e.g. iron deficiency causes characteristic clinical signs and the concentration of iron in the tissues is reduced). In contrast, a Type II response is characterised by reduced growth rate or weight loss in the absence of specific deficiency signs (e.g. zinc deficiency stops growth and is followed by weight loss, but the concentration of zinc in the major tissues remains normal and there are no deficiency signs). Protein and energy derived from carbohydrates and fat are classified as Type II nutrients. This classification emphasises that poor growth is not caused solely by protein-energy malnutrition but can result from other nutrient deficiencies, and it demonstrates that a wide range of nutrients can cause poor growth or weight loss, and the need for a nutritionally balanced diet.

Table 1.16   Summary of functions of macronutrients

Macronutrient	Main functions	Important sources
Carbohydrates	• Conversion to glucose for use by tissues and cells for energy • Maintain correct functioning of the central nervous system, kidneys, brain and muscles • Intestinal health and elimination of body wastes	Starchy roots, tubers and grains and fruits. Other foods, including vegetables, beans and nuts contain lesser amounts
Indigestible carbohydrates (fibre)	• Move wastes from the body (laxation) • Help lower blood cholesterol levels • Decrease risks for coronary heart disease, obesity, and some types of cancers (e.g. colon cancer) • Reduce constipation and formation of haemorrhoids • Assists in maintaining normal blood glucose levels	Fruits, vegetables and whole grain products. Functional fibre synthesised or isolated from plants or animals
Proteins	• Body growth (especially in children and pregnant women) • Tissue repair • Preserving lean muscle mass • Immune function • Formation of essential hormones and enzymes • Energy when carbohydrate is not available See also essential amino acids (section 1.1.1)	Meats, fish, meat substitutes, cheese, milk, nuts, legumes, and in smaller quantities in starchy foods and vegetables
Fats	• Normal growth and development • Energy • Absorbtion of vitamins A, D, E, K, and carotenoids • Providing cushioning for organs • Maintaining cell membranes	Meats, dairy products, fish, margarines, lard and oils, nuts and grain products (see also types of fat in section 1.1.1)
Water	All metabolic processes	Foods and beverages

1.2   Food spoilage, safety and shelf-life

As biological materials, foods deteriorate over time and although this cannot be completely prevented, one aim of food processing is to slow the rate of deterioration by selecting appropriate methods of processing, ingredient formulations, packaging and storage conditions. Spoilage can be defined in a number of ways: one is that a food is no longer acceptable to the consumer due to changes in sensory properties; however, this is not always straightforward because of differences in perceptions of quality by different consumers (a mature mould-ripened cheese may be perceived as high quality and attractive to some consumers and spoiled and repulsive to others). Perceptions of quality attributes are described further in section 1.1.6. Alternatively foods may be said to have spoiled when the numbers and/or activity of pathogenic micro-organisms make them unsafe to eat and may cause illness or even death, or when deterioration of one or more nutrients means that a food no longer has its declared nutritional value (Singh and Anderson 2004). The time taken to reach one of these conditions is the 'shelf-life' of the product and in most countries it is a legal requirement to identify a 'best before', 'sell-by' or 'use-by' date on the package label. Details of labelling legislation vary in different countries but in Europe the Food Labelling Directive (2000/13/EEC) requires most prepacked foods to carry a date of minimum durability. A 'best before' date that indicates when a food will retain its optimum condition is used for most foods. The 'use by' date is used for foods that are microbiologically highly perishable and are therefore likely to cause a danger to health if consumed beyond this date (assuming they have been stored correctly). These foods have to be stored at low temperatures to maintain their safety rather than their quality, and include:

- foods stored at ambient or chill temperatures that are capable of supporting the formation of toxins or growth of pathogens to a level that could lead to food poisoning;
- foods intended for consumption either without cooking or after reheating to a temperature that is unlikely to destroy food poisoning micro-organisms that may be present.

Examples of these foods include: dairy products (milk, cheeses, fromage frais, mousses and products that contain whipped cream); cooked ready-to-eat products (cooked meats, hams, some fermented sausages, poultry, fish, eggs or sandwiches that contain these foods); cooked cereals, pulses and vegetables whether or not they are intended to be eaten without further reheating, smoked or cured fish, uncooked meat products, ready-to-eat prepared vegetable salads, coleslaw, mayonnaise, uncooked or partly cooked pastry dough (including pizzas, sausage rolls) or fresh pasta, containing meat, poultry, fish or vegetables, and foods packed in a vacuum or modified atmosphere and held at chill temperatures.

Foods, such as bread and cakes, that deteriorate in quality rather than safety and chilled foods that do not support the growth of pathogens (e.g. butter and margarines) do not need a use-by date. Some prepacked foods are exempt from the requirement to carry a date mark (e.g. honey, coffee, chocolate, frozen foods) and foods that are sold loose or packed for direct sale. Further details are given by Anon (2003) and Tatham and Richards (1998) and details of methods to evaluate the shelf-life of foods are given by Anon (2004, 1993) and Man and Jones (2000),.

Foods can spoil or become unsafe due to a number of different mechanisms, including physical changes, the effects of chemical or enzymic reactions and the activities of micro-organisms (as well as infestation by insects or other animals or contamination by foreign bodies). A summary of the causes of spoilage of selected foods is given in Table 1.17. The factors that control spoilage are:

- environmental conditions used for storage (temperature, humidity, exposure to oxygen and light);
- composition of the food (pH, moisture content (or more accurately water activity), nutrients available in the food).

Further details are given by Betts and Walker (2004).

1.2.1   Physical changes to foods
Physical damage caused by poor handling is a significant cause of spoilage in crisp products such as extruded, baked, fried or frozen foods (Chapters 15, 16, 19 and 22) and,

Table 1.17 Spoilage mechanisms for selected foods

Food product/category	Type of spoilage
Bakery products (bread)	Moisture migration (staling), starch retrogradation, mould growth
Soft (cakes)	Moisture migration (drying out or softening, staling) starch retrogradation, mould growth
Crisp (biscuits)	Moisture migration (softening), oxidation, breakage
Beers	Oxidation, microbial growth
Cereals	Moisture migration (softening), starch retrogradation, oxidation, breakage
Chocolate	Sugar or fat crystallisation (bloom), oxidation
Coffee/tea	Oxidation, volatile loss
Cooked meats	Microbial growth, moisture loss, oxidation
Crisp fried/extruded foods	Moisture migration (softening), oxidation, breakage
Dairy products	Oxidation, hydrolytic rancidity, bacterial growth, lactose crystallisation
Dried products	Oxidation, browning, moisture pickup, caking
Fresh fruits	Enzymic softening or browning, bruising, moisture loss, yeast or mould growth
Fresh meat/fish/seafood	Bacterial growth, moisture loss, oxidation
Fresh vegetables	Enzymic softening or colour loss, moisture loss, wilting, bacterial or mould growth
Frozen foods	Oxidation, dehydration (freezer burn), texture loss due to ice crystal growth
Sugar confectionery	Moisture migration (stickiness), sugar crystallisation
Wines	Oxidation, microbial growth

Adapted from Singh and Anderson (2004) and Kilcast and Subramanian (2000)

similarly, physical damage to fresh fruits and vegetables causes bruising, which can in turn lead to accelerated microbial growth, enzymic browning reactions and wilting due to moisture loss (Chapter 21). Another example of physical spoilage is the destabilisation or breakdown of emulsions, such as mayonnaise due to freezing, high temperatures or extreme vibration (Depree and Savage 2001).

Moisture migration

Physical changes caused by moisture migration are temperature dependent and involve the water activity (a_w) of the food (section 1.1.2) and glass transitions (section 1.4.1). The glass transition temperature of a food is the temperature at which it changes from, for example, a brittle, glassy state to a softer, rubbery state. An example of glass transition causing spoilage is staling of bakery products, which is due in part to moisture migration from the crumb (high a_w) to the crust (low a_w) (also Chapter 18, section 18.3.1). The loss of moisture in the crumb raises the glass transition temperature to the point where it undergoes a glass transition to become hard and brittle. Conversely, in the crust moisture migration lowers the glass transition temperature and it changes from hard and crisp to become tough and rubbery. Similar changes occur when hard, dry, baked or fried foods (e.g. biscuits or potato crisps) or sugar confectionery, absorb moisture from the atmosphere to become soft or sticky respectively. Jaya *et al.* (2002) report similar changes in food powders, in which moisture absorption causes them to become sticky and form cakes. Labuza and Hyman (1998) describe more complex moisture transfer in multi-component foods, where each component has a different a_w. Retrogradation of starches is caused by recrystallisation of amylose (see section 1.1.1).

Temperature

Temperature is one of the most important factors that influences the rate of spoilage: for example, rates of microbial growth, oxidation of lipids or pigments, browning reactions and vitamin losses are each directly controlled by temperature (Taoukis and Giannakourou 2004). Other forms of spoilage include high temperatures that melt fats and produce unacceptable oil leakage in some foods; respiration of fresh fruits and vegetables is temperature dependent and control of the storage temperature delays ripening and senescence to maximise the post-harvest life; some fruits and vegetables are susceptible to chilling injury, which can cause the development of off-flavours, texture changes, discoloration and accelerated senescence (Chapter 21, section 21.3.1); fluctuating temperatures cause spoilage of frozen foods due to loss of moisture that results in freezer burn and recrystallisation of ice crystals causes grittiness in ice cream and loss of texture in other frozen foods (see Chapter 22, section 22.2.4). Temperature control during production and distribution of chilled and frozen foods is an essential component of hazard analysis and critical control points (HACCP) and details are given in Chapter 21 (section 21.2.4) and Chapter 22 (section 22.2.4).

The rate of deterioration of a food and the prediction of its shelf-life (section 1.3.7) can be made by studying one or more quality index that is characteristic of the food (e.g. loss of a nutrient or characteristic flavour, growth of a target micro-organism, production of an off-flavour or discoloration). The effect of temperature on these indices is measured using kinetic studies based on the Arrhenius equation (Equation 1.16):

$$K = K_A \exp\left(-E_A/R\theta\right) \hspace{4cm} \boxed{1.16}$$

where K = reaction rate, K_A = Arrhenius equation constant, E_A (J mole^{-1}) = activation energy (i.e. the energy barrier that the quality parameter has to overcome for the reaction to proceed), R (8.3144 J mol^{-1} K^{-1}) = universal gas constant (section 1.1.2) and θ (K) = temperature.

The equation constant (K_A) is the value of the reaction at zero K, which is not useful for practical studies and the equation is therefore modified to include a reference temperature (Equation 1.17):

$$K = K_{ref} \exp\left[-E_A/R(1/\theta - 1/\theta_{ref})\right] \hspace{3cm} \boxed{1.17}$$

where K_{ref} (K) = rate constant at reference temperatures of 255 K for frozen foods, 273 K for chilled foods and 295 K for ambient temperature storage.

Values of K are measured at different temperatures and ln K is plotted against $(1/\theta - 1/\theta_{ref})$ on a semilog scale to obtain both the reaction rate and activation energy. Most reactions that cause loss of quality have been classified as either zero order (e.g. Maillard browning) or first order (e.g. microbial growth, vitamin loss and oxidation). Further details are given by Taoukis and Giannakourou (2004) (see also z-values in Chapter 10, section 10.3). The temperature dependence of these reactions is also the basis for accelerated shelf-life testing (section 1.3.7). Methods to validate food spoilage models are described by Betts and Walker (2004).

1.2.2   Biochemical changes

Chemical reactions involving fats, carbohydrates, proteins and micronutrients can each produce changes to the colour, texture or flavour of foods that consumers find unacceptable. The main factors that affect these reactions are the temperature of storage, exposure to light and oxygen, and the a_w and pH of the food.

Oxidation reactions include the development of off-flavours and colour changes due to oxidative rancidity in fats (autoxidation), and lipid oxidation in meat, fish and dairy products (Morrissey and Kerry 2004). In meat, red myoglobin and oxymyoglobin proteins can also become oxidised to brown metmyoglobin. St Angelo (1996) gives details of lipid oxidation and further examples are described in Chapter 19 (section 19.3.1). Autoxidation is a chain reaction that produces free radicals in oils and fats, which in turn decompose to form volatile hydrocarbons, alcohols and aldehydes that produce the characteristic rancid flavour. The reactions are accelerated by metal ions (especially iron and copper), moisture and UV light, and are slowed by antioxidants that scavenge free radicals. Further details of autoxidation are given by Gordon (2004) and lipolysis is described by Davies (2004).

The Maillard reaction (or 'non-enzymic browning') is a complex series of reactions between reducing sugars and amino acids or amino groups on proteins (see section 1.1.1). Depending on the composition of the food and the processing conditions, the Maillard reaction can produce thousands of different compounds. These include volatile pyrazines, pyridines, furans and thiazoles that alter the aroma of foods, low molecular weight compounds that affect the taste, antioxidants and brown melanoidin pigments that alter the appearance of foods (see also Chapter 18, section 18.3.1). Maillard reactions are important to develop the required sensory characteristics in bakery products, fried foods, roasted coffee or cocoa and in soy sauce. The reactions also cause flavour deterioration in products such as fruit juices, dairy products and beer. Further details are given by Arnoldi (2004).

Enzymic reactions
Naturally occurring enzymes in foods catalyse a wide variety of reactions that can adversely affect the flavour, colour and texture of foods during storage. Details are given by Ashie *et al.* (1996). A very large number of extracellular enzymes is also produced by micro-organisms and these are an important cause of food spoilage (section 1.2.3). The factors that affect the rate of enzyme reactions are similar to those that control microbial activity, although enzyme production can take place under conditions that do not support cell growth (Braun *et al.* 1999).

1.2.3   Microbiological changes
In order to predict the changes to foods and their expected shelf-life, it is necessary to understand the types of microbial contamination and the factors that affect microbial growth, activity and destruction. The main factors that control the types of micro-organisms that contaminate foods and the extent of their growth or activity can be summarised as:

- availability of nutrients in the food (e.g. carbon and nitrogen sources, and any specific nutrients required by individual micro-organisms);
- environmental conditions in the food (pH, moisture content or a_w, redox potential (E_h), presence of preservative chemicals);
- storage conditions (temperature, exposure to light or oxygen);
- stage of growth of micro-organisms;
- presence of other competing micro-organisms; and
- the interaction of these factors.

These factors are discussed in greater detail in microbiological texts (for example Adams

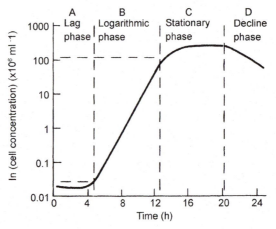

Fig. 1.15   Phases in microbial growth.

and Moss 2007, Jay *et al.* 2005 and Roberts and Greenwood 2002). Growth of micro-organisms follows a characteristic pattern (Fig. 1.15) (Todar 2007):

A *Lag phase*. Immediately after inoculation into a new substrate, cells remain temporarily unchanged. There is no apparent cell division, but cells may be growing in volume or mass, synthesising enzymes and increasing their metabolic activity. The length of the lag phase depends on a wide variety of factors including: the size of the inoculum; the time needed to recover from shock/damage caused by the transfer; time required for synthesis of essential coenzymes or inducible enzymes required to metabolise the substrates in the medium.

B *Logarithmic (or exponential) phase*. All cells are dividing regularly by binary fission, and numbers increase exponentially (i.e. by geometric progression). The cells divide at a constant rate that depends on the composition of the medium and the incubation conditions. The rate of exponential growth is expressed as the 'generation time', or the doubling time of the population. For most micro-organisms, generation times vary from $\approx$15 min to 1 h.

C *Stationery phase*. In a batch culture, cell growth is limited by exhaustion of available nutrients, or accumulation of inhibitory metabolites and the population stops increasing. Some micro-organisms produce secondary metabolites during the stationary phase and spore-forming micro-organisms begin sporulation.

D *Death phase*. The population of viable cells declines exponentially.

For many foods, micro-organisms are the most important, and often the most rapid, causes of spoilage. It is not always the number of micro-organisms in a food that indicates the extent of spoilage, but it is microbial activity that is very important. For example, micro-organisms that are highly active may alter the quality of a food by one or more of the following mechanisms (Sutherland 2003):

• Production of extracellular hydrolytic enzymes (e.g. cellulolytic enzymes, carbohydrases, proteases or pectinases) that alter the structural components of foods and result in softening, or liquefaction.
• Production of enzymes that break down macromolecules to release for example organic acids, hydrogen sulphide and mercaptans. Lipases break down fats to fatty acids and volatile compounds produce off-odours in the food, whereas non-volatile compounds may produce changes to the flavour of foods.

- Production of extracellular polysaccharides that cause sliminess in the food.
- Production of pigments that change the colour of foods.
- Acid production that alters the colour of natural pigments in the food and/or changes the taste.
- Gas production that may cause the product to swell or split, or inflate a package.

Where substantial microbial growth takes place, visible colonies on the food are an indicator of spoilage.

The different types of spoilage micro-organisms are commonly referred to as bacteria, yeasts and moulds and they are described in detail by Blackburn (2006). Technically, micro-organisms are grouped into 'eukaryotes' (which have cells containing a nucleus and complex structures (or 'organelles') such as mitochondria or chloroplasts enclosed by membranes) and 'prokaryotes' (which lack a cell nucleus or other membrane-bound organelles). Prokaryotes are divided into two domains: the bacteria and the archaea. Therefore, Eukaryotes, Archaea and Bacteria are the fundamental classifications in the three-domain system (Woese et al. 1990).

Bacteria

Bacteria are single-celled micro-organisms, 1–5 μm in size that reproduce by binary fission. Most spoilage bacteria cannot grow below $a_w = 0.91$, although halophilic bacteria can grow at $a_w = 0.75$. The optimum pH for growth of most species is 6.0–7.0, but lactic acid bacteria have optimum growth at pH 5.5–5.8 and can grow in foods at pH 4. The redox potential (E_h) at which bacteria grow determines whether they are aerobic (E_h is positive) or anaerobic (E_h is negative). Facultative aerobes can grow under both E_h conditions. Bacteria are also classified according to the optimum temperature for growth into:

- psychrophiles (minimum −10 to 5 °C, optimum = 12–18 °C, range = −10 to 20 °C);
- psychrotrophs (minimum <0–5 °C optimum = 20–30 °C, range = <0–30 °C);
- mesophiles (minimum 5–10 °C, optimum = 30–40 °C, range = 10–45 °C);
- thermophiles (minimum: 20–40 °C, optimum = 55–65 °C, range = 45–75 °C);
- hyperthermophile (minimum ≈ 80 °C, optimum = 90–100 °C, range = 65–120 °C).

Details of the minimum growth temperature (MGT) of micro-organisms are given in Chapter 21 (section 21.5) and bacteria that compromise food safety are described in section 1.2.4 and Appendix B.1.

Fungi

Yeasts are a small group (1%) of fungal species that are single-celled micro-organisms, 3–5 μm in size that reproduce by budding or sporulation. They are widely distributed in soils, on plants (especially fruits), and commonly associated with insects and the hides and feathers of animals. Most yeasts cannot grow below $a_w = 0.88$–0.9, but some osmophilic yeasts can grow at $a_w = 0.6$–0.7. For example *Zygosaccharomyces rouxii* can tolerate high concentrations of sugar and salt. The optimum pH for growth of most yeasts is 4.5–5.5, but some (e.g. *Zygosaccharomyces bailii*) are very tolerant of acetic acid and can grow in foods at pH 1.5. Others have a wide pH tolerance and can grow at pH values between 3 and 10. Yeasts are aerobes but about half of the species can ferment sugars and these are facultative anaerobes. Most yeast species are mesophiles although some psychrotrophic yeasts can grow at several degrees below 0 °C and have optimum growth below 20 °C. However, these are overall data and changes in one environmental factor can influence the response to others. For example, the minimum temperature for growth increases at lower a_w and both the minimum pH and a_w for growth are higher at lower temperatures.

Deak (2004) identified the following groups of foodborne spoilage yeasts.

- *Strongly fermentative yeasts*: the most common spoilage yeasts belong to the *Saccharomyces, Zygosaccharomyces, Torulaspora* and *Kluyveromyces* genera. *K. marxianus* and *K. lactis* both ferment lactose and are common causes of spoilage in dairy products. Yeasts in this group that are useful in food processing are described in Chapter 6 (section 6.4).
- *Weakly fermenting yeasts*: the most important is *Debaryomyces hansenii*, which is salt-tolerant and occurs widely in food products. Others, including *Pichia membranifaciens* and *Issatchenkia orientalis*, develop films on liquid foods.
- *Hyphal yeasts*: these species develop filamentous cells. An example is *Yarrowia lipolytica*, which has strong proteolytic and lipolytic activity and causes spoilage of meat and dairy products.
- *Imperfect yeasts*: the most important genera is *Candida* and *C. tropicalis*, *C stellata* and *C. zeylanoides* are common spoilage yeasts in meat, fish and dried foods.
- *Red yeasts*: these form typical red/orange colonies and include *Rhodotorula* and *Sporobolomyces* spp. They have strong hydrolytic activity and can grow at low temperatures, but do not ferment foods.

Under aerobic conditions yeasts grow on mono- and disaccharide sugars, organic acids, alcohols and amino acids, but most have limited ability to hydrolyse larger molecules such as starch. Exceptions are *Saccharomyces diastaticus, Debaromyces occidentalis* and *Saccharomycopsis fibuligera* which each have amylolytic activity and may cause spoilage of bakery products. Loureiro *et al.* (2004) describe detailed methods for detecting spoilage yeasts.

Moulds
Moulds are a type of fungi that form a mycelium of branched cells, each 30–100 μm in size that reproduce by sexual or asexual sporulation. Most moulds cannot grow below a_w $\approx$ 0.8, but some xerophilic moulds can grow at $a_w \approx 0.65$. The minimum pH for growth of some species (e.g. *Fusarium* spp.) is about 2.0 and others (e.g. *Penicillium* spp.) can grow at $-6\,°C$. Details of growth of spoilage moulds are given by Sautour *et al.* (2002) and moulds that cause food poisoning are described in section 1.2.4. Moulds cause rots in fresh fruits and vegetables (e.g. *Botrytis cinerea* on grapes and strawberries, blue mould rot on tomatoes and oranges caused by *Penicillium* spp. and *Fusarium* spp., and watery soft rot of apples caused by *Sclerotinia sclerotiorum*). Other moulds affect a wide range of foods, including bakery products, nuts, cheese and other dairy products.

The use of multiple preservation techniques can result in a synergistic effect on microbial destruction. Examples described in Chapters 7–9 include the use of ultrasound, irradiation, electric fields and high pressure. When these techniques are used in combination with mild heating there is less product damage in terms of colour and flavour retention and hence enhanced consumer appeal.

1.2.4   Food safety
Injury to consumers by contaminated foods can arise from harmful chemicals or foreign bodies (see Chapter 2, section 2.2), or from micro-organisms. This section outlines the

common microbial pathogens that may contaminate foods. Pathogenic bacteria may also produce toxins in the food in relatively low numbers or grow to sufficient numbers to cause poisoning. Further details are given in food microbiology texts, including Anon (2008d), Fratamico and Bayles (2005), McMeekin (2003) and Lund *et al.* (2000). Details of the taxonomy, pathogenicity, detection and distribution of important pathogens are given by Walker and Betts (2000), Marth (1998) and Anon (1996a). The main types of foodborne pathogens are bacteria, viruses and parasites and a few mould species. No foodborne yeast species are known to cause food poisoning. The main bacterial pathogens and their growth parameters are shown in Appendix B.1. The following section is a brief description of each pathogen.

Aeromonas spp.
Three species of *Aeromonas* are thought to be responsible for gastroenteritis, particularly in immunocompromised people or people suffering from cancer: *A. hydrophilia* HG 1, A, *A. caviae* HG 4 and A, and *A. veronii* HG 8, although their role as an enteropathogen has not been definitively established. In healthy people, children below 5 years old are at greatest risk. Two types of symptoms are mild diarrhoea and a severe illness similar to cholera that may be life-threatening. *A. hydrophilia* is also associated with a range of other illnesses, including inflammation of the gallbladder, septicaemia and meningitis, which can cause death in more than 60% of cases (Motarjemi 2002, Sutherland and Varnam 2002). The pathogen invades cells in the intestine and although it is known to produce toxins, their role has not been definitely established. It has been found in water and a wide range of foods, including vegetables, meat and poultry and fish and shrimps, although it is possible that its presence in these foods is due to contaminated water used in processing.

Bacillus cereus
There are two types of food poisoning caused by *B. cereus*: (1) nausea and vomiting similar to symptoms of *Staphylococcus aureus* poisoning caused by an emetic toxin produced in foods; and (2) diarrhoea and abdominal pain without vomiting, similar to symptoms of *Clostridium perfringens* toxin, caused by diarrhoeagenic toxins produced in the small intestine (Appendix B.1). The minimum cell concentration for both types of illness is approximately 10^5 cells. Because both types of symptoms are relatively mild and short-lived, the incidence of *B. cereus* food poisoning is thought to be significantly under-reported. The emetic type is mostly associated with rice products, where spores survive heating and germinate in the cooling product, producing toxins in the food. The diarrhoeal type of toxin is found in meat, milk and vegetable products where cells survive heat processing, multiply sufficiently before consumption and produce toxins in the intestine. Other species, similar to *B. cereus*, also cause food poisoning including *B. subtilis*, *B. licheniformis* and *B. thuringiensis*. *B. subtilis* causes acute vomiting within 2–3 hours of consumption, followed by diarrhoea, whereas *B. licheniformis* causes diarrhoea within 8 hours, sometimes with vomiting. They grow in foods mainly as a result of temperature abuse and numbers exceeding 10^6 can cause intoxication. *B. thuringiensis* is also known to produce toxins and concern has been expressed over its use as an insecticide used on vegetables (it is lethal to cabbage white caterpillars) (Gibbs 2002).

Brucella spp.
The strains of *Brucella* spp. that cause human brucellosis are *B. abortis* (from cows), *B. melitensis* (from sheep and goats) and *B. suis* (from pigs). People working closely with

farm animals mainly contract the disease but it can also be contracted by consumption of raw milk and unpasteurised dairy products. The acute symptoms include fatigue, weakness, muscle and joint pain and weight loss within two months of infection. It can also result in chronic health problems including inflammation of joints, genitourinary, cardiovascular and neurological conditions and insomnia and depression (Motarjemi 2002).

Campylobacter spp. and *Arcobacter* spp.

Campylobacter jejuni is the most important pathogen in this group, causing up to 90% of reported campylobacteriosis infections, with *C. coli* also causing less common illnesses. Before 1991, two species of *Arcobacter* (*A. butzleri* and *A. cryaerophilus*) were known as aerotolerant *Campylobacter*. They have been associated with enteritis in humans and abortions and enteritis in animals, but have not been linked to widespread outbreaks of food poisoning. Both *Campylobacter* and *Arcobacter* are normal intestinal flora of animals and poultry, and illness is usually caused by eating recontaminated foods or foods that are raw or inadequately cooked. *Campylobacter jejuni* is an important cause of sporadic gastroenteritis because, even though the cells do not survive for long periods in foods, they are highly virulent and only a few hundred cells are needed to cause infection. *C. jejuni* is thermophilic (optimum 37–42 °C), but can survive at low temperatures. It is sensitive to heating, drying, freezing, acidity, oxygen and low concentrations (2%) of salt. Symptoms in healthy people are described in Appendix B.1. The higher incidence of illness in very young children may indicate that there is protective immunity after infection. Immunocompromised people have a severe and prolonged illness, septicaemia or other infections. *C. jejuni* infection is also associated with development of Guillain-Barré syndrome (GBS), an autoimmune disease that causes limb weakness and paralysis that is sometimes fatal. It can also cause chronic arthritis, meningitis, abortion and neonatal sepsis (Motarjemi 2002). *C. jejuni* can form viable but non-culturable (VBNC) cells that are metabolically active but cannot be made to grow by culturing. This can be brought about by sub-lethal environmental conditions such as freeze–thaw injury. However, outbreaks of illness from VBNC cells are possible, indicating that they may recover in the intestines (McClure and Blackburn 2002).

Clostridium botulinum

Although bacteria of the genus *Clostridia* can be anaerobic to aerotolerant, *Clostridium botulinum* is strictly anaerobic and can only sporulate under anaerobic conditions. It is widely distributed in soils and marine and freshwater sediments. There are four phenotypes (numbered I–IV) but only groups I and II produce significant food poisoning. The different toxins produced by *Cl. botulinum* are labelled A–G, with Group I proteolytic bacteria producing A, B and F toxins and Group II non-proteolytic bacteria producing B, E and F toxins. The properties of each group are shown in Appendix B.1.

The toxins are among the most poisonous of natural toxins and block acetylcholine release across nerve synapses to cause muscular paralysis. Symptoms of botulism usually appear within 12–36 hours, but may be delayed by up to ten days. They include vomiting and nausea, quickly followed by double vision, speech impediment and difficulty in swallowing. This is followed by general muscular weakness, lack of coordination and respiratory failure. An antitoxin has been developed which reduces the mortality rate when it is administered quickly, but patients may still require artificial respiration to enable recovery. Because of the severity of the intoxication, particular care is taken by the canning industry to ensure that correct time/temperature combinations are used

(Chapter 13), and incidence of poisoning from commercial products is low. Most cases arise from home vegetable canning (in the USA), improperly cured or under-cooked fish or meats, or inclusion of fresh herbs and spices in cooking oils (Gibbs 2002).

Clostridium perfringens

Different strains (A-E) of *Cl. perfringens* (previously *Cl. welchii)* produce some of the four main types of enterotoxin. Type A *Cl. perfringens* is most common and results in relatively mild poisoning (Appendix B.1) that lasts for about 24 hours. Because of the short duration and relatively mild symptoms this type of poisoning is thought to be severely under-reported (Gibbs 2002). *Cl. perfringens* has a relatively high optimum temperature for growth (43–45 °C) and can grow up to 50 °C, doubling in number every 8–10 minutes under optimum conditions, making it one of the fastest growing food poisoning bacteria. The main sources are cooked meats that suffer temperature abuse or inadequate refrigeration after cooking. Inadequate cooking can also stimulate spore germination on cooling. Refrigeration or freezing kills cells but spores may survive to germinate and grow rapidly on reheating or thawing. Cell growth is inhibited by 6–8% salt and up to 400 μg/kg of nitrite, and properly cured meats are not usually a source of this pathogen.

Enteropathogenic Escherichia coli

Although *E. coli* is the most common non-pathogenic flora in the human intestine and it has long been used as an indicator of faecal contamination of foods, some strains have developed the ability to cause disease and illnesses attributed to *E. coli* have been acknowledged for 100 years. However, in recent years attention has increased because of significant morbidity and mortality in food poisoning outbreaks, particularly those associated with Vero cytotoxin-producing *E. coli* (Vero cytotoxigenic or VTEC). In industrialised countries the focus has been on *E. coli* O157:H7, but others including *E. coli* O26, O103, O111, O118 and O145 may pose an equal or greater threat to public health (Bell and Kyriakides 2002a). Of the pathogenic types of VTEC, there are differences in the virulence genes that result in six types of pathogenicity identified so far (Bell and Kyriakides 2002a, Scaletsky *et al.* 2002, Mulla 1999):

1 Entero-pathogenic *E. coli* (EPEC) – onset within 9–72 hours, duration 6 hours to 3 days. These types invade mucosal cells causing severe diarrhoea, fever, vomiting and abdominal cramps.
2 Entero-toxigenic *E. coli* (ETEC) – onset within 8–44 hours, duration 3–19 days. These adhere to small intestine mucosa and produce toxins that act on mucosal cells causing diarrhoea, cramps, nausea.
3 Entero-invasive *E. coli* (EIEC) onset within 8–72 hours, duration days–weeks. These types invade epithelial cells in the colon causing dysentery, vomiting, fever, headache and abdominal cramps.
4 Entero-haemorrhagic *E. coli* (EHEC) onset within 3–9 days, duration 2–9 days. These serotypes, including *E. coli* O157:H7 attach to mucosal cells and produce potent toxins, causing severe abdominal pain, bloody diarrhoea and vomiting but no fever. In young children and the elderly they may cause acute renal failure, seizures, coma and death.
5 Entero-aggregative *E. coli* (EAEC) onset within 7–48 hours, duration 14 days to weeks. These types bind in clumps to cells of the small intestine and produce toxins that cause persistent diarrhoea but not fever or vomiting.

6 Diffusely adherent *E. coli* (DAEC) onset and duration not yet established. Epidemiology and clinical profiles of the illness not yet established, but may cause diarrhoea in older children.

Methods to identify and control VTEC are described by Bell and Kyriakides (2002a).

Listeria monocytogenes

This is the most important pathogen among six *Listeria* species and it has two serotypes (4b and 1/2a). Listeriosis is rare and in healthy adults only causes mild flu-like symptoms or vomiting and diarrhoea when large numbers of cells are ingested (Appendix B.1). However, if elderly people or people who have compromised immune systems (e.g. patients taking immunosuppressant drugs after organ transplants, HIV/AIDS infection or cancer treatment), become infected they can develop meningitis, encephalitis and/or septicaemia, often with high mortality rates (Motarjemi 2002). Cross-infection in maternity hospitals and foodborne infections are the main sources of transmission. If pregnant women become infected, *Listeria* can cause infection of the uterus, bloodstream or central nervous system, resulting in spontaneous abortion, stillbirth, or infection of the foetus and birth of a premature severely ill baby. *L. monocytogenes* is psychrotrophic and can grow at refrigeration temperatures, and national and international standards as well as food company specifications describe maximum levels of contamination, especially for chilled and ready-to-eat foods (see also Chapter 21). Methods to identify and control *L. monocytogenes* are described by Bell and Kyriakides (2002b).

Mycobacterium paratuberculosis

Now known as *Mycobacterium avium* subsp. *paratuberculosis*, this bacterium causes Johne's disease in cattle and has been associated with the similar Crohn's disease in humans (incurable highly debilitating chronic inflammation of the gastrointestinal tract). It has been found in pasteurised milk but the results of a number of studies of its ability to survive pasteurisation conditions remain inconclusive (Griffiths 2002).

Plesiomonas spp.

A single species, *Pl. shigelloides*, shares similar characteristics to both *Vibrio* spp. and *Aeromonas* spp. It causes three types of diarrhoeal symptoms: secretory and shigella-like, both of which are more common and can vary in severity, and the less common cholera-like symptoms. It can also cause meningitis and has a mortality rate of *ca* 80% (Sutherland and Varnam 2002). It can invade intestinal cells and also has the ability to produce enterotoxins, protease, elastase and haemolysin. People at most risk include young children, the elderly and those suffering from cancer. *Plesiomonas* is found in freshwater and illness is most often caused by drinking contaminated water or eating raw shellfish from such water.

Salmonella spp.

Salmonella spp. are among the most important causes of foodborne disease worldwide and an individual outbreak may affect several thousand people at a time. Symptoms range from mild/severe food poisoning (gastroenteritis) to severe typhoid, paratyphoid and septicaemia (Appendix B.1). These severe conditions produce high rates of morbidity and mortality. There are an estimated 2400 serotypes, mostly designated as the species *S. enterica*, but new strains are evolving, some of which exhibit multiple antibiotic resistance. *Salmonella* spp. cause illness by invading intestinal cells and releasing an enterotoxin that causes inflammation and diarrhoea. They can also enter blood vessels and the lymphatic system to cause the more severe illnesses, including reactive arthritis, pancreatitis, osteomyelitis and meningitis. The numbers of ingested cells needed to cause illness is >10 000, but may be as low as 10–100 cells when infected fatty foods such as cheese or salami are eaten and the fat protects the cells. *Salmonella* may persist in faeces after recovery from illness, making the person a carrier and thus a potential hazard if employed as a food handler. The above data and methods to identify and control *Salmonella* are described by Bell and Kyriakides (2002c). Because of its importance, specified levels of *Salmonella* spp. in foods (negative in 25 g samples) are incorporated into national legislation and international standards. Appendix B.1 indicates the minimum growth conditions for *Salmonella* spp. when other conditions are optimal. However, some serotypes are able to survive for long periods under frozen storage (e.g. 4 months in poultry and minced beef at $-18\,°C$, or 7 years in ice cream at $-23\,°C$) and under conditions of low a_w (e.g. 9 months in chocolate at $a_w = 0.41$ and 6 weeks in peanut butter at $a_w = 0.2$–0.33) (Bell and Kyriakides 2002c).

Shigella spp.
The genus *Shigella* of the Enterobacteriaceae family has four subgroups (*Sh. dysenteriae, Sh. flexneri, Sh. boydii* and *Sh. sonnei*). They infect people mainly by person-to-person transmission and contaminated food and water in areas that have poor hygienic standards. The bacteria multiply in the colon and invade epithelial cells where they cause ulcerative lesions. *Sh. dysenteriae* produces a heat-sensitive cytotoxin (Shiga toxin) which kills the colon cells, and it may also produce an enterotoxin and neurotoxin. Among healthy adults it causes symptoms described in Appendix B.1, which may last for 3–4 days, but is rarely a cause of death. However, it may cause convulsions and delirium and is a common cause of mortality among immunocompromised people and among infants in places where hygiene is poor. *Shigella* is not found in the general environment and most infections are caused either by people who are carriers and infect foods, or by faecal contamination of crops. In practice almost any food could be a source of contamination if hygiene standards are low and there is a high incidence of *Shigella* infection in the general population.

Staphylococcus aureus
This pathogen is a normal part of the flora that is found on the skin and in nasal cavities of humans and animals. It can produce up to 11 enterotoxins when growing in food and intoxication can result from eating as little as 94–184 ng of one of the toxins (Varnam and Evans 1996). However, substantial cell growth is needed to produce sufficient toxin to cause illness and toxin production is more easily inhibited than cell growth by control of a_w and pH. The toxins differ from other enterotoxins in that they do not act directly on cells in the intestine, but act more like a neurotoxin, stimulating nerves that in turn stimulate the vomiting centre in the brain (Sutherland and Varnam 2002). Symptoms described in Appendix B.1 rarely last more than 24 h followed by rapid recovery, and deaths are rare. This relatively mild intoxication may be responsible for significant under-reporting of *St. aureus* illnesses.

   Staphylococcus aureus cells are readily destroyed by normal heating conditions used in processing, but the toxin is not. It is therefore possible for foods such as pasteurised dried milk and salami to contain the toxin without evidence of cellular contamination. Cells are unable to grow below $7\,°C$ and toxins are not produced below $10\,°C$, thus making refrigeration the best method of control for products that are not heated. *St.*

aureus is more resistant to preservatives such as NaCl and nitrite than many pathogens, enabling it to grow in cured meats. It is also able to grow in vacuum packed and MAP foods (Chapter 25, section 25.3), although toxin production may be inhibited. Its relatively high tolerance to low a_w (Appendix B.1) compared with other bacteria allows it to grow during manufacture of dried and intermediate moisture products (Chapter 16).

Vibrio spp.

There are ten *Vibrio* species that cause gastrointestinal illness, the most important being *V. parahaemolyticus* and *V. cholerae*, the latter causing Asiatic cholera. Another (*V. vulnificus*) also causes septicaemia. They are associated with seawater and seafood is the most common source of foodborne infection, although *V. cholerae* is also associated with contaminated freshwater as well as foods. *V. cholerae* has two serotypes (O1 and O139) that produce a number of enterotoxins, including haemolysins and cytotoxins. They cause copious diarrhoea within 6 hours to 3 days, with rapid loss of body fluids and mineral salts causing dehydration and leading to death if not treated by rehydration and salt replacement. Non-toxigenic strains cause less severe gastroenteritis, abdominal cramps and fever. In contrast, *V. parahaemolyticus* is an invasive non-toxigenic pathogen, which causes diarrhoea that lasts for 2–3 days, rarely causing death. There is also a more severe strain that produces Shiga-like cytotoxin and enterotoxin and causes dysentery. It is associated with consumption of raw fish and seafoods, or cooked seafoods that are recontaminated due to poor hygiene. In healthy people, *V. vulnificus* causes mild gastro-enteritis but in people suffering from medical conditions, including hepatitis, cirrhosis or gastric disease, it causes skin lesions and septicaemia, resulting in death in approximately 50% of cases. It destroys body tissues by secreting haemolysins, proteinases, collagenases and phospholipases. It is strongly associated with consumption of raw oysters, particularly in summer months when seawater temperatures are higher (Sutherland and Varnam 2002).

Yersinia enterocolitica

Yersinia enterocolitica and to a lesser extent *Y. pseudotuberculosis* are infectious foodborne pathogens associated mostly with pork. They are members of the family Enterobacteriaceae but unlike most other bacteria in this family, *Y. enterocolitica* is able to grow at 4 °C. In addition to the symptoms in healthy adults caused by infection of the intestinal tract, it may cause autoimmune thyroid disease, liver abscesses, pneumonia, septicaemia tissue infections, conjunctivitis and pharyngitis, especially among elderly and immunocompromised people. Children aged from young infants to young teenagers are most susceptible to infection. *Y. enterocolitica* is not especially heat-resistant and is destroyed at normal processing temperatures. However, recontamination of processed food is important because of its ability to grow at refrigeration temperatures. It is not inhibited by vacuum packing but is sensitive to carbon dioxide used in MAP (Chapter 25, section 25.3) (Sutherland and Varnam 2002).

Mycotoxins

Toxigenic moulds produce a variety of toxins when they grow on cereals (e.g. maize), legumes (e.g. peanuts (or groundnuts)), nuts, spices, pulses and oilseeds. Mould growth may occur on the growing crop, especially if it is subjected to drought stress, but more frequently it is due to inadequate drying of the harvested crop and/or humid storage conditions. In general the toxins do not produce acute food poisoning, as is the case with pathogenic bacteria, but cause chronic toxicity that may result in cancer, liver damage and/or immunosuppression. The most important types are aflatoxins, ochratoxin A, patulin and fumonisins.

Aflatoxins are produced by some strains of *Aspergillus* spp. including *A. parasiticus, A. flavus, A. nomius* and *A. ochraceoroseus*. The optimum temperature for toxin production is 30 °C and it is therefore most commonly found in tropical and sub-tropical regions. The most toxic is aflatoxin B_1, which is possibly carcinogenic and also acutely toxic, often fatally. Aflatoxin B_1 is metabolised in the body to aflatoxin M_1, and is then secreted in mothers' milk. As a result legislative standards for aflatoxin M_1 are more stringent than for aflatoxin B_1 because of the risk of consumption by very young children.

Fumonisins are produced by *Fusarium moniliforme* and other *Fusarium* species, primarily on maize when damp conditions exist during development of the cob, or on insect-damaged maize grains. Fumonisin B_1 causes a number of serious animal illnesses and may be linked to oesophageal cancer in humans.

Ochratoxin is produced in temperate climates by *Penicillium verrucosum* mainly on barley and other cereals, and also a number of *Aspergillus* species, especially *A. ochraceus*, which grows in tropical and sub-tropical regions on a wide variety of crops including cocoa, coffee, grapes and spices (Moss 2002). It is also found in the meat of animals that have consumed contaminated crops. It is carcinogenic, causing urinary tract tumours, and it also causes kidney damage. It remains present after infected coffee beans have been roasted and it has also been found in wine made from infected grapes.

Patulin is produced by *Penicillium* spp., mainly *P. expansum*, and some strains of *Aspergillus* spp. and *Byssochlamys* spp. that cause soft rot in fruits, particularly apples. Where mould growth is evident the fruits would normally be discarded during the sorting operation at the start of processing, but if mould growth takes place in the core of the fruit and whole fruits are used in processing (e.g. in apple juice production), the toxin may pass into the product. Under the acidic conditions in juices it can survive pasteurisation temperatures, but it is destroyed by fermentation during cider making or by treatment with sulphur dioxide.

Other mycotoxins that may be pathogenic include trichothecenes, which are all immunosuppressive, and are produced by several *Fusarium* spp. on cereals, T-2 toxin which is acutely toxic and produced by *F. sporotrichioides*, vomitoxin (deoxynivalenol) produced by *F. graminearum*, citrinin, which causes kidney damage and is produced by some species of *Penicillium*, and sterigmatocystin, produced by *A. versicolor* and found as part of the flora on the surface of cheeses stored at low temperatures for long periods (Moss 2002). Further details of mycotoxins in foods are given by Magan (2004).

Viruses and parasites
Viruses are much smaller than bacteria (22–110 nm) and replicate in living cells, but not in food or water. They may enter the living plant or animal or may contaminate foods due to infected water, food handlers, animals or insects. The most common foodborne pathogenic viruses contaminate food or water via faecal material from infected people, especially food handlers. Contaminated products have a normal colour, odour and taste. There are three groups of foodborne viruses: those that cause gastroenteritis; hepatitis viruses; and those that replicate in the intestine but migrate to other organs to cause illness. The most common cause of viral gastroenteritis is the Norwalk-like calicivirus (NLV), also known as small-round-structured viruses (SRSV). Other less common types include the enteric adenovirus (types 40/41), rotaviruses (groups A–C), sapporo-like caliciviruses (SLV), and astrovirus. In the second group, hepatitis A virus is more common than hepatitis E viruses. It is transmitted by water contaminated by faeces, in

shellfish from such waters, or by infected food handlers. Symptoms of fever, nausea and vomiting appear after 2–6 weeks, followed by hepatitis and damage to the liver (Koopmans 2002). Entroviruses are a less-common cause of illness in the third group. A summary of characteristic viral infections is given in Appendix B.2. There are no antiviral treatments for any of these infections. Most food- or water-borne viruses are relatively resistant to heat, acidity (pH > 3) and disinfection, and can survive for extended periods (days/weeks) on surfaces or in the air. In water some viruses (e.g. hepatitis A and poliovirus) can survive for one year at 4 °C (Biziagos *et al.* 1988). Further information is given by Riemann and Cliver (2006).

Protozoa, including *Giardia duodenalis*, *Cryptosporidium parvum*, *Cyclospora cayetanensis* and *Toxoplasma gondii* are pathogenic intestinal parasites that may con-taminate water and foods. They produce cysts or oocysts that are excreted in faeces and are capable of survival outside of the host for long periods in damp, dark environments. *Giardia* cysts and *Cryptosporidium* oocysts can immediately re-infect people, whereas *Cyclospora* oocysts require a period of maturation before they become infective. Their life cycles are described by Nichols and Smith (2002). Most cases of food poisoning are due to:

- contamination of raw fruits, salads or vegetables by food handlers who have both the infection and poor personal hygiene;
- use of contaminated water for making ice, washing salad vegetables and fruits, or incorporation in other foods that are eaten raw;
- contamination of crops by infected animals, birds, insects or slurry-spraying/manure.

In the short acute phase, giardiasis is characterised by flatulence, abdominal extension, cramps and diarrhoea. Chronic giardiasis causes maladsorption of nutrients resulting in weight loss and general malaise. The infectious dose is 25–100 cysts. Cryptosporidiosis causes acute, self-limiting gastroenteritis in otherwise healthy people, with symptoms starting within 3–14 days after infection and lasting up to two weeks. Symptoms include diarrhoea, severe abdominal pain, nausea, flatulence, vomiting, mild flu-like fever and weight loss. In immunocompromised people or those receiving immunosuppressive drugs, similar symptoms persist and develop into serious weight loss and infections of the gastrointestinal tract, the respiratory tract, gall bladder and pancreas. These symptoms persist until death unless immunosuppressive therapies are discontinued (Nichols and Smith 2002). Cyclosporiasis results in a flu-like fever with diarrhoea, fatigue, nausea, vomiting, abdominal pain and weight loss. Symptoms begin within 2–11 days of ingesting oocysts and can last for more than six weeks in healthy people, but are prolonged in immunocompromised patients. *Toxoplasma gondii* produces oocysts that remain infective in water or soil for up to one year. Although cats are the reservoir for *T. gondii*, cattle, sheep, goats and pigs are intermediate hosts, and infection can occur by eating undercooked or raw meat that contains the oocysts. Symptoms include transplacental infection during pregnancy that may cause stillbirth or perinatal death, or in surviving babies, toxoplasmosis involving damage to sight, hearing or the central nervous system. Some infected babies show symptoms of hydrocephalus and mental retardation.

Other parasites include flatworms (trematodes), which are most commonly found in watercress, salad plants and raw or undercooked freshwater fish and shellfish. They can cause a variety of acute infections including fever, fatigue, diarrhoea, abdominal pain and jaundice. Chronic health problems resulting from infections include damage to the liver, spleen and pancreas, dwarfism and retardation of sexual development (Motarjemi 2002).

Two species of *Taenia* are foodborne parasites: *T. saginata* in cattle and *T. solium* in pigs. *T. solium* causes both intestinal infection by worms contained in undercooked or raw pork, and infection by its eggs (cysticercosis). The eggs may be consumed in foods or water that is contaminated with faeces. They hatch in the intestine and larvae migrate to other organs, including the eye, heart and central nervous system. Symptoms may take between a few days and 10 years to appear and can be very serious, including seizures, psychiatric disturbance and death. Raw or undercooked meat may also contain larvae of *Trichinella spiralis*. Infection is caused when the larvae develop into adults in the intestine, causing nausea, vomiting diarrhoea and fever. The adult worm then produces larvae which migrate through the blood and lymphatic systems to muscles, where they produce rheumatic conditions, followed by damage to the eyes (retinal haemorrhage, photophobia) and then profuse sweating, prostration and cardiac and neurological complications 3–6 weeks later. Death may result from heart failure.

1.3   Types of processing

In general, adequate food safety and shelf-life or the alteration of sensory characteristics cannot be achieved using a single type of processing (such as heating or pH control) and multiple methods (operations) are used. The development of the hurdle concept (section 1.3.1) expanded and developed the idea of multiple preservation factors to inhibit microbial growth and ensure food safety and an adequate shelf-life. In traditionally preserved foods, such as smoked fish or meat, jams and other preserves, there are a combination of factors that ensure microbiological safety and stability of the food, and thus enable it to be preserved for the required shelf-life. In smoked products for example, this combination includes heat, reduced moisture content (a_w) and antimicrobial chemicals deposited from the smoke on to the surface of the food. Some smoked foods may also be dipped or soaked in brine or rubbed with salt before smoking to add a further preservative mechanism. Smoked products may also be chilled or packed in modified atmospheres to further extend the shelf-life. In jams and other fruit preserves, the combined factors are heat, a high solids content (reduced a_w) and high acidity. These preservative factors also strongly influence the sensory characteristics of the product and contribute to important differences in flavour, texture or colour between different products. In vegetable fermentation (Chapter 6, section 6.4.3), the desired product quality and microbial stability are achieved by a sequence of factors that arise at different stages in the fermentation process: the addition of salt selects the initial microbial population, which uses up the available oxygen in the brine. This reduces the redox potential and inhibits the growth of aerobic spoilage micro-organisms and favours the selection of lactic acid bacteria. These then acidify the product and stabilise it. Further treatments may include pasteurisation (Chapter 12) and packaging (Chapter 25) to extend the shelf-life. Similar changes take place during fermentation of milk to yoghurt (Chapter 6, section 6.4.3).

1.3.1   Hurdle concepts
The demand by consumers for high-quality foods having 'fresh' or 'natural' characteristics but with an extended shelf-life has led to the development of foods that are preserved using mild technologies (see minimal processing (Chapter 9) and chilling (Chapter 21)). The concept of combining several factors to preserve foods has been developed by Leistner (1995) and others into the 'hurdle' concept (in which each factor is

a hurdle that micro-organisms must overcome). This in turn has led to the application of hurdle technology (also known as 'combined processes', 'combination preservation' or 'combination techniques'), where an understanding of the complex interactions of temperature, a_w, pH, chemical preservatives, etc. (Table 1.18) is used to design a series of hurdles to control the growth of spoilage or pathogenic micro-organisms and ensure microbiological safety of processed foods. Micro-organisms maintain a stable internal environment (homeostasis), but preservatives that act as hurdles can disturb one or more of the homeostasis mechanisms, thereby causing micro-organisms to remain inactive or even die. Hurdle technology using multiple hurdles has the least effect on product quality, because it permits hurdles of lower intensity to be used, and different hurdles in a food have both an added effect on stability and can act synergistically.

Table 1.18 Examples of hurdles used to preserve foods

Type of hurdle	Examples	Chapters containing additional information
Physical hurdles	Aseptic packaging	13, 25
	Electromagnetic energy (microwave, radio frequency, pulsed magnetic fields, pulsed electric fields)	9, 20
	High temperatures (blanching, pasteurisation, heat sterilisation, evaporation, extrusion, baking, frying)	11–19
	Ionising radiation	7
	Low temperatures (chilling, freezing)	21, 22
	Modified atmospheres	21, 25
	Packaging (including active packaging)	25, 26
	Photodynamic inactivation	9
	Ultra-high pressures	8
	Ultrasonication	9
	Ultraviolet radiation	9
Physicochemical hurdles	Carbon dioxide	6, 21, 25
	Ethanol	1, 6
	Lactic acid	1, 6
	Lactoperoxidase	–
	Low pH	1, 6
	Low redox potential	1, 6
	Low water activity	1, 14–19, 22, 23
	Maillard reaction products	1, 15–19
	Organic acids	1, 6
	Oxygen	1
	Ozone	2
	Phenols	1
	Phosphates	1
	Salt	1, 6, 17
	Smoking	17
	Sodium nitrite/nitrate	1, 17
	Sodium or potassium sulphite, sulphur dioxide	1, 16
	Spices and herbs	–
	Surface treatment agents	1, 9, 27
Microbially-derived hurdles	Antibiotics	6
	Bacteriocins	6
	Competitive flora	6
	Protective cultures	6

Adapted from Leistner and Gorris (1995)

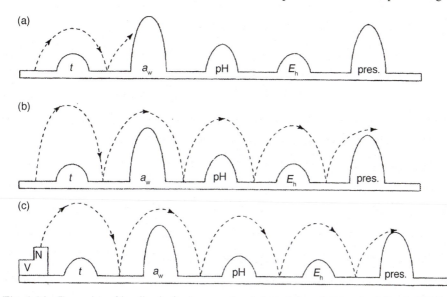

Fig. 1.16   Examples of hurdles in food processing (adapted from Leistner and Gorris 1995). (t = chilling; a_w = low water activity; pH = acidification; E_h = low redox potential; pres. = preservatives; V = vitamins; N = nutrients).

To be successful, the hurdles must take into account the initial numbers and types of micro-organisms that are likely to be present in the food. The hurdles that are selected should be 'high enough' so that the anticipated numbers of these micro-organisms cannot overcome them. However, the same hurdles that satisfactorily preserve a food when it is properly prepared (Fig. 1.16a), are overcome by a larger initial population of micro-organisms (Fig. 1.16b), when for example raw materials are not adequately cleaned (Chapter 2, section 2.2). In this example, the main hurdles are low water activity and chemical preservatives in the product, with storage temperature, pH and redox potential having a smaller effect. Blanching vegetables or fruits (Chapter 11) has a similar effect in reducing initial numbers of micro-organisms before freezing or drying. If in Fig. 1.16 the same hurdles are used with a different product that is richer in nutrients that can support microbial growth (Fig. 1.16c), again the hurdles may be inadequate to preserve it and a different combination may be needed or the height of the hurdles increased. It should be noted that although the hurdles in Fig. 1.16 are represented as a sequence, in practice the different factors may also operate simultaneously or synergistically.

As an example of hurdle technology, fermented sausages (e.g. salami) are produced using a sequence of hurdles: salt and sodium nitrite preservatives inhibit many contaminating bacteria, allowing other bacteria to multiply and use up oxygen, thereby causing the redox potential to fall; this inhibits aerobic micro-organisms and selects lactic acid bacteria, which acidify the meat and increase the pH hurdle; during ripening, the moisture content falls and causes the water activity to decrease. The final product therefore has low water activity as the main hurdle, with lower contributions from nitrite, redox potential and pH, making it stable at ambient temperature for an extended shelf-life.

The combination of hurdle technology and HACCP (section 1.5.1) in process design is described by Leistner (1994). By combining hurdles, the intensity of individual preservation techniques can be kept comparatively low to minimise loss of product quality, while overall there is a high impact on controlling microbial growth. Examples of novel mild processes that retain product quality are described in Chapter 9 ('minimal' processes), Chapter 20, section 20.2 (ohmic heating), Chapter 21 (chilling) and Chapter 25, section 25.3 (modified atmospheres).

1.3.2   Heat transfer

Preservation of foods by heating remains one of the most important methods of processing and the theoretical aspects of heat transfer are described in Chapter 10, with examples of processing methods given in Chapters 11–20. Preservation by removing heat is described in Chapters 21–23.

1.3.3   Moisture management

The transfer of matter is an important aspect of a large number of food processing operations, especially in solvent extraction (Chapter 5, section 5.4), membrane processing (Chapter 5, section 5.5) and distillation (Chapter 14, section 14.2). It is also an important factor in loss of nutrients during blanching (Chapter 11). Mass transfer of gases and vapours is a primary factor in evaporation (Chapter 14), dehydration (Chapter 16), baking and roasting (Chapter 18), frying (Chapter 19), freeze drying (Chapter 23), the cause of freezer burn during freezing (Chapter 22, section 22.2.4) and a cause of loss in food quality in chilled, MAP and packaged foods (Chapters 21 and 25 respectively).

In an analogous way to heat transfer (Chapter 10, section 10.1.2), the two factors that influence the rate of mass transfer are a driving force to move materials and a resistance to their flow. When considering dissolved solids in liquids, the driving force is a difference in the solids' concentration, whereas for gases and vapours it is a difference in partial pressure or vapour pressure. The resistance arises from the medium through which the liquid, gas or vapour moves and any interactions between the medium and the material. An example of materials transfer is diffusion of water vapour through a boundary layer of air in operations such as dehydration and baking. Packaging also creates additional boundary layers which act as barriers to movement of moisture and to heat transfer (Fig. 1.17).

The rate of diffusion is found using:

$$N_A = \frac{D_w}{RTx} \cdot \frac{P_T}{P_{Am}} (P_{w1} - P_{w2}) \hspace{2cm} \boxed{1.18}$$

where N_A (kg s^{-1} or kmol s^{-1}) = rate of diffusion, D_w = diffusion coefficient of water vapour in air, R = the gas constant (= 8.314 kJ kmol^{-1} K^{-1}), T (K) = temperature, x (m) = distance across stationary layer, P_T (kN m^{-2}) = total pressure, P_{Am} (kN m^{-2}) = mean pressure of non-diffusing gas across the stationary layer and $P_{w1} - P_{w2}$ (kN m^{-2}) = water vapour pressure driving force.

Formulae for diffusion of solutes through liquids and for gases dissolving in liquids are given by Toledo (1999d) and Lewis (1990).

Mass balances

The law of conversion of mass states that 'the mass of material entering a process equals the mass of material leaving'. This has applications in, for example, mixing (Chapter 4, section 4.1), fermentation (Chapter 6, section 6.4) and evaporation (Chapter 14, section 14.1).

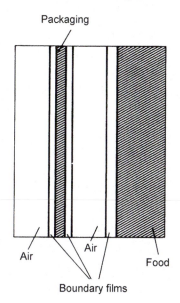

Fig. 1.17   Barriers to mass transfer (and heat flow) in packaged food.

In general a mass balance for a process takes the following form:

Mass of raw materials in = mass of products and wastes out
+ mass of stored materials + losses $\boxed{1.19}$

Mass balances are used to calculate the quantities of materials in different process streams, to design processes, to calculate recipe formulations, the composition after blending, process yields and separation efficiencies. Many mass balances are analysed under steady-state conditions where the mass of stored materials and losses are equal to zero. A typical mass balance is shown in Fig. 1.18.

Here the total mass balance is

$$W + A = \text{moist air} + D \qquad \boxed{1.20}$$

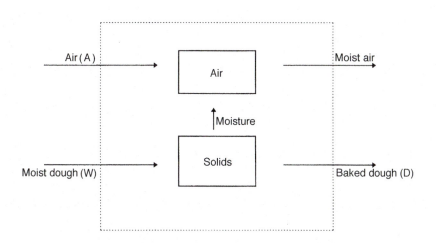

Fig. 1.18   Diagram of material flow during baking in an oven.

The mass balance for air is

$$A + \text{moisture} = \text{moist air} \qquad \boxed{1.21}$$

The mass balance for solids is

$$W = \text{moisture} + D \qquad \boxed{1.22}$$

Examples of mass balance calculations are shown in Chapter 14 (sample problems 14.2 and 14.3).

In applications involving concentration or dilution, the use of mass fraction or weight percentage is often used:

$$\text{Mass fraction A} = \frac{\text{mass of component A}}{\text{total mass of mixture}} \qquad \boxed{1.23}$$

or

$$\text{Total mass of mixture} = \frac{\text{mass of component A}}{\text{mass fraction A}} \qquad \boxed{1.24}$$

If the mass of the component and its mass fraction are known, the total mass of the mixture can be calculated.

Sample problem 1.3

Calculate the total mass balance and component mass balance for mixing ingredients to make 25 kg of beef sausages having a fat content of 30%, using fresh beef meat and beef fat. Typically, beef meat contains 18% protein, 12% fat and 68% water and beef fat contains 78% fat, 12% water and 5% protein.

Solution to sample problem 1.3

Let F = mass of beef fat (kg)

Let M = mass of beef meat (kg)

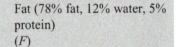

Fat (78% fat, 12% water, 5% protein) (F)

Meat (M)
18% protein, 12% fat and 68% water

Bowl chopper

Sausage (S)
(25 kg, 30% fat)

Total mass balance

$$M + F = 25$$

Fat balance

$$0.12M + 0.78F = 0.3(25)$$

Substitute $M = 25 - F$ into the fat balance

$$0.12(25 - F) + 0.78F = 7.5$$
$$3.0 - 0.12F + 0.78F = 7.5$$
$$= 6.82 \text{ kg}$$

and

$$M = 25 - 6.82$$
$$= 18.18 \text{ kg}$$

A simple method to calculate the relative masses of two materials that are required to form a mixture of known composition is the 'Pearson Square' (Anon 1996b). If, for example, homogenised milk (3.5% fat) is to be mixed with cream (20% fat) to produce a light cream containing 10% fat, the Pearson Square (Fig. 1.19) is constructed with the fat composition of ingredients on the left side, the fat content of the product in the centre. By subtracting diagonally across the square, the resulting proportions of milk and cream can be found (i.e. 10 parts milk and 6.5 parts cream in Fig 1.19). Toledo (1999e) describes that application of mass balances to multi-stage processing and includes a computer programme for the use of mass balances in recipe formulations.

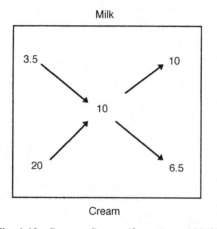

Fig. 1.19   Pearson Square (from Anon 1996b).

1.3.4   Fluid flow
Fluid flow
Many types of liquid food are transported through pipes during processing, and powders and small-particulate foods are more easily handled as fluids (by fluidisation). Gases obey the same laws as liquids and, for the purposes of calculations, gases are treated as compressible fluids. Properties of selected fluids are shown in Table 1.8. The study of fluids is therefore of great importance in food processing. It is divided into fluid statics (stationary fluids) and fluid dynamics (moving fluids).

A property of static liquids is the pressure that they exert on the containing vessel. The pressure is related to the density of the liquid and the depth or mass of liquid in the vessel. Liquids at the base of a vessel are at a higher pressure than at the surface, owing to the

Sample problem 1.4

Use a Pearson Square to calculate the amounts of orange juice (10% sugar content) and sugar syrup (60% sugar content) needed to produce 50 kg of fruit squash containing 15% sugar.

Solution to sample problem 1.4

Orange juice

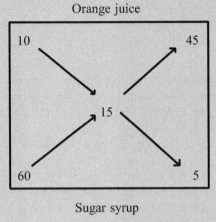

Sugar syrup

The result indicates that 45 kg of orange juice should be mixed with 5 kg of sugar syrup.

weight of liquid above (the 'hydrostatic head'). This is therefore important in the design of holding tanks and processing vessels, to ensure that the vessel is constructed using materials of adequate strength. A large hydrostatic head also affects the boiling point of liquids, which is important in the design of some types of evaporation equipment (Chapter 14, section 14.1.1).

When a fluid flows through pipes or processing equipment (Fig. 1.20), there is a loss of energy and a drop in pressure which are due to frictional resistance to flow. These friction losses and changes in the potential energy, kinetic energy and pressure energy are described in detail in food engineering texts. The loss of pressure in pipes is determined by a number of factors including the density and viscosity of the fluid, the length and diameter of the pipe and the number of bends, valves or other fittings in the pipeline. To overcome this loss in energy, it is necessary to apply power using pumps to transport the fluid. The amount of power required is determined by the viscosity of the fluid (section 1.1.2), the size of the pipework, the number of bends and fittings and the height and distance that the fluid is to be moved.

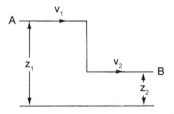

Fig. 1.20   Application of Bernoulli's equation to frictionless fluid flow.

Bernoulli's equation, which is a statement of the conservation of energy, is used to calculate the energy balance when a liquid flows through pipework, the effect of valves or bends on the flow rate, or the pressure developed by a pump.

$$\frac{P_1}{\rho_1} + \frac{v_1^2}{2} = z_1 g = \frac{P_2}{\rho_2} + \frac{v_2^2}{2} + z_2 g \qquad \boxed{1.25}$$

where P (Pa) = the pressure, ρ (kg m^{-3}) = the fluid density, g (= 9.81 m s^{-1}) = acceleration due to gravity, v (m s^{-1}) = the velocity of the fluid and z (m) = the height. The subscript $_1$ indicates the first position in the pipework and the subscript $_2$ the second position in the pipework.

Sample problem 1.5

A 20% sucrose solution flows from a mixing tank at 50 kPa through a horizontal pipe 5 cm in diameter at 25 m^3 h^{-1}. If the pipe diameter reduces to 3 cm, calculate the new pressure in the pipe. (The density of sucrose solution is 1070 kg m^{-3} (Table 1.8).)

Solution to sample problem 1.5

$$\text{Flow rate} = \frac{25}{3600} \text{ m}^3 \text{ s}^{-1}$$

$$= 6.94 \times 10^{-3} \text{ m}^3 \text{ s}^{-1}$$

$$\text{Area of pipe 5cm in diameter} = \frac{\pi}{4} D^2$$

$$= \frac{3.142}{4} (0.05)^2$$

$$= 1.96 \times 10^{-3} \text{ m}^2$$

$$\text{Velocity of flow} = \frac{6.94 \times 10^{-3}}{1.96 \times 10^{-3}}$$

$$= 3.54 \text{ m s}^{-1}$$

$$\text{Area of pipe 3 cm in diameter} = 7.07 \times 10^{-4} \text{ m}^2$$

$$\text{Velocity of flow} = \frac{6.94 \times 10^{-3}}{7.07 \times 10^{-4}}$$

$$= 9.82 \text{ m s}^{-1}$$

Using Equation (1.25),

$$\frac{3.54^2}{2} + \frac{50 \times 10^3}{1070} + 0 = \frac{P_2}{1070} + \frac{9.82^2}{2} + 0$$

Therefore,

$$P_2 = 5120 \text{ Pa}$$

$$= 5.12 \text{ kPa}$$

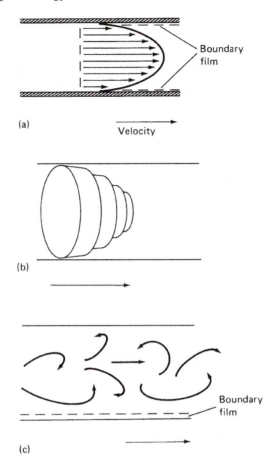

Fig. 1.21   Fluid flow: (a) velocity distribution and boundary layer; (b) streamline flow; (c) turbulent flow.

In any system in which fluids flow, there exists a boundary film of fluid next to the surface over which the fluid flows (Fig. 1.21a). The thickness of the boundary film is influenced by a number of factors, including the velocity, viscosity, density and temperature of the fluid. Fluids which have a low flow rate or high viscosity may be thought of as a series of layers which move over one another without mixing (Fig. 1.21b). This produces movement of the fluid, which is termed 'streamline' (or laminar) flow. In a pipe, the velocity of the fluid is highest at the centre and zero at the pipe wall. Above a certain flow rate, which is determined by the nature of fluid and the pipe, the layers of liquid mix together and 'turbulent' flow is established (Fig. 1.21c) in the bulk of the fluid, although the flow remains streamline in the boundary film. Higher flow rates produce more turbulent flow and hence thinner boundary films.

Fluid flow is characterised by a dimensionless group named the 'Reynolds number' (Re). This is calculated using:

$$Re = \frac{Dv\rho}{\mu}$$

1.26

where D (m) = the diameter of the pipe, v (m s^{-1}) = average velocity, ρ (kg m^{-3}) = fluid density and μ (N s m^{-2}) = fluid viscosity.

A Reynolds number of less than 2100 describes streamline flow and a Reynolds number of more than 4000 describes turbulent flow. For Reynolds numbers between 2100 and 4000, 'transitional' flow is present, which can be either laminar or turbulent at different times. These different flow characteristics have important implications for heat transfer and mixing operations; turbulent flow produces thinner boundary films, which in turn permit higher rates of heat transfer. The implications of this for the design and performance of equipment are discussed in Chapter 10 (section 10.1.2) for liquids through pipes or over metal plates, and Chapter 16 (section 16.1) for air moving over the surface of food or metal. The Reynolds number can also be used to determine the power requirements for pumps and mixers (Chapter 4, section 4.1, and Chapter 27, section 27.1.3). A relevant sample problem is given in Chapter 12 (sample problem 12.2).

In turbulent flow particles of fluid move in all directions and solids are retained in suspension more readily. This reduces the formulation of deposits on heat exchangers and prevents solids from settling out in pipework. Streamline flow produces a wider range of residence times for individual particles flowing in a tube. This is especially important when calculating the residence time required for heat treatment of liquid foods, as it is necessary to ensure that all parts of the food receive the required amount of heat. This aspect is discussed in more detail in relation to aseptic and ohmic heating (Chapter 13, section 13.2, and Chapter 20, section 20.2 respectively). Turbulent flow causes higher friction losses than streamline flow does and therefore requires higher energy inputs from pumps.

Sample problem 1.6
Two fluids, milk and rapeseed oil, are flowing along pipes of the same diameter (5 cm) at 20 °C and at the same flow velocity of $3\,\text{m}\,\text{s}^{-1}$. Determine whether the flow is streamline or turbulent in each fluid. (Physical properties of milk and rapeseed oil are shown in Table 1.8.)

Solution to Sample problem 1.6
For milk from Table 1.8, $\mu = 2.10 \times 10^{-3}\,\text{N s m}^{-2}$ and $p = 1030\ \text{kg m}^{-3}$. From Equation 1.26,

$$\text{Re} = \frac{Dvp}{\mu}$$

Therefore,

$$\text{Re} = \frac{0.05 \times 3 \times 1030}{2.1 \times 10^{-3}}$$

$$= 73\,571$$

Thus the flow of milk is turbulent (Re > 4000).

For rapeseed oil, from Table 1.8, $\mu = 118 \times 10^{-3}\,\text{N s m}^{-2}$ and $p = 900\ \text{kg m}^{-3}$. Therefore

$$\text{Re} = \frac{0.05 \times 3 \times 900}{118 \times 10^{-3}}$$

$$= 1144$$

Thus the flow of oil is streamline (Re < 2100).

Fluid flow through fluidised beds

When air passes upwards through a bed of food, the particles create a resistance to the flow of air and reduce the area available for it to flow through the bed. As the air velocity is increased, a point is reached where the weight of the food is just balanced by the force of the air, and the bed becomes fluidised (e.g. fluidised-bed drying (Chapter 16, section 16.1)). If the velocity is increased further, the bed becomes more open (the 'voidage' is increased), until eventually the particles are conveyed in the fluid stream (e.g. pneumatic separation (Chapter 2, section 2.2.2), pneumatic drying (Chapter 16, section 16.2.1) or pneumatic conveying (Chapter 27, section 27.1.2)). The velocity of the air needed to achieve fluidisation of spherical particles is calculated using:

$$v_f = \frac{(\rho_s - \rho)g}{\mu} \frac{d^2 \epsilon^3}{180(1 - \epsilon)}$$

$\boxed{1.27}$

where v_f (m s^{-1}) = fluidisation velocity, ρ_s (kg m^{-3}) = density of the solid particles, ρ (kg m^{-3}) = density of the fluid, g (m s^{-2}) = acceleration due to gravity, μ (N s m^{-2}) = viscosity of the fluid, d (m) = diameter of the particles, ϵ = the voidage of the bed.

The minimum air velocity needed to convey particles is found using:

$$v_e = \sqrt{\left[\frac{4d(\rho_s - \rho)}{3C_d \rho}\right]}$$

$\boxed{1.28}$

where v_e (m s^{-1}) = minimum air velocity and C_d (= 0.44 for Re = 500–200 000) = the drag coefficient.

Sample problem 1.7

Peas which have an average diameter of 6 mm and a density of 880 kg m^{-3} are dried in a fluidised-bed dryer (Chapter 16). The minimum voidage is 0.4. Calculate the minimum air velocity needed to fluidise the bed if the air density is 0.96 kg m^{-3} and the air viscosity is 2.15×10^{-5} N s m^{-2}.

Solution to sample problem 1.7
From Equation 1.27,

$$v_f = \frac{(880 - 0.96)9.81}{2.15 \times 10^{-5}} \frac{(0.006)^2(0.4)^3}{180(1 - 0.4)}$$

$$= 8.5 \text{ m s}^{-1}$$

1.3.5   Biochemical preservation

Traditional biochemical preservation methods involve either direct microbial destruction using preservative chemicals (e.g. acetic acid and sodium chloride in pickling (Chapter 6, section 6.4.3) or chemicals absorbed from smoke (Chapter 17, section 17.1.1)), or by reducing the pH of the food to a level that prevents the growth of micro-organisms (e.g. milk fermentation to yoghurt (Chapter 6, section 6.4.3)). Other commonly used preservative chemicals include sodium or potassium nitrates and nitrites that are used to preserve meats such as ham and bacon, and sulphites that are used to prevent microbial spoilage (and enzymic browning) of fruits and vegetables (section 1.1.1). In many countries, there is increasing consumer and medical pressure to adopt processing methods that reduce the amounts of preservatives added to foods because of health concerns: salt because of its effects on blood pressure; smoke chemicals (Chapter 17, section 17.3) and nitrites/nitrates because of potential problems of carcenogenicity, and sulphites because of their inducement of asthma, nausea and vomiting in sulphite-sensitive people. Newer antimicrobial compounds, bacteriocins, are used in a wide variety of foods (Chapter 6, section 6.7), either by adding bacteria that produce the bacteriocin, or adding the purified bacteriocin itself to reduce the risk of pathogen growth.

1.3.6   Preservation by modification of environmental conditions

A number of minimal processing methods employ modification of environmental conditions including irradiation (Chapter 7), high-pressure processing (Chapter 8), modification of gas composition (Chapters 21, section 21.1.2, and 25, section 25.3), use of high-intensity light or ultraviolet light (Chapter 9, section 9.4). The mechanisms of operation and methods of processing are described in the relevant chapters.

1.3.7   Calculation of shelf-life

The shelf-life of foods depends on the formulation of ingredients, the method(s) of processing, the type of packaging and the storage conditions. It is necessary to understand the interaction between these four factors and to optimise each to achieve the required shelf-life. The quality indices for measuring shelf-life include chemical changes (e.g. lipid oxidation leading to rancidity), microbiological changes, changes in a sensory attribute or loss of a particular nutrient. Changes in sensory quality can be assessed by chemical analysis (e.g. monitoring production of a particular chemical such as trimethylamine in fish) or by sensory analysis using a trained taste panel. Further details are given by Singh and Cadwallader (2004) and methods of sensory analysis using taste panels are described by Meilgaard et al. (1999).

An alternative to the Arrhenius equation (section 1.2) for calculating shelf-life is the Q_{10} concept. This is the ratio of reaction rate constants at temperatures that differ by $10\,°C$ (equation 1.29) and is a similar concept to the z-value in microbial inactivation (Chapter 10, section 10.3).

$$Q_{10} = K_{(\theta+10)}/K_\theta \qquad \boxed{1.29}$$

$$= t_s(\theta)/t_s\,(\theta + 10) \qquad \boxed{1.30}$$

where K = reaction rate at temperature θ and t_s (days) = shelf-life.

Q_{10} shows the change in shelf-life if a food is stored at a temperature that is $10\,°C$ higher, and when $\ln k$ is plotted against temperature a straight line is obtained (Fig. 1.22). Additional formulae for calculating shelf-life of foods that undergo glass transitions (section 1.4.1) are given by Taoukis and Giannakourou (2004).

For most perishable products shelf-life is based on changes to sensory qualities or microbiological quality that take place within a few weeks or months. However, with long shelf-life products (e.g. frozen foods, biscuits) deterioration may be caused by slow biochemical changes over a year or more. Accelerated shelf-life testing (ASLT) is used to reduce the time needed to assess shelf-life. Studies of the loss of quality indices are made at an elevated temperature (typically $20\,°C$ above normal storage temperature) and the kinetic results are extrapolated to normal storage temperatures. In this way a study that would take a year can be completed within a month. Details of ASLT methodologies are

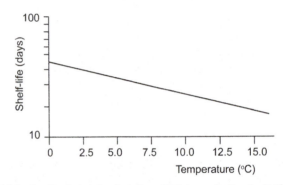

Fig. 1.22   Shelf-life plot for fruit juice based on 50% loss of vitamin C (from Taoukis and Giannakourou 2004).

described by Mizrahi (2004) and Man (2004) and further information is given by Steele (2004), Lees (2003) and Singhal *et al.* (1997).

1.4   Effects of processing

1.4.1   Effects on physical properties of foods
Liquids, gases and some solids (e.g. powders and particulate materials) are termed 'fluids' and can flow without disintegration when a pressure is applied to them. In contrast, solids deform when pressure is applied to them (section 1.1.4). In this section, the properties of fluids and solids that are relevant to both the design of food processes and the quality of processed food are described. More detailed mathematical treatments and derivations of formulae used in food engineering calculations are given in a number of texts including Toledo (1999a), Lewis (1990) and Brennan *et al.* (1990).

Phase transition: properties of liquids, solids and gases
The transition from solid to fluid and back is known as a 'phase transition' and this is important in many types of food processing (e.g. water to water vapour in evaporation and distillation (Chapter 14) and dehydration (Chapter 16); water to ice in freezing (Chapter 22) and freeze drying or freeze concentration (Chapter 23) or crystallisation of fats (section 1.1.1 and Chapter 24, section 24.1.1)). Phase transition takes place isothermally at the phase transition temperature by release or absorption of latent heat, and it can be represented by a 'phase diagram' (Chapter 10, Fig. 10.1).

Glass transition
A second type of transition, known as 'glass transition', takes place without the release or absorption of latent heat and involves the transition of a food to an amorphous glassy state at its glass transition temperature. The transition depends on the temperature of the food, time and the moisture content of the food. Examples of glass transition temperatures are given in Chapter 22 (Table 22.2). When materials change to glasses, they do not become crystalline, but retain the disorder of the liquid state. However, their physical, mechanical, electrical and thermal properties change as they undergo the transition. In their glassy state, foods become very stable because compounds that are involved in chemical reactions that lead to deterioration are immobilised and take long periods of time to diffuse through the material to react together. Details of nine key concepts underlying the relationship between molecular mobility and food stability are

given by Fennema (1996). Processes that are significantly influenced by transition to a glassy state include aroma retention, crystallisation, enzyme activity, microbial activity, non-enzymic browning, oxidation, agglomeration and caking. The relationship between glass transition and water activity (section 1.1.2) and factors that affect glass transition are described in detail by Rahman (1999), Fennema (1996) and Blanshard (1995). Further details are given in Chapter 22 (section 22.1.2).

1.4.2   Effects on the sensory properties of foods

A continuing aim of food manufacturers is to find improvements in processing technology that retain or create desirable sensory qualities or reduce the damage to food caused by processing. Examples of such improvements are described in subsequent chapters.

Changes in texture (section 1.1.4) are caused by loss of moisture or fat, formation or breakdown of emulsions and gels, hydrolysis of polymeric carbohydrates, and coagulation or hydrolysis of proteins. The taste of foods is largely determined by the formulation used for a particular food and is mostly unaffected by processing. Exceptions to this include respiratory changes to fresh foods (Chapter 21, section 21.3.1) that increase sweetness and changes in acidity or sweetness during food fermentations (Chapter 6, section 6.4). Characteristic flavours and aromas may be lost during processing, which reduces the intensity of flavour or reveals other flavour/aroma compounds. Volatile aroma compounds are also produced by the action of heat, ionising radiation, and oxidation or enzyme activity on proteins, fats and carbohydrates. Examples include the Maillard reaction between amino acids and reducing sugars or carbonyl groups and the products of lipid degradation (also Chapters 18 and 19), or hydrolysis of lipids to fatty acids and subsequent conversion to aldehydes, esters and alcohols. The perceived aroma of foods arises from complex combinations of many hundreds of compounds, some of which act synergistically. More detailed descriptions of the production of aroma compounds or loss of naturally occurring aroma compounds are given by Lindsay (1996b) and in subsequent chapters.

Many naturally occurring pigments are destroyed by heat processing, chemically altered by changes in pH or oxidised during storage (Table 1.6). As a result the processed food may lose its characteristic colour and hence its value. Synthetic colourants (Appendix A4) are more stable to heat, light and changes in pH, and they are therefore added to retain the colour of some processed foods. Details of changes to naturally occurring pigments are described in subsequent chapters. Maillard browning is an important cause of both desirable changes in food colour (e.g. in baking or frying (Chapter 18, section 18.3.1, and Chapter 19, section 19.4.2 respectively), and in the development of off-colours (e.g. during canning (Chapter 13, section 13.3) and drying (Chapter 16, section 16.5.1)).

1.4.3   Effects on nutritional value of foods

Many unit operations, especially those that do not involve heat, have little or no effect on the nutritional quality of foods. Examples include mixing, cleaning, sorting, freeze drying and pasteurisation. Unit operations that intentionally separate the components of foods (Chapters 5 and 14), alter the nutritional quality of each fraction compared with the raw material. Unintentional separation of water-soluble nutrients (minerals, water-soluble vitamins and sugars) also occurs in some unit operations (e.g. blanching (Chapter 11), and in drip losses from roast or frozen foods (Chapters 18 and 22)).

Heat processing is a major cause of changes to nutritional properties of foods. For example, improvements in nutritional value due to heat include gelatinisation of starches and coagulation of proteins, which improve their digestibility, and destruction of anti-nutritional compounds (e.g. a trypsin inhibitor in legumes). However, heat also destroys some types of heat-labile vitamin (Fig. 1.9), reduces the biological value of proteins (owing to destruction of amino acids or Maillard browning reactions) and promotes lipid oxidation. The effects of heat on proteins are described in detail by Meade *et al.* (2005) and Korhonen *et al.* (1998).

Oxidation is a second important cause of nutritional changes to foods. This occurs when food is exposed to air (e.g. in size reduction (Chapter 3) or hot-air drying (Chapter 16)) or as a result of the action of heat or oxidative enzymes (e.g. peroxidase or lipoxygenase). The main nutritional effects of oxidation are:

- the degeneration of lipids to hydroperoxides and subsequent reactions to form a wide variety of carbonyl compounds, hydroxy compounds and short chain fatty acids, and in frying oils to toxic compounds (Chapter 19, section 19.3.1); and
- destruction of oxygen-sensitive vitamins (Fig. 1.9).

Changes to lipids are discussed by Nawar (1996), the properties of vitamins and losses during processing are described by Morris *et al.* (2004) and examples of vitamin losses caused by individual unit operations are described in subsequent chapters.

The importance of nutrient losses during processing depends on the nutritional value of a particular food in the diet. Some foods (e.g. bread, potatoes and milk in Western countries, rice in Asia, or maize in Africa and the Caribbean) are an important source of nutrients for large numbers of people. Vitamin losses are therefore more significant in these foods than in those which either are eaten in small quantities or have low concentrations of nutrients. Reported vitamin losses during processing are included in subsequent chapters to give an indication of the severity of each unit operation. However, such data should be treated with caution. Variation in nutrient losses between cultivars or varieties can exceed differences caused by particular methods of processing. Growth conditions, or handling and preparation procedures prior to processing (Chapter 2), also cause substantial variation in nutrient losses. Data on nutritional changes cannot be directly applied to individual commercial operations, because of differences in ingredients, processing conditions and equipment used by different manufacturers. Meade *et al.* (2005) have reviewed the effects of processing on the nutritional quality of proteins in foods.

1.5   Control of processing

A requirement of all food processing is to ensure that products are safe for consumption, and previously quality control systems were based on the inspection of ingredients and end-product testing, with rejection of any batches that did not meet agreed standards. This reactive approach was then recognised to be a waste of resources (money has already been spent on producing the food by the time it is tested and rejection means a financial loss). A more proactive preventative approach to food safety and quality management, termed 'quality assurance' was developed during the 1980s, based on the principles of good manufacturing practice (GMP) (Anon 1998). It aimed to ensure that quality and safety are maintained throughout a process and thus prevent product rejection and financial losses.

1.5.1   Management of quality and food safety

The pressures for this development came from two sources: first, commercial pressures including increasing competition between companies, and the need to conform to international quality legislation in order to access expanding national and international food markets; and secondly, product quality management systems required by major retailers. A major shift in emphasis from national legislation to international legislation occurred in 1994, when a GATT agreement recommended acceptance of hazard analysis and critical control point (HACCP) principles, developed by the Codex Alimentarius Commission, as the required standard for free international movement of food.

HACCP

Hazard analysis is the identification of potentially hazardous ingredients, storage conditions, packaging, critical process points and relevant human factors which may affect product safety or quality. HACCP enables potential hazards in a process to be identified, assessed, and controlled or eliminated. Potential hazards include pathogenic micro-organisms and toxic chemicals (e.g. aflatoxins (section 1.2)), or physical contaminants that could cause harm to consumers (also Chapter 2, section 2.2). A HACCP plan sets tolerances for hazards and defines appropriate control measures, the frequency of their application, sampling procedures, specific tests to be used and the criteria for product acceptance. The system is based on monitoring of 'critical control points' (CCPs; processing factors of which a loss of control would result in an unacceptable food safety or quality risk) and the action to be taken when results of monitoring are outside the preset limits (Fig. 1.23). HACCP is used throughout each stage of a process, and includes raw materials, processing, storage and distribution. It can be used for all potential hazards, including inadequate quality as well as safety, and can identify areas of control where failure has not yet been experienced, making it particularly useful for new operations. Brown (2004) describes methods of microbial risk assessment.

Implementation of a HACCP scheme involves the following stages:

1 Identify the essential characteristics of the product and its use and define the hazards or potential hazards that could threaten the consumer. Included in this is the product formulation, processing, packaging, handling procedures and storage conditions, expected consumer preparation and consumption. Produce a detailed flow diagram of the process, including preparation and transport of raw materials, production methods and schedules, and in-process stock, confirmed by a site visit.
2 Identify critical stages (CCPs) required to control identified hazards to ensure safety. Consider all stages in the process, including realistic process deviations. The judgement of risk is made using one of three methods: probabilistic, comparative or pragmatic, and should be made by people who have a high degree of expertise and experience.
3 Devise target levels and critical limits for each CCP.
4 Establish effective procedures to monitor the CCPs. Monitoring may be by physical, microbiological or chemical tests, or by visual or sensory observations.
5 Establish corrective actions that are needed when there is a deviation from a CCP.
6 Establish procedures to verify that the HACCP system is working correctly.
7 Establish record-keeping systems that fully document the HACCP system and the procedures needed to review it.

Records should include for example, the frequency of monitoring and satisfactory compliance criteria, recipe formulations, cleaning procedures (what is cleaned, how and

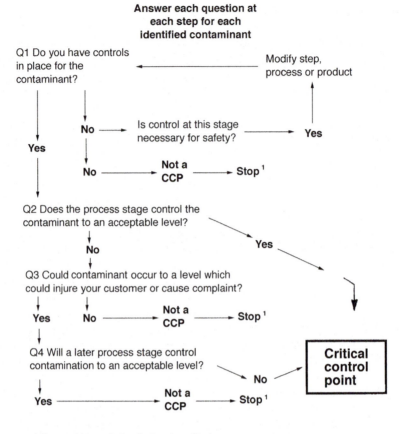

Fig. 1.23   Critical control point decision tree (from Dillon and Griffith 1996).

when it is cleaned, who cleans it and what with), temperatures or other processing conditions, hygienic practices, opportunities for cross-contamination and workers' illness or infections. Further details are given by Mayes (2001) and Shapton and Shapton (2000), and examples of HACCP applications are described by Mortimer and Wallace (1998) and Dillon and Griffith (1996), and for small-scale operations by Fellows et al. (1995).

The requirement for effective food monitoring and control systems to control resources and ensure that safe, high quality products were consistently manufactured led to the concept of total quality management (TQM). The aim of TQM is to define and understand *all* aspects of a process, to implement controls, monitor performance and measure improvements. TQM is therefore a management philosophy that seeks to continually improve the effectiveness and competitiveness of the business as a whole. In outline, a TQM system covers the following areas:

- Raw materials, purchasing and control (including agreed specifications, supplier auditing, raw material storage, stock control, traceability, inspection, investigation of non-conformity to specification).
- Process control (including identification, verification and monitoring of critical control points in a HACCP scheme, hygienic design of plant and layouts to minimise

cross-contamination, cleaning schedules, recording of critical production data, sampling procedures and contingency plans to cover safety issues).

- Premises (including methods of construction to minimise contamination, maintenance, waste disposal).
- Quality control (including product specifications and quality standards for non-safety quality issues, monitoring and verification of quality before distribution).
- Personnel (including training, personal hygiene, clothing and medical screening).
- Final product (including types and levels of inspection to determine conformity with quality specifications, isolating non-conforming products, packaging checks, inspection records, complaints monitoring systems).
- Distribution (to maintain the product integrity throughout the chain, batch traceability and product recall systems).

Rose (2000) has produced overviews of TQM, and Dillon (1999) describes a practical approach to beginning TQM projects in smaller businesses.

'Audits' are the regular systematic collection and feedback of objective information by competent independent personnel. They are used, for example, to monitor suppliers' ability to meet agreed requirements or to monitor production routines. They are an effective tool for monitoring the success of a quality system and are described in detail by Dillon (2001), Mortimer and Wallace (1998) and Dillon and Griffith (1997).

The benefits of a properly implemented TQM system are stated by Rose (2000) as:

- economic (more cost effective production by 'getting it right first time', reduction in wasted materials, fewer customer complaints, improved machine efficiency and increased manufacturing capacity);
- marketing (consistently meeting customer needs, increased customer confidence and sales);
- internal (improved staff morale, increased levels of communication, better trained staff and awareness/commitment to quality, improved management control and confidence in the operation);
- legislative (demonstrating due diligence, providing evidence of commitment to quality and ability to improve).

In the UK, the British Retail Consortium has issued a 'Technical Standard for Companies Supplying Retailer Branded Food Products', which is being used by retailers as the definitive standard for suppliers, and forms the basis of their terms of business (Rose 2000). It covers six key areas: HACCP systems, quality management systems, factory environmental standards, product control, process control and personnel, and it is implemented through third-party inspection bodies (e.g. the European Food Safety Inspection Service (EFSIS)).

The international standard for food safety management systems (ISO 22 000) was developed from an earlier standard (ISO 9001) in 2005. Food safety hazards can be introduced at any stage of the food chain, from ingredient supplies by primary producers, through food manufacturers, transport and storage operators, to retail and food service outlets. Adequate control is therefore needed throughout the food chain by the combined efforts of all who participate. ISO 22 000 specifies the requirements for a food safety management system, which include the following:

- Interactive communication with customers and suppliers to ensure that all relevant food safety hazards are identified and adequately controlled at each step within the food chain.

- System management to ensure that effective food safety systems are established, operated and updated within the framework of a structured management system and incorporated into the overall management activities of the company.
- Combined prerequisite programmes (PRPs), operational PRPs and HACCP principles. PRPs include GMP, good hygiene practice (GHP), good production practice (GPP) and good distribution practice (GDP). Operational PRPs are defined by the hazard analysis and are used to control food safety hazards or contamination of the products. The standard requires that all hazards that may be reasonably expected to occur in the food chain are identified and assessed. This provides the means to document why certain hazards need to be controlled and why others need not, and enables a strategy to be developed to ensure hazard control by combining the PRP(s) and HACCP plan.

The standard takes into account new product development, raw materials and ingredient supplies, production facilities and operations, environmental and waste management (also Chapter 27, section 27.1.4), health and safety in the working environment, as well as validation and verification of the food quality and safety systems. Wallace (2006) describes the establishment of prerequisite programmes and Mayes (2001) describes the implementation of HACCP systems.

References

ADAMS, M.R. and MOSS, M.O., (2007), *Food Microbiology*, Royal Society of Chemistry, Cambridge.

AKOH, C.C. and MIN, D.B., (Eds.), (2002), *Food Lipids: Chemistry, Nutrition, and Biotechnology*, 2nd edn, Marcel Dekker, New York.

ALZAMORA, S.M, TAPIA, M.S. and WELTI-CHANES, J., (2003), The control of water activity, in (P. Zeuthen and L. Bogh-Sorensen, Eds.), *Food Preservation Techniques*, Woodhead Publishing, Cambridge, pp. 126–153.

AMBROSE, D., (2008), Vapour pressure of water at temperatures between 0 and 360C, Kaye and Laby Online, National Physical Laboratory, available at www.kayelaby.npl.co.uk/chemistry/3_4/3_4_2.html.

ANON, (undated), Minerals, information provided by University of California, Davis, available at http://teaching.ucdavis.edu/nut111av/handouts/Complete_Text_Mineral.pdf.

ANON, (1993), Shelf-life of foods – Guidelines for its Determination and Prediction, Institute of Food Science and Technology (UK), 5 Cambridge Court, 210 Shepherds Bush Road, London W6 7NL, www.ifst.org/site/cms/contentCategoryView.asp?category=388.

ANON, (1996a), *Micro-organisms in Foods (5), Characteristics of Microbial Pathogens*, International Committee on Microbiological Specifications for Foods, International Union of Biological Sciences (ICMSF), Blackie Academic & Professional, London.

ANON, (1996b), The Pearson Square – common calculations simplified. *Food Chain*, **17**, (March), Practical Action, Rugby.

ANON, (1998), *Food and Drink Good Manufacturing Practice*, Institute of Food Science and Technology (IFST), London.

ANON (2000), What is colour and how is it measured?, information from Hunter Lab., available at www.hunterlab.com/appnotes/an05_00.pdf.

ANON, (2002), A list of additives currently permitted in food within the European Union and their associated E Numbers, available at www.ukfoodguide.net/enumeric.htm.

ANON, (2003), Use by date guidance notes, Food Standards Agency, available at www.food.gov.uk/foodindustry/guidancenotes/labelregsguidance/usebydateguid.

ANON, (2004), Evaluation of product shelf-life for chilled foods, CCFRA Guideline No. 46, Campden and Chorleywood Food Research Association, Chipping Campden, at www.campden.co.uk/publ/pubfiles/g46.htm.

ANON (2005), Cybercolloids Searchable Library – Pectin, company information from Cybercolloids Ltd, available at www.cybercolloids.net/library/pectin/properties.php.

ANON, (2006), *Dietary Reference Intakes: The Essential Guide to Nutrient Requirements*, Institute of Medicine of the National Academies, Washington, DC.

ANON, (2007a), Current EU approved additives and their E Numbers, UK Food Standards Agency, available at www.food.gov.uk/safereating/chemsafe/additivesbranch/enumberlist.

ANON, (2007b), Approximate pH of Foods and Food Products, FDA/Center for Food Safety & Applied Nutrition, available at http://vm.cfsan.fda.gov/~comm/lacf-phs.html.

ANON, (2008a), USDA National Nutrient Database, Nutrient Data Laboratory, available at www.nal.usda.gov/fnic/foodcomp/search/.

ANON, (2008b), Chemical properties of water, information from Planet Water, available at www.ozh2o.com/h2phys.html.

ANON, (2008c), Physical properties of water, information from the Faculty of Science, University of Waterloo, available at www.science.uwaterloo.ca/~cchieh/cact/applychem/waterphys.html.

ANON, (2008d), *Foodborne Pathogenic Micro-organisms and Natural Toxins Handbook*, Center for Food Safety and Applied Nutrition, US Food and Drug Administration, available at www.cfsan.fda.gov/~mow/intro.html.

ARNOLDI, A., (2004), Factors affecting the Maillard reaction, in (R. Steele, Ed.), *Understanding and Measuring the Shelf-life of Food*, Woodhead Publishing, Cambridge, pp. 111–127.

ASHIE, I.N.A., SIMPSON, B.K. and SMITH, J.P., (1996), Mechanisms for controlling enzymatic reactions in foods, *Critical Reviews in Food Science and Nutrition*, **36** (1/2), 1–30.

ASHURST, P.R. and DENNIS, M.J., (Eds.), (1996), *Food Authentication*, Blackie Academic and Professional, London.

BELITZ, H.-D., GROSCH, W. and SCHIEBERLE, P., (2004), *Food Chemistry*, 3rd edn, Springer Publications, Berlin/Heidelberg.

BELL, C. and KYRIAKIDES, A., (2002a), Pathogenic *Escherichia coli.* in (C. de W. Blackburn and P.J. McClure, Eds.), *Foodborne Pathogens – Hazards, Risk Analysis and Control*, Woodhead Publishing, Cambridge, pp. 279–306.

BELL, C. and KYRIAKIDES, A., (2002b), *Salmonella*, in (C. de W. Blackburn and P.J. McClure, Eds.), *Foodborne Pathogens – Hazards, Risk Analysis and Control*, Woodhead Publishing, Cambridge, pp. 307–335.

BELL, C. and KYRIAKIDES, A., (2002c), *Listeria monocytogenes*, in (C. de W. Blackburn and P.J. McClure, Eds.), *Foodborne Pathogens – Hazards, Risk Analysis and Control*, Woodhead Publishing, Cambridge, pp. 337–361.

BEMILLER, J.N. and WHISTLER, R.L., (1996), Carbohydrates, in (O.R. Fennema, Ed.), *Food Chemistry*, 3rd Edn, Marcel Dekker, New York, pp. 157–223.

BETTS, G.D. and WALKER, S.J., (2004), Verification and validation of food spoilage models, in (R. Steele, Ed.), *Understanding and Measuring the Shelf-life of Food*, Woodhead Publishing, Cambridge, pp. 184–217.

BIZIAGOS, E., PASSAGOT, J., CRANCE, J.M. and DELOINCE, R. (1988), Long term survival of hepatitis A virus and poliovirus type 1 in mineral water, *Applied and Environmental Microbiology*, **54** (11), 2705–10.

BLACKBURN, C., (2006), *Food Spoilage Micro-organisms*, Woodhead Publishing, Cambridge.

BLANSHARD, J.M.V., (1995), The glass transition, its nature and significance in food processing, in (S.T. Beckett, Ed.), *Physico-chemical Aspects of Food Processing*, Blackie Academic and Professional, London, pp. 17–48.

BOURNE, M.C. (2002), *Food Texture and Viscosity: Concept and Measurement*, Academic Press, London.

BOWEN, R., (2006), Dietary polysaccharides: structure and digestion, available at www.vivo.colostate.edu/hbooks/pathphys/digestion/basics/polysac.html.

BRAITHWAITE, E., BURLINGAME, B., CHENARD, C., SELLEY, B. and STUMBO, P., (2006), International Nutrient Databank Directory, produced for the 30th National Nutrient Databank Conference Honolulu, USA, available at www.medicine.uiowa.edu/gcrc/nndc/survey.html.

BRAUN, P., FEHLHABER, K., KLUG, C. and KOPP, K., (1999), Investigations into the activity of enzymes produced by spoilage-causing bacteria: a possible basis for improved shelf-life estimation, *Food Microbiology*, **16** (5), 531–540.

BRENNAN, J.G, BUTTERS, J.R., COWELL, N.D. and LILLEY, A.E.V., (1990), *Food Engineering Operations*, 3rd edn, Elsevier Applied Science, London.

BRENNDORFER, B., KENNEDY, L., OSWIN-BATEMAN, C.O. and TRIM, D.S., (1985), *Solar Dryers*, Commonwealth Science Council, Commonwealth Secretariat, London.

BROWN, M., (Ed.), (2004), *Microbiological Risk Assessment in Food Processing*, Woodhead Publishing, Cambridge.

BUHLER, D.R. and MIRANDA, C., (2000), Antioxidant activities of flavonoids, Department of Environmental and Molecular Toxicology, Oregon State University, available at http://lpi.oregonstate.edu/f-w00/flavonoid.html.

CARDELLO, A.V., (1998), Perception of food quality, in (I.A. Taub and R.P. Singh, Eds.), *Food Storage Stability*, CRC Press, Boca Raton, FL, pp. 1–38.

CARPI, A., (2008), Carbohydrates, available at www.visionlearning.com/library/module_viewer.php?mid=61.

CHAPLIN, M., (2008a), Cellulose, available at http://www.lsbu.ac.uk/water/hycel.html#fun.

CHAPLIN, M., (2008b), Carrageenan, available at http://www.lsbu.ac.uk/water/hycel.html#fun.

CLARK, R.C., (1990), Flavour and texture factors in model gel systems, in (A. Turner, Ed.), *Food Technology International Europe*, Sterling Publications International, London, pp. 271–277.

CLARK, T.J., (2006), Mineral functions in the body, available at www.tjclark.com.au/colloidal-minerals-library/mineral-functions.htm.

COULTATE, T.P., (1984), *Food, the Chemistry of its Components*, Royal Society of Chemistry, London, pp. 102–129.

CUI, S., (ED.) (2008), Food carbohydrates: chemistry, physical properties, and applications, available at www.foodnetbase.com/ejournals/books/book_summary/summary.asp?id=2600.

DAMODARAN, S., (1996), Amino acids, peptides and proteins, in (O.R. Fennema, Ed.), *Food Chemistry*, 3rd edn, Marcel Dekker, New York, pp. 321–429.

DAMODARAN, S., PARKIN, K.L. and FENNEMA, O.R., (2007), *Fennema's Food Chemistry*, 4th edn, CRC Press, Boca Raton, FL.

DAVIES, C., (2004), Lipolysis in lipid oxidation, in (R. Steele, Ed.), *Understanding and Measuring the Shelf-life of Food*, Woodhead Publishing, Cambridge, pp. 142–161.

DAVIS, B.G. and FAIRBANKS, A.J., (2002), *Carbohydrate Chemistry*, Oxford Chemistry Primers, 99, Oxford University Press, Oxford.

DEAK, T., (2004), Spoilage yeasts, in (R. Steele, Ed.), *Understanding and Measuring the Shelf-Life of Food*, Woodhead Publishing, Cambridge, pp. 91–110.

DEPREE, J.A. and SAVAGE, G.P., (2001), Physical and flavour stability of mayonnaise, *Trends in Food Science and Technology*, **12**, 157–163.

DILLON, M., (1999), *Cost Effective Food Control*, The Global Hygiene Forum, Helsinki, Finland.

DILLON, M., (Ed.), (2001), *Auditing in the Food Industry*, Woodhead Publishing, Cambridge.

DILLON, M. and GRIFFITH, C., (1996), *How to HACCP*, 2nd edn, MD Associates, Grimsby.

DILLON, M. and GRIFFITH, C., (1997), *How to Audit: Verifying Food Control Systems*, MD Associates, Grimsby.

EARLE, R.L., (1983), *Unit Operations in Food Processing*, 2nd edn, Pergamon Press, Oxford, pp. 24–38, 46–63.

EASTWOOD, M., (2003), *Principles of Human Nutrition*, 2nd edn, Blackwell Publishing, Oxford.

ESSE, R. and SAARI, A., (2004), Shelf-life and moisture management, in (R. Steele, Ed.), *Understanding and Measuring the Shelf-life of Food*, Woodhead Publishing, Cambridge, pp. 24–41.

FELLOWS, P.J., AXTELL, B.L. and DILLON, M., (1995), *Quality Assurance for Small Scale Rural Food Industries*, FAO Agricultural Services Bulletin 117, FAO, Rome, Italy.

FENNEMA, O.R., (1996), Water and ice, in (O.R. Fennema, Ed.), *Food Chemistry*, 3rd edn, Marcel Dekker, New York, pp. 17–94.

FRATAMICO, P.M. and BAYLES, D.O. (EDS.), (2005). *Foodborne Pathogens: Microbiology and Molecular Biology*, Caister Academic Press, Norwich.

GAONKAR, A.G. and MCPHERSON, A., (Eds.), (2006), *Ingredient Interactions – Effects on Food Quality*, 2nd edn, CRC Press, Boca Raton, FL.

GIBBS, P., (2002), Characteristics of spore-forming bacteria, in (C. de W. Blackburn and P.J. McClure, Eds.), *Foodborne Pathogens – Hazards, Risk Analysis and Control*, Woodhead Publishing, Cambridge, pp. 418–435.

GIBNEY, M. J., VORSTER, H. H. and KOK, F. J., (Eds.), (2002) *Introduction to Human Nutrition*, The Nutrition Society Textbook, Blackwell Science, Oxford.

GORDON, M.H., (2004), Factors affecting lipid oxidation, in (R. Steele, Ed.), *Understanding and Measuring the Shelf-life of Food*, Woodhead Publishing, Cambridge, pp. 128–141.

GORGA, F.R., (2007), Introduction to protein structure, available at http://webhost.bridgew.edu/fgorga/proteins/default.htm.

GREGORY, J.F., (1996), Vitamins, in (O.R. Fennema, Ed.), *Food Chemistry*, 3rd edn, Marcel Dekker, New York, pp. 534–616.

GRIFFITHS, M., (2002), *Mycobacterium paratuberculosis*, in (C. de W. Blackburn and P.J. McClure, Eds.), *Foodborne Pathogens – Hazards, Risk Analysis and Control*, Woodhead Publishing, Cambridge, pp. 489–500.

GUNSTONE, F., (Ed.), (2006), *Modifying Lipids for Use in Food*, Woodhead Publishing, Cambridge.

GUNSTONE, F.D., HARWOOD, J.L. and PADLEY, F.B., (1995), *The Lipid Handbook*, Chapman and Hall, London.

GURR, M.I., (1999), *Lipids in Nutrition and Health: a Reappraisal*, Vol. 11, The Oily Press Lipid Library, PJ Barnes & Associates, Bridgwater.

HALVORSEN, B.L., CARLSEN, M.H., PHILLIPS, K.M., BØHN, S.K., HOLTE, K., JACOBS, D.R. and BLOMHOFF, R., (2006), Content of redox-active compounds (ie, antioxidants) in foods consumed in the United States, *American J. Clinical Nutrition*, **84**, 95–135, available at http://folk.uio.no/runeb/pdf%20filer/Halvorsen%20AJCN%202006.pdf.

HELLER, L., (2007), Food temperature affects taste, 4Hoteliers.com, 21 January, available at www.4hoteliers.com/4hots_fshw.php?mwi=1243.

HOEFLER, A., (2004), *Hydrocolloids*, Eagan Press, St. Paul, MN.

JAY, J.M., LOESSNER, M.J. and GOLDEN, D.A., (2005), *Modern Food Microbiology*, 7th edn, Springer Science and Business Media, New York.

JAYA, S., SUGHAGAR, M. and DAS, H., (2002), Stickiness of food powders and related physico-chemical properties of food components, *J. Food Science and Technology*, **39** (1), 1–7.

KILCAST, D., (1999), Sensory techniques to study food texture, in (A.J. Rosenthal, Ed.), *Food Texture*, Aspen Publishers, Gaithersburg, MD, pp. 30–64.

KILCAST, D., (Ed.), (2004), *Texture in Food, Vol.2 – Solid Foods*, Woodhead Publishing, Cambridge.

KILCAST, D. and MCKENNA, B.M., (2003), *Texture in Food*, Woodhead Publishing, Cambridge.

KILCAST, D. and SUBRAMANIAN, P., (2000), *The Stability and Shelf Life of Food*, Woodhead Publishing, Cambridge.

KOOPMANS, M., (2002), Viruses, in (C. de W. Blackburn and P.J. McClure, Eds.), *Foodborne Pathogens – Hazards, Risk Analysis and Control*, Woodhead Publishing, Cambridge, pp. 440–452.

KORHONEN, H., PIHLANTO-LEPPÄLA, A., RANTAMÄKI, P. and TUPASELA, T., (1998), Impact of processing on bioactive proteins and peptides, *Trends in Food Science and Technology*, **9** (8–9), 307–319.

KUDA, T., SHIMIZU, K. and YANO, T., (2004), Comparison of rapid and simple colorimetric microplate assays as an index of bacterial count, *Food Control*, **15** (6), 421–425.

LABUZA, T.P. and HYMAN, C.R., (1998), Moisture migration and control in multi-domain foods, *Trends in Food Science and Technology*, **9** (2), 47–55.

LAFARGE, C., BARD, M-H., BREUVART, A., DOUBLIER, J-L. and CAYOT, N., (2008), Influence of the structure of cornstarch dispersions on kinetics of aroma release, *J. Food Science*, **73** (2), S104–S109.

LAWLESS, H.T. and HEYMAN, H., (1998), *Sensory Evaluation of Food – Principles and Practice*, Aspen Publishers, Gaithersburg, MD, pp. 379–405.

LEES, M, (Ed.), (2003), *Food Authenticity and Traceability*, Woodhead Publishing, Cambridge.

LEISTNER, L., (1994), Food preservation by combined processes, in (L. Leistner and L.G.M. Gorris, Eds.), *Food Preservation by Combined Processes*, Final Report of FLAIR Concerted Action No 7, Subgroup B, (EUR 15776 EN) Commission of the European Community, Brussels, Belgium.

LEISTNER, L., (1995), Principle and applications of hurdle technology, in (G.W. Gould, Ed.), *New Methods of Food Preservation*, Blackie Academic and Professional, London, pp. 1–21.

LEISTNER, L. and GORRIS, L.G.M., (1995), Food preservation by hurdle technology, *Trends in Food Science and Technology*, **6**, 41–46.

LEWIS, M.J., (1990), *Physical Properties of Foods and Food Processing Systems*, Woodhead Publishing, Cambridge.

LINDSAY, R.C., (1996a), Food additives, in (O.R. Fennema, Ed.), *Food Chemistry*, 3rd edn, Marcel Dekker, New York, pp. 767–823.

LINDSAY, R.C., (1996b), Flavours, in (O.R. Fennema, Ed.), *Food Chemistry*, 3rd edn, Marcel Dekker, New York, pp. 723–765.

LOUREIRO, V., MALFEITO-FERREIRA, M. and CARREIRA, A., (2004), Detecting spoilage yeasts, in (R. Steele, Ed.), *Understanding and Measuring the Shelf-life of Food*, Woodhead Publishing, Cambridge, pp. 233–288.

LUND, B., BAIRD-PARKER, A.C. and GOULD, G.W., (2000), *Microbiological Safety and Quality of Food*, Aspen Publishing, New York.

MAARUF, A.G., CHE MAN, Y.B., ASBI, B.A., JUNAINAH, A.H. and KENNEDY, J.F., (2001), Effect of water content on the gelatinisation temperature of sago starch, *Carbohydrate Polymers*, **46** (4), 331–337.

MACDOUGALL, D., (Ed.), (2002), *Colour in Food*, Woodhead Publishing, Cambridge.

MAGAN, N., (Ed.), (2004), *Mycotoxins in Food*, Woodhead Publishing, Cambridge.

MAN, C.M.D., (2004), Shelf-life testing, in (R. Steele, Ed.), *Understanding and Measuring the Shelf-life of Food*, Woodhead Publishing, Cambridge, pp. 340–356.

MAN, C.M.D. and JONES, A.A., (2000), *Shelf-life Evaluation of Foods*, Aspen Publishing, Gaithersburg, MD.

MARTH, E.H., (1998), Extended shelf-life refrigerated foods: microbiological quality and safety, *Food Technology*, **52** (2), 57–62.

MAYES, T. (ED.), (2001), *Making the Most of HACCP*, Woodhead Publishing, Cambridge.

MCCLURE, P. and BLACKBURN, C., (2002), *Campylobacter* and *Arcobacter*, in (C. de W. Blackburn and P.J. McClure, Eds.), *Foodborne Pathogens – Hazards, Risk Analysis and Control*, Woodhead Publishing, Cambridge, pp. 363–384.

MCKENNA, B.M., (Ed.), (2003), *Texture in Food, Vol. 1: Semi-solid Foods*, Woodhead Publishing, Cambridge.

MCMEEKIN, T., (Ed.), (2003), *Detecting Pathogens in Food*, Woodhead Publishing, Cambridge.

MEADE, S.J., REID, E.A. and GERRARD, J.A., (2005), The impact of processing on the nutritional quality of food proteins, *J. AOAC International*, **88** (3), 904–922.

MEILGAARD, M.C., CIVILLE, G.V. and CARR, B.T., (1999), *Sensory Evaluation Techniques*, 3rd edn, CRC Press, Boca Raton, FL.

MEJIA, L.A., (1994), Fortification of foods: historical development and current practices, *Food and Nutrition Bulletin*, 15 (4) (December 1993/1994), The United Nations University Press, available at www.unu.edu/unupress/food/8f154e/8F154E03.htm#Fortification%20 of%20foods:%20Historical%20development%20and%20current%20practices.

MELLICAN, R.I., LI, J., MEHANSHO, H. and NIELSEN, S.S., (2003), The role of iron and the factors affecting off-color development of polyphenols, *J. Agriculture and Food Chemistry*, **51** (8), 2304–2316.

MILLER, D.R., (1996), Minerals, in (O.R. Fennema, Ed.), *Food Chemistry*, Marcel Dekker, New York, pp. 617–650.

MILSON, A. and KIRK, D., (1980), *Principles of Design and Operation of Catering Equipment*, Ellis Horwood, Chichester.

MIZRAHI, S., (2004), Accelerated shelf-life tests, in (R. Steele, Ed.), *Understanding and Measuring the Shelf-life of Food*, Woodhead Publishing, Cambridge, pp. 317–339.

MOHSENIN, N.N., (1970), *Physical Properties of Plant and Animal Materials*, Vol. 1 *Structure, Physical Characteristics and Mechanical Properties*, Gordon and Breach, London.

MORRIS, A., BARNETT, A. and BURROWS, O-J., (2004), Effect of processing on nutrient content of foods, *Caribbean Food and Nutrition Institute*, **37** (3), 160–164, available at www.ops-oms.org/English/CFNI/cfni-caj37No304-art-3.pdf.

MORRISSEY, P.A. and KERRY, J.P., (2004), Lipid oxidation and the shelf-life of muscle foods, in (R. Steele, Ed.), *Understanding and Measuring the Shelf-life of Food*, Woodhead Publishing, Cambridge, pp. 357–395.

MORTIMER, S. and WALLACE, C., (1998), *HACCP – A Practical Approach*, 2nd edn, Aspen Publishers, Gaithersburg, MD.

MOSS, M., (2002), Toxigenic fungi, in (C. de W. Blackburn and P.J. McClure, Eds.), *Foodborne Pathogens – Hazards, Risk Analysis and Control*, Woodhead Publishing, Cambridge, pp. 479–488.

MOTARJEMI, Y., (2002), Chronic sequelae of foodborne infections, in (C. de W. Blackburn and P.J. McClure, Eds.) *Foodborne Pathogens – Hazards, Risk Analysis and Control*, Woodhead Publishing, Cambridge, pp. 501–513.

MULLA, Z., (1999), *E. coli*: serotypes other than O157:H7, available at www.doh.state.fl.us/Disease_ctrl/epi/htopics/reports/ecoli3.pdf.

NAVE, R., (2008), The C.I.E. Color Space, Dept. Physics and Astronomy, Georgia State University, available at http://hyperphysics.phy-astr.gsu.edu/hbase/vision/cie.html and http://hyperphysics.phy-astr.gsu.edu/hbase/vision/colsys.html#c1.

NAWAR, W.W., (1996), Lipids, in (O.R. Fennema, Ed.), *Food Chemistry*, 3rd edn, Marcel Dekker, New York, pp. 225–319.

NEDDERMAN, R.M., (1997), Newtonian fluid mechanics, in (P.J. Fryer, D.L. Pyle and C.D. Rielly, Eds.), *Chemical Engineering for the Food Processing Industry*, Blackie Academic and Professional, London, pp. 63–104.

NICHOLS, R. and SMITH, H., (2002), Parasites: *Cryptosporidium, Giardia* and *Cyclospora* as foodborne pathogens, in (C. de W. Blackburn and P.J. McClure, Eds.), *Foodborne Pathogens – Hazards, Risk Analysis and Control*, Woodhead Publishing, Cambridge, pp. 453–478.

O'BRIEN, R.D., (2004), *Fats and Oils: Formulating and Processing for Applications*, CRC Press, Boca Raton, FL.

O'NEIL, C.E. and NICKLAS, T.A., (2007), State of the art reviews: relationship between diet/ physical activity and health, *American J. Lifestyle Medicine*, **1** (6), 457–481.

OWUSU-APENTEN, R., (2004), *Introduction to Food Chemistry*, CRC Press, Boca Raton, FL.

PALAV, T. and SEETHARAMAN, K., (2006), Mechanism of starch gelatinization and polymer leaching during microwave heating, *Carbohydrate Polymers*, **65** (3), 364–370.

PARIZA, M.W., (1996), Toxic substances, in (O.R. Fennema, Ed.), *Food Chemistry*, 3rd edn, Marcel Dekker, New York, pp. 825–840.

PELEG, M., (1983), Physical characteristics of food powders, in (M. Peleg and E.B. Bagley, Eds.), *Physical Properties of Foods*, AVI, Westport, CT.

PELEG, M. and BAGLEY, E.B. (EDS.), (1983), *Physical Properties of Foods*, AVI, Westport, CT.

PETSKO, G.A. and RINGE, D., (2004), *Protein Structure and Function*, New Science Press Ltd, London.

PHILLIPS, G. and WILLIAMS, P., (2000), *Handbook of Hydrocolloids*, Woodhead Publishing, Cambridge.

PIDWIRNY, M., (2006), *Physical Properties of Water*, Fundamentals of Physical Geography, 2nd edn, available at www.physicalgeography.net/fundamentals/8a.html.

RAHMAN, M.S., (1999), Glass transition and other structural changes in foods, in (M.S. Rahman, Ed.), *Handbook of Food Preservation*, Marcel Dekker, New York, pp. 75–94.

RIELLY, C.D., (1997), Food rheology, in (P.J. Fryer, D.L Pyle and C.D. Rielly, Eds.), *Chemical*

Engineering for the Food Processing Industry, Blackie Academic and Professional, London, pp. 195–233.

RIEMANN, H.P. and CLIVER, D.O., (EDS), (2006), *Foodborne Infections and Intoxications*, Elsevier, Oxford.

ROBERTS, D. and GREENWOOD, M., (2002), *Practical Food Microbiology*, 3rd edn, Blackwell Publishing, Oxford.

ROE, M.A., FINGLAS, P.M. and CHURCH, S.M., (2002), *McCance and Widdowson's The Composition of Foods*, 6th summary edn, Royal Society of Chemistry, Cambridge.

ROSE, D., (2000), Total quality management, in (M. Stringer and C. Dennis, Eds.), *Chilled Foods*, 2nd edn, Ellis Horwood Ltd., Chichester, Ch. 14.

ROSENTHAL, A.J., (1999), Relation between instrumental and sensory measures of food texture, in (A.J. Rosenthal, Ed.), *Food Texture*, Aspen Publishers, Gaithersburg, MD, pp 1–17.

ROUSEFF, R.L., (1990), Bitterness in Foods and Beverages, Elsevier, New York.

ST ANGELO, A.J., (1996), Lipid oxidation in foods, *Critical Reviews in Food Science and Nutrition*, **36** (3), 175–224.

SAUTOUR, M., SOARES-MANSUR, C., DIVIES, C., BENSOUSSAN, M. and DANTIGNY, P., (2002), Comparison of the effects of temperature and water activity on growth rate of food spoilage moulds, *J. Industrial Microbiology and Biotechnology*, **28** (6), 311–315.

SCALETSKY, I.C.A., FABBRICOTTI, S.H., SILVA, S.O.C., MORAIS, B. and FAGUNDES-NETO, U., (2002), HEp-2-adherent *Escherichia coli* strains associated with acute infantile diarrhea, São Paulo, Brazil. Emergent Infectious Diseases [serial online] 8 August, available from www.cdc.gov/ncidod/EID/vol8no8/01-0492.htm.

SEABOURNE, N.L., (1999), Bitterness sensitivity and its relationship to food preferences, Thesis (MS), University of Illinois at Urbana-Champaign.

SHAPTON, D.A. and SHAPTON, N.F., (Eds.), (2000), *Principles and Practice for the Safe Processing of Foods*, Woodhead Publishing, Cambridge.

SINGH, R.P. and ANDERSON, B.A., (2004), The major types of food spoilage: an overview, in (R. Steele, Ed.), *Understanding and Measuring the Shelf-life of Food*, Woodhead Publishing, Cambridge, pp. 3–23.

SINGH, R.P. and CADWALLADER, K.R., (2004), Ways of measuring shelf-life and spoilage, in (R. Steele, Ed.), *Understanding and Measuring the Shelf-life of Food*, Woodhead Publishing, Cambridge, pp. 165–183.

SINGHAL, R.S., KULKARNI, P.K. and REG, D.V., (Eds.), (1997), *Handbook of Indices of Food Quality and Authenticity*, Woodhead Publishing, Cambridge.

SPILLANE, W.J., (Ed.), (2006), *Optimising Sweet Taste in Foods*, Woodhead Publishing, Cambridge.

STEELE, R., (Ed.), (2004), *Understanding and Measuring the Shelf-life of Food*, Woodhead Publishing, Cambridge.

STRAYER, D. (CHAIR OF 20-MEMBER TECHNICAL COMMITTEE), (2006), *Food Fats and Oils*, Institute of Shortening and Edible Oils, Washington, USA, available at www.iseo.org/foodfatsoils.pdf.

SUTHERLAND, J., (2003), Modelling food spoilage, in (P. Zeuthen and L. Bogh-Sorensen, Eds.), *Food Preservation Techniques*, Woodhead Publishing, Cambridge, pp. 451–474.

SUTHERLAND, J. and VARNAM, A., (2002), Enterotoxin-producing *Staphylococcus, Shigella, Yersinia, Vibrio, Aeromonas* and *Plesimona*, in (C. de W. Blackburn and P.J. McClure, Eds.), *Foodborne Pathogens – Hazards, Risk Analysis and Control*, Woodhead Publishing, Cambridge, pp. 386–415.

SZCZESNIAK, A.S., (1963) Classification of textural characteristics, *J. Food Science*, **28**, 385–389.

TAKEOKA, G.R., GUNTERT, M. and ENGEL, K-H., (2001), *Aroma Active Compounds in Foods: Chemistry and Sensory Properties*, 2nd edn, American Chemical Society, Division of Agricultural and Food Chemistry, Washington, DC.

TAOUKIS, P.S. and GIANNAKOUROU, M.C., (2004), Temperature and food stability: analysis and control, in (R. Steele, Ed.), *Understanding and Measuring the Shelf-life of Food*, Woodhead Publishing, Cambridge, pp. 42–68.

TATHAM, D. and RICHARDS, W., (1998), *ECTA Guide to E.U. Trade Mark Legislation*, European

Communities Trade Mark Association/Sweet and Maxwell, London.

TAYLOR, A.J., (2002), Food Flavour Technology, 2nd edn, CRC Press, Boca Raton, FL.

THIS, H., (2006), Molecular Gastronomy: Exploring the Science of Flavor, Translated by M. B. DeBevoise, Columbia University Press, New York.

TODAR, K., (2007), Growth of bacterial populations, Department of Bacteriology, University of Wisconsin-Madison, available at www.textbookofbacteriology.net/growth.html.

TOLEDO, R.T., (1999a–e), Fundamentals of Food Process Engineering, 2nd edn, Aspen Publishers, Aspen: (a) pp. 109–131, (b) pp. 160–231, (c) pp. 456–506, (d) pp. 548–566, (e) pp. 66–108.

TROLLER, J.A. and CHRISTIAN, J.H.B., (1978), Water Activity and Food, Academic Press, London.

VAN DEN BERG, C., (1986), Water activity, in (D. MacCarthy, Ed.), Concentration and Drying of Foods, Elsevier Applied Science, pp., 11–36.

VARNAM, A.H. and EVANS, M.G., (1996), Foodborne Pathogens: An Illustrated Text, Manson Publishing, London.

VON ELBE, J.H. and SCHWARTZ, S.J., (1996), Colorants, in (O.R. Fennema, Ed.), Food Chemistry, 3rd edn, Marcel Dekker, New York, pp. 651–722.

WALKER, S.J. and BETTS, G., (2000), Chilled foods microbiology, in (M. Stringer and C. Dennis, Eds.), Chilled Foods, 2nd edn, Ellis Horwood Ltd., Chichester, Ch. 6.

WALLACE, C.A., (2006), Safety in food processing, in (J.G. Brennan, Ed.), Food Processing Handbook, Wiley-VCH, pp. 351–372.

WARLDAW, G. M., (2003), Contemporary Nutrition, Issues and Insights, 5th edn, McGraw-Hill Higher Education, Columbus.

WILKINSON, C., DIJKSTERHUIS, G.B. and MINEKUS, M., (2000), From food structure to texture, Trends in Food Science and Technology, 11 (12), 442–450.

WIRAKARTAKUSUMAH, M.A. and HARIYADI, P., (1998), Technical aspects of food fortification, Food and Nutrition Bulletin, 19 (2), UNU Press, United Nations University, available at www.unu.edu/unupress/food/v192e/ch03.htm#b1.

WHITAKER, J.R., (1996), Enzymes, in (O.R. Fennema, Ed.), Food Chemistry, 3rd edn, Marcel Dekker, New York, pp. 431–530.

WOESE, C., KANDLER, O. and WHEELIS, M., (1990), Towards a natural system of organisms: proposal for the domains Archaea, Bacteria, and Eucarya, Proceedings National Academy Science, USA, 87 (12), 4576–4579.

WYETH, L. and KILCAST, D., (1991), Sensory analysis technique and flavour release, in (A. Turner, Ed.), Food Technology International Europe, Sterling Publications International, London, pp. 239–242.

XIE, F., LIU, H., CHEN, P., XUE, T., CHEN, L., YU, L. and CORRIGAN, P., (2006) Starch gelatinization under shearless and shear conditions, International J. of Food Engineering, 2 (5), Article 6, available at www.bepress.com/ijfe/vol2/iss5/art6.

YADA, R., (Ed.), (2004), Proteins in Food Processing, Woodhead Publishing, Cambridge.

ZAMORA, A., (2005), Fats, oils, fatty acids, triglycerides, available at www.scientificpsychic.com/fitness/fattyacids.html.

ZAPSALIS, C. and BECK, R.A., (1985), Food Chemistry and Nutritional Biochemistry, John Wiley, New York, pp. 415–504, 549–579.

Part II

Ambient-temperature processing

Methods used to prepare freshly harvested or slaughtered food for further processing (Chapter 2,) to alter the size of foods (Chapter 3), to mix ingredients (Chapter 4) or to separate components of food (Chapter 5) are each essential unit operations in nearly all food processes. They are used to prepare specific formulations, to aid subsequent processing, or to alter the sensory characteristics of foods to meet the required quality. In each of these unit operations the sensory characteristics and nutritional properties of foods may be changed by removal of components or by the action of naturally occurring enzymes or contaminating micro-organisms, but there is negligible damage to food quality due to heat.

Over recent years, consumer demand has increasingly required processed foods to have a more 'natural' flavour and colour, with a shelf-life that is sufficient for distribution and home storage before consumption. There have been significant developments in processes that do not significantly heat the food and are thus able to retain to a greater extent their nutritional quality and sensory characteristics. Traditionally, fermented foods have many of these characteristics and these have been supplemented more recently with functional foods and probiotic foods (Chapter 6). Irradiation (Chapter 7) has been adopted in some countries as a minimal method of food preservation. There has also been increasing interest in developing other novel methods to achieve mild preservation, and processing using high hydrostatic pressure (Chapter 8) has now been commercialised. Other examples of minimal processing using electric arc discharges, oscillating magnetic fields, pulsed light and UV light, X-rays and ultrasound are described in Chapter 9. The principle underlying the use of mild processing involves the use of combinations of these low-temperature unit operations with refrigerated storage and distribution (Chapter 20) and packaging (Chapters 25 and 26). Each minimal processing method destroys or inhibits microbial growth, and in some cases enzyme activity, but there are no substantial increases in product temperature. There is therefore little damage to pigments, flavour compounds or vitamins and, in contrast to heat processing (Part III), the sensory characteristics and nutritional value of foods are largely retained.

2

Raw material preparation

Abstract: To ensure that raw materials have a uniformly high quality for subsequent processing or for sale in the fresh market sector, it is necessary to cool them to reduce metabolic activity and microbial growth, to remove contaminants, and to sort foods to produce uniform characteristics. This chapter describes the unit operations of cooling, cleaning, sorting, grading and peeling raw materials in preparation for processing.

Key words: hydrocoolers, contaminants, crop cleaning, grading, sorting, machine vision systems, peeling.

Foods require cooling immediately after harvest or slaughter to reduce both metabolic activity and the growth of micro-organisms, and hence reduce changes to organoleptic and nutritional qualities. Most raw materials are also likely to contain contaminants, to have components that are inedible or to have variable physical characteristics (e.g. shape, size or colour). It is not possible to produce high-quality processed foods from substandard raw materials and it is therefore necessary to perform one or more of the unit operations of cooling, cleaning, sorting, grading or peeling to ensure that foods having a uniformly high quality are prepared for subsequent processing or for sale in the fresh market sector. Chilled foods are described in Chapter 21 and other separation operations are described in Chapter 5.

2.1 Cooling crops and carcasses

Rapid cooling of all raw materials slows spoilage by naturally occurring enzymes and contaminating micro-organisms, and therefore extends their shelf-life. However, rapid cooling of animal carcasses immediately after slaughter is avoided to allow the processes of rigor mortis to take place and prevent 'cold shortening' (details are given by Lawrie 1998 and Ranken 2000, and in Chapter 21, section 21.4). Most wholesale buyers now require fresh produce to be cooled before it is transported to distribution warehouses or retailers (also Chapter 27, section 27.3). Produce can be cooled using chilled air or water. Air cooling involves passing refrigerated air over products to cool them without freezing. Vacuum cooling is suitable for leafy vegetables (e.g. cabbage, lettuce and spinach). The surface of the produce is moistened with water and it is placed in a vacuum chamber to induce evaporative cooling. Details are given by Sun and Wang (2001).

'Hydrocooling' uses chilled water to rapidly cool produce, approximately 15 times faster than air cooling (Boyette *et al.* 2006). Unlike in air cooling, produce is not partially dehydrated and slightly wilted produce may be revived. The main limitation of hydrocooling is the sensitivity of a product to wetting and resulting microbial growth. Hydrocooling is particularly suitable for foods that have a large volume in relationship to their surface area (e.g. whole sweetcorn, apples, peaches and other similar fruits). Foods are sprayed or submerged in chilled water at 1.5 °C, produced by a refrigeration unit. Alternatively crushed ice can be used, which reduces the capital cost of the equipment and may be more suitable for growers that have smaller amounts of produce or a short harvest season, provided that there is a reliable source of ice at a reasonable cost. Fresh fruit and vegetable crops are susceptible to postharvest infection when they are stressed (e.g. by too much or too little water) or damaged by bruises or abrasion. To prevent any spread of infection from a few infected items, the recirculated hydrocooling water is chlorinated with ≈ 2–3 mg l^{-1} chlorine. Details of chlorine production and applications are given by Anon (2008a). There is, however, growing concern over the use of chlorine because of its reactions with organic materials to form trihalomethanes, halogenated acetic acids and other chemical residues, which have been linked to a variety of environmental and health problems, including ecosystem damage, fish death, cancer and physiological diseases. Additionally, chlorine has a limited effect in killing surface bacteria at the permitted concentrations.

Xu (1999) describes the use of ozone (O_3) as an alternative to chlorine. It has an oxidation potential 1.5 times stronger than chlorine and the ozone decomposes in water to produce only oxygen. It is more effective than chlorine over a wider spectrum of spoilage and pathogenic bacteria, and can also destroy undesirable flavours produced by bacteria, as well as pesticides and chemical residues in water. Soluble ferrous and manganese salts that discolour water are oxidised to form insoluble salts that can be removed by filtration, and hydrogen sulphide is similarly oxidised to elemental sulphur and removed. Ozone rapidly oxidises bacterial membranes, which weakens the cell wall and leads to cellular rupture and cell death. For example, Kim *et al.* (1999) describe a reduction in microbial numbers on fresh fruits and vegetables that are washed in ozonated water. Ozone is produced by corona discharge in air or oxygen, or by UV lights. Water is collected, filtered and re-treated with ozone before recycling through the washing equipment, and it can be discharged without environmental damage. Details of commercial applications are given by Anon (2008b).

Theory

Heat transfer by conduction is described in Chapter 10 (section 10.1.2). In hydrocooling, the cooling rates of different crops using water at a known temperature in a particular design of cooler are measured (Fig. 2.1) and the data is then collated (Fig. 2.2). This information is used to calculate the time taken to cool the interior of a food from the field temperature to the required temperature for distribution.

The vertical axis in Fig 2.2 is the decimal temperature difference (DTD) which is found using Equation 2.1:

$$\text{DTD} = \frac{\theta_t - \theta_w}{\theta_p - \theta_w} \qquad \boxed{2.1}$$

where: θ_t = target produce temperature (°C), θ_w = chilled water temperature (°C), θ_p = produce starting temperature (°C). Sample problem 2.1 shows the calculation of cooling times for cucumbers to reach different distribution temperatures.

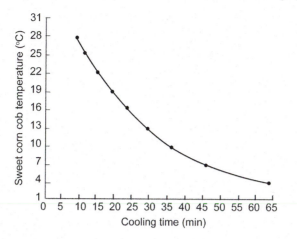

Fig. 2.1 Cooling curve for fresh maize cobs (courtesy of North Carolina Cooperative Extension Service, North Carolina State University) (adapted from Boyette *et al.* 2006).

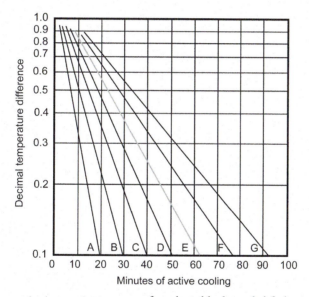

Fig. 2.2 Time–centre temperature response for selected hydrocooled fruits and vegetables (courtesy of North Carolina Cooperative Extension Service, North Carolina State University) (adapted from Boyette *et al.* 2006) (A = kale, green leafy crops, B = peas, asparagus florets, beans, C = radishes, small beets, D = small apples and peaches, E = sweetcorn cobs, apples, and peaches, F = cucumbers, large apples and peaches, G = cantaloupes, large eggplant).

Equipment

Produce is packed into wire-bound wooden crates, mesh bags or perforated bins that have a large amount of open space. For effective heat transfer to the cooling water, the design of containers and the stacking layout on pallets must enable water to flow through and not around the packages.

There are four types of hydrocoolers, which differ in their cooling rates and processing efficiencies: batch, conveyor, immersion and truck hydrocoolers. In batch hydrocoolers,

Sample problem 2.1
Cucumbers with a centre temperature of 30 °C are to be hydrocooled by immersion in water at 1.5 °C. How long will it take to reduce the centre temperature to 10 °C? What is the additional cooling time required for the product to be cooled to 5 °C?

Solution to sample problem 2.1
DTD is calculated by using Equation 2.1:

$$DTD = \frac{(10 - 1.5)}{(20 - 1.5)}$$

$$= 0.30$$

Locating the point on Fig. 2.2 where curve F (for cucumber) intersects the DTD = 0.30 line produces a cooling time of 44 minutes. (Note that this figure has a log scale.)
For the time taken for product to be cooled to 5 °C:

$$DTD = \frac{(5 - 1.5)}{(30 - 1.5)}$$

$$= 0.12$$

Locating the point on Fig. 2.2 where curve F intersects the DTD = 0.12 line produces a cooling time of 70 minutes. Therefore an extra 26 minutes of cooling is required for the cucumbers to be cooled to 5 °C.

palletised bins of produce are loaded into an enclosure with a forklift truck and chilled water is sprayed over the produce, collected, re-cooled and recycled. They are relatively inexpensive and are suitable for growers that have a limited amount of produce or a short harvest season. In another design (Fig. 2.3), a high-capacity fan is used to suck a fine mist of chilled water through the stack of bins (known as 'hydro-air-cooling'), which produces more uniform cooling of foods. Conveyor hydrocoolers, having a length of up to 15 m and a width of up to 2.5 m, pass containers of produce under a shower of chilled water on a conveyor. Water flow rates are typically 750 litres min^{-1} per m^2 of active cooling area, and require circulation of around 30 000 litres of water per minute. The speed of the conveyor is adjusted for different crops or bin sizes. Because of their relatively high cost, they must operate for long periods in a year to be economically justified. In immersion hydrocoolers, crates of produce are moved by a submerged conveyor through a large, shallow tank of recirculated chilled water. This system produces more rapid cooling than other types and is, for example, nearly twice as fast as a conveyor cooler, because the water has greater contact with food surfaces and heat transfer rates are correspondingly higher.

In truck hydrocooling, packaged produce is loaded into a trailer and moved to a cooling system, where perforated pipes are inserted into the trailer above the load. These produce a shower of chilled water at up to 4000 l min^{-1} and the water flowing out of the trailer is collected, re-cooled and recycled. After cooling, the pipes are removed, a layer of crushed ice is placed above the crates and the produce is transported. Truck hydrocoolers can be constructed by a grower at a central location on a farm for a much lower cost than a commercial hydrocooler, but the rates of heat transfer are lower and hence cooling times are longer than other designs (Boyette *et al.* 2006).

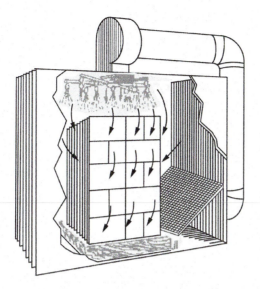

Fig. 2.3   Hydro-air cooler (courtesy of North Carolina Cooperative Extension Service, North Carolina State University) (adaped from Boyette *et al.* 2006).

2.2   Cleaning

Cleaning removes contaminating materials from foods and separates them to leave the food in a suitable condition for sale in the fresh market sector or for further processing. Peeling fruits and vegetables (section 2.4), skinning meat or descaling fish may also be considered as cleaning operations. In vegetable processing, blanching (Chapter 11) also helps to clean the product. The presence of contaminants in processed foods is a major cause of prosecution of food companies (Fig. 2.4) and most retail buyers have detailed specifications of the maximum levels of contaminants that are tolerable. A classification of the type of contaminants found on raw foods is shown in Table 2.1 and details are given by Creaser (1991).

Cleaning should take place at the earliest opportunity in a process both to prevent damage to subsequent processing equipment by for example stones, bone or metal fragments, and to prevent resources from being spent on processing contaminants that are then discarded. The early removal of small quantities of food contaminated by micro-

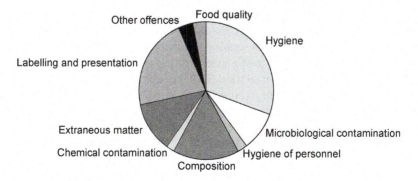

Fig. 2.4   Causes of prosecutions in UK in 1995 (from Jukes 2003 © D.J. Jukes 2003).

Table 2.1   Contaminants found on raw foods

Contaminant	Examples	Possible sources
Ferrous and non-ferrous metals	Filings, nuts, bolts	Mechanised harvesting, handling and processing equipment
Mineral	Soil, engine oil, grease, stones	
Plant	Leaves, twigs, weed seeds, pods, skins	Crops or animals
Animal	Hair, bone, excreta, blood, insects, larvae	
Chemical	Fertilisers, herbicides, insecticides or fungicides	Incorrect application or over-use
Microbial cells	Soft rots, fungal growth, yeasts	Pre-harvest – irrigation water, manure, or animals
Microbial products	Toxins, odours, colours	Post-harvest – wash water, poorly cleaned equipment or cross-contamination

Adapted from Brennan *et al.* (1990)

organisms also prevents the subsequent loss of the remaining bulk by microbial growth during storage or delays before processing. Cleaning is thus an integral part of quality assurance and HACCP systems to protect the consumer (Jongen 2005, and Chapter 1, section 1.5.1), and an effective method of reducing food wastage to improve the economics of processing (Wallin and Haycock 1998, Edwards 2004).

Equipment for cleaning is categorised into wet procedures (e.g. soaking, spraying) and dry procedures (separation by air, magnetism or physical methods). The selection of a cleaning procedure depends on the nature of the product to be cleaned, the types and amounts of contaminant that are likely to be present and the degree of decontamination that is required. In practice, buyers specify an 'acceptable' standard for the particular end use and the least-cost method of achieving this is used by the processor. In general, a combination of cleaning procedures is required to remove the different contaminants found on most foods.

2.2.1   Wet cleaning
Wet cleaning is more effective than dry methods for removing soil from root crops or dust and pesticide residues from fruits or vegetables. It is also dustless and causes less damage to foods than dry methods. Different combinations of detergents and sterilants at different temperatures allow flexibility in operation. Warm water improves cleaning efficiency, especially if mineral oil is a contaminant. However, the use of warm water increases costs and may damage the texture of some foods. It also accelerates bio-chemical and microbiological spoilage unless careful control is exercised over washing times and reducing delays before processing. Furthermore, wet procedures produce large volumes of effluent, often with high concentrations of dissolved and suspended solids (measured as biological oxidation demand (BOD) or chemical oxidation demand (COD)). There is then a requirement both to purchase clean water and to either pay for effluent disposal charges or build in-factory water treatment facilities (also Chapter 27, section 27.1.4). To reduce costs, recirculated, filtered and chlorinated water is used. Concentrations of 100–200 mg l^{-1} chlorine may be used, although its effectiveness for decontamination has been questioned and it is not permitted in some countries (also section 2.1). Further details of cleaning and disinfection are given in Chapter 27 (section 27.1.1) and by Lelieveld and Mostert (2003).

Examples of wet-cleaning equipment include soaking tanks, spray washers, brush washers, drum or rod washers, ultrasonic cleaners and flotation tanks. They are described in detail by Brennan *et al.* (1990) and Grandison (2006). Soaking is a preliminary operation before cleaning heavily contaminated root crops to partially remove soil and stones. Soaking tanks may be fitted with stirrers or paddles to agitate the water. For delicate foods, including strawberries or asparagus, or products that trap dirt internally (e.g. celery), air can be sparged through the water to increase cleaning efficiency. 'Fluming' (carrying foods by water in troughs over a series of weirs) is used to clean small fruits, peas and beans while simultaneously transporting the crops to the next stage in a process. Dewatering screens are routinely used to separate washwater from the clean product.

Spray washing using drum washers or belt washers is widely used for many types of crops. The cleaning efficiency is higher using small volumes of high-pressure water, provided that delicate products are not damaged. The efficiency depends on the volume and temperature of the water and time of exposure to the sprays. Larger food pieces are rotated so that the whole surface is sprayed, and some equipment designs have brushes or flexible rubber discs that gently clean the food surfaces.

Flotation washing is based on differences in density between foods that float (especially fruits or root crops), and contaminating soil, stones or rotten crops that sink in water. 'Froth flotation' is used to separate contaminants from peas, lima beans, corn and other products (Fig. 2.5). For example, peas are dipped in an oil/detergent emulsion and air is blown through the bed of food. This forms a foam that washes away the contaminating material and the cleaned peas are then spray washed. It is able to remove over 95% of light contaminants, handling up to $9000\,kg\,h^{-1}$ (Anon 2007). The simultaneous cleaning and disinfection of fresh crops by short hot water rinse and brushing (HWRB) involves placing crops on rotating brushes and rinsing with hot water for 10–30 s (Orea and Gonzalez Urena 2002).

Fig. 2.5   Froth flotation washer (courtesy of Key Technology; Anon 2007).

2.2.2   Dry cleaning

The main types of equipment used for dry cleaning of foods are air classifiers, magnetic separators and electrostatic separators. Screens, shape sorters and imaging machines (section 2.3) are also used to remove contaminants. Air classifiers (or 'aspiration cleaners') use a fast-moving stream of air to separate contaminants from foods using differences in their densities and the projected areas of particles. A method to calculate the air velocity required for separation is described in Chapter 1 (section 1.3.4). Air classifiers are widely used in harvesting machines to separate denser contaminants (e.g. soil, stones) and less dense contaminants (e.g. leaves, stalks and husks) from grains and legumes. Very accurate separations are possible, but the equipment requires large amounts of energy to produce the air streams. Fast-moving air may also be used to blow loose contaminants from larger foods such as eggs or fruit. Air classifiers are used for products that have high mechanical strength and low moisture content (e.g. grains and nuts). After cleaning, the surfaces remain dry and the concentrated wastes may be disposed of more cheaply than wet effluents. In addition, plant cleaning is simpler and chemical and microbial deterioration of the food is reduced compared to wet cleaning. However, aspiration cleaning tends to be less effective than wet methods in terms of cleaning efficiency. It may also be necessary to prevent or control dust, which not only creates a health and explosion or fire hazard, but also re-contaminates the product.

Magnetic and electrostatic separators

Contamination by metal fragments or loosened bolts from machinery is a potential hazard in all processing. Magnetic cleaning systems for ferrous metals include magnetised drums, magnetised conveyor belts or magnets located above conveyors or in filters or pipework. Permanent magnets are cheaper but require regular inspection to prevent a build-up of metal that can be lost into the food all at once to cause gross recontamination. Electromagnets are easier to clean (by switching off the power supply). Recent developments include auto-cleaning units operated by programmable logic controllers (PLC) that collect and log all contaminants for future reference in case of a customer complaint. Until the 1990s, magnetic separators were made from ceramic magnets, but they now incorporate rare earth materials (e.g. neodymium with iron and boron or samarium with cobalt) that are magnetically six times more powerful than ceramic material. Details of these materials are given by Anon (2000a). Magnetic separators are placed at goods intake points, before and after processing machines and at the end of the process line. Each magnet therefore only has to remove contamination produced since the previous magnet in a processing line, which enables QA staff to readily identify the source of any contamination. It also acts as an early warning of wear in particular equipment that can prevent major machine failure (McAllorum 2005).

Mild steel is magnetic and easy to separate, but non-ferrous metals (e.g. copper and brass) are non-magnetic. Metal detectors (Chapter 26, section 26.7) are therefore used both to protect processing equipment (e.g. mills) that could be seriously damaged by metal fragments, and at the end of a processing line to detect metal contamination in packaged foods as part of a quality assurance and HACCP scheme (Chapter 1, section 1.5.1).

Electrostatic cleaning can be used in a limited number of applications where the surface charge on a raw material differs from that of contaminants. It has been used to remove seeds of similar geometry but having different surface charge from grains, and for cleaning tea leaves. The food is conveyed on a charged belt and foods or contaminants are attracted to an oppositely charged electrode according to their surface charge. There are also developments in electrostatic space charge systems to clean air and reduce the transmission of dust and bacteria (Richardson *et al.* 2003).

Screens and shape sorters

Screens in the form of rotary drums and flatbed designs (section 2.3.2) are size separators that can be used to remove contaminants. Larger contaminants, such as leaves and stalks, can be removed from smaller foods (termed 'scalping') or, alternatively, smaller particles of sand or dust can be removed from larger foods (termed 'sifting' or 'de-dusting'). Screening may produce incomplete separation of contaminants and is often used as a preliminary cleaning stage before other methods. The efficiency of screens to separate materials may be improved by vibrating the screen, or by using abrasive discs or brushes to remove materials from the apertures in the screen.

One type of grain cleaner (Fig. 2.6) that can be used for most varieties of cereal grains, oilseeds and vegetable or legume seeds, has three screen decks, each approximately 1–1.5 m^2. The top screen scalps contaminants with the product falling through the screen. The bottom two screens can be configured in either sifting or scalp-sifting operation, using more than 175 different sizes of perforated metal or wire cloth screens. Weed seeds, foreign material and splits drop through the bottom screens and the product passes over them. Next, the product is passed through air from a bottom blast fan that has adjustable fan velocity. The aspiration removes lightweight materials and dust that may have remained after screening. The equipment removes dust, chaff, straw and small seeds from grains and can also be used for final cleaning to separate dust and fines from seeds before bagging (Anon 2008c). Similar equipment is used in sugar refineries, where a 4.75 mm scalping screen removes sugar contaminants and sugar lumps at up $70\,\mathrm{t\,h^{-1}\,m^{-2}}$ (Anon 2004).

Physical separation of contaminants from foods is also possible when the food has a regular well-defined shape. For example, round foods (peas, blackcurrants and rapeseed) are separated from contaminants by exploiting their ability to roll down an inclined, upward moving conveyor belt. Contaminants, such as weed seeds in rapeseed or snails in

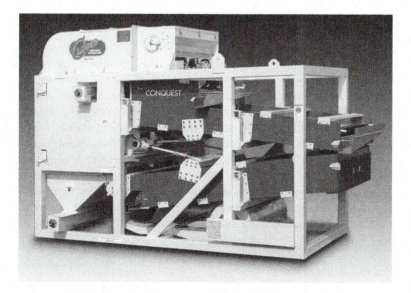

Fig. 2.6   Seed cleaner (courtesy of A.T. Ferrell Company; Anon 2008c).

Fig. 2.7   Open spiral separator (courtesy of Profile Industries Inc., www.profile-ind.com) (Anon 2008d).

blackcurrants, are carried up the conveyor and separated. Similarly, the spiral separator (Fig. 2.7) is used for removing non-round materials (chaff, leaves, seeds, etc.) from round seeds, such as mustard, peas or soybeans. The separator contains a spiral that is fed by gravity from a hopper at the top, using an adjustable feed plate. The seed leaving the hopper runs over a cone divider that spreads the seeds evenly to each of the five flights. Round seeds travel down the flights at a faster speed than non-round materials. Their momentum increases until they run over the edge of the inner spirals, drop into an outer spiral and discharge at the bottom of the machine. The non-round material remains on the inner flights and slides down to a separate chute at the bottom. Since there are no moving parts, no power is required to operate the machine and when the feedrate has been correctly set, the separator continues to operate automatically as long as there is seed in the hopper (Anon 2008d).

Colour and imaging machines
Optical and machine vision systems that have been developed to sort and grade foods are also able to clean foods by removing contaminants. They are described in detail in section

2.3.3. Small-particulate foods are checked for contaminants using microprocessor-controlled colour sorters. The food is illuminated and the reflected light is compared to a pre-set standard. Any contaminants that have a different colour are automatically rejected. Advances in computing power, 'smart' cameras and the use of laser light sources enable contaminants that have the same colour but a different shape to the product to be removed (e.g. green stalks from green beans (section 2.3.3)). Concern over food allergens has increased demand for machine vision systems that not only check ingredients for contamination by known allergens (peanuts, soybeans, shellfish, tree nuts, etc.) but also inspect packages and barcodes to verify that the product is correctly packaged and labelled (Hardin 2005).

X-rays are used to detect metals and other types of solid contaminant in both raw materials and inside packaged foods. The X-rays pass through the food as it passes on a conveyor and are converted to visible light. This is transmitted to an image intensifier and the final image is displayed on a monitor. The system automatically rejects any contaminated items. Developments include the use of solid-state X-ray sensitive elements that collect information about a product as it passes over the sensors. This information is modified by sophisticated computer image processing techniques to create a two-dimensional image of the contaminant. Further details are given by Batchelor *et al.* (2004). Applications include detection of stones, bone fragments, plastics, seafood shells and ceramics or concrete in addition to metals, and it also detects missing or under-filled packages (Greaves 1997). In a system described by Anon (2000b), an X-ray scanner can detect contaminants in cans or jars by size, shape, density and texture, and can discriminate between the container wall and the contaminant, even when they are in close contact.

There are a number of promising technologies for detecting contaminants based on electromagnetic or imaging techniques that are under development but not yet used commercially (Table 2.2), described in detail by Graves *et al.* (1998) and George (2004). Electromagnetic techniques include capacitive systems, impedance spectroscopy and electrical resistance tomography. Microwave reflectance holography using reflected or back-scattered radiation has a number of potential advantages. Many foods differ from

Table 2.2 Summary of techniques used in foreign body food inspection

Technique	Wavelength	Food product	Foreign bodies
Magnetic	N/A	Loose and packaged foods	Metals
Capacitance	N/A	Products <5 mm thick	More research needed
Microwave	1–100 mm	Fruits, possible others but needs more research	Fruit pits
Nuclear magnetic resonance	1–10 mm + magnetic field	Fruits and vegetables	Fruit pits and stones
Infrared	700 nm–1 mm	Nuts, fruits, vegetables	Nut shells, stones and pits
Optical	400–700 nm	Any loose product, fruit and vegetables	Stones, stalks
Ultraviolet	1–400 nm	Meat, fruits, vegetables	Fat, sinews, stones and pits
X-rays	<1 nm	All loose and packaged foods	Stone, plastic, metal, glass, rubber, bone
Ultrasonics	N/A	Potatoes in water, more research needed	Stones

N/A = not applicable.
From Graves *et al.* (1998)

contaminants in their specific microwave impedance and very small contaminants can be detected in three dimensions (Benjamin 2004). Imaging techniques including ultrasound are being studied as cost-effective methods of inspection. Foreign bodies have different acoustic impedances to foods and can be identified by changes in reflection, refraction and scattering of ultrasound waves as they pass through the food (Basir *et al.* 2004). Ongoing research into nuclear magnetic resonance and magnetic resonance imaging aims to shorten the time needed to acquire a full three-dimensional image (Hills 2004). Surface penetrating radar has shown some promising results in laboratory tests, especially in detecting metal foreign bodies in wet, homogeneous materials. Preliminary studies using microwave radar have shown that it can detect small (1 mm) pieces of stone, glass, stainless steel and plastic in homogeneous foods, but the method is not suitable for products in metallic or foil-wrapped containers (Barr and Merkel 2004).

2.3   Sorting and grading

The terms 'sorting' and 'grading' are often used interchangeably, but strictly sorting means 'the separation of foods into categories on the basis of a measurable physical property'; the ones used in relation to foods being size, shape, weight and colour. Grading is 'the assessment of overall quality of a food using a number of attributes'. The distinction was originally made to characterise simple sorting machines (section 2.2.2) from grading carried out by skilled inspectors who are trained to simultaneously assess a number of variables. Examples of grading include examination of carcasses by meat inspectors for disease, fat distribution, bone : flesh ratio and carcass size and shape. Other graded foods include cheese and tea, which are assessed by specialist tasters for flavour, aroma, colour, etc., and eggs, which are visually inspected by operators over tungsten lights ('candling') to assess up to 20 factors, to remove those that are fertilised or malformed or those that contain blood spots or rot. In some cases the grade of food is determined from the results of laboratory analyses. For example, wheat flour is assessed for protein content, dough extensibility, colour, moisture content and presence of insects.

The distinction between sorting and grading is steadily breaking down with the development of sophisticated machine vision-based systems that can simultaneously assess a number of attributes (section 2.3.3). For example in meat inspection, machine vision systems can assess bruising, skin colour and damage on chicken meat (Ade-Hall *et al.* 1996), and machines can now inspect eggs for cracks (Zuech 2006a) as well as size (Anon 2008e) (also section 2.3.2). Previously fruits were graded by operators into different grades using characteristics such as colour distribution, surface blemishes, size and shape of the fruit. Grading is now done by machines that can simultaneously measure 100 characteristics including colour, weight, diameter, sugar content, ripeness, blemishes, or internal characteristics such as 'water core' (Anon 2008f).

The move away from trained inspectors to machine-based systems is driven by a number of changes to the industry:

- Increased pressure from large retailers for uniform fresh-pack products, especially fruits and vegetables, and stricter controls on contaminants in all processed foods.
- The need for processors to improve product tracking and traceability, and also concerns about liability claims. The food industry is moving towards individual product identification so if a problem arises it can be identified, isolated and tracked to affected stocks (Chapter 27, section 27.3). Traceability is made even more important

in many food sectors because products may be sorted in one location, processed in another facility, and labelled, cartoned and palletised in another.

- Increased labour costs for inspectors and operators. People are generally not very effective in performing routine inspection of apparently similar products. After a relatively short time their visual acuity becomes reduced, even if doing a simple task of separating good from substandard products, or foreign materials from products. Grading a product is even more difficult. Application-specific machine vision systems perform significantly better in these sorting operations and incur lower operating costs.
- Reduced costs, easier operation and increased performance of sorters. More information about the product can now be obtained by machine vision systems and this data can also be incorporated by processors into electronic QA and accounting systems (Zuech 2005).

Like cleaning, sorting and grading should be employed as early as possible in a process to ensure a uniform product for subsequent processing and to prevent expenditure on materials that are subsequently discarded as substandard.

Theory
The effectiveness of a sorting or grading procedure is calculated using:

$$\text{effectiveness} = \frac{PX_p\,R(1 - X_r)}{FX_f\,F(1 - X_f)} \qquad \boxed{2.2}$$

where P (kg s^{-1}) = product flowrate, F (kg s^{-1}) = feed flowrate, R (kg s^{-1}) = rejected food flowrate, X_p = the mass fraction of desired material in the product, X_f = the mass fraction of desired material in the feed and X_r = the mass fraction of desired material in the rejected food.

There are three criteria that define the performance of a sorter:

1 *Defect removal efficiency*: the percentage of incoming 'defective product' that the sorting system removes. This is the measure of its ability to remove product that does not meet the specifications for acceptable product.
2 *Recovery efficiency*: the percentage of incoming 'acceptable' product that leaves the system in the 'acceptable' stream. This quantifies the 'false accept/false reject' performance.
3 *Throughput*: the maximum amount of product that the system can sort without losing performance in defect removal efficiency and recovery efficiency.

All three criteria are required to define performance, although this is made more complex when products have more than one type of defect. The main constraint on sorting performance in many instances is now the efficiency of the materials handling system to deliver sufficient product to/from the sorter (Zuech 2006b).

2.3.1   Shape and size sorting
The shape of some foods is important in determining their suitability for processing or their retail value in the fresh-pack sector. For example, for economical peeling, potatoes should have a uniform oval or round shape without protuberances. Cucumbers and gherkins are more easily packaged if they are straight, and foods with a characteristic shape (e.g. pears) have a higher retail value if the shape is uniform. Retailers usually specify the size range of products in the fresh-pack sector and meeting these

specifications has a significant effect on the price received and profitability of the growers' operations. In the processing sector, the size of individual pieces of food is particularly important when a product is heated, dried or cooled, because it in part determines the rate of heat or mass transfer, and any significant variation in size causes over-processing or under-processing. The correct size distribution of small particulate foods such as sugar or powdered ingredients (e.g. starch, colourants, thickeners etc.) is also important to achieve uniform products in mixing and blending operations (Chapter 4, section 4.1).

Theory
Size sorting (termed 'sieving' or 'screening') is the separation of solids into two or more fractions on the basis of differences in size. The particle size distribution of a material is expressed as either the mass fraction of material that is retained on each sieve or the cumulative percentage of material retained (Fig. 2.8). Different mesh sizes are shown in Appendix C. The mean overall diameter of particles (volume or mass mean diameter) is found using:

$$d_v = \frac{\sum d \, m}{\sum m}$$

$$\boxed{2.3}$$

where d_v (μm) = volume or mass mean diameter, d (μm) = the average diameter and m (g) = mass retained on the sieve.

Equipment
Shape sorting is useful where foods contain contaminants that have similar size and weight. Examples of equipment used for shape sorting include the disc sorter and various types of screens. Machine vision systems (section 2.3.3) can also be used to sort foods on

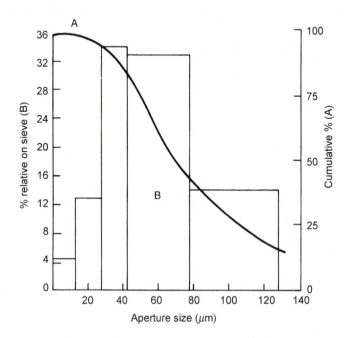

Fig. 2.8   Retention of particles on sieves: (A) cumulative percentage, (B) mass fraction.

Sample problem 2.2
A sieve analysis of powdered sugar showed the following results. Calculated the mass mean diameter of the sample.

Sieve aperture (μm)	Mass retained (%)
12.50	13.8
7.50	33.6
4.00	35.2
2.50	12.8
0.75	4.6

Solution to sample problem 2.2
The cumulative percentages are as follows:

Data plotted for cumulative % in Fig. 2.8

Aperture size (μm)	12.50	7.50	4.00	2.50	0.75
Cumulative percentage	13.8	47.4	82.6	95.4	100.0

To find the mass mean diameter, find d as follows:

Average diameter of particles d (μm)	m (%)	md
0.375	4.6	1.725
1.625	12.8	20.8
3.25	35.2	114.4
5.75	33.6	193.2
10.0	13.8	138.0
Total	100.0	468.125

From Equation 2.3,

$$\text{mass mean diameter} = \frac{468.125}{100}$$

$$= 4.68 \ \mu\text{m}$$

the basis of their shape. The disc sorter has vertical discs that have indentations precisely machined to match the shape of a specific grain. The discs are partly embedded in a mass of grain and selectively lift the required grains, leaving seeds and other materials behind. Different discs can be fitted for each of the common food grains.

Screens with either fixed or variable apertures are used for size or shape sorting. The screen may be stationary or, more commonly, rotating or vibrating. Fixed aperture screens include the flat bed screen (or sieve) and the drum screen (or 'rotary screen', 'trommel' or 'reel'). The multideck flat bed screen is similar to equipment shown in Fig. 2.6 (section 2.1.2). It has a number of inclined or horizontal mesh screens, which have aperture sizes from 20 μm to 125 mm, stacked inside a vibrating frame. Food particles that are smaller than the screen apertures pass through under gravity until they reach a screen with an aperture size that retains them. The smallest particles that are separated commercially are of the order of 50 μm. The capacity of a screen is the amount of food that passes through per square metre per second. The rate of separation is controlled by:

- the shape and size distribution of the particles;
- the nature of the sieve material;
- the amplitude and frequency of shaking; and
- the effectiveness of methods used to prevent blocking (or blinding) of the screens.

These types of screen are widely used for sorting dry foods (e.g. flour, sugar and spices). The main problems encountered are:

- excessive moisture or high humidity, which causes small particles to stick to the screen or to agglomerate and form larger particles, which are then discharged as oversize;
- blinding, particularly if the particle size is close to that of the screen aperture;
- excessively high feed rates, which cause the screens to become overloaded and small particles are discharged with the oversized particles.

Where vibration alone is insufficient to separate particles adequately, a gyratory movement is used to spread the food over the entire sieve area, and a vertical jolting action breaks up agglomerates and dislodges particles that block sieve apertures. Other equipment uses polypropylene balls on the sieve that perform a similar function.

Drum screens are almost horizontal, perforated metal or mesh cylinders that may be concentric (one inside another), parallel (foods leave one screen and enter the next) or series (a single drum constructed from sections with different sized apertures). All types have a higher capacity than flatbed screens and fewer problems associated with blinding. In some designs the screens may be fitted with brushes to reduce blinding. Drum screens are used for size sorting small-particulate foods (e.g. nuts, peas or beans) that have sufficient mechanical strength to withstand the tumbling action inside the screen.

Variable-aperture screens have either a continuously diverging aperture or a stepwise increase in aperture. Both types handle foods more gently than drum screens and are therefore used to sort fruits and other foods that are more easily damaged. Continuously variable screens (Fig. 2.9) have pairs of diverging rollers, cables or conveyor belts. In the expanding roller sorter, foods pass along the machine until the space between the rollers or belts is sufficiently large for them to pass through. The belts or rollers may be driven at different speeds to rotate the food and thus to align it to present the smallest dimension to the aperture. Stepwise increases in aperture are produced by adjusting the gap between driven rollers and an inclined conveyor belt. The food rotates and the same dimension is therefore used as the basis for sorting (e.g. the diameter along the core of a fruit).

2.3.2   Weight sorting

Weight sorting is more accurate than other methods and is therefore used for more valuable foods (e.g. eggs, cut meats and, in industrialised countries, some tropical fruits). Eggs are sorted at up to 12 000 h^{-1} into five to nine categories with a tolerance of 0.5 g. They are first graded by 'candling' and then pass to a weight sorter (Fig. 2.10). This consists of a conveyor that transports the eggs to a series of spring-loaded, strain gauge or electronic weighing devices incorporated into the conveyor. The conveyor operates intermittently and while stationary, tipping or compressed air mechanisms remove heavier eggs, which are discharged into a padded chute. Lighter eggs are replaced on the conveyor to travel to the next weighing device where the procedure is repeated.

Fig. 2.9   Roller grader (courtesy of Vizier Systems, http://vizier.co.za/sizers.htm).

Equipment is computer controlled and can provide data on quantities of eggs and size distributions from different suppliers.

Aspiration and flotation sorting equipment exploits differences in food density to sort foods and is similar in design and operation to machines used for aspiration and flotation cleaning (section 2.1). Grains, nuts and pulses can be sorted by aspiration, and vegetables can be sorted by flotation in brine (specific gravity = 1.1162–1.1362). For example, the density of peas correlates with their tenderness and sweetness, and denser, starchy, over-

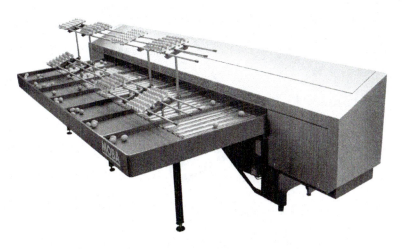

Fig. 2.10   Egg grader (courtesy of Moba BV) (Anon 2008e).

mature pieces sink and are separated from the required product which floats in the brine. Similarly, the density of potatoes directly correlates with solids content, which determines their suitability for crisp manufacture.

The collation into bulk packs of foods that have variable weight (e.g. frozen fish fillets) is time consuming and laborious when performed manually. Operators select pieces of food from a pool of materials and collate them by trial and error to form a pack that is as close as possible to the declared weight. There is frequently a large give-away to ensure compliance with fill-weight legislation. Collation sorting is now performed automatically by a microcomputer-controlled weight sorter. Items of food are weighed and placed in a magazine. Their weights are stored in the computer, which then selects the best combination of items to produce the desired number in a pack with minimum give-away. Other examples of microprocessor-controlled weighing and packaging are described in Chapter 27 (section 27.1.2).

2.3.3   Colour and machine vision sorting and grading

There are two types of colour sorter: machines that use photodetectors to sort small, particulate foods and those that use cameras to sort larger foods such as bakery products and fresh fruits and vegetables. Small particulate foods are sorted at high rates (up to $16\,th^{-1}$) using microprocessor-controlled colour sorting equipment (Fig. 2.11). Particles are fed into the chute one at a time. The angle, shape and lining material of the chute are

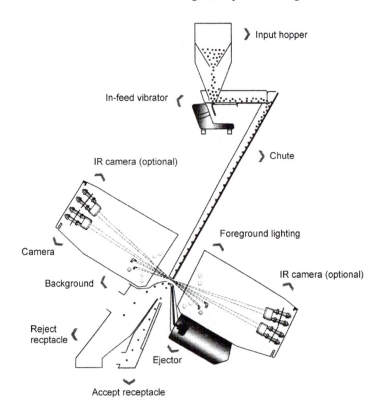

Fig. 2.11   Operation of microprocessor controlled colour sorting equipment (courtesy of Buhler Sortex Ltd., picture © Buhler Sortex Ltd) (Anon 2008g).

adjusted to control the velocity of the pieces as they pass a photodetector. The colour of the background and the type and intensity of the light used for illuminating the food (including infrared and ultraviolet options) are closely controlled for each product. Photodetectors measure the reflected colour of each piece and compare it with pre-set standards, and defective foods are separated by a short blast of compressed air. The microprocessor can store 100 product configurations to enable rapid changeover to different products using an operator touchpad. Typical applications include peanuts, Michigan navy beans (for baked beans), rice, diced carrot, maize kernels, cereals, snackfoods and small fruits. Further details are given by Brimelow and Groesbeck (1993) and Low et al. (2001).

Colour sorters for fruits and vegetables can sort up to $15\,t\,h^{-1}$ of product. For example, vegetables in free-fall are scanned 1000 times per second, as they leave a conveyor belt, using concentrated helium–neon or laser light beams and a high-speed rotating mirror. The machine detects differences in reflectivity between products and unwanted materials (Anon 1990). A different type of equipment employs a sensor located above a conveyor belt, which views products as they pass beneath. The sensor detects up to eight colours and activates an alarm or reject mechanism whenever a pre-selected colour (e.g. burned areas on bakery products) passes the detector beam. It is also able to identify and separate different coloured foods that are to be processed separately.

These earlier colour sorting machines use relatively simple detector arrangements, but in the 1990s machine vision systems began to be developed that could measure colour distribution, shape, size and surface properties, in addition to colour. Improved perform-ance was achieved when colour cameras became lower in price and replaced mono-chrome cameras. More recently, machine vision technology has developed further after:

- the development of higher-resolution digital cameras that can scan both finer line widths and areas of the product;
- advances in microprocessor technology and new software developments that have the computing power to handle the output from these cameras;
- development of lamps that have specific spectral outputs. Many of the systems developed so far are semi-custom; that is, they use a lighting spectrum that is specific to the product being inspected, which enhances sorting of foreign objects or product grading based on appearance (Zuech 2006b).

These application-specific machine vision systems have been developed for both the fresh-pack and the food processing markets (Table 2.3). The components of a machine vision system are as follows (Zuech 2006b, 2006c).

Cameras
Digital cameras, fitted with telecentric optics[1] have light-sensitive cells which produce a voltage that is proportional to the intensity of light received. Advanced CMOS[2]

---

1. Normal lenses have varying magnification of objects at different distances from the lens. This causes problems for machine vision systems because the apparent size of objects changes with their distance from the camera and the apparent shape of objects varies with the field of view (e.g. circles near the centre of the field of view become ellipses at the periphery). Telecentric lenses create images of the same size for objects at any distance and across the entire field of view.
2. A complementary metal oxide semiconductor is a type of digital logic circuit that only uses significant power when its transistors switch between on and off states and therefore use little power, do not produce as much heat as other forms of logic circuits, and allow a high density of logic functions on a chip.

Table 2.3   Applications of machine vision-based technologies

Fresh-pack product applications		Food processing applications	
Apples	Peaches/nectarines	Cereals	Dates
Avocados	Pears	Corn	French fries
Cherries	Persimmons	Lentil	Lima beans
Clementines	Plums	Nuts	Meat cubes
Cranberries	Pomegranates	Oats	Olives
Grapefruit	Tomatoes	Rice	Peas
Kiwi fruit	Cucumbers	Rye	Prunes
Lemons/limes	Onions	Snack foods	Raisins
Mangos	Potatoes	Sunflower seeds	Rutabagas
Melons	Sweet potatoes	Wheat	Shrimp
Oranges/mandarins/	Green beans		
tangerines	Carrots		
Papayas	Spinach		

Adapted from Zuech (2006a)

(complementary metal oxide semiconductor) logic circuits use this voltage to produce a high-resolution image and have the ability to pre-process the image data in real-time. When the camera with a PC and frame grabber (a component of a computer system designed for digitising analog video signals) are housed together, it forms a self-contained machine vision system or 'smart camera'. These cameras have preset programmes for different products that are easily changeable by operators using a video display. Future developments may include wireless cameras and the use of fibre-optic cabling to produce high-resolution images from multiple cameras over long distances.

Lighting

Lighting is the most critical factor in the success of a machine vision system, because it determines which traits the sensors detect on the product. Lighting based on light-emitting diodes (LED) is more efficient and brighter than other types, and it is being used in more machine vision applications. LEDs that have ultraviolet output are used with cameras that have greater sensitivity to ultraviolet light to make more accurate dimensional measurements. Near-infrared cameras have been developed for specific products and these are likely to find wider applications as more thermal and infrared traits are recognised as important sorting and grading parameters. Anon (2001a) describes fluorescent laser sorting technology, which shines diffused laser light onto green vegetables such as spinach, peas and beans. It detects and analyses the amount of fluorescent light emitted by chlorophyll in the vegetables, or by green discoloration on potato products. The chlorophyll appears white on the display whereas everything else, including dead cells that form brown spots on the vegetables, do not transform the laser energy and appear black. Up to 99% of brown spots are detected and the system can also be used to detect up to 99.9% of other contaminants.

Computers and software

Software developments now enable machine vision functions to be performed that could not be considered before because they required too much computing power. Increased computer power using 64-bit processing has enabled more accurate dimensional measurements at lower cost, both in terms of price and the time required for computations. Software has become easier to use and increasingly it is designed so that users can optimise systems for specific applications. A new generation of machine graders uses advanced image processing algorithms and complex decision-making mechanisms based on neural networks and fuzzy logic-based decision-making software (Chapter 27, section 27.2.3) to classify products using more than 100 parameters. Whereas older quality sorting systems classify defects on the basis of a library of samples held in the system, these new types of equipment can be 'trained' by operators to evaluate defects, by simply 'showing' examples to the system. The equipment will then place products into different quality classes according to the type of defects that are detected. In the future these graders will have the ability to better match the performance of human inspectors in applications that require decisions based on subjective analyses (Zuech 2006b).

Equipment

Machine vision graders that are used for fresh-pack foods measure colour (e.g. to reflect ripeness), colour distribution, shape, size, surface conditions (bruises, gouges, insect infestation, etc.). They may also include weighing stations to provide information on size in addition to the physical size derived from the vision system. Cubeddu *et al.* (2002) describe advanced optical techniques that can give information on both surface and internal properties of fruits, including their texture and chemical composition. This can be used to grade fruit according to maturity, firmness or the presence of defects, or the amounts of chlorophyll, sugar or acid in the fruit. A system described by Anon (2008g) is able to simultaneously sort green beans into three categories: acceptable; retaining the stalk (for re-snibbing); and reject discoloured product or contaminants (Fig. 2.12). It can also sort carrots into acceptable, discoloured and misshapen pieces at up to $10 \, \mathrm{t \, h^{-1}}$. The machine can remove 95% of defects that are >10 mm diameter, 85% between 5 and 10 mm and 70% between 3 and 5 mm. Camera sorting equipment using laser light is also used to sort potatoes for French fries after peeling (to check for remaining peel), before cutting (to check for contaminants or rotten potatoes that would damage the knives) and after they have been frozen (to check for discoloration and contaminants). Other applications in the bakery industry include three-dimensional and colour-based sorting of bread, crackers and cookies to remove misshapen products and so avoid jamming automatic packaging machines. Three-dimensional techniques are also important where products are stacked to fit into size-specific packs. By monitoring the height of the

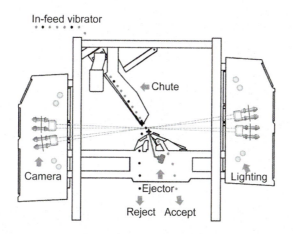

Fig. 2.12   Schematic of SORTEX K colour sorter (courtesy of Buhler Sortex Ltd, picture © Buhler Sortex Ltd) (Anon 2008g).

individual product, the stack height can be calculated to guarantee that it will fit into the pack (Zuech 2006a).

Some machines also have X-ray-based vision systems, which can detect internal flaws or physiological conditions that are not detectable by scanning the surface of the product. Colour/weight/diameter graders are used for grading fragile fruits such as peaches, avocados, mangoes and kiwis, and vegetables including bell peppers and tomatoes. Fruits or vegetables in multiple rows are placed in individual cups or pockets and pass under an arrangement of cameras that capture images of the entire surface of the product. Mechanisms may be incorporated in the pockets to rotate the product under the camera. The images can be used for shape sorting based on maximum/minimum diameters, ratios of sizes, etc., or colour measurement, assessed using simple percentage colour ratios, intensity value histograms or minimum/maximum defined areas. Typically, machines have up to 12 grading lines and a central computer controls the system, including feed conveyor belts, crate-handling equipment, filling and fruit packaging operations. All data can be directly incorporated into accounting and administration systems (Anon 2008f). In other types of graders, individual fruits are picked up gently by grippers, rotated in front of cameras to detect colour, dimensions and quality, and then transported while suspended from the machine to a weighing unit where their weight is recorded to an accuracy of ±2 g. Graders have a capacity of eight grabs per second per line, with different models offering four to ten lines. After the individual fruit has been assessed, the computer activates a switching system that causes the gripper to place it in a selected lane according to its grade, where a belt transports the fruits to a packing machine.

In other methods, near-infrared (NIR) light is used to measure sugar content (Brix) and other quality characteristics of whole fruit or vegetables (e.g. acidity, aroma, ripeness, maturity or internal properties such as water core). NIR light is shone on the fruit and a sensor collects the light that passes through. Sophisticated mathematical models analyse the light absorption at different wavelengths and translate the information into Brix or other quality indicators. The system has a higher accuracy than can be achieved by human taste and also has advantages in being rapid and non-destructive. It is able to measure characteristics across the whole fruit. This is important because, for example, sugar distribution within a fruit is typically not uniform, and the method provides a prediction of the average sugar content and hence sweetness for each piece of fruit (Anon 2008f).

An 'acoustic firmness sensor' enables on-line, non-destructive measurement of firmness and related characteristics of whole fruits. The grader gently taps the product and 'listens' to the vibration pattern. The acoustic signal (or 'resonance attenuated vibration') is characteristic of the overall firmness, juice content, freshness and the internal structure of the product, including for example, tissue breakdown or dehydration. The signal is analysed to create a 'firmness index', which allows the ripeness and other quality aspects to be determined. This gives a more accurate, reliable and consistent result when compared with other destructive methods, regardless of where on the fruit the measurement has been taken (Anon 2008f).

2.4   Peeling

Many fruits and vegetables are peeled to remove unwanted or inedible material, and to improve the appearance of the final product. The main considerations are to minimise costs by removing as little of the underlying food as possible and leaving the peeled surface clean and undamaged; and to reduce energy, labour and material costs to a minimum. The main methods of peeling are flash steam peeling, knife peeling and, at a smaller scale, abrasion peeling. The older methods of caustic peeling and flame peeling are now less commonly used.

In flash steam peeling, foods such as root crops are fed in batches into a pressure vessel that is rotated at 4–6 rpm. High-pressure steam ($\approx$ 1500 kPa) is introduced and all food surfaces are exposed to the steam for a predetermined time by the rotation of the vessel. The high temperatures cause rapid heating of the surface layer (within 15–30 s) but the low thermal conductivity of the product prevents further heat penetration, and the product is not cooked. Texture and colour are therefore preserved. The pressure is then instantly released which causes steam to form under the skin, and the surface of the food 'flashes off'. Most of the peeled material is discharged with the steam, and water sprays are needed only to remove any remaining traces. This type of peeler is popular owing to the lower water consumption, minimum product loss, good appearance of the peeled surfaces, high throughput (up to 4500 kg h^{-1}) with automatic control of the peeling cycle, and the production of a more easily disposable concentrated waste.

Peelers may be linked to a scanner that monitors the quality of the peeled product, automatically adjusts the operating parameters of the peeling equipment to compensate for fluctuations in the raw material, and maintain precise standards for the peel quality that are predetermined by the processor. This can result in annual savings of up to 2% on raw material throughputs, which for large-scale potato processing (e.g. for French fries) represents a substantial saving (Anon 2008h). The scanner control panel continuously displays in graphical form 'product steam time' and the 'peel removal' percentage and compares these with selected parameters in the quality assurance programme.

Specific knife peeling machines have been developed to peel individual crops, including onions, shrimps and fruits. In onion peeling, a blade slits the onion and the outer skin is gently removed using compressed air. This reduces spoilage and onions have a hand-peeled and polished appearance. The machine can process up to 4000 onions per hour ranging in size from 40 to 110 mm (Anon 2008i). The operation of a shrimp peeler is described by Anon (2008j) (Fig. 2.13): (1) a clamp picks up a shrimp; (2) it is carried through a centring guide where a clamp grips the shell; (3) the body shell is broken from the tail segment; (4) the clamp carries the shrimp through a cutter that precisely splits the shell; (5) brushes remove the vein; (6) a fork pulls the shrimp meat cleanly from the shell; (7) the shell is discharged separately. The equipment operates at up to 5000 shrimps h^{-1}.

In fruit peeling, stationary blades are pressed against the surface of rotating fruits or vegetables to remove the skin. Alternatively the blades may rotate against stationary foods. This method is particularly suitable for citrus fruits where the skin is easily removed and there is little damage or loss of fruit. In abrasion peeling, root crops including potatoes, carrots, celeriac, beets, etc. are fed onto silicon carbide or carborundum rollers or placed into a rotating bowl that is lined with carborundum. The abrasive surface removes the skin and it is washed away by water sprays. An example is the peeling of onions where the skin is removed by abrasive rollers at production rates of up to 2500 kg h^{-1}. The advantages of the method include low energy costs as the process operates at room temperature, low capital costs, no heat damage and a good surface appearance of the food. The limitations of the method are:

- a higher product loss than flash peeling (25% compared with 8–18% losses, for vegetables);

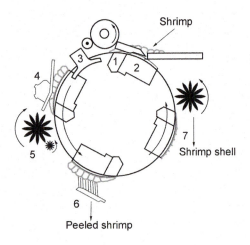

Fig. 2.13   Shrimp peeler (courtesy of Gregor Jonsson Inc.) (Anon 2008j).

- the production of larger volumes of dilute waste which are more difficult and expensive to dispose of;
- relatively low throughputs, as all pieces of food need to contact the abrasive surfaces; and
- irregular product surfaces (e.g. 'eyes' on potatoes) may require hand finishing.

Caustic peeling uses a dilute solution of sodium hydroxide (named 'lye'), which is heated to 100–120 °C. Food is passed through a bath of 1–2% lye, which softens the skin, and it is then removed by high-pressure water sprays. Product losses are of the order of 17%. Although once popular for root crops, this method causes changes in the colour of some products and incurs higher costs, and it is now largely replaced by flash steam peeling. A development of lye peeling is 'caustic' peeling. Food is dipped in 10% sodium hydroxide and the softened skin is removed with rubber discs or rollers. In an example described by Anon (2001b), dry caustic peeling of 72 tonnes of beet per day reduced water consumption by 75% and solid waste by 90% compared with wet peeling. The effluent contained 88% less suspended solids, 94% less COD and 93% less BOD than that from wet peeling.

Flame peeling was developed for onions and consists of a conveyor belt that carries and rotates the food through a furnace at 1000 °C. The outer 'paper shell' and root hairs are burned off, and the charred skin is removed by high-pressure water sprays. Average product losses are 9%.

References

ADE-HALL, A., JEWELL, K., BARNI, M. and MECOCCI, A., (1996), Machine vision for assessing the quality of natural objects, in (R. Gill and C. Syan, Eds.), *Proceedings of the 12th International Conference on CAD/CAM Robotics and Factories of the Future*, 14–16 August, Middlesex University, London, pp. 835–842.

ANON, (1990), Elbicon colour scanning equipment, *Food Trade Review*, Feb., 92.

ANON, (2000a), Neodymium iron boron magnets, company information from Magnet Sales & Manufacturing Company, Inc., available at www.magnetsales.com/Neo/Neo1.htm.

ANON, (2000b), Sorting the good from the bad, *Food Engineering and Ingredients*, Sept., 35–36.

ANON, (2001a), Fluorescent sorting for greens, *Food Engineering and Ingredients*, Feb., 35.

ANON, (2001b), Reducing water and waste costs in fruit and vegetable processing, Envirowise document GG280, available from http://www.envirowise.gov.uk/page.aspx?o=117363.

ANON, (2004), Worldwide screening, *Food Processing*, August, 19.

ANON, (2007), Froth flotation cleaner, company information from Key Technology, available at www.key.net/products/froth-flotation-cleaner/default.html.

ANON, (2008a), Chlorine, company information from Lenntech Water Treatment and Air Purification Holding BV, available at www.lenntech.com/home.htm, search 'chlorine' and select 'Chlorine as a disinfectant for water'.

ANON, (2008b), Ozone information, company information from Ozone Solutions, available at http://www.ozonesupplies.com/Ozone_Info.html.

ANON, (2008c), Clipper Prelude model 526 cleaner, company information from Seedboro Equipment Company, available at http://www.seedburo.com/online_cat/categ08/pr526.asp.

ANON, (2008d), Open spiral separator, company information from Profile Industries Inc., available at www.profile-ind.com.

ANON, (2008e), Egg sorting, company information from Moba, available at http://www.moba.nl/products/products.htm.

ANON, (2008f), Sorting technology, company information from Aweta BV, available at www.aweta.nl/uk/, select 'Technology' and follow links to equipment.

ANON, (2008g), Principles of colour sorting, company information from Buhler Group, available at www.buhlergroup.com/Docs/35753EN.pdf.

ANON, (2008h), Peeling equipment – peel scanning, company information from Odenberg Inc., available at www.odenberg.com/peeling_scanning.htm.

ANON, (2008i), Nakaya onion peeler, company information from Process Plant Network Pty Ltd., available at www.onionpeeler.com.

ANON, (2008j), Jonsson shrimp peeling systems, company information available at http://www.jonsson.com/peeling-systems, or www.jonsson.com and select 'Processors'.

BARR, U-K. and MERKEL, H., (2004), Surface penetrating radar, in (M. Edwards, Ed.), *Detecting Foreign Bodies in Food*, Woodhead Publishing, Cambridge, pp. 172–192.

BASIR, O.A., ZHAO, B. and MITTAL, G.S., (2004), Ultrasound, in (M. Edwards, Ed.), *Detecting Foreign Bodies in Food*, Woodhead Publishing, Cambridge, pp. 204–225.

BATCHELOR, B.G., DAVIES, E.R. and GRAVES, M., (2004), Using X-rays to detect foreign bodies, in (M. Edwards, Ed.), *Detecting Foreign Bodies in Food*, Woodhead Publishing, Cambridge, pp. 226–264.

BENJAMIN, R., (2004), Microwave reflectance, in (M. Edwards, Ed.), *Detecting Foreign Bodies in Food*, Woodhead Publishing, Cambridge, pp. 132–153.

BOYETTE, M.D., ESTES, E.A. and RUBIN, A.R. (2006), Hydrocooling, North Carolina Cooperative Extension Service, available at www.bae.ncsu.edu/programs/extension/publicat/postharv/ag-414-4/index.html.

BRENNAN, J.G., BUTTERS, J.R., COWELL, N.D. and LILLEY, A.E.V., (1990), *Food Engineering Operations*, 3rd edn, Elsevier Applied Science, London, pp. 17–63.

BRIMELOW, C.J.B. and GROESBECK, C.A., (1993), Colour measurement of foods by colour reflectance instrumentation, in (E. Kress-Rogers, Ed.), *Instrumentation and Sensors for the Food Industry*, Woodhead Publishing, Cambridge, pp. 63–96.

CREASER, C., (Ed.), (1991), *Food Contaminants*, Woodhead Publishing, Cambridge.

CUBEDDU, R., PIFFERI, TARONI, P. and TORRICELLI, A., (2002), Measuring fruit and vegetable quality: advanced optical methods, in (W. Jongen, Ed.), *Fruit and Vegetable Processing: Improving Quality*. Woodhead Publishing, Cambridge, pp. 150–169.

EDWARDS, M., (Ed.), (2004), *Detecting Foreign Bodies in Food*, Woodhead Publishing, Cambridge.

GEORGE, R.M., (Ed.), (2004), *Guidelines for the Prevention and Control of Foreign Bodies in Food*, 2nd edn, Guideline #5, Campden and Chorleywood Food Research Association, Chipping Campden.

GRANDISON, A.S., (2006), Postharvest handling and preparation of foods for processing, in (J.G. Brennan, Ed.), *Food Processing Handbook*, Wiley-VCH, Weinheim, Germany, pp. 1–32.

GRAVES, M., SMITH, A. and BATCHELOR, B., (1998), Approaches to foreign body detection in foods, *Trends in Food Science and Technology*, **9**, 21–27.

GREAVES, A., (1997), Metal detection – the essential defence, *Food Processing*, **5**, 25–26.

HARDIN, W., (2005), Search for 'perfectly safe' products pushes machine vision into food industry, Machine Vision Online, available at www.machinevisiononline.org/public/articles/archivedetails.cfm?id=2380, 03/21/2005.

HILLS, B., (2004), Nuclear magnetic resonance imaging, in (M. Edwards, Ed.), *Detecting Foreign Bodies in Food*, Woodhead Publishing, Cambridge, pp. 154–171.

JONGEN, W., (Ed.), (2005), *Improving the Safety of Fresh Fruits and Vegetables*, Woodhead Publishing, Cambridge.

JUKES, D., (2003), UK Food Inspection Statistics 1995, available at www.foodlaw.rdg.ac.uk/uk/INSPEC95.HTM.

KIM, J.G., YOUSEF, A.E. and CHISM, G.W., (1999), Use of ozone to inactivate microorganisms on lettuce, *J. Food Safety*, **19**, 17–33.

LAWRIE, R.A., (1998), *Lawrie's Meat Science*, 6th edn, Woodhead Publishing, Cambridge.

LELIEVELD, H. and MOSTERT, I.T., (Eds.), (2003), *Hygiene in Food Processing*, Woodhead Publishing, Cambridge.

LOW, J.M., MAUGHAN, W.S., BEE, S.C. and HONEYWOOD, M.J., (2001), Sorting by colour in the food industry, in (E. Kress-Rogers and C.J.B. Brimelow, Eds.), *Instrumentation and Sensors for the Food Industry*, 2nd edn, Woodhead Publishing, Cambridge, pp. 117–136.

MCALLORUM, S., (2005), Magnetic separation in process industries, *Food Science and Technology Today*, **19** (1), 43, 45–46.

OREA, J.M. and GONZALEZ URENA, A. (2002). Measuring and improving the natural resistance of fruit, in (W. Jongen, Ed.), *Fruit and Vegetable Processing: Improving Quality*, Woodhead Publishing, Cambridge, pp. 233–266.

RANKEN, M.D., (2000), *Handbook of Meat Product Technology*, Blackwell Science, Oxford.

RICHARDSON, L.J., MITCHELL, B.W., WILSON, J.L. and HOFACRE, C.L., (2003), Effect of an electrostatic space charge system on airborne dust and subsequent potential transmission of microorganisms to broiler breeder pullets by airborne dust, *Avian Diseases*, **47** (1), 128–133.

SUN, D-W. and WANG, L-J., (2001), Vacuum cooling, in (D-W. Sun, Ed.), *Advances in Food Refrigeration*, Leatherhead Publishing, Leatherhead, pp. 264–304.

WALLIN, P. and HAYCOCK, P., (1998), Foreign Body Prevention, Detection and Control: A Practical Approach, Blackie Academic and Professional, London.

XU, L., (1999), Use of ozone to improve the safety of fresh fruits and vegetables, *Food Technology*, **53** (10), 58–62.

ZUECH, N., (2005), What's happening with machine vision in the food industry? Machine Vision Online, available at www.machinevisiononline.org/public/articles/archivedetails.cfm?id=2569.

ZUECH, N., (2006a). Machine vision in the food industry, Machine Vision Online, available at www.machinevisiononline.org/public/articles/archivedetails.cfm?id=790.

ZUECH, N., (2006b). Machine vision trends – 2006, Machine Vision Online, available at www.machinevisiononline.org/public/articles/archivedetails.cfm?id=2684.

ZUECH, N., (2006c). Configurable machine vision systems, Machine Vision Online, available at http://www.machinevisiononline.org/public/articles/archivedetails.cfm?id=2790.

3

Size reduction

Abstract: Size reduction of solid foods and emulsification or homogenisation of liquid foods are used to change their organoleptic characteristics and their suitability for further processing. This chapter describes the theory of size reduction, the equipment used for milling, slicing, dicing, pulping and homogenisation, and the effects of processing on sensory properties of foods.

Key words: size reduction, milling, emulsification, homogenisers, emulsifying agents.

Size reduction or 'comminution' is the unit operation in which the average size of solid pieces of food is reduced by the application of grinding (shearing), compression or impact forces. When applied to the reduction in size of globules of immiscible liquids (e.g. oil globules in water) size reduction is more frequently referred to as homogenisation or emulsification. The size reduction of liquids to droplets (by atomisation) is described in Chapter 16, section 16.1. Size enlargement is achieved by forming (Chapter 4, section 4.2), extrusion (Chapter 15) or agglomeration (Chapter 16, section 16.2.1).

Size reduction has the following benefits in food processing:

- There is an increase in the surface-area-to-volume ratio of the food, which increases the rate of drying, heating or cooling and improves the efficiency and rate of extraction of liquid components (e.g. fruit juice or cooking oil extraction (Chapter 5, section 5.4)).
- When combined with screening (Chapter 2, section 2.3.1), a predetermined range of solid particle sizes is produced which is important for the correct functional or processing properties of some products (e.g. uniform bulk density and flowability of powders, consistent reconstitution of products such as dried soup and cake mixes). A similar range of particle sizes also meets consumer requirements for foods that have specific size requirements and allows more precise portion control and more complete mixing of ingredients (e.g. spices, emulsions, food colourants, icing sugar, flours and cornstarch) to give a uniform colour and taste in products (also Chapter 4, section 4.1).

Size reduction and emulsification are therefore used to improve the organoleptic 感官的. quality or suitability of foods for further processing and to increase the range of products. The operations have little or no preservative effect unless other preservative treatments are employed. Size reduction may actually promote degradation of foods by the release of naturally occurring enzymes from damaged tissues, by increased microbial activity or by oxidation of food components at the larger area of exposed surfaces (section 3.1.4).

Different methods of size reduction can be grouped according to the size range of particles produced:

- Chopping, cutting, slicing, dicing, mincing, shredding and flaking:
  - large to medium pieces (e.g. stewing steak, and sliced fruit for canning),
  - medium to small pieces (bacon, sliced green beans, potatoes for French fries or crisps (US: potato chips), diced carrots for freezing, sliced cheeses and wafer-thin cooked meats, mushrooms or gherkins for pizza toppings, diced meats for pies), and
  - small to granular pieces (minced or shredded meat, flaked cheese, fish or nuts, citrus peels for marmalade, and shredded lettuce or cabbage).
- Milling to powders or pastes of increasing fineness (grated products > spices > flours > fruit pulps or nectars > powdered sugar > starches > smooth pastes, such as peanut butter).
- Emulsification and homogenisation (mayonnaise, milk, essential oils, sauces, butter, ice cream and margarine).

3.1   Size reduction of solid foods

3.1.1   Theory

When stress (force) is applied to a food the resulting internal strains are first absorbed and cause deformation of the tissues. If the strain does not exceed a certain critical level, named the elastic stress limit (E in Fig. 3.1), the tissues return to their original shape when the stress is removed, and the stored energy is released as heat (elastic region, O–E). However, when the strain within a localised area exceeds the elastic stress limit, the food is permanently deformed. If the stress is continued, the strain reaches a yield point (Y), above which the food begins to flow (known as the 'region of ductility', Y–B). Finally, the breaking stress is exceeded at the breaking point (B) and the food fractures along a line of weakness. Part of the stored energy is then released as sound and heat. As little as 1% of applied energy may actually be used for size reduction. It is thought that foods fracture at lower stress levels if force is applied for longer times. The extent of size reduction, the energy expended and the amount of heat generated in the food therefore depend on both the size of the forces that are applied and the time that food is subjected to the forces. As the size of the piece is reduced, there are fewer lines of weakness available, and the breaking stress that must be exceeded increases. When no lines of weakness remain, new fissures must be created to reduce the particle size further, and this requires an additional input of energy. There is therefore a substantial increase in energy required to reduce the size of the particles as they get smaller (see sample problem 3.1). It is important to specify the required particle size distribution in a product to avoid unnecessary expenditure of time and energy in creating smaller particles than are required for a particular application.

When a food is first ground in a mill, the size of the particles varies considerably and is a mixture of all sizes from large particles to dust. As milling continues, the larger particles are reduced in size, but there are fewer changes to the size of the fine particles. For each combination of a particular mill and food, a specific size range of particles becomes the predominant size fraction, however long the grinding continues (see, for example, Chapter 2, Fig. 2.8).

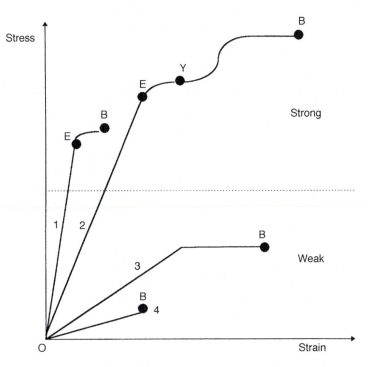

Fig. 3.1 Stress–strain diagram for various foods: E = elastic limit; Y = yield point; B = breaking point; O–E = elastic region; E–Y = inelastic deformation; Y–B = region of ductility; (1) = hard, strong, brittle material; (2) = hard, strong, ductile material; (3) = soft, weak, ductile material; and (4) = soft, weak, brittle material (after Loncin and Merson 1979).

Three types of force are used to reduce the size of foods: compression forces, impact forces and shearing (or attrition) forces. In most size reduction equipment all three forces are present, but often one is more important than the others (e.g. in hammer mills (section 3.1.2) impact forces are more important than shearing forces, and compression forces are least important). The factors that determine the effectiveness of size reduction and influence the selection of equipment are:

- the friability of the food (its hardness and tendency to crack);
- its moisture content; and
- the heat sensitivity of the food.

The amount of energy that is needed to fracture a food is determined by its friability, which in turn depends on the structure of the food. Harder foods have fewer lines of weakness and therefore require a greater energy input to create fractures (Fig. 3.1). Hard foods require a longer residence time in the mill, or larger machines are needed to achieve similar production volumes to those obtained with softer foods. Hard foods are also more abrasive and therefore the materials used to construct the contact parts in a mill need to be harder to resist wear (e.g. manganese steel instead of stainless steel). Compression forces are used to fracture friable or crystalline foods; combined impact and shearing forces are necessary to reduce the size of fibrous foods; and shearing forces are used for fine grinding of softer foods.

The energy required to reduce the size of solid foods is calculated using one of three equations, as follows:

1 Kick's law states that the energy required to reduce the size of particles is proportional to the ratio of the initial size of a typical dimension (e.g. the diameter of the pieces) to the final size of that dimension:

$$E = K_K \ln\left(\frac{d_1}{d_2}\right)$$

3.1

where E (W h^{-1} kg^{-1}) = the energy required per mass of feed, K_k = Kick's constant, d_1 (m) = the average initial size of pieces, and d_2 (m) = the average size of ground particles. d_1/d_2 is known as the size reduction ratio (RR) and is used to evaluate the relative performance of different types of equipment. Coarse grinding has RRs below 8:1, whereas in fine grinding, ratios can exceed 100:1 (Brennan *et al.* 1990).

2 Rittinger's law states that the energy required for size reduction is proportional to the change in surface area of the pieces of food (instead of a change in dimension described in Kick's law):

$$E = K_R\left(\frac{1}{d_2} - \frac{1}{d_1}\right)$$

3.2

where K_R = Rittinger's constant and d_1 and d_2 are as defined in Equation 3.1.

3 Bond's law is used to calculate the energy required for size reduction from:

$$\frac{E}{W} = \sqrt{\left(\frac{100}{d_2}\right)} - \sqrt{\left(\frac{100}{d_1}\right)}$$

3.3

where W (J kg^{-1}) = the Bond Work Index (40 000–80 000 J kg^{-1} for hard foods such as sugar or grain), d_1 (m) = diameter of sieve aperture that allows 80% of the mass of the feed to pass and d_2 (m) = diameter of sieve aperture that allows 80% of the mass of the ground material to pass.

In practice it has been found that Kick's law gives reasonably good results for coarse grinding in which there is a relatively small increase in surface area per unit mass. Rittinger's law gives better results with fine grinding where there is a much larger increase in surface area and Bond's law is intermediate between these two. However, Equations 3.2 and 3.3 were developed from studies of hard materials (coal and limestone) and deviation from predicted results is likely with many foods.

In addition to the friability of foods, the other factors that influence the extent of size reduction, the energy required and the type of equipment selected are the moisture content and heat sensitivity of the food. The moisture content significantly affects both the degree of size reduction and the mechanism of breakdown in some foods. For example, before milling cereals are 'conditioned' to optimum moisture content (e.g. 16% moisture for wheat) in order to obtain complete disintegration of the starchy material. Maize can also be thoroughly soaked and wet milled at approximately 45% moisture content. However, excessive moisture in a 'dry' food can lead to agglomeration of particles that then block the mill.

Excessive dust is created if some types of food are milled when they are too dry, which causes a health hazard and is extremely flammable and potentially explosive (section 3.1.3). Details of separation methods to remove dust from air using cyclone separators are given in Chapter 16, section 16.2.1).

Some foods (e.g. spices, cheese and chilled meats) are sensitive to increases in temperature or oxidation during comminution, and mills are therefore cooled by chilled water, liquid nitrogen or carbon dioxide (section 3.1.2).

Sample problem 3.1
Granulated sugar, having an average particle size of 500 μm, is milled to produce icing sugar having an average particle size of 25 μm using a 12 hp motor. What would be the reduction in throughput if the mill were used to produce fondant sugar having an average particle size of 19 μm? Assume 1 hp = 745.7 W.

Solution to sample problem 3.1
Fine grinding (size reduction ratio = 20), so use Rittinger's Law.
   From Equation 3.2, for grinding icing sugar:

$$12 \times 745.7 = K_R(1/25 \times 10^{-6} - 1/500 \times 10^{-6})$$
$$8948.4 = K_R(40\,000 - 2000)$$
$$K_R = 8948.4/38\,000$$
$$K_R = 0.235$$

For grinding fondant sugar:

$$E = 0.235(1/19 \times 10^{-6} - 1/500 \times 10^{-6})$$
$$E = 0.235(52\,632 - 2000)$$
$$E = 11\,898\,W\ (= 16\,hp)$$

Assuming that the mill fully utilises the 8.9 kW produced by the motor, the throughput is reduced to 8948.4/11 898 $\approx$ 0.75 of the original rate. That is a reduction in throughput of 25%.

3.1.2   Equipment
This section describes selected equipment used to reduce the size of fibrous foods to smaller pieces or pulps, and dry particulate foods to powders. Summaries of the main applications are shown in Table 3.1. Further details of the properties of powders are given by Lewis (1996) and of powders in spray dryers (Chapter 16, section 16.2.1).
   There are three main groups of size reduction equipment for solid foods, grouped in order of decreasing particle size as follows:

- cutting, slicing, dicing, mincing, shredding and flaking equipment;
- milling equipment;
- pulping equipment.

Cutting, slicing, dicing, mincing, shredding and flaking equipment
All types of cutting equipment require blades to be forced through the food with as little resistance as possible. Blades must be kept sharp, to both minimise the force needed to cut the food and to reduce cell rupture and consequent product damage. In moist foods, water acts as a lubricant, but in some sticky products such as dates or candied fruits, food grade lubricants may be needed to cut them successfully. In general blades are not coated with non-slip materials, such as 'Teflon' or polytetrafluoroethylene (PTFE) as these may wear off and contaminate the product, and instead they are mirror-polished during manufacture. Disposable blades are used in some equipment to maintain sharp cutting edges.
   Meats, fruits and vegetables and a wide range of other processed foods are cut during their preparation or manufacture. Cutting using powered knives, cleavers or band saws is

Table 3.1 Applications of selected size reduction equipment

Equipment	Type(s) of force	Type of product				Fineness				Typical products
		Brittle, crystalline	Hard, abrasive	Elastic, tough	Fibrous	Coarse lumps/pieces	Coarse grits	Medium to fine	Fine to ultra-fine	
Ball mills	Impact and shear	✓							✓	Food colourants
Bowl choppers	Impact and shear			✓	✓	✓				Sausagemeat, fruits for mincemeat
Dicers	Impact			✓	✓		✓			Fruits, vegetables, cheese
Disc mills	Shear	✓			✓			✓	✓	Cereals, starch, sugar, spices
Hammer mills	Impact	✓	✓		✓		✓	✓	✓	Sugar, maize, spices, dried vegetables
Mincers	Shear and impact			✓	✓			✓	✓	Fresh meats
Pin and disc mills	Impact and shear	✓	✓		✓			✓	✓	Cocoa powder, starch, spices
Pulpers	Shear or compression				✓			✓	✓	Fruits, oilseeds
Roller mills	Compression and shear	✓			✓			✓	✓	Wheat, sugar cane (Fluted rollers), chocolate refining (smooth rollers)
Shredders	Impact				✓	✓	✓			Fresh vegetables
Slicers	Impact			✓	✓	✓				Cooked and fresh meats, cheese, vegetables

Adapted from: Anon (1986)

an important operation in meat and fish processing. Post-slaughter cutting includes splitting carcases, and removal of offal, excess fat and bones. Meat carcases are further reduced in size to retail joints or prepared for further processing into ham, bacon, sausage, etc. by de-boning, skinning, de-fatting, slicing, mincing or shredding (Anon 2008a). Meats are frozen, or 'tempered' to just below their freezing point (Chapters 22 and 21), to increase their firmness and improve the efficiency of cutting. Fruits and vegetables have an inherently firmer texture and are cut at ambient or chill temperatures.

Slicing equipment has rotating or reciprocating blades that cut the food as it is passed beneath the blades. For many years specifically designed cutters have been used for many individual products. These include bread slicers, where reciprocating vertical sawtooth blades or bandsaws cut the bread, and bacon slicers, in which food is held on a carriage as it travels across a circular rotating blade. These machines continue to be used in food service and retail operations, but have largely been replaced in food processing factories. Increasingly, slicing machines can be easily adapted to cut a wide spectrum of products into a range of sizes. In some designs (Fig. 3.2) food is held against the slicer blades by centrifugal force created by high-speed rotation of the cutting head and each slice falls away freely. This eliminates the problems found in earlier cutters, where multiple knife blades caused compression and damage to the food when it passed between the blades. High-speed cutters are used to slice 'wafer thin' cooked meats and sliced cheeses at up to 2000 slices per minute, and vegetables at up to six tonnes per hour. More sophisticated slicers are able to cut vegetables into a wide variety of shapes including tagliatelle or garland shapes. Machines are computer controlled and can be easily programmed by operators to bulk-slice and stack a range of products including cheeses, meats, mushrooms or vegetables (e.g. for pizza toppings). An 'intelligent' cheese cutter weighs and measures each block to determine the maximum number of portions that can be cut to the required weight with the minimum amount of waste (Sharp 1998). The growth of the chilled sandwich market has stimulated the development of high-speed slicers for both sliced fillings (such as cooked meats, cheese, cucumber, tomato, etc.) that are applied to the sandwich bread, and for cutting bread precisely from corner to corner.

There has been growth in demand for partially processed fresh fruits and vegetables (also known as 'minimally processed', 'lightly processed', 'fresh-processed' or 'pre-prepared' products) that provide convenient fresh products to consumers. Examples include packaged mixed salads, sliced peeled potatoes, shredded lettuce and cabbage, fruit slices, vegetable snacks such as carrot and celery sticks or cauliflower and broccoli florets, diced onions and trays of microwaveable fresh vegetables. They are prepared using a range of size reduction operations including trimming, coring, slicing or shredding.

Harder fruits such as apples are simultaneously sliced and de-cored as they are forced over stationary knives fitted inside a tube. In a similar design (the 'hydrocutter') foods are conveyed by water at high speed over fixed blades. Intermittent guillotine cutters are used to cut confectionery products, such as liquorice and extruded foods (Chapter 15, section 15.3). The blade advances with the product on the conveyor to ensure a square cut edge regardless of the conveyor speed or cut length. Continuous slicers that have a similar design to that in Fig. 3.2 feed the slices to circular knives that produce 'strip cuts' (e.g. flat or crinkle cut potatoes, pepperoni and other cooked meats for pre-prepared salads or pizzas). Diced foods are first sliced and then cut into strips by rotating blades. The strips are fed to a second set of rotating cross-cut knives that operate at right angles to the first set and cut the strips into cubes (Fig. 3.3). Products include all types of frozen tempered fresh meats, cooked meats (e.g. bacon bits, diced beef or poultry for pies and pork skin for fried rinds), diced apricots or pineapple pieces.

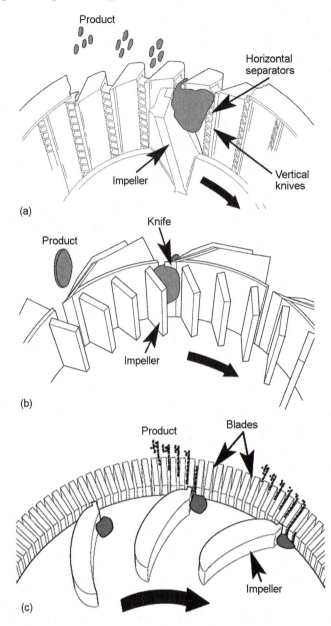

Fig. 3.2 Operation of slicing equipment as used on the Comitrol® Processor by Urschel Laboratories: (a) cutting head, (b) slicing head and (c) microcut head (with permission of Urschel Laboratories Inc.® Comitrol and DiversaCut 2110 are registered trademarks of Urschel Laboratories, Inc. All rights reserved) (Anon 2008c).

Ultrasonic cutters use knife blades or 'horns' (probes) that vibrate at 20 kHz, and have a cutting stroke of 50–100 μm. Details of the component parts and method of cutting are described by Rawson (1998). They have benefits over traditional cutting blades because the required cutting force is significantly reduced, so less sharp blades are needed, the blade is self-cleaning and longer intervals are required between sharpening compared with conventional blades. There is little damage to cells in the product so crumbs and

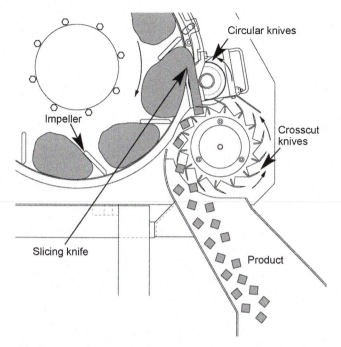

Fig. 3.3 Operation of dicing equipment as used on the DiversaCut 2110® Processor by Urschel Laboratories (with permission of Urschel Laboratories Inc.® Comitrol and DiversaCut 2110 are registered trademarks of Urschel Laboratories, Inc. All rights reserved) (Anon 2008c).

debris are significantly reduced and the cut face has a smooth appearance. The equipment can cut multi-layered products or hard particles contained within a soft food. It is particularly suitable for products that are difficult to cut using other methods (e.g. sticky confectionery, hot bread and soft cake). It is increasingly used for bakery products of all types, frozen pies, ice cream, and fresh meats, fish and vegetables.

Meat mincers have a screw auger that feeds product to a perforated cutting plate and/ or rotating knives at the outlet. They mince by a combination of cutting and shearing as the meat passes through the holes in the cutting plate. Flaking or shredding equipment for flaked fish, cheese, nuts or meat and shredded vegetables is similar to slicing equipment. Adjustment of the blade type and spacing is used to produce the flakes or shreds. Alternatively, shredders can be modified hammer mills in which knives are used instead of hammers to produce the cutting action. A 'squirrel cage' disintegrator has two concentric cylindrical cages inside a casing. They are fitted with knife blades along their length and the two cages rotate in opposite directions. Food is subjected to powerful shearing and cutting forces as it passes between them.

Milling equipment
There are many designs of mills to grind specific types of food. A selection of common types is described in this section and a summary of their applications is shown in Tables 3.1 and 3.2. Substantial amounts of heat are generated in high-speed mills and depending on the heat sensitivity of the food, it may be necessary to cool the mill to maintain the temperature rise within permissible limits. In cryogenic grinding (e.g. spices), liquid nitrogen or solid carbon dioxide is mixed with foods before milling to cool the product and to retain volatiles or other-heat sensitive components. Alternatively, chilled water

Table 3.2   Properties and applications of selected size reduction equipment

Type of equipment	Type(s) of force	Peripheral velocity (m s^{-1})	Typical products
Pin-and-disc mill	Impact	80–160	Sugar, starch, cocoa powder, nutmeg, pepper, roasted nuts, cloves
Wing-beater mill	Impact and shear	50–70	Alginates, pepper, pectin, paprika, dried vegetables
Disc-beater mill	Impact and shear	70–90	Milk powder, lactose, cereals, dried whey
Vertical toothed disc mill	Shear	4–8	Frozen coffee extract, plastic materials
		17	Coarse grinding of rye, maize, wheat, fennel, pepper, juniper berry
Cutting granulator	Impact (and shear)	5–18	Fish meal, pectin, dry fruit and vegetables
Hammer mill	Impact	40–50	Sugar, tapioca, dry vegetables, extracted bones, dried milk, spices, pepper
Ball mill	Impact and shear	–	Food colours
Roller mills	Compression and shear	–	Sugar cane, wheat (fluted rollers) Chocolate refining (smooth rollers)

After Loncin and Merson (1979)

may be circulated through internal channels within mills to cool them. Nibblers, which use a grating rather than a grinding action, have been used instead of mills, and are claimed to reduce problems of increased temperatures, noise and dust (Sharp 1998).

Ball mills
Ball mills have a rotating, horizontal steel cylinder that contains steel or ceramic balls 2.5–15 cm in diameter. At low speeds or when small balls are used, shearing forces predominate. With larger balls or at higher speeds, impact forces become more important. They are used to produce fine powders, such as food colourants. A rod mill has rods instead of balls to overcome problems associated with the balls sticking in adhesive foods.

Disc (or plate) mills
In traditional mills of this type for grinding corn, named Buhr mills, stones were mounted horizontally, but more commonly the steel plates in more modern disc mills are mounted vertically. In each design the feed enters the mill near the centre of the plates and is ground as it passes to the periphery. There are a large number of designs of disc mill, each employing predominantly shearing forces for fine grinding. For example:

- In single-disc mills, food passes through an adjustable gap between a stationary casing and a grooved disc that rotates at high speed.
- Double-disc mills have two discs that rotate in opposite directions to produce greater shearing forces.
- Pin-and-disc mills have intermeshing pins fixed either to a single disc and casing or to double discs (Fig. 3.4). These improve the effectiveness of milling by creating impact forces in addition to shearing forces (also section 3.2.3, colloid mills).

Hammer mills
These mills have a horizontal or vertical cylindrical chamber, lined with a toughened steel 'breaker plate'. A high-speed rotor inside the chamber is fitted with swinging hammers (rectangular pieces of hardened steel) along its length. Food is disintegrated by

Fig. 3.4 Pin-and-disc mill (courtesy of Alpine Process Technology; Anon 1986).

impact forces from repeated hammer impacts as the hammers drive it against the breaker plate, and also particle on particle impacts. Typically, rotor speeds of 3000–7200 rpm are used with flat hammers for fine grinding, and speeds of 1000–3000 rpm are used with sharp-bladed hammers for coarse grinding. The mill can be operated using a free flow of materials through the mill in a single pass, but more commonly a screen restricts the discharge from the mill. The screen is either a perforated metal plate or a bar grate. The perforations in the screens can be round or square holes, diagonal or straight slots, or wire mesh. When the mill operates under these 'choke' conditions (Table 3.3) food remains in the mill until the particles are reduced to a size that can pass through the apertures. Under these conditions, shearing forces also play a part in the size reduction.

However, the size of the screen apertures does not in itself determine the particle size of the product and there is a complex interrelationship between aperture size, shape, rotor speed, screen thickness and total open surface area of the screen. When particles are struck by hammers, they approach the screen at a shallow angle. The higher the rotor speed, the smaller the angle that a particle approaches the screen and hence the smaller the aperture that the particle 'sees' (Fig. 3.5a). Similarly the thickness of the screen determines the particle size that can pass through it, with thicker gauged screens allowing only smaller particles through (Fig. 3.5b). The total open surface area of the screen also affects the size distribution of particles. When a particle that is small enough to pass through the screen approaches it, there is a greater probability that it will pass through an aperture if the screen has a large open surface area. Screens that have rectangular apertures have a higher open surface area than those having circular apertures. Screens that have a smaller open surface area are more likely to cause the particle to bounce back. This results in excessive size reduction and hence creates more 'fines' (undersize particles) in the product. However, there is a balance required between faster processing using a larger open surface area and the greater mechanical strength of screens with a small open area to withstand impacts from particles (Anon 2008b).

Table 3.3   Methods of operating grinding equipment

Type of operation	Advantages	Limitations
Open circuit grinding – product passing straight through the mill.	Simplest method of operation, lower energy consumption.	Wide range of particle sizes because some pass through more quickly than others.
Free crushing – similar to open circuit grinding, with feed material falling under gravity through an 'action zone' where it is crushed.	Residence time kept to a minimum, reduced production of undersized particles, lower energy consumption.	Wide range of particle sizes.
Choke feeding – outlet restricted by a screen, material remains in action zone until small enough to pass through the screen.	Prevents oversize particles in product, large reduction ratios can be achieved.	Long residence times can produce undersize particles and additional energy consumption.
Closed circuit grinding – short residence times, classifier separates oversize particles and recycles them through the mill.	More energy efficient, smaller range of particle sizes.	Higher capital cost of classifier.
Wet milling – mix material with water and mill as a slurry.	No dust created, food can be separated by centrifugation. Useful if a soluble component is to be extracted (e.g. maize milling). Smaller particles produced.	Higher energy consumption, higher capital costs for dewatering equipment.

Adapted from Young (2003)

The screen aperture size and shape, rotor speed and hammer configuration can be changed individually or in any combination to produce the precise particle size required in the product. These mills are widely used for crystalline and fibrous foods including spices, dried cereals and legumes, and sugar.

In another design, sharp knives are arranged in a cylinder and an impeller operating at 2000–12 000 rpm pushes the product over the knives to give a controlled comminution to micro-fine powder. At this speed, products such as rice pass the blades at speeds in excess of $90 \, \mathrm{m \, s^{-1}}$ (320 kph) and are rapidly reduced to flour. Similar equipment is used to produce a wide range of pastes and puréed foods, including peanut butter, mustards, meats for use in gravies or sauces, and fruits, vegetables and meats for baby foods (Anon 2008c) (see also 'pulping equipment' below).

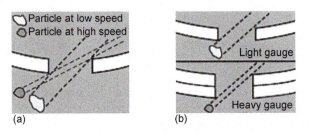

Fig. 3.5   (a) Effect of rotor speed on particle passing through screen. (b) Effect of screen thickness on ability of particle to pass through (courtesy of Fitzpatrick Co. Ltd) (Anon 2008b).

Jet pulverising mills
Jet pulverising (or 'jet energy') mills are used to grind friable or crystalline foods (e.g. sugar, salt, cocoa powder) and ingredients such as food colourants, to fine powders. Food is fed into a central toroidal chamber of the mill and is driven at near sonic velocity around the perimeter of the chamber by multiple air jets. The chamber has replaceable liners made from ceramic, tungsten carbide, hardened steel, polyurethane or rubber. Size reduction is caused solely by high-velocity collisions between particles of food and no grinding mechanism is involved. The particles recirculate in the chamber until they are sufficiently reduced in size to be discharged via a central port. The equipment has no moving parts or screens and no excess heat is produced, making it suitable for heat-sensitive materials. The mill has lower energy costs because it does not require refrigeration to remove heat and it has no grinding surfaces to maintain. Precise metering of the product input and control of air velocity produces a narrow range of particle sizes, from as small as 0.25 μm up to 15 μm (Anon 2008d).

Roller mills
These machines have two or more precisely machined pairs of smooth steel rollers that revolve towards each other and pull particles of food through the 'nip' (the space between the rollers). Overload springs protect the rollers against accidental damage from metal or stones. The main force is compression but if the rollers are rotated at different speeds, or if the rollers are fluted (flutes are shallow ridges along the length of the roller), shearing forces are also exerted on the food. In operation, grains pass between a series of rollers, each having a smaller nip. As the material falls under gravity into the next set of rollers it is crushed into progressively smaller particles (Table 3.4). The total number of rollers depends on the properties of the food and the degree of particle size reduction required, but typically, there are between three and eight pairs of rollers (Fig. 3.6). The low friction results in minimal temperature increases in the product and reduced power requirements. Roller mills are widely used to mill wheat, oats and other cereals to flours and as flaking rolls to produce flaked breakfast cereals. Details are given by Kent and Evers (1994). A mill can be configured for different grains by changing the surface texture of the rollers and controlling the roller speeds and nip settings, to achieve a narrow particle size distribution. This produces less product waste (fines), higher yields and increased milling efficiencies.

Pulping equipment
There are a number of different designs of pulper that use a combination of compression and shearing forces to extract juice from fruits or vegetables, to extract cooking oil from oilseeds or nuts, and to produce a variety of puréed and pulped foods. For example, a

Table 3.4 Particle size and yield of milled maize products

Product	Particle size range (mm)	Yield (% by weight)
Flaking grits	5.8–3.4	12
Coarse grits	2.0–1.4	15
Medium grits	1.4–1.0 ⎫	
Fine grits	1.0–0.65 ⎬	23
Coarse meal	0.65–0.3	10
Fine meal	0.3–0.17	10
Flour	<0.17	5

Adapted from Kent and Evers (1994)

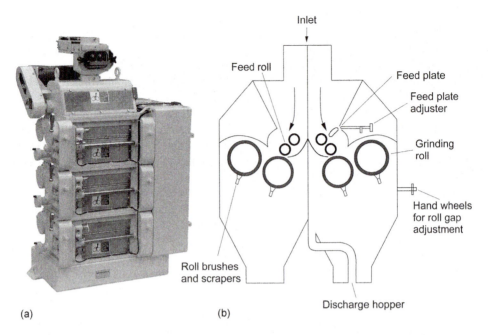

Inlet

Feed roll

Feed plate

Feed plate adjuster

Grinding roll

Hand wheels for roll gap adjustment

Roll brushes and scrapers

Discharge hopper

(a)   (b)

Fig. 3.6   (a) A roller mill (courtesy of Modern Process Equipment Corp. at www.mpechicago.-com/grinders/grinding-equipment.html). (b) Roller mill operation (courtesy of Indopol Food Processing Machinery at www.indopol.com/html/roller_mill_ip-rm04.html).

rotary fruit pulper consists of a cylindrical metal screen fitted internally with high-speed rotating brushes or paddles (Fig. 3.7). These break down grapes, tomatoes, passionfruit or other soft fruits and force the pulp through the perforations in the screen. The size of the perforations determines the fineness of the pulp. Skins, stalks and seeds are discarded from the end of the screen. Other types of pulper may be specially designed for particular fruits, including pineapple crushers, mills for pulping apples for juice or cider production,

Fig. 3.7   Fruit pulper (photo by the author).

Fig. 3.8   Bowl chopper (courtesy of Union Food Machinery Ltd at www.ufm-ltd.co.uk/bowlcutters.html).

or olives for oil extraction. Sugar cane rollers crush the cane to extract juice. They consist of pairs of heavy horizontal fluted cylinders that rotate in opposite directions. The rollers may be driven at the same speed or at different speeds, or only one roll is driven.

Various designs of size reduction equipment are used to prepare nuts and oilseeds prior to pressing or expelling the oil. For example, groundnut decorticators shell groundnuts and separate the husks and kernels prior to milling kernels into groundnut flour (for example, Anon 2008e); sunflower seed crackers and palm nut crackers both break the seeds/nuts to produce a mixture of kernels and husks for crushing (Anon 2008f); and copra cutters break coconut cups into small pieces that are suitable for crushing in an oil expeller. Crushing stones, pin mills or hammer mills are also used to produce olive pulp for pressing to extract the oil. Details of methods of oil and juice extraction are described in Chapter 5, section 5.3.2 and by Rossell (1999).

Bowl choppers (Fig. 3.8) are used to chop meat and harder fruits and vegetables into a pulp (e.g. for sausage meat or mincemeat preserve). A horizontal, slowly rotating bowl moves the ingredients beneath a set of high-speed rotating blades. Food is passed several times beneath the knives until the required degree of size reduction and mixing has been achieved. Solid carbon dioxide may be used to cool meat in the manufacture of sausage meat, or vacuum bowl choppers may be used. In another design a circular set of knives described by the manufacturer as a 'microcut head and impeller assembly' rotates at high speed to reduce the size of products from a maximum of 9.5 mm to a pulp in a single pass. It is used for continuous production of pulped products (e.g. mustard and soy milk (Anon 2008c)). Homogenisers that are used to produce other pastes, purées, sauces, etc. are described in section 3.2.3.

3.1.3   Developments in size reduction technology
The three main factors that drive developments in size reduction technology are the need to (1) have sanitary milling conditions, (2) reduce the costs of cleaning, changeover and maintenance, and (3) comply with legislation to prevent dust explosions. Retail buyers increasingly require improved sanitation standards and food safety, and new equipment

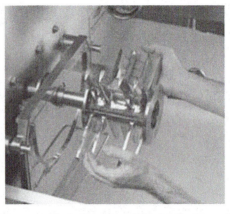

Fig. 3.9   Cantilevered rotor assembly in hammer mill – detail of easy disassembly of FitzMill motor (courtesy of The Fitzpatrick Company Europe N.V.) (Anon 2008b).

designs isolate the food-contact zone from motors, gears, belts and other drive components. This minimises the risk of contamination by oils, dust, etc., and makes the food-contact zone easily accessible for cleaning. Rotors may have a blunt edge and a sharp edge so that they can be quickly reversed to change from impact to knife configuration. The new designs have quick-release parts and components that can withstand high-pressure washing, so that equipment can now be cleaned and changed over to a new product within a few minutes, whereas previously it took several hours to dismantle, clean and reassemble individual components. This has increased throughputs and reduced labour costs. The use of tilt-back machine components on self-supporting hinges including one-piece cantilevered rotors (Fig. 3.9) has also removed the need for staff to lift and carry heavy parts for cleaning (Anon 2008b).

There have also been significant developments in the design of screens for mills. Whereas previously they were simply perforated plates, new designs act more like cutting devices. They improve machine performance and reduce processing times by, for example, preventing the build-up of fibrous materials such as cereals and vegetables in the grinding chamber. Screens are also readily interchangeable to give greater flexibility for the same machine to process a wider range of products. More sophisticated human–machine interfaces (HMIs) and advanced controls such as touch-screen monitors are replacing push-button controls to assist operators in management of product quality, control of machines and fault detection or troubleshooting (Anon 2008b). PC-controllers (Chapter 27, section 27.2.2) also now integrate shredders, dicers and mills with conveyors, tumbling mixers, weighing scales and other equipment.

Depending on the ignition characteristics of the powder being milled, the introduction in 2003 of a European Explosive Atmospheres and Gassy Mines (ATEX) Directive (Anon 2005) requires milling equipment to have features that prevent explosions and fires. These include explosion-proof motors and other safeguards, such as an enclosure for the machine that can withstand a 12 bar explosive pressure. For continuous milling, where an enclosed machine is not feasible, systems have been developed to automatically purge the milling chamber with nitrogen to prevent ignition of dust and an explosion. Older mills may be retro-fitted with nitrogen systems to create an inert environment for ATEX compliance. In a more sophisticated design, a pressure regulator, flowmeter and flow switch are used to automatically send a shutdown signal to the mill controller if the oxygen concentration exceeds a predetermined level. Vacuum conveying systems

(Chapter 27, section 27.1.2) also control dust to reduce the potential for an explosion, and can improve throughput (Higgins 2006).

3.1.4 Effect on foods

Size reduction is used to control the textural or rheological properties of foods (Chapter 1, section 1.1.4), which improves the efficiency of mixing, extraction or heat transfer. There is also an indirect effect on the aroma and flavour of some foods. Dry foods (e.g. grains or nuts) have a sufficiently low a_w (Chapter 1, section 1.3.3) to permit storage for several months after milling without substantial changes in nutritional value, eating quality or microbial safety. However, moist foods deteriorate rapidly if other preservative measures such as chilling, freezing or heat processing are not undertaken.

Sensory characteristics

There are small changes to the colour, flavour and aroma of dry foods during size reduction. For example, oxidation of carotenes bleaches flour and reduces the nutritional value. There is a loss of volatile constituents from spices and some nuts, which is accelerated if the temperature is allowed to rise during milling. In fresh foods size reduction causes physical damage to cells when new surfaces are created, which allows enzymes and substrates to become more intimately mixed. This increases the rates of metabolic processes that cause deterioration (e.g. respiration, ethylene production and other biochemical reactions that cause accelerated deterioration of colour, aroma, flavour and texture). Milling substantially alters the texture of foods, both by the physical reduction in the size of tissues and also by the release of hydrolytic enzymes. The type, duration and temperature of size reduction and the delay before subsequent preservation operations are closely controlled to achieve the desired texture. The relationship between the size of food particles and perceived texture is discussed by Engelen *et al.* (2005a,b) and Steffe (1996).

Nutritional value

The increase in surface area of foods as a result of size reduction may cause a loss of nutritional value due to oxidation of fatty acids, carotenes and heat-sensitive, oxygen-sensitive and light-sensitive nutrients (Chapter 1, sections 1.1.1 and 1.4.3). Losses of vitamin C and thiamin in chopped or sliced fruits and vegetables are substantial (e.g. 78% reduction in vitamin C during slicing of cucumber). Vitamin losses during storage depend on the temperature and moisture content of the food and on the concentration of oxygen in the storage atmosphere.

In dry foods the main changes to nutritional value result from separation of the product components after size reduction. For example, in cereal and legume milling, one objective is to remove the fibrous seed coat from the endosperm, but the nutrient-rich aleurone layer and germ may also be removed, resulting in loss of vitamins (Table 3.5), minerals, proteins and lipids. Because cereals are an important dietary component in most countries, losses of these nutrients in for example polished rice, pearled barley and white wheat flours, can have serious public health consequences. Cereals are also an important source of dietary fibre and this is substantially reduced in white wheat flour (1.5 g per 100 g) compared with wholemeal flour (5.8 g per 100 g) (Anon 2002). However, there are large variations in nutrient losses due to the uneven distribution of nutrients in the grains and the degree of milling. For example, in wheat, there are significant losses of phosphorus, magnesium, chromium, zinc and manganese, which are located in the seed

Table 3.5   Effect of milling on vitamin content of selected grains (reproduced with permission of the United Nations University Press)

Food	Content per 100 g						
	Thiamin (mg)	Riboflavin (mg)	Niacin (mg)	Pantothenic acid (mg)	Vitamin B_6 (mg)	Folate (μg)	Biotin (μg)
Maize							
Whole grain	0.47	0.09	1.62	–	0.54	30.0	7.3
Dehulled	0.44	0.07	1.39	–	0.54	20.0	5.5
Kernel	0.15	0.12	1.7	0.54	0.16	26.8	11.0
Flour	0.20	0.06	1.4	–	0.19	10.0	1.4
Rice							
Grain	0.34	0.05	4.7	1.10	0.55	20.2	12.0
White grain	0.07	0.03	1.6	0.55	0.17	14.1	5.0
Bran	2.26	0.25	29.8	2.8	2.5	150	60
Wheat							
Grain (hard wheat)	0.57	0.12	4.3	1.5	0.4	14.4	12
Wholemeal							
(100% extraction)	0.46	0.08	–	0.8	0.5	25	7
72% extraction[a]	0.31	0.03	1.6	0.3	0.15	14	3
40% extraction[a]	0.32	0.02	1.1	0.3	0.10	5	1
Bran	0.72	0.35	21.0	2.9	0.82	155	49

Data adapted from Anon (2002) and Bauernfeind and De Ritter (1991)
[a] % extraction = weight of flour per 100 parts of flour milled.

coat, but the bio-availability of iron increases (Table 3.6). This is partly because complex polysaccharides, polyphenolic compounds (e.g. tannins) and phytates in the bran that limit iron availability are removed by milling. Similarly by decorticating sorghum, iron availability is increased from 19.6 to 28.7% (Deosthale 1984). During parboiling, water-soluble nutrients migrate from the outer to the inner layers of rice grains and this minimises losses of thiamin, riboflavin and niacin and also the minerals chromium, molybdenum and manganese, which are retained after milling. A description of the properties of vitamins and minerals and the effects of deficiencies in the diet is given in Chapter 1 (section 1.1.1).

Milling also increases the bioavailability of other nutrients by reducing the particle size and hence increasing the surface area of the food available to attack by digestive

Table 3.6   Soluble and ionisable iron content of some unmilled and milled cereals and pulses

Cereal/legume	Iron		
	Iron in grain (mg/100 g)	Soluble (%)	Ionisable (%)
Wheat flour			
Whole	6.1	5.9	4.3
Refined	1.8	13.2	8.2
Sorghum			
Whole	4.2	–	19.6
Pearled	3.7	–	28.7
Chickpea			
Whole	6.0	12.6	2.7
Dhal	4.9	22.6	14.0

Deosthale (1984)

enzymes. Slavin *et al.* (2000) and Alldrick (2002) have reviewed the effects of milling on the nutrient content of cereals (proteins, vitamins, minerals, phytochemicals, β-glucans) and the effects on health (cholesterol, glycaemic index and laxation response).

3.1.5   Effect on micro-organisms

Provided dry foods remain dry after milling there are negligible changes caused by micro-organisms. Berghofer *et al.* (2003) found that the most frequently detected micro-organisms in wheat and wheat flour were *Bacillus* spp., coliforms, yeasts and moulds. The most common moulds isolated were *Aspergillus* spp., *Penicillium* spp., *Cladosporium* spp. and *Eurotium* spp. When wheat grain components become separated during milling, contaminants are concentrated in the bran and wheatgerm (the outer layers of the grain) and the inner endosperm fraction contains lower microbial counts. The flour is microbiologically the most clean. The quality of incoming wheat has an important influence on the microbiological quality of the products, but the authors found higher microbiological counts midway in the milling process, which indicated that contamination from equipment may also contribute to the observed contamination.

The peel on intact fruits and vegetables is a physical and biochemical barrier to most micro-organisms, but the damaged surfaces caused by size reduction release cellular nutrients that provide a suitable substrate for microbiological growth. This can also result in the development of off-flavours and aromas. Additionally, the micro-environment surrounding packed fresh sliced foods provides conditions that increase the numbers and types of micro-organisms that develop, and can result in significant changes to the microflora. For example, the risk of pathogenic bacteria such as *Clostridium* spp., *Yersinia* spp. and *Listeria* spp. growing on minimally processed sliced fruit and vegetable products (e.g. cantaloupe, watermelon, cucumber, lettuce and coleslaw vegetables) may increase as a result of the high relative humidity and lower oxygen concentrations inside film packaging, especially if products are stored above 5 °C. Conversely, low temperatures may select psychrotrophic spoilage micro-organisms such as *Pseudomonas* spp. These aspects are discussed in more detail in Chapter 1 (section 1.2.3) and Chapter 21 (section 21.5). There are also greater opportunities for contamination by pathogens from increased handling of fresh-cut products. For example, Abdul-Raouf *et al.* (1993) reported the growth of *E. coli* O157:H7, psychrotrophic and mesophilic micro-organisms on shredded lettuce, sliced cucumber and shredded carrot stored under different storage temperatures simulating those routinely used in commercial practice. *E. coli* O157:H7 declined on vegetables stored at 5 °C and increased at 12 °C and 21 °C up to 14 days. This is controlled by cooling the product before processing, using stringent sanitary conditions and careful preparation and handling procedures, by removing surface moisture using centrifuges, vibrating screens or air blasts, and strict temperature control after processing.

Sliced cold meat products have a high risk of contamination by pathogenic and spoilage micro-organisms unless food handling guidelines are strictly observed. The slicing operation may increase the microbial load on products via blades, as well as increasing nutrient availability as a result of tissue damage. Even with stringent hygienic conditions and chilling, extensive handling before and after slicing may cause significant contamination of cold meat products. For example, Voidarou *et al.* (2006) found contamination by the bioindicators *E. coli, S. aureus* and *C. perfringens* on sliced turkey, pork ham, smoked turkey and smoked pork ham.

3.2   Size reduction in liquid foods

Emulsification is the formation of a stable emulsion by the intimate mixing of two or more immiscible liquids, so that one (the dispersed phase) is formed into very small droplets within the second (the continuous phase). Homogenisation is the reduction in size (to 0.5–30 μm), and hence the increase in number, of solid or liquid particles in the dispersed phase by the application of intense shearing forces. Both are used to change the functional properties or eating quality of foods and have little or no direct effect on nutritional value or shelf-life.

3.2.1   Theory
The two types of liquid-liquid emulsion are

1   oil in water (o/w) (e.g. milk and cream); and
2   water in oil (w/o) (e.g. butter, low-fat spreads and margarine).

These are relatively simple systems but more complex emulsions are found in such products as ice cream, salad cream and mayonnaise, sausagemeat and cake batters (section 3.2.4).
   The stability of emulsions is determined by the:

- size of the droplets in the dispersed phase;
- viscosity of the continuous phase;
- difference between the densities of the dispersed and continuous phases;
- interfacial forces acting at the surfaces of the droplets; and
- type and quantity of emulsifying agent used.

   Stoke's Law relates factors that influence the stability of an emulsion:

$$v = \frac{d^2 g (\rho_p - \rho_s)}{18 \mu}$$

$\boxed{3.4}$

where v (m s^{-1}) = terminal velocity (i.e. velocity of separation of the phases), d (m) = diameter of droplets in the dispersed phase, g = acceleration due to gravity = 9.81 m s^{-2}, ρ_p (kg m^{-3}) = density of dispersed phase, ρ_s (kg m^{-3}) = density of continuous phase and μ (N s m^{-2}) = viscosity of continuous phase.
   The equation indicates that stable emulsions are formed when droplet sizes are small (in practice between 1 and 10 μm), the densities of the two phases are reasonably close and the viscosity of the continuous phase is high. Stabilisers (Chapter 1, section 1.1.1) are polysaccharide hydrocolloids and proteins that dissolve in water to form viscous solutions or gels (e.g. starch, egg albumin or gelatin). In o/w emulsions, they increase the viscosity and form a three-dimensional network that stabilises the emulsion and prevents coalescence of the oil droplets. Microcrystalline cellulose and related cellulose powders are able to stabilise w/o emulsions. However, there are a large number of complex factors that influence the stability of emulsions, including interfacial properties and electrostatic and van der Waals interactions between molecules in the food. Further details are given by McClements (1999) and Friberg et al. (2004).

3.2.2   Emulsifying agents and stabilisers
Mechanical energy is used to create an emulsion and emulsifying agents (or 'surfactants') reduce the energy required and stabilise the emulsion once it is formed. They act by

Table 3.7   Selected synthetic emulsifying agents used in food processing

Emulsifying agent	HLB value	Function and typical application
Ionic		
Phospholipids (e.g. lecithin)	18–20	Crumb softening in bakery products
Potassium or sodium salts of oleic acid		Reduction in stickiness in pasta, extruded snackfoods and chewing gum
Non-ionic		
Polyoxyethylene sorbitol fatty esters:		Multipurpose water-soluble
Polyoxyethylene sorbitan tristearate (Tween 65)	10.5	emulsifier. Fat crystal
Polyoxyethylene sorbitan trioleate (Tween 85)	11.0	modification in peanut butter
Polyoxyethylene sorbitan monostearate (Tween 60)	14.9	
Polyoxyethylene sorbitan monolaurate	16.7	
Sorbitol esters of fatty acids:		Retardation of bloom in
Sorbitan monolaurate (Span 20)	8.6	chocolate and control of
Sorbitan monopalmitate (Span 40)	6.7	overrun in ice cream
Sorbitan monostearate (Span 60)	4.7	
Sorbitan monooleate (Span 80)	4.3	
Sorbitan sesquioleate (Arlacel 83)	3.7	
Sorbitan tristearate (Span 65)	2.1	
Sorbitan trioleate (Span 85)	1.8	
Propylene glycol fatty acid esters	3.4	
Glycerol monostearate	3.8	Anti-staling and crumb softening in bakery products

Adapted from Lewis (1990) and Anon (2008h)

preventing droplets from coalescing and keeping the phases apart. They are either present in, or added to, a food, and have the ability to bind with both hydrophilic (or polar) and lipophilic (or non-polar) parts of molecules, to form micelles around each droplet in the disperse phase. Emulsifying agents that contain mostly polar groups (e.g. hydrophilic glycerides) bind to water and therefore produce o/w emulsions. Non-polar agents (e.g. lipophilic fatty acids) are adsorbed to oils to produce w/o emulsions (also Chapter 1, section 1.1.2).

There are a large number of emulsifying agents, each having different functional properties, which can be characterised by their hydrophilic/lipophilic balance (HLB) (Table 3.7) and/or ionic charge, to give a guide to their performance. HLB values vary from 0 to 20 which indicates the solubility and relative attraction of an emulsifier to oil or water. A low HLB value (0–6, strongly lipophilic) indicates solubility in oil and these are used in w/o emulsions. High HLB emulsifiers (12–18, strongly hydrophilic) are soluble in water and produce o/w emulsions. Intermediate value emulsifiers act as detergents and solubilisers (Lewis 1990, Anon 2008g). Although the HLB classification is limited to simple emulsions (e.g. salad dressings), it is useful for the selection of the correct emulsifier: that is the higher the percentage of water as the continuous phase, the higher the HLB number (and conversely the higher the proportion of oil, the lower the HLB number).

Polar emulsifying agents are also classified into ionic and non-ionic types. Ionic types, including stearoyl lactylates and diacetyl tartaric acid esters of monoglycerides, have a negative (anionic) charge due to the carboxylic acid group on the ester part of the molecule. They have different surface activities over the pH range, owing to differences in their dissociation behaviour. For o/w emulsions, an ionic emulsifier may provide greater stability because the negative charge on the hydrophilic portion of the molecule repels other oil droplets. The activity of non-ionic emulsifiers is independent of pH. Some emulsifiers, including acetic acid esters, lactic acid esters, polyglycerol esters, propylene glycol esters and sorbitan esters, have one form, whereas others are polymorphic and exist in different crystal forms.

Naturally occurring proteins and phospholipids also act as emulsifying agents (e.g. the surface-active components of egg yolk are lecithin, an o/w emulsifier, and cholesterol, a w/o emulsifying agent). Mustard and paprika are finely divided solids that are able to stabilise emulsions. Commercially, lecithin is produced from soybean oil. It contains a mixture of phospholipids that are hydrophilic, and two fatty acids that are the lipophilic portion of the molecule. The different phospholipids can be separated to give specific types of lecithin having different functional properties (Anon 2003).

There are currently several thousand emulsifying agents commercially available and careful selection of the type of emulsifying agent is needed to create the required emulsion in a given food system, especially if the emulsion is subjected to subsequent heating or freezing (e.g. microwaveable sauces or ready meals that contain emulsified ingredients) (Flick 1990). Methods for calculating the amounts of emulsifier required in a system and HLB values of mixed emulsions are given by Haw (2004). The amount of emulsifier added to a food also depends in part on the energy imparted by the homogenisation equipment (section 3.2.3). For example, whereas 5–10% emulsifier maybe required using a high-speed mixer, this is reduced to 2–5% if a hydroshear homogeniser or colloid mill is used, and further reduced to 0.2% using a high-pressure homogeniser (Anon 1996).

Because of their unique molecular structure, emulsifiers are also used for a number of functional effects to improve the quality of a wide range of processed foods in addition to stabilising emulsions. These include the following:

- *Complexing starch in bakery products*. This may be the most widespread application of emulsifiers. During baking, starch granules absorb water and swell to become gelatinised. During storage the starch molecules gradually associate, forcing out absorbed water and causing the starch to recrystallise (known as starch retrogradation). Emulsifiers slow retrogradation and maintain the softness in bakery products.
- *Protein interaction in bakery products*. In bread making, hydrated gluten forms an elastic network that gives structure to the dough and helps contain the carbon dioxide produced by yeast activity. Emulsifiers, such as the stearoyl lactylates, interact with gluten to form a stronger network.
- *Foam stabilisation in bakery and dairy products*. Emulsifiers improve the way that air is incorporated and retained in baked goods. The correct texture of cake requires air bubbles to be mixed into the cake batter, which then expand during baking to give the characteristic texture of the cake. Efficient aeration is needed to avoid over-mixing the batter, which would make the cake tough. Emulsifiers increase the whipping rate of cake batters by reducing the surface tension and enabling mixer blades to more easily incorporate air. They also help to increase cake volume and to create a more uniform cell structure. In ice cream and whipped toppings, naturally occurring proteins stabilise the emulsion by binding to the triacylglycerides on the surface of fat globules and preventing the fat from agglomerating. However, agglomeration is necessary for creating a foam because fat globules coat the surface of air bubbles and stabilise them. Emulsifiers destabilise this emulsion and promote agglomeration, by displacing the

proteins from the fat globule surface to the aqueous phase. This increases the viscosity of the liquid cream and allows the fat globules to agglomerate. The increased viscosity promotes aeration and the agglomerated fats stabilise the air bubbles when air is incorporated.

- *Crystal modification in spreading fats and chocolate confectionery.* Chocolate is tempered to promote the formation of more stable fat crystals in the cocoa butter (Chapter 24, section 24.1.1). However, if chocolate is exposed to heat and then cooled (e.g. during transport or storage) small amounts of cocoa butter migrate to the surface of the chocolate and appear as a greyish 'bloom'. Adding an emulsifier to chocolate inhibits the conversion of cocoa butter crystals and slows the appearance of fat bloom. Similarly, lipophilic emulsifiers are used to modify the size of fat crystals in margarine to prevent a gritty mouthfeel and produce a smoother product (see section 3.2.4).
- *Instantising powders.* Powdered mixes should disperse rapidly and completely when added to water, but mixes that contain significant amounts of fat may have fat on the surface of the individual particles, which slows or prevents dispersion. An emulsifier such as lecithin applied to the surface of particles aligns the lipophilic portion with the fat, leaving exposed the hydrophilic portion that has a greater affinity for water to aid dispersion. In coffee whiteners, an even dispersion of small fat globules is obtained using an ionic emulsifier such as sodium stearoyl lactylate to prevent agglomeration and give a smooth mouthfeel.
- *Release agent for processing equipment.* Lecithin is used in food lubricants (e.g. bakery pan oils) to prevent products sticking, and acetylated monoglycerides are used for lubricating bread slicer blades (Hegenbart 1995 and section 3.1.2). Emulsifying agents used in cleaning chemicals are described in Chapter 27 (section 27.1.1)

3.2.3   Equipment
The terms 'emulsifiers' and 'homogenisers' are often used interchangeably for equipment used to produce emulsions. The six types of equipment are:

1 high-speed mixers;
2 hydroshear homogeniser;
3 membrane emulsifiers;
4 pressure homogenisers;
5 rotor-stator homogenisers and colloid mills;
6 ultrasonic homogenisers.

The selection of a suitable homogeniser depends mostly on the viscosity of the product and any changes in viscosity that may take place as a result of the shearing action in the homogeniser (see also discussion of Newtonian and non-Newtonian fluids in Chapter 1, section 1.1.2). Other factors include the temperature sensitivity of the material and whether the product requires pumping after homogenisation (high-pressure homogenisers do not require ancillary pumps, whereas other types of equipment do). Further details of homogenisers are given by Pandolfe (1991).

High-speed mixers
High-speed mixers use turbines or propellers to pre-mix emulsions of low-viscosity liquids. They operate at speeds of 6000–50 000 rpm and create a shearing action on the food at the edges and tips of the blades (details in Chapter 4, section 4.1.3).

Hydroshear homogeniser

The hydroshear homogeniser is a double-cone shaped chamber that has a tangential feed pipe at the centre and outlet pipes at the end of each cone. The feed liquid enters the chamber at high velocity and is made to spin in increasingly smaller circles and increasing velocity. The differences in velocity between adjacent layers of liquid causes high shearing forces, which break droplets in the dispersed phase to within a range of 2–8 μm (Anon 1996). This type of homogeniser produces lower pressures and energy levels than high-pressure homogenisers and is therefore not suitable for producing sub-micron emulsions.

Membrane emulsification

Membrane emulsification is a new process, which uses nanoporous membranes (Fig. 3.10) to produce emulsions that have droplet sizes in the range of 1–100 μm diameter with very narrow droplet size distributions (Fig. 3.11). The liquid disperse phase is passed through membrane pores to emerge, one droplet at a time, on the permeate side. These are detached and carried away by the continuous phase flowing across the membrane surface (Nakajima 2001). The pores are uniformly spaced and have a uniform size, leading to a more consistent product. Emulsions with droplets having diameters above 2 μm are produced using circular pores, whereas a sintered porous glass microsieve that has pore diameters down to 0.8 μm is used for droplet sizes below 2 μm (Anon 2008j). Depending on the type of emulsion to be produced, the membrane surfaces are specially treated by coating them with a hydrophilic or hydrophobic surfactant to prevent wetting by the disperse phase. Details are given by Gijsbertsen-Abrahamse (2003).

Compared with conventional emulsification techniques, membrane emulsification has the following advantages:

- Better control of the average droplet diameter. The size distribution of droplets is

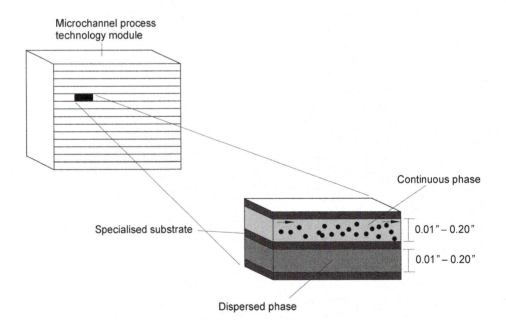

Fig. 3.10 Membrane emulsifier 1 inch = 25.4 mm (courtesy of Velocys Technology Ltd) (Anon 2008i).

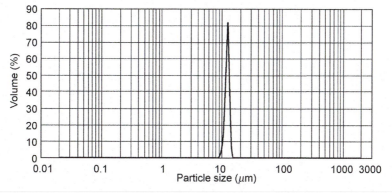

Fig. 3.11   Droplet size distribution of a sunflower oil in water emulsion (courtesy of Nanomi
Emulsification Systems Ltd) (Anon 2008j).

mostly influenced by the size of membrane pores, the viscosity of the continuous
phase, interfacial tension between the phases, the speed of agitation and the type of the
emulsifying agent that is used.

- It produces lower shear stress because the shearing forces are more reproducible and
controllable than the varying shear rates found in homogenisers. As a result the
product quality is usually improved and the technology is suitable for making droplets
from shear-sensitive ingredients or for making double emulsions (below).
- Compared with conventional high-pressure homogenisers the energy consumption is
negligible. As a result the emulsions are not heated during production, which allows
the use of temperature-sensitive ingredients such as proteins. The process has lower
capital and operating costs than more traditional equipment.
- It requires less emulsifying agent due to the absence of the Oswald ripening
phenomenon. (During Oswald Ripening, the diameter of particles in a precipitate
increases due to a decrease in the interfacial free energy between the precipitate and
the continuous phase. Small particles dissolve and the resultant material deposits on
larger particles, causing them to grow.)
- The method produces stable emulsions that have very high disperse phase fractions
without having to recirculate the continuous phase (e.g. 90% for o/w emulsion and
85% for w/o emulsion) (Suzuki and Hagura 2002).
- The process also enables in-line heating of ingredients, cooling to enhance emulsion
stability or in-line cleaning (Anon 2008i).

Owing to the low shearing forces in droplet formation, the technology can also be used to
produce encapsulated (or double) emulsions. Typical examples are water in oil in water
(w/o/w), oil in water in oil (o/w/o) and solid in oil in water (s/o/w) emulsions.
Conventional emulsification methods are used to make composite droplets, and a
homogeneous polydisperse distribution of small droplets is then produced inside each
composite droplet (Fig. 3.12) (Anon 2008j).

Membrane emulsification has so far been used for low-fat spreads and culinary cream
production. Research is focused on reducing clogging of membrane pores because of
droplets adhering to or spreading over the membrane surface, which reduces the performance of the membrane and increases the droplet size and size distribution. Different
membrane materials, including ceramic or stainless steel membranes, and new operating
methods such as rotating membranes, are under development and are reviewed by
Charcosset *et al.* (2004).

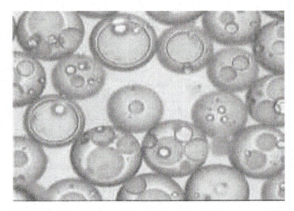

Fig. 3.12   Encapsulated or 'double' emulsion (courtesy of Micropore Ltd at www.micropor-e.co.uk/).

Pressure homogenisers

Pressure homogenisers consists of a high-pressure, positive-displacement pump operating at 10 000–70 000 kPa, which is fitted with a homogenising valve on the discharge side (Fig. 3.13). When liquid is pumped through the small (up to 300 μm) adjustable gap between the valve and the valve seat, the high pressure produces a high liquid velocity (80–150 m s^{-1}). There is then an almost instantaneous drop in pressure and velocity as the liquid emerges from the valve. These extreme conditions of turbulence produce powerful shearing forces that disrupt the droplets in the dispersed phase, which is enhanced by impact forces created by placing a hard surface (a breaker ring) in the path of the liquid. The drop in pressure also creates vapour bubbles in the liquid and when these implode (termed 'cavitation') they produce shock waves that disrupt globules and reduce their size. In some foods, for example milk products, there may be inadequate distribution of the emulsifying agent over the newly formed surfaces, which causes fat globules to clump together. A second similar valve is then used to break up the clusters of globules (Fig. 3.13). Pressure homogenisers are widely used before pasteurisation (Chapter 12) and ultra high-temperature sterilisation (Chapter 13, section 13.2) of milk, and in the production of salad creams, ice cream and some soups and sauces.

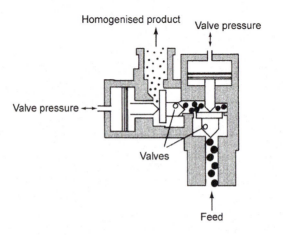

Fig. 3.13   Hydraulic two-stage pressure homogenising valve (courtesy of APV Crepaco Inc. at www.apv.com/us/eng/products/homogenisers/APV+Homogenisers.asp).

Rotor-stator homogenisers and colloid mills

Rotor-stator homogenisers have a high-speed rotor positioned inside a static head or tube (the stator) that contains slots or holes. The food is subjected to intense shearing forces as it is forced through the holes. Anon (2008c) describes an inline, multi-stage rotor-stator design that has both high flowrates and ultra-high shear, producing droplet sizes in the range of 1–5 μm in a single pass. By changing the rotor/stator head, the machine can homogenise a wide range of products at a lower cost than using equipment such as colloid mills. In another design, the rotor and stator have concentric rows of intermeshing teeth. The feed enters at the centre of the homogeniser and moves outward through radial channels cut in the rotor/stator teeth. It is subjected to intense mechanical and hydraulic shear caused by the very high rotor speeds that produce tip speeds of up to 90 m s^{-1} (compared with conventional single-stage rotor/stator mixers that have tip speeds in the range of 15–20 m s^{-1}). Applications include production of mayonnaise and mustard.

Colloid mills are essentially disc mills (section 3.1.2) that have a small clearance (0.05–1.3 mm) between a stationary disc and a vertical disc rotating at 3000–15 000 rpm. They create high shearing forces and are more effective than pressure homogenisers for high-viscosity liquids. With intermediate-viscosity liquids they tend to produce larger droplet sizes than pressure homogenisers do. Numerous designs of disc, including flat, corrugated and conical shapes, are available for different applications. Modifications of this design include the use of two counter-rotating discs or intermeshing pins on the surface of the discs to increase the shearing action. For highly viscous foods (e.g. peanut butter, meat or fish pastes) the discs may be mounted horizontally. The greater friction created in viscous foods may require these mills to be cooled by recirculating water.

Ultrasonic homogenisers

Ultrasonic homogenisers use high-frequency sound waves to cause alternate cycles of compression and tension which shear low-viscosity liquids (also termed 'sonication'). Under the correct conditions, ultrasound also causes the formation of micro air bubbles, which grow and coalesce until they reach a size that resonates with the sound. They then vibrate violently and implode; the cavitation producing a shock wave and liquid jet streams travelling at up to 400 km h^{-1} that shear the liquid to form emulsions with droplet sizes of 1–2 μm (Anon 2008k). The two phases of an emulsion are pumped through the homogeniser at pressures of 340–1400 kPa and the ultrasonic energy is produced by a piezoelectric generator made from lead zirconate titanate crystals, having a power output of between 10 and 375 W. The vibration is transmitted down a titanium horn or probe (Fig. 3.14) that is tuned to make the unit resonate at 15–25 kHz (Yacko 2006). This type of homogeniser is used for the production of salad creams, ice cream, synthetic creams,

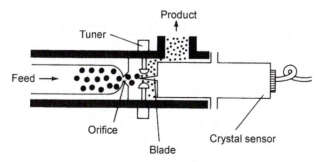

Fig. 3.14   Ultrasonic homogeniser (after Loncin and Merson 1979).

baby foods and essential oil emulsions. It is also used for dispersing powders in liquids (Chapter 4, section 4.1.3). Further details of processing using ultrasound are given in Chapter 9 (section 9.6).

3.2.4   Effect on foods

Viscosity or texture

In many liquid and semi-liquid foods, the desired mouthfeel is achieved by careful selection of the type of emulsifying agent and stabiliser and by control over homogenisation conditions. In milk, homogenisation reduces the average size of fat globules from $4 \, \mu m$ to less than $1 \, \mu m$, thereby giving the milk a creamier texture. The increase in viscosity is due to the higher number of globules and adsorption of casein onto the globule surface. These changes are discussed in detail by Walstra *et al.* (2005), Anon (2001) and Wong *et al.* (1999).

Cream is an o/w emulsion that is mechanically agitated (churned) to cause a partial breakdown of the emulsion when it is made into butter. During this stage, air is incorporated to produce a foam. Liquid fat is released from globules at the surfaces of air bubbles, and this binds together clumps of solid fat to form butter 'grains'. These are then mixed at low speed (worked) to disperse water as fine droplets throughout the mass and to rupture any fat globules remaining from the cream. Although butter is thought of as a w/o emulsion, the complete inversion of the o/w emulsion of cream does not take place. The final product has a continuous phase of 85% fat which contains globules and crystals of solid fat and air bubbles and a dispersed phase (15%) of water droplets and buttermilk, giving the characteristic texture. The stability of butter is mostly due to its semi-solid nature, which prevents migration of bacteria trapped in water droplets. Details of butter production are given by Lane (1992).

Margarine and low-fat spreads are w/o emulsions produced from a blend of oils, which is heated with a solution of skim milk, salt, vitamins and emulsifying agents. The warm mixture is emulsified and then chilled and worked to the desired consistency in a continuous operation. The fats crystallise as they cool, to form a three-dimensional network of long thin needles, which produce the desired smooth texture. Fats are polymorphic and have three forms. It is the β'-form that is required; the β-form is larger and causes a grainy texture; and the α-form rapidly undergoes transition to the β'-form (see also a discussion of fats in chocolate in Chapter 24, section 24.1.1). The fat content of margarine is similar to butter, whereas low-fat spreads have approximately 40% fat. The oils are chosen to have low melting points and these products are therefore spreadable at refrigeration temperatures. Details are given by Lane (1992).

In solid food emulsions the texture is determined by the composition of the food, the homogenisation conditions and post-processing operations such as heating or freezing. Meat emulsions (e.g. sausage and pâté) are o/w emulsions in which the continuous phase is a complex colloidal system of gelatin, proteins, minerals and vitamins, and the dispersed phase is fat globules. The stability of the continuous phase is determined in part by the water-holding capacity (WHC) and fat-holding capacity (FHC) of the meat proteins. The factors that affect WHC and FHC are described by Laurie (1998). The quality of the emulsion is influenced by the ratios of meat and ice water: fat, the use of polyphosphates to bind water, and the time, temperature and speed of homogenisation. The texture of the emulsion is set by heat during subsequent cooking.

In ice cream, the emulsion is formed as a liquid, and the texture of the final product is partly determined by the subsequent freezing. Ice cream is a thick o/w emulsion that has a complex continuous phase consisting of ice crystals, colloidal milk solids, dissolved sugar, flavouring, colouring and stabilisers, together with a solid-air foam. The dispersed phase is milk fat. Air is incorporated into the emulsion during freezing to create a foam having air cells <100 μm in diameter. This increases the softness and lightness of the product and allows it to be easily scooped. The amount of air is measured as the overrun:

$$\% \text{ overrun} = \frac{\text{volume of ice cream} - \text{volume of mix used}}{\text{volume of mix used}} \times 100 \qquad \boxed{3.5}$$

For example, 400 l of ice cream mix produces 780 l of ice cream, so the overrun $= [(780 - 400)/400] \times 100 = 95\%$ overrun. Commercial ice creams have overruns of 60–100%.

Freezing partially destabilises the emulsion to produce a degree of clumping of fat globules, which improves the texture. Commercial ice creams usually have a softer texture than homemade products due to (1) faster freezing, which produces smaller (40–50 μm) ice crystals (Chapter 22, section 22.1.1), (2) a higher overrun, and (3) emulsifiers (e.g. esters of mono- and diglycerides) and stabilisers (e.g. alginates, carrageenan, gums or gelatin (Chapter 1, section 1.1.1)), which cause a larger proportion of the aqueous phase to remain unfrozen. This prevents lactose crystallisation and reduces graininess. As a result, less heat is needed to melt the ice cream and it does not therefore feel excessively cold when eaten. Details of ice cream production are given by Jaspersen (1989) and Andreasen and Nielsen (1992).

Cake batters are also o/w emulsions, in which the continuous phase is a solution of sugar and flavours, colloidal starch and a foam produced during mixing. The dispersed phase is added fats or oils. The texture of the final product is partly determined by subsequent baking, and details of the changes to cake batters during mixing and baking, and the effects of variations in their formulation are described by Mizukoshi (1990). Homogenisation also partially separates the fibres of cellulose to increase its water-binding capacity and changes the size of tomato fibres to produce thicker ketchup (Anon 1996).

Colour, aroma and nutritional value
Homogenisation has an effect on the colour of some foods (e.g. milk) because the larger number of globules causes greater reflectance and scattering of light. Flavour and aroma are improved in many emulsified foods because volatile components are dispersed throughout the food and hence have greater contact with flavour receptors when eaten. The nutritional value of emulsified foods is changed if components are separated (e.g. in butter making), and there is improved digestibility of fats and proteins owing to the reduction in particle size. The nutritional value of other foods is determined by the formulation used and is not directly affected by emulsification or homogenisation. However, the additional unit operations (e.g. chilling, freezing and baking), which are necessary to extend the shelf-life, may also cause changes to nutritional value. In all food emulsions, degradative changes such as hydrolysis or oxidation of pigments, aroma compounds and vitamins are minimised by careful control over the processing, packaging and storage conditions.

3.2.5   Effect on micro-organisms
Microbial growth on the finely dispersed material in emulsions is prevented by hygienic control of production and implementation of HACCP procedures (Chapter 1, section

1.5.1). In many countries, specific regulations are in force to control hygienic standards during preparation of food emulsions (particularly meat and dairy emulsions) owing to the risk of dispersing pathogenic bacteria throughout the food (see Chapter 1, section 1.2.3). This is particularly required for products, such as ice cream, that are not heated before consumption (Anon 2008l), where recontamination with pathogens has been shown to exist (Kanbakan *et al.* 2004, Barbini de Pederiva and Stefanini de Guzman 2000).

References

ABDUL-RAOUF, U.M., BEUCHAT, L.R. and AMMAR, M.S., (1993), Survival and growth of Escherichia coli O157:H7 on salad vegetables, *Applied Environmental Microbiology*, **59** (7), 1999–2006, abstract available at http://aem.asm.org/cgi/content/abstract/59/7/1999.

ALLDRICK, A.J., (2002), The processing of cereal foods, in (C.J.K. Henry and C. Chapman, Eds.), *The Nutrition Handbook for Food Processors*, Woodhead Publishing, Cambridge, pp. 301–313.

ANDREASEN, T.G. and NIELSEN, H., (1992), Ice cream and aerated desserts, in (R. Early, Ed.), *Technology of Dairy Products*, Blackie, London, pp. 197–220.

ANON, (1986), *Alpine Process Technology Technical Literature 019/5e*, Alpine Process Technology Ltd, Runcorn.

ANON, (1996), Processing of emulsions and dispersions by homogenisation, information from the APV Homogenization Group, available at www.geocities.com/grupoindustrialaisa/procemudisp.html.

ANON, (2001), Dairy chemistry and physics, NEM Business Solutions, available at www.cip.ukcentre.com/chem1.htm#top.

ANON, (2002), *McCance and Widdowson's The Composition of Foods*, 6th Summary edn, Compiled by Food Standards Agency and Institute of Food Research, Royal Society of Chemistry Publications, London, pp. 29–82.

ANON, (2003), Lecithin applications, American Lecithin Company information, available at http://americanlecithin.com/leci_appfood.html.

ANON, (2005), European Union guidelines on the application of directive 94/9/EC of 23 March 1994 on equipment and protective systems intended for use in potentially explosive atmospheres (ATEX), Second edn, July.

ANON, (2008a), Cutting, slicing, chopping, mincing, pulping and pressing, Hyforma product information, available at www.hyfoma.com/en/content/processing-technology/size-reduction-mixing-forming/cutting-slicing-chopping-mincing/.

ANON, (2008b), Information from The Fitzpatrick Company, available at www.fitzmill.com/food/size_reduction/theory/theory_sr.html.

ANON, (2008c), Information from Urschel Ltd, available at www.urschel.com/literature.

ANON, (2008d), Micron-Master® Jet Pulverizers, The Jet Pulverizing Company, product information, available at www.jetpul.com/mequip/milloper.htm.

ANON, (2008e), Groundnut decorticator from Raj Kumar Agro Engineers Pvt. Ltd, available at http://rajkumaragromachinery.com/decoraticator.htm.

ANON, (2008f), Oilseeds processing toolkit, Food and Agriculture Organisation of the UN, available at www.fao.org/inpho, followed by search 'oilseeds'.

ANON, (2008g), HLB values, from Pharmaceutical Calculations, available at http://pharmcal.tripod.com/ch17.htm#cal5.

ANON, (2008h), Surfactants – emulsifiers. Presentation of Ohio State University available at http://class.fst.ohio-state.edu/fst621/Lectures/PPT%20presentations/Emulsifiers.ppt.

ANON, (2008i), Information from Velocys Technology available at www.velocys.com/8966.cfm.

ANON, (2008j), Information from Nanomi Emulsification Systems, available at www.nanomi.com/nanomi/Monodisperse_Double_Emulsions.html.

ANON, (2008k), Hielscher – Ultrasound Technology, available at www.ultrasonic-systems.com/ultrasonics/emulsify_01.htm?gclid=CIvuuK70hoYCFRhXEgodBj88hg.

ANON, (2008l), Safe handling and serving of soft ice-cream, Food Safety Authority of Ireland, available at www.fsai.ie/publications/leaflets/ice_cream.pdf.

BARBINI DE PEDERIVA, N.B. and STEFANINI DE GUZMAN, A.M, (2000), Isolation and survival of *Yersinia enterocolytica* in ice cream at different pH values, stored at −18°C, *Brazilian J. Microbiology*, 31, 174–177, available at http://www.scielo.br/pdf/bjm/v31n3/v31n3a05.pdf.

BAUERNFEIND, J.C. and DE RITTER, E., (1991), Cereal grain products, in (J.C. Bauernfeind and P.A. Lachance, Eds.), *Nutrient Addition to Foods*, Food and Nutrition Press, Trumbull, CT.

BERGHOFER, L.K., HOCKING, A.D., MISKELLY, D. and JANSSON, E., (2003), Microbiology of wheat and flour milling in Australia, *International J. Food Microbiology*, 15; 85 (1–2), 137–149, available at www.ncbi.nlm.nih.gov/entrez/query.fcgi?itool=abstractplus&db=pubmed&cmd=Retrieve&dopt=abstractplus&list_uids=12810278.

BRENNAN, J.G., BUTTERS, J.R., COWELL, N.D. and LILLEY, A.E.V., (1990), *Food Engineering Operations*, Elsevier Applied Science, London.

CHARCOSSET, C., LIMAYEM, I. and FESSI, H., (2004), The membrane emulsification process – a review, *J. Chemical Technology & Biotechnology*, 79 (3), 209–218, abstract available at http://www3.interscience.wiley.com/cgi-bin/abstract/107063951/ABSTRACT?CRETRY=1&SRETRY=0.

DEOSTHALE, Y.G., (1984), The nutritive value of foods and the significance of some household processes, in (K.T. Achaya, Ed.), *Interfaces Between Agriculture, Nutrition, and Food Science*, The United Nations University, available at www.unu.edu/unupress/unupbooks/80478e/80478E0j.htm#Milling%20of%20food%20grains.

ENGELEN, L., VAN DER BILT, A., SCHIPPER, M. and BOSMAN, F., (2005a), Oral size perception of particles: effect of size, type, viscosity and method, *J. Texture Studies*, 36 (4), 373, abstract available at www.blackwell-synergy.com/doi/abs/10.1111/j.1745-4603.2005.00022.x.

ENGELEN, L., DE WIJK, R.A., VAN DER BILT, A., PRINZ, J. F., JANSSEN, A. M. and BOSMAN, F., (2005b), Relating particles and texture perception, *Physiology and Behavior*, 86 (1–2), 111–117, abstract available at http://cat.inist.fr/?aModele=afficheN&cpsidt=17137243.

FLICK, E.W., (1990), *Emulsifying Agents – An Industrial Guide*, William Andrew Publishing/Noyes, online version available at: www.knovel.com/knovel2/Toc.jsp?BookID=407&VerticalID=0.

FRIBERG, S.E., LARSSON, K. and SJOBLOM, J., (Eds.), (2004), *Food Emulsions*, 4th edn, Marcel Dekker, New York.

GIJSBERTSEN-ABRAHAMSE, A., (2003), Membrane emulsification: process principles, Thesis, Wageningen University, The Netherlands, available at http://library.wur.nl/wda/dissertations/dis3392.pdf.

HAW, P., (2004), The HLB system – a time saving guide to surfactant selection. Presentation to the Midwest Chapter of the Society of Cosmetic Chemists, 9 March, available at http://lotioncrafter.com/pdf/The_HLB_System.pdf.

HEGENBART, S., (1995), Emulsifier Applications, Food Product Design, available at www.foodproductdesign.com/articles/465/465_1095DE.html.

HIGGINS, K. T., (2006), Size (reduction) matters, Food Engineering, available at www.foodengineeringmag.com/CDA/Archives/1089eb5a01b49010VgnVCM100000f932a8c.

JASPERSEN, W.S., (1989), Speciality ice cream extrusion technology, in (A. Turner, Ed.), *Food Technology International Europe*, Sterling Publications International, London, pp. 85–88.

KANBAKAN, U., CON, A.H. and AYAR, A., (2004), Determination of microbiological contamination sources during ice cream production in Denizli, Turkey. *Food Control*, 15 (6), 463–470.

KENT, N.L. and EVERS, A.D., (1994), *Kent's Technology of Cereals*, 4th edn, Woodhead Publishing, Cambridge, Table 3.3, p. 139.

LANE, R., (1992), Butter, margarine and reduced fat spreads, in (R. Early, Ed.), *Technology of Dairy Products*, Blackie, London, pp. 86–116.

LAWRIE, R.A., (1998), *Lawrie's Meat Science*, 6th edn, Woodhead Publishing, Cambridge.

LEWIS, M.J., (1990), *Physical Properties of Foods and Food Processing Systems*, Woodhead

Publishing, Cambridge, pp. 184–195.

LEWIS, M.J., (1996), Solids separation processes, in (A.S. Grandison and M.J. Lewis, Eds.) *Separation Processes in the Food and Biotechnology Industries*, Woodhead Publishing, Cambridge, pp. 243–286.

LONCIN, M. and MERSON, R.L., (1979), *Food Engineering*, Academic Press, New York, pp. 246–264.

MCCLEMENTS, D.J., (1999), *Food Emulsions: Principles, Practice and Techniques*, CRC Press, Boca Raton, FL.

MIZUKOSHI, M., (1990), Baking mechanism in cake production, in (K. Larssen and S.E. Friberg, Eds.), *Food Emulsions*, 2nd edn, Marcel Dekker, New York, pp. 479–504.

NAKAJIMA, M., (2001), Novel microchannel system for monodispersed microspheres, *RIKEN Review*, **36**, June, available at www.riken.go.jp/lab-www/library/publication/review/pdf/No_36/36_021.pdf.

PANDOLFE, W.D., (1991), Homogenizers, in (Y.H. Hui, Ed.), *Encyclopedia of Food Science and Technology*, John Wiley & Sons, New York, p. 1413.

RAWSON, F.F., (1998), An introduction to ultrasound food cutting, in (M.J.W. Povey and T.J. Mason, Eds.), *Ultrasound in Food Processing*, Blackie Academic and Professional, London, pp. 254–270.

ROSSELL, B., (1999), *Oils and Fats Handbook: Vegetable Oils and Fats*, Woodhead Publishing, Cambridge.

SHARP, G., (1998), At the cutting edge, *Food Processing*, August, 16, 17, 19.

SLAVIN, J.L., JACOBS, D. and MARQUART, L., (2000), Grain processing and nutrition, *Critical Reviews in Food Science and Nutrition*, **40** (4), 309–326.

STEFFE, J.F. (1996), *Rheological Methods in Food Process Engineering*, 2nd edn, Freeman Press, East Lansing, MI.

SUZUKI, K. and HAGURA, Y., (2002), Possibility of the membrane emulsification method to prepare food emulsions with unique properties, *Japan Journal of Food Engineering*, **3** (2), 35–40, available at http://www.jsfe.jp/journal/jjfe-e/jjfe0302e.htm.

VOIDAROU, C., TZORA, A., ALEXOPOULOS, A. and BEZIRTZOGLOU, E., (2006), Hygienic quality of different ham preparations, IUFoST 13th World Congress of Food Sciences Technology, 17/21 September, Nantes, France, available at http://dx.doi.org/10.1051/IUFoST:20060771.

WALSTRA, P., WOUTERS, J.T.M. and GEURTS, T.J., (2005), *Dairy Science and Technology*, 2nd edn, Culinary and Hospitality Industry Publications Services, available at www.vonl.com/chips/dairysi2.htm.

WONG, N.P., JENNESS, R., KEENEY, M. and MARTH, E.H. (1999), *Fundamentals of Dairy Chemistry*, 3rd edn, Springer-Verlag, Heidelberg, Germany.

YACKO, R.M., (2006), *The Field of Homogenizing*, PRO Scientific Company, available at www.proscientific.com/Homogenizing.shtml.

YOUNG, G., (2003), Size reduction of particulate material, available at www.erpt.org/032Q/youc-01.pdf.

4

Mixing and forming

Abstract: Mixing has wide application in most food industries where it is used to combine ingredients to achieve different functional properties or sensory characteristics. Forming is used to increase the variety and convenience of foods such as baked goods, confectionery and snackfoods. This chapter describes the theory of mixing solid and liquid foods. It reviews the wide range of mixers available for the large number of mixing applications and describes moulding and forming equipment that is used for bread, pies, biscuits and confectionery products.

Key words: mixing, forming, non-Newtonian liquids, mixing and moulding equipment.

Mixing (or blending) is a unit operation in which a uniform mixture is obtained from two or more components, by dispersing one within the other(s). The larger component is sometimes called the 'continuous phase' and the smaller component the 'dispersed phase' by analogy with emulsions (Chapter 3), but these terms do not imply emulsification when used in this context. Mixing has no preservative effect and is intended solely as a processing aid or to alter the eating quality of foods. It has very wide applications in most food industries where it is used to combine ingredients to achieve different functional properties or sensory characteristics. In some foods, the correct degree of mixing is used to ensure that the proportion of each component complies with legislative standards (e.g. mixed vegetables, mixed nuts, sausages and other meat products). Extruders (Chapter 15) and some types of size reduction equipment (Chapter 3) also have a mixing action.

Forming (section 4.2) is a size enlargement operation in which foods that have a high viscosity or a dough-like texture are moulded into a variety of shapes and sizes, often immediately after a mixing operation. It is used as a processing aid to increase the variety and convenience of foods such as baked goods, confectionery and snackfoods. It has no direct effect on the shelf-life or nutritional value of foods. Close control over the size of formed pieces is critical (e.g. to ensure uniform rates of heat transfer to the centre of baked foods or to ensure the uniformity of pieces of food and hence to control fill weights). Extrusion also has a forming function.

4.1   Mixing

When food products are mixed there are a number of aspects that differ from other industrial mixing applications:

- Mixing is often used primarily to develop desirable product characteristics, rather than simply ensure homogeneity.
- It is often multi-component, involving ingredients of different physical properties and quantities.
- It may often involve high-viscosity or non-Newtonian liquids (Chapter 1, section 1.1.2).
- Some components may be fragile and damaged by over-mixing.
- There may be complex relationships between mixing parameters and product characteristics that may change as mixing proceeds.

The criteria for successful mixing have been described as first achieving an acceptable product quality (in terms of sensory properties, functionality, homogeneity, particulate integrity, etc.) followed by adequate safety, hygienic design, legality (compositional standards for some foods), process and energy efficiency, and flexibility to changes in processing (Campbell 1995). Lindley (1991a–c) gives a detailed review of mixing operations.

4.1.1   Theory of solids mixing

In contrast with liquids and viscous pastes (section 4.1.2) it is not possible to achieve a completely uniform mixture of dry powders or particulate solids. The degree of mixing that is achieved depends on:

- the relative particle size, shape and density of each component;
- the moisture content, surface characteristics and flow characteristics of each component;
- the tendency of the particles to aggregate; and
- the efficiency of a particular mixer for mixing those components.

In general, materials that are similar in size, shape and density are able to form a more uniform mixture than are dissimilar materials. During a mixing operation, differences in these properties also cause 'unmixing' (or separation) of the component parts. In some mixtures, uniformity is achieved after a given period and then unmixing begins, and it is therefore important in such cases to time the mixing operation accurately. The uniformity of the final product depends on the equilibrium achieved between the mechanisms of mixing and unmixing, which in turn is related to the type of mixer, the operating conditions and the component foods.

If a two-component mixture is sampled at the start of mixing (in the unmixed state), most samples will consist entirely of one of the components. As mixing proceeds, the composition of each sample becomes more uniform and approaches the average composition of the mixture. One method of determining the changes in composition is to calculate the standard deviation of each fraction in successive samples:

$$\sigma_{\mathrm{m}} = \sqrt{\left[\frac{1}{n-1}\sum(c-\bar{c})^2\right]}$$

[4.1]

where σ_{m} = standard deviation, n = number of samples, c = concentration of the component in each sample and $\bar{c}$ = the mean concentration of samples. Lower standard deviations are found as the uniformity of the mixture increases.

Different mixing indices are available to monitor the extent of mixing and to compare alternative types of equipment:

$$M_1 = \frac{\sigma_m - \sigma_\infty}{\sigma_0 - \sigma_\infty} \qquad\qquad - 两种差不多 \tag{4.2}$$

$$M_2 = \frac{\log\sigma_m - \log\sigma_\infty}{\log\sigma_0 - \log\sigma_\infty} \qquad -- 多-少 \tag{4.3}$$

$$M_3 = \frac{\sigma_m^2 - \sigma_\infty^2}{\sigma_0^2 - \sigma_\infty^2} \qquad\qquad - 液体 \tag{4.4}$$

where σ_∞ = the standard deviation of a 'perfectly mixed' sample, σ_0 = the standard deviation of a sample at the start of mixing and σ_m = the standard deviation of a sample taken during mixing. σ_0 is found using:

$$\sigma_0 = \sqrt{[V_1(1 - V_1)]} \tag{4.5}$$

where V = the average fractional volume or mass of a component in the mixture.

In practice, perfect mixing (where $\sigma_\infty = 0$) cannot be achieved, but in efficient mixers the value becomes very low after a reasonable period. The mixing index M_1 is used when approximately equal masses of components are mixed and/or at relatively low mixing rates, M_2 is used when a small quantity of one component is incorporated into a larger bulk of material and/or at higher mixing rates, and M_3 is used for liquids or solids mixing in a similar way to M_1. In practice, all three are examined and the one that is most suitable for the particular ingredients and type of mixer is selected.

The mixing time is related to the mixing index using:

$$\ln M = -Kt_m \tag{4.6}$$

where K = mixing rate constant, which varies with the type of mixer and the nature of the components, and t_m (s) = mixing time.

Sample problem 4.1

During preparation of a dough, 700 g of sugar are mixed with 100 kg of flour. Ten 100 g samples are taken after 1, 5 and 10 min and analysed for the percentage sugar. The results are as follows:

Percentage after	1 min	0.21	0.32	0.46	0.17	0.89	1.00	0.98	0.23	0.10	0.14
Percentage after	5 min	0.85	0.80	0.62	0.78	0.75	0.39	0.84	0.96	0.58	0.47
Percentage after	10 min	0.72	0.69	0.71	0.70	0.68	0.71	0.70	0.72	0.70	0.70

Calculate the mixing index for each mixing time and draw conclusions regarding the efficiency of mixing. Assume that for 'perfect mixing' there is a probability that 99.7% of samples will fall within three standard deviations of the mean composition ($\sigma = 0.01\%$).

Solution to sample problem 4.1
Average fractional mass V_1 of sugar in the mix

$$= \frac{700}{100 \times 10^3}$$
$$= 7 \times 10^{-3}$$

From Equation 4.5:

$$\sigma_0 = \sqrt{[7 \times 10^{-3}(1 - 7 \times 10^{-3})]}$$
$$= 0.083\,37$$
$$= 8.337\%$$

After 1 min

$$\text{mean } \bar{c} \text{ of the samples} = 0.45$$

Using Equation 4.1, after 1 min,

$$\sigma_m = \sqrt{\left[\frac{1}{10-1}\sum(c - 0.45)^2\right]}$$

(that is, subtract 0.45 from c for each of the ten samples, square the result and sum the squares):

$$\sigma_m = \sqrt{(0.11 \times 1.197)}$$
$$= \sqrt{0.131\,67}$$
$$= 0.3629\%$$

After 5 min,

$$\sigma_m = \sqrt{(0.11 \times 0.298\,24)}$$
$$= \sqrt{0.032\,806}$$
$$= 0.1811\%$$

After 10 min,

$$\sigma_m = \sqrt{(0.11 \times 0.001\,41)}$$
$$= \sqrt{0.000\,155}$$
$$= 0.0125\%$$

Using Equation 4.3, after 1 min,

$$M_2 = \frac{\log 0.3629 - \log 0.01}{\log 8.337 - \log 0.01}$$

$$= \frac{-0.44 - (-1.99)}{0.92 - (-1.99)}$$

$$= 1.55/2.91$$

$$= 0.533$$

After 5 min,

$$M_2 = \frac{\log 0.1811 - \log 0.01}{\log 8.337 - \log 0.01}$$

$$= \frac{-0.74 - (-1.99)}{0.92 - (-1.99)}$$

$$= 1.25/2.91$$

$$= 0.429$$

And, after 10 min,

$$M_2 = \frac{\log 0.0125 - \log 0.01}{\log 8.337 - \log 0.01}$$

$$= \frac{-1.90 - (-1.99)}{0.92 - (-1.99)}$$

$$= 0.09/2.91$$

$$= 0.031$$

Interpretation: if the log M_2 is plotted against time, the linear relationship indicates that the mixing index gives a good description of the mixing process and that mixing takes place uniformly and efficiently.

Using Equation 4.6, after 10 min,

$$\ln 0.031 = -k \times 600$$

Therefore,

$$k = 0.0057$$

The time required for $\sigma_m = \sigma_\infty = 0.01\%$ is then found:

$$\ln 0.01 = -0.0057\, t_m$$

$$T_m = 808 \text{ s}$$

Therefore

$$\text{remaing mixing time} = 808 - 600$$

$$= 208 \text{ s}$$

$$\approx 3.5 \text{ min}$$

Note: The means and standard deviations can be found for each of the sampling times using a spreadhseet.

4.1.2   Theory of liquids mixing

The component velocities induced by a mixer in low-viscosity liquids are as follows (Fig. 4.1):

- A radial velocity that acts in a direction perpendicular to the mixer shaft.
- A longitudinal velocity (parallel to the mixer shaft).
- A rotational velocity (tangential to the mixer shaft).

To achieve successful mixing, the radial and longitudinal velocities imparted to the liquid are maximised by baffles, off-centre or angled mixer shafts, or angled blades (Fig. 4.2).

Most liquid foods are non-Newtonian, and the viscosity changes with rate of shear (Chapter 1, section 1.1.2). These properties are described in detail by Lewis (1990). The most common types fall into one of the following categories:

- *Pseudoplastic* (the viscosity decreases with increasing shear rate). Foods such as sauces form a zone of thinned material around a small agitator and the bulk of the food does not move. The higher the agitator speed, the more quickly the zone becomes apparent. Planetary or gate mixers (section 4.1.3) are used to ensure that all food is subjected to the mixing action.
- *Dilatant* (the viscosity increases with shear rate). Foods such as cornflour and chocolate should be mixed with great care. If adequate power is not available in the mixer, the increase in viscosity may cause damage to drive mechanisms and shafts. A folding or cutting action, as for example in some planetary mixers or paddle mixers (section 4.1.3), is suitable for this type of food.
- *Thixotropic* (the structure breaks down and viscosity decreases with increasing shear rate). Foods such as yoghurt exhibit both a shear-thinning viscosity and a time-dependent thixotropic effect. Stirring leads to a reduction in the viscosity of the mixture.

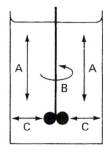

Fig. 4.1   Component velocities in fluid mixing: A, longitudinal; B, rotational; C, radial.

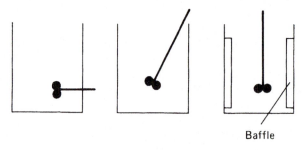

Baffle

Fig. 4.2   Position of agitators for effective mixing of liquids.

- *Viscoelastic* (materials that exhibit viscous and elastic properties including stress relaxation, creep and recoil). Foods such as bread dough require a folding and stretching action to shear the material. Suitable equipment includes twin-shaft mixers and planetary mixers with intermeshing blades (section 4.1.3). The design of equipment should enable thorough mixing without overloading the motor or reducing the mixing efficiency.

The rate of mixing is characterised by a mixing index (section 4.1.1). The mixing rate constant (Equation 4.6) depends on the characteristics of both the mixer and the liquids. The effect of the mixer characteristics on K is given by:

$$K \propto \frac{D^3 N}{D_t^2 z} \qquad \boxed{4.7}$$

where D (m) = the diameter of the agitator, N (rev s^{-1}) = the agitator speed, D_t (m) = the vessel diameter and z (m) = the height of liquid.

The power requirements of a mixer vary according to the nature, amount and viscosity of the foods in the mixer and the position, type, speed and size of the impeller.

Liquid flow is defined by a series of dimensionless numbers: the Reynolds number Re (Equation 4.8, also Chapter 1), the Froude number Fr (Equation 4.9) and the Power number Po (Equation 4.10):

$$\mathrm{Re} = \frac{D^2 N \rho_m}{\mu_m} \qquad \boxed{4.8}$$

$$\mathrm{Fr} = \frac{D N^2}{g} \qquad \boxed{4.9}$$

$$\mathrm{Po} = \frac{P}{\rho_m N^3 D^5} \qquad \boxed{4.10}$$

where P (W) = the power transmitted via the agitator, ρ_m (kg m^{-3}) = the density of the mixture and μ_m (N s m^{-2}) = the viscosity of the mixture. These are related as follows:

$$\mathrm{Po} = K(\mathrm{Re})^n (\mathrm{Fr})^m \qquad \boxed{4.11}$$

where K, n and m are factors related to the geometry of the agitator, which are found by experiment. The Froude number is only important when a vortex is formed in an unbaffled vessel and is therefore omitted from Equation 4.11.

The density of a mixture is found by addition of component densities of the continuous and dispersed phases:

$$\rho_m = V_1 \rho_1 + V_2 \rho_2 \qquad \boxed{4.12}$$

where V = the volume fraction. The subscripts 1 and 2 are the continuous phase and dispersed phase respectively.

The viscosity of a mixture is found using the following equations for baffled mixers and for unbaffled mixers:

$$\mu_m(\text{unbaffled}) = \mu_1^{V_1} \mu_2^{V_2} \qquad \boxed{4.13}$$

$$\mu_m(\text{baffled}) = \frac{\mu_1}{V_1} \left(\frac{1 + 1.5 \mu_2 V_2}{\mu_1 + \mu_2} \right) \qquad \boxed{4.14}$$

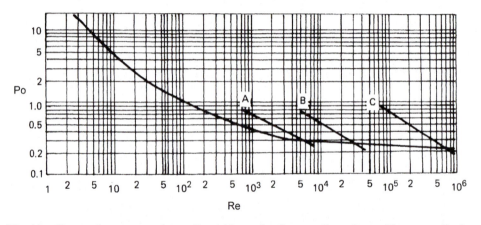

Fig. 4.3   Changes in power number vs Reynolds number for propeller agitator: 30 cm propeller in
137 cm diameter tank, liquid depth 137 cm, propeller 30 cm above base. (A) viscosity =
0.189 N s m^{-2}; (B) viscosity = 0.028 N s m^{-2}; (C) viscosity = 0.109 N s m^{-2}. Propeller speed varied
from 100 to 500 rpm (after Rushton *et al.* 1950).

Characteristic changes in power consumption Po of propellers at different Reynolds
numbers are shown in Fig. 4.3 and further details of power consumption of different
impeller designs are given by Anon (2005a). Similar studies using ribbon mixers are
described by Espinosa-Solares *et al.* (1997).

The specific mechanical energy (SME) (kJ kg^{-1}) can be used to compare the
effectiveness of different mixing systems, and a number of authors report studies of the
effect of SME on dough mixing (e.g. Cuq *et al.* 2006, Icard-Verniere *et al.* 1999, Redl *et al.*
1999). Further details are given in relation to extruder operation (Chapter 15, section 15.1.2).

Historically mixing time has been difficult to accurately predict because of the large
number of variables involved. Different types of computer software have been designed
to predict mixer performance in a particular application. Computational fluid dynamics
(CFD) software is used to simulate fluid flow and heat and mass transfer, involving
turbulent and multiphase flow. Mixing time is the time taken to fully mix the contents of
a vessel after addition of one or more ingredients. CFD enables alteration of, for example,
the type, size, number and location of impellers, the shape of the vessel, the power input,
speed of mixing and limitations on temperature increases for a specific mix. It can be
used to find the optimum performance for a particular mixer in a given application, or to
select an appropriate mixer for a particular food. Other software based on a Eulerian
population balance approach has been developed for modelling particle size distribution
(in this context, particles can also include liquid droplets or bubbles). The software keeps
track of the population of particles as they undergo dispersion, breakage, aggregation or
dissolution. The results of the modelling are used in particular to control and improve the
operation of high-shear mixers (section 4.1.3) in operations such as aeration, liquid-liquid
extraction and emulsion formation (Connelly 2006, Anon 2006a).

4.1.3   Equipment
There is a very wide range of mixers available, owing to the large number of mixing
applications and the historically empirical nature of mixer design and development. The
selection of the correct type of mixer for a particular application depends on the type and
amount of food being mixed and the speed of operation needed to achieve the required

Sample problem 4.2

Olive oil and rapeseed (canola) oil are blended in a ratio of 1 to 5 (by volume) by a propeller agitator 20 cm in diameter operating at 750 rpm in a cylindrical tank 1 m in diameter at 20 °C. Calculate the size of the motor required.

Solution to sample problem 4.2

From Table 1.8 (Chapter 1), the viscosity of olive oil at 20 °C is 0.084 N s m^{-2}, the density of olive oil is 910 kg m^{-3}, the viscosity of rapeseed oil 0.118 N s m^{-2} and the density of rapeseed oil 900 kg m^{-3}.

From Equation 4.13,

$$\mu_m = 0.084^{0.2}0.118^{0.8}$$

$$= 0.110 \text{ N s m}^{-2}$$

From Equation 4.12,

$$\rho_m = 0.2 \times 910 + 0.8 \times 900$$

$$= 902 \text{ kg m}^{-3}$$

From Equation 4.8,

$$Re = (0.2)^2 \frac{750}{60} \frac{902}{0.110}$$

$$= 4100$$

From Fig. 4.3 (curve A) for Re = 4100, Po = 0.4. From Equation 4.10,

$$P = 0.4 \times 902 \left(\frac{750}{60}\right)^3 (0.2)^5$$

$$= 225.5 \text{ J s}^{-1}$$

$$= 0.225 \text{ kW}$$

Therefore the size of the motor required is 0.225 kW.

degree of mixing. Details of different mixers are given by Paul *et al.* (2003) and factors to consider when choosing between batch and continuous mixers are described by Micron (2002). Until a few years ago, mixer selection was relatively straightforward and was made according to factors such as the viscosity of the food, the mixer capacity, the shear rate required and the energy consumption, as well as price and delivery time. Most mixer manufacturers continue to offer selection charts (e.g. Table 4.1) that show the operating characteristics for different types of equipment.

However, developments in design and engineering have made mixers in most categories far more versatile, and there is now a considerable overlap in the capabilities of different types of equipment to mix a particular food. This is especially the case when mixing high-viscosity foods and incorporating powders into liquids. New agitator designs have been developed; new technology has allowed the combination of dissimilar agitators to control flow and shear more effectively; and new auxiliary devices have been

Table 4.1 Factors in mixer selection (courtesy of Charles Ross & Son Company)

Factor	Planetary	Multi-shaft	Ribbon blender	Vertical blender	Disperser	Rotor-stator	Kneader	Motionless
High viscosity (>500 000 cps)	✓	✓					✓	
Medium viscosity (50 000–500 000 cps)	✓	✓		✓	✓	✓		✓
Low viscosity (<10 000 cps)	✓	✓	✓	✓	✓	✓		✓
Emulsification capability		✓✓			✓✓	✓✓✓		
Batch	✓	✓✓	✓	✓	✓	✓✓	✓	
In-line						✓✓		
Solid/solid	✓✓✓		✓	✓				✓
Liquid/liquid	✓✓	✓	✓	✓	✓✓	✓	✓	✓✓✓✓
Low speed/low shear	✓		✓	✓			✓	✓
High speed/high shear		✓✓			✓✓	✓✓		✓✓✓
Vacuum/pressure operation	✓	✓	✓	✓	✓	✓	✓	✓

Anon (2006b)

developed to improve powder wetting and dispersion. Processors now need to evaluate mixing efficiency, costs and product quality using different types of machines and the actual ingredients, volumes and process conditions in their production. The performance of a mixer can be tailored to a particular process by adjusting different combinations of agitators, speeds and shear rates (Ames 2000). Alternatively, the sequence and rate at which ingredients are added in a process can be altered to enhance the capabilities of a mixer. For example, the lowest-viscosity ingredient is added first because it takes less power to mix this than it does to move a viscous mass in order to introduce a less viscous ingredient (Anon 2004a).

There have also been changes in the demands made on mixer performance by processors. For example:

- Mixers must meet new hygiene and safety standards, and increasingly should be sterilisable as well as being fully and easily cleaned. The development of a cantilevered mixer design with a full diameter access door (Fig. 4.4) enables the mixing zone to be isolated from the motor and bearings, and improved hygiene and cleaning (Anon 2001a).
- Increased awareness of allergens requires manufacturers to ensure that mixers are completely cleaned of for example gluten or nut pastes when changing between products.
- Expensive ingredients such as viscosity modifiers require specific mixing techniques to produce the required functionality in the product without over- or under-shear.
- Vitamins must be thoroughly dispersed in vitamin-enriched foods to ensure that the declared vitamin content is guaranteed in the production of some functional foods (Anon 2004b) (also Chapter 6, section 6.3).

Mixers have programmable logic controllers (PLCs; Chapter 27, section 27.2.2) to control electronic weighing of ingredients, pre-programmed for different recipes for rapid

Fig. 4.4   Cantilevered mixer (courtesy of the Kemutec Group at www.kemutec.com/htm/products/
products.htm).

change of products, on-screen diagnostics and management information (e.g. use of ingredients, reconciled with stock levels), and accurate energy control (Anon 2008a). Programmable controllers are used to control the mixing schedules for any number of variously sized tanks from one central location. Touch-screens are routinely used for data input and operator displays. Typically, displays show speed of rotation of the mixer shaft, drive torque, rotation count, mixer load weight, position of loading/unloading valves, blend time and product temperature; and alarms are activated if pre-set parameters are exceeded. There may also be failsafe devices such as a door interlock to stop the machine operating if the loading/unloading door is not properly secured, or an automatic emergency stop if an operator opens a safety gate or gets too close to a machine and breaks a light curtain. Pneumatic control systems that contain no electrical components meet all requirements for mixers operating in hazardous environments.

Mixers can be grouped into types that are suitable for:

- dry powders or particulate solids;
- low- or medium-viscosity liquids;
- high-viscosity liquids and pastes; and
- dispersion of powders in liquids.

A summary of selected types of mixers in each group is described below.

Mixers for dry powders or particulate solids
These mixers have two basic designs: a tumbling action inside rotating vessels and the positive movement of materials in screw-type mixers. They are used for blending cereals or flours, dried foods, flavourings, spices, sugars, instant drink mixes, cake mixes, instant potato and dried soups. Tumbling mixers rotate at speeds of 4–60 rpm and the powders are mixed as they fall through the vessel. The speed of the vessel is optimised for mixing a particular blend of ingredients, but speeds should not exceed the 'critical speed', when centrifugal force exceeds gravity and mixing ceases. Internal baffles or counter-rotating arms can be used to improve the efficiency of mixing. Different designs include drum, double-cone, cube, Y-cone, V-cone (Fig. 4.5), and rotating stainless steel pans, similar to a cement mixer. In the Y-cone mixer, the powders are also divided into two portions each time the arms of the 'Y' are lowered and remixed when they are raised during the next rotation. The process of continually dividing and returning the powder gives very efficient mixing. Most types of tumbling

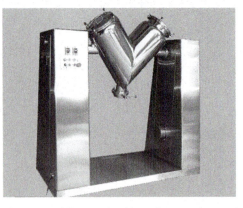

Fig. 4.5   V-cone blender (courtesy of JDA packaging and processing equipment at www.jdapro-gress.com/newribbon.htm).

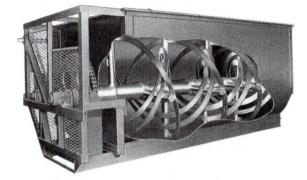

Fig. 4.6   Ribbon mixer (courtesy of S. Howes. at www.showes.com).

mixers have the facility to add sprays of liquid ingredients and some can be operated under partial vacuum. A cryogenic drum mixer, cooled with liquid nitrogen, is described by Anon (2002) for mixing frozen cheese starter cultures. If hazardous products (e.g. ground sugar or cornflour) are mixed, the mixer should be designed to minimise the risk of an explosion or fire (see also Chapter 3, section 3.1.3). Cone mixers are also used as butter churns and a 'dragée' coating pan is used for forming and coating applications (section 4.2.3 and Chapter 24, section 24.2.3).

Screw-type mixers, including U-trough mixers or ribbon mixers (Fig. 4.6), have two or more narrow metal blades formed into helices that counter-rotate in a closed horizontal U-shaped trough. In all designs, the ribbons or paddles are close-fitting to the trough, which ensures that all product is included in the mixing action and there is complete discharge of product on emptying (e.g. Anon 2004b). Improvements in mixer design enable the ribbon assembly to be removed as a single unit for rapid cleaning or changeover to a new product; bearings are sealed to prevent leakage; and the motor and drive mechanism are located outside the mixing vessel to ensure hygienic operation. The diameter, pitch and width of each ribbon are accurately proportioned so that material is moved in a predetermined pattern within the mixer. The movement of food back and forth in a continuous 'figure-of-8' pattern produces a rolling, folding action with vertical and lateral displacement to prevent localised accumulation of the product. There is also a shearing action in the interface between the opposing pitched ribbons. In some designs, a net forward movement of material is produced to convey it through the machine. These continuous mixers have paddle or plough type agitators (see paddle mixers below) instead of ribbons with a series of mixing stages that progressively move and blend ingredients. 'Double reversing' agitators are a hybrid between a paddle and a ribbon mixer. They provide the 'figure-of-8' movement of a paddle mixer with the folding action of a ribbon mixer (Fig. 4.7). This mixing action transfers more energy to the product and gives more efficient mixing. In most designs, mixing of dry, free-flowing materials typically takes 3 to 5 minutes. Amounts as small as 20% of the mixer capacity can be blended with the same accuracy as a full batch, thus allowing flexibility in batch sizes (Anon 2008b). Additive additions as low as 1% may be uniformly mixed and spray nozzles permit the controlled addition of liquid additives.

A vertical-screw mixer is a conical vessel that contains a rotating vertical screw, which orbits around a central axis to mix the contents. This type of equipment is particularly useful for the incorporation of small quantities of ingredients into a bulk material.

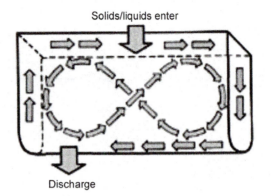

Solids/liquids enter

Discharge

Fig. 4.7 Flow pattern of double reversing agitator (information supplied by Marion Mixers Inc.)
(Anon 2008b).

Mixers for low- or medium-viscosity liquids
To adequately mix low-viscosity liquids, turbulence must be induced throughout the bulk of the liquid to entrain slow-moving parts within faster-moving parts. A vortex should be avoided because adjoining layers of liquid travel at a similar speed and simply rotate around the mixer so mixing does not take place. A large number of designs of agitator are used to mix liquids in baffled or unbaffled vessels. The advantages and limitations of each vary according to the particular application but are summarised in Table 4.2. Details of different designs of impeller mixers are given by Anon (2005c) and on-line software to calculate propeller mixer diameters for different applications is given by Anon (2007).

The simplest paddle agitators are wide, flat blades, which measure 50–75% of the vessel diameter and rotate at 20–150 rpm. The blades are often pitched to promote longitudinal and radial flow in unbaffled tanks. Impeller agitators consist of two or three blades attached to a rotating shaft. Turbine agitators are impeller agitators that have four or more blades mounted together. Their size is 30–50% of the diameter of the vessel and they operate at 30–500 rpm. Impellers that have short blades (less than a quarter of the diameter of the vessel) are known as propeller agitators and these operate at 400–1500 rpm. They are used for blending miscible liquids, diluting concentrated solutions (e.g. tomato paste for sauces or concentrated fruit purées for pastries and puddings), preparing syrups or brines or dissolving other ingredients (e.g. stabilising gums, CMC, guar or gelatine) and rehydrating powdered products (e.g. eggs, milk, whey, potato powder or flours for dips and batters).

To promote longitudinal and radial flow and to prevent vortex formation, the agitator is located in one of the positions shown in Fig. 4.2. Alternatively, baffles are fitted to the

Table 4.2 Advantages and limitations of selected liquid mixers

Type of mixer	Advantages	Limitations
Paddle agitator	Good radial and rotational flow, cheap	Poor perpendicular flow, high vortex risk at higher speeds
Multiple-paddle agitator	Good flow in all three directions	More expensive, higher energy requirements
Propeller impeller	Good flow in all three directions	More expensive than paddle agitator
Turbine agitator	Very good mixing	Expensive and risk of blockage

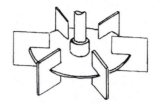

Fig. 4.8 Vaned disc impeller (after Smith 1985).

vessel wall to increase shearing of the liquids and to interrupt rotational flow, but care is necessary in the design to ensure that the vessel may be adequately cleaned (Chapter 27, section 27.1.1).

Blades may also be mounted on a flat disc (the 'vaned disc impeller'; Fig. 4.8), fitted vertically in baffled tanks. High shearing forces are developed at the edges of the impeller blades and they are therefore used for pre-mixing emulsions (Chapter 3, section 3.2.3).

A variety of mixers, including horizontal helical blade mixers (HHBM), paddle mixers (or 'pug mills') and plough mixers, are similar in design to, but more sturdy than, ribbon mixers. They have single or double shafts and blades, paddles or ploughs with adjustable pitch. The paddles lift, tumble, divide and circulate materials in an intense but gentle mixing action. These mixers are used for mixing medium-viscosity foods including chocolate, batters, pastes, slurries and emulsions; to mix dry materials with oils, binders or liquids; and to break down agglomerates. They have capacities ranging from 50 to 10 000 litres and can be heated, cooled or operated under pressure or vacuum.

Pumps also mix ingredients by creating turbulent flow both in the pump itself and in the pipework (Chapter 1, section 1.3.4). There are a large variety of pumps available for handling different fluids and suspensions: the different designs and applications are discussed in Chapter 27 (section 27.1.3).

Mixers for high-viscosity liquids and pastes
In high-viscosity liquids, pastes or doughs, mixing occurs by kneading the material against the vessel wall or into other material, folding unmixed food into the mixed part and shearing to stretch the material (Fig. 4.9). Efficient mixing is achieved by creating and recombining fresh surfaces in the food, but because the material does not easily flow, it is necessary to either move the mixer blades throughout the vessel or to move the food to the mixer blades.

Viscous liquids are mixed using slow-speed vertical-shaft impellers such as multiple-paddle (gate) agitators that develop high shearing forces. The basic design in this group is the 'anchor and gate' agitator. Some complex designs have arms on the gate that intermesh with stationary arms on the anchor to increase the shearing action, whereas others have inclined vertical blades to promote radial movement in the food. This type of equipment is also used with heated mixing vessels, when the anchor is fitted with scraper blades to prevent food from burning onto the hot surface. A development of the basic design has three separate agitators including an anchor, high-speed disperser and a rotor-stator homogeniser. Figure 4.10 shows a mixer with a helical anchor design that is suitable for higher-viscosity foods.

Planetary (or 'orbital') mixers take their name from the path followed by rotating blades that include all parts of the vessel in the mixing action (Fig. 4.11a). Blades rotating at 40–370 rpm may be located centrally in a static bowl, or blades are offset from the

Fig. 4.9   Experimental dough mixing with dye showing mixing action (photo by the author).

centre of a co-currently or counter-currently revolving vessel. In both types there is a small clearance between the blades and the vessel wall. Gate blades are used for mixing pastes, blending ingredients and preparation of spreads; hooks are used for dough mixing, and whisks are used for batter or sauce preparation. Details of different types of equipment are described by Anon (2005b). A patented design (Fig. 4.11b) combines a planetary blade and a high-speed dispersion blade. Both agitators rotate on their own axes and also rotate continuously around the vessel. The planetary blade feeds materials directly into the high-shear zone of the orbiting high-speed disperser. Both agitators have independently variable speeds and this combination of mixing actions replaces the need for multiple mixers (Anon 2006b).

Software used to simulate the action of a planetary mixer shows the trajectories of particles during turns of the blades, and is used to calculate the 'distributive mixing index' (a quantitative measure of mixing efficiency) (Anon 2006a). A more recent

Fig. 4.10   Mixer with multishaft helical anchor design (courtesy of Ross Food Mixers & Blenders Ltd) (Anon 2006b).

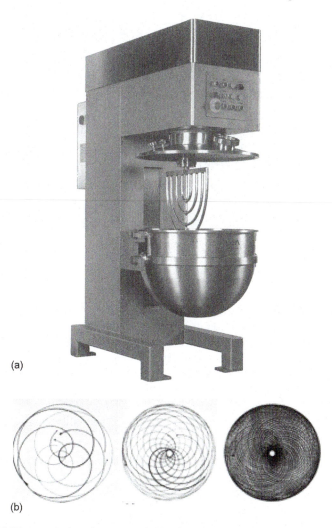

(a)

(b)

Fig. 4.11   (a) Planetary mixer (courtesy of GEA Pharma Systems). (b) Pattern of combined planetary blade and high-speed dispersion blade in planetary mixer operation (courtesy of Ross Food Mixers & Blenders Ltd) (Anon 2008c).

development is double planetary mixers, which are equipped with two vertical mixing blades to give double the rotation motion. Mixing is reported to be 30% faster than with other planetary mixers. They can operate under vacuum and a jacketed mixing tank enables temperature-controlled mixing for heat-sensitive products. All planetary mixers have a sealed gearbox to prevent contamination of foods with oil or grease. Sizes range from 10–1500 litres and they can mix products ranging in viscosity from a few thousand to several million centipoise (Anon 2008c).

The Z-blade (or sigma-blade) mixer (Fig. 4.12) consists of two heavy-duty blades that are mounted horizontally in a metal trough. The blades intermesh and rotate towards each other at either similar or different speeds (14–60 rpm) to produce high shearing forces, both between the blades and between the blades and the close fitting trough. These mixers are primarily used for dough mixing, but are also used for sugar pastes, chewing gum and marzipan. They use a substantial amount of power, which is dissipated in the

Fig. 4.12   Z-blade mixer (courtesy of Winkworth Machinery Ltd at www.mixer.co.uk/zbmix.html).

product as heat unless the walls of the trough are jacketed for temperature control. Special designs for shredding and mixing have serrated blades, and other blade configurations include gridlap, double naben and double claw (McDonagh 1987). Mathematical modelling and numerical simulations to predict the effectiveness of mixing using Z-blade mixers are described by Connelly (2006).

Rotor-stator mixers comprise a high speed (3600–10 000 rpm) centrifugal rotor, closely fitted into a slotted stationary casing. The advantages of high shear rotor-stator batch mixers over conventional stirrers or agitators arise from a four-stage mixing/shearing action when materials are drawn through the specially designed workhead (Fig. 4.13). The four stages are as follows:

(a) High-speed rotor blades develop a low pressure area that draws liquid and solid materials upward from the bottom of the vessel into a precision machined stator head.
(b) Centrifugal force moves materials to the periphery of the head where they are subjected to hydraulic shearing forces, and mechanical shearing between the ends of the rotor blades and the inner wall of the stator.
(c) The materials are forced out at high velocity through holes or slots in the stator and are subjected to intense mechanical shearing at the edges of rotor blades and the slots in the stator, which causes further mixing and particle size reduction.
(d) The materials expelled from the head are projected radially at high speed towards the sides of the mixing vessel and fresh material is continually drawn into the head. The horizontal expulsion and vertical suction into the head creates a circulation pattern that maintains the mixing cycle and minimises disturbance of the liquid surface, so reducing entrained air which would cause aeration.

Work-heads are easily interchangeable and allow a wide range of mixing operations including emulsifying, homogenising, disintegrating, dissolving, dispersing solids into liquids, blending, and breaking down solids and agglomerates. An example of a head is shown in Fig. 4.14(a) for general mixing applications, disintegration of solids and preparation of gels and thickeners, suspensions, solutions and slurries. A slotted head (Fig. 4.14b) is used to disintegrate fibrous materials such as animal and vegetable tissue. Other heads are suitable for the preparation of emulsions and fine colloidal suspensions

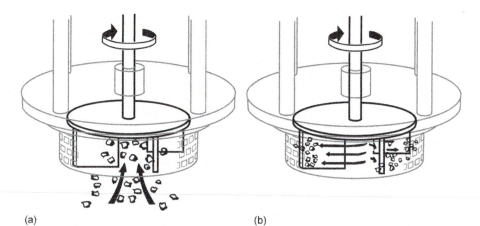

(a)   (b)

(c)   (d)

Fig. 4.13 Operation of rotor-stator mixer (courtesy of Silverson Machines Ltd) (Anon 2008d).

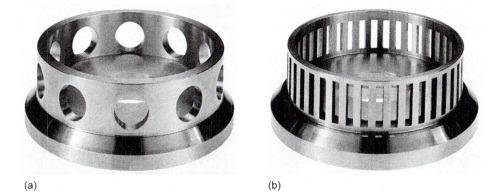

(a)   (b)

Fig. 4.14 Stator head designs: (a) for general mixing applications and (b) slotted head to disintegrate fibrous materials (courtesy of Silverson Machines Ltd) (Anon 2008d).

(Chapter 3, section 3.2.3). Compared with conventional mixers, processing times are reduced by up to 90% and product consistency and process efficiency are improved. They can operate at throughputs from 15 to 200 000 l h^{-1} (Anon 2008d).

In contrast to disc impellers, which rely mostly on hydraulic shear produced by very high tip speeds (24–27 m s^{-1}), the combined hydraulic and mechanical shearing action in rotor-stator mixers requires lower tip speeds (15–18 m s^{-1}) and therefore less power. Some new designs also have a revolving stator, driven by fluid friction, which moves at between one-tenth and one-fifteenth the speed of the rotor. This increases the pumping action of the mixer to increase flowrates while maintaining high shear rates at lower tip speeds (12–15 m s^{-1}), again reducing power consumption. Another development is a fixed rotor and stator combined as a single unit. Material is drawn by impellers into the mixing head from above and below, and it is sheared by grooves on the rotors. The two high-velocity counter-current flows meet in the stator and result in high levels of turbulence and hydraulic shearing. The hydraulic pressure forces the food to the periphery of the stator, where it is subjected to mechanical shear as it passes through sharpened slots. This type of mixer is designed for high-viscosity mixes that cannot be mixed using a conventional agitator but do not require the intense shearing action of a rotor-stator mixer. It can rapidly incorporate large volumes (up to 15 000 kg h^{-1}) of powders into liquids, achieving a consistent homogeneous product – one of the most difficult of all mixing applications. The single-piece mixing head has a hygienic design for cleaning-in-place (CIP) or sterilising-in-place (SIP) and low maintenance because of no wearing parts. It also has reduced power requirements compared with conventional high shear mixers (Ames 2000). Typical operating conditions for these types of mixers are shown in Table 4.3.

Static or 'motionless' mixers have been developed to mix viscous materials and fluids, or to incorporate powders with liquids. These mixers comprise a series of precisely aligned static mixing elements (Fig. 4.15) that are contained within pipework in the processing line. The elements split, rotate and integrate the food material in a precisely defined pattern, according to the type of food to be mixed and the degree of mixing required. In 'low-pressure drop' motionless mixers, semi-elliptical plates are positioned

Table 4.3   Summary of operating conditions for low-shear and high-shear mixers

Condition	Type of equipment	Value
Tip speed – high shear	Open disc impeller	24–27 m s^{-1}
	In-line	27–33.5 m s^{-1}
	Closed rotating rotor-stator	15–18 m s^{-1}
	Rotor-stator with revolving stator	12–18 m s^{-1}
	Fixed rotor and stator	12–18 m s^{-1}
Tip speed – low shear	–	3–9 m s^{-1}
Geometric similarity	Low speed/shear	0.25–0.60
(ratio of diameters of	High speed/shear	0.10–0.20
mixing head and tank)		
Bulk fluid velocity	Slow mixing/high viscosity	0.12–0.18 m s^{-1}
	Vigorous mixing – most applications	0.18–0.24 m s^{-1}
	Vigorous mixing – difficult applications	0.24–0.3 m s^{-1}
	Violent mixing	0.3–0.45 m s^{-1}
Tank turnover	Low viscosity (1–100 cps) < 1100 litres	4–6 per min
	High viscosity (500–5000 cps) < 1100 litres	2–4 per min
	Low viscosity (1–100 cps) 1100–2200 litres	2–4 per min
	High viscosity (500–5000 cps) 1100–2200 litres	1–2 per min

Adapted from Ames (2000), Beaudette (2001) and Anon (2001b)

Fig. 4.15   In-line motionless mixer element (courtesy of Komax Systems Inc. at www.komax.-com/det-fdsanitary.html).

in a tubular housing and are best suited for low-viscosity turbulent flow mixing and blending applications. 'Interfacial surface generator' motionless mixers are used for high-viscosity laminar flow applications. This design generates over 2 million mathematically predictable layers using ten mixing elements (Anon 2006c).

Motionless mixers operate using three mixing actions: radial mixing, flow division and transient mixing. In radial mixing, the food is deflected by the elements through a series of 180° rotations that force it from the centre to the wall of the pipe and back again. In flow division, the material is split into two components by the first mixing element and then rotated through 180° before being split into four streams by the second element and so on past succeeding elements until the required degree of mixing has been achieved. Transient mixing employs spaces between the elements to allow for relaxation of viscous material after successive radial mixings. These mixers eliminate the need for tanks, agitators and moving parts, thus reducing capital costs and maintenance requirements. They have been used in chocolate manufacture for the processing of cocoa mass (Richards 1997) and for blending ingredients to give uniform temperature or concentration profiles. Some types of mixers are more suited to turbulent flow or laminar flow (Chapter 1, section 1.3.4), whereas others operate with a wide range of fluid properties (Anon 2006c). The development and application of motionless mixers is reviewed by Gyenis (2002).

A number of other types of equipment, including bowl choppers, roller mills and colloid mills (Chapter 3, section 3.2.3), are suitable for mixing high-viscosity materials and are used in specific applications, often with simultaneous homogenisation or size reduction. In extrusion (Chapter 15) and butter or margarine manufacture, single or twin screws are used to convey viscous foods and pastes through a barrel and to force it through perforated plates. The small clearance between the screw and the barrel wall causes a shearing and kneading action, which is supplemented by shearing and mixing as the food emerges from the end plate.

4.1.4   Effect on foods and micro-organisms

The main effect of mixing is to increase the uniformity of products by evenly distributing ingredients throughout the bulk. Changes in nutritional value are therefore dependent on the amount and types of ingredients that are added rather than the mixing action per se. The action of a mixer has no direct effect on the shelf-life of a food, but it may have an indirect effect by intimately mixing added components and causing them to react together. The nature and extent of interactions depend on the components involved, but may be accelerated if significant heat is generated in the mixer. In general, mixing has a substantial effect on sensory qualities and functional properties of foods, and these changes are often one of the expected outcomes of the operation (e.g. gluten development in dough mixing to produce the desired texture in the bread). There is little published information on the effect of mixing operations on micro-organisms. It is unlikely that the shearing conditions or temperature in a mixer would reduce the number of contaminating micro-organisms and hence mixing does not have a preservative effect. In some instances, especially where the temperature of the food is allowed to rise during mixing, there may be an increase in numbers of microbial contaminants, caused in part by the greater availability of nutrients as a result of the mixing action. Proper sanitation and HACCP procedures both to control levels of contaminants in ingredients and to adequately clean mixing equipment are therefore essential to maintain the microbial quality of foods after mixing (also Chapter 1, section 1.5.1).

4.2   Forming

There are many designs of forming and moulding equipment that are made specifically for individual products. In this section the equipment used for bread, biscuits, pies, snackfoods and confectionery is described. Other forming operations using extruders are described in Chapter 15 (section 15.3)

4.2.1   Bread moulders

After mixing, bread dough is formed into the shape(s) required to produce the finished loaf. It is first prepared into pieces using a dough divider. An intelligent servo-controlled ram pushes the dough over a knife in a 'dividing box'. Software controls the ram so that it responds to changes in the dough consistency, giving a uniform scaling accuracy (e.g. having a standard weight deviation of 2.5–3.5 g for an 800 g loaf) with outputs varying from 2400 to 14 400 pieces per hour and dough weights from 120 to 1600 g. Servo-control (Appendix D.1) of the weight adjustment mechanism has an interface with a checkweigher to ensure consistent weight of dough pieces (Anon 2008a). The divided dough pieces then pass to either a conical or cylindrical moulder (Fig. 4.16), where they are formed into ball shapes.

A 'moulder panner' shapes dough into cylinders that will expand to the required loaf shape when proofed. It consists of a pre-shaping roller and two to four pairs of sheeting rollers that have successively smaller gaps to roll the dough gently into sheets without tearing it. The pressure is gradually increased to expel trapped air. A variety of designs are used to change the direction of the sheet and roll the trailing edge first. This prevents compression of the dough structure that would cause the moisture content to increase at the trailing end of the sheet. Each pair of rollers has individually controlled variable speeds and gaps to control the size, shape and length-to-width ratio of the dough. The sheet is then curled, rolled into a cylinder and sealed by either a revolving drum, which presses the dough against a pressure plate, or a moulding conveyor with a hinged pressure board. New designs have control software that compensates for dough variations to maintain an even pressure on the dough, and built-in weight control using feedback from checkweighers. The moulded dough pieces are fed into a panner, which deposits them into baking tins. Gentle dough handling retains air cells in the dough structure, which produces bread that has a finer, more resilient crumb structure and better colour. Greater

Fig. 4.16   Conical moulder (courtesy of Baker Perkins Ltd) (Anon 2008a).

gas retention also reduces the costs of ingredients such as improvers and yeast (Anon 2008a).

4.2.2   Pie, tart and biscuit formers

The pastry dough for pie and tart casings is produced by passing it through two or three pairs of chilled 'sheeting' and 'gauge' rolls that are polished to a mirror finish. Casings are formed by depositing pieces of dough into aluminium foil containers, tins or re-usable moulds (Fig. 4.17a). The dough is then pressed to form the pie base using a 'blocking unit' (Fig. 4.17b) and the filling is added into the casing (Fig. 4.17c). A continuous sheet of dough is laid over the top, from which reciprocating blades cut the lids.

Biscuits are formed by one of four methods:

1   A dough sheet is fed from sheeting/gauge rollers and is pressed into shaped cavities in a metal moulding roller (known as die forming) (Fig. 4.18a).
2   Twin cutting rollers cut shapes from a sheet of dough and simultaneously imprint a design on the upper surface of the biscuit using raised characters (Fig. 4.18b).
3   Soft dough is extruded through dies in a wire-cut machine (Fig. 4.18c).
4   A continuous ribbon of dough is extruded from a rout press (similar to a wire-cut machine but without the cutting wires) and the ribbon is then cut to the required length using a reciprocating blade.

There are also numerous designs of equipment for laminating sheets of dough with fat (for croissants and pastries), folding doughs (to form pasties and rolls) and filling doughs (to form sausage rolls, fruit bars such as 'fig rolls', and cakes). These are described by Levine and Drew (1994) and a large number of equipment manufacturers (e.g. Anon 2008e, 2008f). Equipment for forming and encasing balls of dough with other materials is

(a)   (b)   (c)

Fig. 4.17 Stages in pie filling: (a) dough deposited into foil or tins, (b) blocking unit presses dough to form pie base, (c) filling deposited into pie base (courtesy of Rijkaart Ltd) (Anon 2008e).

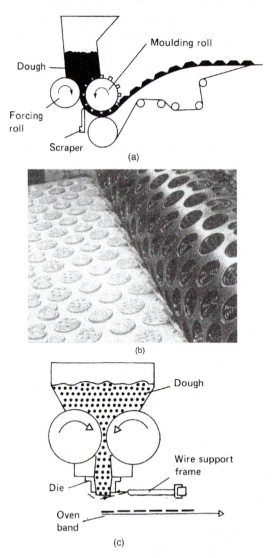

Fig. 4.18 Biscuit formers: (a) rotary moulder, (b) moulding roller and (c) wire-cut machine (courtesy of Baker Perkins Ltd) (Anon 2008a).

described by Hayashi (1989). In this process, the inner material and outer material are co-extruded and then divided and shaped by two 'encrusting discs' (Fig. 4.19a). In contrast to conventional forming techniques, where the size of the product is determined by the size of the feed material, the relative thickness of the outer layer and the diameter of the inner sphere are determined by the flowrate of each material. It is therefore possible to alter the relative thickness of inner and outer layers (Fig. 4.19b) simply by adjusting the flowrates, giving a high degree of flexibility for the production of different products. This equipment was developed in Japan for production of cakes having an outer layer of rice dough and filled with bean paste, but they have found wide application and are used to produce sweetbreads filled with jam, doughnuts, meat pies, hamburgers filled with cheese and fish filled with vegetables.

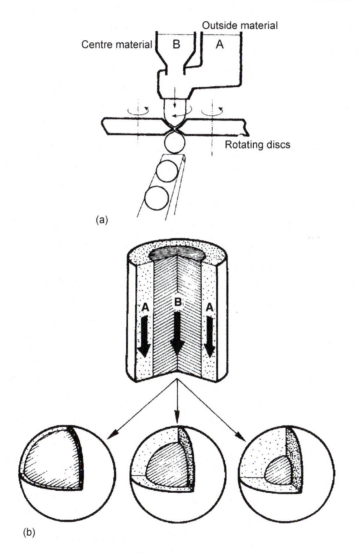

Fig. 4.19   (a) Two revolving crusting discs continuously divide food and shape it into balls. (b) Differences in thickness of outer layer (A) and inner layer (B) result from different material flowrates (courtesy of Hayashi 1989).

4.2.3   Confectionery moulders and depositors

There is a very large range of confectionery products that can be grouped according to differences in their texture into the following:

- High-boiled sweets and lozenges, including brittles (e.g. peanut crunch), humbugs, butterscotch, Edinburgh rock and 'sugar mice'. Boiled sweets may also have a wide variety of centres made from for example fudge, effervescent sherbet powder, chocolate, liquorice, caramel, or pastes made from hazelnut, fruits, spearmint, coconut, almond, etc.
- Toffees and caramels.
- Chocolate products, including solid bars, coins, buttons or chips, multi-coloured or marbled chocolate, fondant crème-filled chocolate, solid chocolate with inclusions such as fruit, nuts, biscuit or puffed rice, hollow goods (e.g. Easter eggs) and aerated products.
- Fudges, fondants, chews, gums, liquorice, pastilles, jellies, marshmallow, nougats and chewing gum or bubble gum.

Confectionery is formed using four systems: those that use moulds to form the shape of the product; those that form a 'rope' of product that is then cut into pieces; depositors that place a measured amount of product onto a flat belt; and sugar panning. Details of coatings used in confectionery production are given in Chapter 24 (section 24.2.3) and the textural characteristics of different confectionery hydrocolloids are shown in Table 4.4 (see also Appendix A.1). Extruded products are described in Chapter 15 (section 15.3).

Moulding equipment

Confectionery moulding equipment consists of individual moulds that have the required size and shape for a specific product, attached to a continuous conveyor (Fig. 4.20). They are carried below a depositor (a type of piston filler (Chapter 26, section 26.1.1)), which accurately deposits the required volume of hot sugar mass into each mould. Food can also be deposited in layers, or centre filled (e.g. liquid centres or chocolate paste in hard-

Table 4.4   Some characteristics of hydrocolloids used in confectionery products

Type	Agar	Gelatin	Gum arabic	Pectin	Starch
Source	Red seaweed	Animal tissues	Acacia tree	Apple pomace or citrus peels	Maize
Characteristic:					
Usage levels in confectionery products (%)	1–2	6–10	20–50	1–2	10–30
Solubilisation temperature (°C)	90–95	50–60	20–25	70–85	70–85
Setting temperature (°C)	35–40	30–35	20–35	75–85	20–35
Setting time (h)	12–16	12–16	24	1	12
Textural characteristics	Short, tender	Elastic, firm	Very firm	Short, tender, clean bite	Soft to firm, chewy
Used in combination with:	Starch, gelatin	Pectin, starch, gum arabic	Starch, gelatin	Starch, gelatin	Gelatin, gum arabic, pectin

From Carr *et al.* (1995)

Fig. 4.20   Confectionery moulding (courtesy of CMI 2003 Ltd at www.c-m-i.org.uk/cdrange.htm).

boiled sweets). The confectionery is then cooled in a cooling tunnel. When it has hardened sufficiently, individual sweets are ejected and the moulds restart the cycle (Fig. 4.21) (Verity 1991). Details of chocolate moulding are given by Perreau (1989).

The three main types of moulding equipment differ in the method of ejection and the material used for the mould:

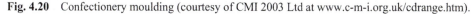

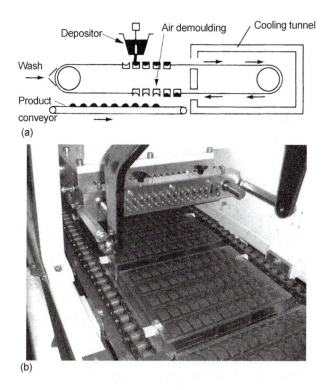

Fig. 4.21   Confectionery moulding: (a) an air demoulding depositor (courtesy of Baker Perkins) (Anon 2008g); (b) depositing chocolate in moulds (courtesy of CMI 2003 Ltd at www.c-m-i.org.uk/cdrange.htm).

1 Metal or polycarbonate moulds fitted with ejector pins are used to produce a wide range of multi-colour and multi-component hard confectionery in up to four colours, in stripes, layers or random patterns.
2 Flexible polyvinyl chloride moulds, which eject the food by mechanical deformation of the mould, are used for soft confectionery (e.g. toffee, fudge, jellies, caramel and fondant).
3 Polytetrafluoroethylene (PTFE)-coated aluminium moulds that have compressed-air ejection are used for jellies, gums, fondant and crèmes. The aluminium moulds optimise heat transfer, cooling times and energy usage.

Moulding gives high dimensional and weight accuracy, negligible waste, efficient wrapping, high levels of hygiene and low maintenance. Each type of equipment is automatically controlled and more than 3000 pieces per minute can be made, giving outputs of up to $1000\,kg\,h^{-1}$ for hard confectionery and up to $720\,kg\,h^{-1}$ for toffee, fondant or fudge (Anon 2008g).

Developments in confectionery moulds using 3D-software and laser technology include polycarbonate moulds, one-shot moulds, one-shot double moulds, frozen cone moulds, moulds with electronic chips, silicon moulds and spinning moulds. A chocolate manufacturer has produced a patented 'Hologram Bar' to show two different messages on chocolate. The intelligent chocolate mould has an electronic chip, which gives processors the possibility of collecting digitised production data on the moulding line (Anon 2008h). Capacities range from 600 to $1500\,kg\,h^{-1}$ (Anon 2008i). Chocolate used for coatings, toppings and fat-based pastes is mechanically treated and cooled by a scraped surface heat exchanger to 22 °C, which is the lowest possible depositing temperature. Depending on the product, no further cooling is required after depositing (Anon 2008j).

Starch moulding can be done by hand at a small scale or it is used in large-scale plants known as 'moguls'. Moulding starch is imprinted with the shape of the required jellies or gums and boiled sugar mass is deposited into the starch to form the shape of the confectionery. Moulding starch contains a small amount of mineral oil, which causes it to hold its shape and prevents dust forming. The moisture content of moulding starch is typically about 6% to ensure that excess moisture is adsorbed from the confectionery product into the starch.

Extruded confectionery 'ropes'
A second type of forming equipment extrudes sugar confectionery and shapes it using a series of rollers to produce a sugar 'rope'. Individual sweets are then cut from the rope and shaped by dies and conveyed to wrapping machines for packaging.

Depositors
Microprocessor-controlled depositors are used to form a wide variety of shapes of confectionery products (and also meringues and cake batters). Typically, a depositor comprises a manifold that has a row of depositing heads over a conveyor (Fig. 4.22). They are controlled by servo-motors, and the three-axis movement (up and down, forward and backward and in a sideways direction) enables a wide range of deposit patterns, fillings and weights of different products to be produced on the same line (e.g. balls, pretzels, Christmas trees, doughnuts and animal shapes such as chickens), as well as jam-filled or other encapsulated products. The depositor is PLC-controlled and can be programmed to produce different products, which are called up by the operator using a touch sensitive screen or keypad. The PLC provides full process visualisation, recipe

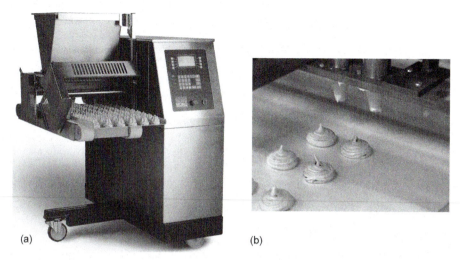

Fig. 4.22 Depositor: (a) machine (courtesy of MONO Equipment at www.monoequip.com); (b) depositing process (courtesy of JRAC Innovation LLC at www.jracinnovation.com/html/products.html).

management, alarms and data logging. The system can be integrated with upstream and downstream equipment, or into a SCADA plant supervisory system (Chapter 27, section 27.2.2).

Sugar panning and coating foods
Panning is the process of forming multiple thin layers of sugar, glucose compound or chocolate onto centres of fondant, fruit, nuts, etc., using a revolving copper or stainless steel 'dragée pan'. Other foods are formed by coating them with sweet or savoury coating materials (see Chapter 24, section 24.2.3).

References

AMES, G., (2000), High shear mixing advances for foods, pharmaceuticals, cosmetics. *Mixing, Blending and Size Reduction Handbook*, pp. 2–4, Admix Inc., available at www.admix.com/pdfs/MixHandbook.pdf.

ANON (2001a), New concept eliminates bearings and seals, *Food Processing*, Feb, p. 42. Further information available at Kemutec Group at www.kemutec.com/htm/products/gardner/mixers/plough.htm.

ANON, (2001b), Mixing scale-up speeds new products to market, *Mixing, Blending and Size Reduction Handbook*, pp. 5–7, Admix Inc., available at www.admix.com/pdfs/MixHandbook.pdf.

ANON, (2002), Cryogenic blending, *Food Processing*, Feb., 18.

ANON, (2004a), Mixing – the shear truth, *Food Processing*, Oct., 17–18.

ANON, (2004b), J R Boone develops special mixer for Adams foods, Processingtalk, available at www.processingtalk.com/news/bon/bon100.html. A similar report appears in *Food Processing*, 2005, Feb., 22.

ANON, (2005a), Power number of the impellers in standard tank geometries, Post Mixing Optimisation and Solutions information, available at www.postmixing.com/mixing%20forum/Macro/Fluid%20Motion/process_intensifier/power_number.htm.

ANON, (2005b), Information from Premier Mill at www.spxprocessequipment.com/sites/premiermill/planetary.asp.

ANON, (2005c), Impellers, Information from Post Mixing Optimization and Solutions, available at www.postmixing.com/mixing%20forum/impellers/impellers.htm.

ANON, (2006a), information supplied by Fluent Inc., available at www.fluent.com/solutions/brochures/size-distrib-broch-SIng-Double-final-06.pdf and at www.fluent.com/solutions/food/index.htm.

ANON, (2006b), PowerMix, company information from Charles Ross and Son Company, available at www.mixers.com/Proddetails.asp?ProdID=81.

ANON, (2006c), Motionless/Static Mixers, from Charles Ross & Son Company, information available at www.mixers.com/company.asp, and follow link to products > static mixers.

ANON, (2007) Information from AJ Design Software, available at www.ajdesigner.com/phpmixing/propeller_mixing_power_laminar_impeller_diameter.php.

ANON, (2008a), Product information from Baker Perkins Ltd available at www.bakerperkinsgroup.com/bakery, select 'Systems and equipment'.

ANON, (2008b), Information from Marion Mixers Inc., available at www.marionmixers.com/.

ANON, (2008c), Planetary Mixers, Information from Charles Ross & Son, available at www.planetarymixers.com/ and follow link to multi-shaft mixers.

ANON, (2008d), Information from Silverson Machines Ltd, available at www.silverson.com/UK/Products/BottomEntry-Parts.cfm.

ANON, (2008e), Product information from Capway Rijkaart Food Processing Equipment, available at www.rijkaart.net/, and follow links to equipment.

ANON, (2008f), Product information from Werner & Pfleiderer Industrielle Backtechnik, available at www.wpib.de/English/Products/Products_e.htm.

ANON, (2008g), Innovative confectionery equipment, Product information from APV Baker Ltd, available at http://www.bakerperkinsgroup.com/confectionery/, and follow link to Innovative confectionery equipment.

ANON, (2008h), Information from Agathon GmbH & Co. KG, available at www.agathon-moulds.com/?rubrika=1139.

ANON, (2008i), company information from S.A. Martin Lloveras, available at www.lloveras.com/scripts/web/home_e.asp, and follow link to machines/equipment.

ANON, (2008j), Information from F.B. Lehmann Maschinenfabrik GmbH available at www.fblehmann.de/frameset_e.htm, and follow link to chocolate.

BEAUDETTE, L., (2001), Successful scale-up of high speed batch and in-line mixers. Company information from Admix Inc., available at www.admix.com/pdfs/MixerScaleUpBooklet.pdf.

CAMPBELL, G.M., (1995), New mixing technology for the food industry, in (A. Turner, Ed.), *Food Technology International Europe*, Sterling Publications International, London, pp. 119–122.

CARR, J.M., SUFFERLING, K. and POPPE, J., (1995), Hydrocolloids and their use in the confectionery industry, *Food Technology*, July, 41–44.

CONNELLY, R., (2006) Looking inside dough mixers, *Fluent News*, Spring, available at ww.fluent.com/about/news/newsletters/06v15i1/a13.pdf.

CUQ, B., DANDRIEU, L.E., CASSAN, D. and MOREL, M.H. (2006), Impact of particles characteristics and mixing conditions on wheat flour agglomeration behaviour. Topical W: Fifth World Congress on Particle Technology, 24–27 April, available at http://aiche.confex.com/aiche/s06/techprogram/P35385.HTM.

ESPINOSA-SOLARES, T., BRITO-DE LA FUENTE, E., TECANTE A. and TANGUY P.A., (1997), Power consumption of a dual turbine–helical ribbon impeller mixer in ungassed conditions, *Chemical Engineering Journal*, **67** (3), 215–219.

GYENIS, J., (2002), Motionless mixers in bulk solids treatments – a review, *KONA*, **20**, 9–23, available at www.kona.or.jp/search/20_009.pdf.

HAYASHI, T., (1989), Structure, texture and taste, in (A. Turner, Ed.), *Food Technology International Europe*, Sterling Publications International, London, pp. 53–56.

ICARD-VERNIERE, C., FEILLET, P., COLONNA, P. and GUILBERT, S., (1999), Mixing conditions and pasta dough development. (Meeting, Montpellier, France, 28–30 September 1998, INRA, Paris, France). *Biopolymer Science: Food and Non-food Applications*, **91**, 229–233. Abstract available at http://cat.inist.fr/?aModele=afficheN&cpsidt=1828370.

LEVINE, L. and DREW, B.A., (1994), Sheeting of cookie and cracker doughs, in (H. Faridi, Ed.), *The Science of Cookie and Cracker Production*, Chapman Hall, New York, pp. 353–386.

LEWIS, M.J., (1990), *Physical Properties of Foods and Food Processing Systems*, Woodhead Publishing, Cambridge.

LINDLEY, J.A., (1991a), Mixing processes for agricultural and food materials: 1 Fundamentals of mixing, *J. Agricultural and Engineering Research*, **48**, 153–170.

LINDLEY, J.A., (1991b), Mixing processes for agricultural and food materials: 2: Highly viscous liquids and cohesive materials, *J. Agricultural and Engineering Research*, **48**, 229–247.

LINDLEY, J.A., (1991c), Mixing processes for agricultural and food materials: 3: Powders and particulates, *J. Agricultural and Engineering Research*, **48**, 1–19.

MCDONAGH, M., (1987), Mixers for powder/liquid dispersions, *The Chemical Engineer*, March, 29–32.

MICRON, H., (2002), Avoiding blender blunders, *Food Processing*, July, pp. 24–25.

PAUL, E.L., ATIEMO-OBENG, V.A. and KRESTA, S.M. (Eds.), (2003), *Handbook of Industrial Mixing: Science and Practice*, John Wiley & Sons, Hoboken, NJ.

PERREAU, R.G., (1989), The chocolate moulding process, in (A. Turner, Ed.), *Food Technology International Europe*, Sterling Publications International, London, pp. 73–76.

REDL, A., MOREL, M.H., BONICEL, J., GUILBERT, S. and VERGNES, B., (1999), Rheological properties of gluten plasticized with glycerol: dependence on temperature, glycerol content and mixing conditions, *Rheologica Acta*, **38** (4), pp. 311–320, abstract available at http://cat.inist.fr/?aModele=afficheN&cpsidt=1969660.

RICHARDS, G., (1997), Motionless mixing efficiency with economy, *Food Processing*, June, 29–30.

RUSHTON, J. N., COSTICH, E.W. and EVERETT, H.S., (1950), *Chemical Engineering Progress*, **46**, 395.

SMITH, T., (1985), Mixing heads, *Food Processing*, Feb., 39–40.

VERITY, R., (1991), Confectionery manufacture and starchless moulding, in (A. Turner, Ed.), *Food Technology International Europe*, Sterling Publications International, London, pp. 97–99.

5

Separation and concentration of food components

Abstract: Extraction or separation of food components is used to prepare products such as fruit juices, cream or cooking oils, to produce sugar or gelatin for use as ingredients in other processes, or to retrieve high-value compounds, including essential oils and enzymes. This chapter describes the theory and equipment that are used for physical separation of food components by centrifugation, filtration, expression, solvent extraction and membrane separation.

Key words: centrifugation, filtration, expression, extraction, supercritical carbon dioxide, reverse osmosis, ultrafiltration, ion exchange membranes.

Foods are complex mixtures of compounds and the extraction or separation of food components is fundamental for the preparation of ingredients to be used in other processes (e.g. cooking oils from oilseeds, sugar from cane or beet, or gelatin from connective tissue); or for retrieval of high-value compounds, such as essential oils and enzymes (e.g. papain from papaya for meat tenderisation or rennet from calf stomachs for cheese making). Other types of separation methods are:

- those used to clean foods by separating contaminating materials (Chapter 2, section 2.2);
- those used to sort foods by separating them into classes based on size, colour or shape (Chapter 2, section 2.3); or
- those used to selectively remove water from foods using heat by evaporation (Chapter 14, section 14.1) or by dehydration (Chapter 16), by crystallisation (described by Brennan *et al.* 1990 and Heldman and Hartel 1997a) or alcohol by distillation (Chapter 14, section 14.2). Osmotic dehydration of fruits and vegetables, by soaking in concentrated solutions of sugar or salt respectively, is described in Chapter 16 (section 16.1.3).

In this chapter, the unit operations that are used for the physical separation of food components by centrifugation, filtration, expression, solvent extraction and membrane separation are described. There are two main categories:

1 Separation of liquids and solids where either one or both components may be valuable (e.g. fruit juices, pectin and coffee solubles), or liquid–liquid separation (e.g. cream and skimmed milk).

2 Separation of small amounts of solids from liquids. Here the main purpose is purification of water or clarification of liquids such as wine, beer, juices, etc. and the solids are not a product.

Each operation is used as an aid to processing and is not intended to preserve the food. Changes to both the organoleptic and nutritional qualities of products are caused by the intentional separation or concentration of food components, but generally the processing conditions cause little damage to these properties of foods.

5.1 Centrifugation

There are two main applications of centrifugation: separation of immiscible liquids and separation of solids from liquids. Separation of solid particles from air by centrifugal action in the 'cyclone' separator is described in more detail in the section describing spay drying in Chapter 16 (section 16.2.1) and by Heldman and Hartel (1997a).

5.1.1 Theory

If two immiscible liquids, or solid particles mixed in a liquid, are allowed to stand they separate due to the force of gravity (F_g) on the components. This can be expressed as:

$$F_g = mg \qquad \boxed{5.1}$$

where m (kg) = mass of the particle and g (9.81 m s^{-2}) = acceleration due to gravity.

This type of separation can be seen for example when yeast collects in the base of a wine fermenter or when cream floats to the surface of milk. However, separation may take place very slowly, especially if the specific gravities of the components are similar, or if forces are holding the components together (e.g. in an emulsion, Chapter 3, section 3.2.2). Commercially, short separation times and greater control of separation are required and this is achieved using centrifugal force in a centrifuge. In calculations, the centrifugal forces are much greater than gravity (up to 10 000 times greater in commercial centrifuges) and the effects of gravity are ignored.

Centrifugal force is generated when materials are rotated; the size of the force depends on the radius and speed of rotation and the density of the centrifuged material. When particles are removed from liquids in centrifugal clarification, the particles move to the bowl wall under centrifugal force. The centrifugal force is calculated using:

$$F_c = mr\omega^2 \qquad \boxed{5.2}$$

where F_c = centrifugal force acting on the particle, r (m) = radius of the path travelled by the particle, and ω (rad s^{-1}) is the angular velocity of the particle.

The angular velocity is related to the tangential velocity of the particle, v (m s^{-1}) by:

$$\omega = v/r \qquad \boxed{5.3}$$

Therefore:

$$F_c = (mv^2/r) \qquad \boxed{5.4}$$

Centrifuge speeds are normally expressed in revolutions per minute (N) (rpm), and Equation 5.3 can also be written as:

$$\omega = 2\pi N/60 \qquad \boxed{5.5}$$

Therefore:

$$F_c = mr(2\pi N/60)^2$$

5.6

or

$$F_c = 0.011\, mrN^2$$

5.7

Sample problem 5.1 shows the increase in acceleration in a centrifuge compared to gravity.

Sample problem 5.1
In a centrifuge operating at 2600 rpm, with a radius of 15 cm, calculate the size of the centrifugal force as a multiple of gravity.

Solution to sample problem 5.1
From Equations 5.1 and 5.7:

$$F_g = mg \quad \text{and} \quad F_c = 0.011\, mrN^2$$

Therefore:

$$F_c/F_g = (0.011rN^2)/g$$

$$= (0.011 \times 0.15 \times 2600^2)/9.81$$

$$= 1137$$

That is the centrifuge operates with a force that is more than 1100 times gravity.

As noted above, in a centrifuge, the force acting on a particle depends on the radius and speed of rotation and the mass of the particle. Because the radius and the speed of rotation have the same effect on all particles, it is differences in the mass of the particles that influence the force acting on them. Hence heavier (i.e. more dense) particles have greater force acting on them than less dense particles. As a result solids are forced to the periphery of centrifuge bowls (or in the case of immiscible liquids, the denser liquid moves to the bowl wall and the lighter liquid is displaced to an inner annulus (Fig. 5.1)).

If liquid flow is streamlined (Chapter 1, section 1.3.4), the rate of movement is determined by the densities of the particles and liquid, and the viscosity of the liquid (Equation 5.8). Earle (1983) describes separation under turbulent flow conditions:

$$Q = \frac{d^2\omega^2(\rho_s - \rho)V}{18\mu \ln(r_W/r_B)}$$

5.8

where Q (m^3 s^{-1}) = volumetric flowrate, d (m) = diameter of the particle, ω ($= 2\pi N/60$) = angular velocity, ρ_s (kg m^{-3}) = density of particles, ρ (kg m^{-3}) = density of liquid, V (m^3) = operating volume of the centrifuge, μ (N s m^{-2}) = viscosity of liquid, r_B (m) = radius of liquid, r_W (m) = radius of centrifuge bowl, N (rev s^{-1}) = speed of rotation.

For a given particle diameter, the average residence time of a suspension equals the time taken for a particle to travel through the liquid to the centrifuge wall:

$$t = \frac{V}{Q}$$

5.9

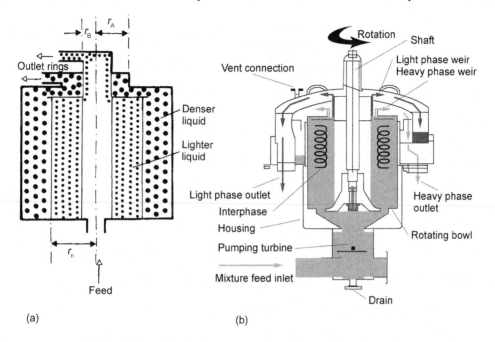

Fig. 5.1   Separation of immiscible liquids in a bowl centrifuge, r_A, radius of dense phase outlet, r_B, radius of light phase output, r_n, radius of neutral zone: (a) principle of operation; (b) components of centrifugal separator (courtesy of Rousselet-Robatel at www.rousselet-robatel.com) (Anon 2008a).

where t (s) = residence time. The flowrate can therefore be adjusted to retain a specific range of particle sizes. Derivations and additional details of these equations are given by Brennan *et al.* (1990) and Earle (1983).

Sample problem 5.2
Beer with a specific gravity of 1.042 and a viscosity of $1.40 \times 10^{-3}\,\mathrm{N\,s\,m^{-2}}$ contains 1.5% solids which have a density of $1160\,\mathrm{kg\,m^{-3}}$. It is clarified at the rate of $240\,\mathrm{l\,h^{-1}}$ in a bowl centrifuge which has an operating volume of $0.09\,\mathrm{m^3}$ and a speed of $10\,000\,\mathrm{rpm}$. The bowl has a diameter of $5.5\,\mathrm{cm}$ and is fitted with a $4\,\mathrm{cm}$ outlet. Calculate the effect on feedrate of an increase in bowl speed to $15\,000\,\mathrm{rpm}$ and minimum particle size that can be removed at the higher speed. All conditions except the bowl speed remain the same.

Solution to sample problem 5.2
From Equation 5.8:

$$\text{Initial flow rate } Q_1 = \frac{d^2(2\pi N_1/60)^2(\rho_s - \rho)V}{18\mu \ln(r_w/r_B)}$$

$$\text{New flow rate } Q_2 = \frac{d^2(2\pi N_2/60)^2(\rho_s - \rho)V}{18\mu \ln(r_w/r_B)}$$

As all conditions except the bowl speed remain the same,

$$\frac{Q_2}{Q_1} = \frac{(2\pi N_2/60)^2}{(2\pi N_1/60)^2}$$

$$Q_1 = 240 \; 1 \, h^{-1} = 0.24 \; m^3 \; h^{-1}$$
$$= 0.24/3600 \; m^3 \; s^{-1}$$
$$= 6.67 \times 10^{-5} \; m^3 \; s^{-1}$$

$$\frac{Q_2}{6.67 \times 10^{-5}} = \frac{(2 \times 3.142 \times 15\,000/60)^2}{(2 \times 3.142 \times 10\,000/60)^2}$$

Therefore:

$$Q_2 = 1.5 \times 10^{-4} \; m^3 \; s^{-1}$$
$$= 540 \; 1 \, h^{-1}$$

To find the minimum particle size from Equation 5.8:

$$d^2 = \frac{Q_2[18\mu \, \ln(r_w/r_B)]}{\omega^2(\rho_s - \rho)V}$$

$$= \frac{Q_2[18\mu \, \ln(r_w/r_B)]}{(2\pi N_2/60)^2(\rho_s - \rho)V}$$

$$= \frac{1.5 \times 10^{-4}(18 \times 1.40 \times 10^{-3} \times \ln(0.0275/0.02)}{(2 \times 3.142 \times (15\,000/60)^2(1160 - 1042) \times 0.09}$$

$$= 4.578 \times 10^{-14} \; m$$

Therefore diameter $= 2.13 \times 10^{-7}$ m $(0.213 \; \mu m)$

In liquid–liquid separations, the thickness of the annular rings in the centrifuge is determined by the density of the liquids, the pressure difference across the layers and the speed of rotation. A boundary region forms between the liquids at a radius r_n at a given centrifuge speed, where the hydrostatic pressure of the two layers is equal. This is termed the 'neutral zone' and is important in equipment design to determine the position of feed and discharge pipes. It is found using:

$$r_n^2 = \frac{\rho_A r_A^2 - \rho_B r_B^2}{\rho_A - \rho_B} \qquad \boxed{5.10}$$

where ρ (kg m^{-3}) = density and r (m) = the radius. The subscripts A and B refer to the dense and light liquid layers respectively.

If the purpose is to remove light liquid from a mass of heavier liquid (e.g. separating cream from milk), the residence time in the outer layer exceeds that in the inner layer. This is achieved by using a smaller radius of the outer layer (r_A in Fig. 5.1) and hence reducing the radius of the neutral zone. Conversely, if a dense liquid is to be separated from a mass of lighter liquid (e.g. the removal of water from oils), the radius of the outer layer (and the neutral zone) is increased.

Sample problem 5.3

A bowl centrifuge is used to break an oil-in-water emulsion (Chapter 4). Determine the radius of the neutral zone in order to position the feed pipe correctly. (Assume that the density of the continuous phase is $1000\,kg\,m^{-3}$ and the density of the oil is $870\,kg\,m^{-3}$. The outlet radii from the centrifuge are 3 and 4.5 cm.)

Solution to sample problem 5.3

From Equation 5.10:

$$r_n = \sqrt{\left[\frac{1000(0.045)^2 - 870(0.03)^2}{1000 - 870}\right]}$$

$$= \sqrt{\left(\frac{2.025 - 0.783}{130}\right)}$$

$$= 0.098 \text{ m or } 9.8 \text{ cm}$$

5.1.2 Equipment

Centrifuges are classified into three groups for:

1 separation of immiscible liquids;
2 clarification of liquids by removal of small amounts of solids (centrifugal clarifiers); and
3 removal of solids (desludging, decanting or dewatering centrifuges).

Separation of immiscible liquids

The simplest type of equipment is the tubular bowl centrifuge. It consists of a vertical cylinder (or bowl), typically 0.1 m in diameter and 0.75 m long, which rotates inside a stationary casing at between 15 000 and 50 000 rpm depending on the diameter. Feed liquor is introduced continuously at the base of the bowl and the two liquids are separated and discharged through a circular weir system into stationary outlets (Fig. 5.1).

Better separation is obtained by the thinner layers of liquid formed in the disc bowl centrifuge (Fig. 5.2). Here a cylindrical bowl, 0.2–1.2 m in diameter, contains a stack of inverted metal cones that have a fixed clearance of 0.5–1.27 mm and rotate at 2000–7000 rpm. The cones substantially increase the area available for separation and have matching holes that form flow channels for liquid movement. Feed is introduced at the base of the disc stack and an inlet zone is used to accelerate the liquid up to the speed of the rotating bowl. This reduces the shear forces on the product, prevents foaming and minimises temperature increases. The denser fraction moves towards the wall of the bowl, along the underside of the discs. The lighter fraction is displaced towards the centre along the upper surfaces. Both liquid streams are removed continuously by either a weir system at the top of the centrifuge (in a similar way to the tubular bowl system) or by a stationary 'paring' disc that decelerates the rotating liquid and transforms the kinetic energy into pressure energy, thus pumping the liquid out of the centrifuge. Changing the disc configuration and shape enables a wide range of liquids or solids to be separated. Disc bowl centrifuges are used to separate cream from milk, to remove gums and traces of water from vegetable oils, to clarify coffee extracts and juices, and to separate high-value citrus oils for use in confectionery and beverages (Anon 2008b). Disc bowl and tubular centrifuges have

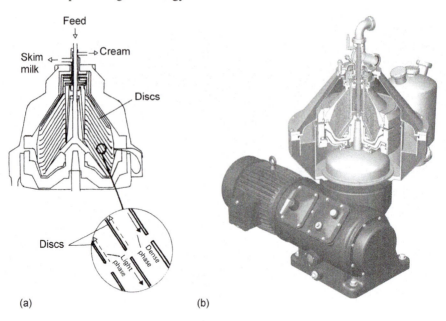

Fig. 5.2   Disc bowl centrifuge: (a) principle of operation (adapted from Hemfort 1983); (b) equipment (courtesy of Alfa Laval Ltd) (Anon 2008b).

capacities of up to $150\,000\,\mathrm{l\,h^{-1}}$. The ranges of particle sizes that can be separated are shown in Table 5.1.

Centrifugal clarifiers

The simplest solid–liquid centrifuge is a solid bowl clarifier, which is a rotating cylindrical bowl, 0.6–1.0 m in diameter. Liquor, with a maximum of 3% w/w solids, is fed into the bowl and the solids form a cake on the bowl wall. When this has reached a predetermined thickness, the bowl is drained and the cake is removed automatically through an opening in the base. Feeds that contain higher solids contents (Table 5.1) are separated using nozzle centrifuges or valve discharge centrifuges. These are similar to disc bowl types, but the bowls have a biconical shape. In the nozzle type, solids are continuously discharged through small holes at the periphery of the bowl and are collected in a containing vessel. In the valve type the holes are fitted with valves that periodically open for a fraction of a second to discharge the accumulated solids. The advantages of the latter design include less wastage of liquor and the production of drier solids. Both types are able to separate feed liquor into three streams: a light phase, a dense phase and solids. Centrifugal clarifiers are used to treat oils, juices, beer and

Table 5.1   Operating characteristics of centrifuges used in food processing

Centrifuge type	Range of feed particle sizes (μm)	Solids content of feed (%)
Disc bowl	0.5–500	<5
Nozzle bowl	0.5–500	5–25
Decanter	5–50 000	9–60
Basket	7.5–10 000	5–60
Reciprocating conveyor	100–80 000	20–75

Adapted from Hemfort (1983)

starches and to recover yeast cells. They have capacities up to $300\,000\,l\,h^{-1}$. Centrifugal extractors are used to separate immiscible liquids that have different densities.

Desludging, decanting or dewatering centrifuges
Feeds with high solids contents (Table 5.1) are separated using desludging centrifuges, including conveyor bowl, screen conveyor, basket and reciprocating conveyor centrifuges. In the conveyor bowl (or 'decanter') centrifuge (Fig. 5.3) a solid bowl rotates up to 25 rpm faster than a screw conveyor contained inside. The slurry is fed through a fixed central pipe into a distributor located in the screw conveyor. The product is then accelerated smoothly and passes through feed ports in the screw to the bowl. Under centrifugal force the solids separate from the liquid and settle against the bowl wall, while the clarified liquid exits over the adjustable overflow weir. The screw rotates at a differential speed to that of the bowl and conveys the separated solids to the conical end. Screws may have a single flight or multiple flights depending on the application and process requirements.

The slower the screw moves in relation to the bowl, the longer the sediment remains in the centrifuge and the greater the liquid–solid separation. The residence time can be adjusted by changing the difference in speed between bowl and conveyor to produce the required liquid–solid separation for a particular slurry and the moisture content in the cake. Three-phase centrifuges separate two liquid phases having different specific gravities while also dewatering solids. Further details are given by Anon (2004a).

The screen conveyor centrifuge has a similar design but the bowl is perforated to remove the liquid fraction. The basket centrifuge has a perforated metal basket lined with a cloth bag or filtering medium, which rotates at up to 2600 rpm in automatically controlled cycles which last 5–30 min, depending on the feed material. The feed liquor first enters the slowly rotating bowl; the speed is then increased to separate solids; and finally the bowl is slowed and the cake is discharged through the base by a blade. The reciprocating conveyor centrifuge is used to separate fragile solids (e.g. crystals from liquor). Feed enters a rotating basket, 0.3–1.2 m in diameter, through a funnel that rotates at the same speed. This gradually accelerates the liquid to the bowl speed and thus minimises shearing forces. Liquid passes through perforations in the bowl wall. When the layer of cake has built up to 5–7.5 cm, it is pushed forwards a few centimetres by a reciprocating arm. This exposes a fresh area of basket to the feed liquor. Capacities of these dewatering centrifuges are up to $90\,000\,l\,h^{-1}$ and they are used to recover animal and vegetable proteins, to separate coffee, cocoa and tea slurries and to desludge oils. Basket centrifuges are also used at lower speeds to (Anon 2008e):

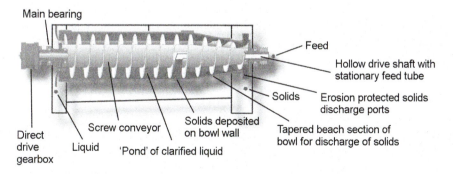

Fig. 5.3   Decanter centrifuge (courtesy of Alfa Laval Ltd) (Anon 2008b).

Fig. 5.4 'Through the wall' centrifuge design (courtesy of Rousselet-Robatel at www.rousselet-robatel.com) (Anon 2008f).

- de-water a range of foods including whole or pre-cut salads or fresh fruits and vegetables after washing, herbs prior to freezing or vegetables after blanching;
- extract honey from honeycombs;
- remove excess oil from fried products;
- remove excess sugar from crystallised fruits;
- clarify fruit juices, vegetable oils and animal fats.

Advances in centrifuge design include cleaning-in-place facilities and a fully opening housing to give access to the food contact area, which reduces the time needed for maintenance and cleaning. They can also be installed with a 'through the wall' design that separates the food contact area from the mechanical parts (Fig. 5.4). Centrifuges are fitted with touch-screen control panels and are controlled by programmable logic controllers (PLCs) that have programmable operation cycles. If the application involves separating solvents from foods (e.g. in solvent extraction of oils), they may also be fitted with gas-tight seals to protect against fire or explosions (i.e. be ATEX-compliant (Anon 2005a). In general centrifugation has advantages over filtration in not requiring disposable components (e.g. membranes or filters) and having more rapid separation.

5.2   Filtration

Filtration is the removal of insoluble solids from a suspension (or 'feed slurry') by passing it through a porous material (a 'filter medium'). The resulting liquor is termed the 'filtrate' and the separated solids are the 'filter cake'. Filtration is used to clarify liquids by the removal of small amounts of solid particles (e.g. from wine, beer, juices, oils and syrups).

5.2.1   Theory

When a suspension of particles is passed through a filter, the first particles become trapped in the filter medium and as a result reduce the area through which liquid can flow. This increases the resistance to fluid flow and a higher pressure difference is needed to maintain the flowrate of filtrate. The rate of filtration is expressed as follows:

$$\text{Rate of filtration} = \frac{\text{driving force (the pressure difference across the filter)}}{\text{resistance to flow}} \quad \boxed{5.11}$$

Assuming that the filter cake does not become compressed, the resistance to flow through the filter is found using:

$$R = \mu r (V_c V / A + L) \quad \text{有文献. 应是 } m^{-1} \quad \boxed{5.12}$$

where R (m^{-2}) = resistance to flow through the filter, μ $(N\,s\,m^{-2})$ = viscosity of the liquid, r (m^{-2}) = specific resistance of the filter cake, V (m^3) = volume of the filtrate, V_c = the fractional volume of filter cake in the feed liquid volume, V, A (m^2) = area of the filter and L = equivalent thickness of the filter and initial cake layer.

For constant rate filtration, the flowrate through the filter is found using:

$$Q = \frac{\mu r V V_c}{A^2 \Delta P} + \frac{\mu r L}{A \Delta P} \quad \Rightarrow \quad Q = \frac{R}{A \cdot \Delta P} \quad \cdot \times \text{这里错了} \quad \boxed{5.13}$$

where Q (V/t) $(m^3\,s^{-1})$ = flowrate of filtrate, ΔP (Pa) = pressure difference and t (s) = filtration time. This equation is used to calculate the pressure drop required to achieve a desired flowrate or to predict the performance of large-scale filters on the basis of data from pilot scale studies.

If the pressure is kept constant, the flowrate gradually decreases as the resistance to flow, caused by the accumulating cake, increases. Equation 5.13 is rewritten with ΔP constant as:

$$\frac{tA}{V} = \frac{\mu r V_c}{2\Delta P A}\frac{V}{A} + \frac{\mu r L}{\Delta P} \quad \boxed{5.14}$$

If $t/(V/A)$ is plotted against V/A, a straight line is obtained (Fig. 5.5). The slope (Equation 5.15) and the intercept (Equation 5.16) are used to find the specific resistance of the cake and the equivalent cake thickness of the filter medium:

$$\text{Slope} = \mu r V_c / 2\Delta P \quad \boxed{5.15}$$

$$\text{Intercept} = \mu r L / \Delta P \quad \boxed{5.16}$$

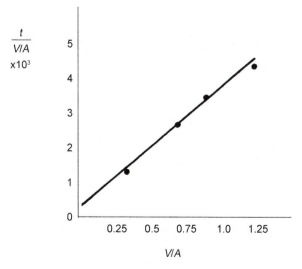

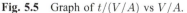

Fig. 5.5   Graph of $t/(V/A)$ vs V/A.

If the filter cake is compressible (that is the specific resistance changes with applied pressure) the term r is modified as follows:

$$r = r'\Delta P^s \qquad \boxed{5.17}$$

where r is the specific resistance of the cake under a pressure difference of 101×10^3 Pa and $s = $ the compressibility of the cake. This is then used in Equation 5.12. Derivations of the above equations and further details are given by Earle (1983) and Toledo (1999a).

Sample problem 5.4
Pulp which contains 15% solids is filtered in a plate and frame fitter press (section 5.2.2) with a pressure difference of 290 Pa. The masses of filtrate are shown below for a 1.5 h cycle. Calculate the specific resistance of the cake and the volume of filtrate that would be obtained if the cycle time were reduced to 45 min. (Assume that the area of the filter $= 5.4 \, m^2$, the cake is incompressible, the viscosity of the filtrate is $1.33 \times 10^{-3} \, Ns \, m^{-2}$ and the density of the filtrate $= 1000 \, kg \, m^{-3}$.)

Time (min)	7.5	30.4	50	90
Mass of filtrate (kg)	1800	3800	4900	6800

Solution to sample problem 5.4

Time (s):	450	1825	3000	5400
$V \, (m^3)$	1.8	3.8	4.9	6.8
V/A	0.33	0.69	0.89	1.24
$t/(V/A)$	1364	2645	3371	4355

Plotting $t/(V/A)$ vs (V/A) (Fig. 5.5):

$$\text{Slope} = 3304 \text{ s m}^{-2}$$
$$\text{Intercept} = 332 \text{s m}^{-1}$$

From Equation 5.15:

$$3304 = \frac{1.33 \times 10^{-3} \times r \times 0.15}{2 \times 290 \times 10^3}$$
$$r = 9.61 \times 10^{12} \text{ m}^{-2}$$

From equation 5.14:

$$tA/V = 2666.7 \, (V/A) + 300$$

For a 45 min (2700 s) cycle:

$$2700 = 3304(V/5.4)^2 + 332(V/5.4)$$

Solving this quadratic equation for $(V/5.4)$

$$\frac{V}{5.4} = \frac{-332 + \sqrt{(332^2 - 4 \times -2700 \times 3304)}}{2 \times 3304}$$
$$= 0.855$$

Therefore $V = 0.855 \times 5.4 = 4.62 \text{ m}^3 = 4620 \text{ l.}$

5.2.2 Equipment

Gravity filtration is slow and finds little application in the food industry. Filtration equipment operates either by the application of pressure to the feed side of the filter bed or by the application of a partial vacuum to the opposite side of the filter bed. The main applications are the processing of beer, cider, corn syrup, fruit and vegetable juices, molasses, soft drinks, sugar, vegetable oils and wine. Filter aids may be applied to the filter or mixed with the food to improve the formation of filter cake. They are chemically inert powders and include:

- 'perlite', a generic name for porous volcanic rock composed mostly of SiO_2 that when heated at 1100 °C expands up to 20 times its original volume (Anon 2008e);
- cellulose, used to retain the colour and aroma of wines or as a fibrous pre-coat filter aid or where special chemical compatibility is required;
- activated carbon, made from hardwoods or coconut shells;
- bone char, used especially to remove heavy metals from production and waste streams;
- pumice;
- bentonites (sodium-calcium bentonite) used for flocculation, adsorption and stabilisation to prevent colloidal and protein hazes in filtration of fermentation must;
- bleaching clay;
- diatomaceous earth (or 'Kieselghur'), which is deposits of fossilised plankton that are predominantly silica, used especially for beer and wine filtration.

Further information on filter aids is given by Anon (2002, 2003, 2004b). Centrifugal filtration using a basket centrifuge is described in section 5.1.2.

Pressure filters

Three types of pressure filters are the plate-and-frame filter press (Fig. 5.6a) (also termed the 'recessed chamber filter press' or 'diaphragm filter press'), the shell-and-leaf pressure filter and the rotary filter press. In the plate-and-frame design, filters made from woven nylon, polypropylene, cloth or paper, are supported on vertical (or, less commonly, horizontal) plates. Synthetic cloths have smooth surfaces that are more easily cleaned than natural fibre cloths or paper, and being more robust they can be re-used more often. Perforated plastic or stainless steel screens may be used where additional strength and rigidity are required (Anon 2008g).

When the stack of plates is pressed together, they form a series of chambers with each filter cloth acting as a gasket between the chambers. The thickness of each supporting frame determines the volume of cake that can be accumulated. Each chamber has holes in the corners that align to form interconnecting discharge manifolds ports connected to the external piping of the press. A centre feed inlet also forms a manifold that connects the individual chambers (Fig. 5.6b). Presses can accommodate up to 120 chambers, providing a filtration area up to 540 m² (Anon 2008d).

In operation, feed liquor is pumped into the press and the pressure is gradually increased until a pre-determined value is reached, typically a maximum of 500–700 kPa. An automatic pressure setting maintains the pressure throughout the filtration cycle. It is the pressure differential between the feed pressure and the gravity discharge that causes filtration to occur. Liquid passes through the filter cloths and flows down the grooved surfaces of the plates to the drain manifold (Anon 2008e).

Once a 1–2 mm layer of solid particles has accumulated on the filter cloths, this 'precoat' layer serves to filter finer particles, yielding a filtrate that has very low turbidity.

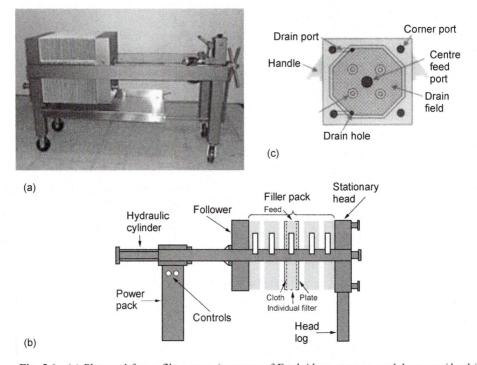

Fig. 5.6 (a) Plate and frame filter press (courtesy of Ertel Alsop at www.ertelalsop.com/depth/ equipment/plateframe.html); (b) principle of operation; (c) detail of a filter element showing feed and discharge ports (courtesy of Seimens AG Water Technologies Corp.) (Anon 2008h).

As filtering continues, solids within the slurry flow to the area of cake that has the lowest pressure differential, resulting in uniform cake build-up over the filter surface. This continues until the spaces between the plates in each chamber are filled by filter cake, typically 35–40 mm. Some designs have a diaphragm in each chamber, and high-pressure water is pumped behind the diaphragms to compress the cake and removed additional filtrate. When filled, the filter chambers are separated, the filter cakes are discharged by gravity and the plates are back-washed with water, ready to begin another cycle.

These filter presses have relatively low capital costs and high flexibility for filtering different foods; they are reliable and easily maintained and are widely used in the production of apple juice, cider, beer, wine and other alcoholic beverages, soft drinks, oils, syrups and gelatin. Older designs were time consuming to use and relatively labour intensive, but large (filter area $600\,m^2$) automatic presses are now available that have PLC-controlled filtration and washing cycles (Anon 2008f). Further information is given by Anon (2008h, 2008i).

The shell-and-leaf pressure filter has been used to overcome the problems of higher labour costs and lack of convenience of plate and frame presses. It consists of mesh 'leaves', which are coated in filter medium and supported on a hollow frame that forms the outlet channel for the filtrate. The leaves are stacked horizontally or vertically inside a pressure vessel, and in some designs they rotate at 1–2 rpm to improve the uniformity of cake build-up. Feed liquor is pumped into the shell at a pressure of approximately 400 kPa. When filtration is completed, the cake is blown or washed from the leaves. This equipment has a higher cost than plate and frame filters and is best suited to routine filtration of liquors that have similar characteristics.

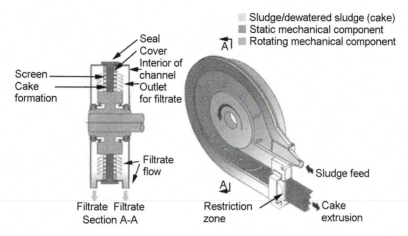

Fig. 5.7   Rotary filter press (courtesy of Fournier Industries Inc.) (Anon 2008j).

In the rotary filter press, food is fed in one or more channels between two parallel revolving stainless steel screens and the filtrate passes through the screens. The material continues to dewater as it travels around the channel, forming a cake as it approaches the outlet to the press. A restriction on the outlet results in the extrusion of a very dry cake (Fig. 5.7).

This type of press is claimed by manufacturers to have advantages over the plate and frame design in having continuous operation, constant low-pressure feed (10–50 kPa), few ancillary screens that have a long life (e.g. 10 years) and uniform cake dryness. It also has simple start-up and shut-down procedures, a wash cycle lasting five minutes per day compared with up to two hours washing per cycle in the plate and frame press, and lower maintenance and operating costs. Anon (2008j) reports energy consumption of 10–20 kW h/dry tonne compared with 171 kW h/dry tonne for centrifugal filtration. The press is PLC controlled, allowing automatic adjustment to accommodate variations in the solids content of the feed and the moisture content of the press-cake. For a multi-channel machine, one or more channels may be removed for maintenance, while the machine continues in operation, whereas maintenance of other types of filtration equipment requires a total machine shut-down and consequent production down-time. This type of press is mostly used in dewatering applications, including bottling wastewater, brewery waste sludge and brewery barley waste.

Vacuum filters
Vacuum filters are limited by the cost of vacuum generation to a pressure difference of about 100 kPa. However, cake is removed at atmospheric pressure and these types of filter are therefore able to operate continuously (pressure filters have batch operation because the pressure must be reduced for cake removal). Two common types of vacuum filter are the rotary drum filter and rotary disc filter.

The rotary drum filter consists of a horizontal cylinder which has the surface divided into a series of shallow compartments, each covered in filter cloth and connected to a central vacuum pump (Fig. 5.8a). As the drum rotates, it dips into a bath of liquor and filtrate flows through the filter and out through channels in the drum. When a compartment leaves the bath, the filter cake is sucked free of liquor, washed with sprays and the vacuum is released. Compressed air is blown from beneath the cloth to loosen the cake, which is removed by a scraper before the individual compartment restarts the cycle

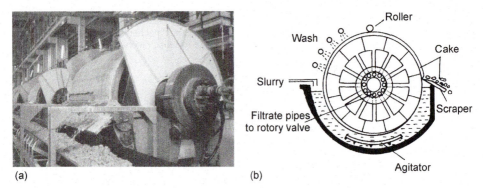

Fig. 5.8 (a) Rotary drum filter (© RPA Process Technologies at www.rpaprocess.com/produtos/
produtos.php?is=2&pid=13); (b) principle of operation (after Leniger and Beverloo 1975).

(Fig. 5.8b). Rotary vacuum disc filters consist of a series of vertical discs that rotate
slowly in a bath of liquor in a similar cycle to drum filters. Each disc is divided into
portions that have outlets to a central shaft. The discs are fitted with scrapers to remove
the cake continuously. They can remove particles within the range 30–200 μm depending
on the filter material and have a large filter area (e.g. 100 m^2) (Anon 2008k). These types
of filter are compact and have low labour costs and high capacity (e.g. 5000 kg m^{-2} h^{-1}).
However, they have a high capital costs and produce cake which has a moderately high
moisture content.

5.3   Expression

5.3.1   Theory

The main applications of expression are the extraction of components of plant materials
(e.g. fruit juices, sugar, vegetable oils and grape juice for wine). These materials are located
within the cell structure of the plants and it is necessary to disrupt the cells in order to
release them. This is achieved either in a single stage, which both ruptures the cells and
expresses the liquid, or in two stages: size reduction to produce a pulp or flour (Chapter 3,
section 3.1), followed by separation in a press. In general the single-stage operation is more
economical, permits higher throughputs and has lower capital and operating costs, but for
some products (e.g. oil extraction from nuts) two-stage expression is more effective. It is
also necessary to increase the pressure slowly to avoid the formation of a dense
impenetrable press cake, as the solid material is easily deformed and blocks the press.

In fruit processing, the press should remove the maximum quantity of juice from fruit
pulp without substantial quantities of solids, and with a minimum amount of phenolic
compounds from the skins, which cause bitterness and browning. This is achieved using
lower pressures and fewer pressings. Details of fruit juice production are given in a
number of texts, including Hui *et al.* (2006), Fellows (1997) and Dauthy (1995). In oil
processing, better extraction is achieved by size reduction of the seeds or nuts to flour,
followed by heating to reduce the oil viscosity, to release the oil from intact cells and to
remove moisture. There is an optimum moisture content for each type of oilseed to obtain
a maximum yield of oil (Table 5.2). Details of oil processing are given by Rossell (1999)
and Anon (1995), and for individual oils by Anon (2008l) for olive oil, Ohler (2006) for
coconut oil, Poku (2002) for palm oil and palm kernel oil, Gunstone (2004), for rapeseed
oil, and Erickson and Wiedermann (1989) and Semon *et al.* (1997) for soybean oil.

Table 5.2 Selected properties of cooking oils

Oil	Moisture content (%)	Oil content (%)	Yield of oil (%)	Melting point (°C)	Iodine value
Coconut (fresh)	40–50	35–40	55–62	25	10
Copra	3–4.5	64–70	60–70	25	10
Cotton seed	5	15–25	13	−1	99–119
Maize (germ)	13	30–50	25–50	–	103–128
Mustard	7	25–45	31–33	−17	96–110
Olive	50–70	35–39	25	−6	75–94
Palm	40	56	11–20	35	54
Palm kernel	10	46–57	36–51	24	37
Peanut (groundnut)	4	28–55	40–42	3	80–106
Rapeseed	9	40–45	25–37	−10	94–120
Sesame	5	25–50	45–50	−3 to −6	104–120
Soybean	13	16–19	14–20	−16	120–143
Sunflower	5	25–50	20–32	−17	110–143

Adapted from Calais and Clark (undated), Anon (2008m) and Anon (2005b)

The factors that influence the yield of juice or oil from a press include:

- maturity and growth conditions of the raw material;
- extent of cell disruption;
- thickness of the pressed solids and their resistance to deformation;
- rate of increase in pressure, the time of pressing and the maximum pressure applied;
- temperatures of solids and liquid, and the viscosity of the expressed liquid.

5.3.2 Equipment
Methods and equipment used to produce flours from oilseeds and nuts, or pulps from fruits are described in Chapter 3 (section 3.1). The following section describes batch and continuous equipment used to express oil or fruit juices from prepared flours or pulps.

Batch presses
Common types of equipment for expressing juices are the tank press and the cage press. The tank press (or bladder press) is used for juice production from soft fruit pulps (e.g. grapes) and consists of a horizontal cylinder that is divided internally by a membrane. During an automatically controlled pressing cycle of approximately 1.5 h, up to 7 t of fruit pulp is fed into the tank on one side of the membrane and compressed air at 200–600 kPa is applied to the opposite side. Juice flows out through channels and, when pressing is completed, the tank is rotated to loosen and discharge the press residue. High yields of good-quality juice (e.g. 650–800 l t^{-1} fruit) are obtained by the gentle increase in pressure, at capacities ranging from 3600 to 25 000 kg (Vine 1987). Phillips (2006) gives details of different designs of tank presses.

The cage press (or basket press) contains up to 2 t of fruit pulp or oilseed flour in a vertical stack, or contained within a perforated or slatted cage, either loose or in cloth bags depending on the nature of the material (Fig. 5.9). In larger presses, ribbed metal layer plates are placed in the press at intervals as the feed is added to reduce the thickness of the individual pulp or flour beds. The pressure is gradually increased on the top plate by a hydraulic or ratchet system, or a motor-driven screw, and liquid is forced through the perforations in the cage and collected at the base of the press. The average pressing cycle

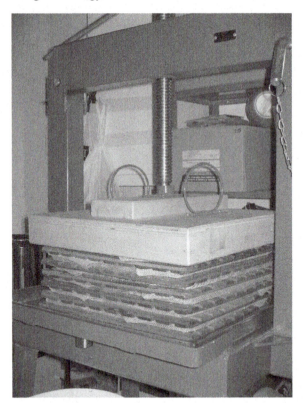

Fig. 5.9   Cage press (photo with Creative Commons License, courtesy of UK Cider at http://ukcider.co.uk).

for a cider press is 20–30 minutes. The equipment allows close control over the pressure exerted on the pulp/flour and may operate semi-automatically to reduce labour costs. This type of press is widely used for apple juice production (Bates *et al.* 2001) or for small-scale expression of cooking oils from oilseeds and nuts (e.g. Anon 2008m, Owolarafe *et al.* 2002). Yields (weight of juice per weight of pulp) are typically > 50% for apple and higher for other fruits, and for oil the yield varies from ≈ 35% using manual presses (Potts and Machell 1993) to ≈ 55% using powered presses.

Continuous presses
There are several types of continuous press used commercially: the belt press for fruit processing, the screw expeller for both fruit processing and oil extraction and the roller press for sugar cane processing. Details of solvent extraction of oils are given in section 5.4.

Belt presses consist of a continuous belt, made from canvas–plastic composite material that passes over a series of rollers (Fig. 5.10). The fruit pulp is distributed uniformly over the lower belt and pressed with a roller to extract the juice. Further press rollers having decreasing diameters then increase compression and shear forces on the pulp to produce a juice yield of ≈ 84%. The pomace is scraped from the belts by adjustable plastic scrapers and conveyed away. Capacities of up to $40\,t\,h^{-1}$ can be achieved (Anon 2008c). In another design, the belt passes under tension over two hollow stainless steel cylinders, one of which is perforated. Pulped fruit is fed onto the inside of the belt and is pressed between the belt and the perforated cylinder. Juice flows through

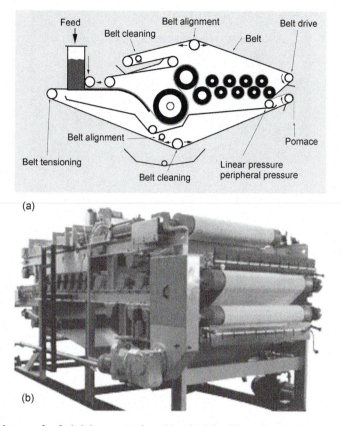

(a)

(b)

Fig. 5.10   Belt press for fruit juice processing: (a) principle, (b) equipment (courtesy of Flottweg AG at www.flottweg.com).

the perforations and the presscake continues around the belt and is removed by a scraper or auger. The presses produce high yields of good-quality juice but have relatively high capital costs.

The demand for more 'natural' cloudy fruit juices is increasing, and light colours and stable turbidity are the main quality characteristics required by customers. To achieve these in apple processing, a fruit mill is fitted directly above the feed to a belt press, which ensures that crushing and juice extraction are completed within 2–3 min. This is followed by immediate heating to inactivate polyphenoloxidases and pectic enzymes, which both prevents browning reactions to produce a light coloured juice and retains pectic materials to produce the required turbidity. The degree of turbidity in the final product may subsequently be adjusted and standardised using a separator. Other juices are standardised using a decanter centrifuge (section 5.1.2).

The screw expeller has a robust horizontal barrel containing a stainless steel helical screw similar to an extruder (Chapter 15, section 15.2.1). The pitch of the screw flights gradually decreases towards the discharge end, to increase the pressure on the material as it is carried through the barrel (Fig. 5.11). The final section of the barrel is perforated to allow expressed liquid to escape and presscake is discharged through the barrel outlet. The pressure in the barrel is regulated by adjusting the diameter of the discharge port. The equipment is mostly used to extract oil from oilseeds where frictional heat reduces the viscosity of the oil and improves the yield. Some types of expellers have supplementary

(a)   Feed

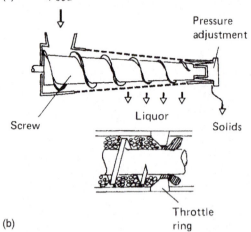

(b)

Fig. 5.11   (a) Oil expeller showing detail of screw barrel (courtesy of De Smet Rosedowns); (b) operation of screw press/oil expeller, showing throttle ring used for oilseed processing (after Brennan *et al.* 1990 and Stein 1984).

heaters fitted to the barrel, or a series of 'throttle rings' located in the barrel to create high shearing forces on the seeds and improve yields. Capacities range from 40–8000 kg h^{-1}. The oilcake has 5–18% residual oil, depending on the type of oilseed and the operating conditions (Barker 1987).

When used for juice extraction, this type of equipment is known as a 'pulper-finisher', which simultaneously pulps soft fruits and expresses the juice (also Chapter 3, section 3.1.2). The equipment may be fitted with a paddle arrangement instead of a screw and the barrel may be cooled to reduce frictional heat, which would have undesirable effects on the flavour and aroma of the product. Further details are given by Anon (2008n).

5.4   Extraction using solvents

Unit operations that involve separation of specific components of foods using a solvent are important in a number of applications, including production of:

- cooking oils or speciality oils from nuts and seeds;
- flavour extracts, herbs, spices and essential oils;
- instant coffee or tea, and decaffeinated coffee and tea.

Once the solvent has been removed from the extracted foods, some may be used directly (e.g. cooking oils) or they may be further processed by concentration (section 5.5 and Chapter 14) and/or dehydration (Chapter 16). Many extraction operations take place close to ambient temperature, but even when elevated temperatures are used to increase the rate of extraction, there is little damage caused by heat and the product quality is not significantly affected. The main types of solvents used for extraction are water, organic solvents or supercritical carbon dioxide. These are described below.

5.4.1   Theory

Solid–liquid extraction involves the removal of a desired component (the solute) from a food using a liquid (the solvent) that is able to dissolve the solute. This involves mixing the food and solvent together, either in a single stage or in multiple stages, holding for a predetermined time and then separating the solvent. During the holding period there is mass transfer of solutes from food material to the solvent, which occurs in three stages: (1) the solvent enters the particle of food and dissolves the solute; (2) the solution moves through the particle of food to its surface; and (3) the solution becomes dispersed in the bulk of the solvent. During extraction, the holding time should be sufficient for the solvent to dissolve sufficient solute, and this depends on the following:

- The solubility of a given solute in the selected solvent.
- The temperature of extraction. Higher temperatures increase both the rate at which solutes dissolve in the solvent and the rate of diffusion into the bulk of the solvent. The temperature of most extraction operations is limited to less than 100 °C by economic considerations; by extraction of undesirable components at higher temperatures; or by heat damage to food components.
- The surface area of solids exposed to the solvent. The rate of mass transfer is directly proportional to the surface area, so reductions in particle size (giving an increase in surface area) increase the rate of extraction up to certain limits.
- The viscosity of the solvent. This should be sufficiently low to enable the solvent to easily penetrate the bed of solid particles.
- Flowrate of the solvent. Higher flowrates reduce the boundary layer of concentrated solute at the surface of particles and thus increase the rate of extraction.

Further details are given in Brennan *et al.* (1990) and Toledo (1999a), and mass balance calculations are described in Chapter 1 (section 1.3.3).

5.4.2   Solvents

The types of solvent used commercially to extract food components are shown in Table 5.3. Extraction using water (leaching) has obvious advantages of low cost and safety, and it is used to extract sugar, coffee and tea solubles. Oils and fats require an organic solvent and as these are highly flammable, great care is needed in operating procedures to ensure that equipment is gas-tight, and that electrical apparatus is isolated from the solvent and/ or spark-proof to comply with ATEX requirements (Anon 2005a).

5.4.3   Supercritical CO$_2$

Carbon dioxide becomes 'supercritical' when it is raised above its critical temperature and critical pressure (i.e. above the critical pressure line (7.386 MPa) and to the right of the critical temperature line (31.06 °C) in Fig. 5.12). As a solvent it acts as a liquid but has

Table 5.3   Solvents used to extract food components

Food	Solvent	Final solute concentration (%)	Temperature (°C)
Decaffeinated coffee	Supercritical carbon dioxide, water or methylene chloride	N/A	30–50 (CO_2)
Fish livers, meat byproducts	Acetone or ethyl ether	N/A	30–50
Hop extract	Supercritical carbon dioxide	N/A	<100–180
Instant coffee	Water	25–30	70–90
Instant tea	Water	2.5–5	N/A
Olive oil	Carbon disulphide	N/A	
Seed, bean and nut oils (e.g. from soybeans, groundnuts, cottonseed, sunflower seed, etc.)	Hexane, heptane or cyclohexane	N/A	63–70 (hexane) 90–99 (heptane) 71–85 (cyclohexane)
Sugar beet	Water	approx. 15	55–85

Adapted from data of Brennan *et al.* (1990) and Clarke (1990)

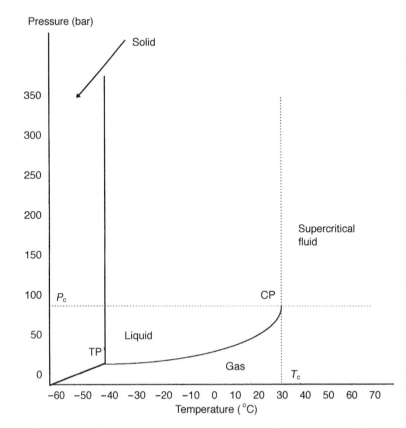

Fig. 5.12   Pressure–temperature diagram for carbon dioxide. TP = triple point; CP = critical point; T_c = critical temperature; P_c = critical pressure (adapted from Brennan *et al.* 1990).

a low viscosity and diffuses easily like a gas. The absence of surface tension means that it rapidly penetrates foods and increases the extraction efficiency. Being highly volatile, it is easily separated without leaving any residues in the food. This is increasingly important as more stringent environmental regulations and the demand for high added-value products have led processors to seek alternatives to chemical solvents. Carbon dioxide is also non-flammable, generally regarded as safe (GRAS), non-corrosive, bacteriostatic and inexpensive. In addition, it has a low critical temperature that reduces heat damage to extracted food components. It is used under conditions that are close to the critical point (near-critical fluid or NCF extraction) for de-odourising applications or for extracting highly soluble solutes. For more complete extraction or for applications that involve less soluble solutes, it is used at higher temperatures and pressures (Rizvi *et al.* 1986). The upper limit for the operating temperature is the heat sensitivity of the food components, and the upper limit for pressure (about 40 MPa) is determined by the cost of pressurised equipment. Del Valle and de la Fuente (2006) describe details of mass transfer during supercritical CO_2 extraction.

Supercritical CO_2 has found increasingly widespread application for removing caffeine from coffee or tea and for producing hop extracts for brewing. It is also used to extract and concentrate flavour compounds from fruits and spices (including pepper, marjoram, nutmeg, cardamom, cloves, parsley, vanilla and ginger), speciality oils from citrus and a variety of nuts and seeds, and high-quality essential oils (Catchpole *et al.* 2003, Mohamed and Mansoori 2002). Bohac (1998) describes the removal of 98% of cholesterol from dried egg yolk using supercritical CO_2 and Mohamed and Mansoori (2002) also report the removal of cholesterol from dairy products using both supercritical ethane/CO_2 extraction and a combined extraction/adsorption process using alumina as an adsorbent. This removed 97% of cholesterol from butter oil and also created oil fractions that had different properties to the original butter oil. Mermelstein (1999) describes the use of supercritical fluid chromatography to extract omega-3 fatty acids for use in neutraceuticals (Chapter 6, section 6.3) and Spilimbergo *et al.* (2002) report research work on microbial inactivation in heat- and pressure-sensitive foods using supercritical CO_2. Gaehrs (1990) describes the economics of extraction and applications to different products, and the topic is reviewed by Gopolan (2003), Rozzi and Singh (2002) and Steytler (1996).

Processes are being developed that involve the rapid expansion of supercritical CO_2 solutions through small nozzles to produce micro- and nano-particles. They include the gas antisolvent process (GAS), rapid expansion of supercritical solutions (RESS) process, the supercritical antisolvent (SAS) process and the gas-saturated solution (PGSS) process. They are being investigated for the production of particles from pure polymers or the production of polymer particles that contain active ingredients. Although these processes have so far been applied to pharmaceutical products, they may find food applications in the future and have been reviewed by Yeo and Kiran (2005).

5.4.4   Equipment
CO₂ extractors
The components of a typical extraction unit that uses near-critical CO_2 solvent are shown in Fig. 5.13. The essential components are an extraction vessel, a separation vessel, a condenser and a pump. CO_2 is stored as a near-critical liquid in the condenser and then pumped to the extraction vessel through a heat exchanger by a high-pressure pump. The state of the CO_2 in the extractor is determined by the pressure, controlled by a pressure

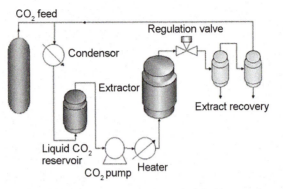

Fig. 5.13   Schematic diagram of continuous supercritical fluid extraction roller mill (from Mohamed and Mansoori 2002).

valve, and the temperature, thermostatically controlled by liquid recirculating through a jacket surrounding the vessel. The material to be extracted is purged with CO_2 gas to remove air and then liquid CO_2 is pumped in at a rate that permits a sufficient residence time for equilibrium conditions to be established. The solution is then passed to the separation vessel in which conditions are adjusted to minimise the solubility of the extracted components (often by decompression). After separation, the CO_2 is returned to the cooled condenser for re-use and the extract is removed from the separation vessel.

Single stage solvent extractors
Extractors are either single stage or multi-stage batch tanks or continuous extractors. A summary of the requirements for batch and continuous solvent extraction of oil is shown in Table 5.4 and details of cooking oil production are given by Salunkhe (1992). Closed tanks are fitted with a mesh base to support the solid particles of food. Solvent percolates down through the particles and is collected below the mesh base, with or without recirculation. They are used to extract speciality oils or to produce coffee or tea extracts. Instant coffee production is described in detail by Clarke (1990). Although they have low capital and operating costs, single stage extractors produce relatively dilute solutions that may require expensive solvent recovery systems for organic solvents or pollution control measures when water is used as the solvent.

Multi-stage solvent extractors
These comprise a series of up to 15 tanks, each similar to single extractors, linked together so that solvent emerging from the base of one extractor is pumped co-currently or counter-currently to the next in the series. A similar arrangement for evaporators is shown in Chapter 14 (Fig. 14.4). These are used at a larger scale to produce oils, tea and coffee extracts, and to extract sugar from beet.

Table 5.4   Requirements for solvent extraction of vegetable oils

Requirements per tonne of oilseed	Batch processing	Continuous processing
Steam (kg)	700	280
Power (kW h)	45	55
Water (m^3)	14	12
Solvent (kg)	5	4

From Bernadini (1976), quoted by Kessler (1985)

Continuous extractors

There are a large number of designs of extractor, each of which may operate counter-currently and/or co-currently. For example, one design has an enclosed, vapour-tight tank containing two vertical bucket elevators (Chapter 27, section 27.1.2) made from perforated buckets that are linked to form a continuous ring. Fresh material is loaded into descending buckets of one elevator and solvent is pumped in at the top to extract solutes co-currently. As the buckets then move upwards, fresh solvent is introduced at the top of the second elevator to extract solutes counter-currently. The solution collects at the base and is pumped to the top of the first elevator to extract more solute, or it is separated for further processing. Other designs of equipment employ perforated screw conveyors instead of bucket elevators, or use a rotating carousel in which segments with perforated bases contain the feed material. Solvent is sprayed onto each segment, collected at the base and pumped to the preceding segment to produce counter-current extraction. In all designs, the material is continuously conveyed into the extractor through a vapour lock that prevents solvent vapours from escaping. They are used to extract oils, coffee, and beet sugar, and for the preparation of protein isolates. Further details of their operation are given by Brennan *et al.* (1990).

In oilseed processing, low-pressure expellers may be used to remove part of the oil from seeds that have high oil contents (e.g. cottonseed, groundnut or sunflower) to reduce processing costs. This is followed by flaking and solvent extraction using hexane. The pre-pressing is important for cottonseed processing because it also reduces antinutritional gossypol in the meal.

In older equipment, solvent recovery took place in a series of steam-jacketed paddle conveyors that evaporated most of the solvent. Steam was then blown through the flakes to remove the remaining solvent. This method was then improved by using super-heated hexane vapour to quickly remove most of the solvent, followed by a steam treatment. However, for soybean flakes neither method eliminates trypsin inhibitors and a toasting stage, in which the flakes are heated to about 125 °C, is required if the flakes are to be fed to non-ruminant animals. More recent methods involve conveying the solvent-wet flakes (containing 35% moisture) to a 'desolventiser-toaster'. This is a vertical set of steam-jacketed kettles with gates that allow the flakes to fall from one kettle into the next while being treated with direct steam. The lower kettles act as dryers to reduce the moisture content to 18%. The dried hot flakes are then cooled by air. In all methods water-cooled tubes are used to condense the solvent/water vapour mixture and the liquid solvent is separated from the water using a decanting centrifuge for recovery and reuse. The hot vapours from the toaster are used to heat the first-stage kettle and create important energy savings (Kessler 1985).

5.5   Membrane concentration

The separation or concentration of food components using membranes is well established, especially in the fruit processing, dairy processing and alcoholic beverage industries. It is also used to purify process water and treat wastewaters in a wide variety of food industries. Membrane emulsification is described in Chapter 3 (section 3.2.3). There are seven types of membrane systems in use in food industries, grouped as follows according to the driving force for transport across the membranes:

- Hydrostatic pressure systems – reverse osmosis, nanofiltration, ultrafiltration, microfiltration and pervaporation.

Table 5.5   Comparison of reverse osmosis and evaporation of whey

Parameter	Reverse osmosis	Evaporation
Steam consumption	0	250–550 kg per 1000 l water removed
Electricity consumption	10 kW h per 1000 l water removed (continuous); 20 kW h per 1000 l water removed (batch)	Approximately 5 kW h per 1000 l water removed
Energy use (kW h)	3.6 (6–12% solids) 8.8 (6–18% solids) 9.6 (6–20% solids)	One effect 387 (6–50% solids) Two effects 90 (6–50% solids) Seven effects 60 (6–50% solids) MVR* 44
Labour	4 h day^{-1}	Normally two operators during whole operation (boiler house and evaporator)
Cooling-water consumption	0–29 300 kJ per 1000 l water removed (continuous); 0–58 600 kJ per 1000 l water removed (batch)	$(5.2–1.2) \times 10^6$ kJ per 1000 l water removed
Economical plant size	6000 l day^{-1} day or more, no upper limit	80 000–100 000 l day^{-1}
Consideration in final product	Maximum 30% total solids. Capacity varies with concentration	Up to 60% total solids

* MVR, mechanical vapour recompression.
Adapted from Madsen (1974)

- Systems where a concentration difference is the driving force – ion-exchange and electrodialysis.

In contrast to evaporation by boiling (Chapter 14) where heat is used to convert water to steam in order to remove it, membranes remove water from foods without a change in phase. This uses energy more efficiently and the lack of heating results in little damage to the organoleptic or nutritional properties of foods, making this unit operation particularly suitable for separation or concentration of heat-sensitive foods, flavourings, colourants and enzyme preparations. Membrane concentration also has lower labour and operating costs than vacuum evaporators. The main limitations of membrane concentration are higher capital costs than some types of evaporation equipment, variation in the product flowrate when changes occur in the concentration of feed liquor, concentration polarisation or fouling of the membranes (section 5.5.1) and a maximum concentration of 30% total solids in the product (Table 5.5).

5.5.1   Theory
Hydrostatic pressure systems
Osmosis is the movement of a solvent from an area of low solute concentration, through a semipermeable membrane, to an area of high solute concentration (e.g. in the uptake of water by cells in plant roots, each cell contains a higher concentration of solutes than its exterior and water flows across the cell membrane causing the cell to expand due to osmotic pressure). Membrane concentration forces a solvent from a feed that contains a high solute concentration through a membrane to a region of low solute concentration by applying a higher pressure than the osmotic pressure (hence the term 'reverse' osmosis).

Osmotic pressure is found for dilute solutions using van't Hoff's equation:

$$\Pi = \frac{cR\theta}{M} \qquad \boxed{5.18}$$

where Π (Pa) = osmotic pressure, c = solute concentration (kg m^{-3}), θ (K) (where K = °C + 273) = absolute temperature, R (J mol^{-1} K^{-1}) = universal gas constant and M = molecular weight.

Gibb's relationship is more accurate over a wider range of solute concentrations:

$$\Pi = \frac{R\theta \ln X_A}{V_m} \qquad \boxed{5.19}$$

where V_m = molar volume of pure liquid and X_A = mole fraction of pure liquid.

Knowing the osmotic pressure of the feed liquid is useful in selecting the correct membrane because the membrane must be able to withstand the pressure difference across the membrane (the trans-membrane pressure) needed to overcome it (see sample problem 5.5).

The driving force for transport across the membrane is the hydrostatic pressure applied to the feed liquid. The types of materials that pass through the membrane (the 'permeate') depend on its chemical composition, physical structure, or the size and size distribution of any pores in the membrane. The solution that is concentrated is termed the 'retentate'. The flowrate of liquid (the 'transport rate' or 'flux') is determined by the solubility and diffusivity of the liquid molecules in the membrane material, and by the difference between the osmotic pressure of the liquid and the applied pressure. This hydrostatic pressure difference (or trans-membrane pressure) is found using:

$$\Delta P = \frac{P_f + P_r}{2} - P_p \qquad \boxed{5.20}$$

where ΔP (Pa) = hydrostatic pressure difference, P_f (Pa) = pressure of the feed (inlet), P_r (Pa) = pressure of the retentate (outlet) (high molecular weight fraction) and P_p (Pa) = pressure of the permeate (low molecular weight fraction).

Sample problem 5.5
Calculate the osmotic pressure of apple juice that contains 15% total solids at 18 °C. Assume that the density of the juice is 1.3500 kg l^{-1}, the predominant sugar affecting the osmotic pressure is glucose (180 kg/kg mol) and the universal gas constant = 8.314 m^3 kPa/(mol K).

Solution to sample problem 5.5
The temperature = (273 + 18) = 291 K.
The density of apple juice = 1350 kg m^{-3} and the concentration of solids = 0.15 kg solids per kg product.
Multiplying the concentration by the density:

$$c = 0.15 \times 1350 = 202.5 \text{ kg m}^{-3}$$

From van't Hoff's equation (Equation 5.18),

$$\Pi = \frac{202.5 \times 8.314 \times 291}{180}$$

$$\Pi = 2721.8 \text{ kPa}$$

Water flux increases with an increase in applied pressure, increased permeability of the membrane and lower solute concentration in the feed stream. It is calculated using:

$$J_w = k_w A (\Delta P - \Delta\Pi) \qquad \boxed{5.21}$$

where J_w (kg h^{-1}) = water flux, K_w (kg m^{-2} h^{-1} kPa^{-1}) = mass transfer coefficient of water through the membrane, which is a function of permeate viscosity and membrane thickness, A (m^2) = area of the membrane, ΔP (kPa) = trans-membrane pressure and $\Delta\Pi$ (kPa) = osmotic pressure difference across the membrane.

Osmotic pressure is found for dilute solutions using:

$$\Pi = (MR\theta)^1 - (MR\theta)^2 \qquad \boxed{5.22}$$

where M (mol m^{-3}) = molar concentration, R (Pa m^3 mol^{-1} K^{-1}) = universal gas constant, θ (K) = absolute temperature, and superscripts 1 and 2 are conditions on each side of the membrane.

The flowrate of solute through a membrane is calculated using:

Sample problem 5.6

Fruit juice containing 9% w/w solids is pre-concentrated at 35 °C by reverse osmosis, prior to concentration in an evaporator. If the operating pressure is 4000 kPa and the mass transfer coefficient is 6.3×10^{-3} kg m^{-2} h^{-1} kPa^{-1}, calculate the area of membrane required to remove 5 t of permeate in an 8-h shift. (Assume that sucrose (molecular weight = 342) forms the majority of solids that contribute to the osmotic pressure of the juice, the density of the juice = 1003 kg m^{-3} and that the universal gas constant is 8.314 Pa m^3 mol^{-1} K^{-1}.)

Solution to sample problem 5.6

The concentration of solids = 0.09 kg solids per kg product. Multiplying the concentration by the density,

$$c = 0.09 \times 1003 = 90.27 \text{ kg m}^{-3}$$

From Equation 5.18,

$$\Pi = \frac{90.27 \times 8.314(273 + 35)}{342}$$

$$= 676 \text{ kPa}$$

Therefore,

$$\text{Required flux} = \frac{5000}{8}$$

$$= 625 \text{ kg h}^{-1}$$

From Equation 5.23,

$$6.25 = 6.3 \times 10^{-3} A(4000 - 676)$$

Thus,

$$A = 29.8 \text{ m}^2 \approx 30 \text{ m}^2$$

Table 5.6 Osmotic pressures of selected solutions

Solution	Concentration	Osmotic pressure (Pa $\times$ 10^5)
Apple juice	15° Brix	2.04
Citrus juice	10° Brix	1.48
Coffee extract	28% TS	3.40
Lactose	1% w/v	0.37
Milk	–	0.69
Salt solution	15% TS	13.8
Sucrose solution	44° Brix	6.9
Sugar syrup	20° Brix	3.41
Tomato paste	33° Brix	6.9
Whey	–	0.69

TS = total solids, w/v = weight per volume.
Adapted from Lewis (1996a)

$$J_s = K_s A \Delta c \hspace{6cm} \boxed{5.23}$$

where J_s (kg h^{-1}) = solute flux, K_s (kg m^{-2} h^{-1} kPa^{-1}) = mass transfer coefficient of solute through the membrane, and Δc (kg m^{-3}) = change in solute concentration across the membrane.

From these equations it can be seen that water flux is influenced by hydraulic pressure difference across the membrane, but this has no effect on the solute flux, which is influenced by the difference in solute concentration across the membrane. Further details are given by Girard and Fukumoto (2000).

Reverse osmosis (or 'hyperfiltration') membranes allow water to pass through, while salts, monosaccharides and aroma compounds are rejected (retained) by the membrane. They have no pores and movement of molecules is by diffusion and not by liquid flow. Water molecules dissolve at one face of a dense polymer layer in the membrane, are transported through it by diffusion and then removed from the other face. Solutes that are rejected either have a lower solubility than water in the membrane material or diffuse more slowly through it. Low molecular weight solutes have a high osmotic pressure (Table 5.6) and it is therefore necessary to apply a high hydrostatic pressure to overcome this and achieve separation.

Ultrafiltration membranes are designed to separate macromolecules from smaller molecular weight solutes and water, and microfiltration membranes separate suspended particles (Fig. 5.14). Membranes for both operations have a porous structure (section 5.5.2). Water and small molecular weight solutes flow through porous membranes under hydraulic (streamline, viscous) flow (Chapter 1, section 1.3.4), and larger solutes become concentrated at the membrane surface. The rate of rejection (Equation 5.26) for ultra-filtration membranes is 95–100% of high molecular weight solutes and 0–10% of low molecular weight solutes (virtually free passage). Other factors that affect the flowrates through the membrane include the resistance of boundary layers of liquid on each side of the membrane, the extent of fouling by physical blocking of pores by solid particles (especially in hollow fibre and spiral wound membranes (section 5.5.3)), by 'adsorptive fouling' (interaction of macromolecules with the membrane surface leading to chemical blocking of pores) or by concentration polarisation.

Concentration polarisation occurs in both reverse osmosis and ultrafiltration when molecules in the retentate accumulate in the boundary layer next to the membrane surface. Their concentration becomes higher than that in the bulk of the feed material and this has a significant effect on the performance of the membrane. In reverse osmosis, the

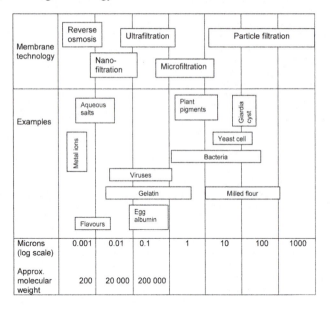

Fig. 5.14   Size separation capabilities of different membrane systems (from Anon 1997).

concentration of low molecular weigh solutes increases the osmotic pressure, which at a constant trans-membrane pressure causes the flux through the membrane to decrease. In addition the concentration gradient causes back diffusion of the solute from the boundary layer into the bulk of the liquid. In ultrafiltration, the larger molecules do not increase the osmotic pressure, but their increased concentration may cause them to precipitate and accumulate as a gel on the membrane surface. This increases the resistance to flow and reduces the flux through the membrane. The effect is described in more detail by Girard and Fukumoto (2000), and Singh and Heldman (2001) describe calculations of the effect of concentration polarisation on the flux through both reverse osmosis and ultrafiltration membranes. D'Souza (2005) has reviewed the mechanisms and methods of cleaning membranes.

The flux in ultrafiltration membranes is controlled by the applied pressure and the solute concentrations in the bulk of the liquid and at the membrane surface:

$$J = KA \ln\left(\frac{c_1}{c_2}\right)$$

[5.24]

where c_1 = concentration of solutes at the membrane and c_2 = concentration of solutes in the liquid.

Other factors that influence the flux include the liquid velocity, viscosity and temperature. A high flowrate is necessary to reduce concentration polarisation. In batch operation the liquid is recirculated until the desired concentration is achieved, whereas in continuous production an equilibrium is established, where the feedrate equals the sum of the permeate and concentrate flowrates. The ratio of this sum determines the degree of concentration achieved. Theoretical aspects of membrane separation are discussed in more detail by Singh and Heldman (2001), Toledo (1999b), Heldman and Hartel (1997b) and Lewis (1996a,b).

The performance of a membrane can be described by membrane retention and rejection values or its conversion percentage. The membrane retention value (R_f) is found using:

$$R_f = \frac{(c_f - c_p)}{c_f}$$

$$\boxed{5.25}$$

where c_f (kg m^{-3}) = concentration of solute in the feed and c_p (kg m^{-3}) = concentration of solute in the permeate.

Similarly, the rejection value (R_j) is found using:

$$R_j = \frac{(c_f - c_p)}{c_p}$$

$$\boxed{5.26}$$

The membrane conversion percentage (Z) is calculated using:

$$Z = \frac{j_p \times 100}{j_f}$$

$$\boxed{5.27}$$

where j_p (kg h^{-1}) = product flux and j_f (kg h^{-1}) = feed flux. Thus a membrane that has a conversion percentage of 65% will produce 65 kg h^{-1} of permeate and 35 kg h^{-1} of retentate from a feedrate of 100 kg h^{-1}.

5.5.2   Equipment and applications

Reverse osmosis
Different types of membranes reject solutes within specific ranges of molecular weight (Fig. 5.14). These molecular weight 'cut-off' points are used to characterise membranes. For reverse osmosis membranes, the cut-off points range from molecular weights of 100 Da at 4000–7000 kPa to 500 Da at 2500–4000 kPa (Table 5.7).

An important commercial food application of reverse osmosis is the concentration of whey from cheese manufacture, either as a pre-concentration stage prior to drying or for use in the manufacture of ice cream. Reverse osmosis is also used to:

- concentrate and purify fruit juices, enzymes, fermentation liquors, vegetable oils, citric acid, egg albumin, milk, coffee, syrups, natural extracts and flavours;
- to extract neutraceuticals from plant materials (Jennings 2000);
- to clarify wine and beer;
- to 'dealcoholise' beers, cider and wines to produce low-alcohol products;
- to recover proteins or other solids from distillation residues, dilute juices, waste water from corn milling or other process washwaters (e.g. Fradin 2001);
- to pre-concentrate coffee extracts and liquid egg before drying, or juices and dairy products before evaporation, so improving the economy of dryers and evaporators;
- to demineralise and purify water from boreholes or rivers or to desalinate seawater. Monovalent and polyvalent ions, particles, bacteria and organic materials with a molecular weight greater than 300 are all removed by up to 99.9% to give high-purity process water for beverage manufacture and other applications.

Table 5.7   Characteristics of membrane processes (reprinted by permission of the publisher, Taylor and Francis Ltd at: http://informaworld.com)

Process	Size range (μm)	Operating pressures (kPa)	Typical flux (1 m^{-2} h^{-1})
Reverse osmosis	0.0001–0.001	1380–6890	3–30
Ultrafiltration	0.001–0.1	345–1380	30–300
Microfiltration	0.1–2	20–345	100–300

Adapted from Girard and Fukumoto (2000)

Table 5.8   Advantages and limitations of different types of membrane for RO and UF

Type of membrane	Advantages	Limitations
Cellulose acetate	High permeate flux Good salt rejection Easy to manufacture	Break down at high temperatures pH sensitive (can only operate between pH 3–6) Broken down by chlorine, causing problems with cleaning and sanitation
Polymers (for example polysulphones, polyamides, poly-vinyl chloride, polystyrene, polycarbonates, polyethers	Polyamides have better pH resistance than cellulose acetate Polysulphones have greater temperature resistance (up to 75 °C), wider pH range (1–13) and better chlorine resistance (up to 50 $\mu g\,kg^{-1}$). Easy to fabricate Wide range of pore sizes	Do not withstand high pressures and restricted to ultrafiltration Polyamides more sensitive to chlorine than cellulose acetate
Composite or ceramic membranes (for example porous carbon, zirconium oxide, alumina)	Inert Very wide range of operating temperatures and pH Resistant to chlorine and easily cleaned	Expensive

Adapted from the data of Heldman and Hartel (1997a)

The molecular structure of reverse osmosis membranes is the main factor that controls the rate of diffusion of solutes. The materials should have high water permeability, high solute rejection and durability. Membranes are made from polyamides, polysulphones and inorganic membranes made from sintered or ceramic materials (Table 5.8). These are able to withstand temperatures up to 80 °C and a pH range from 3 to 11.

A typical reverse osmosis plant operates with a flux of $450\,l\,h^{-1}$ at 4000 kPa up to a flux of $1200–2400\,l\,h^{-1}$ at 8000 kPa. A fourfold concentration of whey typically would have production rates of $80–90\,t\,day^{-1}$. Lewis (1996a) compares 'once-through' and multi-stage recycling of liquid through banks of RO membranes.

Nanofiltration, ultrafiltration and microfiltration
The term nanofiltration (or 'loose reverse osmosis') is used when membranes remove materials having molecular weights in the order of 300–1000 Da (Rosenberg 1995). This compares to a molecular weight range of 2000–300 000 Da for ultrafiltration membranes, although there is overlap with microfiltration (Fig. 5.14). Nanofiltration is capable of removing ions that contribute significantly to the osmotic pressure and thus allows operation at pressures that are lower than those needed for reverse osmosis.

Ultrafiltration membranes have a higher porosity and retain only large molecules (e.g. proteins or colloids) that have a lower osmotic pressure. Smaller solutes are transported across the membrane with the water. Ultrafiltration therefore operates at lower pressures (50–1500 kPa). The most common commercial application of ultrafiltration is in the dairy industry to concentrate milk prior to the manufacture of dairy products, to concentrate whey to 20% solids, or to selectively remove lactose and salts. In cheese manufacture, ultrafiltration has advantages in producing a higher product yield and nutritional value, simpler standardisation of the solids content, lower rennet consumption and easier processing. Other applications include:

- concentration of sucrose and tomato paste;
- treatment of still effluents in the brewing and distilling industries;
- separation and concentration of enzymes, other proteins or pectin;
- removal of protein hazes from honey and syrups;
- treatment of process water to remove bacteria and contaminants (greater than 0.003 μm in diameter); and
- pre-treatment for reverse osmosis membranes to prevent fouling by suspended organic materials and colloidal materials.

The main requirement of ultrafiltration membranes is the ability to form and retain a 'microporous' structure during manufacture and operation. Rigid ceramic or glassy polymers, which are thicker than reverse osmosis membranes (0.1–0.5 μm) are used. They are mechanically strong, durable and resistant to abrasion, heat and hydrolysis or oxidation in water. These materials, which do not creep, soften or collapse under pressure, are described in Table 5.8. The pore size of the inner skin determines the size of molecules that can pass through the membrane; larger molecules being retained on the inside of the membrane. The pores of ultrafiltration membranes are typically 0.01–100 μm. Typical operating pressures in ultrafiltration plants are 70–1000 kPa, at flux rates of up to 40 l min^{-1} per tube.

An extension of ultrafiltration, in which water is added back to the extract during the concentration process is known as 'diafiltration'. This is useful in selectively removing lower molecular weight materials from a mixture, and is described in detail by Lewis (1996b). It offers an alternative to ion exchange or electrodialysis for removal of anions, cations, sugars, alcohol or antinutritional compounds.

Microfiltration is similar to ultrafiltration in using lower pressures than reverse osmosis, but is distinguished by the larger range of particle sizes that are separated (0.01–2 μm) (Fig. 5.14). Whereas ultrafiltration is used to separate macromolecules, microfiltration separates dispersed particles such as colloids, fat globules or cells, and may therefore be thought of as falling between ultrafiltration and conventional filtration (Grandison and Finnigan 1996). Microfiltration membranes are similar to ultrafiltration membranes, having two parts: a macroporous support material and a microporous coating on the surface. The macroporous support materials are produced from sintered materials such as alumina, carbon, stainless steel and nickel, and have a pore diameter of 10 μm or more to allow the permeate to drain away freely. Inorganic materials such as glass and compounds of aluminium, zirconium and titanium are used for the microporous component of membranes. These have good structural strength and resistance to higher temperatures and damage from chemicals or abrasion. Microfiltration membranes have high permeate fluxes initially, but become fouled more rapidly than reverse osmosis or ultrafiltration membranes. They are therefore 'backflushed' (a quantity of permeated is forced back through the membrane) to remove particles from the membrane surface. Other methods of maintaining flux levels are described by Grandison and Finnigan (1996).

An example of the applications of ultrafiltration, microfiltration and nanofiltration for the fractionation of whey is shown in Fig. 5.15. These processes enable new possibilities to tailor the functional properties of milk proteins (e.g. water-holding capacity, fat binding, emulsification characteristics, whippability and heat stability) for specific applications as food ingredients (Rosenberg 1995).

Pervaporation

Pervaporation is a membrane separation technique in which a liquid feed mixture is separated by partial vaporisation through a non-porous, selectively permeable membrane.

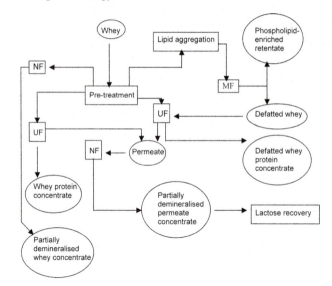

Fig. 5.15   Applications of UF, MF and NF in whey processing (reprinted from Rosenberg (1995), © 1995, with permission from Elsevier).

It produces a vapour permeate and a liquid retentate. Partial vaporisation is achieved by reducing the pressure on the permeate side of the membrane (vacuum pervaporation) or less commonly, sweeping an inert gas over the permeate side (sweep gas pervaporation). There are two types of membrane, which are used in two distinct applications: hydrophilic polymers (e.g. polyvinyl alcohol or cellulose acetate) preferentially permit water permeation, whereas hydrophobic polymers (e.g. polydimethylsiloxane or polytrimethylsilylpropyne) preferentially permit permeation of organic materials. Vacuum pervaporation at ambient temperatures using hydrophilic membranes is used to dealcoholise wines and beers, whereas hydrophobic membranes are used to concentrate aroma compounds, such as alcohols, aldehydes and esters, to up to one hundred times the concentration in the feed material. The concentrate is then added back to a food after processing (e.g. after evaporation (Chapter 14) to improve its sensory characteristics). Reviews of these and other applications of pervaporation are given by Girard and Fukumoto (2000) and Karlsson and Tragardh (1996).

5.5.3   Types of membrane system
The membrane plus support material are termed a 'module'. The design criteria for modules include:

- provision of a large surface area in a compact volume;
- configuration of the membrane to permit suitable turbulence, pressure losses, flowrates and energy requirements;
- no dead spaces and capability for cleaning-in-place (CIP) on both the concentrate and permeate sides;
- easy accessibility for cleaning and membrane replacement (Lewis 1996a).

The two main configurations of membranes are the tubular and flat plate designs. Tubular membranes are held in cylindrical tubes mounted on a frame with associated pipework and controls. The two main types are the hollow fibre and wide tube designs. Hollow

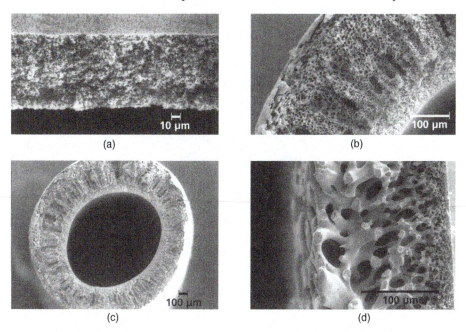

(a)
(b)
(c)
(d)

Fig. 5.16 Membrane structures: (a) asymmetrical membrane cross-section; (b) symmetrical membrane cross-section; (c) hollow-fibre asymmetrical membrane cross-section; (d) flat sheet asymmetrical membrane cross-section (courtesy of Anon 1997).

fibre systems (Fig. 5.16) have 50–1000 fibres, 1 m long and 0.001–1.2 mm in diameter, with membranes of about 250 μm thick. The fibres are attached at each end to a tube sheet to ensure that the feed is uniformly distributed to all tubes. These systems have a large surface area to volume ratio and a small hold-up volume. They are used for reverse osmosis applications such as desalination, but in ultrafiltration applications the low applied pressure and laminar flow limit this system to low-viscosity liquids that do not contain particles. They are also more expensive because an entire cartridge must be replaced if one or more fibres burst. However, they are easy to clean and do not block easily.

In the wide tube design (Fig. 5.17) a number of perforated stainless steel tubes are fabricated as a shell and tube heat exchanger (see Chapter 14, section 14.1.3) and each tube is lined with a membrane. The tubes support the membrane against the relatively high applied pressure. Special end caps connect up to 20 tubes, each 1.2–3.6 m long and 12–25 mm in diameter, in series or in parallel depending on the application. These systems operate under turbulent flow conditions with higher flowrates than hollow fibre systems and can therefore handle more viscous liquids and small particulates. They are less susceptible to fouling and are suitable for CIP, although the higher flowrates mean that pumping costs are correspondingly higher than hollow fibre systems.

Flat plate systems can be either plate-and-frame types or spiral-wound cartridges. The plate-and-frame design is similar to a plate filter press (Fig. 5.6) or plate heat exchanger (Chapter 12, Fig. 12.3), having membranes stacked together with intermediate spacers and collection plates to remove permeate. Flow can be either laminar or turbulent and feed can be passed over plates in either series or parallel. The design allows a high surface area to be fitted into a compact space and individual membranes can be easily replaced (Lewis 1996a).

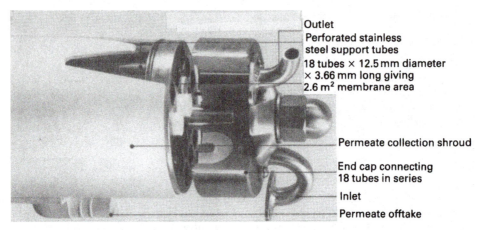

Outlet
Perforated stainless
steel support tubes
18 tubes × 12.5 mm diameter
× 3.66 mm long giving
2.6 m² membrane area

Permeate collection shroud

End cap connecting
18 tubes in series

Inlet

Permeate offtake

Fig. 5.17   Tubular membrane (courtesy of Patterson Candy International Ltd, now part of Black & Veatch at www.bv.com/).

In the spiral wound system (Fig. 5.18), alternating layers of polysulphone membranes and polyethylene supports are wrapped around a hollow central tube and are separated by a channel spacer mesh and drains. The cartridge is approximately 12 cm in diameter and 1 m long. Feed liquor enters the cartridge and flows tangentially through the membrane. Permeate flows into channels and then to the central tube, and the concentrate flows out of the other end of the cartridge. Separator screens cause turbulent flow to maximise the flux, and this, together with the low volume of liquid in relation to the large membrane area, reduce the need for large pumps. These systems are relatively low cost and are gaining in popularity.

Ion exchange and electrodialysis
Ion exchange and electrodialysis are both separation methods that remove electrically charged ions and molecules from liquids. Whereas the driving force for transport across

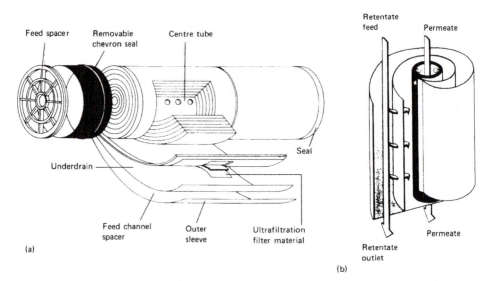

(a)

Feed spacer   Removable   Centre tube
               chevron seal

Underdrain

Feed channel   Outer   Ultrafiltration
spacer   sleeve   filter material

Seal

(b)

Retentate
feed   Permeate

Permeate

Retentate
outlet

Fig. 5.18   Spiral cartridge membrane: (a) components; (b) schematic flow diagram (courtesy of Millipore Ltd at www.millipore.com).

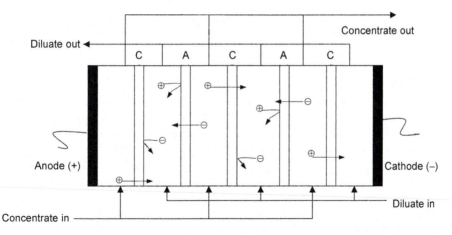

Fig. 5.19 Electrodialysis system: C = cation exchange membrane; A = anion exchange membrane (adapted from Girard and Fukumoto 2000).

reverse osmosis and ultrafiltration membranes is the hydrostatic pressure applied to the feed liquid, in ion exchange and electrodialysis it is the concentration difference of ions in solution. In ion-exchange, solutes such as metal ions, proteins, amino acids and sugars are transferred from a feed material and retained on a solid ion-exchange material by a process of electrostatic adsorption (i.e. attraction between the charge on the solute and an opposite charge on the ion exchanger). They can then be separated by washing them off the ion exchanger. The ion exchanger is either a cation exchanger (having a negative charge) or an anion exchanger (having a positive charge). Electrodialysis is used to separate electrolytes from non-electrolytes and to exchange ions between solutions. A direct current is passed through a solution and, depending on their electrical charge, ions or molecules migrate towards an anode or a cathode. Separation is based on ion-selective membranes (sheets of cation- and anion-exchange materials) that act as barriers to either anions or cations. Therefore anions that migrate towards the anode pass through anion membranes but are rejected by cation membranes and vice versa. The membranes are arranged alternately to form ion-diluting cells and ion-concentrating cells (Fig. 5.19).

The energy consumption in electrodialysis is found using:

$$E = I^2 nRt \qquad \boxed{5.28}$$

where E (J) = energy consumption, I (A) = electric current, n = number of cells, R (Ω) = resistance of the cell and t (s) = time. Further details are given by Bazinet (2005) and Singh and Heldman (2001).

Ion-exchange and electrodialysis equipment is constructed using a porous matrix made from polyacrylamides, polystyrene, dextrans or silica. The applications in food processing include decolorisation of sugar syrups, protein recovery from whey or blood, water softening and dimineralisation, and separation of valuable materials such as purified enzymes. In an example described by Grandison (1996), cheese whey is circulated through ion-diluting cells and brine is circulated through ion-concentrating cells. Mineral ions leave the whey and become concentrated in the brine, thus demineralising the whey. This is a major application of electrodialysis, and demineralised whey is used in infant feeds, drinks, salad dressing, confectionery coatings, ice cream mixes and bakery products. Electrodialysis can also be used to remove potassium and tartaric acid from wines to prevent precipitate formation, to desalinate pickling brines and to de-acidify fruit juices to reduce their sourness. It is also used to purify water and to obtain salt from seawater.

5.5.4   Effects on foods and micro-organisms

Each of the unit operations described in this chapter is intended to remove components of the food and they are therefore used to intentionally alter or improve the sensory properties of the resulting products. The effects on nutritional value are difficult to assess in most operations and are usually incidental to the main purpose of altering eating qualities. However, with the exception of some types of solvent extraction, these operations take place at ambient temperatures and loss of heat-sensitive nutrients is not significant. The main changes occur as a result of the physical removal of food components (Table 5.9). In milk processing for example, the fat-soluble vitamins, retinol, carotene and vitamin D, are removed in the milkfat when it is separated from skimmed milk and concentrated in cream and butter. Conversely, water soluble vitamins and minerals are largely unchanged in skimmed milk, but substantially reduced in cream and butter.

In contrast to concentration by boiling (Chapter 14, section 14.1), most of the separation procedures described in this chapter concentrate foods without heat to produce good retention of sensory and nutritional qualities. For example in membrane concentration of whey, the functional properties (emulsifying ability, foaming ability and solubility) of proteins are retained, and different products that have specified ranges of protein and lactose contents can be produced for use in fortified jams, low-calorie mayonnaise, dips, sauces and skinless sausages, and as alternatives to egg albumin (Smallwood 1986). Table 5.10 shows changes in the composition of orange juice before

Table 5.9   Composition of whole cow's milk before and after ultrafiltration

Component (%)	Raw milk	Concentrate	Permeate
Total solids	13.1	43.3	6.7
Fat	4.1	21.8	0
Protein	3.6	16.1	0.49
Non-protein nitrogen (NPN)	0.19	0.18	0.19
Lactose	4.7	3.2	5.2
Ash	0.7	1.9	0.54

Adapted from Glover (1985) and Jensen (1995)

Table 5.10   Composition of orange juice before and after ultrafiltration (reprinted by permission of the publisher, Taylor and Francis Ltd at: http://informaworld.com)

Component $(mg\,l^{-1})$	Feed	Permeate
Water-soluble pectin	135	27
Mono-galacturonic acid	288	193
Polyphenols	1536	1354
Essential oils	370	20
Anthocyanogens	1077	825
Total nitrogen	1330	1176

Note: Minerals, total acidity and sugars (sucrose, glucose and fructose) had similar concentrations before and after UF.
Adapted from Girard and Fukumoto (2000)

and after ultrafiltration. Both types of membrane retain proteins, fats and larger carbohydrates, but the larger pore size of ultrafiltration membranes allows sugars, vitamins and amino acids to be lost.

Effect on micro-organisms

Centrifugation and filtration are used to remove yeast cells from wines and beers after fermentation. Membrane filtration is used for microbiological control in beverage processing to remove micro-organisms, including yeasts, moulds and bacteria, but not viruses, from foods or process waters, where they are concentrated in the retentate. They reduce the numbers of spoilage micro-organisms in permeates while retaining the sensory and nutritional qualities of the products. Other separation technologies have no direct effect on micro-organisms, but inadequate standards of hygiene or process control could allow contamination of foods during or after these operations. Because heat is not involved, this could result in accelerated spoilage or toxicity if appropriate quality assurance measures are not taken (see also Chapter 1, section 1.5.1).

References

ANON, (1995), Oilseed processing, US Environment Protection Agency, Technology Transfer Network Clearinghouse for Inventories & Emissions Factors, AP 42, Fifth Edition, Volume I Chapter 9: Food and Agricultural Industries, available at www.epa.gov/ttn/chief/ap42/ch09/final/c9s11-1.pdf.

ANON, (1997), Cost-effective membrane technologies for minimising wastes and effluents, Guide GG54, Environmental Technology Best Practice Programme, UK Government.

ANON, (2002), How filteraid works, Information provided by Beaver Chemicals Ltd., available at www.filteraid.com/how_filteraid_works.htm.

ANON, (2003), Product information from General Filtration, available at www.generalfiltration.com/filter_aids.html.

ANON, (2004a), Decanter centrifuges, company information provided by US Centrifuge, available at www.uscentrifuge.com/decanter-centrifuges.htm.

ANON, (2004b), Scios product information, available at www.scios.co.nz/cellulose-filters.htm.

ANON, (2005a), European Union guidelines on the application of Directive 94/9/EC of 23 March 1994 on equipment and protective systems intended for use in potentially explosive atmospheres (ATEX), Second edn, July.

ANON, (2005b), Our industry major commodities, information from Fediol, available at www.fediol.be/2/index.php.

ANON, (2008a) Rousselet-Robatel Model BXP centrifugal separator, company information from Rousselet-Robatel, available at http://www.rousselet-robatel.com/products/bxps.php.

ANON, (2008b), Alfa-Laval disc stack centrifuge technology, Alfa-Laval at http://local.alfalaval.com/en-gb/key-technologies/separation/separators/.

ANON, (2008c), Process and product development information, IDL Process Solutions Inc, available at www.idlconsulting.com/equip.html.

ANON, (2008d), Automatic and frame filter presses for wastewater treatment, power, chemical, mining and food industries, information from Novatek AB, available at www.idswater.com/water/us/novatek/automatic_frame_filter_press/194_0/g_supplier_4.html.

ANON, (2008e), Company information from Nordisk Perlite ApS, available at www.perlite.dk/whats.htm.

ANON, (2008f), Food industry centrifuges, Rousselet-Robatel company information, available at www.rousselet-robatel.com/products/food.php.

ANON, (2008g), Filter presses, information from Petrochemical Services Incorporated, available at www.petrosvs.com/fltrprss.htm.

ANON, (2008h), Information provided by Siemens AG Water technologies, available at http://www.usfilter.com, and follow links to food and beverages then filter presses.

ANON, (2008i), Plate and frame filter presses, Ertel Alsop company information, available from http://www.ertelalsop.com/depth/equipment/plateframe.html.

ANON, (2008j), Information from Les Industries Fournier Inc., available at www.rotary-press.com/equipement-deshydratation/principle_description-and-principle_ang.cfm. Comparative information on other types of filtration equipment at www.rotary-press.com/equipement-deshydratation/advantages_plate-and-frame-press_ang.cfm, www.rotary-press.com/equipement-deshydratation/advantages_belt-filter-press_ang.cfm, www.rotary-press.com/equipement-deshydratation/advantages_centrifuge_ang.cfm and www.rotary-press.com/equipement-deshydratation/advantages_screw-press_ang.cfm.

ANON, (2008k), Product information from Krauss-Maffei Process Technology, available at http://www.kmpt.net/disk_filter_SF.aspx.

ANON, (2008l), The Olive Oil Source, available at www.oliveoilsource.com/mill_and_press_facts2.htm.

ANON, (2008m), Principles of oil extraction, Practical Action, available at http://practicalaction.org/docs/technical_information_service/oil_extraction.pdf.

ANON, (2008n), Extractor, finisher, separator product information from Brown International Corporation, available at www.brown-intl.com/download/Mdl3900c.pdf or www.brown-intl.com/download/MDL202c.PDF.

BARKER, A., (1987), Private communication, Simon Rosedowns Ltd, Hull.

BATES, R.P., MORRIS, J.R. and CRANDALL, P.G., (2001), Tree fruit: apple, pear, peach, plum, apricot and plums, in *Principles and Practices of Small- and Medium-scale Fruit Juice Processing*, FAO Agricultural Services Bulletin, No. 146, Food and Agriculture Organization of the United Nations, Rome, available at www.fao.org/docrep/005/Y2515E/y2515e15.htm.

BAZINET, L., (2005), Electrodialytic phenomena and their applications in the dairy industry: a review, *Critical Reviews in Food Science and Nutrition*, **45**, 307–326.

BERNADINI, E., (1976), Batch and continuous solvent extraction, *J. American Oil Chemists' Society*, **53**, 278.

BOHAC, C.F., (1998), Assessment of methodologies for colorimetric cholesterol assay of meats, *J. Food Science*, **53**, 1642–1644.

BRENNAN, J.G, BUTTERS, J.R., COWELL, N.D. and LILLEY, A.E.V., (1990), *Food Engineering Operations*, 3rd edn, Elsevier Applied Science, London.

CALAIS, P. and CLARK, A.R., (undated), Waste vegetable oil as a diesel replacement fuel, available at www.shortcircuit.com.au/warfa/paper/paper.htm.

CATCHPOLE, O.J., GREY, J.B., PERRY, N.B., BURGESS, E.J., REDMOND, W.A. and PORTER, N.G., (2003), Extraction of chili, black pepper, and ginger with near-critical CO_2, propane, and dimethyl ether: analysis of the extracts by quantitative nuclear magnetic resonance. *J. Agriculture Food Chemistry*, **51** (17), 4853–4860.

CLARKE, R.J., (1990), Instant coffee technology, in (A. Turner, Ed.), *Food Technology International Europe*, Sterling Publications International, London, pp. 137–139.

DAUTHY, M.E., (1995), *Fruit and Vegetable Processing*, FAO Agricultural Services Bulletin No. 119, Food and Agriculture Organization of the United Nations, Rome.

DEL VALLE, J.M. and DE LA FUENTE, J.C., (2006), Supercritical CO_2 extraction of oilseeds: Review of kinetic and equilibrium models, *Critical Reviews in Food Science and Nutrition*, **46**, 131–160.

D'SOUZA, N.M., (2005), Membrane cleaning in the dairy industry: a review, *Critical Reviews in Food Science and Nutrition*, **45**, 125–134.

EARLE, R.L., (1983), *Unit Operations in Food Processing*, 2nd edn, Pergamon Press, Oxford, pp. 143–158.

ERICKSON, D.R. and WIEDERMANN, L.H., (1989), Soybean oil – modern processing and utilization, American Soybean Association, available at www.asa-europe.org/pdf/sboprocess.pdf.

FELLOWS, P.J., (1997), *Guidelines for Small Scale Fruit and Vegetable Processors*, FAO Technical

Bulletin No. 127, Food and Agriculture Organization of the United Nations, Rome.

FOOD STANDARDS AGENCY, (2002), *McCance and Widdowson's the Composition of Foods*, 6th Summary edn, Royal Society of Chemistry, Cambridge.

FRADIN, B., (2001), Recovery of sugar with membranes, *Food Processing*, June, 19.

GAEHRS, H.J., (1990), Supercritical gas extraction: optimising the economics, in (A. Turner, Ed.), *Food Technology International Europe*, Sterling Publications International, London, pp.77–81.

GIRARD, B. and FUKUMOTO, L.R., (2000), Membrane processing of fruit juices and beverages: a review, *Critical Reviews in Food Science and Nutrition*, **40** (2), 91–157.

GLOVER, F.A., (1985), *Ultrafiltration and Reverse Osmosis for the Dairy Industry*, Technical Bulletin No. 5, The National Institute for Research in Dairying, Reading.

GOPOLAN, A.S., (2003), Supercritical carbon dioxide: separations and processes, ACS Symposium #860: American Chemical Society Publication, Washington, DC.

GRANDISON, A.S., (1996), Ion-exchange and electrodialysis, in (A.S. Grandison and M.J. Lewis, Eds.), *Separation Processes in the Food and Biotechnology Industries*, Woodhead Publishing, Cambridge, pp. 155–177.

GRANDISON, A.S. and FINNIGAN, T.J.A., (1996), Microfiltration, in (A.S. Grandison and M.J. Lewis, Eds.), *Separation Processes in the Food and Biotechnology Industries*, Woodhead Publishing, Cambridge, pp. 141–153.

GUNSTONE, F., (Ed.), (2004), *Rapeseed and Canola Oil – Production, Processing, Properties and Uses*, Blackwell Publishing, Oxford.

HELDMAN, D.R. and HARTEL, R.W., (1997a), Other separation processes, in *Principles of Food Processing*, Chapman and Hall, New York, pp. 219–252.

HELDMAN, D.R. and HARTEL, R.W., (1997b), Liquid concentration, in *Principles of Food Processing*, Chapman and Hall, New York, pp. 138–176.

HEMFORT, H., (1983), *Centrifugal Separators for the Food Industry*, Westfalia Separator AG, Germany.

HUI, Y. H., BARTA, J., PILAR CANO, M., GUSEK, T.W., SIDHU, J. and SINHA, N., (Eds.), (2006), *Handbook of Fruits and Fruit Processing*, Blackwell Publishing, Oxford.

JENNINGS, B., (2000), Filtration technology forms food of the future, *Food Processing*, June, 49–50.

JENSEN, R.G., (1995), *Handbook of Milk Composition*, Academic Press, London.

KARLSSON, H.O.E. and TRAGARDH, G., (1996), Applications of pervaporation in food processing. *Trends in Food Science and Technology*, **7** (March), 78–83.

KESSLER, N., (1985), Understanding solvent extraction of vegetable oils, Technical Paper No. 41, Volunteers in Technical Assistance, available at http://journeytoforever.org/biofuel_library/solvextract.html.

LENIGER, H.A. and BEVERLOO, W.A., (1975), *Food Process Engineering*, D. Reidel, Dordrecht, pp. 498–531.

LEWIS, M.J., (1996a), Pressure-activated membrane processes, in (A.S. Grandison and M.J. Lewis, Eds.), *Separation Processes in the Food and Biotechnology Industries*, Woodhead Publishing, Cambridge, pp. 65–96.

LEWIS, M.J., (1996b), Ultrafiltration, in (A.S. Grandison and M.J. Lewis, Eds.), *Separation Processes in the Food and Biotechnology Industries*, Woodhead Publishing, Cambridge, pp. 97–140.

MADSEN, R.F., (1974), Membrane concentration, in (A. Spicer, Ed.), *Advances in Preconcentration and Dehydration of Foods*, Applied Science, London, pp. 251–301.

MERMELSTEIN, N.H., (1999), Commercial-scale separation, *Food Technology*, **53** (2), 82–83.

MOHAMED, R.S. and MANSOORI, G.A., (2002), The use of supercritical fluid extraction technology in food processing, *Food Technology Magazine*, June, available at www.uic.edu/labs/trl/1.OnlineMaterials/SCEinFoodTechnology.pdf#search=%22solvent%20extraction%20equipment%2Bfood%22.

OHLER, J.G., (2006), Modern Coconut Management; palm cultivation and products, available at http://ecoport.org/ep?SearchType=earticleView&earticleId=127&page=1424.

OWOLARAFE, O.K., FABORODE, M.O. and AJIBOLA, O.O., (2002), Comparative evaluation of the digester screw press and a hand-operated hydraulic press for palm fruit processing, *J. Food Engineering*, **52**, 249–255.

PHILLIPS, C., (2006), Product review: choosing the best tank press for your winery, *Wine Business Monthly*, 15 February 2006, available at Wine Busienss.com at www.winebusiness.com/ReferenceLibrary/webarticle.cfm?dataId=42352.

POKU, K., (2002), *Small-scale Palm Oil Processing in Africa*, FAO Agricultural Services Bulletin 148, Food and Agriculture Organization of the United Nations, Rome, available at www.fao.org/DOCREP/005/Y4355E/Y4355E00.HTM

POTTS, K.H. and MACHELL, K., (1993), *The Manual Screw Press for Small-scale Oil Extraction*, Intermediate Technology Publications, London.

RIZVI, S.S.H., DANIELS, J.A., BENADO, A.L. and ZOLLWEG, J.A., (1986), Supercritical fluid extraction: operating principles and food applications, *Food Technology*, **40** (7), 56–64.

ROSENBERG, M., (1995), Current and future applications for membrane processes in the dairy industry, *Trends in Food Science and Technology*, **6**, 12–19.

ROSSELL, B., (1999), *Oils and Fats Handbook: Vegetable Oils and Fats*, Woodhead Publishing, Cambridge.

ROZZI, N.L. and SINGH, R.K., (2002), Supercritical fluids and the food industry, *Comprehensive Reviews in Food Science and Food Safety*, **1**, 33

SALUNKHE, D. K., (1992), *World Oilseeds: Chemistry, Technology and Utilization*. Van Nostrand Reinhold, Amsterdam.

SEMON, M., PATTERSON, M., WYBORNEY, P., BLUMFIELD, A. and TAGEANT, A., (1997), Soybean oil, in oil seed, edible oil, and essential oil extraction student webpage overview, available at www.wsu.edu/~gmhyde/433_web_pages/433Oil-web-pages/Soy/soybean1.html.

SINGH, R.P. and HELDMAN, D.R., (2001), Membrane concentration, in *Introduction to Food Engineering*, 3rd edn, Academic Press, London, pp. 531–556.

SMALLWOOD, M., (1986), Concentrating on natural proteins, *Food Processing*, September, 21–22.

SPILIMBERGO, S., ELVASSORE, N. and BERTUCCO, A., (2002), Microbial inactivation by high-pressure, *J. Supercritical Fluids*, **22**, 55–63.

STEIN, W., (1984), New oil extraction process, *Food Engineering International*, **59**, 61–63.

STEYTLER, D., (1996), Supercritical fluid extraction and its application to the food industry, in (A.S. Grandison and M.J. Lewis, Eds.), *Separation Processes in the Food and Biotechnology Industries*, Woodhead Publishing, Cambridge, pp. 17–64.

TOLEDO, R.T., (1999a), Extraction, in *Fundamentals of Food Process Engineering*, 2nd edn, Aspen Publications, Gaithersburg, MD, pp. 548–566.

TOLEDO, R.T., (1999b), Physical separation processes, in *Fundamentals of Food Process Engineering*, 2nd edn, Aspen Publications, Gaithersburg, MD, pp. 507–547.

VINE, R.P., (1987), The use of new technology in commercial winemaking, in (A. Turner, Ed.), *Food Technology International Europe*, Sterling Publications International, London, pp. 146–149.

YEO, S-D. and KIRAN, E., (2005), Formation of polymer particles with supercritical fluids: a review, *J. Supercritical Fluids*, **34** (3), 287–308.

6

Food biotechnology

Abstract: This chapter describes recent developments in food biotechnology, including genetic modification of foods and micro-organisms, nutritional genomics and the development of functional foods. Fermentation improves the palatability and acceptability of raw materials by producing flavour and aroma components and modifying food texture. This chapter describes the theory of microbial growth in food fermentations and the equipment used to produce fermented foods in submerged cultures and solid substrate fermentations. It outlines the production of microbial enzymes and selected commercially important fermented foods, including alcoholic drinks, bakery and dairy products, and mycoprotein; and concludes with a summary of developments in the production of bacteriocins and antimicrobial ingredients.

Key words: genetic modification, functional foods, fermentation, microbial enzymes, bacteriocins.

The Convention on Biological Diversity (Anon 2000a) defines biotechnology as 'any technological application that uses biological systems, living organisms, or derivatives thereof, to make or modify products or processes for specific use'. It therefore includes all forms of plant and animal breeding and development, traditional food fermentations and waste treatment. The Institute of Food Science and Technology has defined food biotechnology as 'the application of biological techniques to food crops, animals and micro-organisms to improve the quality, quantity, safety, ease of processing and production economics of food' (Anon 2004a). This chapter describes developments in food biotechnology including genetically modified foods, probiotics, prebiotics, microbial enzymes and selected food fermentations. 益生菌

6.1 Genetic modification

Genetic modification is the alteration of the genetic makeup of an organism that can be passed on to its descendants. Strictly, this includes traditional methods of selective breeding of crops or animals for specific attributes by normal reproduction, breeding closely related species or isolating mutants, which have been practised by farmers and breeders for thousands of years. However, the term is now more often used to describe 'genetic engineering' in which techniques in molecular biotechnology are used to manipulate genes, outside the normal reproductive process of a plant, animal or micro-organism, to produce new physiological or physical characteristics in a food. Inter-nationally, the FAO and WHO have defined genetically modified organisms (GMOs) as 'organisms in which the genetic material (DNA) has been altered in a way that does not occur naturally' (Anon 2007). EU legislation defines GM food as 'food containing, consisting of, or produced from, a genetically modified organism' (Anon 2008a).

There are a number of methods for making GM foods, each based on alteration of one or more genes. The production of an individual protein by an organism is specified by a gene (although a single gene can encode for more than one protein) and this can be modified by changing the DNA of the gene. There are a number of methods to do this:

- Chromosomal substitution (all, or part, of a chromosome from a donor organism replaces that of the recipient).
- Gene addition (a functional gene or part of a gene is inserted into an organism, e.g. by recombination).
- Rendering a gene non-functional (e.g. by mutation or the removal of all or part of the gene).
- Induced mutation (the modification of genetic material by mutagenic compounds or irradiation).
- Transfection (the introduction of genetic material into an organism often in the form of a plasmid).

Of these, recombination is important for food crops and the stages involved in producing a recombinant GMO are as follows:

1 Identification and isolation of the piece of DNA containing the targeted gene.
2 Precisely cutting out the nucleotide sequence of the gene using bacterial enzymes. The sequence is identified by a 'promoter' (a DNA sequence that is recognised by RNA polymerase and initiates transcription), and a termination sequence stops transcription of the new gene, producing several copies of the isolated gene using the polymerase chain reaction (PCR) method to replicate nucleotide sequences and help isolate the required DNA.
3 Reintroducing (splicing) the sequence into a different DNA segment in the cell. Where genetic material is transferred between different species the new GMO is known as a 'transgenic' organism. Splicing can be done by exploiting natural pathogens, such as *Agrobacteriaum tumefaciens* or *A. rhizogenes*, that infect cells by injecting genetic material; or by direct methods such as microinjection (also known as the 'gene gun' or 'biolistic transformation') where small particles of gold or tungsten, coated with DNA material, are accelerated to four times the velocity of sound using compressed helium, to penetrate cells without causing significant damage (Wilm 2005). Plant viruses can also be used to insert genetic material into cells, but the technique is limited by the range of hosts that the virus is able to infect.

Transposons (also called 'jumping genes' or 'mobile genetic elements') are useful to alter DNA inside an organism. They are sequences of DNA that can move to different positions within the genome of a cell to cause mutations and change the amount of DNA in the genome (a process called 'transposition'). Transposons can be grouped based on their mechanism of transposition into those that move by being transcribed to RNA and then back to DNA by reverse transcriptase (known as 'retrotransposons'), and those that move directly from one position to another using a transposase to 'cut and paste' them within the genome. Another genetic modification method is to interfere with the expession of a particular gene and 'silence' it using particular fragments of double-stranded RNA (dsRNA) (RNA interference (RNAi) is a mechanism that exists in both plants and animals and has a role in defence against viruses). RNAi and transposon silencing mechanisms are used to create GMOs for new food and animal feed crops.

Whichever method is used, GM plants are regenerated from a single new embryo cell into multicellular organisms. Plants also have a trait known as 'totipotency' in which cells from an adult plant are able to regenerate into new adults, enabling a range of cell types to be used for manipulation. A marker gene, often for antibiotic resistance, is included so that successfully transformed cells can be selected by their ability to grow on media containing antibiotics, whereas those that have not been transformed die. Mammals can also now be cloned from adult cells. Details of GM mechanisms and methods are described in more detail by Heller (2003).

6.1.1   GM food crops

The first commercially grown GM food crop in 1994 was a tomato (the 'FlavrSavr') that carried a gene to reduce the level of polygalacturonase and hence reduce softening and increase the shelf-life. A variant of this was used to produce tomato paste that was first sold in 1996 (Tester 2000). Commercially produced GM foods now include crops that are made resistant to herbicides or insects, new protein ingredients, or production of enzymes by GM micro-organisms (section 6.1.3). Examples of GM food crops include the following:

- Soybeans that tolerate glyphosate herbicides, maize and cotton that are herbicide-tolerant and have insect protection traits (the latter due to a gene for endotoxin production from *Bacillus thuringiensis* – Bt insecticidal protein) and maize with enhanced levels of lysine to improve the quality of protein for animal feeds.
- 'Golden rice', developed to synthesise the precursors of β-carotene in the edible parts of the grain for areas where there is a dietary shortage of vitamin A (not currently (2008) available for human consumption) and transgenic rice containing the milk proteins, lactoferrin and lysozyme, to improve oral rehydration therapy for diarrhoea.
- Insect-protected and herbicide-tolerant rapeseed (canola), oilseeds and oil-bearing plants that have altered fatty acid and lipid compositions (under development), in particular to produce oils that have a predominant fraction of conjugated linoleic acid (Dunford 2001) (also section 6.3).
- Virus resistant sweet potato.
- Plants that are better able to tolerate water and nitrogen limitation, or survive high-salinity, acidic soils or high or low ambient temperatures.

By 2005, more than 80 GM crops had been grown commercially or in field trials in over 40 countries on six continents. The most important GM crops are soybeans, maize, rapeseed and cotton, which have been widely adopted in the USA and in several developing countries (Table 6.1). Brazil in particular has increased cultivation of GM soybeans from 5 million hectares in 2004 to 9.4 million hectares in 2005. In 2004, the share of crops in global trade that contained GM material was estimated to be 90% for soybeans, 80% for maize, 44% for cottonseed and 73% for rapeseed (Brookes *et al.* 2005). The importance of these four crops is seen by their uses in a wide range of food ingredients (Table 6.2).

The development of transgenic farm animals has lagged behind crop development because of higher costs and lower reproductive rates (Anon 2005a). There is also

Table 6.1   Global area of transgenic crops

Country	Global area of transgenic crops			
	2001		2004	
	Area (million ha)	% of total	Area (million ha)	% of total
Argentina	11.8	23	16.2	20
Brazil	–	–	5.0	6
Canada	3.2	7	5.4	6
China	1.5	1	3.7	5
Paraguay	–	–	1.2	2
South Africa	0.2	0.5	0.5	1
USA	35.7	68	47.6	59
Total	52.6	99.5	79.6	99

From Embarek (2005)

widespread lack of public support for genetic modification of animals to improve productivity, with almost three-quarters of global consumers opposed to this (Hoban 2004). By 2006, GM fish were available and in some countries a recombinant version of the bovine growth hormone, somatotropin, was injected into dairy cattle to increase milk production, but there were no foods derived from GM livestock or poultry (Anon 2005a). Possible future applications of GMOs include the production of drugs in food (e.g. fruits that produce antibodies against infectious diseases for human vaccines) (Tester 2000), gluten-free wheat for people suffering from coeliac disease, GM fruit and nut trees that mature more quickly and produce crops years earlier than at present, and plants that produce new types of plastics that have unique properties (Anon 1999).

Table 6.2   Uses of ingredients from the four main GM crops

Crop	Ingredient	Uses
Soybeans	Beans	Soya sauce, soya milk, tofu, miso and tempeh
	Bran	Breakfast cereals
	Full-fat, enzyme-active soyflour	Bread improver
	Defatted flour	Texturised vegetable protein meat analogue or extender
	Soy concentrates/isolates	Infant formula, meal replacement preparations used in weight loss programmes, supplements for athletes' training diets, non-dairy creamer, meat extender
	Hydrolysed vegetable proteins (from acid- or enzymic-hydrolysis of soyflour)	Flavourings with savoury meat-like flavour used in a wide range of processed foods – soups, stock cubes, gravy powders, snack foods and condiments. Flavour enhancers and hypo-allergenic infant foods
	Soya oil (pure oil or hydrogenated)	Direct consumption or used in a wide range of foods and commercial frying oils
	Fatty acids and glyceride esters	Emulsifiers used in a wide range of products
	Lecithin (refined, fractionated, hydrolysed or enzymically modified)	Emulsifier used in bakery products, chocolate and dairy products
	Tocopherols	Antioxidant or vitamin E precursor
	Sterols, stanols and phytosterols	Use in functional foods (section 6.3)
	Isoflavones	Use in functional foods

Table 6.2   Continued

Crop	Ingredient	Uses
Maize	Grains and flours	Direct consumption, flakes, flours for bakery products, breakfast cereals, extruded snackfoods, tortillas, crumb coatings, carriers. Pre-gelatinised for infant foods and bakery products. Feedstock for fermentation to alcohol
	Bran	Breakfast cereals and poultry feeds. Isoflavones may be obtained from bran
	Oil (pure, refined, hydrogenated, enzymically modified)	Direct consumption or used in a wide range of foods and commercial frying oils
	Maize germ oil	Health foods
	Starch and derivatives (maltodextrins, chemically and enzymically modified, glucose syrups, dextrose, high fructose corn syrup	Very wide applications as thickeners, sweeteners, encapsulants, carriers or diluents for other ingredients
	Fermentation or enzymic conversion from starch to:	
	Sugar alcohols (sorbitol, xylitol, etc.)	Used in sugar-free foods and low-calorie drinks
		Flavourings and colourants
	Ethanol	Carrier for flavourings, blended alcoholic beverages
	Ascorbic acid and sodium ascorbate	Antioxidant, widely used in bakery products
	Lactic acid/lactates, citric acid/citrates, acetic acid, gluconic acid/gluconates, glucono-delta-lactone	Very wide range of applications as acidulants, e.g. for the preservation of meat and fish products and in ice cream, sorbets, liqueurs, baking powder, yeast substitutes, preparation of egg products
	Caramels from glucose syrups	Flavourings and colourants
	Sterols and stanols	Use in functional foods
	Monosodium glutamate	Flavour enhancer
	Xanthan gum	Many uses in a variety of food applications for thickening, suspending and gelling in low pH dressings, ice cream, processed cheese dips and spreads and frozen desserts
Rapeseed	Oil	Salad oils, shortening, margarine, coffee whiteners, bakery products, frying oils, anti-dust agent
	Esterified oil (diacylglycerol)	Novel and functional foods (see section 6.3)
	Sterols and vitamin E	Margarine
	Monoglycerides	Emulsifiers in bakery products
	Ammonium phosphatide	Emulsifier in chocolate
Cottonseed	Oil and hydrogenated oil	Cooking oils, salad oils/dressings, frying oils, mayonnaise, shortening, margarines
	Cellulose and chemically modified cellulose gums from linters	Emulsifiers, stabilisers, viscosity modifiers and thickeners in a wide range of products including meat products, sauces, salad dressings, bakery products, low-calorie products, ice cream, yoghurt and high-fibre foods

Adapted from Brookes *et al.* (2005)

6.1.2   Public perceptions of GM foods

Hoban (2004) reviewed the results of surveys of public attitudes towards GM foods conducted from the mid-1990s to 2002. The largest survey of 35 000 respondents in 35 countries, conducted in 2000, found that support for GM foods varied widely in different parts of the world. Support was most evident in the 13 countries surveyed in North and South America, 2 countries surveyed in Africa, 6 of the 8 countries surveyed in Asia and the Pacific, but only 3 of the 8 countries surveyed in Europe. In each case more respondents agreed that 'the benefits of biotechnology outweigh the risks' than those who disagreed. The countries in which GM foods were least supported were Japan and Korea in SE Asia, and France, Germany, Greece, Italy and the UK in Europe, where more respondents disagreed than agreed. In reviewing previous surveys, the author noted that in the USA although there is strong support for GM foods, awareness has fluctuated since the mid-1990s. Few consumers believed that GM foods were in wide use in the food supply, with only 14% believing that 'more than half of US foods contain GM ingredients' (Anon 2001), whereas it is estimated that 60% of all US processed foods contain a GM ingredient (Wagner 2006).

In Europe, consumer support for GM foods in all countries except Spain and Austria declined throughout the mid–late 1990s. Attitudes turned slightly more positive in 2000, except for Italy, France and the Netherlands, which experienced further declines. A later 2002 survey (Anon 2003a) found that in seven countries surveyed, support for 'scientifically altered' (i.e. GM) fruits and vegetables was strongest in the USA and Canada, but some Western Europeans and Japanese were overwhelmingly opposed (89% in France, 81% in Germany and 76% in Japan). This perception may have arisen from a series of food crises in Europe in the 1990s, which increased consumer apprehension over food safety, and weakened consumer trust in government oversight of the food industry and the ability of scientists to guarantee food quality. Although there may be benefits to producers from GM crops, these consumers perceived no evident benefit to them. These aspects are explored in more detail by Santaniello and Evenson (2004a).

There remains substantial controversy between supporters and opponents of GM foods, focused mainly on environmental and health concerns, labelling and consumer choice. Supporters of GM foods claim environmental benefits from reduced pesticide and herbicide use. For example:

- GM cotton has reduced synthetic pesticide use in the USA, Australia and India (Anon 2005b), increased cotton yields by 50% as a result of reduced losses to insects, and insect protected crops can reduce exposure of farmers to synthetic insecticides. Insect protected GM maize can also have lower mycotoxin levels due to reduced insect damage to the crop (Wu 2006). GM crops can be developed for cultivation in soils or climates where they previously could not grow, thus releasing more land for crop production and increasing crop yields per unit area of land to meet the needs of population growth, and thus contribute to sustainable food security. Reduced ploughing or conservation tillage associated with GM crops reduces fuel use and increases carbon sequestration in soils, thus contributing to reduced greenhouse gas emissions from agriculture.

Supporters also point to economic benefits:

- Higher yields increase farm outputs and lower food prices, which benefit farmers and consumers respectively; insect-protected GM crops have greater economic benefits in developing countries because agriculture is often a large part of the economy and reduced crop losses due to insect damage and improved yields benefit large numbers of people. Higher yields of feed crops also meet demand from increased consumption of animal protein.

In relation to safety concerns, GM supporters note that:

- There have been no major health hazards identified since GM foods were introduced, and more than 100 studies have been published indicating their safety (Anon 2004b). There are no significant differences in nutritional value between GM crops and corresponding conventional crops, and no residues of recombinant DNA or novel proteins have been found in organs or tissues of animals fed on GM feeds. They conclude that manufacturers of GM foods are responsible for the safety of their products and are required to demonstrate this before their use is approved. It is in the manufacturers' interests to ensure that safety testing is adequately done.

The main concerns raised by opponents of GM foods are that they threaten unintended, undesirable or unforeseeable consequences to the environment, and create a risk to consumer health and safety. Environmental concerns have arisen due to the following:

- The possibility that unwanted traits could be introduced along with the desired ones, or the risk of cross-contamination of GM genes to traditional crops or wild plants. For example, the uncontrolled spread of GM insecticidal proteins into wild plants could give a competitive advantage to those plants or disrupt the role of insects in natural ecosystems. The use of herbicides substantially reduces weed densities, which can have significant impact on wildlife that consume the weed seeds and hence reduce biodiversity (Tester 2000). For example, the results of a UK farm-scale trial showed that some broadleaf weeds were less numerous in fields containing GM rapeseed and produced one-third fewer seeds for farmland birds to eat at the end of the season, compared with those in a conventional crop. Two years later, there were still fewer seeds, even though the weedkiller had not been applied again, and there were also fewer bees and butterflies in the GM crops (Burke 2005).

Consumer rights groups also emphasise the long-term health risks that GM foods could pose, or that the risks have not yet been adequately investigated. These include the following:

- Exposure of populations to large amounts of novel proteins could cause unpredictable problems with allergenicity that are difficult to detect because symptoms can take several years to develop. Examples include a Bt-maize, which was reported to have caused allergic responses, some life-threatening, in over 200 people and was subsequently banned (Anon 2003b), and the unintentional transfer in 1993 of a gene for an allergenic trait from the Brazil nut into GM soybeans. The intention was to increase the methionine content to improve the protein quality of soybeans for animal feed, but the methionine-rich protein was also the source of Brazil nut allergy and further development was discontinued (Nordlee *et al.* 1996).
- In the USA, the Food and Drug Administration (FDA) reported no evidence for adverse health effects of GM bovine somatotropin. However, there have been reports of the hormone having a negative impact on the ability of cows to conceive, and that cows often developed severe infections requiring treatment with antibiotics (Smith 2003). As a result, some dairies have stopped using the hormone and it is banned in Europe because of health concerns.

Opponents also claim inadequate oversight by regulatory bodies of the safety of GM foods and that their safety relies on the veracity of manufacturers who do the testing. Many believe that the existing regulatory models fail to protect consumers and are subject to influence by the industry. There is concern that the patenting of GM technologies and private funding of developments enables corporations to gain excessive power in the marketplace (Smith 2003). There are also issues over the requirement of farmers, particularly in developing countries, to pay licensing fees for use of modified seeds and preventing them from saving seeds for future planting, especially if crops are developed that contain the so-called 'Terminator Gene', which allows seeds to germinate only once.

Developments in genomics have created a new agricultural biotechnology named marker-assisted selection (MAS) that does not have the potential disadvantages of GM technologies. MAS is a sophisticated method that can accelerate crop and animal breeding by identifying genes that are associated with traits such as yield, pest resistance or increased nutritional or sensory values (known as 'quantitative trait loci' or QTL). When particular genes are identified, related wild varieties of the crop are scanned for the presence of those genes. Instead of using gene splicing technologies to create a GM crop, the wild varieties are cross-bred with the food crop to create the desired traits. This can reduce the time needed to develop new varieties by 50% or more. Crops so far developed include an aphid-resistant lettuce, rice that remains firm after processing, pearl millet that is drought-tolerant and resistant to mildew (Rifkin 1999, 2006) and new varieties of hops, soya and potato. Further details of marker-assisted selection in animals are given by Van der Werf (2006) and an overview of MAS developments in cereal and forage crops, livestock, fruit trees and farmed fish is given by Anon (2003c).

6.1.3   Genetically modified micro-organisms

The use of micro-organisms in food processing is described in sections 6.3.1 and 6.4.3 and 'technical' enzymes produced by micro-organisms are described in section 6.5. This section describes developments in the use of GM micro-organisms (GMMs) and their products, particularly extracellular enzymes. The first GMM food product, bacterial chymosin (rennet) for cheese manufacture, was approved for use in the USA in 1990. Since then there has been widespread and increasing use of enzymes produced by GM bacteria, moulds and yeasts, but this has not given rise to consumer concerns in the same way the GM plants and animals have (section 6.1.2). The enzymes that are commercially produced by GMMs, especially *Bacillus* sp., *Aspergillus* sp. and *Trichoderma* sp. are shown in Appendix B.3.

Enzymes from GMMs and non-GM sources are also incorporated into animal feeds to enhance digestion and complement either the gut microflora of the animal or natural enzymes in the feedstuff. In particular added enzymes are used to hydrolyse antinutritional factors and increase the availability of nutrients in the feed. Examples include the following:

- Phytase produced from GM *Aspergillus niger* or *A. oryzae*, which breaks down phytate to increase the availability of phosphate in the feed. This reduces the amount of added phosphate by up to 30%, so reducing levels of phosphate in manure and contributing to reduced contamination of groundwater.
- Partial reductions in indigestible polysaccharides and oligosaccharides in feeds, including reduced β-glucan in barley feeds by β-glucanase from *Aspergillus* sp. or *Penecillium* sp., reduced arabinoxylan in wheat feeds by xylanase from GM *Asper-*

gillus sp., *Bacillus* sp. or *Trichoderma* sp., and reduced oligosaccharides in vegetable (e.g. soybean) feeds using a GM α-galactosidase from *Aspergillus* sp. In each case the enzymes increase the efficiency of feed utilisation by the animal (Brookes *et al.* 2005).

Examples of food applications of GMMs include GM rennet (chymosin) derived from genetically modified *Escherichia coli*, *Kluyveromyces lactis* and *Aspergillus niger* for cheese production, lactic acid bacteria that are resistant to viruses for the production of safer dairy and meat products (section 6.4.3), GM *Bacillus subtilis* to hydrolyse starch for the production of glucose, citric acid and other products, and microbial production of vitamins B_1 and B_2, aromas, amino acids and flavour enhancers. There are also developments to produce oils and fats that have high concentrations of conjugated linoleic acid (CLA) using micro-organisms that produce linoleic isomerase, thereby converting linoleic acid to CLA (Dunford 2001). CLA has been shown to suppress tumour development and is being studied for its role in cancer prevention (Lee *et al.* 2005).

6.1.4 Legislation and safety testing

A summary of the development of EU legislation is described by Lee and Carson (2004) and the OECD (Anon 2008b). Santaniello and Evenson (2004b) give further details of EU, US and international legislation. Anon (2006) summarises the regulatory status of countries worldwide. There are significant differences in approach between regulators in the USA and EU that are summarised below. In the USA, a legislative policy of 'substantial equivalence' permits GMOs to qualify as traditional or GRAS foods if their nutritional composition is essentially the same (i.e. GM foods that have similar proteins, and toxins are deemed to be no different from conventional food, without further investigation of the effects of any other differences) (Anon 2002). A comprehensive review of the US position on biotechnology was published by the Institute of Food Technologists in 2000 (Anon 2000b–e). Since 2004, EU regulations require all GMO products to be labelled as GM, whether or not they contain GM material. This includes any product derived directly from a GM plant, animal or micro-organism (e.g. maize starch, which includes traces of GM material) or indirectly (e.g. glucose syrup made from maize starch, in which all GM material is removed during processing) (Appendix B.3). Foods that contain less than 0.9% GM material that arises from accidental contamination during cultivation, harvesting and processing need not be labelled.

The EU regulatory system requires authorisation for placing a GM food on the market, in accordance with the precautionary principle that requires a comprehensive pre-market risk assessment and traceability of products at each stage of their production and distribution. Rules for traceability and labelling apply to all GM-derived products except those derived from microbial genetic engineering, To prevent GM crops from contaminating traditional varieties, EU legislation also requires a quarantine zone around GM fields where traditional varieties of the same crop cannot be grown (Anon 2008a).

The Cartagena Protocol is an international agreement on the movement of GMOs from one country to another that seeks to protect biological diversity from the potential risks posed by GMOs. It established procedures that require specific notification to, and agreement of, the importing country before export of a GMO may go ahead. An importing country can declare via the Biosafety Clearing House (an information exchange mechanism) that it wishes to base a decision on risk assessment before agreeing to accept an imported GMO for food, feed or processing.

The differences in approach between regulatory agencies in the EU and the USA have resulted in strong disagreements over the labelling and traceability requirements for GM foods. As a result, the USA claims that restrictions on sales of GM products in the EU violate free trade agreements. This is described in detail by Toke (2004).

One result of the EU legislation and the negative consumer attitudes to GM foods in Europe has been an increase in 'GM avoidance' policies by major EU retailers. This in turn has led to increased demand by food processors for 'Identity Preserved' (GM-free) ingredients, which has increased costs in the supply chain up to but not including the retail sector (Brookes *et al.* 2005). A manufacturer wishing to label products as 'GM-free' must institute systems and records to ensure that only ingredients from non-GM sources are used. Similar considerations also apply to livestock or poultry meat products where feed ingredients contain soya or maize, and to imported finished products from countries that do not segregate GM and non-GM crops. Identity Preserved ingredients with non-GM certification (e.g. soybean lecithin, β-carotene, curcumin, lutein and caramelised sugar) are now becoming available and it is possible that xanthan gum derived from GM-free sugar will be available to replace that from maize derivates because of doubts about the non-GM status of maize.

In 2006, the legal status of foods that contain GM fermentation products was that such products are excluded from regulation if foods are produced with, rather than from, the GMM. If the GMM is present in the final product, whether alive or dead, the food falls within the scope of the regulations. The distinction is based on whether there is genetic material in the food (i.e. if cellular material from GMMs is incorporated) or whether modifications to the product can be demonstrated. The legislation controlling use of enzymes in food processing, whether from GMM or non-GMM sources, categorises them into 'additives' (substances that have a technological function in the food) and 'processing aids' (substances that are added during processing for technical reasons, but do not have a technological function in the final product). At present, lysozyme and invertase are the only two enzymes that are considered as additives. All other enzymes are processing aids that do not have to be declared on the label under GM or food labelling regulations. The position is under review (in 2008) and a proposal to reclassify most enzymes as additives is under discussion. If approved, this would have important implications for processors that operate GM avoidance policies and obtain enzymes from GMM sources.

Safety testing
GM foods are subject to safety assessment for possible changes to allergenic potential or nutritional profile. If the originating material is known to be allergenic (Brazil nuts, peanuts, kiwifruit, eggs, crustaceans, etc.) the testing regime should verify whether allergens are transferred into the host organism. Other factors that are taken into account include the stability to processing and digestion, the final amount of allergen in food and the identity of other non-allergenic proteins. Allergenity is tested *in vitro* using RAST (radioallergosorbent) and ELISA (enzyme-linked immunosorbent assay) and if this is negative, *in vivo* skin prick tests are used. (The RAST is a blood test for 'true allergies' that are immediate IgE reactions such as hives, swelling, rapid heartbeat, whereas the ELISA test is for IgG reactions that are delayed reaction allergies in which symptoms can appear 2–48 h after exposure to the allergen. IgE and IgG are acronyms for the two different kinds of immunoglobulin antibodies produced by the immune system in allergic reactions to food.) Further details are given by Allen (1990).

In 2003, the Codex Alimentarius Commission approved principles for the risk analysis of GM foods and guidelines for the safety assessment of GM plants and micro-organisms (Anon 2003d,e, 2008a). These procedures form the basis of assessments of GM foods and ingredients by national regulatory authorities.

Analytical methods including gas chromatography mass spectrometry (GC-MS), high-performance liquid chromatography (HPLC) and capillary electrophoresis are unable to detect these materials and immunological methods (e.g. ELISA) are used. The DNA is extracted and the promoter sequence and gene are amplified using the polymerase chain reaction (PCR). The products of the PCR can then be made visible using gel electrophoresis. The modified genetic material is often present in very small amounts in foods and the transgenetic protein may be located in non-edible parts of the plant (e.g. Bt toxin is present in leaves but not in maize kernels). Also other compounds, such as polysaccharides, can inhibit the PCR and interfere in the detection, leading to false negative results.

6.2 Nutritional genomics

Nutritional genomics (or 'nutrigenomics') is the application to human nutrition, especially the relationship between nutrition and health, of:

- *genomics* (the study of an organism's genome and use of the genes);
- *proteomics* (the study of structures and functions of proteins and their interactions with the genome and the environment); and
- *metabolomics* (the study of unique chemical fingerprints that specific cellular processes leave behind).

Research focuses on the prevention of disease by understanding nutrient-related interactions at the gene, protein and metabolic levels and their effects on tissues and organs. With progress in genetics research, including the Human Genome Project, genetic variation is known to affect food tolerances and may also influence the dietary requirements of individuals. Nutritionally related biochemical disorders have been linked to genetic origin and the influence of diet on health may therefore depend in part on the genetic makeup of an individual (Stover 2006).

When food is digested, some dietary chemicals are not metabolised and become ligands (molecules that bind to proteins that are involved in turning on certain genes). These dietary chemicals change the expression of genes and even the genome itself. The theory is that a diet that is out of balance will cause gene expressions that may cause chronic illness. This has been observed in the Alaskan Inuit, whose metabolism of high-fat food was suited to high levels of activity in the cold climate and who now show high levels of obesity, diabetes and cardiovascular disease as a result of changes in their lifestyle; those members of the Maasai in East Africa who have abandoned their traditional meat, blood and milk diet for maize and beans, and have since developed new health problems; and differences in lactose tolerance between people of northern European and south-east Asian or African ancestry (Grierson 2003).

Future developments in nutrigenomics may enable identification of people who are genetically predisposed to diet-related diseases, and the development of foods or personalised precisely tailored 'intelligent diets' for disease prevention and health promotion (Hasler 2000, 2002). This is also known as targeted or 'prescription' nutrition, or 'eat right for your genotype'. A few companies have started offering a genetics testing service to produce a preventive health profile and nutritional supplements (Grierson 2003). However, the interactions between diet and genes are complex and poorly understood. There may be genetic subpopulations that would incur different benefits or risks from generalised fortification policies based on genomic criteria. Research is ongoing to understand the molecular mechanisms that underlie gene–nutrient interactions and their modification by genetic variation before dietary recommendations and nutritional interventions can be made to optimise individual health. Another development in genomics, MAS, is described in section 6.1.2.

6.3   Functional foods

The term 'functional foods' (also 'nutraceuticals') was first used in Japan and is used to refer to foods that provide benefits other than the nutrients required for normal health. It can be said to be a concept rather than particular types of products (Roberfroid 2000). It is not defined in law but a number of definitions have been proposed, including 'any substance that is a food or food ingredient that provides medical or health benefit, including the prevention and treatment of disease' (Defelice 1995). Interest in functional foods has grown since the mid-1990s due to a number of factors, including new evidence of the links between diet and health that has resulted in an increasing interest by many consumers in 'healthy eating', compounded by increasing healthcare costs and an ageing population in many industrialised countries. Consumer interest has prompted the commercial development of new foods that are designed to address a range of specific medical conditions or health concerns, including cardiovascular disease, cancer, osteoporosis, neural tube defects, immune deficiencies, intellectual performance, gut health/bowel disorders, ageing and obesity. These are summarised by Henry and Heppell (1998). Research is also focused on functional attributes that are being discovered in many traditional foods, and new food products that are being developed to contain beneficial components.

Functional foods can be grouped into the following four categories:

1   Products in which an ingredient that is normally present is increased or reduced (e.g. breakfast cereals with added vitamins or bran, drinks fortified with antioxidant vitamins or dairy products that have reduced fat).
2   Products in which an ingredient that is not normally present is introduced (e.g. fibre added to fruit juice, folic acid or plant oestrogens added to bread and margarines, and snack bars enriched with stanols to reduce cholesterol absorption).
3   Dairy products, fermented using probiotic bacteria that are selected for their functional benefits to aid digestion and protect against infections (section 6.3.1). Some products have added oligosaccharides to support the growth of these bacteria.
4   Products that are specially formulated to meet a particular nutritional requirement (e.g. sports drinks that give a balanced replacement of fluids lost during exercise or provide additional energy, or cereals that are formulated to slowly release carbohydrates and supply energy over a prolonged period).

Other functional foods may contain herbal extracts that claim to help address a range of problems from premenstrual syndrome to lack of energy. Caffeine stimulation drinks can also be described as functional foods.

A summary of the presently known (2008) functional components of foods is given in Appendix A.5. The addition of functional ingredients to confectionery products is described by Pickford and Jardine (2000), and to spreadable fats by de Deckere and Verschuren (2000). Functional dairy products are described by Mattila-Sandholm and Saarela (2003) and Saarela (2007), and the legislative and health aspects of functional foods are described by Gibson and Williams (2000). Lindsay (2000) describes the use of GM technology to increase the levels of functional components in plant foods.

Some functional ingredients, such as plant extracts, are microencapsulated (section 6.6 and Chapter 24, section 24.3) to mask their taste, and microencapsulation can also be used to improve the nutritional value of products for children (e.g. microencapsulated minerals or vitamins in chewing gum or sweets) or to produce processed foods for targeted nutrition (section 6.2).

6.3.1   Health claims and regulation

New functional foods are being developed at an increasing rate and they are usually accompanied by health claims for marketing purposes. In most countries they are treated as other foods and manufacturers are not allowed to claim that foods can prevent, treat or cure disease. So whereas it is acceptable to state that a food 'provides calcium which is important for strong bones', or 'helps lower blood cholesterol when consumed as part of a low-fat diet', it is illegal to claim that the food 'provides calcium, which helps prevent osteoporosis' or 'helps prevent heart disease' (Anon 2004c). A number of reports, including Heller *et al.* (1999), Jacobson and Silverglade (1999) and Katan and De Roos (2004), reviewed the regulatory and marketing aspects of functional foods in Japan, the USA and the EU. They concluded that some foods were being marketed without sufficient evidence of their health benefits, or that some manufacturers use symbols such as a heart on the packaging that implied health benefits that were not proven, and that regulations should be strengthened. Health claims should take into account the required level of intake within normal dietary patterns and how often/how long it is necessary to consume a product to make any functional impact.

The Japanese Government was the first to develop a specific regulatory approval process for functional foods, named Foods for Special Health Use (FOSHU), and approved foods bear a seal of approval from the Japanese Ministry of Health and Welfare. Over 100 products have been licensed as FOSHU foods in Japan (Hasler 2005). Other countries, including Canada, have specific laws covering the labelling of functional foods. In the USA, claims are regulated by the FDA under a complex set of rules that are described in detail by Schmidl and Labuza (2000). In 1997 in the UK, the Joint Health Claims Initiative (JHCI), a panel of nutritional and health professionals, drew up a Code of Practice on Health Claims on Foods. This sets out the general principles for making a claim, describes precise criteria for substantiating claims, monitors the operation of the code and considers the scientific validity of any particular claim. Food producers may optionally use the JHCI to determine whether their claims are likely to be legally sustainable. In Europe in 2005, the PASSCLAIM (Process for the Assessment of Scientific Support for Claims on Foods) project developed criteria for the scientific substantiation of food claims (Anon 2005c,d). Arvanitoyannis and van Houwelingen-Koukaliaroglou (2005) have reviewed health claims and legislation. Berry Ottaway (2000) describes the EU legislation that relates to different functional ingredients and Schmidl and Labuza (2000) describe US legislation and functional health claims. The issues of guidance to manufacturers and the substantiation of claims are ongoing, and updated information is available at Anon (2008c).

In the following sections, the types of functional foods known as probiotics and prebiotics are examined in more detail.

6.3.2   Probiotic foods

Daily consumption of bioyoghurts and other fermented dairy products that contain probiotic bacteria has been shown to promote gut health. Probiotic bacteria include

Lactobacillus acidophilus, L. casei Shirota, *L. johnsonii, L. rhamnosus* GG, *L. lactis, L. plantarum, Streptococcus thermophilus, Bifidobacterium lactis, B. adolescentis, B. longum, B. breve, B. bifidus* and *B. infantis* (Krishnakumar and Gordon 2001). Their beneficial effects are described by Sanders (1999) and Mattila-Sandholm and Saarela (2000) and include increased immunity to a range of intestinal pathogens, including *E. coli* O157, *Salmonella* spp. and *Shigella* spp. (see Chapter 1, section 1.2.4). The mechanisms by which they do this are described by Isolauri and Salminen (2000). Dairy products that contain *L. acidophilus* can also reduce the incidence of vaginal infections by reducing gut colonisation by *Candida albicans* (Surawicz and McFarland 1996). Other health benefits may include reduction in cancer risk, particularly colon cancer, because lactic acid bacteria can alter the activity of faecal enzymes that may play a role in the development of this disease. They may also have hypocholesterolemic (cholesterol-lowering) and anticarcinogenic actions (Mital and Garg 1995). However, there is not yet a full understanding of the roles that intestinal bacteria play in health and the influence that diet has on their composition.

Their effectiveness depends on the ability of bacteria to survive storage as frozen or freeze dried cultures, processing, and passage through stomach acids to colonise the large bowel and change the composition of its flora. Champagne and Gardner (2005) have reviewed the factors that should be considered to enable probiotic micro-organisms to survive processing and storage. Microencapsulation of probiotics improves their survival rate during processing and digestion and is described in detail by Siuta-Cruce and Goulet (2001) (see also section 6.6). Microencapsulation using calcium-alginate gel capsules, kappa-carrageenan, gelatin or starch (Kailasapathy 2002), and the use of cryoprotectants such as trehalose or gum acacia (Ross *et al.* 2005), have improved the viability of cells and microencapsulated cultures are commercially available. Fermentation technology for the production of probiotic bacteria is described in section 6.4.

6.3.3   Prebiotic foods

Prebiotics are foods that contain ingredients that are not digested but stimulate the growth of probiotic bacteria in the colon. They are complex fermentable carbohydrates including fructo-, gluco-, xylo- and galacto-oligosaccharides, inulin, chitin, dietary fibres, other non-absorbable sugars and sugar alcohols (Roberfroid and Slavin 2000, Playne and Crittenden 1996). Of these, oligosaccharides (Chapter 1, section 1.1.1) have received the most attention, and numerous health benefits have been attributed to them (Gibson *et al.* 2000). They are found naturally in many fruits and vegetables (including banana, garlic, onions, artichokes), milk and honey, but most commercially available prebiotics are manufactured using enzymic reactions. These may involve building the oligosaccharide from sugars using transglycosylation, or by hydrolysis of large polysaccharides. The advantage of prebiotics is that their benefit does not rely on the viability of probiotic bacteria in a given product and they can be added to any food that normally contains carbohydrates. To date they have mostly been added to fermented dairy products, but cereal products, infant formulae, confectionery or savoury pastes and spreads are each potential applications (Rastall *et al.* 2000).

Synbiotics consist of a live probiotic micro-organism and a prebiotic oligosaccharide. These products have advantages in that the prebiotic aids the establishment in the colon of the particular probiotic bacterium, which has known benefits. There is therefore great flexibility in the combination of probiotic micro-organisms and different oligosaccharides to achieve the desired health benefits (Shah 2001, Rastall *et al.* 2000).

6.4   Fermentation technology

Fermented foods are among the oldest processed foods and have formed a part of the diet in almost all countries for millennia. Today they form major sectors of the food processing industry, including fermented cereal products, alcoholic drinks, fermented dairy products and soy products among many others. Mycoprotein has also been developed as a meat-substitute. A summary of commercially important food fermentations is given in section 6.4.3.

During food fermentations, the controlled action of selected micro-organisms is used to alter the texture of foods, preserve foods by production of acids or alcohol, or to produce subtle flavours and aromas that increase the quality and value of raw materials. The preservative effect is supplemented by other unit operations (e.g. pasteurisation, chilling or modified atmosphere packaging (Chapters 12, 21 and 25, section 25.3). These combined effects are examples of hurdle technology, described in Chapter 1 (section 1.3.1). The main advantages of fermentation as a method of food processing are:

- the use of mild conditions of pH and temperature which maintain the nutritional properties and sensory characteristics of the food;
- enrichment of foods with vitamins, proteins, essential amino acids or fatty acids, or detoxification by removal of antinutritional factors (e.g. phytate, trypsin inhibitor) or toxins (e.g. aflatoxins) (Steinkraus 2002);
- the production of foods that have a diversity of flavours, aromas or textures that cannot be achieved by other methods;
- low energy consumption due to the mild operating conditions;
- relatively simple technologies that have low capital and operating costs (section 6.4.2).

6.4.1   Theory

Food fermentations can be grouped into those in which the main products are lactic acid, alcohol and carbon dioxide, and acetic acid. Details of the metabolic pathways that are used to produce these products are readily available (for example Jay *et al.* 2005). Many fermentations involve complex mixtures of micro-organisms or sequences of microbial populations that develop as changes take place in the pH, redox potential or substrate availability. The factors that control the growth of micro-organisms are described in Chapter 1 (section 1.2.3) and selected examples are given in section 6.4.3.

Batch culture

Cell growth during the logarithmic (or exponential) phase is at a constant rate (Chapter 1, Fig. 1.15), which is shown by:

$$\ln c_b = \ln c_0 + \mu t \qquad \boxed{6.1}$$

where c_0 = original cell concentration, c_b = cell concentration after time t, (biomass produced), μ (h^{-1}) = specific growth rate and t (h) = time of fermentation. Graphically, the natural logarithm (ln) of cell concentration versus time produces a straight line, the slope of which is the specific growth rate. The highest growth rate (μ_{max}) occurs in the logarithmic phase.

The rate of cell growth eventually declines owing to exhaustion of nutrients and/or accumulation of metabolic products in the growth medium. If different initial substrate concentrations are plotted against cell concentration in the stationary phase, it is found that an increase in substrate concentration results in a proportional increase in cell yield

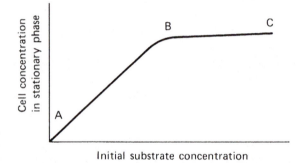

Fig. 6.1   Effect of initial substrate concentration on cell concentration at the end of the logarithmic phase of growth (after Stanbury *et al.* 1995).

(AB in Fig. 6.1). This indicates substrate limitation of cell growth, which is described by:

$$c_b = Y(S_0 - S_r)$$ $\boxed{6.2}$

where c_b = concentration of biomass, Y (dimensionless group) = yield factor, S_0 (mg l^{-1}) = original substrate concentration, S_r (mg l^{-1}) = residual substrate concentration. The portion of the curve BC in Fig. 6.1 shows inhibition of cell growth by products of metabolism.

The reduction in growth rate is related to the residual substrate concentration by Monod's equation:

$$\mu = \frac{\mu_{max} S_r}{(K_s + S_r)}$$ $\boxed{6.3}$

where μ_{max} (h^{-1}) = maximum specific growth rate, K_s (mg l^{-1}) = substrate utilisation constant. K_s is a measure of the affinity of a micro-organism for a particular substrate (a high affinity produces a low value of K_s).

Growth parameters that enable the production of biomass or a specific metabolic product (e.g. ethanol, amino acids or citric acid) can be determined. The yield coefficient (Y) is calculated from the amount of a limiting nutrient, usually the carbohydrate source, that is converted into the microbial product. For biomass, it is found using:

$$c = Y_{c/s}(S - S_r)$$ $\boxed{6.4}$

where c = biomass concentration (g l^{-1}), $Y_{c/s}$ = yield coefficient (g biomass/g substrate utilised), S = initial substrate concentration (g l^{-1}), and S_r = residual substrate concentration (g l^{-1}).

The yield coefficient describes the amount of biomass produced per gram of substrate utilised: the higher the yield coefficient, the greater the amount of substrate that is converted into biomass. For the production of metabolic products, the yield coefficient (Y_p) describes the amount of metabolic product that is produced in relation to the amount of substrate used ($Y_{p/s}$). Yield coefficients are important because the cost of the growth medium, particularly the carbon source, can be a significant proportion of the total production cost.

The specific rate of product formation for primary products varies with the specific growth rate of cells. The rate of production of secondary products (those produced from primary products such as aromatic compounds and fatty acids), which are produced in the stationary growth phase, does not vary in this way and may remain constant or change in more complex ways.

Sample problem 6.1

An inoculum containing 3.0×10^4 cells ml^{-1} of a fast-growing yeast is grown on $50\,g\,l^{-1}$ glucose in a batch culture for 20 h. Cell concentrations are measured at 4 h intervals and the results are plotted (Fig. 1.15, in Chapter 1). The harvested broth contained $0.024\,g\,cells^{-1}$ and $30\,g\,l^{-1}$ of glucose. Calculate the maximum specific growth rate and the yield coefficient of the culture.

Solution to sample problem 6.1

From Fig. 1.15, final cell concentration $= 2 \times 10^8$ cells ml^{-1} after logarithmic growth for 8.5 h.

From Equation 6.1 for the logarithmic phase,

$$\ln 2 \times 10^8 = \ln 3 \times 10^4 + \mu_{max}8.5$$

Therefore,

$$\mu_{max} = \frac{\ln (2 \times 10^8) - \ln (3 \times 10^4)}{8.5}$$

$$= 0.95 \text{ h}^{-1}$$

From Equation 6.4,

$$0.024 = Y_{c/s}(50 - 30)$$

$$Y_{c/s} = 0.024 \times 20$$

$$= 0.48 \text{ or } 48\%$$

The average productivity of a culture (the amount of biomass produced in unit time is found using:

$$P_b = \frac{(c_{max} - c_0)}{t_1 - t_2} \qquad \boxed{6.5}$$

where P_b $(g\,l^{-1}\,h^{-1})$ = average productivity, c_{max} = maximum cell concentration during the fermentation, c_0 = initial cell concentration, t_1 (h) = duration of growth at the maximum specific growth rate, t_2 (h) = duration of the fermentation when cells are not growing at the maximum specific growth rate and including the time spent in culture preparation and harvesting.

Continuous culture

Cultures in which cell growth is limited by the availability of substrate in batch operation have a higher productivity if the substrate is added continuously to the fermenter and biomass or products are continuously removed at the same rate. Under these conditions the cells remain in the logarithmic phase of growth. The rate at which substrate is added under such 'steady state' conditions is found using:

$$D = \frac{F}{V} \qquad \boxed{6.6}$$

where D (h^{-1}) = dilution rate, F $(l\,h^{-1})$ = substrate flowrate and V (l) = volume of the fermenter.

The steady state cell concentration and residual substrate concentration respectively are found using:

$$\bar{c} = Y(S_0 - \bar{S}) \tag{6.7}$$

$$S = \frac{K_s D}{\mu_{max} - D} \tag{6.8}$$

where $\bar{c}$ = steady state cell concentration, Y = yield factor, S = steady state residual substrate concentration, K_s (mg l^{-1}) = substrate utilisation constant.

The maximum dilution rate that can be used in a given culture is controlled by μ_{max} and is influenced by the substrate utilisation constant and yield factor (Fig. 6.2). The overall productivity of a continuous culture (P_c) is found using:

$$P_c = D\bar{c}\left(1 - \frac{t_3}{t_4}\right) \tag{6.9}$$

where t_3 (h) = time before steady state conditions are established and t_4 (h) = duration of steady state conditions. Further details of the above equations are given by Blanch and Clark (1995) and in microbiological texts.

Sample problem 6.2
Brewers' yeast is grown in a fermenter with an operating volume of 12 m^3. A 2% inoculum, which contains 5% yeast cells, is mixed with the substrate and the yeast is harvested after 20 h. Calculate the mass of yeast harvested from the fermenter. (Assume that the yeast has a doubling time of 3.2 h and the density of the broth is 1010 kg m^{-3}.)

Solution to sample problem 6.2

$$\text{Mass of broth} = 12 \times 1010$$
$$= 1.212 \times 10^4 \text{ kg}$$

$$\text{Yeast concentration} = \frac{\text{concentration in the inoculum}}{\text{dilution of inoculum}}$$

$$= \frac{5/100}{100/2}$$

$$= 0.001 \text{ kg kg}^{-1}$$

The doubling time is 3.2 h.
Therefore in 20 h there are 20/3.2 = 6.25 doubling times.
As 1 kg of yeast grows to 2 kg in 3.2 h, 1 kg grows to $1 \times 2^{6.25} = 76$ kg in 20 h.
Therefore,

$$\text{mass of product} = \text{yeast concentration} \times \text{growth} \times \text{mass of broth}$$
$$= 0.001 \times 76 \times (1.212 \times 10^4)$$
$$= 921 \text{ kg}$$

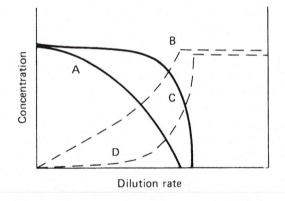

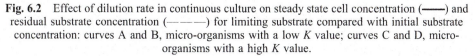

Dilution rate

Fig. 6.2   Effect of dilution rate in continuous culture on steady state cell concentration (———) and residual substrate concentration (———) for limiting substrate compared with initial substrate concentration: curves A and B, micro-organisms with a low K value; curves C and D, micro-organisms with a high K value.

6.4.2   Equipment

Submerged cultures

Liquid substrates are fermented in stainless steel tanks that are fitted with cleaning-in-place (CIP) and sterilising-in-place (SIP) facilities. Figure 6.3 shows a batch fermenter, and continuous fermenters have similar controls but with additional arrangements to

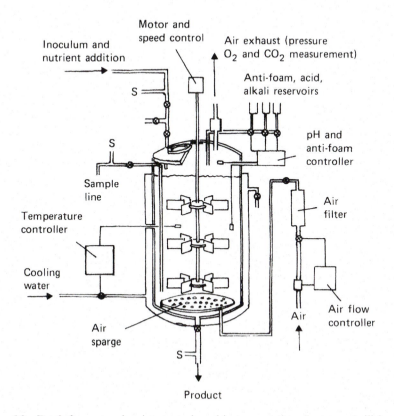

Fig. 6.3   Batch fermenter showing controls and instrumentation: S, steam sterilising points.

continually remove the product. In operation, the inoculum is produced from a freeze-dried stock culture and introduced into a fermentation medium in a sterilised fermentation vessel. The medium contains a carbon source (e.g. glucose syrup) with added nutrients depending on the requirements of the micro-organism for growth and/or metabolite or enzyme production. In aerobic fermentations, air is pumped into the fermenter to initiate a period of growth under constant conditions of temperature, dissolved oxygen (DO) and pH, which are maintained by automatic control mechanisms (types of control mechanisms and systems are described in Chapter 27, section 27.2). In batch operation, the fermenter is emptied at the end of the fermentation, judged by accumulated biomass and/or enzymes or metabolites, or by depletion of nutrients. In continuous fermentations, a feed of nutrients is maintained with the simultaneous removal of fermenter broth so that an excess of nutrients remains in the supernatant. In aerobic fermentations the supply of air (and if necessary supplementary oxygen) is automatically regulated in response to continuous monitoring of the DO concentration in the broth. The concentration of biomass, enzymes or metabolites is controlled by the flow of medium and by monitoring factors such as the carbon dioxide evolution rate using on-line analysis. In anaerobic fermentations, a low DO concentration is maintained using nitrogen or carbon dioxide.

In both batch and continuous fermentations the main factors that require control are temperature, pH, dissolved oxygen, degree of agitation and foaming. Methods of control are described by a number of equipment suppliers including Anon (2008d,e).

Temperature control
There are three ways to control the optimum temperature for cell growth and/or metabolite production: (1) the growth medium is continuously recirculated through a heat exchanger, linked to a thermostatic controller; (2) the fermentation vessel is fitted with a thermostatically controlled water-jacket and water is recirculated through the jacket and a heat exchanger; or (3) the fermenter is fitted with an electric heating blanket, linked to a thermostat.

pH measurement and control
A digital pH probe is linked to a solid state controller that measures and controls pH to an accuracy of ± 0.01 pH units. The microprocessor controls the calibration of the pH probe using standardised buffers and regulates pumps to add acid or alkali. The limits of the working range are established by adjustable high and low set-point controls, and pH values that exceed the set-points in either direction cause the controller to activate the pumps. Adjustable electronic timers determine the length of time that pumps operate to provide a delay between corrective additions, thus minimising overshooting. The set-point limits have an associated alarm to alert the operator to depleted reagent supplies or to signal the need for adjustment or intervention.

Aeration/dissolved oxygen control
A digital DO probe continuously monitors the DO concentration in the medium. It is calibrated by the software in the controller with temperature compensation, and a display indicates DO concentration in a range of 0–100%. The required DO concentration range is maintained using upper and lower set-points, and an alarm indicates when the set-point range is exceeded. The controller is linked to gas flowmeters and the agitator speed monitor. It controls oil-free air pumps or regulators on pressurised gas tanks to add oxygen, air, carbon dioxide or nitrogen to the vessel, and/or to control the speed of

agitation, to maintain the DO concentration according to the needs of process. In-line filters maintain the sterility of incoming and outgoing process air or gas.

Agitation

A controller equipped with digital speed measurement monitors the degree of agitation and adjusts a variable speed motor on the mixer shaft, typically operating between 50 and 1250 rpm.

Foaming

Agitation and aeration may result in excessive foaming, depending on the growth medium and the types of metabolites produced by the fermentation. Foaming is controlled to prevent foam being forced out of the probe ports, thereby risking contamination, and to prevent it inhibiting oxygen transfer. Teflon-coated stainless steel foam-sensing probes are mounted within the fermentation vessel. The sensitivity of the probes is adjustable to eliminate signals caused by occasional splashing. Brown *et al.* (2001) describe an improved method using continuous on-line conductance measurement of foam. When foam contacts the probes they activate a controller that drives a pump to introduce chemical antifoam reagent into the vessel. Antifoam chemicals used in food fermentations include silicone, mineral oils, polyols, fatty acids, alcohols, hydrophobic silica and polyalkylenes. Flowmeters are linked to the controller and adjustable electronic timers determine the length of the antifoam addition and the period of delay between doses.

Medium addition

In continuous fermentations, a solid state electronic controller operates a variable speed metering pump to control the addition of presterilised culture medium to the fermenter, and a medium flowmeter verifies the flowrate. An 'anti-growback' medium inlet tube provides a barrier against potential backgrowth of cultured micro-organisms into the medium feed pipe. Sterile air is directed through the anti-growback inlet tube as a further precaution. Medium and product reservoirs each have an exhaust air vent with autoclavable 0.3 μm air filters and a sampling device.

Fermentation time

The duration of the fermentation is controlled using a programmable timer with automatic shutdown of the fermenter.

Displays and data logging

Data logging and control software enable the fermentation cycle to be programmed from a computer or panel-mounted display screen, and different fermentation cycle programmes can be saved for future reference. A data plot screen (Fig. 6.4) has a graphical display of the entire process and a synoptic screen, updated every few minutes, displays the fermentation parameters and individual controls throughout the fermentation. All data can be printed and saved through a PC interface.

Automatic control of fermenters

Fermentation processes are becoming increasingly sophisticated and 'bioinformatics' is used to automate the entire operation and to obtain traceability of the process. A series of commands, designed to control a fermenter without operator intervention, are known as 'sequences' and they perform the actions and decisions that an operator would take if the fermenter were being operated manually (see also Chapter 27, section 27.2.2). These

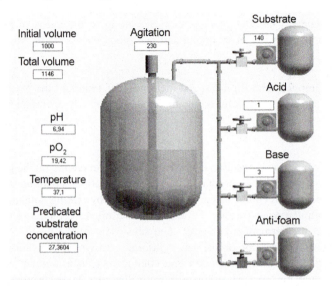

Fig. 6.4   Fermenter control screen (courtesy of Foxy Logic at http://imsb.au.dk/~mha/toverview.htm).

include inoculating or stopping the fermentation process, driving actuators (pumps or valves), or modifying the timing ratios for an actuator to be active/non-active (Andersen 2005). Sequential programming first identifies the measured values that should be logged (e.g. temperature, agitator speed, dissolved oxygen, pH, air flow), and their set-points and alarm settings. Then predefined changes that should take place in these values are selected and these 'profiles' are programmed in as control loops for the fermentation to follow. They may include specified times for media addition, or specified changes in pH as the fermentation proceeds using the elapsed fermentation time as the cue for an action to be taken. All set-points are displayed by numerical settings, controlled by a keyboard or by graphical indicators on a display screen. The controller collects process data and displays it as scrollable graphs that can be printed for traceability and management and accounting procedures (Felezeu 2001).

Fermentations are biological systems that are often non-linear and difficult to model mathematically. More intelligent control is possible with 'fuzzy logic' control or neural network software (Chapter 27, section 27.2.3). Fuzzy logic control is empirically based without any mathematical model. It is equivalent to having the control computer reason as a human from prior experience and acting accordingly to it, but faster. It thus enables the control of systems that would not normally be suitable for automation. It uses imprecise inputs such as 'a little', 'hot' and 'bigger' which are formulated into simple rules. The fuzzy logic controller uses a set of parameters to evaluate these rules and decide what output it should apply to the fermentation process (Park *et al.* 1993). Neural networks can be used to forecast process values in a fermentation on the basis of inputs such as culture volume, pH, DO concentration or substrate feed rate. Before a neural network can be used to control a fermentation, it is 'trained' using historical process data obtained from previous fermentations to produce a parameter file. This is used by the fermenter control program in a network that has modelled the training data. Further information is given by Karim and Rivera (1992) and Linko *et al.* (1997). Neural networks are used in large-scale fermentation systems to aid the control and maximise the product output. For example, Jin *et al.* (1996) describe neural networks and fuzzy logic control of a batch fermentation.

Solid substrate fermentations

Solid substrate fermentation (SSF) involves the growth of micro-organisms (mainly fungi) on moist solid materials that have a large surface area per unit volume (typically 10^3–10^6 m^2 cm^{-3}) (Raimbault 1998). Traditional SSFs include koji, Indonesian tempeh, sake and Indian ragi. SSF is also used to produce enzymes, (e.g. amylases, amyloglucosidase, cellulases, proteases, pectinases, xylanases and glucoamylases) organic acids, aroma compounds and flavours. Substrates include rice, tubers, wheat, millet, barley, beans, maize, sugar beet pulps and soybeans, and also food-processing residues (e.g. wheat bran and soy flakes remaining after oil extraction). The stages in a SSF are first to prepare a substrate by homogenisation, enzymatic hydrolysis, grinding or flaking, and sometimes heating to soften the material and remove contaminants. It is then inoculated and incubated with the micro-organism. Finally the product is either collected or leached from the fermented solids to obtain the required extracts. Some meat products are fermented in plastic or cellulose casings and SSF is also used to treat food and agricultural wastes. Pérez-Guerra *et al.* (2004) describe the advantages of SSF over submerged culture fermentations as:

- simple reactor design and smaller reactors required;
- similar or higher yields of products;
- low energy requirements (sterilisation, mechanical agitation and aeration are not always necessary);
- relatively simple culture media – the substrate usually provides all nutrients required;
- the low moisture availability may favour production of compounds that cannot be produced in submerged culture;
- smaller volumes of polluting effluents produced.

The main disadvantages when compared to submerged culture are that it is only suitable for micro-organisms that can grow at relatively low moisture contents; fermentation cycle times are longer because high inoculum volumes are required and inoculated spores have a lag time before germination, leading to a risk of contamination by undesirable fungi; and because of the nature of the substrate, it is more difficult to monitor process parameters (e.g. moisture content, oxygen and biomass produced).

For micro-organisms to grow on a solid substrate they should be able to tolerate lower water activities, penetrate the substrate and utilise complex mixtures of different polysaccharides. The hyphae of filamentous fungi grow into solid substrates and secrete enzymes that enable the utilisation of available nutrients at low water activity and high osmotic pressure/nutrient concentration. As a result these are the most commonly used micro-organisms and examples in food processing are shown in Appendix B.4. Bacteria (*Bacillus* spp.) are used for enzyme production, and for producing fermented products such as sausages, Japanese natto and fermented soybean paste. Yeasts are used for tapé production, fermented cassava and rice, ethanol production and protein enrichment of agricultural residues for feeds (Appendix B.4).

The factors that control the fermentation are:

- temperature;
- rates of gas and nutrient diffusion; and
- water activity of the substrate.

The complex chemical composition of most substrates has a buffering effect and pH control is usually not necessary. When it is needed, buffering solutions are added.

Temperature control
Metabolic activity increases the temperature in SSFs if the rate of heat removal is inadequate, and this directly affects spore germination, growth and product formation. The temperature increase depends on the type of micro-organism, and the porosity, particle size and depth of the substrate bed. This determines the rate at which heat is transferred from the substrate surface to air, and the rate of diffusion of heated air from the substrate. Temperature control is therefore more difficult than in submerged fermentations and relies on mechanical agitation or forced aeration.

Aeration
Aeration has two functions: to supply oxygen for aerobic metabolism and to remove CO_2, heat and volatiles produced by cell metabolism. The rate of diffusion depends (in addition to the moisture content of the substrate) on the particle size and porosity of the substrate, the substrate depth, and any forced aeration or agitation that is used.

Moisture content
The water activity of the substrate affects biomass growth, metabolic reactions and mass transfer processes. Depending on the application, the optimum substrate moisture content (usually between 30% and 75%) allows maximum cell growth or production of metabolites (enzymes, organic acids, etc.). Lower values induce sporulation of fungi and higher levels can reduce the porosity of the substrate, limit oxygen transfer and increase the risk of bacterial contamination. Moisture loss during the fermentation is prevented by the use of humidifiers and water-saturated air.

Most solid substrates act as both a support material and a nutrient source. They may be supplemented with nutrients (phosphorus, nitrogen, salts) and the pH and/or moisture content adjusted. Less commonly, an inert support material (e.g. sugar cane, bagasse, hemp, inert fibres, resins, polyurethane foam or vermiculite) is impregnated with a liquid medium that contains the required nutrients. This approach enables better monitoring and control of the process, allows higher rates of oxygen and nutrient diffusion, and produces more reproducible fermentations, but it has higher costs.

SSF bioreactor systems are mostly unsophisticated. Batch processes are the most common and different types of bioreactors include stationary packed beds and trays, rotating drums, fluidised beds (similar to fluidised dryers, Chapter 16, section 16.2.1), stirred aerated beds and rocking drums. Filtration of the inlet and outlet air streams may be used to prevent contamination and the uncontrolled release of the process micro-organism into the environment. In agitated bioreactors, shear forces generated by the mixing can damage fungal hyphae. A more sophisticated bioreactor is described by Pérez-Guerra et al. (2004). It consists of a slowly rotating basket that contains a bed of substrate, and it is fitted with slowly rotating helical baffles for mixing the substrate with minimal damage to hyphae. Sterilisation and inoculation are automated. The slow rotation of the basket removes temperature gradients in the substrate bed. Forced aeration with humid air from the base of the basket maintains oxygen levels without evaporative moisture loss. In other designs, microprocessor-controlled chamber bioreactors that have automatic loading, stirring and unloading and capacities of up to 2 tonnes per batch are described by Raimbault (1998).

6.4.3   Commercial food fermentations
Throughout the world a very large number of fermented foods are important components of the diet (Appendix B.4). In this section a brief outline is given of selected commercial fermentations and further details are given by Steinkraus (1995, 2002, 2004).

Alcohol production

Beer wort is produced by boiling malted grains (e.g. barley) to release maltose and other sugars, and in some beers by adding hop extract to produce bitterness (Forster *et al.* 1995). Developments in wort production include the use of dextrose syrups to increase product uniformity, and higher-temperature shorter-time boiling to reduce energy consumption. Other substrates including millet, sorghum and maize are also used where these are the staple crops. Variation in the composition of the wort, the strain of yeast, and the fermentation time and conditions result in the wide range of beers produced. These differences are described in detail by Goldammer (2000). After fermentation, beers are filtered or centrifuged (Chapter 5, sections 5.1 and 5.2) and pasteurised (Chapter 12) before pouring into bottles, cans, kegs or bulk containers (Chapter 26, section 26.1).

Sugars in grape juice (or 'must') are fermented to produce 6–14% ethanol in wines. Cells are removed by filtration or centrifugation and the wine is aged to reduce the acidity and to develop a characteristic bouquet. The main acid in most wines is tartaric acid but, in some red wines, malic acid is present in a high concentration. A secondary malo-lactic fermentation by lactic acid bacteria converts malic acid to lactic acid, which reduces the acidity and improves the flavour and aroma. This is reviewed by Moreno-Arribas and Polo (2005). Details of grape wine production are given by Jackson (2000), Ribereau-Gayon *et al.* (2000), Boulton *et al.* (1996) and Fleet (1993). Other wines are produced throughout the world from many fruits, tree saps, honey and vegetable pods. Distillation of wines to produce spirits is described in Chapter 14 (section 14.2).

Bakery products

Leavened breads are produced in batch operations by first mixing ingredients in a Z-blade or planetary mixer (Chapter 4, section 4.1.3) to form a dough. This is then fermented in a proving cabinet by *Saccharomyces cerevisiae* to produce carbon dioxide, which leavens the dough and makes it anaerobic. Ethanol is produced in minor amounts, which are evaporated during baking. In sour dough fermentations, lactic acid bacteria and yeasts are used to ferment different cereal and legume mixtures using similar equipment. In a continuous liquid fermentation system for doughs, the growth of yeast and *Lactobacillus* sp. are separated and optimised. Yeast is mixed with flour and water and stored until it is needed. It is then activated by addition of dextrose and added to the dough mixer. Similarly a flour and water mixture is seeded with *Lactobacillus* culture and when the pH has dropped to around 3.8, part of the liquor is pumped to a storage vessel, to be used over several weeks' production. The liquor is pumped to a continuous mixer and as it is used, it is replaced by fresh flour/water to allow the fermentation to continue. The computer-controlled process is claimed to greatly improve fermentation efficiency, reduce labour, floor space, eliminate the need for dough-tubs and a fermentation room, and produce more consistent and hygienic doughs (Anon 1998). Other lactic acid fermentations produce leavened pancakes from a variety of cereals and legumes, including idli in India and injera in Ethiopia (Appendix B.4). In all products baking in an oven (Chapter 18, section 18.2) or on a griddle destroys the micro-organisms. Further details of bakery fermentations are given by Hui *et al.* (2004) and Cauvain and Young (2006).

Dairy products

There are a large number of cultured milk products produced throughout the world using lactic acid bacteria. Differences in flavour are due to differences in the concentrations of lactic acid, volatile aldehydes, ketones, organic acids and diacetyl (acetyl methyl carbinol), which give the characteristic 'buttery' aroma to dairy products, that are each produced by the fermentation. Changes in texture are due to coagulation of casein micelles to form characteristic flocs when lactic acid bacteria lower the pH to the isoelectric point. The sequence of lactic acid bacteria in a fermentation is determined mainly by their acid tolerance. In milk, *Streptococcus liquifaciens, Lactococcus lactis* or *Streptococcus cremoris* are inhibited when the lactic acid content reaches 0.7–1.0%. They are then outgrown by more acid-tolerant species including *Lactobacillus casei* (1.5–2.0% acid) and *Lactobacillus bulgaricus* (2.5–3.0% acid). These changes are described in detail by Fox *et al.* (2000), Hui (1993a–c) and Anon (2008f). Modifications to the type of starter culture, incubation conditions and subsequent processing conditions are used to control the size and texture of the coagulated protein flocs, and hence produce the many different textures encountered in fermented dairy products. Preservation is achieved by increased acidity (yoghurt and cultured milks) or reduced water activity (cheese) and by chilling (Chapter 21).

In yoghurt production, skimmed milk is mixed with dried skimmed milk and pasteurised at 82–93 °C for 30–60 minutes. It is inoculated with a mixed culture of *S. thermophilus* and *L. bulgaricus* that act synergistically, with *L. bulgaricus* producing most of the lactic acid and also acetaldehyde and diacetyl, to give the characteristic flavour and aroma in yoghurt. Details of production are described by Tamime and Robinson (2007). Developments in biotechnology have produced lactic acid bacteria that also have stabilising and viscosity modifying properties (Mogensen 1991). These are used in a wide variety of fermented milks, dressings and breads to reduce or avoid the use of synthetic stabilisers and emulsifiers. Other lactic acid bacteria, including *Leuconostoc* sp., *Lactobacillus* sp. and *Pediococcus* sp. produce a range of bacteriocins (section 6.7).

More than 400 types of cheese are produced throughout the world, created by differences in bacteria and fermentation conditions, pressing and ripening conditions, and are described in detail by Fox (1999). The fermentation of cottage cheese is stopped once casein precipitation has occurred and the flocs are removed along with some of the whey, but most other cheeses are pressed and allowed to ripen for several weeks or months. In Cheddar cheese manufacture, rennet (section 6.5) is added and the culture is incubated for 1.5–2 h until the curd is firm enough to cut into small cubes. It is then heated to 38 °C to shrink the curd and expel whey. The curd is recut and drained several times, milled, salted and placed in hoops (press frames). It is pressed to remove air and excess whey, and the cheese is then ripened in a cool room. It is matured for different periods of time, and the flavour gradually changes from a mellow creamy taste after 2–8 months, to a tangy flavour of mature cheese after 8–12 months, and then to a strong, more bitter flavour of vintage cheese after >12 months. The changes are due to enzymes from both the micro-organisms and the cheese (including proteases, peptidases, lipase, decarbox-ylase and deaminases) that produce compounds with characteristic aromas and flavours. Details of cheese production are given by Mullan (2001), Fox *et al.* (2000), Law (1997) and Banks (1992).

Mycoprotein
Fusarium venenatum (PTA-2684) is used to make mycoprotein, a meat substitute that is flavoured and textured to resemble chicken or beef and formed into patties, pie ingredients, sausages, 'deli' slices and cutlets. It is marketed as being a healthy food, a good source of protein and fibre, lower in fat and saturated fat than meat equivalents and with no cholesterol. Manufacturers also claim improved taste and texture over soy-based equivalent products (Wilson 2001). It contains dietary fibre in the form of β-1.3 and -1,6 glucans and chitin (Appendix A.5), which may also act as a prebiotic (section 6.3.3).

Mycoprotein has been shown to significantly reduce total and LDL cholesterol levels and also reduce glycaemia (Turnbull and Ward 1995). An evaluation of the safety and nutritional value of mycoprotein is reported by Miller and Dwyer (2001).

Production of the fungal mycelia by continuous axenic fermentation is described by Rodger (2001). The fermentation medium is continuously removed from the fermenter and is heated rapidly to kill the mycelia, coagulate cellular proteins and cause the cell RNA to leak through cell walls into the supernatant. This reduces the RNA content of the mycelia from 10% to <2%, which is necessary to prevent the formation of serum uric acid when consumed (RNA is metabolised to uric acid). The biomass is collected as a paste that contains approximately 75% moisture using a dewatering centrifuge (Chapter 5, section 5.1.2).

The mycelial hyphae measure 400–700 μm by 3–5 μm in diameter and this high length:diameter ratio means that they are similar in morphology to animal muscle cells. The hyphae are mixed with an egg albumin or milk protein binder, vegetable fat, flavourings and colourants, in a ratio of approximately 90% hyphae:10% added ingredients, and formed into the required shape. The pieces of mycoprotein are then heated to cause the binder to gel and hold the hyphae together in a similar way that connective tissue holds myofibril cells together in meat. The texture of mycoprotein can be adjusted by varying the amounts of binder and/or hyphae, which influence the firmness, chewiness and fibrousness of the product, or by adjusting the amount of added fat to control juiciness. Other applications of mycoprotein, including use as a fat replacer in dairy products, and in extruded cereals and snacks are under development (Roger 2001).

Pickling and curing
Vegetables (e.g. olives, cucumbers) are submerged in 2–6% brine, which inhibits the growth of putrefactive spoilage bacteria. The brine is inoculated with either *L. plantarum* alone or a mixed culture with *P. cerevisiae*. Alternatively a naturally occurring sequence of lactic acid bacteria grows in the anaerobic conditions to produce approximately 1% lactic acid. Nitrogen gas is continuously purged through the vessel to remove carbon dioxide and to prevent cucumbers splitting. Other methods of pickling involve different salt concentrations: for example in 'dry salting' to make sauerkraut or kim-chi, alternate layers of vegetable and granular salt are packed into tanks. Juice is extracted by the salt to form a brine and the fermentation takes place as described above. In each case preservation is achieved by the combination of acid and salt, and, where products are packed into jars, by pasteurisation.

Fermented sausages (e.g. salami, pepperoni and bologna) are produced from a mixture of finely chopped meats, spice mixtures, curing salts (sodium nitrite/nitrate), salt and sugar. The meat is filled into sausage casings, fermented to cause the pH to fall from ≈6.0 to 4.5–4.6, and then pasteurised at 65–68 °C for 4–8 h, dried and/or smoked and stored at 4–7 °C. The fermentation causes flavour development and softening of meat tissues due to proteolysis. The technology of production is described in detail by Campbell-Platt and Cook (1995) and Hutkins (2006). Preservation is due to the anti-microbial action of nitrite–spice mixtures and salt; 0.84–1.2% lactic acid from the fermentation; heat during pasteurisation and/or antimicrobial components in smoke when the product is smoked; reduction in water activity due to salt and drying; and low storage temperature (see also hurdle technologies in Chapter 1, section 1.3.1). In Southeast Asia, small fish, shrimp or waste fish are mixed with dry salt and fermented to produce a range of sauces and pastes (Appendix B.4). Proteins in the fish are broken down by the combined action of bacterial enzymes, acidic conditions and autolytic action of the natural fish enzymes. Dirar (1993) describes the production of similar fermented fish pastes and fermented mullet in Sudan and North Africa. Other fermented porridges are described by Steinkraus (1995, 2002, 2004).

Soy sauce and similar products are made by a two-stage fermentation in which one or more fungal species are grown on a mixture of ground cereals and soy beans. Fungal proteases, α-amylases and invertase act on the soy beans to produce a substrate for the second fermentation stage. The fermenting mixture is transferred to brine and the temperature is slowly increased. Acid production by *P. soyae* lowers the pH to 5.0, and an alcoholic fermentation by *S. rouxii* takes place. Finally the temperature is gradually returned to 15 °C and the characteristic flavour of soy sauce develops over a period of 6 months to 3 years. The process is described in detail by Fukushima (1985). The liquid fraction is separated, clarified, pasteurised and bottled. The final product has a pH of 4.4–5.4 and is preserved by 2.5% ethanol and 18% salt. In the production of tempeh, soy beans are soaked, deskinned, steamed for 30–120 min and fermented to reduce the pH to 4.0–5.0. Fungal enzyme activity softens the beans, and mycelial growth binds the bean mass to form a solid cake. The fermentation changes the texture and flavour of soy beans but has little preservative effect. The product is either consumed within a few days or preserved by chilling.

6.4.4   Effects on foods

The mild conditions used in food fermentations produce few of the deleterious changes to sensory characteristics and nutritional quality that are found with many other unit operations. In fact fermentation can improve food quality indices, including texture, odour, flavour, appearance, nutrition and safety.

Sensory characteristics

Fermenting foods are complex systems that have enzymes from ingredients interacting with metabolic activities of fermentation micro-organisms. Factors such as particle sizes, temperature, and DO concentration also have important effects on biochemical changes that occur during the fermentation. Fermentation improves the palatability and acceptability of raw materials by introducing flavour and aroma components and modifying the food texture. Flavour changes are complex (see Voilley 2006) and in general poorly documented for fermented foods. One example is Chaven and Kadam (1989), who describe flavour compounds formed in fermented seasonings and condiments. Flavour changes during fermentation have been reviewed by McFeeters (2004). The aroma of fermented foods is due to a large number of volatile chemical components (e.g. amines, fatty acids, aldehydes, esters and ketones) and products from interactions of these compounds during fermentation and maturation. In bread, coffee and cocoa, the subsequent unit operations of baking and roasting produce the characteristic aromas. The taste and texture of fermented foods are altered by the production of acidulants and include for example a reduction in sweetness and increase in acidity due to fermentation of sugars to organic acids. The acids in turn alter milk proteins to produce the characteristic yoghurt gel (Sodini *et al.* 2004) or cheese flocs. There is an increase in saltiness in some foods (pickles, soy sauce, fish and meat products) due to salt addition and reduction in bitterness of some foods due to the action of debittering enzymes.

The colour of many fermented foods is retained owing to the minimal heat treatment and/or a suitable pH range for pigment stability. Changes in colour may also occur owing to formation of brown pigments by proteolytic activity in fermented meats, degradation of chlorophyll and enzymic browning in pickles.

Nutritional value

Microbial growth causes complex changes to the nutritive value of fermented foods by changing the composition of proteins, fats and carbohydrates. Some micro-organisms hydrolyse polymeric compounds to produce substrates for cell growth, which may improve the digestibility of proteins and polysaccharides. Fermentation of cereals improves the nutritive value by increasing the amount and quality of proteins (Chavan *et al.* 1988) and available lysine. Micro-organisms also utilise or secrete sugars, vitamins, or essential fatty acids and amino acids, and remove antinutrients, natural toxins and mycotoxins. Changes in the vitamin content of foods vary according to the types of micro-organism and the raw material used, but there is usually an increase in B-vitamins (Chavan and Kadam 1989) and riboflavin and niacin contents may significantly increase (Table 6.3) (Steinkraus 1994). Although in general fermentations do not alter the mineral content of a food, hydrolysis of chelating agents (e.g. phytic acid) during fermentation may improve their bioavailability. The effect of fermentation on toxins and antinutritional components in plant foods is reviewed by Reddy and Pierson (1994). These include reductions in trypsin inhibitor, phytates and flatus producing oligosaccharides. However, fermentation of cereals by *Rhizopus oligosporus* has been reported to release bound trypsin inhibitor, thus increasing its activity (Haard *et al.* 1999).

Safety

The safety of fermented foods is reviewed by Nout (1994) and different pathogens that may infect fermented foods are described in Chapter 1 (section 1.2.4). Fungal and lactic acid fermentations reduce aflatoxin B_1, and lactic acid bacteria inhibit pathogenic micro-organisms by the production of bacteriocins (section 6.7). Cases of foodborne infection, or intoxication due to microbial metabolites such as mycotoxins, ethyl carbamate and biogenic amines may occur due to contaminated raw materials, lack of pasteurisation and inadequately controlled fermentation conditions (Haard *et al.* 1999).

6.5   Microbial enzymes

Advances in biotechnology have had a significant effect on the number and type of new enzymes that are available for use in food processing, especially bakery products, fruit juices, glucose syrups and cheese, or for production of specialist ingredients, which have been reviewed by Pszczola (1999) (GMM enzymes are described in section 6.1.3). There has been rapid growth in the use of enzymes to reduce processing costs; to increase yields of extracts from raw materials; to improve handling of materials; and to improve the shelf-life and sensory characteristics of foods (Appendix B.3, Anon 2004d and Table 6.4). They are separated and purified from microbial cells, or less commonly from animal or plant sources, and are either added to foods as concentrated solutions or powders, or immobilised on support materials in a 'bioreactor'. The main advantages in using enzymes instead of chemical modifications are:

- Enzymic reactions are carried out under mild conditions of temperature and pH, and are highly specific, thus reducing the number of side reactions and by-products.
- Enzymes are active at low concentrations and the rates of reaction are easily controlled by adjustment of incubation conditions.

Table 6.3 Nutritional improvement of selected foods by fermentation

Food	Thiamin (μg g^{-1})		Food	Lysine (mg g^{-1} N)		Food	Riboflavin (mg 100 g^{-1})	
	Unfermented	Fermented		Unfermented	Fermented		Unfermented	Fermented
Finger millet	0.30	0.47	Barley	19	75	Whole milk	0.18	0.50 (cheese)
Maize	0.37	0.86	Maize	18	46	Soybean	0.06	0.37 (soy sauce)
Pearl millet	0.37	0.64	Millet	2	37			0.49 (tempeh)
Sorghum	0.20	0.47	Oats	18	104			
Soybean	0.22	0.88	Rice	4	46			
		(soy sauce)	Wheat	24	65			

Adapted from Chaven and Kadam (1989), Hamad and Fields (1979) and Paul and Southgate (1991)

Table 6.4   Examples of enzymes used in food processing

Enzyme	Operating conditions	
	pH range	Temperature (°C)
Microbial sources		
α-amylases	4.0–5.0	50–70
Amyloglucosidase	3.5–5.0	55–65
Cellulases	3.0–5.0	20–60
Glucoamylases	3.5–5.0	30–60
Glucose isomerase	7.0–7.5	60–70
Glucose oxidase	4.5–7.0	30–60
Hemi-cellulases	3.5–6.0	30–65
Invertase	4.5–5.0	50–60
Pectic enzymes	2.5–5.0	25–65
Proteases (acid)	4.5–7.5	20–50
Proteases (neutral)	7.0–8.0	20–50
Proteases (alkaline)	9.0–11.0	20–50
Pullanase	3.5–5.0	55–65
Plant sources		
Bromelain (from pineapple)	4.0–9.0	20–65
Ficin (from fig)	6.5–7.0	25–60
Papain (from papaya)	6.0–8.0	20–75
Animal sources		
Catalase (from beef liver)	6.5–7.5	5–45
Lipase (from porcine pancreas)	5.5–9.5	20–50
Rennet (from calf stomach)	3.5–6.0	40

- There is minimal loss of nutritional quality at the moderate temperatures employed.
- Lower energy consumption than corresponding chemical reactions.
- They enable the production of new foods, not achievable by other methods.

The use of enzymes in food analysis is also rapidly expanding and is discussed in detail by Pomeranz and Meloan (2000) and Schwedt and Stein (1994).

However, the cost of many enzymes is high and, in some products enzymes must be inactivated or removed after processing, which adds to the cost of the product. Like other proteins, enzymes may cause allergic responses in some people, and they may be microencapsulated or immobilised on carrier materials to reduce the risk of inhalation of enzyme dust by operators. Selection of the precise enzyme for a particular application can be difficult and guidelines on methods to do this are given by West (1988). Details of the factors that influence enzyme activity and reaction rates are described in standard biochemistry texts and by Whitehurst and Law (2002).

Microbial enzymes have optimum activity under similar conditions to those that permit optimum cell growth. They are either secreted by the cells into the surrounding medium ('extracellular' production) or retained within the cell ('intracellular' enzymes). Extracellular enzyme production occurs in either the logarithmic phase or the stationary phase of growth, whereas intracellular enzymes are produced during logarithmic growth but are released into the medium only when cells undergo lysis in the stationary or decline phases. The requirements of commercial enzyme production from micro-organisms are as follows:

- Micro-organisms must grow well on an inexpensive substrate.
- Substrates should be readily available in adequate quantities, with a uniform quality.
- Micro-organisms should produce a constant high yield of enzyme in a short time.

- Methods for enzyme recovery should be simple and inexpensive.
- The enzyme preparation should be stable.

Enzymes are produced by either solid substrate fermentations (e.g. rice hulls, fruit peels, soybean meal or wheat flour) or by submerged culture using liquid substrates in fermenters (e.g. molasses, starch hydrolysate or corn steep liquor (section 6.4.2)). Submerged cultures have lower handling costs and a lower risk of contamination and are more suited to automation than are solid substrates. The success of commercial enzyme production depends on maximising the activity of the micro-organism and minimising the costs of the substrate and incubation and recovery procedures. Extracellular enzymes are recovered from the fermentation medium by centrifugation, filtration (Chapter 5, sections 5.1 and 5.2), fractional precipitation, chromatographic separation, electrophoresis, membrane separation (Chapter 5, section 5.5), freeze drying (Chapter 23, section 23.1) or a combination of these methods. Intracellular enzymes are extracted by disruption of cells in a homogeniser or mill (Chapter 3). Recovery is more difficult and the yield is lower than for extracellular enzymes, because some enzymes are retained within the cell mass. If required, the specific activity of the enzyme is increased by precipitation using acetone, alcohols or ammonium sulphate or by ultrafiltration (Chapter 5, section 5.5.3).

In batch operation the enzyme is mixed with food, allowed to catalyse the required reaction, and then either retained within the food or inactivated by heat. This is used when the cost of the enzyme is low. In continuous operation, immobilised enzymes are either mixed with a liquid substrate and removed by centrifugation or filtration and re-used, or the feed liquor is passed over an immobilised bed of enzyme in a bioreactor. Different methods include the following:

- Micro-encapsulation in membranes that retain the enzyme but permit passage of substrates and products.
- Electrostatic attachment to ion exchange resins.
- Adsorption onto colloidal silica, charcoal, polyacrylamide or glass, and/or cross linking with glutaraldehyde. Chitin and its derivatives are suitable as supports in immobilised enzyme reactors, for example by coating magnetic beads that are easily removed from the reactor (Synowiecki and Al-Khateeb 2003).
- Entrapment in polymer fibres (e.g. cellulose triacetate or starches).
- Covalent bonding to organic polymers or co-polymerisation with maleic anhydride.

Immobilised enzymes should have the following characteristics: short residence times for a reaction; stability to variations in temperature and other operating conditions over a period of time (e.g. glucose isomerase is used for ≈1000 h at 60–65 °C) and suitability for regeneration. The main advantages of enzyme immobilisation are that enzymes are re-used, so reducing costs, and they allow continuous processing and closer control of pH and temperature to achieve optimum activity. Immobilisation is used when an enzyme is difficult to isolate or expensive to prepare. The main limitations are increased costs of carriers, equipment and process control, changes to the reaction kinetics of enzymes, loss of activity and risk of microbial contamination. A summary of the food applications of the main enzyme groups is shown in Appendix B.3 and information on enzymes on the Internet has been compiled by Swadling (2001).

6.5.1   Novel enzyme technologies
Extreme thermophilic and hyperthermophilic micro-organisms (also 'extremophiles') of the *Pyrococcus* spp. and *Thermococcus* spp. produce enzymes that have optimum activity

between 80 and 100 °C and are more resistant than mesophilic equivalents to environmental conditions such as low or high pH. Their reactions can be terminated by simply lowering the temperature and the lower viscosity of concentrated substrates at high temperatures allows improved mixing and pumping. However, commercial applications may be limited because these micro-organisms produce enzymes in amounts that are insufficient for large-scale enzyme production, and they may also produce toxic or corrosive metabolites. Synowiecki *et al.* (2006) describe studies of recombinant α-amylase from *Pyrococcus woesei* and a thermostable α-glucosidase from *Thermococcus thermophilus* that were expressed in an *E. coli* host. α-amylase had optimum activity at pH 5.6 and 93 °C and the α-glucosidase retained 80% of its activity between pH 5.8 and 6.9. The authors concluded that these enzymes were suitable for starch processing. They also reported a recombinant β-galactosidase from *P. woesei* that was active in the pH range 4.3–6.6 with high thermostability. It would be suitable for the production of low-lactose milk or whey at temperatures that restrict microbial growth in continuous-flow immobilised enzyme reactors. Details of the applications of novel enzyme technologies are described by Rastall (2007).

6.6   Microencapsulation and controlled release technologies

Microencapsulation (also Chapter 24, section 24.3) and liposomes have been used to protect enzymes from adverse environmental conditions during processing (e.g. heat or pH) and enable their delayed or controlled activity within the processed food. Liposomes are spherical vesicles, varying in size from <30 to >300 nm, which have one or more bilayer membranes of polar phospholipids. The two layers are formed by the hydrophobic ends of the lipids being oriented towards each other, thus presenting hydrophilic ends of the molecules to both the inside and outside of the vesicle (Fig. 6.5). Liposomes were originally developed for the pharmaceutical industry but have found applications in food processing for the controlled delivery of functional ingredients, including proteins, enzymes and flavours. These applications are reviewed by Gibbs *et al.* (1999). They are

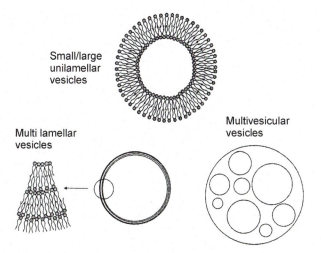

Small/large
unilamellar
vesicles

Multivesicular
vesicles

Multi lamellar
vesicles

Fig. 6.5   Structure of liposome vesicle (reprinted by permission of the publisher, Taylor and Francis Ltd (http://www.tandf.co.uk/journals) (from Taylor *et al.* 2005).

produced by high-intensity ultrasound, high-pressure homogenisation or membrane homogenisation (Chapter 3, section 3.2.3). The functionality of liposomes depends on their interfacial properties, especially the surface charge, the phospholipid concentration, the concentration of entrapped material, as well as the pH, ionic strength and temperature of the medium in which they are contained. Details of the structure and properties of different types of liposomes are reviewed by Taylor *et al.* (2005).

Examples of enzymes that have been microencapsulated or entrapped in liposomes include proteinases that accelerate cheese ripening and increase levels of flavour compounds such as diacetyl, acetoin and methyl ketones, and lipases that reduce the firmness of cheese while increasing the elasticity and cohesiveness and improving the flavour profile (Kheadr *et al.* 2002). The microcapsules or liposomes protect the enzymes from the low pH and high ionic strength in the cheese (el Soda *et al.* 1989). Liposome-entrapped β-galactosidase has been added to dairy products to aid digestion by lactose-intolerant consumers (Rao *et al.* 1995). Other studies indicate that entrapment of enzymes, including α-amylase and glucose oxidase, can restrict their access to substrate and delay their activity, and also protect them from adverse environmental conditions of pH and temperature. Research using liposome-entrapped vitamins has also shown that there is potential to protect these nutrients against degradation and develop new methods for fortifying foods (Taylor *et al.* 2005).

6.7   Bacteriocins and antimicrobial ingredients

Bacteriocins are biologically active peptides that are produced by some stains of lactic acid bacteria and which inhibit the growth of spoilage or pathogenic bacteria. An example is nisin, produced by *Lactococcus lactis* in European cheeses, which prevents growth of *Clostridium tyrobutyricum* and thus prevents off-flavour development and 'blowing' of Swiss-type cheese during ripening. Nisin is also effective against *Listeria monocytogenes* and, although it has been added to cultures in the past, its production by *Lactococcus lactis* is a cheaper and more effective method of removing this pathogen from cheese. *L. lactis* has also been used to inhibit the growth of *Cl. botulinum* in processed cheese and other dairy products, processed vegetables, soups, sauces and beer (Roller *et al.* 1991). Liposomes containing entrapped nisin have been found to withstand the temperatures used in cheese making and remain active against *L. monocytogenes* during storage (Were *et al.* 2003, 2004). The advantage of using bacteriocins such as nisin is the reduction or avoidance of chemical preservatives such as nitrate, sorbic acid and benzoic acid (Mogensen 1991). Other applications of nisin have been reviewed by de Vuyst and Vandamme (1994) and other inhibitory metabolites of lactic acid bacteria are described by Breidt and Fleming (1997) in their application to minimally processed fruits and vegetables. Roller (2003) describes the use of natural antimicrobial chemicals for minimal processing. The use of antimicrobial chemicals to improve the safety of fresh foods is reviewed by Cherry (1999). The action of pediocin PA-1 and its applications in food systems are reviewed by Rodriguez *et al.* (2002) and the use of 'Microgard', reuterin and lactoferrin are described by Mahapatra *et al.* (2005). *Pediococcus acidilactici*, when used in fermented meat, also has the potential to inhibit spoilage bacteria and thus reduce the need for nitrate addition.

Chitin and chitosan

After cellulose, chitin is the most abundant natural polysaccharide, synthesised by insects, crustaceans, molluscs, algae, fungi and yeasts. Chitosans (deacetylated chitin) inhibit phyto-pathogenic fungi and bacteria and induce natural resistance in plants to fungal or viral infections (Pospieszny and Maćkowiak (1997)). Microcrystalline chitosan and its salts also show strong antiviral activity (Synowiecki and Al-Khateeb 2003). Sulphobenzoyl chitin prevents the growth of *Pseudomonas* spp., *Salmonella* spp., *Aeromonas* spp. and *Vibrio* spp. (Chapter 1, section 1.2.4). Treatment of potatoes with chitosan solution protects against pathogens such as those causing soft rot by inactivating polygalacturonase, pectate lyase and pectin methylesterase that are secreted by the pathogens. Chitosan with added calcium ions creates semipermeable membranes on the surface of treated fruits, tomatoes and cucumbers, which alters the rate of oxygen and carbon dioxide permeation and significantly extends the shelf-life (Li and Chung 1997). Chitosan coatings and coated packaging paper also inhibit microbial growth (also Chapter 24, section 24.3, and Chapter 25, section 25.4.1). Wastes from the processing of crabs, lobsters, shrimps, oysters and clams are a rich source of chitin. Chitosan is produced by deacetylation of chitin using concentrated NaOH solution, but this produces variable degrees of deacetylation and loss of nutritionally valuable proteins. Synowiecki and Al-Khateeb (2003) report mild enzymic methods that can deproteinise shrimp shells and deacetylate chitin, and Tharanathan and Kittur (2003) have reviewed the food applications of chitin.

References

ALLEN, J.C., (1990), The value of immunoassys to food analysis, in (J.H. Rittenburg, Ed.), *Development and Application of Immunoassay for Food Analysis*, Elsevier Applied Science, London, pp. 59–80.

ANDERSEN, M.K., (2005), Fermenter control programme, FoxyLogic.com available at http://imsb.au.dk/~mha/bhome.htm.

ANON, (1998), Continuous liquid fermentation, technical information from Baker Perkins Group, available at www.bakerperkinsgroup.com.

ANON, (1999), GM plants offer 'green' plastics; genetically modified oilseed developed by Monsanto Co. International Academy of Life Sciences, available at www.plantpharma.org/ials/index.php?id=63.

ANON, (2000a), Cartagena Protocol on Biosafety to the Convention on Biological Diversity, United Nations Environmental Programme, January.

ANON, (2000b), IFT Expert Report on Biotechnology and Foods – Introduction, *Food Technology*, **54** (8), 124–136.

ANON, (2000c), IFT Expert Report on Biotechnology and Foods – human food safety evaluation of rDNA biotechnology-derived foods, *Food Technology*, **54** (9), 53–61.

ANON, (2000d), IFT Expert Report on Biotechnology and Foods – labeling of rDNA biotechnology-derived foods, *Food Technology*, **54** (9), 62–74.

ANON, (2000e), IFT Expert Report on Biotechnology and Foods – benefits and concerns associated with rDNA biotechnology-derived foods, *Food Technology*, **54** (10), 61–80.

ANON, (2001), Public sentiment about genetically modified food, Pew Initiative on Food and Biotechnology, available at www.pewagbiotech.org.

ANON, (2002), Genetically modified foods – experts view regimen of safety tests as adequate, but FDA's evaluation process could be enhanced, US General Accounting Office report to Congressional Requesters (GAO-02-566), May, 2002 available at www.gao.gov/new.items/d02566.pdf#search=%22gm%20foods%20safety%20testing%22.

ANON, (2003a), Broad opposition to genetically modified foods, Pew Global Attitudes Project, available at http://pewglobal.org/commentary/display.php?AnalysisID=66.

ANON, (2003b), What do the experts say about the potential human health effects of genetically engineered corn?, Friends of the Earth, available at www.humboldt.org.ni/transgenicos/docs/what_experts_says_human_effects.pdf.

ANON, (2003c), Marker assisted selection: A fast track to increase genetic gain in plant and animal breeding? Proceedings of an International Workshop by Fondazione per le Biotecnologie, the University of Turin and FAO, 17–18 October, Turin, Italy. Papers available at the Electronic Forum on Biotechnology In Food And Agriculture, FAO, www.fao.org/Biotech/Torino.htm.

ANON, (2003d), Guideline for the conduct of food safety assessment of foods derived from recombinant-DNA plants (CAC/GL 45-2003), CODEX Alimentarius Commission, available at www.codexalimentarius.net/download/standards/10021/CXG_045e.pdf.

ANON, (2003e), Guideline for the conduct of food safety assessment of foods produced using recombinant-DNA micro organisms (CAC/GL 46-2003), CODEX Alimentarius Commission available at www.codexalimentarius.net/download/standards/10025/CXG_046e.pdf.

ANON, (2004a), IFST Current hot topics: genetic modification and food, Institute of Food Science and Technology, UK.

ANON, (2004b), Safety of genetically engineered foods: approaches to assessing unintended health effects, by Committee on Identifying and Assessing Unintended Effects of Genetically Engineered Foods on Human Health: Board on Life Sciences, Institute of Medicine, Food and Nutrition Board, Board on Agriculture and Natural Resources and National Research Council Of The National Academies, National Academies Press, Washington, DC, available at www.nap.edu/books/0309092094/html/.

ANON, (2004c), Functional foods, British Nutrition Foundation, available at www.nutrition.org.uk/upload/Functional%20Foods%20pdf.pdf.

ANON, (2004d), AMFEP list of commercial enzymes, the Association of Manufacturers and Formulators of Enzyme Products available at www.amfep.org/list.html.

ANON, (2005a), IUFoST Scientific Bulletin on Biotechnology and Food, International Union of Food Science and Technology, available at www.iufost.org/.

ANON, (2005b), GM crops: the global economic and environmental impact – the first nine years 1996–2004, J. Agrobiotechnology Management and Economics, *AgBioForum*, **8** (2&3), 187–196, available at AgbioForum. www.agbioforum.org or www.pgeconomics.co.uk/GM_global_study.htm.

ANON, (2005c), PASSCLAIM Consensus Document, available at http://europe.ilsi.org/ecprojects/PASSCLAIM/passpubs.htm.

ANON, (2005d), Common Guidelines on the nutrition and health claims made on foods, Council of the EU, available at http://ec.europa.eu/food/food/labellingnutrition/claims/common_position05_en.pdf.

ANON, (2006), Genetically Engineered Crops and Foods: Worldwide Regulation and Prohibition, The Centre for Food Safety, Washington DC, available at www.centerforfoodsafety.org/pubs/World_Regs_Chart%20_6-2006.pdf.

ANON, (2007), FAO/WHO Expert Consultations on GM Foods, available at http://www.who.int/foodsafety/biotech/consult/en/index.html.

ANON, (2008a), Genetically modified food, Food Additives and Ingredients Association, available at www.faia.org.uk/gmnew.php.

ANON, (2008b), Harmonisation of regulatory oversight in Biotechnology, OECD, available at www.oecd.org/department/0,2688,en_2649_34387_1_1_1_1_1,00.html.

ANON, (2008c), Functional foods and neutraceuticals, available at www.ffnmag.com/ASP/home.asp.

ANON, (2008d), Omni fermenters, available at www.virtis.com/category/2208.

ANON, (2008e), Cell culture fermentation systems selection guide, Cole-Parmer Technical Library, available at www.coleparmer.com/techinfo/techinfo.asp?htmlfile=SelectingCellCulFerm.htm&ID=660.

ANON, (2008f), Dairy chemistry and physics, available at www.foodsci.uoguelph.ca/dairyedu/chem.html.

ARVANITOYANNIS, I.S. and VAN HOUWELINGEN-KOUKALIAROGLOU, M., (2005), Functional foods: a survey of health claims, pros and cons, and current legislation, *Critical Reviews in Food*

Science and Nutrition, **45**, 385–404.

BANKS, J.M., (1992), Cheese, in (R. Early, Ed.), *Technology of Dairy Products*, Blackie, London, pp. 39–65.

BERRY OTTAWAY, P., (2000), EU legislation and functional foods – a case study, in (G.R. Gibson and C.M. Williams, Eds.), *Functional Foods – Concept to Product*, Woodhead Publishing, Cambridge, pp. 29–42.

BLANCH H.W. and CLARK, D.S., (1995), *Biochemical Engineering*, Marcel Dekker, New York.

BOULTON, R.B., SINGLETON, V.L., BISSON, L.F. and KUNKEE, R.E., (1996), *Principles and Practices of Winemaking*, Springer Science and Business Media, Heidelberg, Germany.

BREIDT, F. and FLEMING, H.P., (1997), Using Lactic acid bacteria to improve the safety of minimally processed fruits and vegetables, *Food Technology*, **51** (9), 44–48.

BROOKES, G., CRADDOCK, N. and KNIEL, B.K., (2005), The Global GM Market – Implications for the European Food Chain, an independent report commissioned by Agricultural Biotechnology Europe, available at www.agbios.com/docroot/articles/05-266-001.pdf#search=%22global%2BGM%2Bmarket%22.

BROWN, A.K.C., GALLAGHER, I.S., DODD, P.W. and VARLEY, J., (2001), An improved method for controlling foams produced within bioreactors, *Food and Bioproducts Processing*, **79**, Issue C2, 114–121.

BURKE, M., (2005), Managing GM crops with herbicides – effects on farmland wildlife, Farmscale Evaluations Research Consortium and Scientific Steering Committee, DEFRA, available at http://www.defra.gov.uk/environment/gm/fse/results/fse-summary-05.pdf.

CAMPBELL-PLATT, G. and COOK, P.E. (Eds.), (1995), *Fermented Meats*, Blackie Academic and Professional, London.

CAUVAIN, S. and YOUNG, L., (2006), *Baked Products*, Blackwell Publishing, Oxford.

CHAMPAGNE, C.P. and GARDNER, N.J., (2005), Challenges in the addition of probiotic cultures to foods, *Critical Reviews in Food Science and Nutrition*, **45**, 61–84.

CHAVAN, U.D., CHAVAN, J.K. and KADAM, S.S., (1988), Effect of fermentation on soluble proteins and *in vitro* protein digestibility of sorghum, green gram and sorghum-green gram blends. *J. Food Science*, **53**, 1574–1575.

CHAVEN, J.K. and KADAM, S.S., (1989), Nutritional improvement of cereals by fermentation, *Critical Reviews in Food Science and Technology*, **28** (5), 349–400.

CHERRY, J.P., (1999), Improving the safety of fresh produce with antimicrobials, *Food Technology*, **53** (11), 54–59.

DE DECKERE, E.A.M. and VERSCHUREN, P.M., (2000), Functional fats and spreads, in (G.R. Gibson and C.M. Williams, Eds.), *Functional Foods – Concept to Product*, Woodhead Publishing, Cambridge, pp. 233–257.

DEFELICE, S. L., (1995), The neutraceutical revolution, its impact on food industry research and development, *Trends in Food Science and Technology*, **6**, 59–61.

DE VUYST, L. and VANDAMME, E.J., (1994), Nisin, an antibiotic produced by *Lactobacillus lactis* subsp. *Lactis*: properties, biosynthesis, fermentation and applications, in (L. De Vuyst and E.J.Vandamme, Eds.), *Bacteriocins of Lactic Acid Bacteria*, Blackie Academic and Professional, Glasgow, pp. 151–221.

DIRAR, H.A., (1993), *The Indigenous Fermented Foods of Sudan*, CAB International, Wallingford, UK, pp. 345–384.

DUNFORD, N.T., (2001), Health benefits and processing of lipid-based nutritionals, *Food Technology*, **55** (11), 38–44.

EL SODA, M., PANNELL, L. and OLSON, N., (1989), Microencapsulated enzyme systems for the acceleration of cheese ripening, *J. Microencapsulation*, **6** (3), 319–326.

EMBAREK, P.K.B., (2005), WHO Workshop on Guiding Public Health Policy in Areas of Scientific Uncertainty, Univ. Ottawa, Canada, 11–13 July, available at www.who.int/peh-emf/meetings/archive/benembarek_ottawajuly05.pdf.

FELEZEU, D., (2001), Intelligent software serving control and supervision of fermentation processes, Samedan Ltd, available at www.samedanltd.com/members/archives/EBR/Spring2002/

DoruFelezeu.htm.

FLEET, G.H., (Ed.), (1993), *Wine Microbiology and Biotechnology*, Taylor and Francis, London.

FORSTER, A., BALZER, U. and MITTER W., (1995), Large scale brewing tests with different hop extracts, available at www.barthhaasgroup.com/cmsdk/content/bhg/research/scientific1/53.html.

FOX, P., (1999), *Cheese: Chemistry, Physics and Microbiology*, Vol 2: *Major Cheese Groups*, Aspen Publishers, Gaithersburg, MD.

FOX, P.F., MCSWEENEY, P., COGAN, T.M. and GUINEA, T.P., (2000), *Fundamentals of Cheese Science*, Aspen Publishers, Gaithersburg, MD.

FUKUSHIMA, D., (1985), Fermented vegetable protein and related foods of Japan and Chine, *Food Review International*, **1** (1), 149–209.

GIBBS, B.F., KERMASHA, S., ALII, L. and MULLIGAN, C.N., (1999), Encapsulation in the food industry: a review. *International J. Food Science and Nutrition*, **50**, 213–224.

GIBSON, G.R. and WILLIAMS, C.M., (Eds.), (2000), *Functional Foods – Concept to Product*, Woodhead Publishing, Cambridge.

GIBSON, G.R., BERRY OTTAWAY, P. and RASTALL, R.A., (2000), *Prebiotics: New Developments in Functional Foods*, Chandos Ltd, Oxford.

GOLDAMMER, T., (2000), *The Brewers' Handbook – The Complete Book to Brewing Beer*, Apex Publishers, Clifton, VA.

GRIERSON, B., (2003), Eat right for your genotype, *The Guardian*, 15 May.

HAARD, N.F., ODUNFA, S.A., LEE, C-H., QUINTERO-RAMÍREZ, R., LORENCE-QUIÑONES, A. and WACHER-RADARTE, C., (1999), *Fermented Cereals. A Global Perspective*, FAO Agricultural Services Bulletin No. 138, Food and Agriculture Organization of the United Nations, Rome.

HAMAD, A.M. and FIELDS, M.L., (1979), Evaluation of the protein quality and available lysine of germinated and fermented cereal, *J. Food Science*, **44**, 456–459.

HASLER, C.M., (2000), The changing face of functional foods, *J. American College of Nutrition*, **19** (5), 499S–506S, also available at http://www.jacn.org/cgi/content/full/19/suppl_5/499S.

HASLER, C. M., (2002), Functional foods: benefits, concerns and challenges – a position paper from the American Council on Science and Health, *Nutrition*, **132** (12), 3772–81 and at http://moodfoods.com/functional-foods.html.

HASLER, C.M., (2005), *Regulation of Functional Foods and Nutraceuticals – A Global Perspective*, Institute of Food Technologists Series, Blackwell Publishing, Oxford.

HELLER, K.J., (2003), *Genetically Engineered Food*, Wiley-VCH, Weinheim, Germany.

HELLER, R., TANIGUCHI, Y. and LOBSTEIN, T., (1999), Functional foods: public health boon or 21st century quackery?, International Association of Consumer Food Organizations, available at www.cspinet.org/reports/functional_foods/.

HENRY, C.J.K. and HEPPELL, N.J., (1998), *Nutritional Aspects of Food Processing and Ingredients*, Aspen Publishers, Gaithersburg, MD, pp. 45–65.

HOBAN, T.J., (2004), *Public Attitudes Towards Agricultural Biotechnology*, ESA Working paper no. 04-09, May, Agricultural and Development division, FAO, Rome, available at www.fao.org/docrep/007/ae064e/ae064e00.htm.

HUI, H., GODDIK, L.M., HANSEN, A.S., JOSEPHSEN, J. and NIP, W-K., (Eds.), (2004), *Handbook of Food and Beverage Fermentation Technology*, Marcel Dekker, New York.

HUI, Y.H., (1993a), *Dairy Science and Technology Handbook. Vol. 1. Principles and Properties*, Wiley-VCH, Hoboken, NJ.

HUI, Y.H., (1993b), *Dairy Science and Technology Handbook. Vol. 2. Product Manufacturing*, Wiley-VCH, Hoboken, NJ.

HUI, Y.H., (1993c), *Dairy Science and Technology Handbook. Vol. 3. Technology and Engineering*, Wiley-VCH, Hoboken, NJ.

HUTKINS, R., (2006) *Microbiology and Technology of Fermented Foods: A Modern Approach*, Blackwell Publishing, Oxford, pp. 207–232.

ISOLAURI, E. and SALMINEN, S., (2000), Functional foods and acute infections, in (G.R. Gibson and C.M. Williams, Eds.), *Functional Foods – Concept to Product*, Woodhead Publishing, Cambridge, pp. 167–180.

JACOBSON, M. and SILVERGLADE, B., (1999), Functional foods: health boon or quackery?, *British Medical Journal*, **319**, 205–206, also available at http://bmj.bmjjournals.com (October 2001, resources updated March 2005).

JACKSON, R.S., (2000), *Wine Science: Principles, Practice, Perception*, Academic Press, San Diego, CA.

JAY, J.M., LOESSNER, M.J. and GOLDEN, D.A., (2005), *Modern Food Microbiology*, 7th edn, Springer Science and Business Media, Heidelberg, Germany.

JIN, S., YE, K., SHIMIZU, K. and NIKAWA, J., (1996), Application of artificial neural network and fuzzy control for fed-batch cultivation of recombinant *Saccharomyces cerevisiae*, *J. Fermentation Bioengineering*, **81**, 412–421.

KAILASAPATHY, K., (2002), Microencapsulation of probiotic bacteria: technology and potential applications, *Current Issues Intestinal Microbiology*, **3** (2), 39–48. Available at http://www.horizonpress.com/ciim/abstracts/v3/05.html.

KARIM, M.N. and RIVERA, S.L., (1992), Artificial neural networks in bioprocess state estimation, *Advances Biochemical Engineering/Biotechnology*, **46**, 1 –33.

KATAN, M.B. and DE ROOS, N.M., (2004), Promises and problems of functional foods, *Critical Reviews in Food Science and Nutrition*, **44**, 369–377.

KHEADR, E.E., VUILLEMARD, J.C. and EL-DEEB, S.A., (2002), Acceleration of Cheddar cheeses lipolysis by using liposome-entrapped lipases, *J. Food Science*, **67**, 485–492.

KRISHNAKUMAR, V. and GORDON, I.R., (2001), Probiotics: challenges and opportunities, *Dairy Industry International*, **66** (2), 38–40.

LAW, B.A., (Ed.), (1997), *Microbiology and Biochemistry of Cheese and Fermented Milk*, 2nd edn, Blackie Academic and Professional, London.

LEE, K.W., LEE, H.J., CHO, H.Y. and KIM, Y.J., (2005), Role of the conjugated linoleic acid in the prevention of cancer, *Critical Reviews in Food Science and Nutrition*, **45**, 135–144.

LEE, R. and CARSON, L., (2004), GM foods – a regulatory history, available at www.ccels.cf.ac.uk/literature/publications/2004/leepaper.pdf#search=%22GM%3Dlegislation%22.

LI, C.F. and CHUNG, Y.C., (1997), The benefits of chitosan postharvested storage and the quality of fresh strawberries, in (A. Domard, G.A.F. Roberts and K.M. Varum, Eds.), *Advances in Chitin Science*, Jacques Andre Publishers, Lyon, pp. 908–913.

LINDSAY, D.G., (2000), Maximising the functional benefits of plant foods, in (G.R. Gibson and C.M. Williams, Eds.), *Functional Foods – Concept to Product*, Woodhead Publishing, Cambridge, pp 183–208.

LINKO, S., LUOPA, J. and ZHU, Y.-H., (1997), Neural networks as 'software sensors' in enzyme production, *J. Biotechnol.*, **52**, 257–266.

MAHAPATRA, A.K., MUTHUKUMARAPPAN, K. and JULSON, J.L., (2005), Applications of ozone, bacteriocins and irradiation in food processing – a review, *Critical Reviews in Food Science and Nutrition*, **45**, 447–461.

MATTILA-SANDHOLM, T. and SAARELA, M., (Eds.), (2000), Probiotic functional foods, in (G.R. Gibson and C.M. Williams, Eds.), *Functional Foods – Concept to Product*, Woodhead Publishing, Cambridge, pp. 287–313.

MATTILA-SANDHOLM, T. and SAARELA, M., (2003) *Functional Dairy Products*, Vol. 1, Woodhead Publishing, Cambridge.

MCFEETERS, R.F., (2004), Fermentation micro-organisms and flavor changes in fermented foods, *J. Food Science*, **69**, 35–37.

MILLER, S.A. and DWYER, J.T., (2001), Evaluating the safety and nutritional value of mycoprotein, *Food Technology*, **55** (7), 42–47.

MITAL, B.K. and GARG, S.K., (1995), Anticarcinogenic, hypocholesterolemic and antagonistic activities of Lactobacillus acidophilus, *Critical Reviews Microbiology*, **21**, 175–214.

MOGENSEN, G., (1991), Replacing chemical additives through biotechnology, in (A. Turner, Ed.), *Food Technology International Europe*, Sterling Publications International, London, pp. 165–168.

MORENO-ARRIBAS, M.V. and POLO, M.C., (2005), Winemaking biochemistry and microbiology: current knowledge and future trends, *Critical Reviews in Food Science and Nutrition*, **45**, 265–286.

MULLAN, W.M.A., (2001), Dairy science and technology, available at http://www.dairyscience.info/manufacture.htm.

NORDLEE, J.A., TAYLOR, S.L., TOWNSEND, J.A., THOMAS, L.A. and BUSH, R.K., (1996), Identification of a Brazil-nut allergen in transgenic soybeans, *New England J. Medicine*, **334** (11), 688–692.

NOUT, M.J.R., (1994), Fermented foods and food safety, *Food Research International*, **27**, 291–298.

PARK, Y.S., SHI, Z.P., SHIBA, S., CHANTAL, C., IIJIMA, S. and KOBAYASHI, T. (1993), Application of fuzzy reasoning to control glucose and ethanol concentrations in baker's yeast culture. *Applied Microbiology Biotechnology*, **38**, 649–655.

PAUL, A.A. and SOUTHGATE, D.A.T. (1991) *McCance and Widdowson's The Composition of Foods*, 5th edn, Royal Society of Chemistry, Cambridge.

PÉREZ-GUERRA, N., TORRADO-AGRASAR, A., LÓPEZ-MACIAS C. and PASTRANA L., (2004), Main characteristics and applications of solid substrate fermentation, available at http://ejeafche.uvigo.es/2(3)2003/001232003F.htm.

PICKFORD, E.F. and JARDINE, N.J., (2000), Functional confectionery, in (G.R. Gibson and C.M. Williams, Eds.), *Functional Foods – Concept to Product*, Woodhead Publishing, Cambridge, pp. 260–286.

PLAYNE, M.J. and CRITTENDEN, R., (1996), Commercially available oligosaccharides, *Bulletin International Dairy Foundation*, **313**, 10–22.

POMERANZ, Y. and MELOAN, C.E., (2000), Enzymatic methods, in *Food Analysis: Theory and Practice*, Springer Publications, Heidelberg, Germany, pp. 506–529.

POSPIESZNY, H. and MAĆKOWIAK, A., (1997), Effect of chitosan derivatives of plants by pathogenic batcteria, in (A. Domard, G.A.F. Roberts and K.M. Varum, Eds.), *Advances in Chitin Science*, Jacques Andre Publishers, Lyon, pp. 759–762.

PSZCZOLA, D.E., (1999), Enzymes: making things happen, *Food Technology*, **53** (2), 74–80.

RAIMBAULT, M., (1998), General and microbiological aspects of solid substrate fermentation, *Electronic J. Biotechnology*, **1** (3), 174–188, available at http://www.scielo.cl/scielo.php?pid=S0717-34581998000300007&script=sci_arttext.

RAO, D.R., CHAWAN, C.B. and VEERAMACHANENI, R., (1995), Liposomal encapsulation of β-galactosidase: comparison of two methods of encapsulation and *in vitro* lactose digestibility. *J. Food Biochemistry*, **33**, 12247–12254.

RASTALL, R., (Ed.), (2007), *Novel Enzyme Technology for Food Applications*, Woodhead Publishing, Cambridge.

RASTALL, R.A., FULLER, R., GASKINS, H.R. and GIBSON, G.R., (2000), Colonic functional foods, in (G.R. Gibson and C.M. Williams, Eds.), *Functional Foods – Concept to Product*, Woodhead Publishing, Cambridge, pp 72–95.

REDDY, N.R. and PIERSON, M.D., (1994), Reduction in antinutritional and toxic components in plant foods by fermentation, *Food Research International*, **27**, 281–290.

RIBEREAU-GAYON, P., GLORIES, Y., MAUJEAN, A. and DUBOURDIEU, D., (2000), *The Handbook of Enology*: Vol. 2, *The Chemistry of Wine Stabilisation and Treatments*, John Wiley and Sons, Chichester.

RIFKIN, J., (1999), *The Biotech Century*, Tarcher Penguin, New York.

RIFKIN, J., (2006), This crop revolution may succeed where GM failed, *The Guardian*, 26 October, p. 38.

ROBERFROID, M.B., (2000), Defining functional foods, in (G. R. Gibson and C. M. Williams, Eds.), *Functional Foods – Concept to Product*, Woodhead Publishing, Cambridge, pp 9–27.

ROBERFROID, M. and SLAVIN, J., (2000), Nondigestible oligosaccharides, *Critical Reviews in Food Science and Nutrition*, **40** (6), 461–480.

RODGER, G., (2001), Mycoprotein – a meat alternative new to the US, *Food Technology*, **55** (7), 36–41.

RODRIGUEZ, J.M., MARTINEZ, M.I. and KOK, J., (2002), Pediocin PA-1, a wide-spectrum bacteriocin from lactic acid bacteria, *Critical Reviews in Food Science and Nutrition*, **42** (2), 91–121.

ROLLER, S., (Ed.), (2003), *Natural Antimicrobials for Minimal Processing of Foods*, Woodhead

Publishing, Cambridge.

ROLLER, S., BARNBY-SMITH, F. and SWINTON, S., (1991), Natural food ingredients from biotechnology, in (A. Turner, Ed.), *Food Technology International Europe*, Sterling Publications International, London, pp. 175–179.

ROSS, R.P., DESMOND, C., FITZGERALD, G.F. and STANTON, C., (2005), A review: overcoming the technological hurdles in the development of probiotic foods, *J. Applied Microbiology*, **98**, 1410–1417. Also available at http://www.blackwell-synergy.com/doi/pdf/10.1111/j.1365-2672.2005.02654.x#search=%22review%20probiotic%20encapsulation%22.

SAARELA, M., (Ed.), (2007), *Functional Dairy Products*, Vol. 2, Woodhead Publishing, Cambridge.

SANDERS, M.E., (1999), Probiotics, *Food Technology*, **53** (11), 67–77.

SANTANIELLO, V. and EVENSON R.E., (EDS), (2004a), *Consumer Acceptance of Genetically Modified Foods*, CABI Publishing, Wallingford.

SANTANIELLO, V. and EVENSON, R.E., (EDS), (2004b), *The Regulation of Agricultural Biotechnology*, CABI Publishing, Wallingford.

SCHMIDL, M.K. and LABUZA, T.P., (2000), US legislation and functional health claims, in (G.R. Gibson and C.M. Williams, Eds.), *Functional Foods – Concept to Product*, Woodhead Publishing, Cambridge, pp. 43–68.

SCHWEDT, G. and STEIN, K., (1994), Immobilized enzymes as tools in food analysis, *Zeitschrift für Lebensmittel-Untersuchung und -Forschung*, **199** (3), pp. 171–182.

SHAH, N.P., (2001), Functional foods from probiotics and prebiotics, *Food Technology*, **55** (11), 46–53.

SIUTA-CRUCE, P. and GOULET, J., (2001), Improving probiotic survival rates, *Food Technology*, **55** (10), 36–42.

SMITH, J.M., (2003), *Seeds of Deception, Yes! Books*, Institute for Responsible Technology, Fairfield, IA, pp. 77–105, available at www.seedsofdeception.com.

SODINI, I., REMEUF, F., HADDAD, S. and CORRIEU, G., (2004), The relative effect of milk base, starter and process on yoghurt texture: a review, *Critical Reviews in Food Science and Nutrition*, **44**, 113–137.

STANBURY, P.F., WHITAKER, A. and HALL, S.J., (1995) *Principles of Fermentation Technology*, Elsevier Science, Oxford.

STEINKRAUS, K.H., (1994), Nutritional significance of fermented foods. *Food Research International*, **27**, 259–267.

STEINKRAUS, K., (1995), *Handbook of Indigenous Fermented Foods*, 2nd edn, Marcel Dekker, New York.

STEINKRAUS, K.H., (2002), Fermentations in world food processing, *Comprehensive Reviews in Food Science and Food Safety*, **1**, 23–32.

STEINKRAUS, K., (2004), *Industrialization of Indigenous Fermented Foods*, 2nd edn, Marcel Dekker, New York.

STOVER, P.J., (2006), Influence of human genetic variation on nutritional requirements, *American J. Clinical Nutrition*, **83** (2), 436S–442S. also available at http://moodfoods.com/nutritional-genomics.html.

SURAWICZ, E.G.W. and MCFARLAND L.V., (1996), Biotherapeutic agents: a neglected modality for the treatment and prevention of selected intestinal and vaginal infections, *J. American Medical Association*, **275** (11), 870–876.

SWADLING, I., (2001), Enzyme information on the Net, *Food Engineering and Ingredients*, April, 49–50.

SYNOWIECKI, J. and AL-KHATEEB, N.A., (2003), Production, properties and some new applications of chitin and its derivatives, *Critical Reviews in Food Science and Nutrition*, **43** (2), 145–171.

SYNOWIECKI, J., GRZYBOWSKA, B. and ZDZIEBLO, A. (2006), Sources, properties and suitability of new thermostable enzymes in food processing, *Critical Reviews in Food Science and Nutrition*, **46**, 197–205.

TAMIME, A.Y. and ROBINSON, R.K., (2007), *Tamime and Robinson's Yoghurt: Science and Technology*, 3rd edn, Woodhead Publishing, Cambridge.

TAYLOR, T.M., DAVIDSON, P.M., BRUCE, B.D. and WEISS, J., (2005), Liposomal nanocapsules in food science and agriculture, *Critical Reviews in Food Science and Nutrition*, **45**, 587–605.

TESTER, M., (2000), The dangerously polarized debate on genetic modification, SCOPE GM Food Controversy Forum (1 December), available at scope.educ.washington.edu/gmfood/commentary/show.php?author=Tester.

THARANATHAN, R.N. and KITTUR, F. S., (2003), Chitin – the undisputed biomolecule of great potential, *Critical Reviews in Food Science and Nutrition*, **43** (1), 61–87.

TOKE, D., (2004), *The Politics of GM Food: A Comparative Study of the UK, USA and EU*, Routledge, Abingdon, Oxon.

TURNBULL, W.H. and WARD, T., (1995), Myco protein reduces glycemia and insulinemia when taken with an oral glucose tolerance test, *American J. Clinical Nutrition*, **61**, 135–140.

VAN DER WERF, J.H.J., (2006), Basics of Marker-Assisted Selection, available at www-personal.une.edu.au/~jvanderw/15_basics_of_marker_assisted_selection.PDF.

VOILLEY, A., (Ed.), (2006), *Flavour in Food*, Woodhead Publishing, Cambridge.

WAGNER, H., (2006), Despite information spate, consumers still in dark about genetically modified foods, Research News, available at http://researchnews.osu.edu/archive/roeGMO2.htm.

WERE, L.M., BRUCE, B.D., DAVIDSON, P.M. and WEISS, J., (2003), Size, stability and entrapment efficiency of phosphlipid nanocapsules containing polypeptide antimicrobials, *J. Agriculture and Food Chemistry*, **51**, 8073–8079.

WERE, L.M., BRUCE, B.D., DAVIDSON, P.M. and WEISS, J., (2004), Encapsulation of nisin and lysozyme in liposomes enhances efficacy against *Listeria monocytogenes*, *J. Food Proteins*, **67**, 922–927.

WEST, S.I., (1988), A systematic approach to enzyme use, in (A. Turner, Ed.), *Food Technology International Europe*, Sterling Publications International, London, pp. 240–242.

WHITEHURST, R.J. and LAW, B.A., (Eds.), (2002), *Enzymes in Food Technology*, Culinary and Hospitality Industry Publications Services, Weimar, TX.

WILM, K. H. (2005), Our Food, available at www.ourfood.com/Genetic_modification_food.htm.

WILSON, D., (2001), Marketing mycoprotein: the Quorn Foods story, *Food Technology*, **55** (7), 48, 50.

WU, F., (2006), Mycotoxin reduction in Bt corn: potential economic, health, and regulatory impacts, *Transgenic Research*, **15** (3), 277–289.

7

Irradiation

Abstract: Ionising radiation, electrons or X-rays destroy pathogenic or spoilage bacteria, disinfest foods, or extend the shelf-life of fresh produce by slowing the rate of germination, ripening or sprouting. Irradiation does not involve heating and the sensory and nutritional properties of foods are largely retained. This chapter describes the theory of irradiation, and methods to measure radiation dose and to detect irradiated foods. It describes irradiation equipment and commercial applications, and concludes with the effects of radiation on micro-organisms and the sensory and nutritional qualities of foods.

Key words: irradiation, γ radiation, X-rays, dosimeters, radiolytic products, food irradiation plant, detection of irradiated foods.

Irradiation preserves foods by the use of ionising radiation (γ-rays from isotopes or, commercially to a lesser extent, from electrons and X-rays). It is used to destroy pathogenic or spoilage bacteria, or to extend the shelf-life of fresh produce by disinfestation and slowing the rate of germination, ripening or sprouting (Table 7.1). It does not involve heating the food to any significant extent and sensory and nutritional properties are therefore largely unchanged (section 7.4.3).

Irradiation is permitted in 53 countries (in 2008), not all of which have processing plants. In 2004 around 230 irradiated foods were legally authorised, some of which were commercially available in about 30 countries (Tada 2004). The EU only authorise irradiation of herbs and spices, but allows the irradiation of certain other foods within member states until the completed EU-wide list of products authorised for irradiation enters into force. European countries have different national approvals ranging from the EU minimum in Germany, seven categories of foods in the UK, and a wide range of products in the Netherlands, France and Belgium, including freshwater shrimps, frogs' legs and de-boned chicken. In the USA irradiation is permitted for disinfestation and treatment of fresh produce, and since 2002 for irradiation of red meat to help combat incidence of food poisoning bacteria, especially *E. coli* O1457:H7 in minced beef (Deeley 2006). Details of irradiation facilities in 34 countries are given by Anon (2008a).

The main advantages of irradiation are shown in Table 7.2 together with concerns over the use of food irradiation that have been expressed by some, including Anon (2008b). In the 1970s and 1980s the Joint FAO/IAEA/WHO Expert Committee on the Whole-

Table 7.1   Applications of food irradiation

Application	Dose range (kGy)	Examples of foods
Low dose (up to 1 kGy):		
Inhibition of sprouting	0.05–0.2	Potatoes, garlic, onions, root ginger, yam
Disinfestation (e.g. arthropods)	0.1–0.5	Fruits, grains, flours, cocoa beans, dry foods (e.g. fruits, meats, fish)
Delay ripening	0.2–1.0	Fresh fruits and vegetables
Inactivation/control of parasites	0.3–1.0	Pork meat, fresh fish
Medium dose (1–10 kGy):		
Extension of shelf-life	1–3	Strawberries, fresh fish and meat at 0–4 °C, mushrooms
Destruction of parasites	1–8	Meats
Control of moulds	2–5	Extended storage of fresh fruit
Destruction of pathogens and spoilage micro-organisms	2.5–10	Spices, raw or frozen poultry, meat, shrimps
High dose (>10 kGy):		
Sterilisation of foods	7–10	Herbs, spices, meat, poultry, seafoods
Decontamination of food additives	10–50	Enzyme preparations, natural gums
Sterilisation of packaging materials	10–25	Wine corks
Sterilised hospital diets	30–50	Ready meals

Adapted from Ley (1987), Guise (1986), Loaharanu (1995), Wilkinsin and Gould (1996), Anon (1999b) and Leek and Hall (1998)

Table 7.2   Advantages and concerns over food irradiation

Advantages	Limitations and concerns
• Improves microbial safety and reduces risk of foodborne illness.	• The process could be used to eliminate high bacterial loads to make otherwise unacceptable foods saleable, substitution for GMP.
• Fresh foods may be preserved without the use of chemical pesticides leading to occupational safety, environmental benefits and reduced chemical contaminants on foods.	• If spoilage micro-organisms are destroyed but pathogenic bacteria are not, consumers will have no indication of the unwholesomeness of a food.
• There is little or no heating of the food and therefore negligible change to organoleptic properties.	• Health hazards if toxin-producing bacteria are destroyed after they have contaminated the food with toxins.
• Changes in nutritional value of foods are comparable with other methods of food preservation.	• The possible development of resistance to radiation in micro-organisms.
• Energy requirements are very low.	• Loss of nutritional value or consumption of radiolytic products such as free radicals that may have adverse health effects.
• Packaged foods may be treated and immediately released for shipping giving minimum stockholding and 'just-in-time' manufacturing.	• Public resistance due to fears of induced radioactivity or other reasons connected to concerns over the nuclear industry.
• Processing has low operating costs.	• High capital cost of irradiation plant.

someness of Irradiated Food (JECFI) addressed safety issues and concluded that the then maximum average dose of 10 kGy 'presents no toxicological hazard and no special nutritional or microbiological problems in foods' (Anon 1977, 1981). This was supported by the Advisory Committee on Irradiated and Novel Foodstuffs (Anon 1986). The JECFI recommendations were then formed into a standard by the Codex Alimentarius Commission who recommended an international code for the operation of radiation facilities (Anon 2003a). It emphasises that irradiation should not displace good manufacturing practice (GMP) and is most effective when used as a final critical control point in a HACCP system (see also Chapter 1, section 1.5.1). The radiation dose administered to a food depends on the objective of the treatment and the resistance of the organisms likely to be present. The maximum recommended dose for foods used to be 15 kGy, with the average dose not exceeding 10 kGy (Anon 1994). However, in 1997 a report of the joint FAO/IAEA/WHO study group concluded that on the basis of the scientific evidence reviewed, 'food irradiated to any dose appropriate to achieve the intended technological objective is both safe to consume and nutritionally adequate'. The group concluded that no upper dose limit need be imposed, and that irradiated foods are deemed wholesome throughout the technologically useful dose range from below 10 kGy to envisioned doses above 10 kGy (Anon 1999a,b).

In countries where irradiation is permitted, labelling regulations place a requirement on manufacturers to indicate that the food or any listed ingredients have been treated by irradiation, using a statement 'Treated with radiation' or 'Treated by irradiation' and include the international 'Radura' logo (Fig. 7.1) on the label. Additionally, wholesale foods are required to be labelled with the phrase 'Treated by irradiation, do not irradiate again'. Applications and details of irradiation technology have been comprehensively reviewed by Wilkinson and Gould (1996) and Molins (2001) and current research is described by Sommers and Fan (2006).

Fig. 7.1   Radura logo.

7.1   Theory

Details of the physical and chemical processes involved in the decay of radioactive materials to produce α-, β- and γ-radiation, X-rays and free electrons are described by many sources (e.g. Ehmann and Vance 1991). The units used in irradiation are shown in Table 7.3. Only γ-radiation, and accelerated electrons (which may also be converted to X-rays) are used in food processing applications because other particles cause induced radioactivity. For this reason electron energy is limited to a legal maximum of 10 MeV and X-rays to 5 MeV for food irradiation. The radioisotope cobalt-60 (^{60}Co) has a half-life of 5.3 years and emits γ-rays at two wavelengths which have energies of 1.17 and 1.33 MeV respectively. Caesium-137 (^{137}Cs) has a half-life of 30.2 years and emits γ-rays with energy of 0.66 MeV. Neither can induce radioactivity in foods.

γ-Rays, electrons and X-rays are distinguished from other forms of radiation by their ionising ability (i.e. they are able to break chemical bonds when absorbed by materials). When they interact with atoms in a food (a process known as 'Compton scattering'), the energy causes ionisation and ejection of electrons from food atoms (known as 'Compton electrons'). The products of ionisation may be electrically charged ions or neutral free radicals. These, together with ejected electrons, then further react to cause changes in an irradiated material known as 'radiolysis'. It is these reactions that cause the destruction of micro-organisms, insects and parasites during food irradiation as well as subtle changes to the chemical structure of the food. In foods that have high moisture contents, water is ionised by the radiation. Electrons are expelled from water molecules and break the chemical bonds. The products then recombine to form hydrogen, hydrogen peroxide, hydrogen radicals (H), hydroxyl radicals (OH) and hydroperoxyl radicals (HO$_2$·) (Fig. 7.2).

Hydroxyl radicals are powerful oxidising agents and react with unsaturated compounds, whereas expelled electrons react with aromatic compounds, especially ketones, aldehydes and carboxylic acids (Diehl 1995). The diffusivity of free radicals depends on the availability of free water in the food, and these reactions are less in dry or frozen foods. The radicals are extremely short lived (less than 10^{-5} s) but are sufficient to destroy microbial cells. Similar radicals are also present in non-irradiated foods owing to:

- the action of enzymes (e.g. lipoxygenases and peroxidases);
- oxidation of fats and fatty acids; and
- degradation of fat-soluble vitamins and pigments.

The presence of oxygen has an important influence on the amount and types of radiolytic changes in a food. Irradiation in the presence of oxygen can lead to the formation of ozone, hydroperoxy radicals and superoxide anions, which can give rise to

Table 7.3   Summary of units used in irradiation

Unit	
Becquerel (Bq)	One unit of disintegration per second
	Curie (Ci) $= 3.7 \times 10^{10}$ Bq
	One million Ci (MCi) $= 14.8$ kW power
Half-life	The time taken for the radioactivity of a sample to fall to half its initial value
Electronvolt (eV)	Energy of radiation (usually expressed as mega-electron volts (MeV)). 1 eV $= 1.602 \times 10^{-19}$ J
Gray (Gy)	Absorbed dose (where 1 kGy is the absorption of 1 kJ of energy per kilogram of food). 4 kGy raises the product temperature by approximately 1 °C. Previously rads (radiological units) were used where 1 rad $= 10^{-2}$ J kg^{-1}. 1 Gy therefore equals 100 rads

$$H_2O \rightarrow H_2O^+ + e^-$$

(a)
$$
\begin{aligned}
e^- + H_2O &\rightarrow H_2O^- \\
H_2O^+ &\rightarrow H^+ + OH\cdot \\
H_2O^- &\rightarrow H\cdot + OH^-
\end{aligned}
$$

(b)
$$
\begin{aligned}
H\cdot + H\cdot &\rightarrow H_2 \\
\text{or } OH\cdot + OH\cdot &\rightarrow H_2O_2 \\
\text{or } H\cdot + OH\cdot &\rightarrow H_2O \\
\text{or } H\cdot + H_2O &\rightarrow H_2 + OH\cdot \\
\text{or } OH\cdot + H_2O_2 &\rightarrow H_2O + HO_2\cdot \\
H\cdot &\rightarrow + O_2HO_2.
\end{aligned}
$$

Fig. 7.2   (a) Ionisation of water (after Robinson 1986); (b) formation of free radicals during irradiation (after Hughes 1982).

hydrogen peroxide, all of which are powerful oxidising agents (Hayashi 1991). Fat-soluble components and essential fatty acids are therefore lost during irradiation and some foods (e.g. dairy products) are unsuitable for irradiation owing to the development of rancid off-flavours. Foods that contain fat (e.g. meats) are irradiated in vacuum packs.

7.1.1   Dose distribution

Penetration of γ-radiation, electrons and X-rays depends on the density of the food as well as the energy of the rays. Because radiation is absorbed as it passes through the food, the outer parts receive a higher dose than do inner parts. The depth of penetration is proportional to the energy and inversely proportional to the density of the food (halving the density approximately doubles the depth of penetration). This can be expressed using Equation 7.1:

$$\text{Penetration (cm)} = \frac{0.524E - 0.1337}{\rho} \qquad \boxed{7.1}$$

where $E =$ energy (MeV) and $\rho =$ density (kg m^{-3}). The penetration of electron beams into different materials is shown in Table 7.4.

For a given food there is thus a limit on both the maximum dose permitted at the outer edge (D_{max}), due to unacceptable organoleptic changes, and a minimum limit (D_{min}) to achieve the desired effects of treatments (section 7.3). The uniformity of dose distribution can be expressed as a ratio of $D_{max} : D_{min}$. This 'overdose ratio' is fundamental to the

Table 7.4   Penetration of electron beams in selected materials

Material	Density (g cm^{-3})	Penetration depth (cm)		
		8 MeV	10 MeV	12 MeV
Air	0.001	3051	3838	4626
Water	1.0	4.0	5.1	6.1
Plastic	1.2	3.3	4.2	5.1
Glass	2.4	1.7	2.1	2.6
Aluminium	2.7	1.5	1.8	2.3
Stainless steel	7.9	0.5	0.6	0.7

From Leck and Hall (1998)

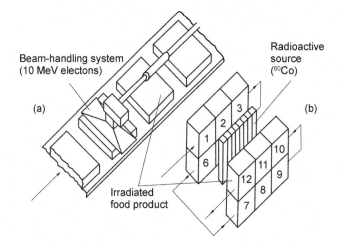

Fig. 7.3   Irradiation plant: (a) electron beam irradiation and (b) ^{60}Co source irradiation. In an electron beam facility, process loads are brought to the radiation field one at a time. In a radionuclide facility, several process loads surround the radiation source and are irradiated together (courtesy of REVISS® Services (UK) Ltd, manufacturers of PURICEC® Irradiation Technologies at: www.puridec.com) (from Anon 2002a).

effectiveness of the process and to the design and economic operation of the irradiation plant. For foods that are sensitive to radiation, such as chicken, this ratio should be as low as possible and not more than about 1.5. Other foods, for example onions, can tolerate a ratio of around 3 without unacceptable changes.

   The product loading configuration controls the distribution of dose received from a γ-radiation source. Good designs produce a minimum overdose ratio and maximise the dose efficiency (the ratio of radiation absorbed to radiation emitted) (Fig. 7.3). Dosimeters are placed at points throughout the package to determine the dose received and to ensure that the correct $D_{max} : D_{min}$ ratio is achieved. High-energy electrons are directed over foods, but they have a lower penetration than γ-rays and are not suitable for bulk foods. They are used for thin packages or for surface treatments. The selection of a radiation source therefore depends on the type of product and its density, the dimensions of the package and the reason for the treatment.

7.2   Equipment

Commercial irradiation equipment consists of an isotope source to produce γ-rays or, less commonly, a machine source to produce a high-energy electron beam. γ-radiation from ^{60}Co is used in most commercial plants but ^{137}Cs is also permitted. The activity of the ^{60}Co or ^{137}Cs sources is rated at $(222–370) \times 10^{10}$ Bq g^{-1} (or 10^{13} Bq kg^{-1}) giving typical dose rates in the order of kGy h^{-1}. The dose rates of electron beams are higher, typically kGy s^{-1}, because they can be highly focused, whereas gamma sources radiate in all directions. Electron beams and X-ray sources are specified by their beam power and 25–50 kW is typical for food irradiation (Anon 2008c). A summary of processing parameters for food irradiation plants is shown in Table 7.5. The processing speed (or residence time of the food) is determined by the dose required, the density of the food and the power output of the source. Because the source power is constant, products that have

Table 7.5 Comparison of typical processing parameters for a food irradiation plant (reproduced by kind permission of REVISS Services (UK) Ltd; full data sheet available from www.puridec.com)

	γ-rays	X-rays	Electron beams
Power source (kW)	52	25	35
Processing speed @ 4 kGy (t h^{-1})	12	10	10
Source energy (MeV)	1.33	5	5–10
Penetration depth (cm)	80–100	80–100	8–10
Dose homogeneity	High	High	Low
Dose rate	Low	High	Higher
Applications	Bulk processing of large boxes or palletised product in shipping containers	Bulk processing of large boxes or palletised product in shipping containers	Sequential processing of primary/secondary packaged products in-line or at-line

From Anon (2007)

a higher density or higher minimum dose requirement need a longer treatment and hence reduce the throughput of the plant.

Radiation is contained within the processing cell by the use of thick (3 m) concrete walls and lead shielding. Openings in the shielding, for entry of products or personnel, must be carefully constructed to prevent leakage of radiation. A dose of 5 Gy is sufficient to kill an operator. It is therefore essential, even at the lowest commercial doses (0.1 kGy), that stringent safety procedures are in place, as specified by the IAEA (Anon 2003b). These include mechanical, electrical and hydraulic interlocks, each functioning independently to prevent the source from being raised when personnel are present and to prevent entry to the building during processing. Entry to the irradiation cell is through a 10 tonne plug door, with no personnel entry via the product conveyor system.

An isotope source cannot be switched off and so it is shielded within a 6 m deep pool of water below the process area when not in use to allow personnel to enter (Fig. 7.4). ^{60}Co has a half-life of 5.26 years and therefore requires the replacement of 12.3% of the activity each year to retain the rated output of the plant. Continuous processing is therefore desirable for economic operation. To process foods the source is raised, and packaged products are loaded onto automatic conveyors and transported through the radiation field, exposing each side to the source. This makes maximum use of the emitted radiation and ensures the correct overdose ratio and a uniform dose distribution (Fig. 7.3).

Isotope sources require a more complex handling system than that used with machine sources may involve multi-pass, multi-level transport systems. Packages suspended from an overhead rail are preferable to powered roller conveyors because conveyors take up more space in the most effective part of the irradiation cell where the dose is at a maximum. Pneumatic or hydraulic systems are used to move products, but hydraulic cylinders must be located outside the irradiation cell for ease of maintenance and to prevent damage by radiation to the oil and seals.

Machine sources are electron accelerators that consist of a heated cathode to supply electrons and an evacuated tube in which a high-voltage electrostatic field accelerates electrons. Either the electrons are used directly on the food, or a suitable target material is bombarded to produce X-rays. The main advantages of machine sources are that they can be switched off and the electron beams can be directed over the packaged food to ensure an even dose distribution. In operation a conveyor carries packaged foods sequentially through two radiation beams that irradiate one or both sides of the package (Fig. 7.3). Handling equipment is therefore relatively simple powered roller conveyors. However,

Fig. 7.4   Isotope source in shielding pool for routine maintenance (courtesy of Puridec Irradiation Technologies) (Anon 2008c).

machine sources are expensive and relatively inefficient in producing radiation. Details of electronic irradiation are given by Brennan (2005).

A programmable logic controller (PLC) is used to control the whole process from goods-in to distribution. It automatically controls the speed at which products pass through the irradiation plant, correcting the cycle time to account for the drop in source activity due to decay; it calculates the dose received, creates dose maps and validates dosimetry results; it monitors control and safety systems to ensure that the plant operates safely, and automatically lowers the source and shuts down the plant if a fault occurs; and it archives product and irradiation process documentation for regulatory compliance and company audits (Comben and Stephens 2000).

7.2.1   Measurement of radiation dose
For many years authorities were concerned about illegally irradiated foods being sold without clearance, but this is no longer the main challenge. Detection of products that are irradiated but not labelled is now possible and it may also become possible to detect products that falsely claim to have been irradiated. Dosimetry is used to enforce good irradiation practice and verification that a product has been irradiated within specified dose limits (Ehlermann 1995).

Dosimeters are devices that, when irradiated, produce a quantifiable and reproducible physical and/or chemical change that can be related to the dose absorbed. These changes

Table 7.6 Useful dose ranges for routine dosimeters

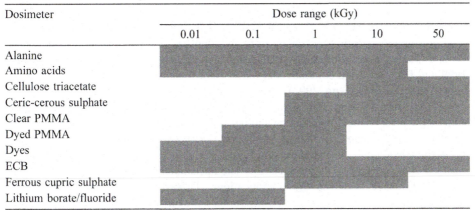

Dosimeter	Dose range (kGy)				
	0.01	0.1	1	10	50
Alanine					
Amino acids					
Cellulose triacetate					
Ceric-cerous sulphate					
Clear PMMA					
Dyed PMMA					
Dyes					
ECB					
Ferrous cupric sulphate					
Lithium borate/fluoride					

PMMA = polymethylmethacrylate, ECB = ethanol-chlorobenzene.
Adapted from the data of Anon (2002a)

can then be measured using analytical instruments. Dosimeters used in production facilities are standardised against reference dosimeters held in national laboratories. Commonly used production dosimeters, formed into pellets, films or cylinders, include dyed polymethylmethacrylate (PMMA), radiochromic and cellulose triacetate (CTA) films, cobalt glass, crystalline alanine dosimeters, ethanol-chlorobenzene (ECB) dosimeters, and thermoluminescent and lyoluminescent dosimeters (Table 7.6). For example, in the operation of an alanine dosimeter, free radicals are produced when it is irradiated and their number is proportional to the radiation dose absorbed by the crystal. The amount of free radicals is determined by electron spin resonance (ESR) spectroscopy. Further details of dosimetry and types of dosimeters are given by Anon (2002a) and Rosenthal (1993).

7.3 Applications

The types of irradiation process can be categorised by the intention of processing and the dose used as described in the following sections (also Table 7.1).

7.3.1 Sterilisation (or 'radappertisation')

Although it is technically possible to sterilise meats and other products using irradiation, the dose required (e.g. 48 kGy for a 12D reduction of *Cl. botulinum* (Lewis 1990)) would make products organoleptically unacceptable. There is thus little commercial interest in sterilisation, with the exception of herbs and spices which are frequently contaminated by heat-resistant, spore-forming bacteria. These products can be sterilised using a dose of 7–10 kGy (Table 7.1), which reduces the microbial load to an acceptable level without significant loss of volatile oils, the main quality characteristic. The main advantage of irradiating spices is the replacement of chemical sterilisation using ethylene oxide, which has been banned by the EU since 1991 as a result of concerns over residues in the product and safety of workers handling the gas. The other main application is sterilisation of hospital meals (Table 7.1).

7.3.2   Reduction of pathogens (or 'radicidation')

Food poisoning bacteria can be destroyed by doses of 2.5–10 kGy (Guise 1986). This is increasingly one of the most important applications of food irradiation as the incidence of food poisoning is steadily increasing in many countries (Loaharanu 1995) (also section 7.5 and Chapter 1, section 1.2.4). Irradiation of red meat has been legal in the USA since 1999 and has been used commercially since 2002, mainly for ground/minced meat for burger patties (Ehlermann 2002). Fresh poultry carcasses irradiated with a dose of 2.5 kGy are virtually free of *Salmonella* and the shelf-life is doubled when the product is held below 5 °C. Higher doses can be applied to frozen poultry or shellfish (at −18 °C) to destroy *Campylobacter* spp., *Escherichia coli* O157:H7 or *Vibrio* spp. (e.g. *V. cholerae*, *V. parahaemolyticus*, *V. vulnificus*) without causing the unacceptable organoleptic changes that would occur in products irradiated at ambient temperatures. Crawford and Ruff (1996) have reviewed these applications. Thayer and Rajkowski (1999) have reviewed the incidence of food poisoning from a wide range of fresh fruit and vegetables and products such as fruit juices, and describe studies that have been made to control pathogens using irradiation. They concluded that irradiation has the capability to improve the safety of fresh produce and that it may be a better alternative to chemical surface treatments (Chapter 2, section 2.2).

7.3.3   Prolonging shelf-life (or 'radurisation')

Relatively low doses (Table 7.1) are needed to destroy yeasts, fungi and non-spore-forming bacteria. This process is used to increase shelf-life by an overall reduction in vegetative cells. Bacteria that survive irradiation are more susceptible to heat treatment and the combination of irradiation with heating is therefore beneficial in causing a greater reduction in microbial numbers than would be achieved by either treatment alone (Gould 1986). Venugopal *et al.* (1999) review the use of radurisation for extending the shelf-life of fresh fish and shellfish.

Some types of fruits and vegetables, such as strawberries and tomatoes, can be irradiated to extend their shelf-life by two or three times when stored at 10 °C. Doses of 2–3 kGy cause a two-fold increase in shelf-life of mushrooms and inhibition of cap opening. A combination of irradiation and modified atmosphere packaging (Chapter 25, section 25.3) has been shown to have a synergistic effect, and as a result a lower radiation dose can be used to achieve the same effect.

7.3.4   Control of ripening

Fruit and vegetable products should be ripe before irradiation because irradiation inhibits hormone production and interrupts cell division and growth, thereby inhibiting ripening. Enzymic spoilage of foods is not entirely prevented by irradiation and a separate heat treatment is required for prolonged storage.

7.3.5   Disinfestation

Grains and tropical fruits and vegetables may be infested with insects, larvae and fungi, which lowers their export potential and requires a quarantine period for disinfestation. Many countries have banned the fumigants ethylene dibromide, ethylene dichloride and ethylene oxide. Methyl bromide, which depletes the ozone layer, was banned in most industrialised countries in 2005 and will be phased out in developing countries by 2015

(Anon 2004). Irradiation avoids the use of fumigants and pesticides or surface treatments that leave chemical residues on foods. Irradiation doses of 0.25–1.0 kGy can control parasitic protozoa and helminths in fresh fish and prevent development of insects in dried fish (Venugopal *et al.* 1999) (e.g. *Toxoplasma gondii* is inactivated at doses of 0.25 kGy and *Trichinella spiralis* at 0.3 kGy (Olson 1998). Low doses, below 1 kGy, are effective for disinfestation and also extend the shelf-life by delaying ripening and preventing sprouting. The use of irradiation to treat dried fruits and nuts is described by Johnson and Marcotte (1999).

7.3.6   Inhibition of sprouting

The technology is effective in inhibiting sprouting of potatoes and in Japan for example, doses of about 150 Gy have been used since 1973 on potatoes intended for further processing (Stevenson 1990). Similar doses are effective in preventing sprouting of onions and garlic.

7.4   Effect on Foods

7.4.1   Induced radioactivity

At recommended doses, ^{60}Co and ^{137}Cs have insufficient emission energies to induce radioactivity in foods. Machine sources of electrons and X-rays do have sufficient energy, but the levels of induced radioactivity are insignificant (2% of the acceptable radiation dose in the worst case and 0.0001% under realistic processing and storage conditions) (Gaunt 1986). These aspects have been comprehensively reviewed by Diehl (1995).

7.4.2   Radiolytic products

The ions and radicals produced during irradiation (section 7.1) are capable of reacting with components of the food to produce radiolytic products. The extent of radiolysis depends on the type of food and the radiation dose employed. However, the majority of the evidence from feeding experiments, in which animals were fed irradiated foods and high doses of radiolytic products, indicates that there are no adverse effects.

7.4.3   Nutritional and sensory properties

At commercial dose levels, ionising radiation has little or no effect on the macronutrients in foods. The digestibility of proteins and the composition of essential amino acids are largely unchanged. At higher dose levels, cleavage of the sulphydryl group from sulphur amino acids in proteins causes changes in the aroma and taste of foods. Depending on the dose received, carbohydrates are hydrolysed and oxidised to simpler compounds and may become depolymerised and more susceptible to enzymic hydrolysis. The physical properties (viscosity, texture, solubility, etc.) of foods that contain high molecular weight carbohydrates, such as pectin, starch, cellulose and gums, is substantially affected by irradiation and the functionality of these foods is changed. However, there is no change in the degree of utilisation of the carbohydrate and hence no reduction in nutritional value.

The effect on lipids is similar to that of autoxidation, to produce hydroperoxides and the resulting unacceptable changes to flavour and odour. The effect is reduced by irradiating foods while frozen, but foods that have high concentrations of lipids are

Table 7.7 Comparison of vitamin contents of heat sterilised and irradiated (58 kGy at 25 °C) chicken meat

Vitamin	Vitamin concentration (mg/kg dry weight)[a]			
	Frozen control	Heat sterilised	γ-irradiated	Electron irradiated
Thiamin HCl	2.31	1.53[b]	1.57[b]	1.98
Riboflavin	4.32	4.60	4.46	4.90[c]
Pyridoxine	7.26	7.62	5.32	6.70
Nicotinic acid	212.9	213.9	197.9	208.2
Pantothenic acid	24.0	21.8	23.5	24.9
Biotin	0.093	0.097	0.098	0.013
Folic acid	0.83	1.22	1.26	1.47[c]
Vitamin A	2716	2340	2270	2270
Vitamin D	375.1	342.8	354.0	466.1
Vitamin K	1.29	1.01	0.81	0.85
Vitamin B_{12}	0.008	0.016[c]	0.014[c]	0.009

[a] Concentrations of vitamin D and vitamin K are given as IU/kg.
[b] Significantly lower than frozen control.
[c] Significantly higher than frozen control.
From Satin (1993).

generally unsuitable for irradiation. Radiolytic products also cause oxidation of myoglobin, leading to discoloration of meat and fish products. These changes are reviewed in a number of publications including Ehlermann (2002) and Venugopal *et al.* (1999).

There is conflicting evidence regarding the effect on vitamins as many studies have used vitamin solutions, which show greater losses than those found in the heterogeneous mixtures of compounds in foods. Water-soluble vitamins vary in their sensitivity to the products of radiolysis of water (section 7.1). The extent of vitamin inactivation (loss of biological activity) also depends on the dose received and the type and physical state of food under investigation. Vitamin C is oxidised to dehydroascorbic acid by γ-radiation but the biological activity is retained and overall there are small changes as a result of the low doses used to irradiate fruits and vegetables. Thiamin in meat and poultry products is the most radiation-sensitive of the B vitamins, with losses that are similar to those in heat sterilisation at doses of 45–68 kGy (Ehlermann 2002). Other vitamins of the B group are largely unaffected (Table 7.7). The order of sensitivity is reported as thiamin > ascorbic acid > pyridoxine > riboflavin > folic acid > cobalamin > nicotinic acid. Fat-soluble vitamins vary in their susceptibility to radiation. Vitamins D and K are largely unaffected whereas vitamins A and E undergo some losses, which vary according to the type of food examined. The order of sensitivity is vitamin E > carotene > vitamin A > vitamin D > vitamin K (Anon 1994). A comparison of irradiated and heat-sterilised chicken meat (Table 7.7) indicates similar levels of vitamin loss.

In summary, the consensus is that, at commercial dose levels, irradiation causes no greater damage to nutritional quality than other preservation operations used in food processing. Changes in nutritional quality are reviewed in detail by Ehlermann (2002) and Diehl (1995).

7.5   Effect on micro-organisms

The reactive ions produced by irradiating foods (Fig. 7.2) injure or destroy micro-organisms immediately, by changing the structure of cell membranes and affecting

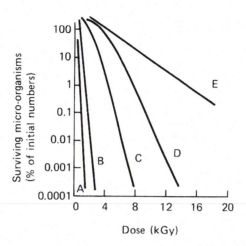

Fig. 7.5 Microbial destruction by irradiation: A, *pseudomonas* sp.; B, *Salmonella* sp.; C, *Bacillus cereus*; D, *Deinococcus radiodurans*; E, typical virus (after Gould 1986).

metabolic enzyme activity. However, a more important effect is on DNA and RNA molecules in cell nuclei, which are required for growth and replication. The effects of irradiation only become apparent after a period of time, when the DNA double helix fails to unwind and the micro-organism cannot reproduce by cell division.

The rate of destruction of individual cells depends on the rate at which ions are produced and inter-react with the DNA, whereas the reduction in cell numbers depends on the total dose of radiation received. The sensitivity of micro-organisms to radiation is expressed as the D_{10}-value (the dose of radiation that reduces the microbial population to 10% of its initial value) by analogy with thermal destruction (Chapter 10, section 10.3). Theoretically a logarithmic reduction in microbial numbers with increasing dose is expected. As in other food preservation methods the rate of destruction varies with microbial species. Some bacterial species contain more than one molecule of DNA and others are capable of repairing damaged DNA. For these the rate of destruction is therefore not linear with received dose (Fig. 7.5).

A simple guide is that the smaller and simpler the organism, the higher the dose of radiation that is needed to destroy it. Viruses are very resistant to irradiation and are unlikely to be affected by the dose levels used in commercial processing. Spore-forming species (e.g. *Clostridium botulinum* and *Bacillus cereus*), and those that are able to repair damaged DNA rapidly (e.g. *Deionococcus radiodurans*) are more resistant than vegetative cells and non-spore-forming bacteria. Gram-negative bacteria, including pathogens such as *Salmonella* spp. and *Shigella* spp., are generally more sensitive than Gram-positive bacteria. Examples of *D*-values for important pathogens are given in Table 7.8.

7.6   Effect on packaging

Radiation is able to penetrate packaging materials and therefore reduces the risk of post-processing contamination and allows easier handling of products. However, packaging materials are themselves subject to changes induced by radiation (Table 7.9). Radiolysis products, including hydrocarbons, alcohols, ketones and carboxylic acids, have the potential to migrate into the product and produce unacceptable flavour taints, particularly

Table 7.8 *D*-values of important pathogens

Pathogen	*D*-value (kGy)	Irradiation temperature (°C)	Suspending medium
A. hydrophilia	0.14–0.19	2	Beef
C. jejuni	0.18	2–4	Beef
E. coli O157:H7	0.24	2–4	Beef
L. monocytogenes	0.45	2–4	Chicken
Salmonella sp.	0.38–0.77	2	Chicken
Staphlococcus aureus	0.36	0	Chicken
Yersinia enterocolitica	0.11	25	Beef
Clostridium botulinum (spores)	3.56	−30	Chicken

Adapted from Olson (1998)

Table 7.9 Changes to packaging materials caused by irradiation

Packaging material	Maximum dose (kGy)	Effect of radiation above maximum dose
Polystyrene	5000	–
Polyethylene	1000	–
PVC	100	Browning, evolution of hydrogen chloride
Paper and board	100	Loss of mechanical strength
Polypropylene	25	Becomes brittle
Glass	10	Browning

Adapted from the data of Guise (1986) and McLaughlin *et al.* (1982)

in fatty foods. Careful choice of packaging materials as well as adhesives, additives and printing materials is necessary to prevent contamination of the food with radiolytic products. Buchalla *et al.* (1993) have reviewed the effects of irradiation on plastic packaging materials.

7.7 Detection of irradiated foods

There has been considerable research since the late 1980s to develop and validate a series of detection methods that can be used by enforcement officers to detect whether a food has been irradiated. A review of the international collaboration to develop standard methods is given by Delincée (1998). Because irradiation produces no major chemical, physical or sensory changes to foods at commercial doses, detection methods focus on minute changes in chemical composition, physical or biological changes to the food. These methods are based on either detecting products formed by irradiation, physical changes such as cell membrane damage or changes to microbial flora. It is unlikely that one method will be suitable for all irradiated foods. The following methods have been validated as EU standards and have been adopted by the Codex Alimentarius Commission (Anon 2007):

- ESR spectroscopy of foods containing (a) bone, (b) cellulose and (c) crystalline sugar.
- Thermoluminescence/photostimulated thermoluminescence.
- Analysis of 2-alkylcyclobutanones by gas chromatography/mass spectrometry (GC-MS) or enzyme-linked immunosorbent assay (ELISA).
- Limulus amoebocyte lysate test with Gram-negative bacteria count (LAL/GNB).

- Direct epifluorescent filter technique with an aerobic plate count (DEFT/APC).
- DNA comet assay.
- Analysis of hydrocarbons by gas chromatography.

7.7.1 Physical methods

ESR spectroscopy

Free radicals produced by irradiation are very short-lived in moist foods and cannot be detected but they are more stable in solid or dry components of foods (e.g. bone, shell or seed) and can be detected by ESR spectroscopy. It has been used to detect irradiated bone-containing meat, fish and shellfish (Venugopal *et al.* 1999), and may have application to a wide range of food products, including foods that contain crystalline cellulose and those that have a low moisture content (e.g pistachio nuts and paprika). It is a non-destructive method that is specific, rapid and relatively simple, although the high cost of ESR spectrometers may limit its application.

Thermoluminescence (TL)

TL is based on the emission of light when electrons trapped in crystalline lattices during irradiation are released by heating the food up to 400 °C. These lattices may be contaminating dust (silicate materials) on spices, minerals from seabeds in the intestines of shellfish, or minerals in fruits and vegetables. The minerals are isolated from the food and when they are heated in a controlled way, the stored energy is released as light and measured by a sensitive photon counter. The method is widely applicable and gives an unequivocal determination that a food has been irradiated, but it has a number of disadvantages: it is laborious because a sufficient amount of minerals (a few mg), free of organic matter, has to be prepared; strict procedures are needed to prevent contamination by dust in the laboratory; and TL analysers are expensive. An adaptation of the method is photostimulated luminescence (PSL) in which pulsed infrared light is used instead of heat to release the stored energy (Anon 2002b). The method does not require isolation of the minerals and a small sample of food can be used directly, to obtain a result within a few minutes. It is also more efficient than TL in detecting an irradiated component in a mixture of materials and may become a 'spot-test' for irradiation. Other methods that detect physical changes are measurement of electrical impedance, changes in viscosity, electric potential, nuclear magnetic resonance and near-infrared spectroscopy (Anon 1994).

7.7.2 Chemical methods

Radiolytic hydrocarbons are produced from fats in foods during irradiation and are detected using gas chromatography (GC). Although these products are formed in other types of food processing, they have a characteristic distribution pattern following irradiation. The use of liquid chromatography coupled to GC increases the sensitivity of the method (Delincée 1998). The method has also been used for low-fat foods including shrimps and oysters (Moorhouse 1996).

2-alkylcyclobutanones (2-CBs)

2-CBs are radiolytic products formed from fatty acids, but are not found as a result of other degradative processes. They are detected using mass spectrometry after separation by GC and can be identified as positive markers to indicate that a food has been irradiated

(Crone *et al.* 1993, Stevenson 1994). The method has been used to detect irradiated poultry meat, pork, beef or lamb meat and liquid whole egg. The test has also been used to detect irradiated exotic fruits, including mango, papaya and avocado (Stewart *et al.* 1998). Other methods that have been used to detect 2-CBs include thin-layer chromatography (TLC), high-pressure liquid chromatography (HPLC) and supercritical fluid extraction/TLC.

Detection of DNA fragmentation by microgel electrophoresis of single cells or nuclei has shown good results but the test is limited to foods that have not been heat treated, as this also results in DNA fragmentation. ELISA tests are simpler and more rapid methods to detect 2-CBs and dihydrothymidine, a DNA breakdown product of irradiation. Details are given by Nolan *et al.* (1998) and Tyreman *et al.* (1998) respectively. In each case the monoclonal antibody used in the assay is raised against the particular marker chemical and subjected to the ELISA assay. The results give both qualitative and quantitative identification of the marker in irradiated samples of food. Microelectrophoresis of single cells (the 'Comet' assay) is a simple and rapid test for DNA damage in irradiated foods. It can be used over a wide dose range and for a variety of products. Details are given by Anon (2008d). Other methods include the detection of trapped hydrogen or carbon monoxide, produced by irradiation, which have been used to identify irradiated shrimps, cod and oysters (Roberts *et al.* 1996).

7.7.3   Biological methods

The LAL/GNB procedure is used to estimate the reduced viability of micro-organisms in a food after irradiation, measuring both dead and live micro-organisms. This method and DEFT/APC give information about the numbers of bacteria destroyed by irradiation. If the DEFT count exceeds the APC by 10^4 or more, it indicates that the food has been irradiated (Anon 1994). These methods also indicate the hygiene status before and after irradiation, and thus help to enforce GMP. They are used for a wide range of irradiated foods, especially to detect irradiated poultry meat. The last method may be limited if initial contamination is low or if low dose levels are used.

Other detection methods are being developed including capillary gel electrophoresis to detect irradiated meat and plant material. Here proteins and peptides are separated on the basis of their size and this was used to detect changes in egg white caused by irradiation (Day and Brown 1998).

References

ANON, (1977), Wholesomeness of irradiated food, Report of the Joint FAO-IAEA-WHO Expert Committee, WHO Technical Report Series No. 604, World Health Organization, Geneva.

ANON, (1981), Wholesomeness of irradiated food, Report of the Joint FAO-IAEA-WHO Expert Committee, WHO Technical Report Series No. 659, World Health Organization, Geneva.

ANON, (1986), The safety and wholesomeness of irradiated foods, Report by the Advisory Committee on the Irradiated and Novel Foods, HMSO, London.

ANON, (1994), Review of the safety and nutritional adequacy of irradiated food, Report of a World Health Organization Consultation, Geneva, 20–22 May 1992, WHO, Geneva.

ANON, (1999a), High-dose irradiation: wholesomeness of food irradiated with doses above 10 KGy, a joint FAO/IAEA/WHO study group. Geneva, Switzerland, 15–20 September 1997, Technical Report Series, No 890, available at www.who.int/foodsafety/publications/fs_management/irradiation/en/.

ANON, (1999b), Facts about food irradiation, International Consultative Group on Food Irradiation, available at www.iaea.org/programmes/nafa/d5/public/foodirradiation.pdf.

ANON, (2002a), Dosimetry for food irradiation, Technical Reports Series No. 409, IAEA, Vienna, STI/DOC/010/409.)), IAEA, A-1400 Vienna, Austria, pp. 32–53, available at www-pub.iaea.org/MTCD/publications/PDF/TRS409scr.pdf.

ANON, (2002b), Foodstuffs – detection of irradiated foods using photostimulated luminescence, Health and Consumer Protection Directorate, European Commission, available at http://ec.europa.eu/food/food/biosafety/irradiation/13751-2002_en.pdf.

ANON, (2003a), Recommended International Code of Practice for Radiation Processing of Food (CAC/RCP 19-1979, Rev. 2-2003), available at www.codexalimentarius.net/web/more_info.jsp?id_sta=18, and General Standard for Irradiated Foods (106-1983, REV. 1-2003 1), available at: www.codexalimentarius.net/web/more_info.jsp?id_sta=16.

ANON, (2003b), Occupational Radiation Protection, International Atomic Energy Agency, available at www-ns.iaea.org/tech-areas/rw-ppss/occupational.htm.

ANON, (2004), Methyl bromide approved for temporary uses. An intergovernmental meeting on the Montreal Protocol on Substances that Deplete the Ozone Layer, Montreal, 26 March, available at www.unep.org/Documents.Multilingual/Default.asp?DocumentID=388&ArticleID=4457&l=en.

ANON, (2007), Information on analytical methods for the detection of irradiated foods standardised by the European Committee for Standardisation (CEN), available at http://ec.europa.eu/food/food/biosafety/irradiation/anal_methods_en.htm.

ANON, (2008a), Authorized Food Irradiation Facilities, IAEA and Food and Agriculture Organization database of authorized facilities by country, available by following the link from www-naweb.iaea.org/nafa/databases-nafa.html.

ANON, (2008b), Food irradiation, The Food Commission, available at www.foodcomm.org.uk/irradiation_probs.htm.

ANON, (2008c), Understanding the key radiation processing parameters – a guide for food processors, company information from Puridec Irradiation Technologies, available at www.puridec.co.uk/puridec/pdfs/Key_radiation_proc.pdf.

ANON, (2008d), Komet^{TM} Imaging Software – DNA Damage & Repair Analysis, company information from Andor Technology, available at www.andor.com.

BRENNAN, C., (2005), *Electronic Irradiation of Foods: An Introduction to the Technology*. Springer Food Engineering series, Springer, New York.

BUCHALLA, R., SCHUTTLER, C. and BOGL, K.W., (1993), Effects of ionising radiation on plastic food packaging materials: a review, *J. Food Proteins*, **56**, 998–1005.

COMBEN, M. and STEPHENS, P. (2000), Irradiation plant control upgrades and parametric release, *Radiation Physics and Chemistry*, **57** (3–6), 577–580.

CRAWFORD, L.M. and RUFF, E.H., (1996), A review of the safety of cold pasteurization through irradiation, *Food Control*, **7** (2), 87–97.

CRONE, A.V.G., HAND, M.V., HAMILTON, J.T.G., SHARMA, N.D., BOYD, D.R. and STEVENSON, M.H., (1993), Synthesis, characterisation and use of 2-tetradecylcyclobutanone together with other cyclobutanones as markers for irradiated liquid whole egg, *J. Science Food Agriculture*, **62**, 361–367.

DAY, L.I. and BROWN, H., (1998), Detection of irradiated liquid egg white by capillary gel electrophoresis, *Food Science Technology Today*, **12** (2), 111–112.

DEELEY, C., (2006), The use of irradiation for food quality and safety, Institute of Food Science and Technology Information Statement, February, available at www.ifst.org/uploadedfiles/cms/store/ATTACHMENTS/Irradiation.pdf.

DELINCÉE, H., (1998), Detection of food treated with ionizing radiation, *Trends in Food Science and Technology*, **9**, 73–82.

DIEHL, J.U.F., (1995), *Safety of Irradiated Foods*, Marcel Dekker, New York.

EHLERMANN, D.A.E., (1995), Dosimetry and identification as a tool for official control of food irradiation, *Proceedings of the 29th International Meeting on Radiation Processing,*

Radiation Physics and Chemistry, **46** (4–6), Part 1, pp. 693–698.

EHLERMANN, D.A.E., (2002), Irradiation, in (C.J.K. Henry and C. Chapman, Eds.), *The Nutrition Handbook for Food Processors*, Woodhead Publishing, Cambridge, pp. 371–395.

EHMANN, W.D. and VANCE, D.E., (1991), *Radiochemistry and Nuclear Methods of Analysis*, Wiley Interscience, New York.

GAUNT, I.F., (1986), Food irradiations – safety aspects, *Proceedings Institute Food Science and Technology*, **19** (4), 171–174.

GOULD, G.W., (1986), Food irradiation – microbiological aspects, *Proceedings Institute Food Science and Technology*, **19** (4), 175–180.

GUISE, B., (1986), Irradiation waits in the wings, *Food Europe*, March–April, 7–9.

HAYASHI, T., (1991), Comparative effectiveness of gamma rays and electron beams in food irradiation, in (S. Thorne, Ed.), *Food Irradiation*, Elsevier Applied Science, London, pp. 169–206.

HUGHES, D., (1982), Notes on ionising radiation: quantities, units, biological effects and permissible doses, *Occupational Hygiene Monograph* No. **5**, Northwood: Science Reviews, Middlesex.

JOHNSON, J. and MARCOTTE, M., (1999), Irradiation control of insect pests of dried fruits and walnuts, *Food Technology*, **53** (6), 46–48, 50, 53.

LEEK, P. and HALL, D., (1998), Portable electron beam systems, L&W Research, Inc., available at http://mbao.org/1998airc/082leek.pdf.

LEWIS, M.J., (1990), *Physical Properties of Foods and Food Processing Systems*, Woodhead Publishing, Cambridge, pp 287–290.

LEY, F.J., (1987), Applying radiation technology to food, in (A. Turner, Ed.,) *Food Technology International Europe*, Sterling Publications International, London, pp. 287–290.

LOAHARANU, P., (1995), Food irradiation: current status and future prospects, in (G.W. Gould, Ed.), *New Methods of Food Preservation*, Blackie Academic and Professional, Glasgow, pp. 90–109.

MCLAUGHLIN, W.L., JARRET, R.D. SNR and OLEJNIK, T.A., (1982), Dosimetry, in (E.S. Josephson and M.S. Peterson, Eds.), *Preservation of Foods by Ionizing Radiation*, Vol. 1, CRC Press, Boca Raton, FL, pp. 189–245.

MOLINS, R.A., (Ed.), (2001), *Food Irradiation: Principles and Applications*, John Wiley & Sons, New York.

MOORHOUSE, K.M., (1996), Identification of irradiated seafood, in (C.H. McMurray, E.M. Stewart, R. Gray and J. Pearce, Eds.), *Detection Methods for Irradiated Foods: Current Status*, The Royal Society of Chemistry, Cambridge, pp. 249–258.

NOLAN, M., ELLIOTT, C.T., PEARCE, J. and STEWART, E.M., (1998), Development of an ELISA for the detection of irradiated liquid whole egg, *Food Science and Technology Today*, **12** (2), 106–108.

OLSON, D.G., (1998), Irradiation of food, *Food Technology*, **52** (1), 56–62.

ROBERTS, P.B., CHAMBERS, D.M. and BRAILSFORD, G.W., (1996), Gas evolution as a rapid screening method for detection of irradiated foods, in (C.H. McMurray, E.M. Stewart, R. Gray and J. Pearce, Eds.), *Detection Methods for Irradiated Foods: Current Status*, The Royal Society of Chemistry, Cambridge, pp. 331–334.

ROBINSON, D.S., (1986), Irradiation of foods, *Proceedings Institute Food Science and Technology*, **19** (4), 165–168.

ROSENTHAL, I., (1993), Analytical methods for post-irradiation dosimetry of foods, *Pure and Applied Chemistry*, **65** (1), 165–172.

SATIN, M., (1993), *Food Irradiation, A Guidebook*, Technomic Publishing Co, Basel, pp. 95–124.

SOMMERS, C.H. and FAN, X., (Eds.), (2006), *Food Irradiation Research and Technology*, Wiley-Blackwell Publishing, Oxford.

STEVENSON, M.H., (1990), The practicalities of food irradiation, in (A. Turner, Ed.), *Food Technology International Europe*, Sterling Publications International, London, pp. 73–76.

STEVENSON, M.H., (1994), Identification of irradiated foods, *Food Technology*, **48**, 141–144.

STEWART, E.M., MOORE, S., MCROBERTS, W.C., GRAHAM, W.D. and HAMILTON, J.T.G., (1998), 2-

Alkylcyclobutanones as markers for exotic fruits, *Food Science and Technology Today*, **12** (2), 103–106.

TADA, M., (2004), Foreword on food irradiation, *Foods Food Ingredients J. Japan*, **209** (12), 1.

THAYER, D.W. and RAJKOWSKI, K.T., (1999), Developments in irradiation of fresh fruits and vegetables, *Food Technology*, **53** (11), 62–65.

TYREMAN, A.L., BONWICK, G.A., BEAUMONT, P.C. and WILLIAMS, J.H.H., (1998), Detection of food irradiation by ELISA, *Food Science and Technology Today*, **12** (2), 108–110.

VENUGOPAL, V., DOKE, S.N. and THOMAS, P., (1999), Radiation processing to improve the quality of fishery products, *Critical Reviews in Food Science and Nutrition*, **39** (5), 391–440.

WILKINSON, V.M. and GOULD, G.W., (1996), *Food Irradiation, A Reference Guide*, Woodhead Publishing, Cambridge.

8

High-pressure processing

Abstract: Processing using high hydrostatic pressures can inactivate micro-organisms and enzymes to effectively pasteurise foods with minimal heating, thus extending the shelf-life of foods while retaining their sensory characteristics and nutritional value. This chapter explains theories of inactivation of microbial cells using high pressures and the equipment used. It concludes with the effects of high pressures on enzymes and foods and describes new combinations of high pressure and other minimal processing techniques.

Key words: high-pressure processing, barosensitivity of micro-organisms, isostatic compression, pressure-assisted freezing and thawing, combined minimal processing.

In high-pressure processing (HPP) (also described as high hydrostatic pressure (HHP) processing or ultra-high pressure (UHP) processing), foods are subjected to pressures between 100 and 1000 MPa for several seconds to minutes. The process is capable of inactivating micro-organisms and enzymes to effectively pasteurise foods with minimal heating while retaining their sensory characteristics and nutritional value for an extended shelf-life. In some foods there is an improvement in functional properties (section 8.1.1).

The first reported use of HPP as a method of food preservation was in 1899 in the USA, where experiments were conducted using high pressures to preserve milk, fruit juice, meat and a variety of fruits. They demonstrated that pressures of 658 MPa for 10 minutes could destroy micro-organisms in these products. In the early years of the twentieth century, other research showed that high pressures could alter the protein structure in egg white. However, these early researchers were constrained by both difficulties in manufacturing high-pressure equipment and inadequate packaging materials to contain the foods during processing, and research was discontinued. Advances in the design of presses together with rapid advances in packaging materials enabled research to begin again on HPP in the 1980s, mainly in Japan.

The process reached the stage of commercial exploitation in 1990 with a range of high-quality pressure-processed jams being sold in Japan, including apple, kiwi, strawberry and raspberry. The jams had a shelf-life of two months under chilled storage, which is required to prevent enzyme activity. Other companies started production of bulk orange and grapefruit juices and other high-acid products including fruit jellies, sauces, fruit yoghurts, purées and salad dressings. These products are suitable for HPP because,

Table 8.1 Examples of commercially available foods produced by HPP

Country/product	Processing conditions	Role of HHP
Japan		
Fruit-based products (jams, sauces, purées, yoghurts	400 MPa, 10–30 min, 20 °C	Pasteurisation, improved gelation, faster sugar penetration, limited residual pectin methylesterase activity
Grapefruit juice,	200 MPa, 10–15 min, 5 °C	Reduced bitterness
Sugared fruits for ice cream/sorbets	50–200 MPa	Faster sugar penetration and water removal
Raw pork ham	250 MPa, 3 hours, 20 °C	Faster maturation, (reduced from 2 weeks to 3 hours), faster tenderisation by endogenous proteases, improved water retention and shelf-life
Fish sausages, terrines and 'pudding'	400 MPa	Gelation, microbial reduction, improved gel texture
Rice wine	–	Yeast inactivated to stop fermentation without heating
Rice cake, hypoallergenic precooked rice	400–600 MPa, 10 min, 45 or 70 °C	Microbial reduction, fresh taste/flavour, enhanced rice porosity and salt extraction of allergenic proteins
Europe		
Fruit juices	400 MPa, room temperature	Inactivation of microflora (up to 10^6 CFU/g), partial inactivation of pectin methylesterase
Sliced processed ham	400 MPa, few min, room temperature	–
Squeezed orange juice	500 MPa, room temperature	Yeast and enzyme inactivation, retained natural taste
USA		
Avocado paste (guacamole, salsa)	700 MPa, 10–15 min, 20 °C	Microbial inactivation and polyphenol oxidase inactivation
Oysters	300–400 MPa, 10 min, room temperature	Microbial inactivation, raw taste/flavour retained, shape and size retained

Adapted from Indrawati *et al.* (2003)

owing to their low pH, they are spoiled by micro-organisms that are relatively sensitive to HPP, and not by pressure-resistant bacterial spores (section 8.3). Similar products later reached the US market (Mermelstein 1997) followed by pressure-treated guacamole, oysters, hummus, chicken strips and fruit 'smoothies'. HPP orange juice and sliced cooked ham are sold in France and Spain respectively (Tewari *et al.* 1999) and HPP orange juice in the UK (Table 8.1). A survey of consumers in three European countries found that high-pressure processing was acceptable to the majority of people that were interviewed (Butz *et al.* 2003). There is increasing interest in using HPP to preserve low-acid foods, including red meats, poultry, seafoods, foie gras, liquid whole egg and cheese (Raso and Barbosa-Canovas 2003).

Commercially produced products include pressure-processed salted raw squid and fish sausages (Hayashi 1995). Pressure-treated oysters have improved microbiological quality, easy opening of the shells which removes the labour-intensive 'shucking' process to open them, and improved yield because some water is absorbed by the oyster during pressure treatment (Ledward 2000). Other possible applications are improved microbiological safety and elimination of cooked flavours from sterilised meats and pâtés. Ready-to-eat meats, including pastrami and Cajun beef, which were pressure treated at 600 MPa at 20 °C for 3 min and stored at 4 °C for 98 days, and showed undetectable or low levels of pathogenic or spoilage micro-organisms. Consumer hedonic ratings for unprocessed and processed meats revealed no difference in acceptability and no deterioration in the sensory quality (Hayman *et al.* 2004). HPP vegetables are reported to have a crisp texture, and high-pressure sterilised main meals, including macaroni cheese, salmon fettuccine, ravioli and beef stroganoff, are each reported to have a freshly prepared flavour, texture and colour (Meyer *et al.* 2000). High pressures also transform cocoa butter into the stable crystal form in chocolate tempering (Chapter 24, section 24.1.1) and have been used to preserve honey and other viscous liquids, and dairy products such as unpasteurised milk and mould-ripened cheese. HPP treatment of pasteurise rice wine resulted in a shelf-stable product that had a taste equivalent to the control, due to the inactivation of spoilage enzymes and micro-organisms (Hara *et al.* 1990). In 2006, about 50 companies were marketing >1500 different HPP systems (Pendrous 2007). Although (in 2008) production volumes are relatively small, HPP is gaining in popularity, not only because of its preservative action and minimal effects on food quality, but also because of its potential for high-pressure freeze/thawing (section 8.2.3) and its ability to alter the functional properties of foods to produce new value-added products (section 8.5). Other advantages of HPP include reduced processing times, minimal energy consumption and virtually no effluents. The main limitations of HPP are the high capital costs and that it cannot be used for dry foods (because water is needed for microbial destruction); or for foods that contain entrapped air, such as strawberries, which would be crushed during the high-pressure treatment.

Details of HPP are given in a number of publications including Doona *et al.* (2007), Indrawati *et al.* (2003), Raso and Barbosa-Canovas (2003), Hendrickx and Knorr (2002), San Martin *et al.* (2002), Palou *et al.* (1999) and Isaacs (1998), and a comprehensive list of research references is given by Anon (2007a).

8.1   Theory

When high hydrostatic pressures are applied to packages of food submerged in a liquid, the pressure is distributed instantaneously and uniformly throughout the food (i.e. it is 'isostatic') so that no pressure gradient exists and all parts receive the same treatment. This application of high pressure to all parts of a food is a significant advantage over other methods of processing because the food is treated evenly throughout, and package size, shape and composition are not factors in determining the process conditions. This overcomes problems of lack of uniformity in processing that are found for example in conductive or convective heating (Chapter 10, section 10.1.2), microwaves and radio frequency heating (due to variation in loss factors (Chapter 20, section 20.1)) or radiant heating (variation in surface properties (Chapter 20, section 20.3)).

Although HPP is considered to be a non-thermal process, the temperature of foods increases at high pressures due to adiabatic heating, generated by compression of water (Fig. 8.1) and other food components (Farid 2005). The temperature rise is approximately 3 °C per 100 MPa depending on the composition of the food, but can be higher for foods that contain significant amounts of fat (e.g. up to 9 °C temperature increase per 100 MPa in soybean oils). The temperature increase is transient and the temperature falls back on

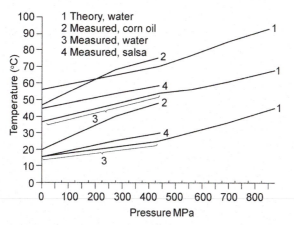

Fig. 8.1 Increase in temperature of water, oil and salsa as a result of adiabatic heating under pressure: 1, water (theoretical); 2, corn oil; 3, water; 4, salsa (each measured) (adapted from Anon 2000 using data from E. Ting, 1999, personal communication, Flow International, Kent, WA).

depressurisation to the initial temperature due to adiabatic cooling (Balasubramaniam 2007).

It is desirable to achieve a uniform (isothermal) temperature increase during compression. To do this, a uniform initial temperature is required in the food, and for isothermal processing the pressure vessel must be held at the same temperature as the final food temperature. However, the temperature distribution in foods can change if there is heat transfer through the walls of the pressure vessel.

As in thermal processing, where D, z and F_o-values are standard processing parameters (Chapter 10, section 10.3), high-pressure process parameters for inactivation of microbial spores are being developed and standardised. The factors that affect process optimisation in HPP include:

- composition and properties of food (pH, a_w, ionic strength and type of ions);
- types (species and strain) of microbial contamination;
- age and stage of growth of micro-organism;
- initial product temperature;
- time to achieve processing pressure;
- operating pressure and temperature (including adiabatic heating), temperature distribution at pressure and holding time at this pressure;
- type of pressure treatment (continuous or cyclic);
- decompression time; and
- packaging integrity.

In pulsed processing (section 8.2.3) additional factors include the pulse waveform, frequency and high- and low-pressure values of the pulses (Anon 2000). Meyer *et al.* (2000) describe the critical variables for achieving sterility in pulsed HPP as two or more pulses, a processing temperature of >105 °C, a uniform initial product temperature and the initial spore load and type. They describe an HPP lethality chart that shows the processing temperatures and pressures needed to achieve sterility. Modelling the thermal behaviour of foods during high-pressure treatments is difficult because of the lack of thermophysical data on foods under pressure. Torrecilla *et al.* (2004) describe the development of a neural network to predict process parameters involved in HPP (see also Chapter 27, section 27.2.3). The neural network was trained using data on applied pressure, the rate of pressure increase, temperatures in the high-pressure vessel and ambient temperature, and was able to accurately predict the time needed to equilibrate the temperature in a food after pressurisation.

8.1.1   Effect on food components

The overriding principle in HPP, known as the Le Chatelier principle, is that reactions resulting in a decrease in volume are accelerated, whereas reactions that produce increases in volume are inhibited. It has been established that high pressures only affect weak linkages such as electrostatic and hydrophobic bonds, which, for example, cause protein molecules to unfold and aggregate. This causes gelation of some food proteins and inactivation of some enzymes. There is increased interest in using HPP to modify enzymic reactions and to cause structural changes in food components to improve functional properties (see section 8.5). However, the effect of high pressures on proteins varies widely because of differences in their hydrophobicity. Up to 100 MPa, hydro-phobic interactions mostly result in a volume increase, but at higher pressures they cause a volume decrease (Suzuki and Taniguchi 1972). High pressures easily disrupt the weak linkages that stabilise the structure of microbial cells (section 8.1.2), but there is almost no effect on covalent bonds (Ledward 2000, Tauscher 1999). Small macromolecules that produce flavour or odour in a food are not changed by pressure and foods that are subjected to HPP at ambient or chill temperatures do not undergo substantial changes to flavour or colour. Likewise the molecular structure of vitamins and availability of minerals is largely unaffected and HPP causes minimal changes to the nutritional value of foods (section 8.5).

Reactions that involve the formation of hydrogen bonds are favoured because they result in a decrease in volume (Hoover et al. 1989). Water is the main component of most foods and although water is almost incompressible compared with gases, high pressures reduce the volume of water by a small amount (Fig. 8.2). High pressures also modify the physicochemical properties of foods, including the density, viscosity, thermal conductivity (up to 40% increase (Balasubramaniam 2007)), ionic dissociation, pH (the pH of water at 25 °C is reduced from 7.00 at 0.1 MPa to 6.27 at 101 MPa (Marquis 1976)) and the freezing point. Ice has different polymorphic forms depending on the pressure at

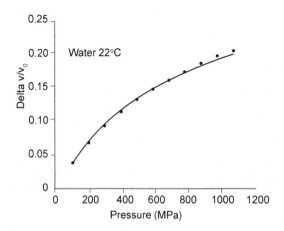

Fig. 8.2   Fractional decrease in water volume under increased pressure (from data of Bridgman 1912, reported by Anon 2000).

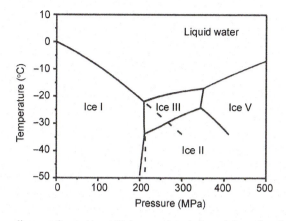

Fig. 8.3 Phase diagram for water at high pressures (adapted from Luscher *et al.* 2004).

which it is frozen. Higher pressures distort the hydrogen bond angles and produce more compact molecular structures. For example, in Fig. 8.3 the melting point of polymorph 'Ice I' decreases with increasing pressure up to 207 MPa and −22 °C. As the pressure is increased further, the ice changes to polymorphs 'Ice III', 'Ice II' up to 'Ice V', each having a higher melting point. Although not shown in Fig. 8.3, the temperature increases to 20 °C at 880 MPa and 30 °C at 1036 MPa. This permits opportunities for fast thawing of foods by pressurising them above their melting point, or freezing foods by decompression below 0 °C (see also section 8.2.3, Chapter 1, section 1.1.1, and Chapter 22, section 22.1.1).

8.1.2 Mechanisms of inactivation of microbial cells

When foods are subjected to HPP the pressure is instantly transmitted through the medium to micro-organisms in the food. It causes microbial inactivation by altering non-covalent bonds in proteins that are responsible for replication, cellular integrity and metabolism in their biologically active form. The irreversible denaturation of one or more of these critical proteins (e.g. membrane-bound ATPases and enzymes involved in DNA replication and transcription and a variety of biochemical changes) results in cell injury or death. Kinetic models are described by Cheftel (1995) based on the formation of an unstable intermediate complex, which decomposes at a rate that is determined by the temperature of the process. Therefore the rate of formation of the activated complex controls the rate of the reaction, whether it is enzyme-catalysed or irreversible protein denaturation. However, further research is needed to predict the effects of HPP on irreversible protein denaturation. Knorr (1995a) describes possible hypotheses for disruption to metabolic processes caused by the effects of high pressures on cellular enzymes and these aspects are reviewed in detail by Palou *et al.* (1999), Earnshaw (1995), Isaacs and Chilton (1995) and Patterson *et al.* (1995).

At increasingly higher pressures, cellular morphology is altered, there are reductions in cell volume, collapse of intracellular vacuoles and damage to cell walls and cyto-plasmic membranes. For example, Perrier-Cornet *et al.* (1995) measured a compression rate of 25% in the cell volume of *Saccharomyces* sp. treated at 250 MPa with partial (10%) reversal of cell compression on return to atmospheric pressure. The loss of cell contents indicated membrane disruption or an increase in permeability. Pressures greater than 500 MPa may cause disruption to the surface of intact cells. Pressure damage to cell membranes includes altered permeability, causing solubilisation and loss of intracellular constituents; leakage of metallic ions (e.g. K^+, Mn^{2+} and Ca^{2+}); permeation of extra-cellular substances into cells; and inhibition of amino acid uptake caused by denaturation of membrane proteins (San Martin *et al.* 2002). Bacteria that have more rigid cell membranes are more susceptible to inactivation by high pressures, whereas those that contain compounds that enhance fluidity in membranes are more resistant to pressure (Russell *et al.* 1995, Smelt *et al.* 1994). Yano *et al.* (1998) isolated barophilic bacteria from the intestines of deep-sea fish retrieved from depths of 3100 and 6100 m, and found that exposure to increasing pressures caused a change from saturated to unsaturated fatty acids in the membranes. These compounds may be an important factor in maintaining membrane fluidity under high pressures. High pressures also sub-lethally stress microbial cells in a similar way to heat injury. The sensitivity to pressure ('barosensitivity') of micro-organisms may be related to the ability of cells to repair leaks after decompression: cells in the exponential growth phase are sensitive to high pressures and cannot repair pressure-damaged membranes; whereas cells in the stationary phase may be able to repair membranes after decompression. Under otherwise optimal conditions, psychrotrophic micro-organisms can recover at refrigeration temperatures, whereas mesophiles require temperatures closer to 37 °C to recover from pressure-induced injury and to resume replication. Refrigerated storage temperatures after pressure treatment prolong the time required for mesophiles to repair membranes (Cheftel 1995). Cell repair after pressure treatment indicates that a critical protein was denatured, but that repair proteins were not damaged. Food acids may inhibit repair of damaged cell proteins and thus make a micro-organism more sensitive to pressure (Cheftel 1995). Since denaturation depends on the protein structure, there is a wide range of tolerances to high pressures by micro-organisms (for example, Smelt (1998) reported a six-fold range in *D*-values among 100 strains of *L. monocytogenes*) (see also section 8.3, Table 8.3).

8.2   Equipment

Companies that make equipment to press metal and ceramic engineering components and quartz crystals for the electronics industry have the necessary expertise to construct high hydrostatic pressure machinery. Palou *et al.* (1999) describe different designs of equipment. The main components of HPP equipment (Anon 2000, Mertens 1995) are:

- a pressure vessel, its end closures and a means of restraining the end closures (e.g. a yoke, threads, or pins);
- a pressure generation system (a low-pressure pump and an 'intensifier' (high-pressure pump) that uses liquid from the low-pressure pump to generate high-pressure fluid for compression);
- a temperature control device;
- a materials handling system; and
- a data acquisition system, controls and instrumentation.

An HPP system can be arranged as a batch process (section 8.2.1) for packaged foods or as a semi-continuous process (section 8.2.2) to treat unpackaged liquid foods. Pressure vessels for commercial operation at pressures of 400–700 MPa are made using two designs: either cylinders made of high-tensile steel alloy 'monoblocs' (forged from a single piece of material) or cylinders are prestressed by winding them with several kilometres of wire to prevent expansion during pressurisation. The wire-wound design is

Fig. 8.4   Horizontal HPP system (courtesy of Avure Technologies at
www.avure.com/systems_hppsystems.htm).

preferred because it allows the equipment to leak and relieve the pressure before a
catastrophic failure of the pressure vessel and explosion could occur. A replaceable liner
is inserted into the cylinder, made from stainless steel so that potable water can be used as
the isostatic compression fluid. Pressure vessels are sealed by either a threaded steel
closure having an interrupted thread so that the closure can be removed more quickly, or
by a retractable prestressed frame that is positioned over the vessel (Fig. 8.4). Pressure is
generated in the vessel by either direct or indirect compression. In direct compression, a
piston moved by hydraulic pressure compresses fluid in the vessel by reducing the
volume. Indirect compression systems use an intensifier to pump the pressurising fluid
directly into the vessel until the desired pressure is reached (Fig. 8.5). Indirect compression systems are preferred for HPP because, compared with direct systems, they have
a lower capital cost for the pressure vessel, simplified materials handling, and require
static pressure seals. Direct compression systems require dynamic pressure seals between
the piston and internal vessel surface, which are more expensive and subject to wear.

The process temperature can be specified from $-20\,°C$ to $>100\,°C$. Temperature
control is achieved by thermostatically controlled electric heating elements wrapped
around the pressure vessel; by pumping a heating/cooling medium through a jacket that
surrounds the vessel; through the top and bottom closures; or through channels between
the wire winding and the process vessel. A PLC controller (Chapter 27, section 27.2.2)

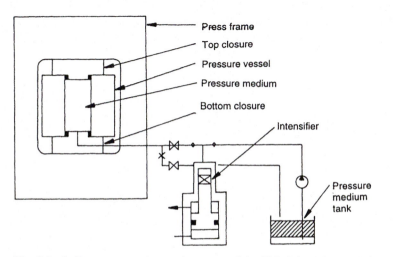

Fig. 8.5   Indirect compression equipment used for HPP (after Mertens 1995).

monitors and controls the process cycle including loading/unloading of the vessel and the time, temperature and pressure conditions during processing.

8.2.1   Batch operation

Batch HPP operation is similar to that of retorts used in heat processing (Chapter 13, section 13.1.2) in that the process cycle involves filling the process vessel with product, closing, bringing the vessel to the required pressure, holding for the required time, decompressing the vessel and removing the product. Presses having vessels of up to 94001 capacity can operate at pressures of 200–500 MPa with operating cycles of as little as 2–3 minutes (Mertens 1995). Palou *et al.* (1999) report a batch press having a throughput of $6001h^{-1}$, operating at 420 MPa, which is used for the commercial production of pineapple juice in Japan. Research is underway to increase the pressures available and the speed of pressurisation by the use of propellants, known as ultra-high-pressure processing using propellants (UHP[3]) (Anon 2007b).

In operation, packaged food is loaded into the pressure vessel, the vessel is sealed and process water is pumped in to displace any air. When the vessel is full, the pressure relief valve is closed and water is pumped in until the required pressure is reached. Expansion of the vessel under pressure increases its volume, and for example a filled 1001 vessel requires an additional 151 water to bring it to a pressure of 680 MPa. The compression time depends on the power of the pump, and for example a 75 kW pump can bring a 501 vessel to an operating pressure of 680 MPa in 3–4 min (Anon 2000). After the specified holding time, the pressure relief valve is opened and the compressed water is allowed to return to atmospheric pressure. The packaged food is removed and is ready for shipment. Foods decrease in volume under pressure and an equal expansion occurs on decompression. There is therefore considerable distortion to the package and stress on the seal when in-container processing is used. The package must be able to accommodate up to 15% reduction in volume and return to its original volume without loss of seal integrity and barrier properties. Suitable packaging materials include glass containers and ethylene-vinyl alcohol copolymer (EVOH) or polyvinyl alcohol (PVOH) films (Chapter 25, section 25.2.4). The effects of HPP on packaging are reviewed by Caner *et al.* (2004).

Materials' handling equipment for in-container processing is similar to that used to load/unload batch retorts (Chapter 13, section 13.1.3). It consists of metal baskets that contain the packaged products, which are transported by conveyors in horizontal HPP units or by cranes in vertical HPP units (Fig. 8.6).

The high cost of pressure vessels, intensifiers and vessel sealing systems means that for economical operation the system should operate as many cycles per hour as possible. Target cycle times are 5–10 min compared with batch thermal processes which may require 60 min to complete a process cycle. The processing cost depends on the holding time, temperature and operating pressure that are used. The capital cost of high-pressure equipment increases exponentially with operating pressure and if moderate heating is used this can reduce the maximum pressure required and significantly reduce the fabrication costs. For example, processing at 400 MPa for 10 min is twice as expensive as processing at 800 MPa with no hold time (Fig. 8.7).

8.2.2   Semi-continuous operation

Semi-continuous systems for treating liquid foods consist of a low-pressure pump to fill the pressure vessel and a freely moving piston to compress the food. As the vessel is

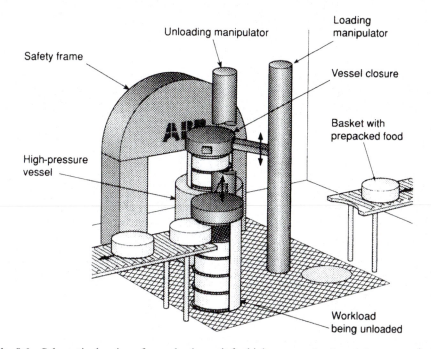

Fig. 8.6   Schematic drawing of a production unit for high pressure treatment at pressures between 400 and 800 MPa (from Olsson 1995).

filled with food, the piston is displaced. When full, the inlet valve is closed and high-pressure water is pumped in behind the piston to compress the food. After the holding time, the high-pressure water is decompressed and the treated food is discharged and aseptically filled into pre-sterilised containers. Materials handling is simpler than batch processing of packaged foods, requiring only pumps, pipework and valves, and several vessels can be operated in parallel using the same high-pressure pump to give a continuous flow of product to the filling line. A comparison of the advantages and limitations of in-container and bulk processing is shown in Table 8.2.

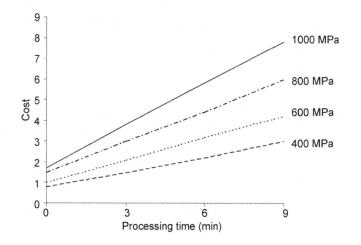

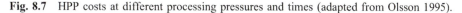

Fig. 8.7   HPP costs at different processing pressures and times (adapted from Olsson 1995).

Table 8.2   Advantages and limitations of in-container and bulk HPP processing

In-container processing	Bulk processing
Advantages	
• Applicable to all solid and liquid foods • Minimal risk of post-processing contamination • No major developments needed for high pressure processing • Easier cleaning	• Simple materials handling • Greater flexibility in choice of container • Maximum efficiency (>90%) in use of high pressure vessel volume • Minimum vessel dead-time (no opening/closing of vessel needed, faster loading/unloading)
Limitations	
• Complex materials handling • Little flexibility in choice of container • Greater dead-time in use of pressure vessel	• Only suitable for pumpable foods • Aseptic filling of containers required – potential post-processing contamination • All pressure components in contact with food must have aseptic design and be suitable for cleaning-in-place (CIP) and sterilising-in-place (SIP)

Adapted from Mertens (1995)

Semi-continuous processing of fruit juices at 4000–6000 l h^{-1} using pressures of 400–500 MPa for 1–5 min at ambient temperature is used by one company in Japan, whereas another uses a similar process operating at 120–400 MPa followed by a short heat treatment before the juice is packaged.

8.2.3   Process developments

Pulsed HPP systems

The use of repeated pressure pulses is more effective for inactivation of spores, vegetative bacteria and yeasts than a single pressurisation of an equivalent time (Meyer *et al.* 2000, Hayakawa *et al.* 1994a,b). The difference in effectiveness varies, and research is continuing to evaluate the benefits against the higher costs of the pressure unit, additional time for each process cycle and possible negative effects on the sensory properties of different products (Anon 2000). Aleman *et al.* (1994, 1996) found that pulsed HPP was more effective than a single pressurisation over similar processing times for inactivation of *S. cerevisiae* in pineapple juice. Repeated pulses of 0.66 s for a total of 100 s gave the same degree of inactivation as one pulse at the same pressure for 5–15 min. However, the type of pulse waveform (ramp, square, sinusoidal, etc.), the pulse frequency and the ratio of time under pressure to time without pressure were critical, and some conditions allowed the total survival of the yeast population. Meyer *et al.* (2000) describe the sterilisation of macaroni cheese at 90 °C using two pressure pulses of 690 MPa for one minute with a pause of one minute at ambient pressure between them. Spore loads of 10^6 g^{-1} of *Clostidium sporogenes* PA3679 and *Bacillus cereus* were destroyed and sterility was achieved.

Freezing and thawing

Foods can have higher quality when frozen or thawed under HPP by taking advantage of the phase changes of water and ice under high pressures (Fig. 8.3 in section 8.1.1). Knorr *et al.* (1998) describe the following types of process:

- Pressure-assisted freezing or thawing – the phase transition takes place by decreasing the temperature (for freezing) or increasing it (for thawing) under constant pressure (i.e. a vertical line in Fig. 8.3).
- Pressure-shift freezing or thawing – the phase transition is caused by a change in pressure. For example, in pressure-shift freezing the change in melting point of ice at higher pressures produces supercooled water (Fig. 8.3), and when the pressure is released there is rapid and uniform nucleation and growth of ice crystals. This produces small ice crystals throughout the food, rather than having an ice front move through the food as in traditional freezing methods (see Chapter 22, sections 22.1.1 and 22.2.5). Further cooling of the food is needed to remove the heat of fusion produced by ice crystal formation to completely freeze it.
- Pressure-induced freezing or thawing – where a phase transition is started by a change in pressure and continues at a constant pressure. For example, in pressure-induced thawing frozen food is pressurised to the liquid area of the phase diagram and heat of fusion is provided to thaw the food.

The advantages of pressure-assisted and pressure-shift freezing or thawing are high freezing or thawing rates (up to two-thirds reduction in the time needed compared to traditional methods (San Martin *et al.* 2002)). In HPP freezing, the smaller ice crystals reduce damage to the cellular structure and consequent enzymic or biochemical changes, resulting in higher-quality foods (e.g. improved texture of frozen vegetables such as cabbage (Fuchigami *et al.* 1998)). Similarly the increased thawing rate in HPP thawing compared with traditional methods reduces the loss of product quality. For example, meat and fish thawed at 50 MPa and −20 °C underwent no colour changes and had reduced drip losses (Chevalier *et al.* 1999, Zhao *et al.* 1998). Luscher *et al.* (2004) describe studies of the effect of pressure-induced ice phase transitions on the inactivation of micro-organisms in frozen suspensions.

Other developments
Another modification to HPP is very rapid (millisecond) controlled decompression. If liquids are rapidly decompressed through a small orifice, the high velocity and turbulent flow increases the shearing forces and causes cavitation in cells and spores that results in physical disruption and death (Earnshaw 1992). Developments in HPP reported by Knorr (1995a) include combined freeze concentration (Chapter 23, section 23.2) and high-pressure blanching. Results suggest that pressure blanched fruits are dried more rapidly than those treated by conventional hot water blanching (Chapter 11).

8.3 Effect on parasites and micro-organisms

Organisms differ in their sensitivity to high pressures and in general more evolutionarily developed life forms are more sensitive to pressure. Although studies of pressure resistance of parasites (e.g. oocysts and spores of *Cryptosporidium* spp. or *Cyclospora* spp. and protozoans *Entamoeba* spp. and *Giardia* spp.) are incomplete, it is likely that they are not as pressure-resistant as bacteria. For example, parasitic *Trichinella spiralis* worms are killed at 200 MPa for 10 min and nematode worms (e.g. *Anisakis simplex*) that occur in cold-water marine fish are killed by processing for 30–60 s at 414 MPa, 90–180 s at 276 MPa or 180 s at 207 MPa (Dong *et al.* 2003).

Micro-organisms vary in their barosensitivity (Table 8.3). There is a minimum critical pressure below which microbial inactivation does not take place regardless of the process

Table 8.3   Effect of HPP on selected micro-organisms

Micro-organism	Treatment conditions			Inactivation (log cycles)	Media
	Pressure (MPa)	Temp (°C)	Time (min)		
Aspergillus awamori	300	N/A	5	5	Satsuma, mandarin juice
Escherichia coli O157	550	20	1	4	Orange juice
Listeria innocua	450	20	10	6.63	Minced beef muscle
Listeria monocytogenes NCTC 11994	375	20	15	2.0	Phosphate buffer saline
L. monocytogenes NCTC 2433	375	20	15	6.0	N/A
Saccharomyces cerevisiae	253	25	10	3	Spaghetti sauce
Salmonella enteriditis	450	20	5	4.04	Liquid whole egg
Salmonella typhimurium	400	20	15	6.2	Phosphate buffer saline
Staphylococcus aureus	400	25	15	1.0	Ovine milk
Vibrio parahaemolyticus	172	23	10	2.5	Phosphate buffer saline
Vibrio vulnificus	200	25	10	2.5	Artificial seawater
Yersinia enterocolitica	275	N/A	15	5	N/A

Adapted from Patterson (1995), Palou *et al.* (1999) and San Martin *et al.* (2002)

time. The following groups of micro-organisms have a decreasing order of barosensitivity: yeasts > Gram-negative bacteria (most pressure sensitive although there are exceptions) > complex viruses > moulds > Gram-positive bacteria > bacterial endospores (most resistant). However, there is much variability in the baroresistance of micro-organisms, even within a single species or genus (Cheftel 1995). The variability in pressure sensitivity in bacteria results from differences in bacterial strains and suspending media, but the range decreases when higher temperatures are used for processing.

8.3.1   Yeasts and moulds
All vegetative cells of heat-resistant moulds *Byssochlamys nivea, B. fulva, Aspergillus fischeri* and *Paecilomyces* spp. are inactivated within a few minutes at 300 MPa at 25 °C (Butz *et al.* 1996), but 600 MPa at 60 °C for 60 min is needed to destroy ascospores except those of *B. nivea*, which required 800 MPa at 70 °C. Low water activity increases the pressure sensitivity of ascospores. Zook *et al.* (1999) observed first-order inactivation curves with *S. cerevisiae* ascospores in orange and apple juice, using pressures ranging from 300 to 500 MPa. Studies of the effect of pulsed HPP on *Saccharomyces cerevisiae* are described in section 8.2.3.

8.3.2   Bacteria
In general, bacteria in the logarithmic phase of growth are more sensitive than cells in the stationary, dormant or death phases. Cocci undergo fewer morphological changes under pressure than rod-shaped bacteria, and are hence more resistant. Inactivation curves for bacteria show an initial lag period, followed by first-order inactivation kinetics during the treatment period and then tailing (possibly due to spores). Sonoike *et al.* (1992), describe pressure-temperature *D*-values for *Lactobacillus casei* and *E. coli* strains based on first-order inactivation kinetics. Biphasic pressure inactivation is found for both vegetative bacteria and spores. An initial rate of inactivation is replaced within a few minutes of pressurisation by a reduced rate, indicating the presence of a small pressure-resistant sub-

population. In this case two *D*-values can be calculated. On occasion the reduced rate can level off, indicating that additional processing has no effect on reducing the remaining microbial population (Anon 2000).

Typically, a pressure of 350 MPa applied for 30 min or 400 MPa applied for 5 min causes a ten-fold reduction in vegetative cells of bacteria, yeasts or fungi (Hoover *et al.* 1989). With the exception of bacterial spores (see below) an increase in processing pressure, time or temperature (to 45–50 °C but also reduced temperatures between −30 and +5 °C), increase the numbers of pathogenic and spoilage micro-organisms that are inactivated. For example, at a pressure of 150 MPa, a decrease in temperature to −20 °C increased inactivation of *Saccharomyces cerevisiae* and *Lactobacillus plantarum* from <1 log up to 7–8 log for each micro-organism (Perrier-Cornet *et al.* 2005).

There is a wide range of pressure sensitivity among pathogenic Gram-negative bacteria. Patterson *et al.* (1995) describe an isolate of *E. coli* O157:H7 that possesses similar pressure resistance to spores, and some strains of *Salmonella* sp. have demonstrated high levels of pressure resistance. Heat-resistant non-spore-forming bacteria are also usually more pressure-resistant than heat-sensitive types, but there are exceptions (e.g. *Salmonella senftenberg* 775W is the most heat-resistant of the *Salmonella* sp. but is more pressure-sensitive than a heat-sensitive strain of *Salmonella typhimurium* (Metrick *et al.* 1989)). *Vibrio parahaemolyticus* is more sensitive to HPP (inactivated within 10 min at 173 MPa) compared with *Listeria monocytogenes* (inactivated within 20 min at 345 MPa at 23 °C) (Mackey *et al.* 1995, Styles *et al.* 1991). Black *et al.* (2007) have reviewed the effects of HPP on pathogens in milk.

The pH of foods is lowered at increased pressure and, for example, Heremans (1995) reported a reduction in the pH of apple juice of 0.5 units per 100 MPa increase in pressure. As pH is reduced, micro-organisms become more susceptible to HPP inactivation, and fewer sub-lethally injured cells recover. Changes in water activity at different pressures have yet to be reported but low water activities due to high salt or sugar concentrations and/or low moisture contents may protect micro-organisms against inactivation. For example, Oxen and Knorr (1993) showed that a reduction in a_w from 0.98–1.0 to 0.94–0.96 increased the survival of micro-organisms suspended in a food. However, recovery of sub-lethally injured cells may be inhibited by low water activity (Anon 2000).

Bacterial spores

In general, bacterial spores are the most pressure-resistant and some can resist 1000 MPa at ambient temperatures. Unless HPP above 800 MPa is used, heat is required to eliminate bacterial spores in low-acid foods and temperatures of 90–110 °C at pressures of 500–700 MPa have been used to inactivate spore-forming pathogenic bacteria such as *Clostridium botulinum*. Spores of *C. botulinum* strains 17B and Cap 9B are the most resistant and can tolerate exposures of 30 min at 827 MPa and 75 °C (Anon 2000). Pulsed pressurisation (section 8.2.3) is more effective in destroying spores than continuous pressure (Heinz and Knorr 2002).

Spores can also be inactivated in two stages: first lower pressures (60–100 MPa) and mild temperatures (e.g. 25–40 °C) are used to induce germination and then germinated spores and vegetative cells are destroyed at a higher pressure (e.g. 600 MPa) and temperature (50–70 °C) (Heinz and Knorr 1998, Seyderhelm and Knorr 1992). However, these effects are not consistent and high pressure with moderate heating can either make micro-organisms more sensitive or more resistant to heat, depending on the type of micro-organism (Galazka and Ledward 1995). Anon (2000) reviews studies of HPP conditions required to inactivate spores of *Bacillus stearothermophilus* (pressures of 200 MPa at 20 °C and 400 MPa at 90 °C). Okazaki *et al.* (1996) report decreasing pressure resistance of spores of *Bacillus subtilis* > *Bacillus coagulans* > *Clostridium sporogenes* PA3679 at pressures of up to 400 MPa and temperatures from 25 to 110 °C. Raso *et al.* (1998a) report that a lower sporulation temperature (20 °C) for *Bacillus cereus* led more pressure-resistant spores than at a higher temperature (37 °C), regardless of the a_w or pH of pressure treatment. High sucrose concentrations protected the spores from pressure inactivation. Wuytack *et al.* (1998) found that exposure to 500 MPa created induced pressure resistance in spores of *B. subtilis* and may also make the spores more resistant to other food preservative methods.

8.3.3   Viruses

Viruses have a wide range of pressure resistances (Smelt 1998) from herpes simplex virus type 1, human cytomegalovirus and bacteriophages that are inactivated at 300–400 MPa, to Sindbis virus which is relatively unaffected by pressures of 300–700 MPa at −20 °C. Human immunodeficiency viruses (HIV) are reduced by 5.5-log viable particles by exposure to 400–600 MPa for 10 min at 25 °C (data from Otake *et al.* 1997, Shigehisa *et al.* 1996, Butz *et al.* 1992, Brauch *et al.* 1990).

8.4   Effect on enzymes

Enzymes that are related to food quality vary in their barosensitivity: some can be inactivated at room temperature by pressures of a few hundred MPa whereas others can withstand >1000 MPa (Cano *et al.* 1997). Other enzymes have their activity enhanced by high pressures. Pressure activation or inactivation depends on pH, substrate composition and temperature. For example, polyphenoloxidase, peroxidase in peas and pectin methylesterase in strawberries can each withstand 1200 MPa (Manvell 1996, Knorr 1993). Lipoxygenase is responsible for off-flavours and colour changes in legumes and can be inactivated at 400–600 MPa at ambient temperature. Hendrickx *et al.* (1998) report differences in polyphenoloxidase barosensitivity from different sources, with potato and mushroom polyphenoloxidase being bororesistant (800–900 MPa required for inactivation), and enzymes from apricot, strawberry and grape being more pressure sensitive (100, 400 and 600 MPa respectively). Inhibition of polyphenoloxidase in avocado by HPP at 500 MPa for 20 min has been used to produce guacamole that retains a natural bright green colour for 4–6 weeks under refrigeration (Ledward 2000).

Pectin methylesterase and polygalacturonase are responsible for cloud destabilisation in juices, gelation of fruit concentrates and loss of consistency in tomato products. They are less resistant than polyphenoloxidase but inactivation pressures for pectin methylesterase vary from 150 to 1200 MPa depending on its origin (Indrawati *et al.* 2003) (e.g. pectin methylesterase from tomato is much more pressure-resistant than that from orange). Inactivation is faster in acidic foods and reduced by high levels of soluble solids. Orange pectin methylesterase is partially (90%) inactivated at 600 MPa at room temperature and does not reactivate during storage (Irwe and Olsson 1994). The situation is further complicated by effects of high pressures on cellular membranes, which when ruptured may permit reactions between released intracellular enzymes and their substrates (Hendrickx *et al.* 1998). The kinetics of pressure-inactivation of enzymes are described by Ludikhuyze *et al.* (2003). Reviews of high pressure processing on enzymes and micro-organisms have been made by Mertens (1995), Hayashi (1993) and Balny *et al.* (1992).

8.5   Effects on foods

HPP causes complex changes to the structure and reactivity of bio-polymers such as starches and proteins. The greater the processing pressure and time, the greater is the potential for changes in the appearance of foods, especially in raw high-protein foods where protein denaturation takes place. Pressures above 300–400 MPa cause unfolding of the molecular structure of proteins and then aggregation and refolding to cause changes to the texture of the food. Gel formation is observed in some proteins, such as soya, gluten, meat, fish and egg albumin. Compared with heat-treated gels, pressure-induced gels have different rheological properties (they are smooth, glossy and soft with greater elasticity) and maintain their natural colour and flavour (Brooker 1999 and Chapter 1, section 1.1.2). The use of pressure, either alone or with different heat treatments, may be used to produce a range of novel textured products (Ledward 2000). For example, *β*-lactoglobulin refolds to a structure that has increased surface hydrophobicity, and hence increased surface activity and improved foaming properties. The effects of high pressures on protein structure are described in more detail by Hendrickx *et al.* (1998) and Heremans (1995), and the effects on protein functionality have been reviewed by Lopez-Fandino (2006), Palou *et al.* (1999) and Messens *et al.* (1997).

Starch molecules are opened and partially degraded by high pressures, to produce increased sweetness and susceptibility to amylase activity. Root vegetables, including potato and sweet potato, became softer, more pliable, sweeter and more transparent whereas the appearance, odour, texture and taste of soybeans and rice did not change after processing (Galazka and Ledward 1995). Starch may also be more sensitive to enzymic modifications after unfolding or gelatinisation during pressure treatment (Brooker 1999).

Meat is tenderised at lower pressures (e.g. 100 MPa at 35 °C for 4 min pre-rigor, or 150 MPa at 60 °C for 1 h post-rigor) by the effect of pressure on myofibrils, but not on connective tissues or collagen (Brooker 1999). Processing at 103 MPa and 40–60 °C for 2.5 min improves the eating quality of meat and reduces cooking losses. The extent of tenderisation depends on all three factors involved: pressure, temperature and holding time. These effects are reviewed by Johnston (1995). High hydrostatic pressures can cause structural changes in structurally fragile foods such as strawberries or lettuce. Cell deformation and cell membrane damage can result in softening and cell serum loss. Usually these changes are undesirable because the food will appear to be processed and no longer fresh or raw. Arroyo *et al.* (1997) processed fresh lettuce and tomatoes at 20 °C for 10 min and 10 °C for 20 min at pressures of up to 400 MPa. Pressures of 300 and 350 MPa reduced Gram-negative bacteria, yeasts and moulds by 1-log cycle but also caused skin to loosen and peel away in tomatoes, and caused lettuce to brown. Other effects on plant tissues are described by Knorr (1995b).

HPP can cause lipid oxidation and hydrolysis, especially in meat and other fatty foods to produce free fatty acids that adversely affect the quality of processed foods (Indrawati *et al.* 2003). The natural colour of fruits and vegetables is mostly unaffected by HPP, with no significant losses of chlorophyll and the colour of processed fruit in jams being retained. The red colour of meat changes when processed at pressures above 300 MPa due to destabilising of myoglobin, but the colour of white meats and cured meats is largely unaffected. The effects of HPP on the Maillard reaction depend on the pH of the food being treated and results of a number of studies are reviewed by Palou *et al.* (1999).

High pressures have little effect on the vitamin contents of foods (e.g. in pressure-processed jams, 95% of vitamin C is retained and 82% of ascorbic acid in peas is retained after treatment at 900 MPa for 5–10 min at 20 °C (Quaglia *et al.* 1996)). Similar high levels of retention for vitamin A, carotene, vitamins B, E and K, are reviewed by Indrawati *et al.* (2003) and Palou *et al.* (1999).

8.6   Combinations of high pressure and other minimal processing technologies

HPP is capable of being used in combination with other types of processing (e.g. antimicrobials (Chapter 6, section 6.7), irradiation (Chapter 7), chilling (Chapter 21) or modified atmospheres (Chapter 25, section 25.3)) to expand the unit operations available to food processors, leading to the development of new products and processes. The resistance of some microbial strains to high pressures and the baroprotective effects of some foods has led to research into combined processes using HPP with compressed or supercritical carbon dioxide (Chapter 5, section 5.4.3), biopolymers (e.g. chitosan), enzymes (e.g. lysozyme), bacteriocins (Chapter 6, sections 6.5–6.7), ultrasound, alternating currents or high-voltage electric field pulses (see Chapter 9, sections 9.1, 9.3 and 9.6, and also hurdle technologies – Chapter 1, section 1.3.1). These combined processes are (in 2008) at the research stage and are not yet used commercially. They have been reviewed by Raso and Barbosa-Canovas (2003), Ledward *et al.* (1995), Knorr (1993) and Balny *et al.* (1992).

Processing foods under low (<15 MPa) pressure with CO_2 is described by Haas *et al.* (1989). Hong *et al.* (1997) describe a CO_2 pressure process of 200 min at 30 °C and a CO_2 pressure of 6.9 MPa to reduce lactobacilli by 5-log cycles in kimchi (fermented Korean vegetables). Ballestra *et al.* (1996) used pressures of 1.2–5 MPa at 25–45 °C to inactivate *E. coli*. The higher temperatures reduced the processing time to 20 min at a pressure of 1.2 MPa. The effectiveness of processing at low pressures is due to the antimicrobial effect of CO_2 which may lower the intracellular pH in cells rather than causing rupture of the cell membranes (section 8.1.2). The relatively long processing times are necessary to allow diffusion of carbon dioxide into microbial cells and these times may limit commercial opportunities for this process.

A combination of HPP and irradiation processing has been used to eliminate *Clostridium sporogenes* in chicken breast (Crawford *et al.* 1996), and Paul *et al.* (1997) used 1.0 kGy of gamma irradiation combined with HPP at 200 MPa for 30 min to destroy 10^4 inoculated staphylococci/g in lamb meat. After storage for three weeks at 0–3 °C, <10^3 staphylococci CFU/ml were detectable.

The effects of combined heat, pressure (at kPa levels) and ultrasound on *Yersinia enterocolitica* were studied by Raso *et al.* (1998b,c). They recorded *D*-values of 1.39 min when heated at 59 °C without other treatment, 1.5 min when treated with ultrasound at 150 decibels (dB) and 20 kHz, and 0.28 min when treated at 300 kPa with 150 dB of ultrasound at 30 °C. A 12 min treatment at 500 kPa and ultrasound at 117 dB and 20 kHz killed 99% of *B. subtilis* spores. The sporicidal effect depended on the pressure, the amplitude of ultrasonic waves and the temperature. A combination of 300 kPa, 20 kHz and 117 dB for 6 min had a synergistic effect on spore inactivation in the range of 70–90 °C.

The combination of HPP with antimicrobial compounds has been studied. Examples include lysozyme and chitosans (Popper and Knorr 1990, Papineau *et al.* 1991) and bacteriocins (Carlez *et al.* 1992). For example, Roberts and Hoover (1996) found that pressures of up to 400 MPa did not reduce the number of viable *Bacillus coagulans* spores, but the combined use of nisin with pressure treatment at 400 MPa at 70 °C for 30 min with nisin at 0.8 IU/ml sterilised samples that contained 2.5×10^6 spore CFU/ml.

Kalchayanand *et al.* (1994) studied electroporation combined with HPP and bacteriocins against Gram-negative and Gram-positive bacteria, and Kalchayanand *et al.* (1998) used HPP (345 MPa at 50 °C for 5 min) in combination with 3,000 activity units ml^{-1} of the pediocin AcH. Of the Gram-negative bacteria in the study, *E. coli* O157:H7 was the most pressure resistant and the Gram-positive bacteria *Leuconostoc sake* and *L. mesenteroides* were the most barotolerant.

Combinations of HPP with other non-thermal processes are described by Raso and Barbosa-Canaovas (2003), Lopez-Caballero *et al.* (2000), Pushpa *et al.* (1997) and Crawford *et al.* (1996). The advantages and limitations of HPP, compared with other minimal processing techniques, are summarised in Chapter 9, Table 9.1.

References

ALEMAN, G.D., FARKAS, D.F., MCINTYRE, S., TORRES, J.A. and WILHELMSEN, E., (1994), Ultra-high pressure pasteurization of fresh cut pineapple, *J. Food Protection*, **57** (10), 931–934.

ALEMAN, G.D., TING, E.Y., MORDRE, S.C., HAWES, A.C.O., WALKER, M., FARKAS, D.F. and TORRES, J.A., (1996), Pulsed Ultra high pressure treatments for pasteurisation of pineapple juice, *J. Food Science*, **61** (2), 388–390.

ANON, (2000), Kinetics of microbial inactivation for alternative food processing technologies – high pressure processing, US Food and Drug Administration, Center for Food Safety and Applied Nutrition, 2 June, available at www.cfsan.fda.gov/~comm/ift-hpp.html.

ANON, (2007a), High pressure references, High Pressure Processing Laboratory, Virginia Tech., available at www.hhp.vt.edu/biblioa-c.html.

ANON, (2007b), New technologies: ultra high pressure processing using propellants (UHP^3), DEFRA Science and Research project UHP-AFM 223 (FT 1519), Dept. for Environment Food and Rural Affairs, UK Government.

ARROYO, G., SANZ, P.D. and PRESTAMO, G., (1997), Effect of high pressure on the reduction of microbial populations in vegetables, *J. Applied Microbiology*, **82**, 735–742.

BALASUBRAMANIAM, V.M., (2007), Safety considerations of high pressure processing of low-acid foods for food industry, project proposal available at www.siu.edu/~foodsafe/PDEHP.html.

BALLESTRA, P., DA SILVA, A.A. and CUQ, J.L., (1996), Inactivation of *Escherichia coli* by carbon dioxide under pressure. *J. Food Science*, **61** (4), 829–836.

BALNY, C., HAYASHI, R., HEREMANS, K. and MASSON, P., (1992), *High Pressure and Biotechnology*, John Libby & Co. Ltd, London.

BLACK, E.P., FOX, P.F., FITZGERALD, G.F. and KELLY, A.L., (2007), Freshly-squeezed milk: effects of high pressure processing on bacteria in milk, FoodInfo, May, *Food Processing*, **76** (5), 4–5, available at www.foodsciencecentral.com/fsc/ixid14744.

BRAUCH, G., HAENSLER, U. and LUDWIG, H., (1990), The effect of pressure on bacteriophages, *High Pressure Research*, **5**, 767–769.

BRIDGMAN, P.W., (1912), Water, in the liquid and five solid forms, under pressure, *Proceedings of the American Academy of Arts and Sciences*, **47**, 441–558.

BROOKER, B., (1999), Ultra-high pressure processing, *Food Technology International*, **59**, 61.

BUTZ, P., HABISON, G. and LUDWIG, H., (1992), Influence of high pressure on a lipid-coated virus, in (R. Hayashi, K. Heremans and P. Masson, Eds.), *High Pressure and Biotechnology*, John Libby & Co., London, pp. 61–64.

BUTZ, P., FUNTENBERGER, S., HABERDITZL, T. and TAUSCHER, B., (1996), High pressure inactivation of *Byssochlamys nivea* ascospores and other heat-resistant moulds, *Lebensmittel-Wissenshaft und Technologie*, **29**, 404–410.

BUTZ, P., NEEDS, E.C., BARON, A., BAYER, O., GEISEL, B., GUPTA, B., OLTERSDORF, U. and TAUSCHER, B., (2003), Consumer attitudes to high pressure food processing, *Food, Agriculture and Environment*, **1** (1), 30–34.

CANER, C., HERNANDEZ, R.J. and HARTE, B.R., (2004), High-pressure processing effects on the mechanical, barrier and mass transfer properties of food packaging flexible structures: a critical review, *Packaging Technology and Science*, **17** (1), 23–29.

CANO, M.P., HERNANDEZ, A. and DE ANCOS, B., (1997), High pressure and temperature effects on enzyme inactivation in strawberry and orange products, *J. Food Science*, **62**, 85.

CARLEZ, A., CHEFTEL, J.C., ROSEC, J.P., RICHARD, N., SALDANA, J.L. and BALNY, C., (1992), Effects of high pressure and bacteriostatic agents on the destruction of *Citrobacter freundii* in minced beef muscle, in (C. Balny, R. Hayashi, K. Heremans and P. Masson, Eds.), *High Pressure and Biotechnology*, John Libby & Co., London, pp. 365–368.

CHEFTEL, J.C., (1995), Review: High-pressure microbial inactivation and food preservation, *Food Science and Technology International*, **1** (2–3), 75–90, available at http://fst.sagepub.com/cgi/content/abstract/1/2-3/75.

CHEVALIER, D., LE BAIL, A. and CHOUROT, J.M., (1999), High pressure thawing of fish (whiting): influence of the process parameters on drip losses, *Lebensmittel-Wissenshaft und Technologie*, **32**, 25–31.

CRAWFORD, Y.J., MURANO, E.A., OLSON, D.G. and SHENOY, K., (1996), Use of high hydrostatic pressure and irradiation to eliminate *Clostridium sporogenes* spores in chicken breast, *J. Food Protection*, **59**, 711–715.

DONG, F.M., COOK, A.R. and HERWIG, R.P., (2003) High hydrostatic pressure treatment of finfish to inactivate *Anisakis simplex*, *J. Food Protection*, 66 (10), 1924–1926.

DOONA, C.J., DUNNE, C. P. and FEEHERRY, F.E., (Eds.), (2007), *High Pressure Processing of Foods*, Blackwell Publishing, Oxford.

EARNSHAW, R.G., (1992), High pressure technology and its potential use, in (A. Turner, Ed.), *Food Technology International Europe*, Sterling Publications International, London, pp. 85–88.

EARNSHAW, R.G., (1995), Kinetics of high pressure inactivation of micro-organisms, in (D.A. Ledward, D.E. Johnson, R.G. Earnshaw and A.P.M. Hasting, Eds.), *High Pressure Processing of Foods*, Nottingham University Press, Nottingham, pp. 37–46.

FARID, M., (2005), Numerical modeling of high pressure food processing using computational fluid dynamics. Paper presented at IFT Annual Meeting, 15–20 July, abstract available at http://ift.confex.com/ift/2005/techprogram/paper_31592.htm.

FUCHIGAMI, M., KATO, N. and TERAMOTO, A., (1998), High-pressure-freezing effects on textural quality of Chinese cabbage, *J. Food Science*, **63** (1), 122–125.

GALAZKA, V.B. and LEDWARD, D.A., (1995), Developments in high pressure food processing, in (A. Turner, Ed.), *Food Technology International Europe*, Sterling Publications International, London, pp. 123–125.

HAAS, G.J., PRESCOTT, H.E., DUDLEY, E., DIK, R., HINTLAN, C. and KEANE, L., (1989), Inactivation of microorganisms by carbon dioxide under pressure, *J. Food Safety*, **9**, 253–265.

HARA, A., NAGAHAMA, G., OHBAYASHI, A. and HAYASHI, R., (1990), Effects of high pressure on inactivation of enzymes and microorganisms in nonpasteurized rice wine (Namazake), *Nippon Nogeikagaku Kaishi*, **64** (5), 1025–1030.

HAYAKAWA, I., KANNO, T., TOMITA, M. and FUJIO, Y., (1994a), Application of high pressure for spore inactivation and protein denaturation, *J. Food Science*, **59** (1), 159–163.

HAYAKAWA, I., KANNO, T., YOSHIYAMA, K. and FUJIO, Y., (1994b), Oscillatory compared with continuous high pressure sterilization on *Bacillus stearothermophilus* spores, *J. Food Science*, **59** (1), 164–167.

HAYASHI, R., (1993), *High Pressure Bioscience and Food Science*, San-Ei Shuppan Co., Kyoto.

HAYASHI, R., (1995), Advances in high pressure processing in Japan, in (A.G. Gaonkar, Ed.), *Food Processing: Recent Developments*, Elsevier, London, p. 85.

HAYMAN, M.M., BAXTER, I., ORIORDAN, P.J. and STEWART, C.M., (2004), Effects of high-pressure processing on the safety, quality, and shelf-life of ready-to-eat meats, *J. Food Protection*, **67**, (8), 1709–1718.

HEINZ, V. and KNORR, D., (1998), High pressure germination and inactivation kinetics of bacterial spores, in (N.S. Isaacs Ed.), *High Pressure Food Science, Bioscience and Chemistry*, The Royal Society of Chemistry, Cambridge, pp. 436–441.

HEINZ, V. and KNORR, D., (2002), Effects of high pressure on spores, in (M.E.G. Hendrickx and D. Knorr, Eds.), *Ultra High Pressure Treatment of Foods*, Kluwer Academic/Plenum Publishers, New York, pp. 77–113.

HENDRICKX, M., LUDIKHUYZE, L., VAN DEN BROECK, I. and WEEMAES, C., (1998), Effects of high pressure on enzymes related to food quality, *Trends in Food Science and Technology*, **9**, 197–203.

HENDRICKX, M.E.G. and KNORR, D., (Eds.), (2002), *Ultra High Pressure Treatment of Foods*, Kluwer Academic/Plenum Publishers, New York.

HEREMANS, K., (1995), High pressure effects on biomolecules, in (D.A. Ledward, D.E. Johnston, R.G. Earnshaw and A.P.M. Hasting, Eds.), *High Pressure Processing of Foods*, Nottingham University Press, Nottingham, pp. 81–97.

HONG, S.I., PARK, W.S. and PYUN, Y.R., (1997), Inactivation of *Lactobacillus* sp. from kimchi by high pressure carbon dioxide, *Lebensmittel-Wissenshaft und Technologie*, **30**, 681–685.

HOOVER, D.G., METRICK, C., PAPINEAU, A.M., FARKAS, D.F. and KNORR, D., (1989), Biological effects of high hydrostatic pressure on food microorganisms, *Food Technology*, **43** (3), 99–107.

INDRAWATI, VAN LOEY, A., SMOUT, C. and HENDRICKX, M., (2003), High hydrostatic pressure technology in food preservation, in (P. Zeuthen and L. Bogh-Sorensen, Eds.), *Food Preservation Techniques*, Woodhead Publishing, Cambridge, pp. 428–448.

IRWE, S. and OLSSON, I., (1994), Reduction of pectinesterase activity in orange juice by high pressure treatment, in (R.P. Singh and F. Oliveira, Eds), *Minimal Processing of Foods and Process Optimisation: An Interface*, CRC Press, Boca Raton, FL, pp. 35–42.

ISAACS, N.S., (1998), *High Pressure Food Science, Bioscience and Chemistry*, Woodhead Publishing, Cambridge.

ISAACS, N.S. and CHILTON, P., (1995), Microbial inactivation mechanisms, in (D.A. Ledward, D.E. Johnson, R.G. Earnshaw and A.P.M. Hasting, Eds.), *High Pressure Processing of Foods*, Nottingham University Press, Nottingham, pp. 65–80.

JOHNSTON, D.E., (1995), High pressure effects on milk and meat, in (D.A. Ledward, D.E. Johnson, R.G. Earnshaw and A.P.M. Hasting, Eds.), *High Pressure Processing of Foods*, Nottingham University Press, Nottingham, pp. 99–122.

KALCHAYANAND, N., SIKES, T., DUNNE, C.P. and RAY, B., (1994), Hydrostatic pressure and electroporation have increase bactericidal efficiency in combination with bacteriocins, *Applied and Environmental Microbiology*, **60**, 4174–4177.

KALCHAYANAND, N., SIKES, A., DUNNE, C.P. and RAY, B., (1998), Interaction of hydrostatic pressure, time and temperature of pressurization and pediocin AcH on inactivation of foodborne bacteria, *J. Food Protection*, **61** (4), 425–431.

KNORR, D., (1993), Effects of high hydrostatic pressure processes on food safety and quality, *Food Technology*, **47** (6), 156.

KNORR, D., (1995a), Hydrostatic pressure treatment of food: microbiology, in (G.W. Gould, Ed.), *New Methods of Food Preservation*, Blackie Academic and Professional, pp. 159–175.

KNORR, D., (1995b), High pressure effects on plant derived foods, in (D.A. Ledward, D.E. Johnson, R.G. Earnshaw and A.P.M. Hasting, Eds.), *High Pressure Processing of Foods*, Nottingham University Press, Nottingham, pp. 123–136.

KNORR, D., SCHLUETER, O. and HEINZ V., (1998), Impact of high hydrostatic pressure on phase transitions of foods, *Food Technology*, **52** (9), 42–45.

LEDWARD, D.A., (2000), Fresher under pressure, *Food Processing*, November, 20, 23.

LEDWARD, D.A., JOHNSTON, D.E., EARNSHAW, R.G. and HASTING, A.P.M., (1995), *High Pressure Processing of Foods*, Nottingham University Press, Nottingham.

LOPEZ-CABALLERO, M.E., PEREZ-MATEOS, M., BORDERIAS, J.A. and MONTERO, P., (2000), Extension of the shelf life of prawns (*Penaeus japonicus*) by vacuum packaging and high pressure treatment, *J. Food Protection*, **63**, 1381–1388.

LOPEZ-FANDINO, R., (2006), Functional improvement of milk whey proteins induced by high hydrostatic pressure, *Critical Reviews in Food Science and Nutrition*, **46**, 351–363.

LUDIKHUYZE, L., VAN LOEY, A., INDRAWATI, I., SMOUT, C. and HENDRICKX, M., (2003), Effects of combined pressure and temperature on enzymes related to quality of fruits and vegetables: from kinetic information to process engineering aspects, *Critical Reviews in Food Science and Nutrition*, **43** (5), 527–586.

LUSCHER, C., BALASA, A., FROHLING, A., ANANTA, E. and KNORR, D., (2004), Effect of high-pressure-induced Ice I-to-Ice III phase transitions on inactivation of *Listeria innocua* in frozen suspension, *Applied and Environmental Microbiology*, **70** (7), 4021–4029, available at http://aem.asm.org/cgi/reprint/70/7/4021.pdf.

MACKEY, B.M., FORESTIERE, K. and ISAACS, N., (1995), Factors affecting the resistance of *Listeria monocytogenes* to high hydrostatic pressure, *Food Biotechnology*, **9**, 1–11.

MANVELL, C., (1996), Opportunities and problems of minimal processing and minimally processed foods, paper presented at EFFoST Conference on Minimal Processing of Foods, November.

MARQUIS, R.E., (1976), High pressure microbial physiology, *Advances Microbial Physiology*, **14** (1), 159.

MERMELSTEIN, N.H., (1997), High pressure processing reaches the US market. *Food Technology*, **51**, 95–96.

MERTENS, B., (1995), Hydrostatic pressure treatment of food: equipment and processing, in (G.W. Gould, Ed.), *New Methods of Food Preservation*, Blackie Academic and Professional, Glasgow, pp. 135–158.

MESSENS, W., VAN CAMP, J. and HUYGEBAERT, A., (1997), The use of high pressure to modify the functionality of proteins, *Trends in Food Science and Technology*, **8**, 107–112.

METRICK, C., HOOVER, D.G. and FARKAS, D.F., (1989), Effects of high hydrostatic pressure on heat-resistant and heat-sensitive strains of Salmonella, *J. Food Science*, **54**, 1547–1564.

MEYER, R.S., COOPER, K.L., KNORR, D. and LELIEVELD, H.L.M., (2000), High-pressure sterilization of foods, *Food Technology*, **54** (11), 67–68, 70, 72.

OKAZAKI, T., KAKUGAWA, K., YAMAUCHI, S., YONEDA, T. and SUZUKI, K., (1996), Combined effects of temperature and pressure on inactivation of heat-resistant bacteria, in (R. Hayashi and C. Balny, Eds.), *High Pressure Bioscience and Biotechnology*, Elsevier Science, Amsterdam, pp. 415–418.

OLSSON, S., (1995), Production equipment for commercial use, in (D.A. Ledward, D.E. Johnson, R.G. Earnshaw and A.P.M. Hasting, Eds.), *High Pressure Processing of Foods*, Nottingham University Press, Nottingham, pp. 167–180.

OTAKE, T., MORI, H., KAWAHATA, T., IZUMOTO, Y., NISHIMURA, H., OISHI, I., SHIGEHISA, T. and OHNO, H., (1997), Effects of high hydrostatic pressure treatment of HIV infectivity, in (K. Heremans, Ed.), *High Pressure Research in the Biosciences and Biotechnology*, Leuven University Press, Leuven, pp. 223–236.

OXEN, P. and KNORR, D., (1993), Baroprotective effects of high solute concentrations against inactivation of *Rhodotorula rubra*, *Lebensmittel-Wissenschaft und Technologie*, **26**, 220–223.

PALOU, E., LOPEZ-MALO, A., BARBOSA-CANOVAS, G.V. and SWANSON, B.G., (1999), High pressure treatment in food preservation, in (M.S. Rahman, Ed.), *Handbook of Food Preservation*, Marcel Dekker, New York, pp. 533–576.

PAPINEAU, A.M., HOOVER, D.G., KNORR, D. and FARKAS, D.F., (1991), Antimicrobial effect of water-soluble chitosans with high hydrostatic pressure, *Food Biotechnology*, **5**, 45–47.

PATTERSON, M.F., QUINN, M., SIMPSON, R. and GILMOUR, A., (1995), Effects of high pressure on vegetative pathogens, in (D.A. Ledward, D.E. Johnson, R.G. Earnshaw and A.P.M. Hasting, Eds.), *High Pressure Processing of Foods*, Nottingham University Press, Nottingham, pp. 47–64.

PAUL, P., CHAWALA, S.P., THOMAS, P. and KESAVAN, P.C., (1997), Effect of high hydrostatic pressure, gamma-irradiation and combined treatments on the microbiological quality of lamb meat during chilled storage, *J. Food Safety*, **16** (4), 263–271.

PENDROUS, R., (2007), Pressure rises on cold solution to food pasteurisation despite high cost barrier, *Food Manufacture*, 2 May, p. 18.

PERRIER-CORNET, J.M., MARECHAL, P.A. and GERVAIS, P., (1995), A new design intended to relate high pressure treatment to yeast cell mass transfer, *J. Biotechnology*, **41**, 49–58.

PERRIER-CORNET, J.M., TAPIN, S., GAETA, S. and GERVAIS. P., (2005), High-pressure inactivation of *Saccharomyces cerevisiae* and *Lactobacillus plantarum* at subzero temperatures, *J. Biotechnology*, 115 (4), 405–412, available at www.avure.com/science_lowtemp.htm.

POPPER, L. and KNORR, D., (1990), Applications of high-pressure homogenization for food preservation, *Food Technology*, **44**, 84–89.

PUSHPA, P., CHAWLA, S.P., THOMAS, P. and KESSAVAN, P.C., (1997), Effects of high hydrostatic pressure, gamma-irradiation and combination treatments on the microbiological quality of lamb meat during chilled storage, *J. Food Safety*, **16**, 263–271.

QUAGLIA, G.B., GRAVINA, R., PAPERI, R. and PAOLETTI, F., (1996), Effect of high pressure treatments on peroxidase activity, ascorbic acid content and texture in green peas, *Lebensmittel-Wissenschaft und Technologie*, **29**, 552–555.

RASO, J. and BARBOSA-CANOVAS, G.V., (2003), Nonthermal preservation of foods using combined processing techniques, *Critical Reviews in Food Science and Nutrition*, **43** (3), 265–285.

RASO, J., BARBOSA-CANOVAS, G.V. and SWANSON, B.G., (1998a), Initiation of germination and inactivation by high hydrostatic pressure of *Bacillus cereus* sporulated at different temperatures, *IFT Annual Meeting: Book of Abstracts*, Atlanta, GA, p. 154.

RASO, J., PALOP, A., PAGAN, R. and CONDON, S., (1998b), Inactivation of *Bacillus subtilis* spores by combining ultrasonic waves under pressure and mild heat treatment. *J. Applied Microbiology*, **85**, 849–854.

RASO, J., PAGAN, R., CONDON, S. and SALA, F.J., (1998c), Influence of temperature and pressure on the lethality of ultrasound, *Applied Environmental Microbiology*, **64** (2), 465–471.

ROBERTS, C.M. and HOOVER, D.G., (1996), Sensitivity of *Bacillus coagulans* spores to combinations of high hydrostatic pressure, heat acidity, and nisin, *J. Applied Bacteriology*, **81**, 363–368.

RUSSELL, N.J., EVANS, R.I., TER STEEG, P.F., HELLEMONS, J., VERHEUL, A. and ABEE, T., (1995), Membranes as a target for stress adaptation, *International J. Food Microbiology*, **28**, 255–261.

SAN MARTIN, M.F., BARBOSA-CANOVAS, G.V. and SWANSON, B.G., (2002), Food processing by high hydrostatic pressure, *Critical Reviews in Food Science and Nutrition*, **42** (6), 627–645.

SEYDERHELM, L. and KNORR, D., (1992), Reduction of *Bacillus stearothermophilus* spores by combined high pressure and temperature treatments. *ZFL, International J. Food Technology, Marketing, Packaging and Analysis*, **43** (4), 17.

SHIGEHISA, T., NAKAGAMI, H., OHNO, H., OKATE, T., MORI, H., KAWAHATA, T., MORIMOTO, M. and UEBA, N., (1996), Inactivation of HIV in bloodplasma by high hydrostatic pressure, in (R. Hayashi and C. Balny, Eds.), *High Pressure Bioscience and Biotechnology*, Elsevier Science, Amsterdam, pp. 273–278.

SMELT, J.P.P., (1998), Recent advances in the microbiology of high pressure processing, *Trends in Food Science and Technology*, **9**, 152–158.

SMELT, J.P.P.M., RIJKE, A.G.F. and HAYHURST, A., (1994), Possible mechanism of high pressure inactivation of microorganisms, *High Pressure Research*, **12**, 199–203.

SONOIKE, K., SETOYAMA, K. and KOBAYASHI, S., (1992), Effect of pressure and temperature on the death rates of *Lactobacillus casei* and *Escherichia coli*, in (C. Balny, R. Hayashi, K. Heremans and P. Masson, Eds.), *High Pressure and Biotechnology*, John Libby & Co., London, pp. 297–301.

STYLES, M.F., HOOVER, D.G. and FARKAS, D.F., (1991), Response of *Listeria monocytogenes* and *Vibrio parahaemolyticus* to high hydrostatic pressure, *J. Food Science*, **56** (5), 1404–1407.

SUZUKI, K. and TANIGUCHI, T., (1972), Effect of pressure on biopolymers and model systems, in (M.A. Sleigh and A.G. Macdonald, Eds.), *The Effect of Pressure on Living Organisms*, Academic Press, New York, pp. 103–124.

TAUSCHER, B., (1999), High pressure and chemical reactions: effects on nutrients and pigments, Emerging Food Science and Technology, Tempere, Finland, 22–24 November, p. 58.

TEWARI, G., JAYAS, D.S. and HOLLEY, R.A., (1999), High pressure processing of foods: an overview, *Sciences Aliments*, **19**, 619–661.

TORRECILLA, J.S., OTERO, L. and SANZ, P.D., (2004), A neural network approach for thermal/pressure food processing, *J. Food Engineering*, **62** (1), 89–95.

WUYTACK, E.Y., BOVEN, S. and MICHIELS, C.W., (1998), Comparative study of pressure-induced germination of *Bacillus subtilis* spores at low and high pressure, *Applied Environmental Microbiology*, **64**, 3220–3224.

YANO, Y., NAKAYAMA, A., KISHIHARA, S. and SAITO, H., (1998), Adaptive changes in membrane lipids of barophilic bacteria in response to changes in growth pressure, *Applied Environmental Microbiology*, **64**, 479–485.

ZHAO, Y., FLORES, R.A. and OLSON, D.G., (1998), High hydrostatic pressure effects on rapid thawing of frozen beef, *J. Food Science*, **63** (1), 272–275.

ZOOK, C.D., PARISH, M.E., BRADDOCK, R.J. and BALABAN, M.O., (1999), High pressure inactivation kinetics of *Saccharomyces cerevisiae* ascospores in orange and apple juice, *J. Food Science*, **64** (3), 533–535.

9

Minimal processing methods under development

Abstract: Minimal processing methods preserve foods without significant heating, while retaining their nutritional qualities and sensory characteristics. This chapter reviews five methods of minimal processing that have been developed more recently. It describes the theory and equipment used for pulsed electric field processing, processing using electric arc discharges, oscillating magnetic fields, pulsed light, UV light and X-rays, and processing using ultrasound. For each it describes the effects on micro-organisms, enzymes and food components.

Key words: pulsed electric field processing, electric arc discharges, oscillating magnetic fields, pulsed light, ultrasound.

Minimal (or 'non-thermal') processing methods are able to preserve foods without significant heating, while substantially retaining their nutritional qualities and sensory characteristics. Minimal processes that are used commercially include fermentation and enzyme technology (Chapter 6, sections 6.4 and 6.5), irradiation (Chapter 7), which has been approved in some countries, and high-pressure processing (Chapter 8). Other processes that are widely adopted to suppress microbial growth with minimal effects on food quality include chilling (Chapter 21), freezing (Chapter 22), controlled or modified atmospheres (Chapters 21, section 21.2.5, and 25, section 25.3) and active packaging systems (Chapter 25, section 25.4.3). These are described in detail by Ohlsson and Bengtsson (2002). This chapter describes research that has been undertaken on other methods of minimal processing that have yet to be widely used at a commercial scale. These include pulsed electric fields, electric arcs, oscillating magnetic fields, high-intensity light, X-rays and ultrasound. The comparative advantages and limitations of these technologies are summarised in Table 9.1, together with examples of potential products that may become commercially important. Barbosa-Canovas *et al.* (2005), Gould (2001), Singh and Yousef (2001) and Morris (2000) have reviewed developments in non-thermal technologies.

There is also increasing interest in developing combinations of existing and novel processing methods to achieve mild preservation. The principle underlying the use of combined techniques to inhibit microbial growth is known as the 'hurdle' concept

Table 9.1 Advantages and limitations of some novel methods of minimal processing

Process	Advantages	Limitations	Examples of commercial applications and products
Pulsed electric fields	• Kills vegetative cells • Colours, flavours and nutrients are preserved • No evidence of toxicity • Relatively short treatment time	• Less effective against enzymes and spores • Only suitable for liquids • Safety concerns in local processing environment • Energy efficiency not yet certain • May be problems with scaling up process	• For liquid foods • Pasteurisation of fruit juices, soups, liquid egg and milk • Decontamination of heat sensitive foods
Oscillating magnetic fields	• Kills vegetative cells • Colours, flavours and nutrients are preserved • Low energy input • Low cost equipment	• No effect on spores or enzymes • Antimicrobial effect is 'patchy' and some vegetative cell growth stimulated • Mode of action not well understood • Poor penetration in electrically conductive materials • Safety concerns in local processing environment • Regulatory issues to be resolved	• Uncertain at present, possibly similar to high-pressure applications
Pulsed light	• Medium cost • Very rapid process • Little or no changes to foods • Low energy input • Suitable for dry foods	• Only surface effects and difficult to use with complex surfaces • Not proven effective against spores • Possible adverse chemical effects • Possible resistance in some micro-organisms • Reliability of equipment to be established • Advantages over high-intensity UV light to be established	• Packaging materials • Baked products • Fresh fruit and vegetables • Meats, seafood and cheeses • Surfaces, water and air

Method	Advantages	Limitations	Applications
Ultrasound	• Effective against vegetative cells, spores and enzymes • Reduction of process times and temperatures • Little adaptation required of existing processing plant • Heat transfer increased • Possible modification of food structure and texture • Batch or continuous operation • Effect on enzyme activity	• Complex mode of action • Depth of penetration affected by solids and air in product • Possible damage by free radicals • Unwanted modification of food structure and texture • Needs to be used in combination with another process (e.g. heating) • Potential problems with scaling up plant	• Mostly in combination with other preservation methods
Photodynamic systems	• Low cost • No additives required • Natural light suitable to activate system • Can be incorporated into packaging or used as a factory process	• Good photosensitisers are not currently 'food-grade' • Oxidation of sensitive foods • Some bacterial resistance • Food constituents can act as quenching agents • Oxygen has limited mobility from immobilising agent • Regulatory issues to be resolved	• Active packaging • Decontamination in washing processes • Water treatment • Sanitation of factory environments by incorporating photosensitisers into paints and plastic surfaces
Gamma radiation	• Well established and reliable • Excellent penetration into foods • Suitable for sterilisation • Suitable for non-microbial applications (e.g. sprout inhibition) • Permitted in some countries • Little loss of food quality • Suitable for large-scale production • Low energy costs • Insecticidal • Suitable for dry foods	• High capital cost • Localised risks from radiation • Poor consumer understanding • 'Politics' of nuclear energy • Changes in flavour due to oxidation	• Fruit and vegetables • Herbs and spices • Packaging • Some meat and fish products

Table 9.1 Continued

Process	Advantages	Limitations	Examples of commercial applications and products
High pressure	• Kills vegetative bacteria (and spores at higher pressures) • No evidence of toxicity • Colours, flavours and nutrients are preserved • Reduced processing times • Uniform treatment throughout food • Desirable texture changes possible • In-package processing possible • Potential for reduction or elimination of chemical preservatives • Positive consumer appeal	• Less effect on enzyme activity, requiring refrigeration of products • Some microbial survival • Expensive equipment • Foods should have approx. 40% free water for antimicrobial effect	• Pasteurisation and sterilisation of fruit products, sauces, pickles, yoghurts and salad dressings • Pasteurisation of meats and vegetables • Decontamination of high risk or high-value heat-sensitive ingredients, including shellfish

Adapted from Earnshaw (1996) and Manvell (1996)

(described in Chapter 1, section 1.3.1) and includes the use of combinations of mild heat, reduced water activity, preservatives, modified atmospheres, redox potential and increased acidity.

In each of the minimal processing methods described in this chapter, processing inactivates micro-organisms, and in some cases enzymes, but there are no substantial increases in product temperature. The methods cause little damage to pigments, structural polymers, flavour compounds or vitamins, and the sensory characteristics and nutritional value of foods are not reduced to a significant extent. The resulting products have higher quality and consumer appeal in markets where the retention of natural taste and colour can command premium prices.

9.1   Pulsed electric field processing

Researchers in the 1960s demonstrated that high voltages (>18 kV) in microsecond pulses enhanced microbial destruction, caused by the electricity itself (in contrast to the use of electric fields to heat foods using ohmic and dielectric heating – Chapter 20). Non-thermal destruction of micro-organisms using electric fields was then demonstrated in model food systems, and this research was extended during the 1980s and 1990s to process a variety of liquid foods, including fruit juices, soft drinks, alcoholic beverages, soups, liquid egg and milk. One such process was termed the 'Elsteril' process for the electric sterilisation or pasteurisation of pumpable conductive foods. The same German company developed the 'Elcrack' process, using similar pulsed electric field (PEF) technology to disrupt vegetable and animal cells to assist in recovery of edible oils and fats. These developments are reviewed by Sitzmann (1995) and Vega-Mercado et al. (1999). PEF was approved for use in the USA in 1995 for antimicrobial treatment of liquid foods (Morris 2000), but problems associated with scale-up of equipment restricted its commercialisation. Developments in solid state switching systems overcame these problems and large-scale equipment can now be manufactured. A commercial-scale PEF plant was installed at Ohio State University in 2002 and the first commercial plant started operating in the US in 2005 (Anon 2007). Processing costs are broadly comparable to conventional thermal processing (Leadley 2003) but there is no scope for heat regeneration. The theory and applications of PEF are reviewed by Picart and Cheftel (2003), Barbosa-Canovas (2001) and Dunn (2001).

9.1.1   Theory

High electric field intensities are achieved through storing energy from a DC power supply in a bank of capacitors, which is then discharged to form high-voltage pulses. When a liquid food is placed between two electrodes and subjected to high electric field strengths (20–$80\,kV\,cm^{-1}$) in short pulses (1–$100\,\mu s$), there is a rapid and significant reduction in the number of vegetative micro-organisms in the food. Two mechanisms have been proposed by which micro-organisms are destroyed by electric fields: electrical breakdown of cells and electroporation (the formation of pores in cell membranes).

In the electrical breakdown mechanism, the microbial membrane can be considered as a capacitor filled with a dielectric. The charge separation across the membrane leads to a normal potential difference of around $10\,mV$, although this is proportional to the radius of the cell. An increase in the membrane potential as a result of an increase in the field strength due to PEF causes a reduction in the cell membrane thickness. Transmembrane pores filled with conductive solution form if the critical breakdown voltage (about 1 V) is reached, which leads to an immediate discharge and decomposition of the membrane. This breakdown is reversible if the pores are small in relation to the total membrane surface. Above the critical field strength and with longer exposure times, larger areas of the membrane break down and the increased size and number of pores cause irreversible mechanical destruction of the cell membrane (Glaser et al. 1988, Zimmermann 1986). The critical field strength for destruction of bacteria with a dimension of approximately 1 μm and critical voltage of 1 V across the cell membrane is in the order of $10\,\mathrm{kV\,cm^{-1}}$ for pulses of 10 μs to 1 ms duration (Anon 2000a).

Electroporation is caused when high electric field pulses temporarily destabilise the lipid bilayer and proteins of cell membranes (Castro et al. 1993). The main effect is to increase membrane permeability due to membrane compression and poration. Pores in the membrane cause the cell to swell and rupture, followed by leakage of cytoplasmic materials and cell death (Vega-Mercado et al. 1996).

Other effects of PEF include disruption to cellular organelles (especially ribosomes); formation of electrolysis products or highly reactive free radicals, produced from components of the food, depending on the type of electrode material used and the chemical composition of the food; induced oxidation and reduction reactions within the cell structure that disrupt metabolic processes; and a certain amount of heat produced by transformation of induced electrical energy. Further details are given by Barbosa-Canovas et al. (2005) and Palaniappan et al. (1990), and mathematical models for microbial inactivation are described by Esplugas et al. (2001), Anon (2000a) and Peleg (1995). Inactivation kinetics indicate approximately straight lines when log survivors are plotted against log treatment time or log number of pulses (Zhang et al. 1995).

The factors that affect microbial inactivation are:

- processing conditions (electric field intensity, pulse waveform and frequency (Hz) and duration, treatment time and temperature);
- type, concentration and growth stage of micro-organisms (section 9.1.3);
- properties of the food (pH, conductivity, ionic strength, presence of anti-microbial compounds) (Anon 2000a, Vega-Mercado et al. 1999).

The degree of inactivation of vegetative cells by PEF is greater at higher electric field intensities and/or with an increase in the number and duration of the pulses. The effect of PEF on enzymes and microbial spores is described in section 9.1.3.

Economically, it is preferable to use higher field strengths and shorter pulses. Electric field pulses may be monopolar or bipolar and the waveform may be sinusoidal (or oscillatory), square or rectangular, or exponentially decaying (Fig. 9.1). Oscillatory pulses are the least efficient for microbial inactivation, and square wave pulses are more energy efficient and more effective at inactivation of micro-organisms than exponentially decaying and bipolar pulses. Bipolar pulses cause greater damage to cell membranes by rapid reversals in the orientation of the electric field, which causes a corresponding change in the direction of charged molecules. This creates stress in cell membranes and enhances their electric breakdown, making bipolar pulses more deadly than monopolar pulses. Bipolar pulses also produce less deposition of solids on electrodes, and cause less electrolysis in foods, which may be organoleptically, nutritionally and toxicologically beneficial (Qin et al. 1994). Bipolar pulses have a relaxation time between pulses whereas instant charge reverse pulses do not. Charge reverse pulses may have a lower critical electric field strength required for electroporation because alternating stresses on cells cause structural fatigue and inactivation. The higher inactivation effectiveness of

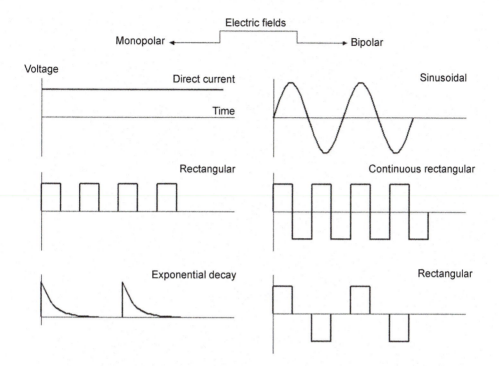

Fig. 9.1   Different types of electric field waveforms (from Ngadi *et al.* 2003) (reprinted from Vega-Mercado *et al.* 1997, © 1997, with permission from Elsevier).

this waveform can save up to 20% of total energy and equipment costs (Anon 2000a). Barbosa-Canovas *et al.* (1999) give further details of electric field pulses and waveforms.

The conductivity of most foods is $0.1–0.5\,\mathrm{S\,m^{-1}}$. Products (e.g. those with added salt) that have a higher ionic strength have a higher electrical conductivity. This reduces the resistance of the chamber and thus requires more energy to achieve a given electrical field strength. Thus, microbial inactivation decreases with increasing conductivity at an equivalent input energy. Also lower food conductivity increases the conductivity driving force between the food and the microbial cytoplasm and causes an increased flow of ions across the membrane, weakening its structure (Anon 2000a).

9.1.2   Equipment
Batch PEF equipment is available but continuous operation is preferable for commercial applications. Major components include a high-voltage (e.g. $40\,\mathrm{kV}/17\,\mathrm{MW}$) repetitive pulse generator, capacitors to store the charge, inductors to modify the shape and width of the electric field pulse, discharge switches to release the charge to electrodes, a fluid handling system for flow control of the product, and a treatment chamber in which the product is subjected to the electric field (Fig. 9.2). Treated food is then fed to an aseptic packaging line. Monitoring and control equipment includes fibre optic temperature sensors, voltage and current monitors and a data acquisition and control microprocessor (Chapter 27, section 27.2.2).

The two designs of treatment chamber used for continuous treatment are coaxial and co-field arrangements. In coaxial chambers (Fig. 9.3), the product flows between inner

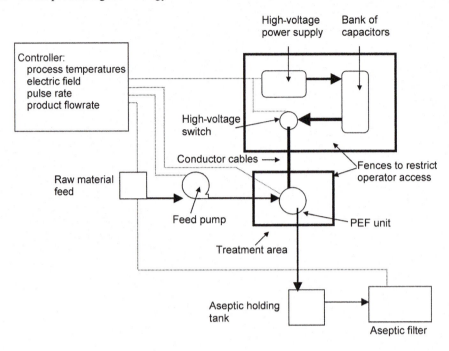

Fig. 9.2   Schematic diagram of PEF equipment (reprinted from Vega-Mercado *et al.* 1997, ©
1997, with permission from Elsevier).

and outer cylindrical electrodes, which are shaped to control the electrical field in the
treatment zone. Co-field chambers have two hollow cylindrical electrodes, separated by
an insulator, forming a tube that the product flows through. The advantages and
limitations of each design and details of equipment and applications are described by
Lelieveld *et al.* (2007).

    Although the process is intended to operate at ambient temperatures, PEF treatment
causes a rise in the product temperature (up to about 30 °C), which depends on the field

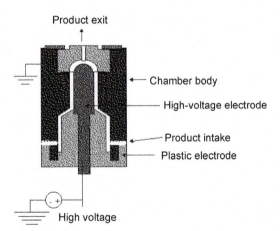

Fig. 9.3   Cross-sectional view of a coaxial PEF treatment chamber (from Anon 2000a).

strength, pulse frequency and number of pulses. Other factors that influence the temperature increase are described by van den Bosch *et al.* (2001) and Lindgren *et al.* (2002). To control the product temperature, equipment is either fitted with refrigeration coils or the food is pumped through heat exchangers before and after treatment. In a single chamber operation the food is recirculated for the required number of times, whereas multi-chamber operations have two or more chambers connected in series, with cooling systems between the chambers. However, the heat produced by PEF is lost in refrigeration systems and cannot be regenerated as it is in thermal pasteurisers (Chapter 12, section 12.2.2), which may affect the economics of commercial processing.

To protect operators from the high voltages, the entire apparatus is contained within a restricted-access area with interlocked gates, and all connections to the chamber including product pipework and refrigeration units, must be isolated and earthed to prevent leakage of energy (Vega-Mercado *et al.* 1999).

The main limitations of PEF processing are as follows:

- Restriction to liquid foods or those with small particles. Dielectric breakdown may occur at particle–liquid interfaces due to differences in their electric constants.
- Restriction to foods that can withstand high electric fields (homogeneous liquids that have low electrical conductivity – see Chapter 20, section 20.2.1). If salt is to be added to foods, it should be done after PEF processing.
- The presence of bubbles in a food causes non-uniform treatment and safety problems. If the electric field exceeds the dielectric strength of the bubbles, it causes discharges inside the bubbles that volatise the liquid and increase their volume. Sparking results if they become big enough to bridge the gap between the two electrodes. Air bubbles must therefore be removed by vacuum degassing.

9.1.3   Effects on micro-organisms, enzymes and food components

Vegetative microbial cells are more sensitive to PEF when in the logarithmic phase of growth than in the stationary phase because cells are undergoing division, during which the cell membranes are more susceptible to the applied electric field. Inactivation increases at higher temperatures, lower ionic strength and lower pH. Bacterial spores and yeast ascospores are considerably more resistant than cells (Sampedro *et al.* 2005). Spores are able to withstand very high-voltage gradients ($>30\,\mathrm{kV\,cm^{-1}}$) (Mertens and Knorr 1992). Gram-negative bacteria are more sensitive than Gram-positive bacteria and yeasts are more sensitive than bacteria due to their larger size.

There are numerous publications on inactivation of vegetative cells, mostly of pathogenic or spoilage micro-organisms, which have been reviewed by Anon (2000a). However, because the lethal effects reported in these studies depend on the specific apparatus and operating conditions used by different researchers, it is difficult to draw direct comparisons between their findings.

Studies in which foods were inoculated with target micro-organisms and processed by PEF have resulted in up to $9D$ reductions of vegetative cells (for *E. coli* in simulated milk using 2 and 3 μs pulses at $70\,\mathrm{kV\,cm^{-1}}$ (Qin *et al.* 1995)). PEF treatment of orange juice at $35\,\mathrm{kV\,cm^{-1}}$ for 59 μs caused 7-log reductions in total aerobic plate count and yeast and mold counts (Yeom *et al.* 2000). Spores of *Bacillus subtilis* required 50 pulses of $16\,\mathrm{kV\,cm^{-1}}$, totalling 12 500 μs to achieve a 4–5D reduction (Pothakamury *et al.* 1995). Other results are shown in Table 9.2 and a review of studies into the inactivation of *E. coli*, *Pseudomonas* sp., *Bacillus* sp., *Staphylococcus aureus*, *Lactobacillus* sp., *Salmonella* sp., *Listeria* sp. and other pathogens is reported by Sampedro *et al.* (2005).

Table 9.2   Inactivation of selected micro-organisms and enzymes by PEF

Micro-organism	Log reduction (D)	Process conditions				Media
		Field intensity (kV cm^{-1})	Temperature (°C)	No. of pulses	Duration of pulses (μs)	
Bacillus subtilis spores ATCC 9372	5.3	3.3 V μm^{-1}, 4.3 Hz, exponential decay	<5.5	30	2	Pea soup
Escherichia coli	3	28.6	42.8	23	100	Milk
Escherichia coli	3.5	5.0 V μm^{-1}, square wave	<30	48	2	Skim milk
Escherichia coli	6	25.8	37	100	4	Liquid egg
Listeria innocua	2.6	50 at 3.5 Hz, exponential decay	15–28	100	2	Raw skim milk (0.2% milkfat)
Listeria monocytogenes	3.0–4.0	30 at 1700 Hz, bipolar pulses	10–50	400	1.5	Pasteurised whole milk (3.5% milkfat), 2% milk (2% milkfat), skim milk (0.2%)
Pseudomonas fluorescens	2.7	50 at 4.0 Hz, exponential decay	15–28	30	2	Raw skim milk (0.2% milkfat)
Saccharomyces cerevisiae	4.0	1.2 V μm^{-1}, exponential decay	4–10	6	90	Apple juice
Salmonella dublin	3.0	15–40	10–50	–	12–127	Skim milk
Salmonella dublin	3	36.7	63	40	100	Milk
Yersinia enterocolitica	6.0–7.0	75	2–3	150–200	500–1300 ns	NaCl solution pH = 7.0
Natural microflora	3	33.6–35.7	42–65	35	1–100	Orange juice
Natural microflora	≈5	6.7	45–50	5	20	Orange juice
	% reduction in activity					
Alkaline phosphatase	65	18–22	22–49	70	0.7–0.8	Raw milk, 2% milk, non-fat milk
Lipase, glucose oxidase	70–85	13–87, instant charge reversal pulses	–	30	2	Buffer solutions
Amylase, peroxidase, phenol oxidase	30–40					

Adapted from Anon (2000a) and Vega-Mercado *et al.* (1997)

Details of studies on specific pathogens and spoilage micro-organisms are given by Fernandez-Molina *et al.* (2001) for *Listeria innocua*, Harrison *et al.* (2001) for *Saccharomyces cerevisiae*, Gongora-Nieto *et al.* (2001) for *Pseudomonas fluorescens* and Jin *et al.* (2001) report studies on the inactivation of *Bacillus subtilis* spores. Details of other studies are given by Picart and Cheftel (2003).

In general it is found that higher electric-field strengths and longer treatment times are more effective at inactivating micro-organisms. Moderate heating of foods (e.g. to 40 °C) significantly increases the lethal effect of PEF, even though the temperature increase itself has no lethal effect. Kalchayanand *et al.* (1994) found that PEF treatment of *E. coli, L. monocytogenes* and *S. typhimurium* made cells more sensitive to the bacteriocins nisin and pediocin (Chapter 6, section 6.7).

The effect of PEF on enzymes is reviewed by Yeom and Zhang (2001). In general, enzymes are more resistant then micro-organisms to PEF processing but different enzymes exhibit a wide range of inactivation. For example, PEF treatment at $35\,kV\,cm^{-1}$ for 59 μs caused about 90% inactivation of pectin methylesterase (Yeom *et al.* 2000). Ho *et al.* (1997) found that lipase, glucose oxidase and α-amylase lost 75–85% of their activity whereas alkaline phosphatase showed a 5% reduction. Inactivation was due to the electric field rather than any changes in temperature or pH and the authors suggest that the differences may be due to the effects of PEF on the secondary or tertiary structure of the enzyme. Yeom and Zhang (2001) summarise the factors that affect enzyme inactivation by PEF as:

- PEF parameters (electric field strength, number of pulses, pulse duration and width, total treatment time);
- enzyme structure (active site, secondary and tertiary structure);
- temperature;
- suspension medium for the enzyme.

There have been numerous studies of the effect of PEF on foods. For example, Qin *et al.* (1994) describes pasteurisation using PEF and Sitzmann (1995) reported a reduction in the ascorbic acid content in milk after PEF treatment (although milk is not a significant source of this vitamin). In general, vitamins are not inactivated to any appreciable extent. Sampedro *et al.* (2005) review studies of PEF-treated milk which showed no physico-chemical or sensory changes compared with untreated products. Studies of PEF-treated liquid whole egg by Qin *et al.* (1995) showed that compared with fresh eggs, PEF treatment reduced the viscosity but increased the colour (β-carotene concentration), but there was no difference between scrambled eggs prepared from the two types of egg. Sponge cakes prepared with PEF-treated egg had a lower volume but sensory evaluation revealed no differences between cakes made using PEF-treated and fresh egg. Studies reported by Vega-Mercado *et al.* (1997) indicate that PEF extended the shelf-life of fresh apple juice to more than 56 days at 22–25 °C with no apparent change in its physicochemical and sensory properties. Zhang *et al.* (1997) assessed the shelf-life of reconstituted orange juice treated with $32\,kV\,cm^{-1}$ pulses at near ambient temperatures. They found that after storage for 90 days at 4 or 22 °C, vitamin C losses were lower and colour was better preserved in PEF-treated juices compared with the heat-treated ones. Similar results were found by Min *et al.* (2003) who compared thermally processed orange juice (at 90 °C for 90 s) and PEF processing at $40\,kV\,cm^{-1}$ for 97 ms. PEF-processed juice retained more ascorbic acid, flavour and colour than thermally processed juice. Other studies of the effect of PEF on foods are reviewed by Anon (2000a).

9.1.4   Combinations of PEF and other treatments

Raso and Barbosa-Canovas (2003) have reviewed research into PEF combined with moderate heating, pH, antimicrobials (Chapter 6, section 6.7) and high hydrostatic pressure (Chapter 8). Results indicate that the lethality of PEF to micro-organisms increases at sub-lethal temperatures, the effects of pH are so far contradictory, and antimicrobials such as nisin and benzoic or sorbic acids have a synergistic effect with PEF to increase microbial inactivation. Sub-lethal HPP treatments (200 MPa for < 1 min) stabilised *B. subtilis* against simultaneous PEF treatment, but a synergistic effect was observed when cells were exposed to HPP (200 MPa for 10 min) followed by PEF treatment immediately before the pressure was released.

9.2   Processing using electric arc discharges

High-voltage arc discharge processing operates by applying rapid voltage discharges between electrodes located below the surface of liquid foods, causing no significant increase in temperature. The arc discharges (e.g. 40 kV for 50–300 μs, using a current of at least 1500 A) create both intense pressure waves and electrolysis, known as 'electro-hydraulic shock' (Anon 2000b). This causes irreversible damage to cell membranes and produces highly reactive oxygen radicals and other oxidising compounds from chemicals in foods, which inactivate micro-organisms and enzymes. Anon (2000b) and Palaniappan *et al.* (1990) have reviewed the effects of electrohydraulic shock on micro-organisms. The chemical action is complex and depends on the applied voltage, the electrode material, the type and initial concentration of micro-organisms and the distribution of chemical radicals.

The process is reported to use less energy than thermal pasteurisation and is more energy efficient than high-pressure processing and pulsed electrical fields for juice processing (Anon 1998). The main drawbacks of this technology that may hinder com-mercialisation are that electrolysis produces a range of unknown compounds that may contaminate the treated food, and shock waves cause disintegration of food particles (Barbosa-Canovas *et al.* 1999).

A different design of equipment may overcome these problems. Anon (1998) describes treatment of oxygenated products that are flowing through an arc plasma chamber. The submerged arc discharge takes place within the gas bubbles and the partial breakdown of gas causes ionisation to produce reactive ozone and UV radiation. However, the kinetics and mechanisms of microbial inactivation are not known and this technology requires validation to reach a stage where commercialisation is likely.

9.3   Processing using oscillating magnetic fields

Oscillating magnetic fields are generated by passing an alternating current through electromagnets. They have high field strengths (5–100 tesla compared with the earth's magnetic field of 25–70 μT) at frequencies of 5–500 kHz, and are applied for between 25 μs and a few milliseconds. They are able to inactivate vegetative cells, but frequencies higher than 500 kHz are less effective for microbial inactivation and also heat the food. From studies conducted to date, there is no effect on spores or enzymes and some types of vegetative cells may be stimulated to grow (Table 9.3). In other types, magnetic fields have no effect on growth. The effects of magnetic fields to inhibit or stimulate microbial growth are not well understood, but may be a result of the magnetic fields themselves or induced electric fields. Possible mechanisms include one or both types of field causing translocation of free radicals, an effect on membrane fluidity or disruption of cell membranes, or breakdown of covalent bonds in DNA molecules. Many cellular metabolic proteins contain ions and an alternative mechanism is that the field(s) could loosen the bonds between ions and proteins (Pothakamury *et al.* 1993). Microbial inactivation may depend on the magnetic field intensity, number and frequency of pulses, and selected

Table 9.3   Effect of oscillating magnetic fields on selected micro-organisms

Micro-organism	Food	Magnetic field strength (T)	Frequency of pulse (Hz)	Effect
Candida albicans	–	0.06	0.1–0.3	Growth simulated; stimulation increases with increase in frequency
Escherichia coli	–	0.15	0.05	Inactivation of cells at 100 cells ml^{-1}
Streptococcus themophilus	Milk	12.0	6000 (1 pulse)	Cells reduced from 25 000 to 970 cells ml^{-1}
Saccharomyces sp.	Yoghurt	40.0	416 000 (10 pulses)	Cells reduced from 3500 to 25 cells ml^{-1}
Saccharomyces sp.	Orange juice	40.0	416 000 (1 pulse)	Cells reduced from 25 000 to 6 cells ml^{-1}
Mould spores	–	7.5	8500 (1 pulse)	Spores reduced from 3000 to 1 spores ml^{-1}

Adapted from Anon (2000c), data from Moore (1979) and Hofmann (1985)

properties of the food (e.g. electrical conductivity and thickness). Foods should have a high electrical resistance (>10–25 Ω cm) to enable the use of larger magnetic field intensities required to achieve microbial inactivation (Anon 2000c).

The high field strengths are achieved using superconducting liquid helium cooled coils or coils that are energised by the discharge of a capacitor. The magnetic fields may be homogeneous or heterogeneous: a homogeneous magnetic field has a uniform field intensity, whereas in a heterogeneous field the field intensity varies, decreasing with the distance from the magnetic coil. Oscillating magnetic fields are applied as pulses, reversing the charge for each pulse (Anon 2000c). The intensity of each pulse varies periodically according to the frequency and type of wave in the magnet and decreases with time to about 10% of the initial intensity (Pothakamury *et al.* 1993). Systems are required to monitor and control the power source, number of pulses and frequencies applied to the food and temperature to ensure consistent treatments.

The process involves subjecting packaged food to 1–100 pulses for a total exposure time of 25–100 ms at 0–50 °C. The temperature of the food increases by about 2–5 °C during treatment. The process has little or no effect on the sensory and nutritional properties of foods and the advantages and limitations of this technology compared with other minimal processing methods are summarised in Table 9.1. There may be safety concerns over the use of powerful magnetic fields in a local processing environment and it is not yet clear whether the technology will be developed into a viable commercial process.

9.4   Processing using pulsed light and UV light

The use of pulsed white light to inactivate vegetative cells and spores on the surfaces of foods and packaging materials was developed by a US company during the 1980s and 1990s, but its commercial application to foods was discontinued in 2002 (Leadley 2003). The technology is used in Chile for commercial processing of fresh grapes exported to the USA (Anon 2000d). The advantages and limitations of pulsed light, compared with other methods of minimal processing, are described in Table 9.1.

The use of UV light to destroy micro-organisms is well documented, particularly in relation to water disinfection, but also as bactericidal lamps used to prevent surface mould growth on bakery products and for air purification. There is also interest in using UV light to reduce microbial contamination in fruit juices (Anon 2000e). Rahman (1999) has reviewed the commercial applications of UV light.

9.4.1 Theory
Pulsed light has a similar spectrum to sunlight, from ultraviolet wavelengths of 170 nm to infrared wavelengths of 2600 nm with peak emissions between 400 and 500 nm (Fig. 9.4). This light is in the 'non-ionising' part of the electromagnetic spectrum (see Fig. 20.1 in Chapter 20) and in contrast to irradiation (Chapter 7), it does not cause ionisation of molecules (Dunn *et al.* 1995).

Pulsed light is produced in short (1 μs to 0.1 s) pulses, typically 1–20 flashes per second that are approximately 20 000 times the intensity of sunlight at sea level. The energy imparted by the light to the surface of a food or packaging material is measured as 'fluence' and is quoted in units of J cm^{-2}. The broad spectrum of pulsed light inactivates micro-organisms by a combination of photochemical and photothermal effects. Inactivation occurs by a number of simultaneous mechanisms, including chemical modifications to proteins, membranes, other cellular material and nucleic acids. The UVC component of light within the 240–265 nm range has a greater photochemical effect than the longer wavelengths, and owing to the high energy and intensity of pulsed light, these UV wavelengths cause microbial inactivation. If most of the energy is in the visual spectrum, the effect is mostly photothermal: that is a large amount of energy is transferred rapidly to the surface of the food, raising the temperature of a thin surface layer sufficiently to destroy vegetative cells.

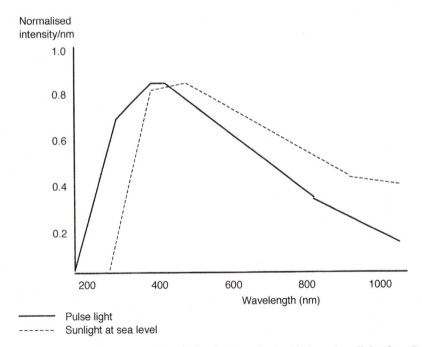

Fig. 9.4 A comparison of the wavelength distribution of pulsed light and sunlight (from Dunn *et al.* 1995).

The wavelengths for UV light are 100–400 nm, subdivided into UVA (315–400 nm), UVB (280–315 nm), UVC (200–280 nm) and the vacuum UV range (100–200 nm). The antimicrobial effects of light at UV wavelengths are due to absorption of the energy by highly conjugated double carbon bonds in proteins and nucleic acids. This causes cross-linking between pyrimidine nucleoside bases in DNA and results in irreversible changes that disrupt cellular metabolism, repair and reproduction (Sizer and Balasubramaniam 1999).

The short treatments with either pulsed light or UV light have little effect on the sensory or nutritional properties of foods. Both penetrate foods to a limited depth and are thus only suitable for surface treatments. In water UV light penetrates further and this allows successful UV water disinfection. Further details are given by Lupal (2007).

9.4.2   Equipment

To generate pulsed light, electricity at normal mains voltage is used to charge capacitors, which then release the power as a high-voltage, high-current, pulse of electricity to lamps filled with inert gas. The electricity passes through the gas in the lamp, causing it to emit an intense pulse of light. Rapid light pulses of tens per second are directed over the surface of the food or packaging material to be treated. Pulsed light systems have ammeters to measure the lamp current for each flash, which determines the light intensity and spectrum, and silicon photodiode detectors to measure the fluence in the UV wavelengths. If an abnormal lamp current is detected or if a pulse has a lower fluence than the minimum threshold, the system is programmed to automatically shut down to avoid under-processing.

Typically, foods are processed using 1–20 pulses that have energy densities of 0.1–50 J cm^{-2} at the food surface (Leadley 2003). The spectral distribution and fluence can be adjusted for different applications. For example, for packaging materials or transparent fluids, UV-rich light is used, having a high proportion of the energy at wavelengths <300 nm. For foods, shorter wavelengths are filtered out to reduce colour loss or lipid oxidation caused by UV, and inactivation is mostly photothermal. The process can be optimised for different foods by altering the number of lamps and frequency of flashes or using simultaneous or sequential flashes for a particular processing speed. As only a few pulses are needed for microbial destruction, this enables high product throughput rates to be achieved.

UV light is produced by low- or medium-pressure xenon lamps, by low-pressure mercury vapour lamps or by lasers. Ordinary glass is not transparent to UV wavelengths and UV lamps are therefore made of special quartz glass that allows 70–90% of UV rays to pass through. Mercury vapour lamps consist of a sealed glass tube that contains an inert gas carrier and a small amount of mercury. When an electric arc is created, it vaporises the mercury, which becomes ionised and gives off UV radiation.

UV treatment is confined to water purification because it is less effective if there is any turbidity in the treated liquid. In a typical operation, water enters the inlet of a UV treatment unit and flows through an annular space between a quartz sleeve that contains the lamp and the outside chamber wall. At a constant UV light intensity, the flowrate is the main controlling factor in microbial destruction (Mone 2007). In the 1990s, UV equipment comprising 24 ultraviolet lamps and reflectors was developed to pasteurise fruit juices. Because 90% of UV light is absorbed within 1 mm of the surface of juices, the equipment uses a transparent helical tube to create a 'continuously renewed surface'. Juice is pumped under turbulent flow, and the coiled tube promotes additional turbulence. This causes a secondary eddy flow effect that results in more uniform velocity and residence time. Details are given by Parisi et al. (2005).

9.4.3   Effect on micro-organisms, enzymes and food components

The lethality of pulsed light to micro-organisms increases with increasing fluence. Studies of pulsed light reported by Dunn *et al.* (1995), using inoculated agar plates, indicate that concentrations of 10^7 CFU g^{-1} of *Staphylococcus aureus* are destroyed by two pulses of 0.75 J cm^{-2}, giving a total fluence of 1.5 J cm^{-2}. Inoculated *Salmonella* sp. on chicken wings and *Listeria innocua* on hot dogs were both reduced by 2-log cycles decimal reductions of 7–9D for other pathogenic bacteria, including *Escherichia coli* O157:H7, *Listeria monocytogenes* and *Bacillus pumilus*, were achieved by using a few pulses at 1 J cm^{-2} per pulse. When water was treated with pulsed light, it was found that oocysts of *Klebsiella* sp. and *Cryptosporidium* sp., which are not affected by chlorination or traditional UV treatments, were reduced by 6–7 logs by either two pulses of 0.5 J cm^{-2} or a single pulse of 1 J cm^{-2}. Other studies are reviewed by Anon (2000d).

When applied to foods, the shelf-life of bread, cakes, pizza and bagels, packaged in clear film, was extended to 11 days at room temperature after treatment by pulsed light. Shrimps had an extension of shelf-life to 7 days under refrigeration and fresh meats had a 1–3 log reduction in counts of total bacterial, lactic acid bacteria, enteric bacteria and *Pseudomonas* sp. (Dunn *et al.* 1995). These lower levels of microbial destruction in foods, compared with those in water or on the smooth surfaces of agar plates or packaging materials, is attributed to the presence of surface fissures in foods that shield some of the micro-organisms from the light. Pulsed light may have applications for sterilising aseptic packaging materials (Chapter 13, section 13.2.3) to replace hydrogen peroxide (Barbosa-Canovas *et al.* 1997).

The UV fluence must be >0.4 J cm^{-2} on all parts of a product to inactivate micro-organisms. UV light causes a sigmoidal reduction curve in microbial numbers (Fig. 9.5) (Anon 2000e). An initial plateau is due to UV light causing injury to micro-organisms. It is followed by a rapid decline in the numbers of survivors because additional UV exposure is lethal after injury has taken place. Finally, tailing of the curve is due to UV resistance of the micro-organisms. Some types of bacteria have a repair mechanism, termed 'photo-reactivation', that is stimulated by visible light, and these photo-reactivated cells have a greater resistance to UV light (Table 9.4). Further details are

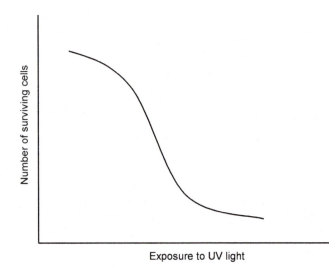

Exposure to UV light

Fig. 9.5   Microbial inactivation by UV light (using data from Anon 2000e).

Table 9.4 Exposure required to UV light at 254 nm to achieve 4-log reductions of micro-organisms

Micro-organism	Exposure required without photo-reactivation ($J\,m^{-2}$)	Exposure required with photo-reactivation ($J\,m^{-2}$)
Enterobacter cloacae	100	330
Enterocolitica faecium	170	200
Escherichia coli	50–110	180–280
Salmonella typhimurium	130	250
Vibrio cholerae	50	210
Yersinia enterocolitica	100	320

Adapted from Hoyer (1998)

given by Anon (2000e). In water treatment the factors that influence the rate of microbial inactivation include the power and wavelength of the UV light, the distance to the product, the transmissivity of the product and the type of product flow (turbulent or laminar – see Chapter 1, section 1.3.4).

9.5   Processing using pulsed X-rays

Linear induction electron acceleration generates ionising radiation (Chapter 7). X-rays that have a broad energy spectrum are produced by targeting the electron beam at a heavy metal converter plate, and then filtering the radiation to produce high-energy, highly penetrating radiation. Compared with electrons, which have a penetration depth of about 5 cm in food, X-rays have significantly higher penetration depths (60–400 cm) depending upon the energy used. The use of high-intensity pulsed X-rays is a new technology that followed the development of a switch that could open for a few nanoseconds and repeatedly deliver ultra-short pulses of power (i.e. up to 1000 pulses per second) in the gigawatt range, at voltages of hundreds of kV. This radiation treatment results in high local concentrations of free radicals and radical–radical recombination reactions (see Chapter 7, section 7.1), although there are thought to be differences in the free radicals formed by electron and X-ray processes. The differences in types of radical produced by X-rays may be the cause of fewer undesirable effects on food quality compared to irradiation.

Microbial inactivation by ionising radiation is described in Chapter 7 (section 7.5) and although the kinetics of microbial inactivation using pulsed X-rays are incompletely understood, it is likely that the mechanisms are similar. However, further research is required to validate the use of pulsed X-rays for food preservation (Anon 2000f).

9.6   Processing using ultrasound

Ultrasound waves have a frequency that is above 16 kHz and cannot be detected by the human ear. In nature, bats and dolphins use low-intensity ultrasound to locate prey, and some marine animals use high-intensity pulses of ultrasound to stun their prey. In food processing, a similar division is made between low-intensity ultrasound ($<1\,W\,cm^{-2}$), which is used as a non-destructive analytical method to assess the composition, structure or flow rate of foods, and high-intensity (or 'power') ultrasound (10–1000 $W\,cm^{-2}$). This is used at higher frequencies (20–100 kHz) to cause physical disruption of tissues, to cut foods (Chapter 3, section 3.1.2), to homogenise foods or create emulsions (Chapter 3, section 3.2.3), and to clean equipment (Chapter 27, section 27.1.1). The applications of high-intensity ultrasound to food processing are reviewed by Anon (2000g), Earnshaw (1998) and McClements (1995).

The antimicrobial effects of ultrasonic waves are well documented (e.g. Suslick 1988) but power ultrasound by itself is not used commercially to preserve foods. This is because the resistance of most micro-organisms and enzymes to ultrasound is so high that the length and intensity of treatment would produce adverse changes to the texture and other physical properties of foods and substantially reduce their sensory characteristics. However, power ultrasound is combined with mild heat treatments and/or pressure to reduce the amount of heating needed for microbial destruction or enzyme inactivation. These processes have not yet been commercialised (Gould 2001), but they may have applications to reduce thermal damage caused by sterilisation or pasteurisation of liquid foods. The potential advantages and limitations of ultrasound as a method of minimal processing are described in Table 9.1.

9.6.1　Theory

During ultrasonication, ultrasonic pressure waves hit the surface of a material and generate a force; if the force is perpendicular to the surface, it results in a compression wave that moves through the food, whereas if the force is parallel to the surface it produces a shearing wave. Both types of wave become attenuated as they move through the food. The depth of penetration and hence the antimicrobial effect depend on the frequency and intensity of the waves and the composition of the food. The very rapid localised changes in pressure cause shear disruption, 'cavitation' (creation of microscopic bubbles in liquid foods), thinning of cell membranes, localised heating and free radical production, each of which have a lethal effect on micro-organisms. Details of cavitation are given by Leighton (1998) and the physics of ultrasound are described by Suslick (1988). The shearing and compression effects of ultrasound cause denaturation of proteins that results in reduced enzyme activity, although short bursts of ultrasound may increase enzyme activity, possibly by breaking down large molecular structures and making the enzymes more accessible for reactions with substrates.

The shearing forces and rapidly changing pressures created by ultrasound waves are more effective in destroying microbial cells when combined with other treatments, including moderate heating, low a_w and low pH (Lillard 1994). There may be future applications for ultrasound to reduce the intensity of conventional heat treatments (e.g. thermosonication using ultrasound with moderate heat as a pasteurisation process) to improve the sensory characteristics and nutritional properties of foods produced by traditional heat processing. Ultrasound combined with slightly raised pressure (tens of megapascals) is termed 'manosonication', and 'manothermosonication' is a combined heat and ultrasound treatment under increased pressure (Sala et al. 1995). Changes to protein structures in enzymes may partly explain the synergistic effect of ultrasound and heat on enzyme inactivation (Sala et al. 1995). Raso and Barbosa-Canovas (2003) have reviewed these treatments.

The microbial death rate in manothermosonication is logarithmic and is thus the same as for heat treatments (Chapter 10, section 10.3). It is likely that ultrasound reduces the heat resistance of micro-organisms by physical damage to cell structures, caused by extreme pressure changes, and disruption of cellular protein molecules that makes them more sensitive to denaturation by heat. Details are given by Raso et al. (1998).

9.6.2 Processing

Sonication equipment consists of an ultrasound generator, typically a piezoelectric transducer (see Chapter 3, Fig. 3.14 in section 3.2.3), which transmits ultrasound to the food via a 'horn' that is submerged in the liquid. This can be placed in heating equipment for thermosonication or in pressurised vessels for manothermosonication. Detailed descriptions of ultrasound generators are given by Zheng and Sun (2004) and Earnshaw (1998). Rahman (1999) has reviewed research into the use of ultrasound to assist drying and diffusion ('acoustic drying'). In some foods (e.g. gelatine, yeast and orange powder), the rates of drying are increased by two to three times. This is thought to be due to both the creation of microscopic channels in solid foods by the oscillating compression waves and by changing the pressure gradient at the air–liquid interface, which increases the rate of evaporation. Acoustic drying has the potential to be an important operation because heat sensitive foods can be dried more rapidly and at a lower temperature than in conventional hot air driers (Chapter 16, section 16.1.1). Additionally, unlike high-velocity air-drying, fragile foods are not damaged by acoustic drying.

Other applications of ultrasound, described by Leadley and Williams (2006), include:

- enhanced rate of crystallisation of fats and sugars, reduced freezing time (Chapter 22);
- degassing fermentation liquors;
- breaking foams and emulsions, creating emulsions (Chapter 3, section 3.2);
- accelerated extraction of solutes;
- humidifying air and possible disinfectant fogging;
- cleaning equipment (Chapter 27, section 27.1.1);
- cutting (Chapter 3, section 3.1);
- enhanced rate of filtration.

9.6.3 Effect on micro-organisms and foods

The mechanism of cell destruction and effects of ultrasound on different micro-organisms have been reviewed by Rahman (1999) and Sala *et al.* (1995). Generally, large cells are more susceptible than small ones, rod-shaped bacteria are more sensitive than cocci, and Gram-positive bacteria are more sensitive than Gram-negative bacteria. Bacterial spores are highly resistant and difficult to disrupt. The use of heat with ultrasound inactivates vegetative micro-organisms, bacterial spores and enzymes (Lopez *et al.* 1994), but the effect is reduced as the temperature is increased.

In manosonication (sonication under pressure) at constant ultrasound amplitude, the D-value of *Yersinia enterolytica* was reduced 8-fold. At constant hydrostatic pressure, microbial inactivation depended on the amplitude of the ultrasound waves and, for example, the D-value of *Y. enterolytica* decreased 11-fold when the amplitude increased from 21 to 150 μm (Raso *et al.* 1998) with lethality having an exponential relationship to the amplitude of the ultrasound waves. Manothermosonication (sonication with heating under pressure) reduces the heat resistance of micro-organisms; the amount depends on the temperature, the micro-organism and its D-value. The lethality of manothermo-sonication is 6–30 times greater than a corresponding heat treatment at the same temperature. Its effectiveness depends on the intensity, amplitude and time of ultrasonication and the applied pressure. The process is also effective in inactivating enzymes, including polyphenoloxidase, lipoxigenase, lipase, protease and pectin methylesterase (Lopez *et al.* (1994)). For example a process that involved heating at 72 °C with ultrasound at 20 kHz and 117 μm under a pressure of 200 MPa increased the inactivation rate of pectin methylesterase by more than 400 times (Vercet *et al.* 1999).

Ultrasound has little effect on smaller molecules responsible for colour and flavour, or on vitamins. It may cause partial denaturation of proteins and changes to other macromolecules. For example, after prolonged exposure of meat to ultrasound, myofibrillar proteins are released to result in tenderisation of meat tissues and improved water binding capacity (McClements 1995).

References

ANON, (1998), Pulse power disinfects fresh juices, extends shelf-life, *Food Engineering*, **10**, 47–50.

ANON, (2000a), Kinetics of microbial inactivation for alternative food processing technologies – pulsed electric fields, US Food and Drug Administration, Center for Food Safety and Applied Nutrition, available at www.cfsan.fda.gov/~comm/ift-pef.html.

ANON, (2000b), Kinetics of microbial inactivation for alternative food processing technologies – high voltage arc discharge, US Center for Food Safety and Applied Nutrition, US Food and Drug Administration, June, available at www.cfsan.fda.gov/~comm/ift-arc.html.

ANON, (2000c), Kinetics of microbial inactivation for alternative food processing technologies – oscillating magnetic fields, US Center for Food Safety and Applied Nutrition, US Food and Drug Administration, June, available at www.cfsan.fda.gov/~comm/ift-omf.html.

ANON, (2000d), Kinetics of microbial inactivation for alternative food processing technologies – pulsed light technology, US Center for Food Safety and Applied Nutrition, US Food and Drug Administration, June, available at www.cfsan.fda.gov/~comm/ift-puls.html.

ANON, (2000e), Kinetics of microbial inactivation for alternative food processing technologies – ultraviolet light, US Center for Food Safety and Applied Nutrition, US Food and Drug Administration, June, available at www.cfsan.fda.gov/~comm/ift-uv.html.

ANON, (2000f), Kinetics of microbial inactivation for alternative food processing technologies – pulsed X-rays, US Center for Food Safety and Applied Nutrition, US Food and Drug Administration, available at www.cfsan.fda.gov/~comm/ift-xray.html.

ANON, (2000g), Kinetics of microbial inactivation for alternative food processing technologies – ultrasound, US Center for Food Safety and Applied Nutrition, US Food and Drug Administration, June, available at http://www.cfsan.fda.gov/~comm/ift-us.html.

ANON, (2007), 2007 IFT Industrial achievement award nomination, Diversified Technologies Inc. and Genesis Juice Corporation, available at http://members.ift.org/IFT/Awards and search past awards.

BARBOSA-CANOVAS, G.V., (Ed.), (2001), *Pulsed Electric Fields in Food Processing*, Woodhead Publishing, Cambridge.

BARBOSA-CANOVAS, G.V., PALOU, E., POTHAKAMURY, U.R. and SWANSON, B.G., (1997), Application of light pulses in the sterilization of foods and packaging materials, in (G.V. Barbosa-Canóvas, U.R. Pothakamury, E. Palou, E. Enrique and B.G. Swanson, Eds.), *Nonthermal Preservation of Foods*, Marcel Dekker, New York, pp. 139–161.

BARBOSA-CANOVAS, G.V., GONGORA-NIETO, M.M., POTHAKAMURY U.R. and SWANSON, B.G., (1999), *Preservation of Foods with Pulsed Electric Fields*, Academic Press, London pp. 1–9, 76–107, 108–155.

BARBOSA-CANOVAS, G.V., TAPIA, M.S. and CANO, M.P., (Eds.), (2005), *Novel Food Processing Technologies*, CRC Press, Boca Raton, FL.

CASTRO, A.J., BARBOSA-CANOVAS, G.V. and SWANSON, B.G., (1993), Microbial inactivation of foods by pulsed electric fields, *J. Food Processing and Preservation*, **17**, 47–73.

DUNN, J., (2001), Pulsed electric field processing – an overview, in (G.V. Barbosa-Canovas and Q.H. Zhang, Eds.), *Pulsed Electric Fields in Food Processing*, Technomic Publishing Co., Lancaster, PA, pp. 1–30.

DUNN, J., OTT, T. and CLARK, W., (1995), Pulsed light treatment of food and packaging, *Food Technology*, Sept., 95–98.

EARNSHAW, R.G., (1996), Non-thermal methods: Realisation of technology – high pressure, irradiation, ultrasonics, high intensity light and pulsed electric fields, Proc. EFFoST conferenceon minimal processing of foods, November, Cologne, Germany.

EARNSHAW, R.G., (1998), Ultrasound – a new opportunity for food preservation, in (M.J.W. Povey and T.J. Mason, Eds.), *Ultrasound in Food Processing*, Blackie Academic and Professional, London, pp. 183–192.

ESPLUGAS, S., PAGAN, R., BARBOSA-CANOVAS, G.V. and SWANSON, B.G., (2001), Engineering aspects of the continuous treatment of fluid foods by pulsed electric fields, in (G.V. Barbosa-Canovas and Q.H. Zhang, Eds.), *Pulsed Electric Fields in Food Processing*, Technomic Publishing Co., Lancaster, PA, pp. 31–44.

FERNANDEZ-MOLINA, J.J., BARKSTOM, E., TORSTENSSON, P., BARBOSA-CANOVAS, G.V. and SWANSON, B.G., (2001), Inactivation of *Listeria innocua* and *Pseudomonas fluorescens* in skim milk treated with pulsed electric fields, in (G.V. Barbosa-Canovas and Q.H. Zhang, Eds.), *Pulsed Electric Fields in Food Processing*, Technomic Publishing Co., Lancaster, PA, pp. 149–166.

GLASER, R.W., LEIKIN, S.L., CHERNOMORDIK, L.V., PASTUSHENKO, V.F. and SOKIRKO, A.V., (1988), Reversible electrical breakdown of lipid bilayers: formation and evolution of pores, *Biochimica et Biophysica Acta*, **940**, 275–281.

GONGORA-NIETO, M.M., SEIGNOUR, L., RIQUET, P., DAVIDSON, P.M., BARBOSA-CANOVAS, G.V. and SWANSON, B.G., (2001), Nonthermal inactivation of *Pseudomonas fluorescens* in liquid whole egg, in (G.V. Barbosa-Canovas and Q.H. Zhang, Eds.), *Pulsed Electric Fields in Food Processing*, Technomic Publishing Co., Lancaster, PA, pp.193–212.

GOULD, G.W., (2001), New processing technologies: an overview, Symposium on Nutritional effects of new processing technologies, London, 21 February, *Proceedings Nutrition Society*, **60** (4), 463–474.

HARRISON, S.L., BARBOSA-CANOVAS, G.V. and SWANSON, B.G., (2001), Pulsed electric field and high hydrostatic pressure induced leakage of cellular material from *Saccharomyces cerevisiae*, in (G.V. Barbosa-Canovas and Q.H. Zhang, Eds.), *Pulsed Electric Fields in Food Processing*, Technomic Publishing Co., Lancaster, PA, pp.183–192.

HO, S.Y., MITTAL, G.S. and CROSS, J.D., (1997), Effects of high field electric pulses on the activity of selected enzymes, *J. Food Engineering*, **31** (1), 69–84.

HOFMANN, G.A., (1985), Deactivation of micro-organisms by an oscillating magnetic field, US Patent 4,524,079.

HOYER, O., (1998), Testing performance and monitoring of UV systems for drinking water disinfection, *Water Supply*, **16** (1/2), 419–442.

JIN, Z.T., SU, Y., TUHELA, L., ZHANG, Q.H., SASTRY, S.K. and YOUSEF, A.E., (2001), Inactivation of *Bacillus subtilis* spores using high voltage pulsed electric fields, in (G.V. Barbosa-Canovas and Q.H. Zhang, Eds.), *Pulsed Electric Fields in Food Processing*, Technomic Publishing Co., Lancaster, PA, pp. 167–182.

KALCHAYANAND, N., SIKES, T., DUNNE, C.P. and RAY, B., (1994), Hydrostatic pressure and electroporation have increased bactericidal efficiency in combination with bacteriocins, *Applied and Environmental Microbiology*, **60**, 4174–4177.

LEADLEY, C., (2003), Developments in non-thermal processing, *Food Science and Technology*, **17** (3), 40–42.

LEADLEY, C.E. and WILLIAMS, A., (2006), Pulsed electric field processing, power ultrasound and other emerging technologies, in (J.G. Brennan, Ed.), *Food Processing Handbook*, Wiley-VCH, Weinheim, Germany, pp. 201–236.

LEIGHTON, T.G., (1998), The principle of cavitation, in (M.J.W. Povey and T.J. Mason, Eds.), *Ultrasound in Food Processing*, Blackie Academic and Professional, London, pp. 151–182.

LELIEVELD, H.L.M., NOTERMANS, S. and DE HAAN, S.W.H., (Eds.), (2007), *Food Preservation by Pulsed Electric Fields*, Woodhead Publishing, Cambridge.

LILLARD, H.S., (1994), Decontamination of poultry skin by sonication, *Food Technology*, **48**, 72–73.

LINDGREN, M., ARONSSON, K., GALT, S. and OHLSSON, T., (2002), Simulation of the temperature increase in pulsed electric field (PEF) continuous flow treatment chambers, *Innovative Food*

Science and Emerging Technologies, **3** (3), September, 233–245.

LOPEZ, P., SALA, F.J., FUENTE, J.L., CONDON, S., RASO, J. and BURGOS, J., (1994), Inactivation of peroxidase, lipoxygenase and polyphenoloxidase by manothermosonication, *J. Agricultural and Food Chemistry*, **42**, 552–556.

LUPAL, M., (2007), UV offers reliable disinfection, Canadian Water Quality Association, available at www.cwqa.com/industry_issues/uv.php.

MANVELL, C., (1996), Opportunities and problems of minimal processing and minimally processed foods, paper presented at EFFoST Conference on Minimal Processing of Foods, November. Cologne, Germany.

MCCLEMENTS, D.J., (1995), Advances in the application of ultrasound in food analysis and processing, *Trends in Food Science and Technology*, **6**, 293–299.

MERTENS, B. and KNORR, D., (1992), Development of nonthermal processes for food preservation, *Food Technology*, **46**, 124–133.

MIN, S. JIN, Z.T., MIN, S.K., YEOM, H. and ZHANG, Q.H., (2003), Commercial-scale pulsed electric field processing of orange juice, *J. Food Science*, **68** (4), 1265–1271.

MONE, J., (2007), Everything you need to know about ultraviolet water purification, available at www.harvesth2o.com/uv.shtml.

MOORE, R.L., (1979), Biological effects of magnetic fields, studies with micro-organisms, *Canadian J. Microbiology*, **25**, 1145–1151.

MORRIS, C.E., (2000), US developments in non-thermal juice processing, *Food Engineering and Ingredients*, **25** (6), 26–27, 30.

NGADI, M., BAZHAL, M. and RAGHAVAN, G.S.V., (2003), Engineering aspects of pulsed electroplasmolysis of vegetable tissues, Agricultural Engineering International, the *CIGR Journal of Scientific Research and Development*, Invited Overview Paper. Vol. V. February, available at http://cigr-journal.tamu.edu/submissions/volume5/Ngadi%20Invited%20Paper%203March2003.pdf.

OHLSSON, T. and BENGTSSON, N., (Eds.), (2002), *Minimal Processing Technologies in the Food Industry*, Woodhead Publishing, Cambridge.

PALANIAPPAN, S., SASTRY, S.K. and RICHTER, E.R., (1990), Effects of electricity on microorganisms: a review, *J. Food Processing Preservation*, **14**, 393–414.

PARISI, B., PATAZCA, E. and KOUTCHMA, T.N., (2005), Validation of UV coiled tube reactor for fresh fruit juices, IFT Annual Meeting, 15–20 July, New Orleans, Louisiana, USA, available at http://ift.confex.com/ift/2005/techprogram/paper_31010.htm.

PELEG, M., (1995), A model of microbial survival after exposure to pulse electric fields, *J. Science Food and Agriculture*, **67** (1), 93–99.

PICART, L. and CHEFTEL, J-C., (2003), Pulsed electric fields, in (P. Zeuthen and L. Bogh-Sorensen, Eds.), *Food Preservation Techniques*, Woodhead Publishing, Cambridge, pp. 360–427.

POTHAKAMURY, U.R., BARBOSA-CANOVAS, G.V. and SWANSON, B.G., (1993), Magnetic-field inactivation of microorganisms and generation of biological changes, *Food Technology*, **47** (12), 85–93.

POTHAKAMURY, U.R., MONSALVE-GONZALEZ, A., BARBOSA-CANOVAS, G.V. and SWANSON, B.G., (1995), High voltage pulsed electric field inactivation of *Bacillus subtilis* and *Lactobacillus delbrueckii*, *Revista española de ciencia y tecnología de alimentos*, **35**, 101–107, abstract available at http://cat.inist.fr/?aModele=afficheN&cpsidt=3616346.

QIN, B., ZHANG, Q., BARBOSA-CANOVAS, G.V., SWANSON, B.G. and PEDROW, P.D., (1994), Inactivation of microorganisms by pulsed electric fields with different voltage wave forms, *IEEE Transactions on Electrical Insulation*, **1**, 1047–1057.

QIN, B.L., POTHAKAMURY, U.R., VEGA-MERCADO, H., MARTIN-BELLOSO, O.M., BARBOSA-CANOVAS, G.V. and SWANSON, B.G., (1995), Food pasteurisation using high-intensity pulsed electric fields, *Food Technology*, **12**, 55–60.

RAHMAN, M.S., (1999), Light and sound in food preservation, in (M.S. Rahman, Ed.), *Handbook of Food Preservation*, Marcel Dekker, New York, pp. 669–686.

RASO, J. and BARBOSA-CANOVAS, G.V., (2003), Nonthermal preservation of foods using combined processing techniques, *Critical Reviews in Food Science and Nutrition*, **43** (3), 265–285.

RASO, J., PAGAN, R., CONDON, S. and SALA, F.J., (1998), Influence of temperature and pressure on the lethality of ultrasound, *Applied Environmental Microbiology*, **64**, 465–471.

SALA, F.J., BURGOS, J., CONDON, S., LOPEZ, P. and RASO, J., (1995), Effect of heat and ultrasound on micro-organisms and enzymes, in (G.W. Gould, Ed.), *New Methods of Food Preservation*, Blackie Academic and Professional, London, pp. 176–204.

SAMPEDRO, F., RODRIGO, M., MARINEZ, A. and RODRIGO, D., (2005), Quality and safety aspects of PEF application in milk and milk products, *Critical Reviews in Food Science and Nutrition*, **42**, 25–47.

SINGH, R.P. and YOUSEF, A.E., (2001), Technical Elements of New and Emerging Non-Thermal Food Technologies, FAO, available at www.fao.org/ag/ags/agsi/Nonthermal/nonthermal_1.htm#_Toc523623838.

SITZMANN, W., (1995), High voltage pulsed techniques for food preservation, in (G.W. Gould, Ed.), *New Methods of Food Preservation*, Blackie Academic and Professional, London, pp. 236–252.

SIZER, C.E. and BALASUBRAMANIAM, V.M., (1999), New intervention processes for minimally processed juices, *Food Technology*, **53** (10), 64–67.

SUSLICK, K.S., (1988), Homogenous sonochemistry, in (K.S. Suslick, Ed.), *Ultrasound. Its Chemical, Physical and Biological Effects*, VCH Publishers, New York, pp. 123–163.

VAN DEN BOSCH, H.F.M., MORSHUIS, P.H.F. and SMIT, J.J., (2001), Temperature distribution in fluids treated by pulsed electric fields (food preservation), Annual Report. Conference on Electrical Insulation and Dielectric Phenomena, 10/14/2001–10/17/2001, Kitchener, Ontario, Canada, pp. 552–555.

VEGA-MERCADO, H., POTHAKAMURY, U.R., CHANG, F.-J., BARBOSA-CANOVAS, G.V. and SWANSON, B.G., (1996), Inactivation of *Escherichia coli* by combining pH, ionic strength and pulsed electric fields hurdles, *Food Research International*, **29** (2), 117–121.

VEGA-MERCADO, H., MARTIN-BELLOSO, O., QIN B., CHANG, F.J., GONGORA-NIETO, M.M., BARBOSA-CANOVAS, G.V. and SWANSON, B.G., (1997), Non-thermal preservation: pulsed electric fields. *Trends in Food Science & Technology*, **8** (May), 151–157.

VEGA-MERCADO, H., GONGORA-NIETO, M.M., BARBOSA-CANOVAS, G.V. and SWANSON, B.G., (1999), Non-thermal preservation of liquid foods using pulsed electric fields, In (M.S. Rahman, Ed.), *Handbook of Food Preservation*, Marcel Dekker, New York, pp. 487–520.

VERCET, A., LOPEZ, P. and BURGOS, J., (1999), Inactivation of heat resistant pectinmethylesterase from orange by manothermosonication, *J. Agriculture Food Chemistry*, **47**, 432–437.

YEOM, H.W. and ZHANG, Q.H., (2001), Enzymic inactivation by pulsed electric fields: a review, in (G.V. Barbosa-Canovas and Q.H. Zhang, Eds.), *Pulsed Electric Fields in Food Processing*, Technomic Publishing Co., Lancaster, PA, pp. 57–64.

YEOM, H.W., STREAKER, C.B., ZHANG, Q.H. and MIN, D.B., (2000), Effects of pulsed electric fields on the activities of microorganisms and pectin methyl esterase in orange juice, *J. Food Science*, **65** (8), 1359–1363.

ZHANG, Q., QIN, B.L., BARBOSA-CANOVAS, G.V. and SWANSON, B.G., (1995), Inactivation of *E. coli* for food pasteurization by high strength pulsed electric fields, *J. Food Processing and Preservation*, **19**, 103–118.

ZHANG, Q.H., QIU, X. and SHARMA, S.K., (1997), Recent development in pulsed electric field processing, *New Technologies Yearbook*, National Food Processors Association, Washington, DC, pp. 31–42.

ZHENG, L. and SUN, D-W (2004), Enhancement of food freezing process by power ultrasound, Inst. Mechanical Engineers, Process Industries Division, available at www.imeche.org.uk/process/pdf/Da%20Wen%20Sun%20Full%20paper%202004.pdf.

ZIMMERMANN, U., (1986), Electrical breakdown, electropermeabilization and electrofusion, *Reviews of Physiology, Biochemistry and Pharmacology*, **105**, 175–256.

Part III

Processing by application of heat

Heat treatment is one of the most important methods used in food processing, not only because of the desirable effects on eating quality (many foods are consumed in a cooked form and processes such as baking produce flavours that cannot be created by other means), but also because of the preservative effect on foods by the destruction of enzymes, micro-organisms, insects and parasites. The other main advantages of heat processing are:

- relatively simple control of processing conditions;
- capability to produce shelf-stable foods that do not require refrigeration;
- destruction of anti-nutritional factors (e.g. trypsin inhibitor in some legumes);
- improvement in the availability of some nutrients (e.g. improved digestibility of proteins, gelatinisation of starches and release of bound niacin).

However, heat can also destroy components of foods that are responsible for their individual flavour, colour, taste or texture and as a result they are perceived to have a lower quality and lower value. Fortunately, the differences in D-values (Chapter 10, section 10.3) between these components and micro-organisms or enzymes can be exploited using higher temperatures for shorter times (HTST) in heat processing. HTST processing can be designed to produce the same level of microbial or enzyme destruction at lower temperatures for longer times, but the sensory characteristics and nutritional value of foods is substantially retained. Developments in blanching (Chapter 11), pasteurisation (Chapter 12), heat sterilisation (Chapter 13), evaporation (Chapter 14, section 14.1) and dehydration (Chapter 16) have each focused on improved technology and better control of processing conditions to achieve higher-quality products. Extrusion (Chapter 15) is by its nature an HTST process and other processes, including dielectric and ohmic heating (Chapter 20, section 20.2) are designed to cause minimal damage to the quality of foods.

Other more severe heat processes, including baking, roasting (Chapter 18) and frying (Chapter 19) are intended to change the sensory characteristics of a product, and preservation is achieved by either further processing (e.g. chilling or freezing, Chapters 21 and 22) or by selection of suitable packaging systems (Chapters 25). In hot smoking

(Chapter 17) a combination of heat, reduced moisture and the antimicrobial chemicals in the smoke preserves the smoked food.

Another important effect of heating is the selective removal of volatile components from a food. In evaporation and dehydration, the removal of water inhibits microbial growth and enzyme activity and thus achieves preservation. In distillation (Chapter 14, section 14.2) either alcohol is selectively removed to produce concentrated spirits, or flavour components are removed, recovered and added back to foods to improve their sensory characteristics.

More recent developments to reduce the severity of heating combine other processing methods that increase the sensitivity of vegetative microbial cells to milder heat. These 'minimal processing' methods were described in Part II and include high-pressure processing (Chapter 8) and thermomansonication (Chapter 9, section 9.6). The reduction in the amount of heating causes less damage to food components responsible for sensory quality and nutritional value and results in higher-quality products.

10

Heat processing

Abstract: This chapter describes the thermal properties of foods and mechanisms of heat transfer, and how these are used to calculate rates of heat transfer in food processing operations. It describes common sources of heat and outlines the operation of heating equipment and methods used to reduce energy consumption. Finally, the effects of heat on micro-organisms, enzymes and food components are outlined.

Key words: thermal conductivity, thermal diffusivity, heat transfer, conduction, convection, heat exchangers, properties of steam, energy saving, heat resistance of micro-organisms, D-value, z-value.

This chapter describes the thermal properties of foods and mechanisms of heat transfer, and how these are used to calculate rates of heat transfer under different conditions. It describes the sources of heat that are commonly used in food processing and outlines the methods of operation of heating equipment. Details of individual types of equipment are given in subsequent chapters (11–20), in which unit operations that involve heating are described. Finally, the effects of heat on micro-organisms, enzymes and food components are outlined, again with details of the effects of each unit operation provided in subsequent chapters. Further details of each of these aspects are given by Richardson (2001).

10.1 Theory

10.1.1 Thermal properties of foods

Three important thermal properties of foods are specific heat, thermal conductivity and thermal diffusivity. Specific heat is the amount of heat needed to raise the temperature of 1 kg of a material by 1 °C. It is found using Equation 10.1 and specific heat values for selected foods and other materials are given in Table 10.1.

$$c_p = \frac{Q}{m(\theta_1 - \theta_2)} \qquad \boxed{10.1}$$

where c_p (J kg^{-1} °C^{-1}) = specific heat of food at constant pressure, Q (J) = heat gained or lost, m (kg) = mass and $\theta_1 - \theta_2$ (°C) = temperature difference.

Table 10.1 Specific heat of selected foods and other materials

Material	Specific heat ($kJ\,kg^{-1}\,°C^{-1}$)	Temperature (°C)
Foods – solid		
Apples	3.59	Ambient
Apples	1.88	Frozen
Bacon	2.85	Ambient
Beef	3.44	Ambient
Bread	2.72	–
Butter	2.04	Ambient
Carrots	3.86	Ambient
Cod	3.76	Ambient
Cod	2.05	Frozen
Cottage cheese	3.21	Ambient
Cucumber	4.06	Ambient
Flour	1.80	–
Lamb	2.80	Ambient
Lamb	1.25	Frozen
Mango	3.77	Ambient
Milk – dry	1.52	Ambient
Milk – skim	3.93	Ambient
Potatoes	3.48	Ambient
Potatoes	1.80	Frozen
Sardines	3.00	Ambient
Shrimps	3.40	Ambient
Foods – liquid		
Acetic acid	2.20	20
Ethanol	2.30	20
Milk – whole	3.83	Ambient
Oil – maize	1.73	20
Oil – sunflower	1.93	20
Orange juice	3.89	Ambient
Water		
Water	4.18	15
Water vapour	2.09	100
Ice	2.04	0
Non-foods – solid		
Aluminium	0.89	20
Brick	0.84	20
Copper	0.38	20
Glass	0.84	20
Glass wool	0.7	20
Iron	0.45	20
Stainless steel	0.46	20
Stone	0.71–0.90	20
Tin	0.23	20
Wood	2.4–2.8	20
Non-foods – gases		
Air	1.005	Ambient
Carbon dioxide	0.80	0
Oxygen	0.92	20
Nitrogen	1.05	0

Adapted from Anon (2005c, 2007a), Singh and Heldman (2001a) and Polley *et al.* (1980)

The specific heat of compressible gases is usually quoted at constant pressure, but in some applications where the pressure changes (e.g. vacuum evaporation (Chapter 14, section 14.1) or high-pressure processing (Chapter 8) it is quoted at constant volume (C_v). The specific heat of foods depends on their composition, especially the moisture content (Equation 10.2). Equation 10.3 is used to estimate specific heat and takes account of the mass fraction of the solids contained in the food:

$$c_p = 0.837 + 3.348M \qquad \boxed{10.2}$$

where M = moisture content (wet-weight basis, expressed as a fraction not percentage),

$$c_p = 4.180X_w + 1.711X_p + 1.928X_f + 1.547X_c + 0.908X_a \qquad \boxed{10.3}$$

where X = mass fraction and subscripts w = water, p = protein, f = fat, c = carbohydrate and a = ash.

Thermal conductivity is a measure of how well a material conducts heat. It is the amount of heat that is conducted through unit thickness of a material per second at a constant temperature difference across the material and is found using Equation 10.4.

$$k = \frac{Q}{t\theta} \qquad \boxed{10.4}$$

where k ($J\,s^{-1}\,m^{-1}\,{}^\circ C^{-1}$ or $W\,m^{-1}\,{}^\circ C^{-1}$) = thermal conductivity and t (s) = time.

Thermal conductivity is influenced by a number of factors concerned with the nature of the food (e.g. cell structure, the amount of air trapped between cells, moisture content), and the temperature and pressure of the surroundings. A formula to predict thermal conductivity based on the composition of foods is shown in Equation 10.5:

$$k = k_w X_w + k_s(1 - X_w) \qquad \boxed{10.5}$$

where k_w ($W\,m^{-1}\,{}^\circ C^{-1}$) = thermal conductivity of water, X_w = mass fraction of water, k_s = ($W\,m^{-1}\,{}^\circ C^{-1}$) = thermal conductivity of solids (assumed to be $0.259\,W\,m^{-1}\,{}^\circ C^{-1}$).

A reduction in moisture content causes a substantial reduction in thermal conductivity. This has important implications in unit operations which involve conduction of heat through food to remove water (e.g. drying (Chapter 16), frying (Chapter 19) and freeze drying (Chapter 23)). In freeze drying the reduction in atmospheric pressure also influences the thermal conductivity of the food.

Ice has a higher thermal conductivity than water and this is important in determining the rate of freezing and thawing (Chapter 22). The importance of thermal conductivity is shown in sample problem 10.1 and sample problem 11.1 (Chapter 11). The thermal conductivities of some materials found in food processing are shown in Table 10.2.

Although, for example, stainless steel conducts heat ten times less well than aluminium (Table 10.2), the difference is small compared with the low thermal conductivity of foods (20 to 30 times lower than steel) and does not limit the rate of heat transfer. Stainless steel is much less reactive than other metals, and is therefore used in most food processing equipment that comes into contact with foods.

Thermal diffusivity is a measure of a material's ability to conduct heat relative to its ability to store heat. It is a ratio involving thermal conductivity, density and specific heat, and is found using Equation 10.6:

$$\alpha = \frac{k}{\rho c_p} \qquad \boxed{10.6}$$

Table 10.2 Thermal conductivity of selected foods and other materials

Material	Thermal conductivity ($W m^{-1} {}^{\circ}C^{-1}$)	Temperature (°C)
Food		
Acetic acid	0.17	20
Apple juice	0.56	20
Avocado	0.43	28
Beef, frozen	1.30	−10
Bread	0.16	25
Carrot	0.56	40
Cauliflower, frozen	0.80	−8
Cod, frozen	1.66	−10
Egg, frozen liquid	0.96	−8
Ethanol	0.18	20
Freeze dried foods	0.01–0.04	0
Green beans, frozen	0.80	−12
Ice	2.25	0
Milk, whole	0.56	20
Oil, olive	0.17	20
Orange	0.41	15
Parsnip	0.39	40
Peach	0.58	28
Pear	0.59	28
Pork	0.48	3.8
Potato	0.55	40
Strawberry	0.46	28
Turnip	0.48	40
Water	0.57	20
Gases		
Air	0.024	0
Air	0.031	100
Carbon dioxide	0.015	0
Nitrogen	0.024	0
Packaging materials		
Cardboard	0.07	20
Glass	0.52	20
Polyethylene	0.55	20
Poly(vinylchloride)	0.29	20
Metals		
Aluminium	220	0
Copper	388	0
Stainless steel	17–21	20
Other materials		
Brick	0.69	20
Concrete	0.87	20
Insulation	0.026–0.052	30
Polystyrene foam	0.036	0
Polyurethane foam	0.026	0

Adapted from Anon (2007a,b), Choi and Okos (2003), Singh and Heldman (2001a) and Lewis (1990)

where α ($m^2 s^{-1}$) = thermal diffusivity and ρ ($kg m^{-3}$) = density. Thermal diffusivity is used to calculate time–temperature distribution in materials undergoing heating or cooling and selected examples are given in Table 10.3.

The thermal diffusivity of foods is influenced by their composition, especially their moisture content, and it can be estimated using Equation 10.7:

Table 10.3   Thermal diffusivity of selected foods

Food	Thermal diffusivity $(\times 10^{-7}\,\mathrm{m^2\,s^{-1}})$	Temperature (°C)
Apples	1.37	0–30
Avocado	1.24	41
Banana	1.18	5
Beef	1.33	40
Cod	1.22	5
Ham, smoked	1.18	5
Lemon	1.07	0
Peach	1.39	4
Potato	1.70	25
Strawberry	1.27	5
Sweet potato	1.06	35
Tomato	1.48	4
Water	1.48	30
Water	1.60	65
Ice	11.82	0

Adapted from Singh and Heldman (2001a) and Murakami (2003)

$$\alpha = 0.146 \times 10^{-6}X_{\mathrm{w}} + 0.100 \times 10^{-6}X_{\mathrm{f}} + 0.075 \times 10^{-6}X_{\mathrm{p}} + 0.082 \times 10^{-6}X_{\mathrm{c}} \quad \boxed{10.7}$$

where X = mass fraction and subscripts w = water, f = fat, p = protein and c = carbohydrate. For example, every 1% increase in the moisture content of vegetables corresponds to a 1–3% increase in their thermal diffusivity (Murakami 2003). Changes in the volume fraction of air can also significantly alter the thermal diffusivity of foods. During heating, the temperature does not have a substantial effect on thermal diffusivity, but in freezing the temperature is important because of the different thermal diffusivities of ice and water.

'Sensible' heat is the heat needed to raise the temperature of a food and is found using Equation 10.4, rearranged from Equation 10.1:

$$Q = m \times c_{\mathrm{p}}(\theta_1 - \theta_2) \quad \boxed{10.8}$$

where Q (J) = sensible heat, m (kg) = mass, c_{p} ($\mathrm{J\,kg^{-1}\,°C^{-1}}$ or $\mathrm{K^{-1}}$) = specific heat of food at constant pressure and θ (°C) = temperature with subscripts 1 and 2 being initial and final values.

Phase changes in water are important in many types of food processing including steam generation for process heating (section 10.2), evaporation by boiling (Chapter 14, section 14.1), loss of water during dehydration, baking and frying (Chapters 16, 18, 19) and in freezing (Chapter 22). 'Latent' heat is the heat used to change phase (e.g. latent heat of fusion to form ice, or latent heat of vaporisation to change water to vapour) where the temperature remains constant while the phase change takes place. A phase diagram (Fig. 23.2 in Chapter 23) shows how temperature and pressure control the state of water (solid, liquid or vapour).

Vapour pressure is a measure of the rate at which water molecules escape as a gas from the liquid. Boiling occurs when the vapour pressure of the water is equal to the external pressure on the water surface (boiling point $=100\,°\mathrm{C}$ at atmospheric pressure at sea level). At reduced pressures below atmospheric, water boils at lower temperatures as shown in Chapter 14 (Fig. 14.1).

The changes in phase can be represented on a pressure–enthalpy diagram (Fig. 10.1) where the bell-shaped curve shows the pressure, temperature and enthalpy relationships

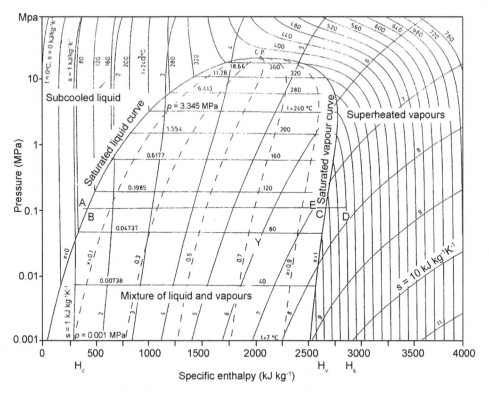

Fig. 10.1   Pressure–enthalpy diagram for water: H_c = enthalpy of condensate; H_v = enthalpy of saturated vapour; H_s = enthalpy of superheated steam (from Straub and Scheibner 1984, with kind permission of Springer Science and Business Media).

of water in its different states. Left of the curve is liquid water, becoming subcooled the further to the left, and right of the curve is vapour, becoming superheated the further to the right. Inside the curve is a mixture of liquid and vapour. At atmospheric pressure, the addition of sensible heat to liquid water increases its heat content (enthalpy) until it reaches the saturated liquid curve (A–B in Fig 10.1). The water at A is at 80 °C and has an enthalpy of 335 kJ kg^{-1} and when heated to 100 °C the enthalpy increases to 418 kJ kg^{-1}. Further addition of heat as latent heat causes a phase change. Moving further across the line (B–C) indicates more water changing to vapour, until at point C all the water is in vapour form. This is then saturated steam that has an enthalpy of 2675 kJ kg^{-1} (i.e. the latent heat of vaporisation of water is 2257 (2675 − 418) kJ kg^{-1} at atmospheric pressure while the temperature remains constant at 100 °C). Within the curve along B–C, the changing proportions of water and vapour are described by the 'steam quality'. For example at point E, the steam quality is 0.9, meaning that 90% is vapour and 10% is water. The specific volume of steam with a quality <100% can be found using Equation 10.9. Further heating (C–D) produces superheated steam. At point D it is at 250 °C and has an enthalpy of 2800 kJ kg^{-1}.

$$V_s = (1 - x_s)V_l + x_s V_v \qquad \boxed{10.9}$$

where V_s (m^3 kg^{-1}) = specific volume of steam, x_s (%) = steam quality, V_l (m^3 kg^{-1}) = specific volume of liquid and V_v (m^3 kg^{-1}) = specific volume of vapour. The data summarised in Fig. 10.1 is also available as steam tables (Keenan *et al.* 1969), and

Table 10.4   Properties of saturated steam

Temperature (°C)	Vapour pressure (kPa)	Latent heat (kJ kg^{-1})	Enthalpy (kJ kg^{-1})		Specific volume (m^3 kg^{-1})	
			Liquid	Saturated vapour	Liquid	Saturated vapour
30	4.246	2431	125.79	2556.3	0.001 004	32.89
40	7.384	2407	167.57	2574.3	0.001 008	19.52
50	12.349	2383	209.33	2592.1	0.001 012	12.03
60	19.940	2359	251.13	2609.6	0.001 017	7.67
70	31.19	2334	292.98	2626.8	0.001 023	5.04
80	47.39	2309	334.91	2643.7	0.001 029	3.41
90	70.14	2283	376.92	2660.1	0.001 036	2.36
100	101.35	2257	419.04	2676.1	0.001 043	1.67
110	143.27	2230	461.30	2691.5	0.001 052	1.21
120	198.53	2203	503.71	2706.3	0.001 060	0.89
130	270.1	2174	546.31	2720.5	0.001 070	0.67
140	316.3	2145	589.13	2733.9	0.001 080	0.51
150	475.8	2114	632.20	2746.5	0.001 091	0.39
160	617.8	2083	675.55	2758.1	0.001 102	0.31
170	791.7	2046	719.21	2768.7	0.001 114	0.24
180	1002.1	2015	763.22	2778.2	0.001 127	0.19
190	1254.4	1972	807.62	2786.4	0.001 141	0.15
200	1553.8	1941	852.45	2793.2	0.001 156	0.13
250	3973.0	1716	1085.36	2801.5	0.001 251	0.05
300	8581.0	1405	1344.0	2749.0	0.001 044	0.02

Adapted from Singh and Heldman (2001b) original data from Keenan, J.H., Keyes, F.G., Hill, P.G. and Moore, J.G., (1969), Steam tables metric units, Wiley, New York, copyright John Wiley & Sons

selected values are shown in Table 10.4 ('steam' is another term for hot water vapour).

When a phase change from water to vapour occurs, there is a substantial increase in the volume of vapour. In some unit operations, such as dehydration, this is not important, but in freeze drying (Chapter 23, section 23.1) and evaporation (Chapter 14, section 14.1) the removal of large volumes of vapour requires special equipment designs.

In steam production using boilers, the vapour produced by the phase change is contained within the fixed volume of the boiler vessel and there is therefore an increase in vapour (or steam) pressure. Higher pressures result in higher-temperature steam (moving further right of the curve in the superheated vapour section of Fig. 10.1). The required pressure and temperature of process steam are controlled by the rate of heating in the boiler (see also section 10.2).

10.1.2   Heat transfer
Energy balances
The first law of thermodynamics states that 'energy can be neither created nor destroyed but can be transformed from one form to another'. This can be expressed as an energy balance (Equation 10.10):

Total amount of heat or mechanical energy entering a process = total energy leaving with the products and wastes + stored energy + energy lost to the surroundings $\boxed{10.10}$

If heat losses are minimised, energy losses to the surroundings may be ignored for approximate solutions to calculation of, for example, the quantity of steam, hot air or refrigerant required. For more accurate solutions, compensation should be made for heat losses. An example of the use of an energy balance is given in sample problem 21.1.

Types of heat transfer
Many unit operations in food processing involve the transfer of heat into or out of a food. There are three ways in which heat may be transferred: by conduction, by convection or by radiation. In the majority of applications more than one type of heat transfer occur simultaneously but one type may be more important than others in particular applications. Further details are given by Hayhurst (1997). Radiation heat transfer is by emission and absorption of electromagnetic waves as, for example, in an electric grill, and is described in detail in Chapter 20 (section 20.3).

Both conduction and convection can take place under 'steady state' or 'unsteady state' conditions. Steady state heat transfer takes place when there is a constant temperature difference between two materials. The amount of heat entering a section of the material equals the amount of heat leaving, and there is no change in temperature of that section of the material. This occurs for example when heat is transferred through the wall of a cold store if the store temperature and ambient temperature are both constant (Chapter 21, section 21.2.3), and in continuous processes once operating conditions have stabilised. However, in the majority of food processing applications the temperature of the food and/or the heating or cooling medium are constantly changing, and unsteady state heat transfer is more commonly found. Calculations of heat transfer under these conditions are complex. They are described by Singh and Heldman (2001a) and Toledo (1999), and a simplified example of unsteady state calculations in heat sterilisation is given in Chapter 11 (section 11.1). The examples below assume steady state conditions, which are simpler to analyse. They are simplified by making a number of assumptions and using prepared charts to obtain useful information for the design and operating conditions of heat processing equipment. Computer models used to give solutions to these calculations are described by Toledo (1999) and Singh and Heldman (2001a).

Conduction
Conduction is the movement of heat by direct transfer of molecular energy within solid materials. Energy transfer is either by movement of free electrons (e.g. through metals) or by vibration of molecules. As molecules gain thermal energy they vibrate with increased amplitude, and this vibration is passed from one molecule to another. Therefore conducted heat moves from an area of higher temperature to an area of lower temperature without actual movement of the molecules through the material. The rate at which heat is transferred by conduction is determined by the temperature difference between the food and the heating or cooling medium, and the total resistance to heat transfer. The resistance to heat transfer is expressed as the thermal conductivity (section 10.1.1). Under steady-state conditions the rate of heat transfer is calculated using Fourier's Law:

$$Q = -kA \frac{(\theta_1 - \theta_2)}{x} \qquad \boxed{10.11}$$

where Q (J s^{-1}) = rate of heat transfer, k (J m^{-1} s^{-1} °C^{-1} or W m^{-1} °C^{-1}) = thermal conductivity, A (m^2) = surface area, $\theta_1 - \theta_2$ (°C) = temperature difference and x (m) = thickness of the material. $(\theta_1 - \theta_2)/x$ is also known as the temperature gradient.

Because the temperature decreases with increasing distance through the food away from the heat source, the negative sign in Equation 10.11 is used to obtain a positive value for heat flow in the direction of decreasing temperature. A calculation based on Equation 10.11 is given in sample problem 10.1 and related problems are given in Chapters 12 and 14.

In contrast to steady state heat transfer, where the temperature varies only with location, in unsteady state conduction, the temperature at a given point within a food depends on the rate of heating or cooling and the position in the food. The temperature therefore changes continuously with both location and time. The factors that influence the temperature change are the:

- temperature of the heating medium;
- thermal conductivity of the food; and
- specific heat of the food.

The basic equation for unsteady state heat transfer in a single direction x is

$$\frac{d\theta}{dt} = \frac{k}{\rho c} \frac{d^2\theta}{dx^2}$$

$$\boxed{10.12}$$

where $d\theta/dt$ = change in temperature with time.

Examples of solutions to this equation for simple shapes (e.g. a slab, cylinder or sphere) are described by Singh and Heldman (2001a) and Toledo (1999).

Convection

When a fluid changes temperature, the resulting changes in density establish natural convection currents. Convection is therefore the transfer of heat by groups of molecules that move as a result of differences in density. Examples include natural-circulation evaporators (Chapter 14, section 14.1.3), air movement in bakery ovens (Chapter 18, section 18.2), and movement of liquids inside cans during sterilisation (Chapter 13, section 13.1). Forced convection takes place when a stirrer or fan is used to agitate the fluid. This reduces the boundary film thickness (see Chapter 1, section 1.3.4) to produce higher rates of heat transfer and a more rapid temperature redistribution. Consequently, forced convection is more commonly used than natural convection in food processing. Examples of forced convection include mixers (Chapter 4, section 4.1.3), fluidised-bed driers (Chapter 16, section 16.2.1), air blast freezers (Chapter 22, section 22.2.1) and liquids pumped through heat exchangers (Chapters 12 and 14).

When liquids or gases are heated in a vessel or a pipe, the rate of heat transfer is complicated because of the motion of the fluid and the presence of boundary layers under laminar flow. A temperature profile (Fig. 10.2) develops in a similar way to a fluid velocity profile, with the fluid nearest to the vessel/pipe wall heating fastest and that at the centre the slowest. This profile depends on the viscosity of the fluid and the type of flow. Calculations of heat transfer are complex and further details are given by Singh and Heldman (2001a), Rotstein *et al.* (1997) and Heldman and Lund (1992) for both Newtonian and non-Newtonian fluids (Chapter 1, section 1.1.2).

Convective heat transfer is found using:

$$Q = h_s A(\theta_b - \theta_s)$$

$$\boxed{10.13}$$

where Q $(J s^{-1})$ = rate of heat transfer, A (m^2) = surface area, θ_s (°C) = surface temperature, θ_b (°C) = bulk fluid temperature and h_s $(W m^{-2} K^{-1})$ = surface (or film) heat transfer coefficient.

Sample problem 10.1

Part 1: In a bakery oven, combustion gases heat one side of a 2.5 cm steel plate at 300 °C and the temperature in the oven is 285 °C. Assuming steady state conditions, and a thermal conductivity for steel of 17 $W\,m^{-2}\,°C^{-1}$, calculate the rate of heat transfer per m^2 through the plate.

Part 2: The internal surface of the oven is 285 °C and air enters the oven at 18 °C. Calculate the surface heat transfer coefficient per m^2, assuming the rate of heat transfer is 10.2 kW.

Solution to sample problem 10.1

Part 1:

From Equation 10.11,

$$Q = -\frac{17 \times 1 \times (300 - 285)}{0.025}$$

$$= 10\ 200\ W$$

Part 2:

From Equation 10.13,

$$h = \frac{10\ 200}{(285 - 18)}$$

$$= 38.2\ W\ m^{-1}\,°C^{-1}$$

This value indicates that natural convection is taking place in the oven.

The surface heat transfer coefficient is a measure of the resistance to heat flow, caused by the boundary film, and is therefore equivalent to the term k/x in the conduction equation (Equation 10.1). It is higher in turbulent flow than in streamline flow. Typical values of h_s are given in Table 10.5. An example of a calculation of heat transfer coefficient is given in Chapter 14 (sample problem 14.2).

The calculations can be simplified by using formulae that relate the physical properties of a fluid (e.g. density, viscosity, specific heat, gravity (which causes circulation due to changes in density)), temperature difference and the length or diameter of the container under investigation. The these factors are expressed as dimensionless numbers as follows:

$$\text{Nusselt number Nu} = \frac{h_c D}{k} \qquad \boxed{10.14}$$

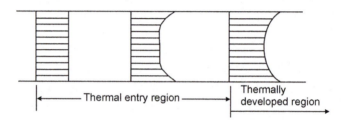

Thermal entry region

Thermally developed region

Fig. 10.2   Temperature profile of liquid being heated in a pipe (reprinted from Singh and Heldman 2001a) © 2001, with permission from Academic Press).

Table 10.5 Values of surface heat transfer coefficients

	Surface heat transfer coefficient ($W\,m^{-2}\,K^{-1}$)	Typical applications
Boiling liquids	2400–60 000	Evaporation
Condensing saturated steam	12 000	Canning, evaporation
Condensing steam		
With 3% air	3500	Canning
With 6% air	1200	
Condensing ammonia	6000	Refrigeration
Liquid flowing through pipes		
Low viscosity	1200–1600	Pasteurisation
High viscosity	120–1200	Evaporation
Moving air ($3\,m\,s^{-1}$)	30	Freezing, baking
Still air	6	Cold stores

Adapted from Delgado and Sun (2003) and Earle (1983)

$$\text{Prandtl number } \mathrm{Pr} = \frac{c_p \mu}{k} \qquad \boxed{10.15}$$

$$\text{Grashof number } \mathrm{Gr} = \frac{D^3 \rho^2 g \beta \Delta\theta}{\mu^2} \qquad \boxed{10.16}$$

where h_c ($W\,m^{-2}\,{}^{\circ}C^{-1}$) = convection heat transfer coefficient at the solid-liquid interface, D (m) = the characteristic dimension (length or diameter), k ($W\,m^{-1}\,{}^{\circ}C^{-1}$) = thermal conductivity of the fluid, c_p ($J\,kg^{-1}\,{}^{\circ}C^{-1}$) = specific heat at constant pressure, ρ ($kg\,m^{-3}$) = density, μ ($N\,s\,m^{-2}$) = viscosity, g ($m\,s^{-2}$) = acceleration due to gravity = $9.81\,m\,s^{-2}$, β ($m\,m^{-1}\,{}^{\circ}C^{-1}$) = coefficient of thermal expansion, $\Delta\theta$ (${}^{\circ}C$) = temperature difference and v ($m\,s^{-1}$) = velocity.

The Nusselt number is found by dividing Equation 10.13 for convection by Equation 10.1 for conduction and replacing the thickness value with the characteristic dimension (D). It can be considered as a measure of the improvement in heat transfer caused by convection over that due to conduction (i.e. if Nu = 1 there is no improvement, whereas if Nu = 4, the rate of convective heat transfer is four times the rate that would occur by conduction alone in stagnant liquid). The Prandtl number relates the boundary layer caused by the fluid velocity (using the fluid viscosity as a measure) with the thermal boundary layer. If Pr = 1 the thickness of the two layers is the same. For gases, Pr = approximately 0.7 and Pr = around 10 for water. The Grashof number is a ratio of the forces that cause lighter liquids to become more buoyant and rise and the viscous forces that slow their movement. It is used for natural convection when there is no turbulence in the fluid to determine whether flow is streamline or turbulent. A further number, the Rayleigh number, is a multiple of the Grashof and Prandtl numbers. Formulae for other types of flow conditions and different vessels are described by Singh and Heldman (2001a) and Toledo (1999).

The Reynolds Number (Chapter 1, section 1.3.4) is used to determine the type of fluid flow. For streamline flow through pipes:

$$\mathrm{Nu} = 1.62 \left(\mathrm{Re}\ \mathrm{Pr}\frac{D}{L}\right)^{0.33} \qquad \boxed{10.17}$$

where L (m) = length of pipe, when Re Pr D/L > 120 and all physical properties are measured at the mean bulk temperature of the fluid.

For turbulent flow through pipes:

$$Nu = 0.023(Re)^{0.8}(Pr)^n$$

10.18

where $n = 0.4$ for heating or $n = 0.3$ for cooling, when $Re > 10\,000$, viscosity is measured at the mean film temperature and other physical properties are measured at the mean bulk temperature of the fluid.

The surface heat transfer data (Table 10.5) indicate that heat transfer through air is lower than through liquids. Larger heat exchangers are therefore necessary when air is used for heating (Chapters 16 and 18) or cooling (Chapters 21 and 22) compared with those needed for liquids. Condensing steam produces higher rates of heat transfer than hot water at the same temperature and the presence of air in steam reduces the rate of heat transfer. This has important implications for canning (Chapter 13, section 13.1) as any air in the steam lowers the temperature and hence lowers that amount of heat received by the food. Both thermometers and pressure gauges are therefore needed to assess whether steam is saturated.

Most cases of heat transfer in food processing involve heat transfer through a number of different materials. For example heat transfer in a heat exchanger from a hot fluid, through the wall of a pipe or vessel to a second fluid, is shown in Fig. 10.3. The heat must first transfer through the boundary film on the hot side, through the metal wall by conduction, and then through the boundary layer of the cold side.

The overall temperature difference is found using:

$$\theta_a - \theta_b = \frac{Q}{A}\left(\frac{1}{h_a}\frac{x}{k} + \frac{1}{h_b}\right)$$

10.19

The unknown wall temperatures θ_2 and θ_3 are therefore not required and all factors to solve the equation can be measured.

The sum of the resistances to heat flow is termed the overall heat transfer coefficient (OHTC) and the rate of heat transfer may be expressed as:

$$Q = UA(\theta_a - \theta_b)$$

10.20

The OHTC is an important term that is used, for example, to indicate the effectiveness of heating or cooling in different types of processing equipment. Examples are shown in Table 10.6.

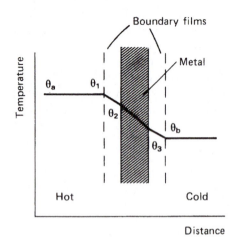

Fig. 10.3   Temperature changes from a hot liquid through a vessel wall to a cold liquid.

Table 10.6   OHTCs in food processing

Heat transfer fluids	Example	OHTC (W m^{-2} K^{-1})
Hot water–air	Air heater	10–50
Viscous liquid–hot water	Jacketed vessel	100
Viscous liquid–hot water	Agitated jacketed vessel	500
Viscous liquid–steam	Evaporator	500
Non-viscous liquid–steam	Evaporator	1000–3000
Flue gas–water	Boiler	5–50
Evaporating ammonia–water	Chilled water plant	500

Adapted from Lewis (1990)

In a heat exchanger, liquids can be made to flow in either the same direction (co-current (or 'concurrent' or 'parallel') flow) (Fig. 10.4a) or in opposite directions ('counter-current' flow) (Fig. 10.4b). In co-current operation, a cold liquid enters the inner pipe at temperature θ_1, flows through the pipe and exits at temperature θ_2. A hot liquid enters at temperature θ_3, flows around the annular space between the inner and outer pipes and exits at temperature θ_4. In the process heat is gained by the cold liquid and lost by the hot liquid. The heat exchanger is insulated to minimise heat losses to the surrounding air and an energy balance (Equation 10.11) can be used to show that the decrease in energy of the hot liquid equals the increase in energy of the cold liquid. This equation is useful to determine the flowrates or temperature changes in a heat exchanger.

$$Q = m_h \times C_{ph}(\theta_3 - \theta_4) = m_c \times C_{pc}(\theta_2 - \theta_1) \qquad \boxed{10.21}$$

where m (kg s^{-1}) = mass flowrate, c_p (kJ kg^{-1} °C^{-1}) = specific heat, both with suffixes 'h' for hot and 'c' for cold, and θ (°C) = temperatures numbered as shown in Fig. 10.4.

Counter-current flow has a higher heat transfer efficiency than co-current flow and is therefore widely used in heat exchangers (e.g. Chapters 11–14). However, the temperature difference varies at different points in the heat exchanger and it is necessary to use a logarithmic mean temperature difference in calculations (sample problem 10.2):

$$\Delta\theta_m = \frac{\Delta\theta_1 - \Delta\theta_2}{\ln(\Delta\theta_1/\theta_2)} \qquad \boxed{10.22}$$

where θ_1 is higher than θ_2.

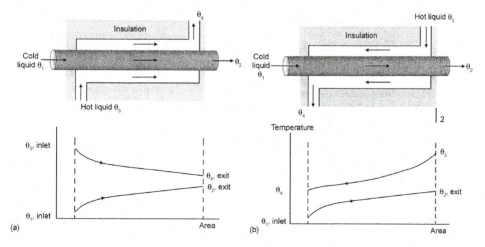

Fig. 10.4   (a) Co-current and (b) counter-current flow through a heat exchanger.

Sample problem 10.2

A heat exchanger is to be used to heat orange juice from 18 °C to 80 °C at a flowrate of 0.5 kg s^{-1}. A counter-current heat exchanger is required and hot water is available at 95 °C to pass through the annular pipe at a flowrate of 1.5 kg s^{-1}. Calculate the required length of the inner juice pipe having a diameter of 8 cm. Assume steady state conditions, no heat losses to the surroundings and an OHTC of 2400 W m^{-2} °C^{-1}. The specific heat of the juice is 3.89 kJ kg^{-1} °C^{-1} and the specific heat of water = 4.18 kJ kg^{-1} °C^{-1} (Table 10.1).

Solution to sample problem 10.2

$$\text{Heat gained by juice} = mc_p(\theta_1 - \theta_2) = 0.5 \times 3.89 \times (80 - 18)$$
$$= 120.59 \text{ kJ}$$

Use a heat balance:

$$\text{Heat gained by juice} = \text{heat lost by water}$$

$$Q = mc_p(\theta_1 - \theta_2) = 0.5 \times 3.89 \times (80 - 18)$$

$$= mc_p(\theta_3 - \theta_4) = 1.5 \times 4.18 \times (95 - \theta_2)$$

The exit temperature of the hot water = $\theta_2 = 76$ °C
From Equation 10.22:

$$\Delta\theta_m = \frac{(95 - 80) - (76 - 18)}{\ln(95 - 80)/(76 - 18)}$$

$$= 31.8 \text{ °C}$$

From Equation 10.20:

$$Q = UA\Delta\theta = u\pi dl\Delta\theta$$

Therefore

$$l = Q/U\pi d\Delta\theta$$

$$= \frac{120.59 \times 1000}{2400 \times 3.142 \times 0.08 \times 31.8}$$

$$= 6.29 \text{ m}$$

Related sample problems are shown in Chapters 11 (sample problem 11.1) and 12 (sample problems 12.1 to 12.3).

The heating time in batch processing is found using:

$$t = \frac{mc}{UA}\ln\left(\frac{\theta_h - \theta_i}{\theta_h - \theta_f}\right)$$ ⟨10.23⟩

where m (kg) = the mass, c_p (J kg^{-1} °C^{-1}) = specific heat capacity, θ_h (°C) = temperature of the heating medium, θ_i (°C) = initial temperature, θ_f (°C) = final temperature, A (m^2) = surface area and U (W m^{-2} °C^{-1}) = OHTC.

Unsteady state heat transfer by conduction and convection
When a solid piece of food is heated or cooled by a fluid the resistances to heat transfer
are the surface heat transfer coefficient and the thermal conductivity of the food. These
two factors are related by the Biot number (Bi):

$$\mathrm{Bi} = \frac{h\delta}{k}$$

10.24

where h $(\mathrm{W\,m^{-2}\,^{\circ}C^{-1}})$ = heat transfer coefficient, δ = the characteristic 'half dimension'
(e.g. radius of a sphere or cylinder, half thickness of a slab) and k $(\mathrm{W\,m^{-1}\,^{\circ}C^{-1}})$ =
thermal conductivity.

At small Bi values (< 0.2) the surface film is the main resistance to heat flow and the
internal resistance of the food is negligible. The time required to heat the solid food is
found using Equation 10.23, using the film heat transfer coefficient h_s instead of U.
However, in most applications the thermal conductivity of the food limits the rate of heat
transfer (Bi = $0.2 - 40$) rather than the surface film resistance. These calculations are
complex, and a series of charts is available to solve the unsteady state equations for
simple shaped foods (Fig. 10.5), known as Gurney–Lurie and Heisler charts. The charts
relate the Bi number (Equation 10.24), the temperature factor (the fraction of the
temperature change that remains to be accomplished (Equation 10.25) and the Fourier
number Fo (a dimensionless number which relates the thermal diffusivity, the size of the
piece and the time of heating or cooling (Equation 10.26)):

$$\frac{\theta_h - \theta_f}{\theta_h - \theta_i}$$

10.25

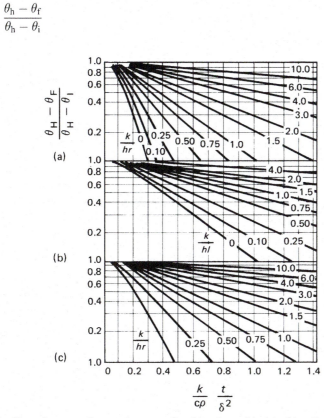

Fig. 10.5 Chart for unsteady state heat transfer: (a) sphere, (b) slab, (c) cylinder (after Henderson
and Perry 1955).

where the subscript 'h' indicates the heating medium, the subscript 'f' the final value and the subscript 'i' the initial value.

$$\text{Fo} = \frac{k \ t}{c\rho \ \delta^2} \qquad \boxed{10.26}$$

Singh and Heldman (2001b) describe computer spreadsheets to perform unsteady state calculations and artificial neural networks have been trained to perform the calculations represented on the charts. They have produced more accurate results than reading from the charts (Pandharipande and Badhe 2004).

An example of an unsteady state calculation is shown in Chapter 11 (sample problem 11.1) and more complex calculations are described by Singh and Heldman (2001b), Toledo (1999), Lewis (1990), Earle (1983) and Jackson and Lamb (1981).

10.2   Sources of heat and methods of application to foods

The cost of energy for heating is one of the major considerations in the selection of processing methods and ultimately in the cost of the processed food and the profitability of the operation. Different fuels have specific advantages and limitations in terms of cost, safety, risk of contamination of the food, flexibility of use, and capital and operating costs for heat transfer equipment. The following sources of energy are the main ones that are used in food processing:

- electricity;
- gas (natural or liquid petroleum gas);
- liquid fuel oil.

In many developing countries, solid fuels are the main source of process heating but in industrialised countries, solid fuels (anthracite, coal, wood and charcoal) are only used to a small extent for heating boilers to generate steam or in specialised applications such as wood chips for food smoking (Chapter 17) or bagasse for sugar boiling. The advantages and limitations of each type of energy source (Table 10.7) can change over time depending on reserves of natural resources and the economics of production in different countries. However, electricity is the preferred source of energy for most applications and gas is widely used for boiler and oven heating. Recent developments include the adaptation of boilers to operate using biomass, including combustible waste materials, which also reduces waste disposal costs.

Table 10.7   Advantages and limitations of different energy sources for food processing

	Electricity	Gas	Liquid fuel	Solid fuel
Energy per unit mass/ volume ($\times 10^3$ kJ kg^{-1})	–	1.17–4.78	8.6–9.3 (fuel oil)	5.26–6.7 (coal) 3.8–5.26 (wood)
Cost per kJ of energy	High	Low	Low	Low
Heat transfer equipment cost	Low	Low	High	High
Efficiency of heating[a]	High	Moderate/ high	Moderate/ Low	Low
Flexibility of use	High	High	Low	Low
Fire/explosion hazard	Low	High	Low	Low
Risk of contaminating food	Low	Low	High	High
Labour and handling costs	Low	Low	Low	High

[a] Efficiency = amount of energy used for heating divided by amount of energy supplied

10.2.1   Direct heating methods

In direct methods the heat and products of combustion from the burning fuel come directly into contact with the food. There is an obvious risk of contamination of the food by odours or incompletely burned fuel and for this reason only gas and, to a lesser extent, liquid fuels are used. Applications include kiln driers (Chapter 16, section 16.2.1) and baking ovens (Chapter 18, section 18.2.1). These direct methods should not be used confused with 'direct' steam injection where the steam is produced in a separate location from the processing plant. Electricity is not a fuel in the same sense as the other types described above. It is generated by steam turbines heated by a primary fuel (e.g. coal, gas or fuel oil) or by hydro-power or nuclear energy. However, electrical energy may also be used directly by pulsed electric field processing (Chapter 9, section 9.1), dielectric heating or ohmic heating (Chapter 20).

10.2.2   Indirect heating methods

Indirect electrical heating uses resistance heaters or infrared heaters. Resistance heaters are nickel–chromium wires contained in solid plates or coils that are attached to the walls of process vessels, in flexible jackets that wrap around vessels, or in immersion heaters that are submerged in the food. These types of heaters are used for localised or intermittent heating. Infrared heaters are described in Chapter 20 (section 20.3).

Indirect heating using fuels requires a heat exchanger to separate the food from the products of combustion. At its simplest an indirect system consists of burning fuel beneath a metal plate and heating by energy radiated from the plate or conducted through it. The most common type of indirect-heating system used in food processing is steam or hot water generated by a heat exchanger (a boiler) located close to the processing area. A second heat exchanger transfers the heat from the steam or water to the food under controlled conditions, or alternatively a heat exchanger transfers heat to air in order to dry foods or to heat them under dry conditions. If steam is directly injected into the food, it is first filtered to remove any traces of condensate and all particles. Details of steam injection are given by Demetrakakes (1997) and in Chapter 13 (section 13.2).

Steam generation

Steam boilers used in food processing are usually the 'water-tube' design in which water is pumped through tubes in the boiler that are surrounded by hot combustion gases from a burner or firebox. An alternative design (the 'fire-tube' boiler) has the combustion gases contained in tubes that pass through water in the boiler vessel. The advantages of the water-tube design include:

- more rapid heat transfer because water is pumped under turbulent flow;
- larger capacities and higher pressures can be obtained;
- greater flexibility of operation;
- safer because steam is generated in small tubes rather than the large boiler vessel (Singh and Heldman 2001b).

To calculate the size of a steam boiler for a particular process, the following steps are taken:

1 Assess the thermal energy requirements of all operations that use steam, including the maximum temperature required. This determines the pressure that the steam is supplied at.
2 Calculate the quantity of steam needed to supply the required energy and using the

specific volume of steam (section 10.1.1) at the given pressure, calculate the size of pipework required to meet the volumetric flowrate.
3 Take account of energy losses (e.g. due to friction losses in pipework (Chapter 1, section 1.3.4) and heat losses through insulated pipes) to calculate the boiler power output required.
4 Take account of the boiler efficiency to calculate the size of boiler needed to meet the process requirements.

Properties of steam are discussed by Brennan *et al.* (1990) and Toledo (1999), and selected properties of saturated steam at different temperatures are shown in Table 10.4.

Anon (2007c) describes the advantages of electrically powered steam generators that can be located alongside processing equipment. In larger plants, steam generation may also be combined with power generation using a steam turbine. Anon (2006) describes the on-site generation of electricity and heat from a single fuel source, known as 'combined heat and power' (CHP) or 'cogeneration'. Waste heat from electricity production can be used to produce process heat, generate steam or heat buildings rather than using additional energy from electricity or gas. The most efficient CHP systems have high thermal loads compared with electric loads and can achieve >80% efficiency. CHP also minimises electricity transmission losses between the generator and end-user. These losses are ≈73% for electricity produced at power stations and transmitted along a grid (comprising about 65% of fuel lost as waste heat during generation and 8% of power lost along transmission lines). Further cost savings can be made if CHP uses waste products (e.g. methane or biomass) instead of fossil fuels. Since electricity is generated on site, processing is also not affected by disruptions in the grid power supply.

Sutter (2007) notes the advantages of hot water compared with steam for heating jacketed vessels: the temperature can be controlled more accurately using hot water, which prevents overheating and product damage; and hot water distributes heat more evenly than steam, which eliminates hot spots that cause product damage. Steam injection is superior to indirect heat exchangers for heating water. It can be programmed to adjust the process temperature at a predetermined rate, giving a rapid response to changing process conditions and ensuring precise temperature control within a fraction of a degree. In contrast to indirect steam heating, where condensate at a relatively high temperature is returned to the boiler with inherent heat losses, the condensate in steam injection has most of the heat extracted, saving up to 17% in fuel costs.

10.2.3 Energy use and methods to reduce energy consumption

In all types of food processing, most of the energy (40–80%) is used for actual processing, and heating, refrigeration and dehydration in particular require significant amounts of energy. For example, Okos *et al.* (1998) report that in the USA, process heating uses approximately 29% of the total energy used by the food industry, while process cooling and refrigeration uses about 16% of total energy inputs. Examples of data from Carlsson-Kanyama and Faist (2000) on energy inputs in different types of food processing are shown in Table 10.8.

The types of process that consume the most energy include wet maize milling, production of sugar from beet, soybean oil mills, production of malted beverages, meat packing plants, canning, and production of frozen foods and bakery products. Less than 8% of the energy consumed by food manufacturing is for non-process uses (e.g. room lighting, heating, air conditioning and on-site transportation). However, in some processes signifi-cant amounts of energy are also used for packaging (11%; range, 15–40%), distribution

Table 10.8 Comparative energy inputs in different types of food processing

Product	Energy (MJ) used per kg product	Notes
Bread	1.53–4.56	
Breakfast cereals	19–66	
Canned fruit and vegetables	2.1–3.8	
Canned meats	5.2–25	
Chocolate	8.6	
Chilled retail display cabinets	0.12	
Coffee instant	50	
Cold storage (e.g. apples)	0.0009–0.017	MJ electricity kg^{-1} per day. Use varies with the size of the cold-room – e.g. $0.0010\,MJ\,l^{-1}$ net volume per day in room of $10\,000\,m^3$ compared with $0.015\,MJ\,l^{-1}$ in a room of $10\,m^3$ (factor of 15 difference)
Drying		Theoretical value for evaporating 1 kg of
Beet pulp (80% to 10% moisture)	6.4	water = 2.60 MJ but actual use is 2–6 times
Soybeans (17% to 11% moisture)	0.47	higher, or 5.2–15.6 MJ (Pimentel and
Potato flakes/granules	15–42	Pimentel 1996)
Freezing	0.3–7.6	A-rated equipment[a] uses 2.7 times less
Ice cream	2.2–3.7	energy per litre of usable volume than older equipment (typically $0.012\,MJ\,l^{-1}$ net volume/day)
Juice from concentrate	1.15	
Juice from fresh citrus fruit	4.6	
Milk processing	0.50–2.6	
Milling wheat flour	0.32–2.58	Electricity the only energy recorded
Oil extraction	0.28–1.5	Energy use allocated between two products (oil and press-cake)
Pasta	0.8–2.4	
Sausages	3.9–36	Large variation according to the extent of processing
Sugar extraction	2.3–26	
Sugar confectionery	6	

[a] EU labelling system for energy efficiency (A-label = most energy efficient, B, C and D labels = descending order of efficiency).
Adapted from data of Carlsson-Kanyama and Faist (2000) and Pimentel and Pimentel (1996)

transport (12%; range, 0.56–30%), cleaning water (15%) and storage (up to 85% of total energy input for deep-frozen foods). Dalsgaard and Abbotts (2003) analyse energy use in different types of food processing and describe methods for improving energy use.

Energy efficiency audits
Energy audits are holistic surveys of a production plant that are undertaken to understand how energy is currently used and to identify areas of potential savings. An energy audit consists of three main parts: understanding energy costs, identifying potential savings and making cost–benefit recommendations. It is used to identify specific areas and equipment within a factory where energy savings can be made. Energy audits can:

- lower energy expenses;
- increase production reliability;
- increase productivity;
- reduce environmental impacts (Anon 2005a).

Details of how to conduct an energy audit are given by Barron and Burcham (2001).

Improvements in energy efficiency make food companies more competitive. Energy saving can be achieved by improving existing plants, developing more energy-efficient processing technologies and creating informed energy policies. Reductions in energy use are possible by changing the type of process technology to a method that uses less energy: for example, supercritical extraction (Chapter 5, section 5.4.3) could be used instead of concentration by boiling; irradiation (Chapter 7) instead of pasteurisation or sterilisation; or drying by vapour recompression supercritical extraction instead of using hot air.

Potentially the main energy savings in food processing are associated with boiler operation, the supply of steam or hot air and re-use of waste heat. Food processors are aware of the potential savings that can be made by reducing energy consumption, and meeting new environmental legislation. Other measures to improve boiler operation are returning condensate as feed water, pre-heating air for fuel combustion, and recovering heat from the flue. Fletcher (2004) reports flue gas loss as a key value in assessing the efficiency of high-pressure boilers. The amount of energy contained in the flue gases that are emitted through the chimney can result in significant energy losses. EU legislation on reduced CO_2 emissions and energy consumption has defined limits and flue gas losses from oil and gas-fired boilers over 50 kW must not now exceed 9%. Flue gas heat exchangers reduce the temperature of flue gas from for example 350 °C to 150 °C to produce a 5% saving in fuel. If a typical factory boiler operates for 8000 hours a year, the efficiency can be increased by up to 15% and fuel cost savings are significant. The substitution of biomass fuels for fossil fuels can reduce hydrocarbon and carbon dioxide emissions. Although their production and use are technologically feasible, there are varying opinions on the economic viability of fuels from renewable resources, with some countries progressing their use more quickly than others. Computer control of boiler operation increases fuel efficiency and is described in detail by Anon (2001).

Energy savings in steam supply to processing areas are achieved by proper insulation of steam and hot-water pipes, minimising steam leaks and fitting steam traps. Individual processing equipment is designed for energy saving and examples include regeneration of heat in heat exchangers (examples in Chapter 11, section 11.2, and Chapter 12, section 12.2), multiple-effect or vapour recompression systems (Chapter 14, section 14.1.2) and automatic defrosting and correct insulation of freezing equipment (Chapter 22, section 22.2). Microprocessor control of processing equipment is widely used to reduce energy consumption.

Recovery of heat from drying air is more difficult than from steam or vapours, but a number of heat exchanger designs are used to recover waste heat from air or gases, described in Chapter 18 (section 18.2) for baking ovens and Chapter 19 (section 19.2.3) for deep-fat fryers. Heat pumps are similar to refrigeration plant (see Chapter 21, Fig. 21.1) but operate by removing heat from a low-temperature source and concentrating it in a heat 'sink' which is then used to heat air or water. Anon (2005b) and Anon (2008) describe the operation of heat pumps.

10.2.4   Types of heat exchangers

Heat exchangers to heat or cool foods are among the most common types of equipment found in food processing operations. Cooling equipment is described in Chapter 21, section 21.1.1. There is a wide variety of heat exchangers used to heat foods (e.g. for blanching, pasteurisation, heat sterilisation, evaporation, drying, frying and baking (Chapters 11–19)). Their design and operation depend on the properties of the foods being processed and the degree of heating required. Equipment can be grouped into direct heating types – steam injection or steam infusion (Chapter 13, section 13.2.3) – and indirect types:

- scraped surface used for evaporation (Chapter 14, section 14.1.3) and freezing (Chapter 22, section 22.2.1);
- tubular used for pasteurisation (Chapter 12, section 12.2) and evaporation;
- shell and tube used for evaporation; and
- plate heat exchangers used for pasteurisation or evaporation.

Details of their operation are given in the chapters as indicated.

10.3   Effect of heat on micro-organisms and enzymes

The preservative effect of heat processing is due to the denaturation of proteins, which destroys enzyme activity and enzyme-controlled metabolism in micro-organisms. The rate of destruction of many micro-organisms is a first-order reaction (see also below for other types): that is when food is heated to a temperature that is high enough to inactivate contaminating micro-organisms, the same percentage die in a given time interval regardless of the numbers present initially. This is known as the 'logarithmic order of death' and is described by a 'death rate curve' (Fig. 10.6).

The time needed to destroy 90% of the micro-organisms (to reduce their numbers by a factor of 10) is referred to as the 'decimal reduction time' or D-value (5 min in Fig. 10.6). D-values differ for different microbial species (Table 10.9) and a higher D-value indicates greater heat resistance.

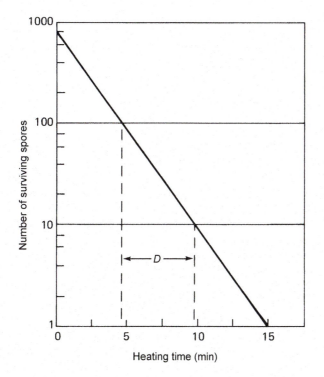

Fig. 10.6   Death rate curve.

Table 10.9   Heat resistance of selected pathogens

Micro-organism	D-value (min)	z-value	Temperature (°C)	Substrate/ typical food
Vegetative cells				
Aeromonas hydrophila	2.2–6.6	5.2–7.7	48	Milk
Bacillus stearothermophilus	3.0–4.0	9–10	–	Vegetables, milk
B. subtilis	0.3–0.76	4.1–7.2	–	Milk products
B. cereus	3.8	36	–	Milk
Campylobacter jejuni	0.62–2.25	–	55–56	Beef/lamb/chicken
Campylobacter jejuni	0.74–1.0	–	55	Skim milk
Clostridium sporogenes	0.7–1.5	8.8–11.1	–	Meats
Cl. thermosaccharolyticum	3.0–4.0	7.2–10.0	–	Vegetables
Escherichia coli O111:B4	5.5–6.6	–	55	Skim/whole milk
E. coli O157:H7	4.1–6.4	–	57.2	Ground beef
E. coli O157:H8	0.26–0.47	5.3	62.8	Ground beef
Listeria monocytogenes	0.22–0.58	5.5	63.3	Milk
L. monocytogenes	1.6–16.7	–	60	Meat products
Staphylococcus aureus	6	–	60	Meat macerate
Staph. aureus	3	–	60	Pasta
Staph. aureus	0.9	9.5	60	Milk
Salmonella senftenberg	276–480	18.9	70–71	Milk chocolate
S. senftenberg	0.56–1.11	4.4–5.6	65.5	Various foods
S. typhimurium	396–1050	17.7	70–71	Milk chocolate
S. typhimurium	2.13–2.67		57	Ground beef
Vibrio cholerae	0.35–2.65	17–21	60	Crab/oyster
V. parahaemolyticus	0.02–2.5	5.6–12.4	55	Clam/crab
V. parahaemolyticus	10–16	5.6–12.4	48	Fish homogenate
Yersinia enterocolitica	0.067–0.51	4–5.78	60	Milk
Spores				
Bacillus subtilis	30.2	9.16	88	0.1% NaCl
Bacillus cereus	1.5–36.2	6.7–10.1	95	Various foods
Clostridium botulinum 62A	0.61–2.48	7.5–11.6	110	Vegetable products
Cl. botulinum B	0.49–12.42	7.4–10.8	110	Vegetable products
Cl. botulinum E	6.8–13	9.78	74	Seafood
Clostridium perfringens	6.6	–	104.4	Beef gravy

Adapted from Anon (2000), Heldman and Hartel (1997) and Brennan *et al.* (1990)

There are two important implications arising from the decimal reduction time: first, the higher the number of micro-organisms present in a raw material, the longer it takes to reduce the numbers to a specified level. In commercial operation the number of micro-organisms varies in each batch of raw material, but it is difficult to recalculate process times for each batch of food. A specific temperature–time combination is therefore used to process every batch of a particular product, and adequate preparation procedures (Chapter 2) are used to ensure that the raw material has a satisfactory and uniform microbiological quality. Secondly, because microbial destruction takes place logarithmically, it is theoretically possible to destroy all cells only after heating for an infinite time. Processing therefore aims to reduce the number of surviving micro-organisms by a predetermined amount. This gives rise to the concept of 'commercial sterility', which is discussed further in Chapter 13 (section 13.1.1).

The destruction of micro-organisms is temperature dependent; cells die more rapidly at higher temperatures. By collating D-values at different temperatures, a semi-logarithmic thermal death time (TDT) curve is constructed (Fig. 10.7). The slope of the TDT curve is termed the z-value and is defined as the number of degrees Celsius required

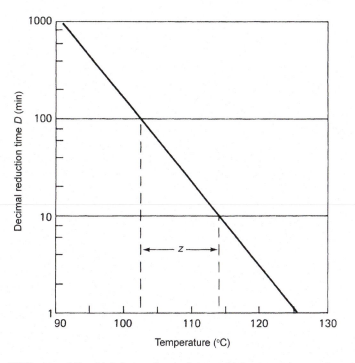

Fig. 10.7   TDT curve. Microbial destruction is faster at higher temperatures (e.g. 100 min at 102.5 °C has the same lethal effect as 10 min at 113 °C.

to bring about a ten-fold change in decimal reduction time (10.5 °C in Fig. 10.7). The *D*-value and *z*-value are used to characterise the heat resistance of a micro-organism and its temperature dependence respectively.

There are a large number of factors that determine the heat resistance of micro-organisms, but general statements of the effect of a given variable on heat resistance are not always possible. The following factors are known to be important:

1   Type of micro-organism: different species and strains show wide variation in their heat resistance (Table 10.9). Spores are much more heat resistant than vegetative cells.
2   Incubation conditions during cell growth or spore formation. These include:
   • temperature (spores produced at higher temperatures are more resistant than those produced at lower temperatures),
   • age of the culture (the stage of growth of vegetative cells affects their heat resistance, and
   • culture medium used (for example mineral salts and fatty acids influence the heat resistance of spores).
3   Conditions during heat treatment. The important conditions are:
   • pH of the food (pathogenic and spoilage bacteria are more heat resistant near to neutrality; yeasts and fungi are able to tolerate more acidic conditions but are less heat resistant than bacterial spores),
   • water activity of the food (Chapter 1, section 1.1.2) influences the heat resistance of vegetative cells; in addition moist heat is more effective than dry heat for spore destruction,
   • composition of the food (proteins, fats and high concentration of sucrose increase the heat resistance of micro-organisms; the low concentration of sodium chloride

used in most foods does not have a significant effect; the physical state of the food, particularly the presence of colloids, affects the heat resistance of vegetative cells), and

- the growth media and incubation conditions used to assess recovery of micro-organisms in heat resistance studies affect the number of survivors observed.

There is growing evidence that survival curves show deviations from the straight semi-logarithmic lines in Figs 10.6 and 10.7 for many vegetative micro-organisms and spores. Peleg and Cole (1998) describe curves that have sigmoidal shapes with shoulders and tails and Geeraerd *et al.* (2004) have characterised eight different types of curve (Fig. 10.8). Peleg (2000, 2002, 2003) has interpreted different shapes in terms of microbial

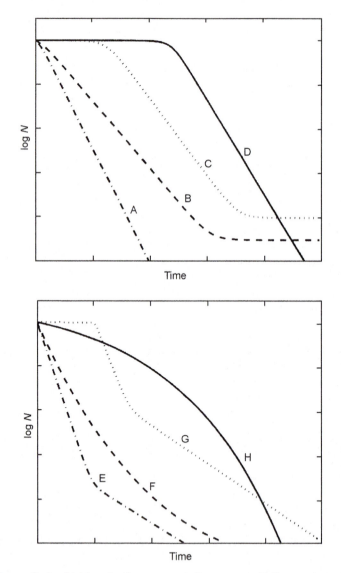

Fig. 10.8   Types of microbial inactivation curves: A, linear curve; B, linear curve with tailing; C, sigmoidal-like curve; D, curve with a shoulder; E, biphasic curve; F, concave curve; G biphasic curve with a shoulder; H, convex curve (adapted from Geeraerd *et al.* 2004).

mortality by using mathematical models based on a 'Weibull' distribution. Further details of the model are described by Peleg (2004), and Geeraerd *et al.* (2004) have reviewed the different models so far developed. Increases in computer power enable these types of mathematical models to be used to re-evaluate the processes based on first-order (linear) inactivation, to predict the survival of newly discovered and heat-resistant strains of pathogens, and to assess whether heat processing conditions are adequate. They can also be used to assess the effects of changes to other factors (e.g. pH, salt content) on microbial survival during heating and so produce a range of heating conditions that can produce a safe product that has satisfactory organoleptic and nutritional properties.

Most enzymes have D and z-values within a similar range to micro-organisms, and factors that influence heat resistance of enzymes are similar to those described for micro-organisms. Enzymes are therefore inactivated during normal heat processing. However, some enzymes are very heat resistant. These are particularly important in acidic foods, where they may not be completely denatured by the relatively short heat treatments and lower temperatures required for microbial destruction. Knowledge of the heat resistance of the enzymes and/or micro-organisms found in a specific food is used to calculate the heating conditions needed for their inactivation. Methods for the calculation of processing time are described in Chapter 13 (section 13.1.1). In practice the most heat-resistant enzyme or micro-organism likely to be present in a food is used as a basis for calculating process conditions. It is assumed that other less heat-resistant enzymes or micro-organisms are also destroyed. Examples of enzymes include peroxidase in vegetables (Chapter 11, section 11.1) and alkaline phosphatase in milk (Chapter 12, section 12.1). Bacteria include *Clostridium botulinum* in meat and vegetable products and *Salmonella seftenberg* in pasteurised liquid egg. Rosnes (2004) has described the thermal destruction of heat-resistant bacteria, bacterial spores, viruses, moulds and prions. (Prions are hypothesised to be proteins that propagate in cells by refolding into a structure that is able to convert normal protein molecules into an abnormally structured form, consisting of tightly packed beta sheets (Chapter 1, section 1.1.1) that are resistant to chemical and physical denaturation. The altered structure is very stable and accumulates in infected tissue, causing tissue damage and cell death. Prions are thought to cause a number of mammalian diseases, including bovine spongiform encephalopathy (BSE) in cattle and Creutzfeldt–Jakob disease (CJD) in humans.)

10.4   Effect of heat on nutritional and sensory characteristics of foods

The destruction of many vitamins, aroma compounds and pigments by heat follows a similar first-order reaction to microbial destruction. Examples of D and z-values of selected vitamins and pigments are shown in Table 10.10. In general z-values are higher than those of micro-organisms and enzymes. As a result, nutritional and sensory properties are better retained by the use of higher temperatures and shorter times during heat processing. It is therefore possible to select particular time–temperature combinations from a TDT curve (all of which achieve the same degree of enzyme or microbial destruction), to optimise a process for nutrient retention or preservation of desirable sensory qualities. This concept forms the basis of individual quick blanching (Chapter 11, section 11.2), high-temperature short-time (HTST) pasteurisation (Chapter 12, section 12.2.2), ultra-high-temperature (UHT) sterilisation (Chapter 13, section 13.2) and HTST extrusion (Chapter 15, section 15.1). The loss of nutrients and changes to sensory quality during individual heat processing operations are reported in Chapters 11–19.

Table 10.10   Heat resistance of selected vitamins and chemicals that contribute to sensory quality of foods in relation to heat-resistant enzymes and bacteria

Component	Source	pH	z-value (°C)	D-value (min)	Temperature range (°C)
Vitamins					
Thiamin	Carrot purée	5.9	25	158	109–149
Thiamin	Lamb purée	6.2	25	120	109–149
Amino acids					
Lysine	Soybean meal	–	21	786	100–127
Pigments					
Anthocyanin	Grape juice	Natural	23.2	17.8[a]	20–121
Betanin	Beetroot juice	5.0	58.9	46.6[a]	50–100
Carotenoids	Paprika	Natural	18.9	0.038[a]	52–65
Chlorophyll *a*	Spinach	6.5	51	13	127–149
Chlorophyll *b*	Spinach	5.5	79	14.7	127–149
Enzymes					
Pectinesterase	Mandarin orange juice	4.0	10.1	3.6	82–94
Pectinesterase	Acidified papaya purée	3.5	14.8	4.8	75–85
Peroxidase	Peas	Natural	37.2	3.0	110–138
Peroxidase	Grape	–	35.4	4.8	65–85
Peroxidase	Strawberry	–	19	5.0	50–70
Polyphenoloxidase	Pear	3.9	–	6.5	75–90
Polyphenoloxidase	Grape	3.3	–	0.45	65–80
Polyphenoloxidase	Apple	3.1	–	0.13	65–80
Bacteria					
Bacillus stearothermilophilus	Various	>4.5	7–12	4.0–5.0	110+
Clostridium botulinum spores	Various	>4.5	5.5–10	0.1–0.3[a]	104
Clostridium butyricum spores	Peach	–	11.5	1.1	90
Bacillus coagulans spores	Tomato paste	4.0	9.0	3.5	75–90
Moulds					
Byssochlamys nivea ascospores	Strawberry pulp	3.0	6.4	193.1	80–93
Neosartorya fischeri ascospores	Apple juice	3.5	5.3	15.1	85–93
Talaromyces flavus ascospores	Strawberry pulp	3.0	8.2	53.9	75–90

[a] *D*-values at temperatures other than 121 °C.
Adapted from Silva and Gibbs (2004), von Elbe *et al.* (1974), Stumbo (1973), Taira *et al.* (1966), Gupta *et al.* (1964), Adams and Yawger (1961), Ponting *et al.* (1960) and Felliciotti and Esselen (1957)

References

ADAMS, H.W. and YAWGER, E.S., (1961), Enzyme inactivation and colour of processed peas, *Food Technology*, **15**, 314–317.

ANON, (2000), Kinetics of microbial inactivation for alternative food processing technologies, overarching principles: kinetics and pathogens of concern for all technologies, Center for Food Safety and Applied Nutrition, US Food and Drug Administration, available at www.cfsan.fda.gov/~comm/ift-over.html.

ANON, (2001), Boiler control systems, United Facilities Criteria document UFC 3-430-11, available at www.wbdg.org/ccb/DOD/UFC/ufc_3_430_11.pdf.

ANON, (2005a), Energy audits, Center for Industrial Research and Service, Iowa State University, available at www.ciras.iastate.edu/energy/energyaudits.asp.

ANON (2005b), Heat pump systems, Energy efficiency and renewable energy, US Dept of Energy, available at www.eere.energy.gov/consumer/your_home/space_heating_cooling/index.cfm/mytopic=12610.

ANON, (2005c), Food and foodstuff – specific heat capacities, The Engineering Toolbox, available at http://www.engineeringtoolbox.com/specific-heat-capacity-food-d_295.html.

ANON, (2006), Industrial combined heat and power, food and beverage growers and processors best practice guide, available at www.fypower.org and follow link to 'Industrial tools' > Products.

ANON (2007a), Thermal properties, Food Resource, Oregon State University, available at http://food.oregonstate.edu/energy/t2.html.

ANON, (2007b), Thermal properties, Food Resource, Oregon State University, available at http://food.oregonstate.edu/energy/t10.html.

ANON, (2007c), Advantages of electric steam, company information from Electrosteam, available at www.electrosteam.com.

ANON, (2008), UK Information Network for Heat Pumps, BRE Sustainable Energy Centre, available at www.heatpumpnet.org.uk/.

BARRON, F. and BURCHAM, J., (2001), Recommended energy studies in the food processing and packaging industry: identifying opportunities for conservation and efficiency, *J. Extension*, **39** (2), available at www.joe.org/joe/2001april/tt3.html.

BRENNAN, J.G, BUTTERS, J.R., COWELL, N.D. and LILLEY, A.E.V., (1990), *Food Engineering Operations*, 3rd edn, Elsevier Applied Science, London.

CARLSSON-KANYAMA, A. and FAIST, M., (2000), Energy use in the food sector – a data survey, Royal Institute of Technology, available at www.infra.kth.se/fms/pdf/energyuse.pdf.

CHOI, Y. and OKOS, M.R., (2003), Thermal conductivity of foods, in *Encyclopaedia of Agricultural, Food, and Biological Engineering*, Taylor and Francis, London, pp. 1004–1010.

DALSGAARD, H. and ABBOTTS, W., (2003), Improving energy efficiency, in (B. Mattsson and U. Sonesson, Eds.), *Environmentally-friendly Food Processing*, Woodhead Publishing, Cambridge, pp. 116–129.

DELGADO, A.E. and SUN, D-W., (2003), Convective heat transfer coefficients, in *Encyclopedia of Agricultural, Food, and Biological Engineering*, Taylor and Francis, London, pp. 156–158.

DEMETRAKAKES, P., (1997), Direct approach (direct steam injection in food processing), *Food Processing*, **58** (6), 83.

EARLE, R.L., (1983), *Unit Operations in Food Processing*, 2nd edn, Pergamon Press, Oxford, pp. 24–38, 46–63, available at www.nzifst.org.nz/unitoperations/matlenerg3.htm.

FELLICIOTTI, E. and ESSELEN, W.B., (1957), Thermal destruction rates of thiamine in puréed meats and vegetables, *Food Technology*, **11**, 77–84.

FLETCHER, A., (2004), Efficient heat exchange for major energy savings, *Food Production Daily.com* – Europe, 28 May, available at www.foodproductiondaily.com/news/ng.asp?id=52442.

GEERAERD, A.H., VALDRAMIDIS, V.P., BERNAERTS, K. and VAN IMPE, J.F., (2004), Evaluating microbial inactivation models for thermal processing, in (P. Richardson, Ed.), *Improving the Thermal Processing of Foods*, Woodhead Publishing, Cambridge, pp. 427–453.

GUPTA, S.M., EL-BISI, H.M. and FRANCIS, F.J., (1964), Kinetics of the thermal degradation of chlorophyll in spinach purée, *J. Food Science*, **29**, 379–382.

HAYHURST, A.N., (1997), Introduction to heat transfer, in (P.J. Fryer, D.L Pyle and C.D. Rielly, Eds.), *Chemical Engineering for the Food Processing Industry*, Blackie Academic and Professional, London, pp. 105–152.

HELDMAN, D.R. and HARTEL, R.W., (1997), Principles of Food Processing, Chapman and Hall, New York, pp. 13–33.

HELDMAN, D.R. and LUND, D.B., (Eds.), (1992), *Handbook of Food Engineering*, Marcel Dekker, New York.

HENDERSON, S.M. and PERRY, R.L., (1955), *Agricultural Process Engineering*, John Wiley, New York.

JACKSON, A.T. and LAMB, J., (1981), *Calculations in Food and Chemical Engineering*, Macmillan, London, pp. 16–57.

KEENAN, J.H., KEYES, F.G., HILL, P.G. and MOORE, J.G., (1969), *Steam Tables – Metric Units*, Wiley, New York.

LEWIS, M.J., (1990), *Physical Properties of Foods and Food Processing Systems*, Woodhead Publishing, Cambridge.

MURAKAMI, E.G., (2003), Thermal diffusivity, in *Encyclopedia of Agricultural, Food, and Biological Engineering*, Taylor and Francis, London, pp. 1014–1017.

OKOS, M., RAO, N., DRECHER, S. RODE, M. and KOZAK, J., (1998), A review of energy use in the food industry, Executive Summary, American Council for an Energy Efficient Economy, available at www.aceee.org/pubs/ie981.htm.

PANDHARIPANDE, S.L. and BADHE, Y.P., (2004), Artificial neural networks for Gurney-Lurie and Heisler charts, *J. Institute of Engineers – India Chemical Engineering Division*, **84**, 65–70, available at www.ieindia.org/publish/ch/0304/mar04ch5.pdf.

PELEG, M., (2000), Microbial survival curves – the reality of flat shoulders and absolute thermal death times, *Food Research International*, **33**, 531–538.

PELEG, M., (2002), A model of survival curves having an 'activation shoulder', *J. Food Science*, **67**, 2438–2443.

PELEG, M., (2003), Calculation of the non-isothermal inactivation patterns of microbes having sigmoidal semi-logarithmic survival curves, *Critical Reviews in Food Science and Nutrition*, **43**, 645–658.

PELEG, M., (2004), Analysing the effectiveness of microbial inactivation in thermal processing, in (P. Richardson, Ed.), *Improving the Thermal Processing of Foods*, Woodhead Publishing, Cambridge, pp. 411–426.

PELEG, M. and COLE, M.B., (1998), Reinterpretation of microbial survival curves, *Critical Reviews in Food Science*, **38**, 353–380.

PIMENTEL, D. and PIMENTEL, M., (1996), *Food, Energy and Society*, revised edition, University Press of Colorado, Niwot, CO.

POLLEY, S.L., SNYDER, O.P. and KOTNOUR, P., (1980), A compilation of thermal properties of foods, *Food Technology*, **36** (11), 76.

PONTING, J.D., SANSHUCK, D.W. and BREKKE, J.E., (1960), Colour measurement and deterioration in grape and berry juices and concentrates, *J. Food Science*, **25**, 471–478.

RICHARDSON, P., (Ed.), (2001), *Thermal Technologies in Food Processing*, Woodhead Publishing, Cambridge.

ROSNES, J.T., (2004), Identifying and dealing with heat-resistant bacteria, in (P. Richardson, Ed.), *Improving the Thermal Processing of Foods*, Woodhead Publishing, Cambridge, pp. 454–477.

ROTSTEIN, E., SINGH, R.P. and VALENTAS, K.J., (Eds.), (1997), *Handbook of Food Engineering Practice*, CRC Press, Boca Raton, FL.

SILVA, F.V.M. and GIBBS, P., (2004), Target selection in designing pasteurisation processes for shelf-stable, high-acid fruit products, *Critical Reviews in Food Science and Nutrition*, **44**, 353–360.

SINGH, R.P. and HELDMAN, D.R., (2001a), Heat transfer in food processing, In *Introduction to Food Engineering*, 3rd edn, Academic Press, London, pp. 208–331.

SINGH, R.P. and HELDMAN, D.R., (2001b), Energy for food processing, in *Introduction to Food Engineering*, 3rd edn, Academic Press, London, pp. 171–205.

STRAUB, U.G. and SCHEIBNER, G., (1984), *Steam Tables in SI Units*, 2nd edn, Springer-Verlag, Berlin.

STUMBO, C.R., (1973), *Thermobacteriology in Food Processing*, 2nd edn, Academic Press, New York.

SUTTER, P.J., (2007), The advantages of hot water vs. steam for jacketed heating, Pick Heaters Inc., available at www.pickheaters.com/hot_water_for_jacketed_heating.cfm.

TAIRA, H., TAIRA, H. and SUKURAI, Y., (1966), Studies on amino acid contents of processed soyabean, Part 8, *Japan J. Nutrition and Food*, **18**, 359.

TOLEDO, R.T., (1999). Heat transfer, in *Fundamentals of Food Process Engineering*, 2nd edn, Aspen Publishers, Gaithersburg, MD, pp. 232–301.

VON ELBE, J.H., MAING, I.Y. and AMUNDSON, C.H., (1974), Colour stability of betanin, *J. Food Science*, **39**, 334–337.

III.A

Heat processing using steam or water

11

Blanching

Abstract: Blanching is a pretreatment that is used to destroy enzymic activity, mostly in vegetables, before unit operations of dehydration or freezing. The chapter describes the theory of blanching, traditional steam and hot water balnchers and developments in microwave blanching. The chapter concludes by discussing the effects of blanching on the sensory and nutritional qualities of foods and on micro-organisms.

Key words: blanching, unsteady state heat transfer, peroxidase, individual quick blanching.

Blanching serves a variety of functions, one of the main ones being to destroy enzymic activity in vegetables and some fruits prior to further processing. As such, it is not intended as a sole method of preservation but as a pre-treatment that is normally carried out between the preparation of the raw material (Chapter 2) and later operations (particularly heat sterilisation, dehydration and freezing (Chapters 13, 16 and 22)). Blanching is also combined with peeling and/or cleaning of foods (Chapter 2) to achieve savings in energy consumption, space and equipment costs.

A few vegetables, for example onions and green peppers, do not require blanching to prevent enzyme activity during storage, but the majority suffer considerable loss in quality if they are not blanched or if they are under-blanched. To achieve adequate enzyme inactivation, food is heated rapidly to a pre-set temperature, held for a pre-set time and then cooled rapidly to near ambient temperatures.

11.1   Theory

Blanching is an example of unsteady state heat transfer (Chapter 10, section 10.1.2), involving convective heating by steam or hot water and conduction of heat within the food. Mass transfer of material into and out of the food tissue is also important (see Chapter 1, section 1.3.3). An example of an unsteady state heat transfer calculation is shown in sample problem 11.1 and further problems are given by Singh and Heldman (2001).

The maximum processing temperature in freezing and dehydration is insufficient to inactivate enzymes and does not substantially reduce the number of micro-organisms in

Sample problem 11.1

Peas which have an average diameter of 6 mm are blanched to give a temperature of 85 °C at the centre. The initial temperature of the peas is 15 °C and the temperature of the blancher water is 95 °C. Calculate the time required, assuming that the heat transfer coefficient is $1200 \, W \, m^{-2} \, K^{-1}$ and, for peas, the thermal conductivity is $0.35 \, W \, m^{-1} \, K^{-1}$, the specific heat is $3.3 \, KJ \, kg^{-1} \, K^{-1}$, and the density is $980 \, kg \, m^{-3}$.

Solution to sample problem 11.1

From Equation 10.24,

$$Bi = \frac{h\delta}{k}$$

$$= \frac{1200(3 \times 10^{-3})}{0.35}$$

$$= 10.3$$

Therefore,

$$\frac{k}{h\delta} = 0.097$$

From Equation 10.25,

$$\frac{\theta_h - \theta_f}{\theta_h - \theta_i} = \frac{95 - 85}{95 - 15}$$

$$= 0.125$$

From the chart for a sphere (Fig. 10.5),

$$Fo = 0.32$$

From Equation 10.26,

$$Fo = \frac{k}{c\rho} \frac{t}{\delta^2}$$

$$= 0.32$$

Therefore, to calculate blanching time (t)

$$t = 0.32 \frac{c\rho\delta^2}{k}$$

$$= \frac{0.32(3.3 \times 10^3)980(3 \times 10^{-3})^2}{0.35}$$

$$= 26.6 \, s$$

unblanched foods. If the food is not blanched, enzymes cause undesirable changes in sensory characteristics and nutritional properties during storage, and micro-organisms are able to grow on thawing or rehydration. In canning, the time taken to reach sterilising temperatures, particularly in large cans, may be sufficient to allow enzyme activity to take place. It is therefore necessary to blanch foods prior to these preservation operations. Under-blanching may cause more damage to food than the absence of blanching does. This is because heat, which is sufficient to disrupt tissues and release enzymes but not inactivate them, causes accelerated damage by mixing the enzymes and substrates. In addition, only some enzymes may be destroyed which causes increased activity of others and accelerated deterioration.

Enzymes that cause loss of colour or texture, production of off-odours and off-flavours, or breakdown of nutrients in vegetables and fruits include lipoxygenase, polyphenoloxidase, polygalacturonase and chlorophyllase. Two heat-resistant enzymes that are found in most vegetables are catalase and peroxidase (Chapter 10, Table 10.10). Although they do not cause significant deterioration during storage, they are used as marker enzymes to determine the success of blanching. Peroxidase is the more heat resistant of the two, so the absence of residual peroxidase activity would indicate that other less heat-resistant enzymes are also destroyed. The factors that control both the rate of enzyme inactivation and the rate of heating at the centre of the product are described in Chapter 10 (sections 10.1.2 and 10.3).

In practice, the time–temperature combinations used for blanching are evaluated for each raw material to achieve a specified temperature at the thermal centre of the food pieces; to achieve a specified degree of peroxidase inactivation; or to retain a specified proportion of vitamin C. The following factors affect blanching conditions:

- the size and shape of the pieces of food;
- the thermal conductivity of the food, which is influenced by the type, cultivar and degree of maturity;
- the blanching temperature and method of heating;
- the convective heat transfer coefficient.

Typical time/temperature combinations vary from 1 to 15 min at 70–100 °C.

The extent of vitamin loss depends on a number of factors including:

- the variety of the food and its maturity;
- methods used in preparation of the food, particularly the extent of cutting, slicing or dicing (Chapter 3, section 3.1);
- the surface-area-to-volume ratio of the pieces of food;
- method of blanching and cooling;
- time and temperature of blanching (lower vitamin losses at higher temperatures for shorter times);
- the ratio of water to food (in both water blanching and cooling).

11.2   Equipment

The two most widespread commercial methods of blanching involve passing food through an atmosphere of saturated steam or a bath of hot water. Both types of equipment are relatively simple and inexpensive. There were substantial developments to blanchers during the 1980s and 1990s to reduce their energy consumption and also to reduce the loss of soluble components of foods. This reduces the volume and polluting potential of effluents (Anon 2008, also Chapter 27, section 27.1.4) and increases the yield of product (the weight of food after processing compared with the weight before processing).

The yield of food from the blanching operation is the most important factor in determining the commercial success of a particular method. In some methods the cooling stage may result in greater losses of product or nutrients than the blanching stage, and it is therefore important to consider both blanching and cooling when comparing different methods. Steam blanching results in higher nutrient retention when cooling is by cold air or cold-water sprays. However, air-cooling causes weight loss of the product due to evaporation, and this may outweigh any advantages gained in nutrient retention. Cooling with running water (fluming) substantially increases leaching losses (washing of soluble components from the food), but the product may gain weight by absorbing water and the overall yield is therefore increased. There are also substantial differences in yield and nutrient retention due to differences in the type of food and differences in the method of preparation.

Recycling of water does not affect the product quality or yield but substantially reduces the volume of effluent produced. However it is necessary to ensure adequate hygienic standards for both the product and equipment to prevent a build-up of bacteria in cooling water. Improved hygiene control may result in additional costs that outweigh savings in energy and higher product yield.

11.2.1 Steam blanchers

The advantages and limitations of steam blanchers are described in Table 11.1. In general this is the preferred method for foods with a large area of cut surfaces as leaching losses are much smaller than those found using hot-water blanchers. At its simplest a steam blancher consists of a mesh conveyor that carries food through a steam atmosphere in a tunnel. The residence time of the food is controlled by the speed of the conveyor and the length of the tunnel. Typically a tunnel is 15 m long and 1–1.5 m wide (Fig. 11.1). The efficiency of energy consumption (i.e. amount of energy used to heat the food divided by the amount of energy supplied) is about 19% when water sprays are used at the inlet and outlet to condense escaping steam. Alternatively, food may enter and leave the blancher through rotary valves or hydrostatic seals to reduce steam losses and increase energy

Table 11.1 Advantages and limitations of steam and hot-water blanchers

Equipment	Advantages	Limitations
Steam blanchers	• Smaller loss of water-soluble components and higher product yield • Smaller volumes of effluent and lower disposal costs than water blanchers, particularly with air cooling instead of water • Better energy efficiency • Better retains product colour, flavour and texture	• Limited cleaning of foods so washers are also required • Uneven blanching if food is piled too high on the conveyor • Some loss of mass in the food • Larger, more complex equipment with higher maintenance costs • More difficult to clean
Hot water blanchers	• Lower capital cost than steam blanchers • More uniform product heating • Use less floorspace	• Large volumes of dilute effluent result in higher costs for both purchase of water and effluent treatment • Risk of contamination of foods by thermophilic bacteria • Turbulence may cause physical damage to some products

Fig. 11.1   Turbo-Flo® steam blancher (courtesy of Key Technologies Inc.).

efficiency to around 27%; or steam may be re-used by passing it through Venturi valves. Energy efficiency is improved to about 31% using combined hydrostatic and Venturi devices (Scott *et al.* 1981).

In older methods of steam blanching, there was often poor uniformity of heating in multiple layers of food. The time–temperature combination required to ensure enzyme inactivation at the centre of the bed resulted in overheating of food at the edges and a consequent loss of quality. Individual quick blanching (IQB) which involves blanching in two stages was developed to overcome this problem. In the first stage the food is heated in a single layer to a sufficiently high temperature to inactivate enzymes. In the second stage (termed 'adiabatic holding') a deep bed of food is held for sufficient time to allow the temperature at the centre of each piece to increase to that needed for enzyme inactivation. This reduces heating times (e.g. 25 s for heating and 50 s for holding 1 cm diced carrot compared with 3 min for conventional blanching, or a reduction from 12 min to 4.5 min for blanching whole sweetcorn cobs) (Anon 2007a). Shorter heating results in improvement in the energy efficiency to 86–91%. The mass of product blanched per kilogram of steam increases from 0.5 kg per kg of steam in conventional steam blanchers to 6–7 kg per kg of steam, when small-particulate foods (e.g. peas, sliced or diced carrots) are blanched. The shorter exposure to heat leads to improvements in colour and flavour, and lower losses of solids and nutrients compared with water-blanched products. There are also substantial savings in steam usage and a reduction in the volume of effluent that has lower solids content.

Nutrient losses during steam blanching are also reduced by exposing the food to warm air (65 °C) in a short preliminary drying operation (termed 'pre-conditioning'). Surface moisture evaporates and the surfaces then absorb condensing steam during IQB. Weight losses and nutrient losses are reduced compared with conventional steam blanching, with no reduction in the yield of blanched food.

The equipment for IQB steam blanching (Fig. 11.2) consists of a heating section in which a single layer of food is heated on a conveyor and then held on a holding conveyor before cooling. Bucket elevators used to load/unload food are located in close-fitting tunnels and the blancher chamber is fitted with rotary valves, both of which minimise steam losses. A cooling section employs a fog spray to saturate the cold air with moisture. This

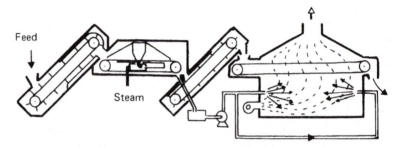

Feed

Steam

Fig. 11.2   Operation of IQB steam blancher (after Timbers *et al.* 1984).

reduces evaporative losses from the food and reduces the amount of effluent produced. The complete inactivation of peroxidase is achieved with a minimum loss in quality, indicated by the retention of 76–85% of ascorbic acid. Anon (2007a) describes a steam blancher having a 1.5 m × 6 m chamber that is fully enclosed and insulated with hydrostatic water seals to prevent evaporation and improve thermal efficiency. It can process $\approx$13 000 kg h^{-1}, creating only 130 l h^{-1} of wastewater, which is 10% of that created by water blanchers. Williams (2007) describes a rotary drum steam blancher that combines the benefits of steam blanching with none of the disadvantages of hot-water drum blanchers, in a compact design that reduces energy and water consumption and has lower capital and maintenance costs than conventional tunnel steam blanchers (Fig. 11.3).

Batch fluidised-bed blanchers operate using a mixture of air and steam, moving at approximately 4.5 m s^{-1}, which fluidises and heats the product simultaneously. The design of the blanching chamber promotes continuous and uniform circulation of the food until it is adequately blanched. Although these blanchers have not yet been widely used at a commercial scale, they have advantages that include: (1) faster, more uniform heating and hence shorter processing times and smaller losses of vitamins and other soluble heat-sensitive components of foods, and (2) substantial reductions in the volumes of effluent.

Fig. 11.3   Clean flow rotary drum blancher (courtesy of Lyco Manufacturing, Inc. at www.lycomfg.com/pdf/csBlancher.pdf).

Steam blanchers may use programmable logic controllers (PLCs) to control belt-speed (residence time) and blanching temperature. The PLC can be accessed or controlled remotely and data can be integrated with factory networks. It can be programmed with different process conditions for specific products and can recall parameters previously used for similar products to fine-tune pre-programmed settings. This reduces operator decision-making and removes the need for manual machine adjustments. This type of control also reduces energy consumption and waste, and processes products up to 60% faster (Anon 2007b, also Chapter 27, section 27.2.2).

11.2.2   Hot-water blanchers

There are a number of different designs of blancher, each of which holds the food in hot water at 70–100 °C for a specified time and then passes it to a dewatering-cooling section. The advantages and limitations of hot-water blanchers are described in Table 11.1. In the reel blancher, food enters a slowly rotating cylindrical mesh drum that is partly sub-merged in hot water. The food is moved through the drum by internal flights. The speed of rotation and length of the drum control the heating time. Pipe blanchers consist of a continuous insulated metal pipe fitted with feed and discharge ports. Hot water is recirculated through the pipe and food is metered in. The length of the pipe and the velocity of the water determine the residence time of food in the blancher. These blanchers have the advantage of a large capacity while occupying a small floor space. In some applications they may be used to simultaneously transport food through a factory.

Developments in hot-water blanchers, based on the IQB principle, reduce energy consumption and minimise the production of effluent. For example the blancher-cooler, has three sections: a pre-heating stage, a blanching stage and a cooling stage. The food remains on a single conveyor throughout each stage and therefore does not suffer physical damage caused by the turbulence of conventional hot-water blanchers. The food is pre-heated with water that is recirculated through a heat exchanger. After blanching, a second recirculation system cools the food. The two systems pass water through the same heat exchanger, and this heats the pre-heat water and simultaneously cools the cooling water. Up to 70% of the heat is recovered. A recirculated water–steam mixture is used to blanch the food and final cooling is by cold air. Effluent production is negligible and water consumption is reduced to approximately $1\,m^3$ per $10\,t$ of product. The mass of product blanched is 16.7–20 kg per kg of steam, compared with 0.25–0.5 kg per kg of steam in conventional hot-water blanchers. Drake and Swanson (1986) report that a counter-current water blancher used 44% less steam to blanch corn than a steam tunnel blancher, producing substantial energy savings with little difference in the yield or quality of products from the two methods.

11.2.3   Newer blanching methods

Microwave heating is described in Chapter 20 (section 20.1). There have been many studies of microwave blanching and most have confirmed that the advantages over conventional blanchers are: (1) more uniform heating, (2) faster heating and reduced energy costs, which lead to (3) reduced processing times and lower nutrient losses. The main disadvantage of microwave blanching is the higher cost of the equipment compared with conventional blanchers. Microwave blanching has been used commercially in Europe and Japan, but not yet widely in the USA. It is reviewed by Dorantes-Alvarez and Parada-Dorantes (2005).

Microwaves are now used commercially for blanching mushrooms to inactivate polyphenoloxidase and prevent browning. Industrial microwave blanchers can be either batch or continuous in operation. However, direct application of microwave energy to entire mushrooms is limited by temperature gradients generated within the pieces, which can vaporise water and damage the texture. A microwave applicator has been developed to modify irradiation conditions and prevent this. Devece *et al.* (1999) used a combined microwave and hot-water bath treatment to produce complete polyphenoloxidase inactivation in a reduced processing time with lower mushroom weight loss and shrinkage. Rodriguez-Lopez *et al.* (1999) found considerable reductions in both the time and temperature required for complete polyphenoloxidase inactivation in whole mushrooms using microwaves compared to conventional hot-water treatments, which resulted in improved product quality. Boyes *et al.* (1997) found that microwave blanching was significantly more efficient at reducing peroxidase activity in sweetcorn kernels when blanched at 85 °C compared with water blanching. They used water-shielding to improve the uniformity of heating and to control the rate of heating by microwaves.

Ramesh *et al.* (2002) compared pulsed microwave blanching at 95 °C with conventional water blanching of spinach, carrot and bell peppers. Microwave blanching reduced nutrient losses and peroxidase inactivation was comparable to water blanching. The absorbed power levels and temperature during microwave blanching were influenced by the quantity and shape of the vegetable, its location in the heater and the microwave power applied. Lin and Brewer (2005) conducted similar studies using peas. Traditional peanut blanching involves heating the nuts in a hot-air, multi-zone oven. The use of a continuous microwave system for peanut blanching was studied by Schirack *et al.* (2007) as a means of reducing processing time and energy costs. They found that the time required to sufficiently heat peanuts to 110 °C and dry them to a final moisture content of <5.5% for acceptable blanching was substantially reduced by microwave heating. Schirack (2006) reported the effects of microwave blanching on flavour attributes in blanched peanuts. Osinboyejo *et al.* (2003) compared microwave and hot-water blanching of green turnip leaves and found that, with the exception of ascorbic acid, nutrient retention was higher in microwave-blanched leaves (25% loss in folic acid, 17% loss in thiamin and 7% loss in riboflavin from microwave blanching compared with 100% loss in each nutrient by hot-water blanching).

Ohmic heating is described in Chapter 20 (section 20.2). It is not yet used commercially for blanching but studies have indicated its potential for mushroom blanching (Sensoy and Sastry 2004). Icier *et al.* (2006) applied ohmic blanching to pea purée and found that peroxidase was inactivated in a shorter time than using water blanching.

High-pressure processing is described in Chapter 8. There have been a number of studies of high pressure blanching (e.g. Van Buggenhout *et al.* 2005, Eshtiaghi and Knorr 1993) which have demonstrated greater nutrient retention than conventional blanching but Cheftel *et al.* (2002) conclude that there is insufficient inactivation of enzymes at high pressure–low temperatures and it is unlikely that this process will replace commercial thermal blanching.

11.3   Effect on foods

Blanching causes cell death and physical and metabolic changes within food cells. Heat damages cytoplasmic and other membranes, which become permeable and result in loss

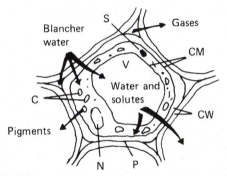

Fig. 11.4   Effect of blanching on cell tissues: S, starch gelatinised; CM, cytoplasmic membranes altered; CW, cell walls little altered; P, pectins modified; N, nucleus and cytoplasmic proteins denatured; C, chloroplasts and chromoplasts distorted.

of cell turgor (Fig. 11.4). Water and solutes pass into and out of cells, resulting in nutrient losses. Heat also disrupts subcellular organelles, and their constituents become free to interact within the cell. Over-blanching can cause excessive softening and loss of flavour in the food, but the heat treatment is less severe than for example in heat sterilisation, and the resulting changes in food quality are less pronounced.

Blanching removes intercellular gases from plant tissues, which together with removal of surface dust, alters the wavelength of reflected light of the food and hence brightens the colour of some vegetables. The time and temperature of blanching also influence changes to food pigments according to their D-value (Chapter 10, section 10.4). Sodium carbonate (0.125% w/w) or calcium oxide may be added to blancher water to protect chlorophyll and to retain the colour of green vegetables, although the increase in pH may increase losses of ascorbic acid. Holding foods such as cut apples and potatoes in dilute (2% w/w) brine prior to blanching prevents enzymic browning. When correctly blanched, most foods have no significant changes to flavour or aroma. Changes in colour and flavour caused by blanching are described in more detail by Selman (1987).

The time–temperature conditions needed to achieve enzyme inactivation may cause an excessive loss of texture in some types of food (e.g. certain varieties of potato) and in large pieces of food. To reduce this, calcium chloride (1–2% w/w) is added to blancher water to form insoluble calcium pectate complexes and thus to maintain firmness in the tissues. In canned foods, blanching softens vegetable tissues, which facilitates filling into containers. The removal of intercellular gases from plant tissues by blanching also assists the formation of a partial vacuum in the head-space of containers. This prevents expansion of air during processing and so reduces strain on the container seams. Removal of oxygen also reduces oxidative changes to the product and corrosion of the can.

The heat received by a food during blanching inevitably causes some changes to sensory and nutritional qualities. Some minerals, water-soluble vitamins and other water-soluble components are lost during blanching. Losses are mostly due to leaching, thermal destruction, and to a lesser extent, oxidation. Fat-soluble components (e.g. β-carotene) are largely retained. Puupponen-Pimiä et al. (2003) studied the effect of blanching on 20 commonly used vegetables. They found that changes were plant species-dependent, but in general that dietary fibre components were either not affected or increased slightly, carotenoids and sterols were not affected, and minerals were stable although there were some leaching losses of soluble minerals. Phenolic antioxidants and vitamins were more heat-sensitive and significant losses of antioxidant activity (20–30%) were found in many

Table 11.2 Effect of blanching method on ascorbic acid losses in selected vegetables

Treatment	Loss (%) of ascorbic acid		
	Peas	Broccoli	Green beans
Water blanch–water cool	29.1	38.7	15.1
Water blanch–air cool	25.0	30.6	19.5
Steam blanch–water cool	24.2	22.2	17.7
Steam blanch–air cool	14.0	9.0	18.6

Differences in both steam versus water blanching and air versus water cooling are significant at the 5% level.
Adapted from Cumming *et al.* (1981)

vegetables. Losses of vitamin C were up to one-third and more than half of the folic acid was lost.

Ascorbic acid is water soluble (so is leached from cells), thermally labile and subject to enzymic breakdown by ascorbic acid oxidase. Losses of ascorbic acid are used as an indicator of the severity of blanching and therefore of food quality (Table 11.2). Typical vitamin losses are 15–20% for riboflavin, 10% for niacin and 10–30% for ascorbic acid (Berry-Ottaway 2002).

11.4 Effect on micro-organisms

The effects of heat on micro-organisms are described in Chapter 10 (section 10.3). Blanching reduces the numbers of contaminating micro-organisms on the surface of foods and hence assists in subsequent preservation operations. This is particularly important in heat sterilisation (Chapter 13), as the time and temperature of processing are designed to achieve a specified reduction in cell numbers. If blanching is inadequate, a larger number of micro-organisms are present initially and this may result in a larger number of spoiled containers after processing. The effect of blanching on micro-organisms has been described by a number of authors including for example, Breidt *et al.* (2000) who found that blanching whole cucumbers for 15 s at 80 °C reduced bacteria by 2–3 log cycles. Vegetative micro-organisms beneath the surface of the cucumber survived blanching.

References

ANON, (2007a), High efficiency blancher/cookers, company information from ABCO Industries Ltd, available at www.abco.ca/foodblancher.html.

ANON, (2007b), New control system adds flexibility to steam blancher/pasteuriser, company information from Key Technology, Inc. in Foodprocessing.com, available at www.foodprocessing.com/vendors/products/2005/226.html.

ANON, (2008), Cleaner Production International LLC, The Food Processing Industry: Improvement of Resource Efficiency and Environmental Performance, available at www.cleanerproduction.com/Directory/sectors/subsectors/FoodProc.html, and follow links to fruit and vegetable processing.

BERRY-OTTAWAY, P., (2002), The stability of vitamins during food processing, in (C.J. Henry and C. Chapman, Eds.), *The Nutrition Handbook for Food Processors*, Woodhead Publishing, Cambridge, pp. 247–264.

BOYES, S., CHEVIS, P., HOLDEN, J. and PERERA, C., (1997), Microwave and water blanching of corn kernels: control of uniformity of heating during microwave blanching, *J. Food Processing and Preservation*, **21** (6), 461–484.

BREIDT, F., HAYES, J.S. and FLEMING, H.P., (2000), Reduction of microflora of whole pickling cucumbers by blanching, *J. Food Science*, **65** (8), 1354–1358.

CHEFTEL, C., THIEBAUD, M. and DUMAY, E., (2002), Pressure-assisted freezing and thawing of foods: a review of recent studies, *High Pressure Research*, **22** (3), 601–611.

CUMMING, D.B., STARK, R. and SANDFORD, K.A., (1981), The effect of an individual quick blanching method on ascorbic acid retention in selected vegetables, *J. Food Processing and Preservation*, **5**, 31–37.

DEVECE, C., RODRIGUEZ-LOPEZ, J.N., FENOLL, L.G., TUDELA, J., CATALA, J.M., DE LOS REYES, E. and GARCIA-CANOVAS, F., (1999), Enzyme inactivation analysis for industrial blanching applications: comparison of microwave, conventional, and combination heat treatments on mushroom polyphenoloxidase activity, *J. Agricultural and Food Chemistry*, **47** (11), 4506–4511.

DORANTES-ALVAREZ, L. and PARADA-DORANTES, L., (2005), Blanching using microwave processing, in (H. Schubert and M. Regier, Eds.), *The Microwave Processing of Foods*, Woodhead, Cambridge, pp. 153–173.

DRAKE, S.R. and SWANSON, B.G., (1986), Energy utilization during blanching (water vs steam) of sweet corn and subsequent frozen quality, *J. Food Science*, **51** (4), 1081–1082.

ESHTIAGHI, M.N. and KNORR, D., (1993), Potato cubes response to water blanching and high hydrostatic pressure, *J. Food Science*, **58** (6), 1371–1374.

ICIER F., YILDIZ, H. and BAYSAL, T., (2006), Peroxidase inactivation and colour changes during ohmic blanching of pea puree, *J. Food Engineering*, **74** (3), 424–429.

LIN, S. and BREWER, M.S., (2005), Effects of blanching method on the quality characteristics of frozen peas, *J. Food Quality*, **28** (4), 350–360.

OSINBOYEJO, M.A., WALKER, L.T., OGUTU, S. and VERGHESE, M., (2003), Effects of microwave blanching vs. boiling water blanching on retention of selected water soluble vitamins in turnip greens using HPLC, Annual Meeting of Inst. Food Technologists, Chicago, 15 July.

PUUPPONEN-PIMIÄ, R., HÄKKINEN, S.T., AARNI, M., SUORTTI, T., LAMPI, A-M., EUROLA, M., PIIRONEN, V., NUUTILA, A.M. and OKSMAN-CALDENTEY, K-M., (2003), Blanching and long-term freezing affect various bioactive compounds of vegetables in different ways, *J. Science of Food and Agriculture*, **83** (14), 1389–1402

RAMESH, M.N., WOLF, W., TEVINI, D. and BOGNAR, A., (2002), Microwave blanching of vegetables, *J. Food Science*, **67** (1), 390–398.

RODRIGUEZ-LOPEZ, J.N., FENOLL, L.G., TUDELA, J., DEVECE, C., SANCHEZ-HERNANDEZ, D., DE LOS REYES, E. and GARCIA-CANOVAS, F., (1999), Thermal inactivation of mushroom polyphenoloxidase employing 2450 MHz microwave radiation, *J. Agriculture Food Chemistry*, **47** (8), 3028–3035.

SCHIRACK, A.V., (2006), The effect of microwave blanching on the flavour attributes of peanuts, PhD thesis, North Carolina State Univ., available at www.lib.ncsu.edu/theses/available/etd-07052006-113250/unrestricted/etd.pdf.

SCHIRACK, A.V., SANDERS, T.H. and SANDEEP, K.P., (2007), Effect of processing parameters on the temperature and moisture content of microwave-blanched peanuts, *J. Food Process Engineering*, **30** (2), 225–240.

SCOTT, E.P., CARROAD, P.A., RUMSEY, T.R., HORN, J., BUHLERT, J. and ROSE, W.W., (1981), Energy consumption in steam blanchers, *J. Food Process Engineering*, **5**, 77–88.

SELMAN, J.D., (1987), The blanching process, in (S. Thorne, Ed.), *Developments in Food Processing*, Vol. 4, Elsevier Applied Science, London, pp. 205–249.

SENSOY, I. and SASTRY, S.K., (2004), Ohmic blanching of mushrooms, *J. Food Process Engineering*, **27** (1), 1–15.

SINGH, R.P. and HELDMAN, D.R., (2001), Unsteady state heat transfer, in *Introduction to Food Engineering*, 3rd edn, Academic Press, London, pp. 280–331.

TIMBERS, G.E., STARK, R. and CUMMING, D.B. (1984), A new blanching system for the food industry, I: design, construction and testing of a pilot plant prototype, *J. Food Processing and Preservation*, **2**, 115–133.

VAN BUGGENHOUT, S., MESSAGIE, I., VAN LOEY, A. and HENDRICKX, M., (2005), Influence of low-temperature blanching combined with high-pressure shift freezing on the texture of frozen carrots, *J. Food Science*, **70** (4), S304–S308.

WILLIAMS, D., (2007), Breakthrough blanching technology combines benefits of steam and rotary drum design, Manufacturing Innovation Insider Newsletter, Lyco manufacturing Inc., available at www.lycomfg.com/pdf/csBlancher.pdf.

12

Pasteurisation

Abstract: Pasteurisation is a mild heat treatment in which food is heated to below 100 °C. It is used to minimise health hazards from pathogenic micro-organisms in low-acid foods and to extend the shelf-life of acidic foods such as fruit juices for several days or weeks by destruction of spoilage micro-organisms and/or enzyme inactivation. This chapter describes the pasteurisation of liquid foods either packaged in containers, or unpackaged using heat exchangers, and reviews the effects on micro-organisms and enzymes and changes caused to the sensory characteristics and nutritive value of foods.

Key words: pasteurisation, phosphatase, D-values and z-values of enzymes and micro-organisms, high-temperature short-time (HTST) processing, concentric tube heat exchanger, plate heat exchanger.

Pasteurisation is a relatively mild heat treatment, in which food is heated to below 100 °C. In low-acid foods (pH > 4.5), such as milk, it is used to minimise public health hazards from pathogenic micro-organisms and to extend the shelf-life of foods for several days or weeks. In acidic foods such as fruit juices (pH < 4.5) it is used to extend the shelf-life for several weeks by destruction of spoilage micro-organisms (yeasts or moulds) and/or enzyme inactivation (Table 12.1). In both types of food, the changes caused to the sensory characteristics or nutritive value are small.

This chapter describes the heat pasteurisation of liquid foods either packaged in containers, or unpackaged using heat exchangers. Processing containers of solid foods that are naturally acidic or made acidic by artificially lowering the pH (e.g. bottled fruits or pickles) is similar to canning (Chapter 13, section 13.1) and is termed pasteurisation to indicate the mild heat treatment employed. Technological developments, including irradiation (Chapter 7), high-pressure processing (Chapter 8) and pulsed electric fields (Chapter 9, section 9.1), have broadened the use of the term pasteurisation and the following definition has been proposed to take these into account: 'Any process, treatment, or combination thereof, that is applied to food to reduce the most resistant micro-organism(s) of public health significance to a level that is not likely to present a public health risk under normal conditions of distribution and storage' (Sugarman 2004). Details of theory and equipment are given by Lewis and Heppel (2000).

Table 12.1   Purpose of pasteurisation for different foods

Food	Main purpose	Subsidiary purpose	Examples of minimum processing conditions[a]
pH < 4.5			
Fruit juice	Enzyme inactivation (pectin methylesterase and polygalacturonase)	Destruction of spoilage micro-organisms (yeasts, moulds)	65 °C for 30 min; 77 °C for 1 min; 80 °C for 10–60 s
Beer	Destruction of spoilage micro-organisms (wild yeasts, *Lactobacillus* sp.) and residual yeasts (*Saccharomyces* sp.)	–	65–68 °C for 20 min (in-bottle); 72–75 °C for 1–4 min
pH > 4.5			
Milk	Destruction of pathogens: *Brucella abortis, mycobacterium tuberculosis, Coxiella burnettii*	Destruction of spoilage micro-organisms and enzymes	63 °C for 30 min; 71.7 °C for 15 s; 88.3 °C for 1 s; 90 °C for 0.5 s
Ice cream, ice milk or eggnog	Destruction of pathogens	Destruction of spoilage micro-organisms	69 °C for 30 min; 71 °C for 10 min; 80 °C for 25 s; 82.2 °C for 15 s
Cream or chocolate milk that have >10% milk fat or added sugar	Destruction of pathogens	Destruction of spoilage micro-organisms	66 °C for 30 min or 75 °C for 15 s
Liquid egg	Destruction of pathogens: *Salmonella seftenberg*	Destruction of spoilage micro-organisms	64.4 °C for 2.5 min; 60 °C for 3.5 min

[a] Followed by rapid cooling to 3–7 °C.
NB: Heat treatment conditions vary from country to country.
Adapted from Anon (2007a) and Dauthy (1995)

12.1   Theory

The effects of heat on enzymes and micro-organisms are described in Chapter 10 and examples of D-values and z-values of enzymes and micro-organisms are given in Table 10.10 in Chapter 10 (section 10.4). The extent of the heat treatment required to pasteurise a food is determined by its pH, which in turn determines whether the target for destruction is the most heat-resistant enzyme, pathogen or spoilage micro-organism that may be present. Among low-acid foods, the pasteurisation process for liquid whole egg is based on a reduction in numbers of *Salmonella seftenberg*, requiring a minimum process of 2.5 min at 64 °C with immediate cooling to 3.3 °C. Pasteurisation of milk is based on a 12D reduction in the numbers of the pathogens *Brucella abortis, Mycobacterium tuberculosis* and *Coxiella burnetii* due to their public health significance. The process is based on $D_{63} = 2.5$ min and $z = 4.1$ °C (i.e. 12 times the D-value of 2.5 min = a holding time of 30 min at 63 °C). Therefore a population of 10 pathogens in a container of raw milk would be reduced to a probability of 10^{-11} by the minimum pasteurisation conditions (alternatively there is a probability of one surviving pathogen in 10^{11} containers).

However, some spoilage bacteria are more heat resistant and may not be destroyed by the minimum heat treatment. Pasteurised milk is therefore stored under refrigeration to maintain the required shelf-life. Formulated liquid milk products (e.g. ice cream, ice milk or eggnog) that contain a high sugar content or high viscosity require higher pasteurisation times/temperatures than the minimum conditions for milk (Table 12.1).

Alkaline phosphatase is a naturally occurring enzyme in raw milk, which has a similar D-value to heat-resistant pathogens. The direct estimation of pathogen numbers by microbiological methods is expensive and time consuming, and a simple test for phosphatase activity is therefore routinely used. A similar test for the effectiveness of liquid-egg pasteurisation is based on residual α-amylase activity. If phosphatase activity is not found, it is assumed that milk has been correctly pasteurised and has also not been contaminated by raw milk. However, alkaline phosphatase can be reactivated in some milk products (e.g. cream or cheese) and micro-organisms used in the manufacture of dairy products may also produce microbial phosphatase that can interfere with tests for residual phosphatase activity. Testing is therefore done immediately after pasteurisation to produce valid results. The amount of alkaline phosphatase in milk also varies widely between different species and within species (e.g. raw cow's milk has a higher activity than goat's milk). Phosphatase is adsorbed on fat globules and the fat content of the product therefore influences the concentration that may be present initially (e.g. $400\,\mu g\,ml^{-1}$ in skimmed cow's milk, $800\,\mu g\,ml^{-1}$ in whole cow's milk and $3500\,\mu g\,ml^{-1}$ in 40% cream) (Anon 2007b). Interpretation of test results therefore needs to take these factors into account.

Pathogens are unable to grow in acidic foods (pH < 4.5) and pasteurisation of these foods is based on heat-resistant enzymes or acid-tolerant spoilage micro-organisms such as lactic acid bacteria, yeasts and moulds. For example, in fruit products pectinesterase is more heat resistant than Gram-positive non-spore-forming bacteria and yeasts, and processing is designed to inactivate this enzyme to enable the required shelf-life to be obtained.

Flavours, colours and vitamins in foods are also characterised by D-values. Different pasteurisation conditions can achieve the same degree of microbial destruction, but the time and temperature can be optimised to retain nutritional values and sensory qualities by the use of high-temperature–short-time (HTST) processing. For example, in milk processing the original lower-temperature longer-time process operating at 63 °C for 30 min (the 'Holder' process) causes larger changes to flavour and a slightly greater loss of vitamins than HTST processing at 71.8 °C for 15 s (Fig. 12.1) (Harper 1976). However, the major nutrients and most vitamins are left unchanged by pasteurisation (section 12.3). Higher temperatures and shorter times (e.g. 88 °C for 1 s, 94 °C for 0.1 s or 100 °C for 0.01 s for milk) are described as 'higher-heat shorter-time' processing or 'flash pasteurisation'. A sample problem showing the calculation of alternative equivalent time–temperature combinations is given in sample problem 12.1. Examples of other methods using the 'lethal rate' for the process are described in Chapter 13 and sample problems (problems 13.1 and 13.2) are given in section 13.1.1.

Pasteurisation involves heating and cooling the food without a change in phase (i.e. sensible heat is added or removed). Sensible heat is calculated using equation 10.8 (Chapter 10, section 10.1). The efficiency of heat transfer in a pasteuriser is measured using heat transfer coefficients, which take account of the properties of the food, the flow characteristics and the equipment used. Heat transfer coefficients are calculated using Equations 10.20–10.24 in Chapter 10 (section 10.1.2). Sample problem 12.2 shows the use of these equations to calculate heat transfer coefficients in a pasteuriser (see also sample problem 12.4 in section 12.2.2).

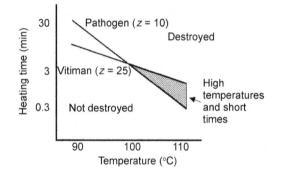

Fig. 12.1   Time–temperature relationships for pasteurisation (courtesy of Professor Douglas Goff, University of Guelph, Canada: www.foodsci.uoguelph.ca/dairyedu/home.html). The shaded area shows the range of times and temperatures used in commercial HTST processing.

Sample problem 12.1

Raw milk containing 4×10^5 bacterial cells ml^{-1} is to be pasteurised at 77 °C for 21 s. The average D-value of the bacteria present in the milk is 7 min at 63 °C and their z-value is 7 °C. Calculate the number of bacteria that will remain after pasteurisation and the processing time required at 63 °C to achieve the same degree of lethality.

Solution to sample problem 12.1

The difference in temperature between 63 °C and 77 °C = 14 °C.

Since the z-value = 7 °C, the D-value at 77 °C is therefore reduced by 2-log cycles from that at 63 °C, i.e. $D = 0.07$ min.

Processing time at 77 °C = 21/60 = 0.35 min,

Therefore processing achieves (0.35/0.07) = 5-log cycle reductions.

Number of bacteria remaining after processing = 4 cells/ml.

At 63 °C, $D = 7$ min, therefore processing time to achieve a 5D reduction = $7 \times 5 = 35$ min.

Sample problem 12.2

Whole milk is cooled in the pipes of a tubular heat exchanger from 30 °C to 10 °C by water at 1 °C. The pipe diameter is 5 cm and the milk flow velocity is 1.0 m s^{-1}. Calculate the heat transfer coefficient for the milk, assuming a specific heat of 3.9 kJ kg^{-1} °C^{-1}, thermal conductivity of 0.56 W m^{-1} °C^{-1} and density of 1030 kg m^{-3}.

Solution to sample problem 12.2

$$\text{Mean bulk temperature} = \frac{30 + 10}{2}$$
$$= 20\,°C$$

From Table 1.8, Chapter 1, for whole milk at 10.5 °C, the viscosity (μ) $= 2.8 \times 10^{-3}$ N s m^{-2}. If Re > 10 000, viscosity should be measured at the mean temperature:

$$\text{Mean film temperature} = \frac{1 + 0.5(30 + 10)}{2}$$

$$= 10.5\,°\text{C}$$

From Equation 1.26 in Chapter 1:

$$\text{Re} = \frac{Dv\rho}{\mu}$$

$$= \frac{0.05 \times 1.0 \times 1030}{2.8 \times 10^{-3}}$$

$$= 18\,393$$

From Equation 10.15 in Chapter 10,

$$\text{Pr} = \frac{c_p \mu}{k}$$

$$= \frac{(3.9 \times 10^3)(2.8 \times 10^{-3})}{0.56}$$

$$= 19.5$$

From Equations 10.14 and 10.18 in Chapter 10,

$$\text{Nu} = \frac{h_c D}{k}$$

$$= 0.023(\text{Re})^{0.8}(\text{Pr})^{0.33}$$

Therefore, the heat transfer coefficient (h_c),

$$h_c = 0.023 \frac{k}{D} (\text{Re})^{0.8} (\text{Pr})^{0.33}$$

$$= 0.023 \frac{0.56}{0.05} (18\,393)^{0.8}(19.5)^{0.33}$$

$$= 0.023 \times \frac{0.56}{0.05} (2581)(2.66)$$

$$= 1768 \text{ W m}^{-2}\,°\text{C}^{-1}$$

The two most important factors to achieve minimum pasteurisation conditions are the temperature to which a product is heated and the time that it is held at that temperature. The controlling factors to establish the correct residence time for a liquid food in a pasteuriser are the velocity of the fastest moving particle in the product and the length of the holding tube. Lethality is based only on the time in the holding tube and, in contrast to canning (Chapter 13, section 13.1.1), the heating and cooling periods are not taken into account. The liquid velocity depends on its flow characteristics (e.g. laminar or turbulent flow (Chapter 1, section 1.3.4) and varies across the diameter of the holding tube. Under laminar flow, the fastest moving particle at the centre of the tube may have twice the average velocity of the liquid, and under turbulent flow it may have 1.2 times the velocity. This information is used to calculate the residence times for the fastest moving particle, and this then determines the length of the holding tube needed to ensure that the minimum holding time is achieved at pasteurisation temperature. A related problem is shown in sample problem 12.3.

Sample problem 12.3
In the counter-current heat exchanger, milk is cooled from 73 °C to 38 °C at the rate of 2500 kg h^{-1}, using water at 15 °C that leaves the heat exchanger at 40 °C. The pipework is 2.5 cm in diameter and constructed from 3 mm thick stainless steel. The surface film heat transfer coefficients are 1200 W m^{-2}°C^{-1} on the milk side and 3000 W m^{-2}°C^{-1} on the water side of the pipe. Calculate the overall heat transfer coefficient (OHTC) and the length of pipe required, assuming a specific heat for milk of 3.9 kJ kg^{-1}°C^{-1}.

Solution to sample problem 12.3
Using Equations 10.19 and 10.20:

$$\frac{1}{U} = \frac{1}{h_a} + \frac{x}{k} + \frac{1}{h_b}$$

$$= \frac{1}{1200} + \frac{3 \times 10^{-3}}{21} + \frac{1}{3000}$$

$$= 1.3 \times 10^{-3}$$

Therefore the OHTC is

$$U = 763.6 \text{ W m}^{-2}\,°\text{C}^{-1}$$

To find the length of pipe required, using Equations 10.20 and 10.22:

$$Q = UA\Delta\theta_m$$

and

$$\Delta\theta_m = \frac{\Delta\theta_1 - \Delta\theta_2}{\ln(\Delta\theta_1/\Delta\theta_2)}$$

$$= \frac{(73 - 40) - (38 - 15)}{\ln[(73 - 40)/(38 - 15)]}$$

$$= 27.7\,°\text{C}$$

Q = heat removed from the milk = $mC_p(\theta_a - \theta_b)$. Therefore,

$$Q = \frac{2500}{3600}(3.9 \times 10^3)(73 - 38)$$

$$= 9.48 \times 10^4 \text{ J}$$

The area of the pipe is:

$$A = \frac{Q}{U\delta\theta_m}$$

$$= \frac{9.48 \times 10^4}{763.6 \times 27.7}$$

$$= 4.48 \text{ m}^2$$

Also

$$A = \pi Dl$$

Therefore the length of the pipe is

$$l = \frac{A}{\pi D}$$

$$= \frac{4.48}{3.142 \times 0.025}$$

$$= 57 \text{ m}$$

12.2   Equipment

12.2.1   Pasteurisation of packaged foods

Some liquid foods (e.g. beers and fruit juices) are pasteurised after filling into containers. Hot water is normally used if the food is packaged in glass, to reduce the risk of thermal shock to the container (fracture caused by rapid changes in temperature). Maximum temperature differences between the container and water are 20 °C for heating and 10 °C for cooling. Metal or plastic containers are processed using steam–air mixtures or hot water, as there is little risk of thermal shock. Containers are cooled to approximately 40 °C to evaporate surface water, which minimises external corrosion to the container or cap, and accelerates setting of label adhesives.

Hot-water pasteurisers may be batch or continuous in operation. The simplest batch equipment consists of a water bath in which crates of packaged food are heated to a pre-set temperature and held for the required length of time. Cold water is then pumped in to cool the product. A continuous version consists of a long narrow trough fitted with a conveyor to carry containers through heating and cooling stages. A second design consists of a tunnel divided into a number of zones (pre-heating, heating and cooling). Very fine (atomised) high-velocity water sprays heat the containers as they pass through each zone on a conveyor, to give incremental rises in temperature until pasteurisation is achieved (Bown 2003). Water sprays then cool the containers as they continue through the tunnel. Savings in energy and water consumption are achieved by recirculation of water between pre-heat sprays, where it is cooled by the incoming food, and cooling zones where it is heated by the hot products. Similar designs of equipment are used to blanch foods (Chapter 11, section 11.2.2). Steam tunnels have the advantage of faster heating, giving shorter residence times, and a smaller space requirement. Temperatures in the heating zones are gradually increased by reducing the amount of air in the steam–air mixtures (Chapter 10, section 10.1.1). Cooling takes place using fine sprays of water or by immersion in a water bath.

12.2.2 Pasteurisation of unpackaged liquids

Open jacketed boiling pans (Chapter 14, section 14.1.3) are used for small-scale batch pasteurisation of some liquid foods. However, large-scale pasteurisation of low-viscosity liquids (e.g. milk, milk products, fruit juices, liquid egg, beers and wines) usually employs continuous equipment, and tube or plate heat exchangers are widely used. The advantages of heat exchangers over in-container processing include:

- more uniform heat treatment;
- lower space requirements and labour costs;
- greater flexibility for different products;
- greater control over pasteurisation conditions;
- greater energy efficiency.

Some products (e.g. fruit juices, wines) require de-aeration prior to pasteurisation to prevent oxidative changes during storage, and milk is homogenised before pasteurisation (Chapter 3, section 3.2.3). They are sprayed into a vacuum chamber and dissolved air is removed by a vacuum pump.

Shell and tube (also 'tube-in-tube' or 'concentric tube') heat exchangers are widely used to pasteurise foods and are particularly suitable for more viscous non-Newtonian foods (e.g. dairy products, mayonnaise, tomato ketchup and baby foods). Different designs comprise a number of concentric stainless steel coils or parallel tubes on a frame, each made from double- or triple-walled tube (Fig. 12.2). Both inner and outer tubes may be corrugated to induce turbulence and reduce fouling. Food passes through the tube, and heating or cooling water is recirculated through the tube walls. Liquid food is passed from one coil to the next for heating and cooling, and heat is regenerated to reduce energy costs.

The plate heat exchanger (Fig. 12.3) was first introduced in 1923 and the compact design and high throughput make it widely used for pasteurisation. It consists of a series of thin vertical stainless steel plates, compressed together in a steel frame that has a fixed plate at one end and a movable pressure plate at the other. The plates form parallel channels for liquid food and heating medium (hot water or steam). The fluids are pumped through alternate channels, usually in a counter-current flow pattern (Fig. 12.4). Each plate is fitted with a synthetic nitrile rubber or neoprene gasket to produce a watertight seal that prevents leakage or mixing of the product and the heating and cooling media.

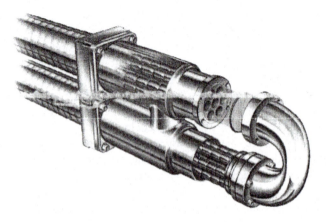

Fig. 12.2 Tubular heat exchanger (courtesy of Tetra Pak).

Fig. 12.3   Plate heat exchanger (courtesy of HRS Heat Exchangers (UK) Ltd at www.hrs.co.uk).

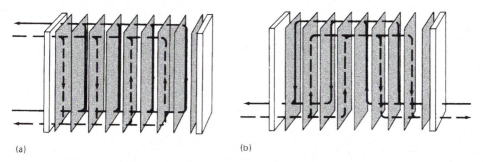

(a)   (b)

Fig. 12.4   Counter-current flow through plate heat exchanger: (a) one pass with four channels per medium; (b) two passes with two channels per pass and per medium (courtesy of HRS Heat Exchangers (UK) Ltd at www.hrs.co.uk).

The plates have corrugations pressed into them (Fig. 12.5), which induce turbulence in the liquids. This, together with the small gap between the plates (≈1–3 mm), and the high velocity induced by pumping, reduces the thickness of boundary films (Chapter 1, section 1.3.4). The small plate thickness (0.3–0.6 mm), high turbulence and low rates of fouling

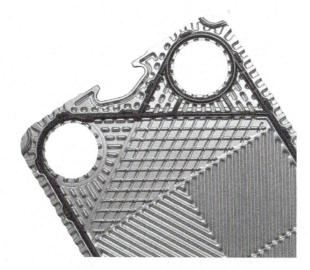

Fig. 12.5   Detail of plate from plate heat exchanger (courtesy of APV Ltd at www.apv.com).

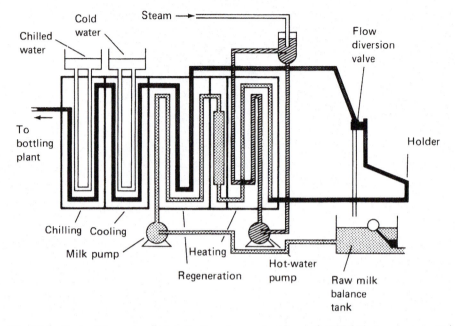

Fig. 12.6 Pasteurising using a plate heat exchanger (courtesy of APV Ltd at www.apv.com).

result in high heat transfer coefficients of the order of 8000–11 500 W m^{-2} °C^{-1}. This compares with traditional shell and tube heat exchangers, in which coefficients are less than 2500 W m^{-2} °C^{-1} (Anon 2007c). The area of plate available for heat transfer varies from 0.02 to 4.6 m^2 and the capacity of the equipment varies according to the size and number of plates, up to 80 000 l h^{-1}.

In operation (Fig. 12.6), food is pumped from a balance tank to a 'regeneration' section, using a positive displacement pump (Chapter 27, section 27.1.3), where it is pre-heated by food that has already been pasteurised. It is then heated to pasteurising temperature in a heating section and passes through a holding tube, retained for the time required to achieve pasteurisation. If the pasteurising temperature is not reached, a temperature sensor activates a flow diversion valve that automatically returns the food to the balance tank to be re-pasteurised. The pasteurised product is then cooled in the regeneration section (and simultaneously pre-heats incoming food) and then further cooled by cold water and, if necessary, chilled water in a cooling section.

The regeneration of heat in this way leads to substantial savings in energy and up to 97% of the heat can be recovered. Heat recovery is calculated using:

$$\text{heat recovery (\%)} = \frac{\theta_2 - \theta_1}{\theta_3 - \theta_1} \times 100 \qquad \boxed{12.1}$$

where θ_1 (°C) = inlet temperature, θ_2 (°C) = pre-heating temperature and θ_3 (°C) = pasteurisation temperature.

After pasteurisation, food is immediately filled into cartons or bottles and sealed to prevent recontamination. Significant levels of spoilage and risks from pathogens can arise from post-pasteurisation contamination, particularly when foods are not re-heated before consumption, and care with cleaning and hygiene is therefore necessary (Chapter 1, section 1.5.1). Further information on methods of pasteurisation is given by Anon (2002).

Sample problem 12.4

Raw whole milk at 7 °C is to be pasteurised at 72 °C in a plate heat exchanger at a rate of 5000 l h^{-1} and then cooled to 4.5 °C. The hot water is supplied at 7500 l h^{-1} at 85 °C and chilled water has a temperature of 2 °C and leaves at 4.5 °C. Each heat exchanger plate has an area of 0.79 m^2. The overall heat transfer coefficients are calculated as 2890 W m^{-2} °C^{-1} in the heating section, 2750 W m^{-2} °C^{-1} in the cooling section and 2700 W m^{-2} °C^{-1} in the regeneration section (see sample problem 12.2). 75% of the heat exchange is required to take place in the regeneration section. Calculate the number of plates required in each section. (Assume that the density of milk is 1030 kg m^{-3}, the density of water is 958 kg m^{-3} at 85 °C and 1000 kg m^{-3} at 2 °C, the specific heat of water is constant at 4.2 kJ kg^{-1} °C^{-1} and the specific heat of milk is constant at 3.9 kJ kg^{-1} °C^{-1}.)

Solution to sample problem 12.4

To calculate the number of plates required in each section, 1 litre = 0.001 m^3; therefore the volumetric flowrate of milk = 5/3600 = 1.39 × 10^{-3} m^3 s^{-1} and the volumetric flowrate of hot water = 7.5/3600 = 2.08 × 10^{-3} m^3 s^{-1}. From Equation 10.13,

$$\text{Heat required to heat milk to 72 °C}$$
$$= 1.39 \times 10^{-3} \times 1030 \times 3900\,(72 - 7)$$
$$= 3.63 \times 10^5\,\text{W}$$

For the regeneration section:

$$\text{Heat supplied} = 75\% \text{ of } 3.63 \times 10^5\,\text{W}$$
$$= 2.72 \times 10^5\,\text{W}$$

and

$$\text{Temperature change of the milk} = 75\% \text{ of } (72 - 7)$$
$$= 48.75\,°\text{C}$$

Therefore the cold milk leaves the regeneration section at 48.75 + 7 °C = 55.75 °C and the hot milk is cooled in the regeneration section to (72 − 48.75) = 23.25 °C. The temperature difference across the heat exchanger plates is (72 − 55.75) = 16.25 °C.

From Equation 10.20 ($Q = UA\,(\theta_a - \theta_b)$),

$$A = \frac{2.72 \times 10^5}{2700 \times 16.25}$$
$$= 6.2\,\text{m}^2$$

As each plate area = 0.79 m^2,

$$\text{Number of plates} = \frac{6.2}{0.79}$$
$$= 7.8 \approx 8$$

In the heating stage,

$$Q = 25\% \text{ of total heat supplied} = 3.63 \times 10^5 \times 0.25$$
$$= 9.1 \times 10^4\,\text{W}$$

From Equation 10.13 for hot water,

$$\theta_a - \theta_b = \frac{9.1 \times 10^4}{2.08 \times 10^{-3} \times 958 \times 4200}$$
$$= 10.85\,^\circ\text{C} = \approx 11\,^\circ\text{C}$$

The temperature of the hot water leaving the heating section is $(85 - 11) = 74\,^\circ\text{C}$. The temperature of the milk entering the heating section is $55.75\,^\circ\text{C}$ and the temperature of the milk after heating is $72\,^\circ\text{C}$.

From Equation 10.22

$$\text{Log mean temperature difference } (\Delta\theta_m) = \frac{(74 - 55.75) - (85 - 72)}{\ln[74 - 55.75)/(85 - 72)]}$$
$$= 15.44\,^\circ\text{C}$$

From Equation 10.20,

$$A = \frac{9.1 \times 10^4}{2890 \times 15.44}$$
$$= 2.04\,\text{m}^2$$

Therefore,

$$\text{number of plates} = \frac{2.04}{0.79}$$
$$= 3$$

For the cooling stage for milk, from Equation 10.13,

$$Q = 1.39 \times 10^{-3} \times 1030 \times 3900\,(23.25 - 4.5)$$
$$= 1.046 \times 10^5\,\text{W}$$

From Equation 10.22 (note that the chilled water leaves at $4.5\,^\circ\text{C}$),

$$\Delta\theta_m = \frac{(23.25 - 4.5) - (4.5 - 2)}{\ln[(23.25 - 4.5)/(4.5 - 2)]}$$
$$= 8.06\,^\circ\text{C}$$

From Equation 10.20,

$$A = \frac{1.046 \times 10^5}{2750 \times 8.06}$$
$$= 4.72\,\text{m}^2$$

Therefore,

$$\text{number of plates} = \frac{4.72}{0.79}$$
$$= 6$$

Novel pasteurisation methods

Research into emerging technologies such as electric arcs (Chapter 7), high-pressure processing (Chapter 8) and pulsed electric fields (PEF) or ultrasound (Chapter 9), together with developments in low-temperature pasteurisation using membrane technology, are intended to replace heat treatments and so reduce changes to organoleptic and nutritional properties. There have been a large number of studies of high-pressure pasteurisation of milk, juices, wines and other products (e.g. Mok *et al.* 2006, Mussa and Ramaswamy 1997). High rates of microbial destruction have been found that are much more rapid than rates of enzyme inactivation or changes to colour and flavour. The process is used commercially for jams and other fruit products and has good potential for pasteurising milk, but further research is required to enable it to become widely used at a commercial scale. High-pressure pasteurisation of juices with 5–20% dissolved CO_2 at 34.5 MPa is reported by Patterson *et al.* (2006) to cause at least 5-log reductions in a range of pathogens without significant changes to sensory characteristics of the juice.

There have also been many studies of PEF pasteurisation; for example PEF treatment of orange juice at $35 \, kV \, cm^{-1}$ for $59 \, \mu s$ caused 7-log reductions in aerobic bacteria, yeasts and moulds, and 90% reduction in pectin methylesterase activity (Yeom *et al.* 2000). Ho and Mittal (2000) reviewed research into PEF pasteurisation and identified further research that is needed to make the process commercially feasible. Ultra-high-pressure homogenisation (UHPH) is being developed as a minimal process to extend the shelf-life and improve the microbial safety of a variety of pasteurised foods (Hayes *et al.* 2005). Pereda *et al.* (2006) compared pasteurisation of milk at 90 °C for 15 s with UHPH treatment at 200 and 300 MPa and 30 or 40 °C. Both processes caused 3–4-log reductions in *Lactococci*. Psychrotrophic bacteria were not detected in pasteurised milk and were reduced by 4-logs in UHPH-treated milk. Coliforms, lactobacilli and enterococci were completely destroyed by both UHPH and heat treatments. However, these methods cannot yet compete with thermal pasteurisation in terms of energy efficiency or scale of operation.

12.3   Effect on foods

Pasteurisation is a relatively mild heat treatment and there are only minor changes to the nutritional and sensory characteristics of most foods. However, the shelf-life of pasteurised foods is usually only extended by a few days or weeks compared with many months with the more severe heat treatment by sterilisation.

Pigments in plant and animal products are also mostly unaffected by pasteurisation. The main cause of colour deterioration in fruit juices is enzymic browning by polyphenoloxidase, and this is prevented by deaeration to remove oxygen prior to pasteurisation. However, Talcott *et al.* (2003) report that reactions, which are independent of oxygen, caused browning of pasteurised passionfruit juice during storage. These are mainly dependent on the temperature of storage. Pasteurisation caused only minor changes to physicochemical attributes, but appreciable juice browning and formation of 5-hydroxymethylfurfural occurred during storage. Fortification with ascorbic acid and sucrose protected carotenoids and retained the colour of the product.

Pasteurised milk is whiter than raw milk but the difference is due to homogenisation (Chapter 3, section 3.2), and pasteurisation alone has no measurable effect. Loss of volatile aroma compounds during pasteurisation of juices causes a reduction in quality and may also unmask other 'cooked' flavours. Jordan *et al.* (2003) reported that deaeration and not pasteurisation of fruit juices caused loss of volatile components. Volatile recovery (Chapter 14, section 14.1.2) may be used to produce high-quality juices but this is not routinely used, owing to the high cost. Loss of volatiles from raw milk removes a hay-like aroma and produces a blander product. Changes to nutritional quality of pasteurised foods are limited to losses of heat-labile vitamins. For example, in milk there is 7% loss of thiamin, 20–25% loss of vitamin C (although milk is not a significant source of this vitamin), losses of 0–10% folate, vitamin B12 and riboflavin, and 5% loss of serum proteins. Further details of vitamin losses in milk are given by Varnam and Sutherland (2001) and Gillis (2005). In fruit juices, losses of vitamin C and carotene are minimised by deaeration.

References

ANON. (2002), Pasteurisation, University of Guelph Dairy Science and Technology, available at www.foodsci.uoguelph.ca/dairyedu/pasteurization.html.

ANON, (2007a), Heat treatments and pasteurisation, Cornell University, available at www.milkfacts.info/Milk%20Processing/Heat%20Treatments%20and%20Pasteurization.htm.

ANON, (2007b), Milk pasteurization basics, Dairy Consultant information, available at www.dairyconsultant.co.uk/pages/Pasteurization.htm.

ANON, (2007c), Plate heat exchangers, Dairy Trade Company information, available at www.dairyconsultant.co.uk/images/DeltaT.doc.

BOWN, G., (2003), Developments in conventional heat treatment, in (P. Zeuthen and L. Bøgh-Sørensen, Eds.), Food Preservation Techniques, Woodhead Publishing, Cambridge, pp. 154–178.

DAUTHY, M.E., (1995), Fruit and vegetable processing, FAO Agricultural Services Bulletins #119, FAO, Rome, available at http://209.85.165.104/search?q=cache:7zA3IRi7DyoJ:www.fao.org/docrep/V5030E/V5030E0o.htm+pasteurization+conditions%2Bjuice&hl=en&ct=clnk&cd=5.

GILLIS, E., (2005), The effect of heat treatment on the nutritional value of milk, California State University, available at www.calstatela.edu/faculty/hsingh2/articles/milk.research.pdf.

HARPER, W., (1976), Processing induced changes, in (W. Harper and C. Hall, Eds.), Dairy Technology and Engineering, AVI, Westport, CT, pp. 539–596.

HAYES, M.G., FOX, P.F. and KELLY, A.L. (2005), Potential applications of high pressure homogenisation in processing of liquid milk, J. Dairy Research, 72 (1), 25–33.

HO, S. and MITTAL, G.S., (2000), High voltage pulsed electrical field for liquid food pasteurisation, Food Reviews International, 16 (4), 395–434.

JORDAN, M.J., GOODNER, K.L. and LAENCINA, J., (2003), Deaeration and pasteurization effects on the orange juice aromatic fraction, Lebensmittel- Wissenschaft und Technologie, 36 (4), 391–396.

LEWIS, M. and HEPPEL, N., (2000), Continuous Thermal Processing of Foods. Pasteurization and UHT Sterilization, Aspen Publishers, Gaithersburg, MD.

MOK, C., SONG, K-T., PARK, Y-S, LIM, S., RUAN, R. and CHEN, P., (2006), High hydrostatic pressure pasteurization of red wine, J. Food Science, 71 (8), M265–M269.

MUSSA, D.M. and RAMASWAMY, H.S., (1997), Ultra high pressure pasteurization of milk: Kinetics of microbial destruction and changes in physico-chemical characteristics, Lebensmittel-Wissenschaft und Technologie, 30 (6), 551–557.

PATTERSON, M.F., LEDWARD, D.A. and ROGERS, N., (2006), High pressure processing, in (J.G. Brennan, Ed.), Food Processing Handbook, Wiley-VCH, Weinheim, Germany, pp. 173–200.

PEREDA, J., FERRAGUT, V., GUAMIS, B. and TRUJILLO, A. J., (2006), Effect of ultra high-pressure homogenisation on natural-occurring micro-organisms in bovine milk, IUFoST, 13th World Congress of Food Science and Technology (DOI: 10.1051/IUFoST:20060250), available at http://iufost.edpsciences.org/ and following link to Proceedings or search authors' names.

SUGARMAN, C., (2004), Pasteurization redefined by USDA committee, definition from the National Advisory Committee on Microbiological Criteria for Foods, reported in *Food Chemical News*, **46** (30), 21.

TALCOTT, S.T., PERCIVAL, S.S., PITTET-MOORE, J. and CELORIA, C. (2003), Phytochemical composition and antioxidant stability of fortified yellow passion fruit (*Passiflora edulis*), *J. Agriculture and Food Chemistry*, **51** (4), 935–941.

VARNAM, A.H. and SUTHERLAND, J.P., (2001), *Milk and Milk Products: Technology, Chemistry, and Microbiology*, Aspen, Gaithersburg, MD.

YEOM, H.W., STREAKER, C.B., ZHANG, Q.H. and MIN, D.B., (2000), Effects of pulsed electric fields on the activities of microorganisms and pectin methyl esterase in orange juice, *J. Food Science*, **65** (8), 1359–1363.

13

Heat sterilisation

Abstract: Heat sterilisation describes both in-container processing (retorting) and ultra-high-temperature (UHT) aseptic processes. Both heat foods to a high temperature to destroy microbial cells, spores and enzymes, thus rendering foods safe and significantly extending their shelf-life at ambient temperatures. This chapter describes the theory of microbial and enzyme inactivation using different time–temperature combinations, the equipment used for in-container and aseptic processing and the effects of heat sterilisation on the safety, sensory characteristics and nutritional value of foods.

Key words: heat resistance of micro-organisms and enzymes, *Cl. botulinum*, heat sterilisation, thermal death time, *F*-value, commercial sterility, canning, ultra-high-temperature (UHT) processing, aseptic processing, heat exchangers.

Heat sterilisation is a unit operation in which foods are heated at a sufficiently high temperature and for a sufficiently long time to destroy vegetative microbial cells, spores and enzymes. As a result, sterilised foods have a shelf-life in excess of six months at ambient temperatures. The foods are also pre-cooked and require minimum heating before consumption, thereby increasing their convenience. However, the severe heat treatment during the older process of in-container sterilisation (canning or bottling) may produce substantial changes in nutritional and sensory qualities of foods. Developments in processing technology aim to reduce the damage to nutrients and sensory components, by either reducing the time of processing in containers, or processing foods before packaging ('aseptic' or 'ultra-high-temperature' (UHT) processing). Reduced processing times can also be achieved by altering the geometry of the container, processing thin layers of product in flexible pouches or trays, or acidification of products. UHT processing typically involves heating foods to a higher temperature (130–150 °C) for a few seconds and then filling the product into pre-sterilised containers. More recent developments, including ohmic heating, are described in Chapter 20 (section 20.2). This chapter describes the optimisation of heat sterilisation processes to achieve the target microbial or enzyme inactivation with minimal effects of food quality, first for in-container heat sterilisation and then for UHT processes. The theory of thermal destruction of micro-organisms and the effect of heat on nutrients and sensory components of foods are described in Chapter 10 (sections 10.3 and 10.4).

Optimisation requires consideration of microbial heat resistance and the rate of heat penetration into foods to achieve the correct design of procedures and equipment. Oliveira (2004) has also described optimisation of heat sterilisation to achieve quality products, taking into account product value and production costs. He notes that the quality gains achieved by UHT processing are only important if there is a higher margin for the products to recover the increased capital investment in the process.

13.1   In-container sterilisation

13.1.1   Theory

The two aims of in-container sterilisation are to inactivate micro-organisms or enzymes and to produce the required sensory and nutritional properties by adequate cooking of the product. The z-values of nutrients and chemicals that contribute to sensory characteristics are four to seven times higher than those of micro-organisms (25–45 °C compared with z-values for micro-organisms of 7–12 °C (Table 13.1)). Therefore, for every 10 °C rise in processing temperature there is approximately a doubling of cooking effect, whereas microbial inactivation is increased ten times (Holdsworth 2004). This has given rise to the concept of a 'cook-value' (or C-value), which is needed to achieve the required change in sensory characteristics (e.g. altering the texture of canned meats or adequate cooking of canned vegetables). The processing time required to achieve a C-value is longer than that needed for sterilisation and Holdsworth (2004) describes equations to calculate C-values at different processing temperatures.

This section focuses on the calculation of processing times that are needed to achieve microbial or enzyme inactivation. The length of time required to sterilise a food is influenced by the:

- heat resistance of micro-organisms or enzymes likely to be present in the food and their numbers, concentration or activity;
- heating conditions;
- pH of the food;
- size and shape of the container; and
- physical state of the food.

To determine the process time for a given food, it is necessary to have information about both the heat resistance of micro-organisms, particularly heat-resistant spores or enzymes that are likely to be present, and the rate of heat penetration into the food.

Table 13.1 z-values for heat-vulnerable components of foods

Component	z-value (°C)
Bacterial spores	7–12
Microbial cells	4–8
Enzymes	3–50
Vitamins	25–30
Proteins	15–37
Sensory factors	
Overall	25–47
Texture-softening	25–47
Colour	24–50

From Holdsworth (1992)

Heat resistance of micro-organisms

The factors that influence heat resistance of micro-organisms or enzymes and their characterisation by *D* and *z*-values are described in Chapter 10 (section 10.3). In low-acid foods (pH > 4.5), the heat-resistant spore-forming micro-organism *Clostridium botulinum,* is the most dangerous pathogen likely to be present. Under anaerobic conditions inside a sealed container it can grow to produce a powerful exotoxin, botulin, which is sufficiently potent to be 65% fatal to humans. *Cl. botulinum* is ubiquitous in soil and it is therefore likely to be found in small numbers on any raw material that has contact with soil, or be transferred by equipment or operators to other foods. Because of the extreme hazard from botulin, the destruction of this micro-organism is therefore a minimum requirement of heat sterilisation. Normally, foods receive more than this minimum treatment as other more heat-resistant spoilage bacteria may also be present (Table 10.9). *Cl. botulinum* cannot grow in more acidic foods (pH 4.5–3.7), and other micro-organisms (e.g. yeasts and fungi) or heat-resistant enzymes are more important causes of food spoilage and are used to establish processing times and temperatures. In acidic foods (pH < 3.7), enzyme inactivation is the main reason for processing and heating conditions are less severe (sometimes referred to as 'pasteurisation').

Thermal destruction of micro-organisms has long been assumed to take place logarithmically at high temperatures (although more recently different thermal destruction kinetics have been found (see Chapter 10, section 10.3)). A logarithmic death rate means that theoretically a sterile product cannot be produced with certainty no matter how long the process time. However, the probability of survival of a single micro-organism can be predicted using details of the heat resistance of the particular microbial strain and the temperature and time of heating. This gives rise to a concept known as 'commercial sterility'. For example, a process that reduces cell numbers by 12 decimal reductions (a 12*D* process) applied to a raw material that contains 1000 spores per container would reduce microbial numbers to 10^{-9} per container, or the probability of one microbial spore surviving in one billion (10^{9}) containers processed. Commercial sterility means in practice that heat processing inactivates substantially all vegetative cells and spores, which if present would be capable of growing in the food under defined storage conditions.

However, if foods that contain more heat-resistant spoilage micro-organisms (Table 10.9) are given a 12*D* process, this would result in over-processing and excessive loss of quality. In practice a 2*D* to 8*D* process is used to give the most economical level of food spoilage consistent with adequate food quality and safety. Because of the comparatively lower heat resistance of *C. botulinum*, the probability of survival remains similar to that obtained in a 12*D* process. The spoilage probability can be expressed as Equation 13.1:

$$\frac{1}{n} = \frac{n_0}{10^{F/D}} \qquad \boxed{13.1}$$

where n = number of containers of processed product, n_0 = initial number of spoilage micro-organisms per container, F = thermal death time required (the process time) and D = decimal reduction time.

The equation can be used in a number of ways to calculate:

- the number of containers that can be processed before there is a probability of one spoiled container;
- an acceptable number of containers that can be processed before one contains a spoilage micro-organism;

- the process time needed to achieve an acceptable level of spoilage;
- the level of spoilage that could be expected from a process that has a known *F*-value (Heldman and Hartel 1997).

For processes to operate successfully, the microbial load on raw materials must be kept at a low level by hygienic handling and preparation procedures (Chapter 2), and in some foods by blanching (Chapter 11). In addition, the correct processing conditions and methods must ensure that all containers receive the same amount of heat. Any failure in these procedures would increase the initial numbers of cells and hence increase the incidence of spoilage after processing.

Rate of heat penetration

In addition to information on heat resistance of micro-organisms and enzymes, it is necessary to collect data on the rate of heat penetration into a food in order to calculate the processing time needed for commercial sterility. Heating containers of food is an unsteady state heat transfer process, which is described in Chapter 10 (section 10.1.2). Heat is transferred from steam or pressurised water through the container and into the food. Generally the surface heat transfer coefficient at the container wall is very high and is not a limiting factor in heat transfer. Table 13.2 describes the main limiting factors on the rate of heat penetration into a food.

A major problem with processing solid or viscous foods is the low rate of heat penetration to the thermal centre. As a result, over-processing reduces the nutritional value and sensory properties of food that is near the walls of the container, in addition to causing long processing times and low productivity. Methods that are used to increase the rate of heat transfer include the use of thinner profile containers and, for viscous foods, agitation of containers. Tucker (2004a) has reviewed the benefits of rotating containers during sterilisation. For example, doubling the speed of end-over-end agitation of cans that contain liquids with particles increases the can-to-fluid heat transfer coefficient by approximately 30% and the fluid-to-particle heat transfer coefficient by 50% (Sablani and Ramaswarmy 1998). An increase in retort temperature would also reduce processing times and protect nutritional and sensory qualities, but this is usually impractical as the higher pressures would require substantially stronger and hence more expensive equipment.

The rate of heat penetration is measured by placing a temperature sensor at the thermal centre of a container (the point of slowest heating or 'critical point') to record temperatures in the food during processing (Fig. 13.1). It is assumed that all other points in the container receive more heat than the thermal centre. In cylindrical containers the thermal centre is approximately one-fifth of the container height above the base for convective heating foods and at the geometric centre of the container for conductive heating foods when the container has a height : diameter ratio between 0.95 and 0.3 (Holdsworth 2004). The convective heating position is also used for products that change from convective heating to conductive heating during the process (foods that contain a high concentration of starch which undergoes a sol-to-gel transition, producing a broken heating curve). A heating curve is produced by plotting temperature versus time on semi-logarithmic graph paper (Fig. 13.2).

Temperature measurement inside containers in static retorts is made using temperature sensors such as copper/constantan thermocouples, platinum resistance thermometers or non-metallic thermistors, each linked by cables to a data logger. In continuous sterilisers, self-contained miniature data acquisition units are fitted inside or outside the

Table 13.2   Factors that influence the rate of heat penetration into a food

Category	Factor	Notes
Product-related factors	Consistency	Liquid or particulate foods have natural convection currents and heat faster than solid foods in which heat is transferred by conduction. They can be grouped into: • most rapid convection heating (e.g. juices, broths, milk) • rapid convection heating (e.g. fruits in syrup, peas in brine) • slower convection/conduction heating (e.g. soups, tomato juice) • conduction heating, water-based foods (e.g. thick purées, rice, spaghetti) • conduction heating, non-water-based foods (e.g. meat pastes and corned beef, high sugar products, low moisture puddings).
	Thermal properties	The low thermal conductivity of most foods is a major limitation to heat transfer in conduction heating foods.
Process-related factors	Temperature of retort	A higher temperature difference between the food and the heating medium causes faster heat penetration.
	Type of heat transfer medium	Saturated steam is most effective. The steam pressure balances the pressure developed inside the container when it is heated. The velocity of water or steam/air mixtures influences the rate of heat transfer (Chapter 10, section 10.1.2).
	Agitation of containers	End-over-end agitation (Fig. 13.6) and, to a lesser extent, axial agitation increases the effectiveness of natural convection currents and thereby increases the rate of heat penetration in viscous or semi-solid foods (e.g. beans in tomato sauce).
	Type of retort	Batch or continuous operation (section 13.1.3).
Package-related factors	Size of containers	Heat penetration to the centre is faster in small containers than in large containers.
	Shape of containers	Tall containers promote convection currents in convective heating foods. Trays, pouches or flat cans are relatively thin and have a higher surface area : volume ratio than cylindrical cans or bottles – each promoting faster heat penetration.
	Container material	Heat penetration is faster through metal than through glass or plastics owing to differences in their thermal conductivities (Chapter 10, Table 10.2).
	Headspace volume	In static retorts, headspace gas insulates the surface of foods and reduces heat penetration. In agitated retorts, movement of headspace gas bubble mixes convective/conductive heating foods.

container and linked to a temperature probe placed at the thermal centre of the container. After processing, the data is downloaded to a data reader connected to a computer. Details of the advantages and limitations of equipment are given by Shaw (2004), and Anon (1997) gives guidelines for conducting heat penetration studies. Developments in non-invasive methods of temperature measurement, including magnetic resonance imaging, microwave radiometry and fibre optic thermometry are described by Nott and Hall (2004).

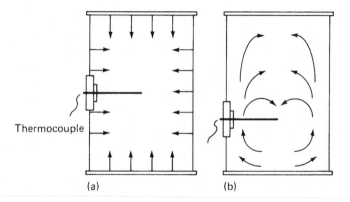

Fig. 13.1   Heat penetration into containers by (a) conduction and (b) convection.

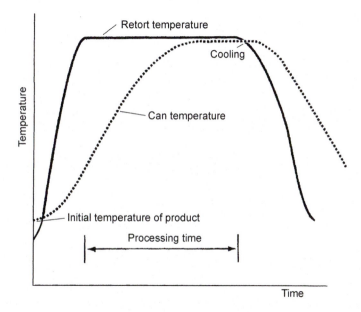

Fig. 13.2   Heat penetration into a container of conductive heating food (from Holdsworth 2004).

Process validation

Oliveira (2004) describes different approaches to creating mathematical models to optimise process times. There are a number of methods to validate the safety of a process, including laboratory simulation trials in which process conditions are replicated and surviving micro-organisms are counted. Another method is mathematical modelling, described in detail by Eszes and Rajko (2004), Bown (2004), Geeraerd *et al.* (2004) and Peleg (2003), and the application of computational fluid dynamics, described by Verboven *et al.* (2004). In canning factories, processes can be validated using time–temperature indicators (TTIs) that involve inoculating an enzyme or a non-pathogenic test micro-organism into the food, or encapsulating it in alginate beads that have similar thermal and physical properties to the food being processed. The test enzyme or micro-organism has similar D and z-values to the target pathogen. After the foods or beads have passed through the process in containers they are recovered and the residual enzyme activity is measured or the numbers of surviving cells are counted. Details are given by Tucker (2004b) and Van Loey et al. (2004).

It is also necessary to conduct studies of the temperature distribution in the retort to locate any 'cold spots', and studies of the rate of heat penetration into the food. Accelerated storage trials on randomly selected cans of food ensure that the level of commercial sterility is maintained before foods are released for retail sale. Some types of spoilage, including microbial spoilage and hydrogen produced by the interaction of acids in the food with the metal of the can, cause swelling. Increasingly severe swelling of a can is termed 'flipper' > 'springer' > 'soft swell' > 'hard swell'. Routine quality assurance measures therefore include observation for swollen or 'bloated' cans (however, swelling is not solely due to spoilage and can also be caused by overfilling, denting, closing after the can has cooled and high storage temperatures or high altitudes). Details of quality assurance and HACCP systems for heat-sterilised foods are given by Shapton and Shapton (1993) and Anon (1993).

Calculation of process times
The thermal death time (TDT), or F-value, is used as a basis for comparing heat sterilisation procedures. It is the time required to achieve a specified reduction in microbial numbers at a given temperature and it thus represents the total time–temperature combination received by a food. It is quoted with suffixes indicating the retort temperature and the z-value of the target micro-organism. For example, a process operating at 115 °C based on a micro-organism with a z-value of 10 °C would be expressed as F_{115}^{10}. The F-value may also be thought of as the time needed to reduce microbial numbers by a multiple of the D-value. It is found using:

$$F = D \left(\log n_1 - \log n_2 \right) \hspace{2cm} \boxed{13.2}$$

where $n_1 =$ initial number of micro-organisms and $n_2 =$ final number of micro-organisms. A reference F-value (F_0) is used to describe processes that operate at 121 °C which are based on a micro-organism with a z-value of 10 °C. Typical F_0 values are shown in Table 13.3.

The slowest heating point in a container may not reach the retort temperature, but once the temperature of the food rises above approximately 70 °C, thermal destruction of vegetative micro-organisms takes place. However, spores are more heat resistant. The processing time is therefore the period that a given can size should be held at a set processing temperature in order to achieve the required thermal destruction of the type of cells or spores that are likely to be present at the slowest heating point in the container. Two methods for calculating process time are described in the following section, the first being a mathematical method based on the equivalent lethality of different time : temperature combinations and the second being a graphical method. It should be noted that these methods were developed at a time when there was limited capability to perform complex calculations. Increased computing power has enabled researchers to develop new more complex models that use actual experimental data on thermal resistance of micro-organisms in specific foods, rather than assuming logarithmic destruction. Details of these new methods are given by Geeraerd et al. (2004), Eszes and Rajko (2004) and Peleg (2003). Holdsworth (1997) compares a number of other mathematical methods.

Table 13.3   Selected commercial F_0 values

Product	Can size (mm) diameter × height	F_0-value
Meat and fish products		
Curried meat and vegetables	73 × 117	8–12
Petfoods	83 × 114	12
Petfoods	153 × 178	6
Frankfurter sausages	73 × 178	3–4
Chicken breast in jelly	73 × 117	6–10
Vegetables		
Celery	83 × 114	3–4
Sweetcorn in brine	83 × 114	9
Peas in brine	83 × 114	4–6
Peas in brine	153 × 178	10
Mushrooms in brine	65 × 101	8–10
Other products		
Baby foods	52 × 72	3–5
Cream soups	73 × 117	4–5
Milk puddings	73 × 117	4–10
Cream	73 × 117	6
Evaporated milk	73 × 117	5

Adapted from Holdsworth (2004)

Formula (or mathematical) method

This method enables calculation of process times for different retort temperatures or container sizes, but it is limited by the assumptions made about the nature of the heating process. The method is based on:

$$B = f_h \log \left(\frac{j_h I_h}{g} \right) \qquad \boxed{13.3}$$

where B (min) = time of heating, f_h (min), the heating rate constant = the time for the heat penetration curve to cover one logarithmic cycle, j_h = the thermal lag factor found by extrapolating the curve in Fig. 13.3 to find the pseudo-initial product temperature (θ_{pih}).

$$j_h = \frac{\theta_r - \theta_{pih}}{\theta_r - \theta_{ih}} \qquad \boxed{13.4}$$

where $I_h = (\theta_r - \theta_{ih})$ (°C) = the difference between the retort temperature and the initial product temperature, g = the difference between the retort temperature and the product temperature at the slowest heating point at the end of heating, θ_r (°C) = retort temperature and θ_{ih} (°C) = initial product temperature. Further details of this method are given by Singh and Heldman (2001).

The heating rate constant varies according to the surface area : volume ratio of the container and therefore depends on the shape and size of the pack. It also depends on whether the product heats by convection or conduction. With the exception of g, the above information can be found from the heating curve (Fig. 13.3). The value of g is influenced by the following factors:

- the TDT of the micro-organism on which the process is based;
- the slope f_h of the heating curve;

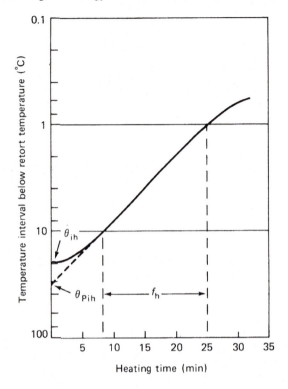

Fig. 13.3   Heat penetration curve on semi-log paper.

- the z-value of the target micro-organism;
- the difference between the retort temperature and the temperature of the cooling water.

To take account of these variables, Ball (1923) developed the concept of comparing the F-value at the retort temperature (denoted F_1) with a reference F-value of 1 min at 121 °C (denoted F). The thermal death time at the retort temperature is described by the symbol U and is related to the reference F-value and F_1 using:

$$U = FF_1 \hspace{5cm} \boxed{13.5}$$

If the reference F-value is known, then it is possible to calculate U by consulting F_1 tables (Table 13.4) or calculating the lethal rate using Equation 13.8. The value of g may then be found from f_h/u and g tables (Table 13.5).

In conductive heating foods, there is a lag before cooling water begins to lower the product temperature, and this results in a significant amount of heating after the steam has been turned off. It is therefore necessary to include a cooling lag factor j_c. This is defined as the time taken for the cooling curve to cover one logarithmic cycle, and is analogous to j_h, the heating lag factor. The cooling portion of the heat penetration curve is extrapolated to find the pseudo-initial product temperature θ_{pic} at the start of cooling, in a similar way to θ_{pih}. j_c is found using:

$$j_c = \frac{\theta_c - \theta_{pic}}{\theta_c - \theta_{ic}} \hspace{4cm} \boxed{13.6}$$

where θ_c (°C) = cooling water temperature and θ_{ic} (°C) = the actual product temperature at the start of cooling. When using Table 13.5, the appropriate value of j_c can then be used

Table 13.4 F_1-values for selected z-values at retort temperatures below 121 °C

$121-\theta_r$ (°C)	z-value					
	4.4 °C	6.7 °C	8.9 °C	10 °C	11.1 °C	12 °C
5.6	17.78	6.813	4.217	3.594	3.162	2.848
6.1	23.71	8.254	4.870	4.084	3.548	3.162
6.7	31.62	10.00	5.623	4.642	3.981	3.511
7.2	42.17	12.12	6.494	5.275	4.467	3.899
7.8	56.23	14.68	7.499	5.995	5.012	4.329
8.3	74.99	17.78	8.660	6.813	5.623	4.806
8.9	100.0	21.54	10.00	7.743	6.310	5.337
9.4	133.4	26.10	11.55	8.799	7.079	5.926
10.0	177.8	31.62	13.34	10.00	7.943	6.579
10.6	237.1	38.31	15.40	11.36	8.913	7.305

Adapted from Stumbo (1973)

Table 13.5 Selected f_h/U and g-values when $z = 10$ and $j_c = 0.4$–2.0

f_h/U	Values of g for the following j_c values					
	0.40	0.80	1.00	1.40	1.80	2.00
0.50	0.0411	0.0474	0.0506	0.0570	0.0602	0.0665
0.60	0.0870	0.102	0.109	0.123	0.138	0.145
0.70	0.150	0.176	0.189	0.215	0.241	0.255
0.80	0.226	0.267	0.287	0.328	0.369	0.390
0.90	0.313	0.371	0.400	0.458	0.516	0.545
1.00	0.408	0.485	0.523	0.600	0.676	0.715
2.00	1.53	1.80	1.93	2.21	2.48	2.61
3.00	2.63	3.05	3.26	3.68	4.10	4.31
4.00	3.61	4.14	4.41	4.94	5.48	5.75
5.00	4.44	5.08	5.40	6.03	6.67	6.99
10.0	7.17	8.24	8.78	9.86	10.93	11.47
20.0	9.83	11.55	12.40	14.11	14.97	16.68
30.0	11.5	13.6	14.6	16.8	18.9	19.9
40.0	12.8	15.1	16.3	18.7	21.1	22.3
50.0	13.8	16.4	17.7	20.3	22.8	24.1
100.0	17.6	20.8	22.3	25.4	28.5	30.1
500.0	26.0	30.6	32.9	37.5	42.1	44.4

Adapted from Stumbo (1973)

to find g. More complex formulae are necessary to calculate processing times where the product displays a broken heating curve.

Finally, in batch retorts, only 40% of the time taken for the retort to reach operating temperature (the 'come-up' time l) is at a sufficiently high temperature to destroy micro-organisms. The calculated time of heating (B) is therefore adjusted to give the corrected processing time:

$$\text{process time} = B - 0.4l \qquad \boxed{13.7}$$

Improved general (graphical) method
This method is based on the fact that different combinations of temperature and time have the same lethal effect on micro-organisms (see Chapter 10, section 10.3). Lethality is therefore the integrated effect of temperature and time on micro-organisms. As the

Sample problem 13.1

A low-acid food is heated at 115 °C using a process based on $F_{121.1}^{10} = 7\,\text{min}$. The following information was obtained from heat penetration data: $\theta_{ih} = 78\,°C$, $f_h = 20\,\text{min}$, $j_c = 1.80$, $f_c = 20\,\text{min}$, $\theta_{pih} = 41\,°C$ and $\theta_{ih} = 74\,°C$. The retort took 11 min to reach process temperature. Calculate the processing time.

Solution to sample problem 13.1

From Equation 13.4:

$$J_h = \frac{115 - 41}{115 - 74}$$

$$= 2.00$$

and

$$I_h = 115 - 78$$

$$= 37\,°C$$

From Table 13.4 (for $121.1 - \theta_r = 6.1$ and $z = 10\,°C$),

$$F_1 = 4.084$$

From Equation 13.5:

$$U = 7 \times 4.084$$

$$= 28.59,$$

$$\frac{f_h}{U} = \frac{20}{28.59}$$

$$= 0.7$$

From Table 13.5 (for $f_h/U = 0.7$, $j_c = 1.80$),

$$g = 0.241\,°C$$

(i.e. the thermal centre reaches 114.76 °C).

From Equation 13.3:

$$B = 20 \log \left(\frac{2.00 \times 37}{0.241} \right)$$

$$= 49.6\,\text{min}$$

From Equation 13.7:

$$\text{Process time} = 49.7 - (0.4 \times 11)$$

$$= 45.2\,\text{min}$$

This gives the process time for $F_0 = 7\,\text{min}$. If the process time had been given, it would be possible to reverse the calculation to find F_0.

Table 13.6 Lethal rates for $z = 10\,°C$

Temperature (°C)	Lethal rate (min[a])	Temperature (°C)	Lethal rate (min)
90	0.001	108	0.049
92	0.001	110	0.077
94	0.002	112	0.123
96	0.003	114	0.195
98	0.005	116	0.308
100	0.008	118	0.489
102	0.012	120	0.774
104	0.019	122	1.227
106	0.031	124	1.945

[a] At 121°C per minute at θ_r.
Adapted from Stumbo (1973)

temperature increases, there is assumed to be a logarithmic reduction in the time needed to destroy the same number of micro-organisms (although more recent studies have shown other destruction kinetics – see Chapter 10, section 10.3). This is expressed as the lethal rate (a dimensionless number that is the reciprocal of TDT) and is shown by the following equation:

$$\text{Lethal rate} = 10^{(\theta - 121)/z} \qquad \boxed{13.8}$$

where θ (°C) = temperature of heating.

The TDT at a given processing temperature is compared to a reference temperature (T) of 121 °C. For example, if a product is processed at 115 °C and the most heat-resistant micro-organism has a z-value of 10 °C,

$$\text{Lethal rate} = 10^{(115 - 121)/10}$$
$$= 0.25$$

As the temperature of a food increases during processing, there is a higher rate of microbial destruction. The initial heating part of the process contributes little towards total lethality until the retort temperature is approached, and most of the accumulated lethality takes place in the last few minutes before cooling begins.

The lethal rate depends on the z-value of the micro-organism on which the process is based and the product temperature, and tables of lethal rate values are available. Table 13.6 is for $z = 10$, the value for most spoilage micro-organisms. This method is preferable in practical situations for determining the impact of a process in terms of equivalent temperature/time relationships.

For convection heating foods, the lethal rate curve (Fig. 13.4) is used to find the point in the process when heating should cease. A line is drawn parallel to the cooling part of the curve so that the total area enclosed by the curve is equal to the required lethality. An example is given in sample problem 13.2 and examples of F_0 values used commercially are shown in Table 13.3.

With conduction heating foods the temperature at the centre of the container may continue to rise after cooling commences, because of the low rate of heat transfer. For these foods it is necessary to determine lethality after a number of trials in which heating is stopped at different times.

Sample problem 13.2

A convective heating food is sterilised at 115 °C to give $F_0 = 7$ min. The come-up time of retort is 11 min and cooling started after 60 min. Calculate the processing time from the following heat penetration data.

Process time (min)	Temperature (°C)	Process time (min)	Temperature (°C)
0	95	35	114.5
5	101	40	114.5
10	108.5	45	114.5
15	111.4	50	114.6
20	113	55	114.6
25	115.5	60	114.6
30	115.5	65	98

Solution to sample problem 13.2

Lethal rates can be found at selected points on a heat penetration curve either by constructing a TDT curve and taking the reciprocal of TDTs (from Fig. 10.7 in Chapter 10) at the selected temperatures, or by consulting the appropriate lethal rate table (Table 13.6). Lethal rates are then plotted against processing time (Fig. 13.4) and the area under the curve is measured by counting squares or using a planimeter.

Process time (min)	Lethal rate	Process time (min)	Lethal rate
0	0.002	35	0.218
5	0.01	40	0.218
10	0.055	45	0.224
15	0.109	50	0.224
20	0.155	55	0.224
25	0.218	60	0.224
30	0.218	65	0.005

For convection heating foods, the lethal rate curve is used to find the point in the process when heating should cease. A line is drawn parallel to the cooling part of the curve so that the total area enclosed by the curve is equal to the required lethality. The area under the curve ACE is 100.5 cm^2. As 1 cm^2 is 0.1 min at 121°C, the area ACE $= 10.05$ min at 121 °C. Therefore by reducing the area under the curve ABD to 70 cm^2 ($F = 7$ min), the process time is 45 min. Thus the process time required for $F_0 = 7$ min is 45 min.

13.1.2   Retorting

The shelf-life of sterilised foods depends in part on the ability of the container to isolate the food completely from the environment. The four main types of heat-sterilisable container are:

- metal cans;
- glass jars or bottles;
- flexible pouches;
- rigid trays.

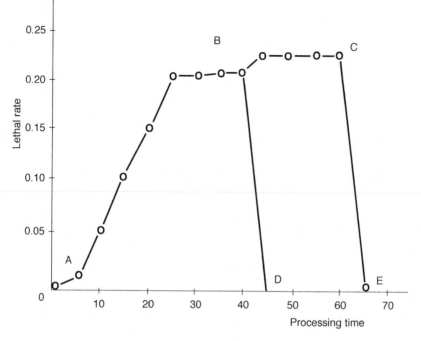

Fig. 13.4   Lethal rate curve.

These materials are described in Chapter 25 (section 25.2) and methods of filling and sealing are described in Chapter 26 (section 26.1).

Before filled containers are processed, it is necessary to remove air by an operation termed 'exhausting'. This prevents air expanding with the heat and therefore reduces strain on the container seals. The removal of oxygen also prevents both internal corrosion of metal cans and oxidative changes in some foods during storage. Containers are exhausted by the following methods:

- Hot filling the food into the container (commonly used as it also pre-heats food which reduces processing times).
- Cold filling the food and then heating the container and contents to 80–95 °C with the lid partially sealed (clinched).
- Mechanical removal of the air using a vacuum pump.
- Steam flow closing, where a blast of steam carries air away from the surface of the food immediately before the container is sealed. This method is best suited to liquid foods where there is little air trapped in the product and the surface is flat and does not interrupt the flow of steam.

Steam replaces the air and on cooling forms a partial vacuum in the headspace. Blanching (Chapter 11) also removes air from vegetables before filling. Filled and sealed containers are then transported into the retort.

Heating by saturated steam
Latent heat is transferred to food when saturated steam condenses on the outside of the container. If air is trapped inside the retort, it forms an insulating boundary film around the cans that prevents the steam from condensing and causes under-processing of the food. It also produces a lower temperature than that obtained with saturated steam

(Chapter 10, Table 10.4). It is therefore important that all air is removed from the retort, and in batch retorts this is done using the incoming steam in a procedure known as 'venting'. Tucker (2004a) describes methods for measuring the temperature distribution in a retort and reducing the incidence of 'cold-spots' caused by air pockets. After sterilisation the containers are cooled by sprays of potable water. Steam is rapidly condensed in the retort, but the food cools more slowly and the pressure in the containers remains high. Compressed air is therefore used to equalise the pressure (pressure cooling) to prevent strain on the container seams. When the food has cooled to below 100 °C, the over-pressure of air is removed and cooling continues to approximately 40 °C, when the crates of containers are removed. At this temperature, moisture on the containers dries rapidly, which prevents surface corrosion and causes label adhesives to set more rapidly.

Heating by hot water
Foods are processed in glass containers or flexible pouches (Chapter 25, sections 25.2.4 and 25.3) using hot water with an over-pressure of air to achieve the required processing temperature. For example, at 121 °C the pressure of saturated steam is 200 kPa so to maintain water as a liquid an overpressure of 100 kPa is created by using a retort pressure of 300 kPa (Bown 2003). Glass containers are thicker than metal cans to provide adequate strength and this, together with the lower thermal conductivity of glass (Chapter 10, Table 10.2), results in a higher risk of thermal shock to the container, slower heat penetration and longer processing times than for cans.

   Foods in rigid polymer trays or flexible pouches heat more rapidly owing to the thinner material, the smaller cross-section of the container and a larger surface area : volume ratio. This enables savings in energy and causes minimal overheating at the outside of the container. The shorter processing cycle time also increases production rates and improves product quality. However, processing polymer trays and flexible pouches is more complex than cans or glass containers because of changes that may occur to the polymer materials during processing. For example, at high temperatures the plastic polymers may stretch or shrink and hence change the container volume; the heat seals may soften and weaken; and the headspace gas pressure may increase sufficiently to cause failure of the seals. The over-pressure should therefore be applied before the headspace gas inflates the container. Liquid or semi-liquid foods may be processed horizontally to ensure that the thickness of food is constant across the pouch. Vertical packs promote better circulation of hot water in the retort, but special frames are necessary to prevent the pouches from bulging at the bottom. Such a change in pack geometry alters the rate of heat penetration to the slowest heating point and hence the lethality achieved. Bown (2003) describes in detail methods used to overcome these variables. An animated description of the sequence of operations for sterilising foods using hot water is available at Anon (2008).

Heating by flames
Sterilisation at atmospheric pressure using direct flame heating of spinning cans at flame temperatures up to 1770 °C produces high rates of heat transfer (Noh *et al.* 1986). The consequent short processing times produce foods of high quality and reduce energy consumption by 20% compared with conventional canning. No brine or syrup is used in the can but high internal pressures (275 kPa at 130 °C) limit this method to small cans. It is used, for example, to process mushrooms, sweetcorn, green beans and cubes of beef.

13.1.3   Equipment

Sterilising retorts may be batch or continuous in operation and Bown (2003) describes the advantages and limitations of batch retorts. Details of retort design and operation are given by May (2001). Batch retorts may be vertical or horizontal (Fig. 13.5); the latter are easier to load and unload and have facilities for end-over-end agitation of containers (Fig. 13.6), but require more floor space. Depending on the scale of production, a battery of

Fig. 13.5   Batch retorts (courtesy of Allpax Products Inc. at www.allpax.com).

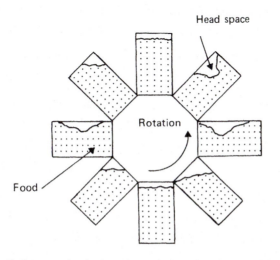

Fig. 13.6   End-over-end agitation of containers (after Hersom and Hulland 1980).

five or more batch retorts is operated in a canning factory. In this situation the process is not truly a batch operation because product is continually loaded and unloaded from individual retorts, thus requiring a continuous supply of food from the preparation section and producing a continuous supply of processed containers to the packing section. Simpson (2004) describes mathematical models to optimise the efficiency of batch retort operation. Bown (2003) notes one of the main advantages of batch operation as being the high degree of flexibility that enables a bank of retorts to be synchronised by computer to act as a continuous processing facility. The use of artificial intelligence to control loading/unloading enables several different products to be automatically processed at one time, using different time–temperature combinations. However, the time required to heat and cool batch retorts in an operating cycle means that they cannot achieve the thermal and operational efficiencies of continuous retorts in which steady-state conditions are maintained. The longer process cycle time contributes to higher processing costs of batch systems compared with continuous sterilisers.

Continuous retorts produce gradual changes in pressure inside cans, and therefore less strain on the can seams compared with batch equipment. They maintain constant conditions of pressure and temperature inside a pressurised chamber and containers are processed as they pass through the chamber. This has advantages in greater thermal efficiency than batch retorts, and because the equipment does not require a heating/cooling cycle, the process efficiency is higher and process times are reduced. The main disadvantages include a higher capital cost than batch equipment and a higher in-process stock that would be lost if a breakdown occurred.

The main types of equipment are cooker-coolers, rotary sterilisers and hydrostatic sterilisers, described by Toledo (1999). Static cooker-coolers carry cans on a roller or chain conveyor through three sections of a tunnel that are maintained at different pressures for pre-heating, sterilising and cooling. Pressure locks allow containers to be transferred between the three sections. Rotary sterilisers consist of a slowly rotating cylindrical cage inside a pressure vessel. Cans are loaded horizontally into the annular space between the cage and the pressure vessel, and as the cage rotates the cans are guided through the steriliser by a static spiral track. The rotation induces forced convection currents and causes the headspace bubble to move through the can, to mix the contents and significantly increase the rate of heat transfer compared to static sterilisers (Tucker 2004a). This type of equipment can process up to 300 containers per min and is mostly designed for a specific container size. It is unsuitable for non-cylindrical containers (e.g. plastic pouches or rectangular cans).

Hydrostatic sterilisers have two columns of water either side of a steam chamber (Fig. 13.7). The height of the water columns (up to 25 m) creates a hydrostatic pressure that balances the steam pressure, and the water seals the steam chamber. Cans are loaded horizontally end to end on carriers that are held between two chains, and these pass through the different sections of the steriliser as shown in Fig. 13.7. These large continuous sterilisers are used for the production of high-volume products (e.g. 1000 cans per min) where there is no requirement to regularly change the container size or processing conditions.

Control of retorts

All types of retorts are fitted with monitoring and control equipment to ensure that they operate at the correct temperature for the required time to achieve the desired lethality with minimum energy expenditure. Process variables that are monitored include:

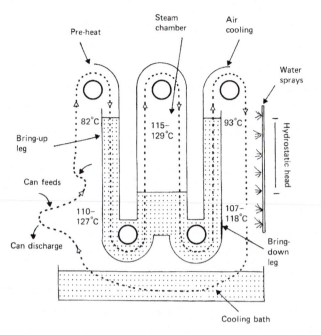

Fig. 13.7   Continuous hydrostatic steriliser.

- temperature and pressure of steam;
- time and temperature of processing;
- temperature of the cooling water;
- heating and cooling rates;
- pressure of compressed air.

Equipment includes continuous time–temperature data loggers and pressure sensors and loggers. Computer control of actuators on steam, water and compressed air valves automatically corrects any deviation from programmed values, and activates alarms if a fault is not corrected (e.g. Anon 2007). More recent developments include artificial intelligence computer control systems that continuously compare the accumulated lethality with the target lethality required for commercial sterility. They detect any process deviation (e.g. low steam pressure or lower than expected rate of product heating) in real time and then calculate the potential risk to public health. The control system can then alter the process variables in real time to correct the deviation and assure product safety. Further details are given by Bown (2004). This type of control means that each process cycle can be optimised to deliver the required lethality rather than using a predetermined time–temperature profile. The use of computer control of retorts is described by Bown (2003) and Simpson et al. (1993), and further details of automatic process control are given in Chapter 27 (section 27.2).

13.2   Ultra-high-temperature (UHT)/aseptic processes

When foods are processed in containers, the outer layers of food thermally insulate the inner layers, and create a resistance to heat transfer that extends the time needed to achieve the required lethality at the slowest heating point. Aseptic processing overcomes this problem by heating foods in thin layers to achieve the required lethality before it is filled into pre-sterilised containers in a sterile atmosphere. This also enables higher processing temperatures (e.g. 130–150 °C) to be used for a shorter time (typically a few seconds) (Fig. 13.8), which improves product quality and process productivity, and reduces energy consumption compared with in-container processing. Aseptic processing has now almost completely replaced in-container sterilisation of liquid foods (Oliveira 2004), including milk, fruit juices and concentrates, cream, yoghurt, salad dressing, liquid egg and ice cream mix. It is also gaining popularity for foods that contain small discrete particles (e.g. cottage cheese, baby foods, tomato products, fruit and vegetables, soups and rice desserts). Aseptic processes for larger-particulate foods have been developed (Manvell 1987), but in-container processing remains the most important method for sterilising solid foods. The high quality of UHT foods compares well with chilled and frozen foods (Chapters 21 and 22) and UHT has an important additional advantage of a shelf-life of at least six months without refrigeration. Lewis and Heppel (2000) and Ramaswarmy *et al.* (1997) have reviewed developments in aseptic processing and details of ohmic heating for UHT processing are given in Chapter 20 (section 20.2).

The advantages of UHT processing compared with canning are summarised in Table 13.7. These include better retention of sensory characteristics and nutritional value, energy savings, easier automation and the use of unlimited package sizes (Sandeep 2004). For example, conventional retorting of A2 cans (selected can sizes are given in Chapter 26, section 26.1.2) of vegetable soup requires 70 min at 121 °C to achieve an F_0-value of 7 min, followed by 50 min cooling. Aseptic processing in a scraped-surface heat exchanger at 140 °C for 5 s gives an F_0-value of 9 min. Increasing the can size to A10 increases the processing time to 218 min, whereas with aseptic processing the sterilisation time is the same. This permits the use of very large containers (e.g. 1 tonne aseptic bags of tomato purée or liquid egg, used as an ingredient for other manufacturing processes).

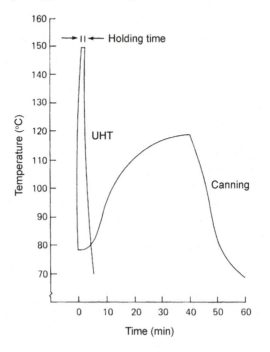

Fig. 13.8   Time–temperature conditions for UHT and canning.

Table 13.7   Comparison of conventional canning and aseptic processing and packaging

Criteria	Retorting	Aseptic processing and packaging
Product sterilisation Process calculations	Unsteady state	Precise, isothermal
Fluids	Routine, convection	Routine
Particulates	Routine, conduction or broken heating	Complex
Low acid particulate processing	Routine	Becoming more common
Other sterilisation required	None	Complex (process equipment, containers, lids, aseptic tunnel)
Energy efficiency	Low	> 30% saving
Sensory quality	Not suited to heat-sensitive foods	Suitable for homogeneous heat-sensitive foods
Nutrient losses	High	Minimal
Value added	Lower	Higher
Stability	Shelf stable at ambient temperatures	Shelf stable at ambient temperatures
Suitability for microwave heating	Only glass and semi-rigid containers	Most semi-rigid and rigid containers (not aluminium foil)
Production rate	600–1000 min^{-1}	≈ 500 min^{-1}
Labour/handling costs	Higher	Low
Downtime	Minimal (mostly caused by seaming and labelling)	Re-sterilisation needed if loss of sterility in filler or steriliser
Flexibility for different container sizes	Need different container delivery equipment and/or retorts	Single filler for different container sizes
Survival of heat resistant enzymes	Rare	Common in some foods (e.g. milk)
Post-process additions	Not possible	Possible (e.g. probiotics added before filling)

Adapted from David (1996)

UHT products may also be fortified with heat-sensitive bioactive components after sterilisation, including probiotics, omega-3 fatty acids and conjugated linoleic acids, phytosterols or fibre (Chapter 6, section 6.3). The main limitations of UHT processing are the high cost and complexity of the plant that arise from the necessity to sterilise packaging materials, associated pipework and tanks, the maintenance of sterile air and surfaces in filling machines. The process also has slower filling speeds than canning, and requires higher skill levels by operators and maintenance staff. In addition, the higher temperatures may cause different changes to the chemistry of some foods (section 13.3).

13.2.1 Theory
For a given increase in temperature, the rate of destruction of micro-organisms and many enzymes increases faster than the rate of destruction of nutrients and sensory components (Fig. 13.9 and Chapter 10, sections 10.3 and 10.4). Food quality is therefore better retained at higher processing temperatures for shorter times. The criteria for UHT processing are the same as for in-container sterilisation (i.e. the attainment of commercial sterility – section 13.1). However, whereas in in-container sterilisation the most lethal effect frequently occurs at the end of the heating stage and the beginning of the cooling stage, UHT processes heat liquid foods rapidly to a holding temperature and the major part of the

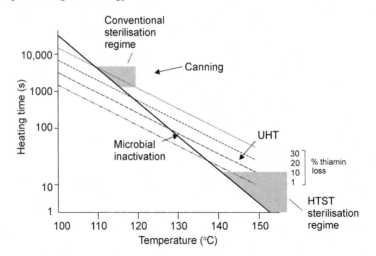

Fig. 13.9   Rates of microbial and nutrient destruction in canning and UHT processing (adapted from Holdsworth 2004 and Killeit 1986).

lethality accumulates at a constant temperature. The sterilising value is calculated by multiplying together the lethal rate at the holding temperature and the holding time. The come-up time and cooling periods are often very short and are not included in calculations, being treated as a safety factor. The flowrate of the fastest moving particle and longest time needed for heat transfer from the liquid to the centre of the particle are together used to determine the time and temperature needed to achieve the required F_0-value. Methods to obtain this data using time–temperature integrators are described by Heppel (2004).

It is important to know both the shortest time that any particle can take to pass through the holding section and the rate of heat transfer from the liquid to the centre of a particle, to ensure that microbial spores cannot survive the process. It is also important to achieve turbulent flow if possible because the spread of residence times is smaller. The slowest heating point in a straight holding tube is at the centre of the tube, but for other designs (e.g. a spiral tube) it is located away from the centre and is found using mathematical modelling. Similarly there should be close control over the particle size range in particulate products. For example if a process is designed to sterilise 14 mm particles to $F_0 = 6$, it can be calculated that the holding tube should be 13 m long. However, if a 20 mm particle passes through under these conditions, it will only reach $F_0 = 0.5$ and will thus be seriously under-processed. Conversely, a 10 mm diameter particle will reach $F_0 = 20$ and will be over-processed (Ohlsson 1992).

It is also necessary to have information on the kinetics of microbial destruction and enzyme inactivation to ensure that the product is adequately sterilised, and information on the kinetics of nutrient destruction and chemical changes such as browning, oxidation and flavour changes, to understand the effects of the process on food quality. These aspects are subject to mathematical modelling and are reviewed by Sandeep and Puri (2001). Typical minimum time–temperature conditions needed to destroy *Cl. botulinum* ($F_0 = 3$) are 1.8 s at 141 °C. The minimum heat treatments for dairy products are 1 s at 135 °C for milk, 2 s at 140 °C for cream and milk-based products and 2 s at 148.9 °C for ice cream mixes (Lewis 1993). Komorowski (2006) has reviewed processing conditions in new dairy hygiene legislation in the EU.

In addition to the use of F_0 to assess microbial destruction, a further two parameters are used in dairy UHT processing: the B^* value, which is used to measure the total integrated lethal effect of a process and the C^* value which measures the total chemical damage taking place during a process. The reference temperature used for these values is 135 °C. A process that is given a B^* value $= 1$ will result in a $9D$ reduction in spores ($z = 10.5$ °C) and would be equivalent to 10.1 s at 135 °C. Similarly a process given a C^* value $= 1$ will cause 3% loss of thiamine and would be equivalent to 30.5 s at 135 °C (Lewis 1993).

Calculation of holding time uses Equations 13.9 and 13.10:

$$B^* = 10^{(\theta-135)/10.5}t/10.1 \qquad \boxed{13.9}$$

$$C^* = 10^{(\theta-135)31.4}t/30.5 \qquad \boxed{13.10}$$

where θ (°C) = processing temperature and t (s) = holding time.

Ideally a process should maximise B^* and minimise C^*, unless for example a specific chemical is to be destroyed (e.g. an enzyme or natural toxin such as trypsin inhibitor) or vegetable tissues are required to be softened.

The calculation of processing times for particulate foods, requires information on the type of fluid flow and the types of heat transfer (convection at the surface of particles and conduction of heat to the slowest heating point in the particle). A sample calculation of UHT processing time is given by Singh and Heldman (2001).

13.2.2   Processing

The process is summarised in Fig. 13.10. Pre-heated food is pumped using a positive displacement metering pump through a vacuum chamber to remove air, to a heat exchanger. Removal of air is important for a number of reasons:

- It ensures that the product has a constant volume in the holding tube. If air in the product expands when heated, there would be a reduction in holding time and possible under-processing.
- It saves energy in heating and cooling.
- It enables a longer shelf-life by reducing oxidative changes during storage at ambient temperatures.

Food is heated in relatively thin layers with close control over the sterilisation temperature, and passed to a holding tube that is of sufficient length to retain the food for the required time. The holding tube is inclined upwards at a shallow angle to ensure that tube is always full of product without air pockets. The sterilised product is cooled either by evaporative cooling in a vacuum chamber or in a second heat exchanger. The pressure required to achieve sterilising temperatures is created by the metering pump and maintained by a back-pressure device. This can be a piston or diaphragm valve or a pressurised tank. The back-pressure device is placed after the cooler to keep the product under pressure both in the holding tube and at the start of cooling, and thus prevent it from boiling. After cooling, the food is temporarily stored in a pressurised sterile 'surge' tank and packaged under sterile conditions. An over-pressure of nitrogen in the surge tank enables the product to be moved to the fillers without pumps. Because containers are not required to withstand the high temperatures and pressures during sterilisation, a variety of materials are suitable and laminated microwaveable cartons (Chapter 25, section 25.2.5) are widely used. Others, including pouches, cups, sachets and bulk packs, are described in Chapter 25 and have been reviewed by Reuter (1989). These packs have considerable economic advantages compared with cans and bottles, in both the cost of the pack and the

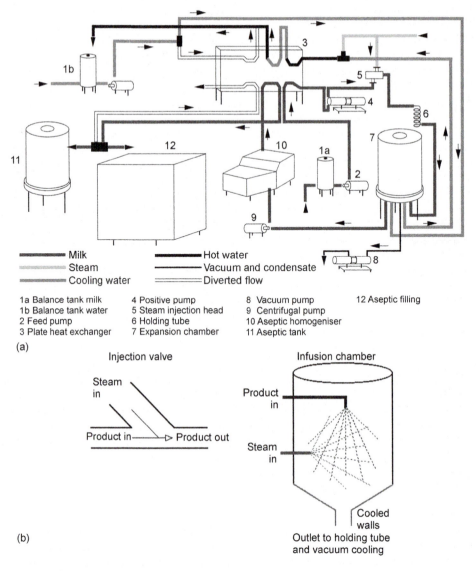

Milk
Steam
Cooling water
Hot water
Vacuum and condensate
Diverted flow

1a Balance tank milk   4 Positive pump   8 Vacuum pump   12 Aseptic filling
1b Balance tank water   5 Steam injection head   9 Centrifugal pump
2 Feed pump   6 Holding tube   10 Aseptic homogeniser
3 Plate heat exchanger 7 Expansion chamber   11 Aseptic tank

(a)

Injection valve

Steam
in

Product in ──────▷ Product out

Infusion chamber

Product
in

Steam
in

Cooled
walls

Outlet to holding tube
and vacuum cooling

(b)

Fig. 13.10   (a) UHT plant (courtesy of Tetra Pak Ltd) (Anon 2003) and (b) injection valve and
infusion chamber.

transport and storage costs. Cartons are pre-sterilised with UV or ionising radiation, with
hydrogen peroxide or heat, and filling machines are enclosed in sterile conditions
maintained by ultraviolet light and a positive air pressure of filtered air to prevent entry of
contaminants. Details of UHT processing of dairy products are given by Anon (2003),
Richardson (2001) and Lewis and Heppel (2000).

The process is successfully applied to liquid and small-particulate foods but has been
less widely developed for foods that contain larger solids. The major difficulties have
been the following:

- Enzyme inactivation at the centre of the pieces of food causes overcooking of the
  surfaces, thus limiting particle sizes.

- Agitation is necessary to improve the rate of heat transfer and to aid temperature distribution, but this causes damage to the product.
- Settling of solids is a problem if the equipment has a holding tube. This causes uncontrolled and overlong holding times and variable proportions of solids in the filled product.

These problems have been addressed for larger particulates (up to 2.5 cm) using a number of processes, including separate treatment of liquid and solid portions in the Jupiter system, the Twintherm system, the single flow 'fraction specific thermal processing' (FSTP) system and the rotaholder (section 13.2.3) or by ohmic heating (Chapter 20, section 20.2). These systems are described in detail by Willhoft (1993). However, these processes that use separate treatments for liquid and solid fractions encountered practical difficulties, and the focus for developments is now on back-pressure devices and pressurised tank systems (Sandeep 2004). The flow of two-phase foods containing particulates through heat exchangers is highly complex and research is continuing to gain a better understanding of control of heat transfer in these systems. Details of methods for validation of aseptic processing, including monitoring methods using miniaturised time–temperature integrators, non-contact real time and subsurface temperature monitoring are given by Palaniappan and Sizer (1997) and Heppel (2004).

13.2.3 Equipment
A theoretically ideal UHT process would heat the product instantly to the required temperature, hold it at that temperature to achieve sterility and cool it instantly to filling temperature. In practice the degree to which this is achieved depends in part on the method used to heat the food and in part on the sophistication of control and hence the cost of equipment. It also depends on the properties of the food (e.g. viscosity, acidity, presence of particles, heat sensitivity and tendency to form deposits on hot surfaces); the potential for fouling; ease of cleaning; and capital and operating costs (Sandeep 2004). With the exception of ohmic heating, equipment used for UHT processing has the following characteristics:

- operation above 132 °C;
- exposure of a relatively small volume of product to a large surface area for heat transfer;
- maintenance of turbulence in the product as it passes over the heating surface;
- use of pumps to give a constant delivery of product against the pressure in the heat exchanger;
- constant cleaning of the heating surfaces to maintain high rates of heat transfer and to reduce burning-on of the product.

Equipment is classified according to the method of heating into:

- direct systems (steam injection and steam infusion);
- indirect systems (plate heat exchangers, tubular heat exchangers (concentric tube or shell-and-tube) and scraped surface heat exchangers);
- other systems (dielectric, ohmic and induction heating (Chapter 20)).

Direct methods
Details of equipment from different manufacturers are reviewed by Ramesh (1999). Steam injection (or 'uperisation') and steam infusion are each used to intimately combine

the product with potable (culinary) steam (Fig. 13.10b, see also Chapter 10, section 10.2.1). In steam injection, steam at ≈ 965 kPa is introduced into a preheated liquid product in fine bubbles by a steam injector, and rapidly heats the product to 150 °C. After a suitable holding period (e.g. 2.5 s) the product is flash cooled in a vacuum chamber to 70 °C, and condensed steam and volatiles in the product are removed. The moisture content of the product therefore returns to approximately the same level as the raw material. The main advantage of this system is that it is one of the fastest methods of heating and the fastest method of cooling, and it is therefore suitable for heat-sensitive foods. However, there are limitations:

- The method is only suitable for low-viscosity liquids.
- There is a requirement for potable steam, which is more expensive to produce than normal processing steam.
- Regeneration of energy is less than 50% compared with more than 90% in indirect systems.
- Flexibility for changing to different types of product is low.

In steam infusion the food is sprayed in a free-falling film into high-pressure (450 kPa) potable steam in a pressurised vessel. It is heated to 142–146 °C in 0.3 s, and is held for 3 s in a holding tube before flash cooling in a vacuum chamber to 65–70 °C. Heat from the flash cooling is used to pre-heat the feed material.

Both systems have computer control of temperature, pressure, level, flowrate, valve operation and the cleaning sequence, at production rates of up to 9000 kg h⁻¹. Steam infusion has advantages over injection methods because the liquid does not contact hotter surfaces and burning-on is therefore reduced. Other advantages include:

- almost instantaneous heating of the food to the temperature of the steam, and very rapid cooling which results in high retention of sensory characteristics and nutritional properties;
- greater control over processing conditions than steam injection;
- lower risk of localised overheating of the product; and
- the method is more suitable for higher-viscosity foods compared to steam injection.

The main disadvantages, in addition to the disadvantages of steam injection, are blockage of the spray nozzles and separation of components in some foods. Further details are given by Lewis and Heppel (2000), Carlson (1996) and Burton (1988).

Indirect systems

Heat exchangers are described in Chapter 10 (section 10.2.4) and a description of the historical development of different types of heat exchangers for indirect UHT heating is given by Carlson (1996). Plate heat exchangers are described in detail for pasteurisation (Chapter 12). In UHT applications, they have a number of limitations due to the higher temperatures and pressures involved (Table 13.8). Tube and shell heat exchangers are described in Chapter 14 (section 14.1.3), in their application to evaporation by boiling. The advantages and limitations of this type of equipment for UHT processing are described in Table 13.8.

There are a large number of designs of tubular heat exchangers, including concentric tube heat exchangers (Fig. 13.11). The double tube design has a corrugated tube concentrically positioned in a larger diameter outer tube. Tube diameters vary according to the required flowrate and the size of particulates, which makes them suitable for heating products with high pulp content, or products that contain particulates that must be

Table 13.8 Comparison of plate and tube-and-shell heat exchangers for UHT processing

Plate heat exchanger		Tube-and-shell heat exchanger	
Advantages	Limitations	Advantages	Limitations
• Relatively inexpensive • Economical in floor space and water consumption • Efficient in energy use ($>90\%$ energy regeneration) • Flexible changes to production rate, by varying the number of plates • Easily inspected by opening the plate stack	• Operating pressures limited by the plate gaskets to approximately 700 kPa • Liquid velocities at relatively low pressure also low ($1.5-2\,\mathrm{m\,s^{-1}}$) • Low flow rates can cause uneven heating and solids deposits on the plates which require more frequent cleaning • Gaskets susceptible to high temperatures and caustic cleaning fluids and are replaced more regularly than in pasteurisation • Limited to low viscosity liquids (up to $1.5\,\mathrm{N\,s\,m^{-2}}$) • Careful initial sterilisation of the large mass of metal in the plate stack is necessary for uniform expansion to prevent distortion and damage to plates or seals • Liable to fouling	• Few seals and easier cleaning and maintenance of aseptic conditions • Operation at higher pressures (7000–10 000 kPa) and higher liquid flow rates ($6\,\mathrm{m\,s^{-1}}$) than plate heat exchangers • Turbulent flow at tube walls due to higher flowrates • Hence more uniform heat transfer and less product deposition	• Difficulty in inspecting heat transfer surfaces for food deposits • Limited to relatively low-viscosity foods (up to $1.5\,\mathrm{N\,s\,m^{-2}}$) • Lower flexibility to changes in production capacity • Larger-diameter tubes cannot be used because higher pressures needed to maintain the liquid velocity and large-diameter pipes have a lower resistance to pressure • Any increase in production rate requires duplication of the equipment

processed with minimal damage. Triple tube designs consist of a centre tube that is concentrically positioned in a middle tube, which is also concentrically located in the outer tube. Stainless steel spacers keep the tubes properly separated. This produces annular spaces in which heat transfer takes place from both sides. An alternative design has smaller tubes enclosed within an outer shell. Counter-current flow and corrugations are used to generate turbulence and hence to increase the rate of heat transfer. This equipment is able to operate at high pressures (up to 2000 kPa) with viscous liquids.

Scraped-surface heat exchangers are also used for freezing (Chapter 22, section 22.2.1), for evaporation by boiling (Chapter 14, section 14.1.3) and for the continuous production of margarine and butter (Chapter 3, section 3.2.3). Their main advantages for UHT processing are their suitability for viscous foods and particulates (<1 cm), and their

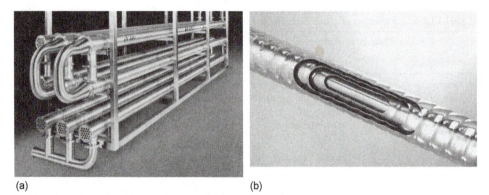

(a)   (b)

Fig. 13.11   (a) Concentric tube heat exchanger; (b) corrugated triple tube heat exchanger (courtesy of APV Ltd at www.apv.com).

flexibility for different products by changing the geometry of the rotor assembly. However, they have high capital and operating costs and heat recovery is not possible. They are used for example to process fruit sauces and fruit bases for yoghurts and pies.

During the 1970s and 1980s, systems were developed to enable greater control over F_0-values that involved separate heat treatments of the liquid fraction and large (2–2.5 cm) particulate components of a food. For example, the 'Jupiter' double-cone heat exchanger combined indirect heating by a rotating jacketed double cone with direct heating by steam or superheated liquor. In a sequence of microprocessor-controlled operations, solid pieces of food are tumbled through steam at ≈ 200 kPa. Liquor, sterilised separately, is added during sterilisation to prevent damage to the solids by the tumbling action. Sterilisation times are 2–4 min to achieve $F_0 = 10$. After sterilisation the product is rapidly cooled with cold water and sterile air, and the condensate-water-stock is removed. The liquid portion of the product is sterilised separately in a plate or tubular system, added to the solids and discharged to an aseptic filler. However, the system had a relatively low capacity, complex operation and relatively high capital costs, and for these reasons has not been widely adopted. In the 'Twintherm' system, particulate food is heated by direct steam injection in a pressurised horizontal cylindrical vessel that rotates slowly. Once sterilised, the particles are cooled with liquid that has been sterilised in conventional UHT heat exchangers, and carried to an aseptic filler. It is claimed to allow more uniform and gentle treatment of particles, compared with continuous processes, and it has been used commercially to produce soups (Ohlsson 1992, Alkskog 1991). The single flow FSTP system employs a cylindrical vessel containing slowly rotating blades on a shaft that form cages to hold particles as they are rotated around the cylinder from inlet to discharge pipes. Liquid moves freely through the cages, giving rapid heat transfer. Again the liquid component is sterilised separately in conventional heat exchangers and is then used to carry the sterilised particles to the filler (Hermans 1991).

More recent developments described by Carlson (1996) and Lewis and Heppel (2000) include tube-in-tube heat exchangers, which use thin-walled tubes that have corrugations and/or twists in the tubes to promote turbulent flow. One tube is inserted inside another, which may be repeated two or more times. Incoming food is heated by sterilised product to regenerate heat and increase energy efficiency. Steam or pressurised hot water is used for the final heating, and after initial cooling by the incoming material, the product is cooled with cold water. This system is relatively low cost and is widely used, although it suffers from some drawbacks: seal integrity is critical to prevent mixing of incoming material, product or heating and cooling media; the seal strength limits the pressure that can be used; and seals may be difficult to clean. Additionally, the equipment is restricted to relatively low-viscosity liquids that do not cause significant fouling (Carlson 1996), but it has been widely used for fruit juices, milk and dairy products.

The problems encountered in straight tube heat exchangers described above are largely overcome by forming a single tube into a continuous helix or coil, which has a carefully defined ratio between the diameter of the coil and the diameter of the tube. The design of the coil promotes secondary flow of liquid within the tube, and this causes turbulence at relatively low flowrates, and high rates of heat transfer (between two and four times the rates in tube-in-tube or shell-and-tube heat exchangers (Carlson 1996)). This enables processing of heat-sensitive products and products that cause fouling of heat exchanger surfaces (e.g. liquid egg). The mixing action in the coil gives a uniform distribution of particles, making the equipment suitable for salad dressings, fruit purées and other foods that contain a range of particle sizes, as well as for high-viscosity liquids such as cheese sauce. Additionally, the continuous tube has no seals and is easily cleaned-in-place, and the simple design is virtually maintenance-free. Indirect systems are automatically cleaned in place after 3–4 h of operation to remove accumulated deposits. The cleaning programme does not involve loss of sterile conditions, and processing resumes immediately afterwards.

Other systems
Three other systems have been developed: the 'Multi-therm', 'Achilles' and ohmic heating. In the first two, the food is heated by a combination of hot liquid and microwave energy. In ohmic heating a conducting fluid is heated directly by electrical energy (Chapter 20, section 20.2). Conversion efficiencies from electrical energy to heat of greater than 90% are claimed, and particulate feeds may be processed without shearing forces associated with some other types of heat exchangers.

13.3   Effect on foods

In this section the changes to foods caused by traditional canning techniques are compared with those caused by UHT processing. Details of the chemical changes to UHT processed foods are given by Neilsen *et al.* (1993).

13.3.1   Canning
The time–temperature combinations used in canning have a substantial effect on most naturally occurring pigments in foods. For example in meats, the red oxymyoglobin pigment is converted to brown metmyoglobin, and purplish myoglobin is converted to red-brown myohaemichromogen. Maillard browning and caramelisation (Chapter 1, section 1.4.2) also contribute to the colour of sterilised meats. However, this is an acceptable change in cooked meats. Sodium nitrite and sodium nitrate are added to some meat products to reduce the risk of growth of *C. botulinum*. The resulting red-pink coloration is due to nitric oxide myoglobin and metmyoglobin nitrite.

In fruits and vegetables, chlorophyll is converted to pheophytin, carotenoids are isomerised from 5,6-epoxides to less intensely coloured 5,8-epoxides, and anthocyanins are degraded to brown pigments. This loss of colour may be corrected using permitted synthetic colourants (Appendix A4). Discoloration of canned foods during storage occurs for example when iron or tin react with anthocyanins to form a purple pigment, or when colourless leucoanthocyanins form pink anthocyanin complexes in some varieties of pears and quinces. In sterilised milk slight colour changes are due to caramelisation, Maillard browning and changes in the reflectivity of casein micelles.

In canned meats there are complex changes to the flavour and aroma. These include pyrolysis, deamination and decarboxylation of amino acids, Maillard reactions and caramelisation of carbohydrates to furfural and hydroxymethylfurfural, and oxidation and decarboxylation of lipids. Interactions between these components produce more than 600 flavour compounds in ten chemical classes. In fruits and vegetables, changes are due to complex reactions, which involve the degradation, recombination and volatilisation of aldehydes, ketones, sugars, lactones, amino acids and organic acids. In milk the development of a cooked flavour is due to denaturation of whey proteins to form hydrogen sulphide, and the formation of lactones and methyl ketones from lipids. In aseptically sterilised foods the changes are less severe, and the natural flavours of milk, fruit juices and vegetables are better retained. Changes to milk are discussed in detail by Burton (1988).

In canned meats, changes in texture are caused by coagulation and a loss of water-holding capacity of proteins, which produces shrinkage and stiffening of muscle tissues. Softening is caused by hydrolysis of collagen, solubilisation of the resulting gelatin, and melting and dispersion of fats through the product. Polyphosphates (Appendix A4) are added to some products to bind water. This increases the tenderness of the product and reduces shrinkage. In fruits and vegetables, softening is caused by hydrolysis of pectic materials, gelatinisation of starches and partial solubilisation of hemicelluloses, combined with a loss of cell turgor. Calcium salts may be added to blancher water (Chapter 11), or to brine or syrup to form insoluble calcium pectate and thus to increase the firmness of canned products. Different salts are needed for different types of fruit (e.g. calcium hydroxide for cherries, calcium chloride for tomatoes and calcium lactate for apples) owing to differences in the proportion of demethylated pectin in each product.

Canning causes the hydrolysis of carbohydrates and lipids, but these nutrients remain available and the nutritional value of the food is not affected. Proteins are coagulated and, in canned meats, losses of amino acids are 10–20%. Reductions in lysine content are proportional to the severity of heating but rarely exceed 25%. The loss of tryptophan and, to a lesser extent, methionine, reduces the biological value of the proteins by 6–9%. Rickman *et al.* (2007a,b) have reviewed recent and classical studies of vitamin losses as a result of canning fruits and vegetables (Table 13.9) and Klein and Kaletz (1997) report a survey of the nutritional value of canned fruits and vegetables. They report that although the thermal treatment can cause losses of water-soluble and oxygen-labile nutrients such as vitamin C and the B vitamins (especially thiamin), these nutrients are relatively stable during subsequent storage owing to the lack of oxygen in the container. Washing, peeling and blanching also cause losses of water-soluble nutrients. Changes to phenolic compounds, which are also water-soluble and oxygen-labile, are more variable in different products. Ascorbic acid is highly sensitive to oxidation and leaching, whereas other vitamins and minerals are more stable and there are for example high levels of retention of vitamin E.

Nutrient losses are highly dependent on the cultivar and maturity of the food, the type of water used in processing (particularly calcium content), the presence of residual oxygen in the container, methods of preparation (peeling and slicing) or blanching (Chapters 2 and 11) and the processing conditions. In some foods, water-soluble vitamins are transferred into the brine or syrup, which is also consumed and there is thus a smaller nutritional loss. Lipid-soluble vitamins are not significantly lost by leaching but are sensitive to oxidation. Thermal processing can also cause isomerisation of the naturally

Table 13.9 Loss of vitamins in canned foods

Food	Losses (%)								
	β-Carotene	Thiamin	Ribo-flavin	Niacin	Vitamin C	Panto-thenic acid	Vit. B$_6$	Folacin	Biotin
Low-acid foods									
Beetroots	–	–	–	–	8–10	–	–	30	–
Broccoli			–	–	84	–	–	–	–
Carrots	+7	67	38–60	32	90	54	80	59	40
Green beans	17	62	54–63	77	63	61	50	57	–
Mushrooms	–	80	32	52	41	54	46	84	54
Peas	22	75	47	71	73	80	69	59	78
Potatoes	–	56	44	7–56	28	–	59	–	–
Spinach	19	66	45	50	62	78	75	35	67
Sweetcorn	0		–	–	0.25	–	–	–	–
Tomatoes	13	53	25	0	30	30	+14–38	54	55
Beef		67	100	100	–	–	–	–	–
Mackerel	4	60	39	29	–	–	46	–	–
Milk	0	35	0	0	50–90	0	50	10–20	–
Salmon	9	73	0	0	–	58	57	–	–
Acid foods									
Apples	4	31	48	–	74	15	0	–	–
Cherries	41	57	64	46	68	–	20	–	–
Peaches	50	49	5–40	7–39	56	71	21	–	–
Pears	–	45	45	0	73	69	18	–	–
Pineapples	25	7	30	0	57	12	–	–	–

Data adapted from Rickman *et al.* (2007a,b), De Ritter (1982), Rolls (1982), Burger (1982) and March (1982)

occurring *trans*-β-carotene into the less biologically active *cis*-β-carotene. Lycopene is more biologically active in its *cis* form and processed tomato products have greater lycopene bioactivity than fresh tomatoes (Dewanto *et al.* 2002, Howard *et al.* 1999). Similarly the level of α-tocopherol increases during tomato processing and vitamin E does not undergo significant losses, although prolonged heating reduces the amount that is present (Rickman *et al.* 2007b). Sterilised soya-meat products may also show an increase in nutritional value owing to a decrease in the stability of the trypsin inhibitor in soybeans.

Although minerals are heat-stable, foods may gain or lose minerals owing to the processing conditions. For example, losses may occur due to leaching into blancher water; or there may be increases, for example in sodium from salt that is added to flavour canned foods, or calcium salts that are added to blancher water to protect the texture of vegetables. Details are given by Martin-Belloso and Llanos-Barriobero (2001). The soluble and insoluble fibre content of fruits and vegetables does not change significantly as a result of canning. The effect of processing conditions on the vitamin content of canned foods is reviewed by Holdsworth (2004) and changes in sterilised milk are discussed by Burton (1988).

13.3.2 UHT processing

In UHT processing, meat pigments change colour, but there is little caramelisation or Maillard browning. Carotenes and betanin are virtually unaffected, and chlorophyll and anthocyanins are better retained. However some enzymes, for example proteases and lipases that are secreted by psychrotrophic micro-organisms in milk, are more heat resistant. These are not destroyed by some UHT treatments and may cause changes to the flavour of products during prolonged storage (David 1996).

Small changes in the viscosity of milk are caused by modification of κ-casein, leading to an increased sensitivity to calcium precipitation and coagulation. Age-gelation may occur during storage of UHT dairy products. This involves a sudden sharp increase in viscosity, gelation and aggregation of casein micelles. The causes may be due to proteolytic breakdown of casein by naturally occurring heat-resistant enzymes, polymerisation of casein and whey proteins by Maillard or other chemical reactions, or formation of κ-casein-β-lactoglobulin complexes.

The relatively long time required for collagen hydrolysis and the relatively low temperature needed to prevent toughening of meat fibres are conditions found in canning but not in UHT processing. Toughening of meat is therefore likely under UHT conditions. The texture of meat purées is determined by size reduction and blending operations (Chapters 3 and 4) and is not substantially affected by aseptic processing. In aseptically processed fruit juices the viscosity is unchanged. The texture of solid fruit and vegetable pieces is softer than the unprocessed food due to solubilisation of pectic materials and a loss of cell turgor but is considerably firmer than canned products.

Aseptically processed meat and vegetable products lose thiamin and pyridoxine but other vitamins are largely unaffected. There are negligible vitamin losses in aseptically processed milk ($\approx$10% losses of thiamin, vitamin B_{12}, folic acid and pyridoxine, compared with 35%, 90%, 50% and 50% respectively in bottled milk) and vitamin C losses are substantially lower than in-container processing ($\approx$ 25% compared with 90%). Ramesh (1999) has reviewed vitamin losses during UHT processing. Denaturation of whey proteins in UHT processing is 60–70% using direct heating and 75–80% using indirect methods, compared with $\approx$87% in bottled milk, and there may be partial precipitation of minerals. β-Lactoglobulin is more affected than α-lactalbumin, but the denaturation does not necessarily affect the nutritive value of the processed milk (Ramesh 1999). Lipids, carbohydrates, riboflavin, pantothenic acid, biotin, nicotinic acid and vitamin B_6 are virtually unaffected by UHT processing.

References

ALKSKOG, L., (1991), Twintherm – a new aseptic particle processing system, paper presented at the News in Aseptic Processing and Packaging seminar, Helsinki, January.

ANON, (1993), Code of hygienic practice for aseptically processed and packaged low-acid foods, CODEX standard CAC/RCP 40-1993, Codex Alimentarius Commission, Rome, available at www.codexalimentarius.net/download/standards/26/CXP_040e.pdf.

ANON, (1997), Guidelines for performing heat penetration trials for establishing thermal processes in batch retort systems, Guideline 16, Campden and Chorleywood Food Research Association, Chipping Campden, UK, available for purchase at www.campden.co.uk/publ/pubfiles/g16.htm.

ANON, (2003), *Dairy Processing Handbook*. Published by Tetra Pak Processing Systems AB, S-221 86 Lund, Sweden, available from www.tetrapak.com/index.asp?navid=250.

ANON, (2007), Animated Steriflow operation, available at www.steriflow.com/flash/principe_en/index.htm.

ANON, (2008), Process and control information, Steriflow thermal processing company information, available at www.steriflow.com/en/14-process-control.

BALL, C.O., (1923), Thermal process time for canned food, *Bulletin of National Research Council*, **7**, No. 37.

BOWN, G., (2003), Developments in conventional heat treatment, in (P. Zeuthen and L. Bøgh-Sørensen, Eds.), *Food Preservation Techniques*, Woodhead Publishing, Cambridge, pp. 154–178.

BOWN, G., (2004), Modelling and optimising retort temperature control, in (P. Richardson, Ed.), *Improving the Thermal Processing of Foods*, Woodhead Publishing, Cambridge, pp. 105–123.

BURGER, I.H., (1982), Effect of processing on nutritive content of food: meat and meat products, in (M. Rechcigl, Ed.), *Handbook on the Nutritive Value of Processed Foods*, Vol. 1, CRC Press, Boca Raton, FL, pp. 323–336.

BURTON, H., (1988), *UHT Processing of Milk and Milk Products*. Elsevier Applied Science, London.

CARLSON, B., (1996), Food processing equipment – historical and modern designs, in (J.R.D. David, R.H. Graves and V.R. Carlson, Eds.), *Aseptic Processing and Packaging of Food*, CRC Press, Boca Raton, FL, pp. 51–94.

DAVID, J., (1996), Principles of thermal processing and optimisation, in (J.R.D. David, R.H. Graves and V.R. Carlson, Eds.), *Aseptic Processing and Packaging of Food*, CRC Press, Boca Raton, FL, pp. 3–20.

DE RITTER, E., (1982), Effect of processing on nutritive content of food: vitamins, in (M. Rechcigl, Ed.), *Handbook on the Nutritive Value of Processed Foods*, Vol. 1, CRC Press, Boca Raton, FL, pp. 473–510.

DEWANTO, V., WU, X., ADOM, K.K. and LIU, R.H., (2002), Thermal processing enhances the nutritional value of tomatoes by increasing total antioxidant activity, *J. Agriculture Food Chemistry*, **50**, 3010–3014.

ESZES, F. and RAJKO, R., (2004), Modelling heat penetration curves in thermal processes, in (P. Richardson, Ed.), *Improving the Thermal Processing of Foods*, Woodhead Publishing, Cambridge, pp. 307–333.

GEERAERD, A.H., VALDRAMIDIS, V.P., BERNAERTS, K. and VAN IMPE, J.F., (2004), Evaluating microbial inactivation models for thermal processing, in (P. Richardson, Ed.), *Improving the Thermal Processing of Foods*, Woodhead Publishing, Cambridge, pp. 427–453.

HELDMAN, D.R. and HARTEL, R.W., (1997), *Principles of Food Processing*, Chapman and Hall, New York, pp. 23–24.

HEPPEL, N., (2004), Optimising the thermal processing of liquids containing solid particles, in (P. Richardson, Ed.), *Improving the Thermal Processing of Foods*, Woodhead Publishing, Cambridge, pp. 481–492.

HERMANS, W., (1991), Single flow fraction specific thermal processing of liquid foods containing particulates, paper presented at the News in Aseptic Processing and Packaging seminar, Helsinki, January.

HERSOM, A. and HULLAND, E., (1980), *Canned Foods*, 7th edn, Churchill Livingstone, pp. 122–258.

HOLDSWORTH, S.D., (1992), *Aseptic Processing and Packaging of Foods*, Elsevier Academic and Professional, London.

HOLDSWORTH, S.D., (1997), Process evaluation techniques, in *Thermal Processing of Packaged Foods*, Blackie Academic and Professional, London, pp. 139–244.

HOLDSWORTH, S.D., (2004), Optimising the safety and quality of thermally processed packaged foods, in (P. Richardson, Ed.), *Improving the Thermal Processing of Foods*, Woodhead Publishing, Cambridge, pp. 3–31.

HOWARD, L.A., WONG, A.D., PERRY, A.K. and KLEIN, B.P., (1999), β-carotene and ascorbic acid retention in fresh and processed vegetables, *J. Food Science*, **64**, 929–936.

KILLEIT, V., (1986), The stability of vitamins, *Food Europe*, March–April, 21–24.

KLEIN, B.P. and KALETZ, R., (1997), A study of canned food nutrition, University of Illinois Department of Food Science and Human Nutrition, available at http://nutrican.fshn.uiuc.edu/explanation.html.

KOMOROWSKI, E.S., (2006), New dairy hygiene legislation, *International J. Dairy Technology*, **59** (2), 97–101.

LEWIS, M.J. (1993), UHT processing: safety and quality aspects, in (A. Turner, Ed.), *Food Technology International Europe*, Sterling Publications International, London, pp. 47–51.

LEWIS, M.J. and HEPPEL, N.J., (2000), *Continuous Thermal Processing of Foods. Pasteurization and UHT Sterilization*, Aspen Publishers, Inc., Gaithersburg, MD.

MANVELL, C., (1987), Sterilisation of food particulates – an investigation of the APV Jupiter system, *Food Science & Technology Today*, **1**, 106–109.

MARCH, B.E., (1982), Effect of processing on nutritive content of food: fish, in (M. Rechcigl, Ed.), *Handbook on the Nutritive Value of Processed Foods*, Vol. 1, CRC Press, Boca Raton, FL, pp. 336–381.

MARTIN-BELLOSO, O. and LLANOS-BARRIOBERO, E., (2001), Proximate composition, minerals and vitamins in selected canned vegetables, *European Food Research Technology*, **212**, 182–187.

MAY, N.S., (2001), Retort technology, in (P. Richardson, Ed.), *Thermal Technologies in Food Processing*, Woodhead Publishing, Cambridge, pp. 7–28.

NEILSEN, S.S., MARCY, J.E. and SADLER, G.D., (1993), Chemistry of aseptically processed foods, in (J.V. Chambers and P.E. Neilsen, Eds.), *Principles of Aseptic Processing and Packaging*, 2nd edn, The Food Processsors Institute, Washington, DC, pp. 87–114.

NOH, B.S., HEIL, J.R. and PATINO H., (1986), Heat transfer study on flame pasteurization of liquids in aluminum cans, *J. Food Science*, **51** (3), 715–719.

NOTT, K.P. and HALL, L.D., (2004), New techniques for measuring and validating thermal processes, in (P. Richardson, Ed.), *Improving the Thermal Processing of Foods*, Woodhead Publishing, Cambridge, pp. 385–407.

OHLSSON, T., (1992), R&D in aseptic particulate processing technology, in (A. Turner, Ed.), *Food Technology International Europe*, Sterling Publications International, London, pp. 49–53.

OLIVEIRA, J.C., (2004), Optimising the efficiency and productivity of thermal processing, in (P. Richardson, Ed.), *Improving the Thermal Processing of Foods*, Woodhead Publishing, Cambridge, pp. 32–49.

PALANIAPPAN, S. and SIZER, C.E., (1997), Aseptic process validated for foods containing particulates, *Food Technology*, **51** (8), 60–68.

PELEG, M., (2003), Modelling applied to processes: the case of thermal preservation, in (P. Zeuthen and L. Bøgh-Sørensen, Eds.), *Food Preservation Techniques*, Woodhead Publishing, Cambridge, pp. 507–523.

RAMASWARMY, H.S., AWUAH, G.B. and SIMPSON, B.K., (1997), Heat transfer and lethality considerations in aseptic processing of liquid/particle mixtures, *Critical Reviews in Food Science and Nutrition*, **37** (3), 253–286.

RAMESH, M.N., (1999), Food preservation by heat treatment, in (M.S. Rahman, Ed.), *Handbook of Food Preservation*, Marcel Dekker, New York, pp. 95–172.

REUTER, H., (ED.), (1989), *Aseptic Packaging of Foods*, Technomic Publishing Co., Lancaster, PA.

RICHARDSON, P., (2001), *Thermal Technologies in Food Processing*, Woodhead Publishing, Cambridge.

RICKMAN, J.C., BARRETT, D.M. and BRUHN, C.M., (2007a), Review: Nutritional comparison of fresh, frozen and canned fruits and vegetables. Part I. Vitamins C and B and phenolic compounds, *J. Science Food and Agriculture*, **87**, 930–944.

RICKMAN, J.C., BARRETT, D.M. and BRUHN, C.M., (2007b), Review: Nutritional comparison of fresh, frozen and canned fruits and vegetables. Part II. Vitamin A and carotenoids, vitamin E, minerals and fiber, *J. Science Food and Agriculture*, **87**, 1185–1196.

ROLLS, B.A., (1982), Effect of processing on nutritive content of food: milk and milk products, in (M. Rechcigl (Ed.), *Handbook on the Nutritive Value of Processed Foods*, Vol. 1, CRC Press, Boca Raton, FL, pp. 383–399.

SABLANI, S. and RAMASWARMY, H.S., (1998), Multi-particle mixing behaviour and its role in heat transfer during end-over-end agitation of cans, *J. Food Engineering*, **38**, 141–152.

SANDEEP, K.P., (2004), Developments in aseptic processing, in (P. Richardson, Ed.), *Improving the Thermal Processing of Foods*, Woodhead Publishing, Cambridge, pp. 177–187.

SANDEEP, K.P. and PURI, V.M., (2001), Aseptic processing of foods, in (J. Irudayaraj, Ed.), *Food Processing Operations Modelling*, Marcel Dekker, New York, pp 37–81.

SHAPTON, D.A. and SHAPTON, N.F., (1993), *H. J. Heinz Company Staff, Principles and Practices for the Safe Processing of Foods*, Woodhead Publishing, Cambridge, pp. 334–357.

SHAW, G.H., (2004), The use of data loggers to validate thermal processes, in (P. Richardson, Ed.), *Improving the Thermal Processing of Foods*, Woodhead Publishing, Cambridge, pp. 353–364.

SIMPSON, R., (2004), Optimising the efficiency of batch processing with retort systems in thermal processing, in (P. Richardson, Ed.), *Improving the Thermal Processing of Foods*, Woodhead Publishing, Cambridge, pp. 50–81.

SIMPSON, R., ALMONACID-MERINO, S.F. and TORRES, J.A., (1993), Mathematical models and logic for the computer control of batch retorts: conduction heated foods, *J. Food Engineering*, **20** (3), 283–295.

SINGH, R.P. and HELDMAN, D.R., (2001), Preservation processes, in *Introduction to Food Engineering*, 3rd edn, Academic Press, London, pp. 333–366.

STUMBO, C.R., (1973), *Thermobacteriology in Food Processing*, 2nd edn, Academic Press, New York.

TOLEDO, R.T., (1999), Thermal process calculations, in *Fundamentals of Food Process Engineering*, 2nd edn, Aspen Publications, Gaithersburg, MD, pp. 315–397.

TUCKER, G.S., (2004a), Improving rotary thermal processing, in (P. Richardson, Ed.), *Improving the Thermal Processing of Foods*, Woodhead Publishing, Cambridge, pp. 124–137.

TUCKER, G.S., (2004b), Validation of heat processes: an overview, in (P. Richardson, Ed.), *Improving the Thermal Processing of Foods*, Woodhead Publishing, Cambridge, pp. 334–352.

VAN LOEY, A., GUIAVARC'H, Y., CLAEYS, W. and HENDRICKX, M., (2004), The use of time-temperature integrators (TTIs) to validate thermal processes, in (P. Richardson, Ed.), *Improving the Thermal Processing of Foods*, Woodhead Publishing, Cambridge, pp. 365–384.

VERBOVEN, P., DE BAERDEMAEKER, J. and NICOLAI, B.M., (2004), Using computational fluid dynamics to optimise thermal processes, in (P. Richardson, Ed.), *Improving the Thermal Processing of Foods*, Woodhead Publishing, Cambridge, pp. 82–102.

WILLHOFT, E.M.A., (1993), *Aseptic Processing and Packaging of Particulate Foods*, Blackie Academic and Professional, London.

14

Evaporation and distillation

Abstract: Evaporation and distillation are processes that use heat to remove water and/or more volatile components from liquid foods by exploiting differences in their volatility. The chapter explains the theory of evaporation, and describes the equipment used and methods to reduce energy consumption. The chapter then discusses the theory of distillation, the equipment used commercially, and the effects of both operations on micro-organisms and food quality are briefly outlined.

Key words: evaporation, distillation, heat and mass balance, multiple effect evaporation, vapour recompression, forced circulation evaporators, azeotropic mixtures, distillation column.

In contrast to other separation operations in which water is removed from a food at ambient temperatures by centrifugation, filtration or membrane separation (Chapter 5), evaporation and distillation use heat to remove water and/or more volatile components from the bulk of liquid foods by exploiting differences in their vapour pressure (their volatility). Evaporation is the partial removal of water from liquid foods by boiling off water vapour. More volatile components are also removed along with the water vapour, and when these contribute to the flavour of a product, they may be recovered and added back to the concentrated food.

Distillation is a unit operation that separates more volatile components in a solution (those that have a higher vapour pressure than water). When vapours are produced from a mixture, they contain the components of the original mixture, but in proportions that are determined by their relative volatilities (i.e. the vapour is richer in components that are more volatile). In fractional distillation, the vapour is condensed and re-evaporated to further separate or purify components. The main uses of distillation in the food industry are to concentrate essential oils, flavours and alcoholic beverages and to deodorise fats and oils. Evaporation and distillation are among the most energy-intensive operations in food processing.

14.1 Evaporation

An important commercial use of evaporation (or concentration by boiling) is to pre-concentrate foods (e.g. fruit juices, milk and coffee extracts) to reduce their weight and volume and hence reduce storage and transport costs. Partly concentrated foods may then be reconstituted at a different location, dried or used as ingredients in other processed foods. Evaporation is also used to increase the solids content of foods (e.g. tomato paste, sugar confectionery, evaporated milk, jams and marmalades) and hence to contribute to their preservation by a reduction in water activity (Chapter 1, section 1.3.3). Concentrated foods are more convenient for the consumer (e.g. fruit cordials for dilution, concentrated soups, garlic pastes) or for the manufacturer (e.g. liquid pectin, fruit concentrates for use in ice cream or baked goods and liquid malt extract for breweries). The operation is also used to concentrate brines and syrups in the production of crystallised salt and sugar respectively.

The simplest method is atmospheric evaporation using an open boiling pan (Section 14.1.3) but this is slow and energy inefficient, and the prolonged exposure to high temperatures would cause unacceptable quality degradation in most foods. Commercially, evaporation of foods is therefore carried out at lower temperatures by heating the product under a partial vacuum. This protects heat-sensitive components of the food and so maintains nutritional quality and sensory properties in the concentrated product.

Evaporation is more expensive in energy consumption than other methods of concentration (especially membrane concentration (Chapter 5, section 5.5) and freeze concentration (Chapter 23, section 23.2)), but it produces a higher degree of concentration than these other methods do (up to ≈85% solids compared with ≈30% solids from membrane concentration) (Heldman and Hartel 1997). Methods that are used to reduce energy consumption in evaporation are described in section 14.1.2.

14.1.1   Theory

Details of types of heat and methods of heat transfer are given in Chapter 10 (section 10.1.1). A phase diagram (Chapter 10, Fig. 10.1) shows how liquid and water vapour are in equilibrium along the vapour pressure–temperature curve. Boiling occurs when the saturated vapour pressure of the water is equal to the external pressure on the water surface (boiling point = 100 °C at atmospheric pressure at sea level). At pressures below atmospheric, water boils at lower temperatures as shown in Fig. 14.1. The heat required

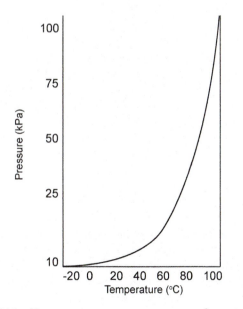

Fig. 14.1   Vapour pressure–temperature curve for water.

to vaporise water (the latent heat of vaporisation) varies according to this temperature. Latent heats are shown in steam tables (Chapter 10, Table 10.4). In the following description of evaporation, water vapour and steam are essentially the same, but the term 'water vapour' is used to describe vapour given off by boiling liquid in an evaporator and 'steam' describes the heating medium.

During evaporation, sensible heat is first transferred from steam to the food to raise the temperature to its boiling point. Latent heat of vaporisation is then supplied by the steam to form bubbles of water vapour that leave the surface of the boiling liquid. The rate of evaporation is determined mostly by the rate of heat transfer into the food. However, the viscosity of some foods increases substantially when their concentration increases and this may reduce the rate of mass transfer of vapour from the food and hence control the rate of evaporation. The rate of heat transfer across evaporator walls and boundary films is found using Equation 10.20 ($Q = UA(\theta_a - \theta_b)$) and a related sample calculation is given in Chapter 10 (sample problem 10.2).

The following factors influence the rate of heat transfer in an evaporator and hence determine processing times and the quality of concentrated products.

Temperature difference between the steam and boiling liquid
Higher temperature differences increase the rate of heat transfer. The temperature difference can be increased by either raising the pressure (and hence the temperature) of the steam (Chapter 10, Table 10.4 and Fig. 10.1) or by reducing the temperature of the boiling liquid by evaporating under a partial vacuum. Very high steam pressures or vacua both require extra strength in equipment and increase its capital cost, and commercial evaporators, therefore operate at pressures of 12–30 kPa, which reduce the boiling point of the liquid to 40–65 °C.

The temperature difference between steam and boiling liquid becomes smaller as foods become more concentrated owing to elevation of the boiling point. Over the temperature range found in commercial evaporators, Dühring's rule states that there is a linear relationship between the boiling temperature of a solution and the boiling point of pure water at the same pressure. This can be represented by Dühring charts for different solutes (Fig. 14.2) and a sample calculation showing elevation of the boiling point in an evaporator is given in sample problem 14.1. Boiling point elevation is more important in foods that contain a high concentration of low molecular weight solutes (e.g. brine or sugar syrups), and causes the rate of heat transfer to fall as evaporation proceeds. It is less important in foods that contain mostly high molecular weight solids (although these may cause fouling problems – see below). In large evaporators, the boiling point of liquid at the base may be slightly raised as a result of increased pressure from the weight of liquid above (the hydrostatic head). In such cases measurement of the boiling point for processing calculations is made half-way up the evaporator.

Boundary films
A film of stationary liquid at the evaporator wall may be the main resistance to heat transfer. The increase in viscosity of many foods as concentration proceeds reduces the flowrate, increases the boundary film thickness, and hence reduces the rate of heat transfer (details are given in Chapter 10, section 10.1.2). More viscous foods are also in contact with hot surfaces for longer periods and as a result suffer greater heat damage. The thickness of boundary films is reduced by promoting convection currents within the food or by mechanically induced turbulence (section 14.1.3).

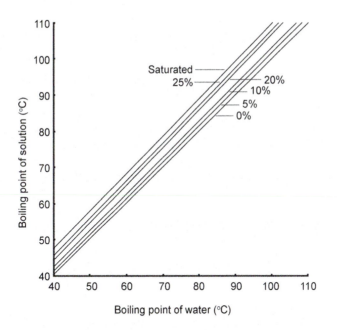

Fig. 14.2   Dühring chart for boiling point of sodium chloride solutions (from Coulson and Richardson 1978).

Sample problem 14.1

A liquid food that has a vapour pressure similar to a 10% salt solution is evaporated to 25% solids in a vacuum evaporator at a pressure of 12.35 kPa. Find the boiling point at the beginning and the end of evaporation and the elevation of boiling point.

Solution to sample problem 14.1

The boiling point of water at 12.35 kPa is found from steam tables (Table 10.4) as 50 °C.

From the Dühring chart for sodium chloride solutions (Fig. 14.2), using the boiling of water at 50 °C, the boiling point of a 10% solution = 52 °C and the boiling point of a 25% solution = 55 °C.

There is therefore a 3 °C elevation of boiling point during the evaporation process.

Deposits on heat transfer surfaces

The 'fouling' of evaporator surfaces reduces the rate of heat transfer. The type of fouling and the rate at which deposits build up depend on the temperature difference between the food and the heated surface and the viscosity and chemical composition of the food. For example, denaturation of proteins or deposition of polysaccharides causes the food to burn onto hot surfaces and the process must then be suspended to clean the equipment. Fouling is reduced in some types of equipment by continuously removing food from the evaporator walls or, for foods that are particularly susceptible to fouling, by maintaining a smaller temperature difference between the food and the heating surface (section 14.1.3).

Metal corrosion on the steam side of evaporation equipment would also reduce the rate of heat transfer, but it is reduced by anti-corrosion chemicals or surfaces.

Heat and mass balances

Heat and mass balances (Chapter 10, section 10.1.2) are used to calculate the degree of concentration, energy use and processing times in an evaporator (see sample problems 14.2 and 14.3). Singh and Heldman (2001) and Toledo (1999) describe similar calculations for multiple effect evaporators (below).

The mass balance states that 'the mass of feed entering the evaporator equals the mass of product and vapour removed from the evaporator'. This is represented schematically in Fig. 14.3. For the water component, the mass balance is given by:

$$m_f(1 - X_f) = m_p(1 - X_p) + m_v \qquad \boxed{14.1}$$

For solutes, the mass of solids entering the evaporator equals the mass of solids leaving the evaporator:

$$m_f X_f = m_p X_p \qquad \boxed{14.2}$$

The total mass balance is $m_f = m_p + m_v$ $\qquad \boxed{14.3}$

Assuming that there are negligible heat losses from the evaporator, the heat balance states that 'the amount of heat given up by the condensing steam equals the amount of heat used to raise the feed temperature to boiling point and then to boil off the vapour':

$$Q = m_s \lambda_s$$
$$= m_f c_p (\theta_b - \theta_f) + m_v \lambda_v \qquad \boxed{14.4}$$

where c_p $(\mathrm{J\,kg^{-1}\,°C^{-1}})$ = specific heat capacity of feed liquor, λ_s $(\mathrm{J\,kg^{-1}})$ = latent heat of condensing steam, λ_v $(\mathrm{J\,kg^{-1}})$ = latent heat of vaporisation of water (Chapter 10, Table 10.4). That is:

Heat supplied by steam = sensible heat + latent heat of vaporisation

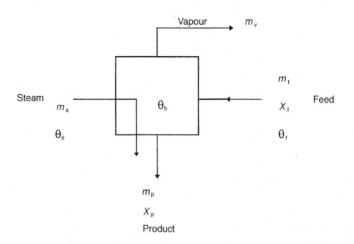

Fig. 14.3   Steady state operation of an evaporator: m_f $(\mathrm{kg\,s^{-1}})$, mass transfer rate of feed liquor; m_p $(\mathrm{kg\,s^{-1}})$, mass transfer rate of product; X_f, solids fraction of feed liquor; X_p, solids fraction of product; m_v $(\mathrm{kg\,s^{-1}})$, mass transfer rate of vapour produced; m_s $(\mathrm{kg\,s^{-1}})$, mass transfer rate of steam used; θ_f (°C), initial feed temperature; θ_b (°C), boiling temperature of food; θ_s (°C), temperature of steam.

For the majority of an evaporation process, the rate of heat transfer is the controlling factor and the rate of mass transfer only becomes important when the liquor becomes highly concentrated.

Sample problem 14.2
A single-effect vertical short tube evaporator (section 14.1.3) is to be used to concentrate syrup from 10% solids to 40% solids at a rate of 100 kg h^{-1}. The feed enters at 15 °C and is evaporated under a reduced pressure of 47.4 kPa (at 80 °C). Steam is supplied at 169 kPa (115 °C). Assuming that the boiling point remains constant and that there are no heat losses, calculate the quantity of steam used per hour and the number of tubes required. (Additional data: the specific heat of syrup is constant at 3.960 kJ kg^{-1} K^{-1}, the specific heat of water is 4.186 960 kJ kg^{-1} K^{-1}, the latent heat of vaporisation of the syrup is 2309 kJ kg^{-1}, latent heat of steam is 2217 kJ kg^{-1} at 115 °C and the overall heat transfer coefficient is 2600 W m^{-2} K^{-1}. The tube dimensions are: length 1.55 m and diameter 2.5 cm.)

Solution to sample problem 14.2

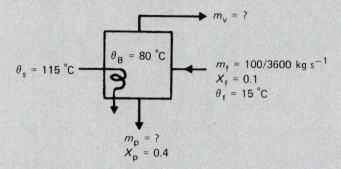

To find the quantity of steam used per hour from Equation 14.2,

$$\frac{100}{3600} \times 0.1 = m_p \times 0.4$$

$$m_p = 0.0069 \text{ kg s}^{-1}$$

From Equation 14.1,

$$\frac{100}{3600}(1 - 0.1) = 0.0069 (1 - 0.4) + m_v$$

$$m_v = 0.0209 \text{ kg s}^{-1}$$

A summary of the mass balance is given in the following table:

	Mass flowrate (kg s^{-1})		
	Solids	Liquid	Total
Feed	0.002 78	0.025	0.0278
Product	0.002 78	0.004 14	0.0069
Vapour			0.0209

From Equation 14.4, the heat required for evaporation is

$$Q = 0.0278 \times 3960 \, (80 - 15) + 0.0209 \times (2309 \times 10^3)$$
$$= 5.54 \times 10^4 \, \mathrm{J \, s^{-1}}$$

$$\begin{array}{c} \text{Heat supplied by 1 kg} \\ \text{steam per second} \end{array} = \text{latent heat} + \begin{array}{c} \text{sensible heat} \\ \text{on cooling at 80 °C} \end{array}$$

$$= (2217 \times 10^3) + 1 \times 4186 \times (115 - 80)$$
$$= 2.36 \times 10^6 \, \mathrm{J \, s^{-1}}$$

Assuming a heat balance in which the heat supplied by the steam equals the heat required for evaporation,

$$\text{Mass of steam} = \frac{5.54 \times 10^4}{2.36 \times 10^6}$$
$$= 0.023 \, \mathrm{kg \, s^{-1}}$$
$$= 84.5 \, \mathrm{kg \, h^{-1}}$$

To find the number of tubes, use Equation 10.20 $(Q = UA(\theta_a - \theta_b))$. Therefore, the total surface area (A) is calculated from:

$$5.54 \times 10^4 = 2600 \times A \, (115 - 80)$$
$$A = 0.61 \, \mathrm{m^2}$$

Now

$$\text{Area of one tube} = (\pi D L) = 0.025 \times 1.55 \times 3.142$$
$$= 0.122 \, \mathrm{m^2}$$

Therefore

$$\text{Number of tubes} = \frac{0.61}{0.122} = 5$$

Sample problem 14.3
Milk containing 3.7% fat and 12.8% total solids is to be evaporated to produce a product containing 7.9% fat. What is the mass of product from 100 kg of milk and what is the total solids concentration in the final product, assuming that there are no losses during the process?

Solution to sample problem 14.3

$$\text{Mass of fat in 100 kg of milk} = 100 \times 0.037 = 3.7 \, \text{kg.}$$

If Y = mass of product:

$$\text{Mass of fat in the evaporated milk} = Y \times 0.079$$

As no fat is gained or lost during the process:

$$0.79 \times Y = 3.7$$

The mass of product $Y = 46.8$ kg

Mass of solids in the milk $= 100 \times 0.128$.

If $Z = \%$ total solids in the evaporated milk

Solids in the product $= 46.8 \times (Z/100)$

i.e.

$$0.4684 \times Z = 12.8$$

$$Z = 27.3$$

Therefore the total solids in the concentrated product $= 27.3\%$.

14.1.2   Improving the economics of evaporation

The main factors that influence the economics of evaporation are high energy consumption and loss of concentrate or product quality (also section 14.1.4).

Reducing energy consumption
A substantial amount of energy is needed to remove water from foods by boiling (2257 kJ per kg of water evaporated at 100 °C). The economics of evaporation are therefore substantially improved by attention to the design and operation of equipment and careful planning of energy use. Evaporator performance is rated on the basis of 'steam economy'. Steam economy is defined as 'the weight (kg) of water evaporated per kilogram of steam used'.

Smith (1997) describes an energy management system used in a sugar refinery that has resulted in substantial reductions in energy consumption. Energy can be saved by re-using heat contained in vapours produced from the boiling food. This is done either by multiple effect evaporation or by vapour recompression.

Multiple effect evaporation
This involves connecting several evaporators (or 'effects') together, with vapour from one effect being used directly as the heating medium in the next. However, the vapour can only be used to boil liquids at a lower boiling temperature, and the effects must therefore have progressively lower pressures in order to maintain the temperature difference between the food and the heating medium.

The number of effects used in a multiple effect system is determined by the savings in energy consumption compared with the higher capital investment required and the provision of increasingly higher vacua in successive effects. As a broad guide, multiple effects are justified when the required rate of evaporation exceeds ≈ 1000 kg h^{-1}. Below this, a single effect evaporator with vapour recompression (below) is more efficient. In a two-effect evaporator, 2 kg of vapour can be evaporated from the product for each 1 kg of steam supplied. The steam economy (mass of vapour produced per mass of steam used) increases as the number of effects increases (Anon 2007). In the majority of applications,

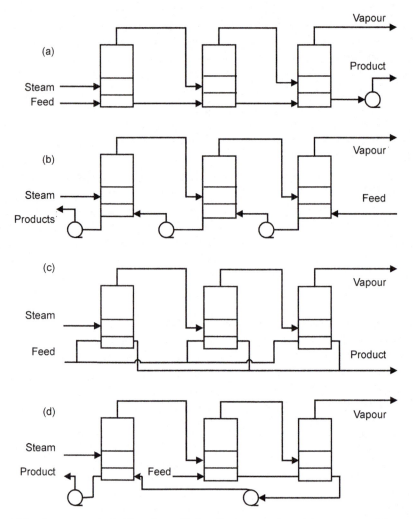

Fig. 14.4   Arrangement of effects in multiple effect evaporation: (a) co-current; (b) counter-current; (c) parallel; (d) mixed (after Brennan *et al.* 1990).

three to six effects are used and steam economies of 6.0–6.5 are found (Heldman and Hartel 1997).

An animation of multiple effect evaporation has been produced by Singh (2001). Different arrangements of multiple effect evaporators are shown in Fig. 14.4 using triple-effect evaporation as an example, and the relative advantages and limitations of each arrangement are described in Table 14.1.

Vapour recompression
In vapour recompression, the pressure (and therefore the temperature) of vapour evaporated from the product is increased and the resulting high-pressure steam is re-used as a heating medium (see also Fig. 10.1 in Chapter 10). Thermo-vapour recompression (TVR) involves passing high-pressure steam through a Venturi-type steam jet and drawing in and compressing a portion of the lower-pressure vapour from the evaporator. The mixture has a higher energy than the evaporator vapour and this method reduces the

Table 14.1 Advantages and limitations of various methods of multiple effect evaporation

Arrangement of effects	Advantages	Limitations
Forward feed	Least expensive, simple to operate, no feed pumps required between effects, lower temperatures with subsequent effects and therefore less risk of heat damage to more viscous product	Reduced heat transfer rate as the feed becomes more viscous, rate of evaporation falls with each effect, best quality steam used on initial feed which is easiest to evaporate. Feed must be introduced at boiling point to prevent loss of economy (if steam supplies sensible heat, less vapour is available for subsequent effects)
Reverse feed	No feed pump initially, best-quality steam used on the most difficult material to concentrate, better economy and heat transfer rate as effects are not subject to variation in feed temperature and feed meets hotter surfaces as it becomes more concentrated thus partly offsetting increase in viscosity	Interstage pumps necessary, higher risk of heat damage to viscous products as liquor moves more slowly over hotter surfaces, risk of fouling
Mixed feed	Simplicity of forward feed and economy of backward feed, useful for very viscous foods	More complex and expensive
Parallel	For crystal production, allows greater control over crystallisation and prevents the need to pump crystal slurries	Most complex and expensive of the arrangements, extraction pumps required for each effect

Adapted from Brennan *et al.* (1990)

amount of fresh steam required. For example, in a multiple effect system using a TVR unit to recompress vapour from the third effect and feed it back to the first effect, the economy improves from 6 to 10 (Heldman and Hartel 1997). However, when high boiling point elevation occurs (e.g. in sugar evaporation) the vapour from the evaporator has a higher temperature and TVR has less economic benefit. The method also results in less control of the evaporation process because the steam jet is set for a particular temperature and quantity of vapour, and when these change as evaporation proceeds (e.g. due to fouling) the efficiency decreases.

Mechanical vapour recompression (MVR) uses a radial fan or centrifugal compressor to compress vapour. Fans achieve lower compression ratios but are more energy efficient than compressors, more reliable and have lower maintenance costs. Once evaporation has started, most of the steam is provided by MVR and only a small amount of fresh steam is needed, giving steam economies of 35–40 (Heldman and Hartel 1997). The use of a MVR evaporator is equivalent to 30–55 effects in a multiple effect system, and combining vapour recompression with multiple effect evaporation is the most thermodynamically efficient method to remove water (Anon 2007). MVR is most suited to products that have a small boiling point elevation and cause little fouling, and where there is little product entrainment and a small temperature difference is required between the steam and product.

Reducing losses of concentrate or product quality
Product losses are caused by 'entrainment', where a fine mist of concentrate is produced by the violent boiling, and is carried out of the evaporator by the vapour. Foods that are liable

to foaming, due to proteins and carbohydrates, may have higher entrainment losses because the foam prevents efficient separation of vapour and concentrate. Most designs of equipment include disengagement spaces to minimise entrainment, or cyclone-type separators (Chapter 16, section 16.2.1) to collect entrained product. Non-ionic surfactants may also be used to control foaming in some applications (e.g. sugar processing) (Anon 2003).

The quality of many products is maintained by using equipment that has a small temperature difference between the steam and product (hence no 'hot-spots' that could cause the product to burn), by reducing the boiling temperature using vacuum evaporation, and by using short residence times. Details of these types of equipment are given in section 14.1.3.

When volatile aroma components are removed along with water vapour, they can be recovered and added back to the concentrate to increase the value of the product. Volatile recovery is based on the lower boiling point of many aroma compounds compared with water. It is achieved by either stripping volatiles from the feed liquor with inert gas or by partial condensation and fractional distillation of vapour from the evaporator (section 14.2). For example in multiple effect juice processing, volatiles can be recovered by heating juice in the first effect and then releasing it into a separator that has a lower pressure. The vapours carry with them some of the volatiles from the liquid (known as the 'first strip'). The temperature at which the first strip takes place depends on the type of juice, and some (e.g. apple) can withstand 80–90 °C without damage to the volatiles, whereas others (e.g. pineapple) require temperatures below 60 °C (Anon 2007). Most of the vapour is used to heat the second effect, but a portion ($\approx$ 10–15%) is diverted to an aroma distillation unit. Here selective stripping and rectification (section 14.2.2) removes more water vapour to leave a concentrated essence, which is chilled and stored. The vapour from the final effect is passed through a 'scrubber' to remove remaining volatiles, and these are similarly concentrated and added to the chilled essence. Flash coolers, in which the food is sprayed into a vacuum chamber, are used to rapidly cool the product and the concentrated essence is mixed in before filling. Because of the reduction in weight of the concentrated product, there is sufficient essence to regain the aroma of the original juice when it is added back.

14.1.3   Equipment

The basic components of an evaporator are:

- a source of steam and a heat exchanger (termed a 'calandria') that transfers heat from steam to the food;
- a feed distributor to uniformly distribute the feed to the heat transfer surface;
- a means of creating a vacuum; and
- a means of separating and condensing the vapours produced.

Other components include control systems, a method for cleaning-in-place and, where required, a mechanical or steam ejector vapour recompression system and/or a volatile recovery system.

Ideally an evaporator should selectively remove water from a food without changing the solute composition, so that the original product is obtained on dilution. This is approached in some equipment, but the closer to the ideal that is achieved the higher the cost. As with other unit operations, the selection of equipment is therefore a compromise between the cost of production and the quality required in the product. The selection of an evaporator should include the following considerations:

- Properties of the product (heat sensitivity, change in viscosity at high concentration, boiling point elevation, volatile content, risk of fouling) in relation to the residence time and temperature of evaporation.
- Operating capacity (as $kg\,h^{-1}$ of water removed) and size of the evaporator in relation to its capacity.
- Degree of concentration required (as % dry solids in the product).
- Required product quality and the need to recover volatiles.
- Ease of cleaning, reliability and simplicity of operation.
- Capital and operating costs in relation to capacity and product quality.

In some applications it may be more cost effective to combine two types of evaporator; for example initial concentration of the bulk liquor (which is less sensitive to heat damage) may be done in a low-cost evaporator that has a high throughput and/or longer residence time, followed by final concentration of the smaller volume of heat-sensitive liquor in a more expensive, but less thermally damaging evaporator as the second effect. Further details are given by Chen and Hernandez (1997).

Evaporator designs can be grouped into those that rely on natural circulation of products and those that employ forced circulation.

Natural circulation evaporators

Today, most evaporators operate continuously but the old method of concentration using batch boiling pans is still used in a few applications (e.g. for concentrating jams that contain whole fruits), for low or variable production rates of small quantities of materials (e.g. preparation of ingredients such as sauces and gravies), or in applications where flexibility is required for frequent changes of product. They are hemispherical pans, similar in appearance to jacketed mixing vessels (Chapter 4, Fig. 4.10). They are heated either by steam passing through internal tubes or an external jacket, or by internal electrical heaters. They have low capital cost and are relatively easy to construct and maintain. However, the food contact area is small and temperature differences between steam and the product must be small to avoid fouling. This results in low heat transfer coefficients (Table 14.2) and a small evaporation capacity. The residence time of products may be several hours and pans are therefore fitted with a lid and operated under vacuum to reduce damage to the product quality. Rates of heat transfer are improved by agitating the liquor using a stirrer or paddle, which reduces the thickness of boundary films and also reduces fouling of the heat transfer surface.

Table 14.2   Comparison of residence times and heat transfer coefficients in selected evaporators

Type of evaporator	Number of stages	Residence time	OHTC ($W\,m^{-2}\,K^{-1}$)	
			Low viscosity	High viscosity
Vacuum boiling pan	1	0.5–2 h	500–1000	<500
Short tube	1	–	570–2800	–
Tubular climbing film	1	10 s–4 min	2000–3000	<300
Tubular falling film	1	5–30 s	2250–6000	–
Plate	3	2–30 s	4000–7000	–
Expanding flow	2	0.5–30 s	2500–3000	–
Wiped/scraped film	1	0.5–100 s	2000–3000	1700
Centrifugal	1	0.6–2 s	8000	–

From Earle (1983), Anon (2007) and Heldman and Hartel (1997)

Short-tube evaporators are shell-and-tube heat exchangers that have a vessel (or shell) that contains a vertical bundle of between 100 and 1600 tubes, depending on the required production rate (Anon 2007). Feed liquor is heated by steam condensing on the outside of the tubes and the vertical tubes promote natural convection currents in the product that increase the rate of heat transfer. Alternatively the tube bundle may be contained in a separate shell outside the boiling vessel (an external calandria). This arrangement has the advantage of higher evaporation capacities because the size of the heater is not dependent on the size of the vessel and the calandria is easily accessible for cleaning. Vapours are removed in a separator and liquor may be recirculated until the desired concentration is achieved. This equipment has in the past been used to concentrate dairy products, syrups, salt, fruit juices and meat extracts, but its most common application is now as a reboiler for distillation columns (section 14.2.2).

Climbing (or rising) film tubular evaporators consist of a vertical bundle of tubes, each up to 5 cm in diameter, contained within a steam shell 3–15 m high. Liquor is heated almost to boiling point before entering the evaporator. It is then further heated inside the tubes and boiling commences, to form a central core of vapour that expands rapidly and forces a thin film of liquor to the walls of the tube (Fig. 14.5). As the rapidly concentrating liquor moves up the wall of the tube, more vapour is formed, resulting in a higher core velocity that produces a thinner, more rapidly moving film of liquor. This produces high heat transfer coefficients and short residence times, which make this type of evaporator suitable for heat-sensitive foods. The concentrate is separated from the vapour and removed from the evaporator or passed to subsequent effects in a multiple-effect system. Vapour is re-used in multiple-effect or vapour recompression systems (section 14.1.2).

The falling film tubular evaporator operates using a similar principle but the feed is introduced at the top of the tube bundle and is distributed evenly to each tube by specially designed plates or nozzles. The force of gravity supplements the forces arising from expansion of the steam to produce very high liquor flowrates (up to 200 m s^{-1} at the end of 12 m tubes) and short residence times (typically 5–30 seconds compared with 3–4 min in a rising film evaporator (Singh and Heldman 2001). Jebson and Chen (1997) studied

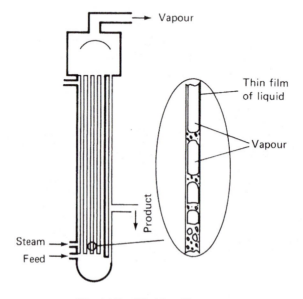

Fig. 14.5   Climbing film evaporator.

the performance of these evaporators and found that heat transfer coefficients varied from 0.3 to 3.0 kW m^{-2} K^{-1} and the steam economy varied from 1 to 4, depending on the number of effects. Other studies are reported by Prost et al. (2006). This type of evaporator is suitable for moderately viscous foods or those that are very heat sensitive (e.g. yeast extracts, dairy products and fruit juices). Falling film evaporators are described in detail by Burkart and Wiegand (1987) and Geankoplis (1993). Anon (1986) describes multiple-effect systems, capable of evaporating 45 000 l h^{-1} of milk. Both types of evaporator have high heat transfer coefficients (Table 14.2) and efficient energy use. Compared with plate evaporators (below) they are suitable for larger-scale production, require less floor space and are capable of handling products that have larger suspended solids.

The number of effects in a climbing film tube evaporator is limited by the temperature difference required to create the rising film. Because the film must overcome gravity to rise up the tube, a temperature difference of >14 °C is required between the steam and product. In an example given by Anon (2007), if the steam temperature is 104 °C and the boiling temperature in the last effect of a multiple effect system is 49 °C, the temperature difference of 55 °C would limit the number of effects to (55/14) ≈ 4 effects. In contrast a falling film evaporator has no requirement to overcome gravity and the temperature difference can therefore be much smaller, permitting up to 10 effects.

In another design, the climbing/falling film tubular evaporator combines the benefits of both types of equipment. Feed liquor enters the climbing film section first and is then fed to the falling film section. The tube bundle is approximately half the height of a climbing or falling film evaporator, thus reducing space requirements.

Forced circulation evaporators
There are a number of designs of forced circulation evaporators that pump food between plates or through tubes, or create thin films of product using scraper assemblies inside cylindrical heat exchangers.

Plate evaporators are similar in construction to heat exchangers used for pasteurisation (Fig. 12.3 in Chapter 12, section 12.2) and ultra-high-temperature (UHT) sterilisation (Chapter 13, section 13.2.3). However, in this application the climbing and/or falling film principle is used to concentrate liquids in thin, rapidly moving films in the spaces between plates, with steam sections arranged alternately between each product section. The number of climbing or falling film sections can be adjusted to meet the production rate and degree of concentration required. In the climbing and falling design, feed liquor enters a climbing film section at a temperature that is slightly higher than the evaporation temperature. This causes the product to flash across the plates and ensures an even distribution of liquor. After passing through the climbing film section, liquor is then passed to a falling film section, and the mixture of vapour and concentrate is separated outside the evaporator. Narrow gaps between the plates and corrugations in the plates cause high levels of turbulence and partial atomisation of the liquor. This generates high rates of heat transfer, short residence times and high energy efficiencies (Table 14.2). Plate evaporators have higher heat transfer coefficients than those found in tubular climbing/falling film evaporators. They are compact, capable of high throughputs (up to 16 000 kg h^{-1} water removed), and are easily dismantled for maintenance and inspection. Falling-film plate evaporators (without the climbing film sections) have higher throughputs (up to 30 000 kg h^{-1} water removed) (Anon 2007). Further details are given by Hoffman (2004). Anon (2007) and Olsson (1988) describe their advantages, compared to tubular falling film evaporators.

Plate evaporators are suitable for heat-sensitive foods of higher viscosity (0.3–0.4 N s m^{-2}) including yeast extract, coffee extract, dairy products (milk, whey protein), pectin and gelatine concentrates, high-solids corn syrups, liquid egg, fruit juice concentrates and purées, and meat extracts. They can also be used as 'finishing' evaporators for fruit purées that are pre-concentrated using other equipment, and to remove solvents during the production of vegetable oils (Chapter 5, section 5.4.2). When arranged as multiple effects and/or multi-stage systems, plate evaporators can achieve high degrees of concentration (up to 98% for sugar solution) in a single pass at operating temperatures from 25 to 90 °C (Anon 2007). The main limitations are products that contain high levels of suspended solids or those that readily foul heat transfer surfaces.

In forced circulation tube evaporators, a pump circulates liquor at high velocity through a calandria, and an over-pressure from a hydrostatic head above the tubes prevents it from boiling. When the liquid enters the separator, which is at a slightly lower pressure, it flashes to a vapour, which is removed and the liquor is recirculated. Dilute feed is added at the same rate that product is removed. This type of evaporator has a higher efficiency than natural circulation tube evaporators, but residence times are longer than climbing/falling film designs. They have been used to concentrate tomato pastes and sugar and are suitable for foods that are prone to crystallisation during concentration (e.g. tartaric acid crystallisation from concentrated grape juice), for crystallisation duties, or for foods that are prone to degrade during heating and deposit solids on the heat transfer surface. The high liquid velocity during recirculation (e.g. 2–6 m s^{-1} compared with 0.3–1 m s^{-1} in natural circulation tube evaporators (Singh and Heldman 2001)) minimises the build-up of crystals on heat exchanger tubes and the temperature difference across the heating surface is very low (2–3 °C) to minimise fouling (Anon 2007). Both the capital and operating costs of this equipment are low compared with other types of evaporators.

The expanding-flow evaporator uses similar principles to the plate evaporator but has a stack of inverted cones instead of a series of plates. Feed liquor flows to alternate spaces between the cones from a central shaft and evaporates as it passes up through channels of increasing flow area (hence the name of the equipment). Steam is fed down alternate channels. The vapour–concentrate mixture leaves the cone assembly tangentially and is separated by a special design of shell that induces a cyclone effect. This evaporator has a number of advantages including compactness, short residence times and a high degree of flexibility achieved by changing the number of cones.

Mechanical (or agitated) thin-film evaporators
Wiped- or scraped-film evaporators are characterised by differences in the thickness of the film of food being processed: wiped-film evaporators have a film thickness of approximately 0.25 mm whereas in scraped-film evaporators it is up to 1.25 mm. Both types consist of a steam or 'hot oil' jacket surrounding a high-speed rotor, fitted with short blades along its length (Fig. 14.6). There are three types: a rigid blade rotor that has a fixed clearance between the blade tip and the heating surface; a rotor with radially moving wipers (with PTFE or graphite elements); and a rotor that has hinged free-swinging metal wiper blades with PTFE tips (Anon 2008a). Feed liquor is introduced between the rotor and the heated surface and evaporation takes place rapidly as a thin film of liquor is swept through the machine by the rotor blades. The blades keep the liquid violently agitated and thus promote high rates of heat transfer (Table 14.2) and prevent the product from burning onto the hot surface. The residence time of the liquor is adjusted between 0.5 and 100 s depending on the type of food and the degree of concentration required. This type of equipment is highly suited to viscous (up to 20 N s m^{-2}) heat-sensitive foods or to those

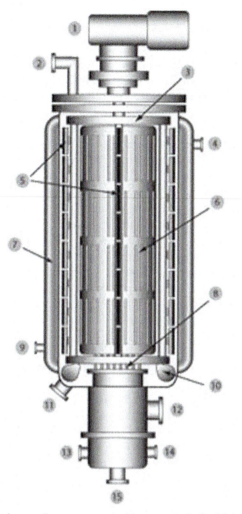

Fig. 14.6   Mechanical short path evaporator: 1, drive motor; 2, feed inlet; 3, distributor plate; 4, hot oil outlet; 5, wiper blades; 6, entrainment separator; 7, hot oil jacket; 8, internal condenser; 9, hot oil inlet; 10, extruder blades; 11, residue discharge; 12, vacuum outlet; 13, cooling water inlet; 14, cooling water outlet; 15, distillate discharge (courtesy of Chem Group Inc.) (Anon 2005).

that are liable to foam or foul evaporator surfaces (e.g. fruit pulps, tomato paste, meat extracts, honey, cocoa mass, coffee and dairy products). However, the capital costs are high owing to the precise alignment required between the rotor and wall. Operating costs are also high as only single effects are possible, which reduces the throughput and gives poor steam economy. It is therefore mainly used for 'finishing' highly viscous products after concentration in other equipment where there is less water to be removed, the product is valuable and there is a substantial risk of heat damage.

A second design of mechanical thin film evaporator is the centrifugal (or rotating cone) evaporator, in which liquor is fed from a central pipe to the undersides of rotating hollow cones. It immediately spreads out to form a layer approximately 0.1 mm thick. Steam condenses on the other side of each cone and rapidly evaporates the liquor. In contrast with the expanding-flow evaporator, in which liquid is moved by vapour pressure, the centrifugal evaporator employs centrifugal force to move the liquor rapidly across the heated surface of the cone. Residence times are 0.6–1.6 s even with concentrated liquors (up to $20\,N\,s\,m^{-2}$), and very high heat transfer coefficients are possible (Table 14.2) (Chen et al. 1997). This is due in part to the thin layers of liquor, but also to removal of droplets of condensed steam that are flung from the rotating cones as fast as they are formed. There is therefore no boundary film of condensate to impede heat transfer. The equipment produces a concentrate which, when re-diluted, has sensory and nutritional qualities that are virtually unchanged from those of the feed material, but it is more expensive and has a lower throughput than other types of evaporator. It is used for concentrating coffee and tea extracts, meat extracts and fruit juices. Further details are given by Chen et al. (1997) and Jebson et al. (2003).

Condensers

The vapour removed from evaporators is condensed using one of two methods: either it is cooled by a surface condenser (a heat exchanger cooled by water) when the condensate is collected for essence recovery; or cooling water is mixed with the condensate. An example of the latter is a barometric condenser, which consists of a sealed chamber above a tall column of water. The water column (or 'barometric leg') seals the chamber and creates a partial vacuum. Vapours enter the chamber from the evaporator and are condensed by water sprays. Toledo (1999) describes this equipment in detail and also gives calculations relating to the operation of barometric condensers.

Control of evaporators

Close control of evaporation conditions enables both savings in energy consumption and production of products that have the required organoleptic properties. In recirculating evaporators, the factors under control are the final solids content of the product, the rate of water removal (or alternatively the liquor feed rate), the level of liquid in the evaporator and the steam pressure. Previously, the solids concentration has been monitored using refractive index, density or viscosity measurements, but mass flow meters that measure flowrate and density are now the standard method. Programmable logic controllers (PLCs) (Chapter 27, section 27.2.2) set the steam pressure and monitor the product density in the recirculation loop. The information is used to control the rate of product removal from the evaporator when it reaches the required solids content and to adjust the flowrate of feed liquor to maintain the correct liquid level in the evaporator. Changes to the throughput of the evaporator are made by changing the steam pressure. In single-pass evaporators it is not possible to delay discharge of the product until it has reached the required density and a different type of control is used. The PLC sets the feed flowrate to the required value and then controls the energy input to achieve the desired degree of concentration. This may involve control of the steam pressure and flowrate or control of the power to a mechanical vapour recompressor.

Evaporators can operate continuously for 6–10 days before being shut down for cleaning, depending on the type of product and the rate at which fouling of heat transfer surfaces takes place. Eggleston and Monge (2007) studied how differences in the time interval between cleaning affects the performance of an evaporator and the amount of product losses. However, low-temperature operation with low-acid foods (e.g. dairy products) risks microbial growth and equipment is therefore cleaned daily. Owing to the high labour costs involved in cleaning evaporators, PLC control of cleaning-in-place in an automated cycle is now standard on most equipment. The PLC also records historical data to optimise the performance of the evaporation system (Anon 2007).

14.1.4   Effect on foods and micro-organisms

Depending on the type of equipment, the operating temperature and residence time, evaporation can produce concentrated foods in which there are few changes to the organoleptic quality or nutritional value, especially when volatile recovery is used. Without volatile recovery, losses of aroma compounds, including esters, aldehydes and terpenes, reduce the flavour of concentrated juices, but in some foods the loss of unpleasant volatiles improves the product quality (e.g. in cocoa liquor and milk). Evaporation darkens the colour of foods, partly because of the increase in concentration of solids, but also because the reduction in water activity promotes chemical changes (e.g. Maillard browning (Chapter 1, section 1.2.2)) and changes to anthocyanins and other pigments in fruit products (Iversen 1999).

In juice concentration there may be substantial losses ($\approx$ 70%) of vitamin C, both as a result of preparation procedures, the evaporation process and, depending on the type of packaging, also during storage. These changes have been studied by Lee and Chen (1998). As a result some juice concentrates are fortified with ascorbic acid (Nindo *et al.* 2007). Vitamin losses in evaporated and sweetened condensed milk vary from insignificant losses (<10%) of thiamin and vitamin B_6 to losses of 80% for vitamin B_{12} and 60% for vitamin C (Porter and Thompson 1976). Vitamins A and D and niacin are unaffected. It is likely that other heat-labile vitamins are also lost during evaporation but little recently published data is available. As these changes are time and temperature dependent, short residence times and low boiling temperatures reduce the extent of these changes and produce concentrates that have a good retention of colour, flavour and vitamins.

Juices are pre-heated because the low temperatures used in evaporation are insufficient to destroy pectic enzymes or contaminating yeasts and lactobacilli. Concentrated juices are preserved to an extent by their natural acidity and increased solids content. Depending on the final product, juices, squashes or cordials may be re-diluted and/or standardised with additional sugar and pasteurised (Chapter 12); 'long-life' juices are UHT sterilised (Chapter 13, section 13.2); and frozen concentrated juices are preserved by freezing (Chapter 22). Similarly, the low temperatures used in vacuum evaporation of milk (40–45 °C) prevent development of a cooked flavour, but they are insufficient to destroy enzymes or any contaminating micro-organisms. Raw milk is therefore pre-heated to 93–100 °C for 10–25 min or 115–128 °C for 1–6 min to destroy osmophilic and thermophilic micro-organisms, and to inactivate lipases and proteases, and to confer heat stability. The milk is concentrated to 30–40% total solids, but remains susceptible to microbial growth. To extend the shelf-life for up to a year, it is sterilised either in cans or under aseptic conditions (Chapter 13). The shelf-life of sweetened condensed milk is extended by the addition of sugar to a concentration $\approx$ 45%. Sugar is added after evaporation to avoid evaporating a high-viscosity liquid. This increases its osmotic pressure and prevents the growth of spoilage or pathogenic micro-organisms. The sweetened evaporated milk is seeded with powdered lactose crystals and cooled while being agitated to promote crystallisation of small lactose crystals. These are necessary to avoid sandiness, a texture defect that affects the mouthfeel. Further information is given by Goff (2007).

14.2   Distillation

Distillation is the separation of more volatile components of a liquid mixture from less volatile components by the application of heat. The volatile-rich vapours are condensed to form a concentrated product. Although common in the chemical industry, distillation in food processing is mostly confined to the production of alcoholic spirits and the preparation of volatile flavour and aroma compounds (e.g. production of essential oils and other flavouring ingredients, or aroma recovery in evaporation).

14.2.1   Theory

In a liquid that contains two components, for example alcohol and water, the molecules are attracted to each other by van der Waals forces. The intermolecular forces (or linkages) that attract similar molecules are greater than those that attract dissimilar molecules. This has two important implications for distillation: first, a dilute alcoholic feed material is easier to distil than a more concentrated feed liquor. This is because in dilute solutions the alcohol molecules are separated by a larger number of water molecules and there are thus fewer of the stronger alcohol–alcohol linkages and more of the weaker alcohol–water linkages. Therefore on heating the weaker alcohol–water linkages are broken more easily and the more volatile alcohol is vaporised. A feed liquor for alcohol distillation is typically 5–7% ethanol rather than the more usual 10–13% found for example in wines (see also fermentation in Chapter 6, section 6.4.3). The second implication of intermolecular attraction is that the distillate contains a high proportion of alcohol molecules, and thus a large number of the stronger alcohol–alcohol linkages. This makes it much more difficult to further concentrate the alcohol by a second distillation. For example, when an ethanol concentration reaches 95.6% w/w the number of alcohol–alcohol linkages is sufficiently high to prevent further separation from water and an alcohol–water equilibrium is established which cannot be changed by further distillation. This is known as an 'azeotropic' (or constant boiling) mixture. Similar equilibria are established for other volatile components of liquors that are distilled to make alcoholic spirits (e.g. other alcohols, aldehydes, ketones) and it is the concentrations of these components that give spirit drinks their individual flavour and aroma characteristics.

In an 'ideal mixture' the intermolecular linkages between different components are the same as those between pure liquids. The vapour pressure of each component of an ideal mixture is related to its concentration by Raoult's Law. This states that 'the partial vapour pressure of a component in a mixture is equal to the vapour pressure of the pure component at that temperature multiplied by its mole fraction in the mixture'. This can be expressed mathematically as:

$$P_A = X_A \cdot P_A^o \qquad \boxed{14.5}$$

and

$$P_B = X_B \cdot P_B^o \qquad \boxed{14.6}$$

where P is the partial pressures of the components A and B, P^o is the partial pressures of the pure components A and B, and X is the mole fraction for components A and B (i.e. the number of moles of a component divided by the total number of moles in the mixture).

Therefore the partial pressure of a component is proportional to its mole fraction at a constant temperature. The total vapour pressure of a mixture is the sum of the partial pressures of the components (Fig. 14.7) (see also sample problem 14.4). In practice food mixtures such as ethanol–water are not ideal mixtures and the relationships are not proportional. This deviation from Raoult's Law results in curves instead of straight lines when vapour pressure is plotted against mole fractions.

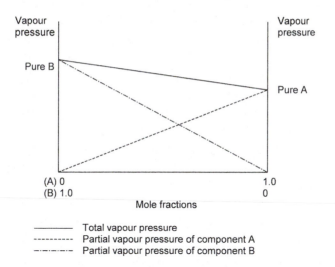

Fig. 14.7   Vapour pressures of a two-component ideal mixture. Note: the partial vapour pressure of pure component B is higher than pure component A, indicating that B is the more volatile liquid (it has a lower boiling point).

Sample problem 14.4
After distillation, an alcohol mixture contains 14 moles of ethanol and 1 mole of methanol. The vapour pressures at the given temperature are 45 kPa for pure ethanol and 81 kPa for pure methanol. Calculate the total vapour pressure of the mixture.

Solution to sample problem 14.4
Assuming the mixture of ethanol and methanol is an ideal mixture, using Raoult's Law:

$$P_{ethanol} = 14/15 \times 45 = 42 \text{ kPa}$$
$$P_{methanol} = 1/15 \times 81 = 5.4 \text{ kPa}$$

Total vapour pressure $= 42 + 5.4 = 47.4 \text{ kPa}$

When a liquid containing components that have different degrees of volatility is heated, those that have a higher vapour pressure and lower boiling points (i.e. they are more volatile components) are separated first. In distillation, the relative vapour pressures of different components govern their equilibrium relationships. The equilibrium curves for a two-component non-ideal vapour–liquid mixture can be shown as a boiling temperature–concentration diagram (Fig. 14.8). For ethanol–water mixtures at atmospheric pressure, the boiling points for pure components are 78.5 °C for ethanol and 100 °C for water at sea level, and the boiling point for the azeotropic mixture is 78.15 °C (89.5 mole% or 95.6% w/w ethanol and 10.5 mole% or 4.4% w/w water).

The horizontal (constant temperature) line in Fig. 14.8 is the boiling temperature, and the point at which it intersects the liquid composition line (x) gives the composition of liquid boiling at this temperature (≈ 0.7 mole fraction of A and ≈ 0.3 mole fraction of B). Similarly where it intersects the vapour composition line (y) shows the composition of

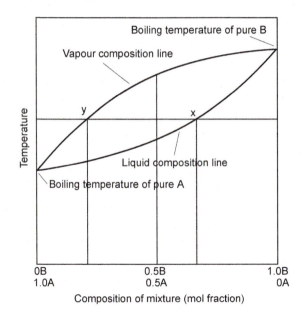

Fig. 14.8   Boiling temperature–concentration diagram.

the vapour. The mole fractions have changed to 0.2 for A and 0.8 for B, indicating the concentration of the more volatile component B has increased in the distillate. The components that have a lower volatility in the remaining liquid are termed 'bottoms' or residues, and they have a lower boiling point than the original mixture. A calculation of the yield of alcohol from a distillation column is given in sample problem 14.5.

In the operation of a multi-stage distillation column that has packing materials to hold condensed liquid at each stage (section 14.2.2), the feed mixture is heated and vapours condense on the first packing that they reach. On condensation the vapours give up latent heat and warm the packing, until the temperature rises sufficiently to vaporise the new mixture. This new mixture has a higher proportion of volatiles than the feed liquor (from Fig. 14.8) and so boils at a lower temperature. This process of repeated vaporisation and re-condensation continues up the column with progressively lower boiling temperatures as the volatile components become separated. In contrast, the boiling temperature of the residue gradually rises as it is left with fewer of the volatile components.

In alcohol distillation, the most volatile alcohol is methanol and this is collected first at the top of the column (known as 'heads') and discarded or used in non-food applications (methanol is toxic, causing permanent blindness and death if 6–10 ml is consumed by adults) (Anon 1993). Ethanol is collected next and then the less volatile alcohols (e.g. propanol) and larger organic molecules (known as 'tails') that contribute distinctive flavours and aromas to individual spirit drinks. Figure 14.9 shows the distillation temperature vs concentration for ethanol. For example, starting with a feed liquor of 10% ethanol boiling at 93 °C (on the liquid curve) and moving horizontally at constant temperature across the diagram to the vapour curve, indicates that the vapour, when condensed, contains 55% ethanol by volume. Similarly, redistilling a 40% spirit produces a condensed vapour containing 80% ethanol. If concentrations of ethanol above the azeotropic concentration of 96.5% are required, it is either distilled under vacuum, mixed with cyclohexane and redistilled at 65 °C to remove the final 3.5% of water, or separated

Sample problem 14.5
A 5% alcohol–water (mass/mass) mixture is distilled at a boiling temperature of 95 °C.
(a) Calculate the mole fraction of alcohol in this mixture. (b) 2.0 mole% of alcohol in
the liquid is in equilibrium with 17 mole% of alcohol in the vapour. Calculate the
concentration of alcohol in the vapour. (Assume an ideal mixture and the molecular
weights of ethanol = 46 and water = 18.)

Solution to sample problem 14.5
(a) In 1 kg of mixture:

$$\text{Number moles alcohol} = 0.05/46 = 1.086 \times 10^{-3}$$
$$\text{Moles water} = 0.95/18 = 0.052$$
$$\text{Therefore mole fraction alcohol} = \frac{1.086 \times 10^{-3}}{1.086 \times 10^{-3} + 0.052}$$
$$= 0.020 \text{ or } 2.0 \text{ mole\%}$$

(b) Let mass fraction of alcohol in vapour $= x$

$$\text{Mass fraction of water} = (1 - x)$$
$$\text{Mole alcohol} = x/46 \text{ and mole water} = (1 - x)/18$$

Since mole fraction of alcohol in vapour $= 0.17$ (17 mole%), then

$$\frac{x/46}{x/46 + (1 - x)/18} = 0.17$$

Thus

$$0.0217x = 0.17(9.452 \times 10^{-3} - 5.763 \times 10^{-3}x)$$
$$0.027\,463x = 9.452 \times 10^{-3}$$
$$x = 0.344$$

Therefore the mass fraction of alcohol in the vapour has been enriched from 5% to
34.4%.

using a molecular sieve (water is a much smaller molecule than ethanol) (Chapter 5,
section 5.5).

Hardy (2007), Petlyuk (2005) and Halvorsen and Skogestad (2000) describe the theory
of continuous distillation and Stichlmair and James (1998) describe material and energy
balance calculations on distillation columns to determine energy use and the degree of
concentration that would be produced under defined conditions. The McCabe–Thiele
graphical method for calculating the size of distillation columns and the analysis of
distillation is described in chemical engineering texts.

14.2.2 Equipment
Although batch distillation in 'pot stills' remains in use in some whisky and other spirit
distilleries (Nicol 1989), most industrial distillation operations use more economical
continuous distillation columns (Fig. 14.10), (Kent and Evers 1994, Panek and Boucher

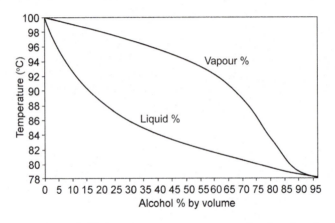

Fig. 14.9   Distillation temperature vs concentration for ethanol.

1989). Heated feed liquor flows continuously through the column and volatiles are produced and separated at the top of the column as distillate. The residue is separated at the base of the column. In order to enhance both the separation of components and equilibrium conditions between the liquid and vapour phases, a proportion of the distillate is added back to the top of the column (reflux) and a portion of the bottoms is vaporised

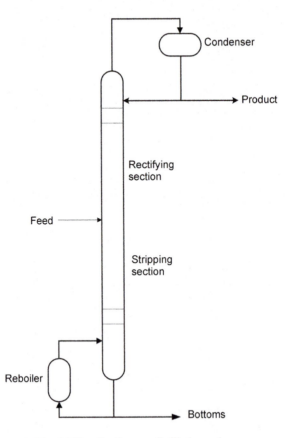

Fig. 14.10   Continuous distillation column.

in a reboiler and added to the bottom of the column. Columns are filled with either a packing material (typically ceramic, plastic or metal rings) or fitted with perforated trays, both of which increase the contact between liquid and vapour phases.

A more recent development is the use of a 'spinning cone column' to remove volatile components from liquids. It is used to recover flavours from beer, coffee, tomato products and fruit juices, to produce low-alcohol wines and beers, and to remove off-flavours. The equipment consists of a column containing a series of rotating inverted cones, which are intermeshed with stationary cones attached to the column wall. Steam or nitrogen is supplied to the base of the column and the feed liquor enters at the top. Thin turbulent films are produced over the large surface area of the cones and rapid separation takes place. The gas passes out of the top of the column and volatile aroma compound are condensed and collected. Because separation is achieved by mechanical energy from the rotating cones, there is less damage to flavours and lower energy consumption. The equipment is also considerably smaller than a packed column having an equivalent throughput. Further information is given by Anon (2008b,c) and Schofield (1995). Wiped film evaporators (section 14.1.3) are also used to distil heat-sensitive foods (Anon 2005).

14.2.3   Effects on food and micro-organisms

Distillation is intended to concentrate volatile components of foods and hence to alter the flavour and aroma of products compared to the feed materials. There are a number of studies of the quality of distilled essences (e.g. Gamarra *et al.* 2006, Kostyra and Baryko-Pikielna 2006), and alcoholic spirits (e.g. Peña y Lillo *et al.* 2005, Bryce 2003), which detail the aromatic compounds and their effects on perceived quality of these products, but these details are beyond the scope of this book. Distillation destroys micro-organisms, partly as a result of heating during the process and partly due to high concentrations of ethanol in alcoholic products, or essential oils that have antimicrobial properties (e.g. terpenes, such as citral, β-pinene and *p*-cymene) (Belletti *et al.* 2004). The antimicrobial properties of essences are reported by Smith-Palmer *et al.* (1998).

References

ANON, (1986), Computer controls evaporation, *Food Processing*, May, 17–18.

ANON, (1993), Methanol toxicity, ATSDR (Agency for Toxic Substances and Disease Registry), US Department for Health and Human Services, *American Family Physician*, **47** (1), 163–171.

ANON, (2003), Pan aid surfactant for sugar beet processing, company information from Penn White Ltd, available at www.pennwhite.co.uk/datasheets/FoamdoctorPanaid_techdata.pdf.

ANON, (2005), Short path and wiped film distillation, company information from Chem Group Inc., available at www.chem-group.com/processing/short-path.tpl.

ANON, (2007), The Evaporator Handbook, APV Americas, Engineered Systems Separation Technologies, available at www.che.utexas.edu/cache/trc/evaporator.pdf.

ANON (2008a), Thin film evaporation, company information by Gooch Thermal Systems, available at www.goochthermal.com/hvethinfilm.html.

ANON, (2008b), Spinning cone column, company information from Conetech, available at www.conetech.com/SpinningConeColumn.html.

ANON, (2008c), What is the spinning cone column, company information from Flavourtech, available at www.flavourtech.co.uk and follow links to 'Flavourtech technologies > spinning cone column'.

BELLETTI, N., NDAGIJIMANA, M., SISTO, C., GUERZONI, M.E., LANCIOTTI, R. and GARDINI, F., (2004),

Evaluation of the antimicrobial activity of citrus essences on *Saccharomyces cerevisiae*, *J. Agriculture Food Chemistry*, **52** (23), 6932–6938.

BRENNAN, J.G., BUTTERS, J.R., COWELL, N.D. and LILLEY, A.E.V., (1990), *Food Engineering Operations*, 3rd edn, Elsevier Applied Science, London, pp. 337–370.

BRYCE, J., (2003), *Distilled Spirits*, Nottingham University Press, Nottingham.

BURKART, A. and WIEGAND, B., (1987), Quality and economy in evaporator technology, in (A. Turner, Ed.), *Food Technology International Europe*, Sterling Publications International, London, pp. 35–39.

CHEN, C.S. and HERNANDEZ, E., (1997), Design and performance evaluation of evaporation, in (K.L. Valentas, E. Rotstein and R.P. Singh, Eds.), *Handbook of Food Engineering Practice*, CRC Press, Boca Raton, FL, pp. 211–251.

CHEN, H., JEBSON, R.S. and CAMPANELLA, O.H., (1997), Determination of heat transfer coefficients in rotating cone evaporators: Part I, *Food and Bioproducts Processing*, **75** C1, 17–22.

COULSON, J.M. and RICHARDSON, J.F., (1978), *Chemical Engineering*, Vol. 2, 3rd edn, Pergamon Press, New York.

EARLE, R.L., (1983), *Unit Operations in Food Processing*, 2nd edn, Pergamon Press, Oxford, pp. 105–115, and available at www.nzifst.org.nz/unitoperations/evaporation.htm.

EGGLESTON, G. and MONGE, A., (2007), How time between cleanings affects performance and sucrose losses in Robert's evaporators, *J. Food Processing and Preservation*, **31** (1), 52–72.

GAMARRA, F.M.C., SAKANAKA, L.S., TAMBOURGI, E.B. and CABRAL, F.A., (2006), Influence on the quality of essential lemon (*Citrus aurantifolia*) oil by distillation process, *Brazilian J. Chemical Engineering*, **23** (1), available at www.scielo.br/scielo.php?pid=S0104-66322006000100016&script=sci_arttext.

GEANKOPLIS, C.J., (1993), *Falling Film Evaporators in the Food Industry, Transport Processes and Unit Operations*, 3rd edn, Prentice Hall, Englewood Cliffs, NJ, pp. 263–267.

GOFF, H.D., (2007), Dairy Science and Technology Series, University of Guelph, available at www.foodsci.uoguelph.ca/dairyedu/concprod.html.

HALVORSEN, I.J. and SKOGESTAD, S., (2000), Distillation theory, in (I.D. Wilson, Ed.), *Encyclopedia of Separation Science*, Academic Press, London, pp. 1117–1134.

HARDY, J.K. (2007), Distillation, information available at http://ull.chemistry.uakron.edu/chemsep/distillation/.

HELDMAN, D.R. and HARTEL, R.W., (1997), Liquid concentration, in *Principles of Food Processing*, Chapman and Hall, New York, pp. 138–176.

HOFFMAN, P., (2004), Plate evaporators in food industry, *J. Food Engineering*, **61** (4), 515–520.

IVERSEN, C.K., (1999), Black current nectar: effect of processing and storage on anthocyanin and ascorbic acid, *J. Food Science*, **64**, 37–41.

JEBSON, R.S. and CHEN, H., (1997), Performances of falling film evaporators on whole milk and a comparison with performance on skim milk, *J. Dairy Research*, **64**, 57–67.

JEBSON, R.S., CHEN, H. and CAMPANELLA, O.H., (2003), Heat transfer coefficients for evaporation from the inner surface of a rotating cone – II, *Transactions Institute Chemical Engineering*, **81**, Part C, December, 293–302.

KENT, N.L. and EVERS, A.D., (1994), Malting, brewing and distilling, in *Kent's Technology of Cereals*, 4th edn, Woodhead Publishing, Cambridge, pp. 218–232.

KOSTYRA, E. and BARYŁKO-PIKIELNA, N., (2006), Volatiles composition and flavour profile identity of smoke flavourings, *Food Quality and Preference*, **17** (1–2), January–March, 85–95.

LEE, H.S. and CHEN, C.S., (1998), Rates of Vitamin C loss and discoloration in clear orange juice concentrate during storage at temperatures of 4–24 °C, *J. Agriculture Food Chemistry*, **46** (11), 4723–4727.

NICOL, D., (1989). Batch distillation, in (J.R. Piggott, R. Sharp and R.E.B. Duncan, Eds.), *The Science and Technology of Whiskies*, Longman Scientific and Technical, Harlow, Essex, pp. 118–149.

NINDO, C.I., POWERS, J.R. and TANG, J., (2007), Influence of refractance window evaporation on quality of juices from small fruits, *LWT – Food Science and Technology*, **40** (6), 1000–1007.

OLSSON, B., (1988), Recent advances in evaporation technology, in (A. Turner, Ed.), *Food Technology International Europe*, Sterling Publications International, London, pp.55–58.

PANEK, R.J. and BOUCHER, A.R., (1989), Continuous distillation, in (J.R. Piggott, R. Sharp and R.E.B. Duncan, Eds.), *The Science and Technology of Whiskies*, Longman Scientific and Technical, Harlow, Essex, pp. 150–181.

PEÑA Y LILLO, M., LATRILLE, E., CASAUBON, G., AGOSIN, E., BORDEU, E. and MARTIN, N., (2005), Comparison between odour and aroma profiles of Chilean Pisco spirit, *Food Quality and Preference*, **16** (1), 59–70.

PETLYUK, F.B., (2005), *Distillation Theory and its Application to Optimal Design of Separation Units*, Cambridge Series in Chemical Engineering, ECT Service, Moscow.

PORTER, J.W.G. and THOMPSON, S.Y., (1976), Effects of processing on the nutritive value of milk, Vol. 1, Proceedings 4th International Conference on Food Science and Technology, Madrid.

PROST, J.S., GONZÁLEZ M.T. and URBICAIN, M.J., (2006), Determination and correlation of heat transfer coefficients in a falling film evaporator, *J. Food Engineering*, **73** (4), 320–326.

SCHOFIELD, T., (1995), Natural aroma improvement by means of the spinning cone, in (A. Turner, Ed.), *Food Technology International Europe*, Sterling Publications International, London, pp. 137–139.

SINGH, R.P. (2001), Animations of figures published in (R.P. Singh and D.R. Heldman, Eds.), *Introduction to Food Engineering*, 3rd edn, Academic Press, London, available at http://rpaulsingh.com/animated%20figures/fig8_2.htm

SINGH, R.P. and HELDMAN, D.R., (2001), Evaporation, in *Introduction to Food Engineering*, 3rd edn, Academic Press, London, pp. 449–472.

SMITH, J., (1997), Energy management, *Food Processing*, March, 16–17.

SMITH-PALMER, A., STEWART, J. and FYFE, L., (1998), Antimicrobial properties of plant essential oils and essences against five important food-borne pathogens, *Letters in Applied Microbiology*, **26**, 118–122.

STICHLMAIR, J. and JAMES, R.F., (1998), *Distillation: Principles and Practice*, Wiley, New York.

TOLEDO, R.T., (1999), Evaporation, in *Fundamentals of Food Process Engineering*, 2nd edn, Aspen Publications, Gaithersburg, MD, pp. 437–455.

15

Extrusion

Abstract: Extrusion cooking is a process that combines the unit operations of mixing, cooking, shaping and forming to produce foods such as breakfast cereals, snackfoods and confectionery. It is a high-temperature short-time process that inactivates micro-organisms and enzymes and reduces the water activity of products, but retains their nutritional value and sensory properties. The first part of the chapter describes the relationship between the properties of ingredients and extruder operating characteristics. It then discusses extrusion equipment and selected applications. The chapter concludes with a summary of the effects of extrusion on micro-organisms and the sensory and nutritional properties of foods.

Key words: extrusion cooking, crispbread, cereals, texturised vegetable protein, twin-screw extruder.

Extrusion is a continuous process that combines several unit operations including mixing, cooking, kneading, shearing, shaping and forming. It is used to produce a wide range of products, including breakfast cereals, snackfoods, biscuits, pasta, sugar confectionery and soya-based meat analogues, as well as petfoods and fish feeds. Extruders consist of either one or two screws contained in a horizontal barrel and are classified according to the method of operation into cold extruders or extruder-cookers. The principles of operation are similar in both types: raw materials are fed into the extruder barrel and the screw(s) convey the food along it. Further down the barrel, the volume is restricted and the food becomes compressed. The screw then kneads the material under pressure into a semi-solid, plasticised mass.

In cold extrusion, the temperature of the food remains below 100 °C. It is used to mix and shape foods without significant cooking or distortion of the food (examples are shown in Table 15.1). The extruder has a deep-flighted screw, which operates at a low speed in a smooth barrel, to knead and extrude the material with little friction (section 15.2). Typical operating conditions are shown in Table 15.3 for low shear conditions.

In this chapter, the focus is on extrusion cooking, where the food is heated above 100 °C. Frictional heat and any additional heating of the barrel cause the temperature to rise rapidly. The food is then subjected to increased pressure and shearing, and finally it is forced through one or more restricted openings (dies) at the discharge end of the barrel. As the food emerges under pressure from the die, it expands to the final shape and cools rapidly as moisture is flashed off as steam. A variety of shapes, including rods, spheres,

Table 15.1 Examples of extruded foods

Type of product	Examples
Cereal-based products	Bases for instant soups and beverages
	Breading
	Crispbreads and croutons
	Expanded snackfoods
	Pasta products
	Pastry doughs
	Precooked and composite flours
	Pregelatinised and modified starches
	Ready-to-eat and puffed breakfast cereals
	Weaning foods
Protein-based products	Caseinates
	Fish pastes
	Processed cheeses
	Sausages, frankfurters and hot dogs
	Semi-moist and expanded pet foods, animal feeds, floating and sinking aquatic feeds
	Surimi
	Texturised vegetable protein (TVP)
Sugar-based products	Chewing gum
	Chocolate, caramel
	Fruit gums
	Fudge
	Hard boiled confectionery (e.g. toffees, caramels, peanut brittle)
	Liquorice
	Nougat, praline

From Heldman and Hartel (1997), Best (1994) and Riaz (2001)

doughnuts, tubes, strips, squirls or shells, can be formed. Typical products include a wide variety of low-density, expanded snackfoods and ready-to-eat puffed cereals (Table 15.1). Extruded products may be subsequently processed further by drying (Chapter 16) or frying (Chapter 19) before packaging (Chapter 25). Many extruded foods are also suitable for coating or enrobing (Chapter 24). Further details of extrusion technology are given by Guy (2001a). Developments using combined supercritical fluid technology (Chapter 5, section 5.4.3) with extruders to produce a range of puffed products, pasta and confectionery are described by Rizvi *et al.* (1995).

Extrusion cooking is a high-temperature short-time (HTST) process that reduces the number of micro-organisms in raw materials and inactivates naturally occurring enzymes. However, the low water activity of products (0.1–0.4) (Chapter 1, section 1.1.2) is the main method of preservation of both hot- and cold-extruded foods, and this is maintained during storage by suitable packaging materials. The HTST process enables many heat-sensitive components to be retained, resulting in good retention of nutritional value and sensory properties.

Extrusion has gained in popularity for the following reasons (Guy, 2001a): it is a versatile process that can produce a very wide variety of products by changing the ingredients, the operating conditions of the extruder and the shape of the dies. Many extruded foods cannot be easily produced by other methods. Extrusion has lower processing costs and higher productivity than other cooking or forming processes. Some original processes (e.g. manufacture of cornflakes and frankfurters) (section 15.3) are more efficient and cheaper when replaced by extrusion. Extruders operate continuously under automatic control and have high throughputs (e.g. production rates of up to $22 \, t \, h^{-1}$

Table 15.2   Advantages and limitations of different types of extruders

Type of extruder	Advantages	Limitations
Single screw	• Lower capital, operating and maintenance costs (capital cost about half of a twin-screw machine). Wet single screw extruders have higher capacity, higher capital costs but lower operating costs than dry extruders.[a] • Less skill required to operate and maintain. • Less complicated assembly of screw configurations. • Wet single screw extruders have greater processing control that produces superior shaped products compared with dry extruders.	• Does not self-clean. There may be problems emptying the extruder barrel if it is allowed to cool and the product solidifies. • Not able to process materials that contain >12–17% fat or >30% moisture due to product slippage in the barrel. • More limited ingredient particle size range (very fine powders and coarse ingredients are not suitable). • Difficult to produce products <1.5 mm in diameter. • Dry extruders[a] require higher motor power and undergo greater wear than other types of extruder. The high exit pressures make it difficult to shape products < 2 mm or process highly viscous materials.
Twin screw	• Can produce intricate shapes and small sizes (<1.0 mm). • Greater flexibility and control than single-screw extruders. • Can handle very viscous, oily (18–27% fat), wet (up to 65% water) or sticky materials (up to 40% sugar compared with 10% in single-screw machines). • Self-wiping screw extruders are easier to clean without dismantling, giving more rapid product changeover. • Can handle very fine ingredients or coarse ingredients directly without pretreatments. • Easier to operate than single-screw extruders.	• More complex than single-screw extruders. • More expensive and have higher maintenance costs.

[a] Wet and dry extruders are described in section 15.2.
Adapted from Riaz (2001), Rokey (2000) and Frame (1994)

for single-screw extruders and up to $14\,\mathrm{t\,h^{-1}}$ for twin-screw extruders (Riaz 2001). The comparative advantages and limitations of single-screw extruders are shown in Table 15.2. Extrusion does not produce process effluents and has no water treatment costs.

Extrusion can be seen as an example of a size enlargement process, in which granular or powdered foods are re-formed into larger pieces. Other examples of size enlargement include forming or moulding (Chapter 4, section 4.2) and agglomeration of powders (Chapter 16, section 16.2.1). Extruders are also used in the plastics industry to produce packaging materials (Chapter 25, section 25.2.4).

15.1   Extrusion cooking theory

Extrusion cooking involves the simultaneous mixing, kneading and heating of ingredients, and results in a large number of complex changes to foods. These include hydration, gelation and shearing of starches and proteins, melting of fats, denaturation or re-orientation of proteins, plasticisation of the material to form a fluid melt, formation of glassy states, and expansion and solidification of food structures when they emerge from the die. For many years, understanding of these interactions developed empirically, but mathematical modelling of fluid flow and heat transfer inside the extruder barrel has led to a greater understanding of the operation and control of extruders. Modelling is described in detail by Elsey *et al.* (1997), Schoner and Moreira (1997), Kulshreshtha *et al.* (1995) and Tan and Hofer (1995).

The factors that influence the quality of extruded products are shown in Fig. 15.1 and can be grouped into those related to the properties of the ingredients and those related to the design and operating conditions of the extruder (including any pretreatments to ingredients). When the relationships between the different variables that produce the required quality have been established for a particular food, the composition of the feed material and the operating variables are standardised to create the required conditions inside the barrel. The variables are maintained within limited tolerances to produce the desired physical and chemical changes to the food and ensure that the extruded product has a consistent quality.

15.1.1   Properties of ingredients

The properties of the feed materials have an important influence on the conditions inside the extruder barrel and hence on the quality of the extruded product. Different types of feed material produce completely different products when the same operating conditions are used in the same extruder. This is because of differences in the type and amounts of starch, proteins, moisture and other added ingredients (e.g. oil or emulsifier), which result in different viscosities and hence different flow characteristics. Mostly, ingredients used in extrusion cooking have low moisture contents (10–40%), and they are transformed into a fluid melt by the shearing action and the high temperature and pressure in the extruder. Under these conditions, ingredients interact with each other to affect the types of transformation that take place. The physicochemical properties of the ingredients are therefore more important than in other food processes (e.g. the hardness, frictional characteristics and particle size of powders or the lubricity and plasticising power of fluids). Similarly, addition of acids to adjust the pH of the feed material causes changes to starch gelatinisation and unfolding of protein molecules. This in turn changes the viscosity and hence the structure and strength of the extruded product. Guy (1994) has characterised ingredients according to their functional roles into:

- structure-forming materials;
- disperse-phase filling materials;
- plasticising or lubricating materials;
- soluble solids;
- nucleating materials;
- colouring materials;
- flavouring materials.

Starch or protein structural components create the texture of extruded foods by forming a three-dimensional matrix that contains the other ingredients. Structural starches are more important in extruded breakfast cereals, snackfoods and biscuits, and these come from cereal or legume flours (e.g. maize, wheat, rice, barley, pea, bean), or from tuber

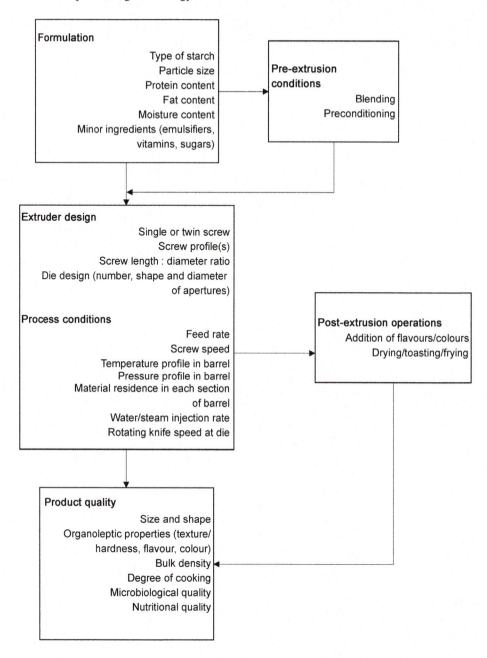

Fig. 15.1   Influence of raw material properties and processing conditions on the quality of extrusion cooked products.

flours (e.g. potato, cassava, tapioca). Details of the differences between different sources of starch and the effects of extrusion on different types of starch granules are described by Guy (2001b). Structural proteins derived from pressed oilseed cake from soybeans, sunflower seeds, fava beans, rapeseed or gluten from wheat, are used to make meat-like products such as texturised vegetable protein (TVP) or pet foods or fish feeds.

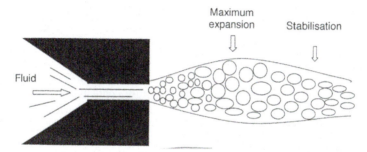

Fig. 15.2   Expansion of extrudate at the die of an extruder, showing bubble growth and stabilisation of the foam (from Guy 2001c).

During extrusion cooking of starch-based foods, any added water is absorbed and causes starch granules to swell and become hydrated. Smaller particles, such as flours or grits, are hydrated and cooked more rapidly than larger particles and this in turn also alters the product quality. The elevated temperature causes starch to gelatinise and form a viscous plasticised fluid melt. This in turn forms the walls of foam bubbles that contain superheated water vapour. When the material leaves the extruder die, the sudden drop in pressure causes these bubbles to expand rapidly, lose moisture by evaporation and simultaneously cool. These changes cause a rapid increase in the viscosity of the material, followed by the formation of a glassy state that cools and sets the cellular structure (Fig. 15.2). The matrices determine the product texture (Blanshard 1995). Changes to starch are described by Guy (1993). The changes in starch solubility under different conditions of temperature and shear rate are monitored by measuring the water absorption index (WAI) and the water solubility index. The WAI of cereal products generally increases with the severity of processing, reaching a maximum at 180–200 °C.

The type of starch has an important influence on the viscosity of the fluid melt and hence on expansion of the foam cells. Starches that have a high proportion of amylose, which is a smaller molecule than amylopectin, produce lower-viscosity fluid melts and hence greater expansion of the food. All starch molecules in raw materials are too large to enable sufficient product expansion, but the shearing action in the extruder barrel reduces the molecular size and provides the correct viscosity to allow expansion. The degree of shearing, the temperature and moisture content in the extruder therefore control the amount of expansion of the product and hence its texture (Guy 2001b). The correct combination of starch type and extrusion conditions is needed to prevent under-expansion as a result of a viscosity that is too high, or collapse of the bubbles after they emerge from the die because the viscosity is too low.

Although proteins are smaller molecules than starches, at high concentrations (>35–40% w/w) and under more moist conditions than those used for starch-based products, the proteins unfold and agglomerate as a result of the shearing action and high temperatures to form high-viscosity, viscoelastic, amorphous complexes. To form a textured structure, the proteins are passed through the extruder die under laminar flow conditions (Chapter 1, section 1.3.4) at 120–130 °C. As the layers of molecules flow together, they polymerise, cross-link and re-orient to form characteristic fibrous structures. The reduction in pressure causes moisture to flash off, both creating bubbles that leave voids in the structure, cooling the product and setting the structure to create the final texture. The nitrogen solubility index is a measure of the extent of protein denaturation, which decreases during extrusion cooking.

Filling materials, including fibrous materials in the ingredients, are dispersed in the starch matrix. At low concentrations they have little effect on product texture, but at concentrations found in wholemeal flours (e.g. 8–9% bran), they reduce the ability of the starch matrix to expand. Plasticisers and lubricants include water and oils. When dry ingredients are extruded, frictional heat is sufficient to raise the temperature of the material to 150 °C. But when ingredients have moisture contents above about 25% the moisture plasticises the material and reduces the amount of heat generated by friction so that barrel heating is required. Oils lubricate both the interacting particles in the material and the particles rubbing against the extruder barrel and screw. Levels as low as 1–2% oil reduce the shearing action on starch molecules and hence reduce the expansion of the product.

Soluble ingredients, including salt and sugar, dissolve in any available water during initial mixing. Depending on their concentration, they may react with starch and/or proteins to reduce the viscosity of the fluid melt and hence affect the degree of expansion of the extruded material. Insoluble ingredients, including bran, powdered calcium carbonate, calcium phosphate or magnesium silicate have been shown to act as nucleating materials that provide surfaces at which bubbles can form. They increase the number of bubbles in the extrudate foam and produce a more expanded product (Guy 2001b). Flavours and colourants may be added during the extrusion process or sprayed onto extruded products. Alternatively, precursor compounds that are present in or added to ingredients are converted to flavours or colours by high-temperature reactions in the extruder (e.g. Maillard browning reactions between added amino acids and reducing sugars (Chapter 1, section 1.2.2)).

15.1.2   Extruder operating characteristics

The most important extruder operating parameters are the temperature and pressure in the barrel, the diameter of the die apertures and the shear rate. The shear rate is influenced by the internal design of the barrel, its length : diameter ratio and the speed and geometry of the screw(s).

Extrusion cooking operates continuously under steady state equilibrium conditions. The amount of heating of feed materials and the rate of heat transfer to the food determines the type and extent of physicochemical changes that take place and hence the quality of the final product. Types of heat transfer are described in Chapter 10 (section 10.1.2) and related sample problems are found in Chapter 10. An energy balance (Equation 15.1) can be used to correlate the energy used in the extruder with the energy transferred to the material.

$$P_{mech} + P_{heat} = P_{cool} + P_{loss} + P_{mat} \hspace{2cm} \boxed{15.1}$$

where P_{mech} = the mechanical power supplied by the motor (for producing frictional heat), P_{heat} = thermal power supplier by barrel heaters, P_{cool} = thermal power absorbed by barrel cooling, P_{loss} = thermal losses to the environment, P_{mat} = thermal power absorbed by the material (Mottaz and Bruyas 2001).

The equation can be used to calculate the amount of heat needed to convert the material as it passes through the extruder, comprising the sensible heat and the change in enthalpy of fusion when the material changes phase from solid to fluid melt. Extruded products may be characterised by the specific mechanical energy (SME), which is the ratio of the energy supplied and the flow of extruded material, having units of kJ kg^{-1} (Equation 15.2).

$$\text{SME} = \frac{\text{Total energy}}{\text{Flow rate}} \qquad \boxed{15.2}$$

If the extruder barrel is not heated, the total energy is frictional heat generated by power from the motor.

The thermally induced changes to feed materials depend on the types of heat transfer within the extruder barrel and can be calculated using Equation 15.3 (Mottaz and Bruyas 2001):

$$dE = dQ + h_{m/b} \times dA_{m/b} \times (\theta_b - \theta_m) + h_{m/s} \times dA_{m/s} \times (\theta_s - \theta_m) \qquad \boxed{15.3}$$

where dE (W) = change in internal energy of the material, which differs depending on its state (powder, melting or molten) and involves thermal properties including sensible heat, melting point and enthalpy of fusion, dQ (W) = heat produced, h (W m^2 K^{-1}) = heat transfer coefficient and dA (m^2) = heat transfer area, both having suffixes $_{m/b}$ indicating heat transfer between the material and the barrel and 'm/s' between the material and the screw, θ (°C) = temperature, having suffixes b, m, s representing the barrel, material and screw respectively.

A simplified model for the operation of an extruder, developed by Harper (1981), assumes that the temperature of the food is constant, fluid flow is Newtonian and laminar, there is no slippage of food at the barrel wall and no leakage of food between the screw and the barrel. With these assumptions, the flow through a single-screw extruder is calculated using Equation 15.4:

$$Q = G_1 N F_d + G_2 \mu \times \Delta P / L \times F_p \qquad \boxed{15.4}$$

where Q (m^3 s^{-1}) = volumetric flowrate in the metering section, N (rpm) = screw speed, μ (N s m^{-2}) = viscosity of the fluid in the metering section, ΔP (Pa) = pressure increase in the barrel, G_1 and G_2 = constants that depend on screw and barrel geometry, L (m) = length of extruder channel and F_d and F_p = shape factors for flow due to drag and pressure respectively.

The first part of the equation represents fluid flow down the barrel caused by pumping and drag against the barrel wall, whereas the second part represents backward flow from high pressure to low pressure, caused by the increase in pressure in the barrel. Clearly, the amount of pressure in the barrel depends in part on the size of the dies: if the barrel is completely open at the die end, there will be no pressure build-up and the extruder will simply act as a screw conveyor. Conversely, if the die end is completely closed, the pressure will increase until backward flow equals drag flow and no further movement will occur. The extruder would become a mixer. In between these two extremes, the size of the die greatly affects the performance of the extruder. The ratio of pressure flow to drag flow is known as the throttling factor (a), which varies from zero (open die hole) to 1 (closed die hole). In practice most extruders operate with a values of between 0.2 and 0.5 (Heldman and Hartel 1997).

The flow through the die is found using Equation 15.5:

$$Q = K' \times \Delta P / \mu \qquad \boxed{15.5}$$

where Q (m^3 s^{-1}) = volumetric flow rate through the die, μ (N s m^{-2}) = viscosity of the fluid in the die, ΔP (Pa) = pressure drop across the die (from inside the barrel to atmospheric pressure) and K' = a flow resistance factor that depends on the number, shape and size of the die holes, usually found experimentally.

The operating conditions for the extruder can be found by calculating the flowrate and die pressure drop that satisfy both equations, which in turn depend on the type of die, depth of flights on the screw, and length and speed of the screw.

It should be noted that the above equations are based on simple models that do not take account of leakage of food between the flights and the barrel, changes in temperature or the effects of non-Newtonian fluids. In practice, modelling is very complex because changes in non-Newtonian fluids are significantly more complicated. The assumptions made in the formulae may therefore limit their usefulness for predicting flow behaviour or operating conditions, but they may be used as a starting point for experimental studies. Most extruder manufacturers use a combination of mathematical modelling and practical experience of the relationships between die shape, extruder construction and characteristics of the product to design their equipment. The situation with twin-screw extruders is even more complex: changes to the degree of intermeshing of the screws (section 15.2) or the direction of rotation, dramatically alter the flow characteristics in the extruder and make modelling equations very complex. Mathematical modelling of extruders is described by Wang et al. (2004), Pomerleau et al. (2003), Mottaz and Bruyas (2001) and Kulshreshtha et al. (1991). Li (1999) describes a model that can simulate and predict extruder behaviour (e.g. pressure, temperature, fill factor, residence time distribution and shaft power) under different operating conditions of feedrate, screw speed, feed temperature/moisture and barrel temperature, which can be used to control extruder operating conditions.

15.2   Equipment

The selection of an appropriate extruder for a particular application should take account of the nature of the ingredients (section 15.1.1), the type of product and its required bulk density, physical and sensory properties, and the expected production rate. The basic design difference is between single- or twin-screw extruders. All extruders should convey the ingredients along the barrel and prevent them spinning with the screw. Twin-screw extruders act like positive displacement pumps, but single-screw extruders require design features that ensure material does not slip and rotate with the screw, which would prevent it moving along the barrel. These features include grooves in the barrel, interrupted flights (spaces intentionally left between the flights on the screw), or restrictions to product flow (known variously as 'throttle rings', 'kneading discs', 'steam locks' or 'shearlocks') or shearing bolts that protrude into the barrel, each of which assists in conveying material along the barrel. In wet extrusion cooking, steam or water may be injected through hollow shearing bolts. There are therefore a large number of design options for extruders, including smooth or grooved barrels, 'dry' (i.e. without addition of steam or water) or 'wet' operation, the size, number, pitch and diameter of the flights on the screw(s), continuous or interrupted flights, and the number and position of shearing bolts or throttle rings.

The greatest wear on extruder barrels and screws occurs at the exit end and these parts require replacement earlier than others. As a result, extruders are constructed in sections that are bolted together. Typically a screw comprises a splined shaft onto which are fitted sections of flights and throttle rings that are arranged in particular configurations for each application. Barrel sections may be fitted with liners made from hard alloys or hardened stainless steel to withstand wear. Worn exit segments can be replaced as required or moved away from the exit to a position where increased clearance between the screw and barrel is less important.

Interchangeable dies have different shaped holes, such as round holes to produce rods, square holes for bars, or slots to produce sheets; or they may have more complex patterns for specially shaped three-dimensional products. Some products require the extruder die to be heated to maintain the viscosity and degree of expansion, whereas others require the die to be cooled to reduce the amount of expansion. Extruders may also be fitted with a special die to continuously inject a filling into an outer shell. This is known as 'co-extrusion' and is used, for example, to produce filled confectionery (section 15.3). Die pressures vary from ≈ 500–$2000\,kPa$ for low-viscosity products to $\approx 17\,000\,kPa$ for expanded snackfoods (Heldman and Hartel 1997). Rotary knives, fitted to the outside of the die, cut the extruded product into required lengths. Details of different extruder designs are given by Riaz (2000) and Mercier *et al.* (1989).

15.2.1 Single-screw extruders

This equipment (Fig. 15.3) is the most widely used design for straightforward cooking and forming applications, when the flexibility of a twin-screw machine is not needed.

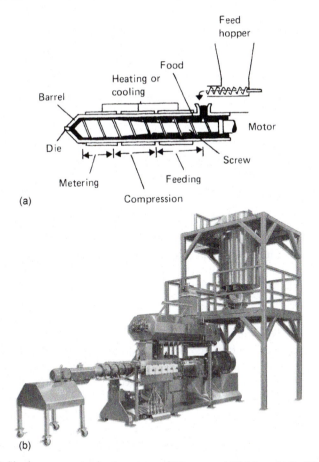

Fig. 15.3 (a) Single-screw extruder (courtesy of Werner and Pfeiderer Ltd); (b) OPTIMA single-screw extruder (courtesy of Extru-Tech Inc., Sabetha, Kansas, USA, a subsidiary of Wenger Manufacturing Inc. at www.wenger.com).

Usually, in wet extrusion the feed material is premixed and preconditioned with steam or water to reduce wear on the components and to improve the product quality. The length : diameter ratio of the barrel is between 2:1 and 25:1 (Hauck 1993). The pitch and diameter of the screw segments, the number of flights and the clearance between the flights and the barrel can each be changed to alter the performance of the extruder. The screw speed is one of the main factors that influences performance; it controls the residence time of the product, the amount of frictional heat generated, heat transfer rates and the shearing forces on the product. Typical screw speeds are 150–600 rpm, depending on the application. Compression is achieved in the extruder barrel by back pressure, created by the die and by:

- increasing the diameter of the screw and decreasing the screw pitch;
- using a tapered barrel with a constant or decreasing screw pitch; or
- placing restrictions to product flow in the barrel.

The screw is driven by a variable speed electric motor that is sufficiently powerful to pump the food against the pressure generated in the barrel.

Single-screw extruders have different degrees of shearing action on the food. High-shear extruders have high speeds and shallow flights to create high pressures and temperatures that are needed to make ready-to-eat breakfast cereals and expanded snackfoods. Medium-shear extruders are used to make breadings, texturised proteins and semi-moist pet foods, and low-shear extruders have deep flights and low speeds to create low pressures for forming pasta, meat products or gums. Operating data for different types of extruder are given in Table 15.3. The temperature and pressure profiles in different sections of a high-shear cooking extruder are shown in Fig. 15.4.

In dry extrusion cooking, much of the energy from the motor generates friction that rapidly heats the food. Throttle rings increase the pressure in the barrel, shearing and heating (up to ≈160 °C). Additional heating can be achieved using a steam-jacketed barrel, a steam-heated screw, or electric induction heating elements around the barrel. An important use of dry extruders is to prepare oilseeds for oil extraction (Chapter 5, section 5.3). Extrusion prior to pressing increases the throughput of an oil expeller and releases antioxidants in oilseeds, thus stabilising the oil. The process produces high-quality oil, similar to a refined, deodorised product, together with an oilcake, containing 50% protein and 90% inactivation of trypsin inhibitors, that is suitable for human consumption.

The advantages and limitations of single-screw extruders are described in Table 15.2 and further details of single-screw extruder designs and operation are given by, for example, Riaz (2001) and Rokey (2000).

15.2.2   Twin-screw extruders

Twin-screw extruders are grouped according to the direction of rotation of the screws (counter-rotating or co-rotating) and the degree to which they intermesh. Non-intermeshing screws act like two single-screw extruders, whereas intermeshing screws produce a positive displacement pumping action to move product through the extruder. The screws rotate within a 'figure of 8' shaped bore in the barrel (Fig. 15.5). Screw length : diameter ratios are between 10:1 and 25:1. One of the main advantages of twin-screw extruders is the greater flexibility of operation to handle a wider range of ingredients and produce different products (Table 15.2). This is achieved by changing the degree of intermeshing of the screws, the number of flights or the angle of the pitch of the screws, or fitting kneading discs to increase the shearing action. In twin-screw

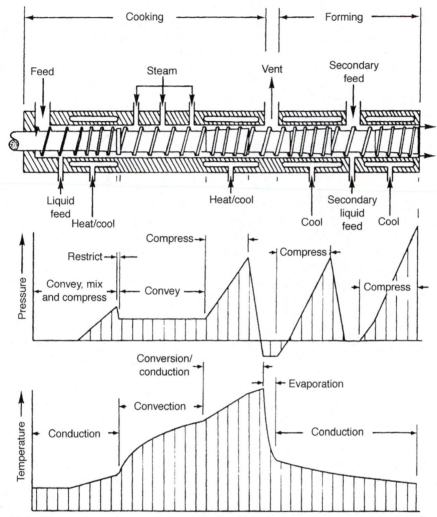

Fig. 15.4   Changes in temperature and pressure in a high-shear, single-screw cooking extruder for expanded food products (from Miller 1990).

Table 15.3   Operating data for different types of single-screw extruders

Parameter	High shear	Medium shear	Low shear
Net energy input to product (kW h kg^{-1})	0.10–0.16	0.02–0.08	0.01–0.04
Barrel length:diameter (*L/D*)	2–15	10–25	5–22
Screw speed (rpm)	> 300	> 200	> 100
Maximum barrel temperature (°C)	110–180	55–145	20–65
Maximum product temperature (°C)	149	79	52
Maximum barrel pressure (kPa)	4000–17 000	2000–4000	550–6000
Product moisture (%)	5–8	15–30	25–75
Product density (kg/m^3)	32–160	160–500	320–800

Adapted from Hauck (1993) and Harper (1979)

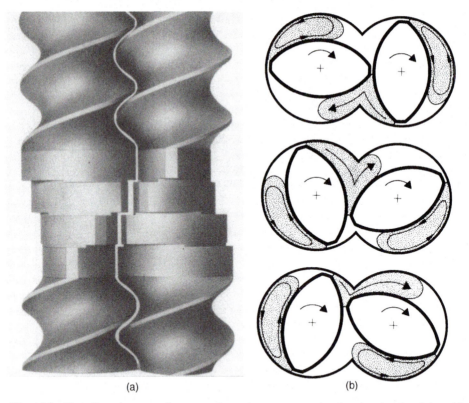

Fig. 15.5 Kneading elements of a co-rotating twin-screw extruder showing dough mixing: (a) sealing profile; (b) movement of material (courtesy of Werner & Pfleiderer at www.wpib.de).

extruder cookers, the spacing between the flights can be adjusted so that large spaces initially convey the material to the cooking section and then smaller spaces compress the plasticised mass before extrusion through an interchangeable die. Their operating characteristics are described by Frame (1994). Co-rotating intermeshing screws, which are self-wiping (the flights of one screw sweep food from the adjacent screw) are most commonly found in food processing applications. Where the barrel is split at the end, products can be directed into two or more channels and different colourants or flavourings can be introduced just before the die to produce two-colour or variegated products. Examples of products from twin-screw machines include co-extruded/filled snackfoods, food gums and jellies, pasta products, TVP, three-dimensional snackfood and confectionery products, marshmallows, cornflakes, chocolate-filled snacks, biscuits and instant rice or noodles. Some products, including sticky caramels and other sweets cannot be made using single-screw extruders, and others, including pet foods that contain up to 30% fresh meat, or ultra-fine and high-fat aquatic feeds, have substantially higher quality using twin-screw machines.

Twin-screw extruders have the following advantages (also Table 15.2):

- Any fluctuations in ingredient feedrate can be accommodated by the positive displacement action of the screws. In contrast, a single screw must be full of material to operate effectively.
- The positive displacement also produces higher rates of heat transfer and better control of heat transfer than a single screw does.

- Twin-screw machines handle oily, sticky or very wet materials, or other products that slip in a single screw. There is therefore greater flexibility to extrude different raw materials.
- Forward or reverse conveying is used to control the pressure in the barrel. For example, in the production of liquorice and fruit gums, the food is heated and compressed by forward conveying, the pressure is released by reverse conveying, to vent excess moisture or to add flavours or colorants, and the food is then recompressed for extrusion.

The main limitations of twin-screw extruders are the relatively high capital and maintenance costs (up to twice the cost of single-screw equipment) (Riaz 2001). The complex gearbox that is needed to drive the twin screws also limits the maximum torque, pressure and thrust that can be achieved.

15.2.3   Control of extruders

A control system for extrusion should encompass the entire process, including formulation of ingredients, preconditioning, extrusion and post-extrusion processing operations (e.g. drying, coating, frying, packaging) in order to obtain the required quality products. There are four main controlled variables for the operation of an extruder:

1 specific mechanical energy (section 15.1);
2 die melt temperature;
3 die pressure;
4 flowrate through the die.

These variables are maintained at predetermined values by controlling the ingredient feedrate, the screw speed, water injection rate, temperature profile of the barrel and the speed of the rotary knife on the die (Chessari and Sellahewa 2001). Powders and granular feed materials are blended with water or steam in a preconditioner to moisten them before feeding into the extruder. This produces a more uniform feed material that can be more accurately metered and provides more uniform extrusion conditions. The residence time in either batch or continuous preconditioners is closely controlled to ensure that each particle is uniformly blended with the liquid, and for steam conditioning, that there is uniform temperature equilibration. Preconditioning with steam or hot water for 4–5 min increases the feed temperature and moisture content, gelatinises starch and denatures proteins. This improves the extruder efficiency, lowers specific energy consumption and reduces equipment wear and maintenance costs (Bailey et al. 1995). Extruders are fitted with feed hoppers that have screw augers or gravimetric vibrating feeders to load material at a uniform rate into the barrel. The weight of product on a feeder (or loss in weight of the hopper) is used to automatically control the feedrate and take account of variations in ingredient bulk density. The feeder is linked to the screw speed to prevent the barrel becoming empty or overfilled.

Liquids are pumped into the extruder barrel by positive displacement pumps, with flow-measuring devices fitted in-line to control the feedrate. The screw speed is controlled via a variable speed motor and the power consumption is monitored by an amp meter in the motor drive circuit. Other instrumentation includes thermocouples to monitor the temperature in the barrel and the fluid melt temperature at the die, and pressure sensors at the die. After material leaves the die, it is cut into the required lengths by a series of knives that rotate across the face of the die. The speed of rotation is adjusted to the throughput to produce the correct length. Alternatively, the product may be transported by conveyor to a separate guillotine for cutting. Further details are given by Chessari and Sellahewa (2001) who also describe novel sensors, including acoustic monitoring of the product as it exits the die, near-infrared spectroscopy, and electronic 'noses' and 'tongues' that are under development to monitor product quality.

Control of extruders by SCADA (supervisory control and data acquisition) software (Chapter 27, section 27.2.2) is now widely used. Typically, a process computer monitors the set-points for the variables that are controlled. This computer controls a programmable logic controller (PLC), which in turn activates local controllers (e.g. water injection pumps or motor speed controllers) to respond to changes in information from the sensor instrumentation. Bailey et al. (1995) describe computerised process control that is able to supervise start-up and shut-down sequences, alarm recognition and storage of formulations. The control system continuously monitors over 100 process alarm conditions, including the product formulation in relation to the operating conditions. It visually alerts the operator if non-specified conditions exist and, in extreme cases, controls an orderly shutdown or an emergency stop. Other systems for control of extruders are described by Chessari and Sellahewa (2001), Moreira (2001), Bailey et al. (1995) and Lu et al. (1993).

The complex nature of extrusion, however, with multiple factors that are non-linear and interact with each other, and product quality attributes that cannot be measured on-line, means that direct feedback control is not feasible, so at present it is not possible to have fully automatic control of extruders. The control systems are used to alert operators to changes from set-point conditions so that adjustments to operating conditions can be made. Because of these difficulties, there has been considerable research into the application of neural networks and fuzzy logic to extruder control, which are described by Fodil-Pacha et al. (2007), Wang et al. (2001) and Zhou and Paik (2000). Inferential control provides a means to the solution of this type of problem without the cost of expensive sensors. The steps to achieve control include identifying the main variables that influence product quality that can be measured on-line, developing a dynamic model between the influential variables and the quality attributes for on-line prediction of product quality, and real time feedback control using these variables (Wang et al. 2001).

15.3   Applications

15.3.1   Confectionery products

HTST extrusion cooking is used to produce gelatinised, chewy products such as fruit gums and liquorice, from a mixture of sugar, glucose and starch. The heat gelatinises the starch and vaporises excess water, which is vented from the machine. Colourings and flavours are added to the plasticised material and, after mixing, it is cooled and extruded. The product texture can be adjusted from soft to elastic by control over the formulation and processing conditions, the shape can be changed by changing the die, and a variety of flavours and colours may be added. These different combinations permit a large range of potential products, including liquorice, toffee, fudge, boiled sweets, creams, and chocolate each produced by the same equipment (Best 1994). Product uniformity is high, no after-drying is required, and there is a rapid start-up and shutdown. Hard-boiled sweets are produced from granulated sugar and corn syrup. The temperature in the extruder is raised to 165 °C to produce a homogeneous, decrystallised mass. Acids, flavours and colour are added to the sugar mass, and the moisture content is reduced to 2% as the product emerges from the die into a vacuum chamber. It is then fed to stamping or forming machines to produce the required shape. Compared with traditional methods which use boiling pans (Chapter 14, section 14.1.3), energy consumption in an extruder operating at 1000 kg h^{-1} is lower (551 compared with 971 kJ per kg of sugar mass), and steam consumption is also lower (0.193 compared with 0.485 kg per kg of sugar mass) (Huber 1984). Further details of extruded confectionery products are given by Jha (2003).

15.3.2   Cereal products

Crispbread

Wheat flour, milk powder, corn starch, sugar and water are mixed and the product is extruded at a high temperature and pressure. The crispbread is then toasted to reduce the moisture content further and to brown the surface. Savings compared with oven baking are up to 66% in energy consumption, as less moisture is removed, and up to 60% in capital costs and floor space, as large ovens are unnecessary (Vincent 1984).

Breakfast cereals

In traditional cornflake manufacture, large maize kernels (grits) were needed, as the size of the individual grit determined the size of the final cornflake. Grits were then pressure cooked, dried, tempered to ensure a uniform moisture distribution, flaked, toasted and sprayed with a vitamin solution. The total processing time exceeded 5 h. Dough pellets are now produced in a low-pressure extruder from any size of maize grit. The size of the pellets determines the size of the cornflakes. They are then flaked, toasted and sprayed as before. The advantages of extrusion cooking are:

- reductions in raw material costs (19.4%), energy consumption (>90%), capital expenditure (44%) and labour costs (14.8%) (Darrington 1987);
- rapid processing to produce cornflakes within minutes of start-up;
- close control over the size and quality of the final product; and
- flexibility to change the product specification easily.

Details of the manufacture of ready-to-eat breakfast cereals are given by Bailey *et al.* (1995) and Bouvier (2001).

Snackfoods

There is a wide variety of extruded snackfoods made from cereal or potato starch doughs. The process involves hydrating the starch and forming a high-temperature (140–180 °C) fluid melt containing superheated water vapour. When this is extruded the material expands to form a foam as water is vaporised, and then cools through the glass transition temperature to form a hard, brittle product. Flavourings and/or colourings are sprayed onto the product after it is extruded. 'Preforms' or 'half-products' are produced from pre-gelatinised doughs. These small, hard, dense pellets are extruded at a lower pressure and slowly dried to a glassy state suitable for extended storage and transport to other processors. The final product is produced by frying or toasting. When half-products are heated rapidly in air or oil, they are softened and develop the necessary physical properties for expansion. The residual moisture (10–12%) in the pellets then turns to steam, to expand the product rapidly to its final shape, which may be up to three times larger than that produced by frying moist dough. Details of extruded snackfood production are given by Guy (2001c).

15.3.3   Protein-based foods

Texturised vegetable protein (TVP)

Extrusion cooking destroys the enzymes present in soybeans, including a urease that reduces the shelf-life, a lipoxidase that causes off-flavours by oxidation of soya oil and also a trypsin inhibitor that reduces protein digestibility. This improves the acceptability, digestibility and shelf-life of the product. Defatted soya flour, soya concentrate or soya isolate are moistened and the pH is adjusted. A lower pH (5.5) increases chewiness in the final product, whereas a higher pH (8.5) produces a tender product and more rapid rehydration. Colours, flavours and calcium chloride firming agent are added, and the material is plasticised in an extruder at 60–104 °C. It is then extruded to form expanded texturised strands, which are cooled and dried to 6–8% moisture content. Details of the production of different texturised soya products are given by Berk (1992).

Weaning foods

Extruded weaning foods are made from a combination of cereals and legumes to produce the correct protein and energy content for growing children. The extruded product may also be fortified with minerals and vitamins. The process produces highly soluble, fully gelatinised flakes or pellets that can be ground to a powder and rehydrated with hot water to form a porridge that is fed to children. The high temperatures used in the extruder ensure that pathogens are destroyed and the products are microbiologically safe. The low water activity ensures a shelf-life in excess of 12 months when packed in moistureproof and airtight packaging. Other weaning foods include ready-to-eat 'rusk' products that resemble aerated biscuits and are designed to dissolve slowly in saliva when eaten by children. Details of extruded weaning food production are given by Kazemzadeh (2001). The process is particularly suitable for production of both commercial weaning foods and those designed as emergency or aid foods in developing countries. Development of these foods is described for example by Milán-Carrillo *et al.* (2007), Plahar *et al.* (2003) and Malleshi *et al.* (1996).

Meat and fish products

Application of extruders to meat and fish products has mostly focused on production of extruded snacks or shelf-stable starch pellets that incorporate previously under-utilised by-products from meat, fish or prawn processing. Additionally, the manufacture of shiozuri surimi from ground, minced fish uses extruders operating with a die temperature of 6–27 °C. A detailed description of surimi processing and the application of the process to red meats and poultry is given by Knight *et al.* (1991).

Other extrusion applications, described by Jones (1990), include the use of extruders as enzymatic reactors. For example:

- Thermostable α-amylase is used to produce modified starches (Chapter 6, section 6.3).
- Caseinates are partially hydrolysed by selected proteases in an extruder and the products are reported to have improved bacteriological quality, colour, flavour and water absorption properties.
- Production of oligosaccharide mixtures from potato starch, without the use of enzymes, that are suitable for dietetic and infant foods.

Extruders are also used for decontamination of spices and for sterilisation of cocoa nibs prior to roasting for chocolate manufacture. In the latter application, extrusion cooking results in a 3*D* reduction in micro-organisms and removal of off-flavours that eliminates the need for a time consuming and expensive conching stage (also Chapter 24, section

24.1.1). Jolly *et al.* (2003) report energy reductions using extruders in chocolate manufacture.

15.4   Effect on foods and micro-organisms

15.4.1   Sensory characteristics

Production of characteristic textures is one of the main features of extrusion cooking technology. The extent of changes to the starch fluid melt (section 15.1) produces the wide range of product textures that can be achieved. The texture depends on the size distribution of air cells in the starch matrix, the cell wall thickness and the nature of the glassy state that starch polymers form on cooling when the product emerges from the die. The relationship between cell structure and texture has been studied by Barrett and Peleg (1992) and Alvarez-Martinez *et al.* (1998). The degree of product expansion also affects the bulk density, which has important implications for filling packs because they are normally filled by weight (if the bulk density is incorrect the packs will be under- or over-filled). Texture is related to bulk density and bulk density is therefore a convenient routine quality assurance check; if the bulk density is within the required limits the product texture will be acceptable.

The HTST conditions in extrusion cooking produce short residence times, so cooked flavours are not produced and there are only minor changes to the natural colour and flavour of foods. In some products, flavour may be produced by Maillard reactions and there has been some research adding specific amino acids as flavour precursors to generate flavours (Breadie *et al.* 1998). Added flavours are mixed with ingredients before or during cold extrusion, but this is largely unsuccessful in extrusion cooking as the flavours are volatilised when the food emerges from the die. Flavours are more often applied to the surface of extruded snackfoods in the form of sprayed emulsions. However, this may cause stickiness in some products and hence require additional drying. Breakfast cereals are toasted after extrusion, which may cause caramelisation of surface sugars and introduce flavours as well as a darker colour.

In many foods the colour of the product is determined by synthetic pigments (Appendix A.4) added to the feed material as water- or oil-soluble powders, emulsions or lakes. Fading of colour due to product expansion, excessive heat or reactions with proteins, reducing sugars or metal ions may be a problem in some extruded foods.

15.4.2   Nutritional value

The shearing and heating conditions in a cooker-extruder reduce the molecular weight of amylose and amylopectin in the starch matrix (section 15.1). These molecules are more rapidly digested and as a result can lead to increases in blood sugar and insulin levels after consumption. Extruded snackfoods, breakfast cereals and biscuits therefore have a relatively high glycaemic index. Camire (2001) describes methods to manipulate extrusion conditions to produce digestion-resistant starch, including adding citric acid to maize meal before extrusion to create increased amounts of polydextrose and oligo-saccharides, adding dietary fibre, or adding protein with removal of fibre. Reducing sugars are lost due to Maillard reactions during extrusion at high temperatures and low moisture contents. Extruded soybean products contain lower levels of flatulence-inducing oligosaccharides, including stachyose and raffinose, than unprocessed soy flours.

Extrusion denatures proteins, which reduces the activity of naturally occurring enzymes, but at lower temperatures it also improves the digestibility of proteins by exposing enzyme access sites. Maillard reactions with amino acids, particularly lysine, which is the limiting amino acid in cereals, reduce protein quality. Destruction of antinutritional components in soya products improves the nutritive value of TVPs.

Lipids may form starch–lipid complexes during extrusion, but these do not affect the nutritive value of the foods. Lipid oxidation does not take place to a significant extent during extrusion, but may occur during storage. Artz et al. (1992) have reviewed the factors that promote oxidation, including metal ions from wear of extruder screws and the increased surface area in expanded products. However, lipolytic enzymes may be inactivated by extrusion and starch–lipid complexes may be more resistant to oxidation. Camire (2001) describes antioxidants used in extruded foods, and foods may also be packaged in nitrogen to reduce oxidation. Extrusion has a variety of effects on the fibre in foods: large insoluble molecules may be partially broken down, which may increase their solubility; the levels of insoluble and soluble non-starch polysaccharides may be either increased or decreased by extrusion depending on the type of raw material. The nutritional implications of these changes are reviewed by Camire (2001).

Vitamin losses in extruded foods vary according to the type of food, the moisture content, the temperature of processing and the holding time. Generally, losses are minimal in cold extrusion. The HTST conditions in extrusion cooking and rapid cooling as the product emerges from the die, cause relatively small losses of most vitamins and essential amino acids. Killeit (1994) has reviewed vitamin retention in extruded foods. For example at an extruder temperature of 154 °C there is a 95% retention of thiamin and little loss of riboflavin, pyridoxine, niacin or folic acid in cereals. However, losses of ascorbic acid and vitamin A are up to 50%, depending on the time that the food is held at the elevated temperatures, and loss of lysine, cystine and methionine in rice products varies between 50 and 90% depending on processing conditions. Many breakfast cereal manufacturers spray vitamin solutions onto products after extrusion to correct perceived deficiencies. Camire (2001) has reviewed the effects of extrusion on vitamins, minerals and antinutrients. Athar et al. (2006) found retention of 44–62% of B-group vitamins during short barrel extrusion of snack foods from different cereal grains. Riboflavin and niacin had the highest stability and pyridoxine was stable in maize, but less so in oats or maize/pea flour ingredients. Thiamin was the least stable vitamin. They concluded that HTST-extruded snacks retained higher levels of heat labile B-vitamins than longer time and lower-temperature methods used in snackfood extruders.

15.4.3   Effect on micro-organisms

Most extrusion-cooked products are microbiologically safe because of both the low water activity of extruded products and the HTST heat treatment (>130 °C for a few seconds) that destroys vegetative cells (Van de Velde et al. 1984). The conditions under which spores are destroyed by extrusion-cooking are not well understood (Chessari and Sellahewa 2001). Likimani et al. (1990) describe a method to calculate D and z-values (Chapter 10, section 10.3) for Bacillus globigii spores. Bulut et al. (1999) studied the effects of extrusion cooking on Microbacterium lacticum and Bacillus subtilis. The results showed a strong correlation between shear stress at the die wall and specific mechanical energy input for destruction of M. lacticum. There were no surviving cells, giving between 4.6 and 5.3 decimal reductions depending on the extruder operating pressures, with the temperature at the extruder die below 61 °C. Bacterial destruction was attributed to heat during extrusion, which weakened cell walls, making them more susceptible to shear forces. A 3.2 decimal reduction in *B. subtilis* spores was obtained using an extruder die temperature below 43 °C, which the authors suggested was a possible 'mechanical germination' inside the extruder. They noted that if shear forces can be optimally combined with thermal forces, an acceptable sterility could be achieved at low temperatures that maximises food quality while minimising process energy requirements. Microbiological safety is a greater concern when extruded products are made from animal by-products that may have high concentrations of pathogens and/or extruded products that have higher moisture contents.

References

ALVAREZ-MARTINEZ, L., KONDURY, K.P. and HARPER, J.M., (1998), A general model for the expansion of extruded products, *J. Food Science*, **53** (2), 609–615.

ARTZ, W.E., RAO, S.K. and SAUER, R.M., (1992), Lipid oxidation in extruded products during storage as affected by extrusion temperature and selected antioxidants, in (J.L. Kokini, C-T. Ho and M.V. Karwe, Eds.), *Food Extrusion Science and Technology*, Marcel Dekker, New York, pp. 449–461.

ATHAR, N., HARDACRE, A., TAYLOR, G., CLARK, S., HARDING, R. and MCLAUGHLIN, J., (2006), Vitamin retention in extruded food products, *J. Food Composition and Analysis*, **19** (4), 379–383.

BAILEY, L.N., HAUCK, B.W., SEVATSON, E.S. and SINGER, R.E., (1995). Ready-to-eat breakfast cereal production, in (A. Turner, Ed.), *Food Technology International Europe*, Sterling Publications International, London, pp. 127–132.

BARRETT, A.H. and PELEG, M., (1992), Extrudates cell structure-texture relationships, *J. Food Science*, **57** (5), 1253–1256.

BERK, Z., (1992), Textured soy protein products, Chapter 7 in *Technology of Production of Edible Flours and Protein Products from Soybeans*, FAO Agricultural Services Bulletin No. 97, Food and Agriculture Organization of the United Nations Rome, available at www.fao.org/docrep/t0532e/t0532e08.htm.

BEST, E.T., (1994), Confectionery extrusion, in (N.D. Frame, Ed.), *The Technology of Extrusion Cooking*, Blackie Academic and Professional, Glasgow, pp. 190–236.

BLANSHARD, J.M.V., (1995), The glass transition, in (S. Beckett, Ed.), *Physico-Chemical Aspects of Food Processing*, Blackie, London.

BOUVIER, J-M., (2001), Breakfast cereals, in (R. Guy, Ed.), *Extrusion Cooking – Technologies and Applications*, Woodhead Publishing, Cambridge, pp. 133–160.

BREADIE, W.L.P., MOTTRAM, D.S. and GUY, R.C.E., (1998), Aroma volatiles generated during extrusion cooking of maize flour, *J. Agric. Food Chem.*, **46**, 1487–1497.

BULUT, S., WAITES, W.M. and MITCHELL, J.R., (1999), Effects of combined shear and thermal forces on destruction of *Microbacterium lacticum*, *Applied Environmental Microbiology*, **65** (10), 4464–4469.

CAMIRE, M.E., (2001), Extrusion and nutritional quality, in (R. Guy, Ed.), *Extrusion Cooking – Technologies and Applications*, Woodhead Publishing, Cambridge, pp. 108–129.

CHESSARI, C.J. and SELLAHEWA, J.N., (2001), Effective process control, in (R. Guy, Ed.), *Extrusion Cooking – Technologies and Applications*, Woodhead Publishing, Cambridge, pp. 83–107.

DARRINGTON, H., (1987), A long-running cereal, *Food Manufacture*, **3**, 47–48.

ELSEY, J., RIEPENHAUSEN, J., MCKAY, B., BARTON, G.W. and WILLIS, M., (1997). Modelling and control of a food extrusion process, *Computers and Chemical Engineering*, **21**, S361–S366.

FODIL-PACHA, F., ARHALIASS, A., AÏT-AHMED, N., BOILLEREAUX, L. and LEGRAND, J., (2007), Fuzzy control of the start-up phase of the food extrusion process, *Food Control*, **18** (9), 1143–1148.

FRAME, N.D. (Ed.), (1994), Operational characteristics of the co-rotating twin-screw extruder, in (N.D. Frame, Ed.), *The Technology of Extrusion Cooking*, Blackie Academic and Professional, Glasgow, pp. 1–51.

GUY, R., (1993), Creating texture and flavour in extruded products, in (A. Turner, Ed.), *Food Technology International Europe*, Sterling Publications International, London, pp. 57–60.

GUY, R., (1994), Raw materials, in (N.D. Frame, Ed.), *The Technology of Extrusion Cooking*, Blackie, London, pp. 52–72.

GUY, R., (Ed.), (2001a), *Extrusion Cooking – Technologies and Applications*, Woodhead Publishing, Cambridge.

GUY, R., (2001b), Raw materials for extrusion cooking, in (R. Guy, Ed.), *Extrusion Cooking – Technologies and Applications*, Woodhead Publishing, Cambridge, pp. 2–28.

GUY, R., (2001c), Snack foods, in (R. Guy, Ed.), *Extrusion Cooking – Technologies and Applications*, Woodhead Publishing, Cambridge, pp. 161–181.

HARPER, J.M., (1979), Food extrusion, *Critical Reviews in Food Science and Nutrition*, **11**, 155–215.

HARPER, J.M., (1981), *Extrusion of Foods*, Vol II, CRC Press, Boca Raton, FL.

HAUCK, B.W., (1993), Choosing an extruder, in (A. Turner, Ed.), *Food Technology International Europe*, Sterling Publications International, London, pp. 81–82.

HELDMAN, D.R. and HARTEL R.W., (1997), *Principles of Food Processing*, Chapman and Hall, New York, pp. 253–283.

HUBER, G.R., (1984), New extrusion technology for confectionery products, *Manufacturing Confectioner*, May, 51, 52, 54.

JHA, M., (2003), *Modern Technology of Confectionery Industries*, Asia Pacific Business Press Inc., Delhi, India.

JOLLY, M.S., BLACKBURN, S. and BECKETT, S.T., (2003), Energy reduction during chocolate conching using a reciprocating multihole extruder, *J. Food Engineering*, **59** (2–3), 137–142.

JONES, S.A., (1990), New developments in extrusion cooking, in (A. Turner, Ed.), *Food Technology International Europe*, Sterling Publications International, London, pp. 54–58.

KAZEMZADEH, M., (2001), Baby foods, in (R. Guy, Ed.), *Extrusion Cooking – Technologies and Applications*, Woodhead Publishing, Cambridge, pp. 182–199.

KILLEIT, U., (1994), Vitamin retention in extrusion cooking, *Food Chemistry*, **49** (2), 149–155.

KNIGHT, M.K., CHOO, B.K. and WOOD, J.M., (1991), Applying the surimi process to red meats and poultry, in (A. Turner, Ed.), *Food Technology International Europe*, Sterling Publications International, London, pp. 147–149.

KULSHRESHTHA, M.K., ZAROR C.A., JUKES D.J. and PYLE, D.L., (1991), A generalized steady state model for twin screw extruders, *Food and Bioproducts Processing*, **69c**, 189–199.

KULSHRESHTHA, M.K., ZAROR, C.A. and JUKES, D.J., (1995), Simulating the performance of a control system for food extruders using model-based set-point adjustment, *Food Control*, **6** (3), 135–141.

LI, C-H., (1999), Modelling extrusion cooking, *Food and Bioproducts Processing*, **77** (C1), 55–63.

LIKIMANI, T.A., SOFOS, J.N., MAGA, J.A. and HARPER J.M., (1990), Methodology to determine destruction of bacterial spores during extrusion cooking, *J. Food Science*, **55** (5), 1388–1393.

LU, Q., MULVANEY, S.J., HSEITH, F. and HUFF, H.E., (1993), Model and strategies for computer control of a twin-screw extruder, *Food Control*, **4** (1), 25–33.

MALLESHI, N.G., HADIMANI, N.A., CHINNASWAMY, R. and KLOPFENSTEIN, C.F., (1996), Physical and nutritional qualities of extruded weaning foods containing sorghum, pearl millet, or finger millet blended with mung beans and nonfat dried milk, *Plant Foods for Human Nutrition*, **49** (3), 181–189.

MERCIER, C., LINKO, P. and HARPER, J., (Eds.), (1989), *Extrusion Cooking*, American Association of Cereal Chemists, St Paul, MN.

MILÁN-CARRILLO, J., VALDÉZ-ALARCÓN, C., GUTIÉRREZ-DORADO, R., CÁRDENAS-VALENZUELA, O.G., MORA-ESCOBEDO, R., GARZÓN-TIZNADO, J.A. and REYES-MORENO, C., (2007), Nutritional properties of quality protein maize and chickpea extruded based weaning food, *Plant Foods for Human Nutrition*, **62** (1), 31–37.

MILLER, R.C., (1990), Unit operations and equipment IV. Extrusion and extruders, in (R.B. Fast and E.F. Caldwell, Eds.), *Breakfast Cereals and How They are Made*, American Association of Cereal Chemists, Inc., St Paul, MN, pp. 135–196.

MOREIRA, R., (2001), *Automatic Control for Food Processing Systems*, Aspen Publications, Gaithersburg, MD, pp. 9–17.

MOTTAZ, J. and BRUYAS, L., (2001), Optimised thermal performance in extrusion, in (R. Guy, Ed.), *Extrusion Cooking – Technologies and Applications*, Woodhead Publishing, Cambridge, pp. 51–82.

PLAHAR, W.A., ONUMA OKEZIE, B. and ANNAN N.T., (2003), Nutritional quality and storage stability of extruded weaning foods based on peanut, maize and soybean, *Plant Foods for Human Nutrition*, **58** (3), 1–16.

POMERLEAU, D., DESBIENS, A. and BARTON, G.W., (2003), Real time optimization of an extrusion cooking process using a first principles model, Control Applications, CCA 2003, Proceedings of 2003 IEEE Conference, Volume 1, 23–25 June, pp. 712–717.

RIAZ, M.N., (2000), *Extruders in Food Applications*, CRC Press, Boca Raton, FL.

RIAZ, M.N., (2001), Selecting the right extruder, in (R. Guy, Ed.), *Extrusion Cooking – Technologies and Applications*, Woodhead Publishing, Cambridge, pp. 29–50.

RIZVI, S.S.H., MULVANEY, S.J. and SOKHEY, A.S., (1995), The combined application of supercritical fluid and extrusion technology, *Trends in Food Science and Technology*, **6**, July, 232–240.

ROKEY, G.J., (2000), Single-screw extruders, in (M.N. Riaz, Ed.), *Extruders in Food Applications*, Technomic Publishing Co., Lancaster, PA, pp. 25–50.

SCHONER, S.L. and MOREIRA, R.G., (1997), Dynamic analysis of on-line product quality attributes of a food extruder, *Food Science and Technology International*, **3** (6), 413–421.

TAN, J.L. and HOFER, J.M., (1995), Self-tuning predictive control of processing temperature for food extrusion, *J. Process Control*, **5** (3), 183–189.

VAN DE VELDE, C., BOUNIE, D., CUQ, J.L. and CHEFTEL, J.C., (1984), Destruction of microorganisms and toxins by extrusion cooking, in (P. Zeuthen and J.C. Cheftel, Eds.), *Proceedings of COST 91 Symposium: Thermal Processing and Quality of Foods*, Applied Science, London, pp. 155–161.

VINCENT, M.W., (1984), *Extruded Confectionery – Equipment and Process*, Vincent Processes Ltd., Shaw, Newbury.

WANG, L., CHESSARI, C. and KARPIEL, E., (2001), Inferential control of product quality attributes – application to food cooking extrusion process, *J. Food Process Control*, **11** (6), 621–636.

WANG, L., GAWTHROP, P., CHESSARI, C., PODSIADLY, T. and GILES, A., (2004), Indirect approach to continuous time system identification of food extruder, *J. Food Process Control*, **14** (6), 603–615.

ZHOU, M. and PAIK, J., (2000) Integrating neural network and symbolic inference for predictions in food extrusion process, in (R. Loganathara, G. Palm and M. Ali, Eds.), *Intelligent Problem Solving. Methodologies and Approaches: 13th International Conference on Industrial and Engineering Applications of Artificial Intelligence*, Prentice Hall, New York, pp. 567–572.

III.B

Heat processing using hot air

16

Dehydration

Abstract: Dehydration using hot air or heated surfaces removes water from foods and reduces their water activity. This inhibits microbial growth and enzyme activity to extend the shelf-life. Drying can cause deterioration of both the eating quality and the nutritional value of foods, and the design and operation of dehydration equipment aim to minimise these changes. The chapter first looks at psychrometrics, the theory of drying and calculation of drying rates. It then summarises the many different types of hot-air and contact drying equipment and methods used to control their operation. The chapter concludes by describing rehydration and the effects of dehydration on foods and micro-organisms.

Key words: dehydration, psychrometrics, drying rate, hot-air dryers, heated-surface dryers, agglomeration of powders, encapsulation, rehydration.

Dehydration (or drying) is 'the application of heat under controlled conditions to remove the majority of the water normally present in a food by evaporation' (or in the case of freeze drying by sublimation). This definition excludes other unit operations that remove water from foods (e.g. mechanical separations and membrane concentration (Chapter 5), evaporation (Chapter 14) and baking (Chapter 18)) as these normally remove much less water than dehydration does. This chapter focuses on dehydration using hot air or heated surfaces. Microwave, radio frequency and radiant dryers are described in Chapter 20 and freeze drying is described in Chapter 23 (section 23.1).

The main purpose of dehydration is to extend the shelf-life of foods by a reduction in water activity (Chapter 1, section 1.1.2). This inhibits microbial growth and enzyme activity, but the processing temperature is usually insufficient to cause their inactivation. Therefore any increase in moisture content during storage, for example due to faulty packaging, can result in rapid spoilage. Drying also causes deterioration of both the eating quality and the nutritional value of the food (section 16.5). The design and operation of dehydration equipment aim to minimise these changes by selection of appropriate drying conditions for individual foods. Details of the chemistry and microbiology of dried foods are given by Hui *et al.* (2007).

The reduction in weight and bulk of dried foods reduces transport and storage costs. Dehydration also provides convenient products that have a long shelf-life at ambient temperature for the consumer, or ingredients that are more easily handled by food processors. Examples of commercially important dried consumer products include baby foods, beans, pulses, nuts, breakfast cereals, coffee whitener, condiments and spices (e.g. garlic, pepper), egg products, flours (including bakery mixes), instant coffee, instant soups, milk, pasta, powdered cheeses, raisins, sultanas and other fruits, sweeteners and tea. Examples of important dried ingredients that are used by manufacturers include dairy products (milk, whey proteins, cheese, buttermilk, sodium caseinate, butter, ice cream mixes), soy powders, soy protein isolate, whole egg, egg yolk and albumen, encapsulated flavourings and colourings, spray-dried meat purées, fruit and vegetable pulps, pastes and juices, maltodextrins (powdered, granulated or agglomerated), lactose, sucrose or fructose powders, enzymes and yeasts (Deis 1997).

16.1   Theory

Dehydration involves the simultaneous application of heat and removal of moisture by evaporation from foods. Factors that control the rates of heat transfer are described in Chapter 10 (section 10.1.2) and mass transfer due to evaporation is described in Chapter 14 (section 14.1.1). There are a large number of factors that control the rate at which foods dry, which can be grouped into categories related to the processing conditions, the nature of the food and the design of dryers. The effects of processing conditions and type of food are described below and differences in dryer design are summarised in section 16.2.

16.1.1   Drying using heated air
Psychrometrics
Psychrometry is the study of inter-related properties of air–water vapour systems. These properties are conveniently represented on a psychrometric chart (Fig. 16.1) and are described mathematically in a number of food engineering textbooks (e.g. Singh and Heldman 2001a, Toledo 1999). There are three inter-related factors that control the capacity of air to remove moisture from a food:

1 the amount of water vapour already carried by the air;
2 the air temperature; and
3 the amount of air that passes over the food.

The amount of water vapour in air is expressed as either absolute humidity (W) (termed 'moisture content' in Fig. 16.1 and also known as the 'humidity ratio'), which equals the mass of water vapour per unit mass of dry air in kg per kg (Equation 16.1), or as relative humidity (RH) (in per cent). This is defined as 'the ratio of the partial pressure of water vapour in the air to the pressure of saturated water vapour at the same temperature, multiplied by 100 and can be represented by Equation 16.2:

$$W = m_w/m_a \hfill \boxed{16.1}$$

where m_w (kg) = mass of water and m_a (kg) = mass of dry air,

$$RH = (\rho_w/\rho_{ws}) \times 100 \hfill \boxed{16.2}$$

where ρ_w (kPa) = partial pressure of water vapour in the air and ρ_{ws} (kPa) = saturated water vapour pressure at the same temperature.

The amount of heat needed to raise the temperature of an air–water vapour mixture is known as the 'humid heat' and corresponds to sensible heat when heating solids or

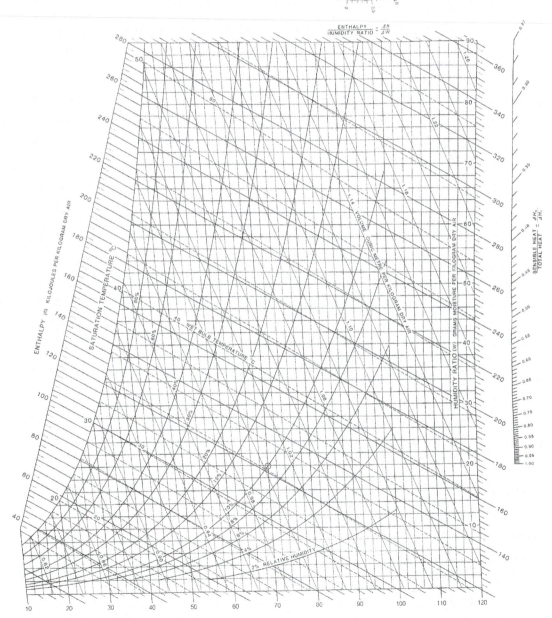

Fig. 16.1   Psychrometric chart based on barometric pressure of 101.3 kPa (courtesy of ASHRAE Inc.).

liquids (Chapter 10, section 10.1.2). Other properties of water vapour are described in Chapter 10 (section 10.1.1) and Chapter 14 (section 14.1.1) and details of the factors that control how much vapour can be carried by air are given by Singh and Heldman (2001a).

Most heat transfer is by convection from the drying air to the surface of the food, but there may also be heat transfer by radiation. If the food is dried in solid trays, heat will also be conducted through the trays to the food. Heat is absorbed by food and both raises the temperature of the food and provides the latent heat needed to evaporate water from the surface. An increase in air temperature, or reduction in RH, causes water to evaporate more rapidly from a wet surface and therefore produces a greater fall in surface temperature. The temperature of the air, measured by a thermometer bulb, is termed the 'dry-bulb' temperature. If the thermometer bulb is surrounded by a wet cloth and air is blown over the cloth, heat is removed by evaporation of water and the temperature falls. This lower temperature is known as the 'wet-bulb' temperature. The difference between the two temperatures is used to find the relative humidity of air on the psychrometric chart (sample problem 16.1).

The dew point is the temperature at which air becomes saturated with moisture (100% RH) and any further cooling from this point results in condensation of the water from the air. Adiabatic cooling lines are the parallel straight lines sloping across the psychrometric chart, which show how absolute humidity decreases as the air temperature increases. The calculations in sample problem 16.1 illustrate how the psychrometric chart is used and further examples are given by Singh and Heldman (2001a) and Toledo (1999).

Sample problem 16.1

Using the psychrometric chart (Fig. 16.1), calculate the following:
1 The absolute humidity of air that has RH = 40% and a dry bulb temperature = 60 °C.
2. The wet bulb temperature under these conditions.
3 The RH of air having a wet bulb temperature = 44 °C and a dry bulb temperature = 70 °C.
4 The dew point of air cooled adiabatically from RH = 30% and a dry bulb temperature = 50 °C.
5 The change in RH of air with a wet bulb temperature = 38 °C, heated from 50 °C to 86 °C (dry bulb temperatures).
6 The change in RH of air with a wet bulb temperature = 35 °C, cooled adiabatically from 70 °C to 40 °C (dry bulb temperatures).
7 Food is dried in a co-current dryer (section 16.2.1) from an inlet moisture content of 0.3 kg moisture per kg product to an outlet moisture content of 0.15 kg moisture per kg product. Air at dry bulb temperature = 20 °C and RH = 40% is heated to the dryer inlet temperature = 110 °C. The dry bulb temperature of the exhaust air from the dryer should be at least 10 °C above the dew point to prevent condensation in pipework. Calculate the exhaust air temperature and RH that meets this requirement and the mass of air required ($kg\,h^{-1}$ (dry basis) per $kg\,h^{-1}$ of dry solids).

Solutions to sample problems 16.1
1 Find the intersection of the 60 °C and 40% RH lines and follow the chart horizontally to the right to read off the absolute humidity (0.0535 kg (or 53.5 g) per kg dry air).

2   From the intersection of the 60 °C and 40% RH lines, extrapolate left parallel to the dotted wet bulb lines to read off the wet bulb temperature (43.8 °C).

3   Find the intersection of the 44 °C and 70 °C lines and follow the sloping RH line upwards to read off the % RH (25%).

4   Find the intersection of the 50 °C and 30% RH lines and follow the dotted wet bulb line left until the RH = 100% (32.6 °C).

5   Find the intersection of the 38 °C wet bulb line and the 50 °C dry bulb line, and follow the horizontal line to the intersection with the 86 °C dry bulb line; read the sloping RH line at each intersection. This represents the changes to the air humidity when it is heated (48 − 9.8%).

6   Find the intersection of the 35 °C wet bulb line and the 70 °C dry bulb line, and follow the wet bulb line left to the intersection with the 40 °C dry bulb line; read the sloping RH line at each intersection. This represents the changes to air as it dries the food; it is cooled and becomes more humid as it picks up moisture from the food (11 − 73%).

7   Find the intersection of the 20 °C dry bulb line and the 40% RH line, and follow the horizontal line left to the dry bulb temperature = 110 °C and read the absolute humidity = 0.006 kg (or 6 g) moisture kg^{-1} dry air.

At 110 °C the wet bulb temperature = 35.5 °C, follow this line left and find the point at which the dry bulb temperature is 10 °C above the dew point. This occurs when the dry bulb temperature = 44 °C and the dew point = 33.8 °C (the dew point is found by reading horizontally left along the absolute humidity line of 0.034 kg (or 34 g) moisture kg^{-1} dry air from the point where the 35.5 °C wet bulb temperature and 44 °C dry bulb temperature intersect to RH = 100%).

At point 35.5 °C wet bulb temperature and 44 °C dry bulb temperature, the RH of the exhaust air = 57%.

The air flowrate is found using a mass balance:

$$\text{Moisture lost per kg dry solids} = 1 \times (0.3 - 0.15) \times 1000 = 150 \, \text{g h}^{-1}$$

This is the moisture gained by the air, which is also equal to

$$M \times (34 - 6) = M \times 28 \text{ g moisture kg}^{-1} \text{ dry air h}^{-1}$$

where M = mass flow of air. Therefore,

$$M = 150/28 = 5.36 \text{ kg air (dry)/kg dry product}$$

Mechanism of drying

The third factor that controls the rate of drying, in addition to air temperature and humidity, is the air velocity. When hot air is blown over a wet food, water vapour diffuses through a boundary film of air surrounding the food and is carried away by the moving air (Fig. 16.2). A water vapour pressure gradient is established from the moist interior of the food to the dry air. This gradient provides the 'driving force' for water removal from the food. The boundary film acts as a barrier to both heat transfer and the removal of water vapour. The thickness of the film is determined primarily by the air velocity; low-velocity air produces thicker boundary films that reduce the heat transfer coefficient. When water vapour leaves the surface of the food, it increases the humidity of the air in the boundary film. This reduces the water vapour pressure gradient and hence slows the rate of drying. Conversely, fast-moving air removes humid air more quickly, reduces the boundary film,

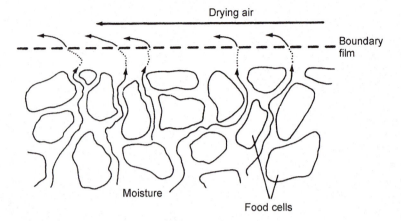

Fig. 16.2   Movement of moisture during drying.

increases the water vapour pressure gradient and hence increases the rate of drying. In summary, the three characteristics of air that are necessary for successful drying when the food is moist are:

1   a moderately high dry-bulb temperature;
2   a low RH; and
3   a high air velocity.

Constant rate period
When food is placed in a dryer, there is a short initial settling down period as the surface heats up to the wet-bulb temperature (A–B in Fig. 16.3a). Drying then commences and, provided that water moves from the interior of the food at the same rate as it evaporates from the surface, the surface remains wet. This is known as the constant-rate period and continues until a certain critical moisture content is reached (B–C in Figs 16.3a and b). The surface temperature of the food remains close to the wet-bulb temperature of the drying air until the end of the constant-rate period, due to the cooling effect of the evaporating water. In practice, different areas of the food surface dry out at different rates

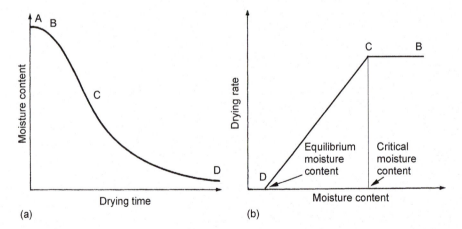

Fig. 16.3   Drying curves. The temperature and humidity of the drying air are constant and all heat is supplied to the food surface by convection.

and, overall, the rate of drying declines gradually towards the end of the 'constant'-rate period. A calculation of the time required to complete the constant rate period is given in sample problem 16.2.

Sample problem 16.2

Diced carrot, having a cube size of 1.5 cm and a moisture content of 88% (w/w basis), is dried in a fluidised bed dryer (section 16.2.1) to a critical moisture content of 38% (w/w basis). During the constant rate period, water is removed at $7 \times 10^{-4}\,kg\,m^{-2}\,s^{-1}$. Calculate the time taken to complete the constant rate period. Assume that the density of fresh carrot is $840\,kg\,m^{-3}$.

Solution to sample problem 16.2

Area of carrot cube available for drying:

$$A = (0.015 \times 0.015) \times 6 \text{ sides (in fluidised bed drying, evaporation of}$$
$$\text{moisture can take place from all sides)}$$
$$= 1.35 \times 10^{-3}\,m^2$$

Drying rate per cube $= 0.0007 \times 0.001\,35$
$$= 9.45 \times 10^{-7}\,kg\,s^{-1}$$

Expressing moisture contents on a dry weight (d/w) basis:

Initial moisture content $= 88\%$ (w/w basis)
$$= 0.88 \text{ kg water per kg product}$$
$$\text{(and therefore 0.12 kg solids per kg product)}$$

Initial moisture content (d/w basis) $= 0.88/0.12$
$$= 7.33 \text{ kg/kg solids}$$

Similarly the critical moisture content (d/w basis) $= (0.38/0.62) = 0.61$ kg/kg solids. The amount of moisture removed during the constant rate period $= 7.33 - 0.61 = 6.72$ kg/kg solids.

The initial mass of one cube $=$ density $\times$ volume
$$= 840 \times (0.015)^3$$
$$= 2.84 \times 10^{-3}\,kg$$

The initial mass of solids of one cube $= (2.84 \times 10^{-3}) \times 0.12$ kg solids per kg product
$$= 3.4 \times 10^{-4}\,kg\,solids$$

Mass of water removed from one cube $= 6.72 \times (3.4 \times 10^{-4})$
$$= 2.28 \times 10^{-3}\,kg$$

$$\text{Time required} = \frac{\text{mass of water removed}}{\text{drying rate}}$$
$$= \frac{2.28 \times 10^{-3}}{9.5 \times 10^{-7}}$$
$$= 2412.7\,s$$
$$\approx 40\,min$$

Falling-rate period

When the moisture content of the food falls below the critical moisture content, the rate of drying slowly decreases until it approaches zero at the equilibrium moisture content (that is, the food comes into equilibrium with the drying air). This is known as the falling-rate period. Non-hygroscopic foods have a single falling-rate period (CD in Figs 16.3a and b), whereas hygroscopic foods have two or more periods. In the first period, the plane of evaporation moves from the surface to inside the food, and water diffuses through the dry solids to the drying air. The second period occurs when the partial pressure of water vapour is below the saturated vapour pressure, and drying is by desorption (see also the discussion of 'bound' and 'free' water, water sorption isotherms and water activity in Chapter 1, section 1.1.2).

During the falling-rate period(s), the rate of water movement from the interior to the surface falls below the rate at which water evaporates to the surrounding air, and the surface therefore dries out (assuming that the temperature, humidity and air velocity are unchanged). If the same amount of heat is supplied by the air, the surface temperature rises until it reaches the dry-bulb temperature of the drying air. Most heat damage to food can therefore occur in the falling rate period. To minimise this, the air temperature is controlled to balance the rate of moisture movement and reduce the extent of heat damage.

The falling-rate period is usually the longest part of a drying operation and, in some foods (e.g. grain drying), the initial moisture content is below the critical moisture content and the falling-rate period is the only part of the drying curve to be observed. During the falling-rate period, the factors that control the rate of drying change. Initially the important factors are similar to those in the constant-rate period and liquid diffusion may be the main mechanism, whereas in later parts, vapour diffusion may be more important. In summary, water moves from the interior of the food to the surface by the following mechanisms:

- Liquid movement by capillary forces, particularly in porous foods.
- Diffusion of liquids, caused by differences in the concentration of solutes at the surface and in the interior of the food.
- Diffusion of liquids that are adsorbed in layers at the surfaces of solid components of the food.
- Water vapour diffusion in air spaces within the food caused by vapour pressure gradients.

Equations for the mechanisms of moisture movement by diffusion of water or movement of water vapour are described by Singh and Heldman (2001b).

The mechanisms that operate in the falling-rate period depend mostly on the temperature of the air and the size of the food pieces. They are unaffected by the RH of the air (except in determining the equilibrium moisture content) and the velocity of the air. In later stages of the falling-rate period, the temperature of the air determines the rate of heat transfer to the plane of evaporation within the food. Heat is transferred by conduction through the food and the rate is limited by the thermal conductivity of the food (see Chapter 10, section 10.1.1). The amount of heat reaching the liquid surface within the food controls the amount of evaporation that takes place and hence the vapour pressure at the liquid surface. The vapour pressure gradient between the liquid surface and the food surface controls the rate at which moisture is removed from the product.

The size of food pieces has an important effect on the drying rate in both the constant-rate and falling-rate periods. In the constant-rate period, smaller pieces have a larger surface area available for evaporation whereas in the falling-rate period, smaller pieces have a shorter distance for heat and moisture to travel through the food. Other factors that influence the rate of drying include the following:

- The composition and structure of the food, which influence the mechanism of moisture removal. For example the orientation of fibres in vegetables (e.g. celery) and protein strands in meat allow more rapid moisture movement along their length than across their structure.
- Moisture is removed more easily from intercellular spaces than from within cells.
- Rupturing cells by blanching (Chapter 11) or size reduction (Chapter 3, section 3.1) increases the rate of drying.
- High concentrations of solutes such as sugars, salts, gums, starches, etc., increase the viscosity of a food and reduce the rate of moisture movement.
- The amount of food placed into a dryer in relation to its capacity influences the drying rate (in a given dryer, faster drying is achieved with smaller quantities of food).

For these reasons the rate at which foods dry may differ in practice from the idealised drying curves described above. Calculation of heat transfer rates in drying systems is often very complex and calculation of drying rates is further complicated if foods shrink during the falling-rate period. Mathematical modelling of dehydration systems is used to address these complexities (section 16.3).

Calculation of drying rate
In commercial operations, the speed of drying is often the limiting factor that controls the production rate that can be achieved. Where simple drying behaviour is found and data on critical and equilibrium moisture contents or thermal properties of foods is known, drying times can be estimated by calculation. However, this data is not available for all foods and pilot-scale drying trials are used to estimate drying times.

The moisture content of a food may be expressed on a wet weight basis (mass of water per unit mass of wet food) or a dry weight basis (mass of water per unit mass of dry solids in the food). In the calculations described below, a dry weight basis is used throughout. The moisture content of foods in other sections is given as wet weight basis. Derivations of the following equations are described by Singh and Heldman (2001c), Barbosa-Canovas (1996) Brennan (1994), and Brennan *et al.* (1990).

The rate of heat transfer is found using:

$$Q = h_s A \left(\theta_a - \theta_s \right) \qquad \boxed{16.3}$$

where Q ($J\,s^{-1}$) = rate of heat transfer, h_s ($W\,m^{-2}\,K^{-1}$) = surface heat transfer coefficient, A (m^2) = surface area available for drying, θ_a (°C) = average dry bulb temperature of drying air, θ_s (°C) = average wet bulb temperature of drying air.

The rate of mass transfer is found using:

$$-m_c = K_g A (H_s - H_a) \qquad \boxed{16.4}$$

Since, during the constant-rate period, an equilibrium exists between the rate of heat transfer to the food and the rate of moisture loss from the food, these rates are related by:

$$-m_c = \frac{h_c A}{\lambda} (\theta_a - \theta_s) \qquad \boxed{16.5}$$

where h_c ($W\,m^{-2}\,K^{-1}$) = surface heat transfer coefficient for convective heating, m_c ($kg\,s^{-1}$) = change of mass with time (drying rate), K_g ($kg\,m^{-2}\,s^{-1}$) = mass transfer coefficient, H_s (kg moisture per kg dry air) = humidity at the surface of the food (saturation humidity), H_a (kg moisture per kg dry air = humidity of air, and λ (J kg^{-1}) = latent heat of vaporisation at the wet bulb temperature.

The surface heat transfer coefficient (h_c) is related to the mass flowrate of air using the following equations. For parallel air flow:

$$h_c = 14.3G^{0.8}$$

$$\boxed{16.6}$$

and for perpendicular air flow:

$$h_c = 24.2G^{0.37}$$

$$\boxed{16.7}$$

where G (kg m^{-2} s^{-1}) = mass flowrate of air per unit area.

For a tray of food, in which water evaporates only from the upper surface, the drying time is found using:

$$-m_c = \frac{h_c}{\rho \lambda x}(\theta_a - \theta_s)$$

$$\boxed{16.8}$$

where ρ (kg m^{-3}) = bulk density of food and x (m) = thickness of the bed of food.

The drying time in the constant-rate period is found using:

$$t = \frac{\rho \lambda x (M_i - M_c)}{h_c(\theta_a - \theta_s)}$$

$$\boxed{16.9}$$

where t (s) is the drying time, M_i (kg per kg of dry solids) = initial moisture content and M_c (kg per kg of dry solids) = critical moisture content.

For water evaporating from a spherical droplet in a spray dryer (section 16.2.1), the drying time is found using:

$$t = \frac{r^2 \rho_1 \lambda}{3h_c(\theta_A - \theta_S)} \frac{M_i - M_f}{1 + M_i}$$

$$\boxed{16.10}$$

where ρ (kg m^{-3}) = density of the liquid, r (m) = radius of the droplet, M_f (kg per kg of dry solids) = final moisture content.

The following equation is used to calculate the drying time from the start of the falling-rate period to the equilibrium moisture content using a number of assumptions concerning for example the nature of moisture movement and the absence of shrinkage of the food:

$$t = \frac{\rho \lambda x (M_c - M_e)}{h_c(\theta_a - \theta_s)} \ln \frac{(M_c - M_e)}{(M - M_e)}$$

$$\boxed{16.11}$$

where ρ (kg m^{-3}) = bulk density of food and x (m) = thickness of the pieces of food, M_e (kg per kg of dry solids) = equilibrium moisture content, M (kg per kg of dry solids) = moisture content at time t from the start of the falling-rate period.

These straightforward equations are suitable for simple drying systems. More complex models are described by Bahu (1997), Turner and Mujumdar (1996) and Pakowski et al. (1991).

Sample problem 16.3

A conveyor dryer (section 16.2.1) is required to dry peas from an initial moisture content of 78% to 16% moisture (wet weight basis), in a bed 10 cm deep that has a voidage of 0.4. Air at 85 °C with a relative humidity of 10% is blown perpendicularly through the bed at $0.9\,\mathrm{m\,s^{-1}}$. The dryer belt measures 0.75 m wide and 4 m long. Assuming that drying takes place from the entire surface area of the peas and that there is no shrinkage, calculate the drying time and energy consumption in both the constant- and falling-rate periods. (Additional data: the equilibrium moisture content of the peas is 9%, the critical moisture content 100% (dry weight basis), the average diameter 6 mm, the bulk density $610\,\mathrm{kg\,m^{-3}}$ and the latent heat of evaporation $2300\,\mathrm{kJ\,kg^{-1}}$.)

Solution to sample problem 16.3

In the constant rate period, from Equation 16.7,

$$h_c = 24.2(0.9)^{0.37}$$

$$= 23.3 \ \mathrm{W \ m^{-2} \ K^{-1}}$$

From Fig. 16.1 for $\theta_a = 85\,^{\circ}\mathrm{C}$ and RH $= 10\%$,

$$\theta_s = 42\,^{\circ}\mathrm{C}$$

To find the area of the peas,

$$\text{Volume of a sphere} = \frac{4}{3}\pi r^3$$

$$= 4/3 \times 3.142\,(0.003)^3$$

$$= 1.131 \times 10^{-7} \ \mathrm{m^3}$$

$$\text{Volume of the bed} = 0.75 \times 4 \times 0.1$$

$$= 0.3 \ \mathrm{m^3}$$

$$\text{Volume of peas in the bed} = 0.3(1 - 0.4)$$

$$= 0.18 \ \mathrm{m^3}$$

$$\text{Number of peas} = \frac{\text{volume of peas in bed}}{\text{volume of each pea}}$$

$$= \frac{0.18}{1.131 \times 10^{-7}}$$

$$= 1.59 \times 10^6$$

$$\text{Area of a sphere} = 4\pi r^2$$

$$= 4 \times 3.142\,(0.003)^2$$

$$= 113 \times 10^{-6} \ \mathrm{m^2}$$

and

$$\text{Total area of peas} = (1.59 \times 10^6) \times (113 \times 10^{-6})$$

$$= 179.67 \ \mathrm{m^2}$$

From Equation 16.5,

$$\text{Drying rate} = \frac{23.3 \times 179.67}{(85 - 42)}$$

$$= 2.3 \times 10^6$$

$$= 0.0782 \text{ kg s}^{-1}$$

From a mass balance,

$$\text{Volume of the bed} = 0.3 \text{ m}^3$$

$$\text{Bulk density} = 610 \text{ kg m}^3$$

Therefore,

$$\text{Mass of peas} = 0.3 \times 610$$

$$= 183 \text{ kg}$$

$$\text{Initial solids content} = 183 \times 0.22$$

$$= 40.26 \text{ kg}$$

Therefore,

$$\text{Initial mass of water} = 183 - 40.26$$

$$= 142.74 \text{ kg}$$

At the end of the constant-rate period, solids remain constant and

$$\text{Mass of water remaining} = 40.26 \times 3$$

$$= 120.78 \text{ kg}$$

Therefore, during the constant-rate period

$$(142.74 - 120.78) = 21.96 \text{ kg water lost}$$

at a rate of 0.026 kg s^{-1}

$$\text{Drying time} = \frac{21.96}{0.026}$$

$$= 844.6 \text{ s} = 14 \text{ min}$$

Therefore,

$$\text{Energy required} = 0.026 \times 2.3 \times 10^6$$

$$= 5.98 \times 10^4 \text{J s}^{-1}$$

$$\approx 60 \text{ kW}$$

In the falling-rate period, from section 16.1.1,

$$M_c = 75/25 = 3$$

$$M_f = 16/84 = 0.19$$

$$M_e = 9/91 = 0.099$$

From Equation 16.11,

$$t = \frac{\rho \lambda x (M_c - M_e)}{h_c (\theta_a - \theta_s)} \ln \frac{(M_c - M_e)}{(M - M_e)}$$

$$= \frac{610 \times 2300 \times 0.1 \, (3 - 0.099)}{23.3 \, (85 - 42)} \ln \frac{(3 - 0.099)}{(0.19 - 0.099)}$$

$$= \frac{140\,300 \times 2.90}{10\,001.9} \times \ln 31.879$$

$$= 406.2 \times 3.4619$$

$$= 1406.35 \text{ s}$$

$$= 23.4 \text{ min}$$

From a mass balance, at critical moisture content, 96.6 kg contains 25% solids = 24.16 kg. After drying in the falling-rate period, 84% solids = 24.16 kg. Therefore

$$\text{Total mass} = 100/84 \times 24.16$$

$$= 28.8 \text{ kg}$$

and

$$\text{Mass loss} = 96.6 - 28.8$$

$$= 67.8 \text{ kg}$$

Therefore,

$$\text{Average drying rate} = \frac{67.8}{5531}$$

$$= 0.012 \text{ kg s}^{-1}$$

and

$$\text{Average energy required} = 0.012 \times (2.3 \times 10^6)$$

$$= 2.76 \times 10^4 \text{ J s}^{-1}$$

$$= 27.6 \text{ kW}$$

16.1.2 Drying using heated surfaces

Slurries of food are deposited on a heated surface and heat is conducted from the hot surface, through the food, to evaporate moisture from the exposed surface. Heat transfer through liquids and solids is described in Chapter 10 (section 10.1.2). The main resistance to heat transfer is the thermal conductivity of the food (Table 10.2 in Chapter 10). Additional resistance arises if the partly dried food lifts off the hot surface to form a barrier layer of air between the food and the hot surface. Knowledge of the rheological properties of the food is necessary to determine the optimum thickness of the layer and the way in which it is applied to the heated surface. Equation 10.20 in Chapter 10 is used to calculate the rate of heat transfer and sample problem 16.4 shows its use.

Sample problem 16.4
A single-drum drier (section 16.2.2) 0.7 m in diameter and 0.85 m long, operates at 150 °C and is fitted with a doctor blade to remove food after three-quarters of a revolution. It is used to dry a 0.6 mm layer of 20% w/w solution of gelatin, pre-heated to 100 °C, at atmospheric pressure. Calculate the speed of the drum required to produce a product with a moisture content of 4 kg of solids per kilogram of water. (Additional data: the density of gelatin feed $= 1020$ kg m^{-3}, the overall heat transfer coefficient $= 1200$ W m^{-2} K^{-1}, and the latent heat of vaporisation of water $= 2.257$ kJ kg^{-1}. Assume that the critical moisture content of the gelatin is 450% (dry weight basis).)

Solution to sample problem 16.4
First,

$$\text{Drum area} = \pi DL$$
$$= 3.142 \times 0.7 \times 0.85$$
$$= 1.87 \, \text{m}^2$$

Therefore

$$\text{Mass of food on the drum} = (1.87 \times 0.75) \, 0.0006 \times 1020$$
$$= 0.86 \, \text{kg}$$

Initially the food contains 80% moisture and 20% solids. From a mass balance,

$$\text{Mass of solids} = 0.86 \times 0.2$$
$$= 0.172 \, \text{kg}$$

After drying, 80% solids $= 0.172$ kg. Therefore,

$$\text{Mass of dried food} = \frac{100}{80} \times 0.172$$
$$= 0.215 \, \text{kg}$$
$$\text{Mass (water) loss} = 0.86 - 0.215$$
$$= 0.645 \, \text{kg}$$

From Equation 16.3,

$$Q = 1200 \times 1.87 \, (150 - 100)$$
$$= 1.12 \times 10^5 \, \text{J s}^{-1}$$
$$\text{Drying rate} = \frac{1.12 \times 10^5}{2.257 \times 10^6} \, \text{kg s}^{-1}$$
$$= 0.05 \, \text{kg s}^{-1}$$

and

$$\text{Residence time required} = \frac{0.645}{0.05}$$
$$= 13 \, \text{s}$$

As only three-quarters of the drum surface is used, 1 revolution should take $(100/75) \times 13 = 17.3$ s. Therefore the speed $= 3.5$ rpm.

16.1.3   Intermediate moisture foods

Intermediate moisture foods (IMFs) having water activities (a_w) between 0.6 and 0.84 are produced by a number of methods:

- Partial drying of raw foods that have high levels of naturally occurring humectants (e.g. dried fruits such as raisins, sultanas and prunes).
- Osmotic dehydration by soaking food pieces in a more concentrated solution of humectant (commonly sugar or salt). Osmotic pressure causes water to diffuse from the food into the solution to be replaced by the humectant (e.g. 'crystallised' or candied fruits using sugar as the humectant, or salt for fish and vegetables). Further details are given by Torreggiani (1993).
- Dry infusion involves drying the food pieces and then soaking in a humectant solution to produce the required water activity.
- Formulated IMFs have food ingredients, including humectants such as glycerol and propylene glycol, sugar or salt, that are mixed to form a dough or paste that is then extruded, cooked or baked to the required water activity (e.g. traditional products such as jams and confectionery, and newer products such as soft, moist snackfoods and petfoods).

IMFs are compact, convenient to consumers, ready-to-eat and are cheaper to distribute because they require no refrigerated transport or storage. Further details are given by Jayaraman (1995).

16.2   Equipment

The following section describes hot-air and heated-surface dryers. Other types of dryers include infrared, radio frequency and microwave dryers (Chapter 20, section 20.1.2) and freeze dryers (Chapter 23, section 23.1.2). Developments in drying technologies are described by Cohen and Yang (1995).

There are a large number of dryer designs and the characteristics of different types of drying equipment and their applications are summarised in Table 16.1. Details of commercial drying operations are given by Greensmith (1998). The relative costs of different drying methods from data by Sapakie and Renshaw (1984) are as follows: forced-air drying, 198; fluidised-bed drying, 315; drum drying, 327; continuous vacuum drying, 1840; freeze drying, 3528.

The selection of a dryer depends largely on the type of product, its intended use and its expected quality. For example, different designs are available for solid and liquid foods, for fragile foods that require minimal handling, and for thermally sensitive foods that require low temperatures and/or rapid drying. Batch dryers are most suited to production rates $150–200\,kg\,h^{-1}$ dried solids, whereas continuous dryers are used for production rates $>1–2\,t\,h^{-1}$. Other considerations include reliability, safety (including protection against fires or explosions for some products), capital and maintenance costs, energy consumption/fuel efficiency and the cost of equipment to ensure that exhaust emissions do not cause dust pollution or nuisance (e.g. from strong odours emitted to the local environment as in garlic or onion drying).

16.2.1   Hot-air dryers

Most commercial-scale dryers use steam to heat the drying air to temperatures <250 °C via fin tube heat exchangers, although electric heaters may be used in small-scale

Table 16.1 Comparison of selected drying technologies

Type of dryer	Characteristics of the food							Drying characteristics				Examples of products
	Batch or continuous	Solid/ liquid	Size of pieces	Initial moisture content	Heat sensitive	Mech- anically strong	Capacity (kg wet food h⁻¹)	Drying rate	Final moisture content	Evaporative capacity (kg h⁻¹)	Labour require- ment	
Bin	B	S	Int	Low		Yes		Low	Low	–	Low	Vegetables
Cabinet	B	S	Int	Mod			300–700	Mod	Mod	75	High	Fruits and vegetables
Conveyor/band	C	S	Int	Mod			2000–5000	Mod	Mod	1800	Low	Breakfast cereals, fruits, biscuits, nuts
Drum	C	S	Sm	Mod			600	Mod	Mod	400	Low	Gelatin, potato powder, infant foods, corn syrup
Foam mat	C	L	–	–	Yes			High	Low	–	Low	Fruit juices
Fluidised bed	B/C	S	Sm	Mod		Yes	–	Mod	Low	900	Low	Peas, grains, sliced/ diced fruits and vegetables, extruded foods, powders
Kiln	B	S	Int	Mod			–	Low	Mod	–	High	Apple rings, hops
Microwave/ dielectric	B/C	S	Sm	Low			–	High	Low	–	Low	Bakery products
Pneumatic/ring	C	S	Sm	Low	Yes	Yes	25 000	High	Low	16 000	Low	Gravy powder, potato powder, soup powder
Radiant	C	S	Sm	Low			–	High	Low	–	Low	Bakery products
Rotary	B/C	S	Sm	Mod	Yes	Yes	–	Mod	Mod	5500	Low	Cocoa beans, nuts
Spin flash	C	L	–	–	Yes		–	High	Low	7800	Low	Pastes, viscous liquids
Spray	C	L	–	–	Yes		30 000	High	Mod	16 000	Low	Instant coffee, milk powder
Sun/solar	B	S	Int	Mod			–	Low	Mod	–	High	Fruits and vegetables
Trough	C	S	Int	Mod			250	Mod	Mod	–	Low	Peas, diced fruits and vegetables
Tunnel	C	S	Int	Mod		Yes	5000	Mod	Mod	–	Mod	Fruits and vegetables
Vacuum band	C	L	–	–	Yes		–	High	Low	150	Low	Chocolate crumb, juices, meat extract
Vacuum tray	B	S,L	–	–	Yes		–	High	Low	–	Mod	Fruit pieces, meat or vegetable extracts

Key: S = solid, L = liquid, B = batch, C = continuous, Sm = small (powders), Int = intermediate to large (granules, pellets, pieces), Mod = moderate.
Adapted from Barr and Baker (1997), Axtell and Bush (1991) and Sapakie and Renshaw (1984), Greensmith (1998) and Heldman and Hartel (1997)

equipment (Brennan 1994). Direct heating using burning gas is used in some applications (e.g. spray drying or pneumatic ring drying), which is more thermally efficient than indirect heating but has two main disadvantages: first, moisture is produced by combustion which increases the humidity of the air and hence reduces its moisture-carrying capacity (section 16.1); and secondly, there may be other products of combustion that could contaminate foods, including nitrogen oxides (NOX) which could increase the levels of nitrites/nitrates in the food, and carcinogenic N-nitrosamines. Low-NOX burners have been developed to reduce these problems (also Chapter 10, section 10.2).

The cost of fuel for heating air is the main economic factor affecting drying operations, and commercial dryers have a number of features that are designed to reduce heat losses or save energy. Examples from Brennan (1992) include:

- insulation of cabinets and ducting;
- recirculation of exhaust air through the drying chamber, provided a high outlet temperature can be tolerated by the product and the reduction in evaporative capacity is acceptable;
- recovering heat from the exhaust air to heat incoming air using heat exchangers or thermal wheels (Chapter 10, section 10.2.3) or prewarming the feed material;
- drying in two stages (e.g. fluidised beds followed by bin drying, or spray drying followed by fluidised bed drying);
- preconcentrating liquid foods to the highest possible solids content using multiple effect evaporation (Chapter 14, section 14.1.2) – energy use per unit mass of water removed in evaporators can be several orders of magnitude less than that required for dehydration;
- automatic control of air humidity by computer control (section 16.3).

Further details of energy efficiency measures in dehydration are given by Heldman and Hartel (1997) and Driscoll (1995).

Bin dryers
Bin dryers are large, cylindrical or rectangular insulated containers fitted with a mesh base. Heated air (40–45 °C) passes up through a bed of food at relatively low velocities (e.g. $0.5\,\mathrm{m\,s}^{-1}$ per m^2 of bin area). These dryers have a high capacity and low capital and running costs, and are mainly used for 'finishing' products such as cut or whole vegetables (from 10–15% to 3–6% moisture content) after initial drying in other types of dryer. They improve the operating capacity of initial dryers by removing the food when it is in the falling-rate period, when moisture removal is most time consuming. The partial pressure of water vapour in the incoming air must therefore be below the equilibrium vapour pressure of dried food at the drying temperature. The long holding time, typically >36 h, and deep bed of food permit variations in moisture content to be equalised and the dryer acts as a store to smooth out fluctuations in the product flow between drying and packaging operations. The dryers may be several metres high and it is therefore important that foods have sufficient strength to withstand compression. This enables spaces between the pieces to be retained and allow the passage of hot air through the bed. This type of dryer, together with fluidised bed, cabinet, conveyor, trough and kiln dryers (below), are examples of 'through-flow' dryers that are described in detail by Sokhansanj (1997).

Cabinet (or tray) dryers
These dryers consist of an insulated cabinet fitted with a stack of shallow mesh or perforated trays, each of which contains a thin (2–6 cm deep) layer of food. Hot air is

blown at $0.5–5\,m\,s^{-1}$ through a system of ducts and baffles to promote uniform air distribution over and/or through each tray. Additional heaters may be placed above or alongside the trays to increase the rate of drying. Tray dryers are used for small-scale production ($1–5\,t\,day^{-1}$) or for pilot-scale work. They have low capital and maintenance costs and have the flexibility to dry different foods. However, they have relatively poor control and produce more variable product quality because food dries more rapidly on trays nearest to the heat source. A low-cost, semi-continuous mechanism which overcomes this problem by periodically moving trays through the stack has been developed (Axtell and Russell 2000, Axtell and Bush 1991).

Conveyor dryers (belt dryers)
Continuous conveyor dryers are up to 20 m long and 3 m wide (Fig. 16.4). Food is dried on a mesh belt in beds 5–15 cm deep. The air flow is initially directed upwards through the bed of food and then downward in later stages to prevent dried food from blowing out of the bed. Two- or three-stage dryers (Fig. 16.5) mix and re-pile the partly dried food into deeper beds (to 15–25 cm and then 250–900 cm in three-stage dryers). This improves the uniformity of drying and saves floor space. For example, potato strips initially piled on a conveyor in a 10 cm deep layer, shrink to a 5 cm layer by the time they reach the end of the first stage. By restacking the material to a depth of 30 cm, the conveyor area needed for the second stage is 20% of that which would have been necessary without restacking. Foods are dried to 10–15% moisture content and then finished in bin dryers. This equipment has good control over drying conditions and high production rates (e.g. up to $5.5\,t\,h^{-1}$). It is used for large-scale drying of fruits and vegetables. Dryers may have computer controlled independent drying zones and automatic loading and unloading to reduce labour costs.

A second variation is trough dryers (or belt-trough dryers) in which small, uniform pieces of food are dried on a mesh conveyor belt that hangs freely between rollers to form the shape of a trough. Hot air is blown through the bed of food and the movement of the conveyor mixes and turns the food to bring new surfaces continually into contact with the

Fig. 16.4   Conveyor dryer (courtesy of Aeroglide Corp. at www.aeroglide.com).

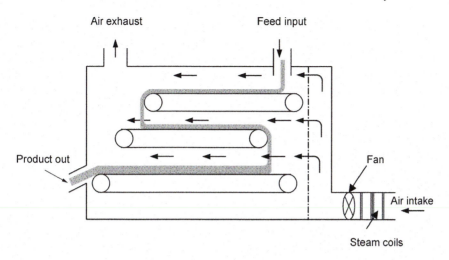

Fig. 16.5   Three-stage conveyor dryer.

drying air. The mixing action also moves food away from the drying air and this allows time for moisture to move from inside the pieces to the dry surfaces. The surface moisture is then rapidly evaporated when the food again contacts the hot air. These dryers have high drying rates (e.g. 55 min for diced vegetables, compared with 5 h in a tunnel dryer), high energy efficiencies, good control and minimal heat damage to the product. They operate in two stages, to 50–60% moisture and then to 15–20% moisture before finishing in bin dryers.

A further application of conveyor dryers is foam-mat drying in which liquid foods are formed into a stable foam by the addition of a stabiliser or edible surfactant (Chapter 1, section 1.1.2 and Appendix A4) and aeration with nitrogen or air. The foam is spread in a thin layer (2–3 mm) on a perforated belt and dried rapidly in two stages by parallel and then counter-current air flows (Table 16.2). Foam mat drying is approximately three times faster than drying a similar thickness of liquid. The thin porous mat of dried food is then ground to a free-flowing powder that has good rehydration properties. Examples

Table 16.2   Advantages and limitations of different arrangements of air flow through dryers

Type of air flow	Advantages	Limitations
Parallel or co-current type: food → air flow →	Rapid initial drying. Little shrinkage of food. Low bulk density. Less damage to food. No risk of spoilage.	Low moisture content difficult to achieve as cool moist air passes over dry food.
Counter-current type: food → air flow ←	More economical use of energy. Low final moisture content as hot air passes over dry food.	Food shrinkage and possible heat damage. Risk of spoilage from warm moist air meeting wet food.
Centre-exhaust type: food → air flow →↑←	Combined benefits of parallel and counter-current driers but less than cross-flow driers.	More complex and expensive than single-direction air flow.
Cross-flow type: food → air flow ↑↓	Flexible control of drying conditions by separately controlled heating zones, giving uniform drying and high drying rates.	More complex and expensive to buy, operate and maintain.

include milk, mashed potatoes and fruit purées. Rapid drying and low product temperatures result in a high-quality product, but a large surface area is required for high production rates, and capital costs are therefore high (see also spray foam drying and vacuum foam mat drying below). The cost of foam mat drying is higher than spray or drum drying but lower than freeze drying.

Explosion puff drying

Explosion puff drying involves partially drying food to a moderate moisture content and then sealing it into a closed, rotating cylindrical pressure chamber. The pressure and temperature in the chamber are increased using superheated steam at ≈250 °C and ≈380 kPa and then instantly released to atmospheric pressure. The rapid loss of pressure causes vaporisation of some of the water and causes the food to expand and develop a fine porous structure. This permits faster final drying (approximately two times faster than conventional methods) particularly for products that have a significant falling-rate period, and also enables more rapid rehydration. Sensory and nutritional qualities are well retained. The technique was first applied commercially to breakfast cereals and now includes a range of fruit and vegetable products.

Fluidised bed dryers

The main features of a fluidised bed dryer are a plenum chamber to produce a homogeneous region of air and prevent localised high velocities, and a distributor to evenly distribute the air at a uniform velocity through the bed of material. Above the distributor, mesh trays contain a bed of particulate foods up to 15 cm deep. Hot air is blown through the bed, causing the food to become suspended and vigorously agitated (fluidised), exposing the maximum surface area of food for drying (Fig. 16.6). A sample calculation of the air velocity needed for fluidisation is described in Chapter 1 (sample

Fig. 16.6   Fluidised bed drying (courtesy of Petrie Technologies at www.petrieltd.com).

problem 1.7). A disengagement or 'freeboard' region above the bed allows disentrainment of particles thrown up by the air. Air from the fluidised bed is fed into cyclones (see Fig. 16.11) to separate out fine particles, which are then added back to the product or agglomerated (Bahu 1997). These dryers are compact and have good control over drying conditions and high drying rates.

In batch operation, the product is thoroughly mixed by fluidisation and this leads to a uniform moisture content. In continuous 'cascade' operation the trays vibrate to move the food from one tray to the next, but there is a greater range of moisture contents in the dried product and bin dryers are therefore used for finishing. The main applications are for small, particulate foods that are capable of being fluidised without excessive mechanical damage, including grains, herbs, peas, beans, coffee, sugar, yeast, desiccated coconut, extruded foods and tea.

In a development of the fluidised-bed dryer, named the 'Toroidal bed' dryer, a fluidised bed of particles is made to rotate around a torus-shaped chamber by hot air blown directly from a burner. The dryer has very high rates of heat and mass transfer and substantially shorter drying times. Larger pieces require a period of moisture equilibration before final drying. The dryer is also suitable for agglomeration and puff drying in addition to roasting, cooking and coating applications (Anon 2005). Brennan *et al.* (2001) studied the effect of drying and puffing conditions in a toroidal bed dryer on the volume of reconstituted dried puffed potato cubes, their rehydration characteristics and texture. They found that the volume of the dried cubes decreased with increase in initial drying time; the shorter the initial drying time the higher the rehydration ratio; and the hardness, springiness, cohesiveness, chewiness and gumminess of the rehydrated cubes all increased with an increase in drying and puffing time.

Another development of the fluidised bed principle is the 'spin flash' dryer (Fig. 16.7), in which a vertical cylindrical drying chamber is fitted with an inverted cone rotor at the base. Hot air from a direct fired gas burner enters tangentially and this, together with the action of the rotor, causes a turbulent rotating flow of air that carries foods up through the chamber. It is used to dry wet cakes or pastes (e.g. filter cakes (Chapter 5, section 5.2) or food pigments). The cake is fed into the drying chamber using a screw conveyor, where the lumps become coated in dry powder. The rotor breaks them into small pieces that are

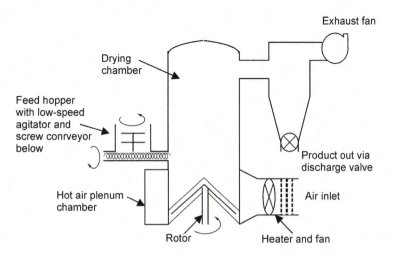

Fig. 16.7   Spin flash dryer (adapted from Anon 2000).

Table 16.3 Comparison of spin flash drying and spray drying, both producing $400 \, kg \, h^{-1}$ of powder at 0.4% moisture using direct heating

Parameter	Spin flash dryer	Spray dryer
Space requirements		
Chamber diameter (m)	0.8	4.25
Floor area (m²)	30	60
Building height (m)	5	14
Building volume (m³)	50	700
Performance		
Feed solids (%)	45	30
Feedrate (kg h⁻¹)	887	1362
Water evaporation (kg h⁻¹)	486	961
Gas consumption (m³ h⁻¹)	62	125
Power consumption (kW h)	30	40

Adapted from Anon (2000)

then fluidised by the drying air. As they dry, the pieces break up and release powder particles that pass up the walls of the dryer, coating new feed as it enters. The particles are therefore removed from the hot air as soon as they are light enough, and the fluidised bed remains at the wet bulb temperature of the drying air, both of which reduce heat damage to the food. At the top of the dryer, the particles pass through a classification screen that is changeable for different product particle size ranges. Dry particles are carried to a cyclone separator by the fluidising air. A comparison of spin flash drying and spray drying is shown in Table 16.3.

The centrifugal fluidised bed dryer is used to pre-dry sticky foods that have a high moisture content, or to dry diced, sliced and shredded vegetables that are difficult to fluidise and/or are too heat-sensitive to dry in conveyor dryers. Food is filled into a drying chamber, which rotates horizontally at high speed. Hot air is forced through the perforated dryer wall and through the bed of food at a high velocity that overcomes the centrifugal force and fluidises the particles (Cohen and Yang 1995).

The 'spouted bed' dryer is used for particles larger than 5 mm that are not readily fluidised in a conventional fluidised bed dryer. The drying air enters a conical chamber at the base and carries particles up through the dryer in a cyclical pattern. This type of dryer produces high rates of mixing and heat transfer and is used for drying heat-sensitive foods.

Fluidised bed dryers are also used to encapsulate solid particles. The particles have an aqueous solution of coating material sprayed onto them, which then dries to form a protective layer when the water is evaporated. Fluidised bed granulators agglomerate particles by spraying a binding liquid into the fluidised bed of granules. The process has benefits in not producing dust and produces powders having particle sizes in the range of 50–2000 μm (Bahu 1997) (see also types of powders in section on spray drying below).

Impingement dryers
Impingement dryers have been used in paper and textile industries for many years and have more recently been used to dry foods such as coffee or cocoa beans, rice and nuts. Dryers have an array of hot air jets that produce high-velocity air, which impinges perpendicularly to the surface of products. The air almost completely removes the boundary layer of air and water vapour (section 16.1) and therefore substantially increases the rates of heat and mass transfer and reduces processing times. Typically, the air temperature is 100–350 °C and the jet velocity is 10–$100 \, m \, s^{-1}$ (Sarkar et al. 2004).

Granular products in particular dry faster and are dried more uniformly because a type of fluidised bed is created by the high air velocity. Moreira (2001) describes the advantages of using superheated steam instead of hot air in impingement dryers, including reduced oxidation of products, improved nutritional value and improved rates of moisture evaporation from the food surfaces.

Kiln dryers

These are two-storey buildings in which a drying room with a slatted floor is located above a furnace. Hot air and the products of combustion from the furnace pass through a bed of food up to 20 cm deep. They have been used traditionally for drying apple rings in the USA, and hops in Europe, but there is limited control over drying conditions and drying times are relatively long. High labour costs are also incurred by the need to turn the product regularly and by manual loading and unloading. However, the dryers have a large capacity and are easily constructed and maintained at low cost.

Pneumatic dryers

In these types of dryer, foods are metered into metal ducting and suspended in hot air. In vertical dryers the air flow is adjusted so that lighter and smaller particles that dry more rapidly are carried to a cyclone separator faster than heavier and wetter particles, which remain suspended to receive the additional drying required. For products that require longer residence times, the ducting is formed into a continuous loop (known as 'pneumatic ring' dryers) and the product is recirculated until it is adequately dried (Fig. 16.8). Humid air is continuously vented from the dryer and replaced with dry air from the heater. High-temperature short-time ring dryers (or 'flash' dryers) have air velocities from 10 to 40 m s^{-1}. Drying takes place within 0.5–3.5 s if only surface moisture is to be removed, or within a few minutes when internal moisture is removed. These dryers are therefore suitable for foods that lose moisture rapidly from the surface, are not abrasive and do not break easily. Evaporative cooling of the particles prevents heat damage to give high-quality products.

Pneumatic dryers have relatively low capital and maintenance costs, high drying rates and close control over drying conditions, which make them suitable for heat-sensitive foods. Outputs range from 10 kg h^{-1} to 25 t h^{-1} (Barr and Baker 1997). They are suitable for drying moist free-flowing particles (e.g. milk or egg powders and potato granules), usually partly dried to less than 40% moisture and having uniform particle size and shape over a range from 10–500 μm. They may be used after spray drying to further reduce the moisture content, and in some applications the simultaneous transportation and drying of the food may be a useful method of materials handling (also Chapter 27, section 27.1.2).

Rotary dryers

An inclined, rotating metal cylinder is fitted internally with flights that cause small pieces of food to cascade through a stream of parallel or counter-current hot air (Table 16.2) as they move through the dryer. The large surface area of food exposed to the air produces high drying rates and a uniformly dried product. The method is especially suitable for foods that tend to mat or stick together in belt or tray dryers. However, the damage caused by impact and abrasion in the dryer restricts this method to relatively few foods (e.g. nuts and cocoa beans). To overcome this problem, a variation of the design, named a 'rotary louvre' dryer, has longitudinal louvres positioned to form an inner drum. Hot air passes through the food particles to form a partially fluidised rolling bed on the base of the drum. Further details are given by Barr and Baker (1997).

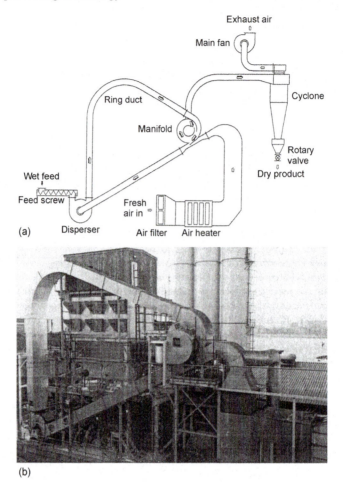

Fig. 16.8 Pneumatic ring dryer: (a) principle of operation (courtesy of Barr Rosin Ltd.); (b) equipment (from Greensmith 1998).

Spray dryers

Spray dryers vary in size from small pilot-scale equipment that can also be used to dry low-volume, high-value products such as enzymes and flavours, to large commercial models capable of producing 10 000 kg h^{-1} of product (Deis 1997). A dispersion of preconcentrated food (40–60% moisture) is first 'atomised' to form a fine mist of droplets that are sprayed into a co- or counter-current flow of hot air (Table 16.2) in a large drying chamber (Fig. 16.9).

One of the following types of atomiser is used:

- *Centrifugal (rotary, disc or wheel) atomiser.* Liquid is fed to the centre of a rotating disc having a peripheral velocity of 90–200 m s^{-1}. Droplets, 50–60 μm in diameter, are flung from the edge to form a uniform spray (Fig. 16.10a) that decelerates to 0.2– 2 m s^{-1} as the droplets fall through the drying camber. It is used for high production rates, unless the feed has a high percentage of fats (e.g. dairy products), when nozzle atomisers are used.

- *Single-fluid (or pressure) nozzle atomiser* (Fig. 16.10b). Liquid is forced at a high pressure (700–2000 kPa) through a small aperture at ≈50 m s^{-1} to form droplet sizes

Fig. 16.9   Spray dryer (from Greensmith 1998).

of 180–250 μm. Grooves on the inside of the nozzle cause the spray to form into a cone shape. The spray angle and spray direction can be varied to use the full volume of the drying chamber. However, nozzle atomisers are susceptible to blockage by particulate foods, and abrasive foods gradually widen the apertures and increase the average droplet size.

- *Two-fluid nozzle atomiser.* Compressed air creates turbulence that atomises the liquid. The operating pressure is lower than the pressure nozzle, but a wider range of droplet sizes is produced. They are used for more viscous or abrasive feeds, or for producing small particle sizes that are not possible with a single-fluid nozzle.
- *Ultrasonic nozzle atomiser.* A two-stage atomiser in which liquid is first atomised by a

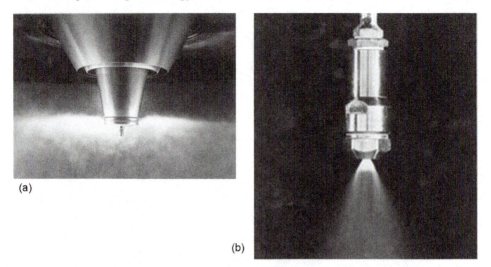

(a)

(b)

Fig. 16.10   (a) Rotary and (b) nozzle atomisers (courtesy of GEA Niro, Soeborg, Denmark at www.niroinc.com).

nozzle atomiser and then using ultrasonic energy to induce cavitation (also Chapter 9, section 9.6).

The viscosity of the feed material and the presence of particles determine which type of atomiser is most suitable. Frequently, both nozzle and disc atomisers are fitted in the same drying chamber to increase the flexibility of the dryer to handle different foods. Further details of the advantages and limitations of different atomisers are given by Masters (1991, 1997). There are a large number of combinations of atomiser, drying chamber design, air heating and powder collecting systems which arise from the different requirements of the very large variety of food materials that are spray dried (e.g. milk, egg, coffee, cocoa, tea, potato, ice cream mix, butter, cream, yoghurt and cheese powder, coffee whitener, fruit juices, meat and yeast extracts, and wheat and corn starch products). Details are given by Deis (1997). Different chamber designs include cylindrical flat bottom types or conical bottom chambers that are designed for the spray pattern produced by disc atomisers. Rapid drying, within 1–30 s, takes place because of the very large surface area of the droplets. The size of the chamber and flow pattern of moving air within the chamber enable the largest droplets to dry before they contact the wall. This prevents the deposition of partially dried product on the wall.

The temperature of drying depends mostly on the type of food and the required powder quality. The factors that are taken into account when selecting the design of a dryer include:

- properties of the feed material, including temperature sensitivity, viscosity, solids content, presence/absence and size of particulates; and
- properties required in the powdered product, including the moisture content, bulk density, explosion hazard and final particle size.

Most of the residence time of particles in a single-stage dryer is used to remove the final moisture, and the outlet temperature must be sufficiently high to do this. Typically, inlet air at 150–300 °C produces an outlet air temperature of 90–100 °C, which corresponds to a wet-bulb temperature (and product temperature) of 40–50 °C. This produces little heat damage to the food, but higher quality is produced using a lower air inlet temperature (e.g. 65–70 °C) (Heldman and Hartel 1997). Co-current air flow is most often used with heat-sensitive materials whereas counter-current air flow gives a lower final moisture content (Table 16.2).

The main advantages of spray drying are rapid drying, large-scale continuous production of powders that have closely controlled properties, low labour costs and relatively simple control, operation and maintenance. The major limitations are high capital costs and the requirement for a relatively high moisture content in the feed to ensure that it can be pumped to the atomiser (Table 16.3). This results in higher energy costs (to remove the moisture) and higher volatile losses. Conveyor-band dryers, spin flash dryers and fluidised bed dryers are gaining in popularity, as they are more compact and energy efficient.

The economics of spray dryer operation are influenced by the temperature of drying and the recycling of drying air. Energy efficiency in spray dryers is increased by raising the inlet air temperature as high as possible (e.g. to ≈220 °C), and keeping the outlet temperature as low as possible (e.g. ≈85 °C) to make maximum use of the energy in the hot air. However, this creates potential risks of heat damage to some products and higher final moisture contents. Therefore a balance is required between the cost of production and product quality. Air recirculation enables up to 25% of the total heat to be reused and air-to-air heat recuperators are also used to recover energy from the exhaust air to further reduce energy costs.

Depending on the design of the dryer, the dry powder and air can be separated in a cyclone separator (Fig. 16.11). Alternatively, the product is separated from exhaust air at the base and removed by either a screw conveyor or a pneumatic system. Fine particles in the air are removed using a cyclone, bag filters, electrostatic precipitators or scrubbers, then bagged or returned for agglomeration.

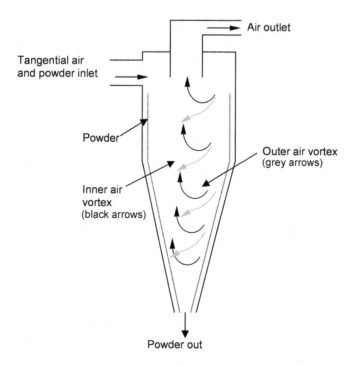

Fig. 16.11   Action of a cyclone separator (adapted from Brennan 1994).

The fluidised spray dryer has a fluidised bed installed in the spray drying chamber. Small particles entrained in the air are recycled from the exhaust system and agglomerated in the fluidised bed. The combination of spray and fluidised bed drying gives efficient use of the drying chamber and produces agglomerated products that have a low bulk density and good instantising characteristics. Further details of these different types of fluidised bed dryers are given by Bahu (1997). A spray dryer with an integrated fluidised bed dryer is an example of a multi-stage dryer. It reduces energy consumption by 15–20% and gives greater control over product quality. Most drying, to 10–15% moisture, takes place in the spray dryer and final drying, to 5% moisture, takes place over a longer time and at a lower temperature in the fluidised bed dryer. The fluidised bed dryer reduces the evaporative load on the spray dryer, which in turn permits a lower outlet air temperature and an increased feedrate to the spray dryer, which leads to higher productivity. As the particles are moist when they enter the fluidised bed, they agglomerate with the dryer particles. An external fluidised bed can also be used for a third cooling stage. Multi-stage drying is also useful for hygroscopic, sticky or high fat products (Deis 1997).

Types of powders
The particle size and bulk density of a powder are important considerations in many applications (Table 16.4). Many powdered foods used as ingredients are required to possess a high bulk density and contain a range of both small and large particles. The small particles fill the spaces between larger ones and thus flow more easily, and also reduce the amount of air in the powder to promote a longer storage life.

There are a number of factors that affect the bulk density of powders:

- Centrifugal atomisers produce smaller droplets and hence higher bulk-density products than nozzle atomisers. Centrifugal atomisers produce more uniform droplet sizes (and hence particle sizes) than nozzle atomisers.
- Aeration of the feed decreases bulk density. Foam spray drying involves making the feed material into a foam using compressed nitrogen or air. Whereas spray-dried particles are hollow spheres surrounded by thick walls of dried material, the foam spray-dried particles have many internal spaces and relatively thin walls, which have typically half the density of spray-dried products.

Table 16.4 Properties of selected powdered foods (reproduced by permission of Elsevier)

Powder	Particle size (μm)	Moisture content (% w/w)	Bulk density (kg m^{-3})	Particle density (kg m^{-3})
Cellulose	43	5	410	1550
Cocoa	7.6	4.4	360	1450
Cornflour	49	9	730	1490
Cornstarch	11.9	10	760	1510
Maltodextrin	55	4.3	600	1390
Salt	12	0.04	1170	2200
Salt	5.8	0.04	870	2210
Soy flour	20.5	6.2	600	1430
Sugar	12	0.06	710	1610
Tea	25	6.6	910	1570
Tomato	320	17.8	890	1490
Wheat flour	51	10	710	1480

Adapted from Fitzpatrick *et al.* (2004)

- Small increases in inlet air temperature decrease bulk density, but an excessively high temperature can also increase bulk density by retaining moisture within case-hardened shells of the particles. During drying in a spray dryer, water evaporates from the surface of particles to form a hard shell, and residual moisture within the shell expands to create porous particles. Maintaining the surface wetness of particles is important for constant-rate drying; if the air temperature is too high, the dried layer at the surface reduces the rate of evaporation.
- If a low bulk density is required, the product should be in contact with the hottest air as it leaves the atomiser, whereas if the outlet air temperature is too low, this increases the moisture content and bulk density of the product.
- Steam injection during atomisation removes air in the centre of the particle and prevents early formation of the shell, to produce a higher bulk density powder.
- More dilute feed material or higher feed temperatures can also increase the bulk density of the powder by forming smaller droplets or by de-aerating them (Deis 1997).
- The nature and composition of the food. Low-fat foods (e.g. fruit juices, potato and coffee) are more easily formed into free-flowing powders than fatty foods such as whole milk or meat extracts.

Fine powders (<50 μm) are difficult to handle, may cause a fire or explosion hazard and are difficult to rehydrate (Brennan 1994). Agglomeration is a size enlargement operation in which an open structure is created when particles of powder are made to adhere to each other. It increases the average size of particles from ≈100 to 250–400 μm and reduces the bulk density of the powder from ≈690 to ≈450 kg m^{-3} (Deis 1997) (Table 16.5). On rehydration, the agglomerated particle sinks below the water surface and breaks apart, allowing the smaller particles to completely hydrate, leading to faster and more complete dispersion of the powder. The characteristics of 'instantised' powders are termed 'wettability', 'sinkability', 'dispersibility' and 'solubility'. For a powder to be considered 'instant', it should complete these four stages within a few seconds. The convenience of instantised powders for retail markets outweighs the additional expense of production, packaging and transport and for processors, agglomeration also reduces problems caused by dust. Details of the properties and handling of powders are given by Lewis (1996) and Barbosa-Canovas et al. (2005).

Powders can be agglomerated by a number of different methods, described by Zemelman and Kettunen (1992). In the 'straight-through' process, fines that are collected in the cyclone are fed back into the feed mist from the atomiser in the spray dryer, where they stick to the moist feed particles to produce agglomerates that are then dried in the dryer. Greater control is achieved using fluidised bed agglomerators: particles are re-

Table 16.5 Differences in reconstitution and physical properties of agglomerated and non-agglomerated skim milk powder

Property	Non-agglomerated powder	Agglomerated powder	
		Integrated spray dryer/ fluidised bed	Re-wetted in fluidised bed
Wettability (s)	>1000	<20	<10
Dispersibility (%)	60–80	92–98	92–98
Insolubility index	<0.1	<0.1	<20
Average particle size (μm)	<100	>250	>400
Bulk density (kg m^{-3})	640–690	450–545	465–500

Adapted from Anon (2000)

moistened under controlled conditions in low-pressure steam, humid air or a fine mist of water and then re-dried in a fluidised bed dryer; or they are discharged from a spray dryer at a slightly higher moisture content (5–8%) onto a fluidised bed dryer. Alternatively, a binding agent (e.g. maltodextrin, gum arabic or lecithin) is used to bind particles together before drying in a fluidised bed. This method has been used for foods with a relatively high fat content (e.g. whole milk powder, infant formulae) as well as fruit extracts, corn syrup solids, sweeteners, starches and cocoa mixes.

Spray drying is the most common means of encapsulation in the food industry. For example, encapsulated flavours are dried by first forming an oil-in-water emulsion (Chapter 3, section 3.2) of the oil-based essence with an aqueous dispersion of a hydrocolloid coating material (e.g. gelatine, dextrin, gum arabic or modified starch). As moisture is evaporated from the aqueous phase, the polymeric material forms a coating around the oily essence. These particles can then be agglomerated to improve dispersability and flowability. Water-soluble materials (e.g. aspartame) are encapsulated by spray coating or fluidised bed coating. Particles are suspended in an upward-moving heated airstream. The coating material is atomised and dries on the particles to coat them uniformly as they are carried up through the dryer several times through a coating cycle. Other ingredients (e.g. powdered shortenings, acidulants, vitamins, solid flavours, sodium bicarbonate or yeast) are encapsulated in a high melting point vegetable fat (e.g. a stearine or wax having melting points of 45–67 °C) and spray-cooled using cold air in the drying chamber to harden the fat. If lower melting point fats (32–42 °C) are used, the material is spray-chilled at lower temperatures (Deis 1997). Further examples of encapsulation are given in Chapter 24 (section 24.3).

Sun and solar drying
Sun drying (without drying equipment) is the most widely practised agricultural processing operation in the world, and more than 250 000 000 t of fruits and grains are dried annually by solar energy. In some countries, foods are simply laid out in fields or on roofs or other flat surfaces and turned regularly until they are dry. More sophisticated methods (solar drying) use equipment to collect solar energy and heat the air, which in turn is used for drying. There are a large number of different designs of solar dryers, described in detail by Brenndorfer *et al.* (1985) and Imrie (1997). These include:

- direct natural-circulation dryers (a combined collector and drying chamber);
- direct dryers with a separate collector; or
- indirect forced-convection dryers (separate collector and drying chamber).

Small solar dryers have been investigated at research institutions, particularly in developing countries, for many years but their often low capacity and insignificant improvement to drying rates and product quality, compared with hygienic sun drying, have restricted their commercial use. Larger solar dryers with photo-voltaic powered fans and having a capacity of 200–400 kg/batch, have been developed by Hohenheim University to a commercial scale of operation. Several hundred dryers are now in use to dry fruit to export standards (Axtell and Russell 2000). Both solar and sun drying are simple inexpensive technologies, in terms of capital and operating costs. Purchased energy inputs and skilled labour are not required, and in sun drying very large amounts of crop can be dried at low cost. The major disadvantages are relatively poor control over drying conditions and lower drying rates than those found in fuel-fired dryers, which result in products that have lower quality and greater variability. In addition, drying is dependent on the time of day and the weather. Extended periods when drying does not occur risks microbial growth on the product. Developments of solar energy include its use to reduce energy consumption in fuel-fired dryers by preheating air and preheating feed water in boilers (Anon 2004). Umesh Hebbar *et al.* (2004) found that a combination of infrared and hot-air drying of carrot and potato reduced the drying time by 48% and consumed 63% less energy compared with drying with hot air alone.

Tunnel dryers

In this equipment, foods are dried on trays that are stacked on trucks, which are programmed to move semi-continuously through an insulated tunnel. Different types of air flow are used depending on the product (Table 16.2). Typically, fruits and vegetables are dried to 15–20% moisture in a 20 m tunnel that contains 12–15 trucks having a total capacity of 5 t of food. The partly dried food is then finished in bin dryers. This ability to dry large quantities of food in a relatively short time made tunnel drying widely used, especially in the USA. However, the method has now been largely superseded by conveyor drying and fluidised bed drying as a result of their higher energy efficiency, reduced labour costs and better product quality.

Ultrasonic and acoustic dryers

High-power ultrasound (Chapter 9, section 9.6) can accelerate mass transfer processes to remove moisture from foods without significant heating. Heat-sensitive foods can therefore be dried more rapidly and at lower temperatures than in conventional hot-air driers without affecting their quality characteristics. The increases in the rate of moisture evaporation are due to pressure variations at air–liquid interfaces. Mulet *et al.* (2003) describe these effects, including 'micro-stirring' at the food interface and rapid alternating contraction and expansion of the material (the 'sponge effect'), which creates microscopic channels that may make moisture removal easier. The high-intensity acoustic waves also produce cavitation of water molecules in the solid food, which may help remove strongly bound moisture (Gallego *et al.* 1999). García-Pérez *et al.* (2006) and de la Fuente-Blanco *et al.* (2006) report studies of fluidised air drying at 40 °C using a high-intensity ultrasonic field at a frequency of 20 kHz, with a power capacity of ≈100 W. They showed that the drying rate is influenced by the air flowrate, ultrasonic power and mass loading of the dryer. However, at high air velocities, the ultrasound acoustic field was disturbed and reduced the effect on drying. Difficulties in the propagation of ultrasound waves in air have led to the development of specially adapted transducers.

Jambrak *et al.* (2007) compared pretreatment of mushrooms, Brussels sprouts and cauliflower by blanching at 80 °C, or treatment using ultrasound with a 20 kHz ultrasound probe or a 40 kHz ultrasound bath before drying. They found that the drying time was shortened for all samples after ultrasound treatment, compared with untreated samples, and that ultrasound-treated samples absorbed more water on rehydration than untreated samples. In ultrasonic spray drying, small droplets are produced by ultrasound and then heated to remove the water. Drying takes place very rapidly (sometimes within seconds) with low-fat solutions, but less well with oily or fatty foods that do not dry easily (Cohen and Yang 1995). Ultrasound has also been used to accelerate mass transfer in osmotic dehydration of apple, and in cheese and meat brining (Mulet *et al.* 2003). An increase in mass transfer is achieved if a threshold power value is achieved for the particular product. However, large-scale commercial development of ultrasound drying has been slow, owing to technical difficulties in transferring acoustic energy from air into the solid material. Although direct contact between the food and the ultrasound transducer significantly increases the drying rate, its application in conventional hot-air driers has proved difficult and further research is required (Gallego 1998).

In acoustic drying, products are atomised and dried at relatively low temperatures (60–90 °C) using intense low-frequency sound waves that promote liquid–solid separation and increase heat and mass transfer coefficients across the boundary layer of the product. Drying rates in these dryers are 3–19 times faster than those of conventional dryers. The dryer requires sound-proofing because of the loud noise produced during drying. Foods that are difficult to dry by conventional methods have been dried successfully in acoustic dryers; for example, liquids containing 5–78% moisture have been dried to 0.5% moisture content. Products with high fat contents (up to 30%) have also been dried in these dryers. Other products that dry well are high-fructose corn syrups, tomato pastes, lemon juice and orange juice. Because the dryer operates at lower temperatures and is relatively fast, degradation of natural colours, flavours and loss of nutritional quality are reduced.

16.2.2   Heated-surface (or contact) dryers

Dryers in which heat is supplied to the food by conduction have three main advantages over hot-air drying:

1   It is not necessary to heat large volumes of air before drying commences and the thermal efficiency is therefore high. Contact dryers are usually heated by steam and typically heat consumption is 2000–3000 kJ per kg of water evaporated compared with 4000–10 000 kJ per kg of water evaporated in hot-air dryers.
2   Dryers produce small amounts of exhaust air and few entrained particles, thus minimising problems and cost of cleaning air before its release to the atmosphere.
3   Drying may be carried out in the absence of oxygen (under vacuum or in a nitrogen atmosphere) to protect components of foods that are easily oxidised.

A comparison of different types of contact dryers is given in Table 16.1.

Ball dryer
In ball drying, a drying chamber is fitted with a slowly rotating screw and contains ceramic balls that are heated by hot air blown into the chamber. Particulate foods are dried mainly by conduction as a result of contact with the hot balls, and are moved through the dryer by the screw to be discharged at the base. The speed of the screw and the temperature of the heated balls control the drying time (Cohen and Yang 1995).

Drum (or roller) dryers
Drum drying involves heat transfer from condensing steam through the metal drums to a layer of product on the outside. This is heated to its boiling point; water is vaporised, the material changes from a liquid to a solid state, and finally the temperature of the product approaches that of the drum. The limiting factor in heat transfer is the thermal resistance of foods caused by their low thermal conductivity (Chapter 10, Table 10.2), which becomes lower as the food dries. Therefore a thin layer of food is needed to conduct heat rapidly without causing heat damage.

Dryers may have a single drum (Fig. 16.12a), double drums (Fig. 16.12b) or twin drums (Fig. 16.12c). In each type of dryer the slowly rotating hollow drums are heated by pressurised steam to 120–170 °C. They are constructed from precision machined, chrome-plated cast iron to provide even heat transfer. The thin layer of food slurry is spread

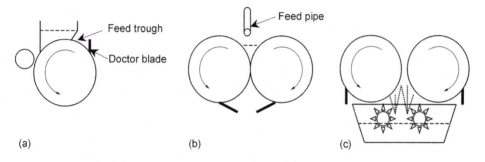

Fig. 16.12   Drum dryers: (a) single drum dryer with top roller feed; (b) double drum dryer with centre feed and bottom discharge; (c) twin drum dryer with splash feed (adapted from Anon 2000).

uniformly over the outer surface by dipping, spraying, spreading, or by auxiliary feed (or applicator) rollers. Feed rollers have advantages when applying wet materials that do not easily form a uniform layer across the drum surface (e.g. starchy materials such as potato pastes). In all types of drum dryers, the dried food is scraped off before the drum has completed one revolution (within 20 s to 3 min) by a 'doctor' blade, which contacts the drum surface uniformly across its width. After removal from the drum, the product is collected by a screw conveyor designed to break up the film or flakes of product into particles. To minimise damage to the surface of the drums, applicator rolls are mounted in spring-loaded assemblies that allow the rolls to lift if a foreign body passes between the rolls and the drum. In twin drum dryers, one drum is fixed and the other is spring loaded to give the same protection. Further information is given by Anon (2008a).

The single drum dryer (Fig. 16.13) is widely used as it has a number of advantages: it has greater flexibility for different products; a larger proportion of the drum area is available for drying; and there is easier access for maintenance. Typical applications include cereal-based breakfast foods, infant foods, pregelatinised starches, potato flakes and fruit pulps. In some designs of double-drum dryers, the drums rotate towards each other and the distance between the drums (the 'nip') determines the thickness of the dried product. Applications include dried yeast and milk products. In twin drum dryers the drums rotate away from each another, and splash feeders apply the wet feed (e.g. gelatine) to the bottom of the drum.

Fig. 16.13   GMF-Gouda single drum dryer showing applicator rolls (manufacturer: GMF-Gouda Processing Solutions at www.gmfgouda.com).

Drum dryers have high drying rates, high energy efficiencies and low labour requirements. They are suitable for slurries in which the particles are too large for spray drying, including a wide range of products that are converted to flakes or powders, which can be quickly rehydrated (e.g. potato flakes, precooked breakfast cereals, some dried soups, fruit purées and whey or distillers' solubles for animal feed formulations). The main disadvantage of drum dryers is that the area available for heat transfer limits the production rate; this is because the surface area : volume ratio decreases with increasing scale, which limits the maximum size of dryers to ≈2 m in diameter and ≈5 m in length (Anon 2007). Maximum production rates are ≈500–600 kg h^{-1}. Drum dryers also have higher capital costs than hot-air dryers owing to the cost of the precisely machined drums. Developments in drum design to improve the sensory and nutritional qualities of dried food include the use of auxiliary rolls to remove and reapply food during drying, the use of high-velocity air to increase the drying rate, or the use of chilled air to cool the product. Drums may be enclosed in a vacuum chamber to dry food at lower temperatures, or in a nitrogen atmosphere for products that are sensitive to oxidation. However, the high capital cost of these adaptations restricts their use to high-value heat-sensitive foods.

Vacuum band and vacuum shelf dryers
In vacuum band drying a food slurry is spread or sprayed onto a steel belt (or 'band') which passes over two hollow drums inside a vacuum chamber at 0.1–10 kPa. The food is dried by the first steam-heated drum and then by steam-heated coils or radiant heaters located over the band. The dried food is cooled by the second water-cooled drum and removed by a doctor blade. Vacuum shelf dryers consist of hollow shelves in a vacuum chamber. Food is placed in thin layers on flat metal trays that are carefully made to ensure good contact with the shelves. A partial vacuum is drawn in the chamber and steam or hot water is passed through the shelves to dry the food. Rapid drying and limited heat damage to the food make both methods suitable for heat-sensitive foods. However, care is necessary to prevent the dried food from burning onto trays in vacuum shelf dryers, and shrinkage reduces the contact between the food and heated surfaces in both types of equipment. They have relatively high capital and operating costs and low production rates.

In vacuum puff drying, a partial vacuum induces foaming in liquid products when dissolved gases are released. The foam is then dried on a heated belt to produce puff-dried foods (see also foam mat drying and foam spray drying using heated air). The increased rate of drying is due to the increase in surface area and the relatively rapid moisture transport through the porous foam structure, compared with the less porous structure of the dried liquid. Heat transfer is less efficient in foam but it is adequate because drying is predominately controlled by the rate of mass transfer. The method reduces damage to heat-sensitive foods such as banana, mango and tomato purées, and the porous structure of the dried foam allows rapid rehydration. Further details are given by Kudra and Ratti (2006).

Vertically agitated dryers
These types of dryers consist of a heated vessel that contains a vertical agitator. The vessel can be either a jacketed cylindrical pan that has a slow-moving paddle to scrape the sides and base, or a jacketed cone that has a vertical screw agitator. The screw rotates about its own axis and also moves around the cone wall on an orbital path. Paddles and agitators provide good mixing of the food and continuously remove food from the wall of the vessel preventing it from over-heating. They are mainly used to dry pastes and slurries, often operating under a partial vacuum (Oakley 1997).

16.3 Control of dryers

The aim of a dryer control system is to produce products that have a uniform moisture content, but this cannot be measured directly on-line and alternative indirect factors are therefore used to control the operation of the dryer. Control of hot-air dryers varies in complexity from simple thermostatic control of air temperature in batch dryers, to more complex feedback control of inlet air and outlet air temperatures and/or humidities. The outlet air temperature is measured using thermocouples or thermistors, air humidity is measured using electronic capacitance or impedance sensors (Pragnell 1993), and air flowrate or product feedrate are measured using flowmeters. A typical control arrangement is to monitor the outlet air temperature and use a programmable logic controller (PLC) (Chapter 27, section 27.2) to adjust the supply of fuel or steam pressure to the air heater. The humidity of the outlet air may also be monitored and used to control vents and dampers to adjust the amount of air that is recirculated. In spray dryers, the outlet air temperature can be used to control the product feedrate to the atomiser (Brennan 1994). Alternatively, the air flow is maintained within a preset range and the inlet air temperature is adjusted to enable the maximum feedrate to be used that produces the required moisture content in the product. In contact dryers the variables are product flowrate, speed of the dryer drum or conveyor, and steam pressure to the dryer. Control of these types of dryer is achieved by monitoring the temperature of the heated surface and using the data to control the steam pressure, keeping the thickness of the feed layer and the drum speed/residence time constant. A process computer linked to a PLC presents real time information to operators, monitors alarms, enables automatic start-up and shut down, and logs data for management and data recording procedures.

Dryers are fitted with alarms to warn operators when a variable exceeds a pre-set condition and interlocks to prevent incorrect operation. Examples of alarm variables include high air inlet or outlet temperatures, high pressure in vacuum dryers and motor failures. Examples of interlocks include stopping the dryer if access doors are opened, preventing feed entering unless the inlet air flow is satisfactory, and preventing the air heater from operating unless the fan is switched on. Further details are given by Gardiner (1997).

The complex nature of drying processes has led to developments in dryer control using neural networks and fuzzy logic control systems. Further details are given in Chapter 27 (section 27.2.3). For example, Qinghua *et al.* (2002) developed an artificial neural network to predict performance indices for rice drying, including energy consumption, final moisture content and moisture removal rate. Inputs to the neural network were four drying parameters: rice layer thickness, hot-air flowrate, hot-air temperature and drying time. A predictive model for heat and mass transfer using a neural network was developed by Hernández-Pérez *et al.* (2004) that can be used for on-line control of drying processes. Other studies of the application of neural networks to dehydration are described by Movagharnejad and Nikzad (2007), Xueqiang *et al.* (2007) and Kaminski *et al.* (1998).

Rodríguez-Jimenes *et al.* (2002) designed an empirically based simulator for a drum dryer and tested an advanced control strategy based on fuzzy logic in order to maintain a constant product moisture content. Zhang and Litchfield (1993) developed a fuzzy logic control system to regulate a continuous crossflow grain dryer. With fuzzy control, similar outlet moisture contents and lower susceptibility of grain to breakage were obtained compared with manually controlled dryer operation. Atthajariyakul and Leephakpreeda (2006) studied optimal conditions for fluidised bed rice drying using fuzzy logic control of the rice moisture content and heat load of the drying air to quantify rice quality and energy consumption. They produced real-time control of moisture content close to the target level with efficient energy consumption. The control system also enables optimum plant utilisation, control over material and energy costs, process diagnostics, and monitoring for safety and environmental/air pollution (Gardiner 1997).

16.4   Rehydration

Rehydration involves three simultaneous processes: imbibition of water into the dried material, swelling, and leaching of soluble materials. The rate and extent of rehydration depend on the extent of disruption to the cellular structure and chemical changes caused by dehydration. Water that is removed from a food during dehydration cannot be replaced in the same way when the food is rehydrated (i.e. rehydration is not the reverse of drying). The reasons include:

- changes to the structure of a food caused by loss of cellular osmotic pressure;
- changes in cell membrane permeability and solute migration;
- crystallisation of polysaccharides (see also glass transition, Chapter 1, Section 1.4.1); and
- coagulation of cellular proteins, which reduces their water-holding capacity.

Heat also reduces the degree of hydration of starch and the elasticity of cell walls. These factors all contribute to texture changes and each is irreversible (Krokida and Maroulis 2001, Rahman and Perera 1999). Most shrinkage (40–50%) occurs in the initial drying stages, and this can be minimised by low-temperature drying to minimise moisture gradients in the product, or creating a porous structure before drying commences to increase mass transfer and the drying rate. Porous products have faster rehydration, but also reduced storage stability because of the increased surface exposure to air.

Volatile losses and heat-induced chemical changes (e.g. Maillard reactions) may also mean that rehydrated food does not have the same flavour and colour compared with the raw material (Section 16.5.1). The rate and extent of rehydration may be used as an indicator of food quality; those foods that are dried under optimum conditions rehydrate more rapidly and completely than poorly dried foods. Studies of the mechanisms in rehydration and the quality of foods after rehydration are reported by for example Agunbiade et al. (2006), Marabi and Saguy (2004) and Lewicki (1998).

16.5   Effect on foods and micro-organisms

16.5.1   Sensory properties

All products undergo changes during drying and storage that reduce their quality compared with the fresh material, and the aim is to minimise these changes while maximising process efficiency. The quality attributes of dried foods are shown in Table 16.6. The main changes to the quality of dried foods are to the texture and flavour or aroma, but changes in colour and nutritional value are also significant in some foods.

Texture
Changes to the texture of solid foods are an important cause of quality deterioration during drying. As a food dries, it becomes more viscous and may pass thorough a series of rubbery and leathery states until a solid state is achieved when most of the water has

Table 16.6 Changes to quality attributes in dried foods

Attribute	Properties
Appearance	Surface morphology, colour, roughness.
Chemical/biochemical	Oxidation, causing oxidative rancidity, loss of colour, Maillard reactions, causing discoloration, change in texture. Enzymatic, e.g. polyphenoloxidase causing enzymic browning, lipoxygenase causing oxidative rancidity, lipase causing lipolytic rancidity, protease causing gelation and flavour and texture changes.
Microbiological	Numbers of surviving pathogens or spoilage micro-organisms.
Nutritional	Vitamin losses, lipid oxidation, protein denaturation.
Physical	Shrinkage, wetting and rehydration characteristics, density changes, alteration of shape and size, reduced solubility. Moisture movement, causing drying and toughening of texture, hydration and softening of texture, aggregation.
Sensory	Changes to colour, aroma, taste and flavour.
Structural	Glass transitions, cellulose crystal structure, bulk density, porosity.
Textural	Hardness, chewiness, stickiness, viscoelasticity, compressive strength, stress needed to fracture.
Thermal	Glass transition temperature, melting point, thermo-mechanical properties.

Adapted from Mujumdar (1997) and Gould (1995)

been removed. At higher viscosities, foods may become sticky and this has important implications for the design of dryers and selection of operating conditions. For example, in spray drying it is important that there is rapid initial moisture loss so that droplets have passed through the high-viscosity sticky state to prevent them from adhering to the dryer walls. The changes to viscosity can be represented by a glass transition curve on a phase diagram (see Fig. 22.3 in Chapter 22, for similar changes to solutes that take place during freezing). Below the curve, the viscosity of a dry food is sufficiently high to prevent it flowing during the timescales that are required for its shelf-life. This has important implications for the packaging and storage conditions for dried foods; any increase in moisture content or temperature during storage would raise the food above the glass transition curve and lead to stickiness and caking.

The nature and extent of pretreatments (e.g. calcium chloride added to blancher water (Chapter 11), peeling (Chapter 2, section 2.4) and the type and extent of size reduction, (Chapter 3, section 3.1)) each affect the texture of rehydrated fruits and vegetables. The loss of texture in these products is caused by the changes to structural polymeric compounds described in section 16.4 and by Khraisheh et al. (2004). Localised variations in the moisture content during drying set up internal stresses, which rupture, crack, compress and permanently distort the relatively rigid cells, to give the food a shrunken, shrivelled appearance (Ratti 1994). Prothon et al. (2003) give a detailed review of this textural collapse. On rehydration, the product absorbs water more slowly and does not regain the firm texture of the fresh material. There are substantial variations in the density, porosity, degree of shrinkage and rehydration with different foods described by Zogzas et al. (1994).

Fully dried meats and fish are commonly produced in some countries (e.g. 'biltong' snack meat in Southern Africa. See also smoked meat and fish in Chapter 17 (section 17.3). Severe changes in texture are caused by aggregation and denaturation of proteins and a loss of water-holding capacity, which leads to toughening of muscle tissue. In general, rapid drying and high temperatures cause greater changes to the texture of foods than moderate rates of drying and lower temperatures.

As water is removed during drying, solutes move from the interior of the food to the surface. The mechanism and rate of movement are specific for each solute and depend on the type of food and the drying conditions used. Evaporation of water causes concentration of solutes at the surface. High air temperatures (particularly when drying fruits, fish and meats) cause complex chemical and physical changes to solutes at the surface that produce a hard impermeable skin. This is termed 'case hardening' and it reduces the rate of drying to produce a food with a dry surface and a moist interior. Later migration of moisture to the surface during storage can then promote mould growth. It is minimised by controlling the drying conditions to prevent excessively high moisture gradients between the interior and the surface of the food.

Flavour and aroma
Lipid oxidation during storage of dried foods causes rancidity, development of off-flavours, and the loss of fat-soluble vitamins and pigments in some foods. Factors that affect the rate of oxidation include the product moisture content, types of fatty acid in the food, oxygen content (related to product porosity), storage temperature and exposure to ultraviolet light, the presence of metals or natural antioxidants, and natural lipase activity.

Heat not only vaporises water during drying but also causes loss of volatile components from the food, and as a result most dried foods have less flavour than the original material. The extent of volatile loss depends on the temperature and moisture content of the food and on the vapour pressure of the volatiles. Volatiles which have a high relative vapour pressure and diffusivity are lost at an early stage in drying. Foods that have a high economic value due to their characteristic flavours (e.g. herbs and spices) are therefore dried at lower temperatures. The rate of flavour loss during storage is determined by the storage temperature, presence of oxygen and the water activity of the food. In dried milk the oxidation of lipids produces rancid flavours owing to the formation of secondary products including δ-lactones. Most fruits and vegetables contain only small quantities of lipid, but oxidation of unsaturated fatty acids to produce hydroperoxides, which react further by polymerisation, dehydration or oxidation to produce aldehydes, ketones and organic acids, causes rancid and objectionable odours. Some foods (e.g. carrot) may develop an odour of 'violets' produced by the oxidation of carotenes to β-ionone. These changes are reduced by:

- vacuum or gas packing;
- low storage temperatures;
- exclusion of ultraviolet or visible light;
- maintenance of low moisture contents;
- addition of synthetic antioxidants (Appendix A4); or
- preservation of natural antioxidants.

The enzyme glucose oxidase (Chapter 6, section 6.5) is also used to protect dried foods from oxidation. A package that is permeable to oxygen but not to moisture, containing glucose and the enzyme, is placed on the dried food inside a container and removes oxygen from the headspace during storage. In modified atmosphere packaging (Chapter 25, section 25.3), milk powders are stored under an atmosphere of nitrogen with 10% carbon dioxide. The carbon dioxide is absorbed into the milk and creates a small partial vacuum in the headspace. Air diffuses out of the dried particles and is removed by re-gassing after 24 h.

Flavour changes caused by oxidative or hydrolytic enzymes are prevented in dried fruits by the use of sulphur dioxide, ascorbic acid or citric acid, by blanching vegetables (Chapter 11) or by pasteurisation of milk and fruit juices (Chapter 12). Other methods that are used to retain flavours in dried foods include:

- recovery of volatiles and their return to the product during drying (see also Chapter 14, section 14.1.2);
- mixing recovered volatiles with flavour fixing compounds, which are then granulated and added back to the dried product (e.g. dried meat powders); and
- addition of enzymes, or activation of naturally occurring enzymes, to produce flavours from flavour precursors in the food (e.g. onion and garlic are dried under conditions that protect the enzymes that release characteristic flavours).

Colour
There are a number of causes of colour changes or losses in dried foods. Drying changes the surface characteristics of a food and hence alters its reflectivity and colour. In fruits and vegetables, chemical changes to carotenoid and chlorophyll pigments are caused by heat and oxidation during drying, but most carotene destruction is caused by oxidation and residual enzyme activity during storage. Loss of the green colour of vegetables during the moist heating conditions of initial drying is due to chlorophyll being converted to olive-coloured pheophytin by losing some of the magnesium in the pigment molecules. Shorter drying times and lower drying temperatures, and blanching or treatment of fruits with ascorbic acid or sulphur dioxide, each reduce pigment losses (Krokida *et al.* 2001).

Enzymic reactions by phenolases (e.g. polyphenoloxidase, cresolase, catecholase, tyrosinase) cause browning of some fruits and vegetables (e.g. banana, apple and potato) on exposure to air by oxidation of phenolic compounds (hydroxybenzenes) to brown melanins. This is a particular problem when fruits are prepared for drying by peeling, slicing etc., or when the tissue is bruised during handling (Krokida *et al.* 2000). It can be inhibited by sulphites metabisulphites or bisulphites that maintain a light, natural colour during storage and also inhibit microbial growth. Only free sulphite is effective in retarding the formation of pigment materials or growth of micro-organisms, and the loss of sulphur dioxide therefore determines the practical shelf-life of the dried product. For moderately sulphured fruits and vegetables the rate of darkening during storage is inversely proportional to the residual sulphur dioxide content. However, sulphur dioxide bleaches anthocyanins, and residual sulphur dioxide is also linked to health concerns. In sulphite-sensitive people, sulphites can provoke asthma and other symptoms of an allergic response such as skin rashes and irritations. Its use in dried products is now restricted in many countries. Packaging under vacuum or nitrogen, or the use of an oxygen scavenger pouch in sealed packs (Chapter 25, section 25.4.3) also reduces browning and flavour changes. The rate of Maillard browning reactions increases at higher temperatures and longer drying times, and at higher solids concentrations (Chapter 1, section 1.2.2). The reaction also makes amino acids unavailable, reducing the nutritional value. The rate of Maillard browning in stored milk and fruit products depends on the water activity of the food and the temperature of storage.

16.5.2 Nutritional value
The loss of moisture as a result of drying increases the concentration of nutrients per unit weight in dried foods compared with their fresh equivalents. Large differences in reported data on the nutritional value of dried foods are due to variations in the composition of raw

materials, differences in preparation procedures, drying temperatures and times, and the storage conditions. In fruits and vegetables, vitamin losses caused by preparation procedures usually exceed those caused by the drying operation. These losses have been studied for many years, and for example Escher and Neukom (1970) showed that losses of vitamin C during preparation of apple flakes were 8% during slicing, 62% from blanching, 10% from puréeing and 5% from drum drying. Other studies are reported by, for example, Erenturk *et al.* (2005).

Vitamins have different solubility in water and as drying proceeds, some (e.g. riboflavin) become supersaturated and precipitate from solution, so losses are small. Others (e.g. ascorbic acid) are soluble until the moisture content of the food falls to very low levels and these react with solutes at higher rates as drying proceeds. Vitamin C is also sensitive to heat and oxidation, and short drying times, low temperatures, and low moisture and oxygen levels during storage are necessary to avoid large losses. Thiamin is also heat-sensitive and losses are ≈15% in blanched tissues, but may be up to 75% in unblanched foods. Lysine is heat-sensitive and losses in whole milk range from 3 to 10% in spray drying and 5–40% in drum drying. Other water-soluble vitamins are more stable to heat and oxidation, and losses during drying rarely exceed 5–10% (excluding preparation and blanching losses).

Oil-soluble nutrients (e.g. essential fatty acids and vitamins A, D, E and K) are mostly contained within the dry matter of the food and they are not concentrated during drying. However, water is a solvent for heavy metal catalysts that promote oxidation of unsaturated nutrients. As water is removed, the catalysts become more reactive, and the rate of oxidation accelerates (Fig. 1.11 in Chapter 1). Fat-soluble vitamins are also lost by interaction with the peroxides produced by fat oxidation. Losses during storage are reduced by lowering the oxygen concentration and the storage temperature, and by exclusion of light. Ultraviolet light (e.g. during sun or solar drying) causes a reduction in carotene and riboflavin content, as well as increasing the rate of darkening.

The biological value and digestibility of proteins in most foods do not change substantially as a result of drying. However, milk proteins are partially denatured during drum drying and this results in a reduction in solubility of the milk powder and loss of clotting ability.

16.5.3   Effect on micro-organisms

Depending on the time–temperature combination used to dry foods, there may be some destruction of contaminating micro-organisms, but the process is not per se lethal, and yeasts, moulds, bacterial spores and many Gram-negative and Gram-positive bacteria can survive in dried foods. Hence most vegetables are blanched (Chapter 11), liquid foods may be pasteurised (Chapter 12) or concentrated by evaporation (Chapter 14) and meat and fish may be treated with salt before drying to inactivate pathogenic bacteria. Dried foods are characterised as having a water activity (a_w) below 0.6, which inhibits microbial growth, provided that the packaging and storage conditions prevent moisture pick-up by the product. If this occurs, xerophilic mould growth at a_w 0.77–0.85 is the most likely form of spoilage, especially by *Eurotium* spp., *Aspergillus* spp., *Penicillium* spp. and *Xeromyces* spp., and also some osmophilic yeasts such as *Saccharomyces rouxii.*. Additionally, some types of *Aspergillus* spp. (*A. parasiticus*, *A. nomius* and *A. niger*) produce a range of mycotoxins, which produce acute symptoms that can be fatal and are potent liver carcinogens (Brown 2006).

Most dried foods are cooked before consumption, which reduces the numbers of surviving micro-organisms, but foods such as dried fruits, nuts, herbs and spices may be consumed uncooked. Pepper, paprika, desiccated coconut and cinnamon in particular have been recognised as posing a particular risk for *Salmonella* contamination. Particular care is needed to ensure that these foods are subject to high standards of hygiene and food handling during both preparation for drying and during post-drying treatments. They are also treated with chemical fumigants, modified atmospheres (Navarro *et al.* 1998) and Chapter 21, section 21.2.5), by irradiation (Chapter 7) or by a combination of methods (Wahid *et al.* 1989). These treatments also destroy contaminating insects. (Note: methyl bromide has been discontinued as a chemical fumigant since 2005 because it is an ozone-depleting gas (Anon 2008b).)

References

AGUNBIADE, S.O., OLANLOKUN, J.O. and OLAOFE, O.A., (2006), Quality of chips produced from rehydrated dehydrated plantain and banana, *Pakistan J. Nutrition*, **5** (5), 471–473.

ANON, (2000), APV dryer handbook, APV Ltd, available at www.che.utexas.edu/cache/trc/dryer.pdf.

ANON, (2004), Solar heat for industrial processes, International Energy Agency Task 33 / Task IV: SHIP, Newsletter No. 1 – December, available at www.iea-ship.org/documents/papersofnewsletterNo1.pdf.

ANON, (2005), The Torbed, company information from Process Research ORTECH Inc, available at www.processortech.com/torbed/default.asp?id=1.

ANON, (2007), Drum dryers – construction and operation, company information provided by R. Simon (Dryers) Ltd., available at www.simon-dryers.co.uk/drum/drum2.htm.

ANON, (2008a), Drum dryer/flaker, company information from JLS International, available at www.jls-europe.de/ and follow the link to 'Drum Dryers/Drum Flakers.

ANON, (2008b), The phase-out of methyl bromide, US Environmental Protection Agency, available at www.epa.gov/ozone/mbr/.

ATTHAJARIYAKUL, S. and LEEPHAKPREEDA, T., (2006), Fluidized bed paddy drying in optimal conditions via adaptive fuzzy logic control, *J. Food Engineering*, **75** (1), 104–114.

AXTELL, B.L.A. and BUSH, A., (1991), *Try Drying It! – Case Studies in the Dissemination of Tray Drying Technology*, Intermediate Technology Publications, London.

AXTELL, B.L.A. and RUSSELL, A., (2000), *Small Scale Drying*, Intermediate Technology Publications, London.

BAHU, R.E., (1997), Fluidised bed dryers. In (G.L. Baker, Ed.), *Industrial Drying of Foods*, Blackie Academic and Professional, London, pp. 65–89.

BARBOSA-CANOVAS, G.V., (1996), *Dehydration of Foods*, Chapman and Hall, New York.

BARBOSA-CANOVAS, G.V., ORTEGA-RIVAS, E., JULIANO, P. and YAN, H., (2005), *Food Powders – Physical Properties, Processing, and Functionality*, Springer Verlag, New York.

BARR, D.J. and BAKER, C.G.J., (1997), Specialized drying systems, in (C.G.J. Baker, Ed.), *Industrial Drying of Foods*, Blackie Academic & Professional, London, pp. 179–209.

BRENNAN, J.G., (1992), Developments in drying, in (A. Turner, Ed.), *Food Technology International Europe*, Sterling Publications International, London, pp. 77–80.

BRENNAN, J.G., (1994), *Food Dehydration – A Dictionary and Guide*, Butterworth Heinemann, Oxford.

BRENNAN, J.G., BUTTERS, J.R., COWELL, N.D. and LILLEY, A.E.V., (1990), *Food Engineering Operations*, 3rd edn, Elsevier Applied Science, London, pp. 371–416.

BRENNAN, J.G., VARNALIS, A. and MACDOUGALL, D.B., (2001), Optimisation of high temperature puffing of potato cubes by response surface methodology, paper presented at IFT Annual Meeting, New Orleans, LA.

BRENNDORFER, B., KENNEDY, L., OSWIN-BATEMAN, C.O. and TRIM, D.S., (1985), *Solar Dryers*, Commonwealth Science Council, Commonwealth Secretariat, London.

BROWN, D., (2006), Mycotoxins, Cornell University poisonous plants informational database, available at www.ansci.cornell.edu/plants/toxicagents/mycotoxin.html and www.ansci.cornell.edu/plants/toxicagents/aflatoxin/aflatoxin.

COHEN, J.S. and YANG, T.C.S., (1995), Progress in food dehydration, *Trends in Food Science and Technology*, **6** (January), 20–25.

DE LA FUENTE-BLANCO, S., RIERA-FRANCO DE SARABIA, E., ACOSTA-APARICIO, V.M., BLANCO-BLANCO, A. and GALLEGO-JUÁREZ, J.A. (2006), Food drying process by power ultrasound, *Ultrasonics*, **44**, Supplement 1, e523–e527.

DEIS, R.C., (1997), Spray-drying – innovative use of an old process, *Food Product Design*, May, available at www.foodproductdesign.com/archive and search for 'spray drying'.

DRISCOLL, R.H., (1995), Energy efficiency in dryers and ovens, *Food Australia*, **47** (7), 310–314.

ERENTURK, S., GULABOGLU, M.S. and GULTEKIN, S., (2005), The effects of cutting and drying medium on the vitamin C content of rosehip during drying, *J. Food Engineering*, **68** (4), 513–518.

ESCHER, F. and NEUKOM, H., (1970), Studies on drum drying apple flakes, *Travaux de Chimie Alimentaire et d'Hygiene*, **61**, 339–348.

FITZPATRICK, J.J., BARRINGEER, S.A. and IQBAL, T., (2004), Flow property measurement of food powders and sensitivity of Jenike's hopper design methodology to the measured values, *J. Food Engineering*, **61**, 399–405.

GALLEGO, J.A., (1998), Some applications of air-borne ultrasound to food processing, in (M.J.W. Povey and T.J. Mason, Eds.), *Ultrasonics in Food Processing*, Thomson Science, London, pp. 127–143.

GALLEGO, J.A., RODRÍGUEZ, G., GÁLVEZ, J.C. and YANG, T.S., (1999), A new high-intensity ultrasonic technology for food dehydration, *Drying Technology*, **17** (3), 597–608.

GARCÍA-PÉREZ, J.V., CÁRCEL, J.A., DE LA FUENTE-BLANCO, S. and RIERA-FRANCO DE SARABIA, E., (2006), Ultrasonic drying of foodstuff in a fluidized bed: parametric study, *Ultrasonics*, **44**, Supplement 1, e539–e543.

GARDINER, S.P., (1997), Dryer operation and control, in (G.L. Baker, Ed.), *Industrial Drying of Foods*, Blackie Academic and Professional, London, pp. 272–298.

GOULD, G.W., (1995), Biodeterioration of foods and an overview of preservation in food and dairy industries, *International Biodeterioration and Biodegradation*, **36** (3), 267–277.

GREENSMITH, M., (1998), *Practical Dehydration*, 2nd edn, Woodhead Publishing, Cambridge.

HELDMAN, D.R. and HARTEL, R.W., (1997), *Principles of Food Processing*, Chapman and Hall, New York, pp. 177–218.

HERNÁNDEZ-PÉREZ, J.A., GARCÍA-ALVARADO, M.A. TRYSTRAM, G. and HEYD, B., (2004), Neural networks for the heat and mass transfer prediction during drying of cassava and mango, *Innovative Food Science and Emerging Technologies*, **5** (1), March, 57–64.

HUI, Y.H., CLARY, C., FARID, M.M., FASINA, O.O., NOOMHORM, A. and WELTI-CHANES, J., (Eds.), (2007), *Food Drying Science and Technology – Microbiology, Chemistry, Applications*, DEStech Publications, Inc., Lancaster, PA.

IMRIE, L., (1997), Solar dryers, in (G.L. Baker, Ed.), *Industrial Drying of Foods*, Blackie Academic and Professional, London, pp. 210–241.

JAMBRAK, A.R., MASON, T.J., PANIWNYK, L. and LELAS, V., (2007), Accelerated drying of button mushrooms, Brussels sprouts and cauliflower by applying power ultrasound and its rehydration properties, *J. Food Engineering*, **81** (1), 88–97.

JAYARAMAN, K.S., (1995), Critical review on intermediate moisture fruits and vegetables, in (J. Welti-Chanes and G. Barbosa-Canovas, Eds.), *Food Preservation by Moisture Control – Fundamentals and Applications*, Technomic Publishing Co., Lancaster, PA, pp. 411–442.

KAMINSKI, W., STRUMILLO, P. and TOMCZAK, E., (1998), Neurocomputing approaches to modelling of drying process dynamics, *Drying Technology*, **16** (6), 967–992.

KHRAISHEH, M.A.M., MCMINN, W.A.M. and MAGEE, T.R.A., (2004), Quality and structural changes in starchy foods during microwave and convective drying, *Food Research International*, **37** (5),

497–503.

KROKIDA, M.K. and MAROULIS, Z.B., (2001), Structural properties of dehydrated products during rehydration, *International J. Food Science and Technology*, **36**, 529–538.

KROKIDA, M.K., KIRANOUDIS, C.T., MAROULIS, Z.B. and MARINOS-KOURIS, D., (2000), Effect of pretreatment on colour of dehydrated products, *Drying Technology*, **18** (6), 1239–1250.

KROKIDA, M.K., MAROULIS, Z.B. and SARAVACOS, G.D., (2001), The effect of the method of drying on the colour of dehydrated products, *International J. Food Science and Technology*, **36**, 53–59.

KUDRA, T. and RATTI, C., (2006), Foam-mat drying: energy and cost analysis, *Canadian Biosystems Engineering*, **48**, 3.27–3.32, available at http://engrwww.usask.ca/oldsite/societies/csae/protectedpapers/c0621.pdf

LEWICKI, P.P., (1998), Some remarks on rehydration of dried foods, *J. Food Engineering*, **36**, 81–87.

LEWIS, M.J., (1996), Solids separation processes, in (A.S. Grandison and M.J. Lewis, Eds.), *Separation Processes in the Food and Biotechnology Industries*, Woodhead Publishing, Cambridge, pp. 243–286.

MARABI, A. and SAGUY, I.S., (2004), Effect of porosity on rehydration of dry food particulates, *J. Science Food and Agriculture*, **84** (10), 1105–1110.

MASTERS, K., (1991) *Spray Drying Handbook*, 5th edn, Longman Scientific and Technical, Harlow, Essex.

MASTERS, K., (1997), Spray dryers, in (G.L. Baker, Ed.), *Industrial Drying of Foods*, Blackie Academic and Professional, London, pp. 90–114.

MOREIRA, R.G., (2001), Impingement drying of foods using hot air and superheated steam, *J. Food Engineering*, **49**, 291–295.

MOVAGHARNEJAD, K. and NIKZAD, M., (2007), Use of artificial neural network models for prediction of drying rates of agricultural products, Special Symposium – Innovations in Food Technology, European Congress of Chemical Engineering – 6, 16–21 Sept., Copenhagen. Abstract available at http://ecce6.kt.dtu.dk/cm/content/abstract/2467/.

MUJUMDAR, A.S., (1997), Drying fundamentals, in (G.L. Baker, Ed.), *Industrial Drying of Foods*, Blackie Academic and Professional, London, pp. 7–30.

MULET, A., CÁRCEL, J. A., SANJUÁN, N. and BON, J., (2003), New food drying technologies – use of ultrasound, *Food Science and Technology International*, **9** (3), 215–221.

NAVARRO, S. DONAHAYE, E., RINDNER, M. and AZRIELI, A., (1998), Disinfestation of Nitidulid beetles from dried fruits by modified atmospheres, 1998 Annual International Research Conference on Methyl Bromide Alternatives and Emissions Reductions, available at www.epa.gov/ozone//mbr/airc/1998/068navarro.pdf.

OAKLEY, D., (1997), Contact dryers, in (G.L. Baker, Ed.), *Industrial Drying of Foods*, Blackie Academic and Professional, London, pp. 115–133.

PAKOWSKI, Z., BARTCZAK, Z., STRUMILLO, C. and STENSTROM, S., (1991), Evaluation of equations approximating thermodynamic and transport properties of water, steam and air for use in CAD of drying processes, *Drying Technology*, **9** (3), 753–773.

PRAGNELL, R., (1993), Relying on humidity, *Process Engineering*, Feb., 41–42.

PROTHON, F., AHRNÉ, L. and SJÖHOLM, I., (2003), Mechanisms and prevention of plant tissue collapse during dehydration: a critical review, *Critical Reviews in Food Science and Nutrition*, **43** (4), 447–479.

QINGHUA, Z., YANG, S.X., MITTAL, G.S. and SHUJUAN Y., (2002), Prediction of performance indices and optimal parameters of rough rice drying using neural networks, *Biosystems Engineering*, **83** (3), 281–290.

RAHMAN, M.S. and PERERA, C.O., (1999), Drying and food preservation, in (M.S. Rahman, Ed.), *Handbook of Food Preservation*, Marcel Dekker, New York, pp. 173–216.

RATTI, C., (1994), Shrinkage during drying of foodstuffs, *J. Food Engineering*, 23, 91–105.

RODRÍGUEZ-JIMENES, G.C., SALGADO-CERVANTES, M.A., GARCÍA-ALVARADO, M.A. and RAYO-GARCÍA, V., (2002), Simulation and control of a drum dryer using a hybrid model and a regulator of fuzzy logic, Annual Meeting Institute Food Technologists – Anaheim, California, abstract available at http://ift.confex.com/ift/2002/techprogram/paper_11904.htm.

SAPAKIE, S.F. and RENSHAW, T.A., (1984), Economics of drying and concentration of foods, in (B.M. McKenna, Ed.), *Engineering and Food*, Vol. 2, Elsevier Applied Science, London, pp. 927–938.

SARKAR, A., NITIN, N., KARWE, M. V. and SINGH, R.P., (2004), Fluid flow and heat transfer in air jet impingement in food processing, *J. Food Science*, **69** (4), CRH113–CRH122.

SINGH, R.P. and HELDMAN, D.R., (2001a), Psychrometrics, in *Introduction to Food Engineering*, 3rd edn, Academic Press, San Diego, CA, pp. 473–495.

SINGH, R.P. and HELDMAN, D.R., (2001b), Mass transfer, in *Introduction to Food Engineering*, 3rd edn, Academic Press, San Diego, CA, pp. 497–528.

SINGH, R.P. and HELDMAN, D.R., (2001c), Dehydration, in *Introduction to Food Engineering*, 3rd edn, Academic Press, San Diego, CA, pp. 557–590.

SOKHANSANJ, S., (1997), Through-flow dryers for agricultural crops, in (G.L. Baker, Ed.), *Industrial Drying of Foods*, Blackie Academic and Professional, London, pp. 31–64.

TOLEDO, R.T., (1999), *Fundamentals of Food Process Engineering*, 2nd edn, Aspen Publications, Gaithersburg, MD, pp. 456–506.

TORREGGIANI, D., (1993), Osmotic dehydration in fruit and vegetable processing, *Food Research International*, **26**, 59–68.

TURNER, I.W. and MUJUMDAR, A.S., (Eds.), (1996), *Mathematical Modelling and Numerical Techniques in Drying*, Marcel Dekker, New York.

UMESH HEBBAR, H., VISHWANATHAN, K.H. and RAMESH, M.N., (2004), Development of combined infrared and hot air dryer for vegetables, *J. Food Engineering*, **65** (4), 557–563.

WAHID, M., SATTAR, A., JAN, M. and KHAN, I., (1989), Effect of combination methods on insect disinfestation and quality of dry fruits, *J. Food Processing and Preservation*, **13** (1), 79–85.

XUEQIANG, L., XIAOGUANG, C., WENFU, W. and GUILAN, P., (2007), A neural network for predicting moisture content of grain drying process using genetic algorithm, *Food Control*, **18** (8), 928–933.

ZEMELMAN, V.B. and KETTUNEN, D.M., (1992), Agglomeration and agglomerator systems, in (Y.H. Hui, Ed.), *Encyclopaedia of Food Science and Technology*, Vol. 1, John Wiley & Sons, Chichester, pp. 11–17.

ZHANG, Q. and LITCHFIELD. B., (1993), Fuzzy logic control for a continuous crossflow grain dryer, *J. Food Process Engineering*, **16** (1), 59–77.

ZOGZAS, N.P., MAROULIS, Z.B. and MARINOS-KOURIS, D., (1994), Densities, shrinkage and porosity of some vegetables during air drying, *Drying Technology*, **12** (7), 1653–1666.

17

Smoking

Abstract: Smoking is used to preserve protein-rich foods such as meat, fish and cheese by the combined action of heat, which destroys micro-organisms and enzymes, and reduces moisture content, and antimicrobial and antioxidant chemicals in the smoke. The chapter describes the constituents of smoke, liquid smoke and the methods of smoking foods. It concludes with the effects of smoking on the sensory characteristics of smoked foods and their microbial safety, and a discussion of health concerns over potentially carcinogenic chemicals in smoked foods.

Key words: smoking, liquid smoke, constituents in smoke, smoking kilns, polycyclic aromatic hydrocarbons, nitrosamines.

Smoking is an ancient process that was used to preserve protein-rich foods for storage at ambient temperatures in times of shortage during winter seasons in temperate climates and during dry seasons in tropical climates. Now, the purpose is to change the flavour and colour of foods, rather than preservation. Smoked foods are preserved by chilling (Chapter 21), and may be packed in modified atmospheres or vacuum packed (Chapter 25, section 25.3) to give the required shelf-life. Smoking is an inexpensive operation that increases the variety of products for consumers, and for processors it adds value to foods. The most commonly smoked foods are fish (e.g. tilapia, mackerel, trout, sable (or black cod), sturgeon and tuna), meats and meat products (e.g. duck, venison, game birds, and pâtés made from these meats, pork, pastrami (pickled, spiced and smoked beef brisket) and beef jerky) and cheeses such as smoked Gouda. Other smoked foods include vegetables such as chipotles (smoked jalapeño peppers), seafoods, nuts, lapsang souchong tea, barley malt used in the manufacture of some types of whisky and ingredients used to make German smoked beer (rauchbier).

There are four types of smoking operations:

1 Cold smoking, in which the food is flavoured and coloured but not cooked. It is typically used for salmon, salamis, kippers, hams and special cheeses.
2 Warm smoking at 25–40 °C is used for bacon, sirloin and some types of sausage.
3 Hot smoking of meats and fish at 60–80 °C, which cooks the food and the heat is sufficient to destroy contaminating micro-organisms. Herring, eel and some sausages are hot-smoked.

4 Dissolving smoke compounds in water to make smoke concentrate or 'liquid smoke' and spraying or coating foods (see also coating, Chapter 24).

Each type of smoking is a surface treatment and smoke chemicals penetrate only a few millimetres into the product.

17.1 Theory

In cold smoking, foods are cured at an air temperature <33 °C for between 6–24 hours and several weeks to produce the required smoked flavour and colour. The texture remains largely unchanged and the products have a milder taste than hot-smoked foods. Micro-organisms are not destroyed and for this reason, cold-smoking is preceded by salt curing. Warm smoking has similar effects and foods are also cured. The preservative action of hot-smoking at 60–80 °C results from a number of factors (see also section 17.4 and hurdle technology (Chapter 1, section 1.3.1)):

- dehydration/reduced moisture content to lower the water activity of the product;
- antioxidant action of some constituent chemicals in the smoke at the surface of foods (e.g. butyl gallate and butylated hydroxyanisole (BHA)) (Brul et al. 2000);
- destruction of micro-organisms and enzymes by heat;
- antimicrobial action of some constituent chemicals in the smoke at the surface of foods (e.g. phenolic compounds, organic acids);
- antimicrobial action of salt pretreatments where these are used.

Fish is the main type of food that is hot-smoked. Details of the process are given by Doe (1998). Goulas and Kontominas (2005) studied the effect of salting and smoking method on the keeping quality of chub mackerel. They found that during the 30 day storage period, salting had a noticeable preservative effect but it was lower than the combined effect of salting and smoking. The heat and humidity cook the products and produce the required smoky taste, golden brown colour, a silky sheen on the skin and uniform weight loss. The products do not require further cooking before consumption.

In hot-smoking, the critical factors to control bacteria are (Fletcher et al. 2003):

- the core temperatures reached by the product during processing;
- control over temperature variations in different parts of the kiln;
- numbers of contaminating bacteria;
- the D-values of the bacteria of interest (see Chapter 10, section 10.3); and
- the amount of smoking.

Although smoke and salt have antimicrobial effects (section 17.4), microbial destruction during hot-smoking is based on the heat treatment received. For example, for control of *Listeria monocytogenes* in smoked salmon, the process should heat the product to a given temperature for a long enough period to ensure that the possibility of any *Listeria* surviving is less than one in a million (10^6). If the highest likely contamination by *L. monocytogenes* is 10^6 g^{-1} of raw fish, the required reduction is from 10^6 to 10^{-6} (12 D-values). For salmon, this requires 12 minutes at 64 °C or 35.1 seconds at 72 °C (Table 17.1). However, if contamination by *L. monocytogenes* is less than 1000 per fish for good quality salmon, then the required reduction is from 10^3 to 10^{-6} (a reduction of 10^9 or a 9D reduction). The D-value for salmon at 64 °C is 58.44 s and the holding time at 64 °C to ensure a *Listeria*-free product is therefore 9×58.44 s = 8.77 minutes. At a core temperature of 72 °C, the time required is 9×2.92 seconds = 26.28

Table 17.1 *D*-values for *L. monocytogenes* and minimum processing times for hot-smoking at different temperatures to ensure 12*D* reductions in selected products

Temperature (°C)	Cod			Salmon			Smoked mussels		
	D-value		Processing time	D-value		Processing time	D-value		Processing time
	min	s		min	s		min	s	
58	4.29		52 min	9.21		1 h 51 min	16.22		3 h 14 min
60	1.97		24 min	4.36		53 min	5.49		1 h 5 min
62		54.02	11 min	2.06		25 min	1.86		22 min
64		24.75	5 min 57 s		58.44	12 min		37.72	7 min 33 s
66		11.34	3 min 16 s		27.64	6 min 32 s		12.76	2 min 33 s
68		5.19	2 min 2 s		13.07	3 min 37 s		4.32	51.8 s
70		2.38	28.6 s		6.18	2 min 14 s		1.46	17.5 s
72		1.09	13.1 s		2.92	35.1 s		0.49	5.9 s
74		0.5	6 s		1.38	16.6 s		0.17	2 s
76		0.23	2.7 s		0.65	7.8 s		0.06	0.7 s
78		0.1	1.3 s		0.31	3.7 s		0.02	0.2 s
80		0.05	0.6 s		0.15	1.8 s		<0.01	0.1 s
82		0.02	0.3 s		0.07	0.8 s		<0.01	<0.1 s
84		0.01	0.1 s		0.03	0.4 s		<0.01	<0.1 s

Adapted from Fletcher *et al.* (2003)

seconds. Further information is given by Bremer and Osborne (1995) and Ben Embarek and Huss (1993).

17.1.1   Constituents in smoke

Mostly, smoke is air and other gases and vapours with a mixture of small hydrocarbon particles of different sizes. Some particles are deposited on the surface of the food, but this is of minor importance for the smoking process. More importantly, the absorption of gases by foods gives the characteristic colour changes and flavour.

Some softwoods, especially pine and fir, contain resins that produce harsh-tasting retene and other components when burned, and these woods are therefore not used for smoking, except for lapsang souchong tea leaves that are smoked and dried over pine or cedar fires. For other foods, hardwoods are considered to produce superior flavours and colours in smoked foods. Hardwood shavings or logs (e.g. oak, beech, chestnut, hickory), dampened with wet sawdust, are burned to produce heat and dense smoke. Sometimes aromatic woods, such as apple, juniper or cherry, or aromatic herbs and spices are also used to produce distinctive flavours.

The important chemical components of smoke are as follows (Anon 1992):

- nitrogen oxides;
- polycyclic aromatic hydrocarbons (PAHs);
- phenolic compounds;
- furans;
- carbonylic compounds;
- aliphatic carboxylic acids;
- tar compounds.

For components that have a high boiling point (e.g. PAHs and phenolic compounds) there is a correlation between their concentration in the smoke and that in the smoked food, whereas more volatile components are not usually found in the food.

When burned, the cellulose and hemicellulose in hardwoods produce sweet, flowery and fruity aromas. Products of pyrolysis of lignin include spicy, pungent phenolic compounds such as guaiacol, responsible for the smoky taste, phenol and syringol, a contributor to a smoky aroma. It also produces sweeter aromas including vanilla-scented vanillin and clove-like isoeugenol (Guillén *et al.* 2006, Hui 2001). Wood also contains small quantities of proteins that contribute roasted flavours. Lesimple *et al.* (1995) identified 62 volatile components in smoked duck fillets, including phenols, alcohols, ethers, aldehydes, ketones, hydrocarbons, acids, esters and terpenes, 34 of which were related to the smoking process. Others (e.g. Anon 1992) report >400 volatile chemical compounds identified in wood smoke. Different species of tree have different amounts of these components and hence their woods impart different flavours to food. The chemical composition and hence the flavours in smoke also depend on the temperature of the fire, the moisture content of the wood, the supply of air to the fire and any water added during burning. High-temperature fires break down flavour molecules into unpleasant tasting compounds. The optimal conditions for producing desirable smoke flavours are lower-temperature, smouldering fires at 300–400 °C. Woods that contain high amounts of lignin burn hotter and a restricted air supply is needed to keep them smouldering, or their moisture content is increased by soaking the pieces in water.

The factors that influence absorption of smoke by food include the density of the smoke and its humidity and temperature. The higher the smoke density, the greater the absorption. In warm- and hot-smoking, the food surface is dried and condensation of the smoke particles is less than on products that are smoked at lower temperatures. If the relative humidity (Chapter 16, section 16.1.1) of the smoke is high, vapour condenses on the food surface and absorption of water-soluble components of the smoke increases. However, if the surface remains moist, colour formation is inhibited as this increases at low moisture contents. If the surface is too dry, there is less penetration of the smoke into the food and hence loss of flavour and preservative action.

A number of wood smoke compounds act as preservatives. Phenol and other phenolic compounds in wood smoke are both antioxidants and antimicrobials. Other antimicrobial chemicals include formaldehyde, acetic acid and other carboxylic acids (section 17.4). These chemicals are also extracted and used as liquid smoke (section 17.1.2). Smoke also contains compounds that have long-term health consequences including PAHs and dioxin-like polychlorinated biphenyls (PCBs) many of which are known or suspected carcinogens (section 17.3). Nitrogen oxides in smoke may also react with amines or amides in the food to form nitrosamines, or with phenols to produce nitro or C-nitroso phenols. Carbonylic compounds and acids react with proteins and carbohydrates in the food. Nitrogen oxides and PAHs are normally only found in small amounts in smoke, but they are of concern because of the potential public health risk of nitrosamines and PAHs (Anon 1992).

'Tasteless smoke' was patented in 1999 as a method of preserving fresh fish. The process involves producing hardwood smoke by burning the chips, and then passing it through filters that remove all particles larger than 1 μm and most of the odour and colour components. The remaining gases (nitrogen, carbon monoxide (CO), carbon dioxide and methane, together with trace amounts of phenolic compounds and hydrocarbons) are applied to tuna fish or other red meats. The product is removed from the smoke chamber, washed in ozonated water to remove any residual smoke odour and kill bacteria, and it is then frozen. The treatment retains the appearance, taste, texture and colour of the fresh seafood after freezing and defrosting (Walsh 2005). The low concentrations of CO alter the colour, when it reacts with meat pigments to produce a cherry-red carboxymyoglobin pigment, but only for a limited time, and eventually the colour diminishes as the fish ages. The process is approved by the EU as a smoking process because, as a component of wood smoke (a GRAS (generally regarded as safe) substance), CO is a legal treatment and is declared on the product label. In contrast, the treatment of fish with industrial carbon monoxide produces colour changes that do not fade. It was banned in the EU in 2003 because it could mislead consumers over the freshness of meat by maintaining a bright red colour. Although the cherry-red colour is different from the oxymyoglobin pigment in fresh meat, it may mislead customers that the product is fresher than it actually is, or enhance the appearance of inferior products. CO treatment can also mask colour changes caused by decomposition, mask other visual evidence of spoilage, and mask potential safety problems such as production of histamines (section 17.4). In Singapore the use of CO is considered a malpractice and Japan bans fish that have an initial CO content $\geq 500\,\mu g\,kg^{-1}$. Currently (2008) in the USA, the Food and Drug Administration (FDA) regards the use of tasteless smoke on tuna as a preservative and it thus needs to be labelled accordingly (Anon 2007a), although this position is challenged.

17.1.2 Liquid smoke

Smoke flavourings or 'liquid smoke' are prepared by condensing smoke derived from burning wood, usually followed by fractionation, purification or concentration. The fractionation steps produce products that have the desired olfactory properties and also reduce the concentration of undesirable by-products of the smoke. For example, as a marker for PAHs (section 17.3) the benzo(a)pyrene content of smoke concentrates is reduced to a maximum limit of $1 \mu g \, kg^{-1}$ of condensate. Much of the tar formed during pyrolysis can also be removed. A typical commercial smoke condensate contains $\approx 70\%$ water, 29% volatile organic compounds and 1% tar (Anon 1992).

Smoke condensates may be used for flavouring food but are more commonly used as the basis for smoke flavouring preparations that contain carrier materials such as salt or dextrose. Their main advantages are that they can be used as a step in continuous processing and reduced production times, compared with batch smoking in kilns. The smoke flavour can be mixed directly into the food, applied as a powder (Chapter 24, section 24.2.2) or sprayed as a fine mist. The use of smoke flavourings also avoids product weight losses that take place during smoking, which affects the product yield and quality. Although the flavour from liquid smoke is almost the same as that produced by traditional smoking, the product does not necessarily acquire the same texture or colour. Smoke flavourings are therefore often combined with a smoking process to get the desired colour from the smoke and the taste from the smoke flavouring.

17.2   Processing

Foods are cured with salt before warm- and cold-smoking and sometimes before hot-smoking. Cold-smoked meats may be cured or precooked before smoking. The most common type of cure in industrialised countries is a wet cure using brine with added sugar and spices. Curing times vary from a few hours to two weeks. Dry curing, in which coarse salt is rubbed directly into fish or meat, is less common because of higher labour costs and less uniform salting, but it is still widely practised in tropical artisan processing. These heavily salted, hot-smoked products can be stored without refrigeration for several weeks, but require boiling in freshwater to make them palatable before consumption. After curing, fish are drained, rinsed and refrigerated for 6–12 h, and then smoked at low temperature (25–33 °C) in a kiln or smokehouse for between half a day to 3 weeks. After removal from the kiln and chilling for 24 h, the food may be sliced or packed as whole pieces.

The following process, described by Bannerman (1984) using fresh chilled or thawed frozen fish, is an example of the manufacture of a hot-smoked product. Frozen fish is preferred for smoking because freezing for at least 30 days kills parasites in the fish. Fish are brined using an 80 brine (211 g salt l^{-1}) to give them flavour and to inhibit the growth of food poisoning micro-organisms without making the product unpleasantly salty. Fish absorb salt more uniformly in weaker brines, but the immersion time is longer, whereas stronger brines can cause salt to crystallise on the surface of the skin after the fish are dried and smoked, creating unsightly white patches. Immersion times in the brine vary according to the size, thickness and fat content of the fish and the aim is to achieve a salt concentration in the finished product of >3%. Yanar et al. (2006) studied the effect of brine concentration on the shelf-life of hot-smoked tilapia and concluded that 5% brine was optimal for a shelf-life of 35 days.

After brining, fish are hung on trolleys in a smoking kiln (section 17.2.1) so that the backs face the flow of smoke. Fillets of fish or meat and small products such as shellfish are arranged on wire mesh trays on the trolleys. The process operates in three stages: first drying at 30 °C for 30–60 min to toughen the skin, then smoking and partial cooking at

50 °C for 30–45 min, and finally cooking at 80 °C for a few minutes up to 30 min for large fish. The total processing time and the times spent at each stage depend on the type of food, its size and fat content, and the type of product required.

In the traditional method for making Arbroath smokies (from haddock) the fish are first salted overnight and then tied in pairs and left overnight to dry. The dried fish are hung in a barrel containing a hardwood fire and sealed with a lid to create a very hot and humid smoky fire without flames. Within an hour of smoking, the intense heat and thick smoke produce the strong smoky taste and aroma that characterises this product and the fish are then ready to eat. In other processes, smoking takes up to 8 h at temperatures of 60–85 °C, with the internal temperature of the fish at 60 °C for at least 30 min to kill pathogenic bacteria.

Cold- and hot-smoked fish are cooled to chill temperatures ($\approx$3 °C) in a chill room before packing. They are commonly packaged into modified atmosphere or vacuum packs (Chapter 25, section 25.3). Muratore and Licciardello (2005) studied the effects of different packaging methods on shelf-life. Products are maintained at chill temperatures during storage, distribution and retail display. Cold-smoked white fish have a longer shelf-life than fatty fish, although it varies greatly with the type of fish, the amount of salting, the extent of smoking and drying, and the storage temperature. Civera et al. (1995) conducted chemical and microbiological analyses of smoked Atlantic and Canadian salmon and found that the effective shelf-life was 40–50 days and 80 days respectively if the fish is stored at 2–3 °C. Smoked products can also be frozen and stored at −30 °C for at least 6 months, or longer when vacuum packed and frozen. Vacuum packaging excludes oxygen and thus slows the development of rancidity, especially in fatty fish products. However, vacuum packing creates an anaerobic environment inside the pack that is suitable for the growth of C. botulinum (section 17.4). Packaging materials are now available that have adequate oxygen transfer to prevent the development of an anaerobic environment within a package (Chapter 25, section 25.4.3). Containers or packages made of materials with an oxygen permeability of 2000 cm^3 m^{-2} per 24 h at 24 °C or higher are permitted for use with products stored at 4 °C or lower for up to 14 days (Anon 1991). Styrofoam trays with a single film over-wrap are permitted if the total permeability of the final package meets the minimum specifications, but these packs should not be stacked in a way that would reduce the oxygen permeability.

Irradiation (Chapter 7) has also been studied as a method to increase the shelf-life of smoked fish. For example, Hammad and El-Mongy (1992) irradiated cold-smoked salmon at 2 and 4 kGy, stored at 2–3 °C. Unirradiated samples reached the maximum accepted mesophilic plate count after one month of storage, while those irradiated reached this level after three and four months at 2 and 4 kGy respectively. No differences in sensory qualities were found between unirradiated samples and those irradiated at 2 kGy, but there was a loss of red colour in samples irradiated at 4 kGy.

17.2.1 Equipment

Smoking equipment should allow the controlled development of flavour and colour in foods, with low levels of carcinogenic or toxic components in the smoke, and low levels of environmental pollution by the smoke. Smoke can either be generated in the kiln or produced in a separate smoke generator. Sawdust and wood should be clean and free from wood preservatives or saw lubricants. Sawdust consumption is $\approx$13 kg h^{-1} in a kiln of 375 kg capacity (Bannerman 1984). Separate smoke generators have advantages in that

Fig. 17.1   Smoking kiln (courtesy of AFOS Ltd at www.intlsmokingsystems.com).

the temperature and humidity in the kiln can be independently controlled so that foods can be dried or cooked before being smoked. They also give better control over the temperature, humidity and density of the smoke. Additionally, the smoke can be filtered, treated with water sprays or by electrostatic precipitation to remove carcinogenic compounds such as benzo(a)pyrene (Ranken 2000) and unwanted particles. However, these treatments also remove some of the components that contribute to the flavour of smoked foods.

Smoking kilns are similar in design to cabinet dryers (Chapter 16, section 16.2.1) or batch ovens (Chapter 18, section 18.2.1) and, in hot-smoking, foods may be dried, cooked and smoked in the same equipment. Kiln heaters are designed to quickly reach and maintain an operating temperature of $\approx$80 °C when fully loaded. Trolleys that have rails or mesh trays are wheeled into the smoking chamber (Fig. 17.1) and the food is smoked in an automatically controlled cycle. Computer control includes management of the smoke temperature, humidity and density via touch-screen or remote controls, a fire protection system and alarms, and an automatic cleaning cycle. The capacity of commercial kilns is 250–2000 kg (Anon 2007b).

17.3   Effect on foods

The main purpose of smoking is to alter the sensory properties of foods, particularly the flavour and colour. Chemicals in smoke that contribute to the flavour and aroma of smoked foods are described in section 17.1.1. Smoking produces a shiny yellow colour, which darkens as smoking time is increased. Earlier studies showed that the colour is produced by interaction of amino groups on proteins in the food with carbonyls in the smoke in a similar way to the Maillard reaction (Ruiter 1979, Gilbert and Knowles 1975). Iliadis *et al.* (2004) studied the effects of pretreatment and hot- and cold-smoking on the chemical, microbiological and sensory quality of mackerel. They found that available lysine was reduced to the same extent (32%) in all hot smoked samples, and that loss of available lysine correlated with colour formation in the cold-smoked products. Nitrogen oxides in smoke can also react with myoglobin to produce a modified colour in smoked meat products. Riha and Wendorff (1993) studied the acceptability of the colour of smoked cheese.

The surface texture of cold-smoked foods is changed by the formation of a firm pellicle, produced by coagulation of proteins by acidic components of the smoke, but the interior of the food is unchanged. In hot-smoked foods, the texture becomes that of cooked foods owing to heat coagulation of proteins.

Salting causes liquid exudates from the flesh of meat and fish, causing losses of water-soluble proteins, vitamins and minerals. Some proteins are also denatured by the salt. Some constituent chemicals in the smoke at the surface of foods (e.g. butyl gallate and BHA) have an antioxidant action (Brul *et al.* 2000). These antioxidant components of smoke reduce oxidative changes to fats, proteins and vitamins. However, hot-smoking also causes nutrient losses due to heat and interaction of the smoke components with proteins. The heat and flow of gases in the smoke cause dehydration of the food and changes to the nutritional and sensory properties described in Chapter 16 (section 16.5) including denaturation of proteins. The loss of moisture also increases the concentration of protein and fat in the food and an increased concentration of salt and other curing agents.

Smoked foods give rise to some health concerns, especially with respect to PAHs and nitrosamines. The potential harmful effects of these smoke components are described by Anon (1992). The PAH component benzo(a)pyrene and 10 other compounds from this group are both mutagenic and carcinogenic (Lawley 1990). High levels of PAHs are found on the surface of products that are smoked for a long time at higher temperatures, but Iliadis *et al.* (2004) found high levels of benzo(a)pyrene, fluoranthene and perylene both in cold- and hot-smoked fish. Witczak and Ciereszko (2006) studied changes in the content of PCBs in mackerel slices during cold and hot smoking. The hot-smoked mackerel showed a decrease in the PCB content, which may have been due to losses with lipid leakage from the product and co-distillation with water vapour. Cold smoking produced an increase in PCB content in the final product compared to the initial raw material.

However, their levels may be reduced by using fire temperatures below 400 °C for smoke generation, by treating smoke (section 17.1.2) and by reduced smoking times. Some products (e.g. frankfurter sausages) are washed after smoking, which also reduces the concentration of these carcinogens. The average intake of PAH components has been calculated to be 1.2 mg per year, but smoked meat and fish contribute only 10% of this, the remainder coming from environmental pollution and tobacco smoking (Anon 1992). The intake of PAHs from food is therefore regarded as of minor public health importance, but some smoked foods can have unacceptably high levels of PAHs due to the smoking process and these are therefore a health concern. In the EU upper limits for smoked food are 5 g of volatile N-nitroso compounds per kg of food and 1 g of benzo(a)pyrene per kg of food, with the exception for smoked fish of 5 g of benzo(a)pyrene per kg. The maximum limit for benzo(a)pyrene in liquid smoke condensates is $10 \, g \, kg^{-1}$, to produce less than $0.03 \, g \, kg^{-1}$ in the food as consumed and for benzo(a)anthracene, the maximum is $20 \, g \, kg^{-1}$, giving less than $0.06 \, g \, kg^{-1}$ in the food (Anon 1992).

Smoked foods may contain N-nitroso compounds, such as N-nitrosodimethylamine. These nitrosamines are among the most carcinogenic substances that have been studied. They are formed by reactions between nitrogen oxides in smoke and amines or amides in the foods, especially in those that have higher concentrations of amines, such as fish and meat. These compounds increase the risk of gastrointestinal cancer in populations where there is a high intake of heavily smoked (and/or salted) foods. Phenolic compounds are important for the taste of smoked food, but where nitrite is used (e.g. in cured smoked meats), phenols may react with the nitrite to form nitro- and nitrosophenols, some of which have been shown to be mutagenic. Some of the phenols may also catalyse the formation of nitrosamines in foods during smoking. Tar compounds in smoke are not well characterised and their effects on health are therefore difficult to evaluate.

17.4   Effect on micro-organisms

The combined effects of salt, antimicrobial chemicals in smoke, heat and partial dehydration during hot-smoking are effective against Gram-negative rods, micrococci and staphylococci. Vegetative bacteria are more susceptible to smoke than bacterial spores and moulds. This means that spoilage by moulds is more likely than bacterial spoilage. However, there is a significant risk of pathogenic bacterial contamination, especially of cold-smoked fish products, and standards for hygienic production and handling are described in the legislation of many countries (e.g. Anon 1979). Sikorski and Kałodziejska (2002) reviewed the incidence of contamination by pathogens on hot-smoked fish. They found low numbers of *Listeria monocytogenes*, *Clostridium botulinum*, *Staphylococcus aureus* and *Vibrio parahaemolyticus* and concluded that the main causes of contamination were unsanitary procedures and airborne micro-organisms during packing of the product. The internal temperature in the fish, which did not exceed 65 °C, and the low concentration of salt were insufficient to inactivate all pathogens or inhibit bacterial growth during storage. Product safety required very fresh fish, handled under hygienic conditions, chilling the product to 2 °C and hygienic handling of the product after smoking. Details of safe handling of chilled foods are given in Chapter 21 (section 21.5).

Listeria monocytogenes poses a health risk for immunocompromised individuals and pregnant women (Chapter 1, section 1.2.4). It is not destroyed during cold-smoking and it can become established in the processing environment and re-contaminate products. It is therefore not possible to produce cold-smoked fish that is consistently free of *L. monocytogenes* (Anon 2001). However, the use of good manufacturing practices (GMP) and good hygienic practices (GHP) (Chapter 1, section 1.5.1) enables production of cold-smoked fish with low levels of *L. monocytogenes* (<1 cell g^{-1}). Growth can also be prevented by freezing, by addition of preservatives (e.g. sodium nitrite) or by use of bioprotective bacterial cultures (Chapter 6, section 6.7). It can also be controlled by limiting the shelf-life (at 4.4 °C) to ensure that not more than 100 cells g^{-1} are present when the food is consumed.

Clostridium botulinum occurs naturally in the aquatic environment and can be present in low numbers on fresh fish. Cases of botulism caused by type E toxin from smoked fish are reported by Fletcher *et al.* (2003) in the anaerobic conditions found in vacuum packs. Korkeala *et al.* (1998) report cases of botulism after eating hot-smoked whitefish, processed from frozen fish. The fish contained botulinum toxin and *Cl. botulinum* was isolated from the fish. They identified safety problems associated with vacuum packed hot-smoked fish and described the product as one of the highest risk industrial foods to cause botulism. They recommended temperature monitoring and the use of time–temperature indicators (Chapter 21, section 21.2.4) to ensure adequately low storage temperatures throughout the processing chain and the use of sodium nitrate and nitrite with a sufficiently high salt concentration to prevent *Cl. botulinum* growth. In the USA, the recommended smoking conditions required to kill *Cl. botulinum* type E are that products achieve a core temperature of 62.5 °C for 30 min and for products stored in air, the fish should contain >2.5% salt in the muscle. Vacuum packaged or modified atmosphere packaged products should contain either >3.5% salt or 3.0% salt plus 100–200 μg g^{-1} of sodium nitrite with storage at chill temperatures <4.4 °C (Anon 1995). Sikorski and Kałodziejska (2002) reviewed studies of other preservative chemicals, including sodium lactate, lactic and sorbic acids, sodium propionate, nisin and lysozyme.

The flesh of some fish species (e.g. tuna, mackerel and mullet) contains high levels of the amino acid histidine, which is converted to the biogenic amine, histamine, by bacteria (e.g. Enterobactericae) growing under suitable conditions on the fish. Histamine causes the symptoms of scombroid poisoning, which are similar to an allergic reaction and include facial swelling, itching of the skin, headache, nausea and vomiting (Fletcher *et al.* 2003). Histamine is not destroyed by subsequent cooking. Iliadis *et al.* (2004) found unacceptable levels of histamine (600 mg kg^{-1}) in unprocessed samples of fish, which increased to levels that would be expected to cause symptoms of scombrotoxin poisoning in both cold- and hot-smoked products (2220 and 2250 mg kg^{-1} respectively). Freezing, salting or smoking may inhibit or inactivate histamine-producing micro-organisms, but growth may take place after thawing before smoking, and post-smoking. Vacuum packaging does not prevent their growth. Handling and processing the fish under sanitary conditions, rapid cooling and continuous refrigeration until consumption each prevent the growth of these bacteria, and hence prevent biogenic amine formation.

Other illnesses that result from consumption of improperly processed smoked foods are caused by *Clostridium perfringens*, *Staphylococcus aureus* and *Salmonella* spp. None of these micro-organisms occurs naturally on raw fish and the sources of infection are food handlers who carry these micro-organisms or shellfish harvested from unsanitary water (see Chapter 1, section 1.2.4). The most important methods to control these bacteria are:

- temperature control during hot-smoking to kill any contaminating micro-organisms;
- holding the product at a sufficiently low temperature before and after smoking to prevent their growth;
- good sanitation and hygienic work practices.

Fletcher *et al.* (2003) describe methods used for the safe production of smoked fish and seafoods using GMP and HACCP (see also Chapter 1, section 1.5.1). Internationally recognised controls require that:

- the core temperature of the fish should be brought to 10 °C or less within 6 h of death and to 4 °C within 24 h;
- chilled fish should not be exposed to temperatures above 4 °C for more than 4 h cumulatively after the initial chilling.
- chilled fish should not be stored for more than 14 days at 0 °C or more than 7 days at 4 °C before smoking.
- frozen fish (stored for 24 weeks or longer) should not be exposed to temperatures above 4 °C for more than 12 h, cumulatively after the initial chilling period, and it should not be exposed to temperatures above 4 °C for more than 6 h of uninterrupted storage (Anon 2001).

References

ANON, (1979), Codex Standards for Smoked Fish, available at www.codexalimentarius.net/download/standards/123/CXP_025e.pdf.

ANON, (1991), Smoked fish: storage conditions, Canadian Food Inspection Agency, available at www.inspection.gc.ca/english/fssa/labeti/retdet/bulletins/smofume.shtml.

ANON, (1992), Health aspects of using smoke flavours as food ingredients, Health protection of consumers, Council of Europe, Publishing and Documentation Service, Strasbourg, available at www.coe.int/t/e/social_cohesion/soc-sp/public_health/flavouring_substances/SMOKE.pdf.

ANON, (1995), Advisory group approves food safety guide for smoked fish, *Food Chemical News*, **37**, 13–14.

ANON, (2001), Processing parameters needed to control pathogens in cold smoked fish – conclusions and research needs, US Food and Drug Administration Center for Food Safety and Applied Nutrition, available at www.cfsan.fda.gov/~comm/ift2list.html.

ANON, (2007a), Selected countries prohibiting carbon monoxide (CO) gas in fresh meat and fresh fish packaging, Carbon Monoxide In Fresh Meat, available at www.co-meat.com/countries.html.

ANON, (2007b), company information from AFOS Ltd, available at www.intlsmokingsystems.com/AFOS.pdf.

BANNERMAN, A.M., (1984), Hot smoking of fish, *Torry Advisory Notes*, **82**, HMSO, available at www.fao.org/wairdocs/tan/x5953e/x5953e00.htm

BEN EMBAREK, P.K. and HUSS, H.H., (1993), Heat resistance of *Listeria monocytogenes* in vacuum packaged pasteurized fish fillets, *International J. Food Microbiology*, **20**, 85–95.

BREMER, P. and OSBORNE, C., (1995), Thermal-death times of *Listeria monocytogenes* in green shell mussels *(Perna canaliculus)* prepared for hot smoking, *J. Food Protection*, **58**, 604–608.

BRUL, S., KLIS, F.M., KNORR, D., ABEE, T. and NOTERMANS, S., (2000), Food preservation and the development of microbial resistance, in (P. Zeuthen and L Bøgh-Sorensen, Eds.), *Food Preservation Techniques*, Woodhead Publishing, Cambridge, pp. 524–543.

CIVERA, T., PARISI, E., AMERIO, G.P. and GIACCONE, V., (1995), Shelf-life of vacuum-packed smoked salmon: microbiological and chemical changes during storage, *Archiv für Lebensmittel-hygiene*, **46** (1), 13–17.

DOE, P.E., (1998), *Fish Drying and Smoking*, Woodhead Publishing, Cambridge.

FLETCHER, G.C., BREMER, P.J., SUMMERS, G. and OSBORNE, C., (2003), Guidelines for the safe preparation of hot-smoked seafood in New Zealand, New Zealand Institute for Crop and Food Research Ltd, available at www.crop.cri.nz/home/research/marine/pathogens/hot-smoked.pdf.

GILBERT, J. and KNOWLES, M.E., (1975), The chemistry of smoked foods – a review, *J. Food Technology*, **10**, 245–261.

GOULAS, A.E. and KONTOMINAS, M.G., (2005), Effect of salting and smoking-method on the keeping quality of chub mackerel (*Scomber japonicus*): biochemical and sensory attributes, *Food Chemistry*, **93** (3), 511–520.

GUILLÉN, M.D., ERRECALDE, M.C., SALMERÓN, J. and CASAS, C., (2006), Headspace volatile components of smoked swordfish (*Xiphias gladius*) and cod (*Gadus morhua*) detected by means of solid phase microextraction and gas chromatography–mass spectrometry, *Food Chemistry*, **94** (1), 151–156.

HAMMAD, A.A.I. and EL-MONGY, T.M., (1992), Shelf-life extension and improvement of the microbiological quality of smoked salmon by irradiation, *J. Food Processing and Preservation*, **16** (5), 361–370.

HUI, Y.H., (2001), *Meat Science and Applications*, Marcel Dekker, New York.

ILIADIS, K.N., ZOTOS, A., TAYLOR, A.K.D. and PETRIDIS, D., (2004), Effect of pre-treatment and smoking process (cold and hot) on chemical, microbiological and sensory quality of mackerel (*Scomber scombrus*), *J. Science Food and Agriculture*, **84** (12), 1545–1552.

KORKEALA, H., STENGEL, G., HYYTIÄ, E., VOGELSANG, B., BOHL, A., WIHLMAN, H., PAKKALA, P. and HIELM, S., (1998), Type E botulism associated with vacuum-packaged hot-smoked whitefish, *International J. Food Microbiology*, **43** (1–2), 1–5.

LAWLEY, P.D., (1990), *N*-Nitroso compounds, in (C.S. Cooper, and P.L. Grover, Eds.), *Handbook of*

Experimental Pharmacology – Chemical Carcinogenesis and Mutagenesis, Springer-Verlag, Berlin, pp. 410–469.

LESIMPLE, S., TORRES, L., MITJAVILA, S., FERNANDEZ, Y. and DURAND, L., (1995), Volatile compounds in processed duck fillet, *J. Food Science*, **60** (3), 615–618.

MURATORE, G. and LICCIARDELLO, F., (2005), Effect of vacuum and modified atmosphere packaging on the shelf-life of liquid-smoked swordfish (*Xiphias gladius*) slices, *J. Food Science*, **70** (5), C359–C363.

RANKEN, M.D., (2000), *Handbook of Meat Product Technology*, Blackwell Science, Oxford, p. 151.

RIHA, W.E. and WENDORFF, W.L., (1993), Evaluation of color in smoked cheese by sensory and objective methods, *J. Dairy Science*, **76**, 1491–1497.

RUITER, A., (1979), Color of smoked foods, *Food Technology*, **33**, 54–63.

SIKORSKI, Z.E. and KAŁODZIEJSKA, I., (2002), Microbial risks in mild hot smoking of fish, *Critical Reviews in Food Science and Nutrition*, **42** (1), 35–51.

WALSH, E., (2005), What is Clearsmoke, *Food and Beverage International*, Spring issue, available at www.fbworld.com/Mag_Spring_2005/advertorial/anova%20-%20clearsmoke/AnovaClearsmoke.html.

WITCZAK, A. and CIERESZKO, W., (2006), Effect of smoking process on changes in the content of selected non-*ortho*- and mono-*ortho*-PCB congeners in mackerel slices, *J. Agriculture and Food Chemistry*, **54** (15), 5664–5671.

YANAR, Y., ÇELIK, M. and AKAMCA, E., (2006), Effects of brine concentration on shelf-life of hot-smoked tilapia (*Oreochromis niloticus*) stored at 4 °C, *Food Chemistry*, **97** (2), 244–247.

18

Baking and roasting

Abstract: Baking and roasting both use hot air to alter the eating quality of foods. The chapter begins with a description of heat and mass transfer during baking followed by a description of batch and continuous baking equipment. The final part looks at the effects of baking on micro-organisms and changes to sensory characteristics and nutritional value of baked foods.

Key words: baking, roasting, continuous ovens, heat and mass transfer.

Baking, like dehydration and smoking, is an ancient process that remains an important industry worldwide. Baking and roasting are essentially the same unit operation in that they both use heated air to alter the eating quality of foods. The terminology differs in common usage; baking is usually applied to flour-based foods or fruits, and roasting to meats, cocoa, and coffee beans, nuts and vegetables. In this chapter the term baking is used to include both operations. A secondary purpose of baking is to preserve foods using heat to destroy micro-organisms and remove moisture to reduce the water activity (a_w) at the surface of the food (also Chapter 1, section 1.1.2). A very large number of baked products are produced commercially, which can be grouped according to their moisture content, a_w and pH (Table 18.1). High moisture content, acidic products have pH values within the range 4.2–4.4, low-acid products between pH >4.6 and <7.0 (the majority of baked products), and a few alkaline products (pH >7.0) made with dough treated with sodium bicarbonate or calcium hydroxide (e.g. crumpets and tortillas). The shelf-life of baked products depends on the composition of the food (acidity, a_w, any chemical preservatives and most importantly the moisture content), its storage temperature and humidity and the type of packaging (see also Hurdle technology, Chapter 1, section 1.3.1).

18.1 Theory

As in dehydration, baking involves the simultaneous transfer of heat into food and removal of moisture by evaporation from the food to the surrounding air (Chapter 16, section 16.1.1). The main difference is the temperature of the heated air, which is higher

Table 18.1   Categories of baked products

Product	Water activity	Examples of products
Low moisture content	0.2–0.3	Crackers, biscuits, wafers, nuts, baked potato crisps
Intermediate moisture content	0.5–0.8	Pastries, cakes, soft cookies, chapattis

High moisture content	0.9–0.99	Alkaline:
		  Crumpets, tortillas
		Low-acid:
		  Breads, rolls, muffins, cheesecake, pizzas, meat pies, sausage rolls, pasties, filled cakes, quiches, baked potatoes, roasted meats
		Acidic:
		  Fruit tarts and pies
		  Sourdough bread

Adapted from Smith *et al.* (2004)

in baking (110–300 °C) than in most dehydration processes. Also in contrast with dehydration, where the aim is to remove as much water as possible with minimal changes to sensory quality, in baking the heat-induced changes at the surface of the food and retention of moisture in the interior of some products (e.g. cakes, breads, meats) are desirable quality characteristics. In other products, such as biscuits and crispbread, loss of moisture from the interior produces the required crisp texture. Heat is therefore used to destroy micro-organisms, to evaporate water, to form a crust, to superheat water vapour (steam) that is transported through the crust, and to superheat the dry crust.

In a hot-air oven, heat is supplied to the food by a combination of:

- infrared radiation (Chapter 20, section 20.3) from the heaters and oven walls that is absorbed into the surface of the food and converted to heat, which is then conducted through the food;
- convection from circulating hot air, other gases and moisture vapour in the oven – the heat is converted to conductive heat at the surface of the food; and
- conduction through the pan or tray on which the food is placed.

Infrared and microwave heating are described in Chapter 20.

The factors that control the rates of heat transfer and equations for the calculation of heat transfer are described in Chapter 10 (section 10.1.2) and sample problem 10.1 (section 10.1.2) is relevant. Mass transfer due to evaporation is described in Chapter 14 (section 14.1.1). The important factors that control heat and mass transfer in hot-air baking are described by Marcotte (2007). These include (1) the baking conditions (particularly the temperature difference between the source of heat and the food, and the velocity of the air in the oven) and (2) the type of food and size of the food pieces.

Baking conditions

A boundary film of air acts as a resistance to heat transfer into the food and to movement of water vapour from the food. The thickness of the boundary layer is determined mostly by the velocity of the air and the surface properties of the food and this partly controls the rates of heat and mass transfer (see also Chapter 16, section 16.1.1). Some designs of hot-air ovens have fans to reduce the thickness of boundary films and increase the rates of heat and mass transfer. Impingement ovens (section 18.2) virtually eliminate boundary films around baking foods. Sakar and Singh (2004) describe heat transfer and fluid flow in impingement ovens.

Type of food

Depending on the type of product that is required, heat must penetrate to the centre of food pieces to adequately cook the food and/or to evaporate moisture to dry the food. Heat passes through most baked foods by conduction (an exception is the convection currents that are established during the initial heating of cake batters (Mizukoshi 1990). The low thermal conductivity of foods (Table 18.2 and Chapter 10, Table 10.2) causes low rates of conductive heat transfer and this is an important influence on baking time.

When a food is placed in an oven, the surface temperature rises to the wet bulb temperature (Chapter 16, section 16.1.1). The heat causes moisture at the surface of the food to evaporate and the low humidity of the hot air creates a moisture vapour pressure gradient, and this in turn creates movement of moisture from the interior of the food to the surface. Moisture movement may be by capillary flow or by vapour diffusion along channels in the food. When the rate of moisture loss from the surface exceeds the rate of movement from the interior, the zone of evaporation moves inside the food, the surface dries out, its temperature rises to the temperature of the hot air and a crust is formed. These changes both enhance eating qualities and retain moisture in the bulk of the food. For example, when baking meats and breads, high oven temperatures are used to produce the required surface crust and a moist interior. As the crust dries, the thermal conductivity falls and further slows the rate of heat penetration (Table 18.2). These changes are similar to those in hot-air drying (Chapter 16, section 16.1.1), but the more rapid heating and higher temperatures in baking cause complex changes to the food components at the surface (section 18.3). Because baking takes place at atmospheric pressure and moisture escapes freely from the food, the internal temperature of the food does not exceed 100 °C.

During storage, moisture slowly migrates from the moist interior to the dry crust. This migration softens the crust, lowers the eating quality and limits the shelf-life of the food (see also glass transition, Chapter 1, section 1.4.1). For products that are required to have uniformly low moisture content throughout (e.g. biscuits and crackers), the temperature

Table 18.2   Thermo-physical properties of doughs and breads

Product	Moisture content (% wet basis)	Density $(kg\,m^{-3})$	Thermal conductivity $(W\,m^{-1}\,K^{-1})$	Thermal diffusivity $(m^2\,s^{-1} \times 10^{-8})$
Bread dough (-16 °C)	46.1	1100	0.980	43.5
Bread dough (19 °C)	46.1	1100	0.500	16.3
Crust	0	417	0.055	7.85
Crumb	44.4	450	0.28	22.2
Bread (8 min) baking	–	307.3	0.72	–
Bread (16 min) baking	–	284.6	0.67	–
Bread (24 min) baking	–	275.1	0.66	–
Bread (32 min) baking	–	263.6	0.64	–
Cake batter	41.5	693.5	0.223	10.9
Cake (centre, $\frac{1}{4}$ baked)	40	815	0.228	8.6
Cake (centre, $\frac{1}{2}$ baked)	39	290	0.195	16.1
Cake (centre, $\frac{3}{4}$ baked)	37.5	265	0.135	16.9
Cake (centre, fully baked)	35.5	300	0.121	14.3
Cake (edge, fully baked)	34	285	0.119	15.0

Adapted from Baik *et al.* (2001)

Table 18.3   Preservation of baked foods

Expected shelf-life	Examples of products	Method of preservation
Short shelf-life (days)	Cereal products	
	Bread	Moisture barrier packaging
	Cakes, pastries	Oxygen and moisture barrier packaging
	Cream filled cakes	Chilling
	Meat pies, pasties, quiches	Chilling
	Meats	
	Sliced ham, beef, chicken	Chilling, modified atmosphere packaging, oxygen scavenging packs
Medium shelf-life (weeks)	Cereal products	
	Bread	Modified atmosphere packaging, oxygen scavenging packs, chemical preservatives
	Cakes, crumpets	Active packaging (oxygen scavenging, ethanol), chemical preservatives (e.g. calcium propionate)
	Meats	
	Meat joints	Chilling, freezing, vacuum packaging
Long shelf-life (months)	Cereal products	
	Biscuits, crackers, snackfoods	Oxygen and moisture barrier packaging
	Cakes	Canning
	Pizzas	Freezing
	Other products	
	Nuts	Oxygen and moisture barrier packaging

of the hot air and hence the rate of heat transfer are reduced to enable moisture to evaporate from the food without forming a surface crust. This is promoted by having uniformly thin pieces of dough. The size of the food pieces is therefore an important factor in baking time as it determines the distance that heat must travel to the centre of the food, and moisture to travel to the surface.

Spoilage of baked products is due to physical changes (moisture loss or gain, staling), chemical changes (rancidity) and microbial growth (section 18.3). Some baked products, such as crackers, crispbread and roasted nuts, have a very low a_w (0.1–0.3) and these have a shelf-life of several months when stored in packs that have high barriers to moisture and oxygen (Chapter 1, section 1.3.7). For other baked products that require an extended shelf-life, a number of different preservation measures are used (Table 18.3).

18.2   Equipment

The technology of bread making is described in detail by Cauvain (2003) and Cauvain and Young (1998, 2001, 2006). The technology of biscuit making is described by Manley (1998, 2001), and the technology of cake making by Bennion and Bamford (1997). Details of the different stages in baking are given by Owens (2001) and food machinery for the production of cereal foods is described by Cheng (1992). A study by Beech (2006) of energy consumption in bread production from receipt of flour at the bakery to delivery of bread to retail outlets showed that the total energy use averaged 6.99 MJ kg^{-1} bread, which increased to 14.8 MJ kg^{-1} bread when wheat growing, flour milling and retailing were included. The author calculated the energy subsidy (primary energy input : food energy output) for the system to be 1.49, which was a factor of five lower than that for other products (e.g. mashed potato, roast beef and reheated canned corn). He concluded that bread is the most energy-efficient staple produced by industrialised food production.

Equipment used to prepare foods prior to baking is described in Chapter 4 (mixing and forming) and post-baking equipment to handle and pack bakery products is described in Chapters 26 and 27 respectively. Microwave and radio frequency ovens are described in Chapter 20 (section 20.1.2) and fuel fired and electric ovens are described in this chapter.

Ovens are chambers or tunnels that are constructed using inner and outer metal walls lined with either firebricks, mineral wool, refractory tiles or other insulating materials to reduce heat losses. The base (or hearth) is constructed from thick steel, ceramic tiles or stone to promote even heat distribution. Ovens can be grouped into direct- or indirect-heating types (also Chapter 10, section 10.2). In directly heated ovens, air and the products of combustion are recirculated over the food by natural convection or by fans. The temperature in the oven is controlled automatically by adjustment of air and fuel flow rates to the burners (section 18.2.3). Liquid petroleum gas (propane or butane) or natural gas are commonly used, but fuel oil or less commonly in industrialised countries, solid fuels are also found. Gas is burned in ribbon burners located above and below conveyor belts in continuous ovens, and at the base of the cabinet in batch ovens. Pressure-relief panels are fitted to the top of ovens to protect personnel should a gas explosion occur (e.g. Anon 2006).

The advantages of direct heating ovens include:

- rapid start-up, as it is only necessary to heat the air in the oven;
- short baking times;
- high thermal efficiencies; and
- good control over baking temperature.

However, care is necessary to prevent contamination of the food by undesirable products of combustion, such as nitrogen oxides (see Chapter 16, section 16.2.1) and gas burners require regular servicing to maintain combustion efficiency.

Electric ovens are heated by induction heating radiator plates or bars. In batch ovens, the walls and base are heated, whereas in continuous ovens, heaters are located above, alongside and/or below a conveyor belt. These radiant ovens have longer start-up times and a slower response to temperature control than direct-heating ovens.

In indirect-heating ovens, the products of combustion do not come into contact with foods. Radiator tubes containing either steam from a remote boiler or combustion gases from burning fuel are used to heat air in the baking chamber. Hot air is commonly recirculated through a heat exchanger. Alternatively, in older designs fuel was burned between a double wall and the combustion products were exhausted from the top of the oven.

18.2.1   Batch and semi-continuous ovens

Batch ovens have inherent disadvantages of higher labour costs and less uniform baking conditions than continuous ovens. There are a number of different designs of batch oven that have features intended to address these disadvantages, including automatic and semi-automatic oven loaders (Fig. 18.1), fans to promote uniform air distribution and different designs that move food through the oven as it bakes.

Among the simplest designs is the deck oven, in which products are placed on shelving inside a heated cabinet. Fans increase the air flow and in some designs, they reverse the direction of flow every few minutes to bake products more evenly and

Fig. 18.1   Pan oven loader (courtesy of Gemini Bakery Equipment at www.geminibe.com).

quickly. This also increases productivity because baking pans do not have to be turned during the baking period. Ovens are fitted with steam injection systems to direct precisely timed blasts of steam mist over the surface of baked products to gelatinise starch and produce a glazed crust on products such as crusty breads, bagels and French baguettes. Deck ovens are highly flexible for different products including breads, rolls, buns, pastries, muffins, cakes, pies, croissants and pizzas. Similar designs are used for cooking hamburgers, bacon and poultry and may be fitted with smoke generators for smoking meats, cheeses and fish (Chapter 17, section 17.2.1). The multi-deck oven (Fig. 18.2) is widely used for baked goods, meats and flour confectionery products. The 'modular' construction allows individual ovens to be simultaneously used for different products, thus increasing the flexibility of operation, and additional modules can be added to

Fig. 18.2   Multi-deck oven (courtesy of Werner and Pfleiderer Ltd at www.wpib.de).

expand production without having to replace the entire plant. As a result, these ovens are popular for small- and medium-sized bakeries. The rack oven is similar in design, with products loaded onto mobile racks that are pushed into the oven, where they may be either rotated or kept stationary during baking.

Rotary-hearth ovens, reel ovens and multi-cycle tray ovens each move foods through the oven on trays, and loading and unloading take place through the same door. Rotary hearth ovens have a similar design to deck ovens, but the racks of food rotate around a vertical spindle in the oven. Reel ovens have hinged trays fitted between slowly rotating wheels. As the wheels turn, the trays of food move vertically through the oven and also horizontally from front to back. Multi-cycle tray ovens move the food through the oven on trays attached to a conveyor chain. The operation of each type is semi-continuous because the oven must be stopped to remove the food. The movement of food through the oven, with or without fans to circulate the air, ensures more uniform heating and permits a larger baking area for a given floor space.

18.2.2 Continuous ovens

Tunnel ovens (Fig. 18.3) consist of a metal tunnel (up to 120 m long and 4 m wide) through which food is conveyed either on steel plates (in a travelling-hearth type oven) or on a solid, perforated or woven metal belt in the band type oven. The oven is divided into heating zones and the temperature and humidity are independently controlled in each zone using heaters and dampers. These retain or remove moisture by adjusting the proportions of fresh and recirculated air in the oven. Vapour and hot air (and in direct-heating ovens, the products of combustion) are extracted separately from each zone and passed through heat recovery systems to remove heat from the exhaust gases and to heat fresh or recirculated air. This gives energy savings of 30% and start-up times can be reduced by 60%. Despite their high capital cost and large floor area, these ovens are widely used for large-scale baking (e.g. 3000 loaves per hour). The main advantages are their high capacity, accuracy of control over baking conditions and low labour costs owing to automatic loading and unloading. Tray ovens have a similar design to tunnel ovens but have metal trays permanently fixed to a chain conveyor. Each tray holds several baking pans and is pulled through the oven in one direction, then lowered onto a second conveyor, returned through the oven and unloaded.

Fig. 18.3   Tunnel oven (courtesy of Werner and Pfleiderer at www.wpib.de).

Impingement ovens have nozzles that direct high-velocity, vertical jets of hot air that impinge the product from top and bottom as it passes through the oven on a conveyor. Each zone in the oven has independently controlled air velocity and temperature. Ovens can be directly or indirectly heated using hot air or steam, depending on the requirements of the product for surface colour or texture development and the required degree of baking (e.g. pizza bases, biscuits, snackfoods and meats). The advantages of impingement heating include high heat transfer rates (5–25 times greater than natural convection), which give advantages in higher speed of baking, product throughput and reduced oven size, and also the ability to bake at lower temperatures, saving fuel costs (Stitley 1995).

18.2.3 Control of ovens

Nearly all oven designs incorporate advanced energy-saving features and microprocessor controls (Chapter 27, section 27.2). Microprocessors control automatic safe shutdown procedures to extinguish burners if abnormal baking conditions arise, and there are interlocks to prevent the oven from being opened during operation. Microprocessors also provide management information of production rates, energy use/efficiency and main-tenance requirements. In tunnel ovens, a programmable logic controller (PLC) controls the belt speed, individual burners, variable speed fans and the position of dampers to automatically adjust the temperature and humidity of baking in each heating zone and produce foods of a predetermined colour or moisture content. Details of automatic colour monitoring of baked products are given in Chapter 2 (section 2.3.3). The PLC also allows operators to select preprogrammed baking cycles for up to 100 individual products using a single touch screen input without the need to manually adjust the oven settings. This enables rapid changeover between products and prevents the use of incorrect baking conditions due to operator error. Additionally, screen touch buttons allow operators to view details of the status of each oven section or pre-programmed operating parameters (Anon 2008). Ovens at different sites can be programmed identically to achieve product uniformity from all factories. Changes to control settings or the introduction of new products can be made from a central computer by direct modem connections to the ovens (also Chapter 27, section 27.2.4).

18.3 Effect on foods and micro-organisms

The purpose of baking is to alter the sensory characteristics of foods, to improve palatability and produce a range of products having different tastes, aromas and textures from similar raw materials. The heat during baking also destroys enzymes and con-taminating micro-organisms and hence extends the shelf-life of products. In this section the effects of baking on sensory and nutritional properties are described, together with likely microbial contamination of baked foods and shelf-life.

18.3.1 Changes to sensory qualities

Texture
Changes in texture are determined by the temperature and duration of heating and the nature of the food; in particular the moisture content and the composition of fats, proteins and the structural carbohydrates: cellulose, starches and pectins (Chapter 1, section 1.1.1). The chemical changes during cereal dough fermentation and baking are described

in detail by Prejean (2007), Sluimer (2005) and Hansen and Schieberle (2005). Gelatinisation of starch begins when the food reaches ≈65 °C in the oven. Starch granules swell by uptake of water, loose their crystalline structure and are transformed into a starch gel. The strength of the gel is determined by the amylose component of the starch, and recrystallisation (retrogradation) of amylose results in the formation of a rigid 3-D gel network which helps stabilise the crumb structure. Pentosans (non-starch polysaccharides, mainly xylans) in wheat flour also contribute to crumb firming. Gelatinisation and dehydration produce the characteristic texture of the impermeable crust, which seals in moisture and fat and protects nutrients and flavour components from degradation.

The use of enzymes in bakery products is reviewed by Hegenbart (1994) (also Chapter 6, section 6.3.1). For example, bacterial amylases are added to bread dough to modify the texture of the crumb. A blend of fungal amylase and protease improves loaf volume and crust colour, and a blend of fungal amylase and hemicellulase offers similar benefits for high-fibre doughs. A blend of glucose oxidase and fungal α-amylase is used instead of chemical oxidants and reduces the use of dough conditioners. Lipases improve dough strength, bread volume and loaf appearance.

The effects of heat on the texture of meat are described in detail by Tornberg (2005), Wattanachant et al. (2005) and Martens et al. (1982). When meat is heated, fats melt and become dispersed as oil through the food or drain out as a component of 'drip losses'. Collagen is solubilised below the surface to form gelatine, and oils are dispersed through the channels produced in the meat by dissolved collagen. Proteins become denatured, lose their water-holding capacity and contract. This forces out additional fats and water and toughens and shrinks the food. The surface dries, and the texture becomes crisper and harder as a porous crust is formed by coagulation, degradation and partial pyrolysis of proteins.

Flavour and aroma
The aromas produced by baking are an important sensory characteristic of baked goods. The number, type and amount of aromas produced during baking depend on:

- the actions of fermenting micro-organisms, especially yeasts and *Lactobacillus* spp. in fermented cereal products;
- the combination of fats, amino acids and sugars present in the surface layers of food;
- the temperature and moisture content of the food throughout the baking period and the time of heating.

Aroma compounds from fermentation by yeasts and *Lactobacilli* include alcohols, aldehydes and organic acids that give flowery, yeasty or malt flavours. Aroma compounds that have a bitter, tallowy or metallic taste are also produced by oxidation of fatty acids by cereal enzymes (e.g. lipoxygenase) during dough mixing. Lactic acid bacteria and yeasts partly inactivate these compounds and fermentation therefore reduces these flavours in dough. The high temperatures and low moisture contents in the surface layers cause Maillard reactions between sugars and amino acids (Chapter 1, section 1.2.2). There are low levels of proteolytic enzymes in wheat flour and amino acids are mostly produced by lactic acid bacteria. Sourdough yeasts assimilate amino acids and there is therefore a lower concentration of amino acids than in dough containing only *Lactobacilli*. Increased amounts of free amino acids (e.g. proline, leucine and phenylalanine) can be achieved by addition of proteases (Chapter 6, section 6.3.1) to improve the flavour of baked bread. The Maillard reaction produces different aromas according to the particular combinations of free amino acids and sugars that are present (Table 18.4).

Table 18.4   Aromas produced by baking or roasting

Food	Predominant amino acids	Selected characteristic aromas after heating with a single sugar
Potato	Asparagine	–
	Glutamine	Caramel, butterscotch, burnt sugar
	Valine	Fruity, sweet, yeasty
	Aminobutyric acid	Caramel, maple syrup, nutty
Peanut	Alanine	Caramel, nutty, malt
	Phenylalanine	Sweet and rancid caramel, violets
	Asparagine	–
	Arginine	Bready, buttery, burnt sugar
Beef	Valine	Fruity, sweet, yeasty
	Glycine	Caramel, smoky, burnt
	Leucine	Toasted, cheesy, malt, bready
Cocoa bean	Leucine	Toasted, cheesy, malt, bready
	Alanine	Caramel, nutty, malt
	Phenylalanine	Sweet and rancid caramel, violets
Valine		Fruity, sweet, yeasty

Adapted from Adrian (1982)

For example, the amino acid proline can produce aromas of potato, mushroom or burnt egg when heated with different sugars and at different temperatures. Odours in bread-crumb include 1-octen-3-one (mushroom odour), 2-phenylethanol (flowery, yeasty odour), 3-methylbutanol (malty, alcoholic odour) and (E)-2-nonenal (stale off-flavour). Important flavour compounds in crusts of wheat and rye bread include Maillard products methional from precursor methionine, 2-acetyl-pyrroline (ACPY) and 6-acetyltetrany-dropyrroline (ACTPY) produced from proline and ornithine (Hille 2007).

Heating also causes caramelisation of sugars and oxidation of fatty acids to aldehydes, lactones, ketones, alcohols and esters, with the more volatile aroma compounds being lost in exhaust air from the oven. Further heating degrades some of the retained volatiles to produce burnt or smoky aromas. Aroma components in roast meat include carbonyls, pyrazines, thiols, thiazoles and other nitrogenous and sulphur-containing compounds. The flavour of roasted coffee and cocoa beans, meats and nuts is one of the main quality characteristics of these products and details of the large number of chemical changes and aromatic compounds in these products are given by Staub (1995) and Czerny et al. (1999) for coffee, Bonvehi (2005) for cocoa, Cerny and Grosch (1992) for meat, and Mason et al. (1966), Johnson et al. (1971) and Buckholtz et al. (1980) for peanuts.

Colour
The characteristic golden brown colour associated with baked cereal foods is due to Maillard reactions; caramelisation of sugars and dextrins (either present in the food or produced by hydrolysis of starches) to furfural and hydroxymethyl furfural; and carbonisation of sugars, fats and proteins. The colour development in roasted red meats is due to the oxidation and denaturation of myoglobin to form brown metmyoglobin.

18.3.2   Changes to nutritional value
During baking, the physical state of proteins and fats is altered, and starch is gelatinised and hydrolysed to dextrins and then reducing sugars. However, these changes do not

substantially affect their nutritional value. The main nutritional changes during baking occur at the surface of foods, and the ratio of surface area to volume is therefore an important factor. In pan bread for example, only the upper surface is affected and the pan protects the bulk of the bread from substantial nutritional changes. However, in biscuits, baked snackfoods and breakfast cereals the thin profile means that thermal destruction of nutrients is greater. Lysine is the limiting amino acid in wheat flour and its destruction by Maillard reactions during baking is therefore nutritionally important (Horvatić and Ereš 2002). Lysine loss is 88% during the manufacture of maize breakfast cereals, which is corrected by fortification. Losses of amino acids in biscuits are: tryptophan, 44%; methionine, 48%; and lysine, 61%. The losses increase as a result of higher temperatures, longer baking times and larger amounts of reducing sugars being present. These in turn depend on the amylase activity of the flour or the use of fungal amylases, addition of sugar to the dough and steam injection to gelatinise the surface starch and improve crust colour. Maillard reactions also lead to the production of carcinogenic acrylomide in bakery products (Becalski *et al.* 2003) and roasted nuts (Lukac *et al.* 2007). Further details of acrylamide are given in Chapter 19 (section 19.4.3).

In meats, nutrient losses are affected by the size of the pieces, the type of joint, the proportions of bone and fat, pre- and post-slaughter treatments and the type of animal. Thiamine is the most important heat-labile vitamin in both cereal foods and meats. In cereal foods the extent of thiamine loss is determined by the temperature of baking and the pH of the food (losses are higher at higher pH values). Loss of thiamine in pan bread is ≈15% but in cakes or biscuits that are chemically leavened by sodium bicarbonate, the losses increase to 50–95%. In meats, thiamine losses range from 30 to 50%, but are considerably higher in pan drippings (80–90%). Vitamin C is also destroyed during baking, but is added to bread dough as an improver. Other vitamin losses are relatively small. In chemically leavened doughs the more alkaline conditions cause the release of niacin, which is bound to polysaccharides and polypeptides, and therefore increase its concentration. The vitamin content of fermented cereal products is determined by the extent of dough fermentation, which increases the amount of B-vitamins. Bread flour has been routinely fortified with B-vitamins for many years and more recently with folic acid in some countries. Neutraceutical bakery products (see Chapter 6, section 6.2) include high-fibre breads, and omega-3 fatty acids, lutein, fructans and oligosaccharides are also now added to bakery products. Baking destroys probiotic bacteria but they can be added to dairy-based fillings and their metabolic products remain in the food.

18.3.3   Effect on micro-organisms

The heat during baking destroys contaminating micro-organisms and inactivates naturally occurring enzymes, and thus contributes to the preservation of the food. Moulds are the most likely contaminants in baked cereal products, but several bakery products, particularly those containing meat or cream fillings, have been implicated in foodborne illnesses involving *Clostridium perfringens, Salmonella* spp., *Listeria monoctyogenes* and *Bacillus cereus*. *Clostridium botulinum* is also a concern in high-moisture bakery products packaged under modified atmospheres (Smith *et al.* 2004). Moulds may contaminate cereals and nuts and there is a risk of some producing mutagenic and/or carcinogenic mycotoxins. Some types, including aflatoxins, fumonisins and zearalenone, undergo 20–50% reduction during dough fermentation and 20% reduction during baking, but others, such as ochratoxin A, are largely unaffected by baking (Marais 2007) (see also Chapter 1, section 1.2.4).

In baking the surface temperature of the food rises to the oven temperature and the internal temperature rises to 100 °C. The baking time and temperature needed to produce the required sensory characteristics in the food are therefore sufficient to destroy vegetative bacterial, yeast or mould cells due to their low thermal resistance (D-values), and baked foods are therefore substantially free of micro-organisms. However, spores of spore-forming bacteria have higher D-values and may be able to survive baking and grow in the product to levels that could cause public health concerns (details of the effect of heat on micro-organisms are given in Chapter 10, section 10.3). The most likely cause of food safety problems is post-baking contamination from the bakery air, slicing machines or bakery staff, and baked foods are therefore cooled and packaged as soon as possible. Particular care is needed when filling pre-baked pastries with dairy products such as cream, as these can be a source of pathogens.

Products may be formulated to contain ingredients that enhance product safety and stability so that they can be distributed without temperature control for limited periods of time. These include preservatives:

- propionic acid/calcium propionate which are effective against moulds;
- sorbic acid/potassium sorbate which are effective against yeasts, moulds and some bacteria;
- acetic acid/acetates against 'rope' bacteria and moulds;
- citric, phosphoric, malic or fumaric acids: or sodium benzoate for fruit fillings (Smith *et al.* 2004).

Other ingredients include:

- humectants (e.g. sugar, glycerine) to reduce a_w;
- water binding agents (gums and starches) to reduce free water; and
- antimicrobial spices (e.g. cinnamon, nutmeg).

Other post-baking treatments for high- or intermediate-moisture baked products include low-dose irradiation (Chapter 7), ultra-high pressures (Chapter 8) and pulsed light and ultraviolet light (Chapter 9, section 9.4).

The a_w of cereal products varies from 0.3 in dry products such as biscuits, to 0.7–0.8 in cakes and ≈0.96–0.98 in bread. Moulds are able to grow at $a_w > 0.6$ and products that are expected to have a longer shelf-life than a few days require additional preservation methods (Table 18.3). The lower a_w of the crust of some baked products acts as a barrier to microbial contamination, but migration of moisture from the interior of the product, or from high-moisture fillings in cakes and pastries, increases the a_w and permits spoilage by moulds. Similarly, if bread is not adequately cooled before packaging, moisture can condense on the inside of the pack to create localised wetting of the crust, leading to mould growth.

Short shelf-life bakery products are sold within a few days, or on the day of production in retail bakeries and thus require only basic packaging to keep them clean. Low-moisture products, including snackfoods and biscuits, have a long shelf-life using packaging that has adequate barriers to moisture and oxygen to prevent oxidative rancidity. Products that contain higher moisture contents are mostly required to have a short shelf-life at ambient temperatures, which may be extended using chilling, freezing, or vacuum- or modified atmosphere packaging to reduce spoilage by moulds. Other products, including cream- or meat-filled pies and pastries and roast meats have the potential to contain pathogens and are stored at chilled or frozen temperatures, sometimes using modified atmospheres.

Active packaging systems (Chapter 25, section 25.4.3) such as oxygen scavenging in food packages are used to control mould growth on bakery products such as cakes, pizza crusts and crumpets for up to 30 days. Ethanol also increases the shelf-life of bread and other baked products when sprayed on to product surfaces prior to packaging or released into the pack from an ethanol releasing system. This uses a small sachet of food-grade ethanol absorbed into a fine inert powder inside the food package. The sachet is permeable to water vapour and when moisture is absorbed from the food, ethanol vapour is released into the package headspace.

References

ADRIAN, J., (1982), The Maillard reaction, in (M. Rechcigl, Ed.), *Handbook of the Nutritive Value of Processed Food*, Vol 1, CRC Press, Boca Raton, FL, pp. 529–608.

ANON, (2006), Hazard alert – Explosion in gas bakery oven, Worksafebc, available at www2.worksafebc.com/i/posters/2006/ha0608_bakery.htm.

ANON, (2008), Control systems, company information from Pladrest Ltd., available at www.pladrest.co.uk/products/2/index.phtml.

BAIK, O.D., MARCOTTE, M., SABLANI, S.S. and CASTAIGNE, F., (2001), Thermal and physical properties of bakery products, *Critical Reviews in Food Science and Nutrition*, **41** (5), 321–352.

BECALSKI, A., LAU, B. P.-Y., LEWIS, D. and SEAMAN, S.W., (2003), Acrylamide in foods: occurrence, sources, and modeling, *J. Agricultural and Food Chemistry*, **51** (3), 802–808.

BEECH, G.A., (2006), Energy use in bread baking, *J. Science Food and Agriculture*, **31** (3), 289–298.

BENNION, E.B. and BAMFORD, G.S.T., (1997), *The Technology of Cake Making*, 6th edn, Blackie Academic and Professional, London.

BONVEHÍ, J.S. (2005), Investigation of aromatic compounds in roasted cocoa powder, *European Food Research and Technology*, **221** (1–2), 19–29.

BUCKHOLTZ, L.L., DAUN, H., STIER, E. and TROUT, R., (1980), Influence of roasting time on sensory attributes of fresh roasted peanuts, *J. Food Science*, **45**, 547–554.

CAUVAIN, S.P., (ED.), (2003), *Bread Making: Improving Quality*, Woodhead Publishing, Cambridge.

CAUVAIN, S.P. and YOUNG, L.S., (1998), *Technology of Breadmaking*, Aspen Publishers, Gaithersburg, MD.

CAUVAIN, S.P. and YOUNG, L.S., (2001), *Baking Problems Solved*, Woodhead Publishing, Cambridge.

CAUVAIN, S.P. and YOUNG, L.S., (2006), *The Chorleywood Bread Process*, Woodhead Publishing, Cambridge.

CERNY, C. and GROSCH, W., (1992), Evaluation of potent odorants in roasted beef by aroma extract dilution analysis, *Zeitschrift für Lebensmitteluntersuchung und -Forschung A*, **194** (4), 322–325.

CHENG, L.M., (1992), *Food Machinery: For the Production of Cereal Foods, Snack Foods and Confectionery*, Woodhead Publishing, Cambridge.

CZERNY, M., MAYER, F. and GROSCH, W., (1999), Sensory study on the character impact odorants of roasted Arabica coffee, *J. Agriculture and Food Chemistry*, **47**, 695–699.

HANSEN, A. and SCHIEBERLE, P., (2005), Generation of aroma compounds during sourdough fermentation: applied and fundamental aspects, *Trends in Food Science and Technology*, **16** (1–3), 85–94.

HEGENBART, S., (1994), Understanding enzyme function in bakery foods, Food Product Design, available at www.foodproductdesign.com/archive/1994/1194DE.html.

HILLE, J.D.R., (2007), Breadmaking: an integral approach of biochemical events, South African Association for Food Science and Technology Expo Bakery Symposium, Durban, 2–5 September, available at www.saafost.org.za/presentations/bakery/Breadmaking_Biochemical.pdf.

HORVATIĆ, M. and EREŠ, M., (2002), Protein nutritive quality during production and storage of dietetic biscuits, *J. Science of Food and Agriculture*, **82** (14) 1617–1620.

JOHNSON, B.R., WALLER, G.R. and BURLINGANE, A.L., (1971), Volatile components of roasted peanuts: basic fraction, *J. Agriculture and Food Chemistry*, **19**, 1020–1024.

LUKAC, H., AMREIN, T.M., PERREN, R., CONDE-PETIT, B., AMADOGRAVE, R. and ESCHER, F., (2007), Influence of roasting conditions on the acrylamide content and the color of roasted almonds, *J. Food Science*, **72** (1), C033–C038.

MANLEY, D.J.R., (1998), *Biscuit, Cookie and Cracker Manufacturing Manuals: 1 – ingredients, 2 – biscuit doughs, 3 – biscuit dough piece forming, 4 – baking and cooling of biscuits, 5 – secondary processing in biscuit manufacturing, 6 – biscuit packaging and storage*, Woodhead Publishing, Cambridge.

MANLEY, D., (2001), *Biscuit, Cracker and Cookie Recipes for the Food Industry*, Woodhead Publishing, Cambridge.

MARAIS, G.J. (2007), The importance of fungi and their mycotoxins in the bakery industry, South African Association for Food Science and Technology Expo Bakery Symposium, Durban, 2–5 September, available at www.saafost.org.za/presentations/bakery.

MARCOTTE, M., (2007), Heat and mass transfer during baking, in (S. Yanniotis and B. Sundén, Eds.), *Heat Transfer in Food Processing*, Wit Press, Southampton, pp. 239–266.

MARTENS, H., STABURSVIK, E. and MARTENS, M., (1982), Texture and colour changes in meat during cooking related to thermal denaturation of muscle proteins, *J. Texture Studies*, **13** (3), 291–309.

MASON, M.E., JOHNSON, B.R. and HAMMING, M.C., (1966), Flavor components of roasted peanuts, *J. Agriculture and Food Chemistry*, **14**, 454–460.

MIZUKOSHI, M., (1990), Baking mechanism in cake production, in (K. Larssen and S.E. Friberg, Eds.), *Food Emulsions*, 2nd edn, Marcel Dekker, New York, pp. 479–504.

OWENS, G. (ED.), (2001), *Cereals Processing Technology*, Woodhead Publishing, Cambridge.

PREJEAN, W., (2007), Baking and baking science, available at www.bakingand bakingscience.com.

SAKAR, A. and SINGH, R.P., (2004), Air impingement heating, in (P. Richardson, Ed.), *Improving the Thermal Processing of Foods*, Woodhead Publishing, Cambridge, pp. 253–276.

SLUIMER, P., (2005), *Principles of Breadmaking*, American Association of Cereal Chemists Press, St. Paul, MN, Chapters 5 and 6.

SMITH, J.P., PHILLIPS DAIFAS, D.P., EL-KHOURY, W. and KOUKOUTSIS, J., (2004), Shelf life and safety concerns of bakery products – a review, *Critical Reviews in Food Science and Nutrition*, **44**, 19–55.

STAUB, C., (1995), Basic chemical reactions occurring in the roasting process, from SCAA Roast Color Classification System, available at www.sweetmarias.com/roast.carlstaub.html.

STITLEY, J., (1995), Heating systems, available at www.tenonline.org/art/fm/9407.html.

TORNBERG, E., (2005), Effects of heat on meat proteins: Implications on structure and quality of meat products, 50th International Congress of Meat Science and Technology, 8–13 August 2004, Helsinki, Finland, *Meat Science*, **70** (3), 493–508.

WATTANACHANT, S., BENJAKUL, S. and LEDWARD, D.A., (2005), Effect of heat treatment on changes in texture, structure and properties of Thai indigenous chicken muscle, *Food Chemistry*, **93** (2), 337–348.

III.C

Heat processing using hot oils

19

Frying

Abstract: Frying is a unit operation in which food is heated in oil to alter its eating quality. The chapter first reviews principles of heat and mass transfer in frying, followed by a description of the equipment used for contact frying and deep-fat frying, and methods to reduce energy consumption and environmental pollution. The criteria for selection of frying oils are described, together with the effects of frying on oil quality. Finally, there is a discussion of the effect of frying on sensory characteristics of foods, changes to their nutritional value, and health concerns over fried foods and methods to reduce their fat contents.

Key words: frying, heat and mass transfer, shallow frying, deep-fat frying, heat recovery, hydrogenated oils, acrylamide, non-volatile decomposition products.

Frying is a unit operation that is mainly used to alter the eating quality of a food by heating it in oil.[1] Frying is an unusual unit operation in that the product of one process (oil extraction, Chapter 5, sections 5.3 and 5.4) is used as the heat transfer medium in another. Fried foods have a characteristic golden colour, a crisp texture, a distinctive mouthfeel and characteristic fried flavours and aromas that are due to both the food and the oil used for frying. The other consideration is the preservative effect of frying that results from thermal destruction of micro-organisms and enzymes, and a reduction in water activity due to dehydration at the surface of the food, or throughout the food, if it is fried in thin slices.

Some foods (Table 19.1) are coated with batter (Chapter 24, section 24.1.2) before frying to create a crisp shell and to hold the food together. Batter coatings add value to fried products, improve their flavour, and reduce moisture loss and weight loss during frying. Other foods that have a skin (e.g. sausage, chicken) can be fried without coatings; the skin holds the food together and retains moisture.

The shelf-life of fried foods is mostly determined by the moisture content after frying: foods that retain a moist interior (e.g. doughnuts, fish and poultry products) have a relatively short shelf-life, owing to moisture and oil migration during storage. These

---

1. Chemically, fats and oils are the same, differing only in their melting point, but commercially many fats (e.g. palm oil and coconut oil) are named oils by custom even though they are solid or semi-solid at ambient temperatures.

Table 19.1   Types of fried foods and frying conditions

Type of food	Frying temperature (°C)	Time (min)
Cereal and legume products		
Bean croquettes	170–175	3–4
Crisp noodles	175–180	2–3
Doughnuts	185–190	3–4
Extruded snackfoods, half products	150	1–5
Fish		
Fillets, batter or breadcrumb coated	175–180	3–5
Goujons	175	2–3
Pre-cooked fish cakes, croquettes, rissoles	175–180	2–3
Scampi	175–180	2–3
Whitebait	190–195	0.5–1
Fruits		
Apple slices, batter coated	175–180	3–4
Other fruits, batter coated	180–185	2–3
Meat and poultry products		
Poultry portions, cooked from raw, breadcrumb coated, large	160–165	8–12
Poultry portions, cooked from raw, batter coated	175–180	4–7
Pre-cooked cutlets	175–180	2–5
Pre-cooked fritters in batter	175–180	
Pre-cooked rissoles	180–185	3–4
Scotch eggs	170–175	5–6
Vegetables		
Carrot, parsnip, sweet potato or beetroot slices, cooked from raw	175–180	2–3
Onion rings, raw and battered	180–185	2–3
Potato chips (French fries), cooked from raw	185	4–6
Potato chips, final browning	190	1–2
Potato crisps (chips in USA), cooked from raw	185	3
Potato puffs, cooked from raw	175	2
Potato puffs, second cooking	190	1
Potato croquettes	190	4–5

Adapted from data of Moreira *et al.* (1999) and Wood *et al.* (1983)

foods are important in food service applications where they are consumed shortly after frying. When they are produced on a larger commercial scale for distribution to retail stores, they are preserved by chilling (Chapter 21) and/or modified atmosphere packaging (Chapter 25, section 25.3) or freezing (Chapter 22). Foods that are more thoroughly dried by frying (e.g. potato crisps (potato chips in the USA) and other potato or maize snackfoods) have a long shelf-life at ambient temperature. The quality is maintained by adequate moisture and oxygen barrier properties of packaging materials (Chapter 25, section 25.1) and correct storage conditions. The production of extruded snackfoods that are subsequently fried is described in Chapter 15 (section 15.3).

There are many types of frying used in food service operations, including sautéing, stir-frying, pan frying and shallow frying, but in the food processing industry deep-fat frying is more common and is described in detail in this chapter. The popularity of frying is due to the short cooking times at the high temperatures employed (Table 19.1) and the sensory qualities of fried foods. However, concerns have arisen over the health implications of eating fried foods, including the production of potential carcinogens and, in industrialised countries, their contribution to obesity and the health problems associated with this (see section 19.4.3). Detailed information on frying is given by Rossell (2001).

19.1   Theory

19.1.1   Heat and mass transfer

Frying involves simultaneous transfer of heat from oil to the food, mass transfer of moisture from the food and subsequent oil absorption by the food (Fig. 19.1). It is similar to dehydration (Chapter 16) and baking (Chapter 18) except that the heating medium is oil rather than hot air. Details of heat transfer are given in Chapter 10 (section 10.1.2) and mass transfer is described in Chapter 14 (section 14.1.1).

Farkas *et al.* (1996) describe different stages in heat transfer during frying. When food is immersed in hot oil, typically at 150–195 °C, convective heat is transferred through a boundary layer of oil and the surface temperature of the food rises rapidly to ≈100 °C within a few seconds. The rate of heat transfer is controlled by the temperature difference between the oil and the food and is described by the surface heat transfer coefficient (Chapter 10, section 10.1.2). Other factors that affect the heat transfer coefficient include the velocity and viscosity of the oil, which affect the thickness of the boundary layer: higher velocities and lower viscosities reduce the thickness of the boundary layer and increase the rate of heat transfer.

In this first stage, heat transfer is by natural convection and the heat transfer coefficient is 250–280 $W\,m^{-2}\,K^{-1}$ (Miller *et al.* 1994). The next stage is boiling, in which moisture at the surface is vaporised and bubbles of steam escape through the oil creating violent turbulence in the oil that reduces the boundary layer of oil surrounding the food. This creates forced convection conditions and the heat transfer coefficient increases. For example, Costa *et al.* (1999) found that the heat transfer coefficient increased two-three-fold to 443–750 $W\,m^{-2}\,K^{-1}$ during escape of vapour bubbles compared with the values measured in the absence of boiling. During the next (falling rate) stage, moisture is progressively removed from the food, a crust is formed and the surface temperature rises. This is characterised by less turbulence and a lower heat transfer coefficient. For example, Budžaki and Šeruga (2005) found that the convective heat transfer coefficient was highest at the beginning of deep-fat frying (197–774 $W\,m^{-2}\,K^{-1}$) at temperatures of 160–190 °C and then fell to 94–194 $W\,m^{-2}\,K^{-1}$. Farinu and Baik (2007) report maximum heat transfer coefficients of 710–850 $W\,m^{-2}\,K^{-1}$ during the first 80–120 s of deep-fat frying sweet potato at 150–180 °C. It then fell to 450–550 $W\,m^{-2}\,K^{-1}$ after 200–300 s.

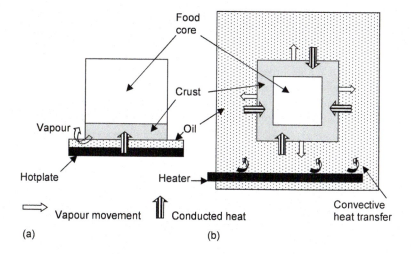

Fig. 19.1   Heat and mass transfer in (a) shallow frying and (b) deep-fat frying.

Other studies using both model systems and different types of foods are reported by Ngadi and Ikediala (2005), Hubbard and Farkas (2000) and Sahin *et al.* (1999a,b).

In shallow (or contact) frying, heat is transferred to the food mostly by conduction from the hot surface through a thin (1–10 mm) layer of oil (Fig. 19.1). In contrast to deep-fat frying, in which all surfaces of the food receive a similar heat treatment, to produce a uniform colour and appearance, the thickness of the layer of oil in shallow frying varies as a result of irregularities in the surface of the food. This, together with the action of bubbles of steam that lift the food off the hot surface, causes temperature variations as frying proceeds and produces the characteristic irregular browning of shallow fried foods. This method is most suited to foods that have a large surface-area-to-volume ratio, and is used for meat products (e.g. bacon, sausages, burgers and other types of pattie), which also contain animal fats that would contaminate the oil in deep-fat fryers. Wichchukit *et al.* (2001) studied heat transfer coefficients during double-sided contact cooking of hamburger patties. They were in the range 250–650 $W m^{-2} K^{-1}$, depending on hotplate temperatures and gap thicknesses between the plates.

In deep-fat frying, moisture in the form of steam leaves the surface of the food through the boundary film of oil, the thickness of which influences the rate of mass transfer. Yildiz *et al.* (2007) report that mass transfer coefficients increased linearly with increasing oil temperature, whereas moisture diffusivity increased exponentially with an increase in frying temperature. Vitrac *et al.* (2003) identified three types of coupled heat and mass transfer during the first minute of frying: first surface vaporisation of moisture having heat flowrates $>100 \, kW \, m^{-2}$; secondly, internal moisture vaporisation with heat flowrates oscillating between 15 and $35 \, kW \, m^{-2}$ controlled by liquid water movement from inside the food, and thirdly a decreasing vaporisation rate after liquid water was removed.

As moisture is lost, the surface begins to dry out in a similar way to that described during baking (Chapter 18, section 18.1). The surface temperature of the food then rises to that of the hot oil, and the internal temperature rises more slowly towards 80–100 °C. The plane of evaporation moves inside the food and a dry crust is formed that has a porous structure, consisting of different-sized capillaries through which steam is removed. The water vapour pressure gradient between the moist interior of the food and the dry oil is the driving force behind moisture loss, in a similar way to hot air dehydration (Chapter 16, section 16.1.1). Costa *et al.* (1999) found that the rate of moisture loss in potato frying increased until the food surface dried and then decreased until the end of frying. Saguy and Pinthus (1995) report a number of studies which show that moisture loss is proportional to the square root of frying time. A number of studies have found that oil is not absorbed while steam is escaping from the product and that oil absorption occurs when the food cools after it is removed from the fryer (Mehta and Swinburn 2001, Moreira *et al.* 1997, Ufheil and Escher 1996). Baumann and Escher (1995) concluded that oil is deposited mainly on the surface and does not penetrate into the food to a significant extent (section 19.4.1). Different theories have been proposed for the mechanisms by which foods absorb oil (Moreira and Barrufet 1998), including:

- condensation of steam in the crust on cooling, which produces a vacuum effect that sucks oil into the crust from the surface;
- a surface phenomenon that involves equilibrium between oil adhesion and drainage;
- capillary forces that draw oil into the crust from the surface of the food.

Bouchon *et al.* (2003) studied oil absorption in deep-fried potato cylinders and identified three types of oil absorption: structural oil (absorbed during frying), penetrated surface oil (sucked into the product as it cooled) and surface oil. They found that little oil was absorbed during frying, suggesting that oil uptake and water removal are not synchronous, and that most of the oil was absorbed by capillary action into the porous crust microstructure as it cooled. Bouchon and Pyle (2005) developed a mathematical model to describe this process. The importance of oil uptake in fried foods is described in section 19.4.1 and oil reduction equipment (section 19.2) has been developed in the light of this research.

Heat is transferred through the food to the thermal centre by conduction, and the rate of heat penetration is controlled by the thermal conductivity of the food (Table 10.2, Chapter 10). Šeruga and Budžaki (2005) report that the thermal conductivity of fried potato dough first increased with temperature to a maximum value $0.60\,\mathrm{W\,m^{-1}\,K^{-1}}$ at $47.5\,^{\circ}\mathrm{C}$ and then decreased to a minimum of $0.47\,\mathrm{W\,m^{-1}\,K^{-1}}$ at $65\,^{\circ}\mathrm{C}$. Mathematical models of deep-fat frying are reviewed by Bouchon (2006) and further details of frying theory are given by Rossell (2001).

19.1.2 Frying time and temperature

The time taken for food to be completely fried depends on the required changes to organoleptic qualities and the time required for the thermal centre to receive sufficient heat to destroy contaminating micro-organisms and become adequately cooked. It is particularly important that comminuted meat products (e.g. sausages or burgers) that are able to support the growth of pathogenic bacteria receive a sufficient time–temperature combination (see Chapter 10, section 10.3). The frying time depends on the type and thickness of the food, the temperature of the oil and the method of frying (shallow or deep-fat frying).

The temperature used for frying is determined by requirements for the development of product sensory attributes and economic considerations: high temperatures reduce processing times and increase production rates, but they also cause accelerated deterioration of the oil (section 19.3.1). This increases the frequency with which oil must be changed and hence increases costs. A second economic loss arises from the vigorous boiling of the food at high temperatures, which causes loss of oil by aerosol formation. Oil is also lost by entrainment in the product and, although deep-fat frying is suitable for foods of all shapes, irregularly shaped food or pieces with a greater surface area : volume ratio entrain a greater volume of oil when it is removed from the fryer (Selman 1989). Methods to reduce these losses are described in section 19.2.3.

19.2   Equipment

19.2.1   Atmospheric fryers

Batch equipment is widely used in food service applications. Shallow fryers have a heated metal surface, covered in a thin layer of oil. Batch deep-fat fryers consist of one or more 5–25 litre stainless steel tanks or 'kettles' of oil that are heated by thermostatically controlled electric resistance heaters immersed in the oil, or by gas burners below the tank. More recent designs use turbojet infrared burners that use 30–40% less energy than gas burners (Anon 2006a, Bouchon 2006). Electric heaters are placed a few centimetres above the base of the tank to create a 'cool zone' below, where food debris can collect, which minimises damage to the oil. New designs of fryer are reported to have up to 69% reduction in energy costs, up to 68% less oil required, 37% quicker heat-up times and double the oil life expectancy of traditional batch fryers (Anon 2007a). No copper-containing bronze or brass fittings are used in valves or heating elements to avoid catalysing oxidation of the oil. In operation the food is suspended in baskets in the bath of hot oil and retained for the required degree of frying. Most foods are fried in deep, narrow baskets that are immersed in the oil and removed manually, although some designs have an automatic basket lifting system that operates when frying is completed. Doughnuts are fried in wide, shallow fryers that are specially designed for this product.

Oil that is absorbed by the food is periodically replaced with fresh oil. The 'oil turnover' represents the time needed to completely replace the oil in the fryer (Equation 19.1 and sample problem 19.1).

$$\text{Oil turnover} = \frac{\text{weight of oil in the fryer}}{\text{weight of oil added per hour}} \qquad \boxed{19.1}$$

In sample problem 19.1, oil turnover is 8 hours and most batch fryers that are operated constantly have turnover rates from 5 to 12 hours, which maintain the oil quality. However, in practice the use of fryers varies, with continuous operation during different parts of the day separated by idle periods. The actual oil turnover for food-service fryers is therefore longer than this, but it should not exceed 20 hours because the oil deteriorates after a few days and has to be completely replaced (Kerr 2006).

Continuous deep-fat fryers consist of a stainless steel mesh conveyor submerged in a thermostatically controlled, insulated oil tank having a capacity of 200–1000 kg of oil (Fig. 19.2) (sample problem 19.2). They are fitted with an extraction system to remove vapour and fumes and an oil filtration system (section 19.2.3). The oil is heated directly by either internal electric heating elements, or pipes that contain the products of gas combustion, or indirectly via a separate gas or electric heat exchanger (Anon 2006b) (see also Chapter 10, section 10.2.4). Some designs have multiple heating zones that can be separately controlled. In this 'zonal flow', multiple oil inlets and outlets are distributed along the length of the frying kettle. The shape of the oil inlets eliminates hot-spots to reduce the formation of acrylamide (section 19.4.3). At the end of each zone, an oil outlet drains the relatively cold oil, taking particles of food with it, and freshly filtered and heated oil is injected. The quantity of oil that is removed is controlled by variable speed pumps to control the oil flow in the kettle and produce uniform heat transfer to the product throughout the kettle (Anon 2007b). This enables the temperature profile to be adjusted to maintain uniform product quality even with variations in raw materials.

Sample problem 19.1
A fryer that holds 700 kg oil can process $1.10 \, \text{t h}^{-1}$ of product. If the product absorbs 8% oil (by mass) during frying, calculate the oil turnover rate.

Solution to sample problem 19.1
The amount of 'make-up' oil that is required to replace oil absorbed by the product is calculated as:

Weight of product fried per hour = 1100 kg

Amount of oil absorbed per hour = make-up oil required

= 1100 × 0.08 = 88 kg

Turnover rate = 700/88 ≈ 8 h

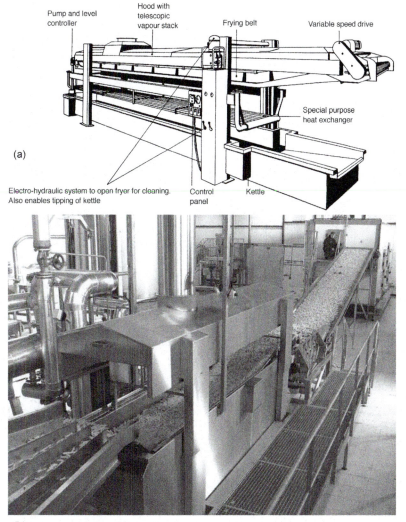

(a)

Pump and level controller

Hood with telescopic vapour stack

Frying belt

Variable speed drive

Special purpose heat exchanger

Electro-hydraulic system to open fryer for cleaning. Also enables tipping of kettle

Control panel

Kettle

(b)

Fig. 19.2 Continuous deep-fat fryers: (a) principle of operation (courtesy of Coat and Fry Ltd at: www.coatandfry.co.uk) and (b) equipment (© Paris Agro-Food Company, Iran, supplied by BMA Nederland BV, the Netherlands) (Anon 2007b).

Sample problem 19.2

A deep-fat fryer tank, measuring 2.8 m long, 1 m high and 1.5 m wide, is constructed from 4 mm stainless steel with 5 mm of fibre insulation on all sides. It is operated at 200 °C for 12 h per day and 250 days per year. Ignoring the resistance to heat transfer caused by boundary films, calculate the annual financial savings arising from reduced energy consumption if the tank insulation is increased to 30 mm of fibre insulation. (Additional data: thermal conductivity of stainless steel $= 21\,W\,m^{-2}\,K^{-1}$, thermal conductivity of fibre insulation $= 0.035\,W\,m^{-2}\,K^{-1}$, average ambient air temperature $= 18\,°C$, and energy cost $= 0.06$ monetary units per kW h.)

Solution to sample problem 19.2
First,

Area of tank insulated with 5 mm insulation

$$= 2(1.51 \times 1.01 + 2.81 \times 1.01 + 2.81 \times 1.51)$$

$$= 17.21 \, \text{m}^2$$

From Equations 10.19 in Chapter 10,

$$(200 - 18) = \frac{Q}{17.21} \left(\frac{0.004}{21} + \frac{0.005}{0.035} \right)$$

$$182 = (Q/17.21) \times 0.143$$

Therefore,

$$Q = 21\,904 \, \text{W}$$

Now

Area of insulated tank with 30 mm insulation

$$= 2(1.56 \times 1.06 + 2.86 \times 1.06 + 2.86 \times 1.56)$$

$$= 18.30 \, \text{m}^2$$

and

$$200 - 18 = \frac{Q}{18.30} \left(\frac{0.004}{21} + \frac{0.03}{0.035} \right)$$

therefore,

$$Q = 3886 \, \text{W}$$

The number of hours of operation per year is 3000, which equals $10.8 \times 10^6 \, \text{s}$.

$$1 \, \text{kW h} = 1000 \, \text{W for } 3600 \, \text{s} = 3.6 \times 10^6 \, \text{J}$$

Therefore,

Cost of energy with 5 mm insulation $= \dfrac{(21\,904)(10.8 \times 10^6)}{3.6 \times 10^6} \times 0.06$

$$= 3943 \, \text{monetary units}$$

and

Cost of energy with 30 mm insulation $= \dfrac{3886(10.8 \times 10^6)}{3.6 \times 10^6} \times 0.06$

$$= 699 \, \text{monetary units}$$

Therefore,

Annual financial saving $= 3943 - 699$

$$= 3244 \, \text{monetary units or } 82.2\% \text{ saving}$$

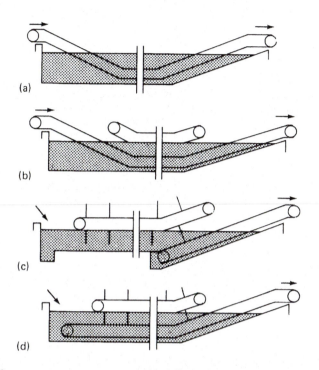

Fig. 19.3   Different conveyor arrangements: (a) delicate non-buoyant products (e.g. fish sticks); (b) breadcrumb-coated products; (c) dry buoyant products (e.g. half-product snackfoods); (d) dual-purpose (e.g. nuts and snackfoods) (courtesy Coat and Fry Ltd) (Anon 2006b).

Food is metered into continuous fryers by slow-moving paddles and it either sinks to a submerged conveyor or, if the food floats, it is held below the surface by a second conveyor (Fig. 19.3). The conveyor speed and oil temperature control the frying time. An inclined conveyor then removes the food and allows excess oil to drain back into the tank. In contrast to batch fryers that may be used intermittently, the oil turnover in continuous fryers has a shorter time, typically 3–8 hours.

An oil reduction system removes surface oil from fried products as they emerge from the fryer. Superheated steam at 150–160 °C is passed through the bed of food and the oil/vapour mixture is filtered and the oil returned to the fryer. This equipment has been developed to address the need to reduce the fat content of fried foods (section 19.4.3) and is reported, for example, to reduce the fat content of potato crisps by 25% (Kochhar 1999). Continuous fryers operate automatically at production rates of up to $25\,\mathrm{t\,h^{-1}}$ of fried product (Bouchon 2006). Products are monitored for consistent colour using computer vision systems (Chapter 2, section 2.3.3) (Gökmen *et al.* 2007).

19.2.2   Vacuum and pressure fryers
Batch vacuum fryers have the kettle enclosed in a vacuum vessel, and in operation the pressure is reduced to ≈3 kPa and frying takes place at 130–140 °C. The pressure is then returned to atmospheric and the product is removed. These are not widely used, although in South East Asia, batch vacuum fryers are used to produce fried fruit chips in small-scale production units. Continuous vacuum fryers have a frying tank installed in a stainless steel vacuum tube and vacuum pumps to reduce the pressure. Product is fed in and removed

through rotary airlocks that maintain the reduced pressure, and a conveyor moves the product through the tank as in continuous atmospheric fryers. Continuous vacuum frying has a number of advantages over atmospheric frying: the lower frying temperature produces a longer frying time for the oil; it can reduce oil absorption by fried foods; it maintains natural colours and flavours in the fried product; and it also reduces the formation of acrylamide (section 19.4.3). For example, Garayo and Moreira (2002) compared vacuum frying at 144 °C and 3.115 kPa with frying at 165 °C at atmospheric pressure. They found that vacuum pressure and oil temperature had a significant effect on the drying rate and oil absorption rate of potato crisps and concluded that vacuum frying could produce potato crisps that had a lower oil content than those fried under atmospheric conditions, and had a desirable colour and texture. However, the process is not yet widely used owing to the higher investment costs compared with atmospheric fryers.

Batch pressure fryers are strong, pressure-resistant sealed containers that are heated either directly or indirectly. The pressure is increased to 163–184 kPa, which increases the temperature to ≈175–184 °C and reduces frying times. It is used in food service applications for products such as chicken, to retain high moisture contents, because the higher pressure prevents moisture loss, and also produces a uniform colour and appearance. Energy consumption per kg of product is reduced by 48% and production rates increased two or three times compared with atmospheric frying (Anon 2007c).

19.2.3   Control of fryer operation, oil filtration and heat recovery

In continuous atmospheric and vacuum fryers, programmable logic controllers (PLCs) (Chapter 27, section 27.2.2) control the oil temperature by regulating the power to electric heaters, or monitor air inlet and exhaust gas temperatures to regulate the gas flow to burners. They also automatically maintain the required oil level in the fryer and control the product feedrate to produce uniformly fried foods (Rywotycki 2003). They can be programmed to select the required settings for different products, which are displayed on a touch-screen menu.

In batch fryers, sediments and food particles are periodically removed from the cool zone and some recent designs also incorporate an oil filter and pump to remove sedimented materials. In continuous fryers, oil is continuously recirculated through external filters (Chapter 5, section 5.1.2), decanter centrifuges or membrane filters (Chapter 5, section 5.5.2). Mechanical filtration systems (Chapter 5, section 5.2.2) have primary screens to remove larger food particles and secondary screens of fine stainless steel or plastic mesh or filter paper that 'polish' recirculated oil to remove particles as small as 10 μm. The most effective filtration systems use a filter powder, such as magnesium silicate or diatomaceous earth. This absorbs oxidised fatty acids, colours, odours and many of the secondary and tertiary by-products of oil degradation, which are then removed along with the filter powder by mechanical filtration. This can increase the frying life of oil by up to 100% by slowing oil degradation, maintain a satisfactory level of free fatty acids in the oil (section 19.3.1), maintain oil colour and minimise the development of off-flavours. Lin et al. (1998, 1999) studied the treatment of frying oils with filter aids and antioxidants to extend their frying life. Combinations of filter aids reduced free fatty acids by 91–94% and total polar components by 6–18%. It is also necessary to prevent carbon build-up on the fryer, which promotes breakdown of the oil and reduces the oil life. This also acts as an insulator between the oil and the heat source, reducing fryer efficiency, oil performance and energy wastage. Further information is given by Phogat et al. (2006) and Bheemreddy et al. (2002a,b).

Table 19.2   Approximate smoke points of some common frying oils

Type of oil	Smoke point (°C)		
	Unrefined	Semi-refined	Refined
Coconut	177	–	–
Cottonseed	–	–	216
Groundnut (peanut)	160	227	232
Maize (corn)	160	–	232
Olive oil (Extra virgin)	160	207	242
Rapeseed (canola)	107	177	204
Palm	–	–	230
Palm olein	–	–	232
Safflower	107	160	232–266
Soybean	160	177	232–257
Sunflower	107	232	232

Adapted from Chu (2004)

Heat recovery systems are used to reduce energy costs (the term 'heat recovery' is also used to describe the time taken for a fryer to regain the operating temperature after cold food is added to the oil). Heat exchangers, mounted in the exhaust hood, recover heat from escaping steam and use it to pre-heat incoming food or oil, or to heat process water.

The smoke point is an important criterion for selecting oils (section 19.3) and frying oils generally have smoke points above 200 °C (Table 19.2). When oil deteriorates the smoke point falls and acrolein, a breakdown product of oil, is produced, which forms a blue haze above the oil and is a source of atmospheric pollution (Anon 2007d). Pollution control systems prevent smoke and other products of oil degradation from being discharged into the atmosphere, by feeding the exhaust air into gas burners used to heat the oil. Oil recovery systems remove entrained oil from the steam and return it to the fryer. Waste frying oil has in the past created additional costs for safe disposal without causing environmental pollution (Paul and Mittal 1997), but in many countries it is now converted to methyl esters to meet the growth in demand for biodiesel (e.g. Kheang *et al.* 2006).

19.3   Types of oils used for frying

The selection of a frying oil for a particular product depends on a number of criteria, but mainly it should be stable to fry for long periods (to give a long fry life) and during product storage (to give the required shelf-life). Factors to take into account include:

- stability against oxidation during both frying and product storage;
- a fatty acid profile that is low in saturated and *trans*-fatty acids;
- low tendency to foam, or polymerise and produce gums;
- high smoke point;
- low viscosity;
- bland flavour;
- low cost.

The melting point of oils is particularly important as it determines the handling and storage methods required, and it affects the sensory quality of products. The type of frying oil used depends on the application: many processors use liquid oils that have been 'brushed' (lightly hydrogenated) for frying chips or nuts, whereas frying doughnuts in liquid oil causes glazes to crack. Doughnuts therefore require oils that have a higher melting point to produce a shell on the surface and impart a less greasy mouthfeel. Other products that are eaten in colder climates may require a lower melting point oil to prevent solidification during product storage and adverse effects on the mouthfeel of the product. Par-fried frozen potatoes are fried in a hardened vegetable oil with a high melting point.

The type of product also affects oil performance: products coated with batters and breadings contain salt and leavening agents that accelerate oil degradation. High-stability frying oils are required for industrial fryers and food service operations. Fifteen years ago, blended animal fats and vegetable oils (e.g. maize, cottonseed, sunflower and groundnut (peanut) oils) were used, but research into the relationship between saturated fats and heart disease resulted in their being replaced with hydrogenated oils. Concerns about both saturated and *trans*-fats have more recently prompted processors to use oils such as high-oleic sunflower that have the required frying properties, despite relatively high levels of monounsaturated fatty acids. In Europe, a blend of refined sesame and rice bran oils is also used (Stier 2003).

Palm oil has a melting point of 35 °C and can be fractionated into liquid palm olein oil (melting point 19–24 °C) and solid palm stearin oil (melting point 44 °C). It can also be double fractionated to produce super olein oil (melting point 13–16 °C), which may be blended with sunflower or groundnut oils to produce liquid frying oil. The advantages of palm oils are high resistance to oxidation and good flavour stability, which produce a long frying life, and product shelf-life. This is due to the higher levels of saturated, mostly palmitic, fatty acids (40–47.5% in palm oil and 38–43% in palm olein oil) balanced by approximately equal levels of unsaturated fatty acids. Olive oil has good frying properties, including resistance to oxidation, but is relatively expensive. Rapeseed (canola), high-oleic sunflower and soybean oils have a higher level of unsaturated fatty acids, which makes them more susceptible to oxidation and off-flavour development. They are partly hydrogenated to reduce the level of linoleic and linolenic acids for use as frying oils (Table 19.3). These fatty acids are susceptible to oxidation and also undergo thermal degradation to produce compounds that increase the viscosity of the oil (section 19.3.1). Hydrogenation improves their oxidative and heat stabilities – especially for foods that require a high ratio of polyunsaturated to saturated fatty acids. However, hydrogenation also removes ≈70% of the natural tocopherol and increases the level of *trans*-fatty acids by ≈40% compared with the original oil. The health concerns over *trans*- and saturated fatty acids are described in section 19.4.3. Manufacturers and foodservice operators are increasingly using healthier, more unsaturated oils such as highly mono-

Table 19.3   Some characteristics of frying oils

Parameter	Palm oil	Palm olein oil	Soybean oil	Hydrogenated soybean oil
Free fatty acids (%)	0.06	0.04	0.03	0.02
Peroxide value (meq kg^{-1})	4.1	1.5	4.0	0.9
Slip melting point (°C)	37.0	20.0	−16	35.8
Smoke point (°C)	214	216	217	219
Trans-oleic fatty acid (%)	Trace	Trace	Trace	39.8
Linoleic fatty acid (%)	9.8	11.3	54.4	1.8
Linolenic fatty acid (%)	0.2	0.2	9.0	0.1

Adapted from Razali and Badri (2003)

unsaturated rapeseed and modified sunflower oils. The composition of different oils is described in Anon (2008b).

Antioxidants (e.g. tertiary butyl hydroquinone (TBHQ), butylated hydroxy anisole (BHA), butylated hydroxy toluene (BHT) and propyl gallate) can enhance both oil life and the shelf-life of fried foods. BHA and BHT are more volatile and have only moderate activity at frying temperatures, and gallates are the preferred antioxidants. TBHQ is very effective but its use is not currently (2008) permitted in all countries. Oils may also include naturally occurring or added tocopherols, which are depleted during the frying process. Lin *et al.* (1998, 1999) studied the treatment of frying oils with antioxidants to extend their frying life. Addition of $50\,\mu g\,kg^{-1}$ of BHT and propyl gallate increased the oxidative stability index value by 49–81%. Food-grade methyl silicone or dimethyl polysiloxane may also be added to reduce foaming (Kerr 2006).

19.3.1   Effect of frying on oils

Changes to foods caused by frying involve both the effect on the oil, which in turn influences the quality of the food, and the direct effect of heat on the fried product (section 19.4). The physical changes to oil caused by frying include a gradual increase in viscosity over time, a reduction in interfacial tension, and an increase in specific heat (Min 2007). Da Silva and Singh (1995) found that when oil was heated, initially the viscosity decreased, and at 200 °C it had fallen by 69% compared with its initial value. They then degraded the oil by frying potatoes for 5 min each hour for 36 h and the viscosity of the oil increased by 28.6%. The increased viscosity lowered the surface heat transfer coefficient and increased the amount of oil entrained by the food.

The high temperatures used in frying, in the presence of air and moisture, cause a complex series of chemical reactions in oil, including hydrolysis, oxidation, polymerisation, isomerisation and cyclisation (Fig. 19.4) (Warner 2007). Oil breakdown products are characterised as volatile decomposition products (VDP) and non-volatile decomposition products (NVDP). Triacylglycerols in the oil are hydrolysed by the steam to form polar diacylglycerols and free fatty acids. Further hydrolysis causes diacylglycerols to break down to monoacylglycerols and free fatty acids and then monoacylglycerols are hydolysed to glycerol and free fatty acids. Finally, glycerol is dehydrated to form acrolein. Triacylglycerols are also oxidised to hydroperoxides. This reaction is limited by the amount of oxygen present in the oil, and initial oxygen is rapidly used up by reaction with natural antioxidants in the oil. Further oxygen can only enter the oil by diffusion from air or from intercellular spaces in the food, and hydroperoxide formation therefore takes place relatively slowly and only after antioxidants have been destroyed. Hydroperoxides are unstable and a series of secondary reactions take place, including decomposition to non-volatile epoxides and hydroxylic compounds, to volatile aldehydes, alcohols, ketones, saturated and unsaturated acids, hydroxy acids, saturated and unsaturated hydrocarbons, esters and lactones, or to higher molecular weight polymeric dimers and trimers (Pokorny 2002, May *et al.* 1983).

Steam produced during frying strips the lower molecular weight VDPs from the oil and they are carried in vapour from the fryer to form the smoke and odour of frying. Analysis of the vapour from fryers has indicated up to 220 different components (Nielsen 1993). Some of these components (e.g. unsaturated lactones) contribute to the flavour of the fried product. The NVDPs formed by oxidation and polymerisation of the oil form sediments on the sides and at the base of the fryer. Polymerisation in the absence of oxygen produces cyclic compounds and high molecular weight polymers that increase the

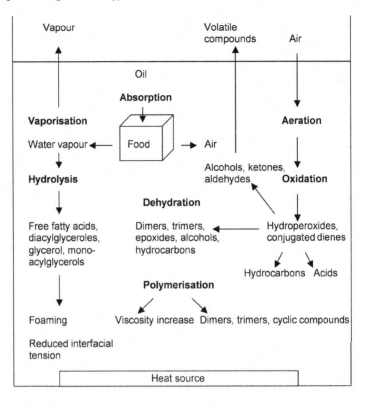

Fig. 19.4   Chemical and physical reactions during deep-fat frying (adapted from Min 2007).

viscosity of the oil (Razali and Badri 2003) and include trilinolenin, trilinolein, triolein and tristearin (May *et al.* 1983). These compounds also affect the sensory, functional and nutritional properties of the oil.

Free-fatty acids and other surfactant materials produced in the oil induce foaming, accelerate oxidation reactions that produce off-flavours, and cause foods to absorb oil. However, they also have beneficial effects in frying: they add flavour to fried foods, they contribute towards the characteristic golden colour and they enable fat retention to give the required mouthfeel. Oil that has been used for a short period gives improved frying compared with fresh oil because polar compounds produced by frying reduce interfacial tension and promote better contact between the oil and the product surface. This results in improved and more uniform heat transfer and flavour absorption (Blumenthal 1991). Frying at lower temperatures (consistent with maintaining product quality) or holding oil at lower temperatures when not in use, reduces thermal degradation of oil. In continuous operation, NVDPs are filtered out (section 19.2.3), thus keeping the oil quality at an optimum level. Optimisation of oil quality is discussed in more detail by Nielsen (1993). The potential toxicity of the products of oil decomposition and other nutritional changes are described in section 19.4.3.

19.4   Effect of frying on foods and micro-organisms

The high temperatures and short processing times used in frying are intended to change the sensory properties of the food and to destroy naturally occurring enzymes and any contaminating micro-organisms. One of the most important considerations is the absorption of oil into the fried product (also section 19.1.1).

19.4.1 Oil absorption

Although absorbed oil contributes to the flavour of fried foods and makes them more palatable, there are a number of problems caused by high rates of oil pick-up: it increases operating costs to replenish lost oil; it may affect the shelf-life and stability of the product; and it may cause an imbalance in the nutritive value between fats and other non-fat components. Many fried foods contain up to ≈40% oil (Table 19.4) (Saguy and Pinthus 1995) and where fried foods form a large part of the diet, excessive fat consumption can be an important source of ill-health (section 19.4.3).

These risks, and consumer trends towards lower-fat products, have highlighted the growing importance of the oil content of fried foods, and have led to considerable research into the factors that affect oil uptake. They have also created pressure on processors to alter processing conditions to reduce the amount of oil absorbed in their products. Research has shown that the type of oil has no effect on the rate of absorption into fried foods (Bouchon 2006, Rimac-Brnčić et al. 2003), but the quality of the oil is very important. This is affected by the:

- temperature and time of frying;
- age and thermal history of the oil, presence of antioxidants;
- interfacial tension between the oil and the product;
- size, surface area, moisture content and surface characteristics of the food;
- design of the fryer and whether frying is continuous or intermittent;
- post-frying treatments.

Each of these factors, together with any pretreatments, such as blanching or partial drying, influences the amount of oil entrained within the food. Rimac-Brnčić et al. (2003) found that pre-frying treatments, especially blanching in 0.5% calcium chloride solution following immersion in 1% solution of carboxymethyl cellulose, reduced oil absorption by 54%.

Bouchon (2006) has reviewed conflicting research into the relationship between frying temperature and oil uptake. For example, Baumann and Escher (1995) found that oil uptake correlated positively to the oil temperature and to the initial dry matter content of the food, and inversely to the slice thickness. Blumenthal (1991) reports that lower temperatures result in the absence of crust formation and greater oil penetration into the food. Mehta and Swinburn (2001) review studies that show 40% more fat being absorbed into French fries that are fried at 10 °C below the recommended 180–185 °C.

Blumenthal (2001) developed a surfactant theory in which several types of surfactants are produced by degradation of the oil or interactions between the food and oil. These act

Table 19.4 Oil content of selected deep-fat fried foods

Product	Oil content (%)
Potato crisps	33–38
Maize crisps	30–38
Tortillas	32–30
Doughnuts	20–25
French fries	10–15

From Min (2007)

as wetting agents that reduce the interfacial tension and cause increased contact between the oil and food, which increases oil absorption. Thanatuksorn *et al.* (2005a) found that both high initial moisture content and surface roughness, created when food is cut or when batter is applied, cause the amount of oil adhering to the surface to increase. As the food cools after frying, the surface oil is absorbed into the food in direct proportion to the surface roughness and the initial moisture content. Pre-drying foods reduces oil uptake. Rubnov and Saguy (1997) confirmed the importance of surface roughness in oil absorption using fractal geometry.

Batters used for coating (also Chapter 24, section 24.1.2) add value to fried products, improve their flavour and reduce moisture loss and weight loss during frying. However, batters can absorb large amounts of oil and hence increase the fat content of the product. Mohamed *et al.* (1998) studied the factors that affect oil absorption by batters, which is related to the porosity of the batter after frying. They found that the amylose content of the batter correlated positively with crispness but negatively with oil absorption. The addition of both ovalbumin and calcium chloride to the batter mix reduced oil absorption and improved the crispness of the fried batter. Thanatuksorn *et al.* (2005b) studied the effect of the moisture content of batter on oil absorption and found that both oil absorption and moisture loss have a linear relation with the square root of the frying time. They suggested that the initial moisture content affected the porous structure of the batter formed by starch gelatinisation during frying, which increased the amount of oil absorbed during frying. Saguy and Pinthus (1995) give a detailed review of studies of oil uptake during frying. Other research (Williams and Mittal 1999, Pinthus *et al.* 1993) into the use of hydrocolloids in batters, including methylcellulose and hydroxypropyl methylcellulose (HPMC), has shown that they create a barrier that reduces oil absorption.

19.4.2   Changes to texture, colour and flavour

The main purpose of frying is the development of a characteristic crisp texture, golden colour, and fried flavours and aromas in foods. These eating qualities are developed by a combination of physicochemical changes in the food and compounds absorbed from the oil. The interactions between frying oils and fried foods are reviewed by Dobarganes *et al.* (2000). Frying produces physicochemical changes to the main food components (proteins, fats and polymeric carbohydrates) that are similar to those produced by baking (Chapter 18). Microstructural changes produce the desirable crispness of fried foods whether in fully dried products such as snackfoods or in the dry, crisp outer crust surrounding a moist cooked core. These changes include starch gelatinisation and dehydration, protein denaturation and changes to protein quality as a result of Maillard reactions with amino acids, evaporation of moisture leading to dehydration of the tissues, reduction in intercellular air, and absorption of oil. Surface cells shrink during frying but do not rupture because, although the starch granules are gelatinised, the rapid dehydration reduces the swelling of granules that is found in moist heating (Costa *et al.* 2000). Aguilera and Gloria (1997) identified the microstructure of fried potato chips as a thin outer layer ($\approx$250 μm thick) formed by cells damaged during slicing, an intermediate layer of shrunken dehydrated cells in the crust that extends to the evaporation front, and a core of hydrated cells containing gelatinised starch.

The golden brown surface colour is formed by Maillard reactions and colour development follows first-order reaction kinetics during frying, and increases as frying time and/or temperature increases (Sahin 2000). Absorbed fat contributes to the flavour and aroma of the crust. Flavour compounds produced in the crust include lactone, hydroxy-nonenoic acids and decadienal (May *et al.* 1983). A small amount of oil oxidation is important to develop the characteristic flavour of fried foods, but as oil breaks down (section 19.3.1) it produces compounds that cause off-flavours and darkening of the oil and may be toxic in high concentrations. A large number of volatiles are produced in the crust by frying (e.g. 2-octenal (tallowy, nutty flavour), dimethyltrisulphide (cabbage flavour) and 2-ethyl-3,6-dimethylpyrazine (roasted, earthy flavour)) (Frankel 2005). Heterocyclic pyrazines and furans have typical fried flavours. Changes to organoleptic quality in the core of the food are caused by steaming so that it retains a moist texture and the flavour of the ingredients. The temperature does not exceed $\approx100\,°C$ and the flavours are not a result of chemical changes caused by high temperatures or fat absorption.

19.4.3   Nutritional changes

The absorption of oil by fried foods (section 19.1.1) is a key nutritional consideration. Excessive fat consumption is an important dietary contributor to obesity, which results in multiple health consequences, including type II diabetes and coronary heart disease. The risk from cancers of the colon, prostate and breast may be increased by high fat intake (Browner *et al.* 1991) and the degree of saturation of oils has a significant influence on the tendency to develop arteriosclerosis and thrombosis. Methods that are used to reduce oil absorption by fried foods are described in section 19.4.1 but many fried foods continue to contain significant levels of fat (Table 19.4). Deep-fat frying using hydrogenated oils causes the production of harmful components such as *trans*-fatty acids, and highly oxidised or polymerised constituents of fatty acids (Vorria *et al.* 2004). Concerns over *trans*-fatty acids in hydrogenated oils and the levels of saturated fatty acids in oils have encouraged the use of alternative, more unsaturated oils (section 19.3). Some manufacturers are now producing baked products (e.g. potato crisps) as an alternative method to frying for reducing their fat content.

Another alternative is to change the type of oil. 'Enova' oil, approved in 2000 as GRAS (generally recognised as safe), is marketed as being able to maintain healthy body weight and body fat. Whereas most vegetable and maize oils used for frying contain triacylglycerides with smaller amounts of diacylglycerides, this oil is processed to increase the concentration of diacylglycerol. The oil has the same calorific value as conventional oils, but it is metabolised differently and more of the oil is used directly as energy by the body rather than being stored as fat. The oil can also reduce post-meal serum triglycerides by 30–50% (Anon 2008a). A sucrose polyester fat substitute named 'Olestra' has also been approved as frying oil. It is made from sucrose and fatty acid methyl esters and has a structure that is resistant to digestive enzymes, and hence has no calorific value. It is used to fry low-fat snackfoods in some countries, but uptake in others is currently (2008) subject to further research. Other lipid-based fat substitutes are described in detail by Sandrou and Arvanitoyannis (2000).

Naturally occurring fats and oils in foods, including highly unsaturated fish oils, are only slightly affected because of the short frying time and limited access of oxygen. The essential fatty acid, linoleic acid, is readily lost and therefore changes the balance of saturated and unsaturated fatty acids in the oil. Starch and non-starch carbohydrates are partially broken down by frying and starch–lipid complexes are formed. Sucrose is hydrolysed to glucose and fructose, which are lost by Maillard reactions and caramelisation during heating, especially in the crust. Proteins are rapidly denatured in the crust and most enzymes are inactivated. Protein availability is reduced and some essential amino acids (e.g. lysine and tryptophan) are destroyed (Pokorny 2002). For example a 17% loss of available lysine is reported in fried fish, although this loss increased to 25% when thermally damaged oil was used (Tooley 1972).

Vitamin losses in fried foods depend on the temperature and time of frying and the condition of the oil. There are also substantial differences between losses in the core of the food and in the crust. Rapid crust formation seals the food surface, which reduces the extent of changes to the bulk of the food, and therefore retains a high proportion of the nutrients. In addition, these foods are usually consumed shortly after frying and there are few losses during storage. For example, vitamin C losses in fried potatoes are lower than in boiling because the vitamin accumulates as dehydroascorbic acid (DAA) owing to the lower moisture content, whereas in boiling, DAA is hydrolysed to 2,3-diketogluconic acid and therefore becomes unavailable. In the crust, there are substantially higher losses of nutrients, particularly fat-soluble vitamins, which results in a loss of nutritional value. Retinol, carotenoids and tocopherols are each destroyed and also contribute to the changes in flavour and colour of the oil. For example, vitamin E that is absorbed from oil by crisps is oxidised during subsequent storage. Bunnel et al. (1965) found 77% loss after 8 weeks at ambient temperature. Oxidation proceeds at a similar rate at low temperatures and French-fried potatoes lost 74% of the vitamin E in a similar period under frozen storage. Heat- or oxygen-sensitive, water-soluble vitamins are also destroyed by frying. However, the preferential oxidation of tocopherols has a protective (antioxidant) effect on the oil.

Some volatile compounds formed during deep-fat frying are known to be toxic (e.g. 1,4-dioxane, benzene, toluene and hexyl-benzene) (May et al. 1983). In 2002, the Swedish National Food Administration announced that high levels of potentially carcinogenic acrylamide had been discovered in fried and baked foods. It is produced by reaction between amino acids and reducing sugars at the high temperatures used in frying, baking and roasting. Acrylamide is considered as a potential carcinogen in animals and may affect humans when consumed in large amounts. Ingested acrylamide is metabolised to a chemically reactive epoxide, glycidamide and there is evidence that exposure to large doses can cause damage to the male reproductive organs of animals (Yang et al. 2005). Acrylamide has been shown to produce various types of cancer in mice and rats, but studies in human populations have so far failed to produce consistent results, possibly because acrylamide is ubiquitous in western diets and it might be difficult to isolate its effects.

There has been considerable research to reduce levels of acrylamide since its discovery in fried foods. The level of reducing sugars in potatoes was found to be the most important parameter for acrylamide formation and cultivars that have low reducing sugar concentrations are now selected for crisp production. Blanching and soaking of potatoes were also identified as important risk reduction processes (Cummins et al. 2006). Soaking uncooked potato products in amino acid solutions reduces acrylamide formation (Kim et al. 2005). The addition of 0.5% glycine to potato snacks reduced acrylamide formation by >70% and soaking potato slices in 3% lysine or glycine produced >80% acrylamide reduction in chips fried at 185 °C. Jung et al. (2003) found that lowering the pH of fried corn chips using 0.2% citric acid reduced the formation of acrylamide by 82%. Dipping cut potato in 1% and 2% citric acid solutions for 1 h before frying produced 73% and 79.7% inhibition of acrylamide formation in French fries, but Wicklund et al. (2006) found blanching had no effect on the concentration of acrylamide in deep-fat fried potato crisps. Other methods to reduce acrylamide formation include storage of potatoes >8 °C, which prevents increases in the fructose content, and lactic acid fermentation of potatoes (Baardseth *et al.* 2006). Kim *et al.* (2005) found that the formation of acrylamide in fried foods depended on the composition of the raw materials, the frying time and temperature. Acrylamide was rapidly formed in potato chips >160 °C, the amount being proportional to the heating time and temperature.

Lower temperatures in vacuum frying have also been shown to reduce acrylamide formation. Vacuum frying sliced potatoes at 118–140 °C reduced acrylamide formation by 94% compared with atmospheric frying at 150–180 °C. The acrylamide content was reduced by 51% when the frying temperature was decreased from 180 to 165 °C during traditional frying and by 63% when the temperature decreased from 140 to 125 °C in vacuum frying (Granda and Moreira 2005, Granda *et al.* 2004). Increased frying time increased acrylamide formation during traditional frying at all temperatures. Chen and Mai Tran (2007) studied pre-drying and vacuum frying of potato crisps to reduce acrylamide formation. Sliced potatoes were blanched and dried to 60% of their initial weight (wet weight basis) and vacuum fried at 120 °C. There was a 92% reduction in acrylamide content and the crisps contained <50% (dry basis) of the oil content compared with crisps that were not pretreated and fried conventionally.

19.4.4   Effect of frying on micro-organisms
There are limited numbers of studies of microbial destruction by frying, but the time and temperature needed to adequately cook the core of fried foods are sufficient to destroy vegetative cells of pathogens and spoilage micro-organisms. For example, Whyte *et al.* (2006) report that shallow frying chicken liver to reach a core temperature of 70–80 °C for 2–3 min was sufficient to inactivate naturally occurring *Campylobacter* spp. Details of thermal destruction of micro-organisms are given in Chapter 10 (section 10.3).

References

AGUILERA, J.M. and GLORIA, H., (1997), Determination of oil in potato products by differential scanning calorimetry, *J. Agriculture and Food Chemistry*, **45**, 781–785.

ANON, (2006a), Energy conservation, frying performance key to frymasters success in food industry, 13 November, available at http://dean.enodis.com/pressrelease.asp?article= EnergyConservation.xml.

ANON, (2006b), Direct heating systems, company information from Coat and Fry Ltd, available at www.coatandfry.co.uk/heatingsystems.

ANON, (2007a), Infinitely Better, information from Falcon Foodservice Equipment, available at www.infinityfryers.com/6.htm.

ANON, (2007b), Potato chips frying systems, company information from BMA Florigo, available at www.bma-nl.com/Potato-chips.888.0.html.

ANON, (2007c), Put some pressure on your competition! – why pressure frying is superior to open frying, company information from The Broaster Company, available at www.broaster.com/ pdfs/lit/pfqa-lr.pdf.

ANON, (2007d), ToxFAQs for Acrolein, Agency for Toxic Substances and Disease Registry, available at www.atsdr.cdc.gov/tfacts124.html#bookmark02.

ANON, (2008a), How Enova oil works, company information available at www.enovaoil.com/about/.

ANON, (2008b), International Food Composition Tables Directory, FAO, Rome, available at www.fao.org/infoods/directory_en.stm.

BAARDSETH, P., BLOM, H., SKREDE, G., MYDLAND, L.T., SKREDE, A. and SLINDE, E., (2006) Lactic acid fermentation reduces acrylamide formation and other maillard reactions in French fries, *J. Food Science*, **71** (1), C28–C33.

BAUMANN, B. and ESCHER, F., (1995), Mass and heat transfer during deep-fat frying of potato slices – I. Rate of drying and oil uptake, *Lebensmittel-Wissenschaft und -Technologie*, **28** (4), 395–403.

BHEEMREDDY, R.M., CHINNAN, M.S., PANNU, K.S. and REYNOLDS, A.E., (2002a), Filtration and filter system for treated frying oil, *J. Food Process Engineering*, **25** (1), 23–40.

BHEEMREDDY, R.M., CHINNAN, M.S., PANNU, K.S. and REYNOLDS, A.E., (2002b), Active treatment of frying oil for enhanced fry-life, *J. Food Science*, **67** (4), 1478–1484.

BLUMENTHAL, M.M., (1991), A new look at the chemistry and physics of deep fat frying, *Food Technology*, **2**, 68–71.

BLUMENTHAL, M.M., (2001), A new look at frying science, *Cereal Foods World*, **46** (8), 352–354.

BOUCHON, P. (2006), Frying, in (J.G. Brennan, Ed.), *Food Processing Handbook*, Wiley-VCH, Weinheim, Germany, pp. 269–290.

BOUCHON, P AND PYLE, D.L., (2005), Modelling oil absorption during post-frying cooling: I: Model development, *Food and Bioproducts Processing*, **83**, Issue C4, 253–260.

BOUCHON, P., AGUILERA, J.M. and PYLE, D.L., (2003), Structure oil-absorption relationships during deep-fat frying, *J. Food Science*, **68** (9), 2711–2716.

BROWNER, W.S., WESTENHOUSE, J. and TICE, J.A., (1991), What if Americans ate less fat? A quantitative estimate of the effect on mortality, *J. American Medical Association*, **265**, 3285–3291.

BUDŽAKI, S. and ŠERUGA, B., (2005), Determination of convective heat transfer coefficient during frying of potato dough, *J. Food Engineering*, **66** (3), 307–314.

BUNNELL, R.H., KEATING, J., QUARESIMO, A. and PARMAN, G.K., (1965), Alpha-tocopherol contents of foods, *American J. Clinical Nutrition*, **17**, 1–10.

CHEN, X.D. and MAI TRAN, T., (2007), Reducing acrylamide formation in fried potato crisps by pre-drying and vacuum frying techniques, paper presented at Chemeca 2007, Victoria, Australia, 23–26 Sept., abstract available at www.chemeca2007.com/abstract/159.htm.

CHU, M. (2004), Smoke points of various fats, available at www.cookingforengineers.com/article/50/Smoke-Points-of-Various-Fats.

COSTA, R.M., OLIVEIRA, F.A.R., DELANEY, O. and GEKAS, V., (1999), Analysis of the heat transfer coefficient during potato frying, *J. Food Engineering*, **39** (3), 293–299.

COSTA, R.M., OLIVEIRA, F.A.R. and BOUTCHEVA, G., (2000), Structural changes and shrinkage of potato during frying, *International J. Food Science and Technology*, **36**, 11–24.

CUMMINS, E., BUTLER, F., GORMLEY, R. and BRUNTON, N., (2006), A methodology for evaluating the formation and human exposure to acrylamide through fried potato crisps, *LWT – Food Science and Technology*, **39** (5), 571–575.

DA SILVA, M.G. and SINGH, R.P., (1995), Viscosity and surface tension of corn oil at frying temperatures, *J. Food Processing and Preservation*, **19** (4), 259–270.

DOBARGANES, C., MÁRQUEZ-RUIZ, G. and VELASCO, J., (2000), Interactions between fat and food during deep-frying, *European J. Lipid Science and Technology*, **102** (8–9), 521–528.

FARINU, A. and BAIK, O-D., (2007), Heat transfer coefficients during deep fat frying of sweetpotato: effects of product size and oil temperature, *Food Research International*, **40** (8), 989–994.

FARKAS, B.E, SINGH, R.P. and MCCARTHY, M.J., (1996), Modeling heat and mass transfer in immersion frying – I. Model development, *J. Food Engineering*, **29**, 211–226.

FRANKEL, E.N., (2005), *Lipid Oxidation*, 2nd edn, The Oily Press, PJ Barnes & Associates, Bridgwater.

GARAYO, J. and MOREIRA, R., (2002), Vacuum frying of potato chips, *J. Food Engineering*, **55** (2), 181–191.

GÖKMEN, V., ŞENYUVA, H.Z., DÜLEK, B. and ÇETIN, A.E., (2007), Computer vision-based image analysis for the estimation of acrylamide concentrations of potato chips and French fries, *Food Chemistry*, **101** (2), 791–798, available at www.aseanfood.info/Articles/11017372.pdf.

GRANDA, C. and MOREIRA, R.G., (2005), Kinetics of acrylamide formation during traditional and vacuum frying of potato chips, *J. Food Process Engineering*, **28** (5), 478–493.

GRANDA, C., MOREIRA, R.G. and TICHY, S.E., (2004), Reduction of acrylamide formation in potato chips by low-temperature vacuum frying, *J. Food Science*, **69** (8), 405–411.

HUBBARD, L.J. and FARKAS, B.E., (2000), Influence of oil temperature on convective heat transfer during immersion frying, *J. Food Processing and Preservation*, **24** (2), 143–162.

JUNG, M.Y., CHOI, D.S. and JU, J.W., (2003), A novel technique for limitation of acrylamide formation in fried and baked corn chips and in French fries, *J. Food Science*, **68** (4), 1287–1290.

KERR, R.M. (2006), Deep-fat frying basics for food services, FAPC 126, Oklahoma Cooperative Extension Service Division of Agricultural Sciences and Natural Resources, available at www.fapc.okstate.edu/news/factsheets.html.

KHEANG, L.S., MAY, C.Y., FOON, C.S. and NGAN, M.A., (2006), Recovery and conversion of palm olein-derived used frying oil to methyl esters for biodiesel, *J. Oil Palm Research*, **18**, 247–252.

KIM, C.T., HWANG, E-S. and LEE, H.J., (2005), Reducing acrylamide in fried snack products by adding amino acids, *J. Food Science*, **70** (5), C354–C358.

KOCHHAR, S.P., (1999), Safety and reliability during frying operations – effects of detrimental components and fryer design features, in (D. Boskou and I. Elmadfa, Eds.), *Frying of Food*, Technomic Publishing, Lancaster, PA, pp. 253–269.

LIN, S., AKOH, C.C. and REYNOLDS, A.E., (1998), The recovery of used frying oils with various adsorbents, *J. Food Lipids*, **5** (1), 1–16.

LIN, S., AKOH, C.C. and REYNOLDS, A.E., (1999), Determination of optimal conditions for selected adsorbent combinations to recover used frying oils, *J. American Oil Chemists' Society*, **76** (6), 739–744.

MAY, W.A., PETERSON, R.J. and CHANG, S.S., (1983), Chemical reactions involved in the deep fat frying of foods, IX: Identification of the volatile decomposition products of triolein, *J. American Oil Chemists' Society*, **60** (5), 990–995, available at: http://class.fst.ohio-state.edu/fst821/Lect/fry.pdf

MEHTA, U. and SWINBURN, B., (2001), A review of factors affecting fat absorption in hot chips, *Critical Reviews in Food Science and Nutrition*, **41** (2), 133–154.

MIN, D.B., (2007), Chemical reactions of deep-fat frying of foods, Dept. Food Science and Technology, Ohio State University, available at www.fst.osu.edu/min/Food%20Lipids%20Chemistry.htm#.

MILLER, K.S., SINGH, R.P. and FARKAS, B.E., (1994), Viscosity and heat transfer coefficients for canola, corn, palm and soybean oil, *J. Food Processing and Preservation*, **18**, 461–472.

MOHAMED, S., HAMID, N.A. and HAMID, M.A., (1998), Food components affecting the oil absorption and crispness of fried batter, *J. Science of Food and Agriculture*, **78** (1), 39–45.

MOREIRA, R.G. and BARRUFET, M.A., (1998), A new approach to describe oil absorption in fried foods: a simulation study, *J. Food Engineering*, **35**, 1–22.

MOREIRA, R.G., SUN, X. and CHEN, Y., (1997), Factors affecting oil uptake in tortilla chips in deep-fat frying, *J. Food Engineering*, **31**, 485–498.

MOREIRA, R.G., CASTELL-PEREZ, M.E. and BARRUFET, M.A., (1999), *Deep Fat Frying – Fundamentals and Applications*, Aspen Publishing, Gaithersburg, MD.

NGADI, M. and IKEDIALA, J.N., (2005), Natural heat transfer coefficients of chicken drum shaped bodies, *International J. Food Engineering*, **1** (3), Article 4, available at: http://www.bepress.com/ijfe/vol1/iss3/art4

NIELSEN, K., (1993), Frying oils technology, in (A. Turner, Ed.), *Food Technology International Europe*, Sterling Publications International, London, pp. 127–132.

PAUL, S. and MITTAL, G.S., (1997), Regulating the use of degraded oil/fat in deep-fat/oil food frying, *Critical Reviews in Food Science and Nutrition*, **37** (7), 635–662.

PHOGAT, S.S., MITTAL, G.S. and KAKUDA, Y., (2006), Comparative evaluation of regenerative capacity of different adsorbents and filters for degraded frying oil, *Food Science and Technology International*, **12** (2), 145–157.

PINTHUS, E.J., WEINBERG, P. and SAGUY, I.S., (1993), Criteria for oil uptake during deep fat frying, *J. Food Science*, **58**, 204–205.

POKORNY, J., (2002), Frying, in (C.J.K. Henry and C. Chapman, Eds.), *The Nutrition Handbook for Food Processors*, Woodhead Publishing, Cambridge, pp. 293–300.

RAZALI, I. and BADRI, M., (2003) Oil absorption, polymer and polar compounds formation during deep-fat frying of French fries in vegetable oils, *Palm Oil Developments*, **38**, 11–15.

RIMAC-BRNČIĆ, S., LELAS, V., RADE, D. and SIMUNDIC, B., (2003), Decreasing of oil absorption in potato strips during deep fat frying, *J. Food Engineering*, **64** (2), 237–241.

ROSSELL, J.B., (ED.), (2001), *Frying: Improving Quality*, Woodhead Publications, Cambridge.

RUBNOV, M. and SAGUY, I.S., (1997), Fractal analysis and crust water diffusivity of a restructured potato product during deep-fat frying, *J. Food Science*, **62** (1), 135–137.

RYWOTYCKI, R., (2003), Food frying process control system, *J. Food Engineering*, **59** (4), 339–342.

SAGUY, I.S. and PINTHUS, E.J., (1995), Oil uptake during deep-fat frying: factors and mechanism, *Food Technology*, **4**, 142–145,152.

SAHIN, S., (2000), Effects of frying parameters on the colour development of fried potatoes, *European Food Research and Technology*, **211** (3), 165–168.

SAHIN, S., SASTRY, S.K. and BAYINDIRLI, L., (1999a), Heat transfer during frying of potato slices, *Lebensmittel-Wissenschaft und -Technologie*, **32** (1), 19–24.

SAHIN, S., SASTRY, S.K. and BAYINDIRLI, L., (1999b), The determination of convective heat transfer coefficient during frying, *J. Food Engineering*, **39** (3), 307–311.

SANDROU, D.K. and ARVANITOYANNIS, I.S., (2000), Low-fat/calorie foods: current state and perspectives, *Critical Reviews in Food Science and Nutrition*, **40** (5), 427–447.

SELMAN, J., (1989), Oil uptake in fried potato products. In *Frying, Symp. Proc. 35*, 25 Feb, 1988, pp. 70–81, British Food Manufacturing Industry Research Association, Leatherhead.

ŠERUGA, B. and BUDŽAKI, S., (2005), Determination of thermal conductivity and convective heat transfer coefficient during deep fat frying of "Krotula" dough, *European Food Research and Technology*, **221** (3–4), 351–356.

STIER, R.F., (2003), Finding functionality in fats and oils, Prepared Foods, available at www.preparedfoods.com/CDA/Archives/5dfe022f62788010VgnVCM100000f932a8c0.

THANATUKSORN, P., PRADISTSUWANA, C., JANTAWAT, P. and SUZUKI, T., (2005a), Effect of surface roughness on post-frying oil absorption in wheat flour and water food model, *J. Science of Food and Agriculture*, **85** (15), 2574–2580.

THANATUKSORN, P., PRADISTSUWANA, C., JANTAWAT, P. and SUZUKI, T., (2005b), Oil absorption and drying in the deep fat frying process of wheat flour–water mixture, from batter to dough, *Japan J. Food Engineering*, **6** (2), 49–55.

TOOLEY, P.J., (1972), The effect of deep-fat frying on the availability of fish lysine, *Nutrition Society Proceedings*, **31**, 2A.

UFHEIL, G. and ESCHER, F., (1996), Dynamics of oil uptake during deep-fat frying of potato slices, *Lebensmittel-Wissenschaft und -Technologie*, **29** (7), 640–643.

VITRAC, O., TRYSTRAM, G. and RAOULT-WACK, A-L., (2003), Continuous measurement of convective heat flux during deep-frying: validation and application to inverse modelling, *J. Food Engineering*, **60** (2), 111–124.

VORRIA, E., GIANNOU, V. and TZIA, C., (2004), Hazard analysis and critical control point of frying – safety assurance of fried foods, *European J. Lipid Science and Technology*, **106** (11), 759–765.

WARNER, K.A., (2007), Frying oil deterioration, in (C. Akoh, Ed.), *Food Lipids: Chemistry, Nutrition and Biotechnology*, Taylor and Francis, Boca Raton, FL, pp. 71–82.

WHYTE, R., HUDSON, J.A. and GRAHAM, C., (2006), *Campylobacter* in chicken livers and their destruction by pan frying, *Letters in Applied Microbiology*, **43** (6), 591–595.

WICHCHUKIT, S., ZORRILLA, S.E. and SINGH, R.P., (2001), Contact heat transfer coefficient during double-sided cooking of hamburger patties, *J. Food Processing and Preservation*, **25** (3), 207–221.

WICKLUND, T., ØSTLIE, H., LOTHE, O., HALVOR KNUTSEN, S., BRÅTHEN, E. and KITA, A., (2006), Acrylamide in potato crisps – the effect of raw material and processing, *LWT – Food Science and Technology*, **39** (5), 571–575.

WILLIAMS, R. and MITTAL, G.S., (1999), Low fat fried foods with edible coatings: modelling and simulation, *J. Food Science*, **64**, 317–322.

WOOD, M.A., HARRIS, K.W., PROUD, D.M. and TREADWELL, D.D., (1983), *Quantity Recipes*, New York State College of Home Economics, New York.

YANG, H-J., LEE, S-H., JIN, Y., CHOI, J-H., HAN, C-H. and LEE, M-H., (2005), Genotoxicity and toxicological effects of acrylamide on reproductive system in male rats, *J. Vetinerinary Science*, **6** (2), 103–109.

YILDIZ, A., PALAZOĞLU T.K. and ERDOĞDU, F., (2007), Determination of heat and mass transfer parameters during frying of potato slices, *J. Food Engineering*, **79** (1), 11–17.

III.D

Heat processing by direct and radiated energy

20

Dielectric, ohmic and infrared heating

Abstract: This chapter describes the uses of dielectric and infrared electromagnetic waves that penetrate food and are absorbed and converted to heat. They are compared with ohmic heating, which uses the electrical resistance of foods to directly convert electricity to heat. Each section explains the theory of heating and the equipment used. The applications of dielectric heating to baking, dehydration, tempering and thawing of foods, and the use of ohmic heating for aseptic processing are described. Finally the effects of each type of heating on both foods and micro-organisms are discussed.

Key words: electromagnetic waves, dielectric heating, dielectric loss factor, radio frequency heating, microwaves, magnetron, dehydration, baking, tempering and thawing, ohmic heating, UHT processing, infrared heating.

Dielectric[1] and infrared (IR or radiant) energy are two forms of electromagnetic energy (Fig. 20.1) (see also high-intensity light/UV light (Chapter 9, section 9.4)). Each has a much lower energy than gamma rays or X-rays (Chapter 7 and Chapter 9, section 9.5) and is a form of non-ionising radiation that only produces thermal effects in foods (see section 20.1.4). Electromagnetic waves penetrate food and are then absorbed and converted to heat. In contrast, ohmic (or resistance) heating uses the electrical resistance of foods to directly convert electricity to heat (see also pulsed electric fields (Chapter 9, section 9.1)).

Dielectric and ohmic heating are direct methods in which heat is generated within the product, whereas infrared heating is an indirect method in which energy is applied to the surface of a food by radiation and then converted to heat. Radiated infrared energy is a component of heat produced by conventional heaters, especially in baking ovens (Chapter 18, section 18.1), but this chapter describes the generation and use of infrared energy as a main means of heating.

Dielectric and ohmic heating are used to preserve foods, whereas infrared radiation is mainly used to alter the eating qualities of foods by changing the surface colour, flavour and aroma. The main applications of these methods are shown in Table 20.1. The advantages of dielectric and ohmic heating over conventional heating (Chapter 10) can be summarised as:

---

1. There are differences in the terminology used to describe dielectric energy and in this chapter the term 'dielectric' is used to represent both radio frequency (RF) and microwave heating.

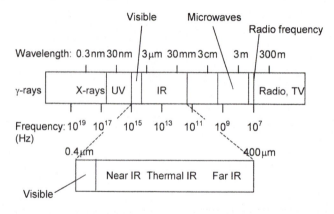

Fig. 20.1   Electromagnetic spectrum (from Keiner 2007).

Table 20.1   Applications of dielectric, ohmic and infrared heating

Method of heating	Applications
Microwave	Cooking, thawing, melting, finish-drying, freeze drying, tempering, pasteurisation, sterilisation, rendering, frying, blanching
Radio frequency	Drying, baking
Ohmic	UHT sterilisation, pasteurisation
Infrared	Drying, baking, frying, thawing, freeze drying, cooking, surface pasteurisation

- rapid heating throughout the food without localised overheating or hot surfaces, which results in minimum heat damage and no surface browning;
- heat transfer is not limited by boundary films and energy conversion efficiencies are high;
- equipment is small, compact and suited to automatic control;
- there is no contamination of foods by products of combustion.

20.1   Dielectric heating

20.1.1   Theory

Microwave and RF energy are transmitted as electromagnetic waves and the depth to which these penetrate foods is determined by both their frequency and the characteristics of the food. Microwave energy has a range of frequencies from 300 MHz to 300 GHz whereas RF energy has lower frequencies, from 1 to 200 MHz. However, because these frequencies are also used for communications and navigation, an international agreement has allocated the following bands for industrial, scientific and medical use:

- Microwaves: 915 MHz (range 902–928 MHz) and 2450 MHz (range 2400–2500 MHz).
- Radio frequency: 13.560 MHz (range 13.553–13.567 MHz), 27.120 MHz (range 26.957–27.283 MHz) and the seldom used 40.68 MHz (range 40.66–40.70 MHz).

The relationship between wavelength, frequency and velocity of electromagnetic waves is shown in Equation 20.1:

$$\lambda = v/f' \qquad \boxed{20.1}$$

where λ (m) = wavelength, v (ms^{-1}) = velocity and f' (Hz) = frequency.

The velocity of electromagnetic waves is $3 \times 10^8 \, \mathrm{m\,s^{-1}}$ (Singh and Heldman 2001) and using Equation 20.1, the calculated wavelength of microwaves is 0.328 m at 915 MHz and 0.122 m at 2450 MHz. Lower-frequency (and longer wavelength (Fig. 20.1)) waves have greater penetration depths. The energy in electromagnetic waves can be considered to be in the form of photons that are discrete, very small quantities of energy. When photons strike a target material, they are either absorbed or they pass through the material. Different atoms in a material have electrons with different allowed atomic energy states. For heating to take place, the energy in the photons must exactly match the energy difference between these atomic energy states. If they are different, the material is transparent to the electromagnetic wave (Ehlermann 2002). This is why, for example, water in foods is heated but plastic and glass are not. The amount of energy absorbed by foods from electromagnetic waves depends on a characteristic known as the 'dielectric loss factor' (ϵ''), a dimensionless number, which relates to the ability of the food to dissipate electrical energy. The higher the loss factor, the more energy is absorbed by the food (Table 20.2). The loss factor depends on the moisture content of the food, its temperature, the presence of salts, and in some foods the structure of the food.

Water has a negatively charged oxygen atom separated from two positively charged hydrogen atoms, which form an electric dipole (Chapter 1, section 1.1.1). When alternating microwave or RF energy is applied to a food, dipoles in the water and other polar components reorient themselves to the direction (or polarity) of the electric field in a similar way to a compass in a magnetic field. Since the polarity rapidly alternates from positive to negative and back again several million times per second (e.g. at the

Table 20.2 Dielectric properties of foods and packaging materials using microwaves at 2450 MHz (materials, except ice, at 20–25 °C).

Material	Dielectric constant (ϵ')	Loss factor (ϵ'')	Penetration depth (cm)
Foods			
Apple	63.4	16	–
Banana (raw)	62	17	0.93
Beef (raw)	51	16	0.87
Bread	4	0.005	1170
Brine (5%)	67	71	0.25
Butter	3	0.1	30.5
Carrot (cooked)	71	18	0.93
Cooking oil	2.6	0.2	19.5
Fish (cooked)	46.5	12	1.1
Ham	85	67	0.3
Ice	3.2	0.003	1162
Potato (raw)	62	16.7	0.93
Strawberry	75.1	36.7	–
Water (distilled)	77	9.2	1.7
Packaging materials			
Glass	6	0.1	40
Paper	4	0.1	50
Polyester tray	4	0.02	195
Polystyrene	2.35	0.001	–

Adapted from Piyasena et al. (2003), Mudget (1982), Buffler (1993) and Mohsenin (1984)

microwave frequency of 2450 MHz, the polarity changes 2.45×10^9 cycles s^{-1}), the dipoles rotate to align with the rapidly changing polarity. The microwaves give up their energy and the molecular movement creates frictional heat that increases the temperature of water molecules. They in turn heat the surrounding components of the food by conduction and/or convection.

The amount of heat absorbed by a food, the rate of heating and the location of 'cold spots' (points of slowest heating) depend on the food composition (e.g. moisture content, ionic strength, density and specific heat), the shape and size of the food, the microwave frequency used and the applicator design. The time that food is heated is also important because its microwave absorption properties and the location of cold spots can change with time.

Water in foods also has a degree of electrical conductivity due to dissolved salts that form electrically charged ions. The charged ions move at an accelerated rate when an electric field is applied (known as 'ionic polarisation') to produce an electric current in the food. Collisions between the ions convert kinetic energy to heat, and more concentrated solutions (which have more collisions) therefore heat more quickly. At RF frequencies, the conductivity of foods and hence the amount of energy absorbed increase at higher temperatures, but at microwave frequencies the loss factor decreases at higher temperatures and so reduces the amount of energy absorbed.

The other important electrical properties of the food, in addition to the loss factor, are (a) the dielectric constant (ϵ'), a dimensionless number that relates to the rate at which energy penetrates a food – in practice most foods are able to absorb a large proportion of electromagnetic energy and heat rapidly; and (b) the loss tangent (tan δ), which gives an indication of how easily the food can be penetrated by electromagnetic waves and the extent to which it converts the electrical energy to heat. These terms are related using Equation 20.2:

$$\epsilon'' = \epsilon' \tan \delta \qquad \boxed{20.2}$$

The dielectric constant and the loss tangent are properties of the food and they influence the amount of energy that is absorbed by the food as shown in Equation 20.3:

$$P = 55.61 \times 10^{-14} f E^2 \epsilon'' \qquad \boxed{20.3}$$

where P (W cm^{-3}) = power absorbed per unit volume, f (Hz) = frequency and E (V cm^{-1}) = electrical field strength. There is therefore a direct relationship between the properties of the food and the energy provided by the dielectric heater. Increasing the electrical field strength has a substantial effect on the power absorbed by the food because the relationship involves a square term.

The depth of penetration of electromagnetic waves is found from the loss factor and the frequency of the waves:

$$x = \frac{\lambda}{2\pi \sqrt{(\epsilon' \tan \delta)}} \qquad \boxed{20.4}$$

where x (m) = the depth of penetration.

The electrical properties of the food also determine how energy is distributed through the food, as represented by the attenuation factor (α') in Equation 20.5:

$$\alpha' = \frac{2\pi}{\lambda} \left[\frac{\epsilon'(\sqrt{1 + \tan^2 \delta} - 1)}{2} \right] \qquad \boxed{20.5}$$

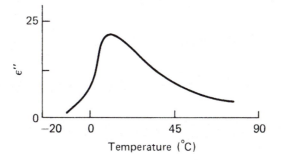

Fig. 20.2   Variation in dielectric loss factor of water and ice (after Lewis 1990).

where α' (m^{-1}) = attenuation factor. Examples of the dielectric constants, loss factors and penetration depths in selected foods are given in Table 20.2 and further details of heat and mass transfer in microwave heating are given by Datta (2001) and Piyasena *et al.* (2003).

It can be seen in Table 20.2 that electromagnetic waves penetrate foods that contain small amounts of water to a much greater depth than moist foods, and that glass, paper and plastic packaging materials have a low loss factor and are almost transparent to microwaves. Microwave penetration increases dramatically when water changes phase to ice (Fig. 20.2), because the molecules are less free to move or absorb energy from the alternating electric field. Ice therefore has a lower loss factor than water and this has important implications for dielectric thawing and tempering applications (section 20.1.3). Continuous metal sheets reflect microwaves and very little energy is absorbed, but small pieces of metal (e.g. wires) and metallised plastic (Chapter 25, sections 25.2.4 and 25.4.3) absorb electromagnetic waves and heat very quickly.

When RF and microwave energy heat water in foods, it increases the vapour pressure and causes movement of moisture from the interior to the surface and rapid evaporation from the surface, therefore making this technology particularly suitable for dehydration as well as heating (section 20.1.3; see also Chapter 16, section 16.1).

In contrast to conventional heating, where the maximum food temperature is that of the heating medium and it is possible to predict the time–temperature history at the slowest-heating point in a food (Chapter 10, section 10.2), this is less straightforward in microwave heating. As food heats, microwave absorption increases, which increases the rate of temperature increase and so further increases the rate of microwave absorption. This 'coupling' continuously generates heat to increase the food temperature and could lead to runaway heating. Microwave equipment therefore needs to be turned on and off (cycled) to keep the temperature within prescribed limits once the target temperature has been reached. Also because microwave absorption is lower at lower temperatures, the waves are able to penetrate further into the food. As it heats, the depth of penetration falls and at higher temperatures, the surface can shield the interior from further heating.

Since heat is generated throughout the food at different rates, the temperature difference between the coldest and hottest points in the food increases with time. This is in contrast to conventional heating, where the coldest point slowly approaches the surface temperature, corresponding to the temperature of the heating medium. Because of their widespread domestic use, there is a popular notion that microwaves 'heat from the inside out'. In fact, the food is heated while the surrounding cold air keeps the surface temperature below that of locations within the food. Surface evaporation from unpackaged

Table 20.3   Summary of process factors in microwave heating

Factor	Examples
Food	Shape, size, composition (e.g. moisture, salt), multiple components (e.g. frozen meals), liquid/solid proportion
Package	Transparency to microwaves, presence of metals (e.g. aluminium foil)
Process	Power level, cycling, presence of hot water or air around the food, equilibration time
Equipment	Dimensions, shape and other electromagnetic characteristics of the oven, wave frequency, agitation of the food, movement of the food by conveyors and turntables, use of stirrers

Adapted from Anon (2000a)

food can further decrease the surface temperature. In some heating applications, such as microwave-heated frozen foods, the surface could be the coldest location.

The shape, volume and surface area of foods can affect the amount and spatial pattern of absorbed microwave energy, leading to overheating at corners and edges and focusing of the energy. For example, a curved shape can focus microwaves and produce a higher internal rate of heating than near the surface.

The moisture and salt contents of a food have a greater influence on microwave processing than in conventional heat processing owing to their influence on the dielectric properties of the food (high salt and moisture contents increase the efficiency of microwave absorption and decrease the depth of penetration). Therefore, the interior part of foods that have high salt or moisture contents is heated less, thus reducing microbial destruction. The composition can also affect thermal properties such as specific heat and thermal conductivity, and change the size and uniformity of temperature increases (e.g. oil that has a low specific heat heats faster than water at the same level of absorbed power). In multi-component frozen dinners, different foods heat at different rates.

Owing to the complexity of the system where the heating pattern depends on a large number of factors (Table 20.3), sophisticated mathematical modelling and computer simulations are used to predict the location of cold spots and the time–temperature history at these locations to develop microbiologically safe processes for specific food and equipment combinations. In calculating process times, the short come-up time in microwave heating is not given the importance that is given in conventional heating (Chapter 13, section 13.1). Software to simulate electromagnetic and heat transfer properties has been described by Dibben (2000) and Zhang and Datta (2000) and mathematical models are described by, for example, Campañone and Zaritzky (2005).

20.1.2   Equipment
Microwave heaters
The components of microwave equipment (Fig. 20.3) are a microwave generator (termed a 'magnetron'), aluminium tubes named waveguides, a stirrer (a rotating fan or 'distributor') and a metal chamber for batch operation, or a tunnel fitted with a conveyor belt for continuous operation. Detailed descriptions of component parts and operation of microwave heaters are given by Buffler (1993) and a number of commercial suppliers (e.g. Anon 2007a).

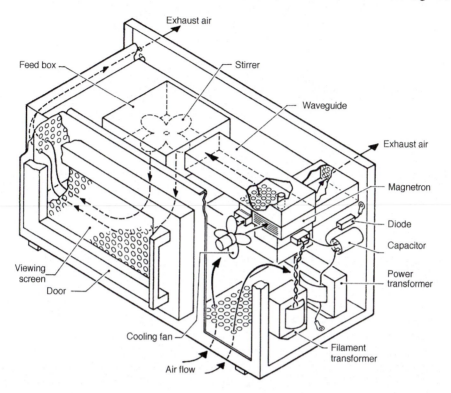

Fig. 20.3   A microwave oven showing the magnetron (from Buffler 1993).

The magnetron is a cylindrical diode ('di' meaning two and 'ode' short for 'electrode'), which consists of a sealed copper tube with a vacuum inside. The tube contains copper plates pointing towards the centre like spokes on a wheel. This assembly is termed the 'anode' and has a spiral wire filament (the cathode) at the centre. When a high voltage ($\approx$4000 V) is applied, the cathode produces free electrons, which give up their energy to produce rapidly oscillating microwaves, which are then directed to the waveguide by electromagnets. The waveguide reflects the electric field internally and thus transfers it to the heating chamber. It is important that the electric field is evenly distributed inside the heating chamber to enable uniform heating of the food. In batch equipment a stirrer is used to distribute the energy evenly throughout the heating chamber, and/or the food may be rotated on a turntable. Both methods reduce shadowing (areas of food which are not exposed to the microwaves). It is important that the power output from the magnetron is matched to the size of the heating chamber to prevent flash-over (unintended electrical discharge around an insulator, or arcing or sparking between adjacent conductors).

In continuous tunnels (Fig 20.4) a different design of distributor is used to direct a beam of energy over the trays or pouches containing the food as they pass on a conveyor. The trays are precisely positioned in the tunnel and power levels of multiple microwave generators are programmed to provide a custom-heating profile for that tray and product. In an alternative design, known as a 'leaky waveguide' applicator, slots are cut in the waveguide to allow the controlled leakage of microwaves to give a uniform power distribution over product widths of up to 3 m. In the 'slotted waveguide' applicator, food passes through a slot running down the centre of the waveguide (Brennan 2006). Automatic control consists of monitors for insufficient power level delivered by a

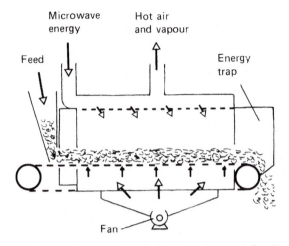

Fig. 20.4   Continuous microwave finish drying equipment (after Decareau 1985).

generator, infrared surface temperature measurement of each tray as they are being transported to the holding area, and monitoring of swelling of the top surface of individual packages. This is due to internal steam generation during heating, and is monitored to ensure that adequate heat has been received to produce enough steam for the package to swell to a predetermined amount. Power settings for individual magnetrons are stored and an alarm warns the operator if the power varies from the set values. Temperature monitoring of microwave processing using thermocouples is difficult because probes reflect and absorb microwaves, and cause electromagnetic field disturbances that change the heating patterns. Instead, fibre-optic temperature probes are used, which are transparent to electric and magnetic fields, accurate and have a fast response time (Anon 2000a).

Because microwaves heat all biological tissues, there is a risk of leaking radiation causing injury to operators. Within limits, the body can absorb microwave energy and the blood flow removes heat to compensate for the temperature increase. However, damage to the eyes is possible at an energy density of $>150\,mW\,cm^{-2}$ because they have insufficient blood flow to provide adequate cooling. The permissible energy density at the surface of microwave equipment is set at a maximum of $10\,mW\,cm^{-2}$ in Europe and the USA (Ehlermann 2002). Chambers and tunnels are sealed to prevent the escape of microwaves, interlocked doors cut the power supply when opened to prevent accidental leakage, and in continuous equipment there are energy trapping devices at the conveyor entry and exit points.

Microwave heaters are very efficient in energy use because moist foods absorb most of the microwave energy, and metals reflect microwaves so that neither the metal of the chamber nor the air is heated. Power outputs of continuous industrial equipment range from 500 W to 15 kW in the 2450 MHz band and 25–120 kW in the 915 MHz band.

Radio frequency heaters
There are several designs of RF applicators (Fig. 20.5):

- The 'through-field' design is the simplest and consists of two electrodes at different voltages that form a parallel plate capacitor, supplied by a high-voltage generator (Fig. 20.5a). Food is placed or conveyed between the plates, and this design is used for

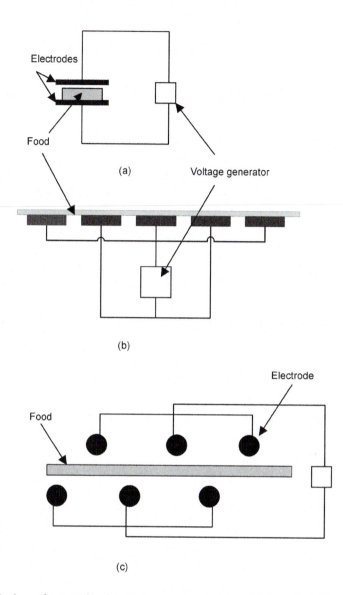

Fig. 20.5 Designs of conventional radio frequency applicators: (a) through-field applicator; (b) fringe-field applicator; (c) staggered through-field applicator (adapted from Jones and Rowley 1997).

relatively thick pieces of food (e.g. in drying chambers or RF units at the end of bakery tunnel ovens). Because the food is an electrical component of the heater, variations in the amount of food passing between the plates, its temperature and moisture content, each cause a variation in the power output of the generator. This is a valuable self-controlling feature: for example, the loss factor of a food falls as the moisture content is reduced and the power output correspondingly falls, so reducing the possibility of burning the food.

- In the 'fringe-field (or 'stray-field') design (Fig. 20.5b), a thin layer of material passes over bars, rods or plates that are connected to either side of the voltage generator and have alternating polarity. The product makes complete contact with the electrodes which ensures that there is a constant electric field in the product between the bars.

- The 'staggered through-field' design (Fig. 20.5c) has bars arranged above and below the product, and is used for foods of intermediate thickness (e.g. biscuits) (Jones and Rowley 1997).

There are two methods of producing and transmitting power to RF applicators: (1) conventional RF equipment in which the applicator is part of the RF generation circuit and may be used to control the amount of power supplied by the generator – the position of the RF applicator plates is adjusted to keep the power within set limits; and (2) 50 ohm (Ω) technology, where the RF generator is separated from the applicator and connected using a high-power coaxial cable. The frequency of the generator is set at 13.56 or 27.12 MHz and an impedance of 50 Ω. An impedance-matching network transforms the impedance of the RF applicator to 50 Ω and is used to adjust the power supplied. The RF applicator can then be designed for optimum performance as it is not itself part of any tuning system. International Electromagnetic Compatibility Regulations (Anon 2005) limit the electromagnetic disturbance that can be emitted by dielectric equipment, and the fixed operating frequency of 50 Ω technology makes it easier to control the frequency to meet the regulations. The use of a matching network also enables advanced process control to give on-line information on the condition of the food (e.g. average moisture content), which can be used to control the RF power, conveyor speed and temperature of air in the applicator. Further information is given by Anon (2007b) and Jones and Rowley (1997).

A simple method to calculate the amount of radio frequency energy needed for a particular process is described by Anon (1999):

$$E = \frac{m(\theta_1 - \theta_2)c_p}{863} \qquad \boxed{20.6}$$

where E (kW) = energy supplied, m (kg h^{-1}), = mass flow rate of product, θ_1 (°C) = final product temperature, θ_2 (°C) = initial product temperature, c_p (kJ^{-1} kg^{-1} K^{-1}) = specific heat.

There are a number of additions to the calculated amount of energy required:

- 1 kW is added for each 1.4 kg of water to be evaporated per hour in drying applications.
- An additional 10–20% of energy required is added to account for surface cooling, depending on the surface area : volume ratio of the product.
- It is assumed that the equipment is 65% efficient in the use of energy supplied, and an additional correction is needed to calculate the actual power requirement.

20.1.3   Applications

The high rates of heating and absence of surface changes have led to studies of dielectric heating of a large number of foods. The most important industrial applications are baking, dehydration, tempering and thawing. Other applications, which involve bulk heating of foods with higher moisture contents (e.g. blanching), are less successful. This is due to the low depth of penetration in large pieces of food and to evaporative cooling at the surface, which results in survival of micro-organisms. Microwave pasteurisation and sterilisation are now used commercially for the production of ready meals. These applications are discussed briefly in this section and details are given by Schubert and

Regier (2005), Piyasena *et al.* (2003), Datta and Anantheswaran (2001), Zhao *et al.* (2000) and Rosenberg and Bogl (1987).

Baking

Conventional ovens operate effectively when products have relatively high moisture contents, but the thermal conductivity falls as baking proceeds and considerable time is needed to bake the centre of the product adequately without causing excessive changes to the surface colour. RF or microwave heaters are located at the exit to tunnel ovens (Chapter 18, section 18.2.2) to reduce the moisture content and to complete baking without further changes in colour. This reduces baking times by 30–50% and hence increases the throughput of the ovens. RF or microwave finishing (removing the final moisture) improves baking efficiency for thin products such as breakfast cereals, babyfoods, biscuits, crackers, bread sheets to be made into breadcrumbs, crispbread and sponge cake. Meat pies, which require a good crust colour in addition to pasteurisation of the filling, can be baked in about one-third of the time required in conventional ovens by combined RF and conventional baking (Jones 1987). Other advantages include: savings in energy, space and labour costs; close control of final moisture contents (typically ±2%) and automatic equalisation of moisture contents as only moist areas are heated; separate baking and drying stages allow control over the internal and external product colour and moisture content; and improved product texture and elimination of 'centre bone' (a fault caused by dense dough in the centre of biscuits). The use of dielectric heating alone is less successful for baking. It causes undesirable qualities in bread, owing to the altered heat and mass transfer patterns and the shorter baking times. These produce insufficient starch gelatinisation, microwave-induced changes to gluten and too-rapid gas and steam production. As a result, microwave-baked breads have no crust and have a tougher, coarser, but less firm texture (Yin and Walker 1995). However, more recently crust-less bread has been produced (i.e. without having to remove the crust), which gives savings of 35% in raw materials. The RF technology also permits automatic control of moisture levels, to produce bread that has lower moisture content (<38%) which increases the shelf-life and reduces evaporation of volatile flavourings. The technology permits significant space savings, and can increase production by 40% because of shorter baking times compared with conventional ovens.

Dehydration

The main disadvantages of hot-air drying are the low rates of heat transfer, caused by the low thermal conductivity of dry foods, and damage to sensory characteristics and nutritional properties caused by long drying times and overheating at the surface (Chapter 16, section 16.1.1). Microwave and RF drying overcome the barrier to heat transfer caused by the low thermal conductivity, by selectively heating moist areas while leaving dry areas unaffected. This improves moisture transfer during the later stages of drying by heating internal moisture and thus increasing the vapour pressure and the rate of drying. For example, the loss factor for free water is higher than that for bound water, and both are higher than the dry matter components. Dielectric heating reduces product shrinkage during the falling rate period, prevents damage to the food surface, and eliminates case hardening. Tohi *et al.* (2002) found a correlation between the capacitance of foods and moisture content, which enables automatic control of drying conditions without sampling the material during the process. Other advantages include energy savings by not having to heat large volumes of air, and minimal oxidation by atmospheric oxygen. However, the use of microwave drying by itself has limitations: the inherent non-uniformity of the microwave electromagnetic field and limited penetration of the microwaves into bulk products compared with RF energy, lead to uneven heating; also microwaves and RF units have higher cost and smaller scales of operation compared with traditional drying methods. Non-uniform field strength can be partly overcome by keeping the food in constant motion to avoid hot-spots (e.g. using a spouted or fluidised bed dryer (Chapter 16, section 16.2.1)) or using pulsed microwaves. However, these factors restrict microwave drying to either finishing of partly dried or low-moisture foods, or their use in 'hybrid' dryers in which microwaves are used to increase the rate of drying in conventional hot-air dryers (Vega-Mercado *et al.* 2001, Garcia and Bueno 1998) (Fig. 20.4). For example, in pasta drying the fresh pasta is pre-dried in hot air to 18% moisture and then in a combined hot-air and microwave dryer to lower the moisture content to 13%. Drying times are reduced from 8 h to 90 min with a reduction in energy consumption of 20–25%, bacterial counts are 15 times lower, there is no case hardening, the drying tunnel is reduced from 36–48 m to 8 m, and clean-up time is reduced from 24 to 6 person-hours (Decareau 1990). In grain finish drying, microwaves are cheaper and more energy efficient than conventional methods and do not cause dust pollution. The lower drying temperature also improves grain germination rates. Zhang *et al.* (2006) have reviewed the advantages and limitations of microwave drying of fruits and vegetables.

Combined microwave–vacuum drying has been used for heat-sensitive products that are difficult to dry using hot air (e.g. fruits that have high sugar contents) but it has high costs owing to the need to maintain the vacuum over long drying periods (Gunasekaran 1999). In conventional freeze drying (Chapter 23, section 23.1) the low rate of heat transfer to the sublimation front limits the rate of drying. Microwave freeze drying overcomes this problem because heat is supplied directly to the ice front, which can reduce the drying time by 50–75% compared to conventional freeze drying (Cohen and Yang 1995). However, careful control over drying conditions is necessary to prevent localised melting of the ice. Because of the difference in loss factors of ice and water (Table 20.2), any water produced by melting ice heats rapidly and causes a chain reaction leading to widespread melting and an end to sublimation. Accelerated freeze drying using microwaves has been extensively investigated but the process remains expensive and is not widely used commercially. It is reviewed by Zhang *et al.* (2006) and further details are given in Chapter 23 (section 23.1.2).

Thawing, melting and tempering
During conventional thawing of frozen foods (Chapter 22, section 22.2.5), the lower thermal conductivity of water, compared with ice, reduces the rate of heat transfer and thawing slows as the outer layer of water increases in thickness. Microwaves and RF energy are used to rapidly thaw small portions of food and for melting fats (e.g. butter, chocolate and fondant cream) (Jones 1987). However, difficulties arise with larger (e.g. 25 kg) frozen blocks, such as egg, meat, fish and fruit juice, that are used in industrial processes. Because water heats rapidly once the ice melts, thawing does not take place uniformly in the large blocks, and some portions of the food may cook while others remain frozen. This is overcome to some extent by reducing the power and extending the thawing period, or by using pulsed microwaves to allow time for temperature equilibration.

A more common application is 'tempering' frozen foods, in which the temperature is raised from around −20 °C to −3 °C and the food remains firm but is no longer hard. After frozen food has been tempered, it is more easily sliced, diced or separated into pieces (Chapter 3, section 3.1.2). Tempering is widely used for meat and fish products, which are more easily boned or ground at a temperature just below the freezing point. If frozen foods are tempered but not allowed to melt, they require much less energy. For example, the energy required to temper frozen beef from -17.7 to $-4.4\,^{\circ}\mathrm{C}$ is $62.8\,\mathrm{J\,g^{-1}}$ whereas $123.3\,\mathrm{J\,g^{-1}}$ is needed to raise the temperature a further $2.2\,^{\circ}\mathrm{C}$ (Decareau 1990). The lower energy cost of tempering gives a good return on investment in dielectric equipment. Production rates range from 1 to $4\,\mathrm{t\,h^{-1}}$ of meat or 1.5–$6\,\mathrm{t\,h^{-1}}$ of butter in equipment that has power outputs of 25–$150\,\mathrm{kW}$. The advantages over conventional tempering in cold rooms include the following:

- Faster processing (e.g. meat blocks are defrosted in 10 min instead of several days). Tempering can also take place when the food is required with little loss or spoilage in the event of a process delay.
- The costs of operating a tempering room are eliminated and savings are made in storage space and labour.
- No drip losses or contamination, which improves product yields and reduces nutritional losses. There is also better control over defrosting conditions and more hygienic defrosting because products are defrosted in the storage boxes, leading to improved product quality.

Other applications

Compared with conventional heating, microwave rendering of fats improves the colour, reduces fines by 95% and costs by 30%, and does not cause unpleasant odours (Decareau 1985). Microwave frying is not successful when deep baths of oil are used, but can be used with shallow trays in which the food is rapidly heated (Chapter 19, section 19.2). There is less deterioration in oil quality and more rapid frying. Pretreating potatoes with microwaves before frying has also been shown to reduce the formation of acrylamide (Belgin *et al.* 2007). Other commercial microwave applications include heating bacon or meat patties in foodservice applications and setting meat emulsions in microwave transparent moulds to produce skinless frankfurters and other sausage products (Decareau 1990).

Microwave blanching has been extensively investigated, but the higher costs, compared with steam blanching (Chapter 11), have restricted its use to products that are more difficult to blanch by conventional methods. Microwave blanching of peanuts causes off-flavours, but control of the processing conditions can limit this (Schirack *et al.* 2006).

Industrial microwave pasteurisation and sterilisation systems have been reported for 30 years starting with batch processing of yoghurt in cups and continuous processing of milk (Anon 2000a) and now focusing on ready-to-eat meals in a few companies worldwide (e.g. Anon 2006). RF pasteurisation and sterilisation are feasible but are not yet used commercially. The lack of widespread commercial operations may be due to the greater complexity, expense and non-uniformity of heating which makes it difficult to ensure sterilisation of the whole package. Pandit *et al.* (2007) describe a computer-vision system (Chapter 2, section 2.3.3) to identify cold-spots in microwave sterilised foods. Microwave and RF heating for pasteurisation and sterilisation require less time to reach the process temperature, especially for solid and semi-solid foods that heating by conduction (Datta and Hu 1992). Other advantages are that equipment can be turned on or off instantly, the product can be pasteurised after packaging so eliminating post-pasteurisation recontamination, and processing systems can be more energy efficient. Microwave pasteurisation of packed complete pasta meals, soft bakery goods and peeled potatoes is reported by Brody (1992).

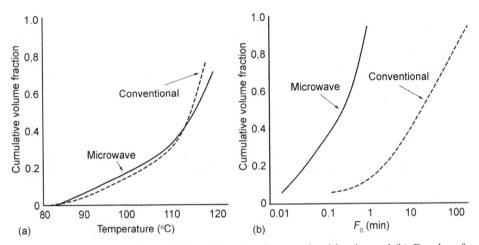

Fig. 20.6   (a) Temperatures reached by dielectric and conventional heating and (b) F_0 values for dielectric and conventional heating for cumulative volume fractions of food (from Anon 2000a). Microwave = 3.5 min at 2 W cm^{-3}, conventional = 40 min at 121 °C.

The temperatures reached by dielectric and conventional heating are similar (Fig. 20.6a), but the F_0 values (time–temperature histories) for cumulative volume fractions of food at each temperature are very different (Fig. 20.6b). Conventional heating shows a larger spread of F_0 values, which indicate non-uniformity of temperatures and long processing times that cause over-processing of the surface parts of the food.

The equipment consists of a pressurised microwave tunnel up to 25 m long, through which food passes on a conveyor in microwave-transparent, heat-resistant, laminated pouches or trays that have shapes specifically adapted for microwave heating. Polypropylene with an ethylene vinyl alcohol (EVOH) barrier or a polyethylene terephthalate (PET) film has been used (Chapter 25, section 25.2.4). Because metal reflects microwaves, packages that have a metal component can change the food temperature distribution. In some applications, metals have been added to the package to redistribute the microwave energy to increase the uniformity of heating. The packs are positioned in the tunnel so that they receive a predetermined amount of microwave energy that is optimised for that type of package.

The process consists of heating, holding for the required period for pasteurisation or sterilisation, and cooling the packs in the tunnel. In the Multitherm process, the microwave-transparent pouches are formed and filled from a continuous reel of film but are not separated. This produces a chain of pouches that passes through a continuous hydrostat system, similar to a small hydrostatic steam steriliser (Chapter 13, section 13.1.3). The pouches are submerged in a medium that has a higher dielectric constant than the product and heating is by microwaves instead of steam.

The design of the equipment can influence the location and temperature of the slowest-heating point in the food, which makes it difficult to predict microbial destruction. The process may therefore include an equilibration stage before holding the heated product to equilibrate the temperatures and avoid non-uniform temperature distribution within the product. Other methods used to improve the uniformity of heating include rotating the packs and using pulsed microwaves. The process operates automatically, with computer control of delivered power, temperature, pressure, conveyor speed and process cycle time. Other processes use a combination of microwaves and hot air at 70–90 °C, followed by an equilibration stage where the slowest heating parts of the packs reach 80–85 °C within 10 min. The packs are then cooled to 1–2 °C and have a shelf-life of ≈40 days at 8 °C. Details of a procedure for the microwave pasteurisation of fruits in syrup to inactivate pectinesterase are reported by Brody (1992).

20.1.4   Effect on foods and micro-organisms

The effects of electromagnetic energy on food components are similar to those found using other methods of heating, although the more rapid heating results in shorter processing times and hence fewer changes to nutritional and sensory properties. The process therefore has the benefits of bacterial destruction with reduced damage to sensory and nutritional properties (Fig. 20.7). These changes are reviewed by Ehlermann (2002) and are described further in Chapter 10 (sections 10.3 and 10.4). As in conventional heating, heat-sensitive vitamins (e.g. ascorbic acid) undergo losses and, for example, Watanabe *et al.* (1998) found that appreciable losses (≈30–40%) of vitamin B_{12} occurred in raw beef, pork and milk after microwave heating.

In pasteurisation and blanching applications, the high rates of heat transfer for a specified level of microbial or enzyme destruction result in reduced losses of heat-sensitive nutrients compared with conventional methods (e.g. there is no loss of carotene in microwave-blanched carrots, compared with 28% loss by steam blanching and 45% loss by water blanching), although Mirza and Morton (2006) found no difference in the colour of carrots that were blanched by four different methods, including microwaves. Ramesh *et al.* (2002) found reduced losses of nutrients after microwave blanching of vegetables, but results for other foods are highly variable and, for these, microwave heating offers no nutritional advantage over steaming. Changes to foods in other types of processing (microwave or RF frying, baking, dehydration, etc.) are similar to conventional methods and are described in the relevant chapters.

Similarly, the energy absorbed from electromagnetic waves can raise the temperature of the food sufficiently to inactivate micro-organisms (e.g. Fujikawa *et al.* 1992). There is disagreement over possible non-thermal effects of electromagnetic energy on micro-organisms (i.e. effects such as ionisation that are not related to lethality caused by heat). Microwaves correspond to an energy range of 1 μeV–1 meV, whereas binding energies of

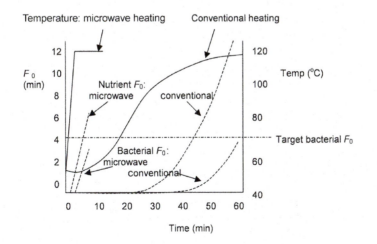

Fig. 20.7   Quality parameters for microwave and conventional heating (F_0 represents accumulated lethality) (from Anon 2000a).

electrons to atoms are >4 eV (Ehlermann 2002) and microwaves are therefore not capable of ionisation. Anon (2000a) has reviewed research into these non-thermal effects and reports that studies are inconclusive and only thermal effects are presumed to exist. All changes are caused by heat alone and microbial inactivation is therefore the same as in conventional heat processing (Chapter 10, section 10.3) (i.e. bacteria are more resistant than yeasts and moulds to thermal inactivation by microwave heating, and bacterial spores are more resistant than vegetative cells). There have been many studies of the effect of microwave heating on pathogenic micro-organisms: *Bacillus cereus*, *Campylobacter jejuni*, *Clostridium perfringens*, pathogenic *Escherichia coli*, *Enterococcus*, *Listeria monocytogenes*, *Staphylococcus aureus*, and *Salmonella* are each reported to be inactivated by microwave heating (Heddleson *et al.* 1994). The effect of microwave pasteurisation on *E. coli* in fruit juices is reported by Cañumir *et al.* (2002). However, non-uniform heating may enable survival of pathogens when measured temperatures indicate that they would be lethal (e.g. survival of pathogens at the surface of poultry due to lower temperatures at the product surface than the measured internal temperature) (Schnepf and Barbeau 1989). The effects of microwaves on micro-organisms and enzymes are described by Anatheswaran and Ramaswarmy (2001).

20.2   Ohmic heating

Also termed 'resistance heating', 'electroconductive heating' or 'Joule heating', this is a process in which an alternating electric current is passed through a food, and the electrical resistance of the food causes the power to be translated directly into heat (see also pulsed electric field processing (Chapter 9, section 9.1)). As the food is an electrical component of the heater, it is essential that its electrical properties are matched to the capacity of the heater. The concept of direct heating in this way is not new, but it was developed into a commercial process during the 1980s–1990s. The process can be used for UHT sterilisation of foods, and especially those foods that contain larger particles that are difficult to sterilise by other methods. It is in commercial use in Europe, the USA and Japan for aseptic processing of high-added-value ready meals, stored at ambient or chill temperatures (see Chapter 13, section 13.2), pasteurisation of particulate foods for hot filling, and preheating products before canning (Fryer 1995).

Ohmic heating has higher energy conversion efficiencies than microwave heating (>90% of the energy is converted to heat in the food). Another important difference is that microwave and radio frequency heating have a finite depth of penetration into a food whereas ohmic heating has no such limitation. However, microwave heating requires no contact with the food, whereas ohmic heating requires electrodes to be in good contact. This means that in practice the food should be liquid or have sufficient fluidity to allow both good contact with the electrodes and to pump the product through the heater.

The advantages of ohmic heating are as follows:

- The food is heated rapidly (>1 °C s^{-1}) throughout the bulk of the food (i.e. volumetric heating) for example, from ambient to 129 °C in 90 s (Ruan and Chen 2002). The absence of temperature gradients results in even heating of solids and liquids if their resistances are the same, which cannot be achieved in conventional heating.
- There are no hot surfaces for heat transfer, as in conventional heating. Therefore heat transfer coefficients do not limit the rate of heating, and there is no risk of surface fouling or damage to heat-sensitive foods by localised over-heating.

- Liquids containing particles are not subject to shearing forces that are found in for example scraped surface heat exchangers (Chapter 13, section 13.2.3), and the method is suitable for viscous liquids because heating does not have the problems associated with poor convection in these materials.
- It has a lower capital cost than microwave heating and it is suitable for continuous processing, with instant switch-on and shutdown.

Further details are given by Ruan et al. (2004), Rahman (1999) and Sastry (1994).

Ohmic heating has been used commercially to pasteurise milk, liquid egg and fruit juices, to process viscous liquids, such as apple sauce and carbonara sauce, to produce high-quality whole fruits for yoghurt, and low-acid products that contain particles, including ratatouille, pasta in tomato or basil sauce, beef bourguignon, vegetable stew, lamb curry and minestrone soup concentrate (Ruan et al. 2004, Ruan and Chen 2002). However, three factors limit the widespread commercial uptake of the process: (1) differences in the electrical conductivities of the liquid and solid components of multi-component foods and variations in conductivity with increasing temperature, which can cause irregular and complex heating patterns and difficulties in predicting the heating characteristics; (2) a lack of data on the critical factors that affect the rate of heating (section 20.2.1); and (3) a lack of accurate temperature-monitoring techniques to profile heat distribution and locate cold-spots during the process. This risks under-processing and the consequent survival of pathogenic spores in low-acid foods. Advances in magnetic resonance imaging (MRI) are being used to address the last issue and are reviewed by Ruan et al. (2004).

20.2.1   Theory

Foods and other materials have a resistance (known as the 'specific electrical resistance') that generates heat when an electric current is passed through them. Electrical 'conductivity' is the inverse of electrical resistance and is measured in a food using a multimeter connected to a conductivity cell. The relationship between electrical resistance and electrical conductivity is found using:

$$\sigma = (1/R)(L/A) \qquad \boxed{20.7}$$

where σ (S m^{-1}) = product conductivity, R (Ω) = measured resistance, L (m) = length of the cell and A (m^2) = area of the cell.

Conductivity measurements are made in product formulation exercises, process control and quality assurance for foods that are heated electrically. Data on electrical conductivity of foods (Table 20.4) are relatively scarce, but it has a much greater range than thermal conductivity (Chapter 10, Table 10.2). For example, it can vary from 10^8 S m^{-1} for copper to 10^{-8} S m^{-1} for an insulating material such as wood. Foods that contain water and ionic salts are more capable of conducting electricity (they have a lower resistance). In composite foods, the conductivity of particles is measured by difference (i.e. the product conductivity minus the carrier medium conductivity).

Unlike metals, where conductivity falls with temperature, the electrical conductivity of a food increases linearly with temperature (Wang and Sastry 1997, Reznick 1996). It can also vary in different directions (e.g. parallel to, or across a cellular structure), and can change if the structure changes (e.g. gelatinisation of starch, cell rupture or air removal after blanching). It can be seen in Table 20.4 that the conductivity of vegetables is lower than muscle tissue, and this in turn is considerably lower than for a

Table 20.4   Electrical conductivity of selected foods at 19 °C

Food	Electrical conductivity $(S\,m^{-1})$
1 Potato	0.037
2 Carrot	0.041
3 Pea	0.17
4 Beef	0.42
5 Starch solution (5.5%)	
(a) with 0.2% salt	0.34
(b) with 0.55% salt	1.3
(c) with 2% salt	4.3

From Kim *et al.* (1996)

sauce or gravy. The salt content of a gravy is typically 0.6–1% and from the data (5b) in Table 20.4 the conductivity of the beef is about a third of that of the gravy. This has important implications for processing of particles (section 20.2.3): if in a two-component food consisting of a liquid and particles in which the particles have a higher conductivity, they are heated at a higher rate. This is not possible in conventional heating owing to the lower thermal conductivity of solid foods, which slows heat penetration to the centre of the pieces (Chapter 10, section 10.1.2) (Fig. 20.8). Ohmic heating can therefore be used to heat sterilise particulate foods under UHT conditions without causing heat damage to the liquid carrier or over-cooking of the outside of particles. Furthermore, the lack of agitation in the heater maintains the integrity of particles and it is possible to process larger particles (up to 2.5 cm) that would be damaged in conventional equipment.

The rate of heating also depends on the density, the pH, thermal conductivity and specific heat capacities of each component, the way that food flows through the equipment and its residence time in the heater, in addition to the electrical conductivity of the components. Each of these may change during processing and hence alter the heating characteristics of the product (Larkin and Spinak 1996).

In two-component foods the heating patterns are not a simple function of the relative conductivities of the particles and liquid carrier. For example, when a particle that has a lower conductivity than the liquid is heated the liquid heats faster, but if the density of the particle is higher, the heating rate may exceed that of the liquid (Sastry and Palaniappan 1992). If two components have similar conductivities, the lower moisture solid portion heats faster than the carrier liquid. The calculation of heat transfer is therefore very complex, involving the simultaneous solution of equations for changes in electrical fields, thermal properties and fluid flow, and is beyond the scope of this book. Details are given by Fryer (1995) and Sastry and Li (1996). Mathematical models are described by Salengke and Sastry (2007), Samprovalaki *et al.* (2007) and Ye *et al.* (2004) and are reviewed by Ruan *et al.* (2004). A simplified theory of heating is given below.

The resistance in an ohmic heater depends on the specific resistance of the product, and the geometry of the heater:

$$R = (R_sL)/A \hspace{3cm} \boxed{20.8}$$

where R (Ω) = total resistance of the heater, R_s $(\Omega\,m^{-1})$ = specific resistance of the product, L (m) = distance between the electrodes and A (m^2) = area of the electrodes.

The resistance determines the current that is generated in the product:

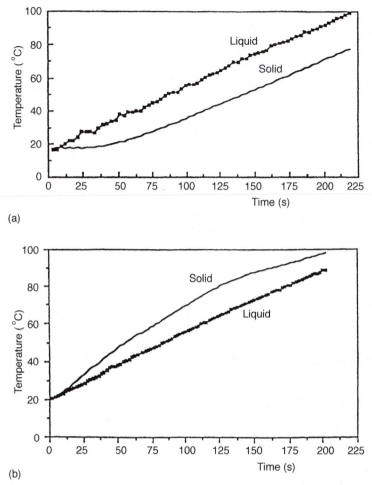

(a)

(b)

Fig. 20.8   Heat penetration into solid pieces of food by (a) conventional heating and (b) ohmic heating (adapted from Fryer 1995).

$$R = \frac{V}{I}$$

[20.9]

where V (V) = voltage applied and I (A) = current.

The available three-phase power sources in most countries have 220–240 V per phase at a frequency of 50 Hz, and to make the best use of the power the geometry of the heater and the resistance of the product have to be carefully matched. If the resistance is too high, the current will be too low at maximum voltage. Conversely, if the resistance is too low, the maximum limiting current will be reached at a low voltage and again the heating power will be too low.

Every product has a critical current density and if this is exceeded there is likely to be arcing (or flash-over) in the heater. The current density is found by:

$$I_d = I/A$$

[20.10]

where I_d (A cm^{-2}) = current density. The minimum area for the electrodes can therefore be calculated once the limiting current density and maximum available current are known. As resistance is determined in part by the area of the electrodes (Equation 20.8), the distance between the electrodes can be calculated. It is important to recognise that the design of the heater is tailored to products that have similar specific electrical resistances and it cannot be used for other products without modification.

The rate of heating is found using Equation 20.11:

$$Q = mC_p\Delta\theta \qquad \boxed{20.11}$$

and the power by:

$$P = VI \qquad \boxed{20.12}$$

and

$$P = RI^2 \qquad \boxed{20.13}$$

Assuming that heat losses are negligible, the temperature rise in a heater is calculated using:

$$\Delta\theta = \frac{V^2\sigma_a A}{Lmc_p} \qquad \boxed{20.14}$$

where $\Delta\theta$ (°C) = temperature rise, σ_a (S m^{-1}) = average product conductivity throughout temperature rise, A (m^2) = tube cross-sectional area, L (m) = distance between electrodes, m (kg s^{-1}) = mass flowrate and c_p (J kg^{-1} °C^{-1}) = specific heat capacity of the product.

In conventional heaters, turbulence is needed to create mixing of the product and maintain maximum temperature gradients and heat transfer coefficients (Chapter 10, section 10.1.2). In ohmic heating, the electric current flows through the product at the speed of light and there are no temperature gradients since the temperature is uniform across the cross-section of flow. The flowrate of product is negligible compared with the velocity of the electric current, but if the flowrate is not uniform across the cross-sectional area, the very high rates of heating mean that slower moving food will become considerably hotter. It is therefore important to ensure that uniform (or 'plug') flow conditions are maintained in the heater (also Chapter 1, section 1.3.4). Kim et al. (1996) give details of experimental studies, which confirm that this takes place. Similarly, the type of pump that is used should provide a continuous flow of material without pulses, as these would lead to increased holding times in the tube and uneven heating. A high pressure is maintained in the heater (up to 400 kPa for UHT processing at 140 °C) to prevent the product from boiling.

20.2.2   Equipment and applications

As described in section 20.2.1, the design of ohmic heaters must include the electrical properties of the specific product to be heated, because the product itself is an electrical component. This concept is only found elsewhere in RF heating and requires more specific design considerations than those needed when choosing other types of heat exchangers.

The factors that are taken into account include the following:

- The type of product, its electrical resistance and change in resistance over the expected temperature rise, its composition, shape size, orientation, specific heat capacity, thermal conductivity and density. For liquid carriers, the additional properties are viscosity and added electrolytes.

- Temperature rise (determines the power requirement) and rate of heating required.
- Flowrate and holding time required.

Early ohmic heater designs used DC power, which caused electrolysis (corrosion of electrodes and product contamination) and also required expensive electrodes. The use of mains power at 50 Hz reduces the risk of electrolysis and minimises the complexity and cost. Alternatively, higher frequencies (>100 kHz) or carbon electrodes may be used to reduce electrolysis. To be commercially successful for aseptic processing, ohmic heaters must have effective control of heating rates and product flowrates that avoids electrolysis or product scorching, and be cost effective. The layout of an ohmic heating system is shown in Fig. 20.9.

The heater consists of a vertical tube containing a series of pure carbon cantilever electrodes (supported from one side) that are contained in a PTFE housing and fit across the tube. The tube sections are made from stainless steel, lined with an insulating plastic such as polyvinyidene fluoride (PVDF), polyether ether ketone (PEEK) or glass. Food is pumped up through the tube and an alternating current flows between the electrodes and through the food to heat it to the required process temperature. The system is designed to maintain the same impedance between the electrodes in each section, and the tube sections therefore increase in length between inlet and outlet because the electrical conductivity of the food increases as it is heated. Food then passes from the heater to a holding tube where it is held for sufficient time to ensure sterility and is then cooled and aseptically packaged (Chapter 13, section 13.2, and Chapter 25, section 25.2.7).

Typically, a heater tube of 2.5 cm diameter and 2 m length could heat several thousand litres per hour (Reznick 1996). Commercial equipment is available with power outputs of

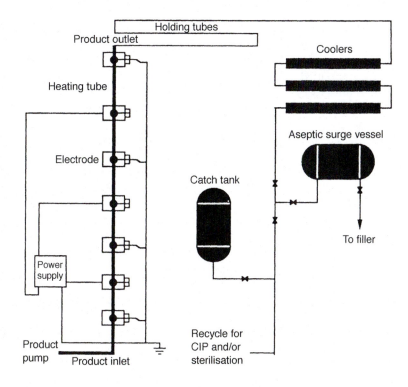

Fig. 20.9   Layout of an ohmic heating system (after Parrott 1992).

75 and 300 kW, which correspond to throughputs of $\approx$750 and 3000 kg h^{-1} respectively (Fryer 1995). The process is automatically controlled via a feed-forward system (Chapter 27, section 27.2), which monitors inlet temperature, product flow rate and specific heat capacity, and continuously adjusts the power required to heat the product (Dinnage 1990).

Ohmic heating has been used to process various combinations of meats, vegetables, pasta and fruits when accompanied by a suitable carrier liquid. It may be necessary to pretreat components of the food to make them more homogeneous (e.g. pregelatinising starch in the carrier liquid, homogenising sauces, especially those that contain fats and heat-sensitive proteins to produce a uniform material, blanching vegetables to expel air and soaking foods in acids or salts to alter the electrical resistance of particles (Zoltai and Swearingen 1996)).

In operation, a small amount of carrier liquid is used to suspend the particles as they pass through the heater. The bulk of the carrier liquid is sterilised by conventional plate or tubular heat exchangers and is then injected into the particle stream as it leaves the holding tube. This has the advantage of reducing the capital and operating costs for a given throughput (Dinnage 1990). The combined product is then aseptically packaged (Chapter 13, section 13.2.3). Ohmic heating costs were found by Allen *et al.* (1996) to be comparable to those for freezing and retort processing of low-acid products. The almost complete absence of fouling in ohmic heaters means that after one product has been processed, the plant is flushed through with a base sauce and the next product is introduced. At the end of processing, the plant is flushed with a cleaning solution.

The process is suitable for foods that contain up to about 60% solids. In contrast to conventional UHT processing of particulate foods, where sufficient amounts of the liquid component are required for heat transfer into the particles, in ohmic heating a high solids content is desirable for two reasons: there is faster heating of low-conductivity particles than the carrier liquid; and a high solids content creates plug-flow conditions in the heater tubes. To obtain high solids concentrations, the particles should be pliable and small, or their geometry is varied to reduce the void spaces between particles. If lower solids concentrations are processed, they require a higher-viscosity carrier liquid to keep the particles in suspension and maintain plug-flow conditions. In all products the viscosity of the sauce or gravy carrier liquid should be carefully controlled, and this may involve using pregelatinised starches to prevent viscosity changes during processing.

The density of the particles should also be matched to the carrier liquid: if particles are too dense or the liquid is not sufficiently viscous, the particles sink in the system and become over-processed. Conversely, if the particles are too light they float, which leads to a variable product composition and the risk of under-processing. It is almost impossible to determine the residence time or heating profiles of particles that float or sink.

To ensure sterility, it is necessary to ensure that the coldest part of the slowest heating particle in the food has received sufficient heat (see Chapter 10, section 10.1.2, and Chapter 13, section 13.1). It is not easy to measure heat penetration into particles, whereas it is relatively easy to measure the temperature of the carrier liquid. The process must therefore demonstrate that solid particles have been heated to an equal or greater extent than the liquid when they enter the holding tube. This is achieved by adjusting the electrical properties of each component (e.g. by control of salt content in the formulation). This is more difficult for non-homogeneous particles such as fatty meat pieces, in which the different components have different electrical resistances. Complexities increase if, for example, salt leaches out of the particles into the surrounding sauce and causes changes to the electrical resistance and hence the rate of heating of both components. The presence of fats or other poorly conductive materials (e.g. pieces of bone, nuts or ice) in particles means that they will not be heated directly, and the slower heating by conduction creates a cold spot within the particle (Larkin and Spinak 1996). If this happens, the surrounding food may also be under-processed and there is a risk of growth of pathogenic bacteria. Further details are given by Anon (2000b).

20.2.3   Effect on foods and micro-organisms

Ohmic heating is an HTST process and therefore has similar benefits to other methods of rapid heating that destroy micro-organisms before there are adverse effects on nutrients or chemicals that produce required organoleptic qualities (see Chapter 10, section 10.3 for details of D- and z-values). It also causes similar changes to foods as does conventional heating, such as starch gelatinisation, melting of fats and coagulation of proteins (Chapter 10, section 10.4). Ohmic heating also increases diffusion of material from solid particles to the carrier liquid, which may be due to electroporation (formation of pores in cell membranes due to electrical potential across the membrane, resulting in leakage), membrane rupture caused by the voltage drop across the membrane, and cell lysis, disrupting internal components of the cell. These effects may contribute to microbial destruction (see also Chapter 9, section 9.1.3). Losses of material from cells only alter the nutritional value if the liquid is not consumed, in for example blanching. A study by Mizrahi (1996) showed that solute losses had a similar pattern in hot-water blanched and ohmic blanched beets, and were proportional to the surface : volume ratio and the square root of processing time. However, it was not necessary to slice the beets before ohmic processing, and this, together with the shorter blanching time, reduced solute losses in ohmic blanching by a factor of ten compared with hot water blanching.

20.3   Infrared heating

The main commercial applications of IR energy are drying low-moisture foods (e.g. breadcrumbs, cocoa, flours, grains, malt, pasta products and tea) and in baking applications (e.g. pizzas, biscuits) or roasting ovens (Chapter 18, section 18.2.1) for products such as coffee, cocoa and cereals. The technology has also been used to fry or thaw foods, and for surface pasteurisation of bread and packaging materials. The main advantages are a reduction in roasting or baking time and savings in energy compared with traditional processes. However, it is not widely used as a single source of energy for drying or baking larger pieces of food because of the limited depth of penetration. Radiant energy is also used in vacuum band driers and cabinet driers (Chapter 16, section 16.2.1), in accelerated freeze driers (Chapter 23, section 23.1.2), in some domestic microwave ovens to brown the surface of foods; and to heat-shrink packaging films (Chapter 26, section 26.3).

20.3.1   Theory

Infrared energy is electromagnetic radiation (Fig. 20.1) that is emitted by hot objects. When it is absorbed, the radiation gives up its energy to heat materials. The rate of heat transfer depends on:

- the surface temperatures of the heating and receiving materials;
- the surface properties of the two materials; and
- the shapes of the emitting and receiving bodies.

The amount of heat emitted from a perfect radiator (termed a 'black body') is calculated using the Stefan–Boltzmann equation:

$$Q = \sigma A T^4 \qquad \boxed{20.15}$$

where Q ($J\,s^{-1}$) = rate of heat emission, σ ($W\,m^{-2}\,K^{-4}$) = Stefan–Boltzmann constant = 5.73×10^{-8}, A (m^2) = surface area, and T ($K = {}^{\circ}C + 273$) = absolute temperature.

This equation is also used for a perfect absorber of radiation, again known as a black body. However, radiant heaters are not perfect radiators and foods are not perfect absorbers, although they do emit and absorb a constant fraction of the theoretical maximum. To take account of this, the concept of 'grey bodies' is used, and the Stefan–Boltzmann equation is modified to:

$$Q = \epsilon \sigma A T^4 \qquad \boxed{20.16}$$

where ϵ = emissivity of the grey body (a number from 0 to 1) (Table 20.5).

Emissivity varies with the temperature of the grey body and the wavelength of the radiation emitted. The amount of absorbed energy, and hence the degree of heating, varies from zero to complete absorption. This is determined by the components of the food, which absorb radiation to different extents, and the wavelength of the radiated energy. The wavelength of infrared radiation is determined by the temperature of the source. Higher temperatures produce shorter wavelengths that have a greater depth of penetration.

Some radiation is absorbed by foods and some is reflected back out of the food. The amount of radiation absorbed by a grey body is termed the 'absorptivity' (α) and is numerically equal to the emissivity (Table 20.5). Radiation that is not absorbed is expressed as the 'reflectivity' ($1 - \alpha$). There are two types of reflection: that which takes place at the surface of the food and that which takes place after radiation enters the food structure and becomes diffuse due to scattering. The net rate of heat transfer to a food therefore equals the rate of absorption minus the rate of emission:

$$Q = \epsilon \sigma A (T_1^4 - T_2^4) \qquad \boxed{20.17}$$

where T_1 (K) = temperature of the emitter and T_2 (K) = temperature of the absorber.

It can be seen from Equation 20.17 and sample problem 20.1 that the temperature of the food has a significant effect on the amount of radiant energy that is absorbed.

Table 20.5   Approximate emissivities of materials in food processing

Material	Emissivity
Burnt toast	1.00
Dough	0.85
Water	0.955
Ice	0.97
Lean beef	0.74
Beef fat	0.78
White paper	0.9
Painted metal or wood	0.9
Unpolished metal	0.7–0.25
Polished metal	< 0.05

From Earle (1983) and Lewis (1990)

Sample problem 20.1

A 12 kW oven operates at 210 °C. It is loaded with a batch of 150 loaves of bread dough in baking tins at 25 °C. The surface of each loaf measures 12 cm × 20 cm. Assuming that the emissivity of dough is 0.85, that the dough bakes at 100°C, and that 90% of the heat is transmitted in the form of radiant energy, calculate energy absorption at the beginning and end of baking and the percentage of radiant energy absorbed by the surfaces of the loaves at the end of baking.

Solution to sample problem 20.1

$$\text{Area of dough} = 150(0.2 \times 0.12)$$
$$= 3.6\,\text{m}^2$$

From Equation 20.17, energy absorbed at the beginning of baking,

$$Q = 0.85 \times (5.73 \times 10^{-8}) \times 3.6 \times (483^4 - 298^4)$$
$$= 8159.8\,\text{W}$$

and the energy absorbed at the end of baking,

$$Q = 0.85 \times (5.73 \times 10^{-8}) \times 3.6 \times (483^4 - 373^4)$$
$$= 6145.6\,\text{W}$$

$$\text{Radiant energy emitted} = 12\,000 \times 0.9\,\text{W}$$
$$= 10\,800\,\text{W}$$

$$\text{Percentage of energy absorbed by the bread} = 6145.9/10\,800$$
$$= 0.57\text{ or }57\%$$

20.3.2   Equipment

Industrial radiant heaters are required to reach operating temperatures quickly to enable good process control and to transfer large amounts of energy (Sköldebrand 2002). Quartz or halogen lamps fitted with tungsten or nichrome electric filaments heated to ≈2200 °C produce near IR radiation with wavelengths of 1.1–1.3 μm and medium wave IR (Table 20.6). Ceramic IR heaters heat up to ≈700 °C and produce far IR radiation. Products are either conveyed through a tunnel or beneath banks of radiant heaters.

20.3.3   Effect on foods and micro-organisms

The rapid surface heating changes the flavour and colour of foods due to Maillard reactions and protein denaturation, and also seals moisture and flavour or aroma compounds in the interior of the food. These changes are similar to those that occur during baking and are described in Chapter 18 (section 18.3). IR heating also has similar effects on micro-organisms to those described during baking.

Table 20.6   Characteristics of infrared emitters

Parameter	Quartz lamp	Quartz tube	Ceramic element
Heated element	Tungsten filament	Nichrome wire	Fe-Cr-Al wire
Type of wave/intensity	Short wave, high intensity	Medium wave, medium intensity	Medium/long wave, medium/low intensity
Operating temperature (°C)	2200–1600	980–760	700–200
Colour of light	Bright white	Cherry red	No visible light
Peak energy wavelength (μm)	1.15–1.6	2.3–2.8	3.2–6
Radiant heat (%)	72–86	40–60	20–50
Convective heat (%)	28–14	60–40	80–50
Heat up/cool down time	1 s	30 s	5 min
Maximum intensity ($kW\,m^{-2}$)	70–1800	15–120	15–60

Adapted from Jackson and Welch (1998)

References

ALLEN, K., EIDMAN, V. and KINSEY, J., (1996), An economic-engineering study of ohmic food processing, *Food Technology*, **50** (5), 269–273.

ANATHESWARAN, R.C. and RAMASWARMY, H.S., (2001), Bacterial destruction and enzyme inactivation during microwave heating, in (A.K. Datta and R.C. Anantheswaran, Eds.), *Handbook of Microwave Technology for Food Applications*, CRC Press, Boca Raton, FL, pp. 191–214.

ANON, (1999), Information from Strayfield Limited, Theale, UK.

ANON, (2000a), Kinetics of microbial inactivation for alternative food processing technologies – microwave and radio frequency processing, US Food and Drug Administration Center for Food Safety and Applied Nutrition, available at http://www.cfsan.fda.gov/~comm/ift-micr.html.

ANON, (2000b) Kinetics of microbial inactivation for alternative food processing technologies – ohmic and inductive heating, US Food and Drug Administration, Center for Food Safety and Applied Nutrition, available at www.cfsan.fda.gov/~comm/ift-ohm.html.

ANON, (2005), The Electromagnetic Compatibility Regulations 2005, available at www.legislation.gov.uk/si/si2005/20050281.htm.

ANON, (2006), Western Europe: Ready Meals 2010, Company Information from RTS Resource Ltd., available at www.rts-resource.com/-pdf/downloads/western%20Europe%20 Ready%20Meals%202010.pdf

ANON, (2007a), RF and microwave components, company information from Spectrum Microwave Inc., available at www.spectrummicrowave.com/.

ANON, (2007b), 50 Ω RF heating equipment, company information from Petrie Technologies Ltd, available at www.petrieltd.com/pages/50Ohm.htm.

BELGIN, E. DU S., PALAZOG, T.K., GOKMEN, V., SENYUVA, H.Z. and EKIZ, H.I., (2007), Reduction of acrylamide formation in French fries by microwave pre-cooking of potato strips, *J. Science Food Agriculture*, **87** (1), 133–137.

BRENNAN, J.G., (2006), Evaporation and dehydration, in (J.G. Brennan, Ed.), *Food Processing Handbook*, Wiley-VCH, Weinheim, Germany, pp. 71–124.

BRODY, A.L., (1992), Microwave food pasteurisation, sterilisation and packaging, in (A. Turner, Ed.), *Food Technology International Europe*, Sterling Publications International, London, pp. 67–71.

BUFFLER, C.R., (1993), *Microwave Cooking and Processing – Engineering Fundamentals for the Food Scientist*. AVI/Van Nostrand Reinhold, New York, pp. 18, 151.

CAMPAÑONE, L.A. and ZARITZKY, N.E., (2005), Mathematical analysis of microwave heating process, *J. Food Engineering*, **69** (3), 359–368.

CAÑUMIR, J.A., CELIS, J.E., DE BRUIJN, J. and VIDAL, L.V., (2002), Pasteurisation of apple juice by using microwaves, *Lebensmittel-Wissenschaft und -Technologie*, **35** (5), 389–392.

COHEN, J.S. and YANG, T.C.S., (1995), Progress in food dehydration, *Trends in Food Science and Technology*, **6** (1), 20–25.

DATTA, A.K., (2001), Fundamentals of heat and moisture transport for microwaveable food product and process development, in (A.K. Datta and R.C. Anatheswaran, Eds.), *Handbook of Microwave Technology for Food Applications*, CRC Press, Boca Raton, FL, pp. 115–172.

DATTA, A.K. and ANANTHESWARAN, R.C., (2001), *Handbook of Microwave Technology for Food Applications*, CRC Press, Boca Raton, FL.

DATTA, A.K. and HU, W., (1992), Quality optimization of dielectric heating processes, *Food Technology*, **46** (12), 53–56.

DECAREAU, R.V., (1985), *Microwaves in the Food Processing Industry*, Academic Press, Orlando, FL.

DECAREAU, R.V., (1990), Microwave uses in food processing, in (A. Turner, Ed.), *Food Technology International Europe*, Sterling Publications International, London, pp. 69–72.

DIBBEN, D., (2000), Electromagnetics: fundamental aspects and numerical modelling, in (A.K. Datta and R.C. Anatheswaran, Eds.), *Handbook of Microwave Technology for Food*, Marcel Dekker, New York, pp. 1–32.

DINNAGE, D.F., (1990), Aseptic processing – use of ohmic heating, in (A. Freed, Ed.), *Changing Food Technology (3)*, Technomic Publishing, Lancaster, PA, pp. 29–42.

EARLE, R.L., (1983), *Unit Operations in Food Processing*, 2nd edn, Pergamon Press, Oxford, pp. 46–63.

EHLERMANN, D.A.E., (2002), Microwave processing, in (C.J.K. Henry and C. Chapman, Eds.), *The Nutrition Handbook for Food Processors*, Woodhead Publishing, Cambridge, pp. 396–406.

FRYER, P., (1995), Electrical resistance heating of foods, in (G.W. Gould, Ed.), *New Methods of Food Preservation*, Blackie Academic and Professional, Glasgow, pp. 205–235.

FUJIKAWA, H., USHIODA, H. and KUDO, Y., (1992), Kinetics of *Escherichia coli* destruction by microwave irradiation, *Applied Environmental Microbiology*, **58** (3), 920–924.

GARCIA, A. and BUENO, J.L., (1998), Improving energy efficiency in combined microwave-convective drying, *Drying Technology*, **16** (1/2), 123–140.

GUNASEKARAN, S., (1999), Pulsed microwave-vacuum drying of food materials, *Drying Technology*, **17** (3), 395–412.

HEDDLESON, R.A., DOORES, S. and ANANTHESWARAN, R.C., (1994), Parameters affecting destruction of *Salmonella* spp. by microwave heating, *J. Food Science*, **59** (2), 447–451.

JACKSON, A.N. and WELCH, D.E., (1998), Industrial applications of electric infrared heating, company information available from Advanced Energy at www.advancedenergy.org.

JONES, P.L., (1987), Dielectric heating in food processing, in (A. Turner, Ed.), *Food Technology International Europe*, Sterling Publications International, London, pp. 57–60.

JONES, P.L. and ROWLEY, A.T., (1997), Dielectric dryers, in (C.G.J. Baker, Ed.), *Industrial Drying of Foods*, Blackie Academic and Professional, London, pp. 156–178.

KEINER, L.E., (2007), The electromagnetic spectrum, available at http://en.wikipedia.org/wiki/Electromagnetic_spectrum.

KIM, H-J., CHOI, Y-M., YANG, T.C.S., TAUB, I.A., TEMPEST, P., SKUDDER, P., TUCKER, G. and PARROTT, D.L., (1996), Validation of ohmic heating for quality enhancement of food products, *Food Technology*, **50** (5), 253–261.

LARKIN, J.W. and SPINAK, S.H., (1996), Safety considerations for Ohmically heated, aseptically processed, multiphase low-acid food products, *Food Technology*, **50** (5), 242–245.

LEWIS, M.J., (1990), *Physical Properties of Foods and Food Processing Systems*, Woodhead Publishing, Cambridge.

MIRZA, S. and MORTON, I.D., (2006), Effect of different types of blanching on the colour of sliced carrots, *J. Science of Food and Agriculture*, **28** (11), 1035–1039.

MIZRAHI, S., (1996), Leaching of soluble solids during blanching of vegetables by ohmic heating, *J. Food Engineering*, **29** (2), 153–166.

MOHSENIN, N.N., (1984), *Electromagnetic Radiation Properties of Foods and Agricultural Products*, Gordon and Breach, New York.

MUDGET, R.E., (1982), Electrical properties of foods in microwave processing, *Food Technology*, **36**, 109–115.

PANDIT, R.B., TANG, J., LIU, F. and MIKHAYLENKO, G., (2007), A computer vision method to locate cold spots in foods in microwave sterilization processes, *Pattern Recognition*, **40** (12), 3667–3676.

PARROTT, D.L., (1992), The use of ohmic heating for aseptic processing of food particulates, *Food Technology*, **46** (12), 68–72.

PIYASENA, P., DUSSAULT, C., KOUTCHMA, T., RAMASWAMY, H. and AWUAH, G., (2003), Radio frequency heating of foods: principles, applications and related properties – a review, *Critical Reviews in Food Science and Nutrition*, **43** (6), 587–606.

RAHMAN, M.S., (1999), Preserving foods with electricity: ohmic heating, in (M.S. Rahman, Ed.), *Handbook of Food Preservation*, Marcel Dekker, New York, pp. 521–532.

RAMESH, M.N., WOLF, W., TEVINI, D. and BOGNAR, A., (2002), Microwave blanching of vegetables, *J. Food Science*, **67** (1), 390–398.

REZNICK, D., (1996), Ohmic heating of fluid foods, *Food Technology*, **50** (5), 250–251.

ROSENBERG, U. and BOGL, W., (1987), Microwave pasteurisation, sterilisation and pest control in the food industry, *Food Technology (USA)*, June, 92–99.

RUAN, R.X.Y. and CHEN, P., (2002), Ohmic heating, in (C.J.K. Henry and C. Chapman, Eds.), *The Nutrition Handbook for Food Processors*, Woodhead Publishing, Cambridge, pp. 407–422.

RUAN, R.X.Y, CHEN, P., DOONA, C. and YANG, T., (2004), Developments in ohmic heating, in (P. Richardson, Ed.), *Improving the Thermal Processing of Foods*, Woodhead Publishing, Cambridge, pp. 224–252.

SALENGKE, S. and SASTRY, S.K., (2007), Models for ohmic heating of solid–liquid mixtures under worst-case heating scenarios, *J. Food Engineering*, **83** (3), 337–355.

SAMPROVALAKI, K., BAKALIS, S. and FRYER, P.J., (2007), Ohmic heating: models and measurements, in (S.Yanniotis and B. Sundén, Eds.), *Heat Transfer in Food Processing*, WIT Press, Southampton, pp. 159–186.

SASTRY, S.K., (1994), Ohmic heating, in (R.P. Singh and F. Oliveira, Eds), *Minimal Processing of Foods and Process Optimisation: An Interface*, CRC Press, Boca Raton, FL, pp. 17–34.

SASTRY, S.K. and LI, Q., (1996), Modeling the ohmic heating of foods, *Food Technology*, **50** (5), 246–247.

SASTRY, S.K. and PALANIAPPAN, S., (1992), Mathematical modelling and experimental studies on ohmic heating of liquid-particulate mixtures in a static heater, *J. Food Process Engineering*, **15** (4), 241–261.

SCHIRACK, A.V., DRAKE, M., SANDERS, T.H. and SANDEEP, K.P., (2006), Impact of microwave blanching on the flavor of roasted peanuts, *J. Sensory Studies*, **21** (4), 428–440.

SCHNEPF, M. and BARBEAU, W.E., (1989), Survival of *Salmonella typhimurium* in roasting chickens cooked in a microwave, convention microwave and conventional electric oven, *J. Food Safety*, **9**, 245–252.

SCHUBERT, H. and REGIER, M., (2005), *The Microwave Processing of Foods*, Woodhead Publishing, Cambridge.

SINGH, R.P. and HELDMAN, D.R., (2001), Heat transfer in food processing – microwave heating, in *Introduction to Food Engineering*, 3rd edn, Academic Press, pp. 306–332.

SKJÖLDEBRAND, C., (2002), Infrared processing, in (C.J.K. Henry and C. Chapman, Eds.), *The Nutrition Handbook for Food Processors*, Woodhead Publishing, Cambridge, pp. 423–432.

TOHI, S., HAGURA, Y. and SUZUKI, K., (2002), Measurement of change in moisture content during drying process using the dielectric property of foods, *Food Science and Technology Research*, **8** (3), 257–260.

VEGA-MERCADO, H., CONGORA-NIETO, M.M. and BARBOSA-CANOVAS, G.V., (2001), Advances in dehydration of foods, *J. Food Engineering*, **49**, 271–289.

WANG, W.C. and SASTRY, S.K., (1997), Changes in electrical conductivity of selected vegetables during multiple thermal treatments, *J. Food Process Engineering*, **20** (6), 499–516.

WATANABE, F., ABE, K., FUJITA, T., GOTO, M., HIEMORI, M. and NAKANO, Y., (1998), Effects of microwave heating on the loss of vitamin B_{12} in foods, *J. Agriculture Food Chemistry*, **46** (1), 206–210.

YE, X. F., RUAN, R., CHEN, P. and DOONA, C., (2004), Simulation and verification of ohmic heating in static heater using MRI temperature mapping, *Lebensmittel-Wissenschaft und -Technologie*, **37** (1), 49–58.

YIN, Y. and WALKER, C.E., (1995), A quality comparison of breads baked by conventional versus nonconventional ovens: a review, *J. Science of Food and Agriculture*, **67** (3), 283–291.

ZHANG, H. and DATTA, A.K., (2000), Electromagnetics of microwave heating: magnitude and uniformity of energy absorption in an oven, in (A.K. Datta and R.C. Anatheswaran, Eds.), *Handbook of Microwave Technology for Food Applications*, Marcel Dekker, New York, pp. 33–68.

ZHANG, M., TANG, J., MUJUMDAR, A.S. and WANG, S., (2006), Trends in microwave-related drying of fruits and vegetables, *Trends in Food Science and Technology*, **17**, 524–534.

ZHAO, Y., FLUGSTAD, B., KOLBE, E., PARK, J.W. and WELLS, J.H., (2000), Using capacitive (radio frequency) dielectric heating in food processing and preservation – a review, *J. Food Process Engineering*, **23** (1), 25–55.

ZOLTAI, P. and SWEARINGEN, P., (1996), Product development considerations for ohmic processing, *Food Technology*, **50** (5), 263–266.

Part IV

Processing by removal of heat

In the unit operations described in this section, a reduction in the temperature of foods slows the biochemical and microbiological changes that would otherwise take place during storage. Preservation by lowering the temperature of foods has important benefits in maintaining their sensory characteristics and nutritional value to produce high-quality and high-value products. As a result these foods have substantially increased in commercial importance during the past 30 years. In particular, rapid expansion of ready-to-eat chilled foods, some packed under modified atmospheres, has been an important development over the past 15 years in many countries (Chapter 21). Many of the developments in minimal processing methods (Chapter 9) as well as storage of fresh foods also rely on chilling as a component of preservation (see also hurdle technology (Chapter 1, section 1.3.1).

In general, the lower the temperature, the longer foods can be stored, and freezing (Chapter 22) continues to be an important method of processing. Micro-organisms and enzymes are inhibited at low temperatures, but unlike heat processing they are not destroyed. Any increase in temperature can therefore permit the growth of pathogenic bacteria or increase the rate of spoilage of foods. Careful control is needed to maintain a low storage temperature and prepare foods quickly under strict hygienic conditions. The need to maintain chill or frozen temperatures throughout the distribution chain is a major cost to producers and retailers, and this area has seen significant developments to improve energy efficiency and reduce costs. Freeze drying and freeze concentration (Chapter 23) are used to produce some high-value products that are stable at ambient temperatures and therefore avoid the costs of a cold distribution chain. However, the high operating costs of these technologies remain significant deterrents to their more widespread adoption.

21

Chilling and modified atmospheres

Abstract: Chilling is a unit operation that is used to extend the shelf-life of foods by reducing their temperature to between $-1\,°C$ and $8\,°C$, which reduces the rates of biochemical and microbiological changes. This chapter first describes the operation of mechanical vapour-compression and cryogenic refrigerators and calculation of the rate of refrigeration. It then describes different types of refrigerants introduced to reduce ozone depletion, chilling and cold storage equipment, methods of temperature monitoring, and modified or controlled atmosphere storage of fresh foods. The chapter concludes by discussing the effects of chilling on pathogenic micro-organisms and food safety, and the effects of chilling on sensory and nutritional qualities of foods.

Key words: refrigeration, vapour-compression refrigerators, properties of refrigerants, coefficient of performance, respiration of fresh fruits and vegetables, cryogenic chilling, CO_2 snow, liquid nitrogen, modified atmosphere storage, critical temperature indicators (CTIs), time–temperature indicators (TTIs), high-risk foods.

Chilling is the unit operation in which the temperature of a food is reduced to between $-1\,°C$ and $8\,°C$ to reduce the rate of biochemical and microbiological changes and hence to extend the shelf-life of fresh and processed foods. It is often used in combination with other unit operations (e.g. fermentation (Chapter 6, section 6.1), pasteurisation (Chapter 12) and minimal processing methods (Chapter 9)) to extend the shelf-life of mildly processed foods.

There is a greater preservative effect when chilling is combined with control of the composition of the storage atmosphere than that found using either unit operation alone. A reduction in the concentration of oxygen and/or an increase in carbon dioxide concentration of the storage atmosphere surrounding a food inhibits microbial and insect growth and also reduces the rate of respiration of fresh fruits and vegetables. When combined with chilling, modified atmosphere packaging (Chapter 25, section 25.3) is an increasingly important method of maintaining high quality in processed foods during an extended shelf-life.

Chilling causes minimal changes to sensory characteristics and nutritional properties of foods and, as a result, chilled foods are perceived by consumers as being high quality, 'healthy', 'natural' 'fresh', convenient and easy to prepare. Since the 1980s there has been substantial product development and strong growth in the chilled food market, particularly for sandwiches, desserts, ready meals, prepared salads, pizza and fresh pasta

(Dennis and Stringer 2000). More recently organic and oriental ready meals have been introduced to markets in industrialised countries. Woon (2007) for example, describes 17% growth in retail sales of organic ready meals in Western Europe between 2005 and 2006, and sales of reduced-fat ready meals increased by 11% over the same period. Prepared salads have been one of the fastest-growing categories, with retail value sales increasing at a compound annual growth rate of 12% between 1998 and 2006. The addition of ingredients that claim active health benefits, such as omega-3, is also contributing to the increase in the range of chilled foods on the market (see Chapter 6, section 6.2). The biggest growth (56%) came from 'Oriental' ready meals, which includes Malaysian, Singaporean, Thai and Chinese dishes. These different developments have made ready meals one of the most dynamic market segments for packaged food.

Chilled foods are grouped into three categories according to their storage temperature range as follows:

1   −1 to +1 °C (e.g. fresh fish, meats, sausages and ground meats, smoked meats and breaded fish);
2   0 to +5 °C (e.g. milk, cream, yoghurt, prepared salads, sandwiches, fresh pasta, fresh soups and sauces, baked goods, pizzas, pastries and unbaked dough);
3   0 to +8 °C (e.g. fully cooked meat and fish pies, cooked or uncooked cured meats, butter, margarine, hard cheese, cooked rice, fruit juices and soft fruits).

Details of the wide range of available chilled foods are given by a number of suppliers including Anon (2006a) and are reviewed by Dennis and Stringer (2000). However, not all foods can be chilled and tropical, subtropical and some temperate fruits, for example, suffer from chilling injury at 3–10 °C above their freezing point (section 21.4).

The successful supply of chilled foods to the consumer depends on sophisticated and relatively expensive distribution systems that involve chill stores, refrigerated transport and retail chill display cabinets (section 21.2.3), together with widespread ownership of domestic refrigerators. Precise temperature control is essential at all stages in the cold chain to avoid the risk of food spoilage or food poisoning. In particular, low-acid chilled foods, which are susceptible to contamination by pathogenic bacteria (e.g. fresh and precooked meats, pizzas and unbaked dough) must be prepared, packaged and stored under strict conditions of hygiene and temperature control. In many countries there is legislation covering the temperature at which different classes of foods should be transported and stored based on an international agreement (the ATP agreement on the Carriage of Perishable Foodstuffs) (Anon 2008a). A summary of GMP and HACCP is given in Chapter 1 (section 1.5.1) and details of legislation that affects temperature control of chilled foods in Europe and North America are given by Anon (2006b), Woolfe (2000) and Goodburn (2000).

21.1   Theory

21.1.1   Refrigeration
There are two methods of chilling foods: mechanical vapour-compression and cryogenics. Mechanical vapour-compression refrigerators (section 21.2.1) have four basic components: an evaporator, a compressor, a condenser and an expansion valve (Fig. 21.1). A refrigerant (section 21.2.1) circulates between these four components, changing state from liquid to gas and back to liquid, with changes in both pressure and enthalpy at each stage. Thermodynamic properties of individual refrigerants are described in

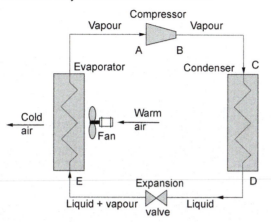

Condenser may be water-cooled or air-cooled

Fig. 21.1   Single stage mechanical (vapour-compression) refrigeration components.

pressure–enthalpy tables (data are available from refrigerant manufacturers and in Granryd 2007), and the properties can also be represented on pressure–enthalpy charts or temperature–entropy charts. Figure 21.2 shows the main components of a pressure–enthalpy chart, with pressure plotted on a logarithmic scale. The area to the left of the bell curve represents subcooled liquid refrigerant, the area under the curve represents mixtures of liquid and vapour, and the area to the right of the curve represents superheated vapour above the saturation temperature of vapour at the corresponding pressure. Within the curve, dryness fraction lines show the proportion of liquid and vapour in the refrigerant. Constant pressure lines are the horizontal lines across the chart and constant temperature lines are vertical in the liquid region of the chart, horizontal under the bell curve and curved downward in the vapour region.

Changes to the refrigerant as it moves through the different components of a vapour-compression cycle can be represented on a pressure–enthalpy diagram (Fig. 21.3) as follows:

1   Refrigerant vapour enters the compressor from the low-pressure side of the cycle (point A in Figs 21.1 and 21.3), having pressure P_1 and enthalpy H_2 and is compressed to a higher pressure P_2 at point B in the superheated region. The outlet pressure from the compressor must be below the critical pressure of the refrigerant (Fig. 21.2) and high enough to enable condensation of the refrigerant by a cooling medium at ambient temperature. During compression, work is done by the compressor, which increases the enthalpy of the refrigerant to H_3 as well as increasing its pressure and temperature. The size of the compressor is selected to pump refrigerant through the system at the required flowrates and pressures. The operating pressure depends on the type of refrigerant being used and the required evaporator temperature.
2   The refrigerant passes to the condenser, where cool air or water flowing through the condenser coils absorbs heat from the hot refrigerant vapour, causing it to condense back to a liquid state. The superheat is first removed (point C) and then the latent heat of condensation (C–D). The enthalpy of the refrigerant falls to H_1 but the pressure remains constant.
3   The liquid refrigerant then passes at a controlled rate through the expansion valve (D–E), which separates the high- and low-pressure parts of the cycle at constant

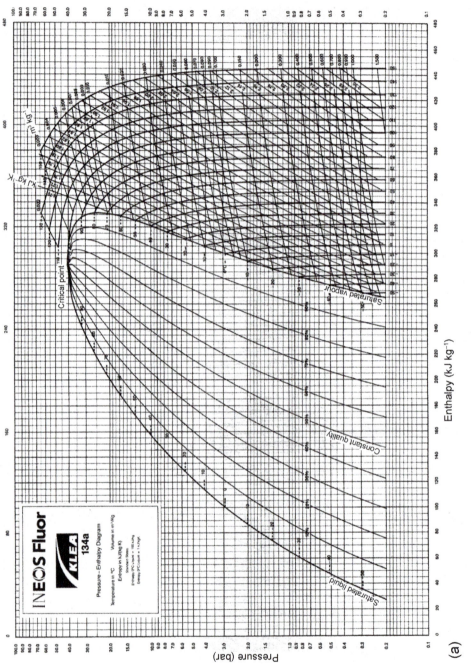

Fig. 21.2 Pressure–enthalpy charts: (a) courtesy of Ineos Fluor, (b) adapted from Singh and Heldman (2001).

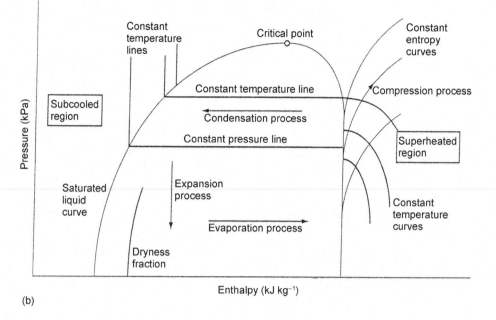

Fig. 21.2 (continued)

enthalpy (H_1). The refrigerant pressure falls to P_1 and some of the refrigerant changes to gas.

4 The gas–liquid mixture passes to the evaporator, where the liquid refrigerant evaporates under reduced pressure, and in doing so absorbs latent heat of vaporisation and cools the freezing medium. The freezing medium can be the relatively warm air in a coldroom, water, brine or food flowing over the evaporator coils. The refrigerant

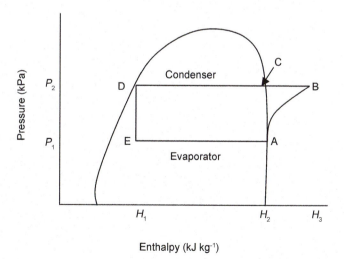

Fig. 21.3 Pressure–enthalpy chart showing vapour-compression cycle (adapted from Singh and Heldman 2001).

evaporates to become a saturated vapour (E–A). The enthalpy of the refrigerant increases from H_1 to H_2 but the pressure remains constant. The refrigerant then passes to the compressor and the cycle continues.

This is an idealised refrigeration cycle and in practice deviations from the idealised cycle including fluid friction, heat transfer losses and component inefficiency, prevent the refrigeration cycle from operating at the optimum performance. These deviations are described by Singh and Heldman (2001). Sun and Wang (2001a) describe other types of refrigeration cycles.

The coefficient of performance (COP) is the ratio of the heat absorbed by the refrigerant in the evaporator and the heat equivalence of energy supplied to the compressor, which is shown in Equation 21.1:

$$COP = \frac{H_2 - H_1}{H_3 - H_2} \qquad \boxed{21.1}$$

where H_1 (kJ kg^{-1}) = enthalpy of refrigerant leaving the condenser, H_2 (kJ kg^{-1}) = enthalpy of refrigerant entering the compressor and H_3 (kJ kg^{-1}) = enthalpy of refrigerant leaving the compressor. The COP is an important measure of the performance of refrigeration systems. For the most common types of refrigeration plant the COP would typically be in the range 3–6 (Heppenstall 2000).

The work done on the refrigerant in the compressor can be calculated from the refrigerant flowrate and the increase in enthalpy using Equation 21.2:

$$q_w = m(H_3 - H_2) \qquad \boxed{21.2}$$

where q_w (kW) = rate of work done on refrigerant and m (kg s^{-1}) = mass flowrate.

Similarly in the condenser the rate of heat removed (q_c (kW)) is found using Equation 21.3:

$$q_c = m(H_3 - H_1) \qquad \boxed{21.3}$$

The difference in enthalpy between the inlet and outlet to the evaporator (known as the 'refrigeration effect') is found using Equation 21.4:

$$q_e = m(H_2 - H_1) \qquad \boxed{21.4}$$

To chill fresh foods it is necessary to remove both sensible heat (also known as 'field heat') and heat generated by respiratory activity. The production of respiratory heat at 20 °C and atmospheric pressure is given by Equation 21.5 and the rate of heat evolution at different storage temperatures is shown in Table 21.1 for a selection of fruits and vegetables:

$$C_6H_{12}O_6 + 6O_2 \rightarrow 6CO_2 + 6H_2O + 2.835 \times 10^6 \text{ J kmol}^{-1} C_6H_{12}O_6 \qquad \boxed{21.5}$$

The processing time required to chill a crop is calculated using unsteady state heat transfer equations (Chapter 10, section 10.1.2 and sample problem 21.1), which are described in detail by Pham (2001). Mathematical models for calculation of heat load and chilling rate are described by, for example, Davey and Pham (1996) and Trujillo and Pham (2003). The calculations are simpler when processed foods are chilled, as respiratory activity does not occur.

The rate of heat removed from a cold store or food is known as the cooling (or refrigeration) load. The refrigerant flowrate can be calculated from the cooling load on the system and the refrigeration effect using Equation 21.6:

Table 21.1   Rates of heat evolved from fruits and vegetables stored at different temperatures

Commodity	Heat evolution (W t^{-1})			
	0 °C	5 °C	10 °C	15 °C
Apples	10–12	15–21	41–61	41–92
Apricots	15–17	19–27	33–56	63–101
Asparagus	81–237	161–403	269–902	471–970
Beans (green)	73–82	101–103	161–172	251–276
Beetroots	16–21	27–28	35–40	50–69
Broccoli	55–63	102–474	–	514–1000
Brussels sprouts	46–71	95–143	186–250	282–316
Cabbage	12–40	28–63	36–86	66–169
Carrots (topped)	46	58	93	117
Cauliflower	53–71	61–81	100–144	136–242
Celery	21	32	58–81	110
Grapes	4–7	9–17	24	30–35
Leeks	28–48	58–86	158–201	245–346
Lemons	9	15	33	47
Lettuce	27–50	39–59	64–118	114–121
Mushrooms	83–129	210	297	–
Onions	7–9	10–20	21	33
Oranges	9–12	14–19	35–40	38–67
Peaches	11–19	19–27	46	98–125
Pears	8–20	15–46	23–63	45–159
Peas (in pods)	90–138	163–226	–	529–599
Plums	6–9	12–27	27–34	35–37
Potatoes	–	17–20	20–30	20–35
Raspberries	52–74	92–114	82–164	243–300
Spinach	–	136	327	529
Strawberries	36–52	48–98	145–280	210–273
Tomatoes (green)	–	21	45	61

From Anon (1978) and Lewis (1990)

$$m_f = \frac{q}{(H_2 - H_1)} \qquad \boxed{21.6}$$

where m_f (kg s^{-1}) = refrigerant flowrate and q (kW) is the cooling load (sample problem 21.2).

Cryogenic chilling
A cryogen is a 'total-loss' refrigerant that cools foods by absorbing latent heat as it changes phase. Cryogenic chillers use solid CO_2, liquefied CO_2 or liquefied nitrogen. Their properties are shown in Table 21.2.

Table 21.2   Selected properties of food cryogens

Property	Liquid nitrogen	Carbon dioxide
Density of liquid (kg m^{-3})	314.9	468
Density of gas (kg m^{-3})	1.2506	1.9769
Boiling point/sublimation temperature at 101.3 kPa (°C)	−195.4	−78.5
Specific heat of vapour (kJ kg^{-1} K^{-1})	1.04	0.85
Heat of vaporisation/sublimation (kJ kg^{-1})	198.3	571.3
Heat removed to freeze food to −18 °C (kJ kg^{-1})	690	565

From Hung (2001), Graham (1984) and Anon (2007a)

Sample problem 21.1

Freshly harvested berries measuring 2 cm in diameter are chilled from 18 °C to 7 °C in a chiller at −2 °C, with a surface heat transfer coefficient of 16 W m⁻² K⁻¹. They are then loaded in 250 kg batches into containers and held for 12 h in a cold store operating at −2 °C, prior to further processing. Each container weighs 50 kg. The cold store holds an average of 2.5 t of berries and measures 3 m high by 10 m × 10 m. The walls and roof are insulated with 300 mm of polyurethane foam and the floor is constructed from 450 mm of concrete. The ambient air temperature in the factory averages 12 °C and the soil temperature 9 °C. An operator spends an average of 45 min per day moving the containers within the store and switches on four 100 W lights when in the store. Calculate the time required to cool the berries in the chiller and determine whether a 5 kW refrigeration plant would be suitable for the cold store. (Additional data: thermal conductivity of the berries = 0.127 W m⁻¹ K⁻¹, thermal conductivity of the insulation = 0.026 W m⁻¹ K⁻¹, thermal conductivity of the concrete = 0.87 W m⁻¹ K⁻¹ (Chapter 10, Table 10.2), specific heat of the berries = 3778 J kg⁻¹ K⁻¹, specific heat of the container = 480 J kg⁻¹ K⁻¹, the density of berries = 1050 kg m⁻³, the heat produced by the operator = 240 W, and the average heat of respiration of the berries = 0.275 J kg⁻¹ s⁻¹.)

Solution to sample problem 21.1

To calculate the time required to cool the berries, from Equation 10.24 for unsteady state heat transfer ($Bi = h\delta/k$) for berries,

$$Bi = \frac{16 \times 0.01}{0.127}$$

$$= 1.26$$

$$\frac{1}{Bi} = 0.79$$

From Equation 10.25 for cooling,

$$\frac{\theta_f - \theta_h}{\theta_i - \theta_h} = \frac{7 - (-2)}{18 - (-2)}$$

$$= 0.45$$

From Fig. 10.5 for a sphere, Fo = 0.38. From Equation 10.26,

$$0.38 = \frac{k}{c\rho} \frac{t}{\delta^2}$$

Therefore,

$$t = \frac{0.38 \times 3778 \times 1050(0.01)^2}{0.127}$$

$$\text{Time of cooling} = 1187 \text{ s}$$

$$= 19.8 \text{ min}$$

To determine whether the refrigeration plant is suitable for the cold store, assume that the berries enter the store at chill temperature.

Total		heat		sensible		heat evolved		heat loss		heat loss
heat	=	of	+	heat of	+	by operators	+	through	+	through
load		respiration		containers		and lights		roof and walls		floor

Now,

$$\text{Heat of respiration} = 2500 \times 0.275$$

$$= 687.5 \text{ W}$$

Assuming that the containers have the same temperature change as the berries and the number of containers is $2500/250 = 10$:

$$\text{Heat removed from containers} = \frac{10 \times 50 \times 480(18 - 7)}{12 \times 3600}$$

$$= 61 \text{ W}$$

and,

$$\text{Heat evolved by operators and lights} = \frac{(240 + 4 \times 100)(45 \times 60)}{24 \times 3600}$$

$$= 20 \text{ W}$$

From Equation 10.11 for a roof and wall area of $60 + 60 + 100 = 220 \text{ m}^2$

$$\text{Heat loss through roof and walls} = \frac{0.026 \times 220[12 - (-2)]}{0.3}$$

$$= 267 \text{ W}$$

Finally,

$$\text{Heat loss through the floor (area} = 100 \text{ m)}^2 = \frac{0.87 \times 100[9 - (-2)]}{0.45}$$

$$= 2127 \text{ W}$$

Therefore

$$\text{Total heat loss is the sum of the heat loads} = 687.5 + 61 + 20 + 2127$$

$$= 2895.5 \text{ W}$$

$$\approx 3 \text{ kW}$$

Therefore a 5 kW refrigeration plant would be suitable.

Sample problem 21.2
A cold store is cooled using R-134a refrigerant in a vapour-compression refrigeration system that has a cooling load of 35 kW. The evaporator temperature is −5 °C and the condenser temperature is 43 °C. Assuming that the compressor efficiency is 80%, calculate the compressor power requirement and the COP of the system.

Solution to sample problem 21.2
Find enthalpies H_1, H_2 and H_3 in Fig. 21.3 using the pressure–enthalpy chart (Fig. 21.2): first draw horizontal line E–A at −5 °C (evaporator temperature) and then line C–D at 43 °C (condenser temperature). Join points D–E (expansion). Extrapolate from point A along the constant entropy curve to meet line D–C that is extended to point B (compression). Read off the enthalpies as follows: H_1 (enthalpy of refrigerant leaving the condenser) = 165 kJ kg^{-1}, H_2 (enthalpy of refrigerant entering the compressor) = 295 kJ kg^{-1} and H_3 (enthalpy of refrigerant leaving the compressor) = 326 kJ kg^{-1}.

From Equation 21.6,

$$\text{Mass flowrate of refrigerant } (m) = \frac{35}{295 - 165}$$
$$= 0.27 \text{ kg s}^{-1}$$

From Equation 21.2,

$$\text{Compressor power requirement } (q_w) = \frac{0.27(326 - 295)}{0.80}$$
$$= 10.46 \text{ kW}$$

and from Equation 21.1,

$$\text{Coefficient of performance} = \frac{(295 - 165)}{(326 - 295)}$$
$$= 4.2$$

Although both nitrogen and CO_2 may be used, CO_2 is preferred for chilling whereas liquid nitrogen is more commonly used for freezing. This is because CO_2 has a higher boiling/sublimation point than nitrogen, and most of enthalpy (heat capacity) is due to the conversion of solid or liquid to gas. Only 13% of the enthalpy from liquid CO_2 and 15% from the solid is contained in the gas itself. This compares with 52% in nitrogen gas (that is, approximately half of the refrigerant effect of liquid nitrogen arises from sensible heat absorbed by the gas). CO_2 does not therefore require gas-handling equipment to extract most of the heat capacity, whereas liquid nitrogen does. The lower boiling point of liquid nitrogen creates a large temperature gradient between the cooling medium and the food, whereas CO_2 has a lower rate of heat removal, which allows greater control in reaching chill temperatures. The main limitation of cryogens is the risk that they can cause asphyxia, particularly by CO_2, and there is a maximum safe limit for operators of 0.5% CO_2 by volume. Excess gas is removed from the processing area by an exhaust system to ensure operator safety, which incurs additional set-up costs. The dangers and detection methods for increased concentrations of CO_2 are described by Henderson (2006) and dangers of asphyxiation by nitrogen are described by Anon (2003). Other hazards associated with liquefied cryogenic gases include cold burns, frostbite and hypothermia after exposure to intense cold.

21.1.2   Modified atmospheres

There remain differences in, and some confusion over, the terminology used to describe different types of modified atmospheres. In this text, modified atmosphere storage (MAS) is the use of gases to replace air around non-respiring foods without further controls during storage. In controlled atmosphere storage (CAS), the composition of gas around respiring foods is monitored and constantly controlled. In commercial operation, CAS and MAS are mostly used for storing apples and smaller quantities of pears and cabbage (see also modified atmosphere packaging, Chapter 25, section 25.3).

The normal composition of air is 78% nitrogen and 21% oxygen by volume, with the balance made up of CO_2 (0.035%), other gases and water vapour. A reduction in the proportion of oxygen and/or an increase in the proportion of CO_2 within specified limits in the atmosphere surrounding a food maintains the original product quality and extends the shelf-life. This is achieved by one or more of the following:

- inhibiting bacterial and mould growth;
- controlling biochemical and enzymic activity to slow ripening and senescence (ageing);
- protecting against insect infestation;
- reducing moisture loss;
- reducing oxidative changes.

For fresh foods that suffer chill injury (section 21.4) the rate of respiration may remain relatively high at the lowest safe storage temperature, and MAS/CAS are used to supplement refrigeration and extend the storage life. The important reaction in respiration is oxidation of carbohydrates (Equation 21.5) and for most products the 'respiratory quotient', defined as the ratio of CO_2 produced to oxygen consumed, is about 1 in air. Reducing the level of oxygen to 3% with or without increasing the level of CO_2 for a particular crop can reduce the rate of respiration to approximately one-third of the rate in air. However, too low an oxygen concentration can cause anaerobic respiration, which produces off-flavours in the product. The lowest oxygen concentrations before the onset of anaerobic respiration vary from 0.8% for spinach and 2.3% for asparagus (Toledo 1999). Typical gas compositions for selected products are shown in Table 21.3. Toledo (1999) also describes calculations of gas composition and flowrate in CAS stores.

The main disadvantages of MAS and CAS are economic: crops other than apples (and to a lesser extent cabbage and pears) have insufficient sales to justify the investment. Short season crops, which increase in price out of season, justify the additional costs of MAS or CAS, but the equipment cannot be used throughout the year. Also plant utilisation cannot be increased by storing crops together, because of the different requirements for gas composition, and the risk of odour transfer. Other limitations of MAS and CAS are as follows:

- The low levels of oxygen, or high levels of CO_2, which are needed to inhibit bacteria or fungi, are harmful to some foods.
- CAS conditions may lead to an increase in the concentration of ethylene in the atmosphere and accelerate ripening and the formation of physiological defects.
- An incorrect gas composition may change the biochemical activity of tissues, leading to development of off-odours, off-flavours, a reduction in characteristic flavours, or anaerobic respiration.

Table 21.3   Controlled atmospheres[a] for selected foods

Product	Carbon dioxide (% by volume)	Oxygen (% by volume)
Fresh crops		
Apples – general	2–5	3
Apples – Bramley's Seedling	8	13
Apples – Cox's Orange Pippin	5	3
Asparagus	5–10	2.9
Broccoli	10	2.5
Brussels sprouts	2.5–5	2.5–5
Cabbage	2.5–5	2.5–5
Green beans	5	2
Lettuce	5–10	2
Pears	5	1
Spinach	11	1
Tomatoes	0	3
Processed foods		
Cheese – mould ripened	0	0
Cheese – hard	25–35	0
Meat – cured	20–35	0
Pasta – fresh	25–35	0

[a] The balance of gases is nitrogen.
Adapted from Toledo (1999) and Day (2000)

- Tolerance to low oxygen and high CO_2 concentrations varies according to type of crop, conditions under which a crop is grown and maturity at harvest.
- Cultivars of the same species respond differently to a given gas composition, and growers who regularly change cultivars are unwilling to risk losses due to incorrect CAS conditions.
- Economic viability may be unfavourable owing to competition from other producing areas that have different harvest seasons, and higher costs of CAS over a longer storage period (twice that of cold storage).

An alternative approach is storage in a partial vacuum, which reduces the oxygen concentration by the same proportion as the reduction in air pressure (i.e. if the pressure is reduced by a factor of 10, then the oxygen concentration is reduced by the same factor). The main advantages are the continuous removal of ethylene and other volatiles from the atmosphere and precise control of air pressure ($\pm 0.1\%$). However, the method is not commonly used owing to the higher costs.

21.2   Equipment

Chilling equipment is designed to reduce the temperature of a product at a predetermined rate to a required final temperature, whereas cold storage equipment is designed to hold foods at a defined temperature, having been cooled before being placed in the store. Chilling equipment is classified by the method used to remove heat into mechanical refrigerators and cryogenic systems. Batch or continuous operation is possible with both types of equipment. All chillers should lower the temperature of the product as quickly as possible through the critical warm zone ($50 \rightarrow 10\,°C$) where maximum growth of micro-organisms occurs (Chapter 1, section 1.2.3). When used in cook–chill applications (section 21.3.2) chillers should be capable of reducing the temperature of 5 cm thick foods from 70 °C to a core temperature of <3 °C within 90 min (Heap 2000).

21.2.1   Mechanical refrigerators

Refrigerants

The refrigerants in mechanical vapour compression refrigerators (Table 21.4) have the following properties:

- A low boiling point and a high critical temperature (Fig. 21.2). At temperatures above the critical temperature, the refrigerant vapour cannot be liquefied.
- A high latent heat of vaporisation to reduce the volume of refrigerant required.
- A dense vapour to reduce the pressure required in the compressor, and hence the size and cost of the compressor.
- Low toxicity and non-flammable.
- Non-corrosive and having low miscibility with oil in the compressor.
- Chemically stable and not environmentally damaging in the event of leakage.
- Low cost.

Refrigerant safety classification consists of two alpha-numeric characters (e.g. A2); the capital letter corresponds to toxicity and the digit to flammability. Refrigerants are divided into two groups according to toxicity:

- Class A: refrigerants for which toxicity has not been identified at concentrations ≤ 400 mg kg^{-1}; and
- Class B: refrigerants for which there is evidence of toxicity at concentrations < 400 mg kg^{-1}.

Table 21.4   Comparison of refrigerants

Refrigerant	R-12[a]	R-22[a]	R-134a[a]	Propane	NH$_3$	CO$_2$
Natural fluid	No	No	No	Yes	Yes	Yes
ODP	0.82	0.055	0	0	0	0
GWP (100 yr) IPCC values	8100	1500	1300	20	<1	1
GWP (100 yr) WMO values	10600	1900	1600	20	<1	1
Critical temperature (°C)	112.0	96.2	101.2	96.7	132.3	31.1
Critical pressure (MPa)	4.14	4.99	4.06	4.25	11.27	7.38
Liquid density at boiling point (kg m^{-3})	1486	523.8	512	582	682	
Enthalpy of liquid at critical temperature (kJ kg^{-1})	183.4	366.6	215.9	425.3	1371[b]	571[b]
Flammable	No	No	No	Yes	Yes	No
Toxic	No	No	No	No	Yes	No
Relative price	–	1.0	4.0	0.3	0.2	0.1

[a] R-12 = Dichlorodifluoromethane, R-22 = monochlorodifluoromethane, R-134a = 1,1,1,2-tetrafluoroethane.
[b] At boiling point.
ODP = ozone depletion potential, GWP = global warming potential.
IPCC = Intergovernmental Panel on Climate Change, 1995 report, Contribution of Working Group I to the Second Assessment Report.
WMO = World Meteorological Organization, 1998 report, Scientific Assessment of Ozone Depletion, WMO Global Ozone Research and Monitoring Project, National Oceanic and Atmospheric Administration, National Aeronautics and Space Administration and the European Commission, Directorate General XII Science, Research and Development.
From Anon (2000a) and Anon (2001)

Refrigerants are divided into three groups according to flammability:

- Class 1: refrigerants that do not burn when tested in air at 21 °C at atmospheric pressure (101 kPa).
- Class 2: refrigerants having a lower flammability limit of $> 0.10 \, \mathrm{kg \, m^{-3}}$ at 21 °C and 101 kPa and a heat of combustion of $<19 \, \mathrm{kJ \, kg^{-1}}$.
- Class 3: refrigerants that are highly flammable $- \leq 0.10 \, \mathrm{kg \, m^{-3}}$ at 21 °C and 101 kPa or a heat of combustion $\geq 19 \, \mathrm{kJ \, kg^{-1}}$ (Anon 2001).

Ammonia has very good properties as a refrigerant and is not miscible with oil, but it is toxic and flammable, and causes corrosion of copper pipes. CO_2 is non-flammable and non-toxic, but can cause asphyxia at relatively low concentrations in the air. It is used for example on refrigerated ships, but it requires considerably higher operating pressures than ammonia. Halogen refrigerants (chlorofluorocarbons or CFCs) are all non-toxic and non-flammable and have good heat transfer properties and lower costs than other refrigerants. However, CFCs remain in the atmosphere and are broken down by UV radiation in the stratosphere to form chlorine radicals. These are thought to interfere with the formation of ozone and deplete the stratospheric ozone layer. The potential adverse health effects of ozone depletion have resulted in an international ban on their use as refrigerants under the 1987 Montreal Protocol. CFC replacements with much lower ozone-depleting potential have been developed, including hydrochlorofluorocarbons (HCFCs) and hydrofluorocarbons (HFCs):

- HCFC-123 (1,1-dichloro-2,2,2-trifluoroethane);
- HCFC-124 (1-chloro-1,2,2,2-tetrafluoroethane);
- HCFC-141b (1,1-dichloro-1-fluoroethane).

Although HCFCs contain chlorine atoms and hence deplete ozone, they are less potent than CFCs (Table 21.4) and have been introduced as temporary replacements for CFCs. Chlorine-free HFCs are compounds containing only hydrogen, fluorine and carbon atoms (Table 21.5):

- HFC-32 (difluoromethane);
- HFC-125 (pentafluoroethane);
- HFC-134a (1,1,1,2-tetrafluoroethane);
- HFC-143a (1,1,1-trifluoroethane);
- HFC-152a (1,1-difluoroethane).

They have weaker carbon–hydrogen bonds that are more susceptible to breaking, and hence have a shorter life in the atmosphere, and they do not deplete the stratospheric ozone layer, but like HCFCs they are greenhouse gases (Heap 1997). R-134a, R407C and R410A are among the currently (2008) widely used refrigerants (Table 21.5).

In contrast to CFCs and HCFCs, ammonia, hydrocarbons and CO_2 all have a zero ozone depletion potential (ODP) and a negligible global warming potential (GWP) (Table 21.4). The ODP of HFCs is zero and their GWP ranges from a few hundred in the case of the flammable R-32 to several thousand in the case of the flammable R-143a and the non-flammable R-125. Although CO_2 has a major impact on global warming ($\approx 63\%$ of the combined effect of all greenhouse gases), its GWP from use as a refrigerant is negligible (Anon 2000a,b).

Chilling equipment
For solid foods, the chilling medium in mechanically cooled chillers may be air, water, brine or metal surfaces. Air chillers (e.g. air-blast chillers) use forced convection to

Table 21.5   Classification and applications of refrigerants

Name	Refrigerant number	Chemical formula	Safety classification	Applications/ properties
Inorganic compounds				
Ammonia	R-717	NH_3	B2	Moderately flammable, toxic
Water	R-718	H_2O	A1	–
Carbon dioxide	R-744	CO_2	A1	Replacement for R-12 and R-22 in refrigerated transport
Organic compounds				
Hydrocarbons				
Propane	R-290	$CH_3CH_2CH_3$	A3	Alternative for R-12 and
		CH_3CH_2	A3	R-22 in air conditioning,
Butane	R-600	CH_2CH_3		highly flammable
Isobutene	R-600a	$CH(CH_3)_2CH_3$	A3	
Propylene	R-1270	CH_3CHCH_2	A3	
Hydrochlorofluorocarbons (HCFCs)				
Dichlorodifluoromethane	R-12	CCl_2F_2	A2	Medium temperature refrigeration
Monochlorodifluoromethane	R-22	$CHClF_2$	A2	Low and medium temperature refrigeration
Hydrofluorocarbons (HFCs)				
Difluoromethane	R-32	CH_2F_2	A2	
Pentafluoroethane	R-125	CHF_2CF_3	A1	
1,1,1,2-Tetrafluoroethane	R-134a	CH_2FCF_3	A1	Replace R-12 in domestic refrigerators, industrial chillers, retail cabinets, refrigerated transport
1,1,1-Trifluoroethane	R-143a	CH_3CF_3	A2	
1,1-Difluoroethane	R-152a	CH_3CHF_2	A2	Replace R-12. Very low global warming potential, but is more flammable
Azeotropic mixtures		Composition (Mass %)		
	R-502	R22/R115 (48.8/51.2)	A1	
	R-507	R125/R143a (50/50)	A1	Used in retail display cabinets, ice machines, refrigerated transport
Zeotropic mixtures				
	R-404A	R125/R143a/ R134a (44/52/4)	A1	Retail display cases, ice machines, alternative to R-502 in refrigerated transport
	R-407C	R32/R125/ R134a (23/25/52)	A1	Replacement for R-22 in air-conditioning and industrial cooling systems, refrigerated transport and cold storage
	R-410A	R32/R125 (50/50)	A1	Used in cold storage, refrigerated transport and industrial chilling

Refrigerants are numbered with an R-, followed by the HFC-number; isomers are identified with lower cases (e.g. R 134a). Inorganic compounds are assigned a number in the 700 series by adding the relative molecular mass of components to 700 (e.g. R717 ammonia has molecular mass = 17). HFC refrigerant blends having the same components but with different compositions are identified with upper case (e.g. R 404A), with R-4 being zeotropic blends of two or more refrigerants and R-5 being azeotropes.
Adapted from Anon (2001) and Sun and Wang (2001a)

circulate air at around -10 to $-12\,^{\circ}C$ at high speed $(4\,m\,s^{-1})$, and thus reduce the thickness of boundary films of air to increase the rate of heat transfer (Chapter 10, section 10.1.2). The two main designs are batch (or static) tunnels, in which trolleys or pallets of food are placed for the required time, and continuous tunnels where the foods are moved through the tunnel at a speed that gives the required residence time for adequate cooling. Details of their design and operation are given by Mascheroni (2001). Larger units have wheeled trolleys that typically each contain up to 45 kg of food on trays. Blast chillers undergo a cycle of loading, chilling and automatic defrosting to remove ice from the evaporator, which may be microprocessor controlled using air temperature probes, product probes (also section 21.2.4) or a timer (e.g. Anon 2007b). They are fitted with alarms for temperature rise/mains failure and trapped personnel inside, and data loggers to record the temperature history of operation and transmit it to a control computer (Anon 2008b). They are also used in refrigerated vehicles, but food should be adequately chilled when loaded onto the vehicle, as the refrigeration plant is only designed to hold food at the required temperature and is not large enough to cool incompletely chilled food.

Other methods of chilling
Eutectic plate systems are another type of cooling that is used in refrigerated vehicles, especially for local distribution. Salt solutions (e.g. potassium chloride, sodium chloride or ammonium chloride) are frozen to their eutectic temperature (i.e. where the water and salt form a single phase at -3 to $-21\,^{\circ}C$) and air is circulated across the plates to absorb heat from the vehicle. The plates are regenerated by re-freezing in an external freezer.

Vacuum cooling of fresh foods (e.g. foods with a large surface area, such as lettuce, mushrooms and broccoli) is described in Chapter 2 (section 2.1). The methods used to vacuum-cool fresh foods, bakery products, liquid foods, such as beer, milk, juices and sauces, are described by Sun and Wang (2001b). The food is placed in a large vacuum chamber and the pressure is reduced to $\approx 0.5\,kPa$. Cooling takes place as moisture evaporates from the surface (a reduction of approximately $5\,^{\circ}C$ for each reduction of 1% in moisture content). Direct immersion in chilled water or brine (also termed 'hydrocooling') is described in Chapter 2 (section 2.1) and by Lucas *et al.* (2001). It is used to remove field heat from fruit and vegetables, for pre-chilling meat and poultry prior to freezing, on-board chilling of fish in refrigerated seawater, and cooling cheese by direct immersion in refrigerated brine. 'Immersion chilling and freezing' (ICF) is described by Lucas *et al.* (2001) and further details are given in Chapter 22 (section 22.2.1).

Recirculated chilled water is also used in plate heat exchangers (Chapter 12, Figs 12.3–12.5) to cool liquid foods after pasteurisation. Liquid and semi-solid foods (e.g. butter and margarine) are cooled by contact with refrigerated metal surfaces in scraped-surface heat exchangers (see examples in their application to heating in Chapters 12, 13 and 14).

21.2.2   Cryogenic chilling
Solid CO_2 can be used in the form of 'dry-ice' pellets, or liquid CO_2 can be injected into air to produce fine particles of solid CO_2 'snow', both of which rapidly sublime to gas. Pellets or snow are deposited onto, or mixed with food in combo bins, trays, cartons or on conveyors (Fig. 21.4). A small excess of snow or pellets continues the cooling during transportation or storage prior to further processing. If products are despatched immediately in insulated containers or vehicles, this type of chilling is able to replace

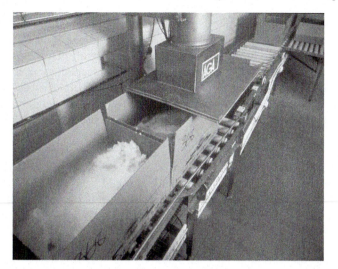

Fig. 21.4   Snow horn dosing cartons with carbon dioxide snow (courtesy of Linde Group at www.linde.com).

on-site cold stores and thus saves space and labour costs. CO_2 snow is replacing dry-ice pellets because it is cheaper and does not have the problems of handling, storage and operator safety associated with dry ice. For example, in older meat processing operations, dry ice pellets were layered with minced meat as it was filled into containers. However, lack of uniformity in distribution of pellets resulted in some meat becoming frozen and some remaining above 5 °C, which permitted bacterial growth and resulted in variable product temperatures for subsequent processing. The use of snow horns to distribute a fine layer of snow over minced meat as it is loaded into combo bins has eliminated these problems and resulted in rapid uniform cooling to 3–4 °C.

Distribution of chilled and frozen food is described by Jennings (1999), in which carbon dioxide 'snow' (section 22.2.4) is added to containers of food, which are then loaded into distribution vehicles. The time that a product can be held at the required chilled or frozen storage temperature can be varied from 4 to 24 hours by adjusting the amount of added snow. Other advantages of the system include greater flexibility in being able to carry mixed loads at different temperatures in the same vehicle, greater control over storage temperature, and greater flexibility in use compared with standard refrigerated vehicles.

Other applications of cryogenic cooling include sausage manufacture, where CO_2 snow removes the heat generated during mixing (Chapter 4, section 4.1.3) and cryogenic grinding where the cryogen reduces dust levels, prevents dust explosions and improves the throughput of mills (Chapter 3, section 3.1.3). In spice milling, cryogens also prevent the loss of aromatic compounds. In the production of multi-layer chilled foods (e.g. trifles and other desserts) the first layer of product is filled and the surface is hardened with CO_2. The next layer can then be added immediately, without waiting for first layer to set, and thus permit continuous and more rapid processing. Other applications include cooling and case-hardening of hot bakery products and chilling flour to obtain accurate and consistent flour temperatures for dough preparation.

Liquid nitrogen is used in freezing (Chapter 22, section 22.2.2) and also in chilling operations. It can be supplied in pressurised containers or made on site as required (e.g. Anon 2008c). For batch chilling, typically 90–200 kg of food is loaded into an insulated stainless steel cabinet, containing centrifugal fans and a liquid nitrogen injector. The liquid nitrogen vaporises immediately and the fans distribute the cold gas around the cabinet to achieve a uniform reduction in product temperature. The chiller has a number of pre-programmed, microprocessor-controlled time–temperature cycles. A food probe monitors the temperature of the product and the control system changes the temperature inside the cabinet as the food cools, thus allowing the same pre-programmed cycle to be used irrespective of the temperature of the incoming food. As with other types of batch equipment, it is highly flexible in operation and it is therefore suitable for low production volumes or where a large number of speciality products are produced.

For continuous chilling, food is passed on a variable speed conveyor to an inclined, insulated, cylindrical barrel having a diameter of 80–120 cm and length 4–10 m. Liquid nitrogen or CO_2 is injected and the barrel rotates slowly and internal flights lift the food and tumble it through the cold gas. The temperature and gas flow rate are microprocessor controlled and the tumbling action prevents food pieces sticking together to produce a free-flowing product. It is used to chill diced meat or vegetables at up to $3 \, th^{-1}$. Controlled temperature liquid nitrogen tumblers are used to improve the texture and binding capacity of mechanically reformed meat products. The gentle tumbling action in a partial vacuum, cooled by nitrogen gas to $-2\,^{\circ}C$, solubilises proteins in poultry meat, which increases their binding capacity and water-holding capacity, thus improving later forming and coating operations.

An alternative design is a screw conveyor inside a 2.5 m long stainless steel housing, fitted with liquid CO_2 injection nozzles. Foods such as minced beef, sauce mixes, mashed potato or diced vegetables are chilled rapidly as they are conveyed through the chiller at up to $1 \, th^{-1}$. It is used to firm foods before portioning or forming operations or to remove heat from previous processing stages.

Details of the hygienic design of chilling plants, cleaning schedules and total quality management (TQM) procedures are discussed in detail by Holah and Thorpe (2000), Holah (2000) and Rose (2000) respectively.

21.2.3   Cold storage

Once a product has been chilled, the temperature must be maintained by refrigerated storage. Chill stores are normally cooled by circulation of cold air produced by mechanical refrigeration units, and foods may be stored on pallets, racks, or in the case of carcass meats, hung from hooks. Transport of foods into and out of stores may be done manually using pallet trucks, by forklift trucks or by computer-controlled robotic trucks (Chapter 27, section 27.3.1). Materials that are used for the construction of refrigerated storerooms are described by Brennan et al. (1990). To meet safety, quality and legal requirements, cold store temperatures should be maintained $<5\,^{\circ}C$. Fresh products may also require control of the relative humidity in a storeroom, and in some cases control over the composition of the storage atmosphere. In all stores it is important to maintain an adequate circulation of air using fans, and foods are therefore stacked in ways that enable air to circulate freely around all sides. This is particularly important for respiring foods, to remove heat generated by respiration (section 21.3.1) or for foods, such as cheese, in which flavour development takes place during storage. Adequate air circulation is also important when high storage humidities are used for fresh fruits and vegetables (Table 21.6) as there is an increased risk of spoilage by moulds if 'dead-spots' permit localised increases in humidity. Hoang et al. (2001) describe computer-aided simulations of air flow, heat transfer and mass transfer in cold stores using computational fluid dynamics

(CFD) to improve their design and operation (CFD is a type of fluid mechanics that uses algorithms to analyse and solve problems that involve fluid flow).

Retail chill storage and display cabinets use chilled air that circulates internally by natural or forced convection. The two most common designs are 'serve-over' or delicatessen cabinets that have food displayed on a chilled base, and vertical multi-deck display cabinets that may be open-fronted or have glass doors. The cost of chill storage is high and to reduce costs, large stores may have a centralised plant to circulate refrigerant to all cabinets. The heat generated by the condenser can also be used for in-store heating. Computer control of multiple cabinets detects excessive rises in temperature and warns of any requirement for emergency repairs or planned maintenance (Cambell-Platt 1987). Other energy-saving devices include plastic curtains or night blinds on the front of cabinets to trap cold air. Details of the design and operation of refrigerated retail display cabinets, chilled distribution vehicles and cold stores are given by Heap (2000).

21.2.4   Temperature monitoring

Temperature monitoring is an integral part of quality management and product safety management throughout the cold chain. Improvements to micro-electronics have produced monitoring devices that can both store large amounts of data and integrate this into computerised management systems (Chapter 27, section 27.2). Woolfe (2000) lists the specifications of commonly used temperature data loggers, which may also be able to sound an alarm if the temperature exceeds a pre-set limit. These are connected to temperature sensors, which measure either air temperatures or product temperatures. There are three main types of sensor that are used commercially: thermocouples, semi-conductors and platinum resistance thermometers (thermistors). The most widely used thermocouples are Type K (nickel–chromium and nickel–aluminium), or Type T (copper and copper–nickel). The advantages over other sensors are lower cost, rapid response time and very wide range of temperature measurement (-184 to $1600\,^{\circ}\text{C}$). Thermistors have a higher accuracy than thermocouples, but they have a much narrower range (-40 to $140\,^{\circ}\text{C}$). Platinum resistance thermometers are accurate and have a temperature range from -270 to $850\,^{\circ}\text{C}$, but their response time is slower and they are more expensive than other sensors. Further details of sensors are given in Chapter 27 (section 27.2.1).

Monitoring air temperatures is more straightforward than product temperature monitoring and does not involve damage to the product or package. It is widely used to monitor chill stores, refrigerated vehicles and display cabinets, and Woolfe (2000) describes in detail the positioning of temperature sensors in these types of equipment. However, it is necessary to establish the relationship between air temperature and product temperature in a particular installation. When air is continuously recirculated through the refrigeration unit and storeroom, cold air is warmed by the incoming products, by lights in a store, vehicles or operators entering. The temperature of the returning air is therefore likely to be the same as the product temperature or slightly higher. The performance of the refrigeration system can be found by comparing the return air temperature with the temperature of the air leaving the evaporator in the refrigeration unit. 'Load tests' are conducted to relate air temperature to product temperature over a length of time under normal working conditions. The operation of open retail display cabinets is sensitive to variations in room temperature or humidity, the actions of customers and staff in handling foods, and lighting to display products. The temperature distribution in the cabinet can therefore change and load testing becomes more difficult. In such situations there is likely to be substantial variations in air temperature, but the mass of the food remains at a more constant temperature, and air temperature

Table 21.6 Optimum storage conditions for selected fruits and vegetables

Product	Optimal storage temperature (°C)	Optimal humidity (%)	Cooling using top ice acceptable	Cooling using water sprinkle acceptable	Ethylene Production	Sensitivity to:	Storage life
Apples	−1–4	90–95	No	No	High	Yes	1–12 months
Apricots	−1–0	90–95	No	No	High	Yes	1–3 weeks
Artichokes, Jerusalem	0–2	90–95	No	No	No	No	4–5 months
Asparagus	–	95–100	No	Yes	No	Yes	2–3 weeks
Blackberries	0–1	90–95	No	No	Very low	No	2–3 days
Broccoli	0	95–100	Yes	Yes	No	Yes	10–14 days
Brussels sprouts	0	90–95	Yes	Yes	No	Yes	3–5 weeks
Cabbage, early	0	98–100	Yes	Yes	No	Yes	3–6 weeks
Cabbage, late	0	98–100	–	–	No	–	5–6 months
Carrots, mature	0	98–100	–	–	No	–	7–9 months
Cauliflower	0–2	90–95	No	No	No	Yes	3–4 weeks
Celery	0	98–100	Yes	Yes	No	Yes	2–3 months
Celeriac	0	97–99	–	–	No	No	6–8 months
Cherries, sweet	0	90–95	No	No	Very low	No	2–3 weeks
Corn, sweet	0	95–98	Yes	Yes	No	No	5–8 days
Cucumbers	10–15	95	No	No	Very low	Yes	10–14 days
Eggplant (aubergine)	7–10	90–95	No	No	No	Yes	1 week
Garlic	0	65–70	No	No	No	No	6–7 months
Grapes	–	85	No	No	Very low	Yes	2–8 weeks
Leeks	0	95–100	Yes	Yes	No	Yes	2–3 months

Commodity							
Lemons	11–13	90–95	No	No	Very low	No	1–6 months
Lettuce	0	98–100	No	Yes	No	Yes	2–3 weeks
Mushrooms	0	95	No	Yes	No	Yes	3–4 days
Nectarines	−0.5–0	90–95	No	No	High	No	2–4 weeks
Okra	–	90–95	No	No	Very low	Yes	7–10 days
Parsnips	0	98–100	Yes	Yes	No	Yes	4–6 months
Peaches	−0.5–0	90–95	No	No	High	Yes	2–4 weeks
Peas, green	0	95–98	–	–	No	–	1–2 weeks
Peppers, hot chilli	–	60–70	No	No	No	Yes	6 months
Peppers, sweet	7–10	90–95	No	No	No	No	2–3 weeks
Plums	−1–0	90–95	No	No	High	Yes	2–5 weeks
Potatoes	3–10	90–95	Yes	No	No	–	2–3 months
Radishes, spring	0	95–100	Yes	Yes	No	Yes	3–4 weeks
Radishes, winter	0	95–100	–	–	No		2–4 months
Rhubarb	0	95–100	No	Yes	No	No	2–4 weeks
Spinach	0	95–100	–	–	No	–	10–14 days
Squashes, summer	0	95	No	No	No	Yes	1–2 weeks
Squashes, winter	0	50–70	No	No	No	Yes	1–6 months
Strawberries	0	90–95	No	No	Very low	No	3–7 days
Tomatoes, mature green	4–10	90–95	No	No	Low	Yes	1–3 weeks
Tomatoes, ripe	4–10	90–95	No	No	Medium	No	4–7 days
Turnips	0	95	Yes	Yes	No	Yes	4–5 months

Adapted from Anon (2005) and Yang (1998)

measurement has little meaning. To overcome this problem the food temperature can be measured using thermocouples, or the air temperature sensor can be electronically 'damped' to respond more slowly and eliminate short-term fluctuations.

In addition to temperature sensors, the temperature history of chilled foods (and also fresh or frozen foods) can be monitored by critical temperature indicators (CTIs) or time–temperature indicators (TTIs), which are widely used in both the chilled food cold chain and the frozen cold chain (Chapter 22, section 22.2.4) (Van Loey *et al.* 1998). They indicate whether a product has been held at the correct storage temperature to give the required shelf-life, or if temperature abuse has occurred so that the product can be moved more rapidly through the cold chain. CTIs show when a product has been exposed to temperatures above a reference temperature for sufficient time to cause a change in the quality or safety of the product. However, they do not show how long the temperature abuse lasted or by how much the critical temperature was exceeded. They are useful for foods that undergo irreversible damage above or below a certain temperature (e.g. freezing of fresh or chilled foods or thawing of frozen foods), or with foods that are susceptible to growth of a pathogen above a certain temperature (section 21.5). TTIs are attached to products and integrate the temperature and the time that a food has been exposed to a particular temperature. These devices are based on irreversible mechanical, chemical or enzymic changes (e.g. melting point temperature, polymerisation, electrochemical corrosion or liquid crystals) (Woolfe 2000, Selman 1995). There are two categories: critical (or 'partial history') time/temperature indicators, and full history time/temperature indicators. Critical TTIs show the cumulative time–temperature exposure above a reference critical temperature. They are useful for indicating the extent of biochemical or enzymic reactions, or microbial growth that can take place only above a certain critical temperature. Full history TTIs produce a continuous integrated time–temperature history of the food as a single measurement that can be correlated with temperature-dependent reactions that result in quality loss. Methods of correlation are described by Le Blanc and Stark (2001).

Examples of indicators include the following:

- Liquid crystal coatings that show the temperature of food and change colour with storage temperature.
- Wax that melts and releases a coloured dye when an unacceptable increase in temperature occurs (temperature abuse).
- A printed label that has an outer ring printed with a stable reference colour and contains diacetylene in the centre of a 'bull's eye'. The diacetylene changes as a function of time and temperature to produce a progressive, predictable and irreversible colour change, and when it matches the reference ring the product has no remaining shelf-life (Fig. 21.5).
- A TTI based on an enzymic reaction which changes the colour of a pH indicator.

A barcode system has been developed that is applied to a pack as the product is dispatched. The barcode contains three sections: (1) a code giving information on the product identity, date of manufacture and batch number, etc. to uniquely identify each container; (2) a second code identifies the reactivity of a TTI; and (3) a section that contains the indicator material. When the barcode is scanned, a hand-held microcomputer display indicates the status and quality of the product with a variety of pre-programmed messages (for example: 'Good', 'Don't use' or 'Call QC'). A number of microcomputers can be linked via modems to a central control computer to produce a portable monitoring system that can track individual containers throughout a distribution chain.

Wessel (2007) describes a prototype TTI that can be attached directly to an RFID

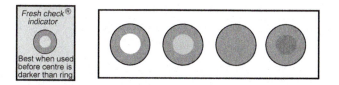

Fig. 21.5   Time–temperature indicator: Expired Lifeline's Fresh-Check® Indicator (from Taoukis and Labuza 2003).

(radio frequency identification) transponder to enable companies to remotely monitor the shelf-life of refrigerated foods based on temperature exposure during shipment. It uses both colour changes and an electrical signal to express the temperature history and it can transfer the electrical signal and temperature information to an active RFID tag (a microchip plus antenna). The tag contains a unique identification number and may have other information, such as the account number for a customer. This type of 'smart label' can have a barcode printed on it, or the tag can be mounted inside a carton or embedded in plastic (see also Chapter 25 (section 25.4.3) and Chapter 27 (section 27.3)). A tag reader interrogates the tag to enable cold-chain operators to calculate the remaining shelf-life of specific goods, based on the temperature information. A detailed description of TTIs is given by Taoukis and Labuza (2003).

21.2.5   Modified and controlled atmosphere storage equipment

In MAS, the store is made airtight, and respiratory activity of fresh foods is allowed to change the atmosphere as oxygen is used up and CO_2 is produced. Individual gases may be added from pressurised cylinders in MAS stores that are not completely gas-tight, to speed up the creation of the required atmosphere rather than relying on the respiratory action of the fruit alone. Gas-tight stores are sealed using metal cladding and carefully sealed doorways. Adjustment to the atmospheric composition is needed in CAS, and solid or liquid CO_2 is used to increase the gas concentration, controlled ventilation is used to admit oxygen, or 'scrubbers' may be used to remove CO_2. Scrubbers absorb CO_2 either by passing air from the store over bags of hydrated calcium hydroxide (lime), under sprays of sodium hydroxide or over activated carbon. The CO_2 content in the atmosphere can be monitored using sensors to measure differences in the thermal conductivity between CO_2 ($0.015\,W\,m^{-1}\,K^{-1}$), N_2 ($0.024\,W\,m^{-1}\,K^{-1}$) and O_2 ($0.025\,W\,m^{-1}\,K^{-1}$) or by differences in infrared absorption. Gas composition is automatically controlled by microprocessors to maintain a predetermined atmosphere using information from the sensors to control air vents and gas scrubbers. MAS and CAS are useful for crops that ripen after harvest, or deteriorate quickly, even at optimum storage temperatures. CA stores have a higher relative humidity than normal cold stores (90–95%) and therefore retain the crispness of fresh foods and reduce weight losses. Details of the atmospheric composition required for different products, building construction, equipment and operating conditions are reviewed by Jayas and Jeyamkondan (2002).

21.3   Applications

21.3.1   Fresh foods

The rate of biochemical reactions in fresh foods caused by naturally occurring enzymes changes logarithmically with temperature (Chapter 1, section 1.2). Chilling therefore

Table 21.7   Respiration rate and storage life of selected foods

Respiration rate		Examples of foods	Typical storage life (weeks at 2 °C)
Class	Rate of CO_2 emission at 5 °C ($mg\ CO_2\ kg^{-1}\ h^{-1}$)		
Extremely high	>60	Asparagus, broccoli, mushroom, pea, spinach, sweetcorn	0.2–0.5
Very high	40–60	Artichoke, snap bean, Brussels sprouts	1–2
High	20–40	Strawberry, blackberry, raspberry, cauliflower, lima bean, avocado	2–3
Moderate	10–20	Apricot, banana, cherry, peach, nectarine, pear, plum, fig, cabbage, carrot, lettuce, pepper, tomato	5–20
Low	5–10	Apple, citrus, grape, kiwifruit, onion, potato	25–50
Very low	< 5	Nuts, dates	>50

Adapted from Saltveit (2004) and Alvarez and Thorne (1981)

reduces the rate of enzymic changes and retards respiration and senescence in fresh foods. The factors that control the shelf-life of fresh crops in chill storage include:

- the type of food and variety or cultivar;
- the part of the crop selected (the fastest-growing parts have the highest metabolic rates and the shortest storage lives (Table 21.7));
- the condition of the food at harvest (e.g. the presence of mechanical damage or microbial contamination, and the degree of maturity);
- the temperature during harvest;
- the relative humidity of the storage atmosphere, which also influences dehydration losses;
- gas composition of storage atmosphere.

These factors are described in more detail by Bedford (2000) and changes to fresh crops and meats are described in section 21.4. Technologies to extend the shelf-life of fresh fruits and vegetables are described by Kader *et al.* (1998).

In CAS of crops, the concentrations of oxygen, CO_2 and sometimes ethylene are monitored and regulated. Oxygen concentrations as low as 0%, and CO_2 concentrations of 20% or higher can be produced in for example grain storage, where these conditions destroy insects and inhibit mould growth. Similarly, the use of CAS for cocoa storage reduces losses due to insect damage and avoids treatments with toxic fumigants (e.g. phosphine, methyl bromide). An additional benefit is that the moisture content of the cocoa stacks can be easily controlled to prevent weight loss (Anon 2008d).

When storing fruits, a higher oxygen concentration is needed to prevent anaerobic respiration, which might produce alcoholic off-flavours. Different types of fruit, and even different cultivars of the same species, require different atmospheres for successful storage and each therefore needs to be independently assessed (see examples in Table 21.3 for Bramley's Seedling and Cox's Orange Pippin at 3.5 °C which produced an increase from 3 months storage in air to 5 months under CAS. This can be further increased to 8 months using a CAS atmosphere of 1% CO_2, 1% O_2 and 98% N_2). Refrigerated storage of winter white cabbage in 5% CO_2, 3% O_2 and 92% N_2 enables the crop to be stored until the following summer (Brennan *et al.* 1990).

21.3.2 Processed foods

The range of retail chilled foods can be characterised by the degree of microbial risk that they pose to consumers as follows:

- Class 1: foods containing raw or uncooked ingredients, such as salad or cheese, ready-to-eat (RTE) foods (also includes low-acid raw foods, such as meat and fish). Some Class 1 products require cooking by the consumer, whereas other cooked–chilled products may be ready to eat or eaten after a short period of re-heating.
- Class 2: products made from a mixture of cooked and low-risk raw ingredients.
- Class 3: cooked products that are then packaged.
- Class 4: products that are cooked after packaging, including ready-to-eat-products-for-extended-durability (REPFEDs) having a shelf-life of 40+ days (the acronym is also used to mean refrigerated-pasteurised-foods-for-extended-durability).

In the above classification, 'cooking' refers to a heat process that results in a minimum reduction in target pathogens (see Chapter 10, section 10.3).

It is essential that foods which rely on chilled storage for their safety are processed and stored below specified temperatures under strict conditions of hygiene. Brown (2000) has reviewed methods to design safe foods using predictive microbial modelling. Gorris (1994) and Betts (1998) describe methods of mild processing to improve the safety of RTE foods (also section 21.5).

The shelf-life of chilled processed foods is determined by:

- the type of food and other preservative factors (e.g. pH, low a_w, use of preservative chemicals);
- the degree of microbial destruction or enzyme inactivation achieved by other unit operations before chilling;
- control of hygiene during processing and packaging;
- the barrier properties of the package; and
- temperatures during processing, distribution and storage.

Each of the factors that contribute to the shelf-life of chilled foods can be thought of as 'hurdles' to microbial growth and further details of this concept are given in Chapter 1 (section 1.3.1). Details of procedures for the correct handling of chilled foods and correct storage conditions for specific chilled products are described by Anon (2004, 1998).

Cook–chill systems

Individual foods (e.g. sliced roast meats) or complete meals are produced by 'cook–chill' or 'cook–pasteurise–chill' processes. An example is 'sous vide' products, which are vacuum packed prior to pasteurisation and chilled storage. These products were developed for institutional catering to replace warm-holding (where food is kept hot for long periods before consumption). The process reduces losses in nutritional value and eating quality and is less expensive. It is described in detail in Ghazala and Trenholm (1998) and Creed and Reeve (1998). Nicolai *et al.* (1994) describe computer-aided design of cook–chill foods.

After preparation, cooked–chilled foods are portioned and chilled within 30 min of cooking. Chilling to 3 °C should be completed within 90 min and the food is stored at 0–3 °C. In the cook–pasteurise–chill system, hot food is filled into a flexible container, a partial vacuum is formed to remove oxygen and the pack is heat sealed. It is then pasteurised to a minimum temperature of 80 °C for 10 min at the thermal centre, followed by immediate cooling to 3 °C. These foods have a shelf-life of 2–3 weeks (Hill 1987).

In addition to normal hygienic manufacturing facilities, the products in Classes 1, 2 and 4 at the beginning of this section require a special 'hygienic area' that is designed to be easily cleaned to prevent bacteria such as *Listeria* spp. from becoming established. RTE products require an additional 'high-care area', which is a physically separated from other areas and is carefully designed to isolate cooked foods during preparation, assembly of meals, chilling and packaging. Such areas have specified hygiene requirements including:

- positive pressure ventilation with micro-filtered air supplied at the correct temperature and humidity;
- entry and exit of staff only through changing rooms;
- 'no-touch' washing facilities;
- use of easily cleaned materials for walls, floors and food contact surfaces;
- only fully processed foods and packaging materials admitted through hatches or air-locks;
- special hygiene training for operators and fully protective clothing (including boots, hairnets, coats, etc.);
- special disinfection procedures and operational procedures to limit the risk of contamination;
- production stopped for cleaning and disinfection every 2 hours.

Detailed descriptions of the special considerations needed for the design, construction and operation of facilities for chilled foods are given by Holah and Thorpe (2000), Brown (2000), Rose (2000) and Anon (1998). Microbiological considerations when producing REPFEDs are described by Gorris and Peck (1998), and Holah (2000) gives details of the special methods needed for cleaning and disinfection of chilling facilities. Creed (2001) describes the production of chilled ready meals, sandwiches, pizzas and chilled deserts.

21.4   Effect on sensory and nutritional qualities of foods

The rate of respiration of fresh fruits is not necessarily constant at a constant storage temperature. Fruits that undergo 'climacteric' ripening (Table 21.8), induced by the plant hormone ethylene, show a short but abrupt increase in the rate of respiration and a significant increase in CO_2 production, which occurs near to the point of optimum ripeness. A climacteric fruit can therefore be picked at full size or maturity but before it is ripe and then allowed to ripen, which increases flavour quality, juice, sugars and other factors. Non-climacteric fruits produce little or no ethylene and no large increase in CO_2 production, and maintain the qualities that they have at harvest.

The production of sensitivity to ethylene in different fruits is shown in Table 21.6, together with control of humidity and cooling methods to achieve the required storage life. Vegetables respire in a similar way to non-climacteric fruits and differences in respiratory activity of selected fruits and vegetables are shown in Tables 21.1 and 21.7. Details of the biochemical action of ethylene are given by Oetiker and Yang (1995) and the ripening processes are described by Saltveit (2004) and reviewed by Giovannoni (2001). Detailed information on crop storage is given by Morris (2001).

Undesirable changes to some fruits and vegetables occur when the storage temperature is reduced below a specific optimum for the individual crop. This is termed 'chilling injury' and results in various physiological changes (Table 21.9) that may be caused by an imbalance in metabolic activity, which results in over-production of metabolites that then become toxic to the tissues (Haard and Chism 1996). Changes in membrane lipid

Table 21.8 Climacteric and non-climacteric ripening fruits

	Climacteric	Non-climacteric
Temperate	Apple	Blueberry
	Apricot	Cherry
	Melon	Cucumber
	Pear	Grape
	Peach	Olive
	Plum	Strawberry
	Tomato	
	Watermelon	
(sub)Tropical	Avocado	Cashew apple
	Banana	Grapefruit
	Breadfruit	Java plum
	Cherimoya	Lemon
	Fig	Lime
	Guava	Litchi
	Jackfruit	Orange
	Kiwifruit	Pepper (green, yellow, red)
	Mango	Pineapple
	Nectarine	Tamarillo
	Papaya	
	Passion fruit	
	Persimmon	
	Soursop	
	Sapote	

Adapted from Harris (1988) and Anon (2008e)

Table 21.9 Chilling injury to selected fruits

Food	Approximate lowest safe temperature (°C)	Chilling injury symptoms
Aubergines	7	Surface scald, *Alternaria* rot
Avocados	5–13	Grey discoloration of flesh
Bananas, green/ripe	12–14	Dull, grey-brown skin colour
Beans, green	7	Pitting, russeting
Cucumbers	7	Pitting, water-soaked spots, decay
Grapefruit	10	Brown scald, watery breakdown
Lemons	13–15	Pitting, membrane stain, red blotch
Limes	7–10	Pitting
Mangoes	10–13	Grey skin, scald, uneven ripening
Melons, honeydew	7–10	Pitting, failure to ripen, decay
Okra	7	Discoloration, water-soaked areas
Oranges	7	Pitting, brown stain, watery breakdown
Papaya	7	Pitting, failure to ripen, off-flavour, decay
Pineapples	7–10	Dull green colour, poor flavour
Potatoes	4	Internal discoloration, sweetening
Pumpkins	10	Decay
Sweet peppers	7	Pitting, *Alternaria* rot
Sweet potato	13	Internal discoloration, decay
Tomatoes, mature green	13	Water-soaked softening, decay
Tomatoes, ripe	7–10	Poor colour, abnormal ripening, *Alternaria* rot
Watermelon	5	Pitting

Adapted from Lutz and Hardenburg (1966)

structure, regulatory enzyme activity and structural proteins result in loss of membrane integrity and leakage of solutes (Brown and Hall 2000).

When operated at optimum conditions for a particular fresh crop, chilling to the correct storage temperature causes little or no reduction in the eating quality or nutritional properties of fresh foods. However, excessive storage times, incorrect temperatures and mechanical damage to crops can cause significant changes, including enzymic browning, wilting and weight loss due to transpiration (evaporation of water from aerial parts of plants). For example, Kidmose and Hansen (1999) studied the effects of storing fresh broccoli florets at 1, 5 or 10 °C for up to 14 days, followed by a short heat treatment and storage for 8 days. They found that storage time and temperature before processing affected the texture, colour, and amount of chlorophyll, vitamin C and β-carotene in the cooked florets. Changes in texture were correlated with water loss during storage of the raw heads. The vitamin C content was significantly affected by the temperature of chill storage of cooked florets and it fell to almost the same level after 3 or 8 days, irrespective of the duration of storage of the raw heads. The β-carotene content of cooked florets was stable when raw heads were stored at 1 and 5 °C, but it fell towards the end of the storage period when heads were stored at 10 °C. After cooking, the β-carotene content remained stable during subsequent chill storage.

Lee and Kader (2000) studied losses of vitamin C in fruits and vegetables and concluded that temperature management after harvest is the most important factor to maintain vitamin C levels. Losses are accelerated at higher storage temperatures and longer storage times, but some chill-sensitive crops show higher losses of vitamin C at lower storage temperatures. Conditions that cause moisture loss after harvest result in a rapid loss of vitamin C especially in leafy vegetables, and losses are also accelerated by bruising and other mechanical injuries, and by excessive trimming. Losses of vitamin C can be reduced by storing fruits and vegetables in atmospheres that contain reduced oxygen and/or CO_2 concentrations $\leq 10\%$, but higher levels of CO_2 can accelerate vitamin C loss. Details of nutrient losses are described by Weatherspoon et al. (2005).

Gil et al. (2006) compared quality indices and nutritional content of fresh-cut and whole fruits (pineapples, mangoes, cantaloupes, watermelons, strawberries and kiwifruits) stored for up to 9 days in air at 5 °C. Losses in vitamin C after 6 days were $\leq 5\%$ in mango, strawberry and watermelon pieces, 10% in pineapple pieces, 12% in kiwifruit slices and 25% in cantaloupe cubes. There were no losses in carotenoids in kiwifruit slices and watermelon cubes, whereas losses in pineapples were the highest at 25% followed by 10–15% in cantaloupe, mango and strawberry pieces after 6 days. No significant losses in total phenolics were found in any of the fresh-cut fruits after 6 days. They concluded that, in general, fresh-cut fruits spoil visually before any significant nutrient loss occurs. The influences of processing and storage on the quality indices and nutritional content of fresh-cut fruits were evaluated in comparison with whole fruits stored for the same duration but prepared on the day of sampling. Fresh-cut pineapples, mangoes, cantaloupes, watermelons, strawberries and kiwifruits and whole fruits were stored for up to 9 days in air at 5 °C. The post-cutting life based on visual appearance was shorter than 6 days for fresh-cut kiwifruit and shorter than 9 days for fresh-cut pineapple, cantaloupe and strawberry. On the other hand, fresh-cut watermelon and mango pieces were still marketable after 9 days at 5 °C. Losses in vitamin C after 6 days at 5 °C were $\leq 5\%$ in mango, strawberry and watermelon pieces, 10% in pineapple pieces, 12% in kiwifruit slices and 25% in cantaloupe cubes. No losses in carotenoids were found in kiwifruit slices and watermelon cubes, whereas losses in pineapples were the highest at 25% followed by 10–15% in cantaloupe, mango and strawberry pieces after 6 days at

5 °C. No significant losses in total phenolics were found in any of the fresh-cut fruit products tested after 6 days at 5 °C. Light exposure promoted browning in pineapple pieces and decreased vitamin C content in kiwifruit slices. Total carotenoid contents decreased in cantaloupe cubes and kiwifruit slices, but increased in mango and watermelon cubes in response to light exposure during storage at 5 °C for up to 9 days. There was no effect of exposure to light on the content of phenolics. In general, fresh-cut fruits visually spoil before any significant nutrient loss occurs.

In animal tissues, aerobic respiration rapidly declines when the supply of oxygenated blood is stopped at slaughter. However, muscles contain glycogen, creatine-phosphate and sugar phosphates that can continue to be used for ATP production by glycolysis. Anaerobic respiration of glycogen to lactic acid causes the pH of the meat to fall from ≈7 to between 5.4 and 5.6. When the supply of ATP ceases, the muscle tissue becomes firm and inextensible, known as rigor mortis. This can take place between 1 and 30 h post-mortem, depending on the type of animal, the physiological condition of the muscle and the ambient temperature. Lactic acid and inosine monophosphate (a breakdown product of ATP) also contribute to the flavour of the meat. The reduced pH of muscle tissues offers some protection against contaminating bacteria, but other non-muscular organs, such as the liver and kidneys, do not undergo these changes and they should be chilled quickly to prevent microbial growth. Provided that there is an adequate supply of glycogen, the rate and extent of the fall in pH are dependent on temperature; the lower the temperature the longer the time taken to reach the pH limit as biochemical reactions are slowed. The reduced pH causes protein denaturation and 'drip losses' and cooling the carcass during anaerobic respiration reduces this and produces the required texture and colour of meat. However, rapid chilling to temperatures below 12 °C before anaerobic glycolysis has ceased causes permanent contraction of muscles known as 'cold shortening', which produce undesirable changes and toughening of the meat.

If animals are exhausted at slaughter, their glycogen reserves are reduced and the production of lactic acid is reduced, leading to a higher pH. Pork that has a pH > 6.0–6.2 produces dark, firm, dry (DFD) meat which is more susceptible to bacterial spoilage. Conversely, if the fall in pH in pork muscle is too rapid or the temperature does not fall sufficiently within the first few hours post mortem, a series of changes produce meat known as pale, soft and exudative (PSE). Soluble sarcoplasmic proteins become denatured and precipitate, to appear as white particles that reflect light and cause paleness in the meat. Changes to membrane-bound myofibrillar proteins cause damage to the cell membranes and as a result they leak intracellular contents to form drip losses and cause cells to soften. The shelf-life of this meat is reduced owing to enhanced microbial growth and oxidation of phospholipids (Brown and Hall 2000). Details of these and other post-mortem changes to meat are described by Lawrie and Ledward (2006), Honikel and Schwagele (2001) and James (2000). Veerkamp (2001) describes chilling of poultry and Neilsen et al. (2001) describe chilling of fish.

Lipid oxidation is a major cause of quality deterioration in chill-stored meat and meat products, which result in adverse changes to flavour, colour, texture and nutritive value, and the possible production of toxic compounds. Jensen et al. (1998) found that pre-slaughter dietary supplementation with vitamin E was effective in reducing lipid oxidation, and improving colour, water-holding capacity and cholesterol oxidation in pig and poultry products. Juncher et al. (2001) reported that the physiological condition of live pigs significantly affects lipid oxidation and the colour and water-holding capacity of chilled pork chops chill-stored for 6 days. After treatments, including exercise and injection of adrenaline (a hormone that increases the supply of oxygen and glucose to the brain and muscles) they noted variations in energy metabolites (glycogen, lactate, creatine phosphate and ATP) and in the final pH of the meat. They concluded that reaching a narrow range of meat pH (pH 5.4–5.8) was the most important factor affecting product quality parameters of colour, lipid oxidation and drip loss, as well as microbiological growth.

Enzyme activity has both positive and negative effects on meat quality: proteases are important to produce loss in muscle stiffness after rigor mortis, known as 'conditioning'. Traditionally, large carcass meat is hung at chill temperatures for 2–3 weeks to become tender, but this occurs faster if the meat is not cooled as the proteases act more quickly. In fish and crustaceans, proteases in the gut weaken the gut wall after death and allow leakage of the contents into surrounding tissues (known as 'belly burst'). It is therefore essential that fish are gutted within hours of being caught and all seafood is chilled quickly to prevent deterioration. The most significant effect of chilling on the sensory characteristics of processed foods is hardening due to solidification of fats and oils. Longer-term chemical, biochemical and physical changes during refrigerated storage may lead to loss of quality, and in many instances it is these changes rather than microbiological growth that limit the shelf-life of chilled foods. These changes include enzymic browning, lipolysis, colour and flavour deterioration in some products, and retrogradation of starch to cause staling of baked products, which occurs more rapidly at refrigeration temperatures than at room temperature.

There have been many studies of the changes in nutrients during cook–chill and sous vide food preparation, largely because of their use in institutional catering and the potential adverse effects of nutrient losses on the health of hospital patients and the elderly. In a review of experimental studies, Williams (1996) found that the greatest loss of vitamins during hot-holding of food (>10% after 2 hours) were vitamin C, folate and vitamin B6, with retinol, thiamin, riboflavin, and niacin being relatively stable. In cook–chill operations, substantial losses of sensitive vitamins occur during each of the chilling, storage and reheating stages. Losses of vitamin C and folate can be >30% when food is reheated after storage for 24 hours at 3 °C. He concluded that vitamin retention is better in conventional foodservice than in cook–chill systems. Nutritional losses in cook–chill systems are reported by Bognar (1980) as insignificant for thiamine, riboflavin and retinol, but vitamin C losses are 3.3–16% per day at 2 °C. The large variation is due to differences in the chilling time, storage temperature, oxidation (the amount of food surface exposed to air and reheating conditions). Vitamin C losses in cook–pasteurise–chill processing are lower than cooked–chilled foods (e.g. spinach lost 66% within 3 days at 2–3 °C after cook–chilling compared with 26% loss within 7 days at 24 °C after cook–pasteurising–chilling).

Lipid oxidation is one of the main causes of quality loss in cook–chilled products, and cooked meats in particular rapidly develop an oxidised flavour termed 'warmed-over flavour', described in detail by Brown (1992). Brown and Hall (2000) have reviewed other effects of lipid oxidation in meats and its control using vacuum packing, modified atmosphere packing or the use of antioxidants either fed to animals pre-slaughter or added to meat products. Lassen et al. (2002) compared simulated warm-holding, conventional cook–chill, modified atmosphere packaging and sous vide meal-service systems for retention of vitamins B_1, B_2 and B_6 in pork roasts. Vitamin B_2 was retained irrespective of the meal-service system and storage period. Vitamins B_1 and B_6 declined by 14% and 21% respectively during 3 h of warm-holding, and by 11% and 19% respectively after 1 day of storage and subsequent reheating (cook–chill, MAP and sous vide). Vitamin B_1 declined by an additional 4% during storage for 14 days in sous vide.

They concluded that conventional and enhanced meal-service systems produced roasts that had similar quality attributes.

Other physicochemical changes in processed foods due to chilling may result in quality deterioration and include: migration of oils from mayonnaise to cabbage in chilled coleslaw; evaporation of moisture from unpackaged chilled meats and cheeses; more rapid staling of sandwich bread at reduced temperatures; and moisture migration from sandwich fillings to the bread, or from pie fillings or pizza toppings into the pastry and crust (Brown 1992). Syneresis in sauces and gravies is due to changes in starch thickeners. In starches that have higher proportions of amylose molecules, the amylose leaches out into solution and form aggregates by hydrogen bonding. These expel water and result in syneresis. Chilled products should therefore use modified starches that have blocking molecules to prevent amylose aggregating, or use starches that have higher proportions of amylopectin (also Chapter 1, section 1.1.1).

21.5   Effect on micro-organisms

As the storage temperature of a food is reduced, the lag phase of microbial growth extends and the rate of growth decreases (see Chapter 1, section 1.2.3). The reasons for this are complex at a cellular level and involve changes to the cell membrane structure, uptake of substrate and enzymic reactions including respiration (Herbert 1989). In chilling, the important factor concerning microbial growth is the minimum growth temperature (MGT), which is the lowest temperature at which an individual micro-organism can grow. Chilling prevents the growth of many mesophilic and all thermophilic micro-organisms that have MGTs of 5–10 and 30–40 °C respectively, but not psychrotrophic or psychrophilic micro-organisms, which have MGTs of 0–5 °C. Psychrotrophs and psychrophiles are distinguished by their maximum growth temperature, which is 35–40 and 20 °C respectively. Most food micro-organisms are psychrotrophs with a few psychrophiles associated with deep-sea fish (Walker and Betts 2000). When food is stored below the MGT of a micro-organism, cells may gradually die, but often the cells can survive and resume growth if the temperature increases. Mechanisms of microbial spoilage of fruits and vegetables are described by Niemira *et al.* (2005).

The effects of CO_2 on microbial growth are discussed by Dixon and Kell (1989) and reviewed by Farber (1991). CO_2 inhibits microbial activity in two ways: it dissolves in water in the food to form mild carbonic acid and thus lowers the pH at the surface of the product; and it has negative effects on enzymic and biochemical activities in cells of both foods and micro-organisms. It is therefore necessary to closely control the degree of atmospheric modification to prevent physiological disorders in the living tissues and secondary spoilage by anaerobic micro-organisms in non-respiring foods.

The most common spoilage micro-organisms in chilled foods are Gram-negative bacteria, which have MGTs of 0–3 °C, some of which may grow well at 5–10 °C. Examples include *Pseudomonas* spp., *Aeromonas* spp., *Acinetobacter* spp. and *Flavobacterium* spp. (Walker and Stringer 1990). They contaminate foods from water or inadequately cleaned equipment or surfaces, and may produce pigments, slime, off-flavours or off-odours, or rots. Yeasts and moulds are able to tolerate chill temperatures but grow more slowly than bacteria and may be out-competed unless other environmental factors limit the growth of bacteria (see also Chapter 1, section 1.2.3, for the influence of other environmental factors such as pH, a_w, preservatives, etc. on microbial growth). If bacterial growth is limited, yeasts may then cause spoilage problems. In addition, many yeasts can grow in the absence of oxygen in modified or controlled atmospheres. Examples of spoilage yeasts include *Candida* spp. *Debaromyces* spp., *Kluveromyces* spp. and *Saccharomyces* spp. Spoilage moulds that affect chilled products include *Aspergillus* spp., *Cladosporium* spp., *Geotrichum* spp., *Penicillium* spp. and *Rhizopus* spp.

Previously it was considered that refrigeration temperatures would prevent the growth of pathogenic bacteria, but it is now known that some species can either grow to large numbers at these temperatures, or are sufficiently virulent to cause poisoning after ingestion of only a few cells. The main microbiological safety concerns with chilled foods are a number of pathogens that can grow slowly during extended refrigerated storage below 5 °C, or as a result of any temperature abuse (Kraft 1992). Examples include *Listeria monocytogenes* (MGT = −0.4 °C), *Clostridium botulinum* types B and F (growth and toxin production 3.3–5 °C), *Aeromonas hydrophilia* (MGT = −0.1–1.2 °C), *Yersinia enterocolitica* (MGT = −1.3 °C) and some strains of *Bacillus cereus* (MGT = 1 °C for cell growth and 4 °C for toxin production) (Walker and Betts 2000, Walker 1992).

Other pathogens are unable to grow at temperatures <5 °C but may grow if temperature abuse occurs and then persist in the food. Examples include *Salmonella* sp. (MGT = 5.1 °C), enteropathogenic *Escherichia coli* (MGT = 7.1 °C), *Vibrio parahaemolyticus* and *Campylobacter* sp. (MGT = >10 °C) (Marth 1998). *E. coli* O157:H7 can cause haemorrhagic colitis after ingestion of as few as ten cells (Buchanan and Doyle 1997). A summary of the sources of these bacteria, types of infection and typical high-risk foods is given in Chapter 1 (section 1.2.4). It is therefore essential that good manufacturing practice (GMP) procedures are enforced as part of the HACCP plan during the production of chilled foods to control the safety of products (Anon 2007c) (see also Chapter 1, section 1.5.1). This includes minimising the levels of pathogens on incoming ingredients and by ensuring that processing and storage procedures do not introduce pathogens or allow their numbers to increase. Brown (2000) has reviewed microbiological hazards in chilled foods, equipment design and decontamination, hygienic design of chilling facilities and process monitoring and control. Sliced cold meat products have a high risk of contamination by pathogenic and spoilage micro-organisms unless food handling guidelines are strictly observed. The slicing operation may increase the microbial load on products via blades, as well as increasing nutrient availability as a result of tissue damage. Even with stringent hygienic conditions, extensive handling before and after slicing may cause significant contamination of cold meat products. For example, Voidarou *et al.* (2006) found contamination by the bioindicators *E. coli*, *S. aureus* and *C. perfringens* on sliced turkey, pork ham, smoked turkey and smoked pork ham.

References

ALVAREZ, J.S. and THORNE, S., (1981), The effect of temperature on the deterioration of stored agricultural produce, in (S. Thorne, Ed.), *Developments in Food Preservation*, Vol. 1, Applied Science, London, pp. 215–237.

ANON, (1978), *Heat Evolution Rates from Fresh Fruits and Vegetables*, American Society of Heating, Refrigeration and Air-conditioning Engineers, Atlanta, Georgia.

ANON, (1998), *Food and Drink Good Manufacturing Practice – A Guide to its Responsible Management*, 4th edn, IFST, London, pp. 67–76.

ANON, (2000a), Carbon dioxide as a refrigerant, 15th Informatory Note on Refrigerants,

International Institute of Refrigeration, Paris, France, available at www.iifiir.org/en/doc/1013.pdf.

ANON, (2000b), Carbon dioxide could replace global-warming refrigerant, Purdue University, reported in Science Daily 4 July, available at www.sciencedaily.com /releases/2000/07/000703091336.htm.

ANON, (2001), Designation and safety classification of refrigerants, American standard ANSI/ASHRAE 34, the American Society of Heating, Refrigerating and Air-Conditioning Engineers, available at Fluorocarbons website at www.fluorocarbons.org/en/applications/refrigeration.html.

ANON, (2003), Hazards of nitrogen asphyxiation, Safety Bulletin 2003-10-B, US Chemical Safety and Hazard Investigation Board, available at www.csb.gov/safety_publications/docs/SB-Nitrogen-6-11-03.pdf.

ANON, (2004), Evaluation of product shelf-life for chilled foods, CCFRA Guideline No. 46, available from Campden and Chorleywood Food Research Association, Chipping Campden.

ANON, (2005), Optimal temperature and humidity conditions for some common fruits and vegetables, available at www.engineeringtoolbox.com/fruits-vegetables-storage-conditions-d_710.html.

ANON, (2006a), Product list from Medallion Chilled Foods, available at www.westphalia.co.uk/docs/productlist.pdf.

ANON, (2006b), *CFA Guidelines for Good Hygienic Practice in the Manufacture of Chilled Foods*, 4th edn, Chilled Foods Association, Peterborough, UK, available at www.chilledfood.org/resources/publications.htm.

ANON, (2007a), Properties of carbon dioxide, Gas Encyclopaedia, information from Air Liquide, available at http://encyclopedia.airliquide.com/Encyclopedia.asp?GasID=26.

ANON, (2007b), Blast chillers: why they are indispensable in commercial kitchens, *Caterer and Hotelkeeper*, 18 July, available at www.caterersearch.com/Articles/2007/07/18/314953/blast-chillers-why-they-are-indispensible-in-commercial-kitchens.html.

ANON, (2007c), *Microbiological Guidance for Produce Suppliers to Chilled Food Manufacturers*, 2nd edn, available from the Chilled Food Association, www.chilledfood.org.

ANON, (2008a), Guide to ATP for road hauliers and manufacturers, Refrigerated Vehicle Test Centre, Cambridge Refrigeration Technology, Cambridge, available at www.crtech.co.uk/pages/ATP/atp-guide.pdf.

ANON, (2008b), Checkpoint wireless temperature system, company information from Omniteam Inc., available at www.omniteaminc.com/documents/checkpoint/checkpoint.pdf.

ANON, (2008c), StirLIN: Stirling liquid nitrogen production plants, company information from Stirling Cryogenics & Refrigeration BV, available at www.stirling.nl/sp/sp3.html.

ANON, (2008d), *GrainPro Newsletter*, October, available at www.grainpro.com/whatsnew.html.

ANON, (2008e), Climacteric and non-climacteric fruit list, information from Quisqualis, available at www.quisqualis.com/Climacteric.html.

BEDFORD, L., (2000), Raw material selection – fruits and vegetables, in (M. Stringer and C. Dennis, Eds.), *Chilled Foods – A Comprehensive Guide*, 2nd edn, Woodhead Publishing, Cambridge, pp. 19–35.

BETTS, G.D., (1998), Critical factors affecting safety of minimally processed chilled foods, in (S. Ghazala, Ed.), *Sous Vide and Cook–Chill Processing for the Food Industry*, Aspen Publications, Gaithersburg, MD, pp 131–164.

BOGNAR, A., (1980), Nutritive value of chilled meals, in (G. Glew, Ed.), *Advances in Catering Technology*, Applied Science, London, pp. 387–407.

BRENNAN, J.G, BUTTERS, J.R., COWELL, N.D. and LILLEY, A.E.V., (1990), *Food Engineering Operations*, 3rd edn, Elsevier Applied Science, London, pp. 465–493.

BROWN, M.H., (1992), Non-microbiological factors affecting quality and safety, in (C. Dennis and M. Stringer, Eds.), *Chilled Foods*, Ellis Horwood, Chichester, pp. 261–288.

BROWN, M.H., (2000), Microbiological hazards and safe process design, in (M. Stringer and C. Dennis, Eds.), *Chilled Foods – A Comprehensive Guide*, 2nd edn, Woodhead Publishing,

Cambridge, pp. 287–339.

BROWN, M.H. and HALL, M.N., (2000), Non-microbiological factors affecting quality and safety, in (M. Stringer and C. Dennis, Eds.), *Chilled Foods – A Comprehensive Guide*, 2nd edn, Woodhead Publishing, Cambridge, pp. 225–255.

BUCHANAN, R.L. and DOYLE, M.P., (1997), Foodborne disease significance of *Escherichia coli*. A scientific status summary of the IFST's expert panel on food safety and nutrition, Chicago III, *Food Technology*, **51** (10), 69–76.

CAMPBELL-PLATT, G., (1987), Recent developments in chilling and freezing, in (A. Turner, Ed.), *Food Technology International Europe*, Sterling, London, pp. 63–66.

CREED, P.G., (2001), Chilling and freezing of prepared consumer foods, in (D-W. Sun, Ed.), *Advances in Food Refrigeration*, Leatherhead Publishing, LFRA, Leatherhead, pp. 438–471.

CREED, P.G. and REEVE, W., (1998), Principles and applications of sous vide processed foods, in (S. Ghazala, Ed.), *Sous Vide and Cook–Chill Processing for the Food Industry*, Aspen Publications, Gaithersburg, MD, pp. 25–56.

DAVEY, L. and PHAM, Q.T., (1996), Construction of a predictive model for product heat load during chilling using an evolutionary method. Proc. Meeting Comm. B1, B2, E1, E2, International Institute of Refrigeration, Melbourne, Feb., available at www.ceic.unsw.edu.au/staff/Tuan_Pham/tanks.pdf.

DAY, B.P.F., (2000), Chilled food packaging, in (M. Stringer and C. Dennis, Eds.), *Chilled Foods – A Comprehensive Guide*, 2nd edn, Woodhead Publishing, Cambridge, pp. 135–150.

DENNIS, C. and STRINGER, M., (2000), Introduction: the chilled foods market, in (M. Stringer and C. Dennis, Eds.), *Chilled Foods – A Comprehensive Guide*, 2nd edn, Woodhead Publishing, Cambridge, pp. 1–16.

DIXON, N.M. and KELL, D.B., (1989), The inhibition by CO_2 of the growth and metabolism of microorganisms, *J. Applied Bacteriology*, **67**, 109–136.

FARBER, J.M., (1991), Microbiological aspects of modified-atmosphere packaging technology – a review, *J. Food Protection*, **54** (1), 58–70.

GHAZALA, S. and TRENHOLM, R., (1998), Hurdle and HACCP concepts in sous vide and cook-chill products, In (S. Ghazala, Ed.), *Sous Vide and Cook–Chill Processing for the Food Industry*, Aspen Publications, Gaithersburg, MD, pp. 294–310.

GIL, M.I., AGUAYO, E. and KADER, A.A., (2006), Quality changes and nutrient retention in fresh-cut versus whole fruits during storage, *J. Agriculture Food Chemistry*, **54** (12), 4284–4296.

GIOVANNONI, J., (2001), Molecular biology of fruit maturation and ripening, *Annual Review of Plant Physiology and Plant Molecular Biology*, **52**, 725–749.

GOODBURN, K., (2000), Legislation, in (M. Stringer and C. Dennis, Eds.), *Chilled Foods*, 2nd edn, Woodhead Publishing, Cambridge, pp. 451–473.

GORRIS, L.G.M., (1994), Improvement of the safety and quality of refrigerated ready-to-eat foods using novel mild preservation techniques, in (R.P. Singh and F.A.R. Oliveira, Eds.) *Minimal Processing of Foods and Process Optimisation – An Interface*, CRC Press, Boca Raton, FL, pp. 57–72.

GORRIS, L.G.M. and PECK, M.W., (1998), Microbiological safety considerations when using hurdle technology with refrigerated processed foods of extended durability, in (S. Ghazala, Ed.), *Sous Vide and Cook–Chill Processing for the Food Industry*, Aspen Publications, Gaithersburg, MD, pp 206–233.

GRAHAM, J., (1984), *Planning and engineering data, 3, Fish Freezing*, FAO Fisheries Circular #771, FAO, Rome.

GRANRYD, E., (2007), *Refrigerant Cycle Data: Thermophysical Properties of Refrigerants for Applications in Vapour-compression Systems*, International Institute for Refrigeration (IIF-IIR), France, available from www.iifiir.org/en/details.php?id=1156.

HAARD, N.F. and CHISM, G.W., (1996), Characteristics of edible plant tissues, in (O.R. Fennema, Ed.), *Food Chemistry*, 3rd edn, Marcel Dekker, New York, pp. 997–1003.

HARRIS, R.S., (1988), *Production is Only Half the Battle – A Training Manual in Fresh Produce Marketing for the Eastern Caribbean*. Food and Agricultural Organization of the United

Nations, Bridgestone, Barbados.

HEAP, R.D., (1997), Environment, law and choice of refrigerants, in (A. Devi, Ed.), *Food Technology International*, Sterling Publications, London, pp. 93–96.

HEAP, R.D., (2000), The refrigeration of chilled foods, in (M. Stringer and C. Dennis, Eds.), *Chilled Foods – A Comprehensive Guide*, 2nd edn, Woodhead Publishing, Cambridge, pp. 79–98.

HENDERSON, R., (2006), Carbon dioxide measures up as a real hazard, Occupational Health & Safety, available at www.ohsonline.com/articles/45034/.

HEPPENSTALL, T., (2000), Refrigeration systems, University of Newcastle upon Tyne, available at http://lorien.ncl.ac.uk/ming/cleantech/refrigeration.htm.

HERBERT, R.A., (1989), Microbial growth at low temperatures, in (G.W. Gould, Ed.), *Mechanisms of Action of Food Preservation Procedures*, Elsevier Applied Science, London, pp. 71–96.

HILL, M.A., (1987), The effect of refrigeration on the quality of some prepared foods, in (S. Thorne, Ed.), *Developments in Food Preservation*, Vol. 4, Elsevier Applied Science, London, pp. 123–152.

HOANG, M.L., VERBOVEN, P. and NICOLAI, B.M., (2001), CFD simulation of cool stores for agricultural and horticultural products, in (D-W. Sun, Ed.), *Advances in Food Refrigeration*, Leatherhead Publishing, LFRA, Leatherhead, pp. 153–192.

HOLAH, J.T., (2000), Cleaning and disinfection, in (M. Stringer and C. Dennis, Eds.), *Chilled Foods – A Comprehensive Guide*, 2nd edn, Woodhead Publishing, Cambridge, pp. 397–428.

HOLAH, J. and THORPE, R.H., (2000), The hygienic design of chilled food plant, in (M. Stringer and C. Dennis, Eds.), *Chilled Foods – A Comprehensive Guide*, 2nd edn, Woodhead Publishing, Cambridge, pp. 355–396.

HONIKEL, K.O. and SCHWAGELE, F., (2001), Chilling and freezing of meat and meat products, in (D-W. Sun, Ed.), *Advances in Food Refrigeration*, Leatherhead Publishing, LFRA, Leatherhead, pp. 366–386.

HUNG, Y-C., (2001), Cryogenic refrigeration, in (D-W. Sun, Ed.), *Advances in Food Refrigeration*, Leatherhead Publishing, LFRA, Leatherhead, pp. 305–325.

JAMES, S.J., (2000), Raw material selection – meat and poultry, in (M. Stringer and C. Dennis, Eds.), *Chilled Foods – A Comprehensive Guide*, 2nd edn, Woodhead Publishing, Cambridge, pp. 63–76.

JAYAS, D.S. and JEYAMKONDAN, S., (2002), PH – Postharvest technology modified atmosphere storage of grains meats fruits and vegetables, *Biosystems Engineering*, **82** (3), 235–251.

JENNINGS, B., (1999), Refrigeration for the new millennium, *Food Processing*, **68** (5), 12–13.

JENSEN, C., LAURIDSEN, C. and BERTELSEN, G., (1998), Dietary vitamin E: quality and storage stability of pork and poultry, Trends in Food Science and Technology, **9** (2), 62–72.

JUNCHER, D., RØNN, B., MORTENSEN, E.T., HENCKEL, P., KARLSSON, A., SKIBSTED, L. and BERTELSEN, G., (2001), Effect of pre-slaughter physiological conditions on the oxidative stability of colour and lipid during chill storage of pork, Meat Science, **58** (4), 347–357.

KADER, A.A., SINGH, R.P. and MANNAPPERUMA, J.D., (1998), Technologies to extend the refrigerated shelf life of fresh fruits and vegetables, in (I.A. Taub and R.P. Singh, Eds.), *Food Storage Stability*, CRC Press, Boca Raton, FL, pp. 419–434.

KIDMOSE, U. and HANSEN, M., (1999), The influence of postharvest storage, temperature and duration on quality of cooked broccoli florets, *J. Food Quality*, **22** (2), 135–146.

KRAFT, A.A., (1992), *Psychrotrophic Bacteria in Foods: Disease and Spoilage*, CRC Press, Boca Raton, FL, pp. 99–112.

LASSEN, A., KALL, M. HANSEN, K. and OVESEN, L., (2002), A comparison of the retention of vitamins B1, B2 and B6, and cooking yield in pork loin with conventional and enhanced meal-service systems, *European Food Research and Technology*, **215** (3), 194–199.

LAWRIE, R.A. and LEDWARD, D., (2006), Biochemical aspects, in *Lawrie's Meat Science*, 7th edn, Woodhead Publishing, Cambridge, pp. 64–71.

LE BLANC, D. and STARK, R., (2001), The cold chain, in (D-W Sun, Ed.), *Advances in Food Refrigeration*, Leatherhead Publishing, LFRA, Leatherhead, pp. 326–365.

LEE, S.K. and KADER, A.A., (2000), Preharvest and postharvest factors influencing vitamin C content of horticultural crops, *Postharvest Biology and Technology*, **20**, 207–220.

LEWIS, M.J., (1990), *Physical Properties of Foods and Food Processing Systems*, Woodhead Publishing, Cambridge.

LUCAS, T., CHOUROT, J-M., RAOULT-WACK, A-L. and GOLI, T., (2001), Hydro/immersion chilling and freezing, in (D-W. Sun, Ed.), *Advances in Food Refrigeration*, Leatherhead Publishing, LFRA, Leatherhead, pp. 220–263.

LUTZ, J.M. and HARDENBURG, R.E., (1966), *The Commercial Storage of Fruits, Vegetables and Forist and Nursery Stocks*, Agricultural Handbook No. 66, USDA, Washington, and available at www.fao.org/docrep/T0073E/T0073E02.htm.

MARTH, E.H., (1998), Extended shelf life refrigerated foods: microbiological quality and safety, *Food Technology*, **52** (2), 57–62.

MASCHERONI, R.H., (2001), Plate and air-blast cooling/freezing, in (D-W. Sun, Ed.), *Advances in Food Refrigeration*, Leatherhead Publishing, LFRA, Leatherhead, pp. 193–219.

MORRIS, S., (2001), Optimal Fresh – the fruit, vegetable and fresh produce expert system, Sydney Postharvest Laboratory, available at www.postharvest.com.au/storage.htm.

NEILSEN, J., LARSEN, E. and JESSEN, F., (2001), Chilling and freezing of fish and fishery products, in (D-W. Sun, Ed.), *Advances in Food Refrigeration*, Leatherhead Publishing, LFRA, Leatherhead, pp. 403–437.

NICOLAI, B.M., SCHELLEKENS, M., MARTENS, T. and DE BAERDEMAEKER, J., (1994), Computer-aided design of cook-chill foods under uncertain conditions, in (R.P. Singh and F.A.R. Oliveira, Eds.), *Minimal Processing of Foods and Process Optimisation – An Interface*, CRC Press, Boca Raton, FL, pp. 293–314.

NIEMIRA, B.A., SOMMERS, C.H. and UKUKU, D.O., (2005), Mechanisms of microbial spoilage of fruits and vegetables, in (O. Lamikanra, S.H. Imam and D. Ukuku, Eds.), *Produce Degradation: Pathways and Prevention*, CRC Press, Boca Raton, FL, pp. 464–482.

OETIKER, J.H. and YANG, S.F., (1995), The role of ethylene in fruit ripening, *Acta Hort. (ISHS)*, **398**, 167–178.

PHAM, Q.T., (2001), Cooling/freezing/thawing time and heat load, in (D-W. Sun, Ed.), *Advances in Food Refrigeration*, Leatherhead Publishing, LFRA, Leatherhead, pp. 110–152.

ROSE, D., (2000), Total quality management, in (M. Stringer and C. Dennis, Eds.), *Chilled Foods – A Comprehensive Guide*, 2nd edn, Woodhead Publishing, Cambridge, pp. 429–450.

SALTVEIT, M.E., (2004), Respiratory metabolism, in (K. Gross, Ed.), *The Commercial Storage of Fruits, Vegetables, and Florist and Nursery Stocks*, Agriculture Handbook No. 66, USDA, ARS, Washington, DC.

SELMAN, J.D., (1995), Time–temperature indicators, in (M.L. Rooney, Ed.), *Active Food Packaging*, Blackie Academic and Professional, London, pp. 215–233.

SINGH, R.P. and HELDMAN, D.R., (2001), Refrigeration, in *Introduction to Food Engineering*, 3rd edn, Academic Press, London, pp. 368–409.

SUN, D-W. and WANG, L-J., (2001a), Novel refrigeration cycles, in (D-W. Sun, Ed.), *Advances in Food Refrigeration*, Leatherhead Publishing, LFRA, Leatherhead, pp. 1–69.

SUN, D-W. and WANG, L-J., (2001b), Vacuum cooling, in (D-W. Sun, Ed.), *Advances in Food Refrigeration*, Leatherhead Publishing, LFRA, Leatherhead, pp. 264–304.

TAOUKIS, P.S. and LABUZA, T.P., (2003), Time–temperature indicators (TTIs), in (R. Ahvenainen, Ed.), *Novel Food Packaging Techniques*, Woodhead Publishing, Cambridge, pp. 103–126.

TOLEDO, R.T., (1999), Refrigeration, in *Fundamentals of Food Process Engineering*, 2nd edn, Aspen Publishers, Gaithersburg, MD, pp. 398–436.

TRUJILLO, F.J. and PHAM, Q.T., (2003), Modelling the chilling of the leg, loin and shoulder of beef carcasses using an evolutionary method, *International J. Refrigeration*, **26** (2), 224–231.

VAN LOEY, A., HAENTJENS, T. and HENDRICKX, M., (1998), The potential role of time-temperature integrators for process evaluation in the cook–chill chain, in (S. Ghazala, Ed.), *Sous Vide and Cook–Chill Processes for the Food Industry*, Aspen Publications, Gaithersburg, MD, pp. 89–110.

VEERKAMP, C.H., (2001), Chilling and freezing of poultry and poultry products, in (D-W. Sun, Ed.),

Advances in Food Refrigeration, Leatherhead Publishing, LFRA, Leatherhead, pp. 387–402.

VOIDAROU C., TZORA A., ALEXOPOULOS A. and BEZIRTZOGLOU E., (2006), Hygienic quality of different ham preparations, IUFoST 13th World Congress of Food Sciences Technology, 17/21 September, Nantes, France, available at http://dx.doi.org/10.1051/IUFoST:20060771.

WALKER, S.J., (1992), Chilled foods microbiology, in (C. Dennis, and M. Stringer, Eds), *Chilled Foods – A Comprehensive Guide*, Ellis Horwood, London, pp. 165–195.

WALKER, S.J. and BETTS, G., (2000), Chilled foods microbiology, in (M. Stringer and C. Dennis, Eds), *Chilled Foods – A Comprehensive Guide*, 2nd edn, Woodhead Publishing, Cambridge, pp. 153–186.

WALKER, S.J. and STRINGER, M.F., (1990), Microbiology of chilled foods, in (T.R. Gormley, Ed.), *Chilled Foods – The State of the Art*, Elsevier Apllied Science, London, pp. 269–304.

WEATHERSPOON, L., MOSHA, T. and NNYEPI, M., (2005), Nutrient loss, in (O. Lamikanra, S.H. Imam and D. Ukuku, Eds.), *Produce Degradation: Pathways and Prevention*, CRC Press, Boca Raton, FL, pp. 223–266.

WESSEL, R., (2007), Chill-on develops prototype RFID-enabled time–temperature indicator, *RFID Journal*, available at http://www.rfidjournal.com/article/articleview/3749/1/1/.

WILLIAMS, P.G., (1996), Vitamin retention in cook/chill and cook/hot-hold hospital foodservices, *J. American Dietetic Association*, **96** (5), 490–498.

WOOLFE, M.L., (2000), Temperature monitoring and measurement, in (M. Stringer and C. Dennis, Eds.), *Chilled Foods – A Comprehensive Guide*, 2nd edn, Woodhead Publishing, Cambridge, pp. 99–134.

WOON, E., (2007), Health drives ready meals in Western Europe, Just Food, available at www.just-food.com/article.aspx?id=100346&lk=s.

YANG, T.C.S., (1998), Ambient storage, in (I.A. Taub and R.P. Singh, Eds.), *Food Storage Stability*, CRC Press, Boca Raton, FL, pp. 435–458.

22

Freezing

Abstract: Freezing reduces the temperature of foods below their freezing point, preserving them by a combination of reduced biochemical, enzymic and microbial activity and reduced water activity, while causing minimal changes to their sensory qualities and nutritional value. This chapter describes mechanisms of ice crystal formation and solute concentration, and methods for the calculation of freezing time. It then describes mechanical and cryogenic freezing equipment, new methods of freezing using pressure, ultrasound and pumpable ice slurries, and methods of thawing foods. The chapter concludes with a discussion of changes that take place to foods during frozen storage and thawing and the effects of freezing on micro-organisms.

Key words: freezing, supercooling, ice crystal formation, eutectic temperature, mechanical freezers, cryogenic freezers, immersion chilling and freezing (ICF), pressure freezing, pumpable ice slurries, ultrasound freezing, magnetic freezing, dehydrofreezing, frozen storage, recrystallisation, thawing.

Freezing is a unit operation that is intended to preserve foods without causing significant changes to their sensory qualities or nutritional value. It involves a reduction in the temperature of a food to below its freezing point, which causes a proportion of the water in the food to undergo a change in state to form ice crystals. The immobilisation of water as ice and the resulting concentration of dissolved solutes in unfrozen water lower the water activity (a_w) of the food (Chapter 1, section 1.1.2). Preservation is achieved by a combination of low temperatures that reduce biochemical changes, enzymic and microbial activity, reduced water activity and, in some foods, pretreatment by blanching. There are only small changes to nutritional or sensory qualities of foods when correct freezing, storage and thawing procedures are followed.

The major groups of commercially frozen foods are:

- baked goods (e.g. bread, cakes, fruit and meat pies);
- fish fillets and seafoods (e.g. cod, plaice, shrimps and crab meat) including fish fingers, fish cakes or prepared dishes with an accompanying sauce;
- fruits (e.g. strawberries, oranges, raspberries, blackcurrants) either whole or puréed, or as juice concentrates);
- meats as carcasses, boxed joints or cubes, and meat products (e.g. sausages, beefburgers, reformed steaks);

- prepared foods (e.g. pizzas, desserts, ice cream, ready meals and cook–freeze dishes);
- vegetables (e.g. peas, green beans, sweetcorn, spinach, sprouts, potatoes).

Rapid increases in sales of frozen foods during the 1970s–1990s were closely associated with increased ownership of domestic freezers and microwave ovens. Frozen foods have an image of high quality and 'freshness' and, particularly in meat, fruit and vegetable sectors, outsell canned or dried products. However, distribution of frozen foods has a relatively high cost, due to the need to maintain a constant low temperature throughout the cold chain and chilled foods (Chapter 21) have gained in popularity over frozen foods in the last 10 years in industrialised countries.

22.1 Theory

There are different stages involved in lowering the temperature of a food below its freezing point. First, sensible heat is removed and in fresh foods, heat produced by respiration is also removed (see Chapter 21, section 21.1). This is termed the 'heat load' and is important in determining the correct size of freezing equipment for a particular production rate. Next, latent heat is removed when water freezes to ice. Most foods contain a large proportion of water (Table 22.1), which has a high specific heat $(4182 \, J \, kg^{-1} \, K^{-1})$ and a high latent heat of crystallisation $(335 \, kJ \, kg^{-1})$ (see Table 1.5, Chapter 1). A substantial amount of energy is therefore needed to remove sensible and latent heat to form ice crystals. The latent heat of other components of the food (e.g. fats) must also be removed before they can solidify but in most foods they are present in smaller amounts and require removal of a relatively small amount of heat for crystal-lisation. Energy for freezing is supplied as electrical energy, which is used to compress refrigerants in mechanical freezing equipment or to compress and cool cryogens (section 22.2). Theoretical aspects of mechanical compression refrigeration are described in Chapter 21 (section 21.1).

22.1.1 Ice crystal formation

The initial freezing point of a food may be described as 'the temperature at which a minute crystal of ice exists in equilibrium with the surrounding water'. However, before an ice crystal can form, a nucleus of water molecules must be present. Nucleation therefore precedes ice crystal formation. There are two types of nucleation: homogeneous nucleation (the chance orientation and combination of water molecules) and hetero-geneous nucleation (the formation of a nucleus around suspended particles or at a cell wall). Heterogeneous nucleation is more likely to occur in foods. All food cells contain solutes such as carbohydrates, salts and other compounds that affect their freezing

Table 22.1 Water contents and freezing points of selected foods

Food	Water content (%)	Freezing point (°C)
Vegetables	78–92	−0.8 to −2.8
Fruits	87–95	−0.9 to −2.7
Meat	55–70	−1.7 to −2.2
Fish	65–81	−0.6 to −2.0
Milk	87	−0.5
Egg	74	−0.5

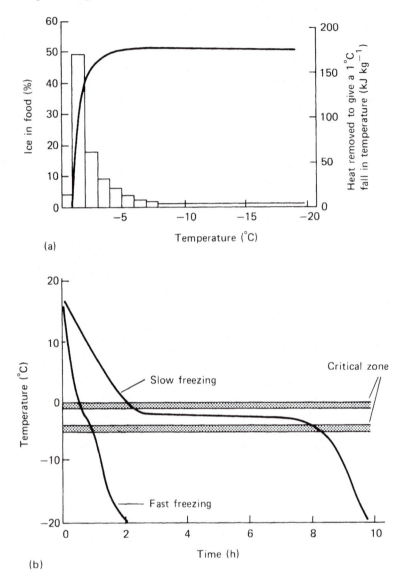

Fig. 22.1 Freezing: (a) ice formation at different freezing temperatures; (b) temperature changes of food through the critical zone (after Leniger and Beverloo 1975).

behaviour. In animal or vegetable tissues, water is both intracellular and extracellular; the extracellular fluids have a lower concentration of solutes, and the first ice crystals are formed there. Higher rates of heat transfer produce larger numbers of nuclei and fast freezing therefore produces a large number of small ice crystals. Energetically it is easier for water molecules to migrate to existing nuclei in preference to forming new nuclei. The time taken for the temperature of a food to pass through the 'critical zone' (Fig. 22.1), i.e. the freezing rate, therefore determines both the number and the size of ice crystals. However, large differences in crystal size are found with similar freezing rates owing to differences in the composition of foods and even in similar foods that have received different pre-freezing treatments.

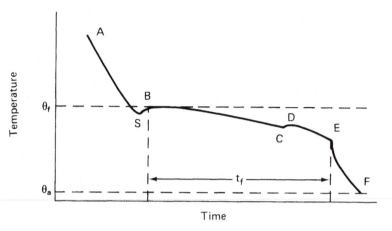

Fig. 22.2   Time–temperature data during freezing.

If the temperature is monitored at the thermal centre of a food (the point that cools most slowly) as heat is removed, a characteristic curve is obtained (Fig. 22.2). The six components of the curve are as follows:

- A–S: the food is cooled to below its initial freezing point (θ_f) which, with the exception of pure water, is always below 0 °C (Table 22.1). At point S the water remains liquid, although the temperature is below the freezing point. This phenomenon is known as 'supercooling', which may be as much as 10 °C below the freezing point, and is the period in which nucleation begins. The length of the supercooling period depends on the type of food and the rate at which heat is removed.
- S–B: the temperature rises rapidly to the freezing point as ice crystals begin to form and latent heat of crystallisation is released.
- B–C: heat is removed from the food at the same rate as before, but it is latent heat being removed as ice forms and the temperature therefore remains almost constant at the freezing point. The freezing point is gradually depressed by the increase in solute concentration in the unfrozen liquor, and as more ice is formed the temperature falls slightly. Once stable nuclei are formed they continue to grow and it is during this stage that the major part of the ice is formed (Fig. 22.1a).
- C–D: one of the solutes becomes supersaturated and crystallises out. The latent heat of crystallisation is released and the temperature rises to the 'eutectic' temperature for that solute (section 22.1.2).
- D–E: crystallisation of water and solutes continues. The total time (t_f) taken for ice crystal growth (the 'freezing plateau') depends on the rate of mass transfer of water from the liquid phase to the nuclei and the rate at which heat is removed. The temperature of the ice–water mixture falls to the temperature of the freezer. A proportion of the water remains unfrozen at the temperatures used in commercial freezing; the amount depends on the type and composition of the food and the temperature of storage. For example at a storage temperature of −20 °C the percentage of frozen water is 88% in lamb, 91% in fish and 93% in egg albumin.
- E–F: if freezing is continued below commercial temperatures, ice formation and solute concentration continue until no more water can be frozen. The temperature falls as sensible heat is removed from the ice. The temperature (θ_a) at point F is known as the 'glass transition temperature' of the amorphous concentrated solution. When a critical, solute-dependent concentration is reached, the physical state of the unfrozen liquid

changes from a viscoelastic liquid to a brittle, amorphous solid glass (see also Chapter 1, section 1.4.1).

For the majority of the freezing plateau the rate of ice crystal growth is controlled by the rate of heat transfer. The rate of mass transfer (of water molecules moving to the growing crystal and of solutes moving away from the crystal) does not control the rate of crystal growth except towards the end of the freezing period when solutes become more concentrated.

22.1.2   Solute concentration

An increase in solute concentration during freezing causes changes to the pH, viscosity, surface tension and redox potential of the unfrozen liquor. As the temperature falls, individual solutes reach saturation point and crystallise out. The temperature at which a crystal of a solute exists in equilibrium with the unfrozen liquor and ice is its 'eutectic temperature' (e.g. for glucose it is −5 °C; for sucrose, −14 °C; for sodium chloride, −21.13 °C; and for calcium chloride, −55 °C). However, it is difficult to identify individual eutectic temperatures in the complex mixture of solutes in foods, and the term 'final eutectic temperature' is therefore used. This is the lowest eutectic temperature of the solutes in a food (e.g. for ice cream it is −55 °C, for meat: −50 to −60 °C and for bread: −70 °C) (Fennema 1996). Maximum ice crystal formation is not possible until this temperature is reached. Commercial foods are not frozen to such low temperatures and unfrozen water is therefore always present in frozen foods.

As food is frozen below point E in Fig. 22.2, the unfrozen material becomes more concentrated and forms a 'glass' that encompasses the ice crystals. This can be represented on a simplified phase diagram for freezing of a solute in water (Fig. 22.3) where:

- A–B: cooling to the freezing point;
- B–C: supercooling;
- C–D: ice crystal growth;
- D–E: the concentration of solutes in the unfrozen phase follows the solubility curve as it is cooled to the eutectic temperature (θ_e);

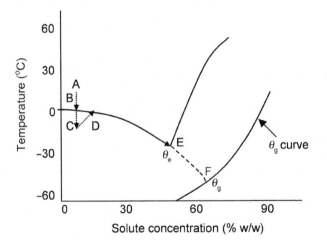

Fig. 22.3   Simplified phase diagram showing the relationship between temperature and solute concentration down to glass transition temperature for a solute in water (adapted from Kennedy 2003).

- E–F: the concentrated phase does not solidify at the eutectic temperature and cooling and concentration continue until the concentration meets the glass transition curve at temperature (θ_g).

Glass transition temperatures for selected foods are shown in Table 22.2. Where the temperature of storage is below this temperature range, the formation of a glass protects the texture of the food and gives good storage stability (e.g. meats and vegetables in Table 22.2). Many fruits, however, have very low glass transition temperatures and as a result suffer losses in texture during frozen storage in addition to damage caused by ice crystals (section 22.3). Further details of glass transition are given by Fennema (1996) and in Chapter 1 (section 1.4.1).

If foods such as ice cream or surimi are formulated to contain maltodextrin, sucrose or fructose, these raise the glass transition temperature, and if this is increased above the storage temperature, the shelf-life of the foods is extended (e.g. Ohkuma *et al.* 2008). In foods that contain a large proportion of water the formation of ice has a dramatic effect on their thermo-physical properties:

Table 22.2   Glass transition temperatures for selected foods

Food	Glass transition temperature t_g (°C)
Dairy products	
Cheddar cheese	−24
Cream cheese	−33
Ice cream	−31 to −37
Ice milk	−30
Fish and meat	
Beef muscle	−12 and −60
Chicken	−16
Cod muscle	−11 and −77
Mackerel muscle	−12
Tuna muscle	−15 and −74
Fruits and fruit products	
Apple	−41 to −42
Apple juice	−40
Banana	−35
Grape juice	−42
Lemon juice	−43
Orange juice	−37.5
Peach	−36
Pear juice	−40
Pineapple juice	−37
Prune juice	−41
Strawberry	−33 to −41
Tomato	−33 to −41
Vegetables	
Broccoli, head	−12
Carrot	−26
Green beans	−27
Maize kernel	−15
Pea	−25
Potato	−12
Spinach	−17

Adapted from Kennedy (2003), Orlien *et al.* (2004) and Fennema (1996)

- The density falls as the proportion of ice increases (Chapter 1, Table 1.5).
- The thermal conductivity increases (the thermal conductivity of ice is approximately four times greater than that of water (Chapter 10, Table 10.2)).
- The enthalpy decreases.
- The specific heat rises substantially as ice is formed and then falls back to approximately the same value as water when the temperature of the food is reduced to $\approx -20\,°C$.
- The thermal diffusivity of the food increases after initial ice formation as the temperature is further reduced.

Further details are given in Singh and Heldman (2001). The changes to thermophysical properties mostly take place as the temperature of the food falls to $\approx -10\,°C$ and then they change more gradually as the temperature falls further to that of frozen storage. Calculation of the freezing point of foods, based on the Clausius–Clapeyron equation and Raoult's Law, and methods to calculate the ice content of foods based on their thermophysical properties are given by Rahman (2001) and further details of the freezing process are given by Sahagian and Goff (1996). Boonsupthip and Heldman (2007) have developed a mathematical model to predict the fraction of frozen water, based on concentrations and molecular weights of specific components of foods.

22.1.3   Calculation of freezing time
Knowledge of the freezing time for a particular food is important for both ensuring its quality and in determining the throughput of a freezing plant. During freezing a moving front inside the food separates the frozen layer from unfrozen food. Heat is generated at the moving front as latent heat of fusion is released. This heat is transferred by conduction through the frozen layer to the surface, and then by convective heat transfer through a boundary film to the freezing medium (see also Chapter 10, section 10.1.2). The factors that influence the rate of heat transfer are:

- thermal conductivity of the food;
- area of food available for heat transfer;
- distance that the heat must travel through the food (size and shape of the pieces);
- temperature difference between the food and the freezing medium;
- insulating effect of the boundary film of air surrounding the food;
- packaging, if present, is an additional barrier to heat flow.

These factors are important for both mechanical and cryogenic freezing (section 22.2).
A number of methods have been developed to calculate freezing time, the earliest of which (Equation 22.1) was developed by Plank in 1941:

$$t_f = \frac{\rho\lambda}{\theta_f - \theta_a}\left(\frac{P'x}{h} + \frac{R'x^2}{k}\right) \qquad \boxed{22.1}$$

where t_f (s) = freezing time, ρ (kg m^{-3}) = density of the food, λ (kJ kg^{-1}) = latent heat of fusion of the food, θ_f (°C) = freezing temperature, θ_a (°C) = temperature of the freezing medium, h (W m^{-2}K^{-1}) = convective heat transfer coefficient at the surface of the food, x (m) = the thickness/diameter of the material, k (W m^{-1}K^{-1}) = thermal conductivity of the frozen food and P' and R' are constants that reflect the shortest distance between the centre and the surface of the food for different shapes. These are $P' = 1/6$ and $R' = 1/24$ for a sphere, $P' = 1/2$ and $R' = 1/8$ for a slab, $P' = 1/4$ and

$R' = 1/16$ for a cylinder. Derivation of the equation is described by Earle (1983) and Singh and Heldman (2001).

Equation 22.1 shows that the freezing time increases with higher food density and increased size of the food, and decreases with higher temperature differences between the food and the freezing medium, thermal conductivity of the frozen food and higher surface heat transfer coefficient. Derivation of the equation involves the following assumptions:

- Freezing starts with all water in the food unfrozen but at its freezing point, and loss of sensible heat is ignored.
- Heat transfer takes place in one direction and is sufficiently slow for steady state conditions to operate.
- The freezing front maintains a similar shape to that of the food (e.g. in a rectangular block the freezing front remains rectangular).
- There is a single freezing point.
- The thermal conductivity and specific heat of the food are constant when unfrozen and then change to a different constant value when the food is frozen.

Its use is limited by not taking into account the removal of sensible heat from unfrozen foods or from foods after freezing, and the lack of accurate data on the density and thermal conductivity of many frozen foods. An example of its use is given in sample problem 22.1.

Sample problem 22.1

Five-centimetre potato cubes are quick frozen in a blast freezer operating at $-40\,°C$ with a surface heat transfer coefficient of $30\,W\,m^{-2}\,K^{-1}$ (Table 22.3). If the freezing point of the potato is measured as $-1.0\,°C$ and the density of potato is $1180\,kg\,m^{-3}$, calculate the expected freezing time for each cube. Also calculate the freezing time for 2.5 cm cubes frozen under the same conditions. (Additional data: thermal conductivity of frozen potato $= 2.5\,W\,m^{-1}\,K^{-1}$ and the latent heat of crystallisation $= 274\,kJ\,kg^{-1}$.)

Solution to sample problem 22.1

To calculate the freezing time for each cube, substitute the data into Equation 22.1, using constants $P' = 1/2$ and $R' = 1/8$.

$$t_f = \frac{1180 \times 274 \times 10^3}{-1 - (-40)} \left(\frac{0.05}{2 \times 30} + \frac{0.05^2}{8 \times 2.5} \right)$$

$$= 8290 \times 10^3 (8.33 \times 10^{-4} + 1.25 \times 10^{-4})$$

$$= 7940\,s$$

$$= 2.2\,h$$

To calculate the freezing time for 2.5 cm cubes,

$$t_f = \frac{1180 \times 274 \times 10^3}{-1 - (-40)} \left(\frac{0.025}{2 \times 30} + \frac{0.025^2}{8 \times 2.5} \right)$$

$$= 8290 \times 10^3 (4.17 \times 10^{-4} + 3.125 \times 10^{-5})$$

$$= 3.715 \times 10^3\,s$$

$$= 1.03\,h$$

Other analytical methods to predict freezing time were developed during the 1960s–1980s to overcome some of the limitations of the Plank equation, which were compared by Cleland (1990). Of these, Pham (1986) developed a simplified Equation 22.2 that included the time taken to lose sensible heat. Derivations and use of the equation are given in Pham (2001).

$$t = \frac{d_c}{E_f h}\left(\frac{\Delta H_1}{\Delta \theta_1} + \frac{\Delta H_2}{\Delta \theta_2}\right)\left(1 + \frac{N_{Bi}}{2}\right) \hspace{2cm} \boxed{22.2}$$

where d_c (m) = a characteristic dimension (radius or shortest distance to the centre), E_f = shape factor (=1 for slab, 2 for a cylinder and 3 for a sphere), ΔH (J m^{-3}) = change in enthalpy with subscripts 1 for precooling unfrozen food and 2 for phase change and cooling of frozen food obtained from Equations 22.3 and 22.4, $\Delta \theta$ (°C) = temperature gradients from Equations 22.5 and 22.6, and N_{Bi} = Biot number.

$$\Delta H_1 = \rho_u c_u (\theta_i - \theta_{fm}) \hspace{2cm} \boxed{22.3}$$

where ρ_u (kg m^{-3}) = density of unfrozen material, c_u (J kg^{-1} K^{-1}) = specific heat of unfrozen material, and θ_i and θ_{fm} (°C) = initial and mean freezing temperatures of the material respectively.

$$\Delta H_2 = \rho_f [\lambda_f + c_f(\theta_{fm} - \theta_c)] \hspace{2cm} \boxed{22.4}$$

where ρ_f (kg m^{-3}) = density of frozen material, λ_f (kJ kg^{-1}) = latent heat of fusion of the food, c_f (J kg^{-1} K^{-1}) = specific heat of frozen material and θ_c (°C) = final temperature at the centre of the food.

The temperature gradients $\Delta \theta_1$ and $\Delta \theta_2$ are found using Equations 22.5 and 22.6:

$$\Delta \theta_1 = \frac{(\theta_i + \theta_{fm})}{2} - \theta_a \hspace{2cm} \boxed{22.5}$$

$$\Delta \theta_2 = \theta_{fm} - \theta_a \hspace{2cm} \boxed{22.6}$$

where θ_{fm} (°C) = mean freezing temperature and θ_a (°C) = temperature of the freezing medium. θ_{fm} is calculated using Equation 22.7:

$$\theta_{fm} = 1.8 + 0.263\theta_c + 0.105\theta_a \hspace{2cm} \boxed{22.7}$$

An example of the use of Equation 22.2 is given in sample problem 22.2 using the same data as sample problem 22.1.

Sample problem 22.2
Using the same data as given in sample problem 22.1, calculate the freezing time for 5 cm potato cubes using Pham's equation. (Additional data: initial product temperature = 12 °C, specific heat of unfrozen potato = 3600 J kg^{-1}, specific heat of frozen potato = 1900 J kg^{-1}, density of frozen potato = 980 kg m^{-3} and the moisture content of the unfrozen potato = 76%.)

Solution to sample problem 22.2
Using Equation 22.7 to find θ_{fm}:

$$\theta_{fm} = 1.8 + [0.263 \times (-18)] + [0.105 \times (-40)]$$
$$= -7.13\,°C$$

Using Equation 22.3 to find ΔH_1:

$$\Delta H_1 = 1180 \times 3600[12 - (-7.13)]$$
$$= 81\,264\,240 \text{ J m}^{-3}$$

Using Equation 22.4 to find ΔH_2:

$$\Delta H_2 = 980\{0.76 \times 274 \times 1000 + 1900[-7.13 - (-18)]\}$$
$$= 224\,315\,140 \text{ J m}^{-3}$$

Using Equation 22.5 to find $\Delta\theta_1$:

$$\Delta\theta_1 = [12 + (-7.13/2)] - (-40)$$
$$= 48.43\,^{\circ}\text{C}$$

Using Equation 22.6 to find $\Delta\theta_2$:

$$\Delta\theta_2 = -7.13 - (-40)$$
$$= 32.87\,^{\circ}\text{C}$$

$$\text{Biot number} = (30 \times 0.025)/2.5$$
$$= 0.3$$

Substituting the data into Equation 22.2, using $E_f = 1$ for a slab:

$$t = \frac{0.025}{1 \times 30}\left[\frac{81\,264\,240}{48.43} + \frac{224\,315\,140}{32.87}\right] \times (1 + 0.3/2)$$

$$= 8.3 \times 10^{-4}(1\,677\,973 + 6\,824\,312) \times 1.15$$

$$= 8115 \text{ s}$$

$$= 2.25 \text{ h}$$

Other mathematical models have been developed by Cleland et al. (1987). Pardo and Niranjan (2006) describe a model developed by Neuman. Fikiin (2003) describes computer programs to model food freezing and the use of computational fluid dynamics software to predict heat and mass flow during freezing.

22.1.4   Thawing

When frozen food is thawed using air or water, surface ice melts to form a layer of water. Water has a lower thermal conductivity and a lower thermal diffusivity than ice and the surface layer of water therefore reduces the rate at which heat is conducted to the frozen interior. This insulating effect increases as the layer of thawed food grows thicker (in contrast, during freezing, the increase in thickness of ice causes heat transfer to accelerate because of the higher thermal conductivity of the ice). Thawing is therefore a substantially longer process than freezing when temperature differences and other conditions are similar.

During thawing (Fig. 22.4), the initial rapid rise in temperature (AB) is due to the absence of a significant layer of water around the food. There is then a long period when

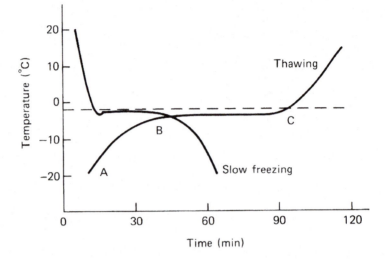

Fig. 22.4 Temperature changes during thawing (after Fennema and Powrie 1964).

the temperature of the food is near to that of melting ice (BC). During this period any cellular damage caused by slow freezing or recrystallisation results in the release of cell constituents to form drip losses (section 22.3.3). Cleland *et al.* (1987) developed an equation to predict thawing time (Equation 22.8):

$$t = \frac{d_c}{E_f h} \frac{\Delta H_{10}}{(\theta_a - \theta_f)} (P_1 + P_2 N_{Bi}) \tag{22.8}$$

where d_c (m) = a characteristic dimension (radius or shortest distance to the centre), E_f = shape factor (= 1 for slab, 2 for a cylinder and 3 for a sphere), h (W m^{-2}K^{-1}) = convective heat transfer coefficient at the surface of the food, θ_f (°C) = freezing temperature and θ_a (°C) = temperature of the freezing medium.

$$P_1 = 0.7754 + 2.2828 N_{Ste} \times N_{Pk}$$
$$P_2 = 0.5(0.4271 + 2.122 N_{Ste} - 1.4847 N_{Ste}^2)$$

and

$$\text{Biot number } (N_{Bi}) = \frac{h d_c}{k_u} \tag{22.9}$$

$$\text{Stephan number } (N_{Ste}) = \rho_u c_u \frac{(\theta_a - \theta_f)}{\Delta H_{10}} \tag{22.10}$$

$$\text{Planck number } N_{Pk} = \rho_f c_f \frac{(\theta_f - \theta_i)}{\Delta H_{10}} \tag{22.11}$$

where ΔH_{10} is the change in enthalpy of the product from 0 to -10°C.

Further details of methods to calculate freezing and thawing times are given by Schwartzberg *et al.* (2007) and the effects of freezing, cold storage and thawing on foods are described in section 22.3.

22.2   Equipment

The selection of freezing equipment should take the following factors into consideration: the rate of freezing required; the size, shape and packaging requirements of the food; batch or continuous operation, the scale of production; range of products to be processed; and not least the capital and operating costs.

Freezers are categorised into:

- mechanical refrigerators, which evaporate and compress a refrigerant in a continuous cycle (see Chapter 21, section 21.2.1) and use cooled air, cooled liquid or cooled surfaces to remove heat from foods; and
- cryogenic freezers, which use solid or liquid carbon dioxide, or liquid nitrogen directly in contact with the food.

In general, mechanical freezers operate at $\approx-40\,°C$ and have higher capital costs than cryogenic freezers, whereas cryogenic freezers operate at -50 to $-70\,°C$ and have higher operating costs because refrigerant is not recirculated and is lost to the atmosphere. Freezers can also be grouped according to the rate of movement of the ice front:

- slow freezers and sharp freezers ($0.2\,cm\,h^{-1}$) – still-air freezers and cold stores;
- quick freezers ($0.5–3\,cm\,h^{-1}$) – air-blast and plate freezers;
- rapid freezers ($5–10\,cm\,h^{-1}$) – fluidised-bed freezers; and
- ultrarapid freezers ($10–100\,cm\,h^{-1}$) – cryogenic freezers.

All types of freezers are constructed from stainless steel and are insulated with expanded polystyrene, polyurethane or other materials that have low thermal conductivity (Chapter 10, Table 10.2). Most freezing equipment has microprocessor control, using programmable logic controllers (PLCs) to monitor process parameters and equipment status, display trends, identify faults and automatically control processing conditions for different products. Riverol *et al.* (2004) describe research into fuzzy logic control of freezers (see also Chapter 27, section 27.2). Larger-scale freezing and frozen storage systems can be monitored through a local area network (LAN – a group of computers that share a common communications line or wireless link) to a central control computer via the Internet. This reduces maintenance costs by simplifying fault analysis and also enables automatic recording of process conditions for HACCP control. Some systems have up to 100 programmable settings that enable operators to set the parameters for cold storage of different products. Additionally, most mechanical refrigeration compressors have solid state controls, so that one compressor can meet the heat load of a number of freezers when the loading of each freezer is done sequentially to keep the load relatively stable. If multiple compressors supply the freezers, computerised load control may be more cost effective. Further details of freezer control are given below and a comparison of selected freezing equipment is shown in Table 22.3.

22.2.1   Mechanical freezers

Cooled air freezers

Chest freezers freeze food in stationary (naturally circulating) air at between -20 and $-30\,°C$. They are not used for commercial freezing owing to long freezing times (3–72 h), which result in poor process economics and loss of product quality (section 22.3). Cold stores are used to freeze carcass meat, to store foods that are frozen by other methods, and as hardening rooms for ice cream. Fans circulate air to produce more uniform temperature distribution, but heat transfer coefficients are low (Table 22.3). A major

Table 22.3   A comparison of freezing equipment

Type of freezer	Typical heat transfer coefficient $(W\,m^{-2}\,K^{-1})$	Typical freezing time to $-18\,°C$ (min)	Examples of foods
Air-blast (5 m s^{-1}) tunnel	10–50	15–20	Unpackaged vegetables
Cryogenic (liquid nitrogen)	1500	0.9	Cake
		2–5	Beefburgers, seafood
		0.5–6	Fruits and vegetables
Fluidised bed	110–160	3–4	Peas, sweetcorn, beans
Immersion			
Freon	500	10–15	Cartons of orange juice
Aqueous solutions	100–950	0.5	Peas
Impingement	350	2–5	Meat patties
Plate	600	120	7.5 kg blocks of fish or minced meat
		25	Cartons of vegetables, seafood, ice cream, hamburgers, fish sticks, bags of product or products in moulds (e.g. fruit pulps, soups, ready meals)
Scraped surface	900	0.3–0.5	Ice cream (1 mm thick layer)
Spiral belt	25	12–19	Fish fingers
Still air (e.g. cold store)	6–9	180–4000	Meat carcass

Data from Mascheroni (2001), Anon (2007) and Arce and Sweat 1980)

problem with cold stores is ice formation on floors, walls and evaporator coils, caused by moisture in the air condensing to water and freezing (e.g. air at 10 °C and 80% relative humidity contains 6 g water per kg of air, so 1000 m³ h^{-1} of air entering the cold store through loading doors would deposit 173 kg of water vapour in the store per day (Weller and Mills 1999). Ice build-up reduces the efficiency of the refrigeration plant, requires frequent defrosting of evaporator coils, uses up energy that would otherwise be used to cool the store, and creates potential hazards from slippery surfaces. A desiccant dehumidifier (e.g. Anon 2008) overcomes these problems by removing moisture from the air as it enters the store, reducing the size of compressors and the energy needed to maintain the store temperature. Details of the operation of cold stores are given by Dellino (1997) and Cano-Muñoz (1991).

Air-blast freezers recirculate air over foods at between −30 and −50 °C at a velocity of 1.5–6.0 m s^{-1}. The high air velocity reduces the thickness of boundary air films surrounding the food and thus increases the surface heat transfer coefficient (Table 22.3). In batch equipment, food is stacked on trays in rooms or cabinets. Continuous equipment has trolleys stacked with trays of food that are dragged through an insulated tunnel (or on conveyor belts – see belt freezers below). Air flow is either parallel or perpendicular to the food and is ducted to pass evenly over all food pieces.

Blast freezing is relatively economical and highly flexible in that foods of different shapes and sizes can be frozen. The equipment is compact and has a low capital cost and a high throughput (up to 20 t h^{-1}). However, moisture from unpackaged food is transferred to the air and builds up as ice on the refrigeration coils, and this necessitates frequent defrosting. The large volumes of recycled air can also cause dehydration losses, freezer burn and oxidative changes to unpackaged or Individually Quick Frozen (IQF) foods (section 22.3). Different designs of blast freezers are described by Mascheroni (2001).

Belt freezers are a variation of continuous air-blast freezers that have a flexible mesh belt passing through an insulated tunnel. Multipass tunnels contain up to 10 separate belts. On the first belt a single layer of product is quickly frozen at the surface to prevent sticking, deformation and weight loss. The products falls to a succession of slower moving belts that hold thicker layers of semi-frozen food. This breaks up any clumps and allows control over the product depth (e.g. a 25–50 mm bed is frozen for 5–10 min and then re-piled to 100–125 mm on the next belt). In 'spiral' freezers (Fig. 22.5a) the belt is formed into a spiral tier that carries food up through a refrigerated chamber. In some designs each tier rests on the vertical sides of the tier beneath (Fig. 22.5b) and the belt is therefore 'self-stacking'. This eliminates the need for support rails and improves the capacity by up to 50% for a given stack height. Cold air moving at $3–8\,\mathrm{m\,s^{-1}}$ is directed

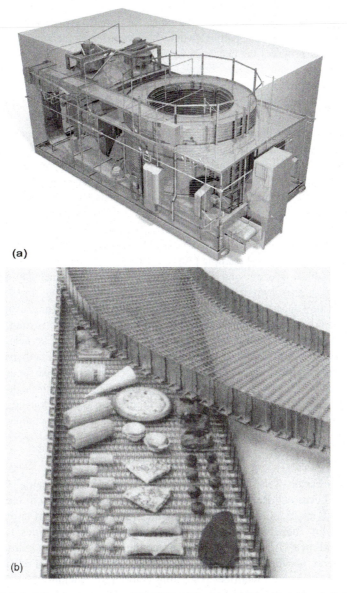

(a)

(b)

Fig. 22.5   (a) Spiral freezer (courtesy of FMC Food Tech) and (b) self-staking belt (courtesy of Frigoscandia Ltd at www.fmctechnologies.com).

through the belt stack in a co-current or counter-current flow. The latter reduces weight losses due to evaporation of moisture. In some designs, baffles and flow dividers are used to direct the air vertically upwards through the lower half of the stack and down through the upper half (known as 'controlled dual flow') (Mascheroni 2001). Spiral freezers require a relatively small floor-space and have high capacity (e.g. a 50–75 cm wide belt in a 32-tier spiral can process up to $3\,t\,h^{-1}$). Other advantages include automatic loading and unloading, low maintenance costs and flexibility to freeze a wide range of foods including pizzas, cakes, pies, ice cream, whole fish and chicken portions.

Impingement freezing involves forcing air at high velocity (e.g. $35\,m\,s^{-1}$) through nozzles to impinge perpendicularly on the food surface. It breaks up the boundary layer of air and increases heat transfer by three to five times that of conventional tunnel freezers (Kennedy 2003). It is most suitable for products that have a high surface area : weight ratio (e.g. meat patties) where heat transfer at the surface is the limiting factor (in thicker products the heat transfer from the core is the limiting factor) (see also impingement cryogenic freezing below, impingement drying in Chapter 16, section 16.2.1, and impingement heating in Chapter 18, section 18.2.2).

Fluidised-bed freezers are modified blast freezers in which air at between −25 and −35 °C is passed at a high velocity (2–$6\,m\,s^{-1}$) through a 2–13 cm bed of food, contained on a perforated tray or conveyor. The products are fluidised (kept in suspension by the high air velocity) and are discharged over a weir, the height of which is adjustable to control the residence time. In some designs ('fluidised bed belt freezers') there are two stages: after initial rapid freezing in a shallow bed to produce an ice glaze on the surface of the food, freezing is completed on a second belt in beds 10–15 cm deep. The formation of a glaze is useful for meat or fruit pieces and other products that have a tendency to stick together. The shape and size of the pieces of food determine the thickness of the fluidised bed and the air velocity needed for fluidisation (a sample calculation of air velocity is given in Chapter 1, sample problem 1.6). Food comes into greater contact with the air than in blast freezers, and all surfaces are frozen simultaneously and uniformly. This produces higher heat transfer coefficients, shorter freezing times (Table 22.3), higher production rates (up to $20\,t\,h^{-1}$) and less dehydration of unpackaged food than blast freezing does. The equipment therefore needs less frequent defrosting. However, the method is restricted to particulate foods (e.g. peas, sweetcorn kernels, cooked rice, shrimps, pasta, Brussels sprouts, mushrooms, strawberries or other berries) or sliced/diced portions of larger foods (e.g. meat cubes, French fries, diced carrot, pepper, onions, sliced fruits, green beans or potatoes). Similar equipment, named through-flow freezers, in which air passes through a bed of food but fluidisation is not achieved, is suitable for larger pieces of food (e.g. fish fillets). Both types of equipment are compact, have a high capacity and are highly suited to the production of IQF foods.

Cooled liquid freezers

In immersion freezers (also known as 'immersion chilling and freezing' (ICF) which is reviewed by Lucas and Raoult-Wack 1998), packaged food is passed through a bath of refrigerant at 0 to −55 °C in perforated containers, in a rotating drum or on a submerged mesh conveyor. Immersion freezing began with the use of brines to freeze fish, vegetables and meat. Aqueous solutions containing soluble carbohydrates (e.g. sucrose, glucose, fructose or other mono- and disaccharides) with additions of ethanol, propylene glycol, salts (e.g. sodium chloride, potassium chloride or calcium chloride) or glycerol have each been studied as possible immersion media. In contrast with cryogenic freezing (section 22.2.2), the refrigerant remains fluid throughout the freezing operation and a change of phase does not occur. The method has high rates of heat transfer (Table 22.3) and capital and operating costs are relatively low. Lucas *et al.* (2001) report that heat transfer coefficients of 210–$290 \, W \, m^{-2} \, K^{-1}$ under natural convection can be increased to 680–$690 \, W \, m^{-2} \, K^{-1}$ using forced convection, to 680–$740 \, W \, m^{-2} \, K^{-1}$ using a rotating drum, and to 550–$900 \, W \, m^{-2} \, K^{-1}$ using fluidisation. Compared with blast freezing, ICF uses less power because it is not necessary to maintain a high fluid velocity (as with high powered fans in blast freezing) and similar processing times can be obtained at higher refrigerant temperatures (Lucas *et al.* 2001). Torreggiani *et al.* (2000) note that the time to freeze small fruits or vegetables from 0 to $-7\,°C$ can be reduced by 4–7 times compared with air-blast freezing. The more rapid freezing better retains the texture of the products and there are lower dehydration losses. It is used commercially to freeze packaged ice cream and concentrated orange juice in laminated card–polyethylene cans, and to pre-freeze film-wrapped poultry before blast freezing.

Commercial development of ICF for freezing unpackaged foods has been hindered by lack of control over mass transfer between the food and freezing medium. Torreggiani *et al.* (2000) describe mass transfer rates of 1–7% w/w water loss and 0.5–1% w/w solute uptake. The extent of mass transfer depends in part on the surface characteristics of the food and the surface area:volume ratio. ICF is best suited to small pieces of food that have smooth non-porous surfaces to minimise entrainment of the refrigerant. Also fatty foods entrain lower amounts of the refrigerant. Surface treatments such as dipping the pre-chilled food in water, create a surface ice barrier when the food comes into contact with the immersion freezing medium. This limits further mass transfer between the food and the freezing medium. Alternatively, Lucas *et al.* (2001) report that adding sucrose to a brine refrigerant reduced sodium chloride uptake in the product by 50%. This was thought to be due to the sucrose forming a concentrated layer at the surface of the food that reduced salt impregnation. However, the method increased water losses from the product (3% compared with 1.2% after 15 min immersion). Although not used commercially, Lucas *et al.* (2001) describe the concept of adding vitamins, flavourings or colourants to the immersion medium to alter the sensory and nutritional properties of the food, or to retain the refrigerant as a protective coating to aid preservation of the food during storage.

Cooled-surface freezers

Plate freezers consist of a vertical or horizontal stack of up to 24 hollow stainless steel or aluminium plates, 2.5–5 cm thick, through which refrigerant is pumped (Fig. 22.6). The liquid refrigerant evaporates inside the plates, absorbing heat and leaving as a mixture of liquid and vapour. Freezers operate at -30 to $-50\,°C$ and may be batch or semi-continuous. Flat, relatively thin foods (e.g. fish blocks (Anon 2007)) are placed in frames as single layers between the plates and a slight pressure is applied by moving the plates together. This improves the contact between surfaces of the food and the plates, and thereby increases the rate of heat transfer. If packaged food is frozen in this way, the pressure prevents the larger surfaces of the packs from bulging. Production rates range from 90 to $3000 \, kg \, h^{-1}$. Advantages of this type of equipment include good economy and space utilisation, high rates of heat transfer (Table 22.3), relatively low operating costs compared with other methods, and little dehydration of the product or weight loss. The main disadvantages are the relatively high capital costs and restrictions on the shape of foods to those that are flat and relatively thin (<8 cm thick).

Surface freezing by a single refrigerated plate is used to form an almost instant crust on foods that are sticky (e.g. pasta, pulped fruits), or prone to lose water (e.g. shrimps,

Fig. 22.6   Horizontal plate freezer (courtesy of Beck Pack Systems AS at www.beck-liner.com).

chicken breasts), or foods that are delicate or prone to lose their shape (e.g. cakes). The frozen crust fixes the shape of the food, prevents it from becoming stuck to conveyor belts and prevents marks from conveyor belts on the product (Mascheroni 2001). After the crust is formed, foods are frozen using a conventional freezer. If the equipment has a provision for producing cold air to freeze the upper surface, it is similar to a continuous belt freezer. Problems may arise due to product adhesion onto metallic belts caused by the adhesive force between superficial ice and the metallic surface. Initially, the adhesive force increases as the temperature of the metal decreases, until this force becomes larger than the strength of the ice. However, as the metal temperature approaches −80 °C, the adhesive force is reduced dramatically and the product can be removed with minimum effort. This temperature can be achieved in cryogenic tunnels and spiral freezers (Estrada-Flores 2002).

Scraped-surface freezers are used for liquid or semi-solid foods (e.g. ice cream). They are similar in design to equipment used for heat sterilisation (Chapter 13, section 13.2.3) and evaporation (Chapter 14, Fig. 14.6) but are refrigerated with ammonia, brine or other refrigerants. In ice cream manufacture, the rotor scrapes frozen food from the wall of the freezer barrel and simultaneously incorporates air. Alternatively, air can be injected into the product. Freezing is very fast and up to 50% of the water is frozen within a few seconds (Jaspersen 1989). This results in very small ice crystals, which are not detectable in the mouth and thus gives a smooth creamy consistency to the product. The temperature is reduced to between −4 and −7 °C and the frozen aerated mixture is then pumped into containers and freezing is completed in a 'hardening room' (see cold stores above). Further details of ice cream production are given in Chapter 3 (section 3.2) and by Steffe (1998) and Drewett and Hartel (2007).

22.2.2 Cryogenic freezers

Freezers of this type use a change of state in the refrigerant (or cryogen) to absorb heat from the freezing food. The heat provides the latent heat of vaporisation or sublimation of the cryogen. The cryogen is in intimate contact with the food and rapidly removes heat from all surfaces to produce high heat transfer coefficients and rapid freezing (Table 22.3). The two most common refrigerants are liquid nitrogen and solid or liquid carbon dioxide. Dichlorodifluoromethane (refrigerant 12 or Freon 12) was also previously used for sticky or fragile foods that stuck together in clumps (e.g. meat paste, shrimps, tomato slices), but its use has been phased out under the Montreal Protocol, owing to its effects on the Earth's ozone layer (further details are given in Chapter 21, section 21.2.1).

The choice of cryogen is determined by its technical performance for a particular product, its cost and availability, environmental impact and safety (Heap 1997). Two advantages of cryogenic freezers compared with mechanical systems are the lower capital cost and flexibility to process a number of different products without major changes to the system (Miller 1998). Others that are described by Estrada-Flores (2002) include short freezing times, reduced dehydration and drip losses, and improved texture of products due to fast freezing and the growth of small ice crystals.

Cryogens

Liquid nitrogen is more commonly used for freezing applications, whereas carbon dioxide refrigerant is more often used for chilling (Chapter 21). Liquid nitrogen is colourless, odourless, non-flammable and inert. It is made by the liquefaction and fractional distillation of air (Hung 2001). The low boiling point creates a large temperature gradient with the food, resulting in high rates of freezing, and the large amount of heat that can be absorbed make it an ideal freezing medium (properties of cryogens are given in Table 21.2 in Chapter 21). Although the method of application of liquid nitrogen and CO_2 is similar, their behaviour during freezing is different: compared with liquid nitrogen, carbon dioxide has a lower enthalpy and a lower boiling point that produces less severe thermal shock. Liquid CO_2 expands and changes to approximately equal parts (by weight) of solid and vapour (or 'snow'). The distribution system creates air/CO_2 currents within the freezer. As solid CO_2 particles contact the food, they instantly sublime to vapour, which draws heat out of the product. Sublimation provides approximately 85% of the refrigeration effect (or freezing capacity). The remaining 15% is a result of contact between the product and the air/gas mixture. To obtain the maximum refrigeration benefit, a typical CO_2 freezer injects liquid CO_2 throughout the length of the freezer (Estrada-Flores 2002). In liquid nitrogen equipment, the liquid nitrogen is sprayed as a very fine mist of droplets using spray nozzles from a distance of $\approx 15\,cm$ to ensure complete coverage of the food and high heat transfer coefficients (Hung 2001). It separates as liquid and vapour, and as droplets touch the product surface the liquid changes to vapour and extracts latent heat from the food. About 50% of the refrigeration effect is supplied by the liquid nitrogen phase change from liquid to vapour. The vapour is recirculated throughout the freezer to achieve optimum use of its freezing capacity and to create convective currents that increase the freezing rate.

Equipment

The designs of cryogenic freezers are similar to mechanical vapour compression freezers and include batch cabinet freezers, continuous tunnel freezers, spiral freezers, fluidised bed freezers, immersion freezers and combined air-blast/cryogenic freezers (or 'cryomechanical' freezers). The different designs and their applications are described by Jha (2005). They have several advantages compared to mechanical freezers:

- Simple continuous operation with relatively low capital costs ($\approx$30% of the capital cost and 5% of the power requirement of mechanical systems because there is no requirement for a compressor or evaporator (Hung 2001)).
- Smaller units for the same production rates because heat exchanger coils are not used.
- Smaller product weight losses due to dehydration (0.5% compared with up to 8.0% in mechanical air-blast systems).
- Rapid freezing produces small ice crystals that cause smaller changes to the sensory and nutritional characteristics of the product.
- Exclusion of oxygen during freezing which reduces oxidative changes to products.
- Rapid start-up and no defrosting time.

The main disadvantage is the relatively high cost of the cryogens, which results in operating costs that are 6–8 times higher than those of mechanical refrigeration systems because cryogens are not recirculated and are lost to the atmosphere. In each design, the very high freezing rates when the cryogen is sprayed onto foods cause a crust to form on the surface, which minimises further loss of moisture and flavours. Freezing the bulk of the food then takes place using the cold gas.

In cryogenic tunnel freezers, packaged or unpackaged food travels on a perforated belt through a tunnel. Earlier designs had liquid nitrogen sprays near the product exit and fans to blow the gas counter-currently over the product as it passed through a tunnel. Newer designs have multiple spray zones that give better process control and do not require fans. Production rates are 45–1550 kg h^{-1} and the newer designs have heat transfer coefficients of 1200 Wm^{-2}K^{-1} (Miller and Butcher 2000). Shaikh and Prabhu (2007a) developed a mathematical model for sizing cryogenic tunnel freezers, which can be used to minimise operating costs by improvements in freezer design and reduce cryogen consumption by up to 30%.

After initial freezing the product temperature is either allowed to equilibrate at the required storage temperature (between -18 and -30 °C) before the food is removed from the freezer, or food is passed to a mechanical freezer to complete the freezing process. Other applications include rigidification of meat for high-speed slicing (Chapter 3, section 3.1), surface hardening of ice cream prior to chocolate coating (Chapter 24, section 24.2) before finishing freezing in mechanical freezers (Londahl and Karlsson 1991). Summers (1998) describes a design of liquid nitrogen freezer that is said to double the output of conventional freezers of the same length, reduce nitrogen consumption by 20% and reduce already low levels of dehydration by 60%. The temperature and belt speed are controlled by microprocessors. Shaikh and Prabhu (2007b) report the development of an improved control mechanism that combines feedback and feed-forward control (Chapter 27, section 27.2) to adjust cryogen consumption and throughput of the tunnel freezers to maintain the product at a pre-set exit temperature, regardless of the heat load of incoming food. The equipment therefore has the same efficiency at or below its rated capacity. This results in greater flexibility and economy than mechanical systems, which have a fixed rate of heat extraction (Tomlins 1995).

The flighted tunnel freezer is used to freeze high-value and often delicate IQF products (e.g. scallops, strawberries, diced fruit, diced poultry, meatballs, sliced mushrooms and other pizza toppings). Individual products are transported through the freezer by flighted conveyors that gently tumble the pieces, keeping them separate and allowing maximum exposure to the cryogen. This reduces product clumping and agglomeration, and the instant crust formation maintains the product moisture level. Automatic temperature and pressure control systems adjust the rate of cryogen injection to compensate for incoming product load and temperature variations. An automatic vapour balance system at the conveyor entrance/exit prevents warm room air from entering the freezer to conserve the cryogen and optimise freezer performance (Anon 2004).

A development of the cryogenic tunnel freezer is the cryogenic impingement freezer. It uses a combination of high-velocity (20–30 m s^{-1}) air jets (air impingement) and cooling by atomised liquid nitrogen that is sprayed onto the surface of the food at the inlet zone of the freezer. The high-velocity impingement jets are applied above and below the product and across the tunnel, which distributes air/gas flow evenly across the length and width of the freezing zone. This is not possible with the velocity profiles of axial-flow fans in a conventional cryogenic freezer. Heat transfer coefficients in the inlet zone are three times higher than in a conventional mechanical impingement freezer and result in a 25% increase in the overall heat transfer coefficient (Morris 2003). The increase is due to disruption of boundary layers of air around the product and the high temperature gradient between product and the nitrogen (typically ≈190 °C). The amount of evaporative cooling with liquid nitrogen can be optimised by controlling the droplet size, spray distribution and gas flowrate along the length of the unit in independently controlled freezing zones. The combined liquid nitrogen and impingement airflow crust-freezes the product almost instantly to minimise dehydration (losses are <0.08% compared with 0.45–0.50% in conventional tunnel freezers). Also the operation at lower temperatures enables high production rates and smaller equipment can therefore be used. Typically it uses one-third of the floor space of a conventional cryogenic freezer for the same production rate. The process is best suited to products having high surface-to-weight ratios with a thickness <20 mm (e.g. hamburger patties, chicken or fish fillets). The impingement jets can also crust-freeze thicker products at the inlet of a mechanical spiral freezer, which reduces product dehydration and ice build-up on evaporator coils, and reduces energy costs because the spiral freezer can operate at higher temperatures (Hung 2001).

Foss et al. (1999) patented a tunnel freezer that uses a mixture of liquid oxygen and liquid nitrogen in a composition similar to that found in normal air. The advantages of the system are related to the safety of the operators, avoiding dangerous build-up of gaseous nitrogen in the surroundings of the freezer and reducing the need for extraction fans. The tunnel has an immersion bath and oxygen sensors to monitor and control the system. Jones et al. (2001) patented a cryogenic freezer that involves dripping flavoured liquid dairy products as droplets into a freezing chamber filled with a mixture of gaseous and liquid cryogenic refrigerant. As they fall through the chamber, they solidify, forming solid beads of flavoured ice cream or yoghurt.

Liquid nitrogen immersion freezers are the fastest method of freezing. Foods (e.g. shrimps, chicken portions and diced meat) are dropped into a bath of liquid nitrogen to crust-freeze them, and a conveyor then lifts the food pieces into a tunnel where the cold gas produced by the immersion of foods continues the freezing process (Fig. 22.7). During immersion the extreme turbulence of the boiling liquid nitrogen prevents pieces of food from sticking together and produces IQF conditions that are suitable for irregular shaped foods. However, the residence time has to be carefully controlled to prevent over-freezing or internal stresses created by the high thermal shock that would cause the food to crack or split. The rapid freezing permits high production rates using small equipment (e.g. a 1.5 m long bath can freeze 1000 kg h^{-1} of small particulate food). An example is given by Anon (2000a).

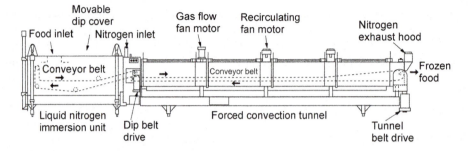

Fig. 22.7   Liquid nitrogen immersion freezer (courtesy of Air Products plc) (Anon 2000a).

22.2.3   New developments in freezing

Although the methods described in this section are not yet used commercially, laboratory and pilot-scale studies have shown that they have the potential to give improved freezing rates and/or higher product quality.

Pressure freezing

An increase in pressure during freezing and thawing influences the ice–water transition and prevents many of the undesirable changes to the texture and sensory properties of foods described in section 22.3. The phase diagram for water shows that when foods are frozen at atmospheric pressure, the ice that is formed is Type I polymorph (Fig. 22.8). As the pressure is increased the freezing point of water is decreased and ice nucleation rate increases. This increases the rate of freezing and produces smaller, more uniform ice

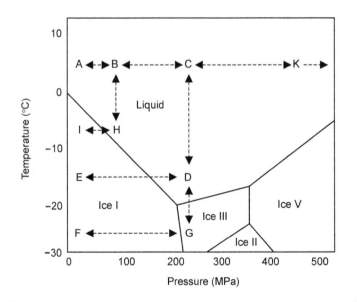

Fig. 22.8   Phase diagram for water at high pressure and effects on phase transitions. A–B–C–D = subzero storage without freezing, A–B–H–I = pressure-assisted freezing and thawing (phase transition at constant pressure), A–B–C–D–E = pressure-shift freezing (phase transition due to pressure change) and pressure-induced thawing (phase transition initiated with pressure change and continued at constant pressure), A–B–C–D–G–F = freezing and thawing to ice III, A–B–C–K to Ice VI (not shown) = freezing above 0 °C (adapted from Fikiin 2003).

crystals. At a pressure of 200 MPa the freezing point falls to $\approx-22\,°C$ (Fikiin 2003). Above ≈200 MPa, higher density Type III ice crystals are formed.

'High-pressure assisted freezing' (HPAF) involves cooling foods to $-21\,°C$ under high pressures (≈300 MPa). Alternatively foods such as meat are cooled under high pressures (to $-3\,°C$ at 24 MPa) without causing freezing. In 'high-pressure shift freezing' (HPSF) the pressure is released and the food instantly freezes, resulting in more uniform ice crystals that are 30–100 times smaller than those produced by blast freezing (Sanz 2005). High-pressure freezing has been reviewed by Sanz (2005), Sanz and Otero (2005) and Fikiin (2008). Otero and Sanz (2003) have reviewed mathematical models of high-pressure processing, including processing at sub-zero temperatures.

A number of studies have shown that foods frozen under high pressure have reduced biochemical damage as a result of reductions in microbial and enzyme activities (e.g. polyphenoloxidase), higher retention of nutrients and reduced damage to the colour, texture and flavour. The homogeneous distribution of small ice crystals throughout the food reduces mechanical damage to cells and there is less drip loss after thawing (section 22.3). For example, Fernández et al. (2006a) report that compared to conventionally frozen broccoli, HPSF broccoli had less cell damage, lower drip losses and better texture. Zhu et al. (2005) measured ice crystal sizes as 91, 73 and 44 μm from HPSF at 100 MPa ($-8.4\,°C$), 150 MPa ($-14\,°C$) and 200 MPa ($-20\,°C$) respectively compared to 145 μm in conventional air freezing at $-20\,°C$ and 84 μm from liquid immersion freezing at $-20\,°C$. Studies have also shown that HPSF is a more advantageous method than HPAF; the degree of supercooling after expansion and the consequent instantaneous freezing of water, together with the temperature drop in the pressure medium, induced short phase transition times (5.9, 8.6, and 13.7 min in HPSF versus 14.8, 14.1 and 23.1 min in HPAF at 0.1, 50 and 100 MPa, respectively) (Fernández et al. 2006b). Chevalier et al. (2002) found that HPSF resulted in smaller and more regular ice crystals in fish (about ten and seven times smaller compared with blast frozen turbot and carp respectively) and produced less thawing and cooking drip losses. Sequeira-Munoz et al. (2005) found similar results when they compared pressure-shift freezing and air-blast freezing of carp.

Pressure freezing influences both proteins and enzymes, and also cellular tissues and micro-organisms. For example, in HPSF of muscle foods, aggregation of myofibrillar proteins causes toughness, which may offset the benefits gained from small ice crystals. Cells in some fruits and vegetables are less damaged by pressure shift freezing than by blast freezing, but Cheftel et al. (2002) report that this may not result in improved appearance, texture or water retention. They also note that there is insufficient enzyme inactivation at high pressure to be able to dispense with blanching, and that microbial inactivation, although enhanced at low temperatures, may not be sufficient for practical applications.

Hydrocolloids (Chapter 1, section 1.1.1) are used in ice-creams and frozen desserts to produce smooth texture and protect the product during storage. Fernández et al. (2007) studied the combination of hydrocolloids with HPAF (at 100 MPa) or HPSF (from 210 MPa to 0.1 and 100 MPa) to determine whether reduced water mobility due to the hydrocolloids affected ice crystal formation. They found that ice crystals were smaller after HPSF than after HPAF, due to greater supercooling following expansion and to shorter phase transition times. Ice crystals were smaller when the hydrocolloids locust bean and xanthan gums were added irrespective of the freezing method. They concluded that formation of a gel-like structure may limit water molecule diffusion and ice crystal growth.

High-pressure thawing also has advantages over conventional techniques, including a 2–5 fold reduction in thawing times, partial destruction or limitation of pathogens, and better retention of organoleptic properties in meats and seafoods (Kennedy 2003). However, Cheftel *et al.* (2002) note that the benefits of pressure thawing in terms of higher rates and better hygiene may not outweigh the higher equipment and packaging costs.

Hydrofluidisation

In freezing by hydrofluidisation a refrigerating liquid is pumped through orifices or nozzles in a vessel to create jets that form highly turbulent liquid, which has extremely high surface heat transfer coefficients. Studies of freezing small fish and vegetables using a sodium chloride solution showed a higher freezing rate than other IQF techniques. At slight or moderate agitation and a refrigerant temperature of $\approx -16\,°C$, fish were frozen from 25 to $-10\,°C$ at the thermal centre in 6–7 min, green beans in 3–4 min, and peas within 1–2 min. The highest surface heat transfer coefficient exceeded $900\,W\,m^{-2}\,K^{-1}$, compared with $378\,W\,m^{-2}\,K^{-1}$ for immersing freezing, $432\,W\,m^{-2}\,K^{-1}$ for spraying refrigerant and $475\,W\,m^{-2}\,K^{-1}$ for immersion with bubbling (Fikiin and Fikiin 2000).

Pumpable ice slurries have been studied as possible replacements for HFC- or HCFC-based refrigerants and may have potential as a refrigerating medium for hydrofluidisation. The minute ice particles in the slurries rapidly absorb latent heat when they thaw on the product surface and produce very high surface heat transfer coefficients (1000–$2000\,W\,m^{-2}\,K^{-1}$), very short freezing times that approach cryogenic freezing, and uniform temperature distribution (e.g. at an ice-slurry temperature of $-25\,°C$ strawberries, apricots and plums can be frozen from 25 to $-18\,°C$ in 8–9 min; raspberries and cherries in 1.5–3 min; and green peas, blueberries and cranberries in ≈ 1 min (Fikiin 2003). The aqueous media used for immersion freezing (section 22.2.1) offer the opportunity to formulate appropriate multi-component refrigerants based on ice slurries. These are less environmentally damaging than CFCs and HCFCs and have low viscosity to enable them to be pumped and produce hydrofluidisation. New products can be produced by using specific freezing media (e.g. fruits frozen in syrup solutions to produce dessert products that retain the characteristic colour, flavour and texture). The immersion media can also include antioxidants, flavourings and micronutrients to extend the shelf-life of products or to improve their nutritional value and sensory properties.

The main advantages of hydrofluidisation over conventional freezing methods include:

- High heat transfer rates with smaller temperature differences between the product and the freezing medium. The evaporator temperature can be maintained at -25 to $-30\,°C$ by a single-stage compressor that has a higher COP (Chapter 21, section 21.2) and nearly two times lower capital and power costs than air fluidisation.
- Hydrofluidisation is achieved at low fluid velocity and pressure, which both saves energy and causes minimal mechanical damage to foods.
- Foods pass quickly through the critical zone for ice crystallisation (section 22.1.1), which ensures formation of small ice crystals that prevent damage to cellular tissues.
- The product surface freezes immediately to form a solid crust that reduces mass transfer to zero and produces an excellent surface appearance.
- The process is continuous, convenient for automation and has low labour costs.
- Frozen products are free-flowing and easily packaged.

Ultrasound freezing

Power ultrasound (Chapter 9, section 9.6) causes cavitation in liquids, which leads to the production of gas bubbles and also 'microstreaming'. During freezing, new nucleation sites are created by the bubbles, which increase the rate of nucleation and crystal growth, and microstreaming accelerates the rates of heat and mass transfer (Zheng and Sun 2006). Within limits that are controlled by the heat produced when ultrasound passes through the medium, a higher ultrasound output power and longer exposure time increases the rate of freezing. For example, Li and Sun (2002a) report an optimum treatment of 2 min with a power output of 15.85 W to achieve the highest freezing rate using potatoes. The high freezing rate resulted in more small intracellular ice crystals and less cell disruption (Sun and Li 2003). Additionally, the alternating acoustic stresses produced by ultrasound cause ice crystals to fracture, leading to foods that contain a smaller size distribution of crystals (e.g. in a scraped surface freezer, where ultrasound can reduce the ice crystal size in ice cream). These and other applications of ultrasonic freezing to produce high-value foods are reviewed by Zheng and Sun (2006).

Magnetic freezing

In conventional refrigeration equipment, undesirable water migration and mass transfer take place within foods as they are undergoing freezing. However, if water could be retained within cells while freezing takes place, the cells would not become dehydrated and the food would retain its original quality characteristics. Magnetic freezing impedes ice crystallisation and allows supercooling below the initial freezing point. Initial studies by Anon (2000b) achieved a temperature reduction from 28 to −1 °C using gadolinium as a magnetic material and the system demonstrated a COP of 4.3. A system for magnetic resonance freezing (MRF) described by Mohanty (2001) comprises a traditional mechanical or cryogenic freezer fitted with a magnetic resonance device. The MRF process has two steps: (1) food is subjected to continuous oscillating magnetic wave vibrations that impede ice crystallisation as the food is supercooled below the initial freezing point; and (2) after a suitable product-specific period of time, the magnetic fields are switched off and the food rapidly undergoes uniform flash-freezing of the entire volume. This produces small ice crystals that do not damage the structure of the food, and there is no water migration and undesirable cellular dehydration. The process has been commercialised as the 'Cells Alive System' (CAS) in Japan and is being used aboard fishing vessels for freezing tuna, and in Alaska to preserve cod milt and roe, products that had previously been impossible to freeze and retain their market value. Suppliers of ingredients for French cuisine use CAS to preserve delicate doughs, foie gras, duck meat and truffles. Other products frozen using CAS include cream, milk, green mangos, sea urchin, sashimi grade seafood and sushi (Anon 2001). A CAS unit costs between 20 and 100% more than some conventional freezers (in 2008) but the system benefits from reduced energy costs and very high product quality retention, which in the case of tuna has allowed the fish to be sold for ≈90% above the market value. A CAS defroster controls the defrosting temperatures at the same level at both the centre and the surface of the food. This prevents a time lag between thawing of the two parts. It is equipped with a humidity control system to defrost foods at their optimal humidity, which thaws foods to their original quality without changes to flavour or colour.

Dehydrofreezing

Partial drying before freezing has been shown to better retain the colour and flavour of foods and reduce drip losses on thawing, as well as producing energy savings because less water is frozen and the weight of food to be transported and stored is reduced. Typically, 50–60% of the moisture is removed by drying or by osmotic dehydration (Chapter 16, section 16.1.3). Browning reactions are inhibited by either blanching

(Chapter 11) or by dipping foods in antioxidant solutions (e.g. ascorbic or citric acid). In osmodehydrofreezing, the incorporation of different solutes raises the glass transition temperature so that the food forms a glass at frozen storage temperatures (section 22.1.2). This reduces the mobility of reactants, slows biochemical deterioration and hence reduces loss of flavours and pigments. For example, chlorophyll in kiwi fruits, vitamin C in apricots and anthocyanins in cherries and strawberries are each better retained after osmotic pre-treatment (Kennedy 2003, Torreggiani *et al.* 2000, 1997). Similarly osmotic pretreatment can protect the texture of frozen fruits and reduce drip losses on thawing (Lazarides and Mavroudis 1995).

Cryoprotectants
Cryoprotectants are compounds that depress the freezing temperature of foods, modify or suppress ice crystal growth during freezing and inhibit ice recrystallisation during frozen storage. They reduce damage to cell membranes and so protect the texture of foods and reduce the loss of nutrients by drip losses. Examples include sugars, amino acids, polyols, methyl amines, carbohydrates and inorganic salts (Kennedy 2003). Cryoprotectant glycoproteins or 'antifreeze proteins' (AFPs) have been isolated from a wide variety of organisms, including bacteria (Kawahara 2002), fungi, plants, invertebrates and fish, such as Antarctic cod and the winter flounder (Payne *et al.* 1994). The organisms have evolved AFPs as mechanisms to protect them against low temperatures and minimise freezing injury. Multiple forms of AFPs are synthesised within each organism, each with a different function. For example, the ice nucleation protein acts as a template for ice formation, whereas the anti-nucleating protein inhibits ice nucleus formation by a foreign particle in the water (section 22.1.1) (Kawahara 2002). In future it may be possible to select an AFP with suitable characteristics and activity for particular food products and introduce it into the food by physical processes, such as mixing or soaking, or by gene transfer (Griffith and Ewart 1995). AFP–ice complexes have interactions with cell membranes and with other molecules present in the solutions. Wang (2000) reviewed studies of cryoprotectants, which show that AFPs have a complex mechanism of action and can display both protective and cytotoxic actions depending on the dose, type, composition and concentration of cryoprotectant, the characteristics of the biological material and the conditions of frozen storage. Li and Sun (2002b) have reviewed developments in methods of freezing (high-pressure freezing, dehydrofreezing and applications of AFPs) and thawing (using high-pressure, microwaves, ohmic and acoustic thawing).

22.2.4   Frozen storage
Once frozen, the temperature of foods should be maintained at or below −18 °C throughout the cold chain. This includes in-factory storage, transport to wholesalers or retailers, wholesale storage and retail displays. Details of the design and construction of frozen food stores and vehicles are given by Le Blanc and Stark (2001) (also Chapter 21, section 21.2.3). Temperature fluctuations are minimised by:

- accurate control of storage temperature (±1.5 °C);
- automatic doors and airtight curtains for stores and for loading refrigerated trucks;
- rapid movement of foods between stores; and
- correct stock rotation and control.

Food handling and distribution logistics are described in Chapter 27 (section 27.3) and the effects of frozen storage on food quality are described in section 22.3.2.

Storage life

There is some confusion and lack of precise information on the storage life of frozen foods, caused in part by the use of different definitions. For example a European Community Directive states that frozen storage must 'preserve the intrinsic characteristics' of foods, whereas the International Institute of Refrigeration defines storage life as 'the physical and biochemical reactions leading to a gradual, cumulative and irreversible reduction in product quality, such that after a period of time the product is no longer suitable for consumption'. Another definition by Bøgh-Sørensen describes practical storage life (PSL) as 'the time the product can be stored and still be acceptable to the consumer' (Evans and James 1993). These definitions differ in the extent to which a product is said to be acceptable and rely heavily on the ability of taste panellists to detect changes in sensory properties that can be used to measure acceptability. For example, PSL is also defined as 'the time that a statistically significant difference ($P < 0.01$) in quality can be established by taste panellists'. These methods therefore measure the period that food remains essentially the same as when it was frozen. This should not be confused with a storage life that is acceptable to consumers as foods may be acceptable for three to six times longer than the PSL. Examples of PSL for selected foods stored at commercial refrigerated storage temperatures are shown in Table 22.4.

Temperature monitoring

The main cause of loss of storage life is fluctuating temperatures (see section 22.3.2). Other factors that affect storage life are discussed in detail by Evans and James (1993)

Table 22.4   Maximum storage times for selected foods at different temperatures

Food	Practical storage life (months)		
	−12°C	−18°C	−24°C
Fish and seafoods			
Clams, oysters	4	6	>9
Oily fish (e.g. herring, salmon, mackerel)	3	4	>9
Prawns, lobster, crab	4	6	>12
White fish (e.g. plaice, sole, cod)	4	8	>12
Meats and meat products			
Bacon	–	2–4	–
Beef joints, steaks	8	12–18	24
Beef mince	6	10	15
Chicken, whole or portioned	9	18	>24
Duck or goose, whole	6	12	18
Lamb joints, chops	12	18	24
Pork joints, chops	6	10	15
Sausages	–	6	
Turkey, whole	8	15	>24
Vegetables			
Broccoli	–	18	>24
Carrots	10	18	>24
Cauliflower	4	15	24
Green beans	–	15	–
Peas	6	18	>24
Potato chips	9	24	>24
Spinach	4	18	>24
Sweetcorn	4	12	>24

Adapted from Singh and Heldman (2001) and Fikiin (2003)

and Fikiin (2003), and include the type of raw material, cultivar or species, pre-freezing treatments (especially the extent of size reduction, blanching or dipping in salt or sugar solutions) and freezing conditions. Temperature fluctuation has a cumulative effect on food quality and the proportion of PSL that is lost can be found by integrating losses over time. Time–temperature tolerance (TTT) and product-processing-packaging (PPP) concepts are used to monitor and control the effects of temperature fluctuations on frozen food quality during production, distribution and storage. These are prepared as recommendations for handling frozen foods throughout the cold chain (e.g. Anon 1999, Fuller 1998), and are incorporated into legislation in many countries (e.g. Sørensen 2002). There are three types of devices that can monitor the temperature of foods during frozen storage and distribution: critical temperature indicators (CTIs), temperature recorders and time–temperature indicators (TTIs). These are described in detail in Chapter 21 (section 21.2.4) and reviewed by Selman (1995).

22.2.5   Thawing

The main considerations in thawing are to avoid overheating the food, to minimise thawing times, and to avoid excessive dehydration of the food. Commercially, foods are thawed in a vacuum chamber by condensing steam, by warm water ($\approx 20\,^\circ C$) or by moist air that is recirculated over the food. Details of types of thawing equipment and methods of operation of are described by Swain and James (2005). More rapid thawing can be achieved using dielectric energy (Chapter 20, section 20.1.3), ohmic heating (Chapter 20, section 20.2.2) and acoustic thawing (Li and Sun 2002a). Thawing under pressure can be achieved at lower temperatures than that at atmospheric pressure (section 22.2.3). The effects of thawing on foods are described in section 22.3.3.

22.3   Effects on foods

22.3.1   Freezing

The freezing process causes negligible changes to pigments, flavours or nutritionally important components. Differences in the variety and quality of raw materials and the degree of control over pre-freezing treatments (Chapters 2 and 3) and blanching (Chapter 11) have a substantially greater effect on food quality than changes caused by correctly operated freezing procedures. Food emulsions (Chapter 3, section 3.2) can be destabilised by freezing and proteins are sometimes precipitated from solution, which, for example, prevents the widespread use of frozen milk. In baked goods a high proportion of amylopectin is needed in the starch to prevent retrogradation and staling during slow freezing.

The main effect of freezing on food quality is damage caused to cells by dehydration; the extent of damage depends on the size of the crystals and hence on the rate of freezing (section 22.1.1). During slow freezing, there is time for cells to lose water by diffusion. Freezing causes an increase in solute concentrations in the unfrozen water surrounding the cells and this creates a water vapour pressure gradient between the cells and the extracellular water. As the cells lose water, they shrink irreversibly until they collapse (e.g. slow freezing causes 3–6% shrinkage of foods). In addition to dehydration, larger ice crystals exert pressure on flexible cell walls, and as ice crystals continue to grow they cause cells to flex. Ice can then grow into the newly created volume and prevent the structure from returning to its original shape. The structural damage to cells leads to rupture and loss of intracellular water, contributing to drip loss during thawing (Fig.

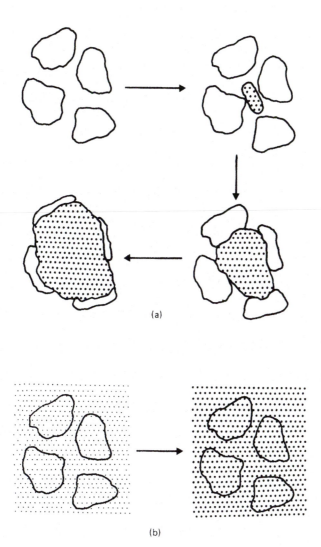

Fig. 22.9 Effect of freezing on plant tissues: (a) slow freezing; (b) fast freezing (after Merryman 1963).

22.9). However, it is incorrect that ice crystals spear through cellular structures; ice crystals grow by adding water molecules to their surfaces and the cellular wall that surrounds an ice crystal makes it difficult for aggregated water molecules to form sharp ice crystals.

Damage to the structure (and texture) of foods during slow freezing is therefore a combination of dehydration and mechanical damage to cells. In contrast, rapid freezing promotes a large number of small ice crystals that are distributed uniformly, both inside and outside the cells. Hence, fast-frozen products suffer less cellular damage or distortion and better retain their texture with less drip loss.

There are important differences in resistance to freezing damage between animal and plant tissues. Meats have a more flexible fibrous structure that separates during freezing, and the texture is not seriously damaged. However, in fish, denaturation and aggregation of proteins can lead to toughening and drying of the muscle and loss of functional properties, especially water-holding capacity, that results in increased drip losses and lower quality. In fruits and vegetables, ice crystals may damage the more rigid cell structure. This results in loss of nutrients in drip losses and loss of texture on thawing. Details of the changes to meats are described by Honikel and Schwägele (2001), Veerkamp (2001) and Devine *et al.* (1996), changes to fish are described by Nielsen *et al.* (2001) and Santos-Yap (1996), and changes to fruits and vegetables are described by Skrede (1996) and Cano (1996).

Volume changes
The volume of ice is 9% greater than that of pure water, and an expansion of foods after freezing would therefore be expected. However, the degree of expansion varies considerably owing to the following factors:

- Moisture content (higher moisture contents produce greater changes in volume).
- Cell arrangement (plant materials have intercellular air spaces which absorb internal increases in volume without large changes in their overall size (e.g. whole strawberries increase in volume by 3.0% whereas coarsely ground strawberries increase by 8.2% when both are frozen to −20 °C (Leniger and Beverloo 1975).
- The concentrations of solutes (high concentrations reduce the freezing point and foods do not freeze – or expand – at commercial freezing temperatures).
- The freezer temperature (this determines the amount of ice and hence the degree of expansion).
- Crystallised components, including ice, fats and solutes, contract when they are cooled and this reduces the volume of the food.
- Rapid freezing causes the food surface to form a crust and prevents further expansion. This causes internal stresses to build up in the food and makes pieces more susceptible to cracking or shattering, especially when they suffer impacts during handling while frozen.

Details of the effect of freezing rate on the cracking resistance of different fruits are described by Sebok *et al.* (1994).

22.3.2   Frozen storage
During storage at normal frozen storage temperatures (≈−18 °C), there is a slow loss of quality owing to both chemical changes and, in some foods, enzymic activity. These changes are accelerated by the high concentration of solutes surrounding the ice crystals, the reduction in water activity (to 0.82 at −20 °C in aqueous foods) and by changes in pH and redox potential. The effects of storage temperature on food quality are shown in Fig. 22.10. If intracellular enzymes are not inactivated, disruption of cell membranes allows them to react to a greater extent with concentrated solutes in the surrounding water.
The main changes to frozen foods during storage are as follows:

- *Degradation of pigments*: chloroplasts and chromoplasts are broken down and chlorophyll is slowly degraded to brown pheophytin even in blanched vegetables. In fruits, changes in pH due to precipitation of salts in concentrated solutions change the colour of anthocyanins.
- *Loss of vitamins*: water-soluble vitamins (e.g. vitamin C, folates and pantothenic acid) are lost at sub-freezing temperatures (Table 22.5) due to oxidation. Vitamin C losses are highly temperature dependent: a 10 °C increase in temperature causes a 6–20-fold increase in the rate of vitamin C degradation in vegetables and a 30–70-fold increase

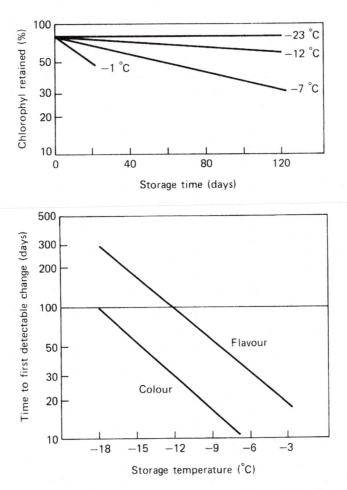

Fig. 22.10   Effect of storage temperature on sensory characteristics (from Jul 1984).

Table 22.5   Vitamin losses during frozen storage

Product	Loss (%) at −18°C during storage for 12 months						
	Vitamin C	Vitamin B$_1$	Vitamin B$_2$	Niacin	Vitamin B$_6$	Pantothenic acid	Carotene
Beans (green)	52	0–32	0	0	0–21	53	0–23
Peas	11	0–16	0–8	0–8	7	29	0–4
Beef steaks[a]		8	9	0	24	22	–
Pork chops[a]		+−18	0–37	+−5	0–8	18	–
Fruit[b]							
Mean	18	29	17	16	–	–	37
Range	0–50	0–66	0–67	0–33	–	–	0–78

+, apparent increase.
[a] Storage for 6 months.
[b] Mean results from apples, apricots, blueberries, cherries, orange juice concentrate (rediluted), peaches, raspberries and strawberries; storage time not given.
Adapted from Burger (1982) and Fennema (1996).

in fruits (Fennema 1996). Losses of other vitamins are mainly due to drip losses, particularly in meat and fish (if the drip loss is not consumed) (Table 22.5).

- *Enzyme activity*: in vegetables that are inadequately blanched, or in fruits, the most important loss of quality is due to polyphenoloxidase activity, which causes browning, and lipoxygenase activity which produces off-flavours and off-odours from lipids and causes degradation of carotene. Proteolytic and lipolytic activity in meats may alter the texture and flavour over long storage periods. In some fish species, enzymic degradation of trimethylamine oxide to dimethylamine and formaldehyde leads to denaturation and loss of solubility of myofibrillar proteins, and loss of quality (Nielsen *et al.* 2001).
- *Oxidation of lipids*: this takes place slowly at $-18\,°C$ and causes off-odours and off-flavours. The storage life of fatty fish is more limited than white fish (Table 22.4) due to oxidative changes to the lipids and associated development of a rancid flavour. Losses of *n*-3 polyunsaturated fatty acids in fatty fish have nutritional significance if these are the main sources in the diet (Fletcher 2002). The effect of lipid oxidation on the quality of frozen foods is reviewed by Erickson (1997).

These changes are discussed in detail by Fennema (1996) and Rahman (1999).

Recrystallisation

Physical changes to ice crystals (e.g. changes in shape, size or orientation of ice crystals) are collectively known as 'recrystallisation' and are an important cause of quality loss in some foods. There are three types of recrystallisation in foods as follows:

1 *Isomass recrystallisation*. This is a change in surface shape or internal structure, usually resulting in a lower surface-area-to-volume ratio.
2 *Accretive recrystallisation*. Two adjacent ice crystals join together to form a larger crystal and cause an overall reduction in the number of crystals in the food.
3 *Migratory recrystallisation*. This is an increase in the average size and a reduction in the average number of crystals, caused by the growth of larger crystals at the expense of smaller crystals.

Migratory recrystallisation is the most important and is largely caused by fluctuations in the storage temperature. When heat is allowed to enter a cold store (e.g. by opening a door and allowing warm air to enter), the surface of the food nearest to the source of heat warms slightly. This causes ice crystals to partially melt and sharper surfaces, which are less stable than flatter ones, become smoother over time; the larger crystals become smaller and the smallest ($<2\,\mu m$) disappear. The melting crystals increase the water vapour pressure, and moisture then moves to regions of lower vapour pressure. This causes areas of the food nearest to the source of heat to become dehydrated. When the temperature falls again, water vapour does not form new nuclei but joins onto existing ice crystals, thereby increasing their size (known as 'Ostwald ripening'). There is therefore a gradual reduction in the numbers of small crystals and an increase in the size of larger crystals, resulting in loss of quality similar to that observed in slow freezing.

In unpackaged foods, moisture leaves the surface of the food to the storage atmosphere and produces areas of visible damage known as 'freezer burn'. This can also occur in cartons of food, where moisture leaves the food and forms ice on the inside of the pack. Areas of freezer burn have a lighter colour due to microscopic cavities, previously occupied by ice crystals, which alter the wavelength of reflected light. This shows that moisture has left the foods and created internal voids, which then become sites for oxidation reactions or cross-linking reactions in biopolymers that cause changes to flavour or texture (section 22.3). Maintaining low constant storage temperatures and packaging in close-fitting, moisture-proof materials can minimise these changes.

Billiard *et al.* (1999) found that air temperatures during defrosting cycles in horizontal open frozen retail display cabinets every 8–12 hours can cause partial thawing of frozen products, with the surface temperature rising to that of the surrounding air. This temperature abuse has a significant effect on product quality and should be taken into account when calculating the actual shelf-life of frozen products (also Chapter 1, section 1.2). Freezer burn is a particular problem in foods that have a large surface area : volume ratio such as IQF foods. The benefits of IQF foods are that they freeze more rapidly; packaged foods may be used by consumers a little at a time without having to thaw the whole pack, and better portion control. However, the low bulk density and high void space causes a higher risk of dehydration and freezer burn. The benefits of fast freezing are lost if frozen products are subjected to temperature abuse and fast-frozen foods can be more sensitive to temperature abuse owing to the delicate matrix of ice crystals formed initially (Estrada-Flores 2002).

22.3.3   Thawing

Details of changes to foods during thawing are described by Fennema (1996). The extent of changes depends on the speed of thawing and the time that food is held close to its freezing point (Fig. 22.4). In the home, food is often thawed using a small temperature difference (e.g. 25–40 °C, compared with 50–80 °C for commercial thawing), which extends the thawing period. The changes to foods also depend on the rate of freezing and the temperature (and temperature abuse) during storage, which determines the damage caused to the cellular structure and hence amount of drip loss on thawing (typically 2–10% of wet weight). For these reasons, thawing is not simply the reverse of freezing.

Drip losses cause loss of water-soluble nutrients: for example in beef losses can be up to 12% thiamine, 10% riboflavin, 14% niacin, 32% pyridoxine and 8% folic acid, and fruits can lose ≈30% of their vitamin C. Drip losses also form substrates for enzyme activity and microbial growth. Microbial contamination of foods before freezing, caused by inadequate cleaning or blanching, has a pronounced effect during this period, permitting growth of psychrotrophic spoilage and pathogenic micro-organisms (section 22.4). Some foods are cooked immediately and are therefore heated rapidly to a temperature that is sufficient to destroy micro-organisms. Others (e.g. cream and frozen cakes) are not cooked and should therefore be consumed within a short time of thawing.

When food is thawed by microwave or radio frequency heaters (Chapter 20, section 20.1.3), heat is generated within the food, and the changes described above do not take place. However, care is needed to control the rate of heating because ice has a lower loss factor than water, which may result in localised overheating and uneven thawing.

22.4   Effect on micro-organisms

Freezing prolongs the shelf-life of products by slowing microbial growth. In general, the lower the temperature of frozen storage, the lower is the rate of microbiological and biochemical changes. However, freezing and frozen storage do not inactivate enzymes and have a variable effect on micro-organisms. The negative effect of freezing on micro-organisms is due to temperature shock, concentration of extracellular solutes, toxicity of intracellular solutes, dehydration and ice formation (Zaritzky 2000). Geiges (1996) reviewed the effect of slow and fast freezing on bacteria and concluded that quick freezing and thawing would result in higher microbial survival rates than those found for slow freezing and thawing. Relatively high storage temperatures (between -4 and $-10\,^{\circ}$C) have a greater lethal effect on micro-organisms than do lower temperatures (between -15 and $-30\,^{\circ}$C). Different types of micro-organism also vary in their resistance to low temperatures; vegetative cells of yeasts, moulds and Gram-negative bacteria (e.g. coliforms and *Salmonella* spp.) are most easily destroyed; Gram-positive bacteria (e.g. *Staphylococcus aureus* and enterococci) and mould spores are more resistant, and bacterial spores (especially *Bacillus* spp. and *Clostridium* spp. such as *Clostridium botulinum*) are virtually unaffected by low temperatures. Other factors that affect the microbiological quality of frozen foods are the physical and chemical characteristics of the product, the pre-freezing microbiological quality and the extent of handling during and after freezing. The majority of vegetables are therefore blanched to inactivate enzymes and to reduce the numbers of contaminating micro-organisms. In fruits, enzyme and microbial activity is controlled by the exclusion of oxygen, acidification or treatment with sulphur dioxide.

References

ANON, (1999), *Control of the Cold Chain for Quick-frozen Foods*, Handbook, IIF-IIR International Institute for Refrigeration, Paris.

ANON, (2000a), Cryo-Quick® Immersion Tunnel Freezer, company information from Air Products, available at www.airproducts.com/Products/equipment/FoodFreezers/PDF_SolutionFinder/Cryo-Quick%20Immersion%20Tunnel%20Freezer.pdf.

ANON, (2000b), Magnetic freezing system developed – a world-first, next-generation freezing technology that uses no CFCs is developed, company information from Chubu Electric Power at www.chuden.co.jp/english/corporate/rd/20001003_1.html.

ANON, (2001), From 'Freezing' to 'CAS-freezing' – The CAS revolution is drastically changing the world of food, company information from Sakura Food Co. Ltd, available at www.sakura.com.vn/cas/feature.htm.

ANON, (2004), Flighted tunnel freezer, company information from Praxair Food Technologies, available at www.praxair.com/praxair.nsf/0/FBF7EF117DEB64A985256C600051F9D9/$file/P-8220B-FlightedInfosheet.pdf.

ANON, (2007), Freezing of standard fish blocks in horizontal plate freezers, company information from Beck Pack Systems A/S, available at www.beck-liner.com/media/Fishblockmanual.pdf.

ANON, (2008), Munters ice-dry®, a simple solution for ice and condensation prevention, company information from Munters Global Food Management Centre Dehumidification Division, available at www.foodprocessing-technology.com/contractors/freezers/munters/.

ARCE, J. and SWEAT, V.E., (1980), Survey of published heat transfer coefficients encountered in food refrigeration processes, *ASHRAE Transactions*, **86** (2), 235–260.

BILLIARD, F., DEFORGES, J., DERENS, E., GROS, J. and SERRAND, M., (1999), *Control of the Cold Chain for Quick-frozen Foods Handbook*, International Institute of Refrigeration Technical Guide, IIR, Paris, pp. 42–43.

BOONSUPTHIP, W. and HELDMAN, D.R., (2007), Prediction of frozen food properties during freezing using product composition, *J. Food Science*, **72** (5), E254–E263.

BURGER, I.H., (1982) Effect of processing on nutritive value of food: meat and meat products, in (M. Rechcigl, Ed.), *Handbook of the Nutritive Value of Processed Food*, Vol. 1, CRC Press, Boca Raton, FL, pp. 323–336.

CANO, M.P., (1996), Vegetables, in (L.E. Jeremiah, Ed.), *Freezing Effects on Food Quality*, Marcel Dekker, New York, pp. 247–298.

CANO-MUÑOZ, G., (1991), Manual on meat cold store operation and management, FAO Animal production and health paper #92, FAO, Rome, available at www.fao.org/DOCREP/004/T0098E/T0098E00.htm#TOC.

CHEFTEL, J.C., THIEBAUD, M. and DUMAY, E., (2002), Pressure-assisted freezing and thawing of foods: a review of recent studies, *International J. High Pressure Research*, **22** (3), 601–611.

CHEVALIER, D., LE BAIL, A., SEQUEIRA-MUNOZ, A., SIMPSON B.K. and GHOUL, M., (2002), Pressure shift freezing of turbot (*Scophthalmus maximus*) and carp (*Cyprinus carpio*): effect on ice crystals and drip volumes, *Progress in Biotechnology*, **19**, 577–582.

CLELAND, A.C., (1990), *Food Refrigeration Processes Analysis, Design and Simulation*, Elsevier, London.

CLELAND, D.J., CLELAND, A.C. and EARLE, R.L., (1987), Prediction of freezing and thawing times for multi-dimensional shapes by simple formulae: 1 regular shapes, *International J. Refrigeration*, **10**, 156–164.

DELLINO, C.V.J., (Ed.), (1997), *Cold and Chilled Storage Technology*, Blackie Academic and Professional, London.

DEVINE, C.E., BELL, R.G., LOVATT, S., CHRYSTALL, B.B. and JEREMIAH, L.E., (1996), Red meat, in (L.E. Jeremiah, Ed.), *Freezing Effects on Food Quality*, Marcel Dekker, New York, pp. 51–84.

DREWETT, E.M. and HARTEL, R.W., (2007), Ice crystallization in a scraped surface freezer, *J. Food Engineering*, **78** (3), 1060–1066.

EARLE, R.L., (1983), *Unit Operations in Food Processing*, 2nd edn, Oxford University Press, Oxford, pp. 78–84.

ERICKSON, M.C., (1997), Lipid oxidation: flavour and nutritional quality deterioration in frozen foods, in (M.C. Erickson and Y-C. Hung, Eds.), *Quality in Frozen Food*, Chapman and Hall, pp. 141–173.

ESTRADA-FLORES, S., (2002), Novel cryogenic technologies for the freezing of food products, *J. Australian Institute of Refrigeration Air Conditioning and Heating, AIRAH*, July, 16–21, available at www.airah.org.au/downloads/2002-07-01.pdf.

EVANS, J. and JAMES, S., (1993), Freezing and meat quality, in (A. Turner, Ed.), *Food Technology International*, Sterling Publications, London, pp. 53–56.

FENNEMA, O.R., (1996), Water and ice, in (O.R. Fennema, Ed.), *Food Chemistry*, 3rd edn, CRC Press, Boca Raton, FL, pp. 17–94.

FENNEMA, O.R. and POWRIE, W.D., (1964), Fundamentals of low temperature food preservation, *Advances in Food Research*, **13**, 219–347.

FERNÁNDEZ, P.P., PRÉSTAMO, G., OTERO, L. and SANZ, P.D., (2006a), Assessment of cell damage in high-pressure-shift frozen broccoli: comparison with market samples, *European Food Research and Technology*, **224** (1), 101–107.

FERNÁNDEZ, P.P., OTERO, L., GUIGNON, B. and SANZ, P.D., (2006b), High-pressure shift freezing versus high-pressure assisted freezing: Effects on the microstructure of a food model, *Food Hydrocolloids*, **20** (4), 510–522.

FERNÁNDEZ, P.P., MARTINO, M.N., ZARITZKY, N.E., GUIGNON, B. and SANZ, P.D., (2007) Effects of locust bean, xanthan and guar gums on the ice crystals of a sucrose solution frozen at high pressure, *Food Hydrocolloids*, **21** (4), 507–515.

FIKIIN, K.A., (2003), Novelties of food freezing research in Europe and beyond, Flair Flow 4 project, British Nutrition Foundation, available at www.nutrition.org.uk/upload/SME%209%20foodfreeze.pdf, pp. 1–36.

FIKIIN, K., (2008), Emerging and novel freezing processes, in (J. Evans, Ed.), *Frozen Food Science and Technology*, Blackwell, Oxford, pp. 101–123.

FIKIIN, K.A. and FIKIIN, A.G., (2000), Individual quick freezing of foods by hydrofluidisation and pumpable ice slurries, in (K. Fikiin, Ed.), *Advances in the Refrigeration Systems, Food Technologies and Cold Chain*, International Institute of Refrigeration Proceedings, Paris, Series: Refrigeration Science and Technology, pp. 319–326.

FLETCHER, J.M., (2002), Freezing, in (C. J. K. Henry and C. Chapman, eds.), *The Nutrition Handbook for Food Processors*, Woodhead Publishing, Cambridge, pp. 331–341.

FOSS, J., MITCHELTREE, M., SCHVESTER, P., RENZ, K., PAGANESSI, J., HUNTER, L., PATEL, R. and BAUMUNK, D., (1999), Liquid air food freezer and method, US Patent 5921091, US Patent Trade Office.

FULLER, R.L., (Ed.), (1998), A practical guide to the cold chain from factory to consumer, Concerted Action Report 1, CT96-1180, available at www.nutrifreeze.co.uk/Documents/THE%20COLD%20CHAIN.pdf.

GEIGES, O., (1996), Microbial processes in frozen food, Advances in Space Research, 18 (12), 109–118.

GRIFFITH, M. and EWART, K.V., (1995), Antifreeze proteins and their potential use in frozen foods, Biotechnology Advances, 13 (3), 375–402.

HEAP, R.D., (1997), Environment, law and choice of refrigerants, in (A. Turner, Ed.), Food Technology International, Sterling Publications, London, pp. 93–96.

HONIKEL, K.O. and SCHWÄGELE, F., (2001), Chilling and freezing of meat and meat products, in (D-W. Sun, Ed.), Advances in Food Refrigeration, Leatherhead Publishing, LFRA, Leatherhead, pp. 366–385.

HUNG, Y-C., (2001), Cryogenic refrigeration, in (D-W. Sun, Ed.), Advances in Food Refrigeration, Leatherhead Publishing, LFRA, Leatherhead, pp. 305–325..

JASPERSEN, W.S., (1989), Speciality ice cream extrusion technology, in (A. Turner, Ed.), Food Technology International, Sterling Publications, London, pp. 85–88.

JONES, M., JONES, C. and JONES, S., (2001), Cryogenic processor for liquid feed preparation of a free-flowing frozen product and method for freezing liquid composition, US Patent 6223542, US Patent Trade Office.

JHA, A.R., (2005), Cryogenic Technology and Applications, Butterworth Heinemann, Oxford.

JUL, M., (1984), The Quality of Frozen Foods, Academic Press, London, pp. 44–80, 156–251.

KAWAHARA, H., (2002), The structures and functions of ice crystal-controlling proteins from bacteria, J. Bioscience and Bioengineering, 94 (6), 492–496.

KENNEDY, C., (2003), Developments in freezing, in (P. Zeuthen and L. Bøgh-Sørensen, Eds.), Food Preservation Techniques, Woodhead Publishing, Cambridge, pp. 228–240.

LAZARIDES, H.N. and MAVROUDIS, N.E., (1995), Freeze/thaw effects on mass transfer rates during osmotic dehydration, J. Food Science, 60 (4), 826–828.

LE BLANC, D. and STARK, R., (2001), The cold chain, in (D-W. Sun, Ed.), Advances in Food Refrigeration, Leatherhead Publishing, LFRA, Leatherhead, pp. 326–365.

LENIGER, H.A. and BEVERLOO, W.A., (1975), Food Process Engineering, D. Reidel, Dordrecht, pp. 351–398.

LI, B. and SUN, D-W., (2002a), Effect of power ultrasound on freezing rate during immersion freezing of potatoes, J. Food Engineering, 55 (3), 277–282.

LI, B. and SUN, D-W., (2002b), Novel methods for rapid freezing and thawing of foods – a review, J. Food Engineering, 54 (3), 175–182.

LONDAHL, G. and KARLSSON, B., (1991), Initial crust freezing of fragile products, in (A. Turner, Ed.), Food Technology International, Sterling Publications, London, pp. 90–91.

LUCAS, T. and RAOULT-WACK, A., (1998), Immersion chilling and freezing in aqueous refrigerating media: review and future directions, International J. Refrigeration, 21 (6), 419–429.

LUCAS, T., CHOUROT, J-M., RAOULT-WACK, A-L. and GOLI, T., (2001), Hydro/immersion chilling and freezing, in (D-W. Sun, Ed.), Advances in Food Refrigeration, Leatherhead Publishing, LFRA, Leatherhead, pp. 220–263.

MASCHERONI, R.H., (2001), Plate and air-blast cooling/freezing, in (D-W. Sun, Ed.), Advances in Food Refrigeration, Leatherhead Publishing, LFRA, Leatherhead, pp. 193–219.

MERRYMAN, H.T., (1963), Food Processing, 22, 81.

MILLER, J., (1998), Cryogenic food freezing systems, Food Processing, 67 (8), 22–23.

MILLER, J. and BUTCHER, C., (2000), Freezer technology, in (C. Kennedy, Ed.), Managing Frozen Foods, Woodhead Publishing, Cambridge, pp. 159–194.

MOHANTY, P., (2001), Magnetic resonance freezing system, Australian Institute of Refrigeration Air Conditioning and Heating (AIRAH) Journal, 55 (6), 28–29.

MORRIS, C.E., (2003), Cryogenic impingement boosts freezer efficiency, Food Engineering, 22

March, available at www.foodengineeringmag.com/CDA/Archives/ 0390f3dcc32f8010VgnVCM100000f932a8c0.

NIELSEN, J., LARSEN, E. and JESSEN, F., (2001), Chilling and freezing of fish and fishery products, in (D-W. Sun, Ed.), *Advances in Food Refrigeration*, Leatherhead Publishing, LFRA, Leatherhead, pp. 403–437.

OHKUMA, C., KAWAI, K., VIRIYARATTANASAK, C., MAHAWANICH, T., TANTRATIAN, S., TAKAI, R. and SUZUKI, T., (2008), Glass transition properties of frozen and freeze-dried surimi products: Effects of sugar and moisture on the glass transition temperature, Food Hydrocolloids, **22** (2), 255–262.

ORLIEN, V., ANDERSEN, M.L., JOUHTIMAKI, S., RISBO, J. and SKIBSTED, L.H., (2004), Effect of temperature and glassy states on the molecular mobility of solutes in frozen tuna muscle as studied by electron spin resonance spectroscopy with spin probe detection, *J. Agriculture and Food Chemistry*, **52** (8), 2269–2276.

OTERO, L. and SANZ, P.D. (2003), Modelling heat transfer in high pressure food processing: a review, *Innovative Food Science and Emerging Technologies*, **4** (2), 121–134.

PARDO, J.M. and NIRANJAN, K., (2006), Freezing, in (J.G. Brennan, Ed.), *Food Processing Handbook*, Wiley-VCH, Weinheim, Germany, pp. 125–145.

PAYNE, S.R., SANDFORD, D., HARRIS A. and YOUNG, O.A., (1994), The effects of antifreeze proteins on chilled and frozen meat, *Meat Science*, **37** (3), 429–438.

PHAM, Q.T., (1986), Simplified equation for predicting the freezing time of foodstuffs, *J. Food Technology*, **21**, 209–219.

PHAM, Q.T., (2001), Cooling/freezing/thawing time and heat load, in (D-W. Sun, Ed.), *Advances in Food Refrigeration*, Leatherhead Publishing, LFRA, Leatherhead, pp. 110–152.

RAHMAN, M.S., (1999), Food preservation by freezing, in (M.S. Rahman, Ed.), *Handbook of Food Preservation*, Marcel Dekker, New York, pp. 259–284.

RAHMAN, M.S., (2001), Thermophysical properties of foods, in (D-W. Sun, Ed.), *Advances in Food Refrigeration*, Leatherhead Publishing, LFRA, Leatherhead, pp. 70–109.

RIVEROL, C., CAROSI, F. and DI SANCTIS, C., (2004), The application of advanced techniques in a fluidised bed freezer for fruits: evaluation of linguistic interpretation vs. stability, *Food Control*, **15** (2), 93–97.

SAHAGIAN, M.E. and GOFF, H.D., (1996), Fundamental aspects of the freezing process, in (L.E. Jeremiah, Ed.), *Freezing Effects on Food Quality*, Marcel Dekker, New York, pp. 1–50.

SANTOS-YAP, E.M., (1996), Fish and seafood, in (L.E. Jeremiah, Ed.), *Freezing Effects on Food Quality*, Marcel Dekker, New York, pp. 109–134.

SANZ, P.D., (2005), Freezing and thawing of foods under pressure, in (G.V. Barbosa-Cánovas, M.S. Tapia and M.P. Cano, Eds.), *Novel Food Processing Technologies*, Marcel Dekker, New York, pp. 233–260.

SANZ, P.D. and OTERO, L., (2005), High-pressure freezing, in (D-W. Sun, Ed.), *Emerging Technologies for Food Processing*, Elsevier Academic Press, San Diego, CA, pp. 627–652.

SCHWARTZBERG, H., SINGH, R.P. and SARKAR, A., (2007), Freezing and thawing of foods – computation methods and thermal properties correlation, in (S. Yanniotis and B. Sundén, Eds.), *Heat Transfer in Food Processing*, WIT Press, Southampton, pp. 61–100.

SEBOK, A., CSEPREGI, I. and BAAR, C., (1994), Causes of freeze cracking in fruits and vegetables, in (A. Turner, Ed.), *Food Technology International*, Sterling Publications, London, pp. 66–68.

SELMAN, J.D., (1995), Time–temperature indicators, in (M.L. Rooney, Ed.), *Active Food Packaging*, Blackie Academic and Professional, London, pp. 215–233.

SEQUEIRA-MUNOZ, A., CHEVALIER, D., SIMPSON, B.K., LE BAIL, A. and RAMASWAMY, H.S., (2005), Effect of pressure-shift freezing versus air-blast freezing of carp (*Cyprinus carpio*) fillets: a storage study, *J. Food Biochemistry*, **29** (5), 504–516.

SHAIKH, N.I. and PRABHU, V., (2007a), Mathematical modeling and simulation of cryogenic tunnel freezers, *J. Food Engineering*, **80** (2), 701–710.

SHAIKH, N.I. and PRABHU, V., (2007b), Model predictive controller for cryogenic tunnel freezers, *J. Food Engineering*, **80** (2), 711–718.

SINGH, R.P. and HELDMAN, D.R., (2001), Food freezing, in *Introduction to Food Engineering*, Academic Press, San Diego, CA, pp. 410–446.

SKREDE, G., (1996), Fruits, in (L.E. Jeremiah, Ed.), *Freezing Effects on Food Quality*, Marcel Dekker, New York, pp. 183–246.

SØRENSEN, L.B., (2002), Frozen food legislation, *Bulletin of the International Institute of Refrigeration*, No. 2002 – 4, available at www.iifiir.org/en/doc/1044.pdf.

STEFFE, J.F., (1998), Ice cream, available at www.egr.msu.edu/~steffe/handbook/icecream.html.

SUMMERS, J., (1998), Cryogenics and tunnel vision, in (A. Turner, Ed.), *Food Technology International*, Sterling Publications, London, pp. 73–75.

SUN, D-W. and LI, B., (2003), Microstructural change of potato tissues frozen by ultrasound-assisted immersion freezing, *J. Food Engineering*, **5** (4), 337–345.

SWAIN, M. and JAMES, S., (2005), Thawing and tempering using microwave processing, in (H. Schubert and M. Regier, Eds.), *The Microwave Processing of Foods*, Woodhead Publishing, Cambridge, pp. 174–191.

TOMLINS, R., (1995), Cryogenic freezing and chilling of food, in (A. Turner, Ed.), *Food Technology International*, Sterling Publications, London, pp. 145–149.

TORREGGIANI, D., FORNI, E. and LONGONI, F., (1997), Chemical-physical characteristics of osmodehydrofrozen sweet cherry halves: influence of the osmodehydration methods and sugar syrup composition, in 1st International Congress on Food Ingredients: New Technologies, Fruits and Vegetables, Allione Ricerca Agroalimentare, SpA, pp 101–109.

TORREGGIANI, D., LUCAS, T. and RAOULT-WACK, A., (2000), The pre-treatment of fruits and vegetables, in (C. Kennedy, Ed.), *Managing Frozen Foods*, Woodhead Publishing, Cambridge, pp. 57–80.

VEERKAMP, C.H., (2001), Chilling and freezing of poultry and poultry products, in (D-W. Sun, Ed.), *Advances in Food Refrigeration*, Leatherhead Publishing, LFRA, Leatherhead, pp. 387–402.

WANG, J-H., (2000), A comprehensive evaluation of the effects and mechanisms of antifreeze proteins during low-temperature preservation, *Cryobiology*, **41** (1), 1–9.

WELLER, M. and MILLS, S., (1999), Frost-free cold stores? *Food Processing*, **68** (8), 21–22.

ZARITZKY, N.E., (2000), Factors affecting the stability of frozen foods, in (C.J. Kennedy, Ed.), *Managing Frozen Foods*, Woodhead Publishing, Cambridge, pp. 111–135.

ZHENG, L. and SUN, D-W., (2006), Innovative applications of power ultrasound during food freezing processes – a review, *Trends in Food Science and Technology*, **17** (1), 16–23.

ZHU, S., RAMASWAMY, H.S. and LE BAIL, A., (2005), Ice-crystal formation in gelatin gel during pressure shift versus conventional freezing, *J. Food Engineering*, **66** (1), 69–76.

23

Freeze drying and freeze concentration

Abstract: Freeze drying and freeze concentration reduce the water activity in foods with minimum damage to nutritional and sensory qualities. This chapter looks at the theory of both freeze drying and freeze concentration, the equipment used in each process and the effects they have on foods and micro-organisms.

Key words: freeze drying, freeze concentration, heat and mass transfer, sublimation, recrystalliser.

The advantages of dried and concentrated foods compared with other methods of preservation are described in Chapters 16 and 14 respectively. The heat used to dry foods, or to concentrate liquids by boiling, removes water and therefore preserves the food by a reduction in water activity (Chapter 1, section 1.1.2). However, the heat also causes a loss of sensory characteristics and nutritional qualities. In freeze drying and freeze concentration a similar preservative effect is achieved by reduction in water activity without heating the food, and as a result these processes cause minimum damage to the nutritional and sensory qualities of heat-labile foods and produce a porous, friable structure that quickly and fully rehydrates (>90% moisture reabsorption). However, both operations are slower than conventional dehydration or evaporation (typically a 4–12 hour drying cycle in freeze drying) (Fig. 23.1). Energy costs for freezing are high and in freeze drying, the production of a high vacuum is an additional expense. This, together with the high capital cost of equipment, results in high production costs for freeze-dried and freeze-concentrated foods. Nijhuis (1998) has reviewed the relative costs of freeze drying and radio frequency drying (Chapter 20, section 20.1.3), and Heldman and Hartel (1997) report the cost of freeze drying as twice that of vacuum band drying and nearly five times the cost of spray drying. Drying using microwave energy at ambient temperature under vacuum creates products that have comparable properties to freeze drying, but in a shorter time and therefore at lower costs (Ahrens *et al.* 2001).

Commercially, freeze drying is more important than freeze concentration. It is used to dry heat-labile foods that have delicate aromas or textures, and have a superior quality and high value. Examples include coffee, mushrooms, herbs and spices, strawberries and raspberries, fruit juices, meats, seafoods or vegetables, and complete meals for hikers and campers, military rations or space flights (e.g. beef stew, beef Stroganoff with noodles, chicken breasts and mashed potatoes with herbs). Newer products include freeze-dried

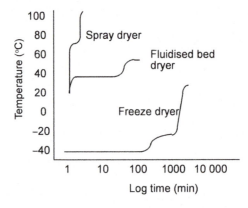

Fig. 23.1   Time–temperature profiles for foods dried by different methods (adapted from Menyhart 1995).

droplets of cream, yoghurt and crème fraiche for use in breakfast cereals, confectionery and as a topping for desserts (Anon 2001). Freeze drying is also used to prepare active enzymes (e.g. for cheese making) and microbial cultures for long-term storage prior to inoculum generation. Freeze concentration is not widely used commercially but has found some applications, including pre-concentrating coffee extract prior to freeze drying, increasing the alcohol content of wine and preparation of fruit juices, vinegar and pickle liquors.

23.1   Freeze drying

Freeze drying differs from other methods of dehydration because water is first frozen and then converted directly to vapour by sublimation, rather than being removed from the food surface by evaporation (Chapter 16 section 16.1.1). Table 23.1 summarises the main differences between freeze drying and conventional hot air drying.

Table 23.1   Differences between conventional drying and freeze drying

Conventional drying	Freeze drying
Successful for easily dried foods (vegetables and grains)	Successful for most foods but limited to those that are difficult to dry by other methods
Meat generally unsatisfactory	Successful with cooked and raw meats
Temperature range 37–93 °C	Temperatures below freezing point
Atmospheric pressures	Reduced pressures (27–133 Pa)
Evaporation of water from surface of food	Sublimation of water from ice front
Movement of solutes and sometimes case hardening	Minimal solute movement
Stresses in solid foods cause structural damage and shrinkage	Minimal structural changes or shrinkage
Slow, incomplete rehydration	Rapid complete rehydration
Solid or porous dried particles often having a higher density than the original food	Porous dried particles having a lower density than original food
Odour and flavour frequently abnormal	Odour and flavour usually normal
Colour frequently darker	Colour usually normal
Reduced nutritional value	Nutrients largely retained
Costs generally low	Costs generally high, up to five times those of conventional drying

23.1.1 Theory

The first stage in freeze drying solid foods is to rapidly freeze small pieces of food to produce small ice crystals, which reduce damage to the cell structure of the food (see Chapter 22, section 22.1.1). In the second stage, the pressure surrounding a food is reduced below 610 Pa, and heat is applied slowly to the frozen food to cause the ice to sublime directly to vapour without melting (Fig. 23.2). The latent of sublimation is either conducted through the food to the sublimation front or produced in the food by microwave or radio frequency heaters. As drying proceeds the sublimation front moves into the frozen food, leaving partly dried food behind it. A water vapour pressure gradient is established because the pressure in the freeze dryer is lower than the vapour pressure at the surface of the ice. Water vapour therefore moves through the dried food into the drying chamber and it is removed by condensing it on refrigeration coils.

In liquid foods that do not have a cellular structure, slow freezing is used to form a lattice of large ice crystals. Channels formed by the sublimed ice allow more rapid removal of vapour than from solid foods (Fig. 23.3).

At commercial freeze drying temperatures, some of the water in foods remains unfrozen in a highly viscous, glassy state (see Fig. 22.3 in Chapter 22 and section 22.1.2). Drying follows the freezing point depression line in Fig. 22.3 to the glass transition line, which depending on the type of food, reduces the moisture content to ≈15% moisture (w/w basis). The remaining unfrozen water is then removed in the third stage by evaporative drying (desorption) to ≈2% moisture (w/w basis). Desorption is achieved by raising the temperature in the dryer to near ambient temperature while retaining the low pressure.

In some liquid foods (e.g. fruit juices and concentrated coffee extract), the formation of a glassy vitreous state on freezing causes difficulties in vapour transfer. Therefore the

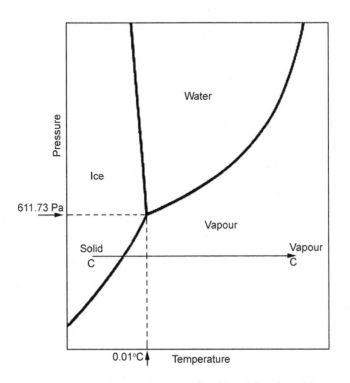

Fig. 23.2 Phase diagram for water showing sublimation of ice.

Fig. 23.3   Porous structure of freeze-dried food.

liquid is either frozen as a foam (vacuum puff freeze drying), or the juice is dried together with the pulp. Both methods produce channels through the food for the vapour to escape. In a third method, frozen juice is ground to produce granules, which both dry faster and allow better control over the particle size of the dried food.

The rate of drying depends mostly on the resistance of the food to heat transfer and to a lesser extent on the resistances to vapour flow (mass transfer) from the sublimation front (Fig. 23.4).

Rate of heat transfer
There are three methods of transferring heat to the sublimation front:

1 *Heat transfer through the frozen layer* (Fig. 23.4a). The rate of heat transfer is controlled by the thickness and thermal conductivity of the ice layer. As drying proceeds, the thickness of the ice is reduced and the rate of heat transfer increases. The heater temperature is limited to avoid melting the ice.
2 *Heat transfer through the dried layer* (Fig. 23.4b). The rate of heat transfer to the sublimation front depends on the thickness and area of the food, the thermal

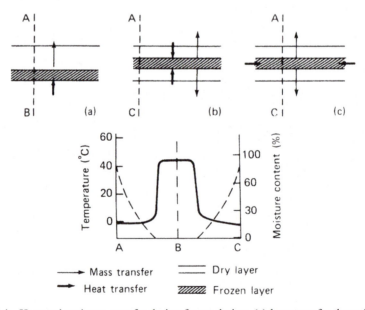

Fig. 23.4   Heat and moisture transfer during freeze drying: (a) heat transfer through the frozen layer; (b) heat transfer from heated surfaces or radiant heaters through the dry layer; (c) heat generated in the ice by dielectric heaters. The graph shows changes in temperature (———) and moisture content (-----) along the line A → B → C through each sample.

conductivity of the dry layer and the temperature difference between the surface of the food and ice front. The dried layer of food has a very low thermal conductivity, similar to insulation materials (Chapter 10, Table 10.2) and therefore offers a high resistance to heat flow. As drying proceeds, this layer becomes thicker and the resistance increases. As in other unit operations, a reduction in the size or thickness of food and an increase in the temperature difference increase the rate of heat transfer (Chapter 10, section 10.1.2). However, in freeze drying, the surface temperature is limited to 40–65 °C, to avoid denaturation of proteins and other chemical changes that would reduce the quality of the food.

3 *Dielectric heating* (Fig. 23.4c). Heat is generated directly at the ice front and the rate of heat transfer is not influenced by the thickness of the dry layer or the thermal conductivity of ice or dry food. However, dielectric heating is less easily controlled, and because water has a higher loss factor than ice (Chapter 20, section 20.1.1), if any ice is melted there is a risk of localised runaway overheating.

Rate of mass transfer

When heat reaches the sublimation front, it raises the temperature and the water vapour pressure of the ice. Vapour then moves through the dried food to a region of low vapour pressure in the drying chamber. One gram of ice forms $2\,m^3$ of vapour at 67 Pa and, in commercial freeze drying, it is therefore necessary to remove several hundred cubic metres of vapour per second through the pores in the dry food. The factors that control the water vapour pressure gradient are:

- the pressure in the drying chamber and the temperature of the vapour condenser, both of which should be as low as economically possible; and
- the temperature of ice at the sublimation front, which should be as high as possible, without melting.

In practice, the lowest economical chamber pressure is ≈ 13 Pa and the lowest condenser temperature is ≈ -35 °C. Theoretically the temperature of the ice could be raised to just below the freezing point. However, above a certain critical 'collapse temperature' (Table 23.2) the concentrated solutes in the food are sufficiently mobile to flow under the forces operating within the food structure. When this occurs, they flow into spaces left by the sublimed ice and there is an irreversible collapse of the food structure. This restricts the rate of vapour transfer and effectively ends the drying operation. The food should

Table 23.2 Collapse temperatures of selected foods in freeze drying

Food	Collapse temperature (°C)
Apple juice (22% moisture)	−41.5
Beef	−12
Cheddar cheese	−24
Coffee extract (25% moisture)	−20
Fish	−6 to −12
Grape juice (16% moisture)	−46
Ice cream	−31 to −33
Potato	−12
Sweetcorn	−8 to −15
Tomato	−41

From Fennema (1996) and Karathanos *et al.* (1996)

therefore stay below the collapse temperature during the sublimation stage of drying and below the glass transition temperature during desorption drying.

When heat is transferred through the dry layer, the relationship between the pressure in the dryer and the pressure at the ice surface is:

$$P_i = P_s + \frac{k_d}{b\lambda_s}(\theta_s - \theta_i)$$

[23.1]

where P_i (Pa) = partial pressure of water at the sublimation front, P_s (Pa) = partial pressure of water at the surface, k_d (W m^{-1} K^{-1}) = thermal conductivity of the dry layer, b (kg s^{-1} m^{-1}) = permeability of the dry layer, λ_s (J kg^{-1}) = latent heat sublimation, θ_s (°C) = surface temperature and θ_i (°C) = temperature at the sublimation front.

The factors that control the drying time are shown in Equation 23.2.

$$t_d = \frac{x^2 \rho (M_1 - M_2)\lambda_s}{8 k_d(\theta_s - \theta_i)}$$

[23.2]

where t_d (s) = drying time, x (m) = thickness of the food, ρ (kg m^{-3}) = bulk density of the dry food, M_1 (dry weight basis) = initial moisture content and M_2 (dry weight basis) = final moisture content in the dry layer. Note that drying time is proportional to the square of the food thickness: doubling the thickness will therefore increase the drying time by a factor of four.

The equations described above are simplified examples of more complex formulae and an example of their use is shown in sample problem 23.1. Toledo (1999) gives derivations of heat and mass transfer equations and additional worked examples of the calculation of drying times. George and Datta (2002) describe heat and mass transfer models for freeze drying.

Sample problem 23.1
Food with an initial moisture content of 400% (dry-weight basis) is poured into 0.5 cm layers in a tray placed in a freeze dryer operating at 40 Pa. It is to be dried to 8% moisture (dry-weight basis) at a maximum surface temperature of 55 °C. Assuming that the pressure at the ice front remains constant at 78 Pa, calculate (a) the drying time and (b) the drying time if the layer of food is increased to 0.9 cm and dried under similar conditions. (Additional data: the dried food has a thermal conductivity of 0.03 W m^{-1} K^{-1}, a bulk density of 470 kg m^{-3}, a permeability of 2.4 × 10^{-8} kg s^{-1}, and the latent heat of sublimation is 2.95 × 10^3 kJ kg^{-1}.)

Solution to sample problem 23.1
Part (a): From Equation 23.1,

$$78 = 40 + \frac{0.03}{2.4 \times 10^{-8} \times 2.95 \times 10^6}(55 - \theta_i)$$

$$= 40 + 0.42(55 - \theta_i)$$

Therefore,

$$\theta_i = -35.7\,°C$$

From Equation 23.2,

$$t_d = \frac{(0.005)^2 470(4 - 0.08)2.95 \times 10^6}{8 \times 0.03[55 - (-35.7)]}$$

$$= 6242 \text{ s}$$

$$\approx 1.7 \text{ h}$$

Part (b): From Equation 23.2,

$$t_d = \frac{(0.009)^2 470(4 - 0.08)2.95 \times 10^6}{8 \times 0.03[55 - (-35.7)]}$$

$$= 20224 \text{ s}$$

$$\approx 5.6 \text{ h}$$

Therefore increasing the thickness of the layer of food from 0.5 to 0.9 cm results in an increase of 3.9 h in the drying time.

23.1.2   Equipment

Freeze dryers consist of a strongly constructed vacuum chamber which contains trays to hold the food during drying, heated shelves or heaters to supply latent heat of sublimation, vacuum pumps and a refrigeration unit. Refrigeration coils inside the chamber are used to condense the vapours directly to ice (i.e. reverse sublimation). They are fitted with automatic defrosting devices to keep the coils free of ice to maximise vapour condensation. This is necessary because the major part of the energy input is used for refrigeration of the condensers, and the economics of freeze drying are therefore determined by the efficiency of the condenser (Equation 23.3):

$$\text{Efficiency} = \frac{\text{temperature of sublimation}}{\text{refrigerant temperature in condenser}} \qquad \boxed{23.3}$$

Vacuum pumps remove non-condensable vapours.

Freeze dryers can be batch, semi-continuous or continuous in operation. Batch or semi-continuous dryers are cylindrical tunnels 1.5–2.5 m in diameter that have air locks through which trolleys containing trays of food enter. In batch dryers the food enters and leaves through the same door, whereas in semi-continuous dryers both ends have airlocks. Fixed heater plates are located in the tunnel and the trays pass between them (Brennan 2006). The dryers may have automatic loading and unloading of product trays, with total tray areas up to 140 m^2 to give batch sizes of up to 2000 kg (Anon undated). The condensers can remove water vapour at rates of up to 300 kg h^{-1}. The product is sealed into the drying chamber, the heater temperature is maintained at 80–120 °C for initial drying and then gradually reduced over a drying period of 4–12 hours. The precise conditions in the drying cycle are determined for individual foods, but the surface temperature of the food does not exceed 60 °C.

In continuous freeze dryers, stacks of trays, having an area of up to 400 m^2, are moved on guide rails through heating zones in a long vacuum chamber (Fig. 23.5). Heater temperatures and product residence times in each zone are pre-programmed for individual foods. The drying cycle is automatically controlled by programmable logic controllers (PLCs) (Chapter 27, section 27.2.2) and the combination of temperature and

Fig. 23.5   Continuous freeze dryer (courtesy of Niro A/S (Copyright)) (Anon undated).

pressure controllers with a cycle sequencer is used to monitor and control process time, temperature and pressure in the chamber, and the temperature at the product surface. The PLC also controls safety interlocks and alarms for equipment malfunctions and product-related deviations from set-points. It can document the conditions in a drying cycle and provide synoptic performance reports that can be imported into management or quality assurance spreadsheets and databases (Anon 2004). Brennan (2006) describes other designs of continuous dryers that are suitable for granular foods, including vacuum chambers that contain multiple conveyors or vibrating decks to move the food and re-pile it as it dries (similar in concept to multi-pass hot air dryers (Chapter 16, section 16.2.1). The food is heated by radiant heaters located above the belts or vibrating decks. He also describes a prototype vacuum spray freeze dryer.

Different types of dryer are characterised by the method used to supply heat to the surface of the food. Conduction and radiation types are used commercially (convection heating is not important in the partial vacuum of the freeze dryer). Microwave freeze drying is also used but is more difficult to control.

Contact (or conduction) freeze dryers
Food is placed on ribbed trays that rest on heated shelves (Fig. 23.6a). This type of equipment dries more slowly than other designs because heat is transferred by conduction to only one side of the food. There is uneven contact between the frozen food and the heated surface, which also reduces the rate of heat transfer. There is a pressure drop through the food that results in differences between the drying rates of the top and bottom layers. The vapour velocity is of the order of $3\,\mathrm{m\,s^{-1}}$ and fine particles of product may be entrained in the vapour and lost. However, contact freeze dryers have higher capacity than other types.

Accelerated freeze dryers
In this equipment, food is held between two layers of expanded metal mesh and subjected to a slight pressure on both sides (Fig. 23.6b). Heating is by conduction but heat is transferred more rapidly into food by the mesh than by solid plates, and vapour escapes more easily from the surface of the food. Both mechanisms produce shorter drying times than contact methods.

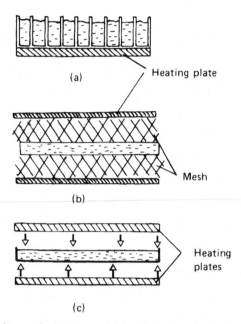

Fig. 23.6 Freeze drying methods: (a) conduction through ribbed tray; (b) expanded mesh for accelerated freeze drying; (c) radiant heating of flat trays (from Rolfgaard 1987).

Radiant freeze dryers
Infrared radiation from radiant heaters (Chapter 20, section 20.3) is used to heat shallow layers of food on flat trays (Fig. 23.6c). Heating is more uniform than in conduction types. The velocity of vapour movement is approximately $1 \, m \, s^{-1}$ and there is little risk of product carryover. Close contact between the food and heaters is not necessary and flat trays are used, which are cheaper and easier to clean than the ribbed trays used in conduction heating.

23.1.3   Effect on foods and micro-organisms
Freeze-dried foods have a very high retention of sensory and nutritional qualities and a shelf-life >12 months when correctly packaged. There are only minor changes to proteins, starches or other carbohydrates. Volatile aroma compounds are not entrained in the water vapour produced by sublimation and are trapped in the food matrix. As a result, aroma retention of 80–100% is possible, giving rehydrated foods an excellent flavour.

The texture of freeze-dried foods is well maintained; there is little shrinkage and no case hardening (Chapter 16, section 16.5.1). The open porous structure allows rapid and full rehydration, but it also makes the food fragile and hygroscopic. It therefore requires protection from moisture pick-up and mechanical damage. The porous structure may also permit oxidative deterioration of lipids, vitamins and pigments to cause a loss of quality. Food is therefore packaged in cartons for mechanical protection and in oxygen- and moisture-barrier films (Chapter 25, section 25.2), and it may be surrounded by an inert gas. These measures increase the cost of packaging freeze-dried foods compared with other dried products.

There are negligible losses of most vitamins, but ascorbic acid losses may range from 8 to 30% and vitamin A losses from 0 to 24% in green vegetables. However, losses of nutrients due to preparation procedures, especially size reduction (Chapter 3, section

3.1.4) and blanching of vegetables (Chapter 11, section 11.3), may be higher than those caused by freeze drying. There have been numerous comparative studies of changes to foods as a result of different drying methods. For example, Lin *et al.* (1998) compared freeze-dried carrot slices to those produced by microwave vacuum drying and hot-air drying. The freeze-dried product had better rehydration, appearance and nutrient retention. Both the microwave vacuum-dried and freeze-dried slices had similar colour, texture, flavour and overall preference. Rehydration rates and α-carotene and vitamin C contents were higher than in carrots prepared by air drying, the density was lower and they had a softer texture. Other studies are described in Chapter 16 (section 16.5).

As in other methods of drying, freeze drying does not necessarily destroy micro-organisms, and unblanched foods may contain pathogens or spoilage micro-organisms that can regrow after rehydration of the product. Details are given in Chapter 16 (section 16.5.3).

23.2   Freeze concentration

Freeze concentration of liquid foods involves the fractional crystallisation of water to ice by freezing and subsequent removal of the ice using wash columns or mechanical separation techniques (Chapter 5). The low temperatures used in the process cause a high retention of volatile aroma compounds and produce little change in nutritional value. However, the process has high refrigeration costs, high capital costs for equipment required to handle the frozen slurries, high operating costs and low production rates compared with concentration by boiling (Chapter 14). The degree of concentration achieved is higher than in membrane processes (Chapter 5, section 5.5), but lower than concentration by boiling (evaporation – Chapter 14, section 14.1). Freeze concentration is used to produce high-value coffee or fruit juice concentrates, and high-quality extracts of fish, meat, vegetables and herbs. It is also used to pre-concentrate these products before conventional drying, which is cheaper than freeze drying and better retains the product quality compared to other methods of pre-concentration (Anon 2008a).

23.2.1   Theory
The factors that control the rate of nucleation and ice crystal growth, including the effect of solute concentration and supercooling, are described by Tähti (2004) and in Chapter 22 (section 22.1.1). In freeze concentration it is desirable for ice crystals to grow as large as economically possible to reduce the amount of concentrated liquor entrained with the crystals. This is achieved in a 'recrystalliser' (section 23.2.2) by slowly stirring a thick slurry of ice crystals and allowing the large crystals to grow at the expense of smaller ones. Calculations of the degree of solute concentration obtained by a given reduction in the freezing point of a solution are used to produce 'freezing point curves' for different products (Fig. 23.7).

The efficiency of crystal separation from the concentrated liquor is determined by the degree of clumping of the crystals, and amount of liquor entrained. It is calculated using:

$$\eta_{sep} = x_{mix} \frac{x_1 - x_i}{x_1 - x_j} \qquad \boxed{23.4}$$

where η_{sep} (%) = efficiency of separation, x_{mix} weight fraction of ice in the frozen mixture before separation, x_1 = weight fraction of solids in the liquor after freezing, $x_i =$

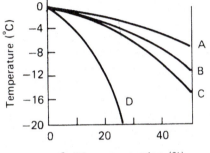

Fig. 23.7   Freezing point curves: A, coffee extract; B, apple juice; C, blackcurrant juice; D, wine (after Kessler 1986).

weight fraction of solids in ice after separation, and x_j = weight fraction of juice before freezing.

23.2.2   Equipment

The basic components of a freeze concentration unit are shown in Fig. 23.8. These are as follows:

- A freezing system (e.g. a scraped surface heat exchanger (Chapters 13 and 14)) to produce ice crystals in the liquid food. Habib and Farid (2006) report that a fluidised bed heat exchanger has lower costs than a scraped surface heat exchanger.
- A mixing vessel (the recrystalliser) to allow the ice crystals to grow. The crystal slurry is recirculated through the heat exchanger to maintain the low temperature.
- A separator to remove the crystals from the concentrated solution, wash them to remove remaining concentrate, melt the crystals and discharge pure water.
- The concentrate is recirculated through the recrystalliser and fresh feed is added to replace the water removed as ice, thus increasing the solids' concentration until the required level is reached.

Wash columns operate by feeding the ice-concentrate slurry from the crystalliser into a vertical enclosed cylinder. The crystal slurry has a solids' concentration of 30–40%. In the wash column, the crystals are compressed to increase the solids to 60–80% and the

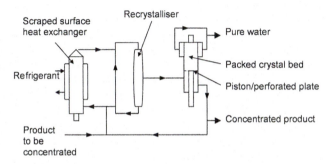

Fig. 23.8   Freeze concentration plant (courtesy of Niro Processing Technology Inc.) (adapted from Anon 2008b).

concentrate is discharged through a filter and recirculated to the recrystalliser. The packed bed of crystals is then washed using melted ice that is forced through the packed bed by increasing the pressure on the column. Detailed descriptions of the process are given by Tähti (2004) and Anon (2008c). Less commonly, separation is achieved by centrifugation, filtration or filter pressing (Chapter 5). Separation efficiencies are 50% for centrifuging, 71% for vacuum filtration, 89-95% for filter pressing and 99.5% for wash columns.

Concentration takes place in either single-stage or, more commonly, multi-stage equipment. Multi-stage concentrators have lower energy consumption and higher production rates and can produce concentrates having up to 45% solids (Kessler 1986).

23.2.3 Effect on foods and micro-organisms

Freeze concentration comes close to the ideal of selectively removing water from foods without changing other components. There are therefore negligible effects on pigments, flavour compounds and nutrients in the food. However, the mild processing conditions also have little or no effect on enzymes and micro-organisms, although the reduced water activity in concentrated products has some effect to inhibit microbial growth. Further details of the effects of freezing and concentration on microbial activity are given in Chapter 22 (section 22.4) and Chapter 14 (section 14.1.4) respectively.

References

AHRENS, G., KRISZIO, H. and LANGER, G., (2001), in (M. Willert-Porada, Ed.), Microwave vacuum drying in the food processing industry, *Advances in Microwave and Radio Frequency Processing*, Report from the 8th International Conference on Microwave and High Frequency Heating, Bayreuth, Germany, 3–7 September, pp. 426–435.

ANON, (undated), Atlas Freeze Drying, company information from Niro Inc., available at www.niroinc.com/html/chemical/cpdfs/freeze_drying.pdf.

ANON, (2001), Freeze dried cream pearls – a new ingredient, *Food Processing*, June, XIV–XV.

ANON, (2004), Freeze dryer control, company information from American Lyophilizer Inc., available at www.freezedrying.com/freeze_dryer_control.htm.

ANON, (2008a), Freeze concentration technology, company information from Niro Process Technology BV, available at www.niro-pt.nl/ndk_website/npt/cmsdoc.nsf/WebDoc/ndkw73ej4s and links to product pages.

ANON, (2008b), ICECON Innovative freeze concentration, company information from Niro Process Technology BV, available at www.niro-pt.nl/ndk_website/npt/cmsdoc.nsf/WebDoc/ndkw73fagh.

ANON, (2008c), Washcolumn separation technology, company information from Niro Process Technology BV, available at www.niro-pt.nl/ndk_website/npt/cmsdoc.nsf/WebDoc/misy739blp.

BRENNAN, J.G., (2006), Evaporation and dehydration, in (J.G. Brennan, Ed.), *Food Processing Handbook*, Wiley-VCH, Weinheim, Germany, pp. 71–124.

FENNEMA, O., (1996), Water and ice, in (O. Fennema, Ed.), *Food Chemistry*, 3rd edn, Marcel Dekker, New York, pp. 18–94.

GEORGE, J.P. and DATTA, A.K., (2002), Development and validation of heat and mass transfer models for freeze-drying of vegetable slices, *J. Food Engineering*, **52** (1), 89–93.

HABIB, B. and FARID, M., (2006), Heat transfer and operating conditions for freeze concentration in a liquid–solid fluidized bed heat exchanger, *Chemical Engineering and Processing*, **45** (8), 698–710.

HELDMAN, D.R. and HARTEL, R.W., (1997), Dehydration, in *Principles of Food Processing*, Chapman and Hall, New York, pp. 177–218.

KARATHANOS, V.T., ANGLEA, S.A. and KAREL, M., (1996), Structural collapse of plant materials during freeze-drying, *J. Thermal Analysis and Calorimetry*, **47** (5), 1451–1461.

KESSLER, H.G., (1986), Energy aspects of food preconcentration, in (D. MacCarthy, Ed.), *Concentration and Drying of Foods*, Elsevier Applied Science, Barking, pp. 147–163.

LIN, T.M., DURANCE, T.D. and SCAMAN, C.H., (1998), Characterization of vacuum microwave, air and freeze dried carrot slices, *Food Research International*, **31** (2), 111–117.

MELLOR, J.D., (1978), *Fundamentals of Freeze Drying*, Academic Press, London, pp. 257–288.

MENYHART, L., (1995), Lyophilization: freeze-drying a downstream process, available at www.rpi.edu/dept/chem-eng/Biotech-Environ/LYO/Fig.2.html.

NIJHUIS, H.H., (1998), Approaches to improving the quality of dried fruit and vegetables, *Trends in Food Science and Technology*, **9**, 13–20.

ROLFGAARD, J., (1987), Freeze drying: processing, costs and applications, in (A. Turner, Ed.), *Food Technology International Europe*, Sterling Publications International, London, pp. 47–49.

TÄIITI, T., (2004), Suspension melt crystallization in tubular and scraped surface heat exchangers, Dissertation zur Erlangung des akademischen Grades Doktor-Ingenieur vorgelegt an der Mathematisch-Naturwissenschaftlich-Technischen Fakultät der Martin-Luther-Universität Halle-Wittenberg available at http://sundoc.bibliothek.uni-halle.de/diss-online/04/04H181/t3.pdf.

TOLEDO, R.T., (1999), Dehydration, in *Fundamentals of Food Process Engineering*, 2nd edn, Aspen Publishers, Gaithersburg, MD, pp. 456–506.

Part V

Post-processing operations

The unit operations described in preceding chapters are used to prepare foods for processing (Part I) or process them to extend their shelf-life and/or alter their sensory characteristics (Parts II to IV). This part describes post-processing operations, including coating food products with chocolate, batter or flavourings (Chapter 24) – which in some cases is part of the process; packaging foods (Chapter 25 and 26) to extend their shelf-life and assist in promotion and marketing; and other ancillary operations, including process control and materials handling within a factory or warehouse, storage and distribution technologies and waste management and disposal (Chapter 27), which are all critical to the success of commercial food processing.

24

Coating

Abstract: Coating foods with batter, breadcrumbs, chocolate or compound coating materials is intended to improve their appearance and eating quality, and to increase their variety. In most cases coatings are not intended to preserve the food. This chapter describes different types of coatings and the equipment used to apply them. The chapter concludes with a description of microencapsulation of ingredients and edible barrier coatings.

Key words: coating, batters, powders and breadcrumbs, chocolate and compound coatings, microencapsulation, pan coating, edible barrier coatings.

Coatings of batter or breadcrumbs are applied to fish, meats or vegetables; chocolate or 'compound' coatings are applied to biscuits, cakes or confectionery; and coatings of salt, sugar, flavourings or colourants are applied to snackfoods, baked goods and confectionery. In each case, the aim is to improve the appearance and eating quality of foods, and to increase their variety. In most cases coatings are not intended to affect the shelf-life or preserve the food, but some (e.g. icing on cakes) may provide a barrier to the movement of moisture and oxygen, or protect the food against mechanical damage. Newer developments in coating particles of food by microencapsulation and edible barrier coatings are described in section 24.3. Coatings are also applied to foods to modify the texture, enhance flavours, improve their convenience and add value to basic products. They change the nutritional value of foods by means of the ingredients contained in the coating material. There are three main methods of coating foods. The selection of an appropriate method depends on the type of coating material to be used and the intended effect of the coating. These are:

1 coating with chocolate, compound coatings or glazes (for sweet foods such as confectionery, ice cream and baked goods), and batters or breadcrumbs (for savoury foods);
2 dusting with spices, flour, sugar, flavourings, colourings, or salt;
3 pan coating with sugar or sugarless coatings.

24.1   Coating materials

24.1.1   Chocolate and compound coatings

The two main types of sweet coatings are chocolate and compound coatings. Different combinations of sugar, cocoa butter, cocoa solids, vanilla and emulsifying agent are used to make the various chocolate coatings, with milk or milk powder added to make milk chocolate and the cocoa solids omitted when making white chocolate. Historically, soya lecithin has been used as an emulsifying agent (Chapter 3, section 3.2.2), but the introduction of genetically modified soya (Chapter 6, section 6.1) has led some manufacturers to use polyglycerol polyricinoleate (PGPR). This is an emulsifier derived from castor oil that also allows producers to reduce the amount of cocoa butter while maintaining the same mouthfeel. It is also used in compound coatings and acts to reduce the viscosity so that coatings can flow more easily when melted.

After blending chocolate ingredients, they are 'conched'. A conche is type of mill, filled with metal beads that grind the ingredients while the chocolate mass is kept liquid by frictional heat. The conching process reduces the size of cocoa and sugar particles so that they are smaller than the tongue can detect, to produce the smooth mouthfeel. High-quality chocolate is conched for $\approx$72 h, but lower grades used for coatings are conched for $\approx$4–6 h. After conching, the chocolate mass is stored in heated tanks at 45–50 °C before 'tempering'. This is the controlled crystallisation of cocoa fat to produce consistently small cocoa butter crystals. Cocoa butter is a polymorphic fat (it crystallises into six different forms or 'polymorphs') (Table 24.1). Uncontrolled crystallisation produces crystals of varying size, some of which are large enough to be seen with the naked eye. This causes the surface of the chocolate to appear mottled without a sheen, produces a gritty mouthfeel, and causes the chocolate to crumble rather than snap when broken. Tempering produces a uniform sheen, a smooth mouthfeel and a crisp bite. This is achieved by controlled heating that causes most of the crystals to form as type V, which are the most stable, so that the appearance and texture do not degrade during storage.

Tempering machines are either tubular or plate heat exchangers (see Chapters 12 and 14) that have accurate temperature control of the heating water. Dark chocolate is first heated to 46–49 °C to melt all six forms of the crystals. Then it is cooled to 28–29 °C,

Table 24.1   Polymorphs of cocoa fat

Polymorph		Melting temperature (°C)	Notes/efffect on chocolate
I	β'_2	17.3	Produces soft, crumbly product that melts too easily.
II	α	23.3	Very unstable, formed by rapid cooling of liquid fat. Produces soft, crumbly product that melts too easily.
III	Mixed	25.5	Produces firm product, poor snap, melts too easily.
IV	β_1	27.3	Unstable and most likely to be present, will slowly change back to form I. Firm product, good snap, melts too easily.
V	β_2	33.8	The stable form, correct tempering should maximise this form. Glossy, firm product, best snap.
VI	β'_1	36.3	The transformation of form V, produces hard product, takes weeks to form. After 4 months at room temperature, leads to white, dusty bloom on chocolate.

Adapted from Elert (2008), Talbot (1999) and Afoakwa *et al.* (2007)

which causes crystal types IV and V to form. The chocolate is stirred to create large numbers of crystal seeds that act as nuclei for small crystals. The chocolate is then heated to 31–32 °C to remove any type IV crystals, leaving only the most stable type V, and it is held at that temperature during production. Milk chocolate is tempered at temperatures ≈2 °C lower than those used for dark chocolate. Subsequent cooling is controlled so that it is not too rapid, to retain only stable fat crystals and avoid the development of a white 'bloom' on the surface of the product. If higher amounts of cocoa butter are used to make chocolate thinner for coating, there is a greater risk of bloom formation and small amounts of emulsifying agent are added to reduce this risk. Further details of chocolate tempering are given by Fryer and Pinschower (2000), Talbot (1999), Loisel et al. (1998) and Beckett (1994). Letourneau et al. (2005) describe a high-pressure tempering process using supercritical CO_2 (Chapter 5, section 5.2.3) to produce type V crystals in cocoa butter.

Owing to the relatively high price of cocoa butter, a number of fats have been developed which are termed 'cocoa butter equivalents' (CBEs) and have similar properties to cocoa butter. They are permitted at levels of up to 5% in many countries. Compound coatings are made from other fats and cannot legally be termed chocolate. These do not require tempering as the fats are not polymorphic. The main ingredients in a compound coating are fat, sugar, corn syrup, flavourings, fat-soluble colourings and emulsifiers, which are mixed in different formulations to achieve the desired properties. Corn syrup and starch are used to reduce the sweetness and cost of coatings. The particle size of the starch has an important effect on the texture and is closely controlled. The thickness of a coating is controlled by the fat content (more fat produces a thinner coating), and the type and amount of emulsifier. The ratios of sugar, starch and fat determine the required flow characteristics for application of the coating material and the desired mouthfeel and taste in the final product. An example of the use of compound coatings is on cakes, such as swiss roll, where the coating is more flexible than chocolate and does not chip off.

24.1.2   Batters, powders and breadcrumbs

Batters are a suspension of flour in water to which various amounts of sugar, salt, thickening agents, flavourings or colourings are added to achieve the required characteristics. They are applied to a wide variety of savoury foods (e.g. fish, poultry and potato products), which may then be pre-fried and chilled or frozen or form part of ready meals (Chapters 21 and 22). The batter is an important contributor to oil pick-up during frying and there are nutritional concerns over the amount and type of fat consumed in some industrialised societies (Chapter 19, section 19.4.1). Fiszman and Salvador (2003) have reviewed research on the formulation of batter mixes to reduce the amount of oil absorbed by these products. Xue and Ngadi (2006) report that the rheological properties of batters affect the quality of batter-coated products and these differ according to the types of flour that are used and their ratios in the batter formulation. For example, the addition of maize flour altered the viscoelastic properties of wheat- and rice-based batters, and salt significantly lowered their viscosity. A single layer of viscous batter (termed 'Tempura') is used for products that are not subsequently breaded. A thinner, adhesive batter is applied to products prior to coating with breadcrumbs.

Examples of powder and crumb coatings are dry mixtures of spices, salt and flavourings applied to savoury foods, or sugar powder on confectionery, biscuits or cakes. Many of the flavour coatings, particularly barbecue dusts, are hygroscopic and require careful storage and handling in the dusting machine to prevent agglomeration and consequent depositing of large granules of dust onto the product. Different types of crumb are available for breading fish, meat or vegetables, including wholemeal wheat or oats, sesame and combinations of wheat, barley and rye. Each are baked and milled to form a crumb that has a known range of particle sizes, and they are flavoured or coloured if required. Maize crumb may be mixed with potato flakes to give a two-tone effect when applied to vegetables. All types of crumb are fragile and require delicate handling. There has been considerable product development in crumb coating materials, including coatings based on Asian or Oriental foods for Western markets that use authentic regional flavourings, herbs and spices to create new products. For example, a specially formulated satay coating is sprinkled onto pieces of meat or poultry, and when grilled produces a hot, spicy, peanut-flavoured sauce. A breadcrumb coating containing coconut and tropical fruit flavours from the Kerala region of India is used on turkey steaks (Anon 2005a).

24.2   Equipment

24.2.1   Enrobers
There are two types of enrober: in the 'submerger' type, food passes through batter on a stainless steel wire conveyor, held below the surface by a second mesh belt. In the second type, foods pass beneath a single or double curtain of hot liquid coating (Fig. 24.1). The coating is applied:

- by passing it through a slit in the base of a reservoir tank;
- over the edge of the tank (spillway enrobers); or
- by coating rollers.

A pan beneath the conveyor collects the excess coating and a pump recirculates it through a heater, back to the enrobing curtain. Excess coating is removed by air knives, shakers, 'licking rolls' and 'anti-tailer' rollers to give a clean edge to the product. Discs, rollers or wires may be used to decorate the surface of the coating. The thickness of the coating is determined by:

- the temperature of both the food and the coating;
- the viscosity of the coating;
- the speed of the air in air blowers; and
- the rate of cooling.

When enrobing products in chocolate, a separate first stage is termed 'pre-bottoming', in which the centres (e.g. peanuts) are passed on a wire belt through the upper surface of tempered chocolate. The pre-bottomed centre then passes over a cooling plate to partially set the chocolate before passing through the enrober curtain. A more detailed description of enrobers is given by Nelson (1994). The type and composition of centres can have a significant effect on the shelf-life of enrobed confectionery. Nut centres, for example, should be sealed to prevent nut oil from seeping into the casing and causing bloom in the chocolate. Ghosh *et al.* (2002) have reviewed the factors that control moisture and fat migration through chocolate coatings. After enrobing, the coating is cooled by recirculated air in a cooling tunnel to prevent fat crystals from re-melting, but not too rapidly to cause overcooling that would produce surface bloom in the chocolate. Products are then held to allow fat crystallisation to continue (e.g. at 22 °C for 48 h).

(a)

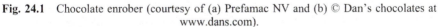

(b)

Fig. 24.1   Chocolate enrober (courtesy of (a) Prefamac NV and (b) © Dan's chocolates at www.dans.com).

Batters are applied either by passing the product on submerged mesh conveyors through a bath of batter, or by passing food pieces through one or more curtains of batter in similar equipment to chocolate enrobers. In both types of equipment, the products then pass beneath air knives to remove excess batter and to control the thickness of the coating.

24.2.2   Dusting or breading

Dusting or breading equipment consists of a hopper fitted with a mesh base, located over a conveyor. The mesh screen is changeable for different types of crumb, seasoning or flavourings. In breading, foods that are coated with a thin batter (section 24.1) pass on a stainless steel wire belt through a bed of breadcrumbs to coat the base and then through a curtain of crumb to coat the upper surface. Excess material is removed by air knives, collected and conveyed back to the hopper by an auger or an elevator (Chapter 27, section 27.1.2). Most coatings and dusts can be readily recirculated, but care is needed when handling crumbs to avoid mechanical damage that would cause changes to the average particle size. Breaded foods are gently pressed between 'tamping' rollers to drive the crumb into the batter and to absorb batter into the material to create a strong bond. After breading, the products are then frozen (Chapter 22) or fried and chilled (Chapters 19 and 21). Similar designs of applicator are used to coat confectionery and baked goods with sugar, flaked nuts or dried fruit pieces.

In another design of coating equipment, a rotating stainless steel drum, slightly inclined from the horizontal, is fitted internally with angled flights or ribs to tumble the food gently and to coat all surfaces with powder (e.g. seasonings). The angle and speed of rotation are adjusted to control the product throughput. Similar equipment is used for spraying products with oil or liquid flavourings. In other designs, flavours or salt are blown directly into the drum by compressed air. Fluidised beds (Chapter 16, section 16.2.1) are also used as flavour applicators.

24.2.3   Pan coating

Panning is the process of building up layers of sugar, sweetener or chocolate onto cores of fondant, fruit or nuts, using a revolving copper or stainless steel pan. This is a type of forming operation (Chapter 4, Section 4.2.3). The process uses a tilted elliptical mixer (Fig. 24.2) that rotates at 15–35 rpm depending on the size of the centre: large nuts, for example, require speeds of 15 rpm and sugar grains ('hundreds and thousands') speeds of 30–35 rpm. Layers of sugar are formed onto the centres as they tumble through sprays of warm sugar solution. Air at 35–65 °C is blown into the pan to give rapid drying of sugar layers, to remove dust and to remove frictional heat. The cycle of coating–drying–coating continues until the required product size and smoothness are formed. During panning, the pieces are periodically removed and sieved to remove waste and break up any clumps. Products are characterised by a smooth, regular surface obtained by the polishing action in the pan. Panning is a slow process involving small batches, but with automatic operation one operator can monitor a bank of ten or more pans. There are three main types of pan-coated products, depending on the type of coating used.

Fig. 24.2   Dragée pan (courtesy of A.M.P. Rose at www.amp-rose.com).

Hard coatings
Centres are coated with a sweetener solution (termed 'wetting') that is added at a rate of 10–15% of the weight of the centres. This crystallises in successive layers and a hard coating is built up around the centres (termed 'engrossing'). If nuts are used, they should first be sealed with a gum arabic/wheat flour mixture to prevent oil seepage during storage. A summary of the properties of hydrocolloids used in confectionery is shown in Chapter 4 (Table 4.4). Coatings are traditionally made from 60–65% sucrose or dextrose syrups. Gum arabic may be added to improve adhesion and reduce brittleness in the product. Sugarless hard coatings are made from sorbitol syrup and are described in detail by Le Bot (1993). Flavouring is added to each charge of wetting syrup and colouring is added in increasing concentration to the last five or six wettings. Cornflour may be added after each wetting to reduce sticking of the pieces but over-use results in agglomeration of the product. Separate glass-lined polishing pans are located away from the humid conditions in a pan room and are kept free of dust. Beeswax, paraffin oil or carnauba wax is used to coat polishing pans and to shine hard pan products. Alternatively, they may be glazed using a mixture of shellac in isopropanol.

Soft coatings
Complex mixtures of liquid glucose syrup and crystalline sucrose are prepared as the centres for soft-coated confectionery such as jelly beans, 'dolly mixtures', etc. Anti-crystallising agents in the liquid phase cause the outer layers only to partially crystallise when added during pan-coating. Successive wettings of 60% glucose syrup are therefore followed by addition of fine castor sugar until the surfaces dry and produce an amorphous soft coating. The hardness of the coating is determined by the ratio of anti-crystallising agents in the syrup. After two or three wettings, the partially coated centres are removed and allowed to dry for 2–3 h. The final coating is dried using icing sugar and the products are then dried for two days at 20 °C in a dust-free room. Sugarless soft coatings made from sorbitol, mannitol, maltitol and xylitol are described by Le Bot (1993).

Chocolate coating
This type of panning is similar to that used for hard coatings. Pans are rotated at around 20 rpm and held at 16 °C for plain chocolate and 14 °C for milk chocolate. Tempered chocolate may be poured or spray-coated onto confectionery centres with successive layers being built up and finished with a hard glaze, or polished with a 50% solution of gum arabic.

24.3   Microencapsulation and edible barrier coatings

Microencapsulation is a process in which minute particles or liquid droplets are coated with a thin film of edible encapsulating material. It is used for a variety of purposes:

- To protect volatile ingredients or those that are heat sensitive or susceptible to oxidation and nutritional loss during processing and storage (e.g. colours, vitamins, flavourings or essential oils). For example, the artificial sweetener aspartame is a dipeptide that is broken down by heat, resulting in a loss of sweetness. Microencapsulated aspartame in bakery products is protected from heat and sweetness is retained in the final product.
- To mask the taste or odour of an ingredient (e.g. fish oils or plant extracts used in functional foods, or incorporation of nutrients and development of nutraceutical foods) (Chapter 6, section 6.3).

- To contain a reactive ingredient (e.g. to separate a food acid from other ingredients to prevent colour and flavour changes in the food) or to enable time-release mechanisms in an ingredient formulation in a product (e.g. microencapsulated sodium bicarbonate in home-baked pizza or bread doughs prevents the early release of bicarbonate and delays the reaction of leavening phosphate until the crust reaches a specific temperature in the oven).
- To control the time and rate of release of enzymes and starter cultures (Chapter 6, section 6.6).
- Transformation of liquids into easily handled solid ingredients.

Common microencapsulated ingredients include sweet flavours (e.g. orange, lemon, butterscotch, bergamot), savoury flavours (garlic, mustard, onion, smoke flavour), herbs and spices (mint, pepper, spice blends) and natural colourants (paprika, annatto, turmeric). The types of coating materials used to produce microcapsules include gums (gum arabic, agar, sodium alginate, caragheenan), celluloses (carboxymethylcellulose, nitrocellulose, cellulose acetate-phthalate) (Carr *et al.* 1995), starch, dextran, sucrose, corn syrup, lipids (paraffin wax, stearic acid, mono- or diglycerides, beeswax, fats), inorganic materials (calcium sulphate, silicates, clays) and proteins (gluten, casein, gelatin, albumin) (Jackson and Lee 1991) (also Chapter 1, section 1.1.1). The material should be insoluble in the entrapped ingredient and not react with it; it should be soluble in the food product and be able to withstand the temperatures used in the microencapsulation process.

A number of methods are used to form microcapsules: spray drying (Chapter 16, section 16.2.1), is commonly used and an emulsion of 25–30% solids (w/w) of the encapsulating material and the entrapped ingredient is formed by adding an emulsifying agent, heating and passing the material through an homogeniser (Chapter 3, section 3.2.3). A 4:1 ratio of encapsulating material to ingredient is used in the emulsion when volatile ingredients are microencapsulated because volatiles are lost at higher levels (Reineccius 1988). The emulsion is then atomised and dried in a spray dryer. The rapid removal of water from the coating material keeps the entrapped ingredient below 100 °C. Spray drying is not suitable for some aromatic ingredients that have a low boiling point and are lost during the drying process (Rosenberg *et al.* 1990). Other disadvantages of spray drying include contamination of the surface of the capsules with ingredient, which can result in oxidation and flavour changes to the encapsulated product. If the feed inlet temperature to the dryer is too low, cracks can form in the surface of the capsules, which may cause loss of volatiles and low-quality ingredients. Incomplete drying results in product sticking inside the drying chamber or forming clumps during storage (Bhandari *et al.* 1992). An alternative method that reduces heat damage to ingredients is spray chilling. A spray of hot encapsulating material is condensed into capsules surrounding the ingredients using refrigerated air in a spray dryer.

Other methods include fluidised-bed coating, extrusion coating and liposome entrapment (Chapter 6, section 6.6). Recent developments in microencapsulation include the encapsulation of probiotics in hydrocolloid beads to improve their survival rate through processing and digestion, and incorporation of omega 3 fatty acids in food products (Chapter 6, section 6.3). Dewettinck and Huyghebaert (1999) and Vistrup (2001) describe these methods in detail and Desai and Park (2005) have reviewed developments in each area.

24.3.1   Edible barrier coatings

Coatings made from a variety of natural materials, including casein, whey protein, carrageenan and chitosan have been developed as a barrier to protect foods from spoilage or contamination, and to retain vitamins and other nutrients to maintain the nutritional value of the food (see also packaging, Chapter 25, section 25.4.3). Coating applications include:

- protecting oxygen-sensitive foods, such as nuts, to reduce the rate of oxidative rancidity and extend their shelf-life, prevent oil migration into surrounding food components and reduce packaging requirements;
- protecting fragile foods (e.g. breakfast cereals and freeze-dried foods (Chapter 23, section 23.1)) by reducing losses due to mechanical damage;
- reducing moisture loss, respiration and colour changes in whole and pre-cut fresh fruits and vegetables to extend their shelf-life;
- adhesives for seasonings in low-fat snack foods (e.g. potato crisps);
- preventing oxidation, moisture migration, or aroma and colour loss in frozen foods. (Krochta 2004, Krochta *et al.* 2002).

For example, glossy coatings for chocolate and other confectionery made from whey protein have been developed as a replacement for ethanol-based shellac and corn zein coatings. Whey protein coatings have also been studied as an oxygen barrier for nuts, and for incorporation of the antimicrobial compound, lactoferrin. Food coatings or glazes made from carrageenan can be applied by spraying, brushing or dipping (Anon 2005b). There is also growing interest in the use of coatings to protect against food poisoning bacteria. For example, Park *et al.* (2004) have combined chitosan from crab and shrimp shells with lysozyme from egg white to create an antimicrobial spray or dip to coat foods. The coating can be formulated with additional nutrients (e.g. vitamin E and calcium) to increase the nutritional value of the food (Park and Zhao 2004, Mei and Zhao 2003). (See also edible protective superficial layers (EPSL) in Chapter 25, section 25.4.1.)

References

AFOAKWA, E., PATERSON, A. and FOWLER, M., (2007), Factors influencing rheological and textural qualities in chocolate: a review, *Trends in Food Science and Technology*, **18** (6), 290–298.

ANON, (2005a), Witwood range offers a true taste of India, company information, available at www.witwoods.co.uk/article7.pdf.

ANON, (2005b), Edible food packaging, company information from Ingredient Solutions Inc., available at www.isinc.to/edible.htm.

BECKETT, S.T., (1994), *Industrial Chocolate Manufacture and Use*, 2nd edn, Blackie Academic and Professional, London.

BHANDARI, B.R., DUMOULIN, E.D., RICHARD, H.M.J., NOLEAU, I. and LEBERT, A.M., (1992), Flavor encapsulation by spray drying: application to citral and linalyl acetate, *J. Food Science*, **57** (1), 217–221.

CARR, J.M., SUFFERLING, K. and POPPE, J., (1995), Hydrocolloids and their use in the confectionery industry, *Food Technology*, July, 41–44.

DESAI, K.G.H. and PARK, H.J., (2005), Recent developments in microencapsulation of food ingredients, *Drying Technology*, **23** (7), 1361–1394.

DEWETTINCK, K. and HUYGHEBAERT, A., (1999), Fluidized bed coating in food technology, *Trends in Food Science and Technology*, **10**, 163–168.

ELERT, G., (2008), Allotropes and polymorphs, The Physics Hypertextbook[TM], available at http://hypertextbook.com/physics/matter/polymorphs/.

FISZMAN, S.M. and SALVADOR, A., (2003), Recent developments in coating batters, *Trends in Food Science and Technology*, **14** (10), 399–407.

FRYER, P. and PINSCHOWER, K., (2000), The materials science of chocolate, *Materials Science in the Food Industry*, **25** (12), 25–29.

GHOSH, V., ZIEGLER, G.R. and ANANTHESWARAN, R.C., (2002), Fat, moisture and ethanol migration through chocolates and confectionery coatings, *Critical Reviews in Food Science and Nutrition*, **42** (6), 583–626.

JACKSON, S.J. and LEE, K., (1991), Microencapsulation and the food industry, *Lebensmittel-Wissenschaft und -Technologie*, **24** (4), 289–297.

KROCHTA, J.M., (2004), Incredible edible films, available at www.ediblefilms.org/applications/.

KROCHTA, J.M., BALDWIN, E.A. and NISPEROS-CARRIEDO, M., (Eds.), (2002), *Edible Coatings and Films to Improve Food Quality*, CRC Press, Boca Raton, FL.

LE BOT, Y., (1993), Stable sugarless coating, in (A. Turner, Ed.,) *Food Technology International Europe*, Sterling Publications, London, pp. 67–70.

LETOURNEAU, J.-J., VIGNEAU, S., GONUS, P. and FAGES, J., (2005), Micronized cocoa butter particles produced by a supercritical process, *Chemical Engineering and Processing*, **44** (2), 201–207.

LOISEL, C., KELLER, G., LECQ, G., BOURGAUX, C. and OLLIVON, M., (1998), Phase transitions and polymorphism of cocoa butter, *J. American Oil Chemists' Society*, **75** (4), 425–439.

MEI, Y. and ZHAO, Y., (2003), Barrier and mechanical properties of milk protein-based edible films incorporated with nutraceuticals, *J. Agriculture and Food Chemistry*, **51** (7), 1914–1918.

NELSON, R.B., (1994), Enrobers, moulding equipment, coolers and panning, in (S.T. Beckett, Ed.), *Industrial Chocolate Manufacture and Use*, 2nd edn, Blackie Academic and Professional, London, pp. 211–241.

PARK, S.I. and ZHAO, Y., (2004), Incorporation of high concentration of mineral or vitamin into chitosan-based films, *J. Agriculture and Food Chemistry*, **52**, 1933–1939.

PARK, S.I., DAESCHEL, M. and ZHAO, Y., (2004), Functional properties of antimicrobial lysozyme-chitosan composite films, *J. Food Science*, **69** (8), M215–M221.

REINECCIUS, G.A., (1988), Spray-drying of food flavours, in (G.A. Reineccius and S.J. Risch, Eds.), *Flavor Encapsulation*, American Chemistry Society, Washington, DC, pp. 55–66.

ROSENBERG, M., KOPELMAN, I.J. and TALMON, Y., (1990), Factors affecting retention in spray drying microencapsulation of volatile materials, *J. Agriculture Food Chemistry*, **38**, 1288–1294.

TALBOT, G., (1999), Chocolate temper, in (S.T. Beckett, Ed.), *Industrial Chocolate Manufacture and Use*, 3rd edn, Blackwell Science, Oxford, pp. 218–230.

VISTRUP, P., (Ed.), (2001), *Microencapsulation of Food Ingredients*, Woodhead Publishing, Cambridge.

XUE, J. and NGADI, M., (2006), Rheological properties of batter systems formulated using different flour combinations, *J. Food Engineering*, **77** (2), 334–341.

25

Packaging

Abstract: Packaging is intended to conveniently contain foods and to protect them against a range of hazards during production, storage and distribution. This chapter describes the barrier properties of different packaging materials and their relationship to the shelf-life of foods. It summarises interactions between packaging materials and foods, and describes in detail the different types of metal, glass, paper and plastic food packaging materials. It then reviews modified atmosphere packaging and developments in edible and biodegradable packaging, active and intelligent packaging, and printing techniques. The chapter concludes with a discussion of the environmental issues in food packaging.

Key words: packaging, types of packaging materials, modified atmosphere packaging (MAP), barrier properties, interactions between packaging and foods, active and intelligent packaging, printing and barcodes, environmental considerations of packaging.

The purpose of packaging is to contain foods and to protect them against a range of hazards during distribution and storage. It may be defined in terms of its protective role as 'a means of achieving safe delivery of products in sound condition to the final user at a minimum cost'. The functions of packaging are:

- containment, to hold the contents and keep them secure for the consumer without leakage until they are used;
- protection against damage caused by micro-organisms, heat, moisture pick-up or loss, oxidation and breakage (section 25.1);
- convenience throughout the production, storage and distribution system, including easy opening, dispensing and re-sealing, and being suitable for easy disposal, recycling or reuse (Paine 1991).

Packaging is an important part of most food processing operations and in some (e.g. canning (Chapter 13, section 13.1)) it is integral to the operation itself. Packaging is one of the most dynamic sectors in food processing and there have been substantial developments in both materials and packaging systems over the past 20 years, which have enabled the development of novel and minimally processed foods (Chapters 8 and 9) and have reduced packaging costs and environmental impacts (section 25.6).

Packaging should not interact with or influence the product (e.g. by migration of toxic compounds, by reactions between the pack and the food or influence the selection of undesirable micro-organisms in the packaged food (e.g. selection of anaerobic pathogens in vacuum packed or modified atmosphere packed products (section 25.3.1)). Other requirements of packaging are smooth, efficient and economical operation on the production line for high speed filling, closing and collating (e.g. ≈ 1000 packs per min); resistance to damage such as fractures, tears or dents caused by filling and closing equipment, loading/unloading or transportation; and, not least, minimum total cost. Methods used to calculate the overall cost of packaging, taking into account the performance of materials on a packing line, are described by Stewart (1995) (also section 25.6).

The other important aspect of packaging is communication, to identify the contents and assist in selling the product. Details of marketing are outside the scope of this book, but the main marketing considerations for a package are:

- advertising the brand image and style of presentation required for the food – it should be aesthetically pleasing and have a functional size and shape;
- flexibility to change the size and design of the containers; and
- compatibility with methods of handling and distribution, and with the requirements of retailers.

Some packages inform the user about method of opening and/or using the contents. Details of the role of packaging as a marketing tool are given by Stewart (1995) and Paine (1991). Developments in 'intelligent' packaging that provide information on the state of contents in a package are given in section 25.4.3. The package design should be attractive and functional (Dale 2000) and also meet legislative requirements concerning the labelling of foods. These vary in different countries and further information is given by Blanchfield (2000).

Packaging materials can be grouped into two main types:

1 *Shipping containers*, which contain and protect the contents during transport and distribution but have no marketing function. They should also inform the carrier about the destination and any special handling or storage requirements. Examples include sacks, corrugated fibreboard cases, shrink-wrapped or stretch-wrapped corrugated fibreboard trays, wooden, plastic or metal cases, crates, barrels, drums, and inter-mediate bulk containers (IBCs) such as combo-bins and large (e.g. 1 tonne) bags made from woven plastic fabric (also Chapter 26, section 26.3). Some types of shipping containers are expensive (e.g. crates, barrels, drums) and are therefore made to be reusable, whereas others (e.g. sacks, or expanded polystyrene trays for fresh fruits) are low cost and used for a single journey.

2 *Retail containers* (or 'consumer units') protect and advertise the food in convenient quantities for retail sale and home storage. Examples include metal cans, glass or plastic bottles and jars, rigid and semi-rigid plastic tubs and trays, collapsible tubes, paperboard cartons and flexible plastic bags, sachets and overwraps.

Section 25.2 describes both shipping and retail containers in categories that reflect their material of construction. Frequently more than one type of material is used to package a single product. A summary of the different packaging materials used for selected food products is shown in Table 25.1.

Table 25.1 Applications of packaging materials for selected foods

Product	Glass Jar	Glass Bottle	Metal Can	Metal Tin/push-on lid	Metal Foil	Metal Tube	Paper Bag/wrap	Paper Pot	Paper Drum/tube	Paper Carton	Plastic Bottle	Plastic Pot	Plastic Tray/overwrap	Plastic Film
Short shelf-life														
Bakery products (e.g. bread, cakes, pies)							✓			✓			✓	✓
Cooked meats													✓	✓
Dairy products (e.g. milk, yoghurt)		✓						✓		✓	✓	✓		
Fresh fruit/vegetables							✓			✓			✓	✓
Fresh meat/fish													✓	✓
Medium/long shelf-life														
Beverages (e.g. juices, wines, beers, carbonated drinks)	✓	✓	✓							✓	✓			✓
Biscuits							✓		✓				✓	✓
Cooking fats				✓								✓		
Cooking oils		✓		✓							✓			
Dairy products (e.g. butter, cheese)					✓		✓	✓				✓	✓	✓
Dried foods (e.g. fruits, cereals, spices, coffee)	✓			✓	✓	✓	✓		✓	✓		✓		✓
Frozen foods							✓	✓				✓		✓
Heat-sterilised foods (canned)	✓		✓											
Pastes and purées (e.g. tomato/garlic paste, peanut butter)	✓		✓			✓								
Preserves (e.g. jams, pickles, chutneys, sauces)	✓	✓									✓	✓		
Snackfoods (fried or extruded)							✓	✓	✓			✓		✓
Sugar							✓			✓				
Sugar confectionery	✓			✓					✓	✓		✓		✓
Syrups, honey	✓	✓								✓	✓	✓		
UHT sterilised foods		✓								✓	✓	✓		✓

25.1   Theory

The shelf-life of packaged foods is controlled by:

1   the properties of the food, (including water activity, pH, susceptibility to enzymic or microbiological deterioration, and the requirement for, or sensitivity to, oxygen, light, carbon dioxide and moisture);
2   environmental factors that cause physical or chemical deterioration of foods (e.g. UV light, moisture vapour, oxygen, temperature changes), contamination by micro-organisms, insects or soils, mechanical forces (damage caused by impact, vibration, compression or abrasion), pilferage, tampering or adulteration;
3   the barrier properties of the package.

Packaging provides a barrier that isolates the food to a predetermined degree from the environment. For some foods, such as low-acid sterilised foods that have a long shelf-life, the packaging provides total protection to prevent recontamination by micro-organisms and oxidative changes, whereas for others, such as respiring fresh fruits, a permeable pack is required to enable the exchange of respiratory gases.

25.1.1   Factors affecting the selection of a packaging material

The properties of foods that affect their shelf-life are described in Chapter 1 (section 1.2). Pfeiffer *et al.* (1999) describe the factors that affect the shelf-life of foods and mathematical models based on these factors that are used to predict shelf-life and optimise packaging. Lyijynen *et al.* (2003) describe methods for scoring packaging options against criteria such as strength, ratio of pack weight : product weight, marketing properties, consumer convenience, cost and disposal options, to select an optimum package for a particular application. This section describes the factors listed in (2) above. Barrier properties of packaging materials are described in section 25.2.

Light
Light transmission is required in packages that are intended to display the contents, but it is restricted when foods are susceptible to deterioration by light (e.g. rancidity caused by oxidation of lipids, loss of nutritional value due to destruction of riboflavin, or changes in colour caused by loss of natural pigments). The amount of light absorbed by food in a package is found using:

$$I_a = I_i T_p \frac{1 - R_f}{1 - R_f R_p}$$
25.1

where I_a (Cd) = intensity of light absorbed by the food, I_i (Cd) = intensity of incident light, T_p = fractional transmission by packaging material, R_p = the fraction reflected by the packaging material and R_f = the fraction reflected by the food.

The fraction of light transmitted by a packaging material is found using the Beer–Lambert law

$$I_t = I_i e^{-\alpha x}$$
25.2

where I_t (Cd) = intensity of light transmitted by the packaging, α = the characteristic absorbance of the packaging material and x (m) = thickness of the packaging material.

The amount of light that is absorbed or transmitted varies with the packaging material and with the wavelength of incident light. Some materials (e.g. low-density polyethylene) transmit both visible and ultraviolet light to a similar extent, whereas others (e.g. polyvinylidene chloride) transmit visible light but absorb ultraviolet light. To reduce light transmission to sensitive products, pigments may be incorporated into glass containers or polymer films, they may be over-wrapped with paper labels, or they may be printed (section 25.5). Alternatively, clear packs may be contained in fibreboard cartons for distribution and storage.

Temperature
The insulating effect of a package is determined by its thermal conductivity (Chapter 10, section 10.1.1) and its reflectivity. Materials that have a low thermal conductivity (e.g. paperboard, polystyrene or polyurethane foams) reduce conductive heat transfer, and reflective materials (e.g. metallised films, aluminium foil) reflect radiant heat. However, control over the temperature of storage to protect foods from heat is more important than reliance on the packaging. In applications where the package is heated (e.g. in-container sterilisation or microwaveable ready meals) the packaging material must be able to withstand the processing conditions without damage and without interaction with the food. Glass containers should be heated and cooled more slowly than metal or plastic containers to avoid thermal shock and the risk of breakage. Similarly, packaging for frozen food should remain flexible and not crack at frozen storage temperatures.

Moisture and gases
Moisture loss or uptake is one of the most important factors that controls the shelf-life of foods. There is a micro-climate within a package, which is determined by the vapour pressure of moisture in the food at the temperature of storage and the permeability of the packaging material. Control of moisture exchange is necessary to prevent microbiological or enzymic spoilage, loss of moisture and drying out of the food (e.g. fresh or cooked meats, cheeses), or freezer burn in frozen foods (Chapter 22, section 22.3.2). Higher permeability is required of packaging for foods such as fresh vegetables and bread to prevent moisture condensation on the inside of packages that would result in mould growth. Similarly, chilled foods require controlled movement of water vapour out of the pack to prevent fogging of the display area if the storage temperature changes. Dried, baked or extruded foods that have a low equilibrium relative humidity require packaging that has a low permeability to moisture to prevent their gaining moisture from the atmosphere. This causes foods, such as biscuits and snackfoods to soften and lose their crispness. If their water activity rises above a level that permits microbial growth they will spoil. Powdered foods (e.g. custard or gravy powder, food colourants, icing sugar) can be highly hygroscopic and if moisture is transmitted through the package they lose their free-flowing characteristics and become caked.

Foods that contain appreciable quantities of lipids or other oxygen-sensitive components can spoil if the package has an inadequate barrier to oxygen. Conversely, fresh foods that are respiring require a high degree of permeability in the material to allow exchange of oxygen and carbon dioxide with the atmosphere, and, in fruits, loss of ethylene from the pack without excessive loss of moisture that would cause weight loss and shrivelling. Fresh red meats require oxygen to maintain the red haemoglobin pigment for their expected shelf-life. Foods that are packaged in modified atmospheres (section 25.3) in which air is replaced by nitrogen and/or carbon dioxide (e.g. cheeses, cooked meats, egg powder and coffee) require materials that have a low permeability to these gases to achieve the expected shelf-life. Packaging should also be impermeable to retain volatile compounds that contribute to desirable odours (e.g. in coffee or snackfoods) or to prevent odour pick-up (e.g. by powders or fatty foods). There should also be negligible odour pick-up from plasticisers, printing inks, adhesives or solvents used in the manufacture of the packaging material (section 25.2.4).

Assuming that a packaging material has no defects (e.g. splits, pinholes or inadequately formed seals in flexible films) and that there is no interaction between the material and the gas or vapour, the permeability of the packaging material is found using Equation 25.3 (usually expressed as either cm^3 (or ml) m^{-2} per 24 h or $g\,m^{-2}$ per 24 h when the inside of the pack is at atmospheric pressure):

$$b = \frac{mx}{A(\Delta P)} \qquad \boxed{25.3}$$

where b = permeability, m = quantity of gas or vapour passing through area A of the material in unit time, x (m) = thickness of the material, A (m^2) = the area of the material and ΔP (Pa) = difference in pressure or concentration of gases between the two sides of the material. Cooksey et al. (1999) describe other equations that are used to predict permeability or transmission rate and further details are given by Massey (2003).

Whereas glass and metal packaging are almost totally impermeable to gases and vapours, plastic films have a wide range of permeabilities, depending on the thickness, chemical composition, and structure and orientation of molecules in the film. Permeance to moisture (or water vapour transfer rate (WVTR)) varies from $<10\,cm^3\,m^{-2}\,day^{-1}$ in low-permeability films to 200–$800\,cm^3\,m^{-2}\,day^{-1}$ in highly permeable films. Permeability to gases varies from $<10\,cm^3\,m^{-2}\,day^{-1}$ in low-permeability films to 100–$25\,000\,cm^3\,m^{-2}\,day^{-1}$ in highly permeable films (Brennan and Day 2006) (Table 25.2). Other units of permeability and conversion factors are described by Singh and Heldman (2001).

The mechanisms of movement of gases, vapours and odour compounds through packaging materials are described by Zobel (1988). Plasticisers and pigments loosen the structure of flexible plastic films and increase their permeability. Permeability is related to both the type of film and the type of gas or vapour, and is not simply a property of the film. For example, the permeability of cellulose, nylon and polyvinyl alcohol films change with variations in humidity owing to interaction of moisture with the film (section 25.2.4).

Permeability is also related exponentially to temperature and it is therefore necessary to quote both the temperature and relative humidity of the atmosphere in which permeability measurements are made (Table 25.2, see also Table 25.15). Methods for testing the permeability and mechanical properties of packaging materials are described by Paine and Paine (1992), Paine (1991), Robertson (1990) and White (1990).

A method to calculate the shelf-life of packaged dry foods, based on the permeability of the pack, the water activity and equilibrium moisture content of the food (Chapter 1, section 1.1.2) is described by Robertson (1993), using the following equation:

$$\ln\left(\frac{M_e - M_i}{M_e - M_c}\right) = b \cdot (A/W_s) \cdot (P_o/x) \cdot (t_s) \qquad \boxed{25.4}$$

where M_e = equilibrium moisture content of the food, M_i = initial moisture content of the food, M_c = critical moisture content of the food, b (g water day^{-1} m^2 (mm Hg)$^{-1}$) = permeability of the packaging material, A (m^2) = area of package, W_s (g) = weight of dry solids in the food, P_o (torr) = vapour pressure of pure water at the storage temperature, x (g H_2O/g solids per unit a_w) = slope of the moisture sorption isotherm (Chapter 1, Fig. 1.12) and t_s (days) = time to the end of the shelf-life.

The calculation of shelf-life where oxygen permeability is the critical factor is found using the following equation:

$$t_s = \frac{Qx}{PA\Delta P}$$

25.5

where Q (ml) = the maximum quantity of oxygen that is permissible in the package and ΔP = difference between the partial pressure of oxygen inside and outside the container (see also Chapter 1, sample problem 1.1).

Grease resistance
Leakage of oils and fats spoils the appearance of a pack. Cooking oils are packaged in metal cans or glass or plastic bottles and cooking fats in plastic tubs, aluminium foil or greaseproof paper. Dry fatty foods (e.g. chocolate) are packed in foil or plastic films, and wet fatty foods (e.g. meat and fish) in treated papers or laminated films and papers.

Sample problem 25.1
Potato crisps having 300 g of dry solids are packaged in a 0.2m^2 sealed bag made from a barrier film that has a water vapour transmission rate of 0.009 ml day^{-1} m^{-2}. From a sorption isotherm from crisps, the equilibrium moisture content = 0.05 g per g of solids, the initial moisture content = 0.015 g per g of solids, the critical moisture content = 0.04 g per g of solids and the slope of the moisture sorption isotherm = 0.04 g H$_2$O/g solids per unit a_w. The crisps are expected to be stored at 20 °C and the vapour pressure of pure water at this temperature = 17.53 torr. Calculate the time for the moisture content to reach the critical moisture content, to find the expected shelf-life using this film.

Solution to sample problem 25.1:
Using Equation 25.4,

$$t_s = \frac{\ln(0.05 - 0.015)/(0.05 - 0.04)}{0.009 \times (0.2/300) \times (17.53/0.04)}$$

$$= \frac{1.2527}{2.6295 \times 10^{-3}}$$

$$= 476.4 \text{ days (approximately 16 months)}$$

Micro-organisms, insects, animals and soils
Packs that are folded, stapled or twist-wrapped are not truly sealed and can become contaminated by micro-organisms. Metal, glass and polymer packaging materials are barriers to micro-organisms, but their seals are a potential source of contamination. The main causes of microbial contamination of adequately processed foods are:

- contaminated air or water drawn through pinholes in hermetically sealed containers as the head space vacuum forms after heat sterilisation (Chapter 13, section 13.1.2);
- inadequate heat seals in polymer films caused by contamination of the seal with product or incorrect heat sealer temperature or time of heating;
- damage such as tears or creases to the packaging material.

Table 25.2　Selected properties of packaging films

Barrier

Film	Thickness (μm)	Yield (m² kg⁻¹)	Moisture vapour transmission rate (ml m⁻² per 24 h) 38 °C 90% RH	23 °C 85% RH	Oxygen transmission rate (ml m⁻² per 24 h) 23 °C 85% RH	25 °C 0% RH	25 °C 45% RH
Cellulose							
Uncoated	21–40	30–18	1500–1800	400–275	25–20		10–8
Nitrocellulose coated	22–24	31–29	12–8	1.8	15–9	10–8	8–6
Polyvinylidene chloride coated	19–42	36–17	7–4	1.7		7–5.5	7–5
Metallised polyvinylidene chloride coated	21–42	31–17	5–4	0.8		3	3–2
Vinyl chloride coated			400–320	80–70			9
Polyethylene							
Low density	25–200	43–5	19–14	3000	120		8000
Stretch-wrap	17–38						
Shrink-wrap	25–200	43–11					
High density	350–1000		6.4				2000–500
Polypropylene							
Oriented	20–30	24	7–5	1.4–1.0	2200–1100		2000–1600
Biaxially oriented	20–40	55–27	7–3	1.2–0.6	1500		
Polyvinylidene chloride coated	18–34	53–30	8–4	1.4–0.6	6–10	13–6	
MG	20–40	55–27	7–4	1.4–0.6	2200–1100	2300–900	
Metallised	20–30	55–36	1.3	0.3–0.2	300–80	300	
Polyester							
Plain	12–23	59–31	40–20	8			110–53
850	12–30	60–24	40–17				120–48
Metallised			2.0–0.8				1.5–0.5
Polyvinylidene chloride coated and metallised			1.3–0.3				0.1
Polyvinylidene chloride	10–50	35–17	4–1	1.7	17–7		2

RH, relative humidity; TP, transparent.
[a]Will not heat seal.

properties				Mechanical properties			Optical properties	
Nitrogen transmission rate (ml m⁻² per 24 h)		Carbon dioxide transmission rate (ml m⁻² per 24 h)		Tensile strength machine direction	Tensile strength transverse direction	Total light transmission	Gloss	Sealing temperature
25°C 0% RH	30°C 0% RH	25°C 0% RH	25°C 45% RH	(MN m⁻²)	(MN m⁻²)	(%)	(%)	(°C)
	28	40–30		33		TP	110	a
		30–20		35		TP	130	90–130
		15		32–60		TP	150	100–130
		20–15		28–60		0	130	90–130
		30		120–130		TP		100–160
	19		40000	16–7				121–170
			8000–7000	61–24				135–170
285		3250		145–200	0.4–0.6		75–85	145
				118–260			80–85	117–124
		30–20				TP		
650–270		7000–3000		210	0.3–0.4	TP	75–85	120–145
85		900		215	0.5–0.6	0.5–3.1		120–145
25–7			500–150			87		100–200
25–10			500–200			88		100–200
1.8	0.0094	20		120–130		90	95–113	100–160

Processes such as irradiation (Chapter 7), pasteurisation (Chapter 12), heat sterilisation (Chapter 13) and ohmic heating (Chapter 20, section 20.2) each rely on packaging to maintain the microbiological quality of the moist processed products. In other processes, low storage temperatures, low moisture contents or the use of preservatives restrict microbial growth, and the role of the package is less critical (see hurdle technology, Chapter 1, section 1.3.1), although protection is still required against contamination by dust and other soils. Metal or glass containers and some of the stronger flexible films and foil laminates are able to resist insect infestation, but only metal and glass containers can protect foods against insects, rodents and birds.

Mechanical strength

The suitability of a package to protect foods from mechanical damage depends on its ability to withstand crushing, caused by stacking in warehouses or vehicles; abrasion caused by rubbing against equipment or during handling; puncturing or fracturing caused by impacts during handling; or by vibration during transport. Some foods (e.g. fresh fruits, eggs, biscuits, etc.) are easily damaged and require a higher level of protection from a package, including cushioning using tissue paper, or from paperpulp or foamed polymer sheets that are formed into shaped containers for individual pieces (e.g egg cartons, fruit trays). For other foods, protection is provided by a rigid container and/or restricted movement by shrink- or stretch-wrapping, or by using plastic films that are tightly formed around the product (see Chapter 26, sections 26.2 and 26.3). Polymer pots, trays and multi-layer cartons (sections 25.2.5 and 25.2.6) also provide protection for specific foods. Strong materials such as metal cans, glass or PET are required to withstand the pressure created by carbonated beverages.

Wooden or metal crates, barrels and drums have long been used as shipping containers as they provide good mechanical protection. These are being replaced by cheaper composite IBCs made from fibreboard and polypropylene (section 25.2.7).

The strength of packaging materials can be assessed by measuring the elongation (strain) that results from an applied force (stress) to give the following from a stress–strain diagram (Fig. 25.1):

- the tensile strength (T);
- Young's modulus (E) (slope of A–B);
- the tensile elongation;
- the yield strength (Y); and
- the impact strength.

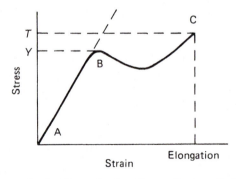

Fig. 25.1   Stress–strain curve for flexible packaging: T = tensile strength; Y = yield strength; slope of curve AB = Young's modulus; C = breaking point (from Briston 1980).

The tensile strength of a material is the maximum amount of tensile (stretching) stress that it can be subjected to before failure. The stretch ratio (also known as elongation) is a measure of the deformation of materials, some of which can have stretch ratios of 3 or 4 before they break. Young's modulus (E) is a measure of stiffness. It is defined as the ratio of the rate of change of stress with strain, and is found from the slope of the stress–strain curve (Fig. 25.1). Metals and glass are isotropic (their mechanical properties are the same in all directions), but plastics are frequently anisotropic (i.e. the Young's modulus and tensile strength differ depending on which direction the force is applied), depending on the orientation of the molecules. Other composite materials have a 'grain' with different mechanical properties when a load is applied in different directions. Young's modulus can be calculated by dividing the tensile stress by the tensile strain (Equation 25.6):

$$E = \frac{F/A_0}{\Delta L/L_0} \qquad \boxed{25.6}$$

where E (MPa) = Young's modulus (modulus of elasticity), F (N) = force applied to the object, A_0 (m) = original cross-sectional area through which the force is applied, ΔL (m) = amount by which the length of the object changes, L_0 (m) = original length of the object.

The yield strength (or 'yield point') of a material is the stress at which it begins to deform. Before the yield point, the material deforms elastically and returns to its original shape when the stress is removed. When the yield point is passed, part of the deformation is permanent and non-reversible. The yield point is therefore the upper limit of the force that can be applied to the material. Each of these factors is influenced by the temperature of the material and the length of time that the force is applied. The molecular structure of polymer films may be aligned in different ways depending on the type of film and method of manufacture. Orientation of molecules in one direction (uniaxial) or in both directions (biaxial) improves the mechanical properties of some films (e.g. polyethylene, polypropylene, polyethylene terephthalate and polystyrene (section 25.2.4)). Each of the properties described in Fig. 25.1 is therefore measured in both the axial (or machine) direction and the lateral (or transverse) direction of the film. Examples of the tensile strength of selected films are given in Table 25.3.

Tamper-evident/resistant features
No package is tamper-proof but tamper-evident or tamper-resistant features can be incorporated into containers. Details are given in Chapter 26 (section 26.4).

Table 25.3   Mechanical properties of polypropylene used either singly or in multi-layer laminates

Type of film	Tensile strength at yield (MPa)		Tensile strength at break (MPa)		Elongation at break (%)		Dart impact (g)
	MD	TD	MD	TD	MD	TD	
Cast polypropylene	20	19	38	34	500	600	300
Laminate: LLDPE/ polypropylene/LLDPE	20	20	65	55	625	750	75
Laminate: polypropylene/ LLDPE/polypropylene	22	20	70	45	600	700	95

MD = machine direction, TD = transverse direction, LLDPE = linear low-density polyethylene.
Data from Anon (2008a)

25.1.2   Interactions between packaging and foods

Any interaction between a packaging material and the food it contains is undesirable for two reasons: the interaction may have toxicological effects on the consumer and/or it may reduce the shelf-life or sensory quality of the food. Details are given by Piringer and Rüter (2000). The migration of oils from foods into plastics is also of concern as this may alter the barrier properties of the material. The materials of concern are mostly flexible films that contain residual monomers from the polymerisation processes (section 25.2.4), and additives to plastics, including nucleating agents, stabilisers, fillers, plasticisers, antifogging agents and pigments. Mercea (2000) describes models for the diffusion of these materials in polymer packaging. Veraart (2008) gives details of the chemicals under investigation and legislation in different countries relating to materials in contact with foods. He also describes methods of testing, mechanisms of migration and status reports on each of the chemicals under consideration. Some types of packaging materials also have volatile compounds that may be absorbed and cause tainting of foods. These may arise from the manufacturing process (e.g. solvents used to make polymer films or containers) or from additives such as wax coating on papers, lacquers and sealing compounds used on cans and closures, printing inks or label adhesives. Materials should therefore be carefully selected to reduce the risk of tainting foods.

In metal containers, interaction of food acids, anthocyanins, sulphur compounds and other components with steel, tin or aluminium are prevented using lacquers and coatings for the metal (section 25.2.2). In tinplate containers, a failure in the lacquer may result in food acids reacting with the tin coating on the steel to form hydrogen, which in extreme cases results in swelling of the can (hydrogen swelling). Glass containers are inert, but materials used in the cap or lid may interact with foods.

25.2   Types of packaging materials

25.2.1   Textiles and wood

Textile containers have poor gas and moisture barrier properties; they are not suited to high-speed filling; have a poorer appearance than plastics; and are a poor barrier to insects and micro-organisms. They are therefore only used as shipping containers for dried foods or in a few niche markets as over-wraps for other packaging. Woven jute sacks (named 'burlap' in USA), which are chemically treated to prevent rotting and to reduce their flammability, are non-slip which permits safe stacking, have a high resistance to tearing, low extensibility and good durability. Jute is a hessian type of weave (plain weave, single yarn); others include tarpaulin (double weave) and twill. They are still used to transport a variety of bulk foods including grain, flour, sugar and salt, although they are steadily being replaced by multi-wall paper sacks, polypropylene sacks or IBCs.

Wooden shipping containers have traditionally been used for a range of solid and liquid foods including fruits, vegetables, tea, wines, spirits and beers. They offer good mechanical protection, good stacking characteristics and a high vertical compression strength-to-weight ratio. However, polypropylene and polyethylene drums, crates and boxes have a lower cost and have largely replaced wood in many applications. The use of wood continues for some wines and spirits because the transfer of flavour compounds from the wooden barrels improves the quality of the product. Wooden tea chests are produced more cheaply than other containers in tea-producing countries and these are still widely used (Fellows and Axtell 2003).

25.2.2 Metal

Metal cans have advantages over other types of container in that they can withstand high-temperature processing and low temperatures; they are impermeable to light, moisture, odours and micro-organisms to provide total protection of the contents; they are inherently tamper-resistant and the metal can be recycled. However, the high cost of metal and relatively high manufacturing costs make cans more expensive than most other containers. They are heavier than other materials, except glass, and therefore incur higher transport costs. The three types of metal used for foods are tinplate, electrolytic chromium-coated steel (ECCS or 'tin-free' steel) and aluminium.

Three-piece cans

Hermetically sealed three-piece 'sanitary' cans, made from tinplate or tin-free steel consist of a can body and two end pieces, and are used to package heat-sterilised foods (Chapter 13, section 13.1). Tinplate cans are made from low-carbon mild steel that contains minor constituents, such as manganese, phosphorus and copper. The strength of the steel depends on the amounts of these constituents, its thickness, and the method of manufacture. There are two manufacturing methods: single (or cold) reduction (CR electroplate) and double reduction (DR electroplate). In both methods, steel is first rolled to a strip 1.8 mm thick, and then dipped into hot dilute sulphuric acid. CR electroplate is then cold-rolled to ≈0.50 mm thick and temper-rolled to ≈0.17 mm. DR electroplate has two cold-rolling stages and produces steel with greater stiffness and so thinner sheet can be used (≈0.15 mm). The tin coating is applied by electrolytic plating to give either the same thickness on each side of the steel or a different coating weight, depending on the requirements of the food. Generally, more acidic foods have a higher coating weight on the inner surface of the can (e.g. D 2.8/1.2 g m^{-2}, where the 'D' indicates the differential coating). At this stage the tinplate has a dull surface, due to the porous finish and it is heated by electric induction (known as 'flow brightening') to slightly melt the tin and improve surface brightness and resistance to corrosion. A mono-layer of edible oil is applied to protect the steel from scratches during the can-making process. Further details are given by Anon (2008d).

The tin may be coated with the following lacquers to prevent interactions with foods (also section 25.1.2):

- Epoxy-phenolic compounds are widely used. They are resistant to acids and have good heat resistance and flexibility. They are used for canned meat, fish, fruit, pasta, vegetables, beer and other beverages. They can also be coated with zinc oxide or metallic aluminium powder to prevent sulphide staining with meat, fish and vegetables.
- Vinyl compounds (vinyl chloride/vinyl acetate copolymers) have good adhesion and flexibility, and are resistant to acids and alkalis, but do not withstand the high temperatures used in heat sterilisation. They are used for canned beers, wines, fruit juices and carbonated beverages and as a clear exterior coating.
- Phenolic lacquers are resistant to acids and sulphide compounds and are used for canned meat or fish products, fruits, soups and vegetables.
- Butadiene lacquers prevent discoloration and have high heat resistance. They are used for beer and soft drinks and with vegetables if they have added zinc oxide.
- Acrylic lacquers are white and are used both internally and externally for fruit and vegetable products.
- Epoxy amine lacquers are expensive, but have good adhesion, heat and abrasion

resistance, flexibility and no off-flavours. They are used for beers, soft drinks, dairy products, fish and meats.

- Alkyd lacquers are low cost and used externally as a varnish over inks. They are not used internally due to off-flavour problems.
- Oleoresinous lacquers are low cost, general purpose, gold coloured coatings, used for beers, fruit drinks and vegetables. They can incorporate zinc oxide ('C' enamel) for use with beans, vegetables, soups, meats and other sulphur-containing foods.

The lacquer is cured by heating at 150–205 °C for ≈10 minutes. The thickness of the lacquer coating can be measured rapidly during production using optical interferometry (Hamilton 2005). Further information is given by Manfredi *et al.* (2005) and the legislative status of lacquers as food contact materials is discussed by Tice (2000). Details of EU legislation are available at Veraart (2008).

If required, the tinplate sheet has an external lithographic decoration applied (section 25.5). The ink is cured in an oven and varnish is applied over the printing and then cured by heat. To make the can body, the steel sheet is first slit by a set of revolving cutters. These cut strips of steel have a width that corresponds to the diameter of the can. A second set of cutters then cuts strips at right angles to the first cut, with the width corresponding to the height of the can. The flat body blank is then rolled into an open-ended cylinder. The two edges are held together, slightly overlapping and a thin copper wire is electrically heated to melt the metal and produce a high-integrity welded seam. This 'lost-wire' welding has a better appearance and greater integrity than the traditional soldered seams. A side-stripe of protective lacquer is applied externally and/or internally to protect the welded area. Alternatively, side seams are bonded by thermoplastic polyamide (nylon) adhesives. The ends of the body are then curled outwards to form a flange that is used to form a double seam (Chapter 26, section 26.1.2) and, if required, the cans are 'beaded' ('beads' are corrugations that are formed in the metal around the can body to increase the strength, or maintain the can strength when using thinner steel). Finally, one can end is stamped out from a tinplate sheet and double seamed onto the can body. The can is then ready for filling and sealing (Chapter 26, section 26.1). Methods of can manufacture are described by Anon (2008b,c).

Tin-free steel is made using a similar process to tinplate, but replacing the tin coating with a $0.15\,g\,m^{-2}$ metallic chromium–chromium oxide coating that is electrolytically deposited onto the surface of CR or DR steel sheet. The production of tin-free steel is described by Anon (2008e). A lacquer is applied to prevent external or internal corrosion (Charbonneau 1997). The discovery that bisphenol A, contained in can lacquer, is an endocrine disrupter (now described as a 'hormonally active agent'), which appears to mimic the female hormone oestrogen (Lyons 2000), has prompted the development of a two-layer PET film (section 25.2.4) that is heat-laminated onto tin-free steel (Yoichiro *et al.* 2006, Anon 2002).

Two-piece cans
Aluminium is the third most abundant element in the Earth's crust, and is most economically recovered from bauxite (40–60% alumina (hydrated aluminium oxide)). One kilogram of aluminium is made from about 4 kg of bauxite by dissolving the bauxite in cryolite (potassium aluminium fluoride) and applying 50 000–150 000 A to electro-lytically reduce the oxide to aluminium and oxygen. The oxygen combines with carbon from the anode to form CO_2 and the aluminium is drawn off into crucibles. Aluminium alloy that contains 1.5–5.0% magnesium is used to make two-piece aluminium cans by either the draw-and-wall-iron (DWI) process or the draw-and-redraw (DRD) process. The

DWI process produces thinner walls than the DRD process does and it is used to produce aluminium cans for carbonated beverages where the gas pressure helps support the container. DRD cans are thicker and are able to withstand the head-space vacuum produced during cooling after heat sterilisation. The advantages of two-piece cans include greater integrity, more uniform lacquer coverage, savings in metal and greater consumer appeal.

In the DWI process (Fig. 25.2a), a disc-shaped blank, 0.3–0.4 mm thick, is cut and formed (drawn) into a cup that has the final can diameter. The cups are then rammed through a series of rings that have tungsten carbide internal surfaces to iron the can walls (reduce the thickness) and to increase the can height. No heat is applied during this process and heat is generated from friction as the metal is thinned. The process permits good control over the wall thickness (≈0.1 mm thick) and therefore saves metal. After forming the can body, the uneven top edge is trimmed to the required height and this is then flanged to accept the can end, which is fitted after the can is filled. If required, the flanged can is then passed through a beading machine to form beads that strengthen the can. Every can is tested by passing it through a light tester that automatically rejects any cans with pinholes or fractures (Anon 2008e).

Modifications to the basic two-piece design include:

- a reduced diameter at the neck of the can which improves the appearance and ability to stack the cans, and saves metal;

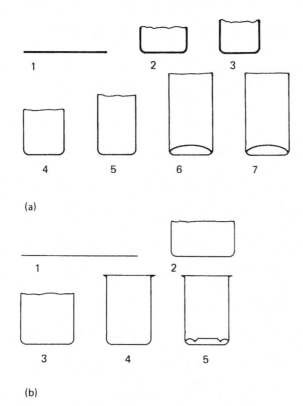

Fig. 25.2   Two-piece can manufacture: (a) DWI cans: 1, body blank; 2 + 3, drawn and redrawn cups; 4–6, three stages of wall ironing and base formation; 7, finished can trimmed to required height; (b) DRD cans: 1, body blank; 2, drawn cup; 3 + 4, stages in redrawn cups; 5, finished trimmed can with profiled base (adapted from Brennan *et al.* 1990).

- ring-pull tabs or full-aperture easy-open ends for greater convenience (Chapter 26, section 26.1.2).

Computer-aided print design and abrasion-resistant inks allow the blank to be printed before the can is formed. The ink is then stretched with the metal during the DWI process, to produce the required design on the finished can.

A lightweight DWI aluminium can in the shape of a bottle, with a spout and screw cap is described by Anon (2000). It is lined with PET and the cap is protected using low-density polyethylene film. The bottle can is designed to compete with PET bottles (section 25.2.5) and is reported to be 65% lighter in weight compared with an equivalent sized bottle, have higher barrier properties and is resealable and recyclable. It is used to pack carbonated drinks, teas, fruit juices, sports drinks and alcoholic beverages.

The DRD process is similar to the initial stages of the DWI process but, instead of ironing to reduce the wall thickness, metal is moved from the base of the container to the wall by reducing the diameter of the container (Fig. 25.2b). This process produces aluminium or tinplate cans that are generally smaller than DWI cans. In both processes, the can end is applied by double seaming and expoxy, phenolic or vinyl-based lacquers are applied internally to prevent interactions between the metal and the product.

Aerosol cans
Aerosol cans are two- or three-piece lacquered tinplate or aluminium cans fitted with a valve through which the product is dispensed. The propellant gas is either mixed with the product or kept separate by a plastic bag or a piston device. The pressure strength of the can should be 1.5 times the maximum vapour pressure of the filled aerosol at 55 °C, with a minimum of 1 MPa. Nitrous oxide propellant is used for UHT sterilised cream, and other gases (e.g. argon, nitrogen and carbon dioxide) are approved for use with foods, including cheese spreads and oil sprays for baking pans.

Other metal packaging
Tinplate and aluminium cans that are fitted with a variety of closures (Chapter 26, section 26.1.2) are used to package powders, syrups and cooking oils. Aluminium is also used for foil wrappers, lids, cups and trays, laminated pouches, collapsible tubes, barrels and closures. Foil is produced by a cold reduction process in which pure aluminium (purity >99.4%) is passed through rollers to reduce the thickness to <0.152 mm and then annealed (heated to control its ductility) to give it dead-folding properties. The advantages of foil include:

- good appearance, no odour to taint products;
- impermeable to moisture, odours, light and micro-organisms, and an excellent barrier to gases; the ability to reflect radiant energy;
- good weight : strength ratio;
- high-quality surface for decorating or printing, and lacquers are not needed because a protective thin layer of oxide forms on the surface as soon as it is exposed to air;
- can be laminated with paper or plastics and compatible with a wide range of sealing resins and coatings for different closure systems.

A potential disadvantage of aluminium is the widely reported incompatibility with use in microwave ovens. Paine (1991) reports a study by the Aluminum Association of Washington and the Aluminum Foil Containers Association of Wisconsin into the effects of aluminium packaging on the performance of microwave ovens. They concluded that in most instances results of food heating were as good as with microwave transparent materials and in many cases heating was more uniform. Foil containers had no effect on the magnetron and in approximately 400 tests, arcing between the foil and oven wall occurred only once. Other tests showed that foil containers did not cause the magnetron to operate outside its allowable ratings and only in the earliest microwaves, before 1969, has any damage to magnetrons occurred.

Foil is widely used for wraps (0.009 mm), bottle caps (0.05 mm) and trays for frozen and ready meals (0.05–0.1 mm). If foil is to be used to contain acid or salty foods it is normally coated with nitrocellulose. Aluminium is also used as the barrier material in laminated films, to 'metallise' flexible films (section 25.2.4) and to make collapsible tubes for viscous products (e.g. tomato purée and garlic paste). Collapsible tubes are supplied pre-formed, with an internal epoxy-phenolic or acrylic lacquer, a sealed nozzle and an open end ready for filling. Aluminium tubes are preferred to polyethylene for food applications because they permanently collapse as they are squeezed, unlike plastic tubes, and thus prevent air and potential contaminants from being drawn into the part-used product. Aluminium packaging is reviewed by Lamberti and Escher (2007).

25.2.3 Glass

Although glass shares some characteristics of the structure of a supercooled liquid, it is generally described as a solid below its glass transition temperature (Chapter 1, section 1.4.1). Glass jars and bottles are made by heating a mixture of sand, the main constituent being silica, soda ash and limestone (Table 25.4), with 30–50% broken glass (or 'cullet'), to a temperature of 1350–1600 °C. Alumina (aluminium oxide) improves the chemical durability of the glass, and refining agents reduce the temperature and time required for melting, and also help remove gas bubbles from the glass. Colourants include chromic oxide (green), iron, sulphur and carbon (amber), and cobalt oxide (blue). Clear (or 'flint') glass contains decolourisers (nickel and cobalt) to mask any colour produced by trace amounts of impurities (e.g. iron). Alternatively glass surfaces may be treated with titanium, aluminium or zirconium compounds to increase their strength and also enable lighter containers to be used.

Table 25.4 Composition and properties of glass

Property	Sodalime glass used for containers
Chemical composition	
(% by weight)	
Silica (SiO$_2$)	70–74
Sodium oxide (Na$_2$O)	12–16
Calcium oxide (CaO)	5–11
Aluminium oxide (Al$_2$O$_3$)	1–4
Magnesium oxide (MgO)	1–3
Potassium oxide (K$_2$O)	≈0.3
Sulphur trioxide (SO$_3$)	≈0.2
Ferric oxide (Fe$_2$O$_3$)	≈0.04
Titanium dioxide (TiO$_2$)	≈0.01
Glass transition temperature, T_g (°C)	573
Coefficient of thermal expansion (ppm/K, ~100–300 °C)	9
Density at 20 °C (g cm^{-3})	2.52
Heat capacity at 20 °C (kJ kg^{-1} K^{-1})	0.49

Adapted from Anon (2008e) and Anon (2008f)

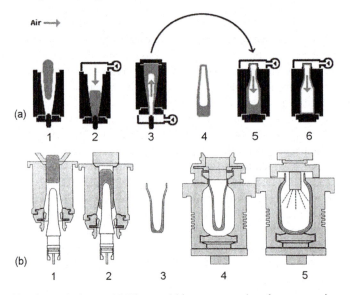

Fig. 25.3 Glass-blowing techniques: (a) Blow and blow process: 1, gob enters parison mould; 2, settle blow to form finish; 3, counter-blow to complete parison; 4, blank formed; 5, blank transferred to blow mould; 6, final shape blown (courtesy of Artech Publishing SRL (www.glassonline.com) (from Anon 2008h). (b) Press and blow process: 1, gob drops into parison mould; 2, plunger presses parison; 3, parison completed; 4, parison transferred to blow mould; 5, final shape blown (from Allaart-Bruin *et al.* 2004).

The molten glass is shaped in a mould by the 'blow-and-blow' process or the 'press-and-blow' process (Fig. 25.3). The 'Independent Section' (IS) machine is now used universally for bottle-making. Narrow neck containers are made by the blow-and-blow process in which a gob of glass at ≈1000 °C is placed in a 'parison' mould where a bubble is formed, and the moulding of the finish (the part that supports the closure) is done. The parison is then inverted and the body is formed by compressed air in the mould. Wide-neck containers are made by the press-and-blow process where the gob is shaped into a parison and the finish is moulded by the upward action of a plunger. It is then transferred for blow moulding as in the blow-and-blow process. Sarwar and Armitage (2003) describe a narrow neck press-and-blow (NNPB) process that was introduced to achieve better control over glass distribution in the container wall and has reduced glass weight by up to 33% without adversely affecting the mechanical performance of the container.

When bottles leave the moulds, the temperature is about 450 °C. If they were allowed to cool on their own, their low thermal conductivity (Chapter 10, section 10.1.1) would cause the interior to cool more slowly than the outside, and the different rates of contraction would cause internal stresses which would make the glass unstable. Glass is therefore annealed at 540–570 °C to remove stresses and then cooled under carefully controlled conditions in an annealing lehr (a long (30 m) tunnel) to prevent distortion or fracturing. Details of glass manufacturing techniques are given by Paine and Paine (1992). Paine (1991) describes improvements in glass-making technology to improve the strength of glass, reduce the risk of fracturing and maintaining container strength with 'light-weighting'. Louis (1998) describes advances in glass-making technology using plasma-arc crucibles to melt raw ingredients. The molten glass can then be co-extruded in a similar way to that currently used for plastic containers to produce jars or bottles of any shape, size or thickness.

Table 25.5   Relative strengths of different shaped glass containers

Container shape	Ratio of relative strengths
Cylindrical	10
Elliptical (2:1)	5
Square with round corners	2.5
Square with sharp corners	1

From Ramsland (1989)

Glass containers have the following advantages:

- Total barrier to moisture, gases, odours and micro-organisms. They are inert and do not react with, or migrate into, food products.
- Rigidity to give good vertical strength and allow stacking without damage to the container.
- Filling speeds comparable to those of cans.
- Suitable for heat processing when hermetically sealed, and transparent to microwaves.
- Marketing advantages include transparency to display the contents; glass is perceived by customers as high value (Nankivell 2001); containers can be decorated, or moulded in a wide variety of shapes and colours, especially for marketing high-value products such as liqueurs and spirits. However, simple cylindrical shapes are stronger and more durable (Table 25.5). Sharp corners and abrasion of glass surfaces weaken the container, and design features include a protruding 'shoulder' (Fig. 25.4) to minimise contact between containers during handling.
- Re-sealable, reuseable and recyclable. An advantage of glass over other recyclable packaging materials is that it can be recycled by simply reheating it until molten and then reforming the container, without loss of container quality or production of by-

Fig. 25.4   Glass container terminology (from Paine 1991).

products. The smooth internal surface of glass is easy to clean and sterilise, so making it reusable as multi-trip containers (e.g. milk and beer bottles).

The main disadvantages of glass include higher weight than other types of packaging, which incurs higher transport costs; lower resistance than other materials to fracturing and thermal shock; more variable dimensions than other containers; and potentially serious hazards from glass splinters or fragments in foods in the event of breakage. Critical faults in glass are broken, cracked or chipped glass, strands of glass stretched across the inside of containers, or bubbles in the glass that make it very thin. Because of the seriousness of potential faults, glass containers are 100% inspected using automated inspection equipment. Glassmaking is also a highly energy-intensive process, although energy consumption is reduced by 20–40% when recycled glass 'cullet' is used.

25.2.4   Flexible films

Flexible packaging describes any type of material that is not rigid, but the term 'flexible film' (or 'web') is usually reserved for non-fibrous plastic polymers (from Greek: *plastikos* = to form), which are less than 0.25 mm thick. The ability to shape plastics is due to long polymer molecules formed by addition reactions (e.g. for polyethylene, the $CH_2=CH_2$ group in ethylene splits at the double bond to form $CH_2–CH_2–CH_2$), or by condensation reactions (e.g. polyethylene terephthalate, where water is eliminated between ethylene glycol and terephthalic acid). Thermoplastic materials are able to undergo repeated softening on heating and hardening again on cooling, whereas thermosetting plastics cross-link the long molecules when heated or treated with chemicals and they do not re-soften.

In general, flexible films have the following advantages:

- A range of barrier properties against moisture and gases, and a range of wet and dry tensile and impact strengths.
- Heat sealable to prevent leakage of contents, and the ability to laminate to paper, aluminium or other plastics.
- Suitable for high-speed filling and ease of handling and printing. Convenient for the manufacturer, retailer and consumer.
- They add little weight to the product and fit closely to the shape of the food, thereby wasting little space during storage and distribution. Relatively low cost.

Ranges of mechanical, optical, thermal and barrier properties are produced for each type of polymer by variation in film thickness, orientation of polymer molecules, amount and type of additives and the type and thickness of coatings. Some films, including polyester, polyethylene, polypropylene and polyamide, are oriented by stretching softened material to cause the molecules to align in either one direction (uniaxial orientation) or two (biaxial orientation) to increase their strength, clarity and flexibility. The process may also increase the moisture and gas barrier properties compared with unoriented films (e.g. polypropylene in Table 25.2). Additives include plasticisers, stabilisers, pigments, antioxidants and slip agents. Plasticisers are added to soften the film and to make it more flexible, especially for use in cold climates or for frozen foods. Pigments may be added to avoid the need for large areas of printing. Films may be used singly, coated with polymer or aluminium, or produced as multi-layered laminates or co-extrusions. There are thus a very large number of possible combinations of polymer and treatment, to meet the varied requirements of foods.

Single films

Most polymer films are made by extrusion, in which pellets of the polymer are melted and extruded under pressure as a sheet or tube (see Chapter 15). Co-extrusion is the simultaneous extrusion of more than one polymer so that they fuse together to form a single film (see coextruded films below). Other methods are 'calendering', where a polymer such as polyvinyl chloride or ethylene vinyl acetate is passed through heated rollers until the required thickness is achieved; or solvent casting, in which a solution of a polymer (e.g. cellulose acetate) and additives is cast through a slot onto a stainless steel belt. The solvent is evaporated by heat to leave a clear sparkling film. Film surfaces are treated with ion beams, flame- or corona-treatments to improve sealability, adhesion, printability or barrier properties. Ozdemir *et al.* (1999) describe a new method of treating film surfaces using cold gaseous plasmas.

The most important types of film for food packaging are described below and a selection of their properties are shown in Table 25.2. The films considered in this section are:

- cellulose;
- ethylene vinyl acetate (EVA) or alcohol (EVOH);
- polyamides (PAs or nylons);
- polyethylene (low density (LDPE) or high density (HDPE));
- polyethylene terephthalate (PET);
- polypropylene (PP);
- polystyrene (PS) and high impact PS (HIPS);
- polyvinyl chloride (PVC);
- polyvinylidene chloride (PVdC);
- rubber hydrochloride.

The properties of single films are reviewed by Driscoll and Patterson (1999), who also quote the relative costs of films as follows: EVOH, 4.00; Nylon, 2.00; PET, 1.00; PP, 0.85; high-impact polystyrene (HIPS), 0.82; high-density polyethylene (HDPE), 0.75; low-density polyethylene (LDPE), 0.70.

Cellulose

Cellulose films are produced by mixing sulphite paper pulp (section 25.2.6) with caustic soda to dissolve it. It is allowed to 'ripen' for 2–3 days to reduce the length of polymer chains and form sodium cellulose. This is then converted to cellulose xanthate by treatment with carbon disulphide, ripened for 4–5 days to form 'viscose', and then cellulose is regenerated by extrusion or casting into an acid–salt bath to form cellulose hydrate. Glycerol is added as a softener and the film is then dried on heated rollers. Higher quantities of softener and longer residence times in the acid–salt bath produce more flexible and more permeable films.

Plain cellulose is a glossy transparent film that is odourless, tasteless, greaseproof and biodegradable within approximately 100 days. It is tough and puncture resistant, although it tears easily. It has low-slip and dead-folding properties and is unaffected by static build-up, which make it suitable for twist-wrapping. However, it is not heat sealable, the dimensions and permeability of the film vary with changes in humidity and it becomes highly permeable when wet. It is used for foods that do not require a complete moisture or gas barrier, including fresh bread and some types of sugar confectionery (Table 25.1). An international code is used to identify the various types of cellulose film (Table 25.6).

Table 25.6   Codes for cellulose films

Code	Explanation
A	Anchored (describes lacquer coating)
/A	Copolymer coated from aqueous dispersion
B	Opaque
C	Coloured
D	Coated one side only
F	For twist wrapping
M	Moisture-proof
P	Plain (non-moisture-proof)
Q	Semi-moisture-proof
S	Heat sealable
/S	Copolymer coated from solvent
T	Transparent
U	For adhesive tape manufacture
W	Winter quality (withstands low temperatures)
X	Copolymer coated on one side
XX	Copolymer coated on both sides

From Driscoll and Paterson (1999)

Cellulose acetate is a clear, glossy transparent, sparkling film that is permeable to water vapour, odours and gases and is mainly used as a window material for paperboard cartons (section 25.2.6). It is one of the biodegradable cellulose films that are being developed further (section 25.4.1). It requires the addition of plasticisers to make into films. It has good printability, rigidity and dimensional stability, and although these films can tear easily, they are tough and resistant to puncturing.

EVA and EVOH
EVA is LDPE polymerised with vinyl acetate. It has high mechanical strength, and flexibility at low temperatures. It has greater resilience than PVC and greater flexibility than LDPE, with higher permeability to water vapour and gases. EVA that contains less than 5% vinyl acetate is used for deep-freeze applications; films with 6–10% vinyl acetate are used in bag-in-box applications and milk pouches; and above 10% vinyl acetate, the material is used as a hot-melt adhesive.

EVOH has a high barrier to oxygen comparable to PVdC, but it is more expensive, hydrophilic and therefore permeable to moisture, and it has a high sealing temperature (185 °C). It is mostly used as a laminate with polypropylene or polyethylene, which provide the moisture barrier properties and heat sealability.

Polyamides (PA or nylons)
PAs are clear mechanically strong films over a wide temperature range (from −60 to 200 °C) that have low permeability to gases and are greaseproof. Water vapour permeability differs among the different types of film. However, the films are expensive to produce, require high temperatures to form a heat seal (≈240 °C), and the permeability changes at different storage humidities. They may be used with other polymers by coating, coextruding or laminating to make them heat sealable at lower temperatures and to improve the barrier properties. They are used to pack meats and cheeses.

Polyethylene (commonly 'polythene')
LDPE film is heat sealable, chemically inert, odour free and shrinks when heated. It is a good moisture barrier but has relatively high gas permeability, sensitivity to oils and poor

odour resistance. Low-slip properties can be introduced for safe stacking, or conversely high-slip properties permit easy filling of packs into an outer container. It is less expensive than most films and is therefore widely used for pouches, bags, coating papers or boards and as a component in laminates. It also has applications in shrink- or stretch-wrapping (Chapter 26, section 26.3). Stretch-wrapping either uses thinner LDPE than shrink-wrapping ($25-38\,\mu$m compared with $45-75\,\mu$m) or linear low-density polyethylene (LLDPE) is used at thicknesses of $17-24\,\mu$m. This material has a highly linear arrangement of molecules and the distribution of molecular weights is smaller than for LDPE. It therefore has greater strength and a higher restraining force. The cling properties of both films are biased on one side, to maximise adhesion between layers of the film but to minimise adhesion to adjacent packages. LDPE is also used as a copolymer in some tubs and trays (section 25.2.5).

HDPE is stronger, thicker, less flexible and more brittle than LDPE and has lower permeability to gases and moisture. It has a higher softening temperature ($121\,°C$) and can therefore be heat sterilised or used in 'boil-in-the-bag' applications. Sacks made from $0.03-0.15\,$mm HDPE have a high tear strength, tensile strength, penetration resistance and seal strength. They are waterproof and chemically resistant and are used instead of multi-wall paper sacks for shipping containers. A foamed HDPE film is thicker and stiffer than conventional film and has dead-folding properties. It can be perforated with up to 80 holes cm^{-1} for use with fresh foods or bakery products. When unperforated, it is used for edible fats. Both types are suitable for shrink-wrapping.

PET

PET is a very strong transparent glossy film that has good moisture- and gas-barrier properties. It is flexible at temperatures from -70 to $135\,°C$ and undergoes little shrinkage with variations in temperature or humidity. There are two types of PET: amorphous and crystalline. Amorphous PET (APET) is clear, and is biaxially oriented to develop the full tensile strength for use in films (e.g. boil-in-bag) or bottles (e.g. for carbonated drinks). It has been described in detail by Turtle (1990). Crystalline PET (CPET) is opaque and is used for microwave trays and semi-rigid containers, such as tubs (section 25.2.5).

PP

PP is a strong film, except at low temperatures when it becomes brittle, and has low permeability to water vapour and gases. It has a high sealing temperature ($\approx170\,°C$) and is therefore coated or laminated with polyethylene or PVdC/PVC to heat seal at a lower temperature. It is used in similar applications to LDPE. Oriented polypropylene (OPP) is a clear glossy film with good optical properties and a high tensile strength and puncture resistance, even at low temperatures. It has moderate permeability to gases and odours and a higher barrier to water vapour, which is not affected by changes in humidity. It is thermoplastic and therefore stretches, although less than polyethylene, and has low friction, which minimises static build-up and makes it suitable for high-speed filling equipment (Chapter 26, section 26.2.1). The coated or laminated forms are used in a wide range of applications including packs for cheese, meat, coffee and biscuits. Biaxially oriented polypropylene (BOPP) has similar properties to oriented PP but is much stronger. PP and BOPP are used for bottles, jars, crisp packets and biscuit wrappers among many other applications. Details are given by Anon (2006a).

PS

PS is a brittle, clear sparkling film that has high gas permeability. It is biaxially oriented (BOPS) to improve the barrier properties and strength, but it still has a relatively high permeability to gases. As a film, it is mainly used for wrapping fresh produce, but it is also used in the form of foam to make cartons or trays for eggs, fresh fruits and takeaway meals. It is also coextruded with EVOH or PVdC/PVC to make semi-rigid containers and blow-moulded bottles. HIPS is used to make rigid/semi-rigid containers and trays that are freezeable, but they are not suitable for use in microwave or conventional ovens or for modified atmosphere packaging (MAP). They are inexpensive and not as brittle as PS trays.

PVC
PVC is a clear transparent, brittle film that can be made by either extrusion or calendering. Plasticisers are used to make the film flexible and the amount and type of plasticiser determine the permeability to water vapour, gases and volatiles. It is greaseproof and can be oriented to make it heat-shrinkable. Highly plasticised films are used in stretch-wrapping and as 'cling film'.

PVdC
Single PVdC film is stiff and brittle and it is therefore used as a copolymer with polyvinyl chloride. This has very low gas and water vapour permeabilies and is heat shrinkable and heat sealable; it is fat resistant and does not melt in contact with hot fats, making it suitable for 'freezer-to-oven' foods. It is very strong and is therefore used in thin films. However, it has a brown tint that limits its use in some applications. The oriented copolymer has greater strength and barrier properties and is heat shrinkable: it is used for shrink-wrapping poultry and meats and as a component of laminates. PVdC is also used as a coating for films and bottles to improve their barrier properties.

Rubber hydrochloride
Rubber hydrochloride is similar to PVC but becomes brittle in ultraviolet light and at low temperatures and is penetrated by some oils. It is not widely used in food applications.

Coated films
Films are coated with other polymers or aluminium to improve their barrier properties or to impart heat sealability. For example nitrocellulose with added waxes and resins is coated on one side of cellulose film to further improve the moisture and gas barrier properties. A nitrocellulose coating on both sides of the film improves the barrier to oxygen, moisture and odours and enables the film to be heat sealed when broad seals are used. A PVdC/PVC coating is applied to both sides of cellulose, using either an aqueous dispersion (MXXT/A cellulose) or an organic solvent (MXXT/S cellulose). MXDT indicates a coating on one side (see Table 25.6). In each case the film is made heat sealable and the barrier properties are improved (Table 25.2). A coating of vinyl acetate on cellulose gives a stiffer film that has intermediate permeability. Sleeves of this material are tough, stretchable and permeable to air, smoke and moisture. They are used for example for packaging meats before smoking and cooking.

A thin coating of aluminium (termed 'metallisation') produces a very good barrier to oils, gases, moisture, odours and light. Metallised film is less expensive and more flexible than foil laminates which have similar barrier properties, and it is therefore suitable for high-speed filling on form–fill–seal equipment (Chapter 26, section 26.2.1). Cellulose, PP or polyester are metallised by depositing vaporised aluminium particles onto the surface of a film under vacuum. The degree of metallisation is expressed in optical density units, up to a maximum of four units. Metallised polyester has higher barrier properties than metallised polypropylene, but PP is used more widely as it is less expensive.

Laminated films

Lamination of two or more films improves the appearance, barrier properties and/or mechanical strength of a package. Commonly used laminates are described in Table 25.7 and recent developments in MAP (section 25.3) have included laminates of nylon–LDPE, nylon–PVdC–LDPE and nylon–EVOH–LDPE for non-respiring products. The nylon provides strength to the pack, EVOH or PVdC provide the correct gas and moisture barrier properties and LDPE gives heat sealability. PVC and LDPE laminates are commonly used for respiring MAP products (Smith *et al.* 1990).

The most versatile method of lamination is adhesive laminating (or 'dry bonding') in which an adhesive is first applied to the surface of one film and dried. The two films are then pressure bonded by passing between rollers. Synthetic adhesives are mostly aqueous dispersions or suspensions of polyvinyl acetate with other compounds (e.g. polyvinyl alcohol, 2-hydroxycellulose ether) to give a wide range of properties. Two-part urethane adhesives, consisting of a polyester or polyether resin with an isocyanate cross-linking agent, are also widely used. Co-polymerised vinyl acetate and ethylene- or acrylic-esters give improved adhesion for producing laminated films. They are also used for case

Table 25.7   Selected laminated films used for food packaging

Type of laminate	Typical food application
Polyvinylidene chloride-coated polypropylene–polyvinylidene chloride-coated polypropylene	Crisps, snackfoods, confectionery, ice cream, biscuits, chocolate confectionery
Polyvinylidene chloride-coated polypropylene–polyethylene	Bakery products, cheese, confectionery, dried fruit, frozen vegetables
Polypropylene–ethylene vinyl acetate	Modified atmosphere packaged (section 25.3) bacon, cheese, cooked meats
Biaxially oriented polypropylene–nylon–polyethylene	Retort pouches
Cellulose–polyethylene–cellulose	Pies, crusty bread, bacon, coffee, cooked meats, cheese
Cellulose acetate–paper–foil–polyethylene	Dried soups
Metallised polyester–polyethylene	Coffee, dried milk, bag-in-box packaging, potato flakes, frozen foods, modified atmosphere packaged (section 25.3) foods
Polyethylene terephthalate aluminium–polypropylene	Retort pouches
Polyethylene–nylon	Vacuum packs for bulk fresh meat, cheese
Polyethylene–aluminium–paper	Dried soup, dried vegetables, chocolate
Nylon–polyvinylidene chloride-polyethylene–aluminium–polyethylene	Bag-in-box packaging
Nylon–medium-density ethylene–butene copolymer	Boil-in-bag packaging

The type of laminate reads from the outside to the inside of the package. All examples of polyethylene are low-density polyethylene.

sealing, spiral tube winding, pressure-sensitive coatings and labelling of plastic bottles. Solvent-based systems have a number of problems including environmental considerations, clean air regulations, higher cost, safety from fire hazards, toxicity and production difficulties, which mean that these are used only when other systems are not suitable, and they are likely to be phased out altogether. Not all polymer films can be successfully laminated; the two films should have similar characteristics and the film tension, adhesive application and drying conditions should be accurately controlled to prevent the laminate from blocking (not unwinding smoothly), curling (edges curl up) or delaminating (separation of the layers).

Coextruded films

Coextrusion is the simultaneous extrusion of two or more layers of different polymers to form a single film. Coextruded films have three main advantages over other types of film: they have very high barrier properties, similar to multi-layer laminates but produced at a lower cost; they are thinner than laminates and closer to mono-layer films and are therefore easier to use on forming and filling equipment; and the layers cannot separate. To achieve strong adhesion, the copolymers used in coextruded films should have similar chemical structures, flow characteristics and viscosities when melted. There are three main groups of polymers:

- Olefins (LDPE, HDPE and PP).
- Styrenes (polystyrene and acrylonitrile–butadiene–styrene (ABS)).
- PVC polymers.

All materials in each group adhere to each other, as does ABS with PVC, but other combinations must be bonded with EVA. There are two main methods of producing coextrusions: blown films and flat-sheet coextrusion. Blown-film coextrusions are thinner than flat-sheet types and are suitable for high-speed form–fill–seal and pouch or sachet equipment (Chapter 26, section 26.2.1). Typically a three-layer coextrusion has an outside presentation layer, which has a high gloss and printability, a middle bulk layer which provides stiffness, strength and split resistance, and an inner layer which is suitable for heat sealing. These films have good barrier properties and are more cost effective than laminated films or wax-coated paper. They are used, for example, for confectionery, snackfoods, cereals and dry mixes. A five-layer coextrusion is used to replace metallised polyester for bag-in-box applications. Flat-sheet coextrusions (75–$3000 \, \mu m$ thick) are formed into pots, tubs or trays (Table 25.8).

Table 25.8 Selected applications of flat-sheet coextrusions

Type of coextrusion	Properties	Applications
High-impact polystyrene–polyethylene terephthalate		Margarine, butter
Polystyrene–polystyrene–polyvinylidene chloride–polystyrene	Ultraviolet and odour barrier	Juices, meats, milk products
Polystyrene–polystyrene–polyvinylidene chloride–polyethylene	Ultraviolet and odour barrier	Butter, cheese, margarine, coffee, mayonnaise, sauce
Polypropylene–saran–polypropylene	Retortable trays	Sterilised foods
Polystyrene–ethylene vinyl acetate–polyethylene	Modified atmosphere packs	Meats, fruits

25.2.5   Rigid and semi-rigid plastic containers

Trays, cups, tubs, bottles and jars are made from single or coextruded polymers. The main advantages, compared with glass and metal, are as follows:

- Lower weight, resulting in savings of up to 40% in transport and distribution costs. Cups, tubs and trays are tapered (a wider rim than base) for more compact stacking for transport and storage.
- Lower production temperature than glass (300 °C compared with 800 °C) and therefore incur lower energy costs. They are produced at relatively low overall cost.
- Precisely moulded into a wider range of shapes than glass, and are tough, unbreakable (impact and pressure resistance) and easy to seal. They have greater chemical resistance than metals.
- Can be easily coloured for aesthetic appeal and UV-light protection.

However, they are not reusable, have a lower heat resistance and are less rigid than glass or metal. Tapered cups, tubs and trays facilitate removal from the mould and are made without sharp corners that would become thin during the moulding process and be a potential source of leakage. In general the height of the container should not exceed the diameter of the rim in order to maintain a uniform thickness of material.

There are seven methods of container manufacture (thermoforming, blow moulding, injection moulding, injection blow moulding, extrusion blow moulding, stretch blow moulding and multi-layer blow moulding). The first is thermoforming in which the film is softened over a mould, and a vacuum and/or pressure is applied (Fig. 25.5a). The six main materials used for thermoforming are PVC, PS, PP, PVC–PVDC, PVC–PVF and PVC–PE–PVDC (Paine 1991). These containers are thin-walled and possess relatively poor mechanical properties. Examples include trays or punnets for chocolates, eggs, or soft fruit, and cups or tubs for dairy products, margarine, dried foods or ice cream.

Blow moulding is similar to glass making (section 25.2.3) and is used in either a single- or two-stage process for producing bottles, jars or pots. Containers are used for example for cooking oils, vinegar, beverages and sauces. In injection moulding, grains of polymer are mixed and heated by a screw in a moulding machine and injected under high pressure into a cool mould. This method is used for wide-mouthed containers (e.g. tubs and jars) and for lids.

In injection blow moulding (Fig. 25.5b), the polymer is injection-moulded around a blowing stick and, while molten, this is transferred to a blowing mould. Compressed air is then used to form the final shape of the container. Injection blow moulding of HDPE, PP and PS bottles gives accurate control of the container weight and precise neck finishes. It is more efficient than extrusion blow moulding and is used for small bottles (<500 ml), but it is not possible to produce containers with handles and capital costs are high (Paine 1991).

In extrusion blow moulding (Fig. 25.5c), a continuously extruded tube of softened polymer is trapped between two halves of a mould and both ends are sealed as the mould closes. The trapped part is inflated by compressed air to the shape of the mould. It is used for >200 ml bottles up to 4500 litre tanks, and can be used to form handles and offset necks. In both types of blow moulding, careful control is needed to ensure uniform thickness in the container wall.

Stretch blow moulding uses a pre-form (or parison), made by injection, extrusion or extrusion blow moulding. It is brought to the correct temperature and rapidly stretched and cooled in both directions by compressed air. The biaxial orientation of the molecules produces a clear container that has increased stiffness, tensile strength, surface gloss,

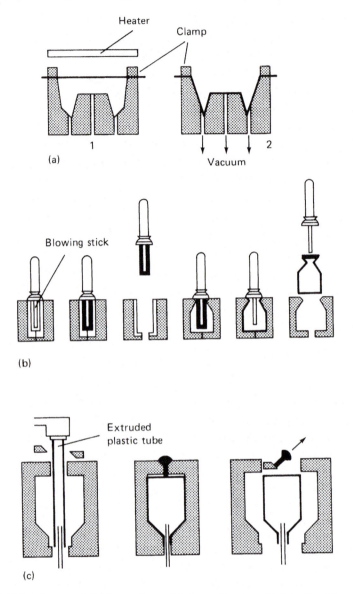

Fig. 25.5   Manufacture of rigid containers: (a) thermoforming; (b) injection blow moulding; (c) extrusion blow moulding (after Briston 1987).

impact resistance, barrier properties to moisture and gases and stability over a wide temperature range. It is mainly used for PET but also for PVC and PP bottles between 450 ml and 1.8 litres.

Multilayer blow moulding using polyethylene vinyl alcohol is high cost but has excellent oxygen barrier properties, and can be used as a thin layer, sandwiched between two layers of PE or PP (HDPE/adhesive/EVOH/adhesive/HDPE).

An important development is PVC blow-moulded bottles for wine, oils and juice concentrates. Previously, this was difficult because thermal degradation of PVC occurs at a temperature a little above the melting point, but the technology has been developed to overcome this problem.

High-nitrile resins (e.g. acrylonitrile–methyl acrylate and acrylonitrile–styrene copolymers) are moulded to form containers that have very good barrier properties and are used to package processed meat, cheese, margarine and peanut butter (Briston 1987).

HIPS and ABS are widely used for thermoformed trays, tubs and cups to contain yoghurt, margarine, cheeses, spreads, ice cream and desserts. In their natural state they are translucent but can be easily coloured. PVC trays, tubs and bottles are made by extrusion- or injection-stretch blow moulding. Food-grade PVC is tough, clear and easy to colour. It has good oil resistance and low gas permeability and is therefore used in trays for meat products and chocolates, in tubs for margarine or jams, and in bottles for edible oils, fruit juices, squashes and concentrates. However, it has lower strength than PET and is not used for carbonated beverages as it cannot withstand the pressure of carbonation. It also has a relatively low temperature resistance (65–70 °C). PP is more expensive than HIPS or PVC and is less widely used for rigid/semi-rigid containers. However, it is resistant to a wider temperature range (deep-freeze temperatures to 120–140 °C) and also provides a good barrier to water vapour and oxygen.

HDPE is at present the most common material used for bottles and jars. It is used for vinegar, milk, syrups, and as drums for salt and bulk fruit juices. PP coextrusion, in which EVA copolymer is a central barrier material, is used for mustard, mayonnaise, jams, tomato ketchup and other sauces to give a shelf-life of 18 months. It is shatterproof, oxygen and moisture resistant, squeezable and suitable for hot filling.

Multi-chamber PET trays have a hygienic smooth white finish. They are fat resistant, heat sealable and lightweight. They are used for example for chilled or frozen ready meals, where the cover is left on during either microwave or conventional cooking and then peeled off to give an attractive table dish. Details of the materials, methods of production and processing are given by Faithfull (1988). Anon (2006b) describes an adhesive patch valve that is fitted over two pre-formed slits in a polyester/polypropylene lid of a microwaveable tray. The pack is intended for microwave steaming foods such as fish fillets, and as the temperature of the product increases, the valve material shrinks to reveal the apertures in the lid and allow the controlled release of steam, while maintaining a positive steam pressure inside the pack. The valve shrinkage is irreversible to prevent it re-closing, which would cause the pack to implode on cooling.

A coextruded five-layer sheet of PP or polycarbonate, with polyvinylidene chloride or EVA barrier layers, is used to form heat-sterilisable trays and pots, by either injection moulding, blow moulding or thermoforming. Plastic cans are made from similar material which is thermoformed or injection blow moulded to form the can body. They are sealed using easy-open aluminium ends and processed on existing canning lines with considerably reduced noise levels. Brody (1992) and May (2004) have reviewed the development of retortable trays, flexible pouches and plastic cans.

25.2.6   Paper and board

Paper pulp is produced by grinding wood chips (mostly spruce) and digesting the pulp by alkaline (sulphate) or acid (sulphite) hydrolysis. This stage dissolves lignin, carbohydrates, resins and gums in the pulp and they are removed by washing to leave mostly cellulose fibres. The alkaline (or 'kraft') process (Swedish for 'strong') involves digesting in sodium hydroxide and sodium sulphate for several hours. The process gives higher yields and process chemicals are more completely and economically recovered (Paine 1991). In the sulphite process, sulphur dioxide and calcium bisulphite are heated with pulp at 140 °C, washed and then bleached with calcium hypochlorite to give very pure cellulose fibres. Both types of pulp then undergo a beating process to split individual cellulose fibres longitudinally. This produces a mass of thin fibrils (termed 'fibrillation') that bind together more strongly to give increased burst, tensile and tear strengths. The extent of beating and the thickness of fibres determine the strength of a paper. Additives are mixed into the pulp to impart particular properties, including:

- fillers ('loading agents') such as china clay, to increase the opacity and brightness of paper and improve surface smoothness and printability;
- binders, including starches, vegetable gums, and synthetic resins to improve the strength (tensile, tear and burst strength);
- resin or wax emulsions (sizing agents) to reduce penetration by water or printing inks;
- pigments to colour the paper and other chemicals to assist in the manufacturing process (e.g. anti-foaming agents).

Two methods are used to produce paper from the pulp: fibres are suspended in water and transferred to a finely woven mesh belt in a 'Fourdrinier' machine. Water is removed by vacuum filtration (Chapter 5, section 5.2) to reduce the moisture content of the fibres to 75–80%, and this is then reduced to 4–8% in the paper by pressing and drying. In the second method, a series of six or more wire mesh cylinders are partly submerged in the pulp suspension. As they rotate, they pick up fibres and deposit them onto a moving felt belt. This absorbs water from the paper and a press reduces the moisture content of the fibres to 60%. The paper is then dried using heated cylinders.

Paper has a number of advantages as a food packaging material: it is produced in many grades and converted to many different forms (below); it is recyclable and biodegradable; and it is easily combined with other materials to make coated or laminated packaging. All types of paper protect foods from dust and soils, but they have negligible water vapour or gas barrier properties and are not heat sealable.

Types of papers
High-gloss machine glazed (MG) and machine finished (MF) papers are produced by passing them between a series of highly polished cylinders (a 'calender stack'), in which one roller is driven and the other is moved by friction with the paper to create a smooth surface. Kraft paper is a strong paper that is used for 25–50 kg multi-wall sacks for powders, flour, sugar, fruits and vegetables. It can be bleached white, printed or used unbleached (brown). It is usually used in several layers or 'plies' to give the required strength. Sack material is described from the outer ply inwards according to the number and weight (or 'substance' in g per m^2) of the layers. For example 2/90, 1/80 means that there are three plies, the outer two having a weight of 90 g m^{-2} and the inner having a weight of 80 g m^{-2}.

Vegetable parchment is produced from sulphate pulp that is passed through a bath of concentrated sulphuric acid to swell and partly dissolve the cellulose fibres, to make them plasticised. This closes the pores and fills voids in the fibre network to make the surface more intact than kraft paper, and thus makes the paper resistant to grease and oils and gives greater wet strength properties. It is used to pack butter, cheese and fresh fish or meat.

Sulphite papers are lighter and weaker than sulphate papers. They are used for grocery bags and confectionery wrappers, as an inner liner for biscuit packs and in laminations. They may be glazed to improve wet strength and oil resistance (MG sulphite paper). Greaseproof paper is made from sulphite pulp in which the fibres are more thoroughly

Table 25.9   Properties of the main types of food papers

Paper	Weight range ($g\,m^{-2}$)	Tensile strength ($kN\,m^{-1}$)	Examples of uses
Kraft	70–300	MD 2.4–11.3 CD 1.2–5.2	Multi-wall sacks, liners for corrugated board
Sulphite	35–300	Variable	Small bags, pouches, waxed papers, labels, foil laminates
Greaseproof	70–150	MD 1.7–4.4 CD 0.85–2.1	Paper for bakery products, fatty foods
Glassine	40–150	MD 1.4–5.2 CD 0.85–2.8	Odour resistant and greaseproof bags, wrappers or liners for boxes, suitable for wax coating to make them water resistant – for dry cereals, potato crisps, dried soups, cake mixes, coffee, sugar
Vegetable parchment	60–370	2.1–14.0	High wet strength and grease-resistant bags, wrappers or liners for boxes used for meat, fish, fats, etc.
Tissue	17–50	–	Soft wrapping paper for bread, fruits, etc.

MD = machine direction, CD = cross direction.
Adapted from Paine (1991)

beaten to produce a closer structure. It is resistant to oils and fats, and although this property is lost when the paper becomes wet, it is widely used for wrapping fish, meat and dairy products. Glassine is similar to greaseproof paper, but is given additional calendering to increase the density and produce a close-knit structure and a high-gloss finish. It is more resistant to water when dry but loses the resistance once it becomes wet. Tissue paper is a soft non-resilient paper used, for example, to protect fruits against dust and bruising. Some properties of papers are given in Table 25.9.

Coated papers
Many papers are treated with wax by coating, dry waxing (in which wax penetrates the paper while hot) or wax sizing (in which the wax is added during the preparation of the pulp). Wax provides a moisture barrier and allows the paper to be heat sealed. However, a simple wax coating is easily damaged by folding or by abrasive foods, and this is overcome by laminating the wax between layers of paper and/or polyethylene. Waxed papers are used for bread wrappers and inner liners for cereal cartons and their benefits are described by Whittle (2000).

The thickness of some plastic films that is needed to give required degree of protection is less than that which can be handled on filling and forming machines. Therefore coating an expensive barrier film onto a thicker, cheaper paper substrate gives the desired strength and handling properties. Coatings can be applied (1) from aqueous solutions (cellulose ethers, polyvinyl alcohol) to make papers greaseproof, (2) from solvent solutions or lacquers, (3) from aqueous dispersions (e.g. polyvinylidene chloride), (4) as hot-melts (e.g. microcrystalline wax, polyethylene and copolymers of ethylene and vinyl acetate) to increase gloss, durability, scuff and crease resistance and permit heat sealability), or (5) as extrusion coatings (e.g. polyethylene). Although they are not affected by temperature, all papers are sensitive to humidity variations, and coated papers in particular may lose moisture from one face and are therefore prone to curling. Smooth papers block if pressed together in a stack. The optimum storage conditions for papers are a temperature of about 20 °C and a relative humidity of approximately 50%.

Paperboard cartons

Paperboard is a generic term covering boxboard, chipboard and corrugated or solid fibreboards. Typically, paperboard has the following structure:

- A top ply of bleached pulp to give surface strength and printability.
- Middle plies of lower grade material.
- An under-liner of white pulp to stop the grey/brown colour of middle plies showing through.
- A back ply of either low grade pulp or better grade pulp if strength or printability are required.

All plies are glued together with hot-melt or aqueous adhesives. Boards are made in a similar way to paper but are thicker to protect foods from mechanical damage. The main characteristics of board are:

- thickness, stiffness and the ability to crease without cracking;
- surface properties, the degree of whiteness and suitability for printing.

White board is suitable for contact with food and is often coated with polyethylene, PVC or wax for heat sealability. It is used for ice cream, chocolate and frozen food cartons. Chipboard is made from recycled paper and is not used in contact with foods. It is used for example as the outer cartons for tea and breakfast cereals. It is often lined with white board to improve appearance and strength. Duplex board has two layers: the liner is produced from bleached woodpulp and the outer is unbleached. It is used for frozen food and biscuit cartons. These paperboards are typically 0.3–1.0 mm thick.

Fibreboard (>0.11 mm thick) is either solid or corrugated. The solid type has an outer kraft layer and an inner bleached board. It is able to resist compression and to a lesser extent impact. Small fibreboard cylinders (or 'composite cans') are made using single ply board, either with or without an aluminium foil layer and a LDPE inner layer. They are spirally wound around a mandrel in a helical pattern and bonded with an adhesive. The correct can length is cut, flanges are formed at each end and they are fitted with plastic or metal caps, which may have an easy-open end or a pouring mechanism. Small tubs or cans are used for frozen juice concentrates, snackfoods, confectionery, nuts, salt, cocoa powder and spices. Larger drums (up to 375 litres) are used as a cheaper alternative to metal drums for powders and other dry foods and, when lined or laminated with polyethylene, for cooking fats. They are lightweight, resist compression and may be water resistant for outside storage. Other products that are handled in drums include frozen fruits and vegetables, peanut butter, sauces and wine. A similar material, made from single-ply board with a moisture-proof membrane below the surface is used to make cases to transport chilled foods. The membrane prevents the board from absorbing moisture and retains its strength throughout the chill chain.

Corrugated board has an outer and inner lining of kraft paper with a central corrugating (or 'fluting') material. This is made by softening the fluting material with steam and passing it over corrugating rollers. The liners are then applied to each side using a suitable adhesive (Fig. 25.6). The board is formed into 'cut-outs' that are then assembled into cases at the filling line (Chapter 26, section 26.1). Corrugated board resists impact, abrasion and compression damage, and is therefore used for shipping containers. Smaller more numerous corrugations (for example 164 flutes m^{-1}, with a flute height of 2.7 mm) give rigidity, whereas larger corrugations (for example 127 flutes m^{-1}, and a flute height of 3.4 mm) or double and triple walls give resistance to impact damage (Table 25.10). Twin-ply fluting, with a strengthening agent between the layers, has the same stacking strength

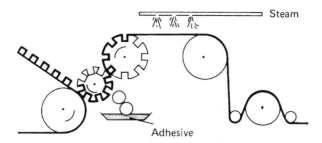

Steam

Adhesive

Fig. 25.6   Corrugated board manufacture.

Table 25.10   Fluting in corrugated board

Flute configuration	No. flutes/m	Flute height (mm)	Minimum flat crush ($N\,m^{-2}$)
A (coarse)	104–125	4.5–4.7	140
B (fine)	150–184	2.1–2.9	180
C (medium)	120–145	3.5–3.7	165
E (very fine)	275–310	1.15–1.65	485

From Paine (1991)

but half the weight of solid board, and a space saving of 30% compared with double-corrugated boards of comparable strength. Corrugated cartons are used as shipping containers for bottled, canned or plastic packaged foods. Boards should be stored in a dry atmosphere to retain their strength and prevent delamination of the corrugated material. Wet foods may be packed by lining the corrugated board with polyethylene, which also reduces moisture migration and tainting (e.g. for chilled bulk meat). Alternatively the liner may be a laminate of greaseproof paper, coated with microcrystalline wax and polyethylene, used for fresh fruit and vegetables, dairy products, meat and frozen foods.

Laminated paperboard cartons are made from combinations of LDPE, paper, aluminium, polyvinylidene chloride or polyamides and are used to make cartons for packaging aseptically sterilised foods (Fig. 25.7) (Chapter 13, section 13.2). There are two systems: in one the material is supplied as individual pre-formed collapsed sleeves with the side seam formed by the manufacturer (Fig. 25.8). The cartons are erected at the filling line, filled and sealed, with the top seal formed above the food. This creates a headspace to both enable the product to be mixed by shaking, and to reduce the risk of spillage on opening. In a second system, the laminate is supplied as a roll, and is formed into cartons on form–fill–seal equipment (Chapter 26, section 26.2.1). The advantages of this system include lower space requirement for storage of the material and easier handling. Details of the production of cartons by both methods are given by Schraut (1989).

These types of carton have the following advantages over metal or glass containers:

- They are very strong and cannot fracture like glass.
- They require no additional capping or labelling.
- They incur lower energy costs in manufacture and give a substantial saving in weight.
- They save on space in distribution, storage and shelf display. Bold graphics on the sides are able to give a 'billboard effect' on display shelves.

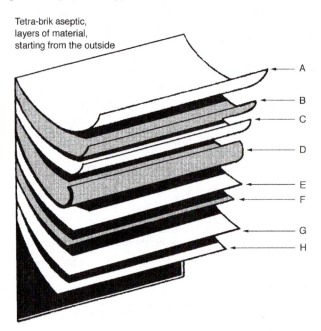

Fig. 25.7 Laminated material used by Tetra-brik aseptic machinery: A, polyethylene; B, printing ink; C, and D, duplex paper; E, polyethylene; F, aluminium foil; G and H, polyethylene (courtesy of TetraPak Inc.) (Anon 2008j).

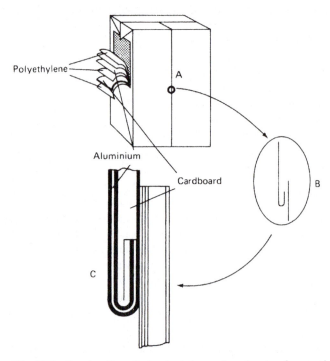

Fig. 25.8 Construction of a laminated paperboard carton for aseptic foods: A, side seam; B, folded to prevent leakage of contents into laminate; C, laminate chamfered to prevent seam bulging (courtesy of Bowater PKL Ltd).

Adhesives

Natural and synthetic water-based adhesives are used to prepare paper and fibreboard containers. They are safe, low cost and have good strength. They are used for seams of bags, corrugated board, labelling cans and bottles and winding paper tubes. Starch pastes and dextrins are used for carton and case closing, tube winding and paperboard laminating and labelling. Latex is blended with acrylic resins and used for cold-seal coatings for confectionery wrappers, carton making, and water-resistant labelling of drums, cans and bottles. Hot melt adhesives contain no water or solvent and are heated to form a solution for application, and then set rapidly on cooling. EVA copolymer is the most widely used and low molecular weight polyethylene is also used for carton and bag sealing.

Moulded paperpulp containers

These are lightweight containers, normally having ≈2.5 mm thick walls, that are able to absorb shocks by distortion and crushing at the point of impact, without transmitting them to the product. They are made from paper pulp using a mould instead of a wire screen, by either pressure injection or suction moulding. Pressure injection uses air at 480 °C to form the pack. It leaves the mould having 45–50% moisture and is then dried. The suction process uses a perforated mould and the pack leaves the mould at 85% moisture before drying. The capital cost of suction moulding is higher than pressure moulding and it is therefore used for higher production rates. The main advantages of pulp, compared with expanded polystyrene and single- or double-wall corrugated board, include low fire risk and no static problems compared with polystyrene and the capacity to make more complex shapes than corrugated board. It is used for egg trays, bottle sleeves and trays for fruits, meats and fish.

25.2.7 Combined packaging systems

Changes in handling and distribution methods (Chapter 27, section 27.3) mean that few foods are now packaged in one material, and an additional one or more shipping containers are used. A common combined packaging system is the use of display cartons to contain multiple packs of food packaged in flexible film. These in turn are either collated and stretch-wrapped, or placed in corrugated board shipping containers and palletised. Other systems have shrink-wrapped trays of cans or glass containers that are palletised and stretch-wrapped, or shrink-wrapped (Chapter 26, section 26.3). Bag-in-box packaging consists of a laminated or coextruded film bag, fitted with an integral plastic tap, and contained within a solid or corrugated fibreboard display case (Briston 1990). These are packed into fibreboard shipping containers. The bag collapses evenly as liquid is withdrawn, which prevents the product from becoming trapped in the folds of the bag and prevents oxidation of the product by air. It is a convenient lightweight secure container for liquid foods (e.g. wines, fruit juices, edible oils, syrups, milk and dairy products, liquid egg, purées and sauces) and is fully recyclable. The design allows savings in weight and space compared with glass (e.g. a saving in weight from 5.13 to 3.24 kg and a saving in volume from 0.011 to 0.004 m^3 for containers of 3 litre capacity, compared with glass) and up to 66% weight reduction compared with bottles. Bag capacities range from 1 to 1400 litres (Anon 2005a).

IBCs, including combo-bins and large bags made from woven plastic fabric, have been introduced to increase handling efficiencies and have largely displaced wooden crates and cases. IBCs have a capacity between that of a bulk road tanker and 220 litre drums (e.g. 1000 litre containers with integral pallet and bottom discharge valve), and are mainly used for powders and liquids. Examples of IBCs include a seamless, 6–9 ply corrugated fibreboard container, capable of withstanding 20 tonnes compression. It can be lined with a multi-ply film for liquids and, because no metal is used in the construction, it is biodegradable and more easily recyclable (Anon 1998). IBCs are claimed to carry 20% more product than drums in a given space, and because they are flat when empty, they use 80% less storage space.

25.3   MAP

The term 'MAP' is used in this book to describe the introduction of an atmosphere, other than air, into a food package without further modification or control. Other terminology that is in use to more specifically designate different operations includes:

- controlled atmosphere packaging (CAP) – continuous monitoring and control of gas composition in bulk containers;
- equilibrium modified atmosphere (EMA) or passive atmosphere modification (PAM) – gas flushing of packs of fresh fruits or vegetables or sealing without gas modification to allow a gas equilibrium to be established as a result of respiration;
- gas-exchange preservation (GEP) – replacing air with a series of gases in quick succession to inhibit enzymes or kill micro-organisms, before packing in nitrogen;
- vacuum packing (VP) – the removal of the majority of air from a pack that has low oxygen permeability, with subsequent changes in gas composition due to metabolic activities of the product or micro-organisms;
- vacuum-skin packaging (VSP) – placing a softened film over the product and applying a vacuum to form a skin (Davies 1995, Church 1994, Wilbrandt 1992; see also controlled atmosphere storage in Chapter 21, section 21.2.5).

MAP is used to extend a product shelf-life to give processors additional time to sell the food without sacrificing quality or freshness. The potential advantages and limitations of MAP are shown in Table 25.11. MAP is used for fresh foods and an increasing number of mildly processed foods, and is gaining in popularity as new applications are developed. Examples of MAP products include raw or cooked meats, poultry, fish, seafood, vegetables, fresh pasta cheese, bakery products, sandwiches, sous vide foods, potato crisps, coffee, tea, prepared salads, part-baked bread croissants, pizzas, peeled fruits and

Table 25.11   Advantages and limitations of MAP

Advantages	Limitations
Increased shelf-life of 50–400%	Added cost
Extended storage results in reduced economic losses and wider distribution radius	Temperature control required
	Different gas compositions for each type of product
Fewer distribution deliveries leads to lower costs	Requirement for special equipment and operator training
Little or no need for chemical preservatives	Increased pack volume has impact on transport and retail display costs
Easier separation of sliced foods (except vacuum packing)	Benefits are lost once the pack is opened or leaks
Good presentation of products	Product safety to be established for some foods

Adapted from Davies (1995), Farber (1991) and Blakistone (1998b)

prepared vegetables with dressing (Davies 1995, Church 1994). Devlieghere (2002) has reviewed the applications of MAP to respiring and non-respiring foods, and the effects of MAP on micro-organisms and the sensory and nutritional quality of foods. Different foods respond in different and sometimes unpredictable ways to modified atmospheres, and each product should therefore be individually assessed to monitor microbial activity, moisture content, pH, texture, flavour and colour changes, in order to determine the optimum gas composition. Care is also needed to prevent temperature abuse during processing and distribution, and high standards of hygienics should be used throughout the production process (also Chapter 21, section 21.2.4).

The atmosphere is not constant in all MAP products and will change according to:

- the permeability of the packaging material;
- microbiological activity; and
- respiration by the food.

Successful MAP requires raw materials to have a low microbiological count and strict temperature control throughout the process (Chapter 21, section 21.2.3).

The three main gases used in MAP are nitrogen, oxygen and CO_2, although others, including carbon monoxide, nitrous oxide, argon, helium and chlorine, have also been investigated. Nitrogen is inert and tasteless, with low solubility in both water and fats. It is used to replace oxygen and thus inhibit oxidation or the growth of aerobic micro-organisms. Details of CO_2 are given in Chapter 21 (section 21.2.2).

For fresh fruits and vegetables, the aim of MAP is to minimise respiration and senescence without causing damage to metabolic activity that would result in loss of quality. However, the effects of low oxygen and raised CO_2 concentrations on respiration are cumulative, and respiration also continually alters the atmosphere in a MA pack. The rate at which oxygen is used up and CO_2 is produced also depends on the storage temperature. In practice, the CO_2 concentration is increased by gas flushing before sealing and a film that is permeable to oxygen and CO_2 is selected to enable respiration to continue. Changes in gas composition during storage depend on the:

- permeability of the packaging material to water vapour and gases;
- external relative humidity, which affects the permeability of some films; and
- surface area of the pack in relation to the amount of food it contains.

Details of MAP for fresh produce are given by Garrett (1998) and, for example, cut lettuce has a 2-week shelf-life at 0–1.1 °C (Brody 1990). The extension of shelf-life of other foods is shown in Table 25.12. In fresh fruits and vegetables, a concentration of 15–20% CO_2 is required to control decay. Some crops can tolerate this level (e.g. broccoli, strawberries and spinach) but most cannot (Table 25.13).

MAP permits a shelf-life for red meat joints of up to 18 days at 0–2 °C, and up to 10 days for ground beef. Oxygen is used to maintain the red colour of oxymyoglobin in unprocessed meats, but in other applications its level is reduced to prevent growth of spoilage micro-organisms and oxidative rancidity. Typically, the shelf-life of fresh red meat is extended by packaging in an 80% O_2/20% CO_2 atmosphere. High oxygen levels may cause development of off-colours in cured meats and bacon, for example, is therefore packed in 35% O_2/65% CO_2 or 69% O_2/20% CO_2/11% N_2. In both atmospheres the oxygen concentration is sufficient to inhibit anaerobic bacteria. Pork, poultry and cooked meats have no oxygen requirement to maintain the colour, and a higher CO_2 concentration (90%) is possible to extend the shelf-life to 11 days. Further details of MAP for fresh produce are given by Day (2003) and Blakistone (1998a).

Table 25.12   Extension of shelf-life using MAP

Product	Shelf-life (days)	
	Air	MAP
Beef[a]	4	12
Bread[b]	7	21
Cake[b]	14	180
Chicken[a]	6	18
Coffee[b]	3	548
Cooked meats[a]	7	28
Fish[a]	2	10
Fresh pasta[a]	2	28
Fresh pizza[a]	6	21
Pork[a]	4	9
Sandwiches[a]	2	21

[a] Refrigerated storage.
[b] Ambient storage.
Adapted from Brody (1990) and Blakistone (1998b)

For processed (i.e. non-respiring) foods, atmospheres should be as low as possible in oxygen and as high as possible in CO_2 without causing the pack to collapse or produce changes to the flavour or appearance of the product. Ground coffee, for example, is protected against oxidation using a CO_2/N_2 mixture or by vacuum packing. A high CO_2 concentration prevents mould growth in cakes and increases the shelf-life to 3–6 months. Other bakery products (e.g. hamburger buns) have the shelf-life increased from 2 days to 3–4 weeks. In MAP of bread, CO_2 inhibits mould growth and the retention of moisture maintains softness. This is not inhibition of staling (a process that involves partially reversible crystallisation of starch) but the effects are similar. Spraying bread with 1% ethanol doubles the ambient shelf-life by retarding mould growth and an apparent inhibition of staling (also Section 25.4). A novel MAP approach to packaging baguettes is to pack them straight from the oven while the CO_2 produced by the fermentation is still being emitted. As they are placed into thermoformed packs the CO_2 expels air and saturates the atmosphere, to give a 3 month shelf-life at ambient temperature. The consumer briefly heats the bread in an oven to create a crust and produce a product that is similar to freshly baked bread (Brody 1990).

CO_2 dissolves in both water and fats in a food and is more soluble in cold water than it is in warm water. Many MAP products are chilled (Chapter 21) and the absorption of CO_2 should therefore be carefully controlled to prevent too great a reduction in gas pressure, which causes collapse of the pack. Nitrogen is often added as a filler gas to prevent pack collapse, although in some products collapse may be advantageous (e.g. hard cheeses), where it forms a tight pack around the product. Additionally, the relative volume of gas and product is important to ensure a sufficiently high gas : product ratio for the gas to have a preservative effect. There should therefore be adequate space between the product and the package to contain the correct amount of gas. Examples of gas mixtures that are used for fresh and processed foods are shown in Table 25.14.

25.3.1   Effect on micro-organisms
Reducing the concentration of oxygen inhibits the development of spoilage micro-organisms, especially *Pseudomonas* spp. (Walker 1992). Other spoilage bacteria that can grow in low oxygen concentrations grow more slowly and so extend the time taken for

Table 25.13 Optimum MAP conditions for selected whole fruits and vegetables

Commodity	Tolerance		Optimum		Recommended storage temperature (°C)
	Maximum CO_2 (%)	Minimum O_2 (%)	CO_2 (%)	O_2 (%)	
Fruits					
Apple	2–5	1–2	1–3	1–2	0–3
Apricot	2	2	2–3	2–3	0–5
Avocado	5	3	3–10	2–5	5–13
Banana	5	2	2–5	2–5	12–15
Cherry (sweet)	15	2	10–12	3–10	0–5
Grapefruit	10	5	5–10	3–10	10–15
Kiwifruit	5	2	3–5	1–2	0–5
Lemon	–	–	0–10	5–10	10–15
Lime	–	–	0–10	5–10	10–15
Mango	5	–	5–8	3–7	10–15
Orange	–	–	0–5	5–10	5–10
Papaya	5	2	5–8	2–5	10–15
Peach	5	2	3–5	1–2	0–5
Pear	2	2	0–1	2–3	0–5
Pineapple	10	2	5–10	2–5	8–13
Vegetables					
Artichoke	2	3	2–3	2–3	0–5
Asparagus	14	5	10–14	air	1–5
Beans, snap	10	2	5–10	2–3	5–10
Bell peppers	2	3	0	3–5	8–12
Broccoli	10	1	5–10	1–2	0–5
Brussels sprouts	5	2	5–7	1–2	0–5
Cabbage	5	2	3–6	2–3	0–5
Carrot	5	5	3–4	5	0–5
Cauliflower	5	2	2–5	2–5	0–5
Chili peppers	2	3	5	3	8–12
Corn, sweet	15	2	10–20	2–4	0–5
Cucumber	10	3	0	3–5	8–12
Lettuce	2	2	0	1–3	0–5
Mushrooms	15	1	5–15	3–21	0–5
Potato	–	–	0	0	4–12
Onion	–	–	0	1–2	0–5
Spinach	15	–	10–20	air	0–5
Tomatoes (mature)	2	3	0	3–5	12–20

Adapted from Anon (2001)

spoilage to occur (e.g. lactic acid bacteria or *Brochothrix thermosphacta,* which cause spoilage by souring) (Nychas and Arkoudelos 1990). Concern has been expressed over potential risks to consumer safety from modified atmospheres or vacuum packaging because they inhibit 'normal' spoilage micro-organisms and thus allow food to appear fresh, while permitting the growth of anaerobic pathogens. Details of pathogens found on chilled foods are given in Chapter 1 (section 1.2.4) and Chapter 21 (section 21.5). Several pathogens including *Clostridium botulinum, Listeria monocytogenes, Yersinia enter-ocolitica, Salmonella* spp., and *Aeromonas hydrophila* are anaerobes or facultative anaerobes (Blakistone 1998b). A large number of studies of the effect of MAP on the microbiology of foods are reported; for example meat poultry and fish (Church 1998) and baked goods (Kotsianis *et al.* 2002). These are reviewed by Devlieghere and Debevere (2003), O'Beirne and Francis (2003), Ooraikul (2003), Davies (1995), Church (1994),

Table 25.14   Recommended gas mixtures for MAP of processed foods

Product	Oxygen (%)	Carbon dioxide (%)	Nitrogen (%)
Bread	0	60–70	30–40
Cakes (dairy)	0	0	100
Cakes (non-dairy)	0	60	40
Cheese (hard)	0	60	40
Cheese (soft)	0	30	70
Cheese (mould ripened)	0	0	100
Cream	0	0	100
Dried/roasted foods	0	0	100
Fish (oily)	0	30–60	40–70
Fish (white)	20–30	40–60	0–30
Kebabs	0–10	40–60	40–60
Meats (cooked/cured)	0	20–35	65–80
Meat (red)	60–85	15–40	0
Meat pies	0	20–50	50–80
Pasta (fresh)	0	50–80	20–50
Pizza	0–10	40–60	40–60
Poultry	0	25	75
Quiche	0	40–60	40–60
Salmon	20	60	20
Sausage	40	60	0
Snackfoods (dry)	0	20–30	70–80

From Parry (1993), Day (1992) and Smith *et al.* (1990)

Ooraikul and Stiles (1991) and Farber (1991). The studies have indicated that growth of pathogens in MAP products is no greater, and frequently lower than in aerobically stored foods. However, for products in which there is a potential safety hazard, it is recommended that one or more of the following criteria are met (see also hurdle technology (Chapter 1, section 1.3.1):

• water activity (Chapter 1, section 1.1.2) is below 0.92;
• pH is below 4.5;
• use of sodium nitrite or other preservative;
• the temperature is maintained below 3 °C.

The application of HACCP techniques (Chapter 1, section 1.5.1) also plays a major role in ensuring the safety of MAP foods. MAP can be combined with other preservation techniques, including irradiation (Chapter 7), pasteurisation (Chapter 12), chilling (Chapter 21), alteration of the product pH or use of salt or other preservatives (Rosnes *et al.* 2003), UV treatment (Chapter 9, section 9.4), or ozone treatment (Lucas 2003).

25.3.2   Packaging materials for MAP

The two most important technical parameters of packaging materials for MAP are gas permeability and moisture vapour permeability (section 25.1.1). Materials are classified according to their barrier properties to oxygen (measured at 90% RH and 23 °C over 24 h) into low barrier (>300 ml m^{-2}) for over-wraps on fresh meat or other applications where oxygen transmission is desirable; medium barrier (50–300 ml m^{-2}); high barrier (10–50 ml m^{-2}); and ultra-high barrier (<10 ml m^{-2}), which protect the product from oxygen to the end of its expected shelf-life. A wide range of different packaging systems is used to produce MAP packs. Typical film materials are single or coextruded films or laminates of

Table 25.15 Permeability of different films used for packaging of MAP produce

Film	Permeability (ml/m^2 day atm) for 25 μm film at 25 °C			Water vapour transmission, g/m^2/day/atm (38 °C and 90% relative humidity)
	Oxygen	Nitrogen	Carbon dioxide	
Ethylene vinyl acetate	12 500	4900	50 000	40–60
Ethylene vinyl alcohol	3–5	–	–	16–18
HDPE	2600	650	7600	7–10
LDPE	7800	2800	42 000	18
Plasticised PVC	500–30 000	300–10 000	1500–46 000	15–40
Polyamide (nylon-6)	40	14	150–190	84–3100
Polypropylene cast	3700	680	10 000	10–12
Polypropylene, oriented	2000	400	8000	6–7
Polypropylene, oriented, PVdC coated	10–20	8–13	35–50	4–5
Polystyrene, oriented	5000	800	18 000	100–125
Polyurethane (polyester)	800–1500	600–1200	7000–25 000	400–600
Polyvinylidene chloride coated (PVdC)	9–15	–	20–30	–
PVdC-PVC copolymer	8–25	2–2.6	50–150	1.5–5.0
Rigid PVC	150–350	60–150	450–1000	30–40

Adapted from Anon (2001)

ethylene vinyl alcohol, polyvinyl dichloride, polyethylene terephthalate, polypropylene, polyethylene, polyester, amorphous nylon (polyamide resin) and nylons, although the last provides only moderate barrier (Table 25.15). Details of types of MAP film and their permeability to moisture and gases are described by Greengrass (1998). Films are usually coated on the inside of the pack with an antifogging agent, typically a silicone or stearate material, to disperse droplets of condensed moisture and permit the food to be visible. Films have been developed that change permeability to moisture and gases under specified temperature conditions, which are designed to match the respiration rate of a fresh product.

In MAP processing, air is removed from the pack and replaced with a controlled mixture of gases, and the package is heat sealed. In batch operation, pre-formed bags are filled, evacuated, gas-flushed and heat-sealed in a microprocessor-controlled sequence. In continuous operation, food is packaged in three ways: in semi-rigid, thermoformed trays covered with film that has the required permeability (e.g. for meats); secondly in pillow pouches (e.g. for fresh salads); and thirdly foods such as baked products are packed in horizontal form–fill–seal equipment or 'flowpacks' (Chapter 26, section 26.2.1).

MAP packs frequently require high oxygen barriers and previously this has only been achievable at reasonable cost using either polyvinylidene chloride (or aluminium foil which does not let the consumer see the product). Developments in transparent low oxygen barrier materials include glass-coated microwaveable pouches, silicon oxide (SiO$_x$) coated films having oxygen transmission rates of <1 ml m^{-2} and moisture vapour transmission rates of <1 g m^{-2}, aluminium oxide coatings and nylon-based co-extrusions having an oxygen permeability of 0.48 ml m^{-2} 24 h^{-1} atm^{-1} (Church 1994). SiO$_x$/PET films, trays and bottles are produced by plasma-enhanced chemical vapour deposition (PECVD), and are likely to become increasingly important as very high barrier, transparent, microwaveable containers. The design of MAP packs for fresh produce is described by Yam and Lee (1995) and the different types of packaging systems are described in detail by Hastings (1998).

25.4   Packaging developments

25.4.1   Edible and biodegradable materials

The widespread and growing concern over the environmental effects of non-biodegradable petrochemical-based plastic packaging materials (section 25.6) has increased interest in the development of non-petroleum-based natural biopolymers (or 'bioplastics') derived from renewable sources, that are biodegradable or compostable (also section 25.2.4). Details of methods of commercial composting are beyond the scope of this book but details are given by Anon (2008i). These natural materials are degraded by the same factors that also cause food spoilage (section 25.1.1). The challenge is therefore to produce bioplastic materials that have sufficient durability to maintain their mechanical and/or barrier properties for the product shelf-life, and then, ideally, biodegrade quickly on disposal. The environmental conditions conducive to biodegradation must be avoided during preparation and use of the package, whereas optimal conditions for biodegradation must exist after disposal. The materials should also function in a similar way to conventional packaging in filling and sealing equipment and have approximately equivalent costs.

These new materials can be grouped according to the method of production into the following three categories:

1 Polymers directly extracted/removed from natural materials (e.g. starch, cellulose, casein and wheat gluten) (see also Chapter 1, section 1.1.1).
2 Polymers produced by chemical synthesis from renewable monomers (e.g. polylactate polymerised from lactic acid monomers).
3 Polymers produced by microbial fermentation (e.g. polyhydroxyalkanoates) (Fig. 25.9).

Starch is widely available and economically competitive with petroleum as a raw material for bioplastics (Liu 2006). Maize (corn in the USA) is currently the most important source of starch for bioplastics, but potato, wheat, rice, barley and oats are also being evaluated as potential starch sources. Bioplastics that contain high concentrations of starch are brittle and do not form films that have the required mechanical properties (flexibility, elongation, tensile strength, etc. (section 25.1.1)). They are made more flexible using biodegradable plasticisers, including glycerol, other low molecular weight polyhydroxy compounds, urea or polyethers. The plasticisers also lower the a_w of the material to limit microbial growth (van Tuil *et al.* 2000). Extruded plasticised materials have good oxygen-barrier properties, but the hygroscopic nature of starch makes them unsuitable for high-moisture products. They can be blended with more hydrophobic polymers, which makes them suitable for injection or blow moulding. Starch-based thermoplastic materials were the first to be commercialised and are used for wrapping, laminating or coating paperboard, producing biodegradable foam laminate foodservice containers, and thermoformed or injection-moulded cups and egg trays. Starch-LDPE polymers have been in use since the 1980s (Arvanitoyannis and Gorris 1999).

Cellulose is the most abundant and inexpensive natural polymer, and like starch it is composed of glucose units. However, unlike starch, the units are joined by β-1,4 glycosidic linkages, which enable cellulose chains to form strong inter-chain hydrogen bonds. The alternating hydroxyl side chains along the cellulose molecule make it hydrophilic, and like starch, cellulose-based packaging materials have poor moisture barrier properties. The side chains also contribute to the highly crystalline structure of cellulose, which produces material that is brittle and has low flexibility and tensile

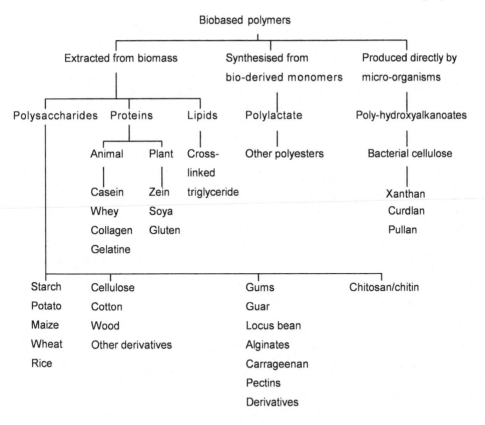

Fig. 25.9   Biobased polymers for food packaging (from van Tuil *et al.* 2000).

strength. Research is therefore aimed at developing materials based on cellulose derivatives that have improved properties. Cellulose and cellulose acetate films (section 25.2.4) have been used for many years and other cellulose derivatives possess excellent film-forming properties. They can be used as wrapping films or injection- or blow-moulded containers, but commercially economic production technologies have not yet been developed and continued research aims to develop more cost-effective methods. Research into protein-based packaging has focused on edible coatings (below) and their sensitivity to humidity has so far limited their applications as packaging films.

Another area of research is in bioplastics synthesised by microbial fermentation of polysaccharides, including polylactic acid (PLA) and poly-hydroxyalkanoates (PHAs). PLAs are biodegradable, thermoplastic polyesters that are derived from lactic acid, which is produced by fermentation of starch or molasses by *Lactobacillus* spp. The lactic acid monomer acid has two forms (D and L) and the properties of PLAs depend on the ratio of the two. For example, 100% L-PLA produces a material that is highly crystalline and has a high melting point, whereas a 90% D/10% L polymer melts more easily and is highly suitable for forming packaging films. The film material has good moisture, oxygen and odour barrier properties, good surface properties for printing, and is suitable for blown-film production, injection moulding and blow- or vacuum-forming processes (section 25.2.5). Typically these films have a glass transition temperature $\approx 60\,^{\circ}\mathrm{C}$ and a melting temperature of $150\,^{\circ}\mathrm{C}$ (de Vlieger 2003). PLAs are similar in appearance and properties to oriented polystyrene films, and can be produced as cast films for wrapping bakery and confectionery products, cast sheets for thermoforming, extrusion processes for coating paperboard, and injection-moulded containers and disposable foodservice tableware (Liu 2006). These bioplastics are made commercially but have yet to be widely adopted because of higher production costs.

PHAs, of which poly-hydroxybutyrate (PHB) is the most common, are linear polyesters produced by bacterial fermentation of sugars or lipids (e.g. by *Alcaligenes eutrophus* or *Ralstonia eutrophus*). There are over 100 monomers that can be combined within this group to produce materials that have a wide range of properties, which depend on the composition of the monomer, the type of micro-organism used in fermentation, and the carbon source for the fermentation process. PHB has a similar melting temperature to polypropylene (175–180 °C) but is stiffer and more brittle. This results in poor impact resistance and, together with its relatively expensive production costs compared with petrochemical plastics, has prevented more widespread use in packaging applications. Other PHAs have lower glass transition and melting temperatures compared with PHB, and low moisture vapour permeability, which is comparable to LDPE. They have been used as cheese coatings, and if cost reductions can be made, the benefits of biodegradability mean that these materials may have potential to replace petroleum-based materials to make bottles, trays and films. In other areas of research, bio-based materials have been developed by blending brittle PLA/PHA with biodegradable polyester, or natural fibres (e.g. kenaf, hemp, pineapple leaf fibre or grasses) are reinforced with bioplastics to improve their performance (Mohanty 2006).

Other novel biodegradable polymers include thermoplastic polyesters, polycaprolactone (PCL), polymethyl-valerolactone, α-amino acids and polyamides, and copolymerisation of lactams and lactones, as each has a low glass transition temperature and low melting point (e.g. -60 and 60 °C respectively for polycaprolactone (de Vlieger 2003)). Petersen *et al.* (1999) describe applications of gluten films, starch/LDPE films and laminated chitosan–cellulose–PCL films. Other bioplastics, including polyurethanes made from castor oil, are described by van Tuil *et al.* (2000).

Cava *et al.* (2006) report studies of the thermal resistance and barrier properties of PET films compared to biodegradable biopolymers, including PCL, PLA, amorphous PLA, PHB copolymer with valeriate and nanocomposites (below). They found that PHB can withstand retorting temperatures and had excellent moisture and aroma (limonene and linalool) barrier properties compared with PET. Resistance to solvents (toluene and ethanol) and oxygen barrier properties were poorer than PET. Nanocomposites of PCL and amorphous PLA were good oxygen barriers but not as good as high oxygen barrier grades of PET. Auras *et al.* (2006) found that PLA showed good aroma barrier properties when they compared sorption and permeability of ethyl acetate and *d*-limonene in PLA with polystyrene, polypropylene, LDPE and PET.

Natural and synthetic polymers may also be blended to improve the film properties while retaining biodegradability. Examples of the time required to compost these polymers are shown in Fig. 25.10. Weber (2000) gives an overview and reviews the status and developments in bio-packaging.

There is also renewed interest in non-petroleum-based edible protective superficial layers (EPSL). These coatings are applied directly on the surface of a food to prevent loss of quality and to protect against microbial spoilage. With fresh fruits and vegetables they are also used to control the rate of respiration and act as moisture barriers (Olivas and Barbosa-Canovas 2005), or as oxygen barriers to prevent enzymic browning. Collagen casing for meat products was one of the first edible films to be used and gelatine, derived from collagen can be formed into films or light foams, although it is very sensitive to

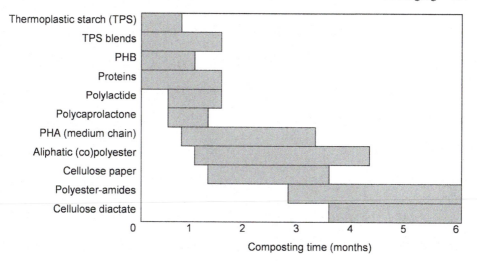

Fig. 25.10 Time required to compost selected bioplastics and synthetic polymers in mechanically turned and actively aerated composting (adapted from van Tuil *et al.* 2000).

moisture. Flexible, hydrophilic EPSLs that have good resistance to breakage and abrasion have also been made from hydrocolloids (e.g. pectin, corn zein (a prolamine derived from corn gluten), wheat gluten, proteins from soybean, peanut, cottonseed, casein and milk whey (Khwaldia *et al.* 2004), chitosan and alginates (Baldwin 1999, Butler *et al.* 1996, Gontard *et al.* 1992a,b, 1993)). Of these plasticised corn zein is used commercially to make films by casting or extrusion and chitosan is widely used for coatings as it also has good gas barrier and antimicrobial properties (also section 25.4.3). Details of other edible lipid coatings, including gums, waxes, oils, resins, fatty acids and acylglycerols, are given by a number of authors, including Morillon *et al.* (2002), Arvanitoyannis and Gorris (1999), Guilbert and Gontard (1995), Baldwin (1994, 1999), McHugh and Krochta (1994) and Debeaufort *et al.* (1998). The properties of some edible films are shown in Table 25.16. Examples of applications of edible and biodegradable materials are shown in Table 25.17 and have been reviewed by Cha and Chinnan (2004). Anon (2006c) describes the first metallised biodegradable cellulose film made from renewable sources of wood pulp.

Active EPSLs that have antimicrobial properties (e.g. using sorbic acid or natural aroma compounds, such as carvacrol (from oregano) or cinnamadehyde (from cinnamon)), or antioxidant properties have been developed (Lopez-Rubio *et al.* 2006). These fix the preservative at the product surface where they are required, and therefore reduce the amounts that are used. An edible coating containing chitosan increases the lag phase of microbial contaminants and could have the potential to preserve dairy products (Coma *et al.* 2003).

25.4.2 Nanotechnology

The inherent problems of natural polymers, including low mechanical strength and poor moisture barrier properties can be improved using nanocomposite technology (Manias 2006). Nanotechnology is the characterisation and manipulation of materials that have dimensions in the range 1–100 nm. It has potential applications in food packaging to improve the properties of existing materials or to develop new materials that have unique properties. Polymer/inorganic nanocomposites use ultra-small inorganic particles to

Table 25.16   Permeability of some edible films compared to petrochemical-based films

Film material	O_2 permeability (ml/m^2 day atm)	CO_2 permeability (ml/m^2 day atm)	Relative humidity (%)
Chitosan	91.4	1553	93
Gluten–beeswax	133	1282	91
Gluten–beeswax and beeswax	<3	13	56
Methylcellulose–beeswax	4	27	42
Methylcellulose–palmitic acid	78.8	–	100
Pectin	57.5	–	87
Sodium caseinate	77	462	77
Sucrose polyester	2.10[a]	–	0.00042[b]
Wheat gluten	190/250	4750/7100	91/94.5
Zein	0.36[a]	2.67[a]	0.116[b]
Film permeability (ml/m^2 day atm for 25 μm film at 25 °C)			
Polyvinylidene chloride coated (PVdC)	9–15	20–30	
LDPE	7800	42 000	
HDPE	2600	7600	
Oriented polypropylene	2000	8000	
Polyester	800–1500	7000–25 000	

[a] Unit of permeability is in fl m/m^2 s Pa ('f' is abbreviation for femto (10^{-15})).
[b] Unit of permeability is ng m/m^2 s Pa ('n' is the abbreviation for nano (10^{-9})).
Adapted from Anon (2001)

change the nature of a polymeric material and these, together with organoclay reinforcement of biopolymers, can improve their mechanical barrier and thermal properties (Mohanty 2006). Polymer/silicate nanocomposites are an emerging class of specialty plastic packaging that have improved properties compared to conventional mineral fillers. These include higher heat resistance, increased flexibility, lower gas permeability, dimensional stability and good surface appearance (Brody 2003).

Typically, a type of silicate, known as 'montmorillonite' (a layered 'smectite' clay) is used. This clay naturally breaks down into nano-sized platelets having thicknesses of 1 nm and surface lengths of 100–1000 nm. It is hydrophilic, which makes dispersion into conventional polymers difficult and it is therefore modified by substitution of its sodium ions with organic ammonium ions to produce an organo-clay complex. Nanocomposites can be prepared by dissolving both the polymer and complex in solvents, or by melting and applying shearing forces to assist dispersion. Nylon-6 is a polar polymer, which combines with the inorganic material. Nylon-6 nanocomposites can achieve oxygen transmission rates four times lower than unfilled nylon-6, with excellent transparency retained in films. Sanchez-Garcia et al. (2007) compared the barrier properties to oxygen, water and limonene of PET nanocomposites made with montmorillonite with those of biopolymers, such as PLA, PHB, PCL and their corresponding nanocomposites.

Chemical vapour deposition is used to coat a nanometre-thick layer of silicon oxide to enhance the barrier properties of polymers. This produces a highly-controlled and consistent coating that is believed to provide an ultra-high gas barrier (OTR < 0.005 ml m^{-2} day^{-1} at 23 °C), crack resistance, clarity (light transmission >93%) using a very thin coating (<1 nm thick). The limitations of the coating are that it is not an effective moisture barrier and its gas barrier properties decrease with increasing relative humidity. The film structure should therefore include a component that has low water vapour permeability. Nanocomposites of natural biopolymers and silicate show increased strength, lower gas permeability and higher moisture barrier properties. These develop-

Table 25.17 Examples of applications of biodegradable or edible packaging materials

Material	Function of packaging	Value added function	Examples of food applications
Alginate	Oxygen and moisture barrier	–	Edible coating for mushrooms
Alginate, carrageenan, cellulose, gelatin, soy protein	Moisture, oxygen, CO_2 barriers, frying oil barrier, batter adhesion, mechanical protection, inhibit microbial growth	Antioxidants, antimicrobial agents	Edible coating for fresh and cured meats, casings, cooked meats
Caseins, whey proteins, lipids, alginate, carageenan	Moisture and oxygen barrier, mechanical protection	Batter adhesion, antioxidants, antimicrobial agents	Edible coating for frozen fish, frozen shrimps
Chitosan/cellulose/polycaprolactone/ protein laminate	Moisture, oxygen, CO_2 barrier, mechanical protection	–	Edible coating for fresh produce (e.g. cut or whole lettuce, broccoli, cabbage, tomatoes)
Corn zein	Moisture, oxygen, CO_2 barrier	–	Coating for fresh pears
Methylcellulose laminated with corn zein and stearic-palmitic acid	Moisture, oxygen, light barrier	–	Pack for potato chips
Paper coated with PLA or starch	Moisture barrier	–	Cups for miscellaneous applications
Paperpulp, starch, PLA and/or PHB trays, with top lids from PLA, cellulose acetate or protein film, starch-based drip pad	High oxygen permeability, low moisture permeability. High moisture absorption pad	Absorption of meat drip	Fresh meat portions
PLA (10%) plus 90% copolyester/ polyamide/starch or PCL	Moisture, light and grease barrier	Antifogging	Pots for yoghurt, butter, margarine
PLA or starch	Moisture barrier	–	Paper coated with bioplastics, window of PLA or starch

Table 25.17 Continued

Material	Function of packaging	Value added function	Examples of food applications
PLA, PHB bottles or paperboard cartons coated with PLA and/or PHB	High moisture, light and gas barrier	Oxygen barrier	Milk
Powdered starch foam	Containment	–	Ready meal containers for French fries, hamburger, chicken
Starch or PHB coated paperboard tray, over-wrapped with starch bag	Moisture barrier	–	Fresh tomatoes
Starch–polyethylene films (0–28% starch)	Oxygen and moisture barrier	–	Ground beef
Sucrose-fatty acid ester, sodium or calcium caseinate-acetylated monoglyceride	Oxygen and moisture barrier	Improve gloss/ appearance, antioxidant and preservative carrier to delay browning and microbial growth	Edible coating for fresh apples
Whey protein and acetylated monoglycerides	Oxygen and moisture barrier	Antioxidant	Edible coating for fish
Whey protein isolate, hydroxypropyl cellulose, corn zein	Oxygen barrier	–	Coating for roasted peanuts

Adapted from data of Haugaard *et al.* (2000)

ments are reviewed by Rhim and Ng (2007) and Lagaron *et al.* (2005). Additionally, biologically active components (e.g. antimicrobials) can be added to the packaging materials. A significant advantage of nanotechnology is that improvements to different properties of materials can be made independently of each other (e.g. composites can have high barrier properties and remain flexible and transparent; or high stiffness and remain ductile and lightweight). Further information is given by Akbari *et al.* (2007).

25.4.3 Active and intelligent packaging

In addition to the functions of packaging materials described in section 25.1, 'active' packaging and 'intelligent' (or 'smart') packaging have additional features. They add functionality to a product, or to consumption, convenience or security. Examples include:

- actively prevent food spoilage and retain product integrity for its shelf-life;
- enhance product attributes (e.g. appearance, taste, flavour, aroma);
- respond actively to changes in the product or the package environment;
- communicate product information, product history or condition to the user;
- confirm product authenticity, and act to counter theft.

Examples of active and intelligent packaging systems are shown in Table 25.18.

There is an important difference between active packaging and intelligent packaging: active packaging senses changes in the internal or external environment of a food package and responds by altering its properties. In intelligent packaging, the package function switches on and off in response to changing external/internal conditions, and includes communication of information on the status of the product to the customer or user. A simple definition of intelligent packaging is 'packaging that senses and informs'. Intelligent packaging signals may be upgraded to active ones to control the environment and maintain the safety or quality of the product contained within the pack.

An early example of active packaging is the 'widget' in cans of beer that releases pressurised gas to produce a foamy 'head' on the product when the can is opened. More recent developments in active packaging are reviewed by Vermeiren *et al.* (1999) and Ahvenainen (2003), and described by Ozdemir and Floros (2004), Labuza (1996), Plaut (1995) and Rooney (1995). Practical applications of active packaging are described by Gill (2003) and Jakobsen and Bertelsen (2003) for meat, by Sivertsvik (2003) for fish, and by Calero and Gomez (2003) for fruits and vegetables.

Moisture control

Much of the research and development in active packaging concerns moisture control. Desiccant sachets or cartridges have been used commercially for decades. Developments include a sachet system that rapidly increases absorption of moisture as the temperature approaches the dew point. It is used to prevent droplets of water from forming on the product, which could promote microbial growth. A similar effect is produced by trapping propylene glycol or diatomaceous earth in a film that is placed in contact with the surface of fresh meat or fish to absorb water and injure spoilage bacteria. These developments are described by Powers and Calvo (2003), Church (1994) and Smith *et al.* (1990).

Gas control

Control of gas composition is a commercially important area of active packaging that has been used since the 1970s and oxygen scavenging is the most commercially developed sub-sector of the active packaging market. Oxygen scavengers prevent oxidative damage

Table 25.18 Examples of active and intelligent packaging technologies

Type of packaging	Mechanism of operation	Purpose	Examples of products
Active packaging			
Antimicrobial release	Films containing organic acids, silver zeolite, spice extracts, allylisothiocyanate, enzymes (e.g. lysozyme) bacteriocins, fungicides, antibiotics. UV-irradiated nylon films to produce surface amines. Films treated with fluorine-based plasmas	Inhibition of spoilage and pathogenic bacteria	Meat, poultry, fish, bread, cheese, fruits and vegetables
Antioxidant release	Films containing BHA, BHT, tocopherol	Inhibition of fat oxidation	Dried and fatty foods
Aroma/flavour release	Films containing flavouring compounds	Masking off-odours, improving flavour	Miscellaneous
Cholesterol removal	Cholesterol reductase immobilised in film	Removal of cholesterol	Dairy products
CO_2 absorbers	Sachets containing calcium, sodium or potassium hydroxide	Removing CO_2 formed during storage	Coffee, dried beef or poultry products
CO_2 production	Sachets of sodium hydrogen carbonate and ascorbate	Inhibition of Gram-negative bacteria and moulds	Fruits, vegetables, fish, meat, poultry
Ethanol emitters	Sachets of ethanol/water absorbed onto silicon dioxide powder to emit ethanol vapour	Inhibition of moulds and yeasts	Bakery products, dried fish
Ethylene scavenging	Sachets containing aluminium oxide and potassium permanganate, activated carbon, or films containing zeolite or clay	Reduction/prevention of ripening	Climacteric fruits – apples, apricots, bananas, mangoes, tomatoes, avocados (Chapter 21, section 21.3.1)
Lactose removal	Lactase immobilised in film	Removal of lactose for lactose intolerant consumers	Dairy products
Microwave heating controllers	'Susceptor' films that have aluminium deposited to create high-temperature treatments	Drying, crisping and browning of microwaved foods	Popcorn, pizzas, ready-to-eat foods
Moisture control	Polyacrylate sheets, propylene glycol films, silica gel or clay sachets	Control of excess moisture in packs, reduction of surface moisture to prevent growth of spoilage micro-organisms	Meat, fish, poultry, bakery products, cut fruit and vegetables
Odour/flavour removal	Cellulose acetate film containing naringinase enzyme, sachets containing ferrous salt and citric acid	Reduction in bitterness of grapefruit juice, improving the flavour of fatty foods	Juices, potato crisps, biscuits and cereal products
Oxygen scavenging	Ferrous compounds, ascorbic acid, glucose oxidase, ethanol oxidase in sachets, labels, films or corks	Inhibition of moulds, yeasts, aerobic bacteria, insects, prevention of oxidation of fats, vitamins, pigments	Cheeses, meat products, ready-to-eat foods, bakery products, coffee, nuts, milk powder

Pesticide emitters	Pesticides absorbed into inner layer of shipping containers	Prevention of growth of moulds, bacteria or pests	Sacks of dried foods (e.g. cereal grains, flour)
Photochromic light protection	Films containing UV-absorbent agent, nylon 6 stabiliser in PET bottles	Reducing light-induced oxidation	Light-sensitive foods such as ham, beer
Sulphur dioxide emitters	Sachets containing sodium metabisulphite in microporous material	Inhibition of moulds	Fruits
Self-cooling packs	Double walled metal containers cooled by ammonium chloride or ammonium nitrate and water	Cooling foods	Non-carbonated drinks
Self-heating packs	Double walled metal containers heated by lime (calcium oxide) and water	Heating/cooking foods	Coffee, tea, ready-to-eat meals
Temperature-sensitive films	Fillers in films control gas permeability at different temperatures	Avoid anaerobic conditions	Fruits and vegetables
Intelligent packaging			
Communications link with appliances or with consumers	Radio frequency identification (RFID) tags, electronic article surveillance tags, magnetic strips	Indicate food type, safety, quality, nutritional attributes, ripeness. Inventory controllers. Location signallers	Miscellaneous
Gas concentration indicators	Redox dyes, pH dyes, enzymes in labels that change colour when a specified level of oxygen or CO_2 is attained	Storage conditions, gas composition, leakage	Foods in modified atmosphere packages
Maximum/extreme temperature indicators	Mechanical, chemical, enzymic	Display temperature abuse conditions	Chilled or frozen foods
Microbial growth/freshness indicators	pH dyes that react with metabolites. Chemical/immunochemical reaction with toxins	Spoilage and pathogen detection. Detection of specific pathogenic bacteria (e.g. E. coli O157)	Fresh meats, poultry, fish
Physical shock indicators	Optically variable films or gas-sensing dyes that irreversibly change colour. Piezoelectric polymers in films that change colour at a stress threshold	Indicate poor handling. Tamper evidence	Miscellaneous
Tamper-evident labels	Change colour when removed or leave a message on the pack that cannot be hidden. 'Self-bruising' closures on bottles or jars	Tamper evidence	Miscellaneous
Time–temperature indicators (TTIs)	Mechanical, chemical, enzymic (Chapter 21, section 21.2.4)	Display loss of shelf-life	Chilled or frozen foods

Oxygen, CO_2, ethylene and moisture scavengers have the most significant commercial use to date.
Adapted from Ahvenainen (2003), Han (2003) and Butler (2001)

to oils, pigments, flavours and vitamins and thus prevent rancidity, loss of colour, taste and nutritional value. They also prevent insect damage instead of using chemical fumigants and prevent the growth of moulds and aerobic bacteria. Older oxygen scavenging systems use sachets that contain iron-based powders plus catalysts that react with oxygen to form a stable oxide. More recent oxygen scavenging technology consists of a polymer, such as PET, an oxygen scavenging/absorbing component (a nylon polymer (MXD6) blended at ≈5% with the PET) and a catalyst (a cobalt salt added at a low concentration ($<200 \, \mathrm{mg \, kg^{-1}}$) that triggers the oxidation of the MXD6). The scavenging system remains active for up to two years to provide protection to oxygen-sensitive products such as beer, wine, fruit juice and mayonnaise throughout their shelf-lives. A four-layer polymer tray, sealed with a high-barrier film is used to pack sliced cured meats. The tray material comprises an inner heat-sealable material, an oxygen absorbing layer, an EVOH oxygen barrier layer and an outside protective layer. Oxygen scavenging adhesive labels are also used with pizzas, baked products, dried foods and coffee. The main advantage of these technologies is that they can reduce oxygen levels to <0.01%, compared with 0.3–3% achieved using MAP (Brennan and Day 2006). Commercially, it is common to remove most oxygen in a pack using MAP (section 25.3) and then use a small scavenger to remove residual oxygen during storage.

Active packaging systems that use both oxygen scavenging and antimicrobial technologies (e.g. sorbate-releasing LDPE film for cheese) extend the shelf-life of perishable foods and reduce the need for preservatives. Other approaches include a film that contains a reactive dye and ascorbic acid, and attachment of immobilised enzymes, including glucose oxidase and alcohol oxidase (Chapter 6, section 6.5) to the inner surface of a film (Brody and Budny 1995). The products of these enzymic reactions also lower the surface pH of the food and release hydrogen peroxide, which extends the shelf-life of, for example, fresh fish. A sachet system containing iron powder and calcium hydroxide scavenges both oxygen and CO_2 and has been used to produce a threefold extension to the shelf-life of packaged ground coffee. Other applications of oxygen scavengers are described by Vermeiren et al. (2003) and Davies (1995), and include bakery products, pre-cooked pasta, cured and smoked meats, cheese, spices, nuts, confectionery, soybean cakes, rice cakes and soft cakes (Table 25.18). Japanese companies have also developed an oxygen-sensitive ink and an indicator that changes from pink to blue when oxygen levels rise from <0.1% to >0.5% (Church 1994), which are used to ensure that the gas composition in MAP is maintained. They may also have applications to non-destructively check the package integrity.

Novel 'breathable' intelligent polymer films that can cope with high respiration rates are in commercial use for fresh cut vegetables and fruits. The films are acrylic side-chain polymers that are made to change phase reversibly at a specified temperature. As the side-chain components melt, gas permeation increases dramatically, and it is possible to tailor the package to adjust the permeation ratios for carbon dioxide:oxygen for individual products. The package is 'intelligent' because it automatically regulates oxygen and CO_2 exchange according to the ambient temperature to maintain an optimum atmosphere around the product during storage and distribution, thereby extending its freshness and quality (Gorris and Peppelenbos 1999). In other situations, low oxygen levels can create favourable conditions for the growth of pathogenic anaerobic bacteria and the film permits a substantial increase in gas permeability to re-oxygenate packs and prevent anaerobic conditions from forming.

Ethylene scavengers are sachets of silica gel containing potassium permanganate, or activated carbon systems that oxidise ethylene to slow the ripening of fruits (Zagory

1995). Other films scavenge off-odours or carbon dioxide. For example, freshly roasted coffee emits CO_2 and this can cause sealed coffee pouches to burst. Dual action oxygen and CO_2 scavengers (sachets containing calcium hydroxide, which is converted by CO_2 to calcium carbonate) and scavenger labels are used to overcome the problem. Dual action oxygen scavengers and CO_2 emitters are used for snackfoods, nuts and long shelf-life cakes.

Antimicrobial packaging and freshness indicators
In solid foods, deteriorative reactions occur mainly at the surface, and lower amounts of antimicrobial compounds are therefore needed to inhibit pathogens when they are incorporated into the packaging than when added to the food, because the compounds are released at the location where they are needed. The slow and controlled release of the antimicrobial provides protection over an extended period of time to extend the shelf-life of the product. Examples of antimicrobials used in this way include the bacteriocins, nisin and pediocin (Chapter 6, section 6.7), organic acids (e.g. propionates, benzoates and sorbates), and natural extracts from grapefruit seed, cloves, cinnamon or horseradish (Han 2003). However, many are not widely used commercially because of cost or regulatory constraints. Ethanol has antimicrobial properties, especially against moulds, and ethanol generators have ethanol adsorbed onto silicon dioxide powder contained in a sachet made from a film that is highly permeable to ethanol vapour (e.g. ethyl vinyl acetate). They have been used to extend the shelf-life of bakery products, cheeses and semi-dried fish products. Similarly a sulphur dioxide-generating film or a film that releases trapped sorbate have been used to extend the shelf-life of grapes by preventing mould growth. Han (2000, 2003) has reviewed antimicrobial packaging and describes models for diffusion of the antimicrobial agent. Antioxidants added to films include BHT, BHA and rosemary extract. For example, the colour of fresh beef is maintained by over-wrapping the meat with BHA-impregnated polyethylene film.

Other intelligent packaging indicates microbial growth to give assurance to consumers over the safety of perishable food products. Visual maximum temperature or tempera-ture–time indicators (TTIs), based on physical, chemical or enzymatic reactions, give a clear, accurate and unambiguous indication of product quality, safety and shelf-life. They are described in detail in Chapter 21 (section 21.2.4) and by Taoukis and Labuza (2003). Freshness indicators directly indicate the quality of a product based on a reaction with microbial metabolites. An example is a small adhesive label on the outside of packaging film that monitors the freshness of seafood products using a colour indicating tag (Fig. 25.11). A barb on the reverse of the tag penetrates the packaging and allows volatile amines, generated by spoilage bacteria on the seafood to pass a chemical sensor that turns

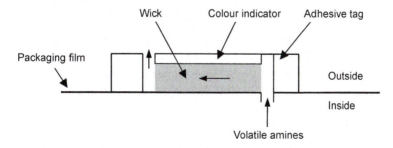

Fig. 25.11   Intelligent packaging to indicate microbial spoilage (adapted from Butler 2001).

progressively bright pink as the seafood ages and can therefore be used to monitor its freshness (Butler 2001). Other freshness indicators described by Brody (2001) include detectors for CO_2, sulphur dioxide, ammonia, hydrogen sulphide, organic acids and toxins. For example, a metmyoglobin-based indicator changes colour from brown to red in the presence of hydrogen sulphide produced by spoilage bacteria on MAP poultry (Smolander et al. 2002). In another type of indicator, a cross-linked polymerised polydiacetylene sensor incorporated into plastic shows a deep blue colour in the presence of E. coli O157:H7 enterotoxin. Smolander (2003) describes other dyes used in freshness indicators, including bromocresol green, cresol red, xylenol blue and methyl red.

Temperature indication and control

In contrast to most microwave packs, which do not heat up in a microwave oven, 'susceptors' absorb a proportion of the microwave energy, become hot (up to 220 °C) and therefore directly affect the rate and type of heating of the food. They are mostly used to impart crispness or browning to the surface of foods, and are used in packs of popcorn, French fries, pizzas, pies and other baked goods. An example of a susceptor is a pet film that is lightly metallised with aluminium and laminated to a paperboard substrate. De-metallised susceptors have areas in which the metallisation is etched off during production. This enables heat to be directed to specific areas of a pack and also prevents carton glues from melting or producing volatiles that could contaminate the product (Woods 1993). Self-venting microwave packs have a vent that opens at a pre-set temperature and re-closes on cooling.

Self-heating containers (e.g. for soup and coffee) and self-cooling containers for beer and soft drinks have been under development for more than 10 years, but have yet to be commercially adopted. One technology uses the latent heat of evaporating water to produce the cooling effect. The water is bound in a gel layer that coats a separate container within the beverage can. When the base of the can is twisted, it opens a valve that exposes the water to a desiccant held in a separate, evacuated chamber. This causes the water to evaporate at room temperature, cooling 300 ml of beverage in a 355 ml can by ≈17 °C in 3 min (Butler 2001). An intelligent self-heating or self-cooling container has a sensor to inform the consumer that it is at the correct temperature. Thermochromic ink dots indicate that a product is at the correct serving temperature following refrigeration or microwave heating. Beer bottle labels can also incorporate thermochromic inks to inform the consumer when the beer has reached the correct temperature to drink after refrigeration.

Tracing, tracking and tamper-evident packaging

Schilthuizen (2000) describes intelligent packaging that has features for tracing and tracking foods through the distribution chain. Earlier examples were barcodes (section 25.5) but methods now include radio frequency identification (RFID) tags, magnetic strips and electronic article surveillance (EAS) tags (these 'security tags' are electromagnetic or radio frequency devices that activate alarms in shops if they are not deactivated at the checkout). These devices allow storage of more data than barcodes and offer protection from fraud. RFID tags contain an electronic chip, a power source and an antenna. Their main advantages are that data can be read from the tag or sent to the tag without having a line of sight to it, and a large number of tags can be identified and addressed. The price of tags is relatively high and they are therefore used on reusable distribution containers as part of the management of logistics chains (Chapter 27, section 27.3). Developments in intelligent labelling are reviewed by Fairley (2006).

Intelligent tamper-evident technologies are also under development. Examples include optically variable films or gas-sensing dyes that irreversibly change colour, or piezoelectric polymers incorporated into packaging materials that change colour at a certain stress threshold. Such 'self-bruising' closures on bottles or jars could indicate that attempts had been made to open them.

Many other active and intelligent packaging concepts are under development, and research is also underway into thin film devices that produce audio and/or visual information, either in response to touch, motion, scanning or activation, to communicate directly to the customer. However, it is not yet clear how applicable or affordable these packs will be. The future success of intelligent packaging requires acceptance by the packaging industry, food manufacturers and consumers, particularly in the light of environmental concerns (section 25.6) over waste and lack of recyclability of disposable packaging. The perception of extra cost, complexity and possible mistrust or unreliability of indicating devices (e.g. showing food to be safe when it is not, leading to potential liability) are factors that could become barriers to acceptance of intelligent packaging. The technologies will also have to comply with food safety regulations concerning possible migration of components from complex packaging materials into products (de Kruijf and Rijk 2003) and environmental controls on recycling features of the packaging. For active and intelligent packaging to be widely adopted it should be inexpensive relative to the value of the product, reliable, accurate, reproducible in its range of operation, environmentally benign and safe to contact foods. For example, Brody (2006) reports that most active packaging technologies have limited commercial application and notes that intelligent self-heating packaging technology was removed from the market in 2006. Excepting moisture control, oxygen scavenging (above) and inventory control (Chapter 27, section 27.3.1) a similar situation may currently exist with other types of intelligent packaging due to a combination of cost, difficulty in meeting legislative controls or perceived need by food manufacturers and retailers. These aspects are reviewed in detail by Butler (2002).

25.5   Printing

Printing inks for films and papers consist of a dye that is dispersed in a blend of solvents, and a resin which forms a varnish. Solvents must be carefully removed after application of the ink to prevent odour contaminating the product and 'blocking' the film during use. The ink should be low in cost and compatible with the film to achieve a high bond strength.

There are five processes used to print films and papers:

1 Flexographic (or 'relief' or 'letterpress') printing in up to six colours, is fast and suitable for lines or blocks of colour. A fast-drying ink is applied to the film by a flexible rubber plate with raised characters. The plate is pressed against an inked roller to cover the raised portions with ink and then against the film or paper (Fig. 25.12a). It is used, for example, for cartons that do not require high print quality.
2 Rotogravure (or 'intaglio') printing is able to produce high-quality detail and realistic pictures. It uses an engraved chromium-plated roller with the printing surfaces recessed in the metal. Ink is applied to the roller and the excess is wiped from all but the recesses. The remaining ink is then transferred to the packaging material (Fig. 25.12b).
3 Offset lithography (or 'planographic') printing is based on the incompatibility of grease and water. A greasy ink is repelled by moistened parts of a printing plate but

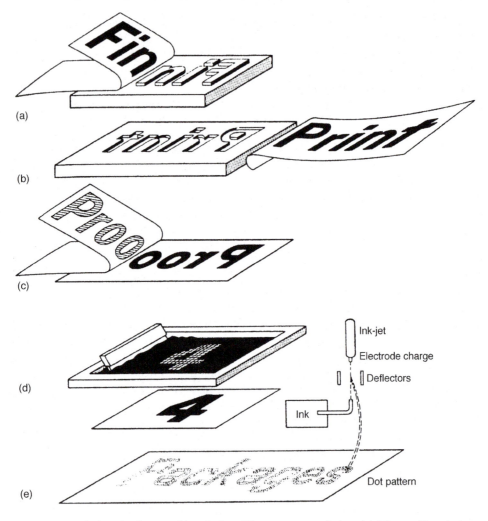

Fig. 25.12   Printing: (a) flexographic printing; (b) rotogravure printing; (c) lithographic printing; (d) screen printing; (e) ink-jet printing (from Lentz 1986).

remains on compatible parts which carry the design. This method produces a print of similar quality to that of rotogravure and is suitable for papers and boards that are too rough for rotogravure printing (Fig. 25.12c).

4   Screen printing in which ink passes through a porous surface of a printing screen (Fig. 25.12d).

5   Ink-jet printing (Fig. 25.12e) in which electrically charged droplets of ink are deflected by charged deflector plates to create the image.

Further details of different method of printing are given by Lentz (1986).

Printing may be on the inside of the film (reverse printing), which has the advantage of producing a high-gloss finish. However, the ink should have negligible odour to prevent contamination of the product. Printing on the outer surface avoids the risk of contact between the ink and product, but the ink must have a high gloss and be scuff-resistant to prevent it from rubbing off during handling. The ink may also be located between two layers of a laminate. This is achieved by reverse printing onto one film and then laminating the two films. Alternatively the ink is overcoated with a polyvinylidene chloride film, which gives a surface gloss, protects the print and contributes to the barrier properties of the film.

Developments in printing for shipping containers include on-line ink-jet printing and 'system laser decoration', in which an ultra-fast laser produces photographic decoration on polymer materials that have special pigments or other additives that selectively change colour under tuned laser light. With microprocessor control, this enables great flexibility in changes to a pack decoration, text language or design, leading to instant decoration, reduced stocks, absence of printing ink residues and enhanced recyclability of materials (Louis 1998, Campey 1987).

UV-curable urethane–acrylate shrink sleeves are increasingly used for PVC, PET and glass bottles, which are cheaper than printing directly on the bottle. They enable greater flexibility in product changeover, and can be removed for recycling PET or glass bottles. The sleeves are designed to have good printability for use with flexographic printing inks, and have good adhesion to the bottle material, excellent scratch and wrinkle resistance and high clarity after shrinking.

25.5.1   Barcodes and other markings

The universal print code (UPC or 'barcode' (Fig. 25.13)) is printed on consumer packs for laser reading at retail checkouts. It avoids the need for individual price labelling of packs and allows itemised bills to be produced for customers. There is no price information

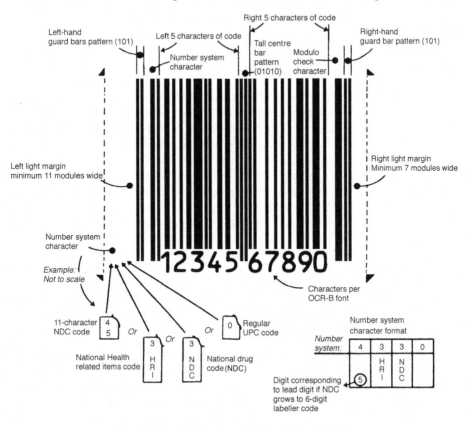

Fig. 25.13   UPC or barcode (from Hirsch 1991).

encoded in a barcode. When the scanner at the checkout scans a product, the cash register sends the UPC number to a central 'point of sale' computer in the store, which identifies the UPC number and returns price of the item at that moment to the cash register. This approach allows the store to change the price of products without having to re-label them. The computer also deducts the item from the store inventory and thus enables faster stock-taking, detection of pilferage and automatic re-ordering. The information is also collated into product sales reports, which can be used by store managers to adjust the shelf space allotment to specific items, or produce data on competitors' sales or the results of promotion and marketing strategies. Shipping containers are also barcoded to inform the carrier about the destination (see also RFID tags, section 25.4.3, and Chapter 27, section 27.3). Further information on UPC codes is given by Adams (2008).

Markings are also required on containers to show the 'sell by' or 'use by' date. A manufacturer's code is printed onto containers to identify the factory, the production line and the shift during which the product was made. Coding lasers are non-contact and produce permanent marks without the use of inks and solvents. They are fully programmable to easily change the characters, and are capable of producing 400–2000 characters per second on paper board metal, glass, plastics and foil. Further details are given by Ravetto (2005).

Developments in labelling, described by Louis (1998), include an information micro-dot on the pack that contains all required label information in a number of languages, leaving the main area of the pack for graphic design and branding. The additional data storage capacity would allow more information to be included on for example, nutritional composition and instructions for use.

25.6   Environmental considerations

The value of packaging materials to protect foods against losses is illustrated by levels of 2–3% food wastage during processing, distribution and storage in industrialised countries, compared with 30–50% in developing countries where sophisticated packaging, storage and distribution are not used. In industrialised countries, around two-thirds of all packaging is used by the food industry and an average household uses around 200 different packaged items (Anon undated). Russotto (1999) reported that 50% of all packaged foods in Europe use plastic packaging, representing 16% of the weight of all packaging. Plastic packaging accounts for ≈11% of municipal waste due to the low rate of recovery compared with the exponential growth in plastics production during the last four decades (Liu 2006). Recycling and reuse are limited because plastic materials increasingly contain multi-layered structures that have been developed to achieve the required barrier properties for particular applications. These multi-layered materials are very difficult to separate into individual layers for recycling. Energy recovery by incineration is practised in some countries and composting biodegradable plastics (section 25.4.1) is under development, but increasing environmental awareness by consumers has targeted food manufacturers and packagers to improve their environmental performance. Retail packs in particular have been subject to criticism by environmental groups on the following grounds: packaging makes excessive use of resources in a wasteful way; it adds to waste disposal problems and uses large amounts of energy, particularly fossil fuels used for plastics. Excessive packaging also adds unnecessarily to the cost of products, limits consumer choice, may be used to deceive consumers, and forms a major part of litter and landfill waste due to poor design for re-use or recycling. A further area of environmental concern is the level of emissions and water pollution during the production of packaging materials. Selke (1994) has reviewed data on airborne emissions of polluting gases and lead, and waterborne emissions of suspended solids, BOD, acids, a range of organic chemicals and heavy metals from the production of a range of packaging materials, including steel, aluminium, plastic films, paper, board and glass.

25.6.1 Packaging costs

The main function of packaging is to enable consumers to receive foods in good condition at the lowest reasonable price. Manufacturers therefore have the responsibility to review the economies of the total production and distribution chain, including reuse and disposal options. Marketing and commercial considerations should be reconciled with economy in the use of materials and energy, and the environmental effects of production and use of materials. It is difficult to compare packaging costs without also including, for example, the costs of shipping containers, depreciation costs of packaging machinery, labour requirements, etc. It is also important to consider the cost of the package in relation to the value of the product. A proportion of the cost of packaging is the energy required for manufacture. The energy needed to make 1 kg of different packaging materials from their raw materials is shown in Table 25.19. It should be noted that these figures are produced on a weight basis, whereas packaging is used on an area or volume basis. The area of a film produced from 1 kg of plastic material is much greater than the area produced from 1 kg tinplate (e.g. one million m^2 of polypropylene film requires 110 tonnes of oil as feedstock and energy, whereas one million 0.33 litre tinplate cans require 75 tonnes of oil). The total environmental impact of production of packaging takes into account the types and sources of raw materials, the energy needed to procure them, to manufacture the packaging material, to convert the material into packages, the amount of wastage during production, the energy needed to handle and use of the packs, and the energy efficiency of the transport methods used.

There are strong economic incentives for manufacturers to reduce the costs of packaging and there have been a number of developments in both the types of materials used for packaging and the methods of handling and distribution of packaged foods that have affected the environmental impact of food packaging (Chiellini 2008). For example,

Table 25.19 Energy needed to make 1 kg of different packaging materials from their raw materials

Packaging material	Total energy	
	$MJ\,kg^{-1}$	Tonnes of oil equivalent (TOE)
Aluminium	293	8.8
Cellulose film	192	4.4
Shrinkwrap film	187	4.3
PET resin	183	4.2
Can sealing compound	180	4.17
Polypropylene film	173	4.0
LDPE resin	104	2.4
Paperboard	99	2.3
Kraft paper	82	1.9
Tinplate	50	1.2
Glass containers	22	0.5

From Paine (1991)

Table 25.20   Levels of recycling of materials for packaging

Material	Weight used in typical container (g and year)	Amount recycled to production (%)	Notes
Aluminium	91 (1950) 17 (1990)	40	Scrap used in aluminium production for all purposes. Plentiful supply of raw material.
Glass	538 (pre 1939) 245 (1990)	35–50	Cullet from glass production and recycled containers from consumer 'bottle banks'. Abundant supplies of raw materials.
Paper and board	N/A	55	Recycled material used in production of all paper and board products. Renewable supply of raw materials.
PET	66 (1983) 42 (1990)	N/A	Scrap used for non-food applications. Non-renewable supplies of oil for raw materials.
PVC	N/A	N/A	Scrap used for non-food applications. Abundant supplies of sodium chloride raw material. Non-renewable supplies of oil raw material but coal, shale or vegetable material could be substituted.
Tinplate	68.9 (1970) 56.6 (1990)	44	Scrap used in ferrous metal production for all purposes. Abundant supplies of all raw materials (iron ore, coal, limestone) except tin. Tin reserves estimated at 25–200 years. 10% of cans produced without tin and percentage increasing.

N/A – data not available.
Adapted from the data of Anon (undated)

100% of scrap from can-making is recycled, which uses 50% less energy compared with making steel from iron ore. Similarly glass-making uses ≈30% recycled glass 'cullet' (section 25.2.3). There is also a move away from materials that require high energy consumption for their manufacture, such as glass and metals, to plastics. This may, however, be a short-term gain because (with the exception of cellulose) plastic materials (section 25.4.1) are produced from non-renewable petrochemical resources, whereas raw materials for glass and metal packaging are relatively abundant and these materials are recyclable. Paper and board packaging is now mostly sourced from renewable forests and the proportion of recycled paper that is used in packaging is steadily increasing (Table 25.20), although there remain substantial variations in the amounts used in different countries. The amount of energy needed to produce a range of packaging materials, including paper, board, aluminium, steel, glass and a variety of plastic films is reviewed by Selke (1994).

Developments that reduce the cost of delivery of raw materials include the use of bulk handling rather than small containers. Ingredients and packaging materials were previously delivered in small unit loads to food manufacturers, which increased both fuel consumption for transportation and the number of packages that were required. Developments in bulk handling of ingredients in which tankers and large (1–2 t) Combo-bins, tanks or bags have replaced 50 kg sacks, have reduced the energy and packaging materials needed to supply ingredients.

The energy needed to transport packaging materials depends in part on the types of packaging: trucks are either filled before they reach their weight limit (e.g. plastics or paper) or they reach their weight limit before they are full (e.g. glass). The energy consumption for a full load is reported by Selke (1994) as ranging from 9.61 MJ/vehicle-km for vehicles below 1 tonne capacity to 1.80–27.5 MJ/vehicle-km for vehicles greater than 18 tonne capacity, including the energy consumed in supplying the fuel. This compares with estimates of 0.58 MJ/tonne-km and 1.09–1.50 MJ/tonne-mile for rail transport and 0.17 MJ/tonne-mile for sea transport. The design of materials, including rolls of film for form–fill–seal machines to replace pre-formed packs, and stackable pots to replace cans and jars, have substantially reduced the volume of packaging materials to be transported and hence the fuel consumed to deliver them.

The weight of packages has also fallen with almost 90% of plastic consumer packs now weighing less than 10 g; up to 80% lighter than in the 1980s (Russotto 1999). 'Lightweighting' of metal, glass, paperboard and plastics has also continuously been adopted for many decades: for example, tinplate thicknesses (section 25.2.2) are half the weight of those used in the 1930s and paperboard for corrugated cases is 30% lighter. Shrink-wrapped trays weighing 80–90 g have largely replaced corrugated fibreboard cartons (weighing 300–350 g) as shipping containers, and even though the total area of flexible packaging material has increased, the weight of materials used has decreased. The development of improved barrier packaging materials has also reduced the weight and cost of retail packaging. For example, in many applications PET has replaced glass bottles; aluminium has replaced tinplate cans; printed co-extruded polymer containers have replaced glass jars with metal lids and printed paper labels; and polymer trays and films have replaced metal or glass containers. Shipping containers are mostly separated and collected at retail outlets and many are recycled or reused. In an audit reported by Russotto (1999), a major European retailer found that it used 20 000 tonnes of shipping containers each year. The company introduced reusable trays and reduced this amount by 40% as part of an efficiency and environmental programme.

25.6.2 Recycling

In contrast to most goods that begin their useful life upon purchase, most packaging materials cease their usefulness at this stage, apart from a period of home storage, and are disposed of. An exception is the use of reusable bottles, which remains common in some countries for milk or beers. Such systems work well when the supplier is relatively close to the consumer or a closed system operates in which the delivery vehicle collects empty bottles. Because reusable bottles are made thicker (and hence heavier) than non-returnable bottles to withstand the reuse, this increases the energy needed for both production of the bottle and transportation. There are environmental benefits only when a bottle has made several journeys, and if it is not returned the initial investment in energy and resources is wasted. In many European countries and in Japan, all new containers have 50–55% recycled glass and some containers are made from 90–100% recycled glass, although in other regions, where collection is less well organised the figures are much lower. Progress has been made in many countries to encourage consumers to separate glass, metal and paper packaging for recycling and there have been significant developments in recycling facilities by local authorities. For example, cans from household waste are separated magnetically, cleaned to remove dirt and treated in a de-tinning plant to extract and reuse the tin.

Twenty years ago nearly all plastic packaging was either incinerated or sent to landfill sites, and only uncontaminated in-house production waste was collected and recycled. However, changes in legislation and pressure from consumers and environmental groups have resulted in more ecologically sound packaging systems, increased levels of post-consumer plastic recycling in many countries, and reduction, reuse and recovery of packaging materials (Anon 2005b, Akre 1991, Griffiths 1991, Hall 1991). The EU directive on packaging waste (Anon 1994) states that by no later than December 2008, 55–80% by weight of packaging waste should be recycled and no later than December 2008 the following targets for materials contained in packaging waste must be attained: 60% by weight for glass, paper and board; 50% by weight for metals; 22.5% by weight for plastics and 15% by weight for wood. Interpretation of European packaging regulations is given by Powell and Steele (1999). The introduction of 'environmental' or 'pollution' taxes, together with potentially large increases in the price of oil-based raw materials, are likely to further change the relative benefits of recyclable materials. These issues are discussed in more detail by Franz and Welle (2003), and White (1994) reviews the options for packaging disposal, and recycling of paper, metals, glass and plastics. Anon (2008j) describes methods for recycling of multi-layer laminated paperboard cartons (section 25.2.6).

The suitability of plastics for recycling depends on their inertness: inert materials absorb smaller amounts of 'contaminants' from the foods they contain and are therefore more suitable for recycling. Franz and Welle (2003) report that PET and PVC are more inert than polystyrene and HDPE and as a result are the most suitable for recycling. The level of contaminants can be controlled off-line using high-performance liquid chromatography (HPLC) or gas chromatography (GC) laboratory equipment, or on-line by sniffing devices. Contaminated containers can then be rejected, which reduces the requirements for cleaning efficiency prior to recycling. A study reported by de Kruijf (1997) concluded that re-using PET and polycarbonate beverage bottles was feasible, without public health risks from contaminated bottles. Modern sorting technologies are capable of providing recycled plastics that are nearly 100% of one polymer type (Franz and Welle 2003). Recycled packs are blended with new polymers in amounts that vary from a few per cent up to 50% in some applications. Alternatively, the recycled material can be used as a core material in multi-layer coextruded plastics, protected from contact with foods by layers made from new polymer (known as a 'functional barrier'). Studies in which this has been successfully achieved for coextruded 3-layer PP yoghurt cups and 3-layer PET bottles and films are reported by Franz et al. (1994, 1996) and Piringer et al. (1998). Many European countries now approve the use of recycled PET as a monolayer for contact with foods, and companies have been established to process 65 000 t of super-clean recycled PET per year for direct food contact applications (Franz and Welle 2003). Further information on plastics recycling is given by Anon (2004).

References

ADAMS, R., (2008), Bar Code 1, information from Adams Communications, available at www.adams1.com/ and follow links to Universal Product Code (UPC) and EAN Article Numbering Code (EAN).

AHVENAINEN, R., (2003), Active and intelligent packaging, in (R. Ahvenainen, Ed.), *Novel Food Packaging Techniques*, Woodhead Publishing, Cambridge, pp. 5–21.

AKBARI, Z., GHOMASHCHI, T. and MOGHADAM, S., (2007), Improvement in food packaging industry with biobased nanocomposites, *International J. Food Engineering*, **3** (4), available at: www.bepress.com/ijfe/vol3/iss4/art3.

AKRE, E., (1991), Green politics and industry, *European Food and Drink Review*, Packaging Supplement, pp. 5–7.

ALLAART-BRUIN, S., HAAGH, G.A.A.V. HEGEN, D., VAN DER LINDEN, B.J. and MATTHEIJ, R.M.M., (2004),

Modelling of the glass press-blow process ECMI 2004, Centre For Analysis, Scientific Computing and Applications, available at www.win.tue.nl/casa/research/casaprojects/allaart-bruin.html.

ANON, (undated), *Packaging and Resources*, The Industry Committee for Packaging and the Environment, London.

ANON, (1994), European Parliament and Council Directive 94/62/EC of 20 December 1994 on packaging and packaging waste, available at http://europa.eu/scadplus/leg/en/lvb/l21207.htm.

ANON., (1998), Seamless corrugated IBCs, in *Food Technology International*, Sterling Publications, London, p. 85.

ANON, (2000), Aluminium can is a resealable bottle, *Food Engineering and Ingredients*, October, 49.

ANON, (2001), Microbiological safety of controlled and modified atmosphere packaging of fresh and fresh-cut produce, analysis and evaluation of preventive control measures for the control and reduction/elimination of microbial hazards on fresh and fresh-cut produce, US Food and Drug Administration Center for Food Safety and Applied Nutrition, September 30, available at www.cfsan.fda.gov/~comm/ift3-6.html.

ANON, (2002), Universal Bright – a new film laminated tin-free steel sheet for food cans, NKK Technical Review No. 87, available at www.jfe-steel.co.jp/archives/en/nkk_giho/87/pdf/87_09.pdf.

ANON (2004), Introduction to plastic recycling, Plastic Waste Management Institute, available at www.pwmi.or.jp/ei/plastic_recycling_2004.pdf.

ANON, (2005a), Bag in Box Packaging for Food, company information from D.S. Smith Plastics Ltd., available at www.rapak.com/packaging-food.asp?ACTRED=1.

ANON, (2005b), Food and drink sector: minimising packaging, Envirowise information, available at www.envirowise.gov.uk/ and follow links to 'select sector – food and drink'.

ANON, (2006a), Polypropylene – BOPP film, company information from Innovene, available at www.innovene.com/genericarticle.do-categoryId=9002717&contentId=7005640.htm#top

ANON, (2006b), Packaging takes the chill off frozen foods, *Food Processing*, **75** (4), 12.

ANON, (2006c), Launch of first metallised NatureFlex, company information from Innovia Films, available at www.packaging-gateway.com/contractors/materials/ubcfilmsplc/press8.html.

ANON, (2008a), Polypropylene, company information from Total Petrochemicals, available at www.totalpetrochemicalsusa.com/brochures/PP_castandblown_film.pdf.

ANON, (2008b), How a three-piece welded food can is made, information from Metal Packaging Manufacturers Association, available at www.mpma.org.uk/pages/pv.asp?p=mpma11&fsize=0.

ANON, (2008c), How a two-piece drawn and wall-ironed food can is made, information from Metal Packaging Manufacturers Association, available at www.mpma.org.uk/pages/pv.asp?p=mpma10&fsize=0.

ANON, (2008d), Tinplate and tin free steel, company information from JFE Steel Corporation, available at www.jfe-steel.co.jp/en/products/sheets/catalog/b1e-006.pdf.

ANON, (2008e), Visypak food packaging, company information from Visy available at http://www.visy.com.au/uploads/steel_can_process.pdf.

ANON, (2008f), Types of glass, company information from British Glass, available at www.britglass.org.uk/aboutglass/typesofglass.html#1.

ANON, (2008g), Glass – an overview, information from CERAM Research Ltd, A–Z of Materials, available at www.azom.com/details.asp?ArticleID=1021#_Sheet_and_Container.

ANON, (2008h), Blow-and-blow process, *Illustrated Glass Dictionary*, Glass Online, available at www.glassonline.com/infoserv/dictionary/80.html.

ANON, (2008i), Composting ca, information from Brent Hansen Environmental Business Consulting & Project Management Ltd., available at www.composting.ca/index.html by following the links to areas of interest.

ANON, (2008j), Recycling, company information from Tetra Pak, available at www.tetrapak.com, and follow links to 'Environment' and 'Recycling and recovery'.

ARVANITOYANNIS, I. and GORRIS, L.G.M., (1999), Edible and biodegradable polymeric materials for food packaging or coating, in (F.A.R. Oliveira and J.C. Oliveira, Eds.), *Processing Foods – Quality Optimisation and Process Assessment*, CRC Press, Boca Raton, FL, pp. 357–368.

AURAS, R., HARTE, B. and SELKE, S., (2006), Sorption of ethyl acetate and *d*-limonene in poly(lactide) polymers, *J. Science of Food and Agriculture*, **86** (4), 648–656.

BALDWIN, E.A., (1994), Edible coatings for fresh fruits and vegetables: past, present and future, in (J.M. Krochta, E.A. Baldwin and M.O. Niosperos-Carriedo, Eds.), *Edible Coatings and Films to Improve Food Quality*, Technomic Publishing, Lancaster, PA, p. 25.

BALDWIN, E.A., (1999), Surface treatments and edible coatings in food preservation, in (M.S. Rahman, Ed.), *Handbook of Food Preservation*, Marcel Dekker, pp. 577–610.

BLAKISTONE, B.A., (1998a), Meats and poultry, in (B.A. Blakistone, Ed.), *Principles and Applications of Modified Atmosphere Packaging of Foods*, 2nd edn, Blackie Academic and Professional, London, pp. 240–284.

BLAKISTONE, B.A., (1998b), Introduction, in (B.A. Blakistone, Ed.), *Principles and Applications of Modified Atmosphere Packaging of Foods*, 2nd edn, Blackie Academic and Professional, London, pp. 1–13.

BLANCHFIELD, J.R., (Ed.), (2000), *Food Labelling*, Wooodhead Publishing, Cambridge.

BRENNAN, J.G. and DAY, B.P.F., (2006), Packaging, in (J.G. Brennan, Ed.), *Food Processing Handbook*, Wiley-VCH, Weinheim, Germany, pp. 291–350.

BRENNAN, J.G., BUTTERS, J.R., COWELL, N.D. and LILLY, A.E.V., (1990), *Food Engineering Operations*, 3rd edn, Elsevier Applied Science, London, pp. 617–653.

BRISTON, J.H., (1980), Rigid plastics packaging, in (S.J. Palling, Ed.), *Developments in Food Packaging*, Vol. 1, Applied Science, London, pp. 27–53.

BRISTON, J.H., (1987), Rigid plastic containers and food packaging, in (A. Turner, Ed.), *Food Technology International Europe*, Sterling Publications, London, pp. 283, 285–287.

BRISTON, J.H., (1990), Recent developments in bag-in-box packaging, in (A. Turner, Ed.), *Food Technology International Europe*, Sterling Publications, London, pp. 319–320.

BRODY, A.L., (1990), Controlled atmosphere packaging for chilled foods, in (A. Turner, Ed.), *Food Technology International Europe*, Sterling Publications, London, pp. 307–313.

BRODY, A.L., (1992), Technologies of retortable barrier plastic cans and trays, in (A. Turner, Ed.), *Food Technology International Europe*, Sterling Publications, London, pp. 241–249.

BRODY, A.L., (2001), What's active about intelligent packaging, *Food Technology*, **55** (6), 75–78.

BRODY, A. L., (2003), Nano, nano food packaging technology, *Food Technology*, **57** (12), 53–54.

BRODY, A., (2006), State of the art of active and intelligent packaging, Institute of Food Technologists Fifth Research Summit, Baltimore, MD, 7–9 May.

BRODY, A.L. and BUDNY, J.A., (1995), Enzymes as active packaging agents, in (M.L. Rooney, Ed.), *Active Food Packaging*, Blackie Academic and Professional, London, pp. 174–192.

BUTLER, B.L., VERGANO, P.J., TESTIN, R.F., BUNN, J.M. and WILES, J.L., (1996), Mechanical and barrier properties of edible chitosan films as affected by composition and storage, *J. Food Science*, **61** (5), 953–956.

BUTLER, P., (2001), Smart packaging – intelligent packaging for food, beverages, pharmaceuticals and household products, *Materials World*, **9** (3), 11–13.

BUTLER, P., (2002), Smart Packaging – Strategic ten-year forecasts and technology and company profiles, available from IDTechEx Ltd, Cambridge.

CALERO, F.A. and GOMEZ, P.A., (2003), Active packaging and colour control: the case of fruit and vegetables, in (R. Ahvenainen, Ed.), *Novel Food Packaging Techniques*, Woodhead Publishing, Cambridge, pp. 416–438.

CAMPEY, D.R., (1987), The application of laser marking, in (A. Turner, Ed.), *Food Technology International Europe*, Sterling Publications, London, pp. 303–304.

CAVA, D., GIMÉNEZ, E., GAVARA, R. and LAGARON, J.M., (2006), Comparative performance and barrier properties of biodegradable thermoplastics and nanobiocomposites versus pet for food packaging applications, *J. Plastic Film and Sheeting*, **22** (4), 265–274.

CHA, D.S. and CHINNAN, M.S., (2004), Biopolymer-based antimicrobial packaging: a review, *Critical Reviews in Food Science and Nutrition*, **44**, 223–237.

CHARBONNEAU, J.E., (1997), Recent case histories of food product–metal container interactions using scanning electron microscopy-x-ray microanalysis, *Scanning*, **19** (7), 512–518.

CHIELLINI, E., (2008), *Environmentally-compatible Food Packaging*, Woodhead Publishing, Cambridge.

CHURCH, N., (1994), Developments in modified-atmosphere packaging and related technologies, *Trends in Food Science and Technology*, **5**, 345–352.

CHURCH, N., (1998), MAP fish and crustaceans – sensory enhancement, *Food Science and Technology Today*, **12** (2), 73–83.

COMA, V., DESCHAMPS, A. and MARTIAL-GROS, A., (2003), Bioactive packaging materials from edible chitosan polymer – antimicrobial activity assessment on dairy-related contaminants, *J. Food Science*, **68** (9), 2788–2792.

COOKSEY, K., MARSH, K.S. and DOAR, L.H., (1999), Predicting permeability and transmission rate for multilayer materials, *Food Technology*, **53** (9), 60–63.

DALE, J., (2000), Practically creative, *Food Processing*, February, 40.

DAVIES, A.R., (1995), Advances in modified-atmosphere packaging, in (G.W. Gould, Ed.), *New Methods of Food Preservation*, Blackie Academic and Professional, Glasgow, pp. 304–320.

DAY, B.P.F., (1992), Chilled food packaging, in (C. Dennis and M. Stringer, Eds.), *Chilled Foods – A Comprehensive Guide*, Ellis Horwood, London, pp. 147–163.

DAY, B.P.F., (2003), Novel MAP applications for fresh-prepared produce, in (R. Ahvenainen, Ed.), *Novel Food Packaging Techniques*, Woodhead Publishing, Cambridge, pp. 189–207.

DEBEAUFORT, F., QUEZADA-GALLO, J-A. and VOILLEY, A., (1998), Edible films and coatings: tomorrow's packagings: a review, *Critical Reviews in Food Science*, **38** (4), 209–313.

DE KRUIJF, N., (1997), Food packaging materials for refilling, in (A. Turner, Ed.), *Food Technology International Europe*, Sterling Publications, London, pp. 85–88.

DE KRUIJF, N. and RIJK, R. (2003), Legislation issues relating to active and intelligent packaging, in (R. Ahvenainen, Ed.), *Novel Food Packaging Techniques*, Woodhead Publishing, Cambridge, pp. 459–496.

DE VLIEGER, J.J. (2003), Green plastics for food packaging, in (R. Ahvenainen, Ed.), *Novel Food Packaging Techniques*, Woodhead Publishing, Cambridge, pp. 519–534.

DEVLIEGHERE, F., (2002), Modified atmosphere packaging (MAP), in (C.J.K. Henry and C. Chapman, Eds.), *The Nutrition Handbook for Food Processors*, Woodhead Publishing, Cambridge, pp. 342–369.

DEVLIEGHERE, F. and DEBEVERE, J., (2003), MAP, product safety and nutritional quality, in (R. Ahvenainen, Ed.), *Novel Food Packaging Techniques*, Woodhead Publishing, Cambridge, pp. 208–230.

DRISCOLL, R.H. and PATTERSON, J.L., (1999), Packaging and food preservation, in (M.S. Rahman, Ed.), *Handbook of Food Preservation*, Marcel Dekker, New York, pp. 687–734.

FAIRLEY, M., (2006), Smart labels offer smart solutions, *Labels and Labelling*, 3 August, available at www.labelsandlabelling.co.uk/scripts/publish/headlines.asp?code=SLF&key=1604.

FAITHFULL, J.D.T., (1988), Ovenable thermoplastic packaging, in (A. Turner, Ed.), *Food Technology International Europe*, Sterling Publications, London, pp. 357–362.

FARBER, J.M., (1991), Microbiological aspects of modified-atmosphere packaging technology – a review. *J. Food Protection*, **54**, 58–70.

FELLOWS, P.J. and AXTELL, B.L., (2003), *Appropriate Food Packaging*, 2nd edn, IT Publications, London, pp. 49–50.

FRANZ, R. and WELLE, F., (2003), Recycling packaging materials, in (R. Ahvenainen, Ed.), *Novel Food Packaging Techniques*, Woodhead Publishing, Cambridge, pp. 497–518.

FRANZ, R., HUBER, M. and PIRINGER, O-G., (1994), Testing and evaluation of recycled plastics for food packaging use – possible migration through a functional barrier, *Food Additives and Contaminants*, **11** (4), 479–496.

FRANZ, R., HUBER, M., PIRINGER, O-G., DAMANT, A.P., JICKELLS, S.M. and CASTLE, L., (1996), Study of functional barrier properties of multiplayer recycled poly(ethylene terephthalate) bottles for soft drinks, *J. Agriculture and Food Chemistry*, **44** (3), 892–897.

GARRETT, E.H., (1998), Fresh-cut produce, in (B.A. Blakistone, Ed.), *Principles and Applications of Modified Atmosphere Packaging of Foods*, 2nd edn, Blackie Academic and Professional, London, pp. 125–134.

GILL, C.O., (2003), Active packaging in practice: meat, in (R. Ahvenainen, Ed.), *Novel Food Packaging Techniques*, Woodhead Publishing, Cambridge, pp. 364–383.

GONTARD, N., GUILBERT, S. and CUQ, J.L., (1992a), Edible wheat gluten films: influence of main process variables on films properties using response surface methodology, *J. Food Science*, **57**, 190–199.

GONTARD, N., GUILBERT, S. and CUQ, J.L., (1992b), Edible films and coatings from natural polymers, in *New Technologies for the Food and Drink Industries*, Campden and Chorleywood Food Research Association, Chipping Campden, available at www.campden.co.uk/publ/product.htm and follow links to relevant publications.

GONTARD, N., GUILBERT, S., AND CUQ, J.L., (1993), Water and glycerol as plasticisers affect mechanical and water vapour barrier properties of an edible wheat gluten film, *J. Food Science*, **58**, 206–211.

GORRIS, L.G.M. and PEPPELENBOS, H.W., (1999), Modified atmosphere packaging of produce, in (M.S. Rahman, Ed.), *Handbook of Food Preservation*, Marcel Dekker, New York, pp. 437–456.

GREENGRASS, J., (1998), Packaging materials for MAP foods, in (B.A. Blakistone, Ed.), *Principles and Applications of Modified Atmosphere Packaging of Foods*, 2nd edn, Blackie Academic and Professional, London, pp. 63–101.

GRIFFITHS, A.J., (1991), Plastics in the environment – the industry's position, *European Food and Drink Review*, Packaging Supplement, pp. 9–12.

GUILBERT, S. and GONTARD, N., (1995), Edible and biodegradable food packaging, in *Foods and Packaging Materials*, Royal Society of Chemistry, Oxford, p. 159.

HALL, M. (1991) Green light at Pakex, *European Food and Drink Review*, Packaging Supplement, pp. 13–16

HAMILTON, D., (2005), Can lacquer coatings accurately controlled with new optical measurement system, company information from Scalar Technologies, available at www.scalartechnologies.com and follow link to 'SG S-can'.

HAN, J.H., (2000), Antimicrobial food packaging, *Food Technology*, **54** (3), 56–65.

HAN, J.H., (2003), Antimicrobial food packaging, in (R. Ahvenainen, Ed.), *Novel Food Packaging Techniques*, Woodhead Publishing, Cambridge, pp. 50–70.

HASTINGS, M.J., (1998), MAP machinery, in (B.A. Blakistone, Ed.), *Principles and Applications of Modified Atmosphere Packaging of Foods*, 2nd edn, Blackie Academic and Professional, London, pp. 39–62.

HAUGAARD, V.K., UDSEN, A-M., MORTENSEN, G., HØEGH, L., PETERSEN, K. and MONAHAN, F., (2000), Food biopackaging, in (C. J. Weber, Ed.), *Biobased Packaging Materials for the Food Industry – Status and Perspectives*, Food Biopack Project, EU Directorate 12, pp. 45–106, available at www.biomatnet.org/publications/f4046fin.pdf.

HIRSCH, A., (1991), *Flexible Food Packaging*, Van Nostrand Reinhold, New York.

JAKOBSEN, M. and BERTELSEN, G., (2003), Active packaging and colour control: the case of meat, in (R. Ahvenainen, Ed.), *Novel Food Packaging Techniques*, Woodhead Publishing, Cambridge, pp. 401–415.

KHWALDIA, K., PEREZ, C., BANON, S., DESOBRY, S. and HARDY, J., (2004), Milk proteins for edible films and coatings, *Critical Reviews in Food Science and Nutrition*, 44, 239–251.

KOTSIANIS, I.S., GIANNOU, V. and TZIA, C., (2002), Production and packaging of bakery products using MAP technology, *Trends in Food Science and Technology*, **13** (9–10), 319–324.

LABUZA, T.P., (1996), An introduction to active packaging of foods, *Food Technology*, **50**, 68–71.

LAGARON, J.M., CABEDO, L., FEIJOO, J.L., GAVARA, R. and GIMENEZ, E., (2005), Improving packaged food quality and safety: (II) Nanocomposites, *Food Additives and Contaminants*, **22** (10), 994–998.

LAMBERTI, M. and ESCHER, F., (2007), Aluminium foil as a food packaging material in comparison with other materials, *Food Reviews International*, **23** (4), 407–433.

LENTZ, J., (1986), Printing, in (M. Bakker and D. Eckroth, Eds.), *The Wiley Encyclopaedia of Packaging Technology*, John Wiley and Sons, New York, p. 554.

LIU, L., (2006), Bioplastics in food packaging: innovative technologies for biodegradable packaging, Packaging Engineering Dept., San Jose University, available at www.iopp.org/files/SanJoseLiu.pdf?pageid=pageid.

LOPEZ-RUBIO, A., GAVARA, R. and LAGARON, J.M. (2006), Bioactive packaging: turning foods into healthier foods through biomaterials, *Trends in Food Science and Technology*, **17** (10), 567–575.

LOUIS, P., (1998), Food packaging in the next century, in (A. Turner, Ed.), *Food Technology International Europe*, Sterling Publications, London, pp. 80–82.

LUCAS, J., (2003), Integrating MAP with new germicidal techniques, in (R. Ahvenainen, Ed.), *Novel Food Packaging Techniques*, Woodhead Publishing, Cambridge, pp. 312–336.

LYIJYNEN, T., HURME, E. and AHVENAINEN, R., (2003), Optimizing packaging, in (R. Ahvenainen, Ed.), *Novel Food Packaging Techniques*, Woodhead Publishing, Cambridge, pp. 441–458.

LYONS, G., (2000), Bisphenol A – a known endocrine disruptor, WWF European Toxics Programme Report, available at http://www.wwf.org.uk/filelibrary/pdf/bpa.pdf.

MANFREDI, L.B., GINÉS, M.J.L., BENÍTEZ, G.J., EGLI, W.A., RISSONE, H. and VÁZQUEZ, A., (2005), Use of epoxy-phenolic lacquers in food can coatings: characterization of lacquers and cured films, *J. Applied Polymer Science*, **95** (6), 1448–1458.

MANIAS, E., (2006), Polymer/inorganic nanocomposites: opportunities for food packaging applications, Institute of Food Technologists Fifth Research Summit, Baltimore, MD, 7–9 May.

MASSEY, L., (2003), *Permeability Properties of Plastics and Elastomers: A Guide to Packaging and Barrier Materials* (Plastics Design Library), William Andrew Publishing, Norwich, NY.

MAY, N., (2004), Developments in packaging formats for retort processing, in (P. Richardson, Ed.), *Improving the Thermal Processing of Foods*, Woodhead Publishing, Cambridge, pp. 138–151.

MCHUGH, T.H. and KROCHTA, J.M., (1994), Permeability properties of films, in (J.M. Krochta, E.A. Baldwin and M.O. Nisperor-Carriedo, Eds.), *Edible Coatings and Films to Improve Food Quality*, Technomic Publishing, Lancaster, PA, pp. 139–188.

MERCEA, P., (2000), Models for diffusion in polymers, in (O. Piringer and A.L. Baner, Eds.), *Plastic Packaging Materials for Food: Barrier Function, Mass Transport, Quality Assurance, Legislation*, Wiley-VCH, Weinheim, Germany, pp. 125–158.

MOHANTY, A.K., (2006), Bio-based materials for a sustainable future in packaging, Institute of Food Technologists Fifth Research Summit, Baltimore, MD, 7–9 May.

MORILLON, V., DEBEAUFORT, F., BLOND, G., CAPELLE, M. and VOILLEY, A., (2002), Factors affecting the moisture permeability of lipid-based edible films: a review, *Critical Reviews in Food Science and Nutrition*, **42** (1), 67–89.

NANKIVELL, B., (2001), Clearly better packaging, *Food Processing*, October, 11–12.

NYCHAS, G.J. and ARKOUDELOS, J.S., (1990), Microbiological and physichemical changes in minced meats under carbon dioxide, nitrogen or air at 3 °C, *International J. Food Science and Technology*, **25**, 389–398.

O'BEIRNE, D. and FRANCIS, G.A., (2003), Reducing pathogen risks in MAP-prepared produce, in (R. Ahvenainen, Ed.), *Novel Food Packaging Techniques*, Woodhead Publishing, Cambridge, pp. 231–275.

OLIVAS, G.I. and BARBOSA-CANOVAS, G.V., (2005), Edible coatings for fresh-cut fruits, *Critical Reviews in Food Science and Nutrition*, **45**, 657–670.

OORAIKUL, B., (2003), Modified atmosphere packaging (MAP), in (P. Zeuthen and L. Bogh-Sorensen, Eds.), *Food Preservation Techniques*, Woodhead Publishing, Cambridge, pp. 339–359.

OORAIKUL, B. and STILES, M.E., (1991), Review of the development of modified atmosphere packaging, in (B. Ooraikul and M.E. Stiles, Eds.), *Modified Atmosphere Packaging of Food*, Ellis Horwood, London, pp. 1–18.

OZDEMIR, M. and FLOROS, J.D., (2004), Active food packaging technologies, *Critical Reviews in Food Science and Nutrition*, **44**, 185–193.

OZDEMIR, M., YURTERI, C.U. and SADIKOGLU, H., (1999), Surface treatment of food packaging polymers by plasmas, *Food Technology*, **53** (4), 54–58.

PAINE, F.A., (1991), *The Packaging User's Handbook*, Blackie Academic & Professional, London.

PAINE, F.A. and PAINE, H.Y., (1992), *A Handbook of Food Packaging*, 2nd edn, Blackie Academic and Professional, London, pp. 53–96.

PARRY R.T., (1993), Introduction, in (R.T. Parry, Ed.), *Principles and Applications of MAP of Foods*, Blackie Academic and Professional, New York, pp. 1–18.

PETERSEN, K., NIELSEN, P.V., BERTELSEN, G., LAWTHER, M., OLSEN, M.B., NILSSON, N.H. and MORTENSEN, G., (1999), Potential of biobased materials for food packaging, *Trends in Food Science and Technology*, **10**, 52–68.

PFEIFFER, C., D'AUJOURD'HUI, M., WALTER, J., NUESSLI, J. and FLETCHER, F., (1999), Optimising food packaging and shelf life, *Food Technology*, **53** (6), 52–59.

PIRINGER, O.G. and RÜTER, M., (2000), Sensory problems caused by food and packaging interactions, in (O.G. Piringer, and A.L. Baner, Eds.), *Plastic Packaging Materials for Food: Barrier Function, Mass Transport, Quality Assurance, Legislation*, Wiley-VCH, Weinheim, Germany, pp. 407–426.

PIRINGER, O-G., HUBER, M., FRANZ, R., BEGLEY, T.H. and MCNEAL, T.P., (1998), Migration from food packaging containing a functional barrier: mathematical and experimental evaluation, *J. Agriculture and Food Chemistry*, **46** (4), 1532–1538.

PLAUT, H., (1995), Brain boxed or simply packed, *Food Processing*, July, 23–25.

POWELL, J. and STEELE, A., (1999), *The Packaging Regulations – Implications for Business*, Chandos Publishing, Witney.

POWERS, T.H. and CALVO, W.J., (2003), Moisture regulation, in (R. Ahvenainen, Ed.), *Novel Food Packaging Techniques*, Woodhead Publishing, Cambridge, pp. 172–185.

RAMSLAND, T., (1989), *Handbook on Procurement of Packaging* (J. Selin, Ed.), PRODEC, Helsinki.

RAVETTO, C., (2005), Making your mark, *Flexible Packaging*, November, available at www.flexpackmag.com/ and search author's name.

RHIM, J-W. and NG, P.K.W., (2007), Natural biopolymer-based nanocomposite films for packaging applications, *Critical Reviews in Food Science and Nutrition*, **47** (4), 411–433.

ROBERTSON, G.L., (1990), Testing barrier properties of plastic films, in (A. Turner, Ed.), *Food Technology International Europe*, Sterling Publications, London, pp. 301–305.

ROBERTSON, G.L., (1993), *Food Packaging - Principles and Practice*, Marcel Dekker, New York.

ROONEY, M.L., (1995), Overview of active food packaging, in (M.L. Rooney, Ed.), *Active Food Packaging*, Blackie Academic and Professional, London, pp. 1–37.

ROSNES, J.T., SIVERTSVIK, M. and SKARA, T., (2003), Combining MAP with other preservation techniques, in (R. Ahvenainen, Ed.), *Novel Food Packaging Techniques*, Woodhead Publishing, Cambridge, pp. 287–311.

RUSSOTTO, N., (1999), Plastics – the 'quiet revolution' in the plastics industry, in (A. Turner, Ed.), *Food Technology International Europe*, Sterling Publications, London, pp. 67–69.

SANCHEZ-GARCIA, M.D., GIMENEZ, E. and LAGARON, J.M., (2007), Novel PET nanocomposites of interest in food packaging applications and comparative barrier performance with biopolyester nanocomposites, *J. Plastic Film and Sheeting*, **23** (2), 133–148.

SARWAR, M. and ARMITAGE, A.W., (2003), Tooling requirements for glass container production for the narrow neck press and blow process, J. Materials Processing Technology, **139** (1–3), 160–163.

SCHILTHUIZEN, S.F., (2000), Communication with your packaging: possibilities for intelligent functions and identification methods in packaging, *Packaging Technology and Science*, **12** (5), 225–228.

SCHRAUT, O., (1989), Aseptic packaging of food into carton packs, in (A. Turner, Ed.), *Food Technology International Europe*, Sterling Publications, London, pp. 369–372.

SELKE, S.E., (1994), Packaging options, in (J.M. Dalzell, Ed.), *Food Industry and the Environment – Practical Issues and Cost Implications*, 2nd edn, Aspen Publishers Inc., New York, pp. 253–290.

SINGH, R.P. and HELDMAN, D.R., (2001), Mass transfer in packaging materials, in *Introduction to Food Engineering*, Academic Press, London, pp. 520–525.

SIVERTSVIK, M., (2003), Active packaging in practice: fish, in (R. Ahvenainen, Ed.), *Novel Food Packaging Techniques*, Woodhead Publishing, Cambridge, pp. 384–400.

SMITH, J.P., RAMASWAMY, H.S. and SIMPSON, K., (1990), Developments in food packaging technology, Part II: storage aspects, *Trends in Food Science and Technology*, November, 111–118.

SMOLANDER, M., (2003), The use of freshness indicators in packaging, in (R. Ahvenainen, Ed.), *Novel Food Packaging Techniques*, Woodhead Publishing, Cambridge, pp. 127–143.

SMOLANDER, M., HURME, E., LATVA-KALA, K., LUOMA, T., ALAKOMI, H-L. and AHVENAINEN, R., (2002), Myoglobin-based indicators for the evaluation of freshness of unmarinated broiler cuts, *Innovative Food Science and Emerging Technologies*, **3** (3), 277–285.

STEWART, B., (1995), *Packaging as an Effective Marketing Tool*, PIRA International, Leatherhead.

TAOUKIS, P.S. and LABUZA, T.P., (2003), Time–temperature indicators, in (R. Ahvenainen, Ed.), *Novel Food Packaging Techniques*, Woodhead Publishing, Cambridge, pp. 103–126.

TICE, P., (2000), EC food contact legislation and how in the future it may be applied to lacquer-coated food and beverage cans, *British Food Journal*, **102** (11), 856–871.

TURTLE, B.I., (1990), PET containers for food and drink, in (A. Turner, Ed.), *Food Technology International Europe*, Sterling Publications, London, pp. 315–317.

VAN TUIL, R., FOWLER, P., LAWTHER, M. and WEBER, C.J., (2000), Properties of biobased packaging materials, in (C.J. Weber, Ed.), *Biobased Packaging Materials for the Food Industry – Status and Perspectives*, Food Biopack Project, EU Directorate 12, pp. 13–44, available at www.biomatnet.org/publications/f4046fin.pdf.

VERAART, R., (2008), Information about materials that come into contact with food, available at www.foodcontactmaterials.com/, and follow link to 'Food contact materials'.

VERMEIREN, L., DEVLIEGHERE, F., VAN BEEST, M., DE KRUIJF, N. and DEBEVERE, J., (1999), Developments in the active packaging of foods, *Trends in Food Science and Technology*, **10**, 77–86.

VERMEIREN, L., HEIRLINGS, L., DEVLIEGHERE, F. and DEBEVERE, J., (2003), Oxygen, ethylene and other scavengers, in (R. Ahvenainen, Ed.), *Novel Food Packaging Techniques*, Woodhead Publishing, Cambridge, pp. 22–49.

WALKER, S.J., (1992), Chilled foods microbiology, in (C. Dennis and M. Stringer, Eds.), *Chilled Foods – A Comprehensive Guide*, Ellis Horwood, London, pp. 165–195.

WEBER, C.J., (Ed.), (2000), Biobased packaging materials for the food industry, Food Biopack Project, EU Directorate 12: The Royal Veterinary and Agricultural University, Frederiksberg, Denmark.

WHITE, R.M., (1990), Package testing and food products, in (A. Turner, Ed.), *Food Technology International Europe*, Sterling Publications, London, pp. 295–299.

WHITE, R., (1994), Disposal of used packaging, in (J.M. Dalzell, Ed.), *Food Industry and the Environment – Practical Issues and Cost Implications*, Blackie Academic and Professional, London, pp. 318–346.

WHITTLE, K., (2000), Wax – the versatile 'green' solution, *Food Processing*, June, 21–22.

WILBRANDT, C.S., (1992), Utilising gases and packaging for quality chilled foods, in (A. Turner, Ed.), *Food Technology International Europe*, Sterling Publications, London, pp. 235–240.

WOODS, K., (1993), Susceptors in microwaveable food packaging, in (A. Turner, Ed.), *Food Technology International Europe*, Sterling Publications, London, pp. 222–224.

YAM, K.L. and LEE, D.S., (1995), Design of modified atmosphere packaging for fresh produce, in (M.L. Rooney, Ed.), *Active Food Packaging*, Blackie Academic and Professional, London, pp. 55–73.

YOICHIRO. Y., HIROKI, I. and TOYOFUMI, W., (2006), Development of laminated tin free steel (TFS), 'Universal Brite' Type F, for food can, *JFE Giho*, **12**, 1–5.

ZAGORY, D., (1995) Ethylene-removing packaging, in (M.L. Rooney, Ed.), *Active Food Packaging*, Blackie Academic and Professional, London, pp. 38–54.

ZOBEL, M.G.R., (1988), Packaging and questions of flavour retention, in (A. Turner, Ed.), *Food Technology International Europe*, Sterling Publications, London, pp. 339–342.

26

Filling and sealing of containers

Abstract: This chapter describes the techniques used to fill and seal rigid and flexible containers and then covers shrink-wrapping and stretch-wrapping. There are descriptions of tamper-evident packaging, labelling, checkweighing and metal detection.

Key words: filling and sealing containers, liquid fillers, form–fill–seal equipment, multi-head weighers, tamper-evident packaging, checkweighing, shrink-wrapping, stretch-wrapping, metal detection.

There have been significant developments in packaging systems during the past ten years, prompted by a number of considerations, for example:

- marketing requirements for different, more attractive packs;
- reductions in pack weight to reduce costs and meet environmental concerns over energy and material consumption (Chapter 25, section 25.6);
- new packaging requirements for minimally processed foods (Chapters 8 and 9) and modified atmosphere packaging (Chapter 25, section 25.3);
- the need for new types of tamper-resistant and tamper-evident packs.

Packaging methods that have been developed to meet these requirements are described in subsequent sections and by Brody and Marsh (1997).

Accurate filling of containers is important to ensure compliance with fill-weight legislation (Tiessen *et al.* 2008) and to prevent product 'give-away' by overfilling (also section 26.6). The composition of some foods (e.g. meat products such as pies and canned mixed vegetables) is also subject to legislation in some countries, and accurate filling of multiple ingredients is therefore necessary. The maintenance of food quality for the required shelf-life depends largely on adequate sealing of containers. Seals are the weakest part of a container and also suffer more frequent faults during production. In this chapter the techniques used to fill and seal rigid and flexible containers are described. By themselves these operations have no effect on the quality or shelf-life of foods, but incorrect sealing has a substantial effect on foods during subsequent storage.

26.1   Rigid and semi-rigid containers

'Commercially clean' metal and glass containers are supplied as palletised loads, which are wrapped in shrink or stretch film (section 26.3) to prevent contamination. They are depalletised and inverted over steam or water sprays to clean them, and they remain inverted until filling to prevent recontamination. Wide-mouthed plastic pots or tubs are supplied in stacks, fitted one inside another, contained in fibreboard cases or shrink film. They are cleaned by moist hot air unless they are to be filled with aseptically sterilised food (Chapter 13, section 13.2), when they are sterilised with hydrogen peroxide. Laminated paperboard cartons are supplied either as a continuous reel or as partly formed flat containers. Both are sterilised with hydrogen peroxide when used to package UHT products.

26.1.1   Filling

All fillers should accurately fill the container ($\pm 1\%$ of the target volume or weight) without spillage and without contamination of the sealing area. They should also have a 'no container, no fill' device and be easily changed to accommodate different container sizes. Except for very low production rates or for difficult products (e.g. bean sprouts), fillers operate automatically to achieve the required filling speeds.

Liquid fillers
No one type of filling machine is suitable for all types of liquids and the selection of equipment depends on the viscosity range, temperature, particulate size and foaming characteristics of the product, and the production rate required. Anon (2006) describes the following types of liquid fillers. Overflow (or 'fill-to-level') fillers are widely used for low- to medium-viscosity and foamy liquids (e.g. bottled water and some dairy products). They are not suitable for products that have a viscosity >25 000 centipoise or have particulates that are larger than ≈ 1.5 mm. The machine fills to a target fill-height in the container rather than volumetrically.

Servo pump fillers are suitable for low-, medium- and high-viscosity liquids, and liquids that have large particulates (e.g. oils, greases, salsa and sauces). Each filling nozzle has a dedicated servo-controlled positive displacement pump, which increases the capital cost of the equipment compared to other types of filler.

Peristaltic liquid fillers are used for high-value, small-volume filling (e.g. essential oils, food dyes) at high ($\pm 0.5\%$) accuracy. In operation, peristaltic pumps make intermittent contact with the outside of flexible product tubes and the product is contained inside the tubing. The tubing is easy to clean or disposable, and prevents product leakage or contamination. Servo drives control the peristaltic pumps and a control computer independently tracks the number of rotations of the pumps to determine precisely how much product has been filled. When the target fill volume is reached, the computer stops the pump (also Chapter 27, section 27.1.3).

Timed gravity fillers are the most economical type of volumetric filling machine, but the range of applications is limited to low-viscosity liquids that do not foam (e.g. bottled water and alcoholic spirits). In operation, the product is pumped into a holding tank above a set of pneumatically operated valves. Each valve is independently timed by a control computer to deliver precise amounts of liquid under gravity into the containers.

Piston fillers (Fig. 26.1) are the oldest and most reliable types of volumetric fillers used in the food industry. They are suitable for viscous products (e.g. pastes or foods containing particulates such as heavy sauces, salsas, salad dressings and creams). In

Fig. 26.1   Carousel piston filler for bottled beer (courtesy of Krones UK).

operation, a piston draws product from a hopper into a cylinder and a rotary valve then changes position so that when the piston returns, the product is filled into containers through a nozzle. The volume of product that is sucked into the cylinder is the volume that is dispensed into the container. This type of equipment is the most cost-effective, accurate and fast way to fill fairly viscous products. Although it is more expensive than overflow and timed gravity machines, it is cheaper than servo pump fillers. The disadvantages of this type of filler are that it is not suitable for low-viscosity products, which can leak between the piston and cylinder; change-overs for different products or container sizes are time consuming; and there is a limited range of fill volumes per piston.

Net weight liquid fillers are suited to liquids that are filled in bulk quantities or to smaller amounts of products that have a high value and are sold by weight. These are not common in food processing but may include enzyme solutions or speciality oils. In operation, the product is pumped to a holding tank above pneumatically operated valves. When the valve is opened, the net weight of product in the container is monitored in real time until the target weight is achieved and the valve is then shut. The advantage of this equipment is high accuracy; the disadvantages are the high cost per filling head and relatively slow rate of filling.

Filling heads can be arranged either in line or in a 'carousel' (or rotary) arrangement. A rotary filler (Fig. 26.2) is used to automatically fill and seal plastic pots or cups with products such as juice, coffee, milk, mineral water, yoghurt or jam at up to 1000 containers h^{-1}.

Rotary fillers for clean and ultra-clean filling of PET, HDPE and PP bottles fill 6000–22 000 bottles h^{-1}, depending on the bottle size, and a compact aseptic machine for filling non-carbonated beverages and liquid dairy products into PET and HPDE bottles has a capacity of 6000 bottles h^{-1} with a filling volume of 200–2000 ml (Anon 2008a). EHEDG (European Hygienic Engineering and Design Group) certification is required for sanitary and aseptic process fillers (Anon 2008b).

Anon (2008c) describes a flowmeter that can be fitted to filling machines to directly measure mass flowrate. Sensor information is collected by a programmable logic controller (PLC) (Chapter 27, section 27.2.2), which directly controls pumps and valves

Fig. 26.2   Rotary pot filler (courtesy of Packaging Automation Ltd, at www.pal.co.uk).

for precise control of filling operations, giving a filling accuracy of better than 0.1%. The same sensor can measure mass flow directly in liquids, pastes or creams that have varying viscosity, temperature, entrained air or suspended solids. Predictive control software uses measurements of each delivery to optimise the accuracy of subsequent fills, thereby reducing product giveaway. A control panel allows an operator to fill a wide range of container sizes, modify the amounts of each ingredient delivered, and select settings from a menu to fill different products on the same machine. This gives productivity benefits from faster product changeover times and eliminating the need for further check-weighing. The mass flow transmitter also produces quality assurance alarms for out of specification values and audit records.

Solids filling
Small particulate solids (e.g. rice, powdered foods) can be filled using similar equipment to liquid fillers or form–fill–seal equipment (section 26.2.1). Larger foods (e.g. confectionery products) can be filled into rigid containers, using photoelectric devices similar to food sorters (Chapter 2, section 2.3) to count individual pieces. Alternatively, a disc fitted with recesses to hold individual items rotates below a holding container and when each recess is filled, the required number is deposited in the pack. Filling solid foods into flexible containers is described in section 26.2. Smith (1999) describes developments in multi-head weighers that are able to weigh different products simultaneously, prior to filling into the same container. Examples include pre-packed mixed salads, mixed nuts and mixed selections of confectionery (Fig. 26.3).

Containers can be filled by weight using a net-weight or gross-weight system. In the former, the product is weighed before filling into the container and sealing, whereas the latter system fills the product and weighs the product plus package before sealing. Both systems have microprocessor control of the rate of filling and final fill-weight using a PLC. A bulk feeder is used to quickly fill a pack to $\approx$90% full, and it is then weighed. The controller calculates the exact weight of material remaining to be filled and activates a fine feed to top up the pack and re-weigh it. The microprocessor also monitors the number of packs and their weights, to produce a statistical record of fill weight variations. New multi-head weighers have rapid weight calculation to produce higher speeds (up to 200 weighings min^{-1} on a 14-head single-discharge machine, with 16-, 20- and 24-head models proportionately faster); and greater accuracy (up to 0.5%), and hence reduced

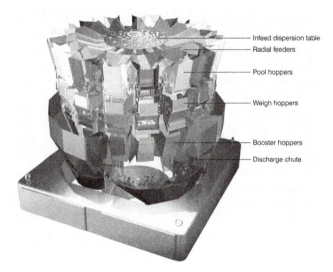

Infeed dispersion table
Radial feeders

Pool hoppers

Weigh hoppers

Booster hoppers
Discharge chute

Fig. 26.3   Multi-head weigher for solid foods (courtesy of Ishida Europe) (Anon 2005a).

product giveaway. Details of this equipment and an animation of the operation of a multihead filler is available at Anon (2005a). The microprocessor-controlled weighers can also be networked across different sites.

There are a wide variety of packing systems to fill fresh fruits and vegetables into distribution cartons or bins, described by Anon (2008d). These include volume filling or loose filling of cartons, which can be the most cost-effective packing system for fruits, such as tomatoes, kiwifruit, citrus fruits and some vegetables (e.g. onions). Fillers range from a simple chute to fully automated filling and carton handling systems (Fig. 26.4). Many packhouses fill distribution containers manually or semi-automatically, and rotary pack tables and belt tables that have padded vinyl surfaces to protect the fruit are the simplest and most economical packing systems. Rotary tables make hand-placement packing more efficient by constantly providing a range of fruits to the packer. Packing belt tables allow multiple packers to access one size/grade of fruit, and are an ergonomic solution to hand placement packing of large quantities of delicate fruits such as stonefruits. Tray-filling conveyors are commonly used for high-volume, gentle filling into trays of pre-sized apples, pears, avocados and kiwifruits. Air bagging heads are used to fill delicate fruit into retail bags. The fruits do not need to be pre-sized and this equipment requires less labour than other manual systems. Bin fillers are used to fill shipment bins evenly and gently with fruits such as apples.

Hermetically sealed glass or metal containers used for heat-sterilised foods are not filled completely. A headspace (or 'expansion space' or 'ullage') is needed above the food to form a partial vacuum. This reduces pressure changes inside the container during processing and oxidative deterioration of the product during storage (also Chapter 13, section 13.1.1). Glass containers and cans have a headspace of 6–10% of the container volume at normal sealing temperatures. When filling solid foods or pastes, it is necessary to prevent air from becoming trapped in the product, which would reduce the headspace vacuum. Viscous sauces or gravies are therefore filled before solid pieces of food. This is less important with dilute brines or syrups, as air is able to escape more easily before sealing.

Fig. 26.4   Filling fruit into bins (courtesy of Durand-Wayland, Inc. (www.durand.wayland.com/packing/fillers/d4.html)).

26.1.2   Sealing

Glass and plastic containers

Glass container terminology is described in Chapter 25 (section 25.2.3). The 'finish' (so-called because it was the last part of the container to be made when glass was hand manufactured) is the most important part of the container when considering sealing. It is the part of the bottle or jar that holds the closure (the lid or cap). It has lugs or threads to secure the cap and a smooth surface to form a seal with the cap. There are many hundreds of different types of closures for glass or plastic containers that can be viewed at suppliers' websites (e.g. Anon 2008e). Each closure has a specific finish with which it is designed to function, and they are not interchangeable. Bottle closures can be grouped into three categories: pressure seals, normal seals and vacuum seals (Fig. 26.5):

Pressure seals are used mostly for carbonated beverages. They include:

- screw-in–screw-out (internal screw), or screw-on–screw-off (external screw);
- crimp-on–lever-off, crimp-on–screw-off, or crimp-on–pull-off;
- roll-on (or spin-on, where the closure is pressed against the finish to form a thread), screw-off (also roll-on-pilfer-proof (ROPP closures)).

Examples include cork or injection-moulded polyethylene stoppers or screw caps, crown caps (pressed tinplate, lined with polyvinyl chloride) or aluminium roll-on screw caps.

Normal seals are used, for example, for non-carbonated beverages (e.g. pasteurised milk or wines). There are many different types of seals, including:

- one or two-piece pre-threaded, screw-on–screw-off;
- lug type screw-on–twist-off;

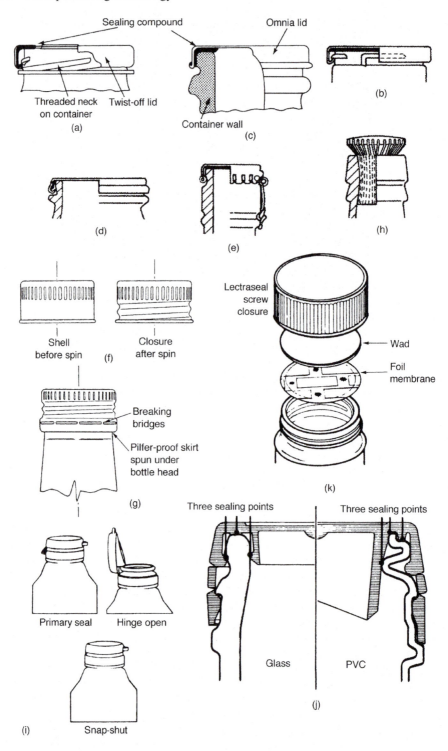

Fig. 26.5 Closures for glass or plastic containers: (a) twist-off; (b) lug cap; (c) Omnia; (d) pry-off; (e) lever cap; (f) screw cap; (g) roll-on-pilfer-proof (ROPP); (h) flanged cork; (i) hinge-open, snap-shut; (j) sealing points on hinge-open, snap-shut lids; (k) pre-threaded closure (after Paine 1991).

- roll-on (or spin-on)–screw-off;
- press-on–prise-off;
- crimp-on–prise-off, or crimp-on–screw-off;
- push-in–pull-out, or push-on–pull-off.

Examples include cork or synthetic cork stoppers fitted with tinned lead, polythene or aluminium capsules, metal or plastic caps, and aluminium foil lids. Plug fittings are made from injection-moulded LDPE and have the required softness and flexibility to form a good seal.

Vacuum seals are used for hermetically sealed containers, preserves or paste jars:

- screw-on–twist-off;
- press-on–prise-off, or press-on–twist-off;
- two-piece screw-on–screw-off, or roll-on–screw-off;
- crimp-on–prise-off.

Of the above, the most common examples of closures for bottles and jars include the following:

- Lug or twist cap (Fig. 26.5a), which consists of a steel cap that may have three to six lugs (inward protrusions from the side of the cap). They are secured by twisting the cap onto the finish, which seats the lugs with the raised threads of glass.
- Plastisol-lined continuous thread cap (Fig. 26.5f). This is a metal cap with a threaded side that is knurled. The cap is applied by screwing the closure onto the glass finish to press the glass into the plastisol gasket.
- Press-on–twist-off caps are most commonly used for baby foods. The steel cap has no lugs and is applied by pressing the cap onto the glass finish after flowing steam over the headspace (steam flow closing). The plastisol gasket covers the top and side of the cap and forms the primary top seal and a secondary side seal where glass threads form impressions in the gasket. Each of these three types of caps is used with hermetically sealed jars, where the partial vacuum helps to maintain the seal (Anon 2008f).

Wide-mouthed rigid and semi-rigid plastic pots and tubs are sealed by a range of different closures, including push-on, snap-on or clip-on lids (e.g. for margarine tubs), and push-on or crimp-on metal or plastic caps (e.g. for snackfoods). Push-on, crimp-on and snap-on lids locate onto a bead at the rim of the container. These closures are not tamper-evident, but the seal is sufficient to retain liquid foods. Membranes made from aluminium foil laminates or thermoplastic films (e.g. for yoghurt pots) are sealed to pots by a combination of pressure and high-frequency activation or heating the heat-seal coating. They are tamper-evident and, depending on the material selected, can be made to provide a barrier to moisture and gases. Where a product is to be used over a period of time, or where additional protection is required for the membrane, a clip-on lid may also be fitted to the pot.

Thermoformed pots or trays are filled and then lidded with a polymer film or foil laminate that is heat sealed to the top flanges. Small containers, such as those used for individual portions for UHT-sterilised milk, honey or jam, are formed, filled and sealed in a single machine at up to 50 000 containers per hour (Guise, 1985). The equipment can also be easily adapted to produce multi-packs of four to six pots.

Plastic jars and bottles are sealed using a variety of closures that can be tamper-evident, recloseable or contain an aperture or pouring spout to dispense the contents. For example, push-on caps or screw-threaded caps have a hinged top that reveals a dispensing opening in the cap when opened. They are used for squeezeable bottles (e.g. for creams, oils, sauces, mustard, mayonnaise or syrups). The caps have a positive 'snap-open, snap-shut' action and a profiled pin to clean the aperture on re-closing to prevent microbial growth and potential contamination of the product. The closure seals both a contact surface with the bottle, and when the lid is closed the protruding pin seals the orifice contact surface. A foil/polymer liner may also be used to seal the bottle and is removed before use. This type of closure is used for liquid products: the lid is opened and the container sides are squeezed to create the internal pressure for dispensing and force liquid out through the orifice. The opening may be fitted with a split membrane valve that helps to control the product flow during dispensing; prevents spillage if the pack is overturned, and prevents the product from dripping. Another type is a 'turret lock' closure which has a dispensing nozzle embedded in the top that is lifted vertically to dispense the product (Anon 2008g).

More rigid containers, including glass, can be used for dry products, because when the lid is opened the contents are poured or shaken from the pack. Other designs of dispensing container include 'disc top' closures (which have a plastic disc that is flipped up to reveal an opening), or 'snap top' closures (similar to a flip top, but it is a non-threaded closure that is pressed onto the finish and secured in place using protruding features on both the finish and on the closure).

In each type of closure the seal is formed by causing a resilient material to press against the rim of the container; the pressure must be evenly distributed and maintained to give a uniform seal around the whole of the cushioning material that is in contact with the rim. Typically, a resilient material is made from PVC, PE, EVA, or stamped out of composite cork or pulpboard sheet, protected with a facing material made from these polymers to prevent any interaction with contents (together these are termed the 'liner'). The tightness with which the cap is fitted to a container is known as the 'tightening torque', and with rolled-on, crimped-on and pressed-on caps, the effectiveness of the seal depends on the pressure exerted on top of cap during the sealing operation. To avoid the need for undue pressure, the width of the sealing edge is kept as narrow as possible. Glass bottles and jars have a narrow round sealing edge, whereas plastic bottles have flat sealing edges. Two other important considerations for caps are the 'thread engagement' (the number of turns of a cap from the first engagement between the cap and rim, and the point where the liner is engaged with the rim). This should be at least one full turn, to allow uniform engagement of liner and rim. The greater the thread engagement, the more effective is the cap-tightening torque in keeping it in place. The 'thread pitch' is the slope or steepness of the thread. The lower the number of turns, the steeper the slope of the thread and the more rapidly the cap will screw on or off (Paine 1991).

Further information on closures for glass and plastic containers is available from suppliers (e.g. Anon 2008h,i) or trade associations (e.g. Anon 2007).

Metal containers
Can lids are sealed by a double seam in a seaming machine (or 'seamer'). The 'first operation roller' rolls the cover hook around the body hook (Fig. 26.6a) and the 'second operation roller' then tightens the two hooks to produce the double seam (Fig. 26.6b). Details are given by Bemand (2001). A thermoplastic sealing compound melts during retorting and fills the spaces in the seam to provide an additional barrier to contaminants. The can seam is the weakest point of the can and the seam dimensions are routinely examined at the manufacturing stage (e.g. using X-ray scanners (Anon 2008h)) and samples are checked by quality assurance staff after filling (see Anon 1997a) to ensure that they comply with specifications (Table 26.1).

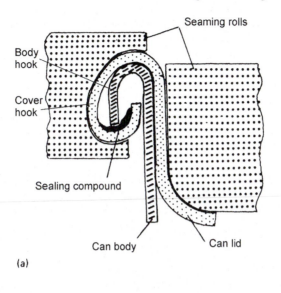

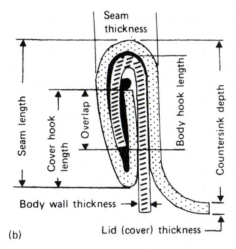

Fig. 26.6 Can seam formation: (a) first operation; (b) double seam (adapted from Anon 2008i, Hersom and Hulland 1980).

Table 26.1 Seam specifications for selected cylindrical cans

Type of can	Dimensions (mm)		Dimensions of seam (mm)		
	Diameter	Height	Length	Thickness	Hooks[a]
A1	65.3	101.6	2.97–3.17	1.40–1.45	1.90–2.16
A2	87.3	115.3	2.97–3.17	1.47–1.52	1.90–2.16
A21/2	103.2	115.3	2.97–3.17	1.52–1.57	1.90–2.16
A10	157.2	177.8	3.10–3.30	1.65–1.70	2.03–2.29
Actual overlap > 1.143 mm, % body hook butting > 70%					

[a] Range of lengths for cover and body hooks.
Adapted from Lock (1969)

The equations used to assess whether these measurements are within specification are:

$$\text{Free space} = \text{seam thickness} - [2(t_b) + 3(t_e)] \qquad \boxed{26.1}$$

$$\text{Percentage body hook butting} = \frac{x - 1.1t_b}{L - 1.1(2t_e + t_b)} \times 100 \qquad \boxed{26.2}$$

$$\text{Actual overlap} = y + x + 1.1t_e - L \qquad \boxed{26.3}$$

where x (mm) = the body hook length, y (mm) = the cover hook length, t_e (mm) = the thickness of the can end, t_b (mm) = the thickness of the can body and L (mm) = the seam length.

Different types of easy-open end are fitted to cans, depending on the product: for example, ring pull closures are used for two-piece aluminium beverage cans, and different designs retain the ring pull within the can after opening to reduce litter problems. Full-aperture ring pull closures are used for meat products, snackfoods and nuts. Both types are produced by scoring the metal lid and coating it with an internal lacquer. In another design, the can body is scored around one end and a metal key is used to unwind a strip of metal to enable the entire end to be removed (e.g. flat cans for fish and meats). Aerosol cans have a pre-sterilised end seamed on, the product is filled through the valve opening and a pre-sterilised aerosol valve is fitted (crimped) onto the can. The propellant is injected under pressure through the valve and pressure checked. Finally, the actuator and tamper-evident dust cap are fitted (Westley 2007).

Aluminium collapsible tubes are sealed by folding and crimping the open end of the tube after filling (polyethylene or laminated plastic tubes are sealed using a heat sealer (section 26.2.1)).

Paperboard cases and cartons
Plain or corrugated cases (Chapter 25, section 25.2.6) are produced as a flat 'blank', which is then cut, creased and folded to form the case or carton. It is important to fit as many blanks as possible to a sheet of paperboard to minimise wastage (Fig. 26.7a). Board is printed, stacked into multiple layers for blanks to be cut out using a guillotine, either on-site or by the case supplier. Each blank is then precisely creased and formed into a carton (Fig. 26.7b), which is either glued or stapled. Different folding carton types are classified by the European Carton Makers Association using an ECMA Code, which also defines how the dimensions should be stated. Details are given by Ramsland (1989). Multiple packs of cans or bottles are held together by paperboard, formed in a similar way to cartons (Fig. 26.7c). These have inter-locking lugs that dispense with the need for staples or glue.

Rigid laminated paperboard cartons for aseptically processed foods have thermoplastic film as the inner layer (Fig. 25.7 in Chapter 25, section 25.2.6). There are two systems for carton production: in one system a continuous roll of laminated material is aseptically formed–filled–sealed (section 26.1.2); in the second system, pre-formed cartons are erected, filled and sealed in an aseptic filler. In the second system the paperboard can be heavier than in form–fill–seal systems, because it does not require the flexibility needed for the forming machine, and as a result the carton is more rigid. UHT cartons (Chapter 25, section 25.2.6, and Chapter 13, section 13.2.3) originally required scissors to open them, and then incorporated a peelable foil strip to reveal a dispensing aperture. Newer

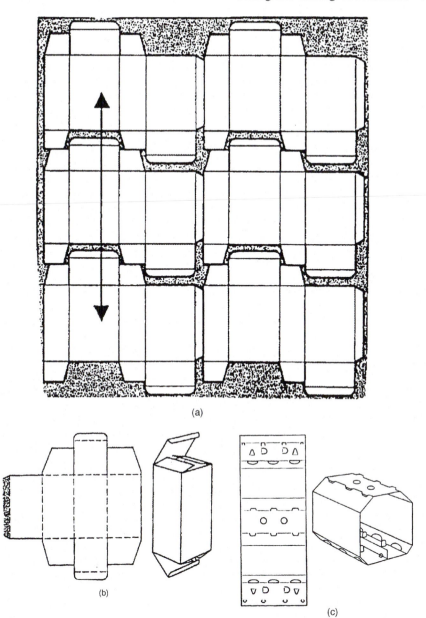

Fig. 26.7 (a) Positioning blanks on a sheet of paperboard to minimise wastage; (b) board is creased and formed into a carton (from Fellows and Axtell 2002); (c) inter-locking lugs which dispense with the need for staples or glue in a bottle carrier carton (from Paine 1991).

developments include a range of screw caps that have different cap sizes, pouring performances and opening styles. Cartons that use this type of screw cap have a hole cut ready to receive the cap and it is applied in the filling machine by an applicator after the carton is filled, and sealed in position by either hot air or hot melt glue. The barrier properties of the carton are fully maintained until the consumer opens the closure by breaking the tamper-evident seal. This type of container is used for fresh and fermented dairy products, ambient juices and wines.

26.2   Flexible containers

Most flexible films are heat sealed but cold seals (adhesive seals) are also used to package heat-sensitive products (e.g. chocolate, chocolate-coated biscuits or ice cream). Thermoplastic materials or coatings become fluid when heated and resolidify on cooling. To seal flexible films, a heat sealer heats the surfaces of two films (or 'webs') until the interface disappears and then applies pressure to fuse the films. The strength of the seal is determined by the temperature, pressure and time of sealing. The seal is weak until cool and should not therefore be stressed during cooling. Three common types of seal are (1) bead seals (Fig. 26.8a), (2) fin seals (Fig. 26.8b) and (3) lap seals (Fig. 26.8c). The bead seal is a narrow weld at the end of the pack. In a lap seal, opposite surfaces are sealed, and both should therefore be thermoplastic, whereas a fin seal has the same surface of a sheet sealed and only one side of the film or one component of a laminate needs to be thermoplastic. Fin seals protrude from the pack and no pressure is exerted on the food during sealing. They are therefore suitable for fragile foods (e.g. biscuits), foods that would be deformed by pressure (e.g. soft bakery goods or sugar confectionery) or heat-sensitive foods.

Hot-wire sealers have a metal wire that is heated to red heat to simultaneously form a bead seal and cut the film, whereas a hot-bar sealer (or jaw sealer) holds the two films in place between heated jaws until the seal is formed. In the impulse sealer, films are clamped between two cold jaws. The jaws are then heated to fuse the films and they remain in place until the seal cools and sets. This prevents shrinkage or wrinkling of the film. The temperature and time of heating using heat sealers should be adjusted to take account of the thickness and melting temperature of each specific type of film because the sealer conducts heat through the film and therefore risks causing heat damage to it or an inadequate seal if incorrect settings are used.

Rotary (or band) sealers are used for higher filling speeds. The centres of metal belts are heated by stationary shoes and the edges of the belts support the unsoftened film. The mouth of a package passes between the belts, and the two films are welded together. The seal then passes through cooling belts that clamp it until the seal sets. Other types of sealer include (1) the high-frequency sealer, in which an alternating electric field (at 1–50 MHz) induces molecular vibration in the film and thus heats and seals it – the film should have a high loss factor (Chapter 20, section 20.1.1) to ensure that the temperature is raised sufficiently by a relatively low voltage; and (2) the ultrasonic sealer which produces high-frequency vibrations (20 kHz), that are transmitted through the film and dissipate as localised heat at the clamped surfaces.

26.2.1   Form–fill–sealing

Form–fill–sealing equipment has been one of the most important developments in food packaging during the past 25 years. The advantages of form–fill–sealing include reduced

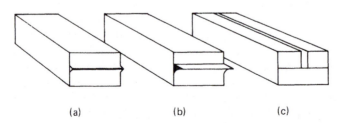

(a)   (b)   (c)

Fig. 26.8   Seals for flexible packaging: (a) bead seal; (b) fin seal; (c) lap seal.

transport, handling and storage costs for materials compared with pre-formed containers, simpler and cheaper package production, lower labour costs and higher outputs (Robinson 1992). The different forms of form–fill–sealing are two types of vertical form–fill–seal (VFFS), known as 'transwrap' or 'flow pack', and horizontal form–fill–seal (HFFS) known as 'pillow pack' or 'flow wrap'.

In vertical transwrap equipment (Fig. 26.9a), a film is pulled intermittently over a forming shoulder by the vertical movement of the sealing jaws. A fin seal is formed at the side. The bottom is sealed by jaw sealers and the product is filled. The second seal then closes the top of the package and also forms the next bottom seal. This type of equipment is suitable for powders and granular products. Filling speeds are $30–120\,\text{packs}\,\text{min}^{-1}$. Films should have good slip characteristics and resistance to creasing or cracking, in order to pass over the forming shoulder, and a high melt strength to support the product on the bottom seal while the seal is still hot.

The vertical flow pack equipment (Fig. 26.9b) differs from the transwrap design in two ways: first a forming shoulder is not used and the film is therefore less stressed; secondly, the action is continuous and not intermittent. The thermoforming machine preheats two plastic films and forms side seams using heaters and crimp rollers; the strips are sealed at the base, filled with product and the top is sealed. The strips are then cut into individual packages or sales units of 5–15 portions. In a development of form–fill–seal, a vertical form–fill–seal machine is combined with a blow-moulding system to produce uniquely shaped, multi-dimensional, single-serve portion packages using PET, HDPE and PP in sizes from 1–150 ml. The process involves blow moulding, filling, capping and labelling. The packs are used for dairy, juice and non-carbonated beverages.

In the HFFS system, products are pushed into the tube of film as it is being formed (Fig. 26.10). The transverse seals are made by rotary sealers, which also separate the

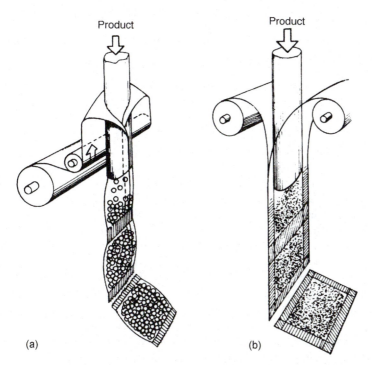

(a)   (b)

Fig. 26.9   Vertical form–fill–seal: (a) transwrap; (b) flow pack machine operations.

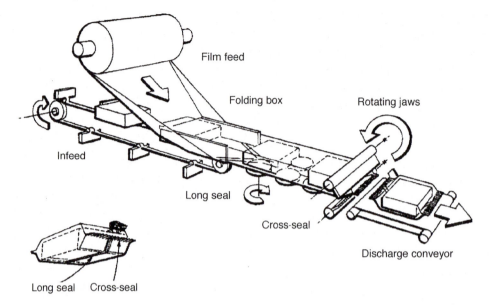

Fig. 26.10   Horizontal form–fill–seal (flow wrap) machine (from Naylor 1992, Wolmsley and Bakker 1986).

packs. Filling speeds are >600 packs min^{-1} and films should therefore be thin and have a high melt strength, to produce a strong seal in a short heating time.

HFFS equipment has gained in popularity owing to its greater speed and flexibility: it can pack single pieces of food or multiple wrapped or unwrapped pieces; packs can accommodate irregular shaped foods that were previously difficult to pack (Martin 1991); it can be used for MAP products (Naylor 1992), (Table 26.2) (also Chapter 25, section 25.3), and the fin seals ensure a good gas and moisture barrier without risk of damage to the product from sealing head pressure.

A modification of this equipment is used to fill laminated cartons aseptically. A web of material is sterilised in a bath of hydrogen peroxide and formed into a vertical tube. An internal heater vaporises any remaining hydrogen peroxide. The tube is then filled, sealed through the product, shaped into a carton and top sealed. The 'ears' on the base of the carton are folded flat and sealed into place. Developments in sterilisation of aseptic packaging using lasers to produce UV light are described by Warriner *et al.* (2004).

In sachet pack machines, either horizontal or vertical packs are formed from single or double sheets of film. Horizontal single-web machines fold the film over a triangular shoulder (or 'plough') and then form two side seams. The sachets are then separated, opened by a jet of compressed air, filled and heat sealed across the top. Horizontal machines have a smaller distance for the product to fall into the package and are therefore more suitable for sticky foods. Vertical machines are similar to the transwrap and flow pack machines, and have lower cost and take up less floor space. A variation on the vertical form–fill–seal machine is the 'chub' pack. Here the film is formed into a tube and filled with product, but instead of heat sealing the individual packs, they are sealed using a metal or plastic clip at each end. This type of packaging is used for liquids and high-viscosity pastes, including cream cheese, sausage meat and other processed ground meats, marzipan and butter. Further details are given by Anon (2008j).

On two-reel machines, one film forms the front and the second forms the back of the pack. Two blanks are cut from a roll of film, brought together and sealed on three sides.

Table 26.2 Packaging systems for MAP products

Type of system	Description	Applications
Thermoforming	Heat forming of semi-rigid and flexible containers, vacuum with gas flushing	Meat, poultry, fish, cooked meats, bakery products, cheese, nuts
Horizontal and vertical form-fill-seal, pillow pack	Single flexible web, gas flushing by lance or tube, venting to atmosphere	Bakery products, snackfoods, cheese, coffee, nuts, meat, fish, salads, fruit, vegetables
Pre-formed tray or bag	Using HDPE, PET or HIPS trays or pre-formed plastic bags, vacuum with gas flushing	Meat, fish, nuts, prepared meals
Thermoforming, composite or pre-formed board/plastic tray	Carton blank and plastic tray form composite tray structure with in-line lidding and vacuum with gas flushing	Meat, poultry, fish, cooked meats, bakery products, cheese, salads, vegetables, prepared meals
Bag-in-box	Barrier or non-barrier bag into corrugated or solid board case, vacuum with gas flushing	Bulk meat, poultry, fruit, cheese, nuts, dried powders
Vacuum skin packs	Multi-layer film top web shrunk over product contained in 'easy-peel' forming bottom web. Gas in headspace before lidding	Meat, fish
Shrink vacuum system	Two webs of film, upper heated and shrunk over product on base web	Meat, fish
Vacuum skin system	Laminate top web shrunk over product on laminated board/film base	Meat, fish
Bag-in-carton	Lined carton, gas flushed	Powders, granular products

Adapted from Hastings (1998)

The package is filled and the final seal is made. Sachet machines are widely used for powders or granules (e.g. coffee, salt and sweeteners), or liquids (e.g. cream, sauces, ketchup and salad cream). Filling speeds are 70–1000 packs min^{-1}, depending on the size of the sachet and the type of product. Sachets are automatically cartoned, and the cartoning machine is therefore an integral part of sachet production (Guise, 1987a,b).

Various devices are used to assist consumers to open flexible packs, particularly where a strong film is used. These include tear-tape applied longitudinally or transversely to the pack, or slits or perforations that are either produced mechanically by the wrapping machine or cut by a laser after wrapping.

26.3   Shrink-wrapping and stretch-wrapping

LDPE is biaxially oriented to produce a range of films that shrink in two directions (Chapter 25, section 25.2.4). The shrink ratios are measured in both the machine direction (MD) and the transverse direction (TD). Films are 'preferentially balanced' (shrink ratios are MD = 50%, TD = 20%), 'fully balanced' (MD = 50%, TD = 50%) or 'low balanced' (MD = 10%, TD = 10%). A small amount of shrinkage is usually required to tighten a loosely wrapped package, but contoured packages require a higher shrink ratio. The size of film required to shrink-wrap a sleeve-wrapped package is calculated using:

$$\text{Width} = A + \tfrac{3}{4}C \hspace{6cm} \boxed{26.4}$$

$$\text{Length} = 2(B + C) + 10\% \text{ shrink allowance} \hspace{2cm} \boxed{26.5}$$

where A (m) = width of the package, B (m) = length of the package and C (m) = height of the package.

The total mass of film used equals the width multiplied by the length, divided by the yield (a measure of film density) (m^2 kg^{-1}). The size of the film required to shrink-wrap an overwrap using centre-fold film is found using:

$$\text{Width} = (B + C) + 10\% \text{ shrink allowance} \hspace{2cm} \boxed{26.6}$$

$$\text{Length} = (A + C) + 10\% \text{ shrink allowance} \hspace{2cm} \boxed{26.7}$$

The total mass of film used equals twice the width (m) multiplied by the length (m) divided by the yield (m^2 kg^{-1}).

The film is shrunk by passing it through a hot-air tunnel or beneath radiant heaters. Alternatively a heat storage gun fires an intermittent pulse of hot air to shrink the film when a package passes beneath. This reduces energy consumption by 70% compared with hot-air tunnels. Shrink-wrapped trays have now largely replaced fibreboard shipping cases for many products (Chapter 25, section 25.2.6).

In stretch-wrapping, LDPE, PVC or LLDPE (Chapter 25, section 25.2.4) is wrapped under tension around collated packages. The main advantages over shrink-wrapping include lower energy use than in shrink tunnels (1.5–6 kW compared with 20–30 kW), and lower film use. In shrink-wrapping, 5–10% extra film is used to allow for shrinkage, whereas stretch film is elongated by 2–5%. Together this gives a 10–15% saving in film.

Other types of flexible wrapper include aluminium foil for unusual shaped foods (e.g chocolate Easter eggs) and twist-wrapped cellulose film for confectionery (Chapter 25, section 25.2.4).

26.4   Tamper-evident packaging

Closures that are designed to enable consumers to use the contents a little at a time require tamper-evident or tamper-proof features. The habit of some consumers of 'grazing' (opening packs, tasting the food and returning it to the shelves), and a number of cases of deliberate poisoning of packaged foods in extortion attempts, caused food manufacturers to modify package designs. Although total protection is not possible, tamper-resistant packaging delays entry into the package and tamper-evident packs indicate whether tampering has been attempted or occurred. The main requirement for tamper evidence is with bottles, pots and jars that need to be re-closable, and examples of tamper-evident or tamper-resistant closures are shown in Table 26.3.

26.5   Labelling

The label is the primary point of contact between a processor and a customer and is therefore an integral part of the marketing strategy for a product. The label is the main method of persuading a purchaser to buy a product without having sampled it, and choosing it rather than a competing brand on a retail shelf. Details of the factors to be taken into account in the design of labels, legislative requirements and the information

Table 26.3 Tamper-evident/resistant packaging

Type of packaging	Tamper-evident or tamper-resistant features
Bottles and jars (glass or plastic)	Foil or membrane seals for wide-mouthed plastic pots and bottles.
	Heat-shrinkable PVC sleeves for bottle necks, or bands or wrappers placed over place over lids. A perforated strip enhances tamper-evidence (must be cut or torn to gain access).
	Breakable caps – rings or bridges to join cap to a lower section on bottles (the container cannot be opened without breaking the bridge or removing the ring and they cannot be replaced).
	ROPP caps for bottles or jars (during rolling, a tamper-evident ring in the cap locks onto a special bead in the neck to produce a seal which breaks on opening and drops slightly).
	A safety button in press-on–twist-off closures for heat-sterilised jars (a concave section, formed in the lid by the head space vacuum, becomes convex when opened).
	A breakable plastic strip that gives a visual signal that a jar has not been previously opened.
	Child-resistant closures (clic-lok, squeeze-lok, ringuard, pop-lok, etc.). These are not normally used for food containers.
Flexible films	Film wrappers that must be cut or torn to gain access.
	Blister or bubble packs that give visible evidence of backing material separating from blister. Each compartment must be broken, cut or torn to gain access.
	Laminated plastic/foil pouches must be cut to gain access.
Tubes (aluminium or plastic)	Foil membrane over tube mouth that has to be punctured to gain access.
Cans	Three-piece, two-piece and aerosol cans are inherently tamper-resistant.
	Composite containers in which the ends are joined to the walls and cannot be pulled apart.

Adapted from Paine (1991) and Anon (2007)

required on a label are beyond the scope of this book and are described by Blanchfield (2000). The following information is the minimum required in most countries:

- Name of the product.
- List of ingredients (in descending order of weight).
- Name and address of the producer.
- Net weight or volume of product in the package.
- A 'use-by', 'best-before' or 'sell-by' date.
- Storage information or instructions on storage after opening.
- Any special instructions for preparing the product.
- A barcode (Chapter 25, section 25.5.1).

Labels are made from paper, plastic film, foil or laminated materials, pre-printed by either lithographic or rotogravure techniques (Chapter 25, section 25.5). A wide variety of label combinations is possible and some of the main types of label used for foods are as follows:

- *Glued-on labels* – the adhesive is applied at the time of labelling or the label is pre-glued and wetted for application. Cans and glass bottles are usually labelled using a hot-melt adhesive at >500 containers min^{-1}.
- *Thermosensitive labels* – heat is applied at time of application (e.g. biscuit and bread wrappers). These are more expensive but they can also be used as a closure.
- *Pressure-sensitive labels* – self-adhesive labels that are pre-coated with adhesive, mounted on a roll of release paper and removed before application.
- *Insert labels* – inserted into transparent packs.
- *Heat transfer labels* – the design is printed onto paper or polyester substrate and transferred to the package by application of heat.
- *In-mould labels* – involves thermoforming the container and labelling at the same time. A printed paper label, that has a heat-activated coating on the reverse side, is placed in the thermoforming mould before the parison is inserted (Chapter 25, section 25.2.5). When air is injected to blow the package shape, the heat activates the coating. A combination of heat, air pressure and the cold surface of the mould secures the label to the pack and sets the adhesive. The label also contributes to the strength of the pack and reduces polymer use by 10–15%.
- *Shrink sleeve decoration* – used for glass and plastic containers. An axially oriented PVC or PP sleeve is made larger than the container and heat-shrunk to fit it. Alternatively, an LDPE sleeve is made smaller than the container and stretched to fit it. In both methods, the sleeve is held in place by the elasticity of the film, and no adhesives are used. When shrunk over the necks of containers sleeves also provide a tamper-evident closure.
- *Stretchable inks* – applied before or during bottle manufacture for labelling plastic bottles.

Labels and containers are coded with a variety of information, including a production batch number, codes for international traceability and quality assurance, and best-before or use-by dates, using either ink-jet printers or laser coding equipment. Both types of equipment are non-contact and can be used to mark most types of packaging materials as well as uneven or delicate surfaces (e.g. flexible films and eggshells). In ink-jet coding the print head sprays a dot-matrix pattern of ink onto the package as it passes beneath. The ink is broken into individual droplets by ultrasonic pressure waves from a vibrating rod as they leave the print nozzle. Each droplet is given an electrostatic charge and the flight of the droplet (and hence its position on the package) is controlled by the size of the charge as it passes through an applied electrostatic field. Laser coders operate by firing a dot-matrix laser to build up each character in the code. The laser either removes coloured ink from a label to reveal a white surface beneath, or etches the surface of a container to leave a mark (e.g. on glass and PET bottles) (Wallin and Chamberlain 2000). Advantages of laser coding over ink-jet printing are the lower operating costs because no ink is required and no moving parts. Coding equipment is also used to mark shipment containers and thus removes the need for different labels and cases for each product, so reducing the inventory costs and simplifying warehouse management (Anon 2003). Other labelling equipment includes microprocessor-controlled print-and-apply labellers, print barcodes or RFID tags onto cartons, cases or palletised loads (Chapter 25, section 25.5.1, and Chapter 27, section 27.3) and weigh-price-labellers that accurately weigh products, and print and apply labels, even to irregular shaped products (e.g. Anon 2005b).

26.6   Checkweighing

Checkweighers are incorporated in all production lines to ensure compliance with fill-weight legislation (average weight or minimum weight legislation) and to minimise product give-away. They are pre-set to the required weight for individual packs and any that are below this weight are automatically removed from the production line. They are microprocessor controlled and are able to weigh up to 450 packs min^{-1} and automatically calculate the standard deviation of pack weights and the total amount of product that is given away. This data is collated by control computers and prepared into reports for use in process management and control procedures (Chapter 27, section 27.2.2). Checkweighers can also be linked by feedback controls to filling machines, which automatically adjust the fill-weight to increase filling accuracy. Anon (2008k) describes a checkweigher that is directly connected to slicing machines in German sausage manufacturing, where any deviation from the net pack weight is used to automatically adjust the slice thickness and eliminate over- and under-filling. Details of the design, operation and statistical control of checkweighers are given by Anon (1997b).

26.7   Metal detection

Details of contaminants in foods and methods of removing them before processing are described in Chapter 2 (section 2.2). Contamination with metal fragments can also occur during processing as a result of wear or damage to equipment, and metal detection is therefore an important component of HACCP systems in all food processing plants as well as a requirement to prove due diligence (Chapter 1, section 1.5.1). The basic components of a metal detection system are:

- a detection head that is correctly matched to the product and set to its optimum sensitivity;
- a handling system that conveys the product under the detection head;
- a reject system that is capable of rejecting all contaminated products into a locked container;
- an automatic fail-safe system if any faults arise in the detection equipment (Greaves 1997).

X-ray detectors are described in Chapter 2 (section 2.3.3). There are two other types of metal detectors: the more common are based on the 'balanced coil system' (also known as 'three-coil' detectors). These detectors are made from a coil of wire that conducts a high voltage to produce a high-frequency magnetic field, with two receiver coils placed either side. The voltages induced in the receiver coils are adjusted to exactly cancel each other out when the magnetic field is not disturbed. When an electrical conductor (e.g. a ferrous or non-ferrous metal contaminant or metal-impregnated grease) passes through the detector, it changes the amplitude and/or the phase of the electrical signal induced in the coils (Graves *et al.* 1998). This change is detected by the electronic circuitry, which activates an alarm and a mechanism to reject the pack (Mayo 1984). The detector is adjusted for each particular product to take account of differences in electrical conductivity of different foods. The second type is a ferrous-only type that is used for products in aluminium containers or foil. Here coils of wire are wrapped around a former that contains a number of magnets. The passage of ferrous metal through the magnetic field causes a voltage to be generated in the coil windings.

Various reject systems are available, including air-blast, conveyor stop, pusher arms for items up to 50 kg, or a retracting section of conveyor that allows the product to fall into a collection bin underneath. Details of the operation of metal detectors are given by Anon (1991) and Patel (2002). Microprocessor control enables the characteristics of up to 100 products to be stored in the detector memory, automatic set-up to compensate for product effects, automatic fault identification and production of printed records to show the number of detections and when they were found. Further information is given by Bowser (2004).

References

ANON, (1991), *Lock International Metal Detection Handbook*, Lock International, Tampa, FL.

ANON, (1997a), Low-cost semi-automatic double seam inspection, company information from SEAMetal, available at http://prev.qbyv.com/seametal9000m.htm.

ANON, (1997b), *Principles of Checkweighing – A Guide to the Application and Selection of Checkweighers*, 3rd edn, Hi-Speed Checkweigher Company Inc., available at www.foodandbeveragepackaging.com/ and search for 'Principles of Checkweighing'.

ANON, (2003), Cutting costs with lasers, *Food Processing*, January, 21.

ANON, (2005a), Multihead weighers, company information from Ishida, available at www.ishidaeurope.com/our_products/weighing_solutions/multi_head_weighers/r_series/.

ANON, (2005b), Weigh-price-labellers, company information from Ishida, available at www.ishidaeurope.com/our_products/weighing_solutions/weigh_price_labellers/.

ANON, (2006), Liquid filler for small and medium sized operations, company information from InLine Filling Systems Inc., available at www.liquidfiller.com/.

ANON, (2007), Technical Bulletin 2: Tamper Evidence, Technical Bulletin 3: Vacuum Closures – Metal, Plastic and Composite for Various Containers, Technical Bulletin 5: Dispensing Closures, Technical Bulletin 10: Types of Closures in Relation to Glass Containers, available from the Closure Manufacturers Association at http://closuremanufacturers.org/.

ANON, (2008a), Carton and plastic based packaging systems, aseptic liquid food packaging, hygienic filling machines, plastic bottle systems, bottle filling machines, dairy and juice packaging solutions and other carton systems, company information from Elopak A/S, available at www.idspackaging.com/packaging/us/elopak/packaging_solutions_systems/398_0/g_supplier.html.

ANON, (2008b), European Hygienic Engineering and Design Group, available at www.ehedg.org/.

ANON, (2008c), The new Micro Motion® Model 1500 filling and dosing transmitter, company information from Emerson Process Management, available at www.frflow.com/home/news/pr/512_micromotion1500.html.

ANON, (2008d), Packing systems, company information from Compac Sorting Equipment Ltd., available at www.compacsort.com/wa.asp?idWebPage=14568&idDetails=121.

ANON, (2008e), Company information from O.BERK Company, available at www.oberk.com and follow link to 'closures'.

ANON, (2008f), Caps for plastic bottles and glass bottles, company information from e-Bottles, available at www.ebottles.com/showbottlefamilys.asp?type=2&mat=closures.

ANON, (2008g), Dispensing closures, company information from Bottle Solutions, available at www.bottlesolutions.com/product/category-8c8bdd72-13dd-4069-9c21-f9e5e370e3de.aspx.

ANON, (2008h), X-ray technology for can seam inspection, company information from InnospeXion ApS, available at www.innospexion.dk/web/systems/pdf/web_canseamscanner_01.pdf.

ANON, (2008i), Double seam formation, terminology & glossary, company information from Dixie Canner Co., available at http://www.dixiecanner.com/1120.htm.

ANON, (2008j), KartridgPak, company information from Oystar Packaging Technologies, available at www.oystar.packt.com/1543.html and follow links to 'Product lines' > KartridgPak > Products > Chubmaker series.

ANON, (2008k), Checkweigher takes over control tasks, *Food Processing*, **77** (5), 36–37.

BEMAND, D., (2001), Can seaming, *International Bottler and Packer*, July, 31–37, available at http://www.angelusmachine.com/PDF/CanSeaming-IntBPJul01.pdf.

BLANCHFIELD, J.R., (Ed.), (2000), *Food Labelling*, Woodhead Publishing, Cambridge.

BOWSER, T., (2004), Metal detectors for food processing, Food Technology Factsheet, Oklahoma State University, available at http://osuextra.okstate.edu/pdfs/FAPC-105web.pdf.

BRODY, A.L. and MARSH, K.S., (Eds.), (1997), *The Wiley Encyclopedia of Packaging Technology*, 2nd edn, Wiley-Interscience, New York.

FELLOWS, P.J. and AXTELL, B.L.A., (2002), *Appropriate Food Packaging*, ITDG Publishing, London, pp. 53–56.

GRAVES, M., SMITH, A. and BATCHELOR, B., (1998), Approaches to foreign body detection in foods, *Trends in Food Science and Technology*, **9**, 21–27.

GREAVES, A., (1997), Metal detection – the essential defence, *Food Processing*, May, 25–26.

GUISE, B., (1985), Shrinking to style, *Food Processing*, **54**, 23–27.

GUISE, B., (1987a), Filling an industry need, *Food Processing*, **56**, 31–33.

GUISE, B., (1987b), Spotlight on sachet packaging, *Food Processing*, **56**, 35–37.

HASTINGS, M.J., (1998), MAP machinery, in (B.A. Blakistone, Ed.), *Principles and Applications of Modified Atmosphere Packaging of Foods*, 2nd edn, Blackie Academic and Professional, London, pp. 39–64.

HERSOM, A.C. and HULLAND, E.D., (1980), *Canned Foods*, 7th edn, Churchill Livingstone, London, pp. 67–102, 342–356.

LOCK, A., (1969), *Practical Canning*, 3rd edn, Food Trade Press, London, pp. 26–40.

MARTIN, A.V., (1991), Advances in flexible packaging of confectionery, *European Food and Drink Review*, Packaging Supplement, Winter, pp. 17–20.

MAYO, G., (1984), Principles of metal detection, *Food Manufacture*, Aug., 27, 29, 31.

NAYLOR, P., (1992), Horizontal form-fill and seal packaging, in (A. Turner, Ed), *Food Technology International Europe*, Sterling Publications, London, pp. 253–255.

PAINE, F.A., (1991), *The Packaging User's Handbook*, Blackie Academic & Professional, London.

PATEL, H., (2002), Metal detectors uncovered, *Food Science and Technology*, **15** (4), 38–41.

RAMSLAND, T., (1989), *Handbook on Procurement of Packaging*, PRODEC, Helsinki.

ROBINSON, C.J., (1992), Form, fill and seal technology, in (A. Turner, Ed.), *Food Technology International, Europe*, Sterling Publications, London, pp. 250–251.

SMITH, S., (1999), Multi-head marvels, *Food Processing*, January, 16–17.

TIESSEN, J., RABINOVICH, L., TSANG, F. and VAN STOLK, C., (2008), Assessing the impact of revisions to the EU horizontal food labelling legislation, Prepared for the European Commission by the RAND Corporation, P.O. Box 2138, Santa Monica, CA, available at www.rand.org/pubs/technical_reports/2008/RAND_TR532.pdf.

WALLIN, R. and CHAMBERLAIN, S., (2000), Code on coding, *Food Processing*, October, 31–32.

WARRINER, K., MOVAHEDI, S. and WAITES, W.M., (2004), Laser-based packaging sterilisation in aseptic processing, in (P. Richardson, Ed.), *Improving the Thermal Processing of Foods*, Woodhead Publishing, Cambridge, pp. 277–303.

WESTLEY, C., (2007), How aerosols are filled, available at www.yorks.karoo.net/aerosol/link3.htm.

WOLMSLEY, T.E. and BAKKER, M., (1986)), *Encyclopaedia of Packaging Technology*, John Wiley and Sons, New York.

27

Materials handling and process control

Abstract: This chapter describes improvements in handling technologies for solid materials and liquid foods, ingredients and packaging materials, from suppliers through production and distribution, to the consumer. It includes hygienic design and methods of cleaning equipment, waste treatment, automatic process control and human–machine interfaces, neural networks, fuzzy logic and robotics. It concludes with developments in warehousing operations and distribution logistics.

Key words: materials handling, solids handling, pumps and valves, process control, sensors, RFID tags, cleaning-in-place (CIP), water and waste management, automatic control systems, programmable logic controller (PLC), robotics, 'Supervisory Control and Data Acquisition' (SCADA) software, Enterprise Resource Planning software, distribution logistics.

Correct handling of foods, ingredients and packaging materials, from suppliers, through the production process and during distribution to the consumer is essential to optimise product quality and minimise costs. Improvements in materials handling technologies have led to substantial increases in production efficiencies, and are used at all stages in a food manufacturing process, including:

- harvest and transportation to raw material stores;
- preparation procedures (Chapter 2) and movement of food through a process or within a factory;
- collection and disposal of process wastes;
- collation of packaged foods and movement to finished product warehouses;
- distribution to wholesalers and retailers;
- presentation of products for sale.

Advances in computer software, and reduced costs and increased power of micro-electronics hardware, have had substantial effects on both the equipment and control systems that are used to handle materials. This, together with developments in bulk handling systems, have led to substantial improvements to the efficiency of handling both foods and wastes. In this chapter, an outline of methods that are used to handle solid and liquid materials is given, together with a description of developments in process control, distribution of foods and waste management.

27.1   Materials handling

Efficient materials handling is the organised movement of materials in the correct quantities, to and from the correct place, accomplished with a minimum of time, labour, wastage and expenditure, and with maximum safety. Some advantages of correct materials handling are summarised in Table 27.1.

Important techniques identified in Table 27.1 are the use of continuous methods of handling, unit loads and bulk handling, the use of a systems approach to planning a handling scheme, and automation (section 27.2). A systems approach applied to raw materials, ingredients, in-process stock and finished products creates optimum flows of materials, in the correct sequence throughout the production process, and avoids bottlenecks or shortages. Further details are given by Tersine (1994) and Shafer and Meredith (1998). In summary, a systems approach should ensure that:

- raw materials, ingredients and packaging materials arrive at the factory at the correct time, in the correct quantities and in the required condition (section 27.3);
- the space, facilities and equipment layout in a factory enable efficient handling and movement of materials without a risk of cross-contamination (Fig. 27.1);
- storage facilities are sufficient for stocks of materials and can maintain the quality of materials for the required time (section 27.3.1);
- handling equipment and staff levels are sufficient to move materials in the required amounts;
- distribution vehicles are sufficient in number and capacity, and journeys are scheduled to optimise fuel consumption and drivers' time, particularly minimising journeys with empty vehicles (section 27.3.2).

Table 27.1   Advantages of correct materials handling techniques and methods of achieving greater efficiency in materials handling

Advantages	Methods of achieving efficiency
• Savings in storage and operating space • Better stock control • Improved working conditions • Improved product quality • Lower risk of accidents • Reduced processing time • Lower costs of production • Less wastage of materials and operator time	• Only move materials when necessary and minimise all movements by placing related activities close together • Handle materials in bulk • Package or group materials for easier handling • Use continuous handling techniques and minimise manual handling • Automate where possible • Combine operations to eliminate handling between them • Use a systems approach to optimise material flows and make paths as direct as possible • Use all layers of a building's height • Use handling equipment that can be adapted to different applications • Use gravity wherever possible

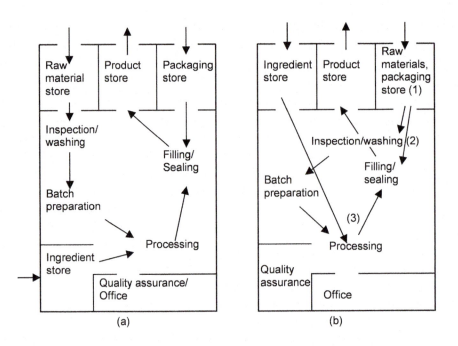

Fig. 27.1 Simplified examples of (a) correct and (b) incorrect factory layouts for small scale vegetable processing. Faults include (1) raw materials and packaging in same store; (2) adjacent filling of prepared food and washing raw material; (3) confused and excessive materials handling (adapted from Fellows 1997).

27.1.1   Hygienic design and cleaning

It is important that all equipment that is used to handle foods is designed to protect them from physical, chemical and biological hazards (see Chapter 1, section 1.2) and allow easy and effective cleaning. Hygienic design is particularly important for machines used to process foods that have a high safety risk to consumers (e.g. liquid filling, especially dairy products, cooked meat slicers, machines that handle short shelf-life chilled or cook–chill foods, conveyor systems for unpacked products such as salads, cold fill cook-in sauce lines, sandwich making equipment). Best practice guidelines on hygienic design are described for example by the European Hygienic Equipment Design Group (EHEDG) (Anon 2008a) and by the FDA and US Food Safety and Inspection Service (Anon 2001a).

Surfaces of food equipment can be divided into those that are (1) in direct contact with foods, or where food residues can drip, drain or diffuse on to them; if contaminated, these surfaces can directly result in product contamination; and (2) non-product contact surfaces (e.g. equipment legs, supports, housings). Non-product contact surfaces are included in sanitary design because contamination of these surfaces can cause indirect contamination of products. All food contact materials and surfaces should be:

- smooth, impervious and free of cracks and crevices;
- non-porous, non-absorbent, non-contaminating and do not transfer odours or taints to the product;
- non-reactive, non-toxic and corrosion resistant;
- durable, maintenance free, and either accessible for cleaning and inspection, or able to

be easily disassembled. If cleaned without disassembly (cleaning-in-place (CIP)), the results should be similar to manual cleaning (Schmidt and Erickson 2005).

Stainless steel is the preferred metal for food contact surfaces because of its corrosion resistance and durability in most applications. Corrosion resistance varies with the chromium content, and structural strength varies with the nickel content. For example, 300 series stainless steel is termed '18/8' (i.e. 18% Cr and 8% Ni) or sanitary standard stainless steel 316 (18/10) are used for most food contact surfaces. During production, surfaces are treated with nitric acid or another strong oxidising agent to create a non-reactive oxide film on the surface. Ground or polished stainless steel surfaces are required to meet a specified smoothness or 'finish', assessed by the 'roughness average' (or R_a value) determined using a 'profilometer' that measures peaks and troughs in the surface. Titanium is used in stainless steel alloys for equipment used to process foods with high acid and/or salt contents (e.g. citrus juice, tomato products). Copper is used for some brewing and distilling equipment (Chapter 14, section 14.2), but not for processing acid products because of reactions that produce copper residues in the products. Aluminium has poor corrosion resistance and can become pitted and cracked by cleaning and sanitising chemicals. In food contact applications, it is coated with a plastic material such as polytetrafluorethylene (PTFE). Cast iron is only used for frying and baking surfaces.

Plastics and rubber-like materials are used for gaskets and membranes that are food contact surfaces. They have poorer corrosion resistance and durability compared with metals and require more frequent examination for deterioration. Other non-metal contact materials include ceramics (e.g. in membrane filtration systems (Chapter 5, section 5.5)) and break-resistant or heat-resistant glass (e.g. Pyrex®) for sight glasses in vessels.

Equipment should be designed and fabricated so that contact surfaces do not have sharp corners and crevices. Standards specify a defined radius for internal angles depending on the application (e.g. an internal angle <135° has a minimum radius of 6.35 mm). The design should avoid dead spaces or bends in pipework that allow the product to accumulate; and threaded bolts, screws or rivets should be avoided in or above food contact areas. Welded joints on stainless steel surfaces should be butt-type, continuous and flush, and ground to a smooth finish. Vessels and tanks should be self-draining and pipework, including CIP pipework, should slope to a drain if it is not routinely dis-assembled. Pipes, gauges or probes in contact surfaces should not create a dead end or an area where food can accumulate that is not accessible for cleaning. Shafts, bearings and seals should be self- or product-lubricated, or use food grade lubricants, and should be attached so that the food contact area is sealed against contamination by lubricants. Components should be accessible and removable for cleaning and disinfection. Openings on tanks and other vessels should be lipped and covered with an overhanging lid, and the rims of equipment should be constructed so that contaminants cannot collect there (Fig. 27.2).

Some keypads and touch screens (or 'human–machine interfaces' (HMIs)) used on food processing equipment are difficult to clean and may be a source of contamination by operators, especially when processing high-risk foods (Hayward 2003). Also they may not tolerate moisture, oil and other hazards in a factory environment, or have membranes that can be damaged by use. A new interactive holographic control technology that uses infrared emitters and detectors, operates by passing a finger though holographic images of what would otherwise be keys of keyboards or icons on touch screens, but are floating in the air at convenient locations for operators. Because there is nothing to touch, operation is hygienic, and problems of wear and hazards are avoided (Anon 2008b).

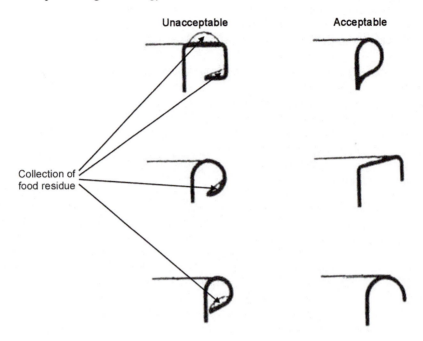

Fig. 27.2   Correct and incorrect top rims of food equipment.

A hygiene risk assessment should be used to identify and resolve any conflict between safety and hygiene (e.g. where frequent access is required for cleaning, hinged guards with an interlocked magnetic switch should be used to prevent operation of equipment) (Anon 2001b).

Non-product contact surfaces should be constructed to be cleanable, corrosion resistant and maintenance free. For example, the legs of equipment should be sealed at the base and not be hollow to prevent them from harbouring micro-organisms, insects or rodents. Tubular steel frameworks should not have bolts or studs and attachments should be welded to the surface, not attached using drilled holes.

Cleaning
Methods for cleaning raw materials are described in Chapter 2 (section 2.2). Adequate cleaning routines for equipment and facilities are part of good manufacturing practice (GMP) (Anon 1998) and form an integral part of management systems needed to implement quality assurance and HACCP programmes (Chapter 1, section 1.5.1). Software to assist managers to devise adequate cleaning schedules is described by Dillon and Griffith (1999) and detailed information on plant cleaning is described by Anon (2002a), Dillon and Griffith (1999) and Quartly-Watson (1998).

The layout of equipment should allow sufficient space for cleaning and avoid harbouring insects and rodents. Unless fitted to a wall, equipment should be at least 10 cm from walls, and floor mounted equipment should be >15 cm above the floor or sealed to the floor or a pedestal. Table-mounted equipment should be sealed to the table or have a space >10 cm underneath for cleaning.

There are two distinct stages in a cleaning operation: removal of soils and disinfection (killing or injuring contaminating micro-organisms). The ability of a disinfectant (or sanitiser) to inactivate or kill micro-organisms depends partly on the presence of soils, and they should be removed before applying a sanitiser. Although water is a good cleaning agent if sufficient energy is applied in the form of heat and force, cleaning compounds decrease the amount of energy required to penetrate, dislodge and remove surface contamination (or 'soils') (see also emulsions (Chapter 3, section 3.2)). The four stages in cleaning are as follows:

1 Applying a detergent solution with good wetting and penetrating properties to make intimate contact with the soil to be removed. Detergent molecules have a lipophilic region of long-chain fatty acids and a hydrophilic region of either a sodium salt of carboxylic acid (soapy detergents or the sodium salt of an alkyl or aryl sulphonate (anionic detergents)). Anionic detergents are not affected by hard water, whereas soapy detergents form a scum in hard water. Non-ionic detergents, which have alcohols, esters or ethers as the hydrophilic component, produce little foam and are easily rinsed off.
2 Displacing soils from the surface by saponifying and emulsifying fats, hydrolysing proteins and dissolving minerals. Enzymes may be added to detergents to remove proteins.
3 Dispersing the soil in water.
4 Preventing re-deposition of the soil onto the clean surface by rinsing.

A cleaning agent should also effectively soften the water, dissolve quickly and completely, be non-corrosive and non-toxic, non-caking and dust-free, and be economical to use and stable during storage (Harper and Spillan undated). Other ingredients may include polyphosphates (to soften water and keep dirt in suspension), sodium sulphate or sodium silicate (to make detergent powder free-flowing) and sodium perborate (bleaching agent). The selection of a cleaning compound depends on the nature of the soil to be removed, the characteristics of the available water and the method of application of the cleaning compound.

The proportions of water-soluble, alkali-soluble and acid-soluble soils deposited on equipment vary according to the composition of the food and the processing conditions at the surface on which the soil is deposited (Table 27.2). The quality of water also varies: it may contain suspended materials and magnesium and calcium salts (water hardness),

Table 27.2   Soil characteristics

Component on surface	Solubility characteristics	Ease of removal	Changes induced by heating soiled surface
Sugar	Water soluble	Easy	Caramelisation, more difficult to clean
Fat	Water insoluble, alkali soluble	Difficult	Polymerisation, more difficult to clean
Protein	Water insoluble, alkali soluble, acid soluble	Very difficult	Denaturation, much more difficult to clean
Salts:			
Monovalent	Water soluble, acid soluble	Easy	None
Polyvalent (e.g. $CaPO_4$)	Water insoluble, acid soluble	Difficult	Interactions with other constituents, more difficult to clean

From Harper and Spillan (undated)

both of which reduce the effectiveness of cleaning agents and form surface deposits on cleaned equipment. Soluble iron and manganese salts can also cause coloured deposits. Sequestering agents in cleaning compounds remove small amounts of water hardness salts but suspended material and soluble iron and manganese salts require water treatment before it can be used for cleaning.

Different components of cleaning compounds are combined to achieve the following functions:

- *Sequestering* – to prevent mineral salts from being deposited on cleaned surfaces;
- *Wetting* – to lower the surface tension of water to increase its ability to penetrate soils;
- *Emulsifying and suspending* – to emulsify fat and suspend other solids in solutions;
- *Dissolving* – to form a solution of inorganic and/or organic solids;
- *Saponifying* – to saponify fats into soaps;
- *Peptising* – to break down and disperse proteins;
- *Dispersion* – to disperse and flocculate mineral films and prevent their re-deposition on clean surfaces;
- *Rinsing* – to separate soils from the surface when flushed with fresh water;
- *Defloculation* – to break up soil flocs on surfaces to improve their removal.

The chemicals used in cleaning compounds can be grouped into (1) alkalis – for emulsifying, saponifying and peptising functions, (2) complex phosphates – for emulsifying and peptising, dispersion of soils, water softening and prevention of soil re-deposition, (3) surfactants – for wetting and dispersion of soils and prevention of re-deposition, (4) chelating compounds – for water softening, control of mineral deposits, peptising and prevention of soil re-deposition, and (5) acids – for control of mineral deposits and water softening.

Alkaline cleaners have a pH ≈ 11–12 and consist of an alkali (e.g. caustic soda, trisodium phosphate or sodium metasilicate), polyphosphates (e.g. sodium hexametapho-sphate, sodium tetraphosphate) and wetting agents (e.g. anionic sulphated alcohols, alkyl aryl sulphonates or cationic quaternary ammonium compounds). Many cleaning com-pounds have reduced the level of polyphosphates to 0.5–1.5% because of environmental legislation that limits the amount of phosphates in waste waters, and chelating agents such as sodium salts of ethylenediaminetetraaceticacid (EDTA) are used to reduce water hardness. Chlorine compounds (e.g. chlorinated trisodium phosphate, hypochlorides) increase the efficiency of alkaline detergents. The chlorinated alkalis do not have bactericidal activity because of their high pH, which also minimises the corrosive activity of the chlorine. Proteolytic enzymes may also be used in combination with alkaline cleaners on equipment that is heavily soiled with proteins (e.g. membrane filters). Details of the advantages and limitations of different chemical components in cleaning agents are described by Harper and Spillan (undated).

Acidic cleaners are blends of organic acids (acetic, lactic, hydroxyacetic, citric, levulinic or tartaric acids), inorganic acids (sulphuric, nitric or phosphoric acids), or acid salts with a wetting agent, and have a pH < 2.5. They are used to remove minerals, including milkstone from high temperature dairy heat exchangers, in can washing and for neutralising alkaline cleaning agents.

Methods of cleaning

In foam cleaning, foam is produced by the introduction of air into a detergent solution as it is sprayed onto the surface to be cleaned. The foam increases the contact time of the cleaning solution. High-pressure cleaning uses increased mechanical force to remove soils. A detergent solution is sprayed at high pressure, often at an increased temperature, to make soil removal more effective. In CIP of interior surfaces of tanks and pipelines, a cleaning solution is circulated and returned to a reservoir to allow reuse. CIP reduces the time spent cleaning, the cost of cleaning because fewer staff are required, and the equipment downtime, so that it is available for production for longer periods. CIP systems have nozzles that are built into equipment together with associated pipework, fittings and instrumentation. The CIP operation is controlled by a programmable logic controller (PLC) (section 27.2.2), which controls pumps, valves, and the duration and sequence of each cleaning operation. This allows optimum timing for efficient cleaning of all parts of the plant. Different cleaning programs may be selected from a menu in the PLC depending on the types of product being processed. A typical CIP operation involves first pumping water through a heater to the CIP nozzles; the first part of the liquid flush that contains high concentrations of product is sent to a drain or to a separate collecting tank. The remaining washwater is filtered and returned to the rinsing tank for reuse. The washing procedure is repeated, with the length of the washing period controlled by the PLC. In the next cycle, a caustic cleaning solution is pumped through the equipment and solids are separated from the liquid to enable the cleaning solution to be reused. There then follows a rinsing cycle with water, and an acid cleaning cycle to neutralise any alkali remaining on the equipment surfaces. Finally, there is a clean water rinsing cycle, and if equipment is required to be dry before start-up (e.g. dehydration equipment), air heaters and fans force warm air through the plant. CIP has been reviewed by a number of authors including Wilson (2002).

Mechanical cleaning using a manual brush or a scrubbing machine (e.g. a floor scrubber) uses mechanical force and detergent to remove soils. Cleaning-out-of-place (COP) involves dismantling of parts of equipment and placing them in a cleaning tank containing an agitated, heated cleaning solution. Ultrasonic cleaning may be used to improve the effectiveness of COP or other types of cleaning/soaking methods. Ultrasonic cleaning is the oldest industrial application of power ultrasound and is usually combined with other cleaning operations such as presoaking or rinsing using different cleaning solutions. The main advantage of ultrasonic cleaning is due to cavitation (see Chapter 9, section 9.6), which is evenly distributed throughout the volume of the liquid and is capable of reaching otherwise inaccessible places. Ultrasonic cleaning equipment normally operates in the range 20–50 kHz, and the high-pressure waves created by implosion of cavitational bubbles loosen the soils. It is best suited to hard materials (metals, glass, ceramics, plastics) that reflect sound and do not absorb it. Gibson *et al.* (1999) studied the effectiveness of different methods of cleaning for removal of bacterial biofilms and further details are given by Marriott and Gravani (2006).

There are two classes of sanitisers used on food processing equipment: halogens (chlorine, iodine or bromine) and quaternary ammonium compounds. Phenols are not widely used because they risk tainting foods. More recently, ozone has been used as a plant sanitiser (Pehanich 2006) and for water treatment (Mahapatra *et al.* 2005).

The factors that influence the effectiveness of disinfection include the following:

- The numbers and characteristics of the microflora present.
- The temperature of use (chlorine, for example, is less effective as the temperature is increased).
- The amount of organic material or mineral deposits present. If the surface has a large amount of organic matter, the sanitiser combines with this rather than the smaller micro-organisms.

- The pH of the sanitiser (chlorine is more effective at lower pH values).
- The germicidal action of the sanitiser (i.e. its selectivity and concentration).
- The length of time the sanitiser is in contact with the surface.
- The phenol coefficient (a comparison of sanitiser activity with pure phenol). This is defined as the highest dilution of sanitising solution that kills all micro-organisms in 10 minutes divided by the highest dilution of phenol giving the same results. The higher the phenol coefficient, the more effective the sanitiser is in killing micro-organisms (Schuler *et al.* 2005).

All cleaning compounds used on product contact surfaces should be approved by regulatory authorities. Methods of disinfection include dipping, soaking and high-pressure spraying.

27.1.2 Solids handling methods
Mechanised handling systems for fresh crops and other raw materials are now routinely used to produce washed and graded crops for processors and retailers (see Chapter 2 and Chapter 26, section 26.1). They include pea viners and combine harvesters, mobile crop washing, destoning and grading equipment, gentle-flow box tippers that transport and unload crops with minimal damage, and automatic cascade fillers for large (1 t or more) intermediate bulk containers (IBCs) (e.g. boxes and 'combo' bins or bags). The bulk movement of particulate and powdered food ingredients by road or rail tanker and their storage in large silos has been common practice in large plants for many years. Increasingly, IBCs are used to ship ingredients, to move food within a production line, and to move part-processed foods between production sites (Chapter 25, section 25.2.7).

Advances in microelectronics have been applied to monitoring and control of storage silos (e.g. fill-level, humidity and temperature). Sensors (section 27.2.1) can be used to detect the loss in weight from a storage tank or silo as it is emptied and calculate the weight of ingredient used. Alternatively sensors on a mixing vessel can detect the increase in weight as different ingredients are added. The information from sensors is used by microprocessors to control pumps, create pre-programmed recipe formulations (e.g. Anon (2004a) and record data for production costing and stock control (sections 27.2 and 27.3)). Multi-ingredient batch weighing and metering systems (Chapter 26, section 26.1.1), using PLCs are an integral part of ingredient or raw material handling.

Continuous handling equipment is an essential component of continuous processes and it also improves the efficiency of batch processing. The most important types of solids handling equipment are conveyors and elevators. Other types of equipment, including chutes, cranes and trucks are described by Anon (2008c) and Brennan *et al.* (1990) and are summarised in Table 27.3.

Conveyors and elevators
In general, conveyors and elevators are best suited to high-volume movement, where the direction of flow of materials is fixed and amounts moved are relatively constant. They can also be used as a reservoir of work-in-progress. Conveyors are widely used in all food processing for the movement of solid materials, both within unit operations, between operations and for inspection of foods (Table 27.3). There are a large number of conveyor designs, produced to meet specific applications, but all types can only cover a fixed path of operation. Details of materials of construction and different designs of conveyors are given by manufacturers (e.g. Anon 2008d,e) and their operation is reviewed by Perera and Rahman (1997). Common types include belt conveyors; an endless belt that is held

Table 27.3   Applications of materials handling equipment

	Conveyors	Elevators	Cranes and hoists	Trucks	Pneumatic equipment	Water flumes
Direction						
Vertical up		*	*		*	
Vertical down		*	*		*	
Incline up	*	*			*	
Incline down	*	*			*	*
Horizontal	*			*	*	
Frequency						
Continuous	*	*			*	*
Intermittent			*	*		
Location served						
Point	*	*			*	*
Path	*				*	*
Limited area			*			
Unlimited area				*		
Height						
Overhead	*	*	*		*	
Working height	*			*	*	*
Floor level	*		*	*		*
Underfloor	*				*	*
Materials						
Packed	*	*	*	*		
Bulk	*	*	*	*	*	
Solid	*	*	*	*	*	*
Liquid				*	*	*
Service						
Permanent	*	*	*		*	*
Temporary			*	*		

From Brennan *et al.* (1990)

under tension between two rollers, one of which is driven. The belts may be stainless steel mesh or wire, synthetic rubber or a composite material made of canvass, steel and polyurethane or polyester. Trough-shaped flexible belts are used to move small particulate materials. Flat belts are used to carry packed foods, and may be inclined up to 45° if they are fitted with cross-slats or raised chevrons to prevent the product from slipping. Spiral conveyors are used to raise or lower products (Fig. 27.3).

Roller conveyors and skate wheel conveyors are usually unpowered, but roller conveyors may also be powered. The rollers or wheels are either horizontal, to allow packed foods to be pushed along, or slightly inclined (e.g. a fall of 10 cm in a length of 3 m) to allow packs to roll under gravity. Rollers are stronger than skate wheels and therefore able to carry heavier loads, but their greater inertia means that they are more difficult to start and stop, and they are more difficult to use around corners. 'Air-cushioned' conveyors overcome these problems by carrying materials on a film of air blown into tubular trough sections by a fan.

Chain conveyors are used to move churns, barrels, crates and similar bulk containers by placing them directly over a driven chain, which has protruding lugs at floor level.

Fig. 27.3   Mass flow spiral conveyor (courtesy of Ryson International Ltd at www.ryson.com).

Monorail conveyors are used for moving meat or poultry carcasses on an overhead track. Screw conveyors comprise a rotating helical screw inside a metal trough. They are used to move bulk foods such as flour and sugar, or small-particulate foods including peas and grains. Their main advantages are the uniform, easily controlled flowrate, the compact cross-section without a return conveyor and total enclosure to prevent contamination. They may be horizontal or inclined, but are generally limited to a maximum length of 6 m, as above this high friction forces result in excessive power consumption.

Vibratory conveyors cause a small vertical movement in particulate foods or powders, which raises pieces a few millimetres off the conveyor, and a forward movement to transport them along the conveyor. The amplitude of vibration is adjusted to control the speed and direction of movement and this precise control makes vibratory conveyors useful as feed mechanisms for processing equipment. Other types of feeders are described by Wheeldon (2002).

Pneumatic conveyors consist of a system of pipes through which powders or small-particulate foods are suspended in recirculated air and transported at 20–35 m s^{-1}. The air velocity is critical: if it is too low, the solids settle out and block the pipework; if it is too high there is a risk of abrasion damage to the food or internal pipe surfaces. The calculation of air velocity needed to suspend foods is described in Chapter 1 (section 1.3.4) and similar equipment is used to classify foods (Chapter 2, section 2.2), and to dry foods (Chapter 16, section 16.2.1). Generation of static electricity by movement of foods is a potential problem that could result in a dust explosion when conveying powders, and this is prevented by earthing the equipment, venting, or explosion containment and suppression techniques (Anon 2007a). The advantages of pneumatic conveyors are that they cannot be overloaded, have few moving parts, low maintenance costs and require only a supply of high-velocity air. Low-velocity dense phase pneumatic conveying systems (Anon 2008f) have lower turbulence, which produces advantages such as reduced attrition of food particles, less abrasive wear on equipment and reduced energy consumption. The selected velocity depends on the physical characteristics of the material, the capacity and conveying distance. Magnetic conveyors are used to hold cans in place with minimal noise, and are able to invert empty cans for washing.

Conveying foods in water using shallow inclined troughs (or 'flumes') and pipes is used for simultaneous washing and transporting of small particulate foods; the main advantage is reduced power consumption as water flows under gravity, especially at factory sites located on hillsides. Water is recirculated to reduce costs and is filtered and chlorinated to prevent a build-up of micro-organisms.

There are many other types of conveyors including sorters, singulators, diverters and descramblers. Singulating conveyors separate products as they travel along them. Accumulating conveyors hold products in place until they are given a signal to release them (e.g. for accumulating the correct number of items such as biscuits for filling into packs, or feeding products to a case filler at a specified rate).

Developments in conveyors include antimicrobial belts (e.g. Anon 2008g) and 'intelligent' conveyors. Rather than having to start and stop conveyors, these systems divide a process into intelligent zones that 'know' when they should operate. Photo-electric sensors feed information on the position of products to a microprocessor that controls the movement of the conveyors using servo motors. In each zone, the micro-processor collates information on products upstream and downstream from it and controls the conveyors by 'deciding' whether to run, accumulate or 'sleep', based on inputs from the surrounding zones (McTigue Pierce 2005). Conveyors can be programmed for control of forward or return velocity, the rate of acceleration or deceleration, and time of operation to optimise the flow of product (Anon 2007b). Details of developments in conveying are given by Sharp (1998) and Anon (2008h).

There are many designs of elevator, but a common type is the bucket elevator: it consists of metal or plastic buckets fixed between two endless chains. They have a high capacity for moving free-flowing powders and particulate foods. The shape and spacing of the buckets and the speed of the conveyor ($15–100\,\mathrm{m\,min^{-1}}$) control the flowrate of materials. Further details of elevators are given by Brennan *et al.* (1990).

27.1.3 Liquid handling methods

Pumps, valves and associated pipework are the usual method of handling liquid foods, or cleaning fluids and there is a very wide range of designs available, often for specific applications. Details of their selection are given by Bowser (undated). The selection of a particular pump is based on the following factors:

- Type of product, particularly its viscosity and shear sensitivity.
- Product flowrate, suction and discharge pressures and temperature.
- Continuous operation/frequent start-stop.

Centrifugal pumps have a rotating impellor inside a stationary casing. They are widely used in food processing and have different types of impeller, depending on the application (Hogholt 1998). They are suited to applications that involve variable flowrates, either by adjusting the speed of the impeller or throttling the flow using an adjustable valve in the discharge pipework. Positive displacement pumps have a cavity that expands and contracts. Liquid flows into the expanding cavity on the suction side and is expelled from the discharge side as the cavity collapses. Unlike centrifugal pumps, these produce the same flowrate at a given speed, independent of the discharge pressure. They are suitable for viscous or shear-sensitive products, or for high pressures and accurate flow control. Examples include rotary lobe pumps (Hammelsvang 1999), rotary piston pumps, gear pumps and diaphragm pumps. Johnson (2001) and Anon (2002b) describe the advantages of peristaltic pumps for gentle product handling. Lee (2000) describes progressing cavity pumps, in which an enclosed helical rotor turns within a stator to form sealed cavities that contain the product. As the rotor turns the cavities move from the suction side to the discharge side, so pumping the product with a non-pulsating, positive displacement action.

There are a large number of different types of valves that are used in pipelines, each of which is suitable for CIP and automatic operation using compressed air and/or electricity. All valves can be fitted with proximity switches to detect and transmit the position of the spindle, and pneumatic or electro-pneumatic actuators to position the valve spindle accurately. In large diameter pipes, or applications involving very high pressures, the force required to close a conventional single seat valve is too great and a double seat design is used. This has two valve plugs on a common spindle, and two valve seats (Fig. 27.4b). Valves can also be grouped into two-port and three-port valves. Two-port valves throttle (restrict) the flow of fluid passing through them and three-port valves can be used to mix or divert liquids into different steams.

Two-port valves that have a linear spindle movement include globe valves and slide valves, and valves with a rotary spindle movement include ball valves, butterfly valves and plug valves (Fig. 27.4a,c). Slide valves have two different designs: 'wedge gate' and 'parallel slide' types. Both are used to isolate fluid flow because they produce a leakproof shut-off. Ball valves comprise a spherical ball that has a hole to allow fluid to pass through, and is located between two sealing rings in the valve body. Rotating the ball through 90° opens and closes the flow. They give a tight closure for many fluids including steam at temperatures up to 250 °C.

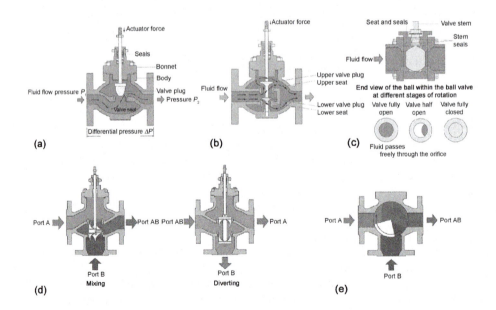

Fig. 27.4   Examples of different types of valves used in fluid handling: (a) flow through a single seat, two-port globe valve; (b) flow through a double seat, two-port valve; (c) ball valve (shown in a fully open position); (d) globe type three-port valves; (e) rotating shoe on a mixing application (courtesy of Spirax-Sarco Ltd) (Anon 2008i).

'Butterfly' valves consist of a disc that rotates inside the valve on bearings. In the open position the disc is parallel to the pipe wall, allowing full flow through the valve. It is rotated to the closed position against a seat, perpendicular to the pipe wall. They are similar to ball valves, but the different sealing arrangement means that ball valves can operate against higher differential pressures than equivalent butterfly valves.

There are three basic types of three-port valve: piston, globe plug and rotating shoe types. Piston valves have a hollow piston that is moved up and down by an actuator, covering and correspondingly uncovering two ports, so that the cumulative cross-sectional area is always equal (e.g. if one port is 30% open, the other port is 70% open and vice versa) (Fig. 27.4d). In globe type three-port valves, the actuator pushes a disc or pair of valve plugs between two seats, increasing or decreasing the flow through the two ports in a corresponding manner. In the rotating shoe three-port valve (Fig. 27.3e), a rotating shoe can be positioned for mixing applications to allow different proportions of fluids from two inlet pipes to be mixed and exit from the third port. An example of its use as a mixing/diverting valve is shown in Fig. 27.5. When the heat exchanger requires maximum heat, port A is fully open and port B fully closed. All water from the boiler is passed through the heat exchanger and through the valve ports AB and A. When the heat load is reduced, port A is fully closed and port B fully open, and the water from the boiler bypasses the heat exchanger and passes through the valve via ports B and AB.

Other types of valve include safety valves to prevent excess pressure in pressure vessels; vacuum valves to protect vessels or tanks from collapse under unwanted vacuums; modulating valves to permit the exact control of product throughputs; non-return valves; and sampling valves to allow bacteriologically safe samples to be taken from a production line without the risk of contaminating the product. Diaphragm valves consist of a polymer membrane or stainless steel bellows that prevent the product from having contact with the valve shaft. They are used for sterile applications up to product pressures of 0.4 MPa.

Further details of the use of valves in process control are given in section 27.2.2. The design of pipelines for handling liquid foods is described by Steffe and Singh (1997).

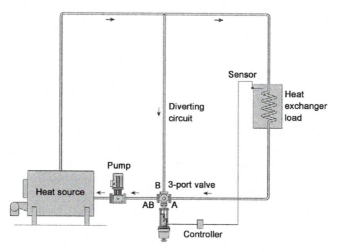

Fig. 27.5   Three-port mixing valve installed on return pipework in a water heating application (courtesy of Spirax-Sarco Ltd) (Anon 2008i).

27.1.4   Water and waste management
Water supplies and water treatment

Variation in the quality of processed foods can arise from the quality of water used in the process as well as from raw materials and processing conditions. Processors should therefore ensure that water supplies are reliable and consistently meet the water quality standards described by local regulations. Water sources may be the public mains supply, a private well or surface water, depending on the location of the factory and the cost of the water plus water treatment. In most countries, public water supplies are treated and tested to ensure that they meet safe (potable) drinking water standards for microbiological, chemical and radiological contaminants.

Water has different uses within a processing operation: for washing raw materials, blanching, cooling, steam generation and process heating, cleaning, or direct use in the product. The categories can be described as: general-purpose, process, cooling and boiler feed water. General-purpose water is all water that is used for washing and sanitising raw materials and equipment and in-plant chlorination is usually the only treatment required. Process water is used as a cooking medium or it is added directly to the product, and it should have sufficiently high quality so that it does not alter the product quality (including freedom from dissolved minerals that would make the water excessively hard and affect the texture of raw materials or affect the taste of a product). Treatment of process water may include softening, reverse osmosis and deionisation. Cooling water that does not contact food products or their containers does not have to be potable, but may be treated to prevent accumulation of scale in pipes and equipment. However, water used to cool thermally processed, low-acid foods in hermetically sealed containers should have residual chlorine or other sanitiser (also Chapter 13, section 13.1.2). Boiler feed water is softened to remove hardness and prevent scale formation, and feed water for high-pressure boilers is demineralised to remove all dissolved solids.

Water from wells or surface sources may contain impurities, including suspended material, micro-organisms, organic matter, colour, taste or odour chemicals, and dissolved gases, and it should be treated to lower these to acceptable levels. Suspended materials are removed by screening, sedimentation, coagulation and flocculation, and filtration. Water must be microbiologically safe from bacteria, viruses, protozoan cysts and worms. Methods used for disinfection include chemical, thermal, ultraviolet radiation and ultrasonic treatment for cell disruption. Chemical treatment using chlorine or its derivatives is the least expensive and most common, and also permits reuse, to conserve water and reduce wastewater volumes. The chlorine 'dose' is the total amount of chlorine added to water, expressed as $mg\,l^{-1}$ (or parts per million (ppm)). Residual chlorine is the effective chlorine left in the water after initial combination with organic matter and micro-organisms, which can be tested using *N,N*-diethyl-phenylenediamine test kits. Organic contaminants that cause colour, taste or odour problems can be removed by coagulation, settling and filtration, or by activated carbon. Dissolved carbon dioxide, oxygen, nitrogen and hydrogen sulphide are naturally found in water. These gases can cause corrosion and reduced heat transfer, or iron pick-up, and are removed by boiling and venting dissolved air, oxygen scavengers or forced draught aeration followed by chlorination to remove hydrogen sulphide.

All water contains dissolved minerals: bicarbonates of magnesium and calcium create temporary hardness because they precipitate on heating to form scale. Permanent hardness is due to magnesium and calcium sulphates, chlorides and nitrates, and may cause hardening of vegetables during blanching and canning. Hard water also causes alkaline detergents to produce scum or deposits during washing. Water softening

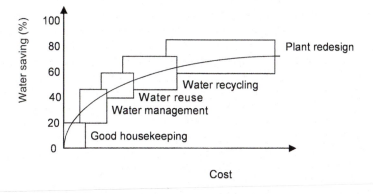

Fig. 27.6 Key areas for water optimisation and associated costs (courtesy of Food Processing at www.foodprocessing.com) (from Holmes 2000).

processes include precipitation, cation exchange, and demineralisation. Further details of water treatment are given by Flores *et al.* (1994).

Holmes (2000) describes methods to save water by improved 'housekeeping' (e.g. repairing leaking pipes, turning off water cooling equipment when not in use), better water management, reusing or recycling water (e.g. water that has been used to clean slightly soiled equipment can be used in the first cleaning stage of heavily soiled equipment), or redesigning the factory to improve water utilisation and minimise effluent production (Fig. 27.6). Recycled water may be purified using membrane filters (Chapter 5, section 5.5) before reuse.

Effluent treatment

With the exception of a few processes (e.g. extrusion (Chapter 15) or baking (Chapter 18)), solid wastes and liquid effluents are produced in large quantities by food processing operations (Table 27.4). They arise due to cleaning and preparation of raw materials (Chapter 2), spillages, cleaning of equipment and floors, and emptying vessels at the end of production or when changing over to different products.

The nature of wastes varies according to the type of food being processed: fruit and vegetable processing effluents have high concentrations of sugars, starch and solid matter

Table 27.4 Volumes and composition of wastes from selected food processing operations

Commodity	Volume (litres per unit)	BOD (mg l^{-1})	Suspended solids (mg l^{-1})
Fruit products[a]	90–450	1000–5000	100–2000
Meat packing[b]	9000–36 300	600–1600	400–720
Milk processing[c]	12–23	20–650	30–360
Mushrooms[b]	30 000	80–850	50–240
Potato chips[b]	18 000	730–1800	800–2000
Poultry packing[d]	6.8	725–1150	770–1750
Sauerkraut	14–80	1400–6300	60–630
Vegetable products[e]	90–1260	500–11 000	30–4000

[a] Per case of product (data from apples, apricots, citrus, pumpkin and tomatoes).
[b] Per ton of product.
[c] Per litre of milk.
[d] Per chicken.
[e] Per case of product (data from beans, carrots, corn, peas, peppers, spinach).
Data adapted from Guthrie (1988), Potter (1986) and Dalzell (1994)

such as peelings, whereas meat and dairy processing effluents contain a higher proportion of fats and proteins. Nearly all processors also produce dilute wastewater from washing equipment, and solid wastes such as discarded packaging materials or office paper. Sources of production of effluents and their treatment methods are described by Waldron (2007), Wilbey (2006) and Tebbutt (1992). Wheatley (1994) describes the composition of typical food industry wastes and methods that can be used to minimise waste production. Anon (1996) describes methods for auditing wastes and reducing their costs (see also Chapter 25, section 25.6).

In large processing plants or those located in unpopulated areas, effluent treatment may be carried out on-site in purpose-built facilities, or effluents are pre-treated to remove solids, fats and greases, or by pH adjustment, before treatment by municipal authorities or private water utilities (Anon 2002c). The direct costs of both purchasing water from utility suppliers and effluent treatment have risen sharply in most countries due to legislation to improve the quality of potable water and to reduce the environmental impact of effluents. The cost of effluent treatment is based on a combination of the volume of effluent and its polluting potential, as measured by chemical oxidation demand (COD) or biological oxidation demand (BOD) and the amount of suspended solids (in $mg\,l^{-1}$). High concentrations of sugars, starches and oils have very high polluting potential (CODs from 500–4000 $mg\,l^{-1}$ compared with domestic sewage at 200–500 $mg\,l^{-1}$) because as micro-organisms utilise these materials they remove dissolved oxygen from water, which may kill fish and aquatic plants. Charges are therefore considerably higher for treatment of these effluents. The cost of effluent treatment in some countries is calculated using the Mogden formula (Anon 2007c):

$$C = R + (V \text{ or } VB \text{ or } VM \text{ or } M) + B(O_t/O_s) + S(S_t/S_s) \qquad \boxed{27.1}$$

where C = total charge per m^3 of trade effluent, R = reception and transport charge per m^3, V = volumetric and primary treatment charge per m^3 in effluent treatment works that do not have biological treatment, VB = volumetric and primary treatment charge per m^3 in effluent treatment works that have biological treatment, VM = treatment and disposal charge per m^3 at non-designated sea outfalls, M = treatment and disposal charge per m^3 at designated sea outfalls, O_t = COD ($mg\,l^{-1}$) of trade effluent after 1 h settlement, B = biological oxidation charge per m^3 of settled sewage, S_t = total suspended solids ($mg\,l^{-1}$), S = treatment and disposal charge per m^3 of primary sludge, O_s = mean strength (COD) of settled sewage at treatment works and S_s = mean suspended solids at treatment works.

In many processes it is possible to reduce treatment costs by separating concentrated waste streams from more dilute ones (e.g. washings from confectionery boiling pans that contain high concentrations of sugar can be isolated from general factory washwater). Anon (2001c) describes the savings achieved using 'pigging'. This involves sending a rubber block that is shaped to fit closely into pipework, to both clean the pipes and recover product at the end of production or before a product change-over. Other means of reducing both polluting potential and wastewater treatment charges include:

- recycling water;
- recovering fats and oils by aeration flotation for sale as by-products;
- storing concentrated effluents and blending them over a period of time with dilute wastes to produce a consistent moderately dilute effluent;
- removing solids using screens, flocculating suspended solids using a chemical coagulant (e.g. lime or ferrous sulphate) or removing suspended solids directly by

sedimentation, filtration or centrifugation (Chapter 5) and discharging them as solid waste to commercial waste disposal companies or for composting;

- treating effluents using a biological method (e.g. trickling filters, activated sludge processes, lagoons, ponds, oxidation ditches, spray irrigation or anaerobic digesters) or low pressure membranes (Chapter 5, section 5.5) (Clark 2000);
- treating effluents with ozone (Cove 2006);
- fermenting waste materials to produce more valuable products (e.g. organic acids, vitamins).

Descriptions of the methods used to treat effluents and details of the advantages and limitations of these treatment methods are given by Tebbutt (1992) and Brennan *et al.* (1990). Solid wastes, packaging and office waste materials are collected in some countries by municipal authorities and in others by private waste management and recycling companies. They are usually disposed of in landfill sites, but increasing shortages of suitable sites and steadily increasing costs of collection have stimulated incentives and opportunities for recycling and reuse, especially for paper, metals and some types of plastics. These developments are described in Chapter 25 (section 25.6). An environmental management system (EMS) is a structured management approach to controlling wastes and pollution, reducing energy consumption, and more generally reducing the environmental impact of a processing operation. Anon (2008j) describes how to set up an EMS to meet international legislation (ISO 14001) or the EU's Eco-management and Audit Scheme (EMAS), and methods for efficient management of energy and water are described by Smith *et al.* (2008).

27.2   Process control

The purpose of process control is to reduce the variability in final products so that legislative requirements and consumers' expectations of product quality and safety are met. It also aims to reduce wastage and production costs by improving the efficiency of processing. Simple control methods (e.g. reading thermometers, noting liquid levels in tanks, adjusting valves to control the rate heating or filling) have always been in place, but the move away from controls based on the operators' skill and judgement to technology-based control systems has taken place as the scale and complexity of processing have increased. First, manually operated valves were replaced by electric or pneumatic operation; and measurements of process variables (e.g. levels of liquids in tanks, pressure, pH, temperature (Table 27.5)) were no longer taken at the site of equipment, but were sent by transmitters to control panels. Within the past 20 years, changes in process control technology to microelectronic control have been widespread and applied in almost every sector of the industry. The impetus for these changes has come from both escalating labour costs and raw material costs, and increasingly stringent regulations for standardised, safe foods following international harmonisation of legislation and standards. For some products, regulations require monitoring, reporting and traceability of all batches produced (Atkinson 1997) which has further increased the need for more sophisticated process control.

All of these requirements have caused manufacturers to upgrade the effectiveness of their process control and management systems. Advances in microelectronics and developments in computer software technology, together with reductions in the cost of computing power, have led to the development of very fast data processing. This has in

Table 27.5 Examples of measured parameters and types of sensors used in food processes

Parameter	Sensor/instrument type	Examples of applications
Bulk density	Radiowave detector	Granules, powders
Caffeine	Near infrared detector	Coffee processing
Colour	Ultraviolet, visible, near infrared light detector	Colour sorting, optical imaging to identify foods or measure dimensions (Chapter 2, section 2.3.3)
Conductivity	Capacitance gauge	Cleaning solution strength
Counting food packs	Ultrasound, visible light	Most applications
Density	Mechanical resonance dipstick, gamma-rays	Solid or liquid foods
Dispersed droplets or bubbles	Ultrasound	Foams
Fat, protein, carbohydrate content	Near infrared, microwave detectors	Wide variety of foods
Fill level	Ultrasound, mechanical resonance, capacitance	Most processes
Flowrate (mass or volumetric)	Mechanical or electromagnetic flowmeters, magnetic vortex meter, turbine meter, ultrasound	Most processes
Foreign body detection	X-rays, imaging techniques, electromagnetic induction (for metal objects)	Most processes
Headspace volatiles	Near infrared detector	Canning, MAP
Humidity	Hygrometer, capacitance	Drying, freezing, chill storage
Interface – foam/liquid	Ultrasound	Foams
Level	Capacitance, nucleonic, mechanical float, vibronic, strain gauge, conductivity switch, static pressure, ultrasound	Automatic filling of tanks and process vessels
Packaging film thickness	Near infared detector	Packaging, laminates
Particle size/ shape distribution	Radiowave detector	Dehydration
pH	Electrometric	Most liquid applications
Powder flow	Acoustic emission monitoring	Dehydration, blending
Pressure or vacuum	Bourdon gauge, strain gauge, diaphragm sensor	Evaporation, extrusion, canning
Pump/motor speed	Tachometer	Most processes
Refractometric solids	Refractometer	Sugar processing, preserves
Salt content	Radiowave detector	Pickle brines
Solid/liquid ratio	Nuclear magnetic resonance	In development
Solute content	Ultrasound, electrical conductivity	Liquid processing, cleaning solutions
Specific micro-organisms	Immunosensors	Pathogens in high-risk foods
Specific sugars, alcohols, amines	Biosensors	Spoilage of high-risk foods
Specific toxins	Immunosensors	High-risk foods
Suspended solids	Ultrasound	Wastewater streams
Temperature	Thermocouples, resistance thermometers, near infrared detector (remote sensing and thermal imaging), fibre-optic sensor	Most heat processes and refrigeration
Turbidity	Absorption meter	Fermentations
Valve position	Proximity switch	Most processes
Viscosity	Mechanical resonance dipstick	Dairy products, blending
Water content	Near infrared detector, microwaves (for powders), radiowaves, NMR	Baking, drying, etc.
Water quality	Electrical conductivity	Beverage manufacture
Weight	Strain gauge	Weighing tank contents, checkweighing

Adapted from Hamilton (1985), McFarlane (1988, 1991), Medlock and Furness (1990), Anon (1993), Kress-Rogers (1993, 2001)

turn led to efficient, sophisticated, inter-linked, more operator-friendly and affordable process control systems being made available to manufacturers. These developments are now used at all stages in a manufacturing process, including:

- ordering and supplying raw materials automatically using just-in-time (JIT) and material resource planning (MRP) software (Wallin 1994);
- production planning and management of orders, recipes and batches;
- controlling process conditions and the flow of product through the process – individual processing machines are routinely fitted with microprocessors, to monitor and control their operation, product quality and energy consumption;
- collation and evaluation of process data and product data;
- control of CIP procedures;
- packaging, warehouse storage and distribution control.

This section describes the principles of process control and automation with selected examples of equipment. Their use in management is described in section 27.2.4 and in logistics in section 27.3.2. Details of mathematical models used in computer control of canning, aseptic processing, evaporation, fermentation and dehydration are described by Teixeira and Shoemaker (1989). Other examples of computer control are reported by Pyle and Zarov (1997) and Hollingsworth (1994) and specific uses of microprocessors within individual processes are described in the relevant preceding chapters.

The advantages of automatic process control can be summarised as (1) more consistent product quality and greater product stability and safety because variations in processing conditions are minimised; (2) more efficient operation through better use of resources (e.g. raw materials, energy, labour), reduced effluents or more uniform effluent loads and increased production rates (e.g. through optimisation of equipment utilisation); and (3) improved safety using rapid automatic fail-safe procedures with operator warnings and verification that operators have made correct inputs. The main disadvantages of automation relate to the social effects of reduced employment when fewer operators are required; higher set-up and maintenance costs; and reliance on accurate sensors to precisely measure process conditions, and the increased risks, delays and costs if an automatic system fails.

The components of an automatic control system are as follows:

- Sensors to detect process conditions and give information on the status of process variables, with transmitters to send this information to a controller.
- A controller to monitor and control a process. Output signals from the controller are sent to actuators.
- Actuators (e.g. motors, solenoids or valves) to make changes to the process conditions.
- A system of communication between a controller and actuators; for example signals from motors or valves, which indicate that the component has been switched on, and input signals from sensors, which indicate that a required process condition has been reached (e.g. maximum or minimum temperature, flowrate or pressure), and an 'interface' for operators to communicate with the control system.

Control systems may also have facilities for collating information and producing analyses of performance or production statistics for management reports.

27.2.1   Sensors
Sensors are instruments that measure and transmit specified process variables, and can be grouped into:

- primary measurements (e.g. temperature, weight, flowrate and pressure);
- comparative measurements obtained by comparison of primary measurements (e.g. specific gravity);
- inferred measurements, where the value of an easily-measured variable is assumed to be proportional to a parameter that is difficult to measure (e.g. hardness as a measure of texture); and
- calculated measurements, found using qualitative and quantitative data from analytical instruments or mathematical models (e.g. biomass growth in a fermenter (Chapter 6, section 6.4.2)). Some process variables are used indirectly as indicators of complex biochemical changes that take place during processing (e.g. the time–temperature combination needed to destroy micro-organisms (Chapter 10, section 10.3). It is therefore necessary to know the precise relationship between the measured variable and the changes that take place in order to be able to exercise effective control.

Solid-state electronic sensors have largely replaced older mechanical or chemical types, owing to their greater reliability, greater accuracy and precision, and faster response times. Examples of the types of sensors used in food processing are shown in Table 27.5. The main requirements of sensors are as follows:

- A hygienic sensing head that is free of contaminants (i.e. contains no reagents or micro-organisms that could contaminate foods) and does not cause potential hazards from foreign bodies (e.g. no glass components) or react with foods.

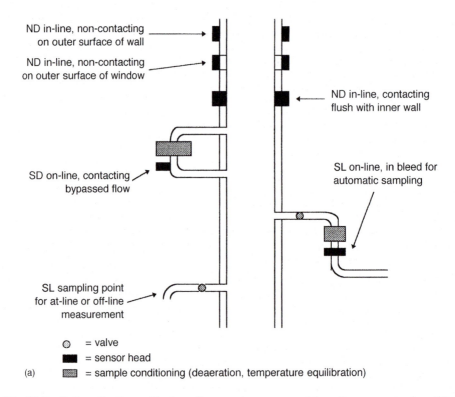

ND in-line, non-contacting on outer surface of wall

ND in-line, non-contacting on outer surface of window

ND in-line, contacting flush with inner wall

SD on-line, contacting bypassed flow

SL on-line, in bleed for automatic sampling

SL sampling point for at-line or off-line measurement

⊙ = valve
■ = sensor head
(a) ▨ = sample conditioning (deaeration, temperature equilibration)

Fig. 27.7 Options for the positioning of sensors in a process: (a) continuous processing; (b) on conveyors; (c) in batch processing (ND = non-destructive measurement, SD = slight damage (e.g. deaeration of sample), SL = sample lost) (adapted from Kress-Rogers 1993).

- Robust to withstand processing temperatures, pressures, food components or effluents, and cleaning-in-place chemicals, or having cheap, easily replaced, disposable sensing heads.
- Reliable with good reproducibility, even when exposed to moisture, steam, dust or fouling by food components, resistant to damage from mechanical vibration or electromagnetic interference in some applications (e.g. microwave or ohmic heaters (Chapter 20)).
- Low maintenance requirement and low total cost (capital, operating and maintenance costs) in proportion to the benefits gained (Kress-Rogers 2001).

Options for the positioning of sensors in a process are shown in Fig. 27.7a–c, and may be summarised as 'in-line,' 'on-line', 'at-line' and 'off-line' (the last being used in analytical laboratories using samples taken from sampling points). On-line and in-line sensors are widely used because of their rapid response time, and selected applications are described by Anon (1993).

Non-contact, at-line sensors have advantages in many applications. These sensors include those using electromagnetic waves, light, infrared radiation (Skjoldebrand and Scott 1993), microwave or radio frequency waves (Chapter 18), gamma-rays or ultrasound. For example, Ritchie (2004) describes the advantages of infrared sensors and

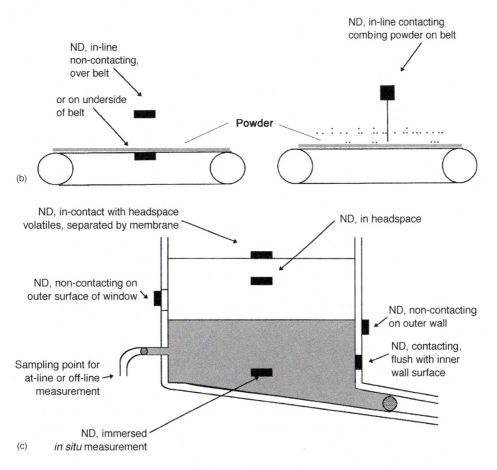

Fig. 27.7 Continued

Jennings (1998) describes a colour sensor, which is able to withstand impact and severe environmental conditions. It is able to assess surface brightness, texture, colour and surface mapping at up to 10 000 measurements per second. It has found application in labelling and sensing registration marks on packaging materials. It is likely that non-contact sensors will continue to be further developed, together with bio-sensors, and those based on nuclear magnetic resonance (NMR) or electron spin resonance (ESR) (e.g. Rohn and Kroh 2005, Kim *et al.* 1999).

Other applications that are being met by new developments in sensors include measurement of chemical composition, complex rheological properties, the size and distribution of particles, droplets or bubbles and volatiles produced by drying, baking or frying (Kress-Rogers 2001). Sensors that measure sound from process machinery or from the product itself, termed 'acoustic emission' are described by Roberts and Wiltshire (1990), with applications to measurement of powder flowrate, CIP, solids content in solid/liquid mixes and cooker-extruder performance. Amrani (2005) describes an 'intelligent pipe' that monitors the electrical properties of liquids as they pass through the pipe. It consists of a transmitting electrode and a receiving electrode that are located either side of a non-conducting section of pipe, inserted into the processing line. The electrical properties of liquids change as their formulation changes, and the intelligent pipe measures these changes using impedance spectroscopy, to produce a characteristic 'signature' of a liquid within 2 s. It can be used, for example, to monitor mixing of ingredients in real time and verify that mixing is complete; to verify that a product is within specification or detect non-conformity; to monitor the stages in CIP or product changeover, leading to savings in products that would otherwise be discarded. It is non-invasive, highly sensitive, hygienic and produces reliable results.

'Electronic noses' and 'electronic tongues' are common names for sensors that detect flavour or odour (i.e. volatiles) or taste (soluble materials). Previously, mass spectro-meters or gas/liquid chromatographs were used to produce a 'fingerprint' of the material being analysed. These have been replaced an array of up to 25 simple sensors that measure changes in voltage or frequency, and are linked to software that enables pattern recognition. Each odour or taste leaves a characteristic pattern on the sensor array, and an artificial neural network (section 27.2.3) is trained using known flavour mixtures, so that it can distinguish between and 'recognise' odours and tastes. The sensors are intended to simulate a sensory response to a specific flavour, or to sourness, sweetness, saltiness and bitterness. Further information is given by Holm (2003), Tan *et al.* (2001) and Bartlett *et al.* (1997). Other new developments in sensors, including time-resolved diffuse reflect-ance spectroscopy (TDRS) in the near infrared/visible range, and surface plasmon resonance (SPR) used in biosensors involving antibodies or enzymes, are described by Johnston (2006), Holm (2003) and Rand *et al.* (2002).

27.2.2   Controllers and PLCs
The information from sensors is used by controllers to make changes to process conditions. Whereas previously operators monitored and manually controlled a process, in all but the smallest scales of food processing control is now achieved using automatic controllers. These operate using logic in a similar way to the decision-making logic demonstrated by human thought (Table 27.6).

The increasingly widespread use of microprocessor-based process controllers is due to their flexibility in operation, their ability to record (or 'log') data for subsequent calculations, and substantial reductions in their cost. They can store and analyse data and

Table 27.6   Comparison of manual and automatic control sequences for changing the supply silo for a sugar ingredient

Question	Action/decision by operator	Action by automatic controller
How much sugar is left in silo 2?	Read from level gauge	Information from level sensors indicates approach to set limit for empty
How long before the change to the new silo?	From experience, about 5 min	Calculation using information from a sugar flowrate transmitter
Which silo?	From knowledge of number of silos in factory	Information from level sensors indicating the level status in other silos
Shall I choose silo 3?	Decide against using – knowledge that discharge valve is under repair	Information from on-line engineering report of ongoing maintenance work
Shall I choose silo 4?	Affirm after checking level reading to confirm that silo 4 is not full	From level sensors indicating the level status in silo 4 (empty)
Is the current silo now empty?	Read from level gauge	Information from a sugar flowrate transmitter to calculate the delay before initiating a pre-programmed sequence of signals to change motorised valves and switch to silo 4
Which valves to open?	From knowledge of plant layout	
Decide to alter valves	Open valve 24, close valve 22	

be connected to printers, communications devices, other computers and controllers throughout a plant. They can also be easily re-programmed by operators to accommodate new products or process changes.

There are two basic operational requirements of a controller: a method of keeping a specified process variable at a predetermined set-point, and a method to control the sequence of actions in an operation. To meet the first requirement, sensors measure a process variable and then a transmitter sends a signal to the controller, where it is compared with a set-point. If the input deviates from the set-point (the error), the controller alters an actuator to correct the deviation and hold the variable constant at the set-point. The type of signals that are sent can be either 'on or off' (e.g. a motor is operating or not, or a valve is open or closed) or continuously variable (e.g. a temperature transmitter or the speed of a pump). Niranjan *et al.* (2006) describe different types of control systems, including on/off, proportional, integral and derivative controllers.

In 'closed-loop' control, there is a continuous flow of information around an electrical loop: information from sensors is used by the controller to produce changes to the actuator; these changes are then registered by the sensor and the loop starts again. A common example is feedback control (e.g. a thermostat automatically controls a steam valve (the actuator) to maintain the steam flow in response to information from a thermocouple sensor measuring product temperature (Fig. 27.8)).

A second type is 'feed-forward' control, in which process parameters are monitored and compared with a model system that anticipates the required process conditions. If the operating parameters deviate from the model, the controller alters them via the actuators. Feed-forward control is preferable because the error can be anticipated and prevented, rather than waiting for the error to be detected and then apply compensation to remove it (McFarlane 1991). However, this is not always possible because it is necessary to know

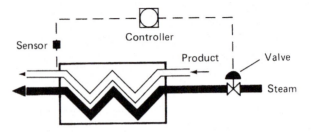

Fig. 27.8   Feedback control of product heating.

in advance what changes may take place in a product before corrective action can be taken. In these situations feedback control may be preferable.

The second requirement of process controllers is the proper sequencing of actions in a process. Previously this was achieved using hard-wired relay circuits (a type of electrical switch), controlled by timers and counters that switched different circuits on or off at predetermined times or in a specified sequence. In microprocessor control, the control loops in a system use 'sequence control' where the completion of one operation signals the controller to start the next operation, with or without a time delay. An example of a simple control sequence is described in Table 27.6 for the operations needed to change the supply from a sugar silo.

Two additional functions of controllers are to monitor a process for faults (self-diagnosis) and to provide management information. When a fault is detected or a required condition is not met, the controller activates electronic interlocks to prevent the process from continuing (e.g. automatic cleaning schedules are prevented until a tank level signal indicates that it is empty). The controller automatically restarts a process when a fault has been corrected. Monitoring also enables the collation and analysis of management information at pre-set intervals (e.g. each shift, each day or month). This information is used to prepare cost analyses or maintenance programmes (e.g. using information on how many hours a machine has operated or a valve has opened/closed since it was last serviced).

Programmable logic controllers (PLCs)
A significant development in process control during the 1980s was the introduction of PLCs. They are based on microcomputers and have the same functions as relays, but with vastly greater flexibility and the power to develop other functions, including recipe storage, data transfer and communications with higher level computers (Marien 1989). PLCs have a fixed program stored in two modes in their memory: the first (teach) mode allows instructions to be programmed into the memory by an operator via a keyboard or touchscreen (or HMI). In the second (run) mode, the program is executed automatically in response to data received from sensors. This is achieved using software building blocks termed 'algorithms', which construct control sequences for each action and allow the operator to program the system. Each algorithm carries out a specific function, and the operator simply defines the series of algorithms and the data on which they are to operate, in response to a series of questions displayed on a screen. A display panel provides information to the operator on the status or progress of the controls in operation and a printed record gives a summary of the processing conditions that have been used. PLCs are highly reliable, relatively low cost and easy to install and use. An important advantage is the ease and speed with which they can be reprogrammed by factory staff who do not have sophisticated computing knowledge. This allows great flexibility in

being able to modify process conditions or change product formulations. PLCs also monitor a process and if a process parameter exceeds a pre-set limit, an alarm is activated to attract the attention of the operator, or the PLC may automatically correct any deviation from the specified limits. Similarly the PLC can constantly monitor the status of equipment and issue an alarm if a malfunction occurs, which greatly speeds up fault tracing and repairs.

Batching and blending

Manufacturers are now required to produce a greater range of products, which requires more frequent changes to product formulations and processing conditions. This is complex and time consuming if performed manually, but is well suited to PLC control. A 'batcher' has a microprocessor which stores information on the types and weights of ingredients for all products made in a factory (e.g. up to 2600 recipes and 30 different mixing cycles). Each formulation is assigned a number and when a number is entered by the operator, the PLC automatically controls the flow of ingredients from storage silos through automatic weighers to mixing vessels. This type of control is widely used in the production of baked goods, snackfoods, confectionery, soft drinks and ice cream. A similar system is used to control raw materials that have a variable composition but produce a final product in which the composition is subject to legislation or trade standards. For example, the alcoholic strength of beer has a pre-set specification, and beer from fermentation tanks that has a higher strength is mixed with deaerated water in the correct proportions. The operator enters the target specific gravity of the product and the PLC receives data from sensors on the specific gravity of the beer and the specific gravity of the diluent. It then calculates the required ratio of the two fluid streams and automatically adjusts one until the measured ratio meets the product specification.

In another application, a PLC determines the least-cost formulation needed to produce a target product specification from different combinations of raw materials. For example, the quality of meat products such as burgers and sausages, and the profitability of their production, are each determined by the fat content of the meat and accurate proportioning of meat and non-meat ingredients. Data on the composition of the raw materials is fed into the PLC, which simulates possible formulations and selects the one that has the lowest cost. PLC control of the formulation also produces the exact lean-to-fat ratio and meat-to-non-meat ratio required in the product, whatever the composition of the raw meat. This system is also used in buying departments to assess the implications of purchasing particular raw materials. Other examples of control by PLCs are described in relation to specific equipment in preceding chapters.

27.2.3 Neural networks, fuzzy logic and robotics

Where complex relationships exist between a measured variable and the process or product, it was previously difficult to automate the process, but developments in 'expert systems' or 'neural networks' have been able to solve such problems. These are able to deduce complex relationships and also to quickly 'learn' from experience. Wallin and May (1995) describe the use of neural networks for a number of applications including automatic control of highly complex extrusion cookers (Chapter 15) by 'intelligent' interpretation of the range of variables in raw materials, processing conditions and final product quality. They are also being used for applications in production line robotics.

Neural networks are applied to biological systems that are too complex to mathematically model because of their inherent non-linearity. Since they can be used to describe non-linear functions, they can be used for example in fermentation processes (Chapter 6, section 6.4), to forecast process values such as substrate concentration on the basis of inputs such as culture volume, pH, dissolved oxygen and substrate feed rate. Before a neural network can be used in process control, the network needs to be 'trained' with historical process data obtained from previous fermentation runs (Andersen 2004). Neural networks are used in large-scale fermentation systems to maintain the quality of the batch and to maximise the yield of product (e.g. Rani and Rao 1999, Linko *et al.* 1997).

Neural networks are also used for crop inspection using a vision system to check fruits and vegetables for defects such as mis-shapes, poor colour, undersize items and foreign bodies (see Chapter 2, section 2.3.3). Other applications include monitoring the correct position and proportion of ingredients on prepared foods such as pizzas; correct positioning of labels on packs, inspection of containers for damage or leakage and 'intelligent' automation of chemical analyses. These systems offer reductions in labour costs and wastage, higher production rates and generation of improved information for more accurate production scheduling and sales forecasting.

Fuzzy logic

Fuzzy logic control mimics how a person would make decisions, and is the equivalent of having computers reason like humans, although much faster. Normally a computer output is 'on/off', 'true or false' or 'yes/no', but fuzzy logic does not need these precise inputs to generate usable outputs. Instead, the output from a microprocessor controller can be based on imprecise inputs (e.g. 'little', 'not so large', 'large', 'larger'). This has strengths over conventional control algorithms because it is empirically based and does not use mathematical models. Fuzzy logic is therefore suited to control of biological systems that are difficult or impossible to model mathematically and would not otherwise be possible to automate; and it can be easily modified and fine-tuned during operation. It can be used in systems ranging from simple micro-controllers to large, networked PC-based data acquisition and control systems.

The essence of fuzzy logic is the use of rules based on simple language such as 'if X = A and Y = B, then Z = C', where inputs (X and Y) are variables (pH, temperature, etc.) and A and B are linguistic variables representing the input values ('little', 'not so large', etc.). Instead of rigid control algorithms in a mathematical model, this small number of rules provide more flexible control. To gather and organise information about the relationship between these inputs and corresponding linguistic output values, a rule matrix is created. By using control words such as 'positive large', 'negative medium' and 'zero' the controller 'knows' what actions to take when a rule is applicable. Terms such as 'IF (process is too cool) AND (process is getting colder) THEN (add heat to the process)'; or 'IF (process is too hot) AND (process is heating rapidly) THEN (cool the process quickly)' are used. The terms are imprecise but descriptive of what actually happens. Fuzzy logic has found particular application in bioprocesses such as microbial fermentations (Chapter 6, section 6.4) (e.g. Park *et al.* 1993). Further information is given by Kaehler (1998) and applications of fuzzy logic in unit operations are described in, for example, Chapter 2 (section 2.2.2), Chapter 6 (section 6.4.2), Chapter 15 (section 15.2.3) and Chapter 22 (section 22.2.2).

Robotics

Advances in neural networks, together with vision systems, pressure-sensitive grippers and laser guidance systems have been applied to robotics. The main components of a robotic machine are the microelectronic processor that uses information from a vision system for product recognition, finding its orientation, and inspection, and servo motors to move arms and grippers. Earlier problems that were associated with maintaining hygienic robotic surfaces and avoiding contamination of foods with grease or lubricants have largely been solved, and the robotics industry foresees many more potential applications in the food industry. Uptake at present (2008) is relatively limited, but production line robots have so far been used for picking and placing items, primary packaging (placing foods into the first layer of packaging) or secondary packaging (inserting the wrapped food into a box, case, carton or tray) (Fig. 27.9), carton erection and palletising, meat deboning or cutting, and cake decoration (Wallin 1994). Pick and place examples include high-speed robotic picking of individual chocolates and loading them into trays. Palletising applications include building mixed load pallets. Robots are also used in bakeries for lidding and de-lidding baking tins, cutting dough, de-panning bread and filling crates (Anon 2008k).

In robotic meat cutting, an infrared scanner is used to make a three-dimensional image of the body of a pig or cow prior to butchering, and software algorithms in the processor determine the precise position of the carcass and where the robotic arm should make the cuts. As the large pieces of meat passes beneath the robot on a conveyor, a vision system using a laser scanner determines the topography of the meat and the controller directs the arm to cut slices from it (Brumson 2007, 2008).

Fig. 27.9   A robotic 'work cell' boxing bread sticks from a moving conveyor (courtesy of Flexicell Inc. at www.flexicell.com).

El Amin (2006) describes a prototype robotic sausage-making line, which is staffed by two workers who feed the line with sausagemeat, casings and packaging. The sausage-meat is automatically filled into casings under aseptic conditions and sausages are passed on a conveyor at up to 200 per minute under a laser light. The light feeds information to a computer on the location of each sausage and the computer controls two crab-like robots, each having three arms. By 'knowing' the location on the conveyor, the robots pick individual sausages and place them five at a time into trays. The trays are film-sealed with a modified atmosphere and are conveyed to a labeller and weigher. Another robot picks the trays four at a time and places them in cartons. When the eight-pack cartons are full, a larger robot places them on a pallet, and when there are 50 cartons in place another robot places the pallet in a shrink-wrapping machine. The wrapped pallet is then conveyed to a delivery vehicle.

27.2.4   Production control

Types of control systems

Developments in hardware and software that enable PLCs and larger computers to be linked together in an integrated control system are described by Teixeira and Shoemaker (1989). In summary, these include the following:

- *Dedicated control systems*, in which PLCs are dedicated to the control of a single unit operation, such as controlling the temperature of a pasteuriser. They do not communicate information to other computers but receive instructions from a central control panel.
- *Centralised control systems* have a mainframe computer located in a centralised control room, which monitors and controls the process in specified zones. For example, a fully automated milk processing plant is described by Anon (1987). The central computer checks the positions of valves, fluid levels, flowrates, densities and temperatures in the plant, and if a fault occurs, it sounds an alarm and produces a printout of the faulty equipment for operators in the control room. Larger systems are able to control 200 plant actuators and monitor 5000 inputs at a rate of 2000 per second. Although centralised control systems have been used for many decades, their major disadvantage is that any failure in the central computer could cause a total plant shutdown, unless an expensive standby computer of equal capacity is available to take over. For this reason, distributed or integrated control systems that do not have this disadvantage are now more commonly installed.
- *Distributed control systems* have each area of a process independently controlled by a PLC, and the PLCs are both linked together (process interlocking), and linked to a central computer via a communications network (Lasslett 1988). Although the capital cost and programming costs are higher than the other systems, these control systems are highly flexible in being able to change processing conditions and do not have the vulnerability to total plant shutdown if one component fails. Further information is given by Anon (2007d), Laduzinsky (1997) and Dahm and Mathur (1990).
- *Integrated control systems* are distributed control systems in different sections of a factory or in different factories that are linked to form a larger management information system. A central computer has mass data storage, sophisticated data manipulation and communications with other management computers. This allows it to be used for other functions, including marketing, quality assurance and plant maintenance, in addition to process control. Each distributed control system is independent and if one component fails, the others maintain full control of their

process areas. An example of an integrated system having three layers of control is given by Selman (1990).

- *Process control systems* are hybrids of PLC and distributed control systems (Williams 1998). They enable complex control combinations to give complete automation using a single highly flexible program, and are able to collect and process data to show the performance of the process plant (e.g. length and causes of stoppages, energy consumption). Factory engineers can use the data to design more effective maintenance programmes and locate faults more quickly, and production staff use data on processing costs and production efficiencies to improve materials handling and scheduling procedures, control stocks, and compare actual production levels with targets.

Developments in the use of computer networks to control manufacturing in different locations via the Internet or via an internal 'Ethernet' (computer networking technologies for local area networks (LANs)) are described by Sierer (2001).

Software developments

One of the most important software developments in the 1980s was the 'Supervisory Control and Data Acquisition' (SCADA) software. This collects data from a PLC that is controlling a process and displays it to plant operators in real-time as animated graphics (Fig. 27.10). Thus, for example, an operator can see a tank filling or a valve change colour as it opens or closes. The graphics are interactive to allow the operator, for example, to start a motor or adjust a process variable. Alarm messages are displayed on the screen when a pre-set condition is exceeded. However, a major limitation of SCADA systems was their inability to analyse trends or recall historical data. This was corrected by new software in the 1990s, based on Microsoft's 'Windows' operating system, which allows data to be transferred between different programs and applications, using a 'dynamic data exchange' (DDE). This enables analysis and reporting of simple trends and historical data using spreadsheets, and is linked in real time to office computer systems. With this software, the office software can be used for real time process control to adjust recipes, schedule batches, produce historical information or management reports. In an example described by Atkinson (1997) this type of system is used in a large ice cream factory to reduce costs by close monitoring and control of the refrigerator's compressors, to display real time running costs and to produce records of refrigeration plant performance.

Developments to enable different software systems to communicate with each other include the 'open database connectivity' (ODBC) standard. These have enabled different information databases to be linked together and be accessed by anyone who is connected to a network. Information such as master recipes, production schedules and plant status can be incorporated into company business systems. Additionally, the development of 'object linking and embedding' for process control (known as OPC) greatly simplifies the linking together of different software. More recently, a system termed 'Common Object Resource Based Architecture' (CORBA) has led to development of manufacturing execution systems (MES). These act as an information broker that not only links PLC process control systems, SCADA systems, OPC and office business systems, but also exchanges information from barcode readers, checkweighing systems, formulation programs, equipment controllers and laboratory information systems. The systems have spreadsheet and Internet programs for producing and accessing information, often linked by Ethernet. They can be used to connect many disparate types of control systems as described by Williams (1998) to 'provide a means to link up the islands of automation

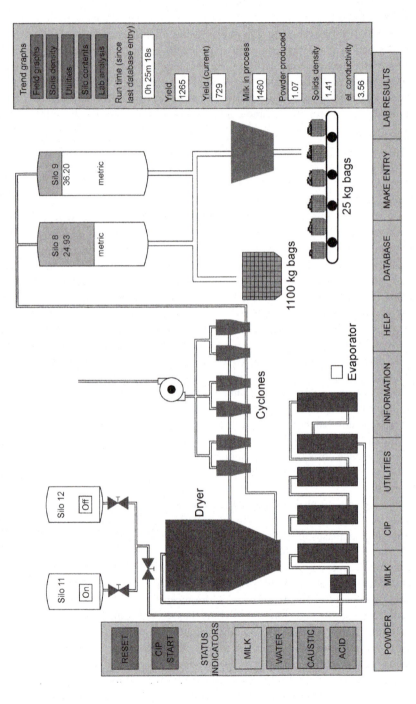

Fig. 27.10 The HMI for the production of dried skim milk (from Cleary *et al.* 1999).

that exist in most food plants and create cohesive systems in which information is freely available'. Bunyan (2003) describes how MES can be used to monitor energy consumption, volumes of steam or process water to determine overall plant efficiency in real time and Hilborne (2006) describes its application to process control.

Managers can use the information to increase productivity, reduce downtime, and track and eliminate wastage. This system can also be used to track ingredients and products throughout the whole manufacturing process and establish a product 'genealogy'. If a problem occurs with a product, the information can be used to quickly back-track through each stage of the process to establish the cause. Where for example, a problem is identified with an ingredient the MES system can use the 'family tree' of product genealogies to establish which products it has been used in, and hence decide the extent of a product recall based on factual information.

Other management systems including enterprise resource planning (ERP) and materials requirement planning (MRP) are described by Anon (2004b,c). MRP systems coordinate decisions on ordering, stock levels, work-in-progress, storage and distribution of finished products. This enables them to calculate the amounts of materials that are needed at a particular time; to generate customer orders and sales forecasts; determine the physical distribution routes for products; and keep records of available inventory. The information can then be used to produce purchase orders for raw materials and work orders for production staff (Johnston *et al.* 1997). The software also monitors production by each department against orders received (either 'customer make-to-order' orders or 'production make-to-inventory' orders). Production is tracked against these orders and variances may be viewed in real time. Operators may view variances on production monitors or on palm computers. As orders are processed, they change status and their colour code on screen changes in real time to identify their current status. If an order has multiple products, the status of each item is also updated in real time.

The concept of MRP has expanded to integrate other parts of the business and has become 'manufacturing resource planning' (MRPII). This is a single integrated system, containing a database that can be accessed by all parts of the company, including engineering departments, sales and marketing departments, finance and accounting departments as well as production managers. Information from sales can therefore be used directly in for example production scheduling, buying and plant maintenance. Aspects of the use of computers in supply chain management are discussed by Bourlakis and Weightman (2004) and Eastham *et al.* (2001), and further information is given by Tersine (1994).

Manufacturing control systems also track the movement of each ingredient and work-in-progress (WIP) during production and into stores, and maintain accurate inventories. The production date and time are recorded for each unit of product, which is used to keep track of the age of products and to ensure first-in-first-out (FIFO) rotation of both final product and WIP inventories. Similar control over ingredients prevents the use of out-of-date materials. This provides an audit trail of all ingredients and products on a daily basis, and collation of this information is used to generate detailed yield reports by department and product. Each case or pallet is labelled with a barcode label or RFID tag and a record is created in a production database. This record has a unique serial number and contains information such as lot number, pallet number, production date, shift, time, individual checkweighing scale identification, gross weight and packaging tare. A barcode label is produced for the case or pallet, which may include any of this information, and is used for all further transactions related to the item. The system also controls placement of products from palletising to storage and picking orders to reduce handling costs. Pallets or individual cases are dispatched using hand-held barcode or RFID scanners. The operator can pick products for orders, scan cases or make up pallet loads against an order and receive product returns. The system stores the composition and weight of each order dispatched, the date and origin of the product (Anon 2008l).

27.3   Logistics and control of storage and distribution

Storage of raw materials, ingredients and products can take place under ambient conditions or under controlled conditions of temperature, humidity or atmospheric composition. Storage of chilled and frozen foods is described in Chapters 21 and 22 respectively, and controlled or modified atmosphere storage is described in Chapter 25 (section 25.3). In this section, storage under ambient conditions is described for a range of representative foods.

In general, manufacturers reduce the amount of stored ingredients and products to a minimum for the following reasons: (1) financial – large amounts of stored materials adversely affect the cashflow of a company because money is tied up in materials that have been paid for, or in final products that have incurred the costs of production; (2) loss of quality – chemical or biochemical changes to foods and deterioration of some types of packaging materials may occur during storage which reduce their quality and value, or render them unusable; (3) risk of pilferage for some high-value products; and (4) high cost of warehousing space. However, because of the seasonality of supply of some raw materials and, for some products a seasonal demand, it is necessary for processors to maintain stocks of ingredients, packaging materials and final products. The JIT methodologies of materials supply that are found in some other industries (e.g. Johnston et al. 1997) are less common in the food manufacturing sector, although they are increasingly used by retailers.

Stored goods (or 'inventory') may be grouped into raw materials, work-in-progress and finished goods. However, they can be categorised more usefully by their role in the production system (Johnston et al. 1997) as follows:

- Buffer (or safety) inventory, to compensate for uncertainties in supply or demand.
- Cycle inventory. This occurs because a processor chooses to produce in batches that are greater than the immediate demand.
- Anticipation inventory. This is created where seasonal demand or supply fluctuations are significant but predictable. It is used especially for supply of seasonal fruits and vegetables, or for products that have a specific seasonal demand (for example, Easter eggs, Christmas cakes).
- Pipeline (or in-transit) inventory, for materials that are in the process of being moved from a point of supply to a point of demand.

Decisions on the size of different inventories depend on the balance between two sets of costs: the cost of buying and the cost of storage. One way of controlling inventory costs is to rank individual materials by their usage value (their rate of usage multiplied by their individual value) into three classes (Johnston et al. 1997):

1 Class A – the 20% of high-value materials that account for 80% of the total usage value.
2 Class B – the next 30% of medium value materials that account for 10% of the usage value.

3   Class C – the lowest value materials that are stocked, comprising 50% of the total, that account for 10% of the usage value.

Class A products are then given inventory preference over Class B and in turn over Class C.

27.3.1   Warehousing operation

Storerooms and warehouses should be constructed to prevent access by rodents, insects and birds, and inspected on a regular basis to ensure that preventative measures are effective. Details of materials used for construction of storerooms are given by Brennan *et al.* (1990). In summary, windows are screened against insects; doors are fitted with screens or air curtains and rooms are equipped with electric insect killers; drainage channels and power cable ducting are fitted with devices to prevent entry by rodents; ceilings and walls are designed to prevent insects, rodents and birds from gaining entry; and floors are covered with vinyl-based coatings to prevent cracks that could harbour insects and micro-organisms.

For ambient temperature storage, the storeroom should be cool with a good ventilation to maintain a flow of air. Fresh foods are only stored for short periods, but the storeroom temperature should be low and humidity sufficiently high to prevent wilting or drying out (see Chapter 21, section 21.2.3). Ingredients such as sugar, salt and powdered flavourings or colourants are hygroscopic. Where this is likely to result in loss of quality or function, the humidity in a storeroom should be controlled to equal the equilibrium relative humidity (ERH) of the stored product (see Chapter 1, section 1.1.2). For example, white sugar which has an ERH of 60% will form a cake if it picks up moisture, and it is therefore stored below 60% humidity. Fats and oils are particularly susceptible to odour pick-up, and spices are likely to contaminate other ingredients with their odour. Both types of ingredient are therefore stored separately from other foods.

Packages which are grouped into larger (or 'unitised') loads require less handling when they are moved through storage and distribution networks. Pallets are commonly used to move unitised loads of cases or sacks by fork-lift or stacker trucks. Fibreboard slipsheets reduce the volume occupied by pallets in vehicles and warehouses. Products are secured to the pallet or slipsheet by shrink-film or stretch-film (Chapter 25, section 25.2.4, and Chapter 26, section 26.3). Sacks or cartons of food are stored on pallets or racks, with space to clean behind the stack. Daily cleaning routines are used as part of a HACCP plan (Chapter 1, section 1.5.1) to prevent dust or spilled food from accumulating which would encourage insects or rodents. They should be stacked to the recommended height for each material to prevent crushing of packs, or collapse and injury to operators. Lighting should be bright and at a high level to reduce shadowing caused by stacked pallets.

Warehouse management systems are used to monitor material movements into and out of the stores, check stock levels, stock rotation, use of materials in the process and the destinations for delivery of products. Developments in software and expanded use of RFID tagging of containers and/or palletised loads have enabled significant improvements in warehouse management. Automation of the entire purchase process includes receipt of loads, stacking in warehouses, assigning, picking and shipping of foods to and from each warehouse, with each load having information on vendors to improve product tracking. Large warehouses use computerised truck-routeing systems, which store information on stock levels, their location in a warehouse and the warehouse layout. Automated guided vehicles (AGVs) follow fixed routes guided either by wires buried in the warehouse floor or coloured lines painted on the floor. These are being replaced by 'free-path' AGVs in which a computer assigns an optimum route for each vehicle. Packaged goods are palletised and each pack and pallet is coded with a barcode or RFID tag that is read by a microprocessor. The coded stock is allocated a storage location by the computer, which compiles both a map of the warehouse and current stock levels in its memory. The progress of each AGV in retrieving or replacing stock is monitored and controlled using information transmitted by an odometer in the vehicle and by barcode directions that are displayed throughout the warehouse, which are read by a laser mounted on the truck. Different designs of AGVs include outrigger, reach, counter-balance, tug/tow, conveyor deck, lift deck and narrow aisle machines (Anon 2008m) and developments in robotic handling and picking in warehouses and other areas of food processing are described by Murphy (1997a,b).

27.3.2 Distribution logistics

The links between harvesting and production of a processed food and purchase by the customer is known as the 'distribution chain' and the different systems involved in distribution are termed 'logistics'. The main factors in an efficient distribution chain are to provide the customer with products at the right place, at the right time and in the right amount; to reduce the cost to a minimum (distribution is an expense but does not add value to a product) and to maintain the product quality throughout the distribution chain. For example, a case study that describes the handling and distribution of peas from harvest to sale of frozen product is described by Chambers and Helander (1997).

Consumer pressure for a greater variety of foods that have higher quality, freshness and year-round availability has resulted in a substantial increase in the volume and range of foods that are handled by the major food retailers. The development of information and communication technologies has also been responsible for the introduction of global production and distribution since the 1980s. This coincided with a sharp increase in competitiveness among large food retail businesses in the main industrialised countries, and spurred a rapid development of global value chains (GVCs). Growth in the buying power of major retail companies, particularly in Europe and the USA, enabled them to drive down prices paid to food processors. As a result, transnational processing companies (TNCs) have adopted a series of strategies to increase their competitiveness, including mergers and acquisitions with food manufacturers in other countries. TNCs have established tightly integrated global-scale systems in which operations take place in widely separated locations. Developments in technology, particularly microelectronic controls and automation in food processing, have assisted globalisation and reduced the need for highly skilled, highly paid workforces. This makes it possible for companies to move their operations to new countries, often in the developing world, where unskilled and lower-paid workers can be employed. These developments enable food production to be coordinated between distant sites, and suppliers can now be called upon to transfer goods across the world at short notice (Shepherd et al. 2007).

There are also higher legal standards for storage and transportation of foods, especially temperature control of chilled and frozen foods in some countries (Chapter 21, section 21.2.4) and shorter lead times for deliveries. These developments have caused retailers to change their methods of storage and distribution. Previously, products from a food manufacturer were transported to a relatively large number of small distribution depots that each handled a single product. Delivery volumes were low and it was not economic to deliver every day. In addition, foods that required temperature-controlled transport had to be carried on separate vehicles, some of which were owned by contractors who operated their own distribution policies and delivery schedules. Each of these aspects increased the cost of distribution and reduced both product quality and distribution efficiency. During the 1980s and 1990s, retailers began to use mathematical models and simulations to improve the logistics of food supply to reduce costs and distribution times by:

- combining distribution streams of various suppliers;
- combining transport of fresh, chilled and frozen foods;
- changing the method and frequency of ordering; and
- redesigning and reorganising warehouses (Koster undated).

These developments resulted in a smaller number of large regional (or 'composite') distribution centres (RDCs) that can handle a wide range of products. A composite depot may cover $23\,000\,m^2$ and can typically handle more than 30 million cases of food per year and serve up to 50 retail outlets. The depot is divided into five temperature zones (ambient, semi-ambient ($+10\,°C$), chill ($+5\,°C$), chilled ($0\,°C$) and frozen ($-25\,°C$)) to handle the range of short- and long shelf-life products found in most large stores (Harrison 1997). Delivery vehicles use insulated trailers that are fitted with movable bulkheads and refrigeration units to create three different temperature zones.

Primary distribution is from the manufacturer to RDCs and secondary distribution is from the RDCs to retail stores. Short shelf-life products are received into distribution depots during the afternoon and evening, and are delivered to retail stores before trading starts the next day (termed the 'first wave' delivery). Longer shelf-life and ambient products are taken from stock and formed into orders for each retail store over a 24 hour period, and are delivered in a 'second wave' between 8 am and 8 pm at scheduled times that are agreed with each store. Many larger retailers now use electronic data interchange (EDI) to automatically order replacement products, directly in response to consumer purchases (also Chapter 25, section 25.5.1). This results in more frequent deliveries of smaller amounts of product, in order to minimise stock levels in stores. Further details are given by Winston (2000), Krcmar et al. (1995) and O'Callaghan and Turner (1995). However, these developments have caused a significant increase in distribution and handling costs for processors. Many smaller- and medium-scale processors now cooperate in logistics, often subcontracting logistics to a third part logistics contractor to gain cost savings and more efficient distribution from the larger volumes that are handled (Webster 2001). Cooperation and sharing of costs enable processors to invest in automatic order picking systems that would be unaffordable for individual companies (Koster undated). The introduction of 'factory gate pricing' by retailers, where the price for foods does not include transport costs (i.e. these are borne by retailers) has enabled retailers or contract hauliers to improve logistics efficiency by 'backhauling': that is, vehicles may be full when delivering from RDCs to stores, but empty on return journeys, giving a 50% efficiency; whereas if retailers use the vehicles to collect goods from manufacturers on their return to RDCs, the overheads are reduced and efficiency is increased (Webster 2002).

References

AMRANI, H., (2005), Non invasive process monitoring using an intelligent pipe, company information from Kaiku Ltd, available at www.strath.ac.uk/Other/cpact/presentations/2005/pdfs/T07_Amrani.pdf

ANDERSEN, M.H. (2004), Fermenter control program, by FoxyLogic, available at www.foxylogic.com/.

ANON, (1987), A fresh approach, *Food Processing*, November, 26, 27.

ANON, (1996), Cutting costs by reducing waste, Envirowise, available at www.envirowise.gov.uk/ EN343.

ANON, (1998), *Food and Drink Good Manufacturing Practice*, 4th edn, Institute of Food Science and Technology, London, available at www.ifst.org and follow link to publications.

ANON, (1993), Sensors in the food industry, in (A. Turner, Ed.), *Food Technology International Europe*, Sterling Publications International, London, pp. 107–110.

ANON, (2001a), Current Good Manufacturing Practices (cGMP's) 21 CFR 110, Food and Drug Administration (FDA) Regulations, US Department of Agriculture Food Safety and Inspection Service, available at www.fsis.usda.gov/OPPDE/rdad/FRPubs/00-014R/FDA-GMPRegs.htm.

ANON, (2001b), Food Information Sheet No. 24, Health and Safety Executive of the UK Government, available at www.hse.gov.uk/pubns/fis24.pdf.

ANON, (2001c), Effluent treatment, *Food Processing*, October, 32.

ANON, (2002a), Fouling, cleaning and disinfection in food processing, proceedings of a conference at Jesus College, Cambridge, UK, 3–5 April, Department of Chemical Engineering, University of Cambridge, Cambridge.

ANON, (2002b), Peristaltic answers, *Food Processing*, June, 30.

ANON, (2002c), Disposing with waste problems, *Food Processing*, February, 31.

ANON, (2004a), Recipe system aids Irn-Bru, *Food Processing*, July, 22.

ANON, (2004b) Saving £300k – not pie in the sky, *Food Processing*, March, 23.

ANON, (2004c), Ready to respond to demand, *Food Processing*, April, 25.

ANON, (2007a), Guidelines on the Application of Directive 94/9/EC of 23 March 1994 on equipment and protective systems intended for use in potentially explosive atmospheres (ATEX), European Union, available at http://ec.europa.eu/enterprise/atex/guide/index.htm.

ANON (2007b), TNA's ROFLO® offers unique gateless distribution and accumulation system, company information from TNA, available at www.tnasolutions.com/ index.php?op=distribution&lang=en.

ANON, (2007c), Mogden formula, in Tracking Water Use to Cut Costs, Envirowise Good Practice Guide GG152R and calculation tool available at www.envirowise.gov.uk/ mogdenformulatool.

ANON, (2007d), PLC brings redundant options to food processing, News Release from Mitsubishi Electric Automation Systems, Edited by the Engineeringtalk Editorial Team on 6 March, available at www.engineeringtalk.com/news/mib/mib234.html.

ANON (2008a) Best practice guidelines by the European Hygienic Equipment Design Group, available at www.ehedg.org/guidelines.htm and follow links to relevant guidelines.

ANON, (2008b), Next generation interfaces for food processing equipment, company information from Holo Touch Inc., in Food Manufacturing, available at www.foodmanufacturing.com/ scripts/ShowPR~RID~4037.asp.

ANON, (2008c), SHAPA Product Selector, information from the Solids Handling and Processing Association, available at www.shapa.co.uk/handling-processing-products.php with links to different types of equipment.

ANON, (2008d), HabaFLOW conveyor belts, company information from Habasit AG., available at www.habasit.com/link.asp?FI=2791.html.

ANON, (2008e), Products, company information from European Conveyor Systems, available at www.europeanconveyors.co.uk/products/index.php.

ANON, (2008f), Young Industries' provides rugged and reliable multi-phase pneumatic conveyors, company information from Young Industries, available at www.younginds.com/PDF_files/ ProductBulletins/201-200.pdf.

ANON, (2008g), HyGUARD® Europe antimicrobial conveyor and processing belts for the food industry, company information available from Habasit at www.habasit.com/1976.html or www.habasit.com/, and search 'HyGuard'.

ANON, (2008h), A moveable feast, *Food Processing*, **77** (1), 32–33.

ANON, (2008i), Control valves, company information from Spirax-Sarco Limited, available at www.spiraxsarco.com/resources/steam-engineering-tutorials/control-hardware-el-pn-actuation/control-valves.asp#head38.

ANON, (2008j), Setting up an environmental management system in the food and drink industry, Envirowise publication GG344, available at www.envirowise.gov.uk/GG344 and publication GG374 environmental management accounting at www.envirowise.gov.uk/GG374.

ANON, (2008k), Bakery automation systems, lamination and dough processing equipment, bread pans, trays and baking forms, company information from Capway Systems, available at www.foodprocessing-technology.com/contractors/materials/capway/.

ANON, (2008l), CAT^2 Software for the food industry – primary processing tool series, company information from CAT^2 available at www.catsquared.com/pptsmodules.htm.

ANON, (2008m), Automatic guided vehicle applications in the food industry, company information from FMC Technologies, available at www.fmcsgvs.com/content/sales/food.htm.

ATKINSON, M., (1997), Following the 'open' road, *Food Processing* (supplement), July, pp. S11–12.

BARLETT, P.N., ELLIOTT, J.M. and GARDNER, J.W., (1997), Electronic noses and their application in the food industry, *Food Technology*, **51** (12), 44–48.

BOURLAKIS, M. and WEIGHTMAN, P., (Eds.), (2004), *Food Supply Chain Management*, Blackwell Publishing, Oxford.

BOWSER, T.J., (undated), Sanitary pump selection, Food Technology Factsheet FAPC-108, Oklahoma Cooperative Extension Service, Division of Agricultural Sciences and Natural Resources, Oklahoma State University, available at www.fapc.okstate.edu/files/factsheets/ and select factsheet 'FAPC-108'.

BRENNAN, J.G., BUTTERS, J.R, COWELL, N.D. and LILLEY, A.E.V., (1990), *Food Engineering Operations*, 3rd edn, Elsevier Applied Science, London, pp. 561–616.

BRUMSON, B., (2007), Food Robotics, available at *Robotics Online*, 19 March at www.robotics online.com/public/articles/details.cfm?id=2829

BRUMSON, B., (2008), Food for thought: robotics in the food industry, available at www.robotics.org/

BUNYAN, P., (2003), How to make MES work, *Food Processing*, Sept., 13.

CHAMBERS, S. and HELANDER, T., (1997), Peas, in (R. Johnson, S. Chambers, C. Harland, A. Harrison and N. Slack, Eds.), *Cases in Operations Management*, Pitman Publishing, London, pp. 310–323.

CLARK, R., (2000), Pouring money down the drain?, *Food Processing*, November, 42.

CLEARY, F., MCGRATH, M. J. GAISER, M., O'CONNOR, J. F. EVERARD, M. and HOLLAND, J., (1999), Development of a low cost data acquisition system for milk powder production line monitoring, *J. Dairy Science*, **82** (9), 2039–2048.

COVE, N., (2006), Treating organically loaded wastewater, *Food Processing*, **75** (11), 14.

DAHM, M. and MATHUR, A., (1990), Automation in the food processing industry: distributed control systems, *Food Control*, **1** (1), 32–35.

DALZELL, J.M., (1994), *Food Industry and the Environment – Practical Issues and Cost Implications*, Blackie Academic and Professional, London.

DILLON, M. and GRIFFITH, C., (1999), *How to Clean – A Management Guide*, MD Associates, Grimsby, UK.

EASTHAM, J., SHARPLES, L. and BALL, S., (Eds.), (2001), *Food Supply Chain Management*, Butterworth Heinemann, Oxford.

EL AMIN, A., (2006), Robotics: the future of food porcessing, Food Production Daily.com, 5 April, available at www.foodproductiondaily.com/news/ng.asp?n=66874-k-robotix-robotics-anuga

FELLOWS, P.J., (1997), *Guidelines for Small-scale Fruit and Vegetable Processors*, FAO Agricultural Services Bulletin 127, Food and Agriculture Organization, Rome.

FLORES, R.A., ARAMOUNI, F.M. and POWELL, G.M., (1994), Water supply for food and beverage processing operations, paper MF-1122, Department of Biological and Agricultural Engineering, Cooperative Extension Service, Kansas State University, available at www.oznet.ksu.edu/library/, search 'MF-1122'.

GIBSON, H., TAYLOR, J.H., HALL, K.E. and HOLAH, J.T., (1999), Effectiveness of cleaning techniques used in the food industry in terms of removal of bacterial biofilms, *J. Applied Microbiology*, **87**, 41–48.

GUTHRIE, R.K., (1988), *Food Sanitation*, 3rd edn, AVI, Westport, CT, pp. 157–166.

HAMILTON, B., (1985), The sensor scene, *Food Manufacture*, September, **39**, 41–42.

HAMMELSVANG, L., (1999), Bringing displacement pumps into the next millennium, *Food Technology International*, pp. 55–56; *World Pumps*, **398**, November, 28–30 (3).

HARPER, W.J. and SPILLAN, M., (undated), Cleaning compounds: characteristics and functions, Department of Food Science and Technology, The Ohio State University, available at http://class.fst.ohio-state.edu/FST401/Information/Cleaning%20and%20Sanitizing.doc.

HARRISON, A., (1997), Tesco composites, in (R. Johnson, S. Chambers, C. Harland, A. Harrison and N. Slack, Eds.), *Cases in Operations Management*, Pitman Publishing, London, pp. 359–367.

HAYWARD, T., (2003), Award winning hygienic control devices and indicator lights, *Food Processing*, September, 6–7.

HILBORNE, M., (2006), Take control, *Food Processing*, **75** (5), 20.

HOGHOLT, S., (1998), Centrifugal pump technology in the food industry, in (A. Turner, Ed.), *Food Technology International*, Sterling Publications, London, pp. 59–62.

HOLLINGSWORTH, P., (1994), Computerised manufacturing trims production costs, *Food Technology*, **48** (12), 43–45.

HOLM, F., (2003), Food quality sensors, a component of Flair-Flow 4, the European Commission, available at www.nutrition.org.uk/upload/FF4foodqualitysensors.pdf.

HOLMES, M., (2000), Optimising the water supply chain, *Food Processing*, November, 38, 40.

JENNINGS, B., (1998), Non-contact sensing, *Food Processing*, November, 9–10.

JOHNSON, D., (2001), A ready pump for ready meals, *Food Processing*, June, 21.

JOHNSTON, K., (2006), Sensors and sensing, *Food Processing*, **75** (12), Faraday Winter supplement, pp. 4–5.

JOHNSTON, R., CHAMBERS, S., HARLAND, C., HARRISON, A. and SLACK, N., (1997), *Introduction to Planning and Control. Cases in Operations Management*, Pitman Publishing, London, pp. 203–222.

KAEHLER, S.D., (1998), Fuzzy logic tutorial, in *Encoder – Newsletter of Seattle Robotics Society*, available at www.seattlerobotics.org/encoder/mar98/fuz/fl_part1.html#WHAT%20IS%20FUZZY%20LOGIC?.

KIM, S-M., CHEN, P., MCCARTHY, M.J. and ZION, B., (1999), Fruit internal quality evaluation using on-line nuclear magnetic resonance sensors, *J. Agricultural Engineering Research*, **74** (3), 293–301.

KOSTER, A.C., (undated). 1) Logistic redesign. 2) Joint Distribution. Briefing papers from Agrotechnological research Institute ATO-DLO, PO Box 17, 6700 AA Wageningen, Netherlands.

KRCMAR, H., BJORN-ANDERSEN, N. and O'CALLAGHAN, R., (1995), *EDI in Europe – How it Works in Practice*, John Wiley and Sons, Chichester.

KRESS-ROGERS, E., (1993), Instrumentation for food quality assurance, in (E. Kress-Rogers (Ed.), *Instrumentation and Sensors for the Food Industry*, Woodhead Publishing, Cambridge, pp. 1–36.

KRESS-ROGERS, E., (Ed.), (2001), *Instrumentation and Sensors for the Food Industry*, 2nd edn, Woodhead Publishing, Cambridge.

LADUZINSKY, A.J. (1997), Distributed control marches modular – distributed control systems, Prepared Foods, Nov., available at http://findarticles.com/p/articles/mi_m3289/is_n12_v166/ai_20216374.

LASSLETT, T., (1988), Computer control in food processing, in (A. Turner, Ed.), *Food Technology International Europe*, Sterling Publications International, London, pp. 105–106.

LEE, D., (2000), Selecting progressing cavity pumps, *Food Processing*, June, 51.

LINKO, S., LUOPA, J. and ZHU, Y.-H., (1997), Neural networks as software sensors in enzyme production, *J. Biotechnology*, **52**, 257–266.

MAHAPATRA, A.K., MUTHUKUMARAPPAN, K. and JULSON, J.L., (2005), Applications of ozone, bacteriocins and irradiation in food processing: a review, *Critical Reviews in Food Science and Nutrition*, **45**, 447–461.

MARIEN, M., (1989), Automation to meet the food industry's needs, in (A. Turner, Ed.), *Food Technology International Europe*, Sterling Publications International, London, pp. 127–133.

MARRIOTT, N.G. and GRAVANI, R.B., (2006), *Principles of Food Sanitation*, 5th edn, Springer Science, New York.

MCFARLANE, I., (1988), Advances in sensors that benefit food manufacture, in (A. Turner, Ed.), *Food Technology International Europe*, Sterling Publications International, London, pp. 109–113.

MCFARLANE, I., (1991), The need for sensors in food process control, in (A. Turner, Ed.), *Food Technology International Europe*, Sterling Publications International, London, pp. 119–122.

MCTIGUE PIERCE, L., (2005), Conveyors: the missing link to a smooth-running line: new levels of flexibility, easy-to-clean/sanitize designs and better package handling are among the advances in conveyor technology today, *Food & Drug Packaging*, 1 October, available at http://findarticles.com/p/articles/mi_m0UQX/is_/ai_n15861206.

MEDLOCK, R.S. and FURNESS, R.A., (1990), Mass flow measurement, *Measurement and Control*, **23** (5), 100–113.

MURPHY, A., (1997a), The future for robotics in food processing, *European Food and Drink Review*, Spring, 31–35.

MURPHY, A., (1997b), Robotics in food processing – towards 2001, paper presented at EFFoST conference on Minimal Processing of Food: A Challenge for Quality and Safety, 6–9 November, Cologne, Germany.

NIRANJAN, K., AHROMRIT, A. and KHARE, A.S., (2006), Process control in food processing, in J.G. Brennan (Ed.), *Food Processing Handbook*, Wiley-VCH, Weinheim, pp. 373–384.

O'CALLAGHAN, R. and TURNER, J.A., (1995), Electronic data interchange – concepts and issues, in (H. Krcmar, N. Bjorn-Andersen, and R. O'Callaghan, Eds.), *EDI in Europe – How it Works in Practice*, John Wiley and Sons, Chichester, pp. 1–20.

PARK, Y.S., SHI, Z.P., SHIBA, S., CHANTAL, C., IIJIMA, S. and KOBAYASHI, T., (1993), Application of fuzzy reasoning to control glucose and ethanol concentrations in baker's yeast culture, *Applied Microbiology and Biotechnology*, **38**, 649–655.

PEHANICH, M., (2006), Cleaning without chemicals, *Food Processing*, available at www.foodprocessing.com/articles/2006/052.html?page=1.

PERERA, C.O. and RAHMAN, M.S., (1997), Can clever conveyors become more intelligent? *Trends in Food Science and Technology*, **8** (3), 75–79.

POTTER, N.N., (1986), *Food Science*, 4th edn, AVI, Westport, CT.

PYLE, D.L. and ZAROV, C.A., (1997), Process control, in (P.J. Fryer, D.L. Pyle and C.D. Rielly, Eds.), *Chemical Engineering for the Food Industry*, Blackie Academic and Professional, London, pp. 250–294.

QUARTLY-WATSON, T., (1998), The importance of ultrasound in cleaning and disinfection in the poultry industry – a case study, in (M.J.W. Povey and T.J. Mason, Eds.), *Ultrasound in Food Processing*, Blackie Academic and Professional, London, pp. 144–150.

RAND, G.A., YE, J., BROWN, C.W. and LETCHER, S.V., (2002), Optical biosensors for food pathogen detection, *Food Technology*, **56** (3), 32–39.

RANI, Y.K. and RAO, R.V.S., (1999), Control of fermenters – a review, *Bioprocess Engineering*, **21**, 77–88.

RITCHIE, F.S., (2004), Don't just fit and forget, *Food Processing*, July, 19–20.

ROBERTS, R.T. and WILTSHIRE, M.P., (1990), Sensing with sound waves, in (A. Turner, Ed.), *Food Technology International Europe*, Sterling Publications International, London, pp. 109–115.

ROHN, S. and KROH, L.W., (2005), Electron spin resonance – a spectroscopic method for determining antioxidative activity, *Molecular Nutrition and Food Research*, **49** (10), 898–907.

SCHMIDT, R.H. and ERICKSON, D.J., (2005), Sanitary design and construction of food equipment, University of Florida, Institute of Food and Agricultural Sciences, available at http://edis.ifas.ufl.edu/FS119.

SCHULER, G.A., NOLAN, M.P., REYNOLDS, A.E. and HURST, W.C., (2005), Cleaning, sanitizing, and pest control in food processing, storage and service areas, extension food science, The Cooperative Extension Service, the University of Georgia College of Agricultural and Environmental Sciences, available at http://pubs.caes.uga.edu/caespubs/pubcd/b927-w.html.

SELMAN, J.D., (1990), Process monitoring and control on-line in the food industry, *Food Control*, January, 36–39.

SHAFER, S.M. and MEREDITH, J.R., (1998), *Operations Management*, John Wiley & Sons, New York.

SHARP, G., (1998), The conveyor collection, *Food Processing*, **67** (9), 28–29.

SHEPHERD, A., MEMEDOVIC, O. and FELLOWS, P.J., (2007), *Agrifood Value Chains and Poverty Reduction*, UNIDO Publications, Vienna.

SIERER, B., (2001), Share technical data with network technologies, *Food Processing*, March, 38–39.

SKJOLDEBRAND, C. and SCOTT, M., (1993), On-line measurement using NIR spectroscopy, in (A. Turner, Ed.), *Food Technology International Europe*, Sterling Publications International, London, pp. 115–117.

SMITH, R., KLEMES, J. and KIM, J-K., (2008), *Handbook of Water and Energy Management in Food Processing*, Woodhead Publishing, Cambridge.

STEFFE, J.F. and SINGH, R.P., (1997), Pipeline design calculations for Newtonian and non-Newtonian fluids, in (K.J. Valentas, E. Rotstein, and R. Singh, Eds.), *Handbook of Food Engineering Practice*, CRC Press, New York, pp. 1–36.

TAN, T., SCHMITT, V. and ISZ, S., (2001), Electronic tongue: a new dimension in sensory analysis, *Food Technology*, **55** (10), 44–50.

TEBBUTT, T.H.Y., (1992), *Principles of Water Quality Control*, 4th edn, Pergamon Press, Oxford.

TEIXEIRA, A.A. and SHOEMAKER, C.F., (1989), Computer control in the food processing plant, in *Computerized Food Processing Operations*, pp. 51–100, and *On-line Control of Unit Operations*, pp. 101–134, AVI, New York.

TERSINE, R.J., (1994), *Principles of Inventory and Materials Management*, 4th edn, Prentice Hall, London.

WALDRON, K. (Ed.), (2007), *Handbook of Waste Management and Co-Product Recovery in Food Processing*, Vol. 1, Woodhead Publishing, Cambridge.

WALLIN, P.J. (1994), Robotics and computers in the food industry, in (A. Turner, Ed.), *Food Technology International Europe*, Sterling Publications International, London, pp. 89–94.

WALLIN, P. and MAY, J., (1995), The developing appetite for neural computing, *Food Processing Hygiene Supplement*, October, XII.

WEBSTER, J., (2001), The logistics of food delivery, *Food Processing*, December, 20.

WEBSTER, J., (2002), The food industry and FGP, *Food Processing*, December, 27–28.

WHEATLEY, A.D., (1994), Water pollution in the food industry: sources, control and cost implications, in (J.M. Dalzell, Ed.), *Food Industry and the Environment – Practical Issues and Cost Implications*, Blackie Academic and Professional, London, pp. 137–258.

WHEELDON, J., (2002), Weighing up feeders, *Food Processing*, March, 49.

WILBEY, R.A., (2006), Water and waste treatment, in (J.G. Brennan, Ed.), *Food Processing Handbook*, Wiley-VCH, Weinheim, pp. 399–428.

WILLIAMS, A. (1998), Process control – the way ahead, *Food Processing Supplement*, September, S4–S5.

WILSON, I., (Ed.), (2002), Fouling, cleaning and disinfection, *Food and Bioproducts Processing*, **80** Special issue: C4, 221–339.

WINSTON, J., (2000), Driving down delivery costs, *Food Processing*, December, 18.

Part VI

Appendices

Appendix A.1

Selection of hydrocolloids

Hydrocolloid	Source	pH range	Solubility at 21°C	65°C	Effect of ion/salt	Gel character	Acid stability	Functionality	Applications
Alginates	Seaweed	3.5–10	Yes	Yes	Ca^{++} required for gel	Rigid/cutable/cohesive/thixotropic	Gels at pH 3.5 with calcium present	Forms non-melting gels and thickening, secondary emulsifier in beverages and in foam stability	Desserts, pie fillings, ice cream, beverage emulsions, confectionary gels, French dressing, meat sauces, syrups, toppings, purées, frozen fruit, icings, gravies, soups
Carrageenan	Red seaweed								
Kappa		4–10	No	Yes	K^+ gels	Rigid/cutable Cohesive/thixotropic	Solutions undergo hydrolysis at pH 3.5; gel is stable	Stabiliser, gelation, improves water-holding capacity of meat emulsions, freeze/thaw stable cream, syneresis control, thickens	Ice cream, evaporated milk, meats, desserts, bakery products, freeze–thaw stable whipped cream, beverages, puddings
Iota		4–10	No	Yes	C^{++} gels				
Lambda		4–10	Yes	Yes	None	None (thickener)			
Carboxyl methyl cellulose (CMC)	Derived from cellulose	4–10	Yes	Yes	Decreases viscosity	None	Stable at pH 7–9 < viscosity at pH 5 precipitates in milk (<pH 3; > 6)	Thickens, suspends, film former	Retards ice crystal growth in frozen desserts, thickener and mouthfeel improver in sauces/spreads, lubricant in extruded products, batter thickener and humectant in cake mixes, moisture binder and retarder of ice crystallisation/syneresis in icings, syrup thickener

Gelatin	Bones	4.5–10	Yes	Yes	None	Cutable/elastic	Stable	Gelation	Dessert gels, capsule
Gellan gum (tetrasaccharide – 2 glucose + glucuronic acid and rhamnose)	Fermentation product of *Pseudomonas elodea*	1–13	No	Yes	Required for gel	Rigid/cutable/cohesive	Stable in acid conditions	Gelation, fluid gels, suspension	Confections, fruit jellies, jams, non-standard jams, fillings, beverages
Guar gum	Guar seed	4–10	Yes	Yes	None	None	Gradual decline at pH 3.5–10	Thickens, improves mouthfeel, softens texture produced by carrageenan, prevents ice crystal growth to give freeze–thaw stability, binds water	Sauces, dairy products, dressings, ice cream, bakery products
Gum arabic	Acacia tree	2–10	Yes	Yes	None	None	Stable	Emulsifier and emulsion stabiliser, low viscosity at high concentrations	Beverages, prevents sucrose crystallisation and emulsifies fats in sugar confectionery, used in flavour powders and emulsions
Locust bean gum	Locust bean (carob seed)	4–10	No	Yes	None	None (gels with xanthan gum)	Stable at pH 5–8	Rigid gels with xanthan and carrageenan, rarely used alone, moisture control, heat shock resistance	Ice cream – smooth melting and texture improvement, cream cheese
Micro-crystalline cellulose (MCC)	Derived from cellulose	3–10	Yes	Yes	None	None	Insoluble	Stabilises foams and emulsions	Reduced fat products, prepared foods, sauces

Hydrocolloid	Source	pH range	Solubility at 21°C	65°C	Effect of ion/salt	Gel character	Acid stability	Functionality	Applications
Methyl cellulose (MC) and hydroxypropyl methylcellulose (HPMC)	Derived from cellulose	3–10	Yes	No	None	Rigid/cutable	Stable in acid conditions	MC provides fat-like characteristics, reduces fat absorption in fried foods, imparts creaminess, gas retention and moisture control in bakery products to impart tenderness and increase shelf-life. HPMC provides foam stability, freeze–thaw stability, prevents phase separation	Bakery products, whipped toppings, sauces, beverages, extruded foods, vegetarian burgers
Pectins	Citrus peel, apple pomace	2–7	No	Yes	HM[a]; none LM[a]; Ca^{++} gels	Rigid/cutable Cohesive/ thixotropic	Stable gels formed in presence of sugar and acid or with Ca^{++}	Gelation, thickening, stabilisation, prevents sucrose crystallisation	Jelly, jams, marmalades, beverages, acid milk drinks, confectionery
Xanthan	Fermentation by *Xanthomonas campestris* bacteria,	1–13	No	Yes	None	None (gels with locust bean gum)	Precipitation in milk (pH < 4.5)	Pseudoplastic, high-viscosity solutions, thickener/emulsion & suspension stabiliser, freeze–thaw stability, syneresis control	Bakery products, dressings, sauces, frozen foods, beverages

[a] High methoxyl and low methoxyl pectins.

Adapted from:

ANON (2005), Hydrocolloids, company information from Univar USA Inc., available at www.univarcanada.com/pdfdoc/food/Texture%20Modification_Sept05.pdf page 2

BEMILLER, J.N. and WHISTLER, R.L., (1996), Carbohydrates, in (O.R. Fennema, Ed.), *Food Chemistry*, 3rd edn, Marcel Dekker, New York, pp. 157–223.

Appendix A.2

Nature, properties, functions and sources of vitamins

Vitamin	Nature, properties and functions	Main sources	Deficiency diseases
Water-soluble vitamins			
Vitamin C (ascorbic acid)	Occurs as both ascorbic acid and dehydroascorbic acid. A major function is synthesis of hydroxyproline, a component of collagen and connective tissues. Essential for growth of cartilage, bones, teeth and for wound healing. It is an antioxidant, helps absorb iron and maintain healthy tissues.	Nearly all fresh fruits and vegetables, especially berries, citrus fruits, green vegetables, tomatoes and leafy green vegetables.	Scurvy, anaemia, capillary wall rupture/ bruising, nosebleeds, bleeding gums, low infection resistance, muscle degeneration, poor digestion, weakened cartilages and slow healing wounds.
Thiamin (vitamin B_1)	Thiamine pyrophosphate is a coenzyme for several enzymes involved in decarboxylation reactions (e.g. decarboxylation of pyruvate) for metabolism of carbohydrates into energy. Essential for correct heart function and healthy nerve cells.	Meats, dairy products, leafy green vegetables, grains and legumes.	Polyneuritis, encephalopathy, infantile beriberi, oedema, abdominal pain, problems with muscle tone in the gastrointestinal tract, vomiting, nervous system disorders, seizures, palpitations, tachycardia, circulatory collapse, and immune system problems.
Riboflavin (vitamin B_2)	Precursor to the coenzymes flavin adenine dinucleotide (FAD) and flavin mononucleotide (FMN), which are hydrogen carriers in a number of oxidation–reduction reactions within mitochondria. Important for body growth and production of red blood cells.	Wide variety of foods, including milk, meats and grains.	Ariboflavinosis with similarities to the niacin deficiency (sore throat with swelling of the mouth and throat mucosa, cracking of lips and mouth, moist scaly skin, and decreased red blood cell count with normal cell size and haemoglobin content).
Niacin (nicotinamide)	Precursor to the coenzymes nicotinamide adenine dinucleotide (NAD) and nicotinamide adenine dinucleotide phosphate (NADPH), which are hydrogen carriers in processes such as glycolysis, Kreb's cycle and oxidative phosphorylation. Cholesterol-lowering effects.	Meats, fish, dairy products, leafy green vegetables, potatoes and nuts. Can be synthesised in the body in small amounts from tryptophan.	Pellagra, which affects the skin, central nervous system and gastrointestinal tract (dementia, dermatitis and diarrhoea).

Vitamin	Nature, properties and functions	Main sources	Deficiency diseases
Pyridoxine (vitamin B_6)	Occurs in three forms: pyridoxine, pyridoxal and pyridoxamine. The precursor to pyridoxal phosphate, a coenzyme for several reactions involving protein metabolism, including transamination reactions necessary for synthesis of amino acids. Helps form red blood cells and maintain brain function.	Pyridoxal and pyridoxamine in meat, milk, fish and poultry, and pyridoxine, pyridoxal in a number of vegetables including potatoes and tomatoes.	Deficiency is uncommon in adults. In infants: dermatitis, abdominal pain, vomiting, ataxia and seizures.
Pantothenic acid	The precursor to coenzyme A, an enzyme critical to the oxidation and/or synthesis of carbohydrates and fatty acids during metabolism. Plays a role in the production of hormones and cholesterol.	A variety of foods, including grains, brassicas, legumes, egg yolk, fish, yeast and meat. Also synthesised by intestinal bacteria.	Very rare and has been observed only in cases of severe malnutrition.
Biotin	Functions as a coenzyme for several enzymes that catalyse carboxylation, decarboxylation and deamination reactions (e.g. pyruvate carboxylase in the Kreb's cycle) for the metabolism of proteins and carbohydrates, and in the production of hormones and cholesterol.	Egg yolk, legumes, nuts, potatoes and liver. Also synthesised by intestinal bacteria.	Rarely, if ever, occurs in individuals who consume a regular diet but includes severe dermatitis, loss of hair, and lack of muscular coordination
Cyanocobalamin (vitamin B_{12})	A cobalt-containing coenzyme involved in numerous metabolic pathways. Important for metabolism and helps form red blood cells and maintain the central nervous system.	Animal products: meat, cheese, fish and seafood, eggs, kidney, liver, milk and milk products (essentially absent from plant products), microbial synthesis.	Defect in red blood cell formation (pernicious anaemia), diminished reflex responses, memory impairment, hallucinations, muscle weakness, problems with digestion, absorption and metabolism of carbohydrates and fats, impaired fertility, growth and development.
Folic acid (folate)	Occurs in various forms expressed as pteroglutamic equivalents. A coenzyme in the synthesis of several amino acids, as well as DNA purines and thymine. Required for energy production, protein metabolism, formation of red blood cells and vital for normal growth and development.	Beans, pulses, meats, organ meats, bran, cheese, chicken, dates, dark-green green leafy vegetables, milk, oranges, root vegetables, fish, whole grains, yeast, eggs. Also synthesised by intestinal bacteria.	Depression, anxiety, fatigue, growth failure and anaemia and birth defects (e.g. spina bifida) in pregnant women.

Fat-soluble vitamins

Vitamin A (retinol)	Required for a range of functions, including production of vision pigments, resistance to infectious agents, maintenance of epithelial cells in soft tissue, mucous membranes, and skin, formation and maintenance of healthy teeth and bones. Disease results from both deficiency and excess.	Animal tissues, especially fish, liver and dairy products. $Trans$-β-carotene in green plants and yellow fruits and vegetables can be converted to vitamin A following ingestion.	Itching and burning eyes, eye sties, night blindness, blindness, dry hair or skin, allergies, loss of appetite, loss of smell, fatigue, insomnia, impaired growth, steroid synthesis reduction; decreased immune system function, cancer susceptibility.
Vitamin D (cholecalciferol)	A steroid hormone that facilitates absorption of calcium from the intestine, and in maintaining blood calcium homeostasis and for healthy teeth and bones. Receptors are present in most cells and it may have many additional effects.	Synthesised in the skin when exposed to UV light. Also present at low concentrations in some foods (fish, eggs, dairy products, oatmeal, sweet potatoes) and added as ergocalciferol in fortified foods (e.g. margarine).	Brittle, soft and fragile bones (osteoporosis and osteopenia) or teeth, rickets, hypocalcaemia, poor metabolism, diarrhoea, insomnia, irregular heartbeat, myopia, nervousness, pale skin, sensitivity to pain.
Vitamin E (tocopherol)	A group of eight compounds – four tocopherols and four tocotrienols, each having a different potency expressed as α-tocopherol equivalents. Function as antioxidants, particularly to prevent oxidation of unsaturated fatty acids and maintain the integrity of cell membranes. A role in red blood cell formation and use of vitamin K.	Vegetable oils, herring, whole grains/unrefined cereals, wheat germ, oatmeal, dark leafy green vegetables, nuts, eggs, milk, organ meats, sweet potatoes, soybeans.	Enlarged prostate gland, gastrointestinal disease, hair loss, impotency or sterility, miscarriages, muscular wasting, or weakness, reduced circulation, slow tissue healing, increased susceptibility to cancer, cardiovascular disease and cataracts
Vitamin K	Necessary for formation of several blood-clotting factors. Essential role in blood clotting and bone formation.	Dark green leafy vegetables, soybeans, egg yolks, liver, oatmeal, rye, and wheat, majority synthesised by bacteria in large intestine.	Brittle or fragile bones (osteoporosis), low blood platelet count and poor blood clotting, high blood glucose levels.

Adapted from:
BOWEN, R., (2003), Vitamins: introduction, available at www.vivo.colostate.edu/hbooks/pathphys/misc_topics/vitamins.html
HIGDON, J., (2004), Pantothenic acid, information from Micronutrient Information Center, Linus Pauling Institute, available at http://lpi.oregonstate.edu/infocenter/vitamins/pa/ or / biotin

MCGEE, W., (2007), Vitamins, information from Medical Encyclopaedia, available at www.nlm.nih.gov/medlineplus/ency/article/002399.htm

Appendix A.3

Properties, functions and sources of minerals

Mineral	Properties and functions	Main sources	Action in foods	Deficiency diseases
Calcium	Maintaining blood electrolyte balance, structure and rigidity of bones and teeth; it is involved in blood clotting, transmission of impulses from nerves to muscles, regulation of the parathyroid gland and heartbeat, functioning of muscles, skin, soft tissues and circulatory, digestive, enzymatic and immune systems. Vitamin D is essential for calcium absorption and utilisation.	Dairy products, molasses, nuts, cereals, fruits, tofu, seafoods, green leafy vegetables, seaweeds.	Texture modifier: forms gels with negatively charged macromolecules such as alginates, low-methoxyl pectins, soy proteins, caseins, etc. Firms canned vegetables when added to canning brine.	Eczema, fragile bones, heart palpitations, hypertension, muscle cramps, osteomalacia, osteoporosis, osteopenia, periodontal disease, rickets, tooth decay, slowed nerve impulse response, decreased muscle growth, arthritis.
Chromium	Essential for absorption and metabolism of glucose for energy production and synthesis of cholesterol, fats and protein. It appears to increase the effectiveness of insulin and its ability to regulate blood sugar levels. It may also be involved in protein synthesis. Correct functioning of adrenal glands, brain, blood circulatory system, heart, immune system, liver and white blood cells.	Fruits, black pepper, calf liver, meat, whole grains, maize and maize oil, dairy products, mushrooms, potatoes, beer, oysters, legumes.	—	Incorrect amino acid metabolism, increased serum cholesterol, myopia, protein/ calorie malnutrition, susceptibility to infection, lowered or escalated blood sugar levels, coronary artery disease.
Copper	A number of proteins and enzymes contain copper, some of which are essential for the utilisation of iron. It is involved in protein metabolism, respiration, healing processes, maintenance of hair colour, and formation of myelin sheaths that protect nerve fibres, and it is a blood antioxidant. It combines with zinc and vitamin C to form elastin and formation of bones and red blood cells.	Avocado, barley, cauliflower, nuts, lamb, oranges, organ meats, raisins, salmon and seafood (oysters), legumes, green leafy vegetables, soybeans.	Catalyst for lipid peroxidation, ascorbic acid oxidation, non-enzymatic oxidative browning. Colour modifier: May cause black discoloration in canned, cured meats. Enzyme cofactor for polyphenoloxidase. Texture stabiliser: stabilises egg-white foams	Elevated serum cholesterol, fractures and bone deformities, osteoporosis, impaired respiration, skin sores, slowed healing processes, poor hair and skin colouring, loss in taste sensitivity.

Mineral	Function	Sources	Food chemistry role	Deficiency symptoms
Iodine	Trace amounts are required for correct functioning of immune system, brain, and thyroid gland which regulates the body's production of energy and metabolic rate. It is involved in conversion of carotene to vitamin A, protein synthesis and synthesis of cholesterol.	Iodised salts, seafood, seaweeds, asparagus, fish, garlic, beans, mushrooms, sesame seeds, soybeans, spinach.	KIO_3 is dough improver, improves baking quality of wheat flour.	Thyroid dysfunction, goitre (enlarged thyroid) and hypothyroidism, leading to slowed mental and physical development.
Iron	Required for production of haemoglobin and myoglobin, some enzymes, oxygenation of red blood cells, healthy growth and resistance to disease, healthy immune system and energy.	Liver, red meat, dark green leafy vegetables, eggs, seafood, fish, brewer's yeast, dates, legumes, peaches, pears, pumpkins, raisins, rice and wheat bran, sesame seeds and soybeans.	Catalyst for Fe^{2+} and Fe^{3+} and lipid peroxidation in foods. Colour modifier – colour of fresh meat depends on valence of Fe in myoglobin and haemoglobin: Fe^{2+} is red, Fe^{3+} is brown. Forms green, blue, or black complexes with polyphenolic compounds. Reacts with S^{2-} to form black FeS in canned foods. Enzyme cofactor: lipoxygenase, cytochromes, ribonucleotide reductase.	Iron deficiency anaemia, breathing difficulties, brittle nails, hair loss, dizziness, constipation.
Magnesium	Essential for enzyme activity and protein synthesis, assists in calcium and potassium uptake to maintain blood electrolyte balance, important role in bone formation, carbohydrate and mineral metabolism, functioning of arteries, bones, cells, digestive, immune and reproductive systems, heart, nerves, and teeth. Extracellular magnesium is critical for maintaining electrical potentials of nerve and muscle membranes and for transmission of nerve impulses to muscles. Use of diuretics, laxatives, vomiting and diarrhoea can significantly contribute to magnesium requirement.	Most foods, especially dairy products, fruits, fish, meat, seafood, garlic, lima beans, sesame seeds, tofu, green leafy vegetables. whole grains.	Colour modifier – removal of Mg from chlorophyll changes colour from green to olive-brown.	Decreased blood pressure and body temperature, coronary heart disease, interference with transmission of nerve and muscle impulses, causes disorientation, hyperactivity, noise sensitivity, insomnia, muscle weakness or tremors; disruption of pH balance, calcium deposits in kidneys, blood vessels and heart, digestive disorders and many other symptoms.

Mineral	Properties and functions	Main sources	Action in foods	Deficiency diseases
Manganese	An antioxidant, activates enzymes and has roles in protein, carbohydrate and fat metabolism, blood sugar regulation, healthy nerves and immune system, sex hormone production, skeletal development.	Whole grains, fruits, vegetables.	Enzyme cofactor: pyruvate carboxylase, superoxide dismutase.	Poor reproductive performance, growth retardation and congenital malformations, abnormal formation of bone and cartilage and impaired glucose tolerance.
Molybdenum	Essential for several metabolic enzymes, for oxidation of fats and metabolism of calcium, magnesium, copper and nitrogen. It may also be an antioxidant.	Plant foods.	–	Impaired reproduction and weight gain, increased rate of heartbeat and breathing, visual problems and shortened life expectancy.
Nickel	Activator for some enzymes and involved in hormone and lipid metabolism and cell membrane integrity.	Plant foods.	Catalyst for hydrogenation of vegetable oils – finely divided, elemental Ni is most widely used for this process.	Reduced growth, impaired liver function, changes in skin colour and reproductive problems.
Phosphorus	A balance of magnesium, calcium and phosphorus is required for these minerals to be used effectively. Essential component of bone and to utilise vitamins to metabolise food and maintain electrolyte balance, correct functioning of brain cells, circulatory and digestive systems, eyes, liver, muscles, nerves and teeth/bones.	Most foods, especially asparagus, maize, dairy products, eggs, fish, fruits, garlic, sunflower seeds, meats, wheat bran and whole grains.	Acidulent – H_3PO_4 in soft drinks. Leavening acid – $Ca(HPO_4)_2$ is a fast-acting leavening acid. Moisture retention in meats – sodium tripolyphosphate improves moisture retention in cured meats. Emulsification aid – phosphates are used to aid emulsification in comminuted meats and in processed cheeses. Stabiliser in evaporated milk.	Bone loss, weakness, anorexia. High levels of phosphorus can inhibit calcium uptake and may result in osteoporosis, arthritis, pyorrhea, rickets and tooth decay. Nervous disorders, heart and kidney problems.

Mineral	Function	Sources	Food application	Effects of deficiency/excess
Potassium	Important in a balance with sodium for cellular metabolism and regulating transfer of nutrients to cells, maintaining blood pressure and electrolyte balance, transmitting electrochemical impulses. Correct functioning of blood, endocrine/ digestive and nervous systems, heart, kidneys, muscles and skin. Use of diuretics, laxatives, vomiting and diarrhoea can significantly contribute to potassium requirement.	Dairy products, fish, fruit, meat, poultry, vegetables, whole grains, beans, nuts, potatoes, wheat bran and yams.	KCl may be used as a salt substitute, but may cause bitter flavour. Leavening agent – potassium acid tartrate.	Interference with transmission of nerve and muscle impulses, heart palpitations and arrhythmias, heart attack, stroke. Decreased blood pressure, salt retention, oedema, increased cholesterol, muscle weakness, respiratory distress, weak reflexes.
Selenium	An antioxidant that protects all membranes, reduces cancer risk, enhances immune system and protects against heart disease. Required with vitamin E for production of antibodies, binding of toxic metals, amino acid metabolism and promotion of growth and fertility. Production of prostaglandins which affect blood pressure and platelet aggregation.	Seafoods, organ meats, cereals (depending on levels in soils).	Enzyme cofactor for glutathione peroxidase.	Premature ageing, heart attack, muscular dystrophy, cystic fibrosis, infertility and increased risk of cancer. Keshan disease (endemic cardio-myopathy in China) was associated with selenium deficiency.
Sodium	With potassium for maintaining cellular water balance and blood electrolyte balance/pH and needed for stomach, lymphatic system, nerve and muscle function. Excessive salt increases blood pressure.	Virtually all foods, especially celery, eggs, meat, dairy products, miso, poultry, processed foods, salt, seafood and seaweeds.	Flavour modifier – NaCl elicits the classic salty taste in foods. Preservative – NaCl may be used to lower water activity in foods. Leavening agents – many leavening agents are sodium salts, e.g. sodium bicarbonate, sodium aluminium sulphate, sodium acid pyrophosphate.	Cramps, decreased resistance to infection, muscle shrinkage or weakness, low blood sugar levels, dehydration, heart palpitations and heart attack, arthritis, rheumatism, neuralgia, short attention span and mental confusion.

Mineral	Properties and functions	Main sources	Action in foods	Deficiency diseases
Sulphur	Essential for resisting bacterial infection, aids oxidation reactions, stimulates bile secretions in the liver and protects against toxic substances. Required for correct functioning of skin, immune system, blood, liver.	Brussels sprouts, beans, cabbage, eggs, garlic, fish, meats and onions.	Browning inhibitor – sulphur dioxide and sulphites inhibit both enzymatic and non-enzymatic browning. Used in dried fruits. Antimicrobial – prevents/controls microbial growth. Widely used in wine making.	Poor resistance to bacterial infections.
Zinc	Needed to maintain required functioning of immune system, thymus and spleen, maintaining concentrations of vitamin E in the blood, correct functioning of blood, bones, eyes, heart, joints, liver and prostate gland. A component of insulin and many enzymes, including those involved in metabolism of phosphorus, proteins and alcohol. Levels can be decreased by diarrhoea, kidney disease, cirrhosis of the liver, diabetes.	Fish, meats, seafood, whole grains, liver, egg yolks, beans, mushrooms, nuts, pumpkin seeds, sardines, soybeans, wheat germ.	ZnO is used in the lining of cans for protein-rich foods to lessen formation of black FeS during heating. Zn can be added to green beans to help stabilise the colour during canning.	Poor memory, decreased learning ability, delayed sexual maturity, sterility, growth retardation and dwarfism, loss of taste and smell, poor circulation, susceptibility to infections and prolonged wound healing, retarded growth, prostate and immune system disorders, liver damage.

Adapted from:
ANON, (undated), Minerals, information provided by University of California, Davis, available at http://teaching.ucdavis.edu/nut11av/handouts/Complete_Text_Mineral.pdf
CLARK, T.J. (2006), Mineral functions in the body, available at www.tjclark.com.au/colloidal-minerals-library/mineral-functions.htm
MILLER, D.R., (1996), Minerals, in (O.R. Femnema, Ed.), *Food Chemistry*, Marcel Dekker, New York, pp. 617–650.

Appendix A.4

Additives currently permitted in food within the EU and their associated E numbers

Colours

E100 Curcumin
E101 (i) Riboflavin, (ii) Riboflavin-5'-phosphate
E102 Tartrazine
E104 Quinoline yellow
E110 Sunset Yellow FCF; Orange Yellow S
E120 Cochineal; Carminic acid; Carmines
E122 Azorubine; Carmoisine
E123 Amaranth
E124 Ponceau 4R; Cochineal Red A
E127 Erythrosine
E128 Red 2G
E129 Allura Red AC
E131 Patent Blue V
E132 Indigotine; Indigo Carmine
E133 Brilliant Blue FCF
E140 Chlorophylls and chlorophyllins
E141 Copper complexes of chlorophyll and chlorophyllins
E142 Green S
E150a Plain caramel
E150b Caustic sulphite caramel
E150c Ammonia caramel
E150d Sulphite ammonia caramel E151

Brilliant Black BN; Black PN
E153 Vegetable carbon
E154 Brown FK
E155 Brown HT
E160a Carotenes
E160b Annatto; Bixin; Norbixin
E160c Paprika extract; Capsanthian; Capsorubin
E160d Lycopene
E160e Beta-apo-8'-carotenal (C30)
E160f Ethyl ester of beta-apo-8'-carotenoic acid (C30)
E161b Lutein
E161g Canthaxanthin
E162 Beetroot Red; Betanin
E163 Anthocyanins
E170 Calcium carbonate
E171 Titanium dioxide
E172 Iron oxides and hydroxides
E173 Aluminium
E174 Silver
E175 Gold
E180 Litholrubine BK

Preservatives

E200 Sorbic acid
E202 Potassium sorbate
E203 Calcium sorbate
E210 Benzoic acid
E211 Sodium benzoate
E212 Potassium benzoate
E213 Calcium benzoate
E214 Ethyl *p*-hydroxybenzoate
E215 Sodium ethyl *p*-hydroxybenzoate
E216 Propyl *p*-hydroxybenzoate
E217 Sodium propyl *p*-hydroxybenzoate
E218 Methyl *p*-hydroxybenzoate
E219 Sodium methyl *p*-hydroxybenzoate
E220 Sulphur dioxide
E221 Sodium sulphite
E222 Sodium hydrogen sulphite
E223 Sodium metabisulphite
E224 Potassium metabisulphite
E226 Calcium sulphite
E227 Calcium hydrogen sulphite

E228 Potassium hydrogen sulphite
E230 Biphenyl; diphenyl
E231 Orthophenyl phenol
E232 Sodium orthophenyl phenol
E234 Nisin
E235 Natamycin
E239 Hexamethylene tetramine
E242 Dimethyl dicarbonate
E249 Potassium nitrite
E250 Sodium nitrite
E251 Sodium nitrate
E252 Potassium nitrate
E280 Propionic acid
E281 Sodium propionate
E282 Calcium propionate
E283 Potassium propionate
E284 Boric acid
E285 Sodium tetraborate; borax
E1105 Lysozyme

Antioxidants

E300 Ascorbic acid
E301 Sodium ascorbate
E302 Calcium ascorbate
E304 Fatty acid esters of ascorbic acid
E306 Tocopherols
E307 Alpha-tocopherol
E308 Gamma-tocopherol
E309 Delta-tocopherol

E310 Propyl gallate
E311 Octyl gallate
E312 Dodecyl gallate
E315 Erythorbic acid
E316 Sodium erythorbate
E320 Butylated hydroxyanisole (BHA)
E321 Butylated hydroxytoluene (BHT)

Emulsifiers, stabilisers, thickeners and gelling agents

E322 Lecithins
E400 Alginic acid
E401 Sodium alginate
E402 Potassium alginate
E403 Ammonium alginate
E404 Calcium alginate
E405 Propane-1,2-diol alginate
E406 Agar
E407 Carrageenan
E407a Processed eucheuma seaweed
E410 Locust bean gum; carob gum
E412 Guar gum
E413 Tragacanth
E414 Acacia gum; gum arabic
E415 Xanthan gum
E416 Karaya gum
E417 Tara gum
E418 Gellan gum
E425 Konjac
E432 Polyoxyethylene sorbitan monolaurate; Polysorbate 20
E433 Polyoxyethylene sorbitan mono-oleate; Polysorbate 80
E434 Polyoxyethylene sorbitan monopalmitate; Polysorbate 40
E435 Polyoxyethylene sorbitan monostearate; Polysorbate 60
E436 Polyoxyethylene sorbitan tristearate; Polysorbate 65
E440 Pectins
E442 Ammonium phosphatides
E444 Sucrose acetate isobutyrate
E445 Glycerol esters of wood rosins
E460 Cellulose
E461 Methyl cellulose
E463 Hydroxypropyl cellulose
E464 Hydroxypropyl methyl cellulose
E465 Ethyl methyl cellulose
E466 Carboxy methyl cellulose

E467 Sodium carboxy methyl cellulose
E468 Cross-linked sodium carboxy methyl cellulose
E469 Enzymatically hydrolysed carboxy methyl cellulose
E470a Sodium, potassium and calcium salts of fatty acids
E470b Magnesium salts of fatty acids
E471 Mono- and diglycerides of fatty acids
E472a Acetic acid esters of mono- and diglycerides of fatty acids
E472b Lactic acid esters of mono- and diglycerides of fatty acids
E472c Citric acid esters of mono- and diglycerides of fatty acids
E472d Tartaric acid esters of mono- and diglycerides of fatty acids
E472e Mono- and diacetyltartaric acid esters of mono- and diglycerides of fatty acids
E472f Mixed acetic and tartaric acid esters of mono- and diglycerides of fatty acids
E473 Sucrose esters of fatty acids
E474 Sucroglycerides
E475 Polyglycerol esters of fatty acids
E476 Polyglycerol polyricinoleate
E477 Propane-1,2-diol esters of fatty acids
E479b Thermally oxidised soya bean oil interacted with mono and diglycerides of fatty acids
E481 Sodium stearoyl-2-lactylate
E482 Calcium stearoyl-2-lactylate
E483 Stearyl tartrate
E491 Sorbitan monostearate
E492 Sorbitan tristearate
E493 Sorbitan monolaurate
E494 Sorbitan monooleate
E495 Sorbitan monopalmitate
E1103 Invertase

Sweeteners

E420 (i) Sorbitol, (ii) Sorbitol syrup
E421 Mannitol
E953 Isomalt
E965 (i) Maltitol, (ii) Maltitol syrup
E966 Lactitol
E967 Xylitol

E950 Acesulfame K
E951 Aspartame
E952 Cyclamic acid and its Na and Ca salts
E954 Saccharin and its Na, K and Ca salts
E957 Thaumatin
E959 Neohesperidine DC

Others (Acids, acidity regulators, anti-caking agents, anti-foaming agents, bulking agents, carriers and carrier solvents, emulsifying salts, firming agents, flavour enhancers, flour treatment agents, foaming agents, glazing agents, humectants, modified starches, packaging gases, propellants, raising agents and sequestrants)

E170 Calcium carbonates
E260 Acetic acid
E261 Potassium acetate
E262 Sodium acetate
E263 Calcium acetate
E270 Lactic acid
E290 Carbon dioxide
E296 Malic acid
E297 Fumaric acid
E325 Sodium lactate
E326 Potassium lactate
E327 Calcium lactate
E330 Citric acid
E331 Sodium citrates
E332 Potassium citrates
E333 Calcium citrates
E334 Tartaric acid (L-(+))
E335 Sodium tartrates
E336 Potassium tartrates
E337 Sodium potassium tartrate
E338 Phosphoric acid
E339 Sodium phosphates
E340 Potassium phosphates
E341 Calcium phosphates
E343 Magnesium phosphates
E350 Sodium malates
E351 Potassium malate
E352 Calcium malates
E353 Metatartaric acid
E354 Calcium tartrate
E355 Adipic acid
E356 Sodium adipate
E357 Potassium adipate
E363 Succinic acid
E380 Triammonium citrate
E385 Calcium disodium ethylene diamine tetra-acetate; calcium disodium EDTA
E422 Glycerol
E431 Polyoxyethylene (40) stearate
E450 Diphosphates
E451 Triphosphates
E452 Polyphosphates
E459 Beta-cyclodextrin
E500 Sodium carbonates
E501 Potassium carbonates
E503 Ammonium carbonates
E504 Magnesium carbonates
E507 Hydrochloric acid
E508 Potassium chloride
E509 Calcium chloride
E511 Magnesium chloride
E512 Stannous chloride
E513 Sulphuric acid
E514 Sodium sulphates

E515 Potassium sulphates
E516 Calcium sulphate
E517 Ammonium sulphate
E520 Aluminium sulphate
E521 Aluminium sodium sulphate
E522 Aluminium potassium sulphate
E523 Aluminium ammonium sulphate
E524 Sodium hydroxide
E525 Potassium hydroxide
E526 Calcium hydroxide
E527 Ammonium hydroxide
E528 Magnesium hydroxide
E529 Calcium oxide
E530 Magnesium oxide
E535 Sodium ferrocyanide
E536 Potassium ferrocyanide
E538 Calcium ferrocyanide
E541 Sodium aluminium phosphate
E551 Silicon dioxide
E552 Calcium silicate
E553a (i) Magnesium silicate, (ii) Magnesium trisilicate
E553b Talc
E554 Sodium aluminium silicate
E555 Potassium aluminium silicate
E556 Aluminium calcium silicate
E558 Bentonite
E559 Aluminium silicate; kaolin
E570 Fatty acids
E574 Gluconic acid
E575 Glucono delta-lactone
E576 Sodium gluconate
E577 Potassium gluconate
E578 Calcium gluconate
E579 Ferrous gluconate
E585 Ferrous lactate
E620 Glutamic acid
E621 Monosodium glutamate
E622 Monopotassium glutamate
E623 Calcium diglutamate
E624 Monoammonium glutamate
E625 Magnesium diglutamate
E626 Guanylic acid
E627 Disodium guanylate
E628 Dipotassium guanylate
E629 Calcium guanylate
E630 lnosinic acid
E631 Disodium inosinate
E632 Dipotassium inosinate
E633 Calcium inosinate
E634 Calcium 5′-ribonucleotides
E635 Disodium 5′-ribonucieotides
E640 Glycine and its sodium salt
E650 Zinc acetate

E900 Dimethylpolysiloxane
E901 Beeswax, white and yellow
E902 Candelilla wax
E903 Carnauba wax
E904 Shellac
E905 Microcrystalline wax
E912 Montan acid esters
E914 Oxidised polyethylene wax
E920 L-Cysteine
E927b Carbamide
E938 Argon
E939 Helium
E941 Nitrogen
E942 Nitrous oxide
E943a Butane
E943b Iso-butane
E944 Propane
E948 Oxygen
E949 Hydrogen
E999 Quillaia extract

E1200 Polydextrose
E1201 Polyvinylpyrrolidone
E1202 Polyvinylpolypyrrolidone
E1404 Oxidised starch
E1410 Monostarch phosphate
E1412 Distarch phosphate
E1413 Phosphated distarch phosphate
E1414 Acetylated starch
E1420 Acetylated Starch
E1422 Acetylated distarch adipate
E1440 Hydroxyl propyl starch
E1442 Hydroxy propyl distarch phosphate
E1450 Starch sodium octenyl succinate
E1451 Acetylated oxidised starch
E1452 Starch aluminium octenyl succinate
Polyethylene glycol 6000
E1505 Triethyl citrate
E1518 Glyceryl triacetate; triacetin
E1520 Propan-1,2-diol; propylene glycol

From:
ANON, (2002), A list of additives currently permitted in food within the European Union and their associated E Numbers, available at www.ukfoodguide.net/enumeric.htm

Appendix A.5

Examples of functional components of foods

Component	Source	Potential benefit
Carotenoids		
Beta-carotene	Carrots, various fruits	Neutralises free radicals which may damage cells; bolsters cellular antioxidant defences.
Lutein, zeaxanthin	Kale, spinach, maize, eggs, citrus fruits	May contribute to maintenance of healthy vision.
Lycopene	Tomatoes, processed tomato products	Role in cancer risk reduction including breast, digestive tract, cervix, bladder, skin and possibly lung.
Dietary (functional and total) fibre		
Insoluble fibre	Wheat bran	May contribute to maintenance of a healthy digestive tract.
Beta glucan	Oat bran, rolled oats, oat flour	Cholesterol-lowering, can reduce total and low-density lipoprotein (LDL) cholesterol, thereby reducing the risk of coronary heart disease (CHD).
Soluble fibre	Psyllium seed husk	May reduce risk of CHD.
Whole grains	Cereal grains	May reduce risk of CHD and cancer; may contribute to maintenance of healthy blood glucose levels.
Fatty acids		
Monounsaturated fatty acids	Tree nuts	May reduce risk of CHD by altering blood cholesterol levels.
Polyunsaturated fatty acids (PUFAs) – omega-3 fatty acids – ALA	Walnuts, flax	May contribute to maintenance of mental and visual function.
PUFAs – omega-3 fatty acids – DHA/EPA	Salmon, tuna, marine and other fish oils	Can lower triglyceride fats in the blood. May reduce risk of CHD, may contribute to maintenance of mental and visual function. The cardioprotective effect of fish consumption observed in some investigations, but not in others. n-3 fatty acids lower triglycerides levels but increase LDL cholesterol.
PUFAs – conjugated linoleic acid (CLA)	Beef and lamb; some cheeses, flaxseed oil	May contribute to maintenance of desirable body composition and healthy immune function. Anticarcinogenic. CLA increases in processed foods (significant because many mutagens and carcinogens have been identified in cooked meats). Effective in suppressing tumours in mice, may be unique mechanism(s) by which CLA modulates tumour development.

Component	Source	Potential benefit
Flavonoids		
Anthocyanidins	Berries, cherries, red grapes	Bolster cellular antioxidant defences; may contribute to maintenance of brain function.
Flavanols	Tea, cocoa, chocolate, apples, grapes	May contribute to maintenance of heart health. Cancer chemopreventive effects in animals but inconclusive epidemiological studies. Benefits may be restricted to high intakes in high-risk populations. May also reduce risk of CVD but not yet conclusive.
Catechins, epicatechins, procyanidins		
Flavanones	Citrus fruits	Neutralise free radicals that may damage cells; bolster cellular antioxidant defences. Protective against a variety of cancers. Evidence accumulating of cancer preventative effect of limonene – suggested as a good candidate for human chemoprevention. A metabolite of limonene, perrillyl alcohol, undergoing Phase I clinical trials in patients with advanced malignancies.
Flavonols	Onions, apples, tea, broccoli	Neutralise free radicals which may damage cells; bolster cellular antioxidant defences.
Proanthocyanidins	Cranberries, cocoa, apples, strawberries, grapes, wine, peanuts, cinnamon	Inhibits adherence of *E. coli* to uroepithelial cells. Contributes to maintenance of urinary tract health and heart health.
Isothiocyanates		
Sulphoraphane	Cruciferous vegetables: cauliflower, broccoli, cabbage, kale, horseradish	Decreased cancer risk. May enhance detoxification of undesirable compounds and bolster cellular antioxidant defences. Enzyme myrosinase catalyses glucosinolates to isothiocyanates and indoles. Indole-3 carbinol being studied for cancer chemopreventive properties, particularly the mammary gland. Sulphoraphane is an isothiocyanate in broccoli.
Phenols	Red wine	Growing evidence that phenolic content in red wine can reduce risk of CVD. Phenolic substances prevent oxidation of LDL in the process of atherogenesis. Alcohol-free wine shown to increase total plasma antioxidant capacity. Grape juice is effective in inhibiting the oxidation of LDL. Red wine is also a source of *trans*-resveratrol, shown to have estrogenic properties which may explain cardiovascular benefits, and also shown to inhibit carcinogenesis *in vivo*.
Caffeic acid, ferulic acid	Apples, pears, citrus fruits, some vegetables	May bolster cellular antioxidant defences; may contribute to maintenance of healthy vision and heart health.
Plant stanols/sterols		
Free stanols/sterols	Maize, soy, wheat, wood oils, fortified foods and beverages	May reduce risk of CHD.

Component	Source	Potential benefit
Stanol/sterol esters	Fortified table spreads, stanol ester dietary supplements	Blocks absorption from the gut of cholesterol in bile and cholesterol ingested in food. Can lower total cholesterol and LDL cholesterol levels, raising or not changing HDL-cholesterol in the blood and having no reported side effects. May reduce risk of CHD.
Polyols		
Sugar alcohols – xylitol, sorbitol, mannitol, lactitol	Some chewing gums and other food applications	May reduce risk of dental caries.
Prebiotics		
Inulin, fructo-oligosaccharides Polydextrose	Whole grains, onions, some fruits, garlic, honey, leeks, fortified foods and beverages	May improve gastrointestinal health; may improve calcium absorption.
Probiotics		
Lactobacilli, bifidobacteria	Yoghurt, other dairy and non-dairy applications	May improve gastrointestinal health and systemic immunity.
Phytoestrogens		
Isoflavones Daidzein, genistein	Soybeans and soy-based foods	May reduce risk of CHD but exact mechanism of hypocholesterolaemic effect not fully elucidated. Thought to play preventive and therapeutic roles in CVD, cancers of the breast, prostate and bowel, osteoporosis, and the alleviation of menopausal symptoms. Because isoflavones are weak oestrogens, they may act as antioestrogens competing with endogenous oestrogens for binding to oestrogen receptor. This may explain why populations that consume significant amounts of soy have reduced risk of oestrogen-dependent cancer but epidemiological are inconsistent. No published clinical trials on the role of soy in reducing cancer risk.
Lignans enterodiol and enterolactone, formed in intestinal tract by bacterial action on plant lignan precursors	Flaxseed, rye, some vegetables	May contribute to maintenance of heart health and healthy immune function. Enterodiol and enterolactone similar to oestrogens and possess weakly estrogenic and antiestrogenic activities and may play a role in prevention of oestrogen-dependent cancers, but no epidemiological data and relatively few animal studies to support this hypothesis.
Soy protein		
Soy protein	Soybeans and soy-based foods	May help reduce risk of CHD.

Component	Source	Potential benefit
Sulphides/thiols		
Diallyl sulphide, allyl methyl trisulphide	Garlic, onions, leeks	Cancer chemopreventive, antibiotic and cholesterol-lowering properties. May enhance detoxification of undesirable compounds; may contribute to maintenance of healthy immune function. Cardioprotective effects may be due to cholesterol-lowering effect but garlic component responsible is unclear. Sulphoxide, alliin, is converted by allinase into allicin which decomposes to form numerous sulphur-containing compounds, some of which may have chemopreventive activity. Inconclusive results may be due to variation in organosulphur compounds in garlic products. Allium vegetables may have protective effect against gastrointestinal tract cancers. Garlic advocated for prevention of CVD, possibly through antihypertensive properties but insufficient evidence to recommend it as a routine clinical therapy.
Dithiolthiones	Cruciferous vegetables	Contribute to maintenance of healthy immune function.

Note: CHD = coronary heart disease, CVD = cardiovascular disease, HDL = high-density lipoproteins, LDL = low-density lipoproteins, DHA = docosahexaenoic acid, EPA = eicosapentaenoic acid.

Adapted from:

ANON, (2006), Functional foods, International Food Information Council, available at www.ific.org/nutrition/functional/index.cfm

ARVANITOYANNIS, I.S. and VAN HOUWELINGEN-KOUKALIAROGLOU, M., (2005), Functional foods: a survey of health claims, pros and cons, and current legislation, *Critical Reviews in Food Science and Nutrition*, **45**, 385–404.

ASCHERIO, A., RIMM, E.B., STAMPFER, M.J., GIOVANNUCCI, E.L. and WILLETT, W.C., (1995), Dietary intake of marine *n*-3 fatty acids, fish intake, and the risk of coronary disease among men, *New England J. Medicine*, **332**, 977–982.

CLINTON, S.K., (1998), Lycopene: chemistry, biology, and implications for human health and disease, *Nutrition Review*, **56**, 35–51.

FRANKEL, E.N., KANNER, J., GERMAN, J.B., PARKS, E. and KINSELLA, J.E. (1993), Inhibition of oxidation of human low-density lipoprotein by phenolic substances in red wine, *The Lancet*, **341**, 454–457.

GOULD, M.N., (1997), Cancer chemoprevention and therapy by monoterpenes, *Environmental Health Perspectives*, **105**, 977–979.

HASLER, C.M., (1998), Functional foods: their role in disease prevention and health promotion, a publication of the Institute of Food Technologists Expert Panel on Food Safety and Nutrition, *Food Technology*, **52** (2), 57–62.

JANG, M., CAI, J., UDEANI, G., SLOWING, K.V., THOMAS, C.F., BEECHER, C.W.W., FONG, H.H.S., FARNSWORTH, N.R., KINGHORN, A.D., MEHTA, R.G., MOON, R.C. and PEZZUTO, J.M., (1997), Cancer chemopreventive activity of resveratrol, a natural product derived from grapes, *Science*, **275**, 218–220.

KATIYAR, S.K. and MUKHTAR, H., (1996). Tea in chemoprevention of cancer: epidemiologic and experimental studies (review), *International J. Oncology*, **8**, 221–238.

KOHLMEIER, L., WEERINGS, K.G.C., STECK, S. and KOK, F.J., (1997). Tea and cancer prevention – an evaluation of the epidemiologic literature, *Nutr. Cancer*, **27**, 1–13.

LAWSON, L.D., WANG, Z-Y.J. and HUGHES, B.G. (1991), Identification and HPLC quantitation of the sulfides and dialk(en)ylthiosulfinates in commercial garlic products, *Planta Med.*, **57**, 363–370.

LI, Y., ELIE, M., BLANER, W.S., BRANDT-RAUF, P. and FORD, J. (1997), Lycopene, smoking and lung cancer, *Proceedings American Association Cancer Research*, **38**, 113 (abstract #758).

MESSINA, M., BARNES, S. and SETCHELL, K.D.R., (1997), Phytooestrogens and breast cancer, *The Lancet*, **350**, 971–972.

SCHMIDT, D.R. and SOBOTA, A.E., (1988), An examination of the anti-adherence activity of cranberry juice on urinary and nonurinary bacterial isolates, *Microbios.*, **55**, 173–181.

TIJBURG, L.B.M., MATTERN, T., FOLTS, J.D., WEISGERBER, U.M. and KATAN, M.B., (1997), Tea flavonoids and cardiovascular diseases: a review, *Critical Reviews in Food Science and Nutrition*, **37**, 771–785.

VERHOEVEN, D.T.H., VERHAGEN, H., GOLDBOHM, R.A., VAN DEN BRANDT, P.A. and VAN POPPEL, G., (1997), A review of mechanisms underlying anticarcinogenicity by brassica vegetables. *Chemical Biochemical Interactions*, **103**, 79–129.

Appendix B.1

Sources and symptoms of pathogenic bacteria

Pathogen	Sources	Minimum growth conditions			D-value (min)	z-value (°C)	Symptoms of infection	Typical high-risk foods
		Temperature (°C)	pH	a_w				
Aerobacter spp.	Warm-blooded animals	15	5.5	–	–	–	Abdominal pain, diarrhoea within 24–72 h	Poultry
Aeromonas hydrophila	Fresh or brackish water	1–5	6.5	–	$D_{60} = 3–7$	–	Diarrhoea, vomiting, fever within 12–36 h	Water, vegetables, cheese, raw milk, poultry, lamb, fish or shellfish
Bacillus cereus	Soil, surfaces of cereals, vegetables and meats	4 (psychrotrophic strains) 35 (mesophilic strains)	4.3	0.92	$D_{100} = 0.3–27$ $D_{121} = 0.02–0.06$	10	2 types: emetic nausea within 1–5 h or diarrhoea within 8–16 h	Reheated rice and products containing cereals or spices (emetic type), meat products, sauces, milk products (diarrhoeal type)
Brucella abortis	Farm animals	–	–	–	$D_{63} = 2.5$	4.1	Acute joint and muscle pain, fatigue. Chronic genitourinary, cardiovascular and neurological conditions	Raw milk, unpasteurised dairy products
Campylobacter jejuni	Warm-blooded animals	>25	4.9	0.98	$D_{60} = 6 s$	5	Vomiting within 1–14 h, diarrhoea within 4–16 h, headache, abdominal pain, within 24–72 h	Poultry products, raw milk and to lesser extent meat and milk products, water, shellfish

Pathogen	Sources	Minimum growth conditions			D-value (min)	z-value (°C)	Symptoms of infection	Typical high-risk foods
		Temperature (°C)	pH	a_w				
Clostridium botulinum Group I (proteolytic)	Ubiquitous, especially soil and water	10	4.6	0.94	$D_{100} = 25$ and $D_{121} = 0.2$ (spores)	10	7 types of toxin: vomiting, blurred vision, progressive difficulty in swallowing, respiratory failure within 12–36 h. Up to 70% fatal	Canned vegetables and other low-acid foods, smoked fish
Group II (non-proteolytic)		3.3	5.0	0.97	$D_{80} = 1.0$, $D_{100} = 0.1–0.2$ and $D_{121} \leq 0.001$ (spores)	10		
Clostridium perfringens	Ubiquitous – soil, raw, dried and cooked foods	12	5.5	0.95	$D_{95} = 1–3$ (heat-sensitive spores) and 18–64 (heat-resistant spores)	10.3 (cells), 16.8 (spores)	Acute diarrhoea, flatulence within 8–18 h, but little nausea fever or vomiting	Cooked and raw meats, poultry, fish, dairy products, dried foods (e.g. soups, spices, pasta)
Enteropathogenic *Escherichia coli*	Intestinal tract of humans and warm-blooded animals	7–8 (all types)	4.4	0.93	$D_{64} = 0.36–0.46$	5	6 types of illness including acute bloody diarrhoea, intestinal haemorrhage, bladder/kidney infections, septicaemia, blood clots on brain within 7h–4 days[a]. Some types fatal	Meat, poultry, fish vegetables, soft cheeses, water, alfalfa shoots
		6.5 (O157:H7)	4.0	0.95	$D_{63} = 0.5$			
Listeria monocytogenes	Ubiquitous – soil, most foods, healthy humans or animals, surfaces	−0.4	4.6	0.92	$D_{60} = 4.25–5.49$	1.97–7.46	Adults: gastroenteritis within 24–48 h. Newborn babies: meningitis and death. Unborn foetus: spontaneous abortion, stillbirth or meningitis. Individuals having compromised immune systems: meningitis or septicaemia.	Milk, seafoods, smoked or marinated fish, sandwiches, raw vegetables/salads, coleslaw, raw or cooked meats/pâté, poultry products, soft cheeses

Organism	Reservoir/source				D value		Disease/symptoms	Foods
Mycobacterium tuberculosis *Mycobacterium avium* subsp. *paratuberculosis*	Milk	25	5.5	—	$D_{71} = 12$ s (in milk)	8.6	Associated with Crohn's disease	Milk and milk products
Plesiomonas shigelloides	Freshwater	8	4.0	—	No survival at 60 °C for 30 min.	—	Diarrhoea, abdominal pain, nausea within 24–48 h	Water or shellfish
Salmonella enterica	Poultry, cattle, pigs, other animals	5.2[b]	3.8[c]	0.94	0.28–10 s (in milk) 0.36 min (in ground beef) 0.55–9.5 min (in liquid egg), 4.5–6.6 h (in chocolate)	—	Gastroenteritis (within 12–48 h, duration 2–7 days): nausea, vomiting, high fever, abdominal pain, may be fatal. Enteric fever (within 7–28 days, duration 14 days): high fever, nausea, abdominal pain. Carrier for months/years. Septicaemia: high fever, abdominal and thoracic pain.	Eggs, poultry, milk, cooked meats, salami, cheeses
Shigella spp.	Infected people, food or water	6.1	4.8	0.96	$D_{60} = 1.0$ $D_{80} = 10$ s	—	Dysentery, severe abdominal pain, fever within 12–50 h	Salads, milk, soft cheese, poultry, cooked rice
Staphylococcus aureus	Animals, human skin and nasal cavity, surfaces	6.5 (10 for toxin production)	4.0	0.83 (growth) 0.85 (toxin production)	$D_{77} =$ 0.001–0.01	8–12	Nausea, vomiting, sometimes diarrhoea, within 2–4 h	Recontaminated heat processed foods, cheese, salami, cooked meats, dried milk, sandwiches
Vibrio parahaemolyticus	Inshore marine waters	5	4.8	0.94	$D_{50} =$ 2.4–9.9	—	Gastroenteritis, abdominal cramps, nausea, fever, within 4–24 h. Cholera (*V. cholerae*) within 6 h–3 days.	Raw, improperly cooked or recontaminated fish and shellfish, water
V. cholerae	Inshore marine waters	10	5.0	0.97	—	—		

Pathogen	Sources	Minimum growth conditions			D-value (min)	z-value (°C)	Symptoms of infection	Typical high-risk foods
		Temperature (°C)	pH	a_w				
Yersinia enterocolitica	Pigs	−1.3	4.2	0.945	$D_{60} = 1.0$		Fever, diarrhoea, severe abdominal pain, vomiting, joint pain within 24–36 h	Pork, milk, tofu, chitterlings (pork intestine), sausages

[a] See section 1.2.4 for details.
[b] Most serotypes do not grow <7 °C.
[c] Most serotypes do not grow below pH 4.5.

Data adapted from:
ANON, (2001), Fish and fisheries products hazards and controls guidance, US Food & Drug Administration Center for Food Safety & Applied Nutrition, Third Edition, June, available at http://vm.cfsan.fda.gov/~comm/haccp4.html, and follow link to Appendix 4.

ANON, (2008), Foodborne disease profiles, Food Safety for Nutritionists and other Health Professionals, Industry Council for Development, available at www.icd-online.org/an/html/ coursesfood.html and follow links to 'Food safety' and 'Foodborne disease profiles'.

BELL, C. and KYRIAKIDES, A.. (2002), Pathogenic *Escherichia coli*, in (C. de W. Blackburn and P.J. McClure, Eds.), *Foodborne Pathogens Hazards, Risk Analysis and Control*, Woodhead Publishing, Cambridge, pp. 279–306.

BELL, C. and KYRIAKIDES. A.. (2002), *Salmonella*, in (C. de W. Blackburn and P.J. McClure, Eds.), *Foodborne Pathogens Hazards, Risk Analysis and Control*, Woodhead Publishing, Cambridge, pp. 307–335.

BELL, C. and KYRIAKIDES, A.. (2002), *Listeria monocytogenes*, in (C. de W. Blackburn and P.J. McClure, Eds.), *Foodborne Pathogens Hazards, Risk Analysis and Control*, Woodhead Publishing, Cambridge, pp. 337–361.

GIBBS, P.. (2002), Characteristics of spore-forming bacteria, in (C. de W. Blackburn and P.J. McClure, Eds.), *Foodborne Pathogens Hazards, Risk Analysis and Control*, Woodhead Publishing, Cambridge, pp. 418–435.

GRIFFITHS. M. (2002), *Mycobacterium paratuberculosis*, in (C. de W. Blackburn and P.J. McClure, Eds.), *Foodborne Pathogens Hazards, Risk Analysis and Control*, Woodhead Publishing, Cambridge, pp. 489–500.

MCCLURE, P. and BLACKBURN, C.. (2002), *Campylobacter* and *Arcobacter*, in (C. de W. Blackburn and P.J. McClure, Eds.), *Foodborne Pathogens Hazards, Risk Analysis and Control*, Woodhead Publishing, Cambridge, pp. 363–384.

PARK, R.W.A., GRIFFITHS, P.L. and MORENO, G.S., (1991), Sources and survival of campylobacters – relevance to enteritis and the food industry, *J. Applied Bacteriology*, **70**, S97–S106.

SUTHERLAND, J. and VARNAM, A.. (2002), Enterotoxin-producing *Staphylococcus*, *Shigella*, *Yersinia*, *Vibrio*, *Aeromonas* and *Plesimona*, in (C. de W. Blackburn and P.J. McClure, Eds.), *Foodborne Pathogens Hazards, Risk Analysis and Control*, Woodhead Publishing, Cambridge, pp. 386–415.

Appendix B.2

Characteristics of viral foodborne infections

Virus type	Incubation time	Symptoms	Age groups at risk	Duration of illness	Severity
Adenovirus	7–14 days	Diarrhoea	Children <5 years	Days/weeks	Mild
Astrovirus	24–48 hours	Diarrhoea	Children <10 years	Days	Mild
Calicivirus	24–48 hours	Vomiting/ diarrhoea	All (especially institutionalised or hospitalised people)	Days	Mild
Enterovirus	1–2 weeks	Diarrhoea, meningitis, encephalitis, paralytic illness	Children <15 years	Days/weeks, can be life-long	Can be severe
Hepatitis A	Up to 50 days	Hepatitis	All if endemic	Weeks	Severity increases with age of first infection
Hepatitis E	Up to 70 days	Hepatitis	All	Weeks	Mild except for pregnant women
Rotavirus	24–48 hours	Vomiting, diarrhoea, fever	Children <5 years	Days	Major cause of death in developing countries

Adapted from:
KOOPMANS, M., (2002), Viruses, in (C. de W. Blackburn and P.J. McClure, Eds.), *Foodborne Pathogens – Hazards, Risk Analysis and Control*, Woodhead Publishing, Cambridge, pp. 440–452.

Appendix B.3

Enzymes from GM micro-organisms used in food processing

Enzyme	Host organism	Donor organism	Application
α-Acetolactate decarboxylase	Bacillus amyloliquefaciens or subtilis	Bacillus sp.	Soft drinks, beers, wines
	Saccharomyces cerevisiae	Enterobacter sp.	Soft drinks, beers, wines
Aminopeptidase	Trichoderma reesei or longibrachiatum	Aspergillus sp.	Cheese, egg, meats, milk, spices & flavours
α-Amylase	Bacillus amyloliquefaciens or subtilis	Bacillus sp.	Bakery products, soft drinks, beers, wines, cereal & starch processing
		Thermoactinomyces sp.	Bakery products
	Bacillus licheniformis	Bacillus sp.	Soft drinks, beers, wines, cereal & starch processing, sugar & honey processing
Arabinofuranosidase	Aspergillus niger	Aspergillus sp.	Wast
Catalase	Aspergillus niger	Aspergillus sp.	Soft drinks, beers, wines
Cellulase	Trichoderma reesei or longibrachiatum	Trichoderma sp.	Cereal & starch processing
Cyclodextrin glucanotransferase	Bacillus licheniformis	Thermoanaerobacter sp.	Soft drinks, beers, wines
α-galactosidase	Saccharomyces cerevisiae	Guar plant	Soft drinks, beers, wines
β-glucanase	Bacillus amyloliquefaciens or subtilis	Bacillus sp.	Cereal & starch processing
	Trichoderma reesei or longibrachiatum	Trichoderma sp.	Soft drinks, beers, wines, fruits & vegetables, cereal & starch processing
Glucoamylase or amyloglucosidase	Aspergillus niger	Aspergillus sp.	Cereal & starch processing
Glucose isomerase	Streptomyces lividans	Talaromyces sp.	Cereal & starch processing
	Streptomyces rubiginosus	Actinoplanes sp	Cereal & starch processing
Glucose oxidase	Aspergillus niger	Streptomyces sp.	Bakery products, cheese, egg, milk
	Aspergillus oryzae	Aspergillus sp.	Bakery products
Hemicellulase	Bacillus amyloliquefaciens or subtilis	Aspergillus sp.	Bakery products, cereal & starch processing
Hexose oxidase	Hansenula polymorpha	Bacillus sp.	Bakery products, cheese, fats & oils, milk, soups & broths, cereal & starch processing
		Chordrus sp.	
Inulase	Aspergillus oryzae	Aspergillus sp.	Cereal & starch processing

Laccase	*Aspergillus oryzae*	*Myceliopthora* sp.	Soft drinks, beers, wines
		Polyporus sp.	Soft drinks, beers, wines
Lactase or β-galactosidase	*Aspergillus oryzae*	*Aspergillus* sp.	Cheese, dietary foods, edible ice, milk
	Kluyveromyces lactis	*Kluyveromyces* sp.	Edible ice, milk
Lipase triaclyglycerol	*Aspergillus oryzae*	*Candida* sp.	Fats & oils
		Fusarium sp.	Bakery products, fats & oils
		Rhizomucor sp.	Cheese, fats & oils, spices & flavours
		Thermomyces sp.	Bakery products, fats & oils
Lipoxygenase	*Escherichia coli*	Pea	Bakery products, spices & flavours
Maltogenic amylase	*Bacillus amyloliquefaciens* or *subtilis*	*Bacillus* sp.	Bakery products, cereal & starch processing
endo-1,4-beta-Mannanase	*Trichoderma reesei* or *longibrachiatum*	*Trichoderma* sp.	Cereal & starch processing
Pectin lyase	*Aspergillus niger* var. *awamori*	*Aspergillus* sp.	Soft drinks, beers, wines, cocoa, chocolate, coffee & tea, fruits & vegetables
Pectin methyl-esterase or Pectinesterase	*Aspergillus niger*	*Aspergillus* sp.	Soft drinks, beers, wines, fruits & vegetables
	Trichoderma reesei or *longibrachiatum*	*Aspergillus* sp.	Soft drinks, beers, wines, cocoa, chocolate, coffee & tea, fruits & vegetables
Pentosanase	*Aspergillus niger*	*Aspergillus* sp.	Soft drinks, beers, wines, cocoa, chocolate, coffee & tea, fruits & vegetables
Phospholipase A	*Aspergillus oryzae*	*Aspergillus* sp.	Soft drinks, beers, wines, fruits & vegetables
	Trichoderma reesei or *longibrachiatum*	*Aspergillus* sp.	Soft drinks, beers, wines, cocoa, chocolate, coffee & tea, fruits & vegetables
Phospholipase B	*Bacillus amyloliquefaciens* or *subtilis*	*Bacillus* sp.	Bakery products
	Aspergillus oryzae	*Fusarium* sp.	Bakery products
Polygalacturonase or Pectinase	*Trichoderma reesei* or *longibrachiatum*	*Aspergillus* sp.	Bakery products, fats & oils
	Trichoderma reesei or *longibrachiatum*	*Aspergillus* sp.	Bakery products, cereal & starch processing
	Aspergillus niger	*Aspergillus* sp.	Soft drinks, beers, wines, cocoa, chocolate, coffee & tea, fruits & vegetables
	Trichoderma reesei or *longibrachiatum*	*Aspergillus* sp.	Soft drinks, beers, wines, cocoa, chocolate, coffee & tea, fruits & vegetables

Enzyme	Host organism	Donor organism	Application
Protease (incl. milk clotting enzymes)	*Aspergillus niger* var. *awamori*	Calf stomach	Cheese
	Aspergillus oryzae	*Rhizomucor* sp.	Cheese, meat
	Bacillus amyloliquefaciens or subtilis	*Bacillus* sp.	Bakery products, soft drinks, beers, wines, cheese, fish, meat, milk, cereal & starch processing
	Bacillus licheniformis	*Bacillus* sp.	Fish, meat
	Cryphonectria or *Endothia parasitica*	*Cryphonectria* sp.	Cheese
	Kluyveromyces lactis	Calf stomach	Cheese
Pullulanase	*Bacillus licheniformis*	*Bacillus* sp.	Cereal & starch processing
	Bacillus subtilis	*Bacillus* sp.	Soft drinks, beers, wines, cereal & starch processing
	Klebsiella planticola	*Klebsiella* sp.	Soft drinks, beers, wines, cereal & starch processing
Xylanase	*Trichoderma reesei* or *longibrachiatum*	*Hormoconis* sp.	Bakery products
	Aspergillus niger	*Aspergillus* sp.	Bakery products, soft drinks, beers, wines
	Aspergillus niger var. *awamori*	*Aspergillus* sp.	Bakery products
	Aspergillus oryzae	*Aspergillus* sp.	Cereal & starch processing
		Thermomyces sp.	Bakery products
	Bacillus amyloliquefaciens or subtilis	*Bacillus* sp.	Bakery products, soft drinks, beers, wines, cereal & starch processing
	Bacillus licheniformis	*Bacillus* sp.	Cereal & starch processing
	Trichoderma reesei or *longibrachiatum*	*Trichoderma* sp.	Soft drinks, beers, wines, cereal & starch processing

Adapted from:
ANON, (2004), AMFEP list of commercial enzymes, the Association of Manufacturers and Formulators of Enzyme Products, available at www.amfep.org/list.html

Appendix B.4

Examples of food fermentations

Type of fermentation	Raw material	Fermented foods and examples of countries/areas	Micro-organisms	Typical incubation conditions	
				Temp (°C)	Time (h)
Lactic acid	Cabbage, radish, red peppers	Sauerkraut (Europe, USA), kimchi (Korea)	Leuconostoc mesenteroides, Lactobacillus brevis, L. plantarum		
	Vegetables, cucumber, olive, mango, lime	Pickles (Middle East, Europe, USA, India, Korea, Thailand, China)	Lactobacillus mesenteroides, L. brevis, L. plantarum, Penecillium cerevisiae	Amb[a]	48–260
	Milk	Yoghurt (worldwide)	Streptococcus thermophilus, Lactobacillus bulgaricus	40–45	2–3
		Fermented milks and creams (e.g. kefir (Russia), liban (Iraq), dahi (India), laban (Egypt))	Lactococcus lactis subsp. cremoris, Lactococcus lactis ssp. lactis, Acetobacter orientalis, Lactobacillus acidophilus, L. delbruecki ssp. bulgaricus, L. salivarius,		
		Cheeses (western world)	Streptococcus cremoris, S. diacetylactis, S. Lactis, Penecillium sp.	32	14–16[b]
	Milk/wheat	Kishk (Egypt), trahanas (Greece, Turkey)			
	Tubers (e.g. cassava)	Kenkey (Ghana), gari (Nigeria), pozol (Mexico)	Corynebacterium sp., Geotrichum sp.	Amb	96
	Rice/shrimps	Balao balao, burong dalag (Philippines)	Bacillus pumilus, B. licheniformis	Amb	24–72
	Wheat, rice, maize, lentils	Sourdough breads (Western world), idli (India), hoppers (Sri Lanka) injera (Ethiopia), kisra (Sudan), puto (Philippines)	Leuconostoc mesenteroides, Streptococcus faecalis		

Type of fermentation	Raw material	Fermented foods and examples of countries/areas	Micro-organisms	Typical incubation conditions	
				Temp (°C)	Time (h)
	Maize	Non-alcoholic beverages (e.g. mahewu (Southern Africa))	*L. delbrueckii*	45	
	Meat	Fermented sausage (Western world, Thailand)	*Pediococcus cerevisiae, Lactobacillus plantarum, L. curvatus*	15–26	24
Acetic acid	Wines, coconut water	Vinegars (Western world), kombucha Europe, Indonesia, Japan), nata de coco (Philippines)	*Acetobacter aceti*	25	72–120
Alcoholic	Grapes, other fruits, honey, palm sap, sugarcane, rice	Wines (worldwide except Muslim countries)	*Saccharomyces cerevisiae, S. cerevisiae var. ellipsoideus, S. carbajali, S. oviformis, S. chevalieri, Saccharomycopsis fibuliger, Kloeckera apiculata, Zymomonas mobilis, Amylomyces rouxii, S. sake, Zymomonas sp.*	22–30	100–360[c]
	Cereals (e.g. rice, maize)	Beers (worldwide except Muslim countries)	*Saccharomyces cerevisiae* (top yeast), *S. carlsbergensis* (bottom yeast), *Leuconostoc mesenteroides*	20 12–15	120–240
	Wheat and other cereals	Leavened breads (Western countries, Middle East)		26	0.5–1
Mixed (lactic acid and/or yeasts/moulds)	Fish, soybeans	Fish sauces (e.g nuocmam (Vietnam), shoyu and miso (Japan), patis (Philippines), budu (Malaysia)), and pastes (e.g. bagoong (Philippines), belachan (Malaysia), mam (Vietnam), prahoc (Cambodia) and soy sauce (China))	1st stage: *Aspergillus oryzae, A. soyae, Mucor* sp., *Rhizopus* sp. 2nd stage: *Pediococcus soyae, Saccharomyces rouxii*	30 15–25	48–72 3–6 months

Cheese	Roquefort, Stilton, Gorgonzola	*Penicillium roquefort*, *Penecillium notatum*		
Cassava, rice	Tapé (Indonesia)	*Amylomyces rouxii*		
Soybeans	Koji (Japan), citric acid; Textured products – tempeh (Indonesia)	*Aspergillus oryzae*, *Rhizopus oligosporus*		
Cocoa beans	Cocoa (West Africa, South America)	*Lactobacillus plantarum, L. mali, L. fermentum, L. collinoides, Acetobacter rancens, A. aceti, A. oxydans*	Amb	144
Coffee beans	Coffee (East Africa, South Asia, South America)	*Leuconostoc* sp., *Lactobacillus* sp., *Bacillus* sp., *Erwinia* sp. *Aspergillus* sp., *Fusarium* sp.	Amb	20–100
Alkaline	Dawadawa, iru, ogiri (West Africa), kenima (India), natto (Japan)	*Bacillus subtilis*	Amb	
Locust bean, melon seeds, sesame seeds, castor beans				

[a] Ambient.
[b] Fermentation of ripened cheeses continues for 1–12 months.
[c] Wines are also aged from 1–5 years.

Adapted from:

RAIMBAULT, M., (1998), General and microbiological aspects of solid substrate fermentation, *Electronic J. Biotechnology*, **1** (3), 174–188, available at http://www.scielo.cl/scielo.php?pid=S0717-34581998000300007&script=sci_arttext

STEINKRAUS, K.H., (2002), Fermentations in world food processing, *Comprehensive Reviews in Food Science and Food Safety*, **1**, 23–32.

Appendix C.1

Units and conversions

Units

Quantity	Unit name	SI unit
Length	metre	m
Mass	kilogram	kg
Temperature	Celsius	°C
Thermodynamic temperature	kelvin	K (273 + °C)
Time	second	s
Derived units with a special name		
Absorbed dose (ionising radiation)	gray (Gy)	$J\,kg^{-1}$
Amount of substance	mole	M
Electric current	ampere (or amp)	A
Energy	joule (J)	$N\,m$
Force	newton (N)	$kg\,m\,s^{-2}$
Frequency	hertz	Hz
Luminous intensity	candela	Cd
Power	watt (W)	$J\,s^{-1}$
Pressure	pascal (Pa)	$N\,m^{-2}$
Derived units without a special name		
Acceleration		$m\,s^{-2}$
Area		m^2
Density		$kg\,m^3$
Dynamic viscosity		$N\,s\,m^{-2}$
Enthalpy		$J\,kg^{-1}$
Heat transfer coefficient		$W\,m^{-2}\,K^{-1}$ (or $W\,m^{-2}\,°C^{-1}$)
Kinematic viscosity		$m^2\,s^{-1}$
Momentum		$kg\,m\,s^{-1}$
Specific gravity		none
Specific heat		$J\,kg^{-1}\,K^{-1}$ (or $°C^{-1}$)
Thermal conductivity		$W\,m^{-1}\,K^{-1}$ (or $W\,m^{-1}\,°C^{-1}$)
Velocity		$m\,s^{-1}$
Volume		m^3

Conversions

Conversions between Imperial and metric units can be made using a number of websites, including:
http://www.simetric.co.uk/
http://www.sengpielaudio.com/calculator-millimeter.htm
http://www.ajdesigner.com/phpnaturallog/natural_log_equation_y.php
http://www.onlineconversion.com/temperature.htm
http://www.convert-me.com/en/

Mesh sizes (adapted from information from Glen Mills Inc. available at www.glenmills.com)

Micrometres	Tyler mesh	US standard mesh	Inches
4760	4	4	0.185
3360	6	6	0.131
2380	8	8	0.093
1680	10	12	0.065
1190	14	16	0.046
840	20	20	0.0328
590	28	30	0.0232
420	35	40	0.0164
297	48	50	0.0116
250	60	60	0.0097
210	65	70	0.0082
177	80	80	0.0069
149	100	100	0.0058
105	150	140	0.0041
74	200	200	0.0029
62	250	230	0.0024
53	270	270	0.0021
44	325	325	0.0017
37	400	400	0.0015

Appendix D.1

Glossary

Absorption	uptake of moisture by dry foods
Acid food	a food with a pH of less than 4.6 and a water activity (a_w) equal to or greater than 0.85
Active packaging	packaging that senses changes in the internal or external environment of a food package and responds by altering its properties
Actuator	a mechanical device for moving or controlling a mechanism or system
Additives	chemicals added to food to improve their eating quality or shelf-life
Adiabatic	changes to the humidity and temperature of air without loss or gain of heat (in drying)
Adiabatic process	processing in which no heat is added or removed from a system
Adulterants	chemicals that are intentionally added to food which are forbidden by law
Aflatoxins	toxins produced by toxigenic strains of *Aspergillus* spp.
Agglomeration	the production of granules from powder particles
Algorithms	software building blocks used to construct control sequences in computerised process control
Alkaline phosphatase	an enzyme in raw milk having a similar D-value to heat-resistant pathogens, used to test for effectiveness of pasteurisation
Annealing	heating to control the ductility of a material
Archaea	prokaryotic micro-organisms. Eukaryotes, Archaea and Bacteria form the three-domain classification system
Aseptic processing	heat sterilisation of foods before filling into pre-sterilised (aseptic) containers
Atomiser	a device to form a fine droplets of food (e.g. in a spray dryer)
Bacteriocins	naturally produced peptides that inhibit other micro-organisms, similar in effect to antibiotics
Baroresistance	resistance to high pressure
Barosensitivity	sensitivity to high pressure
Bingham or Casson plastic fluids	fluids that flow after a critical shear stress is exceeded (e.g. tomato ketchup)
Biological oxidation demand (BOD)	a measure of the oxygen requirement by micro-organisms when breaking down organic matter, used as a measure of the polluting potential of materials in water but is no longer used to calculate effluent treatment charges
Bioplastics	non-petroleum-based natural biopolymers
Black body	a theoretical concept for a material that can either absorb all the heat that lands on it or radiate all of the heat that it contains
Blancher	equipment used to blanch foods
Blanching	heating foods, especially vegetables to below 100 °C for a short time, to both inactivate enzymes which would cause a loss of quality during storage and to remove air and soften the food
Blinding	blocking of a sieve by food particles
Bloom	a thin layer of unstable forms of cocoa fat that crystallise at the surface of a coating to produce dullness or white specks
Botulin	an exotoxin produced by *Cl. botulinum*, able to cause fatal food poisoning
Bound moisture	liquid physically or chemically bound to a solid food matrix that exerts a lower vapour pressure than pure liquid at the same temperature

Boundary film (or surface film)	film of fluid next to the surface over which a fluid flows that causes a resistance to heat transfer
Breading	the application of pre-prepared bread crumbs to the surface of a food
C-value	'cook-value' – processing needed to achieve the required change in sensory characteristics (in heat processing)
Calandria	heat exchanger section in an evaporator
Calendering	passing film through heated rollers
Carborundum	an abrasive material made from silicon and carbon
Case hardening	formation of a hard impermeable skin on some foods during drying, which reduces the rate of drying and produces a food with a dry surface and a moist interior
Cavitation	production of bubbles in foods by ultrasound and their rapid expansion/ contraction
Centrifugation	the separation of immiscible liquids, or solids from liquids by the application of centrifugal force
Chelating agents	chemicals that sequester trace metals
Chemical oxidation demand (COD)	a chemical method used to measure the polluting potential of materials in water
Chilling	reduction in the temperature of a food to between -1 and $8\,°C$
Chilling injury	physiological changes to some types of fruits and vegetables caused by low temperatures that result in loss of eating quality
Chiral	describes atoms that can exist in two different spatial configurations that are the mirror image of each other
Choke	restriction of the outlet to a mill to retain particles until sufficiently small (or restriction of the outlet in an extruder)
Climacteric	abrupt increase in respiration rate in some fruits during ripening
Clinching	partial sealing of can lids
Coating	a generic term to describe the application of a viscous covering (such as batter, chocolate starch/sugar mixtures) to the surface of a food
Coextrusion	(1) the simultaneous extrusion of two or more packaging films to make a co-extruded film; or (2) the extrusion of two foods in which a filling is continuously injected into an outer casing in an extruder
Cold shortening	undesirable changes to meat caused by cooling before rigor mortis has occurred
Collapse temperature	the maximum temperature of a frozen food before solute movement causes a collapse of the food structure and prevents movement of water vapour during freeze drying
Commercial sterility	a term used in heat sterilisation to indicate that processing inactivates substantially all micro-organisms and spores which, if present, would be capable of growing in the food under defined storage conditions
Common Object Resource Based Architecture (CORBA)	computer software that acts as an information broker to link process control systems with other computerised company information
Compound coating	a coating material in which cocoa solids and hardened vegetable oils are used to replace cocoa butter
Conduction	the movement of heat by direct transfer of molecular energy within solids
Conductivity	(1) (electrical) physical property of a food material that determines its ability to conduct electricity (expressed in siemens per m (Sm^{-1}). It enables heating to occur in ohmic heating; (2) (thermal) physical property of a food material that determines its ability to conduct heat (in watts per metre $°C$).
Constant-rate drying	the drying period in which the rate of moisture loss is constant when surface moisture is removed
Continuous phase	the medium that contains the dispersed phase in an emulsion
Convection	the transfer of heat in fluids by groups of molecules that move as a result of differences in density or as a result of agitation
Critical control point (CCP)	a processing factor of which a loss of control would result in an unacceptable food safety or quality risk

Critical moisture content	the amount of moisture in a food at the end of the constant-rate period of drying
Cross-field	an ohmic heating system where the electric field is aligned across the product flow path
Crumb	pre-prepared breadcrumbs used to cover food pieces, or the porous inner part of baked foods
Crust	hard surface layer on baked foods
Cryogen	a refrigerant that absorbs latent heat and changes phase from solid or liquid to a gas (e.g. subliming or evaporating carbon dioxide or liquid nitrogen)
Cryogenic freezers	equipment that uses subliming or evaporating carbon dioxide or liquid nitrogen directly in contact with food to freeze it
Cryogenic grinding	mixing liquid nitrogen or solid carbon dioxide with food to cool it during grinding
Cryoprotectant	compounds that depress the freezing temperature of foods, modify or suppress ice crystal growth during freezing and inhibit ice recrystallisation during frozen storage (e.g. 'antifreeze proteins')
Cryostabiliser	compounds that produce increase the viscosity of solutions and restrict the mobility of water. They may also restrict ice crystal growth by absorption to nuclei but do not depress the freezing point or increase osmolality as do cryoprotectants
Dead-folding	a crease or fold made in a material that will stay in place
Decimal reduction time (D-value)	the time needed to destroy 90% of micro-organisms (to reduce their numbers by a factor of 10)
Dehydrofreezing	partially drying food before freezing
Depositor	machine for placing an accurate amount of food onto a conveyor or into a mould
Desorption	removal of moisture from a food
Detergents	chemicals that reduce the surface tension of water and hence assist in the release of soils from equipment or foods
Dew point	temperature at which an air–water vapour mixture becomes saturated with moisture, marking the onset of condensation
Diafiltration	a process to improve the recovery of solutes by diluting the concentrate during reverse osmosis or ultrafiltration
Die	a restricted opening at the discharge end of an extruder barrel
Dielectric constant	the ratio of the capacitance of a food to the capacitance of air or vacuum under the same conditions
Dielectric heating	a generic term that includes heating by both microwave and radio frequency energy
Dilatant material	food in which the viscosity increases with shear rate (e.g. liquid chocolate and cornflour suspension)
Direct heating ovens	those in which products of combustion are in contact with the food
Dispersed phase	droplets in an emulsion
Dosimeter	a device that qualitatively or quantitatively measures the dose of irradiation
Dry bulb temperature	temperature measured by a dry thermometer in an air–water vapour mixture
Effect	an evaporator
Effective freezing time	the time required to lower the temperature of a food from an initial value to a predetermined final temperature at the thermal centre
Electrical conductivity	the capacity of a material to conduct electricity
Electrodialysis	the separation of electrolytes into anions and cations by the application of a direct electrical current and the use of ion-selective membranes
Electroheating	see ohmic heating
Electroporation	the formation of pores in cell membranes by applied electric fields
Emulsification	creation of an emulsion by the dispersion of one immiscible liquid (dispersed phase) in the form of small droplets in a second immiscible liquid (continuous phase)
Emulsifying agent	chemical that forms micelles around each droplet in the dispersed phase

of an emulsion to reduce interfacial tension and prevents droplets from coalescing

Encapsulation	the coating of microscopic particles with another material
Enrobing	the unit operation in which food pieces are coated with chocolate or other materials
Enthalpy	heat content, a description of thermodynamic potential of a system
Entrainment	(1) oil droplets that are carried over in steam produced by vigorously frying foods, leading to loss of oil; or (2) loss of concentrated droplets of product with vapour during evaporation by boiling
Entropy	in a closed thermodynamic system, a quantitative measure of the amount of thermal energy not available to do work
Equilibrium moisture content	the moisture content of a food at which it neither gains nor loses moisture to its surroundings (at a given temperature and pressure, the food is in equilibrium with the air–vapour mixture surrounding it)
Equilibrium relative humidity	relative humidity of the storage atmosphere in equilibrium with the moisture content of food
Eutectic temperature	the temperature at which a crystal of an individual solute exists in equilibrium with a solution and (in freezing) with the unfrozen liquor and ice
Exhausting	removal of air from a container before heat processing
Expeller	a horizontal barrel, containing a helical screw, used to extract oil from seeds or nuts
Expression	the separation of liquids from solids by applied pressure
Extractors	equipment used to extract food components using solvents
Extruder	one or more screws rotating in a barrel with restricted apertures at the discharge end, used for producing extruded foods
Extrusion	a process that involves the combination of several unit operations including mixing, cooking, kneading, shearing, shaping and forming to produce extruded foods
F-value	the time needed to reduce microbial numbers by a multiple of the D-value
Falling-rate drying	the drying period in which the rate of moisture loss declines
Feedback control	automatic control of a process using information from sensors to adjust processing conditions
Feed-forward control	comparison of processing conditions with a model system, used in automatic process control
Field heat	heat within crops when they are harvested
Filter cake	solids removed by filtration
Filter medium	porous material through which food are filtered
Filtrate	the liquor remaining after solids are removed by filtration
Filtration	the separation of solids from liquids by passing the mixture through a bed of porous material
Final eutectic temperature	(in freezing) the lowest eutectic temperature of solutes in equilibrium with unfrozen liquor and ice
Finish	the part of a jar or bottle that holds the closure
Flash pasteurisation	also known as higher-heat shorter-time processing, involving heat treatment greater than 72 °C for 15 s for milk
Flash-over	arcing of electricity between electrodes without heating taking place
Fluidisation	suspension of powders and small-particulate foods in air so that they can be more easily handled as fluids
Fluence	energy imparted by light to the surface of a material
Flux	flow of liquid through reverse osmosis or ultrafiltration membranes
Foam	a colloidal system with a liquid or solid continuous phase and a gaseous dispersed phase
Focusing	concentration of electromagnetic waves inside a food due to its curved surface (as a lens focusing light waves), which leads to enhanced heating of the interior
Forming	moulding of doughs and other materials into different shapes
Fouling	deposits of food or limescale on surfaces of heat exchangers

Free moisture	moisture in excess of the equilibrium moisture content at a particular temperature and humidity, and so free to be removed
Freeze concentration	concentration of liquid foods by freezing water to ice and removal of ice crystals
Freeze drying	dehydration of food by freezing water to form ice, followed by removal of ice by sublimation
Freezing plateau	the period during freezing when the temperature of a food remains almost constant as latent heat of crystallisation is removed and ice is formed
Friability	the hardness of a food and its tendency to crack
Fuzzy logic	a control system that mimics human decision making, which is empirically based without a mathematical model
Glass transition temperature	the temperature at which a supersaturated solution (amorphous liquid) converts to a glass
Glassy state	a substance existing as an amorphous, non-crystalline solid
Grading	the assessment of a number of attributes to obtain an indication of overall quality of a food
Grey body	a concept used to take account of the fact that materials are not perfect absorbers or radiators of heat
Half-life	the time taken for an isotope to lose half of its radioactivity
Hazard analysis	the identification of potentially hazardous ingredients, storage conditions, packaging, critical process points and relevant human factors which may affect product safety or quality
Headspace	the space in a container between the surface of a food and the underside of the lid
Heat sterilisation	destruction of the majority of micro-organisms in a food by heating
Hermetically sealed container	a package that is designed to be secure against entry of micro-organisms and maintain the commercial sterility of its contents after processing
Heterofermentative micro-organisms	micro-organisms that produce more than one main metabolic product
Homofermentative micro-organisms	micro-organisms that produce a single main by-product
Homogenisation	the reduction in size and increase in number of solid or liquid particles in the dispersed phase
Humectants	chemicals (e.g. salt, sugar, glycerol) that are able to lower the water activity in a food by depressing the vapour pressure
Humidity	partial pressure of water vapour in air
Hydrocooling	immersion of fruits and vegetables in chilled water
Hydrophile–lipophile balance (HLB value)	the ratio of hydrophilic to hydrophobic groups on the molecules of an emulsifier
Hydrogenisation	increasing the number of saturated bonds in oils by treatment with hydrogen
Hydrostatic head	the pressure resulting from the weight of a column of liquid
Hygroscopic foods	foods in which the partial pressure of water vapour varies with the moisture content
Hyperfiltration	reverse osmosis
Impact strength	the force required to penetrate a material
Impingement	(in heating or cooling) air jets that produce high-velocity air, which impinges perpendicularly to the surface of products
Indirect heating ovens	ovens in which heat from combustion is passed through a heat exchanger to heat air which is then in contact with the food
Intelligent packaging	a package that has a function that switches on and off in response to changing external/internal conditions and communicates information on the status of the product to the customer or user
Inventory	the stored accumulation of materials in an operation
Ion exchange	the selective removal of charged molecules from a liquid by electrostatic adsorption, followed by their transfer to a second liquid using an ion-exchange material
Ionisation	breakage of chemical bonds (e.g. during irradiation)

Irradiation	the use of γ-rays to preserve foods by destruction of micro-organisms or inhibition of biochemical changes
Isostatic	uniform pressure throughout a food
Isotope	a source of γ-rays, from a radioactive material such as cobalt[60] or caesium[137]
Just-in-time	a management system in which goods are ordered as they are required and stocks are not held in warehouses
Kinetic energy	energy due to motion
Lacquer	internal protective coating in cans
Lamination	bonding together of two or more packaging films, papers or foods
Latent heat	heat taken up or released when a material undergoes a change of state
Leaching	washing out of soluble components from the food
Lethality	integrated effect of heating temperature and time on micro-organisms
Logistics	systems involved in food distribution
Loss factor	(also termed the 'dielectric loss' or 'loss tangent') a measure of the amount of energy that a material will dissipate when subjected to an alternating electric field (in microwave and dielectric heating)
Low-acid food	a food with a pH greater than 4.6 and a water activity (a_w) equal to or greater than 0.85
Lyophilisation	freeze drying
Magnetic flux density	force that an electromagnetic source exerts on charged particles (measured in tesla)
Magnetron	The component of a microwave system that generates microwaves
Maillard reactions	a series of biochemical reactions between reducing sugars and amino acids that cause non-enzymic browning, flavour development and nutrient losses
Manosonication	ultrasound treatment combined with slightly raised pressure
Manothermosonication	a combined heat and ultrasound treatment under increased pressure
Manufacturing resource planning	computer-based systems used to control distribution networks by using forecasted demand for and actual orders to assist management decisions
Material requirement planning	a single integrated computer system, containing a database that can be accessed by all parts of the company for management planning
Mechanical refrigerators	equipment which evaporates and compresses a refrigerant in a continuous cycle, using cooled air, cooled liquid or cooled surfaces to freeze foods
Metallisation	a thin coating of aluminium on plastic packaging
Microencapsulation	*see* encapsulation
Microfiltration	a pressure-driven membrane process using membranes with a pore size of 0.2–2 μm at lower pressures than ultrafiltration
Microwaves	electromagnetic waves produced commercially at specified frequencies for industrial heating
Mimetics	low-calorie fat substitutes
Mimic panel	a graphical electronic display of a process
Moulders	machines that form dough or confectionery into different shapes
Multiple effect	(in evaporation) the reuse of vapour from boiling liquor in one evaporator as the heating medium in another evaporator operating at a lower pressure
Nanocomposite	a polymer containing ultra-small inorganic molecules, used for food packaging
Nanofiltration	a membrane process to separate particles with molecular weights from 300 to 1000 Da, using lower pressures that reverse osmosis
Neural networks	computer systems that can analyse complex relationships in a process and 'learn' from experience
Neutraceutical	(or 'functional foods') foods that provide benefits other than the nutrients required for normal health
Newtonian fluid	food in which the viscosity does not change with rate of shear
Nip	the gap between rollers in a mill or a moulding/forming machine
Nominal freezing time	the time between the surface of the food reaching 0 °C and the thermal centre reaching 10 °C below the temperature of the first ice formation

Non-hygroscopic foods   foods that have a constant water vapour pressure at different moisture contents

Non-Newtonian liquid   food in which the viscosity changes with rate of shear

Non-thermal effects   effects of a process that are not of thermal origin, and cannot be explained by measured temperature changes

Nucleation   the formation of a nucleus of water molecules that is required for ice crystal formation

Ohmic heating   direct electrical heating of foods

Ostwald ripening   a reduction in the numbers of small ice or fat crystals and an increase in the size of larger crystals

Overall heat transfer coefficient (OHTC)   the sum of the resistances to heat flow due to conduction and convection

Panning   the process of building up thin layers of sugar, sweetener or other coatings in a controlled way onto solid cores of nuts, fruit, etc.

Parison   a mould used to form glass or plastic bottles

Pasteurisation   a relatively mild heat treatment, in which food is heated to below 100 °C to preserve it without substantial changes to sensory characteristics or nutritional value. In low-acid foods, the main reason for pasteurisation is destruction of pathogens

Pinholes   small holes in can seams or flexible packaging

Plasticiser   chemicals added to plastic films to make them more flexible

Polymorphic fat   a fat that can crystallise into more than one form

Potential energy   energy due to position of an object

Power cycling   turning a microwave source on and off

Prebiotic   foods that contain ingredients that are not digested but stimulate the growth of probiotic bacteria in the colon

Preconditioning   adjustment of the moisture content of a food (e.g. before extrusion)

Pre-forms   small dense pellets made by extrusion from pre-gelatinised cereal dough, suitable for extended storage until they are converted to snackfoods by frying, toasting or puffing (also known as 'half-products')

Press cake   solid residue remaining after extraction of liquid component from foods

Probiotic   foods that contain probiotic bacteria that promote gut health

Process inter-locking   linking different parts of a process so that one cannot operate until a second is correctly set up

Programmable logic controller (PLC)   a microprocessor that is used in process control to replace electrical relays and to collect and store process data

Pseudoplastic material   food in which the viscosity decreases with increasing shear rate (e.g. emulsions, fruit purées)

Psychrometrics   the study of inter-related properties of air–water vapour systems

Radiation   the transfer of heat by electromagnetic waves

Radio frequency energy   electromagnetic waves produced commercially at frequencies of 13.56, 27.12 or 40.68 MHz for industrial heating

Radiolysis   changes to a food material caused by ionising radiation to produce chemicals that destroy micro-organisms, etc.

Recrystallisation   physical changes to ice crystals (changes in shape, size or orientation), which are an important cause of quality loss in some frozen foods

Redox potential   oxidation/reduction potential of a food or microbial substrate

Refrigerant   a liquid that has a low boiling point and high latent heat of vaporisation so that it can change phase and absorb or lose heat in a refrigerator

Refrigerators   equipment that evaporates and compresses a refrigerant in a continuous cycle, using cooled air, cooled liquid or cooled surfaces to freeze foods

Relative humidity   the ratio of the partial pressure of water vapour in air to the pressure of saturated water vapour at the same temperature, multiplied by 100

Respiration   metabolic activity of living animal or plant tissues

Retort   a pressurised vessel, used to heat foods above 100 °C during canning

Reverse osmosis   unit operation in which small molecular weight solutes are selectively removed by a semi-permeable membrane under high pressure

Rheopectic material   a food in which the viscosity increases with shear stress (e.g. whipping cream)

Runaway heating	a cycle of increasing temperature in food causing an increasing rate of microwave/ohmic energy absorption that further increases the rate of temperature rise. More prominent in foods undergoing phase change from ice to water and in foods containing significant amounts of salt and other ions
Sanitiser	a disinfectant
Screen	a sieve
Sensible heat	heat used to raise the temperature of a food or removed during cooling, without a change in phase
Sequence control	a type of process control in which the completion of one operation signals the start of the next
Servo	a device used to provide control of a desired operation through the use of feedback (e.g. servo motor)
Soils	a generic term used for all types of contaminating materials on foods or equipment
Sorption isotherm	a curve produced from different values of relative humidity plotted against equilibrium moisture content
Sorting	the separation of foods into categories on the basis of a measurable physical property
Specific electrical resistance (or 'resistivity')	the property of a food by which it resists the flow of electricity, usually measured in ohms and equal to the ratio of the voltage to the current: it is the reciprocal of conductance
Specific growth rate	the slope of the curve when the natural logarithm of microbial cell concentration is plotted against time
Specific heat	the ability of a material to store heat. The amount of heat that causes a unit change in temperature by a unit mass of material
Stabilisers	hydrocolloids that dissolve in water to form viscous solutions or gels
Steady-state heat transfer	heating or cooling when there is no change in temperature at any specific location
Sterilants	chemicals that inactivate micro-organisms
Streamline flow	(or laminar) flow of liquids in layers without significant mixing between layers
Sublimation	a change in state of water, directly from ice to water vapour without melting
Substrate	a growth medium for micro-organisms
Supercooling	a phenomenon in which water remains liquid although the temperature is below its freezing point.
Supercritical carbon dioxide	liquid CO_2 used to extract food components
Supervisory Control and Data Acquisition (SCADA)	a type of computer software that collects data from programmable logic controllers and displays it as graphics to operators in real-time
Surface heat transfer coefficient	a measure of the resistance to heat flow, caused by a boundary film of liquid
Susceptor	a packaging material that is used to create a localised high temperature in microwave ovens, usually made from lightly metallised polyethylene terephthalate
Syneresis	separation of a liquid from a gel
Tempering	(1) cooling food to close to its freezing point; or (2) a process of re-heating, stirring and cooling chocolate to remove unstable forms of polymorphic fats
Tempura	thick batter
Tensile elongation	a measure of the ability to stretch
Tensile strength	the force needed to stretch a material
Tesla	unit of magnetic flux density (B) (1 tesla (T) $= 10^4$ gauss)
Thermal centre	the point in a food that heats or cools most slowly
Thermal conductivity	the ability of a solid material to conduct heat
Thermal death time (TDT)	the time required to achieve a specified reduction in microbial numbers at a given temperature

Thermal diffusivity	the ratio of thermal conductivity of a product to the specific heat of the product, multiplied by the density
Thermal shock	(1) (heating) fracture to a glass container caused by rapid changes in temperature; (2) (freezing) a rapid reduction in temperature that causes foods to fracture
Thermophysical properties	properties that influence the heating rate of a material (e.g. thermal conductivity, specific heat, density)
Thixotropic material	food in which the viscosity decreases with continued shear stress (e.g. most creams)
Ullage	headspace in a container
Ultrafiltration	unit operation in which solutes having molecular weights in the range of 1–200 kDa are selectively removed using a semipermeable membrane operating at lower pressure than reverse osmosis
Ultra-high temperature (UHT) processing	heat sterilisation at above 135 °C for a few seconds
Ultrasonication	treatment of foods using ultrasound
Unitised loads	grouping of packages into larger loads
Unsteady state heat transfer	heating or cooling where the temperature of the food and/or the heating or cooling medium are constantly changing
Usage value	the rate of usage of individual materials in an inventory multiplied by their individual value
Venting	removal of air from a retort before heat processing
Viscoelastic material	food materials that exhibit viscous and elastic properties including stress relaxation, creep and recoil
Voidage	the fraction of the total volume occupied by air (the degree of openness) of a bed of material in fluidised bed drying
Water activity	the ratio of vapour pressure of water in a solid to that of pure water at the same temperature
Waveguide	The component of a microwave system that guides microwaves from the magnetron to the cavity where the food is heated
Web	a packaging film
Wet bulb temperature	temperature measured by a wet thermometer in an air–water vapour mixture
Yield	weight of food after processing compared with weight before processing
Young's modulus	(also modulus of elasticity) a measure of the hardness of a material = stress/strain
z-value	the number of degrees Celsius required to bring about a ten-fold change in decimal reduction time of micro-organisms

Appendix D.2

List of acronyms

ABS	Acrylonitrile–butadiene–styrene
AFP	Antifreeze protein
AGV	Automatically guided vehicle
APET	Amorphous polyethylene terephthalate
AQL	Acceptable quality limit
ASLT	Accelerated shelf-life testing
ATEX	European Explosive Atmospheres and Gassy Mines Directive
ATP	Adenosine triphosphate
BET	Brunauer–Emmett–Teller
BHA	Butylated hydroxy anisole
BHT	Butylated hydroxy toluene
BOD	Biological oxidation demand
BOPP	Bioriented polypropylene
BOPS	Bioriented polystyrene
BSE	Bovine spongiform encephalopathy
CAP	Controlled atmosphere packaging
CAS	Controlled atmosphere storage
CBE	Cocoa butter equivalent
CCP	Critical control point
CFC	Chlorofluorocarbon
CFD	Computational fluid dynamics
CHD	Coronary heart disease
CHP	Combined heat and power
CIE	Commission on illumination
CIP	Cleaning in place
CJD	Creutzfeld–Jakob disease
CMC	Carboxymethyl cellulose
COD	Chemical oxidation demand
COP	Cleaning out of place
COP	Coefficient of performance
CORBA	Common Object Resource Based Architecture
CPET	Crystalline polyethylene terephthalate
CTA	Cellulose triacetate
CTI	Critical temperature indicator
CVD	Cardiovascular disease
DAA	Dehydroacscorbic acid
DAEC	Diffusely-adherent *E. coli*
DC	Direct current
DCS	Distributed control systems
DDE	Dynamic data exchange
DE	Dextrose equivalent
DFD	Dark, firm, dry
DNA	Deoxyribonucleic acid
DO	Dissolved oxygen
DRD	Draw and redraw (cans)
DTD	Decimal temperature difference
DWI	Drawn and wall Ironed (cans)
EAEC	Entero-aggregative *E. coli*
EAS	Electronic article surveillance
ECB	Ethanol chlorobenzene

ECCS	Electrolytic chrome coated steel (tin-free steel)
EDI	Electronic data interchange
EDTA	Ethylenediaminetetraaceticacid
EFA	Essential fatty acid
E_h	Oxidation–reduction (redox) potential
EHEC	Entero-haemorrhagic *E. coli*
EIEC	Entero-invasive *E. coli*
ELISA	Enzyme-linked immunosorbant assay
EMA	Equilibrium modified atmosphere
EMS	Environmental management system
EPEC	Entero-pathogenic *E. coli*
EPSC	Edible protective superficial coating
EPSL	Edible protective superficial layer
ERH	Equilibrium relative humidity
ERP	Enterprise resource planning
ESR	Electron spin resonance
ETEC	Entero-toxigenic *E. coli*
EU	European Union
EVA	Ethylene vinyl acetate
EVOH	Ethylene vinyl alcohol
FAD	Flavin adenine dinucleotide
FAO	Food and Agriculture Organization
FDA	Food and Drug Administration
FFS	Form–fill–seal
FHC	Fat holding capacity
FIFO	First-in-first-out
FLA	Four letter acronym
FUSHU	Foods for specific health use
GAS	Gas anti-solvent process
GATT	General Agreement on Tariffs and Trade
GBS	Guillan–Barré syndrome
GC	Gas chromatograph/chromatography
GDP	Good distribution practice
GEP	Gas exchange preservation
GHP	Good hygienic practice
GM	Genetic modification/genetically modified
GMM	Genetically modified micro-organism
GMO	Genetically modified organism
GMP	Good manufacturing practice
GPP	Good production practice
GRAS	Generally recognised as safe
HACCP	Hazard analysis critical control point
GVC	Global value chain
HCFC	Chlorofluorohydrocarbon
HDL	High-density lipoprotein
HDPE	High-density polyethylene
HFC	Fluorohydrocarbons
HFFS	Horizontal form–fill–seal
HHBM	Horizontal helical blade mixer
HIPS	High-impact polystyrene
HLB	Hydrophile–lipophile balance
HMI	Human–machine interface
HPAF	High-pressure assisted freezing
HPLC	High-pressure liquid chromatography
HPMC	Hydroxypropylmethyl cellulose
HPP	High-pressure processing
HPSF	High-pressure shift freezing
HTST	High-temperature short-time
HWRB	Hot water rinse and brushing

IBC	Intermediate bulk container
ICF	Immersion chilling and freezing
IMF	Intermediate moisture food
IQB	Individual quick blanching
IQF	Individual quick frozen/freezing
IR	Infrared
JIT	Just in time
ISO	International Organization for Standardization
LAN	Local area network
LDL	Low-density lipoprotein
LDPE	Low-density polyethylene
LED	Light-emitting diode
LLPDE	Linear low-density polyethylene
LN	Liquid nitrogen
MAP	Modified atmosphere packaging
MAS	Modified atmosphere storage
MAS	Marker assisted selection
MCC	Microcrystalline cellulose
MD	Machine direction
MDPE	Medium-density polyethylene
MES	Manufacturing execution systems
MF	Machine finished (paper)
MG	Machine glazed (paper)
MGT	Minimum growth temperature
MRI	Magnetic resonance imaging
MRP	Material resource (or requirement) planning
MUFA	Monounsaturated fatty acid
MVR	Mechanical vapour recompression
NIR	Near infrared
NLV	Norwalk-like virus
NMR	Nuclear magnetic resonance
NNPB	Narrow-neck press and blow
NOX	Nitrogen oxides
NVDP	Non-volatile decomposition products
ODBC	Open database connectivity
OECD	Organization of Economic Co-operation and Development
OHTC	Overall heat transfer coefficient
OLE	Object linking and embedding
OPC	Object linking and embedding for process control
OPP	Oriented polypropylene
ORP	Oxidation reduction potential
OTR	Oxygen transmission rate
PAH	Polycyclic aromatic hydrocarbon
PAM	Passive atmosphere modification
PCB	Polychlorinated biphenols
PCL	Polycaprolactone
PCS	Process control system
PECVD	Plasma enhanced chemical vapour deposition
PEEK	Polyether ether ketone
PEF	Pulsed electric field
PET	Polyethylene terephthalate
PG	Propyl gallate
PGPR	Polyglycerol polyricinoleate
PGSS	Gas saturates solution process
PHA	Polyhydroxyalkanoates
PHB	Polyhydroxybutyrates
PLA	Polylactic acid
PLC	Programmable logic controller
PMMA	Polymethylmethacrylate

PP	Polypropylene
PPP	Product processing packaging
PRP	Prerequisite programme
PS	Polystyrene
PSE	Pale, soft, exudative
PSL	Photostimulated luminescence
PSL	Practical storage life
PTFE	Polytetrafluoroethylene
PUFA	Polyunsaturated fatty acid
PVC	Polyvinyl chloride
PVdC	Polyvinylidene chloride
PVDF	Polyvinylidene fluoride
QA	Quality assurance
QDA	Quantitative descriptive analysis
QTL	Quantitative trait loci
RDA	Recommended dietary (or daily) allowance
RDC	Regional distribution centre
REPFED	Ready-to-eat-products-for-extended-durability (or refrigerated-pasteurised-foods-for-extended-durability)
RESS	Rapid expansion supercritical solutions
RF	Radio frequency
RFID	Radio frequency identification
RNA	Ribonucleic acid
ROPP	Roll-on-pilfer-proof
RR	Reduction ratio
RTE	Ready-to-eat
RVP	Relative vapour pressure
SAS	Supercritical antisolvent process
SCADA	Supervisory Control and Data Acquisition
SFA	Saturated fatty acid
SG	Specific gravity
SIP	Sterilise in place
SLV	Sapporo-like virus
SME	Specific mechanical energy (or small and medium-scale enterprise)
SRSV	Small-round-structured virus
SSF	Solid substrate fermentation
TBHQ	Tertiary butyl hydroquinone
TD	Transverse direction
TDT	Thermal death time
TL	Thermoluminescence
TLA	Three letter acronym
TLC	Thin layer chromatography
TNC	Trans-national corporation
TQM	Total quality management
TTI	Time–temperature indicator
TTT	Time–temperature tolerance
TVP	Texturised vegetable protein
TVR	Thermal vapour recompression
UHPH	Ultra-high-pressure homogenisation
UHT	Ultra-high temperature
UPC	Universal print code (barcode)
UV	Ultraviolet
VBNC	Viable but not culturable
VDP	Volatile decomposition products
VFFS	Vertical form–fill–seal
VOC	Volatile organic compounds
VP	Vacuum packaging
VSP	Vacuum skin packaging
VTEC	Vero cytotoxigenic

WAI	Water absorption index
WHC	Water-holding capacity
WHO	World Health Organization
WIP	Work in progress
WOF	Warmed over flavour
WSI	Water solubility index
WTO	World Trade Organization
WVP	Water vapour pressure
WVTR	Water vapour transmission rate

Appendix D.3

Symbols

A	Area
a	Thermal diffusivity
a	Throttling factor (extrusion)
a_w	Water activity
B	Time of heating (canning)
Bi	Biot number
b	Permeability
b	Slope of sorption isotherm
C_d	Drag coefficient (fluid dynamics)
c	Concentration
c	Internal seam length (canning)
c	Specific heat capacity
c_p	Specific heat at constant pressure
D	Diameter (pipe, vessel)
D	Dilution rate (fermentation)
D	Decimal reduction time
D	Diffusion coefficient
d	Diameter (sphere, size of sieve aperture)
d	Differential operator
dB	Decibel
E	Electrical field strength
E	Energy
E	Young's modulus
F	Feed flowrate (sorting, fermentation)
F	F-value
F	Force
F	Shape factors (extrusion)
F_p	Shape factor (freezing)
Fr	Froude number
f	Slope of heat penetration curve (canning)
f	Frequency
G	Geometric constants (extrusion)
G	Air mass flowrate (dehydration)
g	Acceleration due to gravity ($9.81\,\mathrm{m\,s^{-2}}$)
G	Retort temperature minus product temperature (canning)
H	Humidity
H	Enthalpy
H	Heat transfer coefficient
h_c	Convective heat transfer coefficient
h_s	Surface heat transfer coefficient
I	Light intensity
I	Electrical current
I_h	Retort temperature minus product temperature (canning)
J	Flux (membrane concentration)
J	Heating/cooling factor (canning)
K	Mass transfer coefficient (dehydration, membrane concentration)
K	Constant
K	Reaction rate
K_k	Kick's constant (size reduction)
K_R	Rittinger's constant (size reduction)
K_s	Substrate utilisation constant (fermentation)
k	Thermal conductivity

L	Length
L	Equivalent thickness of filter cake
l	Come-up time (canning)
M	Moisture content
M	Molar concentration
m	Mass
m	Mass flowrate
N	Speed
N	Rate of diffusion
Nu	Nusselt number
n	Number
P	Pressure
P	Product flowrate (sorting)
P	Power
P	Productivity (fermentation)
P_o	Power number (mixing)
P_o	Vapour pressure of pure water
Q	Rate of heat transfer
Q	Volumetric flowrate
q	Cooling load/rate of work done (refrigeration)
q_p	Specific rate of product formation (fermentation)
R	Universal gas constant
R	Reject flowrate (sorting)
R	Rejection value (membrane separation)
R	Resistance to flow through a filter
R	Fraction of reflected light (packaging)
R	Electrical resistance
Re	Reynolds number
r	Radius
r	Specific resistance to flow through a filter
S	Substrate concentration (fermentation)
s	Compressibility of filter cake
T	Absolute temperature
T	Fractional transmission of light (packaging)
T	Time
T	Metal thickness (canning)
U	Overall heat transfer coefficient
U	Thermal death time at retort temperature
V	Volume
V	Voltage
V_c	Fractional volume of filter cake
v	Velocity
v_c	Air velocity needed to convey particles
v_f	Air velocity needed for fluidisation
W	Absolute humidity
W	Work index (size reduction)
X	Thickness, depth
X	Mass fraction
x	Average
x	Direction of heat flow
x	Slope of moisture sorption isotherm (packaging)
y	Cover hook length (canning)
Y	Yield or yield factor (fermentation)
z	Height
z	z-value
Z	Conversion percentage (membrane concentration)
α	Absorbance, absorptivity
α	Thermal diffusivity
β	Coefficient of thermal expansion

Δ	Difference, change
δ	Half dimension
$\tan \delta$	Loss tangent (dielectric)
ϵ	Porosity
ϵ	Voidage of fluidised bed
ϵ	Emissivity (infrared radiation)
ϵ'	Dielectric constant
ϵ''	Loss factor (dielectric)
θ	Temperature
λ	Latent heat
λ	Wavelength
μ	Viscosity
μ	Specific growth rate (fermentation)
Π	Osmotic pressure
π	Constant $= 3.142$
ρ	Density
Σ	Sum
σ	Standard deviation
σ	Electrical conductivity
σ	Stephan–Boltzman constant (infrared radiation)
ω	Angular velocity

Index

CPSIA information can be obtained at www.ICGtesting.com
Printed in the USA
BVOW09s1644220215

388711BV00002B/6/P